Edible Films
and
Coatings

Fundamentals and Applications

FOOD PRESERVATION TECHNOLOGY SERIES

Series Editor
Gustavo V. Barbosa-Cánovas

Edible Films and Coatings: Fundamentals and Applications
Editors: María Pilar Montero García, M. Carmen Gómez-Guillén, M. Elvira López-Caballero, and Gustavo V. Barbosa-Cánovas

Introduction to Food Process Engineering
Editors: Albert Ibarz and Gustavo V. Barbosa-Cánovas

Shelf Life Assessment of Food
Editor: Maria Cristina Nicoli

Cereal Grains: Laboratory Reference and Procedures Manual
Sergio O. Serna-Saldivar

Advances in Fresh-Cut Fruits and Vegetables Processing
Editors: Olga Martín-Belloso and Robert Soliva-Fortuny

Cereal Grains: Properties, Processing, and Nutritional Attributes
Sergio O. Serna-Saldivar

Water Properties of Food, Pharmaceutical, and Biological Materials
Editors: Maria del Pilar Buera, Jorge Welti-Chanes, Peter J. Lillford, and Horacio R. Corti

Food Science and Food Biotechnology
Editors: Gustavo F. Gutiérrez-López and Gustavo V. Barbosa-Cánovas

Transport Phenomena in Food Processing
Editors: Jorge Welti-Chanes, Jorge F. Vélez-Ruiz, and Gustavo V. Barbosa-Cánovas

Unit Operations in Food Engineering
Albert Ibarz and Gustavo V. Barbosa-Cánovas

Engineering and Food for the 21st Century
Editors: Jorge Welti-Chanes, Gustavo V. Barbosa-Cánovas, and José Miguel Aguilera

Osmotic Dehydration and Vacuum Impregnation: Applications in Food Industries
Editors: Pedro Fito, Amparo Chiralt, Jose M. Barat, Walter E. L. Spiess, and Diana Behsnilian

Pulsed Electric Fields in Food Processing: Fundamental Aspects and Applications
Editors: Gustavo V. Barbosa-Cánovas and Q. Howard Zhang

Trends in Food Engineering
Editors: Jorge E. Lozano, Cristina Añón, Efrén Parada-Arias, and Gustavo V. Barbosa-Cánovas

Innovations in Food Processing
Editors: Gustavo V. Barbosa-Cánovas and Grahame W. Gould

Edible Films
and
Coatings
Fundamentals and Applications

Edited by
María Pilar Montero García
M. Carmen Gómez-Guillén
M. Elvira López-Caballero
Gustavo V. Barbosa-Cánovas

CRC Press
Taylor & Francis Group
Boca Raton   London   New York

CRC Press is an imprint of the
Taylor & Francis Group, an **informa** business

CRC Press
Taylor & Francis Group
6000 Broken Sound Parkway NW, Suite 300
Boca Raton, FL 33487-2742

Printed on acid-free paper
Version Date: 20160627

International Standard Book Number-13: 978-1-4822-1831-2 (Hardback)

Visit the Taylor & Francis Web site at
http://www.taylorandfrancis.com

and the CRC Press Web site at
http://www.crcpress.com

Printed and bound in the United States of America by Publishers Graphics, LLC on sustainably sourced paper.

Contents

Section III   Strategies to Optimize Coating and Film Functionality

Section IV   Encapsulation and Controlled Release in Films and Coatings

Section V   Applications of Films and Coatings in Foodstuffs

Section VI   Coatings and Films: Drawbacks and Challenges

Preface

In recent decades, there have been many documents referring to research in the development of edible and nonedible films and coatings made from biodegradable materials. Before continuing, it may be useful to clarify what we mean by films and coatings, as there may be separate words for these two concepts in some languages while one word is used to refer to both in others. In both cases, they are materials that cover food, and they may also be described as edible packaging. Film is a type of packaging that is made separately and then applied to food as a covering. Coatings are solutions applied on the surface of food that remain attached to the food; often they are extremely thin and transparent, making them imperceptible to the human eye.

The advances in this area have been breathtaking, and, in fact, their implementation in the industry is already a reality, especially with regard to coatings, as their application in the food industry is relatively simple, despite the fact that there are always obstacles to be overcome. However, the development of biodegradable films is also a fact, although it is still in its infancy. One of the most typical cases is the manufacture of biodegradable bags made from potato peel or from starch obtained from various sources, generally cross-linked with polylactic acid.

Even so, there is still a need for broad development in various fields and from various perspectives. A particularly novel application is their use as edible coatings or films in the design of gourmet foods, where fantasy and imagination provide a lot of scope for innovation in terms of shapes, textures, and colors—in other words, sensory appearance in every sense. With new advances in the incorporation of bioactive compounds, the possibilities are numerous, as there are not only a wide range of compounds but also the ways in which they are incorporated and even loaded in carriers that control their release and activity present challenges in which further advances are constantly being made. Once again, there are two possibilities, depending on the form that the "activity" takes, whether it is directed toward the preservation of food or to the effect after ingestion. Films and coatings also have the potential for applications in agronomy, as yet little explored and exploited, which could provide considerable advances in the preservation and quality of food.

All these matters, and also the management of any waste produced, are rigorously treated in this book from a critical viewpoint, for we feel it is important not only to describe the scientific advances that have been made but also to comment on the weaknesses and gaps that may remain. We hope that this book proves to be a real advancement in comparison to previous publications, making it genuinely worth reading.

We acknowledge the collaboration of all those who have participated in the publication of this book and express our most sincere gratitude for their dedication, effort, and contribution.

The book is produced under the auspices of CYTED (Latin American Science and Technology Development Program), as part of the project "Obtainment of additive materials from plant by-products from the region and their application in the development of biodegradable packaging for agro-food and nutraceutical use" (Action 309AC0382), within the area of Promotion of Industrial Development. Accordingly, special emphasis has been placed on its applicability.

We are very fortunate to be performing work that continues to fill us with enthusiasm every day and that enables us to enjoy the studies in which we are involved—and we have found much enjoyment over the years of working together. We hope that you too will find much to enjoy in the pages of this book.

This is a great opportunity for the CYTED-Agrobioenvase group, providing a way for us all to come to a common understanding of the advances in this field. We will undoubtedly be enriched by these chapters with their different points of view and by the knowledge of all those who are involved. For this reason, we did not want the book to be produced only by the CYTED-Agrobioenvase group, feeling it necessary to include other groups of researchers who could provide new vitality and different viewpoints and thus enrich this work. Accordingly, we conducted an examination and selection of participants of great prestige, who are experts in their specific subject.

The 29 chapters are arranged into sections, making it easier for the reader to find information. We hope that the material presented will be of interest and that you will enjoy reading it.

Editors

María Pilar Montero García is a research professor at the Spanish National Research Council (CSIC) in the Institute of Food Science, Technology and Nutrition (ICTAN), Madrid, Spain. She is author and coauthor of numerous publications in the field of food science and technology, especially valorization of products and development of functional foods, and films and coatings with special emphasis on their application in various fields of science. She earned a PhD in biological sciences from the Complutense University of Madrid (Spain) in 1988.

M. Carmen Gómez-Guillén is a research scientist at the Spanish National Research Council (CSIC) in the Institute of Food Science, Technology and Nutrition (ICTAN), Madrid, Spain. She earned a PhD in veterinary science from the Complutense University of Madrid (Spain) in 1994. She has published extensively in the field of Food Science and Technology, particularly on the development and valorization of seafood products and by-products, fish gelatin, edible films, and bioactive peptides.

M. Elvira López-Caballero is a tenured researcher at the Institute of Food Science, Technology and Nutrition of the Spanish National Research Council (ICTAN-CSIC). She is author and coauthor of numerous professional publications in the field of Food Science and Technology, including food preservation by hurdle technologies. She earned a PhD in veterinary science from the Complutense University of Madrid (Spain) in 1998.

Gustavo V. Barbosa-Cánovas is a professor of food engineering at the Washington State University, Pullman, Washington. His areas of research interest in the food domain include nonthermal processing, dehydration, physical properties, edible films, and water activity. He earned a BS in mechanical engineering from the University of Uruguay and a MS and PhD in food engineering from the University of Massachusetts as a Fulbright scholar.

Contributors

Jessica Acosta-García
School of Industrial Design and Engineering
Department of Mechanical, Chemical and
  Industrial Design
Technical University of Madrid
Madrid, Spain

Véronique Aguié-Béghin
Department of Fractionation of Agricultural
  Resources and Environment
University of Reims Champagne-Ardenne
Reims, France

Teresa Agüinaco-Castro
School of Industrial Design and Engineering
Department of Mechanical, Chemical and
  Industrial Design
Technical University of Madrid
Madrid, Spain

Izabela D. Alvim
School of Applied Sciences
University of Campinas
Campinas, Brazil

María Cristina Añón
Faculty of Exact Sciences
Center for Research and Development in Food
  Cryotechnology
National Scientific and Technical Research
  Council
National University of La Plata
La Plata, Argentina

Gustavo V. Barbosa-Cánovas
Department of Biological Systems Engineering
Washington State University
Pullman, Washington

Ann H. Barrett
Food Processing, Engineering and Technology
  Team
U.S. Army Natick Research Development and
  Engineering Center
Natick, Massachusetts

Valerio Bifani
Department of Chemical Engineering
Universidad de La Frontera
Temuco, Chile

Soottawat Benjakul
Department of Food Technology
Prince of Songkla University
Hat Yai, Thailand

Maria A. Bertuzzi
Faculty of Engineering
National University of Salta
Salta, Argentina

Branko Bugarski
Faculty of Technology and Metallurgy
University of Belgrade
Belgrade, Serbia

Kezban Candoğan
Department of Food Engineering
Ankara University
Ankara, Turkey

Emine Çarkcıoğlu
Department of Food Engineering
Ankara University
Ankara, Turkey

Ximena Carrión-Granda
Department of Food Technology
Public University of Navarra
Pamplona, Spain

Brigitte Chabbert
Department of Fractionation of Agricultural
  Resources and Environment
University of Reims Champagne-Ardenne
Reims, France

Amparo Chiralt
Department of Food Technology
Technical University of Valencia
Valencia, Spain

Talita A. Comunian
Department of Food Engineering
University of Sao Paulo
Sao Paulo, Brazil

María Cecilia Condés
Faculty of Exact Sciences
Center for Research and Development in Food
    Cryotechnology
National Scientific and Technical Research
    Council
National University of La Plata
La Plata, Argentina

Rajinder Kumar Dhall
Department of Vegetable Science
Punjab Agricultural University
Punjab, India

Verica Djordjević
Faculty of Technology and Metallurgy
University of Belgrade
Belgrade, Serbia

Ignacio Echeverría
Faculty of Exact Sciences
Center for Research and Development in Food
    Cryotechnology
National Scientific and Technical Research
    Council
National University of La Plata
La Plata, Argentina

María José Fabra
Department of Food Quality and Preservation
Institute of Agricultural Chemistry and Food
    Technology
Spanish National Research Council
Valencia, Spain

Carmen S. Favaro-Trindade
Department of Food Engineering
University of Sao Paulo
Sao Paulo, Brazil

Idoya Fernández-Pan
Department of Food Technology
Public University of Navarra
Pamplona, Spain

Mariana S.L. Ferreira
Department of Food Technology
Federal University of Rio de Janeiro State
Rio de Janeiro, Brazil

Silvia Karina Flores
Department of Industry
University of Buenos Aires
Buenos Aires, Argentina

Carmen Fonseca-Valero
School of Industrial Design and Engineering
Department of Mechanical, Chemical and
    Industrial Design
Technical University of Madrid
Madrid, Spain

M. Alejandra Garcia
Center for Research and Development in Food
    Cryotechnology
La Plata, Argentina

Lia Noemi Gerschenson
Department of Industry
University of Buenos Aires
Buenos Aires, Argentina

Begoña Giménez
Department of Food Science and Technology
University of Santiago de Chile
Santiago, Chile

Joaquín Gómez-Estaca
Department of Products
Institute of Food Science, Technology and
    Nutrition
Spanish National Research Council
Madrid, Spain

M. Carmen Gómez-Guillén
Department of Products
Institute of Food Science, Technology and
    Nutrition
Spanish National Research Council
Madrid, Spain

Carlos R.F. Grosso
School of Applied Sciences
University of Campinas
Campinas, Brazil

Mónica Ihl
Department of Chemical Engineering
Universidad de La Frontera
Temuco, Chile

Bojana Isailović
Faculty of Technology and Metallurgy
University of Belgrade
Belgrade, Serbia

Adriana Izquier
Department of Food Technology
University of Lleida
Lleida, Spain

Alberto Jiménez
Department of Food Technology
Technical University of Valencia
Valencia, Spain

Hande Kaya-Celiker
Department of Biological Systems Engineering
Virginia Polytechnic Institute and State
    University
Blacksburg, Virginia

José María Lagarón
Department of Food Quality and Preservation
Institute of Agricultural Chemistry and Food
    Technology
Spanish National Research Council
Valencia, Spain

Joao Borges Laurindo
Department of Chemical and Food Engineering
Federal University of Santa Catarina
Florianopolis, Brazil

Steva Lević
Department of Food Technology and Biochemistry
Faculty of Agriculture, University of Belgrade
Belgrade, Serbia

Renata Linhares
Department of Food Technology
Federal University of Rio de Janeiro State
Rio de Janeiro, Brazil

M. Elvira López-Caballero
Department of Products
Institute of Food Science, Technology and
    Nutrition
Spanish National Research Council
Madrid, Spain

P. Kumar Mallikarjunan
Department of Biological Systems Engineering
Virginia Polytechnic Institute and State
    University
Blacksburg, Virginia

Bianca C. Maniglia
Department of Chemistry
University of Sao Paulo
Sao Paulo, Brazil

Milena Martelli
Department of Chemistry
University of São Paulo
Sao Paulo, Brazil

Olga Martín-Belloso
Department of Food Technology
University of Lleida
Lleida, Spain

Juan Ignacio Maté
Department of Food Technology
Public University of Navarra
Pamplona, Spain

Adriana Noemí Mauri
Faculty of Exact Sciences
Center for Research and Development in Food
    Cryotechnology
National Scientific and Technical Research
    Council
National University of La Plata
La Plata, Argentina

Florencia Cecilia Menegalli
School of Food Engineering
University of Campinas
Campinas, Brazil

Jelena Milanović
Faculty of Technology and Metallurgy
University of Belgrade
Belgrade, Serbia

Michael Molinari
Department of Fractionation of Agricultural
    Resources and Environment
University of Reims Champagne-Ardenne
Reims, France

María Pilar Montero García
Department of Products
Institute of Food Science, Technology and
    Nutrition
Spanish National Research Council
Madrid, Spain

Silvia Moreno
Instituto Leloir Foundation
National Scientific and Technical Research
    Council
Buenos Aires, Argentina

Hugo Mújica-Paz
School of Biotechnology and Food
Monterrey Institute of Technology and Higher
    Education
Monterrey, Mexico

Muralidharan Nagarajan
Department of Food Technology
Prince of Songkla University
Hat Yai, Thailand

Viktor Nedović
Department of Food Technology and Biochemistry
Faculty of Agriculture, University of Belgrade
Belgrade, Serbia

Cristina Nerín
Department of Analytical Chemistry
University of Zaragoza
Zaragoza, Spain

Almudena Ochoa-Mendoza
School of Industrial Design and Engineering
Department of Mechanical, Chemical and
    Industrial Design
Technical University of Madrid
Madrid, Spain

María José Ocio
Department of Food Quality and Preservation
Institute of Agricultural Chemistry and Food
    Technology
Spanish National Research Council
Valencia, Spain

Lauren O'Conner
Food Processing, Engineering and Technology
    Team
U.S. Army Natick Research Development and
    Engineering Center
Natick, Massachusetts

Mariana S. de Oliveira
Department of Food Engineering
University of Sao Paulo
Sao Paulo, Brazil

Gabriel Paës
Department of Fractionation of Agricultural
    Resources and Environment
University of Reims Champagne-Ardenne
Reims, France

Ana S. Prata
School of Applied Sciences
University of Campinas
Campinas, Brazil

Thummanoon Prodpran
Department of Material Product Technology
Prince of Songkla University
Hat Yai, Thailand

Michelle J. Richardson
Food Processing, Engineering and
    Technology Team
U.S. Army Natick Research Development and
    Engineering Center
Natick, Massachusetts

Ana María Rojas
Department of Industry
University of Buenos Aires
Buenos Aires, Argentina

Pablo Rodrigo Salgado
Faculty of Exact Sciences
Center for Research and Development in Food
    Cryotechnology
National Scientific and Technical Research
    Council
National University of La Plata
La Plata, Argentina

Gloria Sánchez
Department of Biotechnology
Institute of Agricultural Chemistry and Food
    Technology
Spanish National Research Council
Valencia, Spain

Milla G. dos Santos
Department of Food Engineering
University of Sao Paulo
Sao Paulo, Brazil

Andrea Silva-Weiss
Department of Food Science and Technology
Universidad de Santiago de Chile
Santiago, Chile

Anibal M. Slavutsky
Faculty of Engineering
National University of Salta
Salta, Argentina

Robert Soliva-Fortuny
Department of Food Technology
University of Lleida
Lleida, Spain

Volnei B. Souza
Department of Food Engineering
University of Sao Paulo
Sao Paulo, Brazil

Maria S. Tapia
Department of Food Technology
University of Lleida
Lleida, Spain

Delia R. Tapia-Blácido
Department of Chemistry
University of Sao Paulo
Sao Paulo, Brazil

Aurora Valdez-Fragoso
School of Biotechnology and Food
Monterrey Institute of Technology and Higher
    Education
Monterrey, Mexico

Vito Verardo
School of Biotechnology and Food
Monterrey Institute of Technology and Higher
    Education
Monterrey, Mexico

Noemí E. Zaritzky
Center for Research and Development in Food
    Cryotechnology
La Plata, Argentina

Section I

Preparation, Properties, and Characterization of Coatings and Films

1

Standard and New Processing Techniques Used in the Preparation of Films and Coatings at the Lab Level and Scale-Up

Maria A. Bertuzzi and Anibal M. Slavutsky

CONTENTS

1.1 Introduction

Ongoing research on biodegradable/edible films is being made, and there is a great interest to make this knowledge more widely available and used. In the last 30 years, considerable progress has been made in developing these materials driven by the increasing consumer demand for safe, high-quality, convenient food with long shelf lives, along with an ecological awareness of the limited natural resources and the environmental impact of packaging waste (Janjarasskul and Krochta 2010; Kester and Fennema 1986). Nevertheless, the use of edible films or coatings to extend storage life of food is an ancient human practice of food preservation. Some examples of those old practices are fruit waxing used since the twelfth century in China, meat larding used in England since the sixteenth century (Kester and Fennema 1986), and the production of soy films from soy milk (yuba) traditionally employed in the Orient to wrap and shape ground meats or vegetables (Gennadios and Weller 1991).

Although edible films and coatings cannot totally replace synthetic packaging, they can advantageously substitute it in several applications due to their biodegradability. Edibility gave these coatings additional functions where plastics were not useful, such as providing a barrier between components in a heterogeneous food or giving individual protection to small pieces or portions of food materials (Guilbert 1986). Besides acting as a passive barrier, edible films and coatings have many other applications such as controlled release of active ingredients, immobilization of drug or enzyme, encapsulation of microorganisms or food ingredients, and establishment of modified atmosphere to protect food.

Edible films and coatings are defined as thin layers of material that can be eaten by the consumer and provide a barrier to mass transfer (moisture, oxygen, and solute movement) within the food itself or between the food and the environment. Films are formed as stand-alone sheets of materials, whereas coatings are directly formed on the product (Bourlieu et al. 2009; Guilbert 1986).

Edible packagings are prepared by using edible compounds derived from diverse renewable sources. Polysaccharide, protein, or lipid materials are used in various forms (simple or composite materials, single-layer or multilayer films) to prepare edible films and coatings. Due to their low affinity with water, lipids are generally considered as the most effective moisture barrier. On the other hand, because of their nonpolymeric characteristic, they are usually opaque and present poor mechanical properties (Rhim and Shellhammer 2005). Conversely, due to their hydrophilic nature, polysaccharide and protein films exhibit limited water vapor barrier, but they present good mechanical and gas barrier properties at low relative humidity (Janjarasskul and Krochta 2010). A detailed description of the characteristics of films and coatings made of lipids, proteins, and polysaccharides or their combinations are presented in other chapters of this book.

1.2 Historical Progresses

Throughout the time that edible films have been studied and developed, different strategies have been used to improve the functional properties of these materials. In the early stages, films based on pure hydrocolloids or lipids were studied and numerous publications reported physicochemical, mechanical, and barrier properties of these materials (Janjarasskul and Krochta 2010; Krochta 1992; Liu 2005; Rhim and Shellhammer 2005). As an alternative to improve barrier and mechanical properties of biopolymer-based films, the polymer network was modified through the formation of intramolecular and intermolecular covalent cross-links, for example, by applying thermal treatments or by adding chemical cross-linkers (Kim et al. 2002; Micard et al. 2000). Another approach consisted of combining oppositely charged hydrocolloids such as chitosan with alginate, carboxymethylcellulose, or many other negatively charged polysaccharides or proteins (Arzate-Vázquez et al. 2012; Li et al. 2011b). Lipids have also been used in combination with hydrocolloids as emulsion-based films or bilayer films in order to reduce water vapor permeability of hydrophilic films (Debeaufort and Quezada-Gallo 2000; Pérez-Gago and Krochta 2005; Wu et al. 2002).

In the last years, the progress in the application of nanotechnology to edible films has been reviewed by some authors (Azeredo 2009; Chivrac et al. 2009; Rhim et al. 2013). Nanotechnology has been applied to edible films through the incorporation of nanoreinforcements into the film matrix or the preparation of nanolaminated films. Bionanocomposites consist of a biopolymer matrix reinforced with organic or inorganic particles (nanoparticles) having at least one dimension in the nanometer range (1–100 nm). These materials exhibit improved thermal, mechanical, and barrier properties compared to the base biopolymers due to the high aspect ratio and high surface area of nanoparticles. A nanolaminated film consists of two or more layers of material with nanometric dimensions that are physically or chemically bonded to each other (Fabra et al. 2013). All these developments resulted, to some extent, in improved functional properties of biopolymer-based films. Nevertheless, there is still a big challenge ahead, in order to obtain competitive materials for the food packaging market.

1.3 Film and Coating Formulation

The formulation of edible films and coatings involves using at least one component capable of forming a continuous and cohesive matrix, a characteristic feature of polymeric materials (Guilbert and Gontard 2005). Films may be homopolymeric or heteropolymeric depending on whether they are prepared from a single polymer type or from blended polymers. Composite films are mixtures of hydrocolloids (polysaccharides and proteins), lipids, and additives where film formulation is tailor-made to suit to the needs of a specific product or required functional properties. Such heterogeneity may provide uniquely tailored properties, for example, the decrease of glass transition temperature, the prevention of crack propagation through the system, and the improvement of water resistance (Wu et al. 2002).

Film-forming materials used in edible films and coatings are polysaccharides, proteins, lipids, and resins. Polysaccharides are polymeric carbohydrate structures, formed by repeating units (either mono- or disaccharides) joined together by glycosidic bonds forming long chains. Polysaccharides have good film-forming properties with a wide range of solution viscosities. Generally, polysaccharide films are formed by disrupting interactions among long-chain polymer segments during dissolution and by forming new intermolecular hydrophilic and hydrogen bonding upon evaporation of the solvent to create a film matrix. The main advantages of polysaccharide-based films are their mechanical resistance, their efficient barrier properties against oil and lipids, their gas barrier properties, and, particularly, their selective permeability against oxygen transmission. However, due to the hydrophilic nature of these materials, humidity greatly affects their functional properties (Lacroix and le Tien 2005; Liu 2005; Nisperos-Carriedo 1994).

Proteins are also hydrophilic materials consisting of one or more long chains of amino acid residues. Depending on the amino acid sequence, native protein structures are random—coil, fibrous, or globular. Usually, formation mechanism of protein films involves the denaturation of the protein by the application of heat, solvents, or change in pH. The film matrix is formed when the extended peptide chains associate through new intermolecular interactions, depending on the protein, treatment, elaboration, and drying conditions. Generally, protein-based films have good mechanical and optical properties. They present more flexibility but less tensile strength than polysaccharides based on their strong intra- and intermolecular linkages. On the other hand, their disadvantages are the potential allergic responses by some individuals and the detrimental effect that humidity has on their functional properties due to their hydrophilic characteristics (Gennadios et al. 1994; Lacroix and Cooksey 2005).

Many lipid materials, mainly waxes and resins, have been used as protective coatings against moisture transfer. They have also been used to add gloss in spite of their being opaque and brittle since they are not polymers. They are fragile and do not form cohesive, self-supporting film structures. Owing to their low polarity, lipids have been incorporated into film-forming materials to provide a moisture barrier within composite films. Disadvantages of lipids films are related to their waxy taste and texture, greasy surface, and potential rancidity (Hernandez 1994; Morillon et al. 2002; Rhim and Shellhammer 2005).

Film additives are incorporated to enhance structural, mechanical, and handling characteristics or to provide active functions to the film matrix. Plasticizers are nonvolatile, high boiling point, nonseparating substances that modify certain physical and mechanical properties of the material. Plasticizers commonly used in edible films are mono-, di-, and oligosaccharides (e.g., glucose, fructose, and sucrose), polyols (e.g., glycerol, sorbitol, and polyethylene glycols), and lipids and their derivates (e.g., phospholipids, fatty acids, and surfactants). They reduce tensile strength and glass transition temperature by decreasing the intermolecular forces along the polymer chains (reduction in cohesion), which generally produce a decrease in brittleness, improve flow, impart flexibility, and increase toughness as well as tear and impact resistance (Sothornvit and Krochta 2005). Water is the universal plasticizer of hydrophilic materials, and the influence of water content on functional properties of protein- and polysaccharide-based films must be considered.

Film formulation based on emulsions requires surface-active compounds with both polar and nonpolar character capable of modifying interfacial energy at the interface in immiscible systems. Surfactants perform critical functions in the film-forming process, including emulsion stabilization, and antifoaming during solution preparation. They also serve as wetting agents to ensure adequate adherence to the substrate, leveling agent to minimize surface defects in the film, and releasing agents to ensure easy release of the dry film from the support where it was formed. Some common surfactants or emulsifiers of food grade are acetylated monoglyceride, lecithin, glycerol monostearate, sodium lauryl sulfate, sorbitan monooleate, and many proteins due to their amphiphilic nature. The amount usually incorporated into the formulation is 1% or less, based on the weight of biopolymer (Rossman 2009).

Some compounds are added to the film-forming solution as preservatives. Antimicrobials such as methylparabens, propylparabens, and sodium benzoate and antioxidants such as butylated hydroxytoluene, butylated hydroxyanisole, and ascorbic acid are used to protect the solution and the dry film during storage from deterioration and spoilage.

Active compounds are also incorporated into the film matrix in order to add functions to the passive barrier generated by the film or coating. Films and coatings formulated with active ingredients are known as active films. Incorporation of antimicrobials such as plant extracts, bacteriocins, organic acids, and enzymes have been reported in the literature (Ibarguren et al. 2010; Sánchez-González et al. 2011). Oxidation of food can be reduced by the use of oxygen scavengers and antioxidant agents in the packaging. Antioxidants are compounds that delay oxidation reactions and can be incorporated into active films and coatings to introduce this function. Some examples of antioxidants incorporated in edible films are phenolic compounds, tocopherols, ascorbic acid, and tartaric acid (Gómez-Estaca et al. 2009; Pereira de Abreu et al. 2010).

The addition of dispersed particles (macro-, micro-, or nanosized) in the polymeric matrix has been used as a way to modify some physical and functional properties such as gas and vapor permeability, thermal stability, tensile strength, and abrasion resistance (Avella et al. 2005).

A composite material is made by combining two or more materials that work together in the structure to impart the material unique properties. Usually, the combined materials have very different properties. Composite films are prepared by incorporation of a filler compound into a polymeric matrix. Composites are used to reinforce film structure, to limit the dependence of some film properties on moisture, to reduce film permeation, and to restrain molecular rearrangement. Typical fillers include talc, clays, silica, lipids, biopolymers, and cellulose fibers particularly microcrystalline cellulose and cellulose nanofibers (Azeredo 2009).

In general, macroscopic fillers present poor interaction at the interface, while nanofillers present better interaction between both phases because of their large surface/volume ratio. In the last years, many studies have been dedicated to organic–inorganic systems and, in particular, to those in which layered silicates are dispersed in a polymeric matrix (Avella et al. 2005). The properties of nanocomposites are governed by the extent of the dispersion of nanoparticles in the biopolymeric matrix and by the interaction between nanoparticles and biopolymer (Ray and Bousmina 2005). Thermal stability, gas barrier properties, strength, and low melt viscosity are among the properties that can be achieved by nanoreinforcements of biodegradable polymers. A uniform dispersion of nanoparticles leads to a very large matrix/filler interfacial area that changes the nanostructure, molecular mobility, the relaxation behavior, and the tortuosity of the permeate pathway through the film (Janjarasskul and Krochta 2010). Thus, selection of appropriate techniques to form the nanoparticle dispersion in the film matrix is a critical factor in their performance.

Some composite films were developed to combine complementary functional properties of hydrocolloids and lipids to overcome their individual shortcomings. The edible films based on polar biopolymers, such as polysaccharides and proteins, are sensitive to humidity and their properties are negatively affected by moisture. The efficiency of the lipid material in composite films and coatings depends on the nature of the used lipid, its chemical structure, hydrophobicity, state of aggregation (i.e., solid or liquid), and the interaction with the hydrocolloid (Rhim and Shellhammer 2005).

1.4 Forces Involved in Film Formation

When a polymer film is formed on a substrate, a set of attractive forces operates between the molecules of the same material, on the one hand (cohesion), and between the surface molecules of different substances, such as the film and the substrate surface, on the other hand (adhesion) (Banker 1966). Cohesion refers to the ability of contiguous surfaces of the same material, at molecular or supermolecular level, to form strong bonds that prevent or resist separation at the contact point. Therefore, cohesion depends on polymer structure, that is, molecular length, geometry, molecular weight distribution, and type, and position of lateral groups. Cohesion requires high attractive forces between molecules and coalescence of contiguous surfaces of the particles of film material. Compatibility between components is required to obtain strong cohesion and film strength. Coalescence or the disappearance of boundary layers or surfaces occurs by the diffusion of individual macromolecules between and within these layers. Only high-molecular-weight polymers, owing to their molecular structure, combine sufficient cohesive strength and capacity for coalescence to produce film structures, deposited from appropriate solvents (Banker 1966; Guilbert et al. 1996).

Molecular shape exerts a strong influence on film cohesion, since it largely determines both diffusivity of the macromolecule and the strength of its intermolecular forces. Branched structures hinder diffusion and reduce the interlacing areas. Cohesion is affected by numerous parameters like the presence, concentration, location, and relative polarity of polar groups along the chain, molecular weight distribution, branching, and regularity of the chain. Intermolecular forces that promote cohesion also promote crystallinity. Most polymers used in films and coatings present amorphous or semicrystalline structures. The relative degree of molecular chain order in a film varies according to such factors as film formulation, method and conditions of film formation, and storage time and conditions. Film crystallinity affects physical and functional properties such as film strength, flexibility, permeability, and solubility (Liu 2005).

Plasticizers cause a decrease in cohesion by reducing intermolecular forces of film-forming polymers. As a result, the degree of cohesion in film structures is fundamental to film properties. The increase in cohesion of the polymeric structure raises film density and compactness, may decrease permeability and flexibility, and probably, may intensify brittleness (Guilbert et al. 1996).

Furthermore, adhesive forces are the attractive forces between the film-forming material and the substrate. They are important in the design of the production process because they are related to the spreadability of the film-forming solution and the easy release of the dry film from the substrate in the production system, as well as the premature peeling of the coating from the coated surface (Guilbert et al. 1996; Han and Gennadios 2005).

A large difference between surface energy of film-forming solution and the substrate lessens the work of adhesion, resulting in the incomplete coating of the substrate or the easy peel-off of the film from the substrate. Surface-active agents can be added to the film-forming solution to reduce the surface tension of the solution and, thus, to increase the work of adhesion (Han and Gennadios 2005).

Some operational factors increase film cohesion by promoting molecular diffusion like increasing temperature, contact time, degree of polymer solvation, or decreasing viscosity (Banker 1966). Not all of them are readily controllable in the film formation process.

As the temperature rises, the molecular movement increases due to the dependence of diffusion rate on temperature and the decrease on film density. An increased temperature generally facilitates adhesion between polymer film and substrate. Nevertheless, excessive heat may cause film defects such as curling, blisters, or fat edges (Rossman 2009).

Contact time is the period during which the polymer molecules are capable of diffusion and orientation. An inappropriate high evaporation rate may produce noncohesive films due to the premature immobilization of polymer molecules as well as the poor diffusion between polymeric particles into the film (Banker 1966).

Viscosity is of great importance to the film cohesion, and consequently polymer solvation and polymer concentration are also relevant for matrix cohesion. At low viscosity or high solvation levels, molecular diffusion will be promoted. Furthermore, under these conditions, most film-forming solution will be very dilute; the drying time will be excessively long, and some components could be separated from the bulk of the previously homogeneous system. Consequently, an intermediate viscosity will usually result in the highest cohesive strength. It will be a function of the solvent used, the rate of desolvation, and the stereochemical displacement of the polymer from the solvent during evaporation (Banker 1966).

1.5 Film Formation Process

Film formation is a process where an ordered solid phase is obtained from a liquid phase. The liquid phase may be either a melt or a solution with an appropriate viscosity for the application method. After application on a substrate, the liquid is converted into a solid film. The chemical and physical changes that occur in this process are critical to the final appearance and performance of the films.

Guilbert and Gontard (2005) classified the methods used for film preparation into two general process pathways:

1. The "casting" or "wet process," based on solution or dispersion of the film-forming material in a suitable solvent and the subsequent removal of the solvent
2. The "dry process," based on thermoplastic properties of certain biopolymers when they are subjected to high temperatures and pressures in processes such as extrusion or compression molding

1.5.1 Casting or Wet Process

In general, during the film formation in the wet process, a phase transition from a polymer-in-water (or other solvent) system to a water-in-polymer system occurs (Han and Gennadios 2005). This can be accomplished through different mechanisms depending on material nature.

1.5.1.1 Film Formation Mechanisms

During the dissolution of a macromolecular substance, the cohesive forces between the solute molecules are neutralized by unions with the solvent molecules (solvation). The more crystalline polymer, the greater the intermolecular cohesive forces, and therefore, the more difficult the polymer dissolution. The polymers used in edible films and coatings are generally aliphatic polymeric nonpolar chains with polar substituents along the chain. Solvation conditions depend on the functional group characteristics. As the functional groups on a linear polymer become ionized during dissolution, the charged groups will repel each other, producing a stretching of the polymer chain. If the polymer is soluble in the solvent, there is an attractive interaction between the polymer and the solvent and the net interaction between the polymer segments is repulsive. As a rule, maximum polymer solvation and polymer chain extension will produce the greatest cohesiveness in the film structure (Banker 1966). Since solvation and polymer chain extension is reflected in the viscosity of solution, viscosity provides a useful control parameter during formulation and application steps (Rossman 2009).

The film formation mechanism depends on whether the polymer is in the dissolved or the dispersed state. Film formation from polymer solutions occurs as the solvent evaporates, since the polymer chains are intimately mixed. The polymer chains interpenetrate, going through a gel state, and finally the membrane is formed during drying (Felton 2013).

The mechanism of film formation from polymeric dispersions is more complex than that from solutions. The discrete polymer particles in polymeric dispersions must coalesce and the polymer chains interpenetrate to form a continuous film. In this case, temperature plays an important role in particle coalescence and film formation. In contrast to polymer solutions, dispersed polymer systems require the polymer particles to deform and fuse together to form the film. Thus, these systems exhibit a minimum film-forming temperature (MFFT). Below this temperature, a polymeric dispersion will form an opaque, discontinuous material upon solvent evaporation, whereas a clear continuous film will be formed at temperatures above the MFFT (ASTM 2010), since drying at temperatures above the MFFT provides sufficient capillary force for coalescence. The MFFT depends on the polymer nature, polymer particle diameter, and the coating formulation. In addition, these systems generally require postcoating storage in temperature and humidity-controlled environments to ensure complete polymer coalescence. Conversely, polymer solutions do not exhibit an MFFT and will form a film at room temperature (Steward et al. 2000). Therefore, a previous treatment of polymeric dispersions is convenient in order to turn them into colloidal solutions. This simplifies and facilitates the film formation process. Gelatinization of starch granules before the film formation is a typical example of this treatment.

According to thermodynamic principles, a solid phase is precipitated from a solution if the chemical potential of the polymer in the solid phase is less than that of the material in the solution. In the equilibrium state of a saturated solution, the chemical potential of a component is the same in both, solid and solution phases. Thus, film formation occurs if the equilibrium concentration of the component of interest is exceeded by some supersaturation method, including (Myers 1999)

- Solvent evaporation by heating or vacuum
- Decrease of material solubility by cooling or heating the solution, depending on the enthalpy of the solution
- Adding to the solution another solvent that is miscible with the primary solvent, but is a poor solvent for the material
- Salting out by the addition of substances that may contain a common ion with the polymeric substance, thereby reducing its solubility

- Chemical reaction in the solution changing a soluble substance into an insoluble one
- Changes in other factors that affect the ability of the solvent to solvate the material

These phenomena and their combinations are utilized for film formation according to the nature of film-forming material.

1.5.1.1.1 Biopolymeric Films

Kester and Fennema (1986) classified the mechanisms of film formation from hydrocolloid solutions as simple coacervation, complex coacervation, and thermal gelation or precipitation. Coacervation is the formation of macromolecular aggregates due to phase separation in initially homogeneous polymer solution.

Simple coacervation results from the addition of a substance to a polymer solution that causes phase separation. For example, in aqueous solution, the addition of water-miscible solvent (e.g., ethanol) or a suitable electrolyte that causes the separation into polymer-rich and polymer-poor phases. Coacervation is a reversible process; thus, the system may be further stabilized by changing the pH, by heating, by irradiation, or by chemical cross-linking (Arvanitoyannis and Dionisopoulou 2010; Mu et al. 2012).

Complex coacervation is driven by electrostatic interactions between two or more macromolecules of water-soluble oppositely charged polymers that form a new liquid phase consisting of macromolecular aggregates, for example, chitosan and alginate (Arzate-Vázquez et al. 2012) or arabic gum and gelatin (Green and Schleicher 1957).

Gelation or thermal coagulation occurs during denaturation of proteins or thermal gelation of macromolecules caused by heating or cooling the biopolymer solution, for example, ovalbumin, gelatin, or agar (Giménez et al. 2013; Jerez et al. 2007).

1.5.1.1.2 Lipid Coatings

Morillón et al. (2002) sorted the different ways in which edible coatings based on solid fats, waxes, or resins can be formed as follows:

- *Melting and solidification*: The material is heated above the lipid melting point, homogenized, and cast or applied while melted, followed by cooling in order to produce solidification.
- *Solubilizing in an organic solvent*: The solution is applied in a thin layer (e.g., spraying or brushing) and then the solvent is evaporated.
- *Preparing an emulsion in water*: The emulsion is poured and then the water is evaporated. The addition of a surfactant is recommended to stabilize the emulsion.

1.5.1.1.3 Composite Films

Composite films combine materials with different properties in order to improve film functional properties. Their preparation method depends on the characteristics of film components and the desired structure and properties of the resulting film. Some composite films combine hydrocolloids, as the structural matrix, with lipids, either as a stable emulsion in the hydrocolloid or forming a bilayer.

1.5.1.1.3.1 Emulsion-Based Films   A film based on a stable emulsion is formulated by adding a lipid material and surfactants to a biopolymer solution. The mixture is heated above the lipid melting point, homogenized, and cast or directly applied while melted or after cooling (Krochta 2002). Food grade solvents for edible packaging are generally limited only to water and ethanol. The casting solution can be spread in a thin layer on a support that will release the film after drying to produce a preformed stand-alone film or applied to form a coating directly on food or pharmaceutical products (Baldwin et al. 1997; Krochta 2002). Drawbacks of emulsion films are related to the low lipid melting temperature, the solvent volatilization from the structural network, and the strong effect of emulsion droplet size and distribution, polarity, degree of saturation, and polymorphism of lipid components on water barrier properties and mechanical properties of films and coatings (Perez-Gago and Krochta 2005).

1.5.1.1.3.2 Bilayer Films   They consist of a second distinguishable layer film of lipid laminated over a preformed hydrocolloid film. Two different methods are used to prepare bilayer films:

- *Lamination method*: A molten or solubilized lipid is dispersed or laminated on a polysaccharide- or protein-based film.
- *Emulsion method*: A lipid is dispersed in the biopolymeric solution as an unstable emulsion before casting. The bilayer is formed during the drying process due to a phase separation.

Cracking and delamination frequently occur in bilayer films, ruining their excellent water barrier properties that depend on layer continuity. The preparation process requires two casting and drying steps and high temperature for lipid fusion or organic solvents for lipid dissolution (Perez-Gago and Krochta 2005). Nevertheless, bilayer films have 10–1000 times better barrier efficiency against water transfer than emulsion-based films (Debeaufort and Quezada-Gallo 2000).

1.5.1.1.4 Nanostructured Films

Alternatively, improved film properties have been tackled by controlling the micro- and nanostructure. Thus, the use of nanotechnology in biopolymer films and coatings permitted the incorporation of nanofillers into hydrocolloid matrixes and the development of nanolaminated films.

1.5.1.1.4.1 Nanocomposite Films   Bionanocomposites are composite materials based on the incorporation of nanosized fillers into the biopolymeric matrix. Different nanofillers can be added to biopolymers, but in the last years, most studies have focused on layered silicates such as montmorillonite (MMT) due to its availability, versatility, low cost, and its being environmentally friendly (Chivrac et al. 2009). MMT is a 2:1 layered smectite clay mineral with a platey structure and hydrophilic behavior (Poole and Owens 2007). MMT is characterized by a moderate negative surface charge. The limiting aspect of nanocomposite formulation using nanoclays is the difficulty to disperse them homogeneously into a polymeric matrix due to their preferred face-to-face stacking in agglomerated tactoids (Tunc et al. 2007). Depending on the process conditions and on the polymer/nanofiller affinity, different morphologies can be obtained: microcomposites, intercalated nanocomposites, and exfoliated nanocomposites. In microcomposites, the polymer and the clay remain immiscible (phase separation), resulting in agglomeration of the clay into the matrix and poor macroscopic properties of the final material. Interaction between the layered silicates and polymer chains may produce two types of nanoscale composites: intercalated and exfoliated. The intercalated nanocomposites result from the penetration of polymer molecules into the interlayer region of the clay. As a consequence, an ordered multilayer structure with alternating polymer/inorganic layers at a repeated distance of a few nanometers is formed. On the other hand, the exfoliated nanocomposites involve extensive polymer penetration, with the clay layers delaminated and randomly dispersed in the polymeric matrix (Azeredo 2009; Weiss et al. 2006). Microcomposite morphologies are common in commercially available nanoclay reinforced material. Although a significant amount of work has been performed on hydrocolloid nanocomposites, more researches are still needed to understand the complex plasticizer–matrix–nanofiller interaction and its influence on the resulting morphology and material properties. Many authors have reported different techniques of MMT incorporation into the polymeric matrix. They obtained different types of composite structures (exfoliated, intercalated, or phase separation) depending on the nature of film components and the preparation method (Chivrac et al. 2010; Slavutsky and Bertuzzi 2012; Tunç and Duman 2010).

Some organic nanofillers were also studied in the formulation of bionanocomposites such as cellulose nanocrystals in starch or alginate film matrix (Huq et al. 2012; Slavutsky and Bertuzzi 2014). The authors reported strong molecular interactions between cellulose nanocrystals and alginate or starch film matrix. It was found that the reinforcement with nanofiller concentrations lower than 5% into the film structure increases film tensile strength and thermal stability and reduces water vapor permeability and surface hydrophilicity. Usually, the changes in functional properties of bionanocomposites films are significant, even with very low level of nanofiller incorporation (≤5% w/w), due to the strong interaction between the biopolymer and the nanofiller, which results from the high aspect ratio of nanoparticles.

1.5.1.1.4.2 Nanolaminated Films A nanolaminated film consists of two or more layers of material with nanometric dimensions that are physically or chemically bonded to each other (Rubner 2003). Layer-by-layer assembly allows the preparation of a desired hierarchically organized composite material with a wide variety of different components as constituents of the multilayer films. The versatility of the layer-by-layer technique has allowed the deposition of a broad range of materials (e.g., polyelectrolytes, charged lipids, DNA, nanoparticles, and dye molecules) and the application to solvent accessible surface of almost any kind and shape (biological cells, colloids, fruits, etc.). This technique is based on the alternating deposition, on a solid substrate, of different materials achieved not only by electrostatic interactions but also by hydrogen bonding, hydrophobic interactions, covalent bonding, and complementary base pairing (Decher 2012; Weiss et al. 2006). The type of substances used to create each layer, the total number of layers incorporated into the multilayered film, the sequence of the different layers, and the preparation conditions used to form each layer will determine the functional properties (gas and vapor permeability, mechanical properties, swelling and wetting characteristics) of the final film (Weiss et al. 2006). Furthermore, nanolaminated films can be used to incorporate active compounds, act as controlled release systems, and can coat food products such as fruits and vegetables. There has been an increasing interest in nanolayered systems due to their potential applications in a wide range of areas including pharmaceutical, biomedical, and food packaging. The fabrication of nanolayered films by the layer-by-layer assembly at lab scale can be made by using simple and low-cost operations such as dipping and washing (Weiss et al. 2006).

1.5.1.2 Film Drying

Water is the main solvent used in edible films due to the hydrophilic nature of polysaccharides and proteins. However, there are some disadvantages in using waterborne systems, since the high latent heat of evaporation of water leads to long drying times and high energetic costs.

Usually, once the polymer is dissolved at a concentration required for application, generally with a viscosity in the range 0.05–1 Pa·s, the film-forming solution is applied on a substrate, and the solvent is allowed to evaporate. In the first stage of solvent evaporation from the film, the rate of evaporation is essentially independent of the polymer presence. The evaporation rate depends on the vapor pressure at that temperature, the surface area/volume ratio, and the rate of airflow over the surface. This first stage is the longest and lasts until the polymer has reached approximately 60%–70% volume fraction. Initially, particles move with Brownian motion, but this ceases when molecules undergo a significant interaction once a critical volume of solvent has evaporated (Steward et al. 2000; Wicks et al. 2007).

As solvent evaporates, viscosity increases, glass transition temperature increases, free volume decreases, and the rate of loss of the solvent becomes dependent on how rapidly solvent molecules can diffuse to the surface of a film so that they can evaporate. The remaining solvent leaves the film initially via any remaining interparticle channels and then by diffusion through the fused polymer skin, but the rate of evaporation eventually slows down to the asymptotically approach that of diffusion alone. It is during this final stage that a wet film becomes more homogeneous and gains its mechanical properties as polymer chain interdiffusion occurs, increasing cohesion. Reducing the rate of evaporation can lead to better quality films by allowing the molecules more time to pack into an ordered structure. In contrast, casting at high temperatures gives the particles sufficient energy to overcome their mutual repulsion and the films are formed before the molecules are fully ordered (Wicks et al. 2007). The solvent evaporation rate in hydrocolloid solutions can be reduced by adding hydrophobic or inert additives that increase the diffusion path length or by adding hydrophilic materials that generate polar interactions.

1.5.1.3 Film or Coating Application Methods

Typical methods for application and distribution of the film-forming material in a liquid form onto a substrate include spreading, enrobing, and spraying.

Hand spreading involves the spreading of the film-forming solution with a paint brush or roller onto the substrate or food. The film or coating requires setting or solidifying at ambient temperature or by heating (Debeaufort and Voilley 2009).

Spraying consists in applying the film coating solution onto the substrate with a spray system that ensures consistent and uniform coating with minimal waste. However, this method requires that the product be turned to expose the bottom surface for subsequent coating. Spray-coating is preferred for products with a large surface area and for obtaining thinner films (Dagaran et al. 2005) and for film-forming solution with low viscosity. Spraying is also used when a second application of a cross-linking promoter is required, for example, a calcium solution on alginate (Donhowe and Fennema 1994). Cunning and Caulkins (1965) patented an apparatus for spray-coating wax or similar materials on agricultural products (fruit and vegetables), while they are moving along a conveyor.

Enrobing involves application of a coating layer onto a substrate by falling film enrobing or by dipping and subsequent dripping. The final thickness and coverage may be less uniform and thicker than with other coating methods. This method is suitable for items with irregular shapes and sizes and for highly viscous solution. Dipping is widely used for coating meat and vegetable product in laboratory assays (Donhowe and Fennema 1994). Earle (1968) patented a process to preserve fresh food (meat, seafood, and poultry) consisting of a first step of product immersion in an alginate dispersion, draining off the solution excess, and then a second step of product immersion in an aqueous gelling solution of calcium ions.

1.5.1.4 Film and Coating Production Technologies

The most common types of equipment used to convert the film-forming solution in edible films are preexisting technologies that were adapted to the conditions and requirements of biopolymer materials.

1.5.1.4.1 Wet Casting

The casting technique is the most commonly used for film preparation at laboratory scale. It consists of pouring a film-forming solution or suspension onto a leveled flat surface (e.g., Petri dishes, acrylic plates, etc.), controlling the average thickness of the resulting films through the mass or volume of suspension poured onto the plate. Smooth rods or wire-wound rods are usually used to spread uniformly the coating solution onto the plate. The wire-wound rod creates many narrow ridges on the casting surface that flow and level out during drying. In general, solvents used for edible films production include water, ethanol, or a combination of both (Rossman 2009). Usually, dilute solution of polymer (5%–10%) and plasticizers (up to 60% polymer weight) are used. Most published works report that the drying of film-forming solution takes place at room temperature or in a forced air circulation drying oven at moderate temperatures, requiring drying times up to 24 h (Dagaran et al. 2009; Rhim et al. 2002; Slavutsky and Bertuzzi 2012). However, some local irregularities are usually inevitable such as variations in drying rate or final film thickness related to geometric and drying conditions. For these reasons and due to the long drying time required, this methodology is not suitable for preparing larger films.

1.5.1.4.2 Pan Coating

This is used to apply either thin or thick layer on solid, almost spherical particles in a rotating drum. The solution is ladled or sprayed into the rotating pan, and the particles are tumbled within the drum to distribute the solution over their surface. Forced hot air is used to dry the coating. This method is utilized by pharmaceutical and confectionery industries. Edible wax and shellac are used in confectionery to provide a brilliant surface and a moisture-barrier coating. The drawbacks of using ethanol as solvent of these materials are the potential explosion hazards and the environmental problems caused by solvent emissions during processing. Besides, ethanol may cause nondesirable bitterness and off-flavors on lipid-based coatings such as chocolate. Whey protein–based coatings applied by pan coating resulted in transparent, high-gloss, and flexible coating when cast from aqueous solution (Krochta et al. 2005). Arnold (1968) patented a method of roasting pecan nutmeats by coating with a protective film of arabic gum, salt, and spices in a rotating pan. The coated nutmeats are deposited on an open mesh-type moving belt and moved through an infrared tunnel, where infrared heat is applied from above and below, to roast the nuts.

1.5.1.4.3 Fluidized-Bed Coating

The equipment is comprised of a spray system that atomizes the coating solution that is applied counter-currently to the fluidized particles of the substrate until a desired thickness is achieved. This method is used by the pharmaceutical industry to coat tablets. Fluidized bed reduces the formation of clusters of coated product, a problem commonly found in pan coating. Compared with pan coating, fluidized bed requires a greater amount of coating solution because of the loss on the column wall during spraying. However, it requires a shorter processing time, provides a complete coverage, and minimizes cluster formation. Lin and Krochta (2006) reported fluidized-bed systems as a viable alternative for peanut coating with a whey protein–based coating, considering its high oxygen barrier.

1.5.1.4.4 Belt Conveyors

Film-forming solution is cast or spread uniformly on one end of a continuous belt that passes through a drying chamber to remove excess water. The belt rotates around large drums situated at either end of the line. The dry film is stripped from the belt and is wound into mill rolls. Another option consists in using a carrier web, such as polyester films or paper films. The coated substrate rides on support rollers as it enters and passes through a drying chamber to remove the water and finally is wound into rolls. For edible films, these conveyor lines are typically 1.5–4.0 m in length and 0.5–1.5 m in width. Belts are usually made of stainless steel or silicone. If necessary, a release coating can be applied to the belt to ensure uniform release of the film (Rossman 2009).

A shear-thinning behavior is appropriate for the film-forming solution with low viscosities at high shear rates to ensure adequate flow conditions and high viscosities at low shear rates to prevent outpouring, undesired flow, sedimentation, and phase separation. During the drying process, a decreasing contact angle between the solid substrate and the liquid is observed, the film-forming solution is concentrated, its viscosity is increased, and its surface energy is decreased due to the removal of the solvent (water or ethanol). The work of adhesion increases by the approach of the surface energy values of both phases (substrate surface and coating solution). However, if there is a very high adhesion between film and the substrate (belt surface), the film separation of the belt will be difficult. Conversely, this is a desirable phenomenon for the direct coating onto food surfaces, in order to avoid the peeling problem of the coated layer (Han and Gennadios 2005). The rheological properties and surface tension of film-forming solution may be regulated through biopolymer selection and additives incorporation. They are the control parameters of the film formation process when a film without defects and suitable adhesion to the substrate is desired. The main advantages of this method are the film uniformity, the film functional properties reproducibility, and the operation and maintenance simplicity (Rossman 2009).

The solution can be applied onto the belt by using some conventional coating methods like knife coating or slot die coating that are briefly described.

1.5.1.4.4.1 Knife Coating

A stationary, rigid knife doctors a precise layer of solution onto a web moving under the knife. A pool of film-forming solution is held behind the knife. Micrometer adjustments are provided to set the height of the knife above the web and to control the film thickness. These coaters produce smooth and uniform films; they are simple devices with little maintenance and they can be adapted to small or large productions. The film-forming solution may be spread on the belt conveyor (e.g., steel belt), on a disposable substrate (e.g., release paper), or on a dry preformed film forming a second layer. The spreading solution is dried, in a short time, by heat conduction, heat convection through countercurrent hot air circulation, or infrared emission. The final thicknesses of dry films ranges from 20 μm to 1 mm and depend on biopolymer concentration, spreading speed, and the height of the doctor blade gap (De Moraes et al. 2013; Rossman 2009).

1.5.1.4.4.2 Slot Die Coating   A coating liquid is forced out of a reservoir through a slot by pressure, and transferred to a moving substrate. The slot is significantly smaller in section than the reservoir and is oriented perpendicular to the direction of substrate movement. Slot die coating has many variations that can be distinguished from each other by the design of the die itself, the orientation of the die relative to the substrate, the distance from the die to the substrate (slot die coating, extrusion coating, and curtain coating), the coating structure (pattern coating, continuous coating), and the method used to generate the pressure that forces liquid out of the die. The device has a sealed steel chamber containing the fluid, which reduces its contamination. All of the coating fluid is applied to the substrate via the positive displacement pump and the slot in the die exit. Film thickness is achieved by adjusting the gap between the die lips or by increasing the speed of the belt. Film thickness ranged between 20 nm and 100 μm can be obtained. With this device, it is possible to maintain constant fluid temperature, to distribute uniformly the fluid, and to define the coating width. Dies can be designed to simultaneously lay down multiple layers of wet film (Rossman 2009).

1.5.1.4.5 Nanostructured Multilayers

Nanotechnology has many potential and novel applications within the food and packaging industry for producing materials with new or improved properties. However, many of these developments are either too expensive or too impractical to implement industrially. Layer-by-layer assembly and electrospinning are examples of technologies commercially available for edible film fabrication.

1.5.1.4.5.1 Layer-by-Layer Assembly   Nanolayered films prepared by layer-by-layer assembly consist of successive electrostatically adsorbed layers of oppositely charged polyelectrolytes. This deposition process involves the submersion of a substrate having an intrinsic surface charge (e.g., preformed film based on a charged biopolymer) into a series of polyelectrolyte solutions, and water rinse bath to remove any physically entangled or loosely bound polyelectrolyte. Layer-by-layer dipping technique is carried out by computer-controlled slide-stainer due to the long time required to complete a multilayer. Thereby, the solvent is limited to those with relatively low vapor pressure, such as water, to avoid evaporation and compounds concentration during extended dipping periods. In order to eliminate rinse water contamination, robotic modifications have been made to dipping systems. It involves spraying the sample with water, which immediately drains away. Since the contaminated water drains away and is not left in the rinse bath, the possibility of contamination is avoided. Polyelectrolyte solutions can also be applied on the substrate by spraying. Spray deposition is less time consuming than dipping method (Izquierdo et al. 2005). The quality of the films obtained by both techniques was compared and it was found that multilayers prepared by dipping are thicker, denser, and less rough than films having the same number of layers, that is, having the same number of deposition cycles, obtained by spraying (Kolasinska et al. 2009).

Krogman et al. (2012) developed an automated apparatus capable of spray depositing polyelectrolytes via the layer-by-layer mechanism onto a vertically oriented substrate. To counteract the effects of irregular spray patterns, the substrate can be slowly rotated about a central axis. The droplet pathway can be forced by using vacuum. In this way, a thicker or three-dimensional substrate can be coated, such as rolls of textile.

1.5.1.4.5.2 Electrospinning   Electrospinning is a simple, versatile, and efficient manufacturing technology capable of producing thin, solid polymer strands from solution by applying a strong electric field to a spinneret with a small capillary orifice. Generally, electrospun polymer fibers can range in size from 10 to 1000 nm in diameter and may exhibit unusual functionalities with respect to their mechanical, electrical, and thermal properties (Fabra et al. 2013). Because of the high aspect ratio, the fibers have shown to be ideal materials to produce high oxygen barrier structures by combining layers of biopolymers with an intermediate electrospun fiber layer of other biopolymer to enhance the barrier properties while serving as natural tie or adhesive layer (Lagarón Cabello et al. 2013). The challenge for the development of multilayer bio-based structures is in ensuring proper adhesion between the layers. Intermediate electrospin fiber layers of biopolymers have demonstrated to have the ability of acting as natural adhesives (Busolo et al. 2009). Besides, the electrospinning technique allows tight control of the thickness of interlayers and improves mechanical and barrier properties without affecting optical properties due to the nanometric size of the fibers.

MECC Co. Ltd (Japan) successfully fabricated a nanofibrous membrane with better alignment that exhibited silklike reflection using this technology. They also succeeded in scaling up the process of fabricating aligned-fiber membranes using a modified drum collector system.

The spinning process was modified to form films from casein, replacing the spinneret with a plate die to form flat films. Protein solutions were extruded into a coagulating bath and then collected onto a roller (Frinault et al. 1997).

A list of providers around the world of electrospinning and electrospraying equipment for laboratory and industrial level is provided by ElectrospinTech (http://electrospintech.com). From the commercial perspective, the application of electrospin nanofibers will be in high-performance or high-value-added products. Most likely, the limited application of nanotechnology to the food industry will change as nanofabrication technologies become more cost-effective.

1.5.1.4.5.3 Plasma   Plasma technology is used to alter chemical and physical properties of polymeric surfaces without affecting their bulk properties. This process consists of complex reactions between charged and neutral plasma species, between plasma and surface species, and between surface species (Kaplan 2004). Deposition of thin films by plasma polymerization offers several advantages over conventional coating techniques; the process may occur in a single reaction step, good adhesion between film and substrate is generally achieved, the deposition of the film is fairly uniform over the entire surface, and problems with residual solvents are avoided (Chan et al. 1996). Andrade et al. (2005) used plasma polymerization to modify the surface of maize starch films in order to reduce their water affinity. They deposited hydrogenated-carbon coatings of 20–230 nm in thickness through exposing films to low-pressure glow plasma generated with 1-butene gas. They found that the plasma coating process reduced significantly the hydrophilic character of starch films. A good adhesion between the starch substrate and the deposited coating was observed. Nevertheless, the efficiency of the plasma-polymerized coating as a water barrier increased with thickness up to a critical value due to the presence of microcracks and delamination in deposited layers of larger thickness. This may be caused by stress accumulation in the coating due to the high cross-linking density (Chan et al. 1996).

1.5.1.5 Drying Technology

Film formation requires solvent evaporation from the film-forming solution. During drying process, polymer molecules aggregate forming a gel structure and finally a solid film is formed by progressive evaporation of the volatile solvent. Drying requires the application of heat to evaporate the solvent up to reach the desired moisture content in the dry film, usually between 5% and 15%, without damaging the film characteristics. The rate at which drying is accomplished is governed by the rate at which the heat is transferred to the film-forming solution and by the internal solvent transfer to the surface of the film and its subsequent evaporation. Industrial dryers differ in type and design, depending on the principal method of heat transfer employed (Rossman 2009). The dryers commonly used in edible films production processes are described as follows:

- *Indirect dryers*: They are equipment in which the heating medium (e.g., steam, hot gas, and thermal fluids) does not come into contact with the product being dried. Instead, wet material is dried by contact with a heated surface; heat transfer to the wet material is mainly by conduction from this surface. For example, steam can be applied to the bottom surface of the belt conveyor, and the heat is transferred from the belt to the film-forming solution, which minimizes the tendency of the wet film to develop a dry skin (Devahastin and Mujumdar 2006).

- *Impingement drying*: This system involves blowing hot air in a series of slots or nozzles at high velocity against the wet film. This may cause movements of the film-forming solution creating defects in the dry film (Rossman 2009). Since impingement yields very high heat and mass transfer rates, it is a popular system for convective drying when rapid drying or small equipment is desired. Nevertheless, it has high capital and operating costs because of the more complex fabrication and increased air-handling requirements. Impinging jet drying is recommended only if a major fraction of the moisture to be removed is unbound, making it suitable for the first stage of the biopolymer solutions drying. If the drying rate is internal diffusion controlled, as in the final stage of the film drying, the high heat transfer rates of the impingement system can often result in film degradation if the material is heat-sensitive (Mujumdar 2006).

* *Conveyor dryers*: They are the most versatile dryers available. Hot dry air is blown into the drying chamber and flows counter to the movement of the substrate. Hot air can be introduced above the web, below the web, or on both sides simultaneously (Rossman 2009). Typically, the dryer is separated into independent zones, which have their own heat source and fans. Air temperature and velocity can be controlled as the film progresses through the dryer (Poirier 2006). Air can be heated by electric heating or indirectly with heat exchangers. Infrared heating can also be used as heat source in the drying chamber.

1.5.2 Dry Process

Dry processes are based on heat application to the film-forming material to increase its temperature and to allow its flow. In these processes, biopolymers are plasticized and heated above their glass transition temperature to form a uniform melt by using heat, pressure, and shear (Guilbert and Gontard 2005). The soft and rubbery melt can be shaped into specific forms upon cooling. Cooling reconverts the rubbery product in glassy material with solid characteristics (Hernandez-Izquierdo et al. 2008).

Dry process is based on the thermoplastic properties of some biopolymers (mainly starch and proteins) in low water content conditions (Cuq et al. 1997). Thermoplasticity of the film-forming materials is modified by addition of plasticizers and other additives. Plasticizers reduce the glass transition temperature due to the decrease in cohesion in the polymeric structure. Plasticizers such as glycerol, sorbitol, and polyethylene glycol, in an amount of 10%–60% of polymer weight, are added to modify the thermoplastic properties of used materials. As water also acts as a plasticizer, its addition in concentrations up to 50% polymer weight also reduces polymer melt temperature (Ciannamea et al. 2014; Krishna et al. 2012).

Thermoplastic processes include casting via compression molding, extrusion, and injection molding (Verbeek and van den Berg 2010). Preexisting technology used in the plastics industry has been adapted to edible film production.

Edible films based on biopolymers have been obtained by several authors using thermoplastic processes, such as extrusion of starch (Li et al. 2011a), soy protein (Zhang et al. 2001), zein (Chen et al. 2013), whey protein (Hernández-Izquierdo et al. 2008), and gelatin (Krishna et al. 2012); compression molding of whey protein (Sothornvit et al. 2007), soybean protein (Ciannamea et al. 2014), and wheat gluten (Türe et al. 2012); injection molding of egg albumen (Fernández-Espada et al. 2013); and blown extrusion of sodium caseinate (Belyamani et al. 2014), and gelatin (Andreuccetti et al. 2012).

Most biopolymers are not thermoplastics and cannot be liquefied to stable melts. Accordingly, thermomechanical methods require a previous stage of biopolymer thermoplasticization in order to obtain mixtures with enough fluidity to be processed. Without added water and/or plasticizers, the strong intermolecular interactions would lead to thermal degradation of the material before the film can be formed (Hernandez-Izquierdo and Krochta 2008; Li et al. 2011a).

During heat processing of protein, protein structures disaggregate, denature, dissociate, unravel, and align in the direction of the flow. These changes allow the protein molecules to recombine and cross-link through specific linkages. The cross-linking reactions can result in a high glass transition temperature and high melt viscosity, which require addition of plasticizers to increase free volume and mobility of the molecules and to avoid degradation. As temperature increases above the glass transition, the plasticized proteins turn into a soft, rubbery material that can be shaped into desired forms. Upon cooling, the matrix network gets fixed into the desired structure through hydrogen, ionic, hydrophobic, and covalent interactions (Hernandez-Izquierdo and Krochta 2008; Verbeek and van den Berg 2010).

Process variables such as temperature, pressure, time, and plasticizer contents determine the degree of conformational changes, protein aggregation, and chemical cross-linking that take place during processing. The cross-linking through S–H and S–S groups is significantly dependent on temperature; therefore, the increase in processing temperature will result in films with higher density, decreased solubility, and improved mechanical and barrier properties (Kim et al. 2002).

Starch cannot be thermally processed without a plasticizer or gelatinization agent, since its decomposition temperature is lower than its glass transition temperature. When plasticizers, such as water and glycerol, are added during thermomechanical processing, granules suffer structural degradation, glucose chains gain mobility, glass transition temperature decreases, and starch melts and flows. Under these conditions, starch-based materials can present multiphase transitions, such as gelatinization, melting, decomposition, and recrystallization (Li et al. 2011a; Liu 2005; Xie et al. 2014).

Starches with different amylose and amylopectin contents behave differently during phase transition, have different rheological properties during extrusion, and develop different mechanical properties in the resulting film (Della Valle et al. 2007). Higher temperatures and moisture contents are necessary for processing high-amylose starches, due to their higher gelatinization temperature, originated by the strong interaction forces between the long linear amylose chains. Li et al. (2011a) studied the effects of different amylose/amylopectin ratios on the functional properties of starch-based film obtained by extrusion. They observed that films with higher amylose content exhibit better mechanical properties. In addition, these films retained some unaffected crystalline zones that act as reinforcements.

1.5.2.1 Extrusion Process

Extrusion is a continuous, efficient, high-performance, and low-cost process, advantageous for large-scale production (Krishna et al. 2012). It is used extensively for the production of conventional commercial plastic packaging. Extrusion requires shorter times and lower energy consumption to remove water than casting methods (Hernandez-Izquierdo and Krochta 2008). This process uses one or two rotating screws fitted in a barrel in order to progressively increase the pressure and push forward and mix the ingredients through a die of desired shape where expansion may take place (Nur Hanani et al. 2012). Prior to extrusion, the polymer is blended with plasticizers, fillers, stabilizers, lubricants, and other additives, to produce the desired product property profile. Once the ingredients are mixed, the formulation is fed to the extruder, where it is melted, mixed, and delivered to the die. In general, the extruder barrel can be subdivided into three processing zones: (1) the feeding zone, where the granular, low-density raw material is introduced into the barrel and slightly compressed; (2) the kneading zone, with further compression and increasing pressure, temperature, and material density; and (3) the heating zone, where the highest shear rates, temperatures, and pressures are achieved along with the final product texture, density, and functional properties (Hauck and Huber 1989). After exiting the die, the product is pulled away from the extruder at constant velocity to attain the appropriate cross section and cooled and solidified in the desired shape.

The successful application of extrusion must ensure complete melting and sufficient mixing, while avoiding degradation to obtain a homogenous melt. This can only be achieved by careful control of the key processing parameters. The main process variables include feed rates, screw speed, screw configuration, screw length-to-diameter ratio, barrel temperature profile, and die size and shape (Hernandez-Izquierdo and Krochta 2008). Then, formulation of film-forming material and process variables are important parameters that determine the extent of conformational changes, aggregation, and chemical cross-linking in the final extruded material (Rhim and Ng 2007).

The thermoplastic materials obtained by extrusion can be shaped employing standard equipment used for synthetic polymers, such as extrusion, blowing, injection, and compression molding. This methodology combines two steps: (1) extrusion to obtain pellets of thermoplastic material and (2) blowing, injection, or compression molding to obtain films.

1.5.2.1.1 Blowing

Blown film extrusion is the most common technology to make plastic films, especially for the packaging industry. The process involves a single screw extruder, in which the thermoplastic pellets are successively compacted and melted to form a continuous viscous liquid. This molten plastic is then forced, through an annular die. Air is blown into the center of the die, and the pressure causes the extruded melt to expand in the radial direction, forming a bubble. The bubble is pulled continually upward from the die and a cooling ring blows air onto the film. The film can also be cooled from the inside using internal cooling, and it reduces the temperature and maintains the bubble diameter. After solidification, the film moves into a set of nip rollers that collapse the bubble to maintain the air pressure inside and collect the film. Blown extrusion has been used to produce edible films based on sodium caseinate or starch (Belyamani et al. 2014; Melo et al. 2014).

1.5.2.1.2 Injection Molding

A heated barrel feeds the molten polymer into a prefabricated mold via an extrusion method. The material is introduced into the heated barrel through a feed hopper. The screw melts the polymer and also acts as a ram during the injection phase. The polymer is injected into a mold tool that defines the shape of the molded part. When the material is cooled, the mold is opened and a mechanism is used to push the product out of the mold (Fernández-Espada et al. 2013).

Injection molding is the most commonly used manufacturing process for the fabrication of plastic parts. A wide variety of products are manufactured using this process, which vary greatly in their size, complexity, and application. Jane and Wang (1996) have patented a thermoplastic composition based on soy protein and polysaccharides that has a high degree of flowability for processing by extrusion and injection molding into solid articles as well as films that are biodegradable with a high degree of tensile strength and water resistance.

1.5.2.2 Compression Molding

A specific quantity of raw material is placed into a heated mold that is closed, and pressure is applied to force the molten material to contact all areas of the mold, giving to the material the desired shape. Thus, it operates in a discontinuous manner. The combination of high temperatures, high pressures, short times, and low moisture contents in compression molding causes the transformation of biopolymer–plasticizer mixtures into viscoelastic melts (Hernández-Izquierdo and Krochta 2008).

Compression molding is usually assisted by a prior intensive mixing of all components: biopolymer, plasticizers, and additives. Extrusion can be used to perform the intensive mixing of the film-forming materials prior to compression molding, where, through the addition of a suitable combination of plasticizers, temperature, pressure, and shear, a malleable mixture capable to be molded is obtained. Other blenders used for the prior mixing of the film-forming materials were kitchen aid mixer or food processors (Sothornvit et al. 2007; Türe et al. 2012).

Compression-molded soy protein isolate–glycerol films were produced at an optimum temperature of 150°C, a pressure of 10 MPa, and a dwell time of 2 min (Cunningham et al. 2000). Thermogravimetric analysis indicated that soy protein degraded at temperatures above 180°C (Ogale et al. 2000). Krishna et al. (2012) combined extrusion and compression molding to obtain gelatin-based films with appropriate thickness, water barrier, and mechanical and thermal properties. They observed significant differences in functional properties upon comparing extruded films with solution cast films.

1.6 Challenges and Future Trends

There exists in the literature a lot of information regarding the effect of the preparation process variables on the characteristics and functional properties of the resulting films and coatings based on biomaterials. This knowledge allowed important progress in the development of suitable technologies for the fabrication of films and coatings. In most of the cases, these technologies emerged from the adaptation of previously existing ones used in plastics, to fit the requirements of these new materials. Nowadays, there are a wide variety of available products obtained by wet casting, extrusion, or compression molding of different biopolymeric materials, such as disposable bags, sausage casings, vessels, forks, and knives. Nevertheless, more studies to further improve film and coating properties are needed.

The development of appropriate technologies must support the research efforts focused on films and coatings with improved performance, active films, or intelligent packaging materials. An example of this promising trend is the application of nanotechnology techniques in the production of nanocomposites and nanolaminated films. Great advances in this matter can be envisaged, to the extent that certain technologies and developments can be applied to a larger scale in an economically profitable and competitive manner.

REFERENCES

Andrade, C.T., R. Simão, R.M.S.M. Thiré, and C. Achete. 2005. Surface modification of maize starch films by low-pressure glow 1-butene plasma. *Carbohydrate Polymers* 61(4) (September): 407–413.

Andreuccetti, C., R.A. Carvalho, T. Galicia-García, F. Martinez-Bustos, R. González-Nuñez, and C.R.F. Grosso. 2012. Functional properties of gelatin-based films containing *Yucca schidigera* extract produced via casting, extrusion and blown extrusion processes: A preliminary study. *Journal of Food Engineering* 113(1): 33–40.

Arnold, F.W. 1968. Infrared roasting of coated nutmeats. US Patent 3,383,220. Filed: March 15, 1965. May 14, 1968.

Arvanitoyannis, I.S. and N.K. Dionisopoulou. 2010. Irradiation of edible films of plant and animal origin. In: *Irradiation of Food Commodities*, ed. I.S. Arvanitoyannis, pp. 609–634. London, U.K.: Academic Press.

Arzate-Vázquez, I., J.J. Chanona-Pérez, G. Calderón-Domínguez et al. 2012. Microstructural characterization of chitosan and alginate films by microscopy techniques and texture image analysis. *Carbohydrate Polymers* 87(1): 289–299.

ASTM D2354-10/1. Standard test method for minimum film formation temperature of emulsion vehicles. American Society for Testing Materials, Philadelphia, PA, 2010.

Avella, M., J.J. de Vlieger, M.E. Sabine Fischer, P. Vacca, and M.G. Volpe. 2005. Food chemistry biodegradable starch/clay nanocomposite films for food packaging applications. *Food Chemistry* 93: 467–474.

Azeredo, H.M.C. 2009. Nanocomposites for food packaging applications. *Food Research International* 42(9): 1240–1253.

Baldwin, E.A., M.O. Nisperos-Carriedo, R.D. Hagenmaier, and R.A. Banker. 1997. Use of lipids in coatings for food products. *Food Technology* 51(6): 56–64.

Banker, G.S. 1966. Film coating theory and practice. *Journal of Pharmaceutical Sciences* 55: 81–92.

Belyamani, I., F. Prochazka, and G. Assezat. 2014. Production and characterization of sodium caseinate edible films made by blown-film extrusion. *Journal of Food Engineering* 121(January): 39–47.

Bourlieu, C., V. Guillard, B. Vallès-Pamiès, S. Guilbert, and N. Gontard. 2009. Edible moisture barriers: How to assess of their potential and limits in food products shelf-life extension? *Critical Reviews in Food Science and Nutrition* 49(5) (May): 474–499.

Busolo, M., S. Torres-Giner, and J.M. Lagaron. 2009. Enhancing the gas barrier properties of polylactic acid by means of electrospun ultrathin zein fibers. In: *ANTEC, Proceedings of the 67th Annual Technical Conference*, Chicago, IL, pp. 2763–2768.

Chan, C.-M., T.-M. Ko, and H. Hiraoka. 1996. Polymer surface modification by plasmas and photons. *Surface Science Reports* 24: 1–54.

Chen, Y., R. Ye, X. Li, and J. Wang. 2013. Preparation and characterization of extruded thermoplastic zein–poly(propylene carbonate) film. *Industrial Crops and Products* 49(August): 81–87.

Chivrac, F., H. Angellier-Coussy, V. Guillard, E. Pollet, and L. Avérous. 2010. How does water diffuse in starch/montmorillonite nano-biocomposite materials? *Carbohydrate Polymers* 82(1): 128–135.

Chivrac, F., E. Pollet, and A. Luc. 2009. Progress in nano-biocomposites based on polysaccharides and nanoclays. *Materials Science* 67: 1–17.

Ciannamea, E.M., P.M. Stefani, and R. Ruseckaite. 2014. Physical and mechanical properties of compression molded and solution casting soybean protein concentrate based films. *Food Hydrocolloids* 38(July): 193–204.

Cunning, T.G. and J.M. Caulkins. 1965. Method for spray coating fruit and vegetables. US Patent 3,192,052 A. Filed: July 6, 1962. June 29, 1965.

Cunningham, P., A. Ogale, P. Dawson, and J. Acton. 2000. Tensile properties of soy protein isolate films produced by a thermal compaction technique. *Journal of Food Science* 65(4): 668–671.

Cuq, B., N. Gontard, and S. Guilbert. 1997. Thermoplastic properties of fish myofibrillar proteins: Application to biopackaging fabrication. *Polymer* 38(16): 4071–4078.

Dagaran, K., P.M. Tomasula, and P. Qi. 2009. Structure and function of protein-based edible films and coatings. In: *Edible Films and Coatings for Food Applications*, eds. M.E. Embuscado and K.C. Huber, pp. 25–56. New York: Springer.

De Moraes, J.O., A.S. Scheibe, A. Sereno, and J.B. Laurindo. 2013. Scale-up of the production of cassava starch based films using tape-casting. *Journal of Food Engineering* 119(4): 800–808.

Debeaufort, F. and J.-A. Quezada-Gallo. 2000. Lipid hydrophobicity and physical state effects on the properties of bilayer edible films. *Journal of Membrane Science* 180: 47–55.

Debeaufort, F. and A. Voilley. 2009. Lipid-based edible films and coatings. In: *Edible Films and Coatings for Food Applications*, eds. M.E. Embuscado and K.C. Huber, pp. 135–168. New York: Springer.

Decher, G. 2012. Layer-by-layer assembly (putting molecules to work). In: *Multilayer Thin Films: Sequential Assembly of Nanocomposite Materials*, eds. G. Decher and J.B. Schlenoff, pp. 1–21. Weinheim, Germany: Wiley-VCH.

Devahastin, S. and A.S. Mujumdar. 2006. Indirect dryers. In: *Handbook of Industrial Drying*, ed. A.S. Mujumdar, pp. 137–148. Boca Raton, FL: Taylor & Francis.

Donhowe, G.I. and O.R. Fennema. 1994. Edible films and coatings: Characteristics, formation, definitions, and testing methods. In: *Edible Coatings and Films to Improve Food Quality*, eds. J.M. Krochta, E.A. Baldwin, and M.O. Nisperos-Carriedo, pp. 1–24. Lancaster, PA: Technomic Publishing.

Earle, R.D. 1968. Method of preserving foods by coating same. US Patent 3,395,024. July 30, 1968.

Fabra, M.J., M.A. Busolo, A. López-Rubio, and J.M. Lagaron. 2013. Nanostructured biolayers in food packaging. *Trends in Food Science & Technology* 31(1): 79–87.

Felton, L. 2013. Mechanisms of polymeric film formation. *International Journal of Pharmaceutics* 457(2) (December 5): 423–427.

Fernández-Espada, L., C. Bengoechea, F. Cordobés, and A. Guerrero. 2013. Linear viscoelasticity characterization of egg albumen/glycerol blends with applications in material moulding processes. *Food and Bioproducts Processing* 91(4) (October): 319–326.

Frinault, A., D.J. Gallant, B. Bouchet, and J.P. Dumont. 1997. Preparation of casein films by a modified wet spinning process. *Journal of Food Science* 62: 744–747.

Gennadios, A., T.H. McHugh, C.L. Weller, and J.M. Krochta. 1994. Edible coatings and film based on proteins. In: *Edible Coatings and Films to Improve Food Quality*, eds. J.M. Krochta, E.A. Baldwin, and M.O. Nisperos-Carriedo, pp. 201–77. Lancaster, PA: Technomic Publishing.

Gennadios, A. and C.L. Weller. 1991. Edible films and coatings from soymilk and soy protein. *Cereal Foods World* 36: 1004–1009.

Giménez, B., A.L. de Lacey, E. Pérez-Santín, E.M. López-Caballero, and P. Montero. 2013. Release of active compounds from agar and agar–gelatin films with green tea extract. *Food Hydrocolloids* 30(1): 264–271.

Gómez-Estaca, J.L.B., C.M. Gómez-Guillén, A. Alemán, and P. Montero. 2009. Antioxidant properties of tuna-skin and bovine-hide gelatin films induced by the addition of oregano and rosemary extracts. *Food Chemistry* 112(1): 18–25.

Green, B.K. and L. Schleicher. 1957. Oil-containing microscopic capsules and method of making them. US Patent 2,800,457 A. Filed: June 30, 1953. July 23, 1957.

Guilbert, S. 1986. Technology and application of edible protective films. In: *Food Packaging and Preservation*, ed. M. Mathlouthi, pp. 371–394. New York: Elsevier.

Guilbert, S. and N. Gontard. 2005. Agro-polymers for edible and biodegradable films: Review of agricultural polymeric materials, physical and mechanical characteristics. In: *Innovation in Food packaging*, ed. J.H. Han, pp. 263–276. London, U.K.: Academic Press.

Guilbert, S., N. Gontard, and L.G.M. Gorris. 1996. Prolongation of the shelf life of perishable food products using biodegradable film and coatings. *Lebensmittel-Wissenschaft & Technologie* 29: 10–17.

Han, J.H. and A. Gennadios. 2005. Edible films and coatings: A review. In: *Innovations in Food Packaging*, ed. J.H. Han, pp. 301–307. London, U.K.: Academic Press.

Hauck, B.W., and G.R. Huber. 1989. Single screw vs twin screw extrusion. *Cereal Food World* 24: 930–939.

Hernandez, E. 1994. Edible coating from lipids and resins. In: *Edible Coatings and Films to Improve Food Quality*, eds. J.M. Krochta, E.A. Baldwin, and M.O. Nisperos-Carriedo, pp. 279–303. Lancaster, PA: Technomic Publishing.

Hernandez-Izquierdo, V.M. and J.M. Krochta. 2008. Thermoplastic processing of proteins for film formation—A review. *Journal of Food Science* 73(2) (March): R30–R39.

Hernandez-Izquierdo, V.M., D.S. Reid, T.H. McHugh, J.D.J. Berrios, and J.M. Krochta. 2008. Thermal transitions and extrusion of glycerol-plasticized whey protein mixtures. *Journal of Food Science* 73(4): E169–E175.

Huq, T., S. Salmieri, A. Khan et al. 2012. Nanocrystalline cellulose (NCC) reinforced alginate based biodegradable nanocomposite film. *Carbohydrate Polymers* 90(4): 1757–1763.

Ibarguren, C., L. Vivas, M.A. Bertuzzi, M.C. Apella, and M.C. Audisio. 2010. Edible films with anti-*Listeria monocytogenes* activity. *International Journal of Food Science and Technology* 45(7): 1443–1449.

Izquierdo, A., S.S. Ono, J.C. Voegel, P. Schaaf, and G. Decher. 2005. Dipping versus spraying: Exploring the deposition conditions for speeding up layer-by-layer assembly. *Langmuir* 21(16): 7558–7567.

Jane, J.-L. and S. Wang. 1996. Soy protein based thermoplastic composition for preparing molded articles. US Patent 5,523,293. Filed: May 25, 1994. June 4, 1996.

Janjarasskul, T. and J.M. Krochta. 2010. Edible packaging materials. *Annual Review of Food Science and Technology* 1: 415–448.

Jerez, A., P. Partal, I. Martínez, C. Gallegos, and A. Guerrero. 2007. Egg white-based bioplastics developed by thermomechanical processing. *Journal of Food Engineering* 82(4): 608–617.

Kaplan, S. 2004. Plasma processes for wide fabric, film and non-womens. *Surface and Coatings Technology* 186: 214–217.

Kester, J.J. and O.R. Fennema. 1986. Edible films and coatings: A review. *Food Science and Technology* 40(12): 47–59.

Kim, K.M., C.L. Weller, M.A. Hanna, and A. Gennadios. 2002. Heat curing of soy protein films at selected temperatures and pressures. *LWT: Food Science and Technology* 35(2): 140–145.

Kolasinska, M., R. Krastev, T. Gutberlet, and P. Warszynski. 2009. Layer-by-layer deposition of polyelectrolytes: Dipping versus spraying. *Langmuir* 25(2): 1224–1232.

Krishna, M., C.I. Nindo, and S.C. Min. 2012. Development of fish gelatin edible films using extrusion and compression molding. *Journal of Food Engineering* 108(2): 337–344.

Krochta, J.M. 1992. Control of mass transfer in food with edible-coatings and films. In: *Advances in Food Engineering*, eds. R.P. Singh and M.A. Wirakartakusumah, pp. 517–537. Boca Raton, FL: CRC Press.

Krochta, J.M. 2002. Proteins as raw materials for films and coatings: Definitions, current status and opportunities. In: *Protein-Based Films and Coatings*, ed. A. Gennadios, pp. 1–41. Boca Raton, FL: CRC Press.

Krochta, J.M., K.L. Dangaran, and S.-Y. Lin. 2005. Methods and formulations for providing gloss coatings to foods and for protecting nuts from rancidity. US Patent 20,050,191,390 A1. Filed: March 1, 2004. September 1, 2005.

Krogman, K.C., P.T. Hammon, and N.S. Zacharia. 2012. Automated layer by layer spray technology. US Patent 8,234,998 B2. Filed: September 5, 2007. August 7, 2012.

Lacroix, M. and K. Cooksey. 2005. Edible films and coatings from animal origin proteins. In: *Innovations in Food Packaging*, ed. J.H. Han, pp. 301–307. London, U.K.: Academic Press.

Lacroix, M. and C.L. Tien. 2005. Edible films and coatings from non-starch polysaccharides. In: *Innovations in Food Packaging*, ed. J.H. Han, pp. 338–355. London, U.K.: Academic Press.

Lagarón Cabello, J.M., M. Martínez Sanz, and A. López Rubio. 2013. Procedimiento de obtención de una película multicapa con alta barrera. Patent application ES 201,131,336. Filed: August 1, 2011. April 23, 2013.

Li, M., P. Liu, W. Zou et al. 2011a. Extrusion processing and characterization of edible starch films with different amylose contents. *Journal of Food Engineering* 106(1) (September): 95–101.

Li, X.Y., X.G. Chen, Z.W. Sun, H.J. Park, and D.-S. Cha. 2011b. Preparation of alginate/chitosan/carboxymethyl chitosan complex microcapsules and application in lactobacillus casei ATCC 393. *Carbohydrate Polymers* 83(4) (February 1): 1479–1485.

Lin, S.-Y. and J.M. Krochta. 2006. Fluidized-bed system for whey protein film coating of peanuts. *Journal of Food Process Engineering* 29(530): 532–546.

Liu, Z. 2005. Edible films and coatings from starches. In: *Innovations in Food Packaging*, ed. J.H. Han, pp. 318–337. London, U.K.: Academic Press.

Melo, C.D., P.S. Garcia, M.V.E. Grossmann, F. Yamashita, L.H. Dall'Antônia, and S. Mali. 2014. Properties of extruded xanthan-starch-clay nanocomposite films. *Brazilian Archives of Biology and Technology* 54(6): 1223–1333.

Micard, V., R. Belamri, M.H. Morel, and S. Guilbert. 2000. Properties of chemically and physically treated wheat gluten. *Journal of Agricultural and Food Chemistry* 48(7): 2948–2953.

Morillon, V., F. Debeaufort, G. Blond, M. Capelle, and A. Voilley. 2002. Factors affecting the moisture permeability of lipid-based edible films: A review. *Critical Reviews in Food Science and Nutrition* 42(1): 67–89.

Mu, C., J. Guo, X. Li, W. Lin, and D. Li. 2012. Preparation and properties of dialdehyde carboxymethyl cellulose crosslinked gelatin edible films. *Food Hydrocolloids* 27(1): 22–29.

Mujumdar, A.S. 2006. Impingement drying. In: *Handbook of Industrial Drying*, ed. A.S. Mujumdar, pp. 385–394. Boca Raton, FL: Taylor & Francis.

Myers, D. 1999. *Surfaces, Interfaces, and Colloids: Principles and Applications*. New York: John Wiley & Sons, Inc.

Nisperos-Carriedo, M.O. 1994. Edible coatings and films based on polysaccharides. In: *Edible Coatings and Films to Improve Food Quality*, eds. J.M. Krochta, E.A. Baldwin, and M.O Nisperos-Carriedo, pp. 305–335. Lancaster, PA: Technomic Publishing.

Nur Hanani, Z.A., E. Beatty, Y.H. Roos, M.A. Morris, and J.P. Kerry. 2012. Manufacture and characterization of gelatin films derived from beef, pork and fish sources using twin screw extrusion. *Journal of Food Engineering* 113(4): 606–614.

Ogale, A.A., P.L. Cunningham, P.L. Dawson, and J.C. Acton. 2000. Viscoelastic, thermal, and microstructural characterization of soy protein isolate films. *Journal of Food Science* 65: 672–679.

Pereira de Abreu, D.A., P. Paseiro Losada, J. Maroto, and J.M. Cruz. 2010. Evaluation of the effectiveness of a new active packaging film containing natural antioxidants (from barley husks) that retard lipid damage in frozen Atlantic salmon (*Salmo salar* L.). *Food Research International* 43(5): 1277–1282.

Pérez-Gago Maria, B. and J.M. Krochta. 2005. Emulsion and bi-layer edible films. In: *Innovations in Food Packaging*, ed. J.H. Han, pp. 384–402. New York: Elsevier.

Poirier, D. 2006. Conveyor dryers. In: *Handbook of Industrial Drying*, ed. A.S. Mujumdar, pp. 411–422. Boca Raton, FL: Taylor & Francis.

Poole, C.P. and F.J. Owens. 2007. *Introducción a la Nanotecnología*. Madrid, Spain: Editorial Reverté.

Ray, S.S. and M. Bousmina. 2005. Biodegradable polymers and their layered silicate nanocomposites: In greening the 21st century materials world. *Progress in Materials Science* 50: 962–1079.

Rhim, J.W., A. Gennadios, C.L. Weller, and M.A. Hanna. 2002. Sodium dodecyl sulfate treatment improves properties of cast films from soy protein isolate. *Industrial Crops and Products* 15: 199–205.

Rhim, J.W. and P.K.W. Ng. 2007. Natural biopolymer-based nanocomposite films for packaging applications. *Critical Reviews in Food Science and Nutrition* 47(4): 411–433.

Rhim, J.-W., H.-M. Park, and C.-S. Ha. 2013. Bio-nanocomposites for food packaging applications. *Progress in Polymer Science* 38: 1629–1652.

Rhim, J.W. and T.H. Shellhammer. 2005. Lipid-based edible films and coatings. In: *Innovations in Food Packaging*, ed. J.H. Han, pp. 362–380. San Diego, CA: Academic Press.

Rossman, J.M. 2009. Commercial manufacture of edible films. In: *Edible Films and Coatings for Food Applications*, eds. M.E. Embuscado and K.C. Huber, pp. 367–390. New York: Springer.

Rubner, M.F. 2003. pH-controlled fabrication of polyelectrolyte multilayers: Assembly and applications. In: *Multilayer Thin Films: Sequential Assembly of Nanocomposite Materials*, eds. G. Decher and J.B. Schlenoff, pp. 133–154. Weinheim, Germany: Wiley-VCH.

Sánchez-González, L., M. Vargas, C. Gonzalez-Martinez, A. Chiralt, and M. Cháfer. 2011. Use of essential oils in bioactive edible coatings. *Food Engineering Reviews* 3(1): 1–16.

Slavutsky, A.M. and M.A. Bertuzzi. 2012. A phenomenological and thermodynamic study of the water permeation process in corn starch/MMT films. *Carbohydrate Polymers* 90(1): 551–557.

Slavutsky, A.M. and M.A. Bertuzzi. 2014. Water barrier properties of starch films reinforced with cellulose nanocrystals obtained from sugarcane bagasse. *Carbohydrate Polymers* 110(September 22): 53–61.

Sothornvit, R. and J.M. Krochta. 2005. Plasticizers in edible films and coatings. In: *Innovations in Food Packaging*, ed. J.H. Han, pp. 403–428. New York: Elsevier.

Sothornvit, R., C.W. Olsen, T.H. McHugh, and J.M. Krochta. 2007. Tensile properties of compression-molded whey protein sheets: Determination of molding condition and glycerol-content effects and comparison with solution-cast films. *Journal of Food Engineering* 78(3) (February): 855–860.

Steward, P.A., J. Hearn, and M.C. Wilkinson. 2000. An overview of polymer latex film formation and properties. *Advances in Colloid and Interface Science* 86(3): 195–267.

Tunc, S., H. Angellier-Coussy, Y. Cahyanab, P. Chalier, N. Gontard, and E. Gastaldi. 2007. Functional properties of wheat gluten/montmorillonite nanocomposite films processed by casting. *Journal of Membrane Science* 289: 159–168.

Tunç, S. and O. Duman. 2010. Applied clay science preparation and characterization of biodegradable methyl cellulose/montmorillonite nanocomposite films. *Applied Clay Science* 48(3): 414–424.

Türe, H., M. Gällstedt, and M.S. Hedenqvist. 2012. Antimicrobial compression-moulded wheat gluten films containing potassium sorbate. *Food Research International* 45(1) (January): 109–115.

Verbeek, C.J.R. and L.E. van den Berg. 2010. Extrusion processing and properties of protein-based thermoplastics. *Macromolecular Materials and Engineering* 295(1): 10–21.

Weiss, J., P. Takhistov, and J.D. McClements. 2006. Functional materials in food nanotechnology. *Journal of Food Science* 71(9): 107–116.

Wicks, Z.W., F.N. Jones, P.S. Pappas, and D.A. Wicks. 2007. *Organic Coatings: Science and Technology.* Hoboken, NJ: Wiley.

Wu, Y., C.L. Weller, F. Hamouz et al. 2002. Development and application of multicomponent edible coatings and films: A review. *Advances in Food and Nutrition Research* 44: 347–394.

Xie, F., P. Luckman, J. Milne, L. McDonald, C. Young, C.Y. Tu, T.D. Pasquale, R. Faveere, and P.J. Halley. 2014. Thermoplastic starch: Current development and future trends. *Journal of Renewable Materials* 2(2): 95–106.

Zhang, J., P. Mungara, and J. Jane. 2001. Mechanical and thermal properties of extruded soy protein sheets. *Polymer* 42: 2569–2578.

2

Transport Phenomena in Films and Coatings
Including Their Mathematical Modeling

M. Alejandra Garcia and Noemí E. Zaritzky

CONTENTS

Abstract

In the present chapter, two examples related to transport phenomena in films and coatings are discussed.

One of them represents the heat and mass transfer process in fried foods that were covered with an edible coating based on methylcellulose (MC). This is an alternative to reduce oil uptake (OU) in fried foods due to its lipid-barrier properties. The following aspects are discussed: (1) mathematical modeling of heat and moisture transfer during the deep-fat frying of food, (2) experimental validation of the mathematical model with regard to the temperature profiles and the water losses from the food product, (3) analysis of the relationship between the OU measurements and microstructural changes developed, and (4) performance of applying an edible coating based on MC on a food model dough system.

The mathematical model of the frying process based on the numerical solution of the heat and mass transfer differential equations under unsteady-state conditions was proposed and solved. It allows simulating satisfactorily the experimental data of temperature and water content during the different frying stages. OU was also linearly correlated with water loss at the initial frying stage. A simple equation for OU as a function of frying times was proposed, considering the microstructural changes developed during the frying process. The presence of MC coating reduced the OU, modifying the wetting properties and also becoming a mechanical barrier to the oil.

The second example represents the mathematical modeling of potassium sorbate release from a starch biodegradable active film to a model food system represented by a gel in contact with the active film. Mass transfer partial differential equations in nonstationary conditions were numerically solved using the finite element method. The model assumes a constant initial mass of antimicrobial in the active film that diffuses through the film penetrating in the food system.

The numerical solution allowed the determination of the diffusion coefficients of the antimicrobial agent in both, the film and the gel. Concentration profiles were simulated to predict the time period in which the antimicrobial concentration can be maintained above the critical inhibitory concentration in the packaged food.

Experimental data of sorbate diffusion from active films and from a liquid solution to the semisolid medium were compared with the predicted concentration profiles. The model allows the simulation of nonstationary diffusion of different additives incorporated to polymeric matrixes, taking into account the preservative concentration in the film and the dimensions of the semisolid food system.

2.1 Introduction

The food industry is focusing its efforts toward offering to the consumers safer, more nutritious, and high-quality products. The use of edible films and coatings in food protection and preservation has recently increased since they offer several advantages over synthetic materials, such as being biodegradable and environmentally friendly (Tharanathan, 2003). Biopolymer-based materials are well recognized as efficient barriers against O_2 and CO_2, although they exhibited poor water vapor barrier properties (Lacroix, 2009). Likewise, the ability of edible films to retard moisture, oxygen, aromas, and solute transport may be improved by including additives such as antioxidants, antimicrobials, colorants, and flavors, fortifying nutrients in film formulation (Pranoto et al., 2005).

Besides, polysaccharide-based films have been described as oil barriers. The application of hydrocolloid coatings allows the reduction of oil content in deep-fat fried products due to their lipid-barrier properties. The most widely studied are cellulose derivatives, which exhibit thermogelation (Bertolini-Suárez et al., 2008).

In addition, studies dealing with edible films with antimicrobial properties are enhanced in the last years. These films could prolong the shelf life and ensure the safety of foods by preventing the growth of pathogenic and spoilage microorganisms as a result of their lag-phase extension and/or their growth rate reduction (Quintavalla and Vicini, 2002).

Hydrophilic matrices have been used as controlled delivery systems in the last decades, due to their low cost and the simple technology associated. This technology has been widely studied in the pharmacological area. Although numerous investigations on drug release from hydrophilic matrices have been developed, the mechanisms that drive the mass transfer process is still a matter of debate. The incorporation of active agents into polymeric systems results in a variety of release profiles with different stages. In some cases, the active compound release has been described as a simple matrix diffusion process, with degradation occurring at a later stage, post–active compound release (Gallagher and Corrigan, 2000).

In the present chapter, two examples related to transport phenomena in films and coatings are discussed.

One of them corresponds to the heat and mass transfer process in fried foods that were covered with an edible coating. The second example represents the mathematical modeling of potassium sorbate (PS) release from a starch biodegradable active film to a model food system represented by a gel in contact with the active film.

2.2 Edible Coatings to Control Oil Uptake in Fried Foods

The great volume of production of fried foods and its influence on the consumption of lipids enhanced the study of the frying process because of its strong economic and nutritional impact. From the economical point of view, higher oil contents increase production costs. The reduction of lipid content in fried foods is required mainly owing to its relation with obesity and coronary diseases. An alternative to reduce oil uptake (OU) in fried foods is the use of edible films or coatings.

The application of hydrocolloid coatings allows to reduce oil content of deep-fat fried products due to its lipid-barrier properties; the most widely studied are gellan and cellulose derivatives (Williams and Mittal, 1999a,b; Farid, 2002; Mellema, 2003). Cellulose derivatives, including methylcellulose (MC) and hydroxypropyl methylcellulose, exhibit thermogelation, when suspensions are heated, they form a gel that reverts below the gelation temperature, and the original suspension viscosity is recovered (Farid and Chen, 1998). These cellulose derivatives reduce oil absorption through film formation at temperatures above their gelation point, or they reinforce the natural barrier properties of starch and proteins, especially when they are added in dry form (Balasubramaniam et al., 1995).

With regard to the modeling of deep-fat frying of coated foods, Mallikarjunan et al. (1997) and Huse et al. (1998) demonstrated the effectiveness of various edible coatings in reducing oil absorption in starchy products. Williams and Mittal (1999a,b) working on frying of foods coated with gellan gum determined the effectiveness of edible films to reduce fat absorption in cereal products and developed a mathematical model for the process.

2.2.1 Case Study: Heat and Mass Transfer in Frying Processes

Frying process is considered one of the oldest cooking methods. It is employed to treat foods thermally and to confer unique textures and tastes; the organoleptic characteristics depend on frying conditions. Frying phenomena occur during the immersion of the product in oil at a temperature of 150°C–200°C, where a simultaneous heat and mass transfer takes place (Singh, 1995; Aguilera and Hernández, 2000).

Heat conduction occurs in the core of the solid food, and it is strongly influenced by the physical properties of the food that are continuously changing during the frying process (Campañone et al., 2010). Simultaneously, convection takes place between the oil and the surface of the food. When the formation of bubbles begins, the heat transfer is accelerated because bubbles contribute to the turbulence of the frying medium; however, the formation of foam in the oil produces a significant decrease in the heat transfer rate (Singh, 1995). Then, the amount of vapor decreases due to the reduction of water content (WC) in the core of the food. The rate of heat transfer toward the food core is influenced by the thermal properties and viscosity of the frying medium and the agitation conditions. During the production of bubbles on the surface, forced convection is the controlling regimen, and the determination of heat transfer coefficients becomes difficult, mainly if bubbles remain attached to the surface.

Water transfer is produced by surface boiling and when evaporation decreases; diffusion of water from the food core toward the surface is the dominant physical phenomenon. Water leaves the product as bubbles of vapor and internally migrates by means of different mechanisms (Singh, 1995; Aguilera, 1997; Aguilera and Hernández, 2000). After an initial time, when surface moisture is evaporated, a dehydrated zone begins to form on the surface of the food. Under these conditions, surface temperature is closer to that of the frying medium, while moisture is strongly reduced reaching values close to the bound water value. Thus, an important gradient of moisture between the core and the surface is established.

The mechanism of oil penetration is a subject of controversy (Pinthus and Saguy, 1994; Kassama, 2003; Mellema, 2003). Absorption of oil on the surface of the fried product occurs when samples are removed from the frying medium; the oil that remains on the piece surface enters the product (Pinthus and Saguy, 1994; Ufheil and Escher, 1996; Moreira and Barrufet, 1998; Aguilera and Hernández, 2000; Yamsaengsung and Moreira, 2002a,b; Bouchon et al., 2003; Mellema, 2003; Bouchon and Pyle, 2005a,b; Campañone et al., 2010). According to these authors, oil does not invade the product itself, and OU during frying is negligible. Conditions at which products are removed from the frying oil seem decisive for the uptake of oil; this would be related to the adhesion of oil to the surface and draining phenomena.

Several models have been developed for heat, moisture, and fat transfer during frying of foods (Ni and Datta, 1999; Yamsaengsung and Moreira, 2002a,b). Ateba and Mittal (1994) developed a model for heat, moisture, and fat transfer in deep-fat frying of beef meatballs. Rice and Gamble (1989) used a simplification of Fick's equation to predict moisture loss and the oil absorption during potato frying. Farkas et al. (1996a,b) developed a model for frying considering heat and moisture transfer, and Singh (2000) treated the crust as a moving boundary.

In the present chapter, the following aspects with reference to the use of edible films to control OU are discussed: (1) the mathematical modeling of the heat and moisture transfer during the deep-fat frying process of food systems as a function of operating conditions, (2) the relationship between OU measurements after the frying process with the effect of frying time on sample microstructural changes and the growth of the dehydrated zone, and (3) the performance of applying an edible coating based on MC on a fried-food model dough system by analyzing moisture content, OU, surface microstructure, and quality attributes (texture and color).

2.2.2 Mathematical Modeling of Heat and Mass Transfer during Frying Process

A mathematical model of the frying process was proposed based on the numerical solution of heat and mass transfer partial differential equations under unsteady-state conditions. The frying process was modeled for food discs (Bertolini-Suárez et al., 2008; Campañone et al., 2010). Discs were considered as infinite slabs and unidirectional; heat and mass transfer equations were solved for $0 \leq x \leq L$ (L=half the thickness of the sample). In the proposed model, different stages were considered.

Stage 1: Coupled equations represent the initial heating of the product in which energy and mass transfer between the product and the frying oil occurs. The model considered an initial uniform temperature distribution in the sample (T_{ini}).

To get the temperature profiles, the following microscopic energy balance was solved:

$$\rho c_p \frac{\partial T}{\partial t} = \nabla(k \nabla T) \tag{2.1}$$

Thermal properties corresponded initially to the nonfried product and changed with temperature and moisture content.

The initial and boundary conditions were

$$t = 0 \quad T = T_{ini} \quad 0 \leq x \leq L \tag{2.2}$$

$$x = 0 \quad \frac{\partial T}{\partial x} = 0 \quad t > 0 \tag{2.3}$$

$$x = L \quad -k\frac{\partial T}{\partial x} = h_1(T - T_{oil}) + L_{vap}\, m_{vap,1} \quad t > 0 \tag{2.4}$$

where

$m_{vap,1}$ is the water flux from the product
h_1 is the convective heat transfer coefficient

The heat transfer mechanism was considered governed by natural convection. The value of the heat transfer coefficient (h) for natural convection was calculated for horizontal plates using two different equations:

$$\text{Nu} = \frac{0.108 \text{Gr}^{4/11}\, \text{Pr}^{9/33}}{(1 + 0.44\, \text{Pr}^{2/3})^{4/11}}, \text{(Mallikarjunan et al., 1997)} \tag{2.5}$$

$$\text{Nu} = 0.55\, (\text{Gr}\cdot\text{Pr})^{1/4}, \text{(Huse et al., 1998)} \tag{2.6}$$

where

Nu is the Nusselt number
Gr is the Grashof number
Pr is the Prandtl number

$$\text{Gr} = \frac{L^3 \rho_{oil}^2 g \beta \Delta T}{\mu^2}; \quad \text{Pr} = \frac{Cp\mu}{k}$$

Table 2.1 shows the physical properties of raw and fried dough system fed to the mathematical model and the properties of the used oil.

The obtained average value of the natural convection heat transfer coefficient for horizontal plates was $h_1 = 50$ W/m^2 K.

TABLE 2.1

Physical Properties of Sunflower Oil and the Dough System (Core and Dehydrated Zones) Used in the Model

Property	Core Zone	Dehydrated Zone	Sunflower Oil
Initial water content	0.42	—	—
Density (kg/m^3)	623[b]	579[a]	876[c]
Thermal conductivity (W/[m K])	0.6[b]	0.05[a]	0.6[c]
Specific heat (J/kg °C)	2800[b]	2310[a]	2033[c]
Viscosity (Pa·s)			2.04×10^{-3c}
Porosity	—	0.42[b]	—
Tortuosity	—	8.8[b]	—
Diffusion coefficient of vapor in air (dehydrated zone) (m^2/s)		$\dfrac{5.68 \times 10^{-9}(T + 273.16)^{1.5}}{1.789 - 2.13 \times 10^{-3}(T + 273.16)}$[b]	
Moisture diffusion coefficient in dough (core zone) (m^2/s)		2×10^{-9b}	

[a] Rayner et al. (2000).
[b] Campañone et al. (2010).
[c] Moyano and Pedreschi (2006).

In order to evaluate water profiles, the microscopic mass balance was solved:

$$\frac{\partial C_w}{\partial t} = \nabla(D_w \nabla C_w) \tag{2.7}$$

The following are the initial and boundary conditions:

$$t = 0 \quad C_w = C_{w,ini} \quad 0 \le x \le L \tag{2.8}$$

$$x = 0 \quad \frac{\partial C_w}{\partial x} = 0 \quad t > 0 \tag{2.9}$$

$$x = L \quad -D_w \frac{\partial C_w}{\partial x} = m_{vap,1} \quad t > 0 \tag{2.10}$$

$m_{vap,1}$ is calculated from Equation 2.10 with $C_w = 0$ at the sample/oil interface, because it was considered that water evaporates and immediately leaves the frying medium.

For each time interval, the model calculated the average value of moisture content as follows:

$$C_{w,ave} = \frac{\int_0^L C_w \, dV}{V} \tag{2.11}$$

The second stage starts when the surface temperature of the product reaches that of the water vaporization.

Stage 2: Water boiling is produced on the surface, and a dehydrated layer is formed in the external zone of the product, having different thermophysical and transport properties from those of the wet core (Table 2.1). Several authors (Farkas et al., 1996a,b; Ni and Datta, 1999; Bouchon et al., 2003; Bouchon and Pyle, 2005b) used the denomination of crust for the dehydrated zone. In the present work, we considered that the crust is only the portion of the dehydrated zone containing oil.

When water on the surface is no longer available, vaporization occurs inside the product forming a front that moves toward the inner zone. During this phase of the frying process, the vapor being released from the product surface impedes oil penetration into the food (Singh, 1995; Aguilera and Hernández, 2000; Kassama, 2003). Microscopic-energy balances given by Equation 2.1 were solved for both (core and dehydrated) zones to obtain temperature profiles, considering the corresponding thermal properties.

The initial temperature profile in this stage was given by the final profile obtained at the end of Stage 1. The following boundary conditions were applied:

$$x = 0 \quad \frac{\partial T}{\partial x} = 0 \quad t > 0 \tag{2.12}$$

$$x = x_1 \quad T = T_{vap} \quad t > 0 \tag{2.13}$$

$$x = L \quad -k_d \frac{\partial T}{\partial x} = h_2(T - T_{oil}) \quad t > 0 \tag{2.14}$$

where x_1 is the variable position of the vaporization moving front (measured from the center of the slab). Thus, the thickness of the dehydrated zone (d_Z) measured from the sample/oil interface can be calculated as $d_Z = L - x_1$ (Figure 2.1).

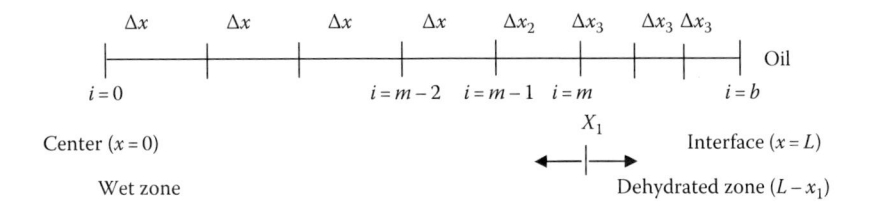

FIGURE 2.1   Scheme of the variable grid employed in the numerical solution (Stage 2).

Values of the heat transfer coefficient (h_2) higher than 250 W/m² K were found in literature for the whole frying process. Costa and coworkers (1998) reported a value of $h_2 = 650 \pm 7$ W/m² K for potato chips fried at 180°C; Dagerskog and Sorenfor (1978) reported values of h_2 ranging between 300 and 700 W/m² K for meat hamburgers fried at 190°C; Moreira et al. (1995) obtained values of 285 and 273 W/m² K for fresh and used soy oil, respectively. These high-heat transfer coefficients allowed us to assume a constant temperature at the product–oil interface. Moisture content decreased due to water vaporization; vapor diffused through the dehydrated zone, which was considered as a porous medium.

The following mass balance was solved in the dehydrated zone:

$$m_{vap,2} = -D_v \frac{\partial C_v}{\partial x} \tag{2.15}$$

D_v was calculated through the following relationship:

$$D_v = D_{va} \frac{\varepsilon}{\tau} \tag{2.16}$$

where
  ε is the porosity of the product
  τ is the tortuosity

The following relationships were considered:

$$x = x_1 \quad C_v = C_{v,equi} \tag{2.17}$$

$$x = L \quad C_v = 0 \tag{2.18}$$

where $C_{v,equi}$ is the equilibrium vapor concentration at T_{vap}; it was calculated using the Clausius–Clapeyron equation, assuming ideal gas behavior for the water vapor.

Water vapor concentration at the product–oil interface was assumed negligible according to Equation 2.18. To calculate the position of the vaporization front (x_1) as a function of time, the following equation was proposed:

$$-C_{w,ave} \frac{dx_1}{dt} = -D_v \frac{\partial C_v}{\partial x} \tag{2.19}$$

where $C_{w,ave}$ is the average value of WC in the core calculated at the end of Stage 1, because it represents the volumetric water distribution inside the food.

Equation 2.19 considers that the amount of water available for vaporization in the core is the same as the humidity travelling in the dehydrated zone; then,

$$\frac{dx_1}{dt} = \frac{D_v}{C_{w,ave}} \frac{\partial C_v}{\partial x} \tag{2.20}$$

When the vaporization front reaches the center of the fried product, the dehydrated food system increases its temperature until the process is completed. Water transfer is absent, and the moisture content corresponds to the bound water. To estimate temperature profiles, the microscopic energy balance (Equation 2.1) with the boundary conditions (2.3) and (2.14) was solved.

Stage 3: The final step corresponds to the cooling of the fried product when it is removed from the hot-oil medium. The energy microscopic balance (Equation 2.1) was solved with the boundary condition given by Equation 2.3, and considering a natural convection heat transfer coefficient at the food–air interface, as follows:

$$x = L \quad -k_d \frac{\partial T}{\partial x} = h_3(T - T_a) \quad t > 0 \tag{2.21}$$

h_3 can be calculated using the following equation:

$$h = 1.32 \left(\frac{\Delta T}{H} \right)^{0.25} \quad \text{for plates with } H = \text{diameter} \left[35 \right] \tag{2.22}$$

with ΔT as the positive temperature difference between the surface of the fried food and the surrounding air. In this step, the OU is produced.

The mathematical model was solved for the different stages. The microscopic energy and water vapor mass balances led to a system of coupled nonlinear partial differential equations.

2.2.3 Numerical Solution

The mathematical model was solved for the different frying stages. Details of the numerical solution based on the finite differences method were reported in Bertolini-Suárez et al. (2008).

The domain has been discretized (Figure 2.1), where 0 indicates the food center, m the moving front (the point m moves with time), and b the surface. The set of equations allowed to calculate temperature profiles as a function of frying times. The number of nodes in each zone changed as long as the vaporization front progressed. At each time step, the position of the moving front was calculated.

2.2.4 Validation of the Mathematical Model: Frying Process of a Coated Dough Disc

2.2.4.1 Description of the Food System and Selection of the Frying Conditions

The food system was a wheat flour dough; discs 60 mm in diameter and 7 mm thick were cut and immediately utilized. For each experiment, 6 discs were fried. The mathematical model was also validated on dough discs that are 5 mm thick (Bertolini-Suárez et al., 2008).

Samples were fried in a controlled temperature deep-fat fryer containing commercial sunflower oil. Different constant frying temperatures were tested to select the working frying conditions according to sample characteristics; these temperatures ranged between 150°C and 170°C. Frying times between 5 and 15 min were analyzed, and an initial temperature of 20°C was considered. Optimum time–temperature frying conditions were determined by a nontrained sensory panel of 6 members; panelists judged color, flavor, texture, and overall appearance of the samples as described in a previous work (García et al., 2002).

Frying conditions determine sensory characteristics and consumer acceptability of fried products. Sensory characterization was performed to select time–temperature conditions. Sensory analysis (color, flavor, texture, and overall appearance) determined that 12 min at 160°C ± 0.5°C were the best frying conditions for dough discs. Besides, this was the time required to complete starch gelatinization in dough discs (confirmed by microscopy observation), and consequently, fried samples could be considered cooked. Accordingly, this temperature was selected for further determinations and the selecting frying times ranged between 3 and 15 min.

2.2.4.2 Application of a Hydrocolloid Coating to Reduce Oil Uptake during Frying

In order to analyze the effect of the hydrocolloid coating to reduce OU in fried products, coating formulations were prepared using 1% (w/w) MC (A4M, Methocel) aqueous solution and 0.75% (w/w) sorbitol (Merck, United States) as a plasticizer. Dough samples were dipped in the coating suspensions for 30 s and immediately fried at the selected conditions, as previously described (Bertolini-Suárez et al., 2008).

2.2.4.3 Heat Transfer Validation: Thermal Histories of Coated and Uncoated Samples during the Frying Process

Sample temperatures of coated and uncoated samples as a function of frying time were measured using copper–constantan thermocouples (Omega). Temperature measurements in the center of the dough were performed by introducing the thermocouple carefully through the disc from the border toward the center; besides, the temperatures of an internal sample point and the temperature of the frying medium were also recorded.

Temperatures were measured and recorded each 5 s using type T thermocouples linked to a data acquisition and control system.

Figure 2.2 shows the measured values (symbols) at the center and in an inner point (1.6 mm from the center). After 120 s, a constant temperature of 100.5°C was reached at the core. The core temperature remained constant due to water evaporation, while the temperature in the dehydrated zone increased up to the end of the process (15 min). Similar results on thermal histories were reported by Costa et al. (1998), working on sliced potatoes fried in sunflower oil at 140°C–180°C.

A cooling period could be observed when the product was removed from the fryer. In the same figure, simulated temperatures using the proposed model of samples 5 or 7 mm thick are shown (lines), and a good agreement with the experimental thermal histories is observed.

Experimental thermal histories of coated and uncoated samples did not differ significantly ($p > 0.05$). These results can be explained taking into account that the average coating thickness determined by scanning electron microscopy (SEM) was approximately 10 μm (García et al., 2002, 2004) leading to a negligible heat transfer resistance.

2.2.4.4 Water Transfer

WC was determined measuring weight loss of fried products, upon drying in an oven at 110°C until constant weight (García et al., 2002). At different frying times, the relative variation of water retention % (WR) in the coated product relative to the uncoated one was calculated as follows:

$$\mathrm{WR} = \left(\frac{\mathrm{WC\ coated}}{\mathrm{WC\ uncoated}} - 1 \right) \times 100 \qquad (2.23)$$

where WC is the water content of the samples (dry basis). For each frying time condition, results were obtained using all the samples from at least two different batches. The equilibrium WC was defined as the humidity reached at long frying times (1080 s).

Experimental values of WC for coated and uncoated samples are shown in Table 2.2; nonsignificant differences ($p > 0.05$) were detected between coated and uncoated discs. This could be attributed to the poor water vapor barrier of MC films (Donhowe and Fennema, 1993; Debeaufort and Voilley, 1997; Krochta and De Mulder-Johnston, 1997). The equilibrium WC values, which correspond to bound water, were 0.157 and 0.1695 g water/g dry solid for uncoated and coated samples, respectively.

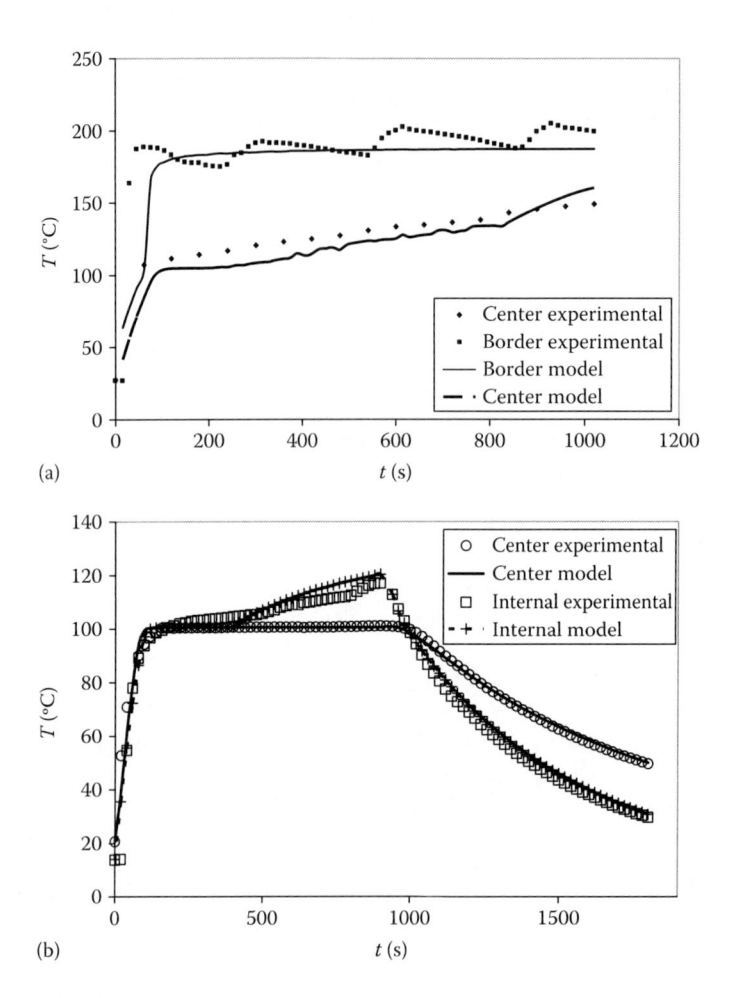

FIGURE 2.2 Thermal histories of the fried product at the center and in an internal point (1.6 mm from the border) along the different frying stages. Experimental data (symbols) and numerical simulations (lines). Sample thickness: (a) 5 mm and (b) 7 mm.

TABLE 2.2

Lipid and Water Contents of Uncoated and Coated Dough Discs as a Function of Frying Time

Time (s)	OU of Uncoated Samples (g Oil/g Dry Solid)	OU of Coated Samples (g Oil/g Dry Solid)	WC of Uncoated Samples (g Water/g Dry Solid)	WC of Coated Samples (g Water/g Dry Solid)	WR, Water Retention Relative Variation (%)	OUR (%)
0	0.00	0.00	0.721 ± 0.043[a]	0.786 ± 0.059	9.09	—
180	0.053 ± 0.010	0.045 ± 0.010	0.492 ± 0.017	0.477 ± 0.013	−3.06	14.07
360	0.058 ± 0.016	0.049 ± 0.001	0.356 ± 0.013	0.352 ± 0.012	−1.05	15.36
540	0.073 ± 0.008	0.054 ± 0.001	0.300 ± 0.013	0.296 ± 0.007	−1.34	26.36
720	0.074 ± 0.016	0.053 ± 0.006	0.253 ± 0.037	0.200 ± 0.006	−21.12	29.07
900	0.082 ± 0.017	0.062 ± 0.013	0.142 ± 0.016	0.159 ± 0.016	11.43	24.62
1080	0.089 ± 0.005	0.063 ± 0.005	0.157 ± 0.005	0.170 ± 0.004	8.03	29.91

[a] Value ± standard deviation.

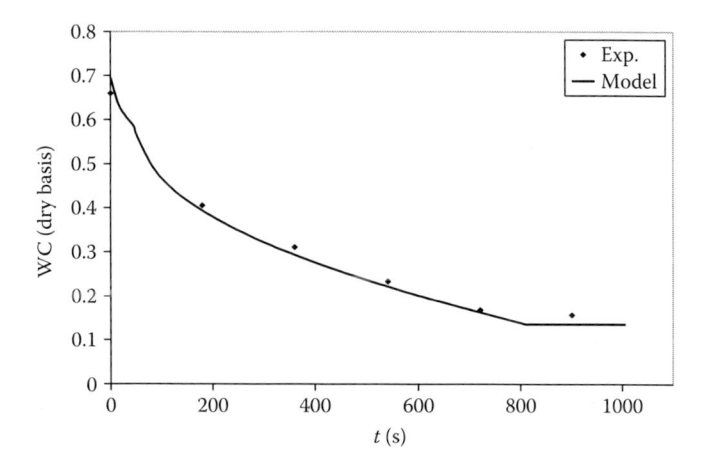

FIGURE 2.3 Water content along the different frying stages. Experimental data (symbols) and numerical simulations (lines).

The presence of the coating did not change the moisture content of the samples during frying because the hydrophilic MC coating became a negligible barrier to water transfer.

Simulated and experimental water concentrations versus frying times are shown in Figure 2.3 for samples that are 5 or 7 mm thick, and a good agreement was achieved.

Figure 2.4a shows the predicted position of the vaporization front as a function of frying time. At 15 min of process, the vaporization front reached the center of the product, and the humidity corresponded to the bound water. WC correlated linearly with the vaporization position x_1 that corresponds to the thickness of the humidity core (Figure 2.4b). Considering $d_Z = L - x_1$, the following equation was obtained:

$$WC = 0.14893 + 158.4(L - d_Z) \quad r^2 = 0.9966 \tag{2.24}$$

2.2.4.5 Temperature and Water Content Predictions Using the Mathematical Model

Once the mathematical model was validated, it was used to predict temperature and WC profiles as shown in Figure 2.5a and b at different frying times.

As can be observed, a sample temperature of the dehydrated zone increased during the frying process reaching values higher than the equilibrium water vaporization temperature.

2.2.5 Oil Uptake of Coated and Uncoated Products

OU of fried products was determined measuring the lipid content of dried samples using a combined technique of successive batch and semicontinuous Soxhlet extractions. The first batch extraction was performed with petroleum ether/ethylic ether (1:1) followed by a Soxhlet extraction with the same mixture and another with n-hexane. Oil uptake relative variation % (OUR) in the coated product relative to the uncoated one was calculated as follows:

$$OUR = \left(1 - \frac{OU \text{ coated}}{OU \text{ uncoated}}\right) \times 100 \tag{2.25}$$

For each frying time condition, results were obtained using all the samples from at least two different batches.

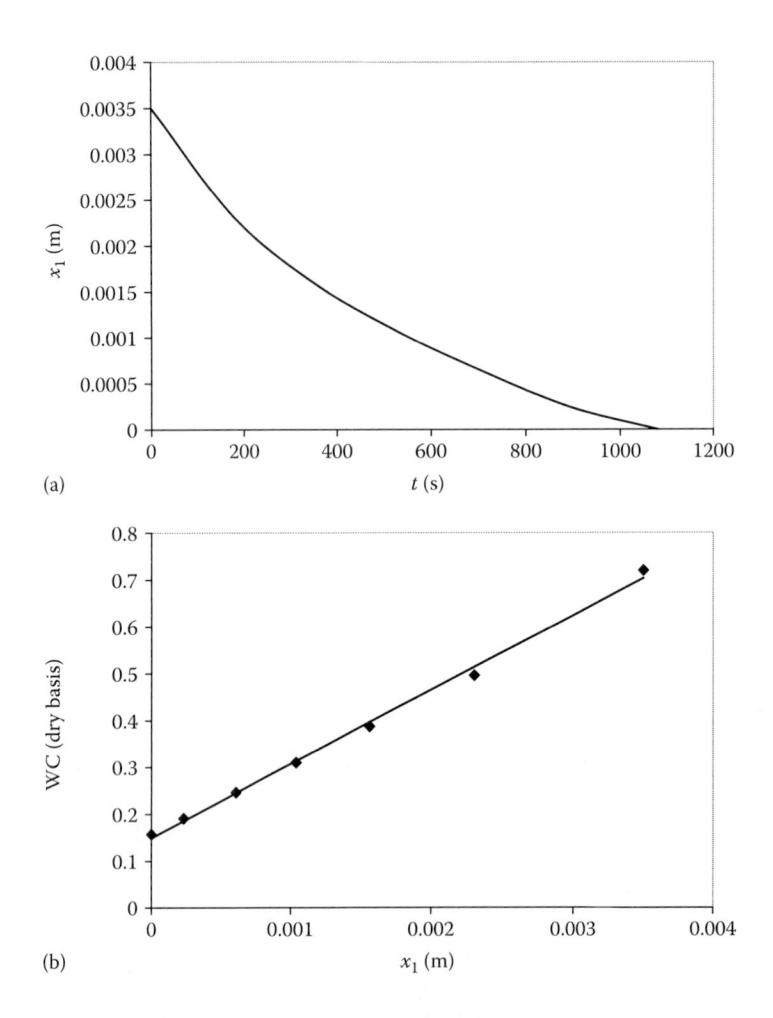

(a)

(b)

FIGURE 2.4 (a) Position of the vaporization front as a function of frying time and (b) water content of fried dough discs as a function of the position of the vaporization front.

The OU values of coated dough discs at each frying time are shown in Table 2.2. The final lipid content, at long frying times, was 0.0894 g oil/g dry solid for uncoated samples. As can be observed, MC coating reduced the lipid content of dough discs significantly ($p < 0.05$); the OU was 30% lower than that corresponding to uncoated samples (Table 2.2). Similarly, Williams and Mittal (1999a,b) reported that MC coatings reduced the lipid content of potato spheres by 34.5% compared to the control samples. This result was attributed to the presence of hydrocolloid films acting as lipid barriers, particularly MC due to its thermal gelation properties. The final lipid content values, at long frying times, for coated samples was 0.0626 g oil/g dry solid; thus, the ratio between the lipid content (db) of uncoated and coated samples was 1.4 at 1080 s, verifying that the MC coating acted as an effective oil barrier.

Many factors affect OU in deep fat frying, such as oil quality, frying temperature, residence time, product shape and size, product composition (initial moisture and protein content), pore structure (porosity and pore size distribution), and prefrying treatments (drying, blanching, surface coating). Interfacial tension was also reported by Pinthus and Saguy (1994) to have a significant influence on OU after deep-fat frying. The mechanism of oil penetration is a subject of controversy (Kassama, 2003). Many researchers have suggested that fat absorption is primarily a postfrying phenomenon (Perkins and Erikson, 1996; Williams and Mittal, 1999a; Yamsaengsung and Moreira, 2002; Bouchon et al., 2003); thus, fat transfer occurs in food product during the cooling stage. Moreira and Barrufet (1998) reported that 80% of the total oil content was absorbed in tortilla chips during the cooling period.

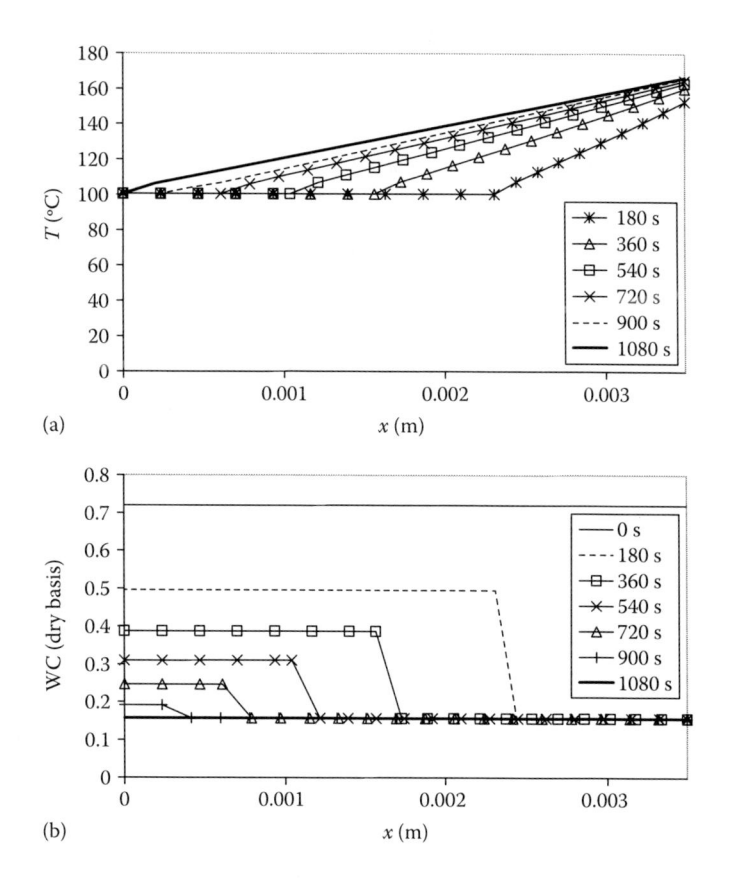

FIGURE 2.5 Profiles of (a) temperature and (b) water content of fried dough disc predicted by the proposed mathematical model.

They explained that as the product temperature decreases, the interfacial tension between gas and oil increases, raising capillary pressure. This sucks the surface oil into the porous medium, thus increasing the final oil content.

Ufheil and Escher (1996) also suggested that the absorption theory was based on surface phenomena, which involve the equilibrium between adhesion and the drainage of oil during cooling. Fat absorption in chips due to surface adherence was the result of steam condensation, as was also suggested by Rice and Gamble (1989). OU depends on structural changes during the process; differences in the starting food microstructure can be expected to be important determinants in the evolution of the characteristics of the end product. Meat products are mainly an arrangement of protein fibers, which may relax and denature upon heating (Aguilera and Stanley, 1999), while flour-based products, such as doughnuts and tortilla, have a different microstructure.

2.2.6 Quality Attributes of Coated and Uncoated Fried Samples

To evaluate whether MC coating application affected the quality attributes, color and texture parameters of coated and uncoated fried dough samples were analyzed.

Colorimetric measurements: The CIELab scale was used, lightness (L^*) and chromaticity parameters a^* (red–green) and b^* (yellow–blue) were measured. L^*, C^* (chroma), H° (hue), and color differences (ΔE) were also calculated as

$$\Delta E = \sqrt{(\Delta L^*)^2 + (\Delta a^*)^2 + (\Delta b^*)^2} \tag{2.26}$$

where

$$\Delta L^* = L_0^* - L_t^*, \quad \Delta a^* = a_0^* - a_t^*, \quad \Delta b^* = b_0^* - b_t^* \tag{2.27}$$

being L_t^*, a_t^*, and b_t^* as the color parameter values of samples fried at different frying times. L_0^*, a_0^*, and b_0^* were selected as the color parameters of samples fried for 3 min at 160°C and not the color of the raw sample in order to analyze the effect of frying time. Samples were analyzed in triplicates, recording four measurements for each sample.

Nonsignificant ($p > 0.05$) differences were observed in the analyzed color parameters of coated and uncoated dough discs. Chromaticity parameter b^* increased significantly ($p < 0.05$) with frying time, while lightness L^* and chromaticity parameter a^* were independent of frying time. Besides, chroma parameter and color differences (ΔE) increased and hue decreased significantly ($p < 0.05$) as a function of frying time for $t < 720$ s (Figure 2.6a). During overcooking of the samples, all parameters remained constant.

Texture analysis: The breaking force of samples was measured by a puncture test using a texture analyzer. Samples were punctured with a cylindrical plunger (2 mm diameter) at 0.5 mm/s. Maximum force at rupture was determined from the force-deformation curves. At least 10 samples

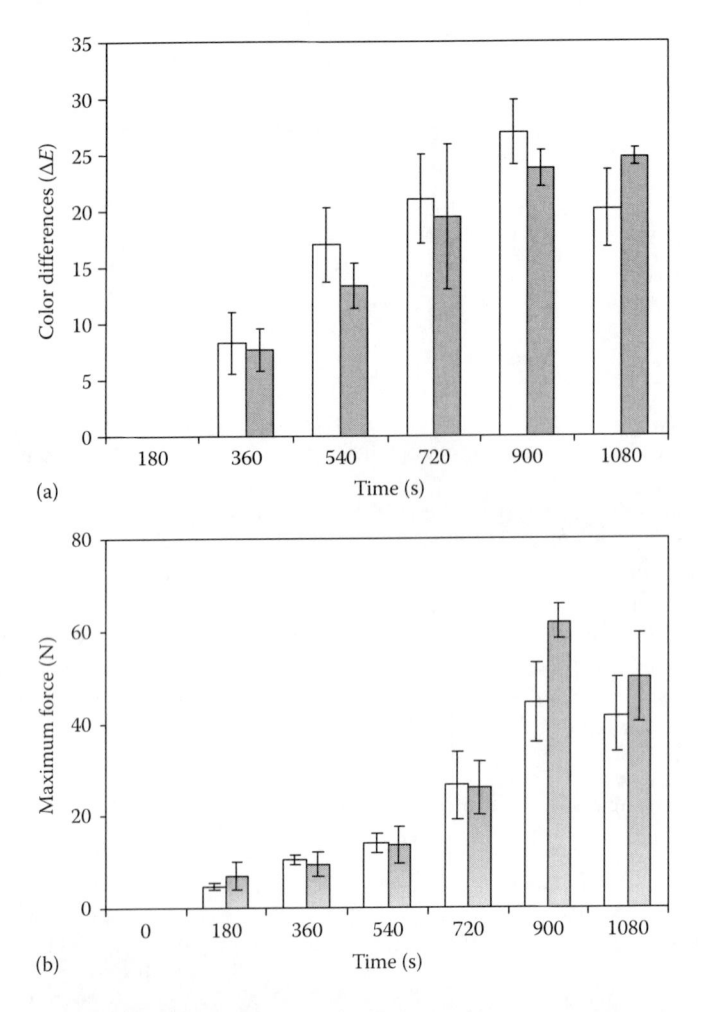

FIGURE 2.6 Quality attributes of coated (filled bars) and uncoated (empty bars) fried dough discs as a function of frying time: (a) color differences and (b) firmness.

were measured for each assay. Samples were allowed to reach room temperature before performing the tests. During the frying process (for $t < 720$ s), similar values of maximum force were obtained for coated and uncoated fried samples. In all cases, the maximum force to puncture the samples increased as a function of frying time, due to the formation of a dehydrated zone (Figure 2.6b). At longer frying times when discs were considered overcooked, the highest maximum forces for coated samples were observed.

With regard to the quality attributes during the cooking process, (for $t < 720$ s) differences between the color parameters and firmness values of coated and uncoated samples were not significant; besides, the panelists could not distinguish between control and coated samples.

2.2.7 Microstructure Analysis of the Fried Product

SEM and ESEM techniques were used to observe the structure of the coated and uncoated fried samples (Bertolini-Suárez et al., 2008; Campañone et al., 2010).

The effects of frying time and coating application on the structure were studied. Figure 2.7 shows that MC coating was getting dehydrated during the frying process and remained attached to the surface of the product, explaining the lower lipid content of the coated product. The thickness of the coating was measured on the micrographs obtaining values ranging between 9 and 24 μm after 12 min of frying.

Figure 2.7 also shows the integrity of the MC layer and the good adhesion of this coating to the fried product. The addition of a plasticizer (sorbitol) to MC coatings was necessary to achieve coating integrity (García et al., 2002). Formation of a uniform coating on the surface of the sample is essential to limit mass transfer during frying (Huse et al., 1998). Sorbitol addition improved barrier properties of coatings by decreasing oil content compared to coated samples without a plasticizer. Similarly, Rayner et al. (2000) reported that the performance of a soy protein film applied to dough discs increased by the addition of glycerol as a plasticizer, reducing the fat uptake by the food.

The coating did not prevent the formation of a dehydrated zone on the surface of the dough. The release of water vapor led to the formation of blisters at the outer surface of the crust being smaller in size and number in the coated dough than in the uncoated one. In the first period of frying, the MC coating loses its water, the dough keeps its humidity, and the starch gelatinizes with higher WC than in the uncoated dough, leading to a more compact network (Figure 2.8).

During frying in both coated and uncoated systems, the core was progressively dehydrated, and starch gelatinization was completed at 12 min. The dehydrated zone showed holes due to the water vapor release; the size of the holes increased along the frying process.

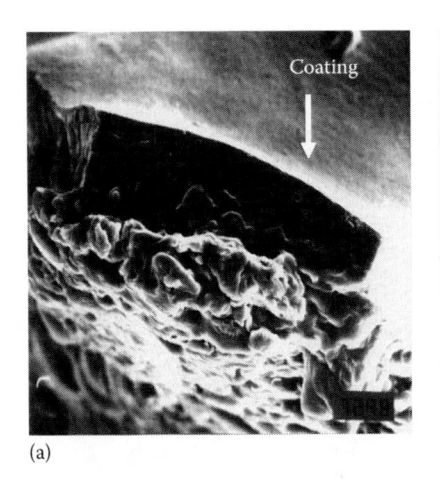 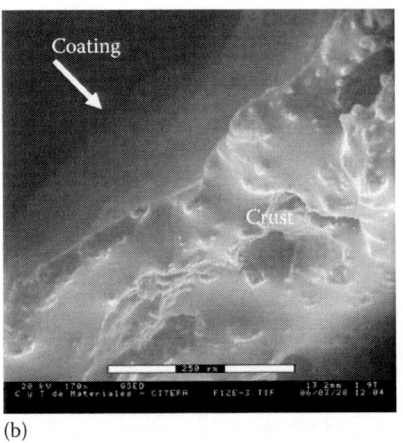

(a) (b)

FIGURE 2.7 Cross-sectional micrographs of a fried dough disc coated with 1% methylcellulose plasticized with 0.75% sorbitol: (a) scanning electron microscopy, magnification 100 μm between marks, and (b) ESEM.

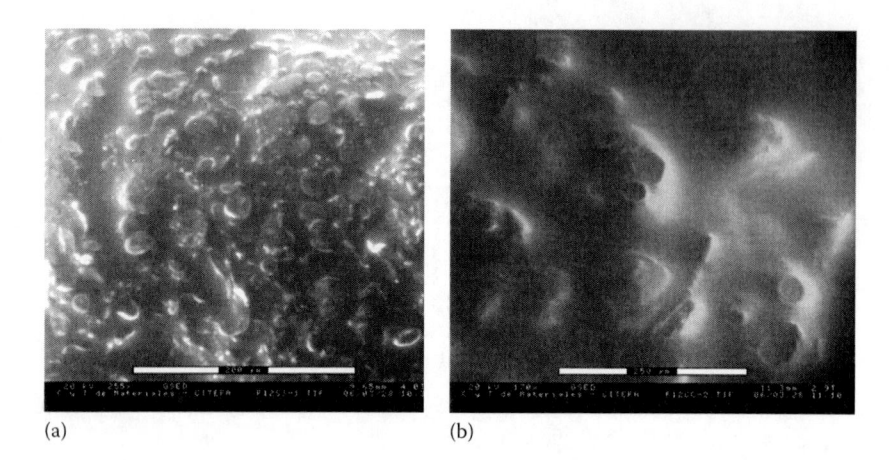

(a)   (b)

FIGURE 2.8 ESEM micrographs of a dough disc fried during 12 min: (a) surface of the uncoated sample and (b) surface of the coated sample.

2.2.8 Relationship between Oil Uptake and Microstructure Changes during the Frying Process

OU depends on structural changes during the process; differences in the starting food microstructure can be expected to be important determinants in the evolution of the characteristics of the end product. Microstructural changes were produced during frying time; the dehydrated zone increased allowing the oil retained by the surface to penetrate into the pores formed by water evaporation.

The mathematical model solved in the present work allowed to estimate the thickness of the dehydrated zone ($d_Z = L - x_1$).

Figure 2.9 shows OU versus frying time for both coated and uncoated samples. Simultaneously, the same figure shows the growth of the dehydrated zone with frying time. The predicted dehydrated zone thickness curve as a function of time is the same for coated and uncoated samples, according to the proposed mathematical model. A simple equation was proposed to interpret experimental results:

$$OU = a(1 - e^{-bt})$$ (2.28)

where
  a is the oil concentration at long times
  b is the coefficient that takes into account the structural changes as a function of frying time

The parameters of Equation 2.28 were estimated by a nonlinear regression showing high values of the correlation coefficient r^2 (Table 2.3).

Results of Figure 2.9 allowed to correlate OU with the thickness of the dehydrated zone (d_Z) as shown in Figure 2.9b. A linear behavior of OU versus dehydrated zone thickness was maintained up to OU values of 0.031 g/g dry solid and 0.071 g/g dry solid for both coated and uncoated samples, respectively.

Values of OU for coated samples were lower than those of uncoated ones. For low d_Z values, the oil retained by the surface could be incorporated into the dehydrated zone, when the sample was removed from the frying medium (linear relationship Figure 2.9b). When the dehydrated zone was large, a deviation from the linear behavior was observed. This could be attributed to the fact that the amount of oil retained at the sample surface is limited, being the oil surface wetting the property related to the interfacial tension, governing these phenomena (Mellema, 2003).

The presence of MC coating with thermal gelation properties modified the surface wetting and also became a mechanical barrier leading to a decrease in the OU of the coated samples.

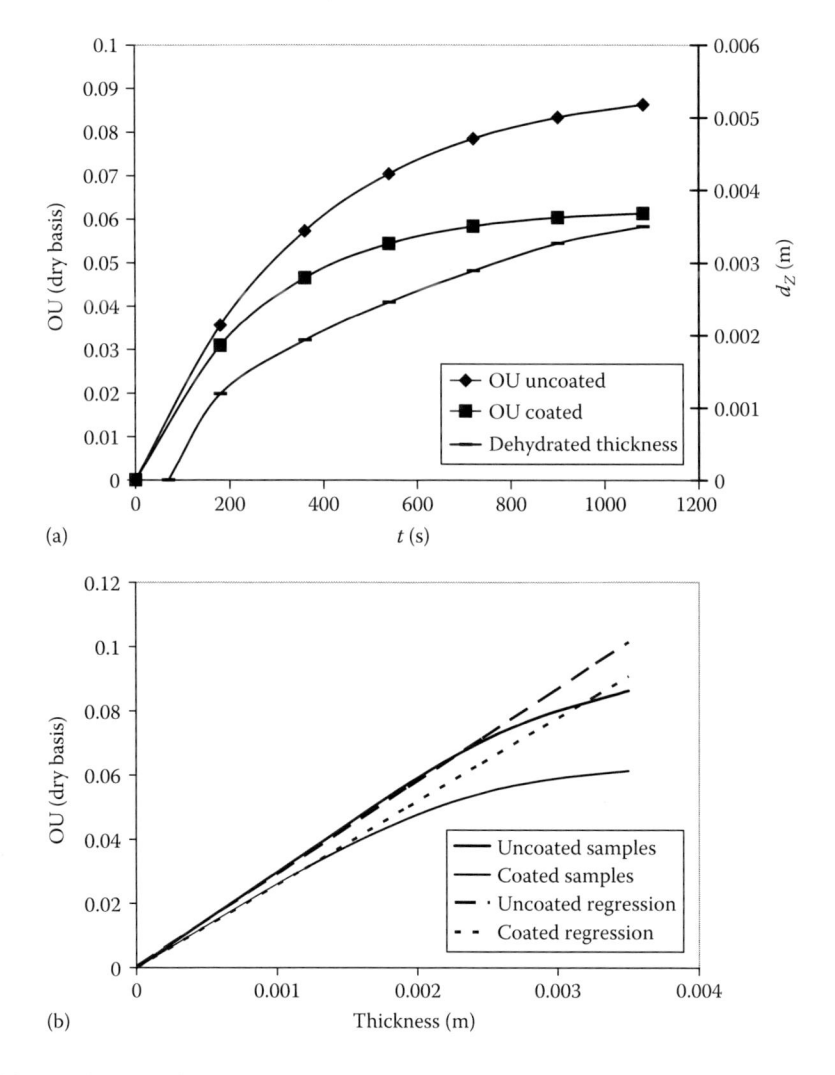

FIGURE 2.9 (a) Oil uptake of uncoated and coated samples and dehydrated zone thickness (d_T) as a function of frying time. d_T curve is the same for coated and uncoated samples. (b) Oil uptake of uncoated and coated samples as a function of dehydrated zone thickness.

TABLE 2.3

Gaseous Permeability (O_2 and CO_2) of Films Based on Native Corn Starch (N), Acetylated Starch (A), and a Mixture of Both in Equal Proportions (50N+50A), with 1.5% p/p Glycerol (1.5G) and 0.3% p/p Potassium Sorbate (0.3KS)

Film Composition	O_2 Permeability $\times 10^{10}$ (cm³/m s Pa)	CO_2 Permeability $\times 10^{9}$ (cm³/m s Pa)	Selectivity Coefficient (β) P_{CO_2}/P_{O_2}
N-1.5G	3.97 ± 0.05	5.37 ± 0.27	13.51
N-1.5G-0.3KS	6.36 ± 0.12	8.30 ± 0.88	13.10
50N+50A-1.5G	3.40 ± 0.04	4.72 ± 0.27	13.87
50N+50A-1.5G-0.3KS	5.86 ± 0.17	7.81 ± 0.56	13.33
A-1.5G	3.12 ± 0.05	4.03 ± 0.02	12.92
A-1.5G-0.3KS	5.50 ± 0.10	7.62 ± 0.12	13.85

The OU was also correlated with the WC showing a negative slope. At high WC, the results were linearly correlated; however, deviations were observed at WC=0.3 and 0.35 (dry basis) for uncoated and coated samples, respectively. Besides considering that WC is a function of d_z, linear equations were obtained for the initial frying periods.

Following the OU criteria introduced by Pinthus and Saguy (1994), the ratio between OU and WC was obtained from Table 2.2, being 0.17 for uncoated and 0.14 for coated samples; these results agree with the findings reported by Moyano and Pedreschi (2006).

2.2.9 Final Considerations about the Modeling of Coated Fried Food

A mathematical model of the frying process based on the numerical solution of the heat and mass transfer differential equations under unsteady-state conditions was proposed and solved using measured physical and thermal properties. It allowed to simulate satisfactorily the experimental data of temperature and WC during the different frying stages.

The model allowed the prediction of the position of the vaporization front and the thickness of the dehydrated zone as a function of frying time. WC correlated linearly with the vaporization front position corresponding to the thickness of the humidity core.

OU that occurs when the sample is removed from the frying medium was correlated with the thickness of the dehydrated zone; a linear behavior was held for the initial frying period. However, deviations were observed when the thickness of the dehydrated zone increased. This could be attributed to the fact that the amount of oil retained at the sample surface is determined by the surface tension property. OU was also linearly correlated with water loss at the initial frying stage. A simple equation for OU as a function of frying times was proposed, considering the microstructural changes developed during the frying process in which the dehydrated zone increases allowing the oil retained by the surface to penetrate into the pores left by water evaporation.

The presence of MC coating reduced the OU due to the thermal gelation behavior, modifying the wetting properties and also becoming a mechanical barrier to the oil.

The application of MC coatings reduced significantly ($p < 0.05$) the oil content of dough discs with regard to control ones, reaching a decrease of 30%. However, the coating did not modify the WC of the samples; this was attributed to the hydrophilic characteristics of the films that led to poor water vapor barrier properties, besides thermal histories of coated and uncoated samples were similar.

Scanning electron microscopy techniques showed the integrity of the MC coating and good adhesion to the food product after its dehydration during the deep-oil-frying process.

2.3 Case Study: Release of an Antimicrobial Agent from a Biodegradable Active Film to a Food System

One of the important emerging functions of edible films and coatings is their use as carrier of antimicrobial and antifungal agents to increase shelf life of foods. This section discussed edible films as delivery matrices for antimicrobial preservatives.

Microbial contamination of food products causes serious diseases and consequent economic losses (Türe et al., 2012). The addition of bactericidal agents or growth inhibitors into food formulation by spraying or immersion methods has been employed to overcome food contamination. However, direct application of the antimicrobial agents has some limitations since they could be neutralized or evaporated or reduce their effective concentration on the surface due to a rapid and inadequate diffusion into the food bulk (Quattara et al., 2000; Pranoto et al., 2005). To overcome these problems, the incorporation of additives to film matrices has been proposed, improving food packages functionality (Cagri et al., 2001; Li et al., 2006). The active agent is slowly released from the film to the food surface, where it remains at high concentration (Gennadios et al., 1997; Quattara et al., 2000; Chollet et al., 2009).

The selection of an antimicrobial agent depends on its activity against a target microorganism. The growth of potential microorganisms that can spoil food products depends on its characteristics such as pH, water activity, composition, as well as storage conditions. The direct incorporation of preservatives in packaging films is a convenient method by which antimicrobial activity can be achieved. Taking into consideration that antimicrobial activity mainly depends on the diffusion of the preservative, it is important to evaluate the release from the polymer matrix. The mathematical modeling provides information about the mechanisms that control the mass transfer process.

Within the natural polymers used in active packaging, starch is one of the most used materials due to its abundance, availability, low cost, and biodegradability (López et al., 2008; García et al., 2009). Corn starch–based films are homogeneous and transparent (López et al., 2008, 2010, 2011). The addition of glycerol as a plasticizer improved material flexibility, avoiding cracks and decreasing the water vapor permeability. The use of acetylated corn starch into formulations led to the development of more resistant films that are less permeable to water vapor (López et al., 2008, 2010). Films based on native and acetylated corn starch showed a good heat sealing capacity making them appropriate materials to develop food packages (López et al., 2011). The scaling up of starch-based materials has been demonstrated since they could be processed by blown extrusion as well as thermocompression (López et al., 2013a).

Sorbic acid and its salts are widely used food preservatives; they are efficient and versatile inhibiting the most common microorganisms (fungi, molds, and yeasts) that can deteriorate foods (Kristo et al., 2008). Flores et al. (2007) studied the performance of tapioca starch–based films as carriers of sorbate and established that films were effective in controlling the growth of *Z. bailii* population, acting as a preservative release agent or as a barrier for external yeast contamination. On the other hand, Türe et al. (2012) developed films based on wheat gluten containing PS demonstrating the antifungal properties of these active materials. The antifungal effectiveness of PS incorporated in guar gum and pea starch coatings was also reported (Mehyar et al., 2011). Pranoto et al. (2005) informed that chitosan films with PS presented antimicrobial activity against *Staphylococcus aureus*, *Listeria monocytogenes*, and *Bacillus cereus*. Sayanjali et al. (2011) applied an antimicrobial coating based on CMC with PS to control the growth of molds, inhibiting substantially *Aspergillus* species.

Despite many works present in the literature concerning active films with PS, it is still relevant to study the diffusion of the additive from the polymeric matrices to the product and its diffusion profile inside the food bulk. The mathematical modeling of this phenomenon could allow to determine active agent diffusion coefficients in both, the film and the product.

In the present section, the effectiveness of the addition of PS in corn plasticized starch films to prevent microbial growth on a food product is discussed. The sorbate release from the polymeric matrix to a semisolid model food system was evaluated, and a mathematical model of the diffusion process was proposed.

2.3.1 Active Film Preparation and Characterization

Starch aqueous suspensions (5% w/w) of native and acetylated corn starch (acetylation degree = 2.2%) as well as a combination in equal proportions were used to prepare active films (López et al., 2010). These suspensions were gelatinized at 90°C during 20 min; then, glycerol as a plasticizer (1.5% w/w) and PS as an antimicrobial agent (0.1%–0.5% w/w) were added. The pH values of the filmogenic suspensions ranged between 5.55 and 5.98. Filmogenic suspensions with pH adjusted at 4.5 were also prepared in order to increase preservative effect of PS, considering that the undissociated form is the effective one.

Active starch films were obtained by the casting method; they were dried at 50°C, removed from the plates, and stored at 20°C and 65% RH. PS incorporation did not modify the filmogenic capacity of the studied formulations; its concentration in the film was quantified using the AOAC (1977) official method based on absorbance measurement at 260 nm (López et al., 2013b). PS concentration in the active starch films decreased approximately 21% after 60 days storage at 20°C and 65% RH due to sorbate degradative oxidation.

2.3.2 Effect of PS on Optical and Barrier Properties of the Active Films

Surface colorimetric measurements of films containing 0.3% w/w PS were recorded in terms of the Hunter parameters (L^*, a^*, b^*) according to the CIE scale showing an increase in color differences.

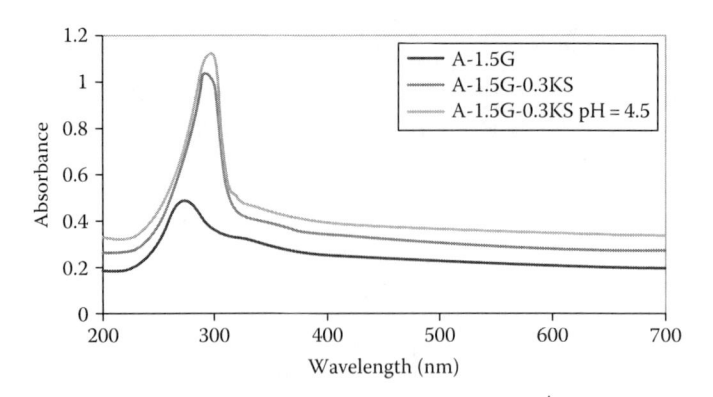

FIGURE 2.10   UV–visible spectra of acetylated corn starch films (A) containing 1.5% p/p glycerol (1.5G) and 0.3% p/p potassium sorbate (0.3KS) with and without pH adjusted to 4.5.

This result was attributed mainly to the increase of parameter b, indicating that active films resulted in more yellow than the control samples due to sorbate oxidative browning (Gerschenson and Campos, 1995; Famá et al., 2006). However, despite this, color modification did not affect the film acceptability.

PS addition increased the UV-barrier capacity of starch-based films, since the absorbance peak located around 270–300 nm increases its intensity (Figure 2.10). In general, the filmogenic suspension pH adjusted to 4.5 that led to an increase in the UV barrier capacity.

Water vapor barrier properties were not significantly ($p > 0.05$) affected by the addition of PS in the starch-based film. However, PS incorporation significantly increases ($p < 0.05$) film gaseous permeability, maintaining their selective permeation to CO_2 and O_2 (Table 2.3).

2.3.3 Experiments to Determine PS Release from the Active Film to a Model Food System

The effectiveness of the polymeric matrices to retain the PS was evaluated through the active agent diffusion from the starch film to a semisolid medium that simulated a food product (agar gel). This gel was prepared using an aqueous agar solution (2% w/w) that was molded in cylinders of 2 cm height and 2.5 cm diameter and solidified by cooling.

Film discs of 2.5 cm diameter containing PS were weighted and deposited onto the top surface of the agar gel cylinders and stored at 4°C. At different times, two cylinders were taken to evaluate the sorbate release using the official method for sorbic acid (AOAC, 1977). Each cylinder was cut in four slices of 0.5 cm thickness, and PS concentration in each slice was determined at different times. Agar gel cylinders without film were used as controls (blanks) for the sorbate spectrophotometric determination.

2.3.4 Mathematical Modeling of PS Diffusion Process

PS diffusion from a medium M_1 constituted by a corn starch film containing PS to a medium M_2 (in contact with M_1) constituted by an agar gel, which represents a model system of an intermediated a_w food product, was analyzed.

Fick's second law was applied; Equation 2.29 corresponds to the unidirectional diffusion under the nonstationary state:

$$\frac{\partial C_i}{\partial t} = \frac{\partial}{\partial x}\left(\frac{D_i \partial C_i}{\partial x}\right) \tag{2.29}$$

where

$i = 1, 2$ corresponds to the medium M_1, M_2, respectively

C_i and D_i correspond to the PS concentration and the sorbate diffusion coefficient in medium i

t is the time

x is the position

A constant diffusion coefficient was considered in each medium. In the corn starch film, x varied between 0 and L_1, while in the agar gel x varied between L_1 and $L_1 + L_2$, being L_2 the length of the gel.

The initial conditions ($t = 0$) are described by

$$C_1 (x, 0) = C_{10} \quad \text{in } 0 \leq x \leq L_1 \tag{2.30}$$

$$C_2 (x, 0) = 0 \quad \text{in } L_1 \leq x \leq L_1 + L_2 \tag{2.31}$$

Boundary conditions at $t > 0$ are represented by

$$\text{(Continuity of the material flux)} \tag{2.32}$$

$$C_2 (L_1, t) = K_{12} C_1 (L_1, t) \text{ (equilibrium at the interface } x = L_1) \tag{2.33}$$

$$\frac{\partial C_2}{\partial x} = 0 \quad \left(\text{Null flux in } x = L_1 + L_2 \right) \tag{2.34}$$

The software COMSOL Multiphysics Finite Element Analysis Simulation Software (USA, 2007) was used to model mathematically the experimental data.

2.3.5 Diffusion Coefficients Determination

Figure 2.11 shows as an example the experimental results of PS diffusion from a polymeric matrix based on native corn starch, in contact with the semisolid medium (agar gel of $a_w \sim 0.8$). When active films were tested, it was observed that 80% of the sorbate contained in the film matrix was released after 36 h, and 91% after 60 h. The results demonstrated that during this period active films maintained their antimicrobial action on the food product surface.

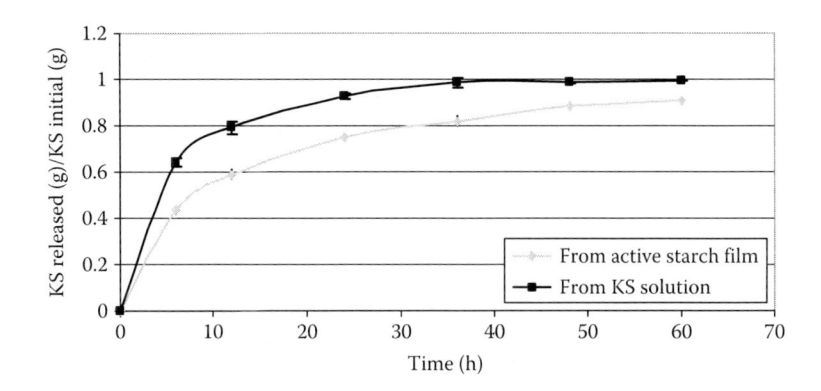

FIGURE 2.11 Potassium sorbate release from an active native starch film and from a sorbate solution to a semisolid medium as a function of time.

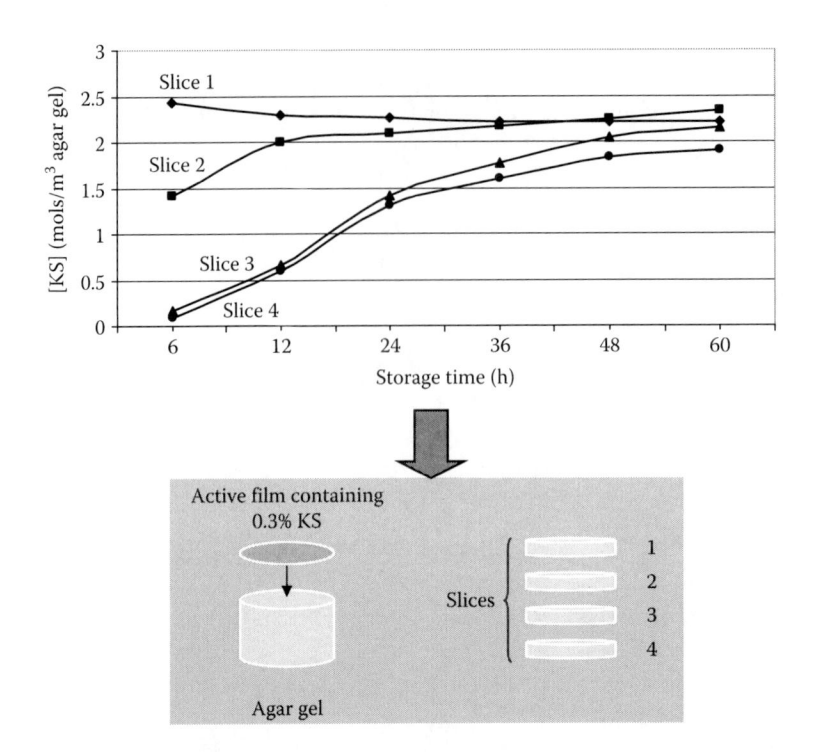

FIGURE 2.12 Potassium sorbate diffusion from the active starch films: experimental sorbate concentration profiles in the agar gel and sampling scheme.

The theoretical curves were fitted to the experimental PS concentrations to estimate diffusivity values. Calculations were performed to determine the best-fit diffusion coefficients by minimizing the sum of squares of the differences between experimental and calculated PS concentrations in the gel.

Experimental data of PS concentration in the different agar gel sections when the preservative diffused from the active starch films are shown in Figure 2.12. Experimental profiles showed that after 60 h PS concentration in the agar slice that was in contact with the film (slice 1) maintained a constant value.

Most of the studies found in the literature describe and model the diffusive process of different additives from films to a liquid medium (Crank, 1975; Vojdani and Torres, 1989; Redl et al., 1996; Ozdemir and Floros, 2001; Flores et al., 2007). However, these systems do not represent a real system in which the active film is in contact with a solid food product.

Besides, modeling the preservative diffusion process from hydrophilic matrices is of great interest not only for the food industry but also for pharmaceutical applications, since these materials are also used for controlled-release drugs (Peppas et al., 2000; Siepmann and Siepmann, 2008). From the numerical simulations, the PS diffusion coefficient in the agar gel was 6×10^{-10} m^2/s, and in the starch film matrix, it was 0.3×10^{-13} m^2/s (López et al., 2013b).

Nonsignificant effects of starch type and controlled pH film formulation on PS diffusivity values were observed. In the literature, there is a wide variation among sorbate diffusion coefficient values reported for biodegradable films. Choi and coworkers (2005) reported a value of 2.6×10^{-13} m^2/s for films of κ-carrageenan. Redl and coworkers (1996) informed sorbate diffusion coefficients in gluten films and in bilayer systems with beeswax of 7.6×10^{-12} and 2.7×10^{-16} m^2/s, respectively. In whey protein films, the diffusion coefficients varied between 4.1 and 9.3×10^{-11} m^2/s, depending on the formulation (Hasan et al., 2006).

Buonocore and coworkers (2003) developed a complex model that describes the release of kinetic of different antimicrobial agents (lysozyme, nisin, and sodium benzoate) from a highly hydrophilic matrix to an aqueous medium. A more realistic model should include the swelling of the film when it is in contact with high-humidity systems.

2.3.6 Application of Active Films in Dairy Products: PS Minimum Inhibitory Concentration

Each preservative that is applied in foodstuffs requires a different minimum concentration in the system to show antimicrobial effect. A dairy product was selected as a food system to determine the minimum inhibitory concentration (MIC) of PS.

A commercial cheese with high humidity content (55%), elaborated with full cream milk acidified with lactic acid bacteria, was placed in contact with the active film.

The method proposed by Fajardo et al. (2010) was used to determine the MIC. Approximately 25 g of cheese was deposited onto starch films containing different PS concentrations (0.1%–0.5% w/w). Samples were stored at 20°C during 6 days; visual observation and photographic record were done daily. The lowest PS concentration that did not allow visual observation of fungi growth on the cheese surface was considered as MIC. Figure 2.13 shows the photographs of the cheese samples, placed on native corn starch films with glycerol (1.5% w/w) and containing PS at different concentrations (0.1%–0.5% w/w), stored at 20°C and 65% RH during 6 days.

PS MIC was 0.3% w/w because in these samples visible fungi growth was not detected during the storage period. The pH of the film formulation did not affect the obtained MIC value.

The absence of the pH effect could be attributed to the proximity of the sorbic acid pK_a (4.76) to the assayed dairy product pH (4.7–5.5); thus the concentration of the nondissociated sorbate form in the interface film product would be similar in both cases.

López et al. (2013b) demonstrated that starch-based active films were able to inhibit the growth of *Candida* spp., *Penicillium* spp., *S. aureus*, and *Salmonella* spp., which are microorganisms having a negative effect on quality or pathogenic bacteria responsible for some foodborne diseases.

2.3.7 Antimicrobial Performance of the Active Film on a Dairy Product

Starch films formulated with glycerol (1.5% w/w) and with PS at the MIC were formulated with and without adjusting pH to 4.5.

Approximately 25 g of the commercial cheese described previously was deposited onto active starch films, and samples were placed into bags of thermo-sealable synthetic films (PD 141 [CRYOVAC®] that is a polyethylene multilayer material 75 μm thickness).

Fresh cheese is characterized to be a slightly fermented product with low acidity (pH ≈ 5), high water activity, low salt content (<3%), and with an electronegative redox potential (oxygen absence). These conditions led to the development of many microorganisms, specially molds and yeasts. Samples were stored

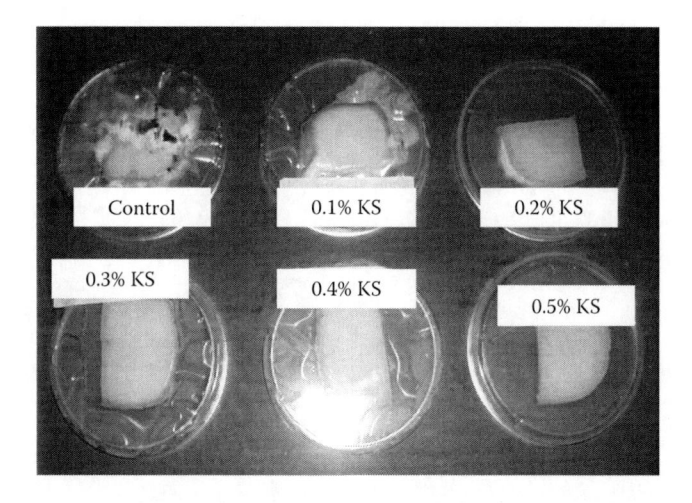

FIGURE 2.13   Cheese samples stored at 20°C and 65% RH placed onto native corn starch film with glycerol (1.5% w/w) and potassium sorbate at different concentrations (0.1%–0.5% w/w).

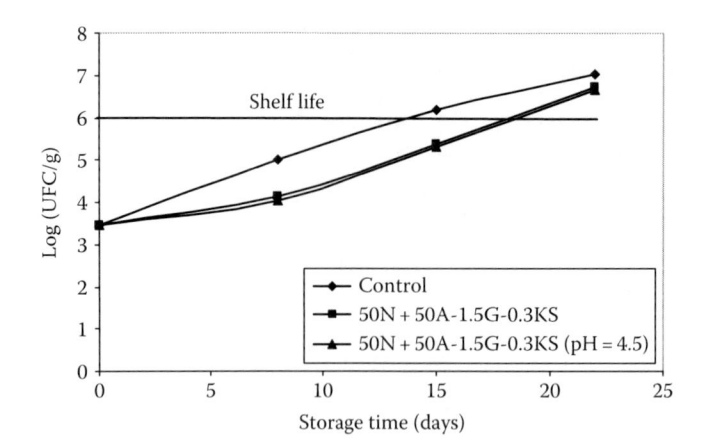

FIGURE 2.14   Yeast and mold growth on cheese samples deposited onto corn starch films based on mixtures of equal proportions of native and acetylated corn starch (50N+50AS) plasticized with 1.5% glycerol (1.5G) containing 0.3% potassium sorbate (0.3KS) with (pH=4.5) and without pH adjustment. Samples were stored at 4°C.

at 4°C; at 7, 15, and 22 days, yeast and mold counts were determined using YGC medium (Yeast extract Glucose Chloramphenicol, Merck) incubating 8 days at 30°C. Viable microorganisms were determined by counting the number of formed colonies, expressing the results in CFU/g cheese.

Figure 2.14 shows the microbial counts in cheese samples placed onto active films containing 0.3% PS. Shelf life of fresh cheese was defined as the time necessary to reach 10^6 CFU/g in the sample. Several authors have stressed that when microbial counts exceed 10^6 CFU/g toxic substances may be produced (Howard and Dewi, 1995; García et al., 2001). In the case of the control sample (packaged without the active film), the observed shelf life was 14 days; while for cheese samples stored on active films, it was extended to 17 days (Figure 2.14).

Cerqueira and coworkers (2010) stressed that galactomannan coatings were effective to prolong shelf life of regional cheese. Several authors have used chitosan-based films to extend the shelf life of different types of cheese under refrigeration or modified atmosphere storage (Gammariello et al., 2008; Del Nobile et al., 2009; Fajardo et al., 2010). On the other hand, other biodegradable materials can be used for this purpose; for example, PLA was used to package Dambo cheese maintaining its quality attributes during 84 days (Holm et al., 2006).

2.3.8 Final Considerations about the Modeling of the Preservative Release in Active Films

The release of PS from active films was studied using native or acetylated corn starch films. The addition of PS to the starch films increased their UV barrier capacity without modifying the water vapor permeability values. Active films resulted more yellowness and less transparent than films without PS, but these characteristics did not affect their acceptability. The PS MIC in active starch films resulted in 0.3% w/w, regardless of the corn starch type and the formulation pH. The tested active films were effective to extend the shelf life of refrigerated cheese samples, from 14 to 17 days. The active agent release from films based on corn starch to a semisolid medium was mathematically modeled, and diffusion coefficients were determined. The PS inclusion in the starch polymeric matrices maintained a high antimicrobial concentration in the product surface, where its action is required. The developed mathematical model assumed a constant initial mass of antimicrobial that diffuses along both media that are in contact; its resolution allows to simulate diffusion processes of different additives incorporated to polymeric matrixes, taking into account only the preservative concentration and the dimensions of the film and the semisolid medium. Finally, further research should consider evaluating the eventual swelling of the film on the controlled release of active compounds.

REFERENCES

Aguilera, J.M. (1997). Fritura de alimentos. In J.M. Aguilera (ed.), *Temas de tecnología de alimentos*. México, D.F.: Programa Iberoamericano CYTED, Instituto Politécnico Nacional, Vol. 1, pp. 185–214.

Aguilera, J.M. and Hernández, H.G. (2000). Oil absorption during frying of frozen parfried potatoes. *Journal of Food Science*, 65(3), 476–479.

Aguilera, J.M. and Stanley, D.W. (1999). *Microstructural Principles of Food Processing and Engineering*. Gaithersburg, MD: Aspen Publishers, Inc.

AOAC. Official Method 974.08. Sorbic acid in wines spectrophotometric method. Final action (1977). Catalá, R. and Gavara, R. (2001). Nuevos envases: De la protección pasiva a la defensa activa de los alimentos envasados. *Arbor CLXVIII*, 661, 109–127.

Ateba, P. and Mittal, G.S. (1994). Modelling the deep-fat frying of beef meatballs. *International Journal of Food Science & Technology*, 29, 429–440, 1994.

Balasubramaniam, V.M., Mallikarjunan, P., and Chinnan, M.S. (1995). Heat and mass transfer during deep-fat frying of chicken nuggets coated with edible film: Influence of initial fat content. *CoFE 1995—Conference on Food Engineering*, Chicago, IL, November 2–3, 1995, pp. 103–106.

Bertolini-Suárez, R., Campañone, L.A., García, M.A., and Zaritzky, N.E. (2008). Comparison of the deep frying process in coated and uncoated dough systems. *Journal of Food Engineering*, 84, 383–393.

Bouchon, P., Aguilera, J.M., and Pyle, D.L. (2003). Structure oil-absorption relationships during deep-fat frying. *Journal of Food Science*, 68(9), 2711–2716.

Bouchon, P. and Pyle, D.L. (2005a). Modelling oil absorption during post-frying cooling, I: Model development. *Trans IChemE, Part C, Food and Bioproduct Processing*, 83(C4), 1–9.

Bouchon, P. and Pyle, D.L. (2005b). Modelling oil absorption during post-frying cooling, II: Solution of the mathematical model, model testing and simulations. *Trans IChem, Part C, Food and Bioproduct Processing*, 83(C4), 1–12.

Buonocore, G.G., Del Nobile, M.A., Panizza, A., Bove, S., Battaglia G., and Nicolais, L. (2003). Modeling the lysozyme release kinetics from antimicrobial films intended for food packaging applications. *Journal of Food Science*, 68(4), 1365–1370.

Cagri, A., Ustunol, Z., and Ryser, E.T. (2001). Antimicrobial edible films and coatings. *Journal of Food Science*, 66(6), 865–870.

Campañone, L.A., García, M.A., and Zaritzky, N.E. (2010). Modelling of heat and mass transfer during deep frying process. In M. Farid (ed.), *Mathematical Analysis of Food Processing*. Boca Raton, FL: CRC Press, Chapter 12, pp. 331–356.

Cerqueira, M.A., Sousa-Gallagher, M.J., Macedo, I., Rodriguez-Aguilera, R., Souza, B.W.S., Teixeira, J.A., and Vicente, A.A. (2010). Use of galactomannan edible coating application and storage temperature for prolonging shelf-life of "Regional" cheese. *Journal of Food Engineering*, 97(1), 87–94.

Choi, J.H., Choi, W.Y., Cha, D.S., Chinnan, M.J., Park, H.J., Lee, D.S., and Park, J.M. (2005). Diffusivity of potassium sorbate in κ-carrageenan based antimicrobial film. *Food Science and Technology*, 38(4), 417–423.

Chollet, E., Swesi, Y., Degraeve, P., and Sebti, I. (2009). Monitoring nisin desorption from a multi-layer polyethylene-based film coated with nisin loaded HPMC film and diffusion in agarose gel by an immunoassay (ELISA) method and a numerical modeling. *Innovative Food Science and Emerging Technologies*, 10, 208–214.

Costa, R.M., Oliveira, F.A.R., Delaney, O., and Gekas, V. (1998). Analysis of heat transfer coefficient during potato frying. *Journal of Food Engineering*, 39, 293–299.

Crank, J. (1975). *The Mathematics of Diffusion*, 2nd ed. Oxford, U.K.: Oxford University Press.

Dagerskog, M. and Sorenfor, P. (1978). A comparison between four different methods of frying meat patties: Heat transfer, yield, and crust formation. *LWT: Food Science and Technology*, 11, 306–311.

Debeaufort, F. and Voilley, A. (1997). Methylcellulose-based edible films and coatings: 2. Mechanical and thermal properties as a function of plasticizer content. *Journal of Agricultural and Food Chemistry*, 45, 685–689.

Del Nobile, M.T., Gammariello, D., Conte, A., and Attanasio, M. (2009). A combination of chitosan, coating and modified atmosphere packaging for prolonging Fior di latte cheese shelf life. *Carbohydrate Polymers* 78(1), 151–156.

Donhowe, I.G. and Fennema, O.R. (1993). The effects of plasticizers on crystallinity, permeability, and mechanical properties of methylcellulose films. *Journal of Food Processing and Preservation*, 17, 247–257.

Fajardo, P., Martins, J.T., Fuciños, C., Pastrana, L., Teixeira, J.A., and Vicente, A.A. (2010). Evaluation of a chitosan-based edible film as carrier of natamycin to improve the storability of Saloio cheese. *Journal of Food Engineering*, 101, 349–356.

Famá, L., Flores, S., Gerschenson, L., and Goyanes, S. (2006). Physical characterization of cassava starch biofilms with special reference to dynamic mechanical properties at low temperatures. *Carbohydrate Polymers*, 66, 8–15.

Farid, M. (2002). The moving boundary problems from melting and freezing to drying and frying of food. *Chemical Engineering and Processing*, 41, 1–10.

Farid, M.M. and Chen, X.D. (1998). The analysis of heat and mass transfer during frying of food using a moving boundary solution procedure. *Heat and Mass Transfer*, 34, 69–77.

Farkas, B.E., Singh, R.P., and Rumsey, T.R. (1996a). Modeling heat and mass transfer in immersion frying. I. Model development. *Journal of Food Engineering*, 29, 211–226.

Farkas, B.E., Singh, R.P., and Rumsey, T.R. (1996b). Modeling heat and mass transfer in immersion frying. II. Model solution and verification. *Journal of Food Engineering*, 29, 227–248.

Flores, S.K., Famá, L., Rojas, A.M., Goyanes, S., and Gerschenson, L. (2007). Physical properties of tapioca-starch edible Films: Influence of filmmaking and potassium sorbate. *Food Research International*, 40, 257–265.

Gallagher, K.M. and Corrigan, O.I. (2000). Mechanistic aspects of the release of levamisole hydrochloride from biodegradable polymers. *Journal of Controlled Release*, 69, 261–272.

Gammariello, D., Chillo, S., Mastromatteo, M., Di Giulio, S., Attanasio, M., and Del Nobile, M.A. (2008). Effect of chitosan on the rheological and sensorial characteristics of apulia spreadable cheese. *Journal of Dairy Science*, 91(11), 4155–4163.

García, M.A., Ferrero, C., Bértola, N., Martino, M., and Zaritzky, N. (2002). Edible coatings from cellulose derivatives to reduce oil uptake in fried products. *Innovative Food Science and Emerging Technologies*, 3, 391–397.

García, M.A., Ferrero, C., Bértola, N., Martino, M., and Zaritzky, N. (2004). Methylcellulose coatings reduce oil uptake in fried products. *Food Science and Technology International*, 10(5), 339–346.

García, M.A., Martino, M., and Zaritzky, N. (2001). Composite starch-based coatings applied to strawberries (*Fragaria × ananassa*). *Nahrung/Food*, 45(4), 267–272.

García, M.A., Pinotti, A., Martino, M., and Zaritzky, N. (2009). Composite edible films and coatings. In M. Embuscado and K. Huber (eds.), *Edible Films and Coatings for Food Applications*. Pondicherry, India: Springer Science, Chapter 6, pp. 169–210.

Gennadios, A., llanna, M.A., and Kurth, L.B. (1997). Application of edible coatings on meats, poultry and seafoods: A review. *Lebensmittel-Wissenschaft & Technologie*, 30, 337–350.

Gerschenson, L.N. and Campos, C.A. (1995). Sorbic acid stability during processing and storage of high moisture foods. In G. Barbosa Canovas and J. Welti Chanes (eds.), *Food Preservation by Moisture Control*. Lancaster, PA: Technomic Publishing, pp. 761–790.

Hasan, S., Deniz, S., and Murat, O. (2006). A mathematical model for potassium sorbate diffusion through whey protein films. *Drying Technology: An International Journal*, 24(1), 21–29.

Holm, V.K., Mortensen, G., Vishart, M., and Agerlin Petersen, M. (2006). Impact of poly-lactic acid packaging material on semi-hard cheese. *International Dairy Journal*, 16, 931–939.

Howard, L.R. and Dewi, T. (1995). Sensory, microbiological and chemical quality of mini-peeled carrots as affected by edible coating treatment. *Journal of Food Science*, 60(1), 142–144.

Huse, H.L., Mallikarjunan, P., Chinnan, M.S., Hung, Y.C., and Phillips, R.D. (1998). Edible coatings for reducing oil uptake in production of akara (deep-fat frying of cowpea paste). *Journal of Food Processing and Preservation*, 22, 155–165.

Kassama, L.S. (2003). Pore development in food during deep-fat frying. Department of Bioresource Engineering, Macdonald Campus of McGill University, Sainte-Anne-de-Bellevue, Quebec, Canada.

Kristo, E., Koutsoumanis, K.P., and Biliaderis, C.G. (2008). Thermal, mechanical and water vapor barrier properties of sodium caseinate films containing antimicrobials and their inhibitory action on Listeria monocytogenes. *Food Hydrocolloids*, 22, 373–386.

Krochta, J.M. and De Mulder-Johnston, C. (1997). Edible and biodegradable polymers films: Challenges and opportunities. *Food Technology*, 51(2), 61–74.

Lacroix, M. (2009). Mechanical and permeability properties of edible films and coatings for food and pharmaceutical applications. In M.E. Embuscado and K.C. Huber (eds.), *Edible Films and Coatings for Food Applications*. New York: Springer.

Li, Y.H., Liu, B., Zhao, Z.B., and Bai, F.W. (2006). Optimized culture medium and fermentation conditions for lipid production by Rhodosporidium toruloides. *Chinese Journal of Biotechnology*, 22(4), 650–656.

López, O., García, M., and Zaritzky, N. (2008). Film forming capacity of chemically modified corn starches. *Carbohydrate Polymers*, 73, 573–581.

López, O.V., Giannuzzi, L., Zaritzky, N.E., and García, M.A. (2013b). Potassium sorbate controlled release from corn starch films. *Materials Science and Engineering C*, 33, 1583–1591.

López, O.V., Lecot, C.J., Zaritzky, N.E., and García, M.A. (2011). Biodegradable packages development from starch based heat sealable films. *Journal of Food Engineering*, 105(2), 254–263.

López, O.V., Zaritzky, N.E., and García, M.A. (2010). Novel sources of edible films and coatings. *Stewart Postharvest Review*, 6(3), 1–8.

López, O.V., Zaritzky, N.E., Grossmann, M.V.E., and García, M.A. (2013a). Acetylated and native corn starch blend films produced by blown extrusion. *Journal of Food Engineering*, 116, 286–297.

Mallikarjunan, P., Chinnan, M.S., Balasubramaniam, V.M., and Phillips, R.D. (1997). Edible coatings for deep-fat frying of starchy products. *LWT: Food Science and Technology*, 30, 709–714.

Mehyar, G.F., Al-Ismail, K., Han, J.H., and Chee, G.W. (2011). Characterization of edible coatings consisting of pea starch, whey protein isolate, and carnauba wax and their effects on oil rancidity and sensory properties of walnuts and pine nuts. *Journal of Food Science*, 77(2), 52–59.

Mellema, M. (2003). Mechanism and reduction of fat uptake in deep-fat fried foods. *Trends in Food Science and Technology*, 14, 364–373.

Moreira, R.G. and Barrufet, M.A. (1998). A new approach to describe oil absorption in fried foods: A simulation study. *Journal of Food Engineering*, 31, 485–498.

Moreira, R.G., Palau, J., Sweat, V.E., and Sun, X. (1995). Thermal and physical properties of tortilla chips as a function of frying time. *Journal of Food Processing and Preservation*, 19, 175–189.

Moyano, P.C. and Pedreschi, F. (2006). Kinetics of oil uptake during frying of potato slices: Effect of pretreatments. *LWT: Food Science and Technology*, 39(3), 285–291.

Ni, H. and Datta, A.K. (1999). Moisture, oil and energy transport during deep-fat frying of food materials. *Trans IChemE, Part C, Food Bioproduct Processing*, 77, 194–204.

Ozdemir, M. and Floros, J.D. (2001). Analysis and modeling of potassium sorbate diffusion through edible whey coating films. *Journal of Food Engineering*, 47(2), 149–155.

Peppas, N.A., Bures, P., Leobandung, W., and Ichikawa, H. (2000). Hydrogels in pharmaceutical formulations. *European Journal of Pharmaceutics and Biopharmaceutics*, 50, 27–46.

Perkins, E.G. and Erikson, M.D. (1996). *Deep Frying: Chemistry, Nutrition, and Practical Applications*. Champaign, IL: AOCS Press.

Pinthus, E.J. and Saguy, I.S. (1994). Initial interfacial tension and oil uptake by deep-fat fried foods. *Journal of Food Science*, 59(4), 804–807.

Pranoto, Y., Rakshit, S.K., and Salokhe, V.M. (2005). Enhancing antimicrobial activity of chitosan films by incorporating garlic oil, potassium sorbate and nisin. *LWT: Food Science and Technology*, 38, 859–865.

Quattara, B., Simard, R.E., Piette, G., Begin, A., and Holley, R.A. (2000). Diffusion of acetic and propionic acids from chitosan-based antimicrobial packaging films. *Journal of Food Science*, 65, 768–773.

Quintavalla, S. and Vicini, L. (2002). Antimicrobial food packaging in meat industry. *Meat Science*, 62(3), 373–380.

Rayner, M., Ciolfi, V., Maves, B., Stedman, P., and Mittal, G.S. (2000). Development and application of soy-protein films to reduce fat intake in deep-fried foods. *Journal of the Science of Food and Agriculture*, 80, 777–782.

Redl, A., Gontard, N., and Guilbert, S. (1996). Determination of sorbic acid diffusivity in edible wheat gluten and lipid film. *Journal of Food Science*, 61, 116–120.

Rice, P. and Gamble, M.H. (1989). Modeling moisture loss during potato slice frying. *International Journal of Food Science & Technology*, 24, 183.

Sayanjali, S., Ghanbarzadeh, B., and Ghiassifar, S. (2011). Evaluation of antimicrobial and physical properties of edible film based on carboxymethyl cellulose containing potassium sorbate on some mycotoxigenic Aspergillus species in fresh pistachios. *LWT: Food Science and Technology*, 44(4), 1133–1138.

Siepmann, J. and Siepmann, F. (2008). Mathematical modeling of drug delivery. *International Journal of Pharmaceutics*, 364, 328–343.

Singh, R.P. (1995). Heat and mass transfer in foods during frying. *Food Technology*, 49(4), 134–137.

Singh, R.P. (2000). Moving boundaries in food engineering. *Food Technology*, 54(2), 44–53.

Tharanathan, R. (2003). Biodegradable films and composite coatings: Past, present and future. *Critical Review in Food Science and Technology*, 14, 71–78.

Türe, H., Gällstedt, M., and Hedenqvist, M.S. (2012). Antimicrobial compression-moulded wheat gluten films containing potassium sorbate. *Food Research International*, 45, 109–115.

Ufheil, G. and Escher, F. (1996). Dynamics of oil uptake during deep-fat frying of potato slices. *LWT: Food Science and Technology*, 29(7), 640–644.

Vojdani, F. and Torres, J.A. (1989). Potassium sorbate permeability of methylcellulose and hydroxypropyl methylcellulose coatings: Effect of fatty acids. *Journal of Food Processing and Preservation*, 13, 417–430.

Williams, R. and Mittal, G.S. (1999a). Water and fat transfer properties of polysaccharide films on fried pastry mix. *LWT: Food Science and Technology*, 32, 440–445.

Williams, R. and Mittal, G.S. (1999b). Low-fat fried foods with edible coatings: Modeling and simulation. *Journal of Food Science*, 64(2), 317–322.

Yamsaengsung, R. and Moreira, R.G. (2002a). Modeling the transport phenomena and structural changes during deep-fat frying: Part I Model development. *Journal of Food Engineering*, 53, 1–10.

Yamsaengsung, R. and Moreira, R.G. (2002b). Modeling the transport phenomena and structural changes during deep-fat frying: Part II Model solution and validation. *Journal of Food Engineering*, 53, 11–25.

3

Barrier Properties of Films

João Borges Laurindo

CONTENTS

3.1 Introduction

The primary functions of food packaging are containment, protection, information, and convenience. Protection is often considered the main function, maintaining the food properties during its transport, stocking, and commercialization. In this way, the package must protect the food against insects, dirt, moisture, and gases, with special attention to the oxygen, carbon dioxide, and ethylene. The moisture barrier property of a polymeric film must be carefully chosen to protect dehydrated foods against air humidity, while the film gas barrier properties must be considered to design a modified atmosphere packaging system. Other potentially important barrier properties, susceptible to be analyzed in films with food packaging applications, are the barrier capacities to odors, light, and fat (Hatzidimitriu et al., 1987; Nelson and Fennema, 1991; Blanco-Pascual et al., 2014).

This chapter is focused on the barrier properties of flexible films to water vapor, oxygen, and carbon dioxide.

3.2 Fundamental Aspects of Water Vapor and Gas Permeation

The barrier properties to water vapor and gases of a package depend on its integrity (good sealing properties and absence of fractures, microholes, and micropores) and on the film permeability. The packaging integrity can be solved using adequate film thickness and sealing procedure, but film permeability is never null for classical polymeric films (Sarantópoulos et al., 2002). This chapter is an introduction to the barrier properties of nonporous polymeric films.

The permeability of films to gases and vapors depends on the solubility of the permeating species and on its mobility in the polymeric film. The permeation occurs in three steps, that is, (1) sorption and solubilization of a permeating species on the film surface in contact with a gas mixture with the higher concentration of the given species, (2) diffusion of the species through the polymeric film due to a concentration gradient, and (3) desorption (and evaporation) of the permeating species from the film

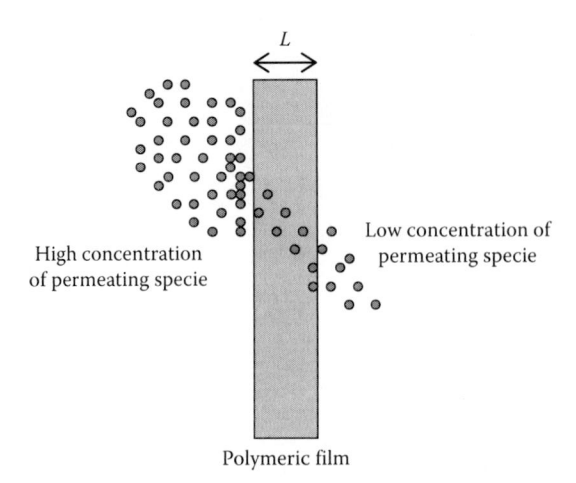

$$L$$

High concentration
of permeating specie

Low concentration of
permeating specie

Polymeric film

FIGURE 3.1 Sketch of the permeation of a species through a polymeric film (thickness L) from the gas mixture with the higher concentration of the permeating species (left side) to the gas mixture with the lower concentration of the permeating species (right side).

surface to the gaseous mixture phase with the lower species concentration. These steps are sketched in Figure 3.1.

The sorption phenomenon depends on the chemical affinity between the permeating species and the polymer. Diffusion occurs mainly in the amorphous regions of the polymer, due to the void spaces created by the thermal agitation of the polymer chains.

In oil-based polymers, the sorption rate of the permeating species on the film surface is supposed to be higher than its diffusion through the film. Therefore, the diffusion phenomenon controls the permeation rate in these hydrophobic materials. The diffusion step is represented by Fick's law, which is given by

$$J_{Az} = -D_{Ap} \frac{dC_A}{dx} \tag{3.1}$$

where
J_{Az} is the flux of the *species A* through the polymeric film (mol/m²·s, g/m²·s)
D_{Ap} is the diffusion coefficient of the *species A* in the polymeric film (m²/s)
dC_A/dx is the gradient of concentration of A through the film

For O_2, N_2, and CO_2, the values of the diffusion coefficient are supposed to be independent of the gas concentration in the polymeric matrix. This consideration is based on their low interactions with the polymer. Nevertheless, this is not the case for the water vapor permeability (WVP) of a hygroscopic film (e.g., starch or protein-based film), for which the solubility coefficient and the diffusion coefficient depend on the film moisture, which directly interacts with the polymer. This difference will be discussed in the following section.

For Fick's law application, the concentrations of the *diffusing species A* must be known at both sides of the film, because this law is only applicable to a same phase, that is, to the polymeric film in the present case. If the concentrations (partial vapor pressures, molar fraction) of the *species A* are known in the gas phases (with higher and lower concentrations), they can be used to calculate their concentrations at the film interfaces. As the solubility of the main permeating species (O_2, CO_2, ethylene, water vapor) in petrol-based polymers is not high, the assumption of a linear partition law to calculate the concentration of the permeating species at the solid interface is a usual assumption. This supposition is represented by

$$C_A = k_s p_A \qquad (3.2)$$

where

C_A is the concentration of the permeating species (A) on the polymer surface
p_A is the partial pressure of the species A in the gaseous phase
k_s is the partition coefficient of the species A between gas and solid phases

This partition coefficient is also called solubility coefficient of a given species in a material (liquid or solid). In many situations, this coefficient depends on the polymer and chemical species interactions and on pressure and temperature.

3.2.1 Water Vapor Permeability

Typically, we are interested in determining the permeation of certain species through a package in steady-state conditions, which represents the situation observed all long the product shelf life. The transient period is normally neglected, because it is very short if compared to the product shelf life, as sketched in Figure 3.2.

This figure represents the water vapor permeation through a starch-based film, at three relative humidity (RH) ranges (11%–33%, 33%–64%, and 64%–90%), and shows how the steady-state situation was reached approximately 2 h after the permeation experiment was started. The water sorption phenomenon in the beginning of the experiment explains the nonlinear behavior of the vapor transfer, sketched in the detail inserted in Figure 3.2.

During the adaptation period (transient period), the film surface can change due to the adsorption of the given species. For example, the water vapor adsorption on the surface of a hygroscopic film causes swelling, as shown in Figure 3.3a, which presents topographic images of a starch film submitted to a high RH atmosphere (close to 100%). The images have the same size and were obtained on the same surface region, with an atomic force microscope.

These images depict how film surface roughness significantly increases in time. The time evolution of the film roughness was determined using Equation 3.3 and represented in Figure 3.3b. This figure shows

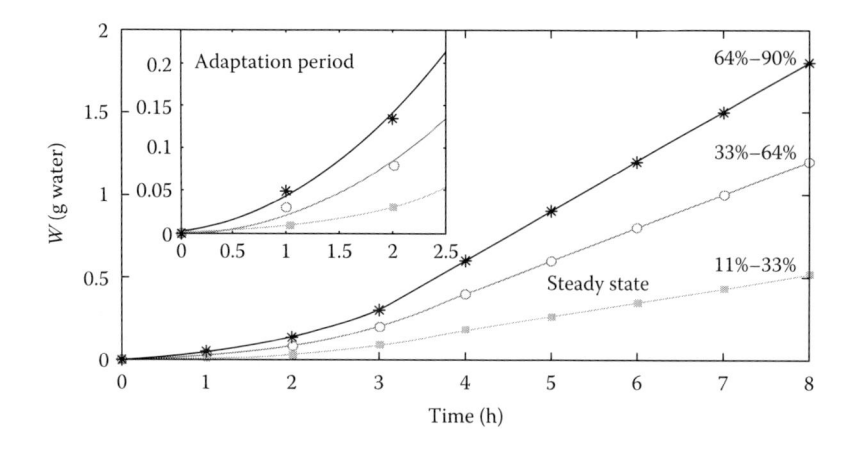

FIGURE 3.2 Experimental data on the permeation of water vapor through a starch-based film.

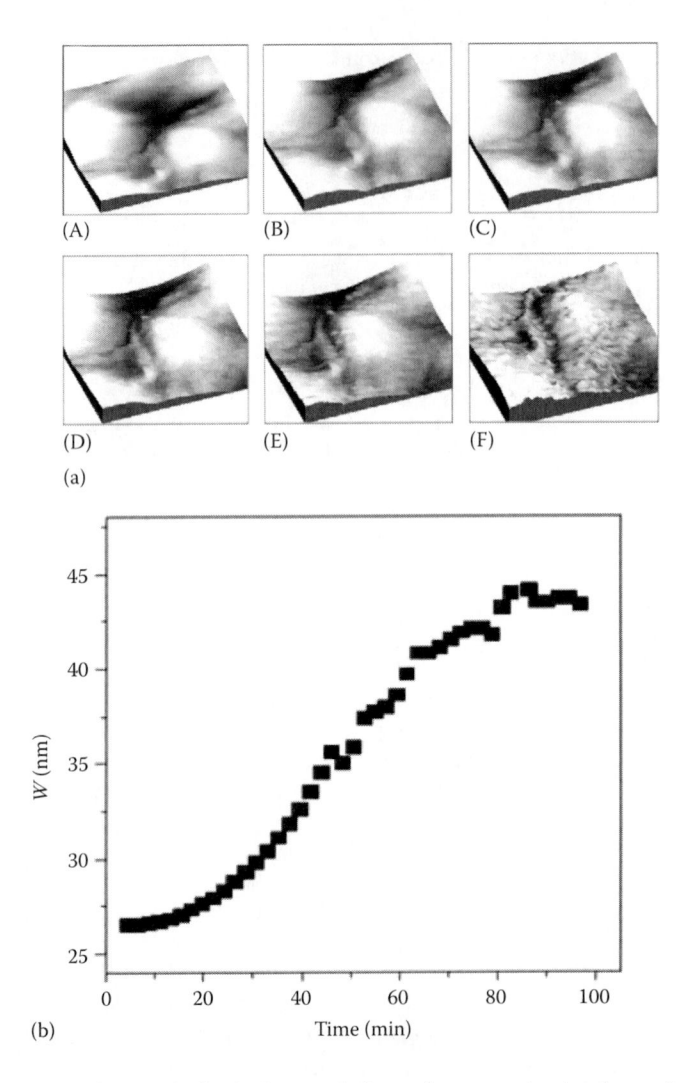

FIGURE 3.3 (a) Topographic images obtained using atomic force microscope, showing the roughening of the film surface exposed to high relative humidity (close to 100%). The film roughness increased from approximately 27 nm (initially dry film, $t = 0$ min) to 38 nm ($t = 55$ min). (A) 0 min, (B) 11 min, (C) 22 min, (D) 33 min, (E) 44 min, (F) 55 min. (b) Time evolution of the film roughness (W).

the surface roughness, $m(l)$, of the layers. The root-mean-square (RMS) deviation of the surface height h from its mean value is $m(l)$, that is, the RMS of the surface height (h) fluctuations, written as

$$m(l) = \sqrt{\langle (h - \langle h \rangle)^2 \rangle} \tag{3.3}$$

where l is the size of the region over which m is measured.

Figure 3.3b shows that the film surface roughness increased from 26 nm to approximately 45 nm in 80 min of exposure time. The surface roughness increased rapidly until 70 min of exposure, following a sigmoidal behavior, but did not significantly change after this period. The roughness increase is probably a result of film heterogeneity at the nanoscale. Literature results reported that the cavities distribution (nanopores) of starch films is influenced by the RH (Stading et al., 2001). These results suggest that the characteristic time needed to reach equilibrium between the film surface and the high RH of the air might not be shorter than 90–100 min.

Considering the situation sketched in Figure 3.1, of a film with thickness L, in which the diffusion coefficient of the permeating species (water vapor) is independent of its concentration in the film, the following equation can represent the mass transfer through the film in steady-state condition:

$$J_{Wz} = \rho_s D_{eff} \frac{X_1 - X_2}{L} \tag{3.4}$$

where

J_{Wz} is the steady-state mass flux through the film with thickness L
X_1 and X_2 are the surface moisture contents on dry weight basis (kg water/kg dry solid) (with $X_1 > X_2$)
ρ_s is the specific gravity of the dried film (kg dry solid/m^3)
D_{eff} is the effective water diffusion in the film (m^2/s)

As discussed before, the mass transfer process depends on the affinity between the film material and water and on the effective diffusivity of water through the film. The values of X (film moisture surface) are related to the air water activity by the film sorption isotherm. Usually, the isotherms of starch and protein-based materials present an initial linear part, for low water activities, and another nonlinear part, for higher water activities. For the linear part, the material moisture can be related to the water activity (a_w), which is given by

$$k_{sw} = \tan(\theta) = \frac{X}{a_w} \tag{3.5}$$

where k_{sw} (g water/kg dry solid · Pa) is the solubility coefficient of water in the solid material.

From Equations 3.4 and 3.5, the equation representing the water flux through the film, in terms of water activity differences, is given by

$$J_{Wz} = (\rho_s k_{sw} D_{eff}) \frac{a_{w1} - a_{w2}}{L} p_s \tag{3.6}$$

The term in parenthesis of this equation is the WVP (K_w) through the film, given by

$$K_w = \rho_s k_{sw} D_{eff} \tag{3.7}$$

which can be represented, for example, in the units (g·µm/m^2·Pa·h).

For oil-based films, the film properties do not change with the concentration of the permeating species. However, it is not the case for hygroscopic films (as starch- and protein-based films), because these films can swell at high moisture, changing its specific mass. Besides, the solubility and the diffusion coefficient can significantly increase with moisture. Thus, at steady-state condition, the water transfer rate (W, g/h) through a film with area of permeation S and thickness L is given by

$$W = K_w S \frac{a_{w1} - a_{w2}}{L} p_s \tag{3.8}$$

This can be used for the experimental determination of K_w from the mass variation over time of a diffusion cell. On the other hand, the film water sorption isotherm can be represented by the extensively used Guggenheim–Anderson–de Boer (GAB) model, which is given by

$$X = \frac{CkX_0 a_w}{[(1 - ka_w)(1 - ka_w + Cka_w)]} \tag{3.9}$$

where C, X_0, and k are constants, with X_0 representing the monolayer moisture content, on a dry basis. This equation allows the determination of the moisture content at the film surfaces submitted to permeation test, which makes possible to assess the diffusion coefficient of water through the film, as reported by Romera et al. (2012).

3.2.1.1 Experimental Determination of the Water Vapor Permeability of Hygroscopic Films

The experimental determination of the *WVP* of films is very important to define their possible applications for food preservation. If the film will be used for packaging dehydrated foods, it must provide a suitable barrier property to prevent moisture gain that leads to the deterioration of the product properties.

The simplest procedure to determine the WVP of films is the gravimetric method, which is based on the increase of the weight of a hygroscopic material present in a permeation cell, on which the top area of a film sample is carefully placed and hermetically fixed. The film sample must have a permeation area of at least 30 cm² and to be hermetically sealed on the cell edges. In some permeation cells, this sealing can be done with a mixture of paraffin and microcrystalline wax, which is not permeable to water vapor and does not absorb water. However, it is much more practical to use permeation cells with O-rings to avoid vapor leaks. An example of permeation cell with O-rings is shown in Figure 3.4. The cell is placed in a container with constant temperature and RH, and its weight increase at different time intervals is used to determine the WVP.

This kind of test is normally used for films utilized as packaging of dehydrated foods, which must be protected against ambient humidity (ASTM, 2000). The permeation cell must be made of noncorrosive material, with low weight and size adequate to the laboratory balances, impermeable, and resistant to handling during the experiments.

A variation of this procedure can be applied to determine the *WVP* of films intended for packages to protect high water activity foods. In this case, distilled water is placed inside the permeation cell, which is put into a container with low RH, and the weight loss measured in time is used for the *WVP* determination.

Another possibility for *WVP* assessment is the use of a thermosealed package itself, instead of a permeation cell. For this purpose, the package is filled with a desiccant, thermosealed, and placed in a container with higher RH, as previously described in the permeation cell method. This procedure can be used to determine the *WVP* of a film only if the film is hermetically sealed in the cell. Otherwise, it can be used to determine the WVP of the package itself, that is, the permeability through the film and the seal.

FIGURE 3.4 Permeation cells with O-rings, used to determine the water vapor permeability of films by the gravimetric method.

The film samples to be tested must be representative of the film and do not present defects or malformation, folds, creases, or microholes. For polysaccharides (e.g., starch) and protein films, which are generally prepared at laboratory scale by the casting technique, it is important to check if the thickness is homogeneous. Films prepared by casting have different surface roughness, because a film surface remains in contact with the support, while the other remains exposed to the air inside the drying oven. It is important to choose and to fix the surface that will be in contact with the higher RH, by which permeation will begin.

The use of a chamber with controlled temperature and RH is very important for the accuracy of the *WVP* determination. Both the chamber temperature and RH should not vary more than 1°C ($T \pm 1°C$) and 3% (RH $\pm$ 3%), respectively. Constant climate chambers specifically designed with controlled RH and temperature exist, but the easier way to maintain fixed RHs inside the permeation cell is with saturated salt solutions with excess of salt, which is able to absorb the water vapor that permeates through the cell (ASTM, 2001a). In order to maintain homogeneous conditions inside this chamber, a circulation air system must be provided, and the air must be able to circulate through the perforated shelves that support the permeation cells.

The WVP test does not reflect the actual conditions to which the film will be submitted in real scenarios. Normally, the real scenario is not taken into account for petrol-based plastic films (polyethylene, polypropylene, PVC, among others), because due to their hydrophobic properties the solubility coefficient of water in these materials does not depend on the RH of the atmospheres inside and outside the package. However, the literature reported the importance of the RH range on the WVP of hygroscopic films, like starch-based films. For the same difference of RH between the inside and the outside of the package, hygroscopic films are more permeable to water vapor at higher RH values (Müller et al., 2008). These authors presented a gravimetric procedure, adapted from the ASTM standard method, to determine the water vapor permeabilities (K_w) of hygroscopic films at different RH ranges, as shown in Figure 3.5. For that purpose, film samples (film discs of 90 mm diameter) are previously conditioned under three RHs, at constant temperature, for 24 h. After this preconditioning period, the samples are placed in the permeation cells with the saturated saline solution responsible for the lower a_w. The permeation cells (in triplicate) are placed in a chamber with the saturated saline solution providing the higher a_w. Small fans, installed inside the chamber, promote circulation of internal air and maintain a homogeneous a_w (Figure 3.5). The chamber with the permeation cells must be placed in a constant temperature environment, which can be accomplished by using different oven models or experimental isolated rooms. An example of the different experimental conditions that can be used to determine the WVP of hygroscopic films at different a_w ranges are presented in Table 3.1.

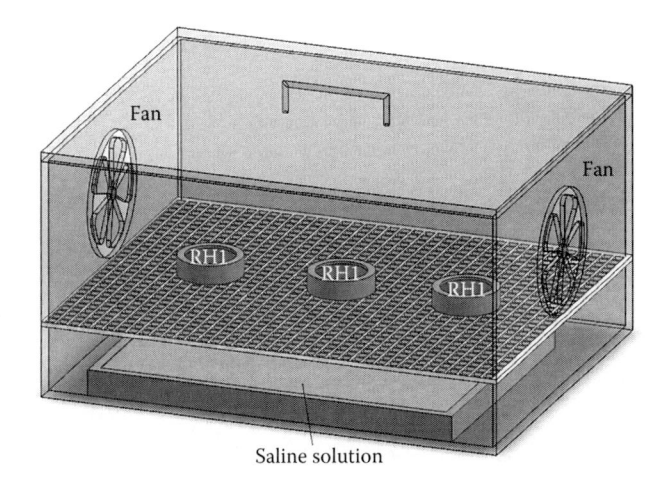

FIGURE 3.5 Sketch of the gravimetric procedure used to determine the water vapor permeability (K_w) of hygroscopic films.

TABLE 3.1

Examples of Possible Experimental Conditions in Order to Determine the
Water Vapor Permeability of Hygroscopic Films at Different a_w Ranges

a_w	Interior of Diffusion Cell	Exterior of Diffusion Cell (Chamber)
0.002–0.33	Calcium chloride ($CaCl_2$)	Magnesium chloride ($MgCl_2$)
0.33–0.62	Magnesium chloride ($MgCl_2$)	Sodium nitrite ($NaNO_2$)
0.62–0.90	Sodium nitrite ($NaNO_2$)	Barium chloride ($BaCl_2$)

As previously described, the water vapor transfer through each sample can be measured by the weight gain of the permeation cell as a function of time. The cells must be weighed at the beginning of the experiment and at fixed time intervals, until observing constant mass transfer rate, which is a characteristic of the steady-state condition. From a practical standpoint, one can consider that the steady state is reached when constant weight gain is observed after six consecutive weighings. It is very important to avoid manual contact with the permeation cell and film sample, which can impair the accuracy of the WVP results. The intervals between two weighings depend on the film WVP. For films with low WVP, the intervals can range from 12 to 24 h (as many petrol-based plastics), while in films, much more permeable to water vapor (as starch- or protein-based films) the interval can be 2 h. The permeation cells must remain outside the chamber only during the period necessary for temperature equilibration and weighing. This time and weighing procedure must be kept the same for all replicates.

In this way, for essays carefully performed as presented earlier, the water vapor transfer rate, in (g/h), is calculated by

$$W = \frac{\Delta m}{\Delta t} \tag{3.10}$$

where $\Delta m/\Delta t$ is the capsule mass variation over time, under the steady-state condition, calculated by the linear regression of the experimental mass transfer data, considering at least five consecutive data. With the values of W, the WVP (K_w, [g·mm/h·m²·Pa]) is calculated by

$$K_w = \frac{WL}{p_s S(a_{w1} - a_{w2})} \tag{3.11}$$

where
  L is the average thickness of the sample
  p_s is the saturation pressure of the vapor at the experimental temperature
  a_{w1} ($RH_1/100$) and a_{w2} ($RH_2/100$) are the water activities in the interior of the chamber and in the interior of the capsule, respectively

Table 3.2 presents water vapor permeabilities (K_w) determined by the steady-state method. These data show that K_w is highly dependent on the air RH and considerably increases at high RH.

TABLE 3.2

Water Vapor Permeabilities (K_w) of Films Determined by the Steady-State Method, under
Different Relative Humidity Ranges (at Equilibrium, RH = $100 \times a_w$)

	ΔRH (%)	Water Vapor Permeability ($\times 10^7$ g/m²·pa·h)
Starch film with 30 g glycerol/100 g of starch	11–33	1.8 ± 0.5
	33–64	6.2 ± 1.0
	64–90	8.4 ± 1.0

TABLE 3.3

Conversion Factors for Water Vapor Permeability

	$\dfrac{g \cdot \mu m}{m^2 \cdot s \cdot Pa}$	$\dfrac{g \cdot \mu m}{m^2 \cdot s \cdot mmHg}$	$\dfrac{g \cdot \mu m}{m^2 \cdot s \cdot bar}$	$\dfrac{g \cdot \mu m}{m^2 \cdot day \cdot Pa}$
$\dfrac{g \cdot \mu m}{m^2 \cdot s \cdot Pa}$	1	133.3224	1×10^5	86,400
$\dfrac{g \cdot \mu m}{m^2 \cdot s \cdot mmHg}$	0.007501	1	750.0617	648.0533
$\dfrac{g \cdot \mu m}{m^2 \cdot s \cdot bar}$	1×10^{-5}	0.001333	1	0.864000
$\dfrac{g \cdot \mu m}{m^2 \cdot day \cdot Pa}$	1.15741×10^{-5}	1.543083×10^{-3}	1.157407	1

There are some automatic systems for *WVP* determinations that promise an easy to use interface, simplicity in operation, with less operator-dependent test quality, and fast and accurate test results. These devices have automatic temperature control and equilibrium detection, which simplifies the determinations of *WVP*, and can be useful when many samples need to be tested.

The values of WVP can be presented in different units, such as $g \cdot \mu m/m^2 \cdot s \cdot Pa$, $g \cdot \mu m/m^2 \cdot s \cdot mmHg$, $g \cdot \mu m/cm^2 \cdot s \cdot Pa$, $g \cdot \mu m/m^2 \cdot s \cdot bar$, $g \cdot cm/m^2 \cdot dia \cdot mmHg$, or $g \cdot \mu m/m^2 \cdot dia \cdot mmHg$. In order to be able to compare different *WVP* values, results must be represented in the same units.

Table 3.3 provides some useful information to make easier *WVP* units conversion. To convert a permeability value from the units given in the first column (on the left) to the units given in the first row, we multiply the conversion value given in the box that corresponds to the intersection known–unknown permeability values.

3.2.2 Determination of the Film Permeability to Gases

As previously discussed, the barrier properties of a film to water vapors and gases depend on the solubility and diffusion coefficients of the permeating species. It is important to remark that a chemical species can have higher permeability through a film (compared to other species), even if its diffusion coefficient through this film is lower. However, this increased permeability occurs only if the solubility coefficient of the species A in the film material is higher enough to compensate its lower diffusivity. The comparison of sizes of the molecules of CO_2, O_2, and N_2 shows that the carbon dioxide is the biggest molecule, which implies that it has the lower diffusion coefficient through a given polymeric film. However, the CO_2 permeability in polymeric films is much higher than those of O_2 and N_2, which can be explained by its much higher solubility coefficient in the polymers (Sarantópoulos et al., 2002).

It is important to take into account the permeability of flexible films to oxygen and carbon dioxide in order to choose the film application in food packaging. Food with components susceptible to oxidation (fat, vitamins, and pigments) needs to be protected against the oxygen present in the ambient, which represents approximately 20% of the air volume (or approximately 23% of the air mass). The permeability rate of a film to oxygen is defined by the ASTM (2001b) as the amount of oxygen that crosses a unit area of a film, at each time unit, under constant temperature, RH, and gradient of oxygen vapor pressure difference between both sides of the film sample. The permeability is calculated by multiplying the permeability rate ($cm^3/m^2 \cdot dia$) by the film thickness and dividing by the oxygen partial pressure difference. Hence, films permeability can be expressed, for example, in $cm^3 \cdot \mu m/m^2 \cdot dia \cdot mmHg$ or $cm^3 \cdot \mu m/m^2 \cdot dia \cdot Pa$. Furthermore, as previously discussed for *WVP*, this test should be carried out with homogeneous and representative film samples, without defects, folds, creases, or microholes. During the film's gas permeability test, the temperature must be kept constant due to the fact that the solubility coefficient of the permeating species (O_2, CO_2) in the polymeric matrix and its diffusion coefficient through the film are extremely dependent on it. Besides, the temperature can change the RH, which consecutively influences the gas permeability. The film moisture (which depends on the RH for polysaccharide- and protein-based films) influences the permeability of hygroscopic films to gases. The water adsorbed and absorbed by the polymer acts as a plasticizer, increasing the polymer free volume and thus the film permeability to gases. Therefore, the conditioning of the film samples at given RH and temperature is very important in gas permeability tests of hygroscopic films (Sarantópoulos et al., 2002).

The determination of permeability to gases can be performed with the equal pressure method, at steady-state conditions, in which the film sample is fixed in the middle of a diffusion cell to separate two small chambers at atmospheric pressure. Nitrogen circulates in one chamber, while oxygen (or a mixture of oxygen with other gases) circulates in the other chamber. The oxygen that permeates the sample is carried by the nitrogen toward a coulometric detector, in which an electric current is generated. This current is proportional to the amount of oxygen that reaches the detector at each unit time. The RH at both chambers can be controlled to simulate the real conditions the film will be applied, considering, for example, the external ambient and the water activity of the food product (Sarantópoulos et al., 2002). The oxygen permeation through the film sample, under steady-state condition, can be reached after 30–60 min to 24–48 h, depending on the film. The oxygen permeability rate of a film is commonly presented in cm^3 (standard temperature and pressure [STP])/m²·dia, at given RH and STP. If the equipment used for the permeability determination does not have a coupled barometer, the ambient pressure needs to be measured to perform the correction of the gas permeability rate to the STP conditions. The permeability coefficient can be determined by multiplying the permeability rate by the film thickness and dividing by the difference of the oxygen partial pressure between the chambers. In fact, this difference is the oxygen partial pressure in the chamber where it circulates, as nitrogen is the only gas circulating in the other chamber.

Permeability results must be presented specifying the experimental conditions, such as the RH at both sides, the difference of oxygen partial pressure, the permeation area, and the number of replicates (Sarantópoulos et al., 2002).

The devices based on coulometric sensors are very practical and easy to use, but they are expensive. Nevertheless, an alternative method based on the concentration variation inside a chamber of a diffusion cell (permeation cell) can be also used for film gas permeability tests. This method has been used in tests of film permeability rates to oxygen and carbon dioxide. The diffusion cell is formed by three small chambers, two external and one in the middle, formed by the fixation of two film samples. The external chambers have inlet and outlet connections to gas circulation, while the internal chamber has both inlet and outlet connections to gas, and a microsyringe (500 µL) system adapted to collect gas samples during the permeation time. The chamber assembly is done with neoprene junctions, which define the permeation area (100 cm²) and the chamber volumes (approximately 50–70 mL). The gas flux needs to be measured and controlled in the range of 5–100 mL/min. Before the test, the film sample needs to be conditioned for 48 h at the RH it will be used. If the test is performed with dry film, it should be previously conditioned in a desiccator with silica gel or calcium chloride. For specific RH, the film sample can be conditioned in desiccator with water or with a saturated salt solution. For hygroscopic materials (e.g., starch- and protein-based films) susceptible to be used for packing foods with high water activity, the samples might be conditioned at RH of 75% or higher. High RH conditions can be easily obtained in a desiccator with sodium chloride saturated (75%) or barium chloride (90%) solutions, among others. Likewise, the RH of the permeating and carrying gases can be fixed by circulating these gases in glass containers with water or saturated salt solutions. As previously discussed, it is also important to keep constant temperature in the diffusion cell, which can be controlled by placing the permeation cell system inside a thermostatic chamber. Any possible leakage from the permeation cell can be detected by using an impermeable film (e.g., laminate with aluminum) as sample and carefully maintaining a constant internal chamber composition.

3.2.2.1 Influence of Temperature and Pressure on Film Permeability to Water Vapor and Gases

Pressure and temperature can influence the permeability of a given species through a film, because both variables can modify the diffusion and solubility coefficients. If the gas or water vapor does not interact with the polymeric chain, the permeability of the species does not depend on the pressure and temperature. For instance, this approximation is done for calculating the WVP through petrol-based films, and it mainly occurs because the solubility of water in these polymers is very low. On the other hand, if the permeating species interacts with the polymer, the permeability tends to increase at higher pressures, which can be explained by the positive influence of the pressure on the solubility and diffusion coefficients.

The influence of the temperature on the values of the diffusion coefficient is frequently represented by an Arrhenius type function, which is given by

$$D = D_o \exp\left(-\frac{E_D}{RT}\right)$$

where
    D is the diffusion coefficient at a given temperature (T)
    D_o is the diffusion coefficient at a reference temperature (T_o)
    R is the ideal gas constant
    E_D is the diffusion activation energy

Similarly, the following equations represent the influence of the temperature on the values of the permeability of a film to a given species (gas or water vapor) and the ability of the solubility coefficient of a species to affect a film material:

$$K = K_o \exp\left(-\frac{E_K}{RT}\right) \tag{3.13}$$

where
    K is the permeability at a given temperature (T)
    K_o is the diffusion coefficient at a reference temperature (T_o)
    R is the ideal gas constant
    E_K is the permeability activation energy

$$k_s = k_{s0} \exp\left(-\frac{E_s}{RT}\right) \tag{3.14}$$

where
    k_s is the solubility coefficient at a given temperature (T)
    k_{s0} is the solubility coefficient at a reference temperature (T_o)
    R is the ideal gas constant
    E_s is the activation energy for the solubility coefficient

REFERENCES

American Society for Testing and Materials (ASTM). Standard test method for water vapor transmission of materials. In *Standard E96-00. Annual Book of American Standard Testing Methods*, 8pp. Philadelphia, PA: ASTM, 2000.

American Society for Testing and Materials (ASTM). Standard practices for maintaining constant relative humidity by means of aqueous solutions. In *Standard E104-85. Annual Book of American Standard Testing Methods*, 3pp. Philadelphia, PA: ASTM, 2001a.

American Society for Testing and Materials (ASTM). Standard test method for oxygen gas transmission rate through plastic film and sheeting using a coulometric sensor. In *Standard D3985-95. Annual Book of American Standard Testing Methods*, 6pp. Philadelphia, PA: ASTM, 2001b.

Blanco-Pascual, N., Fernández-Martín, F., and Montero, P. Jumbo squid (*Dosidicus gigas*) myofibrillar protein concentrate for edible packaging films and storage stability. *LWT: Food Science and Technology* 55(2) (2014): 543–550.

Hatzidimitriu, E., Gilbert, S.G., and Loukakis, G. Odor barrier properties of multi-layer packaging films at different relative humidities. *Journal of Food Science* 52(2) (1987): 472–474.

Müller, C.M.O., Yamashita, F., and Laurindo, J.B. Evaluation of the effects of glycerol and sorbitol concentration and water activity on the water barrier properties of cassava starch films through a solubility approach. *Carbohydrate Polymers* 72 (2008): 82–87.

Nelson, K.L. and Fennema, O.R. Methylcellulose films to prevent lipid migration in confectionery products. *Journal of Food Science* 56(2) (1991): 504–509.

Romera, C.O., de Moraes, J.O., Zoldan, V.C., Pasa, A.A., and Laurindo, J.B. Use of transient and steady-state methods and AFM technique for investigating the water transfer through starch-based films. *Journal of Food Engineering* 109 (2012): 62–68.

Sarantópoulos, C.I.G.L., de Oliveira, L.M., Padula, M., Coltro, L., Alves, R.M.V., and Garcia, E.E.C. In C.I.G.L. Sarantópoulos, L.M. de Oliveira, M. Padula et al. (eds.), *Embalagens Plásticas Flexíveis: Principais Polímeros e Avaliação de Propriedades*. CETEA, ITAL, Campinas, Brazil, 2002.

Stading, M.T., Rindlav-Westling, Å., and Gatenholm, P. Humidity-induced structural transitions in amylose and amylopectin films. *Carbohydrate Polymers* 45(3) (2001): 209–217.

Section II

Traditional and Alternative Sources for Biopolymeric Film and Coating Matrices

4

Films and Coatings from Vegetable Protein

**Adriana Noemí Mauri, Pablo Rodrigo Salgado,
María Cecilia Condés, and María Cristina Añón**

CONTENTS

For several years, plant proteins have been the interest of both researchers and industrialists. The diversity, differences in physicochemical and nutritional properties, contributions to the health of consumers, low cost, and advances in production and processing technologies have made the storage proteins of grains and seeds an extremely attractive potential from the commercial point of view (Moure et al., 2006).

During the past decades, studies have been intensified to replace animal protein by proteins from other sources, including the plant storage proteins. The main sources of such proteins are cereals (e.g., wheat, corn, rice), legumes (e.g., peas, lentils, beans), pseudocereals (e.g., amaranth, quinoa), and those proteins found in grains and seeds rich in oil (e.g., sunflower, soybean, rapeseed, peanut, cotton). The protein content of all of these plant sources is wide, ranging from 35% to 40% for soybean and 7% to 9% for rice. Table 4.1 shows typical protein contents of major cereals, legumes, pseudocereals, and oilseeds.

These proteins fall into the category of sustainable biopolymers and have attracted considerable attention as potential substitutes for existing petroleum-based synthetic polymers in at least some applications, owing to the easy availability from renewable resources of these proteins and their ready biodegradability (Guilbert and Cuq, 2005). This chapter deals with different aspects related to the formation, characteristics, and applications of materials based on plant proteins.

4.1 Structural and Physicochemical Characteristics of Plant Storage Proteins

Proteins are heteropolymers, composed of about 20 α-amino acids that in addition to the amino and carboxyl groups involved in the formation of the peptide bonds contain side chains: these can be electrically charged or uncharged and either hydrophobic or polar, conferring on each amino-acid residue a

TABLE 4.1

Typical Protein Contents of Major Cereals,
Legumes, Pseudocereals, and Oilseed Sources

Source		Protein Content[a]
Cereals	Wheat	8%–15%
	Corn	9%–12%
	Rice	7%–9%
Legumes	Pea	20%–30%
	Lentil	20%–30%
	Bean	20%–25%
Pseudocereals	Amaranth	14%–20%
	Quinoa	12%–23%
Oilseed	Soybean	35%–40%
	Sunflower	15%–27%
	Rapeseed	17%–26%
	Peanut	25%–30%

[a] Data from Day (2013) and FAO (2014).

distinctive character (Damodaran, 1997). Most of these vegetable proteins contain 100–500 amino acids. Depending on their amino-acid sequence (i.e., their primary structure, it being specific to each polypeptide chain), the peptide backbone will assume different spatial orientations on its axis (i.e., the secondary structure—it is primarily stabilized by hydrogen bonding). The next level of protein architecture—the tertiary structure—reflects the three-dimensional organization of the polypeptide chain (it is based on hydrogen bonding, van der Waals forces, electrostatic and hydrophobic interactions, and covalent disulfide linkages) to form globular, fibrous, or randomly coiled protein conformations. Finally, the quaternary structure occurs as a result of the interaction among different polypeptide chains, whether or not they are identical, through any of the aforementioned types of noncovalent bonds to give the final native protein structure (Kannan et al., 2012).

According to Osborne (1924), plant proteins can be grouped into four categories: first, the water-soluble *albumins*; second, the *globulins*, those soluble in saline solutions; third, the *prolamins*—the group characteristic of cereal proteins—soluble in 60%–70% (v/v) aqueous alcohol; and fourth, the *glutelins*, soluble in neither water, saline, nor alcohol solutions, but extractable with alkali.

Today, this classification has been replaced by another based on the structure of the genes, the sequence homology of the constituent amino acids, and the mechanism of accumulation of the proteins in the plant storage bodies (Fukushima, 1991). On the basis of these criteria, two families of storage proteins have been characterized: the *globulins* and the *prolamins*.

The globulin family. Globulins are reserve proteins and constitute the majority of the legume proteins, though also being present in the monocotyledons, dicotyledons, and gymnosperms (Casey, 1999).

Within the globulin fraction, two kinds of proteins having sedimentation coefficients between 7 to 9S and 11 to 12S have been characterized, respectively, designated as the *vicilins* and the *legumins*. Both types of globulins share certain structural characteristics: their subunits consist of two domains (the N- and the C-terminal) of an equivalent structure, which suggests that the two were derived from a common ancestor, though some differences have been found in their overall structures that would be attributable to posttranslational modifications occurring during processing (Adachi et al., 2001, 2003; Argos et al., 1985; Shewry et al., 1995).

Structural analysis of the 11S globulins, the *legumins*, by various techniques confirmed a complex quaternary structure organized in hexamers (of *ca.* 300–360 kDa), composed of two trimers joined by hydrophobic interactions. All those subunits (of *ca.* 50–70 kDa) are held together by noncovalent interactions and are formed by an acidic polypeptide (polypeptide A, of *ca.* 30 kDa) and a basic polypeptide (polypeptide B, of *ca.* 20 kDa) linked by a disulfide bond whose position has been highly conserved

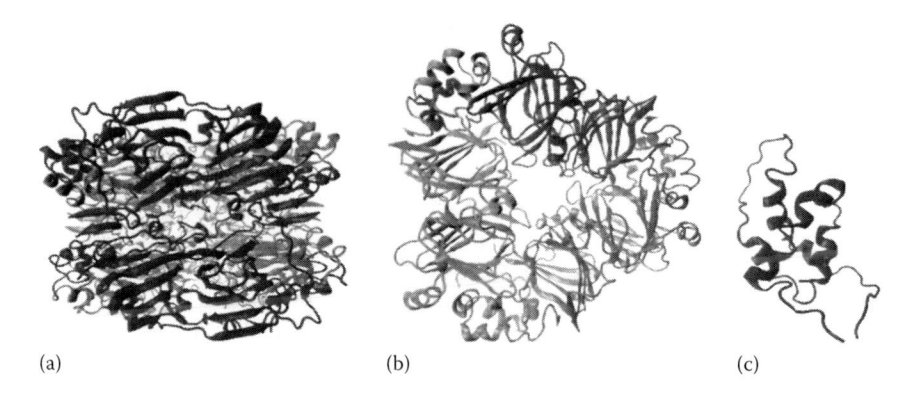

(a)   (b)   (c)

FIGURE 4.1 Three-dimensional molecular structures of: (a) glycinin, the 11S legumin of soybean (PDB: 1oD5); (b) β-conglycinin, the 7S vicilin of soybean (PDB: 1IPK); and (c) 2S albumin of sunflower (PDB: 1S6D).

among the different 11S globulins. Several legumins have been well characterized—such as the soybean glycinin (Figure 4.1a), the sunflower helianthinin, the bean and pea legumin, and the amaranthine of amaranth. These proteins exhibit a significant molecular heterogeneity originating in a polymorphism of the genes encoding these globulins. Similar physicochemical properties characterize the legumins— such as an average hydrophobicity, denaturation temperature, and enthalpy, their dissociation mechanisms, and other features. These proteins are soluble in salt solutions of high ionic strength and neutral pH (Adachi et al., 2001, 2003; González-Pérez and Vereijken, 2007; Marcone, 1999; Molina et al., 2004; Quiroga et al., 2009).

The *vicilins*—characterized by a sedimentation coefficient of between 7 and 9S—are structurally organized as trimers (of *ca.* 150 and 200 kDa), typically composed of two types of subunits (of *ca.* 70–80 and 50 kDa) that share a strong sequence homology and differ in the degree of glycosylation along with the presence or absence of processing sites in the amino-acid sequence that lead to the reduction in size. Unlike what was described earlier for the legumins, these subunits are not stabilized by disulfide bonds and have an isoelectric point of 5.5, but they are also soluble in saline solutions. Figure 4.1b shows the structure of β-conglycinin-7S vicilin of soybean (Maruyama et al., 2002; Quiroga et al., 2010).

The 2S storage proteins were initially identified on the basis of their sedimentation coefficient. The 2S albumins—heterodimeric proteins consisting of two polypeptide chains of 4 and 9 kDa that remain linked by four disulfide bonds—are widely distributed in dicotyledonous seeds. As with other storage proteins, they possess a high degree of polymorphism, making them quite varied in their structure and properties among different plant species. These proteins are water soluble, and at least some are structurally related to the prolamin superfamily (Anisimova et al., 1994; Shewry et al., 1995). Figure 4.1c shows the three-dimensional molecular structure of sunflower 2S-albumin.

The prolamin superfamily. Prolamins constitute the largest group among the storage proteins of cereals and the members of the grass family and include proteins of all cereals belonging to the tribes *Triticeae* (i.e., barley, rye, and wheat) and *Panicoideae* (i.e., corn, sorghum, and millet). Certain exceptions exist, such as rice and oats, in which the major storage proteins are 11S globulins, but also occasionally along with a lower proportion of prolamins (Shewry and Halford, 2002; Shewry and Tatham, 1990).

Most prolamins share two common structural features: (1) the presence of domains with different conformations and (2) sequences of amino acids enriched in specific residues, such as methionine, which are repeated along the chain. These features are responsible for the high proportion of glutamine, proline, and other amino acids (e.g., Phe, Met, Gly, His) that are specific to certain groups of prolamins (Kreis et al., 1985; Shewry and Halford, 2002).

The prolamins of the *Triticeae* tribe can be classified—according to the tribe's amino-acid sequence and composition—into three different categories: those that are sulfur rich and sulfur poor and of high molecular weight. These prolamins are highly polymorphic mixtures whose components have molecular weights ranging between 30 and 90 kDa (Shewry and Halford, 2002).

The sulfur-rich prolamins constitute approximately 70%–80% of the prolamin fraction. These proteins (of *ca.* 30–50 kDa) consist of both polymeric components (joined by interchain disulfide cross-linking) and monomers (containing intrachain disulfide bonds). In each species, at least two families can be distinguished: the β- and γ-hordeins in barley, two types of γ-secalins in rye, and the α- and γ-gliadins plus the low-molecular-weight-subunit glutenins in wheat. The structure of these proteins is characterized by the presence of two separate domains: an N-terminal with repeat sequences of one or two short peptide motifs rich in proline and glutamine and a C-terminal with nonrepeated sequences that possesses most or all of the conserved cysteine residues (Shewry et al., 1995).

The sulfur-poor prolamins, constituting approximately 10%–20% of the total prolamins, include the C-hordein in barley and the ω-secalins in rye and wheat. In all instances, the amino-acid sequence is characterized by the repeat of an octapeptide motif flanked by twelve residues at the N-terminus and by a unique short sequence of six residues in the C-hordeins or of four in the ω-secalins at the C-terminus. These prolamins—of molecular masses reported as ranging from 30 to 80 kDa—contain high levels of glutamine, proline, and phenylalanine but, lacking cysteine residues, cannot form polymers (Kasarda et al., 1983).

Prolamins of high molecular weight are minor components, constituting approximately 10% of the total prolamin fraction of wheat. These proteins have a central domain characterized by extensive repeat sequences containing different peptide motifs flanked by N- and C-terminal nonrepetitive domains. The variation in the length of the repeat domain is responsible for the variability in molecular size of the high-molecular-weight subunits, with their molecular masses ranging from 65 to 90 kDa (Lindsay and Skerritt, 1999).

4.2 Functional Properties of Plant Storage Proteins

The storage protein of grains and seeds described earlier do not exhibit the biologic activities characteristic of other proteins, but instead act as a source of the carbon and nitrogen required for germination of the seedling. Beyond that role, like other proteins, the globulins and prolamins possess different functional properties that contribute to the desirable organoleptic attributes of food. Kinsella (1979) defined protein functional properties as "Those physical and chemical properties which affect the behavior of proteins in food systems during storage, processing, preparation, and consumption." These properties can in principle be divided into three groups depending on the following characteristics (Pilosof and Bartholomai, 2000):

1. *The protein–water interactions (or hydration properties).* This category comprises water sorption and retention, swelling, wettability, solubility, dispersability, and viscosity.
2. *The surface properties.* The ability of a protein to function as an emulsifier or foaming agent depends on its rate of migration to the interface and the ability to form a stable interfacial film.
3. *The protein–protein interactions.* This category includes flocculation, coagulation, gelation, and the assembly into structures such as fibers, doughs, extruders, films, and coatings. The formation and characteristics of these films and coatings will be discussed in the following sections.

These properties are affected both by intrinsic characteristics (such as shape, size, amino-acid composition and sequence, protein structures [secondary, tertiary, and quaternary], distribution of net charges, the hydrophobicity-to-hydrophilicity ratio, and molecular flexibility or rigidity along with the protein's capacity to interact with other components in the food system) and by extrinsic conditions (such as pH, temperature, moisture, chemical additives, mechanical processing, enzymes, and ionic strength) (Damodaran and Paraf, 1997; Zayas, 1996).

4.3 Relationship between Protein Structure and Functional Properties

Because of the complex structure of proteins (primary, secondary, tertiary, and quaternary) and the close relationship between structure and function, improving the functionality of proteins is possible through subjecting them to chemical, physical, or enzymatic treatments that tend to cause any one or all of the following alterations:

1. *Changes in the molecular mass of the protein* through the dissociation of the quaternary structure (e.g., by mild heating, high hydrostatic pressure, reduction of interchain disulfide bridges, decreasing the ionic strength, acidification), through the hydrolysis of the polypeptide chains (enzymatically or chemically), or through aggregation (e.g., by heating at high protein concentration, variations in the ionic strength and acidity of the medium, the addition of specific ions such as Ca, the action of enzymes).

2. *Unfolding and/or denaturation of the protein* (e.g., by heating, variations in pH and/or ionic strength, pressure treatment, limited hydrolysis). The modifications of the bonds that stabilize the native structure of a given protein generally are accompanied by changes in molecular flexibility.

3. *Changes in the net charge of the protein molecule* (e.g., by variations in ionic strength, change in pH, chemical reactions such as deamidation or acetylation). A difference in the overall charge alters the degree of interaction with other protein molecules, with water, and with other ionic compounds.

4. *Modifications in the surface or exposed hydrophobicity* (e.g., by heating, high hydrostatic pressure, treatments with acids and bases, variation in the ionic strength). These changes affect the amphiphilic nature of proteins and their ability to act as surfactants and interact with water and other compounds.

Undoubtedly, these changes in protein's structural properties induced by physical, chemical, or enzymatic treatments constitute powerful alternatives for varying protein functionality—both positively and negatively—and in many instances for broadening the spectrum of functions that the protein from a particular source can exhibit (Damodaran and Paraf, 1997; Foegeding and Davis, 2011; Hall, 1996; Moure et al., 2006; Zayas, 1996).

4.4 Plant Protein Recovery to Prepare Edible Films

Numerous plant proteins—such as soybean, corn zein, wheat gluten, sunflower, peanut, cotton, rapeseed, rice, pea, sorghum, barley, amaranth, quinoa, and triticale—have been studied as potential film-forming agents (Cuq et al., 1998; Gennadios, 2002; Hernández-Izquierdo and Krochta, 2008; Zhang and Mittal, 2010). Extraction and isolation of these proteins involve solubilization and precipitation techniques based on the solubility and the isoelectric point of the protein molecules. Plant proteins are usually available in the form of meals, concentrates, or isolates (Moure et al., 2006), which differ in their protein content (on a dry-weight basis). The term protein *concentrate* or *isolate* is used when the concentration reaches 70% or 90%, respectively (FAO, 2014). Protein concentrates are usually obtained through the application of techniques involving the extraction of the nonproteinaceous material in the meal—consisting essentially in carbohydrates—to increase the concentration of the protein fraction in the resulting product. Those extraction techniques involve the use of 60%–80% (v/v) aqueous alcoholic solutions or water at a pH near the proteins' isoelectric point in order to minimize protein loss into the fraction extracted (Vioque et al., 2001). On the other hand, protein isolates are obtained through the solubilization of proteins in saline solutions and/or at a pH far from their isoelectric point, thus enabling the concentration of proteins by ultrafiltration, through diafiltration membranes, or by isoelectric precipitation. Thereafter, operations involving solid–liquid separations—such as centrifugation and (ultra)filtration—must be conducted in order to insure that the final drying step results in a pure protein powder (Vioque et al., 2001). These physicochemical treatments involved in the recovery of plant proteins can affect the nutritional value of the final products as well as their functional properties (Day, 2013).

Currently, a particular interest has arisen in those proteins derived from agroindustrial wastes or by-products for the production of biodegradable materials such as corn zein from ethanol production, soybean and sunflower proteins from oilcakes, because of their wider availability and lower cost compared to other protein sources and also to avoid competition for food resources, revalorize agricultural by-products and reduce environmental impacts as well as waste-disposal costs (Leceta et al., 2014; Rouilly and Rigal, 2002; Song et al., 2011; Zhang and Mittal, 2010). Nowadays, soybean-protein isolates and concentrates prepared from oilcake are commercially available. During the oil extraction process, the oilcake is obtained as a secondary product. Although it is mainly used for animal feed, it is used for the preparation of protein-enriched products, giving a greater commercial value to this agricultural by-product (Day, 2013; Garrido et al., 2014). From these considerations, the food industry would appear to be interested in extrapolating the experience obtained with soybean oilseeds to other protein sources such as sunflower and rapeseed.

4.5 Characteristics of Protein-Based Films and Coatings

Protein-based films can be prepared by two types of technology: a "wet process" based on protein dissolution or dispersion and a "dry process" based on the thermoplastic properties of the proteins under low hydration conditions (Cuq et al., 1998). Regardless of the processing technique used, the resulting protein material is essentially a three-dimensional reorganized matrix gel, with lower water content, where interactions between proteins and other components of the formulation are highly favored.

The type and number of interactions involved in the stabilization of the protein matrix (disulfide bonds and hydrogen bonding plus electrostatic and hydrophobic interactions) are determined by the amino-acid composition along with the molecular weight of the proteins under consideration (both of which characteristics can vary significantly for different proteins), the degree of protein denaturation, the experimental parameters used in film preparation, and the additives used. All of these conditions would determine the degree of cross-linking and the hydrophilic–hydrophobic character of the resulting protein network and would be reflected in its mechanical and barrier properties and its susceptibility to water (Gennadios, 2002; Mauri and Añón, 2012).

The most beneficial characteristics of protein materials reside in their edibility along with their inherent biodegradability: indeed, these protein matrices can be degraded either naturally or in compost and as such are environmentally friendly. The biodegradation kinetics of these materials depends on the type of protein (i.e., molecular weight, structure) as well as on the type of the additives used (e.g., plasticizers, cross-linkers). For example, materials formulated on a gluten base and obtained by casting or thermomolding become totally degraded after 36 days of aerobic fermentation and after 50 days in cultivated soil (Domenek et al., 2004). In addition, soybean-protein film was found to have disintegrated after 20 days in a simulated farmland-soil environment (Park et al., 2000). These latter values place proteins among those polymers exhibiting the most rapid degradation. Although no toxic effects were observed for either protein films or their metabolites in microbial growth-inhibition assays, in order to insure that such materials are environmentally safe, the toxicity and environmental friendliness must be evaluated through the use of authorized techniques (de Viegler, 2003). As indicated earlier, proteins can form edible films or coatings if all the constitutive ingredients are food grade and the processes used for their manufacture are suitable for food processing. This qualification, in the case of proteins, implies that the only permissible processing changes would be affected through heating, shifts in pH, the addition of salts, the action of enzymes, and/or removal of water. These edible materials could be included in the food product or serve as a covering of it and as such could furthermore enhance the nutritional value of the protected food through the addition of nutritional supplements (Krochta, 2002). In the case of plant protein containing endogenous antinutritional components such as protease inhibitors, glucosinolates, phytic acid, saponins, and gossypol, it is necessary to consider their separation or inactivation before or during the formation of the protein material. Moreover, in the case of soy, peanuts, and wheat proteins, which are common food allergens, their presence should be clearly labeled to provide consumers the corresponding information according to food legislation.

Because of their chemical nature, proteins have the capacity of forming materials with characteristics radically differing from those prepared from other types of polymers, particularly the synthetic ones widely used in the plastics industry—such as the low-density polyethylene (LDPE) and high-density polyethylene (HDPE), polypropylene (PP), polyvinylidene chloride (PVDC), and polyesters like polystyrene (PS). In comparison to films made from these purely synthetic materials, protein films normally exhibit excellent barrier capabilities with respect to oxygen, oils, and aromas along with moderate mechanical properties, though typically present a high permeability to water vapor (Gennadios, 2002). Moreover, protein-based films are highly sensitive to environmental conditions and most notably to the relative humidity (RH), because of their hygroscopic nature. Table 4.2 shows a comparison of the tensile

TABLE 4.2

Mechanical Properties and Water Vapor Permeability of Some Synthetic Polymer Films and Plant-Protein-Based Films

Films		Thickness (μm)	Tensile Strength (MPa)	Elongation (%)	T (°C)	RH (%)	Water Vapor Permeability ($\times 10^{-10}$ g H_2O/m·s·Pa)	T (°C)	ΔRH (%)	References
Synthetic	LDPE	25	13	500	—	—	0.0085	38	90	Gennadios et al. (1993)
	HDPE	25	26	300	—	—	0.0025	38	90	Gennadios et al. (1993)
	PP	—	33	300	—	—	0.005	—	—	Plackett (2011)
	PVDC	25	93	30	—	—	0.0016	38	90	Gennadios et al. (1993)
	PS	—	52	47	—	—	0.05	—	—	Plackett (2011)
	Polyesters (PET)	25	178	85	—	—	0.014	38	90	Gennadios et al. (1993)
Protein based[a]	Soybean	75	4.5	95	20	58	1.50	20	75	Salgado et al. (2010)
	Sunflower	75	4.0	25	20	58	1.45	20	75	Salgado et al. (2010)
	Rapeseed	61	3.7	20	25	50	25.5	20	50	Jang et al. (2011)
	Peanut	115	4.3	105	20	50	63.7	37	50	Jangchud and Chinnan (1999)
	Wheat gluten	88	0.9	260	25	50	6.20	20	50	Aydt et al. (1991)
	Corn zein	81	0.4	5	25	50	4.20	20	50	Aydt et al. (1991)
	Pea	—	0.5	75	20	60	2.90	20	48	Gueguén et al. (1998)
	Triticale	200	0.6	135	25	52	0.56	25	68	Aguirre et al. (2013)
	Lentil	150	4.2	58	25	50	3.10	25	68	Bamdad et al. (2006)
	Kidney bean	75	6.2	7	22	50	3.32	22	100	Ma et al. (2013)
	Amaranth	53	1.3	75	20	58	0.56	20	75	Condés et al. (2013)

[a] All protein-based films were plasticized by glycerol and were obtained by casting.

strength, elongation at break, and water vapor permeability (WVP) of synthetic polymers along with selected films formed from vegetable proteins.

Barrier properties to moisture, gases, flavor, aroma, radiation, and lipids are of particular interest in food packaging. In protein-based materials, these properties depend on the nature and density of the macromolecular network and most particularly on the proportion and distribution of the polar *versus* nonpolar amino acids (Cuq et al., 1998). In general, with protein materials, these permeabilities increase with the temperature and RH of the ambient so that the barrier properties of a protein film can only be compared under identical measuring conditions (Gennadios, 2002).

Protein films, because of their polar nature and inherent hydrophilicity, exhibit high WVPs similar to those of polysaccharide films (Krochta, 1997; McHugh and Krochta, 1994). Therefore, these films can be used as protective barriers against moisture exchange for only short periods or in low-moisture foods. They could also be used with certain other types of food products, such as meat pies and high-moisture cakes, whose packaging requires films that are highly permeable to water vapor (Gennadios et al., 1993). Furthermore, the application of protein coatings to food products before osmotic dehydration can prevent the loss of valuable water-soluble ingredients as a result of diffusion into the dehydration fluids, and conversely, the penetration of the dehydrating agent into the food itself (Dabrowska and Lenart, 2001). In this regard, among the vegetable-protein matrices, those of triticale and amaranth protein form films with the lowest WVPs (Aguirre et al., 2013; Condés, 2013).

The permeability of the protein-based materials to gases (oxygen, carbon dioxide, and ethylene) under low and intermediate RH conditions is considerably lower than that reported for LDPE and HDPE films (Guilbert and Cuq, 2005). The low permeability to oxygen is valuable because these materials would therefore provide protection against lipid oxidation (Kester and Fennema, 1986). Furthermore, whereas the barrier properties in those synthetic materials are stable and not affected by the RH of the environment, in protein materials, they are highly sensitive to both the ambient temperature and the RH (Cisneros-Zevallos and Krochta, 2002). For example, the permeability to oxygen and carbon dioxide of the latter materials at high RHs can become as much as a thousand times greater than at a level of 0%RH, though that effect of ambient moisture is far greater for the so-called "hydrophilic" gases such as CO_2 than for the "hydrophobic" ones such as O_2 (Guilbert and Cuq, 2005). This differential permeability accordingly results in an increase in the coefficient of selectivity for those two gases (i.e., CO_2/O_2) with temperature and RH that must be taken into account when designing packagings for foods in modified atmospheres or using these proteinaceous formulations as edible coatings. Atmospheres low in O_2 and high in CO_2 have been used to extend the shelf life of fresh-cut fruits and vegetables by reducing respiration, product perspiration, and ethylene production and have also been shown to effectively control the enzymatic browning, loss of firmness, and decay of those products (Rojas-Graü et al., 2009).

Likewise, because of their hydrophilic nature, protein films possess excellent barrier properties with respect to aromas and oils (Miller and Krochta, 1997; Trezza and Vergano, 1994), which feature constitutes an extremely useful characteristic for the packaging of ready-to-eat foods or products to be fried. Protein coating can accordingly decrease the absorption of oil by those products during frying, thus both reducing the fat content and calories and improving the nutritional quality of the final fried food (Mallikarjunan et al., 1997). Albert and Mittal (2002) reported that soybean-protein-isolate and wheat-gluten coatings provided the best reduction in fat uptake ($\approx99\%$) when used on pastry mixes and were very promising as coating materials for low-fat fried foods. These findings imply notable health benefits in the use of such food coatings given the relationship between fat consumption and the incidence of obesity and coronary disease (Varela and Fiszman, 2011).

When used as coatings, protein-based materials can enhance the organoleptic attributes of a food product, including the visual quality (i.e., color, glossiness) and tactile features (i.e., surface smoothness, nongreasiness, and/or stickiness). For example, zein is one of a few proteins used as a commercially successful finishing agent to impart surface gloss while also acting as an O_2, oil, and/or moisture barrier for nuts, candies, confectionery products, and other foods (Krochta and Mulder-Johnston, 1997).

Protein materials furthermore offer interesting perspectives in their application to the retention of food aromas or flavors when used as encapsulating wrappings for the appropriate types of food products. In fact, at this time, not many reports have appeared in the literature regarding the use of plant protein materials for such purposes in the way that carrageenans and milk proteins have (Madene et al., 2006).

In general, protein films are weaker and have shorter elongation than synthetic films (Hernández-Izquierdo and Krochta, 2008). Nevertheless, the mechanical behavior of protein materials is sufficient to permit their use in various applications such as coatings, wrappings, or packagings (Krochta, 2002). The production of protein films and coatings requires the addition of a plasticizer to reduce the extensive interactions among the constituent polymer molecules and thus decrease the material brittleness with a consequent increase in its flexibility and handling (Sothornvit and Krochta, 2005). The nature and concentration of the plasticizer used will also influence the properties of the resulting protein materials.

The mechanical properties of protein-based materials depend partially on the distribution and intensity of the inter- and intramolecular interactions generated in the formation of the three-dimensional network (Guilbert and Cuq, 2005), both of which parameters are greatly influenced by the degree of denaturation of the constituent proteins before forming the film. Since the interconnection of protein molecules during the drying process leads to the formation of the film matrix, the extension or unfolding of those proteins could favor that intermolecular linkage and more greatly influence where the junctions could be formed (Hoque et al., 2010). In this regard, films produced from soya proteins that had become completely denatured (and were mainly stabilized by disulfide bonds) possessed a higher tensile strength and greater elongation at the breakpoint than did the native soya-protein films or those composed of partially denatured sunflower proteins—both of these latter being stabilized by mainly hydrogen bonding and hydrophobic interactions (Denavi et al., 2009; Salgado et al., 2010). These mechanical properties were also extremely dependent on the ambient temperature and RH at which the measurements were made.

Protein-based materials prove soluble in water when the energy of the intermolecular bonds is less than those formed during solvolysis between water and the polar groups not involved in the network. The presence of the so-called physical nodes—the occurrence of intermolecular covalent bonds and/or a high matrix density for water penetration and interaction—is sufficient to produce films that are either partially or completely insoluble (Guilbert and Cuq, 2005). Therefore, solubility could be modified according to the processing conditions of film formation and the protein structure. Denavi et al. (2009) reported that the solubility of films prepared with native soybean protein by casting was affected by the processing drying conditions (temperature and RH), while the solubility of films prepared with totally denatured soybean proteins remained relatively constant ($\approx 40\%$) regardless of drying temperatures and humidities. And the higher solubility of native soybean-protein films was attributed to their greater hydrophilicity as a result of their higher amount of exposed polar groups for solubilization and their lower degree of cross-linking. Condés et al. (2013) reported that the unfolded conformation of thermally treated amaranth proteins, upon partial or total denaturation, favored the interactions between the constituent polypeptide chains during film formation. These intermolecular linkages—through mainly disulfide and hydrogen bonds—led to the more extensive cross-linking that became reflected in the greater tensile strength and lower water solubility of those films.

The solubility in water of a protein-based material also determines its potential application. Thus, solubility or the sensitivity to water of such a film is an advantage, for example, in the formulation of quick-dissolving sachets for instant foods, in the development of active material in which the water-imbibing capacity is used to induce a drastic change in the properties of the associated material, or in the liberation of some active compound (such as an antioxidant or antimicrobial agent) upon its delivery to a moist or wet target (Janjarasskul and Krochta, 2010; Ortiz et al., 2013). In contradistinction, if the continued integrity of the film is required to protect a food, then a water-insoluble protein formulation is needed (Krochta, 2002).

At the present time, the investigations on these protein-based materials have adopted two strategies:

1. To seek out applications for which the properties of these materials render them adequately: The thought here is not to replace their synthetic counterparts, but rather to constitute a substitute in specific applications—particularly in circumstances where the rate of use is high and a prolonged shelf life is not needed, such as with packaging for foods or the plastics used in agriculture. These latter types of application would appear both engaging and promising.
2. To improve or enhance the functionality of the protein materials: This possibility is dealt with in the following section.

4.6 Approaches to Improving or Enhancing the Functionality of Plant-Protein-Based Films

Numerous research efforts have been made to enhance the functionality of these materials. A special interest has been taken in improving their mechanical and barrier properties with an aim at widening their application, whether it be to replace those synthetic polymers within their own fields of use or to seek novel roles that are more specific and innovative. These efforts can be grouped according to the following strategies (Mauri and Añón, 2012).

4.6.1 Application of Physical and/or Chemical Treatments

As mentioned earlier, the protein structure influences the type of interactions and the degree of cross-linking of the final materials. In general, an increase in such linkages produces denser matrices with lower water solubilities and greater mechanical resistances, while the types of interactions involved in the stabilization of the matrix influence the magnitude of the change in each property (Mauri and Añón, 2006, 2008; Salgado et al., 2010). Different film-forming dispersions or admixtures prepared with proteins from soybean, amaranth, gluten, peanut, and other sources were heated, treated with acids or alkalis, hydrolyzed, or exposed to high pressure or irradiation before beginning the process of film formation (Condés et al., 2012, 2013; Guerrero and de la Caba, 2010; Lui et al., 2004; Mauri et al., 2006; Roy et al., 1999). Those interventions produced changes in the initial structure of the proteins that usually became reflected more significantly in the mechanical properties and the water solubility than in the permeability to water vapor. The most significant improvements occurred when the formation of sulfhydryl bridges were favored in the stabilization of the protein matrices.

The processing methodologies utilized in the formation of protein materials, just as the operational variables employed, can result in products having differing properties. Thus, for example, the conditions of drying used (e.g., temperature, RH) for obtaining soya-protein films through casting or thermocompression affected the drying speed of the film-producing admixture or dispersion, in which the parameter, in turn, determined the type and proportion of the interactions between the protein chains and as a consequence the film's mechanical properties, solubility, and permeability to water vapor (Denavi et al., 2009; Foulk and Bunn, 2001; Guerrero et al., 2010). In the processing of proteins by extrusion, the parameters most influential for modifying the resulting protein structure are the temperature and the shearing force. These variables produce a major restructuring of the protein material through the unfolding of the polymer chains, their breakage as a result of the mechanical stress, and their elevated reaction rate at higher temperatures (Kumar et al., 2010; Rouilly et al., 2006; Wang and Padua, 2003). In contrast to the gluten-protein films produced by casting, the protein materials obtained from gluten by processes similar to those employed to transform thermoplastic polymers by thermomechanical treatments exhibited changes in the profiles of molecular-weight distribution (Redl et al., 1999).

Physical and chemical treatments of the preformed films in order to modify the properties of the resulting materials are also possible. A thermal treatment after the formation of the films (a curing treatment) affects film properties noticeably: in general, the resistance to stretching and the coloration are increased, while at the same time, the film's elongation and moisture content are decreased, with the permeability to water vapor tending to differ upon variation of the protein source (Gennadios et al., 1996; Liu et al., 2004; Micard et al., 2000; Rhim et al., 2000). In addition, ultraviolet and gamma radiation can affect the film's constituent proteins through the induction of conformational changes, the oxidation of amino acids, the cleavage of covalent bonds, the formation of free radicals, and the generation of recombination and polymerization reactions (Bourtoom, 2009; Rhim et al., 1999). In this regard, Lee et al. (2004, 2005) using gamma rays and Gennadios et al. (1998) employing ultraviolet light succeeded in enhancing the resistance to stretching of soy-protein-based films.

4.6.2 Incorporation of Additives

One of the advantages of plant protein films and coatings is that they can vehiculize additives into the matrix to enhance functionality. This may include additives capable of

1. Improving or modifying the basic functionality of the materials—for example, plasticizers, emulsifiers, and cross-linkers
2. Enhancing the material functionality by increasing the quality, stability, and safety of the product the film or coating protects, especially foods—for example, antioxidants, antimicrobial agents, nutraceuticals, flavors, and/or coloring agents (Han, 2005)

The first additives were incorporated in order to modify the mechanism of film formation. To that end, in the majority of the investigations, low-molecular-weight hydrophilic plasticizers—for example, glycerol, sorbitol, sugars, and other polyols, ethylene glycol, and ethanolamine and their respective derivatives— were added to formulations based on plant proteins (Hernández-Izquierdo and Krochta, 2008; Song et al., 2011; Vieira et al., 2011). The physical properties of protein films were found to be strongly influenced by the type and amount of plasticizer used. The addition of the aforementioned compounds—usually at concentrations varying between 15% and 40% of the weight of the matrix biopolymer—not only enhanced the film's flexibility and resilience, but also resulted in a decrease in the film's strength and its ability to act as a barrier to moisture, oxygen, aroma, and oils, and increased the susceptibility of the matrix to environmental humidity (Vanin et al., 2005). The inclusion of hydrophobic plasticizers, such as citrate esters as an alternative to reduce these drawbacks, has also been investigated (Andreuccetti et al., 2009). The addition of plasticizers also causes a substantial diminution in the glass transition temperature (Tg) and influences the viscoelastic and fluidity properties of the material-forming dispersions: with the combination of plasticizers and elevated temperature, the proteins pass through the glass transition to produce a rubbery mass that can be shaped and then stabilized as such by cooling and/or eliminating volatile plasticizers. The correct selection of the appropriate plasticizers requires a consideration of the issues of plasticizer compatibility, efficiency, permanence, and economics in order to optimize the mechanical properties of a given film with a minimum increase in its permeability (Sothornvit and Krochta, 2005).

Emulsifiers are amphiphilic compounds that in addition to serving as plasticizers can modify surface energy to control the adhesion and wettability of the coating surfaces (Krochta, 2002). The following ones have been widely used in the food industry to develop plant-protein-based films: the Tweens, the Spans, fatty-acid salts, and phospholipids such as lecithin (Andreuccetti et al., 2010, 2011).

Another strategy is to use chemical cross-linkers (generally, low-molecular-weight aldehydes such as formaldehyde, glyoxal, and glutaraldehyde) that react with the amino and sulfhydryl moieties of the proteins to form covalent intra- and intermolecular bridges. But as aldehydes are potentially toxic, their use is limited to nonfood applications (Marquie et al., 1995). In contrast, enzymatic cross-linking—for example, with transglutaminase, lipoxidase, lysyl oxidase (the protein–lysine-6 oxidase), polyphenol oxidase, and peroxidase—by not involving such restrictions, would offer the greater potentiality for general application. In both examples, an increase in the degree of cross-linking produces matrices that are more mechanically resistant and less soluble, though at times also less permeable to water vapor (Bourtoom, 2009). For example, Liu et al. (2004) reported that the inclusion of formaldehyde and glutaraldehyde caused a significant increase in the tensile strength of peanut-protein films and reduced the permeability to water vapor and oxygen. Stuchell and Krochta (1994) found that peroxidase-cross-linked soybean-protein films were extremely brittle. Although those films had a significantly higher elastic modulus than that of the non-cross-linked films, the moisture barriers of the two films were similar. In another study, Yildirim and Hettiarachchy (1998) observed that soybean 11S-globulin films cross-linked by transglutaminase had higher mechanical strength than the non-cross-linked controls, but also exhibited a significantly increased permeability to water vapor.

The ability of edible wrappings to carry and control the release of active compounds is a promising upcoming food packaging and medical function attracting intense current research interest and much attention from both the food and the pharmaceutical industries. Other additives have been incorporated to confer specific functionalities on films with an aim at improving the safety, nutritional value, quality, and/or appearance of foods, such as antioxidants, antimicrobials, vitamins, microorganisms, probiotics, flavors, and pigments. Synthetic antioxidants (BHA, BHT, and n-propyl gallate) and antimicrobials (propionic, ascorbic, benzoic, and sorbic acids and their salts) were investigated initially for the formation of active materials for food packaging (Han, 2005). Nowadays, the current trend in such materials is to incorporate natural additives such as vegetable extracts (Jang et al., 2011; Sivarooban et al., 2008) or essential oils with their inherent antioxidant and antimicrobial properties (Atarés et al., 2010; Emiroğlu et al., 2010; Salgado et al., 2012, 2013) along with phenolic compounds and protein hydrolysates (Salgado et al., 2011); the last of which will be treated in the following sections. All of those compounds, in addition to increasing the functionality of the protein films, can modify the physicochemical properties of the resulting materials because they may also act as a cross-linking agents or plasticizers (Orliac et al., 2002; Ou et al., 2005; Salgado et al., 2010, 2011). Therefore, the retention and/or release of those additives will depend mainly on the interactions they can establish with the protein matrix and the other films' components. Finally, regardless of the promising properties that these incorporated substances might exhibit *in vitro*, the relevance to their usefulness in the protection of foodstuffs must still be determined. Therefore, it is very important to move on applications in real systems.

4.6.3 Formation of Composite and Nanocomposite Materials

At the present time, the formation of composite and nanocomposite materials represents one of the most widely studied and auspicious approaches to improving the properties of vegetable-protein-based materials. These proteins can form composites consisting of two or more layers—that is, bilayered or multilayered—in an interpenetrated network formed by two continuous phases (i.e., involving blends of proteins and other polymers) or, alternatively, a continuous protein matrix with a dispersed discontinuous filler (e.g., discrete particles, fibers, sheets) such that, depending on the filler's dimensions, the overall material is referred to as a composite, microcomposite, or nanocomposite (Vinson and Sierakowski, 2008). Figure 4.2 shows the structural scheme of possible protein-based composites.

Thus, for example, the formation of the following composite materials has been studied:

- *Materials obtained from admixtures of vegetable proteins with other polymers.* These materials have been formulated with an aim at combining the good selective permeability to gases of protein-based films with the superior mechanical properties and/or resistance to moisture characteristic of other types of polymers. Films have been obtained with vegetable proteins laminated or blended with other biopolymers such as chitosan (Ma et al., 2013), alginate (Jia et al., 2009), or carrageenans (Sanchez-García et al., 2010), starch (Rhim et al., 1999), carboxymethylcellulose (Su et al., 2010), and other plant or animal proteins (Denavi et al., 2009), with synthetic biodegradable polymers such as polyvinyl alcohol (Su et al., 2007), and even with nonbiodegradable polymers such as polyethylene and PP (Lee et al., 2008).

- *Materials obtained from admixtures of vegetable proteins and lipids.* The incorporation of lipids into protein films has the potential to improve the moisture barrier. Lipids can be added in

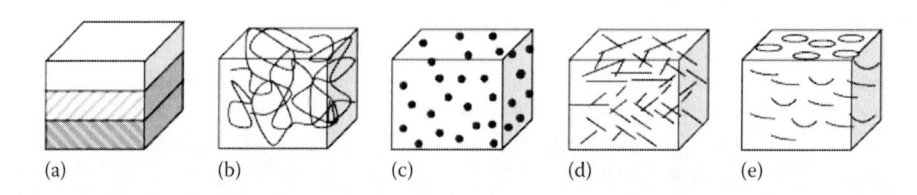

(a)   (b)   (c)   (d)   (e)

FIGURE 4.2   Different types of protein composite materials. A continuous protein matrix plus another continuous phase: (a) multilayered or (b) interpenetrated. A continuous protein matrix with a dispersed discontinuous filler: (c) discrete particles, (d) fibers, and (e) sheets.

the form of a continuous layer over the protein phase or can be dispersed within the hydrophilic matrix (Morillon et al., 2002). In both such bilayer and emulsion films, the WVP depends on mainly the polarity, the degree of saturation of the lipid type, and the nature of the interactions that become established between the lipids and the protein matrix. Among lipids, waxes were the most effective to improve the properties of the protein materials (Debeaufort et al., 2000; Quezada-Gallo et al., 2000). Fatty acids and essential oils have also been studied for the formation of composite protein films (Atarés et al., 2010; Monedero et al., 2009). Bilayer films usually provide a higher moisture barrier than those containing emulsions, but the delamination and the requirement of several additional steps in the formulation of bilayer films make emulsion films more suitable for food applications (Debeaufort et al., 2000; Gontard et al., 2007; Quezada-Gallo et al., 2000).

- *The incorporation of reinforcing elements into protein formulations.* The reason for the inclusion of a reinforcement in the formulation of protein-based materials is that those components could provide technical characteristics to the protein matrix—that is, an enhanced rigidity and consequent structural stability, a resistance to mechanical impact, a decrease in the permeabilities to gases and liquids, and a change in electrical properties. These added features depend on the physicochemical structure and function of the reinforcements along with the inherent properties of the protein matrix, the percent composition of the filler in the formulation, the nature of the interface between the filler and the protein matrix, and the final morphology of the system (Vinson and Sierakowski, 2008). By these means, the mechanical properties of films composed of proteins from zein and/or gluten have been improved by the incorporation of glass fiber (Beg et al., 2005), while those from soya through reinforcement with long, short, or microcrystalline cellulose fibers or with lignocellulose or lignin (Wang et al., 2003). Protein-based materials reinforced with natural fibers always continue to provide the advantages of low cost, biodegradability, and sustainability.

The recent incorporation of nanoreinforcements—fillers having at least one dimension within the 1–100 nm range—has become popular as an approach to improving the properties of the materials. The resulting nanocomposite materials can often exhibit advantageous mechanical and barrier properties, greater heat stability, and chemical resistance, along with an improved outward appearance, even at low incorporation levels (i.e., ≤5%–10% by weight), owing to their high aspect ratio and the extensive nanoparticle surface area that allows a strong interaction with the polymer matrix (Arora and Padua, 2010; Rhim and Ng, 2007; Zhao et al., 2008).

Improvements in nanoreinforced films can be achieved if and when the fillers can be dispersed uniformly within the protein matrix, in which distribution depends on the reinforcement concentration and the chemical affinity of the fillers for the protein matrix. The addition of a variety of nanofillers, such as nanoclays, silica nanoparticles, carbon nanotubes, cellulose nanofibers, and starch nanocrystals, has been found to improve diverse physical and barrier properties of the resulting films (de Azeredo, 2009). For example, the addition of montmorillonite improved the mechanical behavior and permeability to water vapor and gases of soya- and gluten-protein films constructed by different processing techniques (Echeverría et al., 2014; Tunc et al., 2007), the inclusion of microcrystalline cellulose bettered the mechanical properties of soya-protein films (Wang et al., 2013), and the incorporation of nanocrystals of starch strengthened the matrices in soya- and amaranth-protein films (Condés, 2012). These nanocomposites can be used as vehicles for the delivery of active agents and systems for their controlled liberation (Mascheroni et al., 2010). In this regard, the presence of differing proportions of reinforcements can modulate the transfer characteristics of active compounds through the film because of differences produced in protein network by the presence of these fillers and as a result of the consequent variable degree of tortuosity incurred along the mean free path of liberation (Tunc et al., 2007). In spite of the substantial benefits that nanomaterials can confer on the technology of protein-based films, the present lack of information regarding the possible toxicity of such nanoparticles to human health and to the environment must nevertheless be borne in mind. This uncertainty would underscore the necessity for some form of proactive legislation to regulate the manufacture, usage, and disposal of nanomaterials until such time as they are proven safe.

4.7 Contribution of the Presence of Nonprotein Compounds Coextracted with Proteins to the Biologic Activity and Color of Films

During the process of extracting vegetable proteins, certain compounds associated with them—such as the isoflavones in soybean and the phenolic compounds in sunflower—become coextracted (Salgado et al., 2010; Speroni et al., 2010). The presence of those components gives a certain characteristic chromatic to the proteins that is reflected in the resulting films, unlike what happens with the proteins extracted from animal sources—such as gelatin, whey, and myofibrils—where the coextracted components, and thus the resulting films, are colorless. For example, soybean-protein films are light yellow (Denavi et al., 2009), and those of amaranth protein are light brown (Condés et al., 2013), whereas sunflower-protein films could reach a dark-green coloration due to the oxidation of phenolic compounds to o-quinones during the extraction of the protein in alkaline medium (Salgado et al., 2010). The occurrence of an intense color limits certain potential applications of these materials in food packaging. For example, such packaging could not be used for products that should be readily inspected within the packaging (such as minimally processed vegetables) because the impaired visualization could reduce the acceptability to potential consumers. In contrast, such films could be used—if their properties were proven adequate—for applications in which color is irrelevant or in those uses in which a form of chromatic filter may be an advantage, as in the situation with packaging products that are sensitive to a part of the visible-light spectrum (Salgado et al., 2010). In addition, phenolic compounds can exhibit both antioxidant and antimicrobial activities that would, presumably, persist in the resulting films. Thus, for example, the phenols coextracted with sunflower proteins (mainly chlorogenic acid) conferred significant antioxidant properties on the resulting films in a concentration-dependent function, whether free or protein-bound (Salgado et al., 2012). Although Salgado et al. (2010) saw no modification of the mechanical and barrier properties of those sunflower-protein-based films, Orliac et al. (2002) reported that other phenols such as gallic acid could also act as a cross-linking agent in other protein matrices.

4.8 Peptides and Protein Hydrolysates as Sources of Active Compounds for Film Production

Although certain food proteins act directly in their intact form to elicit physiologic effects, the peptides resulting from protein digestion, hydrolysis, or fermentation are considered the most relevant, and these peptides frequently display a greater bioactivity than the parent protein (Korhonen and Pihlanto, 2006). The beneficial health effects of bioactive peptides include antihypertensive, antiobesity, immunomodulatory, antidiabetic, hypocholesterolemic, anticancer, antioxidative, and antimicrobial activities (Udenigwe and Aluko, 2012). Numerous plant proteins—such as amaranth (Tironi and Añón, 2010), soya (Gibbs et al., 2004), sunflower (Megías et al., 2004), pea, chickpea, and lentil (Roy et al., 2010)—constitute a source of hydrolysates and peptides with certain of these health-promoting activities. Those protein-derived preparations could be used to activate protein films. The antioxidant and antimicrobial activities, however, would be the most appropriate for use in food packaging. The incorporation of the increasing percentages of gelatin hydrolysate into squid-skin-gelatin films and the addition of bovine-plasma-protein hydrolysate to soybean- and sunflower-protein-based films have both been reported to result in increased antioxidant activities (measured by the ferric-reducing ability of plasma and the 2,2′-azino-*bis*-(3-ethylbenzothiazoline-6-sulfonic acid) [ABTS] assays, respectively). But those supplements also produced a detriment in the films' mechanical properties and WVPs because the added peptides plasticized the two protein networks, though without significantly altering the appearance of the films (Giménez et al., 2009; Salgado et al., 2011). Moreover, the other biological activities may be very interesting to develop patches or materials for medicine.

4.9 Legislation, Trends, and Applications

Plant proteins (categorized as sustainable biopolymers) can be used to form materials having unique properties because of their complex structure—and, in turn, the variability of the latter depending on the protein source employed—in combination with the method of processing used and the environmental conditions. Accordingly, considerable attention has been given recently to the possibility of using plant proteins as potential substitutes for existing petroleum-based synthetic polymers in at least some short-term applications and/or when a rapid degradation offers an advantage, such as with food packaging. It occurs because of the low cost of those proteins, their easy availability from reproducible resources, and their facile biodegradability. Furthermore, plant proteins used in this field could be an interesting alternative to increase the use of agricultural by-products.

The gradual implementation of laws on the use of biodegradable materials along with the consumer pressure in response to environmental problems generated from the massive use of synthetic plastics and the uncertain future supply of petroleum would drive the real interest in this area.

While a very important development of these materials, in terms of improving its properties and search for specific applications (as discussed earlier), has been reached at research level, this development has not been successfully transferred to industry level with the same intensity yet, and thus these materials are not used massively as would. In this sense, it is necessary to advance in the industrial production of these materials, and for this, it is necessary to strengthen the interaction between the scientific community and industry.

The main challenge confronting the use of these biodegradable protein-based materials in food packaging is the necessity of being certain to maintain their functionality with efficiency during the required time and then become degraded only thereafter. To that end, investigations are focusing on the development of protein-based materials with better barrier properties, in which the difference would imply thinner and lighter protein films as well as a reduction in the cost of processing, production, transport, and waste disposal.

The limitless possibilities of modifying or amplifying the functionality of these protein-based materials will eventually determine their massive implementation. Because of the strict structure–function relationship inherent in the protein materials, it is possible to produce reasonable modifications that can successfully be used in certain applications. Furthermore, it is necessary to evaluate these possible applications in real systems, such as the effectiveness of a given packaging will be influenced by interactions with the internally packaged product and the immediate external ambient. These studies will further optimize the development of materials according to the required functionality.

Finally, no single film or coating is appropriate, or even adequate, for all of the applications involving foods, agriculture, or nutraceuticals. On the contrary, each protein-based material—be it in the form of a film or a coating—possesses properties compatible with a given concrete application, hence the diversity of these systems.

REFERENCES

Adachi, M., Kanamori, J., Masuda, T., Yagasaki, K., Kitamura, K., Mikami, B., and Utsumi, S. Crystal structure of soybean 11S globulin: Glycinin A3B4 homohexamer. *Proceedings of the National Academy of Science*, 100(72), (2003): 7395–7400.

Adachi, M., Takenaka, Y., Gidamis, A.B., Mikami, B., and Utsumi, S. Crystal structure of soybean proglycinin A1aB1b homotrimer. *Journal of Molecular Biology*, 305, (2001): 291–305.

Aguirre, A., Borneo, R., and León, A.E. Properties of triticale protein films and their relation to plasticizing–antiplasticizing effects of glycerol and sorbitol. *Industrial Crops and Products*, 50, (2013): 297–303.

Albert, S. and Mittal, G.S. Comparative evaluation of edible coatings to reduce fat uptake in a deep-fried cereal product. *Food Research International*, 35, (2002): 445–458.

Andreuccetti, C., Carvalho, R.A., Galicia-García, T., Martínez-Bustos, F., and Grosso, C.R.F. Effect of surfactants on the functional properties of gelatin-based edible films. *Journal of Food Engineering*, 103, (2011): 129–136.

Andreuccetti, C., Carvalho, R.A., and Grosso, C.R.F. Effect of hydrophobic plasticizers on functional properties of gelatin-based films. *Food Research International*, 42, (2009): 1113–1121.

Andreuccetti, C., Carvalho, R.A., and Grosso, C.R.F. Gelatin-based films containing hydrophobic plasticizers and saponin from *Yucca schidigera* as the surfactant. *Food Research International*, 43, (2010): 1710–1718.

Anisimova, I., Fido, R.J., Tatham, A.S., and Shewry, P.R. Heterogeneity and polymorphism of sunflower seed 2S albumins. *Biotechnology & Biotechnological Equipment*, 7, (1994): 63–69.

Argos, P., Narayana, S.V., and Nielsen, N.C. Structural similarity between legumin and vicilin storage proteins from legumes. *EMBO Journal*, 4, (1985): 1111–1117.

Arora, A. and Padua, G.W. Revier: Nanocomposites in food packaging. *Journal of Food Science*, 75, (2010): R43–R49.

Atarés, L., De Jesús, C., Talens, P., and Chiralt, A. Characterization of SPI-based edible films incorporated with cinnamon or ginger essential oils. *Journal of Food Engineering*, 99, (2010): 384–391.

Aydt, T.P., Weller, C., and Testin, R.F. Mechanical and barrier properties of edible corn and wheat protein films. *Transactions of the ASAE*, 34, (1991): 207–211.

Bamdad, F., Goli, A.H., and Kadivar, M. Preparation and characterization of proteinous film from lentil (*Lens culinaris*). *Food Research International*, 39, (2006): 106–111.

Beg, M.D.H., Pickering, K.L., and Weal, S.J. Corn gluten meal as a biodegradable matrix material in wood fibre reinforced composites. *Materials Science and Engineering*, 412, (2005): 7–11.

Bourtoom, T. Edible protein films: Properties enhancement. *International Food Research Journal*, 16, (2009): 1–9.

Casey, R. Distribution and some properties of seed globulins. In P.R. Shewry and R. Casey (eds.), *Seed Proteins*, pp. 159–169. Norwell, MA: Kluwer Academic Publishers, 1999.

Cisneros-Zevallos, L. and Krochta, J.M. Internal modified atmospheres of coated fresh fruits and vegetables: Understanding relative humidity effects. *Journal of Food Science*, 67, (2002): 1990–1995.

Condés, M.C. Películas compuestas y nanocompuestas, biodegradables y/o comestibles, en base a proteínas de amaranto y almidones de distinto origen botánico. PhD Thesis. UNLP, Buenos Aires, Argentina, 2012.

Condés, M.C., Añón, M.C., and Mauri, A.N. Amaranth protein films from thermally treated proteins. *Journal of Food Engineering*, 119, (2013): 573–579.

Cuq, B., Gontard, N., and Guilbert, S. Proteins as agricultural polymers for packaging production. *Cereal Chemistry*, 75, (1998): 1–9.

Dabrowska, R. and Lenart, A. Influence of edible coatings on osmotic treatment of apples. In P.F. Maupoei (ed.), *Osmotic Dehydration and Vacuum Impregnation—Applications in Food Industries*, pp. 43–49. Lancaster, PA: Technomic Publishing Company, 2001.

Day, L. Proteins from land plants—Potential resources for human nutrition and food security. *Trends in Food Science & Technology*, 32, (2013): 25–42.

Damodaran, S. Amino acids, peptides and proteins. In O.R. Fennema (ed.), *Food Chemistry*, pp. 321–431. New York: Marcel Dekker, 1997.

Damodaran, S. and Paraf, A. *Food Proteins and their Applications*. New York: Marcel Dekker, 1997.

de Azeredo, H.M.C. Nanocomposites for food packaging applications. *Food Research International*, 42, (2009): 1240–1253.

de Viegler, J.J. Green plastics for food packaging. In R. Ahvenainen (ed.), *Novel Food Packaging Techniques*, Chapter 24, Cambridge, U.K.: Woodhead Publishing Limited, 2003.

Debeaufort, F., Quezada-Gallo, J.A., Delporte, B., and Voilley, A. Lipid hydrophobicity and physical state effects on the properties of bilayer edible films. *Journal of Membrane Science*, 180, (2000): 47–55.

Denavi, G., Tapia Blácido, D.R., Añón, M.C., Sobral, P.J.A., Mauri, A.N., and Menegalli, F.C. Effects of drying conditions on some physical properties of soy protein films. *Journal of Food Engineering*, 90, (2009): 341–349.

Denavi, G.A., Pérez-Mateos, M., Añón, M.C., Montero, P., Mauri, A.N., and Gómez-Guillén, M.C. Structural and functional properties of soy protein isolate and cod gelatin blend films. *Food Hydrocolloids*, 23, (2009): 2094–2101.

Domenek, S., Feuilloley, P., Gratraud, J., Morel, M.-H., and Guilbert, S. Biodegradability of wheat gluten based bioplastics. *Chemosphere*, 54, (2004): 551–559.

Echeverría, I., Eisenberg, P., and Mauri, A.N. Nanocomposites films based on soy proteins and montmorillonite processed by casting. *Journal of Membrane Science*, 449, (2014): 15–26.

Emiroğlu, Z.K., Yemiş, G.P., Coşkun, B.K., and Candoğan, K. Antimicrobial activity of soy edible films incorporated with thyme and oregano essential oils on fresh ground beef patties. *Meat Science*, 86, (2010): 283–288.

FAO. Food and Agriculture Organization. http://www.fao.org/docrep/t0532e/t0532e06.htm and http://www.fao.org/docrep/t0532e/t0532e07.htm (last accessed, October 2014).

Foegeding, E.A. and Davis, J.P. Food protein functionality: A comprehensive approach. *Food Hydrocolloids*, 25, (2011): 1853–1864.

Foulk, J.A. and Bunn, J.M. Properties of compression-molded, acetylated soy protein films. *Industrial Crops and Products*, 14, (2001): 11–22.

Fukushima, D. Structures of plan storage proteins and their functions. *Food Reviews International*, 7(3), (1991): 353–381.

Garrido, T., Etxabide, A., Leceta, I., Cabezudo, S., de la Caba, K., and Guerrero, P. Valorization of soya by-products for sustainable packaging. *Journal of Cleaner Production*, 64, (2014): 228–233.

Gennadios, A. *Protein-Based Films and Coatings*. Boca Raton, FL: CRC Press, 2002.

Gennadios, A., Brandenburg, A.H., Weller, C.L., and Testin, R.F. Effect of pH on properties of wheat gluten and soy protein isolate films. *Journal of Agricultural and Food Chemistry*, 41, (1993): 1835–1839.

Gennadios, A., Ghorpade, V.M., Weller, C.L., and Hanna, M.A. Heat curing of soy protein films. *Transactions of the ASAE*, 39, (1996): 575–579.

Gennadios, A., Rhim, J.W., Handa, A., Weller, C.L., and Hanna, M.L. Ultraviolet radiation affects physical and molecular properties of soy protein films. *Journal of Food Science*, 63, (1998): 225–228.

Gibbs, B.F., Zougman, A., Masse, R., and Mulligan, C. Production and characterization of bioactive peptides from soy hydrolysate and soy-fermented food. *Food Research International*, 37, (2004): 123–131.

Giménez, B., Gómez-Estaca, J., Alemán, A., Gómez-Guillén, M.C., and Montero, M.P. Improvement of the antioxidant properties of squid skin gelatin films by the addition of hydrolysates from squid gelatine. *Food Hydrocolloids*, 23, (2009): 1322–1327.

Gontard, N., Marchesseau, S., Cuq, J.L., and Guilbert, S. Water vapour permeability of edible bilayer films of wheat gluten and lipids. *International Journal of Food Science & Technology*, 30, (2007): 49–56.

González-Pérez, S. and Vereijken, J.M. Sunflower proteins: Overview of their physicochemical, structural and functional properties. *Journal of the Science of Food and Agriculture*, 87, (2007): 2173–2191.

Gueguén, J., Viroben, G., Noireaux, P., and Subirade, M. Influence of plasticizers and treatments on the properties of films from pea proteins. *Industrial Crops and Products*, 7, (1998): 149–157.

Guerrero, P. and de la Caba, K. Thermal and mechanical properties of soy protein films processed at different pH by compression. *Journal of Food Engineering*, 100, (2010): 261–269.

Guerrero, P., Retegi, A., Gabilondo, N., and de la Caba, K. Mechanical and thermal properties of soy protein films processed by casting and compression. *Journal of Food Engineering*, 100, (2010): 145–151.

Guilbert, S. and Cuq, B. Material formed by proteins. In C. Bastioli (ed.), *Handbook of Biodegradable Polymers*, pp. 339–384. Shawbury, U.K.: Rapra Technology Limited, 2005.

Hall, G.M. *Methods of Testing Protein Functionality*. Suffolk, VA: Chapman & Hall, 1996.

Han, J.H. *Innovations in Food Packaging*. San Diego, CA: Elsevier Academic Press, 2005.

Hernández-Izquierdo, V.M. and Krochta, J.M. Thermoplastic processing of proteins for film formation—A review. *Journal of Food Science*, 73, (2008): R30–R39.

Hoque, M.S., Benjakul, S., and Prodpran, T. Effect of heat treatment of film-forming solution on the properties of film from cuttlefish (*Sepia pharaonis*) skin gelatine. *Journal of Food Engineering*, 96, (2010): 66–73.

Jang, S.A., Lim, G.O., and Song, K.B. Preparation and mechanical properties of edible rapeseed protein films. *Journal of Food Science*, 76, (2011): C218–C223.

Jang, S.A., Shin, Y.J., and Song, K.B. Effect of rapeseed protein–gelatin film containing grapefruit seed extract on 'Maehyang' strawberry quality. *International Journal of Food Science & Technology*, 46, (2011): 620–625.

Jangchud, A. and Chinnan, M.S. Properties of peanut protein film: Sorption isotherm and plasticizer effect. *Lebensmittel-Wissenschaft und -Technologie*, 32, (1999): 89–94.

Janjarasskul, T. and Krochta, J.M. Edible packaging materials. *Annual Review of Food Science and Technology*, 1, (2010): 415–448.

Jia, D., Fang, Y., and Yao, K. Water vapor barrier and mechanical properties of konjac glucomannan–chitosan–soy protein isolate edible films. *Food and Bioproducts Processing*, 87, (2009): 7–10.

Kannan, A., Hettiarachchy, N., and Marshall, M. Food proteins-peptides: Chemistry and structure. In N.S. Hettiarachchy (ed.), *Food Proteins and Peptides*, pp. 1–24. Boca Raton, FL: CRC Press, 2012.

Kasarda, D.D., Autran, J.C., Lew, E.J.L., Nimmo, C.C., and Shewry, P.R. N-terminal amino acids sequences of w-gliadinas y w-secalinas: Implication for the evolution of prolamin genes. *Biochimica et Biophysica Acta*, 747, (1983): 138–150.

Kester, J.J. and Fennema, O.R. Edible films and coatings: A review. *Food Technology*, 40, (1986): 47–59.

Kinsella, J.E. Functional properties of soy proteins. *Journal of the American Oil Chemists' Society*, 56, (1979): 242–258.

Korhonen, H. and Pihlanto, A. Bioactive peptides: Production and functionality. *International Dairy Journal*, 16, (2006): 945–960.

Kreis, M., Forde, B.G., Rahman, S., Miflin, B.J., and Shewry, P.R. Molecular evolution of the seed storage proteins of barley, rye and wheat. *Journal of Molecular Biology*, 183, (1985): 499–502.

Krochta, J.M. Edible protein films and coatings. In S. Damodaran and A. Paraf (eds.), *Food Proteins and Their Applications*, pp. 529–549. New York: Marcel Dekker, 1997.

Krochta, J.M. Proteins as raw materials for films and coatings: Definitions, current, status, and opportunities. In A. Gennadios (ed.), *Protein-Based Films and Coatings*, pp. 1–42. Boca Raton, FL: CRC Press, 2002.

Krochta, J.M. and Mulder-Johnston, C.D. Edible and biodegradable polymer films: Challenges and opportunities. *Food Technology*, 51, (1997): 61–74.

Kumar, P., Sandeep, K.P., Alavi, S., Truong, V.D., and Gorga, R.E. Preparation and characterization of bio-nanocomposite films based on soy protein isolate and montmorillonite using melt extrusion. *Journal of Food Engineering*, 100, (2010): 480–489.

Leceta, I., Etxabide, A., Cabezudo, S., de la Caba, K., and Guerrero, P. Bio-based films prepared with by-products and wastes: Environmental assessment. *Journal of Cleaner Production*, 64, (2014): 218–227.

Lee, J.W., Son, S.M., and Hong, S.I. Characterization of protein-coated polypropylene films as a novel composite structure for active food packaging application. *Journal of Food Engineering*, 86, (2008): 484–493.

Lee, M., Lee, S., and Song, K.B. Effect of γ-irradiation on the physicochemical properties of soy protein isolates films. *Radiation Physics and Chemistry*, 72, (2004): 35–40.

Lee, S.L., Lee, M.S., and Song, K.B. Effect of gamma-irradiation on the physicochemical properties of gluten films. *Food Chemistry*, 92, (2005): 621–625.

Lindsay, M.P. and Skerritt, J.H. The glutenin macropolymer of wheat flour doughs: Structure–function perspectives. *Trends in Food Science & Technology*, 10, (1999): 247–253.

Liu, C.C., Tellez-Garay, A.M., and Castell-Perez, M.E. Physical and mechanical properties of peanut protein films. *Lebensmittel-Wissenschaft und -Technologie*, 37, (2004): 731–738.

Ma, W., Tang, C.H., Yang, X.Q., and Yin, S.W. Fabrication and characterization of kidney bean (*Phaseolus vulgaris* L.) protein isolate-chitosan composite films at acidic pH. *Food Hydrocolloids*, 31, (2013): 237–247.

Madene, A., Jacquot, M., Scher, J., and Desobry, S. Flavour encapsulation and controlled release—A review. *International Journal of Food Science & Technology*, 41, (2006): 1–21.

Mallikarjunan, P., Chinnan, M.S., Balasubramaniam, V.M., and Phillips, R.D. Edible coatings for deep-fat frying of starchy products. *Lebensmittel-Wissenschaft und -Technologie*, 30, (1997): 709–714.

Marcone, M.F. Biochemical and biophysical properties of plant storage proteins: A current understanding with emphasis on 11S globulins. *Food Research International*, 32, (1999): 79–92.

Marquie, C., Aymard, C., Cuq, J.L., and Guilbert, S. Biodegradable packaging made from cottonseed flour: Formation and improvement by chemical treatments with gossypol, formaldehyde, and glutaraldehyde. *Journal of Agricultural and Food Chemistry*, 43, (1995): 2762–2767.

Maruyama, N., Ramlan, M., Salleh, M., Takahashi, K., Yagasaki, K., Goto, H., Honatani, N., Nakagawa, S., and Utsumi, S. Structure–physicochemical function relationships of soybean β-conglycinin heterotrimers. *Journal of Agricultural and Food Chemistry*, 50 (2002): 4323–4326.

Mascheroni, E., Chalier, P., Gontard, N., and Gastaldi, N. Designing of a wheat gluten/montmorillonite based system as carvacrol carrier: Rheological and structural properties. *Food Hydrocolloids*, 24, (2010): 406–413.

Mauri, A.N. and Añón, M.C. Effect of solution pH on solubility and some structural properties of soybean protein isolate films. *Journal of the Science of Food and Agriculture*, 86, (2006): 1064–1072.

Mauri, A.N. and Añón, M.C. Mechanical and physical properties of soy protein films with pH-modified microstructures. *Food Science and Technology International*, 14, (2008): 2119–2125.

Mauri, A.N. and Añón, M.C. Proteínas como envases alimentarios. In G.I. Olivas Orozco, G.A. González-Aguilar, O. Martín-Belloso, and R. Soliva-Fortuny (eds.). *Películas y Recubrimientos Comestibles: Propiedades y aplicaciones en alimentos*, pp. 95–124. México: CLAVE, 2012.

McHugh, T.H. and Krochta, J.M. Water vapor permeability properties of edible whey protein-lipid emulsion films. *Journal of American Oil Chemists Society*, 71, (1994): 307–312.

Megías, C., Yust, M.M., Pedroche, J., Lquari, H., Girón-Calle, J., Alaiz, M., Millán, F., and Vioque, J. Purification of an ACE inhibitory peptide after hydrolysis of sunflower (*Helianthus annuus* L.) protein isolates. *Journal of Agricultural and Food Chemistry*, 52, (2004): 1928–1932.

Micard, V., Belamri, R., Morel, M.H., and Guilbert, S. Properties of chemically and physically treated wheat gluten films. *Journal of Agricultural and Food Chemistry*, 48, (2000): 2948–2953.

Miller, K.S. and Krochta, J.M. Oxygen and aroma barrier properties of edible films: A review. *Trends in Food Science & Technology*, 8, (1997): 228–237.

Molina, M.I., Petruccelli, S., and Añón, M.C. Effect of pH and ionic strength modifications on thermal denaturation of the 11S globulin of sunflower (*Helianthus annuus*). *Journal of Agricultural and Food Chemistry*, 52, (2004): 6023–6029.

Monedero, F.M., Fabra, M.J., Talens, P., and Chiralt, A. Effect of oleic acid–beeswax mixtures on mechanical, optical and water barrier properties of soy protein isolate based films. *Journal of Food Engineering*, 91, (2009): 509–515.

Morillon, V., Debeaufort, F., Blond, G., Capelle, M., and Voilley, A. Factors affecting the moisture permeability of lipid-based edible films: A review. *Critical Reviews in Food Science and Nutrition*, 42, (2002): 67–89.

Moure, A., Sineiro, J., Domínguez, H., and Parajó, J.C. Functionality of oilseed protein products: A review. *Food Research International*, 39(9), (2006): 945–963.

Orliac, O., Rouilly, A., Silvestre, F., and Rigal, L. Effects of additives on the mechanical properties, hydrophobicity and water uptake of thermo-moulded films produced from sunflower protein isolate. *Polymer*, 43, (2002): 5417–5425.

Ortiz, C.M., Mauri, A.N., and Vicente, A. Use of soy protein based 1-methylcyclopropene-releasing pads to extend the shelf life of tomato (*Solanum lycopersicum* L.) fruit. *Innovative Food Science and Emerging Technologies*, 20, (2013): 281–287.

Osborne, T.B. *The Vegetable Proteins*. London, U.K.: Longmans, Green and Company, 1924.

Ou, S., Wang, Y., Tang, S., Huang, C., and Jackson, M.G. Role of ferulic acid in preparing edible films from soy protein isolate. *Journal of Food Engineering*, 70, (2005): 205–210.

Park, S.K., Hettiarachchy, N.S., and Were, L. Degradation behavior of soy protein–wheat gluten films in simulated soil conditions. *Journal of Agricultural and Food Chemistry*, 48, (2000): 3027–3031.

Plackett, D. *Biopolymers—New Materials for Sustainable Films and Coatings*. Cornwall, U.K.: John Wiley & Sons, 2011.

Pilosof, A.M.R. and Bartholomai, G.B. *Caracterización funcional y estructural de proteínas*. Buenos Aires, Argentina: EUDEBA, 2000.

Quezada-Gallo, J.A., Debeaufort, F., Callegarin, F., and Voilley, A. Lipid hydrophobicity, physical state and distribution effects on the properties of emulsion-based edible films. *Journal of Membrane Science*, 180, (2000): 37–46.

Quiroga, A.V., Martínez, N.E., Rogniaux, H., Geairon, A., and Añón, M.C. Globulin-p and 11S-globulin from amaranthus hypochondriacus: Are two isoforms of the 11S-globulin. *The Protein Journal*, 28, (2009): 457–467.

Quiroga, A.V., Martínez, N.E., Rogniaux, H., Geairon, A., and Añón, M.C. Amaranth (*Amaranthus hypochondriacus*) vicilin subunit structure. *Journal of Agricultural and Food Chemistry*, 58, (2010): 12957–12963.

Redl, A., Morel, M.H., Bonicel, J., Guilbert, S., and Vergnes, B. Rheological properties of gluten plasticized with glycerol: Dependence on temperature, glycerol content and mixing conditions. *Rheologica Acta*, 38, (1999): 311–320.

Rhim, J.W., Gennadios, A., Fuc, D., Weller, C.L., and Hanna, M.A. Properties of ultraviolet irradiated protein films. *LWT: Food Science and Technology*, 32, (1999): 129–133.

Rhim, J.W., Gennadios, A., Handa, A., Weller, C.L., and Hanna, M.A. Solubility, tensile, and color properties of modified soy protein isolate films. *Journal of Agricultural and Food Chemistry*, 48, (2000): 4937–4941.

Rhim, J.W. and Ng, P.K.W. Natural biopolymer-based nanocomposite films for packaging applications. *Critical Reviews in Food Science and Nutrition*, 47, (2007): 411–433.

Rhim, J.W., Wu, Y., Weller, C.L., and Schnepf, M. Physical characteristics of a composite film of soy protein isolate and propyleneglycol alginate. *Journal of Food Science*, 64, (1999): 149–152.

Rojas-Graü, M.A., Oms-Oliu, G., Soliva-Fortuny, R., and Martín-Belloso, O. The use of packaging techniques to maintain freshness in fresh-cut fruits and vegetables: A review. *International Journal of Food Science & Technology*, 44, (2009): 875–889.

Rouilly, A., Mériaux, A., Geneau, C., Silvestre, F., and Rigal, L. Film extrusion of sunflower protein isolate. *Polymer Engineering & Science*, 46, (2006): 1635–1640.

Rouilly, A. and Rigal, L. Agro-materials: A bibliographic review. *Journal of Macromolecular Science, Part C: Polymer Reviews*, 42, (2002): 441–479.

Roy, F., Boye, J.I., and Simpson, B.K. Bioactive proteins and peptides in pulse crops: Pea, chickpea and lentil. *Food Research International*, 43, (2010): 432–442.

Roy, S., Weller, C.L., Gennadios, A., Zeece, M.G., and Testin, R.F. Physical and molecular properties of wheat gluten films cast from heated film-forming solutions. *Journal of Food Science*, 64, (1999): 57–60.

Salgado, P.R., Fernández, G.B., Drago, S.R., and Mauri, A.N. Addition of bovine plasma hydrolysates improves the antioxidant properties of soybean and sunflower protein-based films. *Food Hydrocolloids*, 25, (2011): 1433–1440.

Salgado, P.R., López-Caballero, M.E., Gómez-Guillén, M.P., Mauri, A.N., and Montero, M.P. Exploration of the antioxidant and antimicrobial capacity of two sunflower protein concentrate films with naturally present phenolic compounds. *Food Hydrocolloids*, 29, (2012): 374–381.

Salgado, P.R., López-Caballero, M.E., Gómez-Guillén, M.P., Mauri, A.N., and Montero, M.P. Sunflower protein films incorporated with clove essential oil have potential application for the preservation of fish patties. *Food Hydrocolloids*, 33, (2013): 74–84.

Salgado, P.R., Molina Ortiz, S.E., Petruccelli, S., and Mauri, A.N. Biodegradable sunflower protein films naturally activated with antioxidant compounds. *Food Hydrocolloids*, 24, (2010): 525–533.

Sanchez-García, M.D., Hilliou, L., and Lagaron, J.M. Nanobiocomposites of carrageenan, zein, and mica of interest in food packaging and coating applications. *Journal of Agricultural and Food Chemistry*, 58, (2010): 6884–6894.

Shewry, P.R. and Halford, N.G. Cereal seed storage proteins: Structures, proteins and role in grain utilization. *Journal of Experimental Botany*, 53, (2002): 947–948.

Shewry, P.R., Napier, J.A., and Tatham, A.S. Seed storage proteins: Structures and biosynthesis. *Plant Cell*, 7, (1995): 945–956.

Shewry, P.R. and Tatham, A.S. The prolamin storage proteins of cereal seeds: Structure and evolution. *Biochemical Journal*, 267, (1990): 1–12.

Sivarooban, T., Hettiarachchy, N.S., and Johnson, M.G. Physical and antimicrobial properties of grape seed extract, nisin, and EDTA incorporated soy protein edible films. *Food Research International*, 41, (2008): 781–785.

Song, F., Tang, D.L., Wang, X.L., and Wang, Y.Z. Biodegradable soy protein isolate-based materials: A review. *Biomacromolecules*, 12, (2011): 3369–3380.

Sothornvit, R. and Krochta, J.M. Plasticizers in edible films and coatings. In J.H. Han (ed.), *Innovations in Food Packaging*, pp. 403–433. San Diego, CA: Elsevier Academic Press, 2005.

Speroni, F., Milesi, V., and Añón, M.C. Interactions between isoflavones and soybean proteins: Applications in soybean-protein–isolate production. *LWT: Food Science and Technology*, 43, (2010): 1265–1270.

Stuchell, Y.M. and Krochta, J.M. Enzymatic treatments and thermal effects on edible soy protein films. *Journal of Food Science*, 59, (1994): 1332–1337.

Su, J.F., Huang, Z., Liu, K., Fu, L.L., and Liu, H.R. Mechanical properties, biodegradation and water vapor permeability of blend films of soy protein isolate and poly (vinyl alcohol) compatibilized by glycerol. *Polymer Bulletin*, 58, (2007): 913–921.

Su, J.F., Huang, Z., Yuan, X.Y., Wang, X.Y., and Li, M. Structure and properties of carboxymethyl cellulose/soy protein isolate blend edible films crosslinked by Maillard reactions. *Carbohydrate Polymers*, 79, (2010): 145–153.

Tironi, V.A. and Añón, M.C. Amaranth proteins as a source of antioxidant peptides: Effect of proteolysis. *Food Research International*, 43, (2010): 315–322.

Trezza, T.A. and Vergano, P.J. Grease resistance of corn zein coated paper. *Journal of Food Science*, 59, (1994): 912–915.

Tunc, S., Angellier, H., Cahyana, Y., Chalier, P., Gontard, N., and Gastaldi, E. Functional properties of wheat gluten/montmorillonite nanocomposite films processed by casting. *Journal of Membrane Science*, 289, (2007): 159–168.

Udenigwe, C.C. and Aluko, R.E. Food protein-derived bioactive peptides: Production, processing, and potential health benefits. *Journal of Food Science*, 77, (2012): R11–R24.

Vanin, F.M., Sobral, P.J.A., Menegalli, F.C., Carvalho, R.A., and Habitante, A.M.Q.B. Effects of plasticizers and their concentrations on thermal and functional properties of gelatin-based films. *Food Hydrocolloids*, 19, (2005): 899–907.

Varela, P. and Fiszman, S.M. Hydrocolloids in fried foods: A review. *Food Hydrocolloids*, 25, (2011): 1801–1812.

Vieira, M.G.A., da Silva, M.A., dos Santos, L.O., and Beppu, M.M. Natural-based plasticizers and biopolymer films: A review. *European Polymer Journal*, 47, (2011): 254–263.

Vinson, J.R. and Sierakowski, R.L. *The Behavior of Structures Composed of Composite Materials.* Dordrecht, the Netherlands: Kluwer Academic Publishers, 2008.

Vioque, J., Sánchez-Vioque, R., Pedroche, J., Yust, M.M., and Millán, F. Production and uses of protein concentrates and isolates. *Grasas y aceites*, 52, (2001): 127–131.

Wang, Y. and Padua, G.W. Tensile properties of extruded zein sheets and extrusion blown films. *Macromolecular Materials and Engineering*, 288, (2003): 886–893.

Wang, Z., Sun, X.X., Lian, Z.X., Wang, X.X., Zhou, J., and Ma, Z.S. The effects of ultrasonic/microwave assisted treatment on the properties of soy protein isolate/microcrystalline wheat-bran cellulose film. *Journal of Food Engineering*, 114, (2013): 183–191.

Yildirim, M. and Hettiarachchy, N.S. Properties of films produced by cross-linking whey proteins and 11S globulin using transglutaminase. *Journal of Food Science*, 63, (1998): 248–252.

Zayas, J.F. *Functionality of Proteins in Food.* Berlin, Germany: Springer-Verlag, 1996.

Zhang, H. and Mittal, G. Biodegradable protein-based films from plant resources: A review. *Environmental Progress & Sustainable Energy*, 29, (2010): 203–220.

Zhao, R., Torley, P., and Halley, P.J. Emerging biodegradable materials: Starch- and protein-based bio-nanocomposites. *Journal of Materials Science*, 43, (2008): 3058–3071.

5

Films and Coatings from Animal Protein

Joaquín Gómez-Estaca, M. Carmen Gómez-Guillén, and María Pilar Montero García

CONTENTS

5.1 Introduction

Edible films and coatings are used to improve food appearance and shelf life by preventing chemical, physical, or biological damage (Fabra et al., 2009). Furthermore, edible films formed as coatings or placed between food components provide possibilities for improving the quality of heterogeneous foods by limiting the migration of moisture, lipids, flavors/aromas, and colors between food components, carrying food ingredients (antimicrobials, antioxidants, nutraceuticals, etc.), and/or improving the mechanical integrity or handling characteristics (Krochta, 1992). Edible films are generally good barriers to oxygen at low and intermediate relative humidity (RH) and present selective permeability to gases (high CO_2/O_2 permeability relationship in comparison with other synthetic polymers). The main drawback of edible films and coatings is their great sensitivity to water, which causes their mechanical properties, their oxygen barrier, and even their integrity to be seriously compromised (Hernandez-Munoz et al., 2005). Besides the mere use as packaging, edible films and coatings may be a design or form part of the food, contributing to product development with great imagination and fantasy. Thanks to its versatility and adaptability, as well as to the possibility of including a large number of compounds of various natures (aromas, colorants, etc.).

Apart from the aforementioned uses of edible films and coatings to improve food quality and shelf life, their production and utilization have also benefits from an environmental point of view. On the one hand, to the degree that an edible film or coating acts as efficient moisture, oxygen, or aroma barrier, the amount and/or complexity of traditional and nonbiodegradable packaging can be reduced. On the other hand, edible films and coatings are commonly formulated from biopolymers extracted from by-products or coproducts from the agrofood industry, thus contributing to give value to such biomass and reducing environmental impact and costs related to their elimination.

5.2 Important Properties of Proteins for the Production of Edible Films and Coatings

Proteins are heteropolymers whose monomer units are α-amino acids. The combinations of the 20 existing amino acids provide an almost unlimited number of different polymer chains, with diverse physicochemical properties. Proteins differ from polysaccharides in that the latter consist of a very limited number of monomers. For example, cellulose and starch, which are the two most common polysaccharide biopolymers, consist of the repetition of a single monomer, glucose (Hernandez-Izquierdo and Krochta, 2008). Moreover, proteins contain a great variety of functional groups, which make it possible to alter them enzymatically, chemically, or physically, varying the properties of the materials obtained in order to adjust them to the specific needs of each application (Hernandez-Munoz et al., 2005).

Preparation of protein-based films requires formation of a continuous low-moisture and more or less ordered macromolecular network containing numerous and uniformly distributed interactions between polymer chains. Such kind of interactions may be hydrophobic, electrostatic, hydrogen bonding, or covalent bonding. The probability of forming intermolecular bonds mainly depends on protein shape (fibrous *vs.* globular) and on physicochemical conditions during processing. High-molecular-weight fibrous proteins generally can easily form films with good mechanical properties, whereas globular proteins generally need to be unfolded previously to film formation (Cuq, 2002). Commonly, unfolding entails the exposition of buried functional groups that allow the establishment of protein–protein interactions that otherwise could not be produced. Special attention is to be paid to disulfide bonds in the case of proteins containing the amino acid cysteine (e.g., β-lactoglobulin, keratin, muscular proteins, or ovalbumin), as the films stabilized by such kind of strong covalent bond generally present improved mechanical and barrier properties (Balaguer et al., 2014; Lagrain et al., 2010).

There are two main methods for producing edible films from proteins: wet processing and dry processing. Edible coatings are produced by wet processing, as it is mandatory for the preparation of a film-forming solution containing the polymer and other additives (plasticizers, antioxidants, antimicrobials, etc.), in which the product is dipped or that is sprayed over the product, and then the solution dried over its surface. By casting and drying the film-forming solution, an edible film is obtained. For industrial scale, the spread coating is a suitable system; in this case, it is necessary to increase the viscosity of the filmogenic solution (De Moraes et al., 2013). The dry processing is based on the viscoelastic behavior of most proteins in the presence of plasticizers at low-moisture levels and high temperatures and with pressure or shear forces, which allows them to be shaped for the production of all kinds of materials, including edible films (Hernandez-Izquierdo and Krochta, 2008). An example of dry processing of proteins is extrusion, which has the advantage of continuous processing of film. Whichever method is employed to produce the edible film, the addition of one or more plasticizers is mandatory to overcome film brittleness caused by extensive intermolecular forces, thus improving film elasticity; however, plasticizers also increase film permeability (Khwaldia et al., 2004). Polyalcohols, saccharides, or fatty acids such as glycerol, sorbitol, glucose, and oleic acid are the most commonly employed plasticizers; however, other low-molecular-weight molecules such as peptides coming from the partial hydrolysis of the polymer have been also shown to possess a plasticizing effect (Blanco-Pascual et al., 2014; Giménez et al., 2009).

Table 5.1 summarizes the main proteins from animal origin with the potential to produce edible films and coatings, as well as some remarks about relevant biochemical properties and some key properties of the resulting materials. These proteins are milk proteins (caseins, whey proteins), muscle proteins (myofibrillar, sarcoplasmic), egg albumin, and keratins from feathers and wool. The ability of collagen/gelatin to form edible films and coatings as well as their properties and applications will be deeply studied in other chapter of this book.

5.3 Edible Films and Coatings from Milk Proteins

Cow's milk contains about 33 g of protein/L, and its proteins are classified in two groups: caseins and whey proteins. In addition to the nutritional value of milk, some of each protein has several key

TABLE 5.1

Key Biochemical Properties of Proteins from Animal Origin with Potential to Produce Edible Films and Coatings and Some Key Properties of the Corresponding Films

Source of Protein	Group of Proteins	Molecular Weight/Structure	Amino Acid Composition	Key Film Properties
Milk proteins	Caseins/caseinates	19–25 kDa/random coil	Poor in cysteine, rich in proline	• Water soluble • Good emulsifying properties • Stable to a wide range of temperature, pH, and salt concentration
	Whey proteins (α-lactalbumin, β-lactoglobulin, and bovine serum albumin, among others)	14–66 kDa/globular	Rich in cysteine	• Water insoluble • Stable to thermal treatment • Very good O_2 and CO_2 barrier
Muscle proteins	Sarcoplasmic	≤50 kDa/globular	Contain cysteine	• Partially water soluble
	Myofibrillar (mainly actin and myosin)	Myosin heavy chain: 200 kDa/fibrous Actin: 42 kDa/globular	Contain cysteine	• Water insoluble
	Stromal proteins (mainly collagen/gelatin)	<20–>300/random coil	Rich in glycine and proline, poor in cysteine	• Water soluble • Good emulsifying properties • High elongation
Wool/feather	Keratin	10–60 kDa/fibrous	Very rich in cysteine	• Water insoluble • High mechanical strength • Low elongation
Egg	Ovalbumin, ovotransferrin, ovomucoid, lysozyme, and others	14–78 kDa/mainly globular structure	Rich in cysteine	• Water insoluble

physicochemical characteristics for effective performance in edible films, such as their solubility in water and ability to act as emulsifiers (Khwaldia et al., 2004). Furthermore, considerable interest exists in finding new uses for milk proteins due to their industrial surplus.

5.3.1 Caseins and Caseinates

Casein comprises 80% of milk protein and consists of three principal components (α, β, and κ-casein) that form colloidal micelles that are stabilized by calcium–phosphate bridging (Kinsella, 1984; Walstra, 1990). Caseins have molecular weights between 19 and 25 kDa and are poor in cysteine, resulting in a few disulfide cross-linkages and thus, an open random structure. They have a high content of proline, being this related with good emulsifying properties (Khwaldia et al., 2004).

Caseinates are manufactured by precipitating the casein from milk by lowering the pH to the isoelectric point (4.6). The casein curd is then washed with water and dissolved in the appropriate alkali to increase the pH to about 7.0, and the dispersion is then spray-dried. Depending on the alkali employed to adjust the pH, different caseinates are formed; the most common are sodium caseinate and calcium caseinate; however, others such as potassium and magnesium are also available commercially. Because of the high sensitivity of casein to calcium ions, calcium caseinate contains aggregates of varying sizes that affect not only the emulsifying properties of casein but also the properties of the films (Fabra et al., 2010; Khwaldia et al., 2004).

Caseinates easily form films from aqueous solutions due to their random coil nature and their ability to form extensive intermolecular hydrogen, electrostatic, and hydrophobic bonds, resulting in an increase of interchain cohesion (Avena-Bustillos and Krochta, 1993; Chen, 2002). Caseinate films are transparent, tasteless or with a slight milky flavor, and water soluble and exhibit resistance to thermal denaturation

TABLE 5.2

Some Physicochemical Properties of Cow Milk Protein Films

Protein	Film Production	Mechanical Properties (TS (MPa)/EB (%))	Water Vapor Permeability ($\times10^{-10}$ $g \cdot m \cdot m^{-2} \cdot s^{-1} \cdot Pa^{-1}$)	Water Solubility (%)	References
Sodium caseinate	Casting/13% protein/25% glycerol	TS = 12.8 ± 0.6 EB = 24 ± 11	0.84 ± 0.01	—	Khwaldia et al. (2004)
Sodium caseinate	Casting/2.5% protein/32% glycerol	TS ≈ 5 MPa EB ≈ 7.9%	≈4.3	≈95%	Longares et al. (2005)
Whey protein isolate	Casting/2.3% protein/ thermal treatment 90°C 20 min/37% glycerol	TS ≈ 6 MPa EB ≈ 3.3%	≈4.3	≈25%	Longares et al. (2005)
Whey protein isolate	Casting/6% protein/ thermal treatment 95°C 15 min/66% glycerol	TS = 4.7 ± 0.4 EB = 114 ± 14	4.48 ± 0.4	28 ± 2	Ustunol and Mert (2004)
Whey protein isolate	Casting/10% protein/ thermal treatment 90°C 30 min/43.5% glycerol	TS = 13.9 EB = 30.8	—	—	McHugh and Krochta (1994)

(CaCas is less heat stable than NaCas) and/or coagulation, which means that the protein films remain stable over a wide range of pH, temperature, and salt concentrations (Kinsella, 1984). In consequence, these films may be much appropriated for applications in which a thermal treatment is to be applied to the food product and the film is intended to maintain its integrity. Due to the highly hydrophilic nature of caseinates, they are not good moisture barriers; however, their good emulsifying properties make them much suitable for the formulation of compound films containing lipid components, in order to reduce the water vapor permeability (Fabra et al., 2008). These kinds of formulations are much appropriate to retard weight loss of fresh fruits and vegetables and frozen foods (Khwaldia et al., 2004). Regarding the mechanical properties, works from literature show a wide range of tensile strength (TS) and maximum elongation; however, in most of the works, these properties range from ≈5 to ≈15 MPa and ≈8% to ≈30% (Chen, 2002) (Table 5.2). Some other modifications, apart from compounding with lipids, may be made in order to improve the physicochemical performance of caseinate films. These are adjusting the pH of the film to the isoelectric point, thus reducing its water solubility and improving mechanical properties, as well as cross-linking with calcium ions, transglutaminase, or γ-irradiation (Khwaldia et al., 2004; Ressouany et al., 1998; Vachon et al., 2000).

5.3.2 Whey Proteins

Whey protein is the protein fraction that remains soluble after casein precipitation at pH 4.6. Indeed, whey is the main coproduct obtained from the cheese production. However, the most commonly employed methods today for the production of whey protein concentrates are ultrafiltration and diafiltration from milk surplus (Morr and Ha, 1993). Whey proteins are a mix of several proteins, majority of which are α-lactalbumin, β-lactoglobulin, and bovine serum albumin (Kinsella, 1984), all with a globular structure and molecular weight ranging from 14 to 66 kDa. β-Lactoglobulin stands up because it contains one free thiol group and two disulfide groups per monomer, and four hydrophobic groups are located inside the globular structure (Khwaldia et al., 2004). Although there are some works dealing with the properties of the films made from individually isolated whey proteins (Maté and Krochta, 1996a; Nicolai et al., 2011), the high purification and separation costs made this economically unviable for the industry.

The application of a thermal treatment has been found to be a fundamental step in order to obtain convenient edible films from whey protein. Thus, films produced from native whey proteins could not be obtained, as they cracked upon drying (McHugh et al., 1994) or showed poor mechanical properties, high oxygen permeability, and high water solubility (Pérez-Gago and Krochta, 1999, 2001). The application of denaturation treatments such as heating induces oxidation of free sulfhydryls and reduction of disulfide bonds, as well as exposition of buried hydrophobic residues, giving rise to the establishment of new bonds (disulfide, hydrophobic) among adjacent protein units, contributing to the increase of the polymer molecular weight and thus forming water-insoluble edible films (Khwaldia et al., 2004). On the contrary, other film properties such as the water vapor permeability are not affected by protein denaturation and cross-linking (Pérez-Gago et al., 1999); this property seems to be mainly governed by the plasticizer content or the amino acid composition. Some physicochemical properties of whey protein films are listed in Table 5.2.

Regarding the practical applications of whey protein films to foodstuffs, as they pose selectivity barrier to the transport of CO_2 and O_2 at low and intermediate RH (Khwaldia et al., 2004), these films are a good choice to retard lipid oxidation (appetizers, frozen fatty fish) or to extend shelf life of fruits and vegetables by reducing respiration rate and moisture loss (Khwaldia et al., 2004; Maté and Krochta, 1996b; Mehyar et al., 2012; Reinoso et al., 2008; Rodriguez-Turienzo et al., 2012). Furthermore, their good mechanical properties make them very useful to improve mechanical resistance of fragile foods (e.g., nuts) during transportation and storage.

5.4 Edible Films and Coatings from Seafood Muscle Proteins

The main source of muscle proteins that have been used for the production of edible films is fish (Benjakul et al., 2008; Chinabhark et al., 2007; Cuq et al., 1995; García and Sobral, 2005; Hamaguchi et al., 2007; Shiku et al., 2004), including species such as sardine (*Sardina pilchardus*), tilapia (*Oreochromis niloticus*), round scad (*Decapterus maruadsi*), blue marlin (*Makaira mazara*), bigeye snapper (*Priacanthus tayenus*), and Alaska pollock (*Theragra chalcogramma*). Recently, other muscle protein sources such as squid (*Todarodes pacificus*) and giant squid (*Dosidicus gigas*) mantles, as well as shrimp (*Litopenaeus vannamei*) muscle, have been also shown as suitable raw materials for the production of films (Blanco-Pascual et al., 2013; Gómez-Estaca et al., 2014; Leerahawong et al., 2011). The production of edible films and coatings from muscle protein may not necessarily compete with protein supply for humans, as they may be produced with underutilized fish species or excess catches, or with fish batches that had been spoiled after long chilling or frozen storage (Benjakul et al., 2008; Hamaguchi et al., 2007; Shiku et al., 2004). Muscle protein edible films and coatings can be also produced from other animal species such as beef, pork, or chicken (Nemet et al., 2010); however, it is very much uncommon than from seafood species.

The protein content of muscle varies between species, but it is about 20% of the total muscle weight (wet basis). Muscle proteins are classified into three major groups according to their solubility: myofibrillar, sarcoplasmic, and stromal proteins. Myofibrillar proteins are soluble in moderate salt solutions and the main proteins are myosin (200 kDa) and actin (42 kDa). Myofibrillar proteins, which are majorly fibrous, can form films with good mechanical properties, whereas sarcoplasmic proteins are water-soluble low-molecular-weight globular proteins that generally need to be denatured previously to film formation (Iwata et al., 2000). The most abundant stromal protein is collagen, a protein soluble in acid whose denatured derivative, gelatin, has been extensively studied as film-forming biopolymer (Gomez-Guillen et al., 2009).

5.4.1 Myofibrillar Proteins

The fibrous structure and high molecular weight of myosin make myofibrillar proteins a good candidate to develop edible films; this is the reason why, among muscle-soluble proteins, myofibrillar proteins were the ones that initially focused the main attention of researchers to develop edible films (Cuq et al., 1995, 1996). In order to isolate myofibrillar proteins, muscle is previously minced and washed with cold water or very diluted salt solutions to remove sarcoplasmic proteins, lipids, and impurities and then passed through a screen (Cuq et al., 1995) or solubilized at acidic or basic conditions and centrifuged, to remove stromal proteins (Blanco-Pascual et al., 2014; Tongnuanchan et al., 2013). Obviously, the protein concentrate obtained will differ in composition depending on the purification method. Thus, when the proteins are obtained by acidic or alkaline solubilization and isoelectric precipitation, a protein concentrate of higher purity is obtained, significantly lowering the lipid content as well as pigments, heme iron, and other impurities (Tongnuanchan et al., 2011). This is going to affect the properties of the resulting films; the lower the impurity content, the better the film properties (Artharn et al., 2007; Benjakul et al., 2008; Tongnuanchan et al., 2011).

The pH is a key parameter when producing edible films from myofibrillar proteins by casting. This is because they are soluble at acidic or alkaline pHs and precipitate at pH 5–6 (isoelectric point) (Hultin and Kelleher, 1999, 2000), with some variations depending on the seafood species used, as reviewed by López-Caballero et al. (2013). For this reason, films from muscle proteins can be obtained in a pH range of 2–3 and 7–12, whereas at pHs between 4 and 7, films cannot be generally formed due to the proximity to the protein isoelectric point (Cuq et al., 1995; Hamaguchi et al., 2007; Shiku et al., 2003). The properties of the films from myofibrillar proteins were similar when produced at pH 2, 3, 11 or 12, and better when produced at pH 7, 8, 9, and 10 (Shiku et al., 2003). In other works, Tongnuanchan et al. (2011) also found scarcely differences among films produced at pH 3 or 11. However, Blanco-Pascual et al. (2014), working with edible films from giant squid myofibrillar proteins, observed that films prepared from alkali-solubilized proteins showed higher water resistance and TS and lower water vapor permeability and transparency than those produced from acid solubilized proteins.

The films made from myofibrillar proteins pose good optical and mechanical properties (Cuq et al., 1995), and they are mainly stabilized by electrostatic, hydrogen, and hydrophobic bonding, as well as disulfide bridges, especially when applying thermal treatment, as myosin heavy chain contains approximately 40 SH groups (Cuq, 2002). As for other protein films, comparison of the properties of the films among different works is very difficult because of the large amount of variables involved in the preparation (fish species, protein extraction method, pH, plasticizer type and concentration, protein concentration, thickness, drying temperature, etc.) and testing of the films. The TS of the films from myofibrillar proteins may range from 6.6 to 17 MPa and elongation at break (EB) from 20% to 150% (Table 5.3)

TABLE 5.3

Some Physicochemical Properties of Fish Muscle Protein Films

Protein	Film Production	Mechanical Properties (TS (MPa)/EB (%))	Water Vapor Permeability ($\times 10^{-10}$ $g \cdot m \cdot m^{-2} \cdot s^{-1} \cdot Pa^{-1}$)	Water Solubility (%)	References
Myofibrillar proteins from blue marlin	Casting/pH 1–13/1% protein/50% glycerol	pH 3 (TS $\approx$ 15/ EB $\approx$ 28) pH 11 (TS $\approx$ 14/ EB $\approx$ 30)	pH 3 (0.66 $\pm$ 0.03 at pH 4) pH 11 (0.86 $\pm$ 0.02 at pH 11)	pH 3 (12.7% $\pm$ 1.2%) pH 11 (18.5 $\pm$ 0.8)	Shiku et al. (2003)
Myofibrillar proteins from round scad	Casting/pH 3/2% protein/50% glycerol	TS $\approx$ 6.5–8.5 MPa EB $\approx$ 110%–150%	$\approx$ 0.9–1	$\approx$ 45–63	Artharn et al. (2007), Benjakul et al. (2008)
Myofibrillar proteins from sardine	Casting/pH 2.5–4/2% protein/35% glycerol	TS = 17.1 MPa EB = 22.7 MPa	0.63	$\approx$ 35	Cuq et al. (1995, 1997)
Sarcoplasmic proteins from blue marlin	Casting/pH 4 or 11/thermal treatment 70°C for 15 min/3% protein/50% glycerol	pH 4 (TS $\approx$ 5/ EB $\approx$ 45) pH 10 (TS $\approx$ 5/ EB $\approx$ 60)	1.18 at pH 4 1.195 at pH 10	—	Iwata et al. (2000), Tanaka et al. (2001)
Myofibrillar and sarcoplasmic proteins from round scad	Casting/pH 3/2% protein/50% glycerol	TS $\approx$ 2.5–4.5 EB $\approx$ 130–150	$\approx$ 1–1.2	$\approx$ 48–55	Artharn et al. (2007), Benjakul et al. (2008)
Myofibrillar and sarcoplasmic proteins from blue marlin	Casting/pH 1–12/2% protein/50% glycerol	pH 3 (TS = 3.1 $\pm$ 0.6/ EB = 89.5 $\pm$ 9.4) pH 11 (TS = 2.98 $\pm$ 0.35/ EB = 74.3 $\pm$ 8.1)	1.55 $\pm$ 0.08 at pH 3 1.50 $\pm$ 0.09 at pH 11	—	Hamaguchi et al. (2007)

(Artharn et al., 2007, 2008; Benjakul et al., 2008; Chinabhark et al., 2007; Cuq et al., 1995; Shiku et al., 2003; Tongnuanchan et al., 2011). Film water solubility may range between 30% and 60% (Table 5.3) and it is greatly affected, as for other protein films, by plasticizer content (Cuq, 2002). Seafood myofibrillar protein films' properties are also susceptible to be improved through cross-linking with thermal or chemical treatments (Gómez-Estaca et al., 2014).

There are very few studies dealing on the aging of the films, despite the great importance of this topic. Blanco-Pascual et al. (2014) studied the aging of giant squid myofibrillar protein films obtained by acid or alkali solubilization over a period of 4 months. In the films obtained by acidic solubilization, an evident yellowing, as well as protein aggregation resulting in a lower protein release and higher mechanical resistance, was observed. The films obtained by alkali solubilization became brittle and lost their transparency with storage time.

5.4.2 Sarcoplasmic Proteins

In industrial *surimi* manufacturing process, minced flesh is repeatedly washed with chilled water to remove sarcoplasmic proteins and other impurities (ammonia, volatile bases, peptides, free amino acids, lipids, etc.) to produce a tasteless and odorless product composed mainly by purified myofibrillar proteins. As a result of washing, approximately 40–50 g/100 g of minced fish solids (containing primarily water-soluble proteins) are lost in the process and considered as waste, but it has the potential for recovery (Bourtoom et al., 2006). It was estimated that 5000 tons (dry weight) of fish water-soluble proteins are discarded annually in the waste water of *surimi* processing plants in Japan (Iwata et al., 2000). Isolated sarcoplasmic proteins also have the ability to form films; however, they generally pose worse properties than those produced from myofibrillar proteins (Iwata et al., 2000). It is mandatory to previously denature the sarcoplasmic proteins in order to form more extended structures; Iwata et al. (2000) found that films could not be formed with heating temperatures below 50°C and that 70°C was the temperature at which the maximum TS was afforded. The pH of the FFS, as occurred for the myofibrillar proteins, is a key factor. The films could not be formed at pH ranging from 7 to 9 (the protein precipitated) and the best TS was obtained for films produced at pH 10 and 4 (≈5 MPa).

5.4.3 Whole Muscle Protein Films

From an industrial point of view, the production of edible films and coatings using the whole muscle arises as a feasible option, avoiding tedious washing and separation processes and achieving the best use of muscle proteins (Paschoalick et al., 2003). Taking into account that the most feasible source of protein to produce edible films and coatings from seafood muscle is low-value species or coproducts, avoiding separation steps that increase the cost is much desirable. However, the properties of the films prepared with whole muscle are generally worse than those of the isolated myofibrillar protein counterparts (Artharn et al., 2007); in spite of this, these may be suitable depending on the application. Thus, the TS and EB of the films made from whole muscle proteins ranged from 3 to 5 MPa and from 80% to 160%, respectively (Artharn et al., 2007; Benjakul et al., 2008; García and Sobral, 2005; Hamaguchi et al., 2007; Sobral et al., 2005), in contrast with the results obtained for films from myofibrillar proteins (Table 5.3). The works by Benjakul et al. (2008) and Artharn et al. (2007) are especially interesting in order to elucidate the effect of muscular components different from myofibrillar proteins on some properties of the films, as they evaluated the properties of films made from both washed and unwashed fish mince. As shown in Table 5.3, it is clear that there is a reduction of TS from ≈6.6–8.5 MPa to 4.2–4.5 MPa for films prepared from washed and unwashed mince, respectively; on the contrary, the effect on EB was not so evident. The worsening of the film's properties is likely to be due to the interference of muscle components other than myofibrillar proteins on the formation of the protein network, such as lipids or sarcoplasmic proteins. Artharn et al. (2008) studied the effect of myofibrillar/sarcoplasmic protein ratios on the properties of round scad muscle protein–based film and found that TS of films decreased with increasing sarcoplasmic protein content. Another explanation may be the proteolysis of myofibrillar proteins under certain pHs (depending on the fish and seafood species), resulting in the reduction of the average molecular weight of the polymer solution, indeed, worsening some film properties as the water solubility and the mechanical properties (Blanco-Pascual et al., 2013; Chinabhark et al., 2007; Gómez-Estaca et al., 2014). Due to the fact that proteolytic activity differs among seafood species, specific studies on the best extraction procedures and film formation conditions are needed. For example, Blanco-Pascual et al. (2013) working with giant squid (*D. gigas*) muscle found higher proteolytic activity and thus worse film properties at acidic pH than at alkaline one, whereas Gómez-Estaca et al. (2014), working with shrimp (*L. vannamei*) muscle, found contrary results.

Protein films in general and those from seafood proteins specifically show high barrier to UV light (200–280 range). This is a very important property of protein films, as compared to those from conventional polymers, as it will contribute to reduce light-induced lipid oxidation when applied to foods (Blanco-Pascual et al., 2013; Gómez-Estaca et al. 2014). The main drawbacks of such edible films and coatings may be the microbial instability and a potential fishy odor; however, they should not be a problem when the films are intended to be used in seafood packaging applications.

5.5 Edible Films and Coatings from Egg White Proteins

According to Gennadios et al. (1996), egg yolk has a greater number of applications in the food industry compared to egg white, leading to a surplus of egg albumen in the egg-breaking industry of North America. Egg white is a complex protein system made up of a solution of globular proteins containing ovomucin fibers (Lim et al., 2002). The protein content of chicken egg white is ≈10%, ovalbumin constituting more than a half of the total. This protein stands up because it contains free sulfhydryl groups, which can be involved in the formation of disulfide bridges. Other proteins such as ovotransferrin, ovomucoid, and lysozyme also contain sulfur though in the form of S–S bridges (Mine, 1992).

As occurs for other globular proteins, it is necessary to unfold egg proteins and rearrange them in a form of a continuous matrix to form a film. For this reason it is mandatory to unfold the protein by dissolving it at alkaline pH (10.5–12) and heating (40°C–45°C for 20–30 min) (Gennadios et al., 1996); this process involves the reduction of S–S bonds to SH groups, as demonstrated by Handa et al. (1999) by measuring surface sulfhydryl groups. These sulfhydryl groups are further oxidized during gelation and drying of the films to intra- and intermolecular S–S bonds (Gennadios et al., 1996) that strengthen the film. Under these conditions, and in the presence of plasticizers, smooth and homogeneous egg white films are obtained. Gennadios et al. (1996) evaluated the effect of different plasticizers, namely, polyethylene glycol, glycerol, and sorbitol (30%–60% w/w of protein), the former producing the films with the best mechanical properties. In addition to S–S bonding, hydrogen bonding, and hydrophobic and electrostatic interactions are important in dictating the physical properties of egg white films (Lim et al., 2002). Cross-linking treatments to improve the physicochemical performance of films from egg white films have been also applied, including treatment with transglutaminase (Lim et al., 1998), dialdehyde starch (Gennadios et al., 1998), and ultraviolet radiation films (Rhim et al., 1999). Mixing with other polymers, such as gelatin, has been also conducted, in order to produce materials with improved properties for carrying and releasing active compounds (Giménez et al., 2012). Egg albumen has been applied alone or in combination with other biopolymers (milk protein, starch) as a coating to improve the quality of hen shell eggs, meat products, or pizza with good results (Lim et al., 2002). In all the cases, the application of thermal treatment was essential for the formation of an effective coating. An example of the main physico-chemical properties of white egg films is shown in Table 5.4.

5.6 Edible Films and Coatings from Wool and Feather Proteins

Keratins are fibrous proteins found in hair, wool feathers, nail, horns, and other epithelial coverings. The most referred sources of keratin to produce films are chicken feathers and wool, with the molecular weight of extracted keratin ranging between 10 and 60 kDa (Barone et al., 2005, 2006; Fraser and Parry, 2008; Yamauchi and Yamauchi, 2002; Zoccola et al., 2012). Keratin is characteristically abundant in cysteine residues; indeed, 7%–20% of the total amino acid residues are cysteine (Yamauchi and Yamauchi, 2002). Keratin fibrils are formed by intra- and intermolecular S–S bonds of cysteine residues

TABLE 5.4

Some Physicochemical Properties of Hen Egg White and Keratin Films

Protein	Film Production	Mechanical Properties (TS (MPa)/EB (%))	Water Vapor Permeability ($\times 10^{-10}$ g·m·m^{-2}·s^{-1}·Pa^{-1}) s	Water Solubility (%)	References
White egg	Casting/9% protein/ thermal treatment 45°C 20 min/pH 11/60% polyethylene glycol	TS = 5 ± 0.1 EB = 62 ± 2	0.23 ± 0.001	48.4 ± 0.6	Handa et al. (1999)
Wool keratin	Thermopressing at 70°C–160°C for 5 min at 5 MPa and 5 min at 10 MPa/ water as plasticizer	120°C (TS = 27.8 ± 2.9/EB = 3.5 ± 0.8) 160°C (TS = 17.7 ± 1.7/EB = 3.4 ± 0.6)	—	—	Katoh et al. (2004)
Feather keratin	Thermopressing at 160°C for 2–8 min at 88,964 N/15%–80% glycerol	2 min thermopressing and 30% glycerol (TS ≈ 10 MPa/ EB ≈ 50%)	—	—	Barone et al. (2005)

between monomeric keratins, these bonds being responsible for the mechanically strong 3D linked network of keratin fiber (Katoh et al., 2004). Traditionally, keratins are classified as either "soft" or "hard." The soft keratins, with a low content of disulfide bonds, are found in the *stratum corneum* and in the callus, whereas the hard keratins are found in epidermal appendages such as feathers, hair, nails, and hoofs and have high disulfide content (Martelli et al., 2006).

In order to produce films from keratin fibers, it is necessary to previously dissolve keratin. Zoccola et al. (2012) reviewed the main extraction methods for wool keratin; according to these authors, there are two main solubilization methods of keratin: The first is hydrolysis and extraction via reduction, oxidation, or sulfitolysis. The second method is the one that accumulates more data in the literature and is related to the cleavage of disulfide bonds. The reductive extraction implies solubilization of keratin with urea and SDS (cleavage of hydrogen and hydrophobic bonds) and treatment with 2-mercaptoethanol to cleave S–S bonds. With a view to applying the films for food applications, after keratin extraction, mercaptoethanol and urea should be removed, for example, by dialysis, although keratins tend to aggregate; partial carboxymethylation of keratins was proposed as a method to ameliorate this effect (Schrooyen et al., 2000). Treatment with methyl iodide or α-iodoacetic acid has been also proposed for the reductive extraction of keratin. The oxidative extraction is an alternative method for preparing soluble proteins from wool or feathers, and it involves treatment with either peracetic or performic acid, which oxidizes cysteine to cysteic acid residues. Successively, the oxidized keratins can be dissolved in alkali. Sulfitolysis and oxidative sulfitolysis describe the cleavage of a disulfide bond by sulfite to give a thiol and an S-sulfonate anion (Barone et al., 2006; Jin et al., 2011). Keratin solubilization can be also achieved via hydrolysis. Soluble keratin with different hydrolysis degrees may be obtained depending on the severity of the acidic or alkaline extraction. Special attention is being recently paid to green processes for keratin extraction, including hydrolysis in superheated water and enzymatic treatments. However, keratin extracted under some of these green processes usually shows low molecular weight and low amount of cysteine, so it is not suitable to produce materials with good mechanical properties.

After keratin extraction, the films may be produced by both wet and dry processes. Although it is possible to obtain films by wet processing without thermal treatment, these films are water soluble and present worse mechanical properties than those in which thermal treatment is applied (Barone et al., 2005; Katoh et al., 2004; Martelli et al., 2006). This is because thermal treatment promotes the oxidation of sulfhydryl groups to disulfide bridges that impair a lot of resistance of the films, improving solubility and mechanical properties. The importance of the thermal treatment to improve the physicochemical performance of keratin films is the reason why its thermoplastic properties have been studied in detail

(Barone et al., 2005, 2006), the films having been obtained by compression molding or extrusion showing TSs as high as 28 MPa, as compared to values of 5 MPa for films produced by casting without thermal treatment (Katoh et al., 2004; Martelli et al., 2006) (Table 5.4). Such a maximum strength could be also obtained by applying cross-linking treatment with ethylene glycol diglycidyl ether in solution or further casting (Tanabe et al., 2004). In another work, Barone et al. (2005) processed by extrusion poultry feathers without reduction or oxidation agents, easily obtaining semitransparent and cohesive films plasticized with glycerol. In spite of the aforementioned work, the generally tedious extraction processes to isolate keratin and the presence of solvents make it difficult to find examples of practical applications of keratin films and coatings in food products in the literature. The use of keratins, however, has a great potential in food packaging applications, specially forming part of multilayer structures, thanks to its good mechanical and barrier properties.

5.7 Conclusions

The selection of the type of protein to produce an edible film or coating will depend on many factors such as the availability of the raw materials, the available technology to process them, or the desired application. The wide range of biochemical characteristics of proteins; the possibility of their modification thanks to the variety of functional groups that they possess; the utilization of different additives such as plasticizers, cross-linkers, and fillers; and the selection of the processing technology (wet or dry processing) and processing parameters (temperature, pH, ionic strength, etc.) allow us to produce edible films and coatings with a wide range of optical, mechanical, barrier, and sensory properties, which may be fitted to each specific food application.

REFERENCES

Artharn, A., Benjakul, S., and Prodpran, T. (2008). The effect of myofibrillar/sarcoplasmic protein ratios on the properties of round scad muscle protein based film. *European Food Research and Technology*, *227*(1), 215–222.

Artharn, A., Benjakul, S., Prodpran, T., and Tanaka, M. (2007). Properties of a protein-based film from round scad (*Decapterus maruadsi*) as affected by muscle types and washing. *Food Chemistry*, *103*(3), 867–874.

Avena-Bustillos, R. J., and Krochta, J. M. (1993). Water vapor permeability of caseinate-based edible films as affected by pH, calcium crosslinking and lipid content. *Journal of Food Science*, *58*(4), 904–907.

Balaguer, M. P., Gomez-Estaca, J., Cerisuelo, J. P., Gavara, R., and Hernandez-Munoz, P. (2014). Effect of thermo-pressing temperature on the functional properties of bioplastics made from a renewable wheat gliadin resin. *LWT: Food Science and Technology*, *56*(1), 161–167.

Barone, J. R., Schmidt, W. F., and Gregoire, N. T. (2006). Extrusion of feather keratin. *Journal of Applied Polymer Science*, *100*(2), 1432–1442.

Barone, J. R., Schmidt, W. F., and Liebner, C. F. E. (2005). Thermally processed keratin films. *Journal of Applied Polymer Science*, *97*(4), 1644–1651.

Benjakul, S., Artharn, A., and Prodpran, T. (2008). Properties of protein-based film from round scad (Decapterus maruadsi) muscle as influenced by fish quality. *LWT: Food Science and Technology*, *41*(5), 753–763.

Blanco-Pascual, N., Alemán, A., Gómez-Guillén, M. C., and Montero, M. P. (2014). Enzyme-assisted extraction of κ/ι-hybrid carrageenan from Mastocarpus stellatus for obtaining bioactive ingredients and their application for edible active film development. *Food and Function*, *5*(2), 319–329.

Blanco-Pascual, N., Fernández-Martín, F., and Montero, M. P. (2013). Effect of different protein extracts from Dosidicus gigas muscle co-products on edible films development. *Food Hydrocolloids*, *33*(1), 118–131.

Blanco-Pascual, N., Fernandez-Martin, F., and Montero, P. (2014). Jumbo squid (*Dosidicus gigas*) myofibrillar protein concentrate for edible packaging films and storage stability. *LWT: Food Science and Technology*, *55*(2), 543–550.

Bourtoom, T., Chinnan, M. S., Jantawat, P., and Sanguandeekul, R. (2006). Effect of select parameters on the properties of edible film from water-soluble fish proteins in surimi wash-water. *LWT: Food Science and Technology, 39*(4), 405–418.

Chen, H. (2002). Formation and properties of casein films and coatings. In A. Gennadios (ed.), *Protein-Based Films and Coatings*, pp. 181–211. Boca Raton, FL: CRC Press.

Chinabhark, K., Benjakul, S., and Prodpran, T. (2007). Effect of pH on the properties of protein-based film from bigeye snapper (*Priacanthus tayenus*) surimi. *Bioresource Technology, 98*(1), 221–225.

Cuq, B. (2002). Formation and properties of fish myofibrillar protein films and coatings. In A. Gennadios (ed.), *Protein-Based Films and Coatings*, pp. 213–232. Boca Raton, FL: CRC Press.

Cuq, B., Aymard, C., Cuq, J. L., and Guilbert, S. (1995). Edible packaging films based on fish myofibrillar proteins: Formulation and functional properties. *Journal of Food Science, 60*(6), 1369–1374.

Cuq, B., Gontard, N., Cuq, J. L., and Guilbert, S. (1996). Stability of myofibrillar protein-based biopackagings during storage. *Food Science and Technology—Lebensmittel-Wissenschaft & Technologie, 29*(4), 344–348.

Cuq, B., Gontard, N., Cuq, J. L., and Guilbert, S. (1997). Selected functional properties of fish myofibrillar protein-based films as affected by hydrophilic plasticizers. *Journal of Agricultural and Food Chemistry, 45*(3), 622–626.

De Moraes, J. O., Scheibe, A. S., Sereno, A., and Laurindo, J. B. (2013). Scale-up of the production of cassava starch based films using tape-casting. *Journal of Food Engineering, 119*(4), 800–808.

Fabra, M. J., Talens, P., and Chiralt, A. (2008). Tensile properties and water vapor permeability of sodium caseinate films containing oleic acid-beeswax mixtures. *Journal of Food Engineering, 85*(3), 393–400.

Fabra, M. J., Talens, P., and Chiralt, A. (2009). Microstructure and optical properties of sodium caseinate films containing oleic acid-beeswax mixtures. *Food Hydrocolloids, 23*(3), 676–683.

Fabra, M. J., Talens, P., and Chiralt, A. (2010). Influence of calcium on tensile, optical and water vapour permeability properties of sodium caseinate edible films. *Journal of Food Engineering, 96*(3), 356–364.

Fraser, R. D. B. and Parry, D. A. D. (2008). Molecular packing in the feather keratin filament. *Journal of Structural Biology, 162*(1), 1–13.

García, F. T. and Sobral, P. J. D. A. (2005). Effect of the thermal treatment of the filmogenic solution on the mechanical properties, color and opacity of films based on muscle proteins of two varieties of Tilapia. *LWT: Food Science and Technology, 38*(3), 289–296.

Gennadios, A., Handa, A., Froning, G. W., Weller, C. L., and Hanna, M. A. (1998). Physical properties of egg white-dialdehyde starch films. *Journal of Agricultural and Food Chemistry, 46*(4), 1297–1302.

Gennadios, A., Hanna, M. A., and Weller, C. L. (1996). Glycerine-plasticized egg albumen films. *Inform, 7*(10), 1074–1075.

Giménez, B., Gómez-Estaca, J., Alemán, A., Gómez-Guillén, M. C., and Montero, M. P. (2009). Improvement of the antioxidant properties of squid skin gelatin films by the addition of hydrolysates from squid gelatin. *Food Hydrocolloids, 23*(5), 1322–1327.

Giménez, B., Gómez-Guillén, M. C., López-Caballero, M. E., Gómez-Estaca, J., and Montero, P. (2012). Role of sepiolite in the release of active compounds from gelatin-egg white films. *Food Hydrocolloids, 27*(2), 475–486.

Gómez-Estaca, J., Montero, P., and Gómez-Guillén, M. C. (2014). Shrimp (Litopenaeus vannamei) muscle proteins as source to develop edible films. *Food Hydrocolloids, 41*, 86–94.

Gomez-Guillen, M. C., Perez-Mateos, M., Gomez-Estaca, J., Lopez-Caballero, E., Gimenez, B., and Montero, P. (2009). Fish gelatin: A renewable material for developing active biodegradable films. *Trends in Food Science & Technology, 20*(1), 3–16.

Hamaguchi, P. Y., Weng, W., Kobayashi, T., Runglertkreingkrai, J., and Tanaka, M. (2007). Effect of fish meat quality on the properties of biodegradable protein films. *Food Science and Technology Research, 13*(3), 200–204.

Hamaguchi, P. Y., WuYin, W., and Tanaka, M. (2007). Effect of pH on the formation of edible films made from the muscle proteins of Blue marlin (*Makaira mazara*). *Food Chemistry, 100*(3), 914–920.

Handa, A., Gennadios, A., Froning, G. W., Kuroda, N., and Hanna, M. A. (1999). Tensile, solubility, and electrophoretic properties of egg white films as affected by surface sulfhydryl groups. *Journal of Food Science, 64*(1), 82–85.

Handa, A., Gennadios, A., Hanna, M. A., Weller, C. L., and Kuroda, N. (1999). Physical and molecular properties of egg-white lipid films. *Journal of Food Science*, 64(5), 860–864.

Hernandez-Izquierdo, V. M. and Krochta, J. M. (2008). Thermoplastic processing of proteins for film formation: A review. *Journal of Food Science*, 73(2), R30–R39.

Hernandez-Munoz, P., Kanavouras, A., Lagaron, J. M., and Gavara, R. (2005). Development and characterization of films based on chemically cross-linked gliadins. *Journal of Agricultural and Food Chemistry*, 53(21), 8216–8223.

Hultin, H. O. and Kelleher, S. D. (1999). Isolated from an animal muscle tissue which comprises myofibrillar proteins substantially free of animal membrane lipids, said proteins capable of being formed into a gel. Google Patents.

Hultin, H. O., and Kelleher, S. D. (2000). High efficiency alkaline protein extraction. Google Patents.

Iwata, K., Ishizaki, S., Handa, A., and Tanaka, M. (2000). Preparation and characterization of edible films from fish water-soluble proteins. *Fisheries Science*, 66(2), 372–378.

Jin, E., Reddy, N., Zhu, Z., and Yang, Y. (2011). Graft polymerization of native chicken feathers for thermoplastic applications. *Journal of Agricultural and Food Chemistry*, 59(5), 1729–1738.

Katoh, K., Shibayama, M., Tanabe, T., and Yamauchi, K. (2004). Preparation and physicochemical properties of compression-molded keratin films. *Biomaterials*, 25(12), 2265–2272.

Khwaldia, K., Banon, S., Desobry, S., and Hardy, J. (2004). Mechanical and barrier properties of sodium caseinate-anhydrous milk fat edible films. *International Journal of Food Science and Technology*, 39(4), 403–411.

Khwaldia, K., Ferez, C., Banon, S., Desobry, S., and Hardy, J. (2004). Milk proteins for edible films and coatings. *Critical Reviews in Food Science and Nutrition*, 44(4), 239–251.

Kinsella, J. E. (1984). Milk proteins: Physicochemical and functional properties. *Critical Reviews in Food Science and Nutrition*, 21(3), 197–262.

Krochta, J. M. (1992). Control of mass transfer in foods with edible-coatings and films. In R. P. Singh and M. A. Wirakartakusumah (eds.), *Advances in Food Engineering*, pp. 517–538. Boca Raton, FL: CRC Press.

Lagrain, B., Goderis, B., Brijs, K., and Delcour, J. A. (2010). Molecular Basis of Processing Wheat Gluten toward Biobased Materials. *Biomacromolecules*, 11(3), 533–541.

Leerahawong, A., Arii, R., Tanaka, M., and Osako, K. (2011). Edible film from squid (*Todarodes pacificus*) mantle muscle. *Food Chemistry*, 124(1), 177–182.

Lim, L. T., Mine, Y., Britt, I. J., and Tung, M. A. (2002). Formation and properties of egg white films and coatings. In A. Gennadios (ed.), *Protein-Based Films and Coatings*, pp. 233–252. Boca Raton, FL: CRC Press.

Lim, L. T., Mine, Y., and Tung, M. A. (1998). Transglutaminase cross-linked egg white protein films: Tensile properties and oxygen permeability. *Journal of Agricultural and Food Chemistry*, 46(10), 4022–4029.

Longares, A., Monahan, F. J., O'Riordan, E. D., and O'Sullivan, M. (2005). Physical properties of edible films made from mixtures of sodium caseinate and WPI. *International Dairy Journal*, 15(12), 1255–1260.

López-Caballero, M. E., Giménez, B., Gómez-Guillén, M. C., and Montero, P. (2013). Valorization and integral use of seafood by-products. In J. A. Teixeira and A. A. Vicente (eds.), *Engineering Aspects of Food Biotechnology*, pp. 359–403. Boca Raton, FL: CRC Press.

Martelli, S. M., Moore, G. R. P., and Laurindo, J. B. (2006). Mechanical properties, water vapor permeability and water affinity of feather keratin films plasticized with sorbitol. *Journal of Polymers and the Environment*, 14(3), 215–222.

Maté, J. I. and Krochta, J. M. (1996a). Comparison of oxygen and water vapor permeabilities of whey protein isolate and β-lactoglobulin edible films. *Journal of Agricultural and Food Chemistry*, 44(10), 3001–3004.

Maté, J. I. and Krochta, J. M. (1996b). Whey protein coating effect on the oxygen uptake of dry roasted peanuts. *Journal of Food Science*, 61(6), 1202–1206.

McHugh, T. H., Aujard, J. F., and Krochta, J. M. (1994). Plasticized whey-protein edible films -water-vapor permeability properties. *Journal of Food Science*, 59(2), 416.

McHugh, T. H. and Krochta, J. M. (1994). Sorbitol- vs glycerol-plasticized whey protein edible films: Integrated oxygen permeability and tensile property evaluation. *Journal of Agricultural and Food Chemistry*, 42(4), 841–845.

Mehyar, G. F., Al-Ismail, K., Han, J. H., and Chee, G. W. (2012). Characterization of edible coatings consisting of pea starch, whey protein isolate, and carnauba wax and their effects on oil rancidity and sensory properties of walnuts and pine nuts. *Journal of Food Science*, 77(2), E52–E59.

Mine, Y. (1992). Sulfhydryl groups changes in heat-induced soluble egg-white aggregates in relation to molecular size. *Journal of Food Science, 57*(1), 254–255.

Morr, C. V. and Ha, E. Y. (1993). Whey protein concentrates and isolates: Processing and functional properties. *Critical Reviews in Food Science and Nutrition, 33*(6), 431–476.

Nemet, N. T., Šošo, V. M., and Lazić, V. L. (2010). Effect of glycerol content and pH value of film-forming solution on the functional properties of protein-based edible films. *Acta Periodica Technologica, 41*, 57–67.

Nicolai, T., Britten, M., and Schmitt, C. (2011). β-Lactoglobulin and WPI aggregates: Formation, structure and applications. *Food Hydrocolloids, 25*(8), 1945–1962.

Paschoalick, T. M., Garcia, F. T., Sobral, P. J. A., and Habitante, A. M. Q. B. (2003). Characterization of some functional properties of edible films based on muscle proteins of Nile Tilapia. *Food Hydrocolloids, 17*(4), 419–427.

Pérez-Gago, M. B. and Krochta, J. M. (1999). Water vapor permeability of whey protein emulsion films as affected by pH. *Journal of Food Science, 64*(4), 695–698.

Pérez-Gago, M. B. and Krochta, J. M. (2001). Denaturation time and temperature effects on solubility, tensile properties, and oxygen permeability of whey protein edible films. *Journal of Food Science, 66*(5), 705–710.

Pérez-Gago, M. B., Nadaud, P., and Krochta, J. M. (1999). Water vapor permeability, solubility, and tensile properties of heat-denatured versus native whey protein films. *Journal of Food Science, 64*(6), 1034–1037.

Reinoso, E., Mittal, G. S., and Lim, L. T. (2008). Influence of whey protein composite coatings on plum (*Prunus domestica* L.) fruit quality. *Food and Bioprocess Technology, 1*(4), 314–325.

Ressouany, M., Vachon, C., and Lacroix, M. (1998). Irradiation dose and calcium effect on the mechanical properties of cross-linked caseinate films. *Journal of Agricultural and Food Chemistry, 46*(4), 1618–1623.

Rhim, J. W., Gennadios, A., Fu, D. J., Weller, C. L., and Hanna, M. A. (1999). Properties of ultraviolet irradiated protein films. *Food Science and Technology—Lebensmittel-Wissenschaft & Technologie, 32*(3), 129–133.

Rodriguez-Turienzo, L., Cobos, A., and Diaz, O. (2012). Effects of edible coatings based on ultrasound-treated whey proteins in quality attributes of frozen Atlantic salmon (*Salmo salar*). *Innovative Food Science and Emerging Technologies, 14*, 92–98.

Schrooyen, P. M. M., Dijkstra, P. J., Oberthü, R. G., Bantjes, A., and Feijen, J. (2000). Partially carboxymethylated feather keratins. 1. Properties in aqueous systems. *Journal of Agricultural and Food Chemistry, 48*(9), 4326–4334.

Shiku, Y., Hamaguchi, P. Y., Benjakul, S., Visessanguan, W., and Tanaka, M. (2004). Effect of surimi quality on properties of edible films based on Alaska pollack. *Food Chemistry, 86*(4), 493–499.

Shiku, Y., Hamaguchi, P. Y., and Tanaka, M. (2003). Effect of pH on the preparation of edible films based on fish myofibrillar proteins. *Fisheries Science, 69*(5), 1026–1032.

Sobral, P. J. A., Dos Santos, J. S., and García, F. T. (2005). Effect of protein and plasticizer concentrations in film forming solutions on physical properties of edible films based on muscle proteins of a Thai Tilapia. *Journal of Food Engineering, 70*(1), 93–100.

Tanabe, T., Okitsu, N., and Yamauchi, K. (2004). Fabrication and characterization of chemically crosslinked keratin films. *Materials Science & Engineering. C, Biomimetic Materials, Sensors and Systems, 24*(3), 441–446.

Tanaka, M., Iwata, K., Sanguandeekul, R., Handa, A., and Ishizaki, S. (2001). Influence of plasticizers on the properties of edible films prepared from fish water-soluble proteins. *Fisheries Science, 67*(2), 346–351.

Tongnuanchan, P., Benjakul, S., and Prodpran, T. (2011). Roles of lipid oxidation and pH on properties and yellow discolouration during storage of film from red tilapia (*Oreochromis niloticus*) muscle protein. *Food Hydrocolloids, 25*(3), 426–433.

Tongnuanchan, P., Benjakul, S., Prodpran, T., and Songtipya, P. (2011). Characteristics of film based on protein isolate from red tilapia muscle with negligible yellow discoloration. *International Journal of Biological Macromolecules, 48*(5), 758–767.

Tongnuanchan, P., Benjakul, S., Prodpran, T., and Songtipya, P. (2013). Properties and stability of protein-based films from red tilapia (*Oreochromis niloticus*) protein isolate incorporated with antioxidant during storage. *Food and Bioprocess Technology, 6*(5), 1113–1126.

Ustunol, Z. and Mert, B. (2004). Water solubility, mechanical, barrier, and thermal properties of cross-linked whey protein isolate-based films. *Journal of Food Science, 69*(3), FEP129–FEP133.

Vachon, C., Yu, H. L., Yefsah, R., Alain, R., St-Gelais, D., and Lacroix, M. (2000). Mechanical and structural properties of milk protein edible films cross-linked by heating and γ-irradiation. *Journal of Agricultural and Food Chemistry, 48*(8), 3202–3209.

Walstra, P. (1990). On the stability of casein micelles. *Journal of Dairy Science, 73*(8), 1965–1979.

Yamauchi, A. and Yamauchi, K. (2002). Formation and properties of wool keratin films and coatings. In A. Gennadios (ed.), *Protein-Based Films and Coatings*, pp. 253–273. Boca Raton, FL: CRC Press.

Zoccola, M., Aluigi, A., Patrucco, A., and Tonin, C. (2012). Extraction, processing and applications of wool keratin. In R. Dullart and J. Mousqus (eds.), *Keratin: Structure, Properties and Applications*, pp. 36–62. Hauppauge, NY: Nova Science Publishers, Inc.

6

Films and Coatings from Collagen and Gelatin

Soottawat Benjakul, Muralidharan Nagarajan, and Thummanoon Prodpran

CONTENTS

Abstract

Films and coatings are thin layers, which can be used to protect foodstuffs by covering/wrapping or direct coating on the food surface. Films are generally preformed material and coatings are liquids, in which foods could be dipped. Biodegradable packaging has been known as an important eco-friendly approach to reduce plastic wastes. Nevertheless, films from synthetic polymers generally have superior properties to biodegradable films. To tackle this problem, the appropriate modifications of biodegradable films are required. Collagen and gelatin are commonly obtained from bovine hides, bones, and pigskins. Nowadays, gelatin and collagen from aquatic animals are gaining increasing attention due to health issues and religious constraints of mammalian counterpart. The industrial use of collagen and gelatin obtained from nonmammalian source is therefore rising significantly. Collagen films/coatings have numerous applications in the field of medical, pharmaceutical, and cosmetics. Due to their film-forming abilities, films/coatings from gelatin can be used to preserve foodstuffs and extend their shelf life. Mechanical properties of films and barrier properties of films/coatings mainly depend on the molecular weight distribution and amino acid composition. Additionally, other components, especially plasticizer, and technology used in conjunction with film and coating formation are of prime factors determining the properties of film or coating. Properties of film/coating can be improved by several techniques to bring about the film/coating with desirable property. Furthermore, the incorporation of active compounds like antioxidants/antimicrobials, essential oils, fatty acids, and plant extracts can make the film/coating become multifunctional and known as "active packaging." The present review covers the up-to-date information on biopolymeric film/coating matrices based on collagen and gelatin and active packaging prepared from those materials. Blend and nanocomposite films based on collagen/gelatin are also addressed.

6.1 Introduction

Biodegradable packaging has been known as an important eco-friendly alternative or replacer for plastic counterpart since the starting materials used for packaging are abundant, inexpensive, and renewable (Vartiainen et al., 2010). Most synthetic films are nonbiodegradable and may cause environmental and ecological problems (Gomez-Guillen et al., 2009). Biodegradable films are generally made from biological materials such as proteins, lipids, and polysaccharides (Tharanathan, 2003). Biodegradable polymers are defined as polymers degraded by the action of naturally occurring microorganisms such as bacteria, fungi, and algae (ASTM, 1999). Proteins are superior to polysaccharides in their ability to form films with greater mechanical and barrier properties (Cuq et al., 1998). Proteins are very different, depending on their origin, structures, and amino acid composition. Among protein-based biopolymers, collagen and gelatin have been known for their film-forming ability and their uses as an outer covering/coating for food products.

Collagen is an acid-soluble protein obtained from animal connective tissues (Benjakul et al., 2012a). Gelatin is water-soluble protein derived from collagen via thermal denaturation or partial hydrolysis (Benjakul et al., 2012b). Collagen and gelatin are generally obtained from bovine hides, bones, and pigskins. The industrial use of collagen and gelatin obtained from nonmammalian source, especially those from fish processing by-products, is increasing as packaging or coating materials (Karim and Bhat, 2009). Films and coatings are thin layers, which can be used to protect foodstuffs by covering/wrapping or coating on food surface. Films are the sheet, which can be used for wrapping or making the bag or pouch via sealing. Coating has been used on the food surfaces by introducing the liquid, which can form the film after drying (Krochta, 2002). Due to the film-forming abilities of gelatins, their films/coatings can be used to protect foodstuffs or extend the shelf life of several food commodities (Gomez-Guillen et al., 2009). Collagen films/coatings have been extended for medical, pharmaceutical, and cosmetic applications (Senaratne et al., 2006). Nevertheless, those films or coatings still have the poorer properties than synthetic films, especially in terms of water vapor barrier property (Tharanathan, 2003). Several technologies have been implemented for the improvement of film properties. Additionally, active or green packagings have been produced from gelatin or collagen, in which a variety of active substances are incorporated (Bower et al., 2006; Gomez-Estaca et al., 2007; Hoque et al., 2011a; Jongjareonrak et al., 2008; Li et al., 2014; Limphisophon et al., 2010; Rattaya et al., 2009; Tongnuanchan et al., 2013; Wu et al., 2013). Recently, nanotechnology has been introduced for packaging materials as a potential means to bring about the desirable properties. Nanocomposite films/coatings developed from biopolymers known as "bio-nanocomposites" show the better properties, compared with polymer alone or microscale composites (Martucci and Ruseckaite, 2010a; Sothornvit et al., 2009). Polymer nanocomposites usually have much better polymer/nanofiller interaction than conventional composites (Bae et al., 2009a).

6.2 Films and Coatings

6.2.1 Definition

Films are thin layers preformed separately from the products and can be used for covering or wrapping the products. Films are prepared by wet casting or dry molding methods (Krochta, 2002; Tharanathan, 2003). Coatings are thin layers formed directly on the products using either liquid film-forming solutions (FFSs)/dispersions or molten compounds, and they become part of the product (Krochta, 2002; Tharanathan, 2003). Coating of products can be carried out by spraying and dipping methods, etc.

6.2.2 Traditional Starting Materials for Protein Films and Coatings

Biodegradable films and coatings can be obtained from different sources. Among them, edible films and coatings from proteins are supposed to provide nutritional value and also have the impressive mechanical and gas barrier properties (Ou et al., 2004). The most distinctive characteristics of proteins compared to other polymers are conformational denaturation, electrostatic charges, and amphiphilic nature (Han et al., 2005). In addition, stronger intermolecular binding potential via covalent bonds is found in protein-based films and not in films from homopolymer polysaccharides (Cuq et al., 1995). Proteins used as film-forming materials are derived from both animal and plant sources, such as animal tissues, milk, egg, grain, and oilseeds (Krochta, 2002). Several proteins have been used for film preparation and coating. Those include myofibrillar protein (Tongnuanchan et al., 2011), soy protein (Rhim et al., 2006), corn zein (Arcan and Yemenicioglu, 2011), wheat gluten (Gennadios et al., 1994), milk protein (McHugh and Krochta, 1994), collagen (Alizadeh and Behfar, 2013; Sionkowska, 2006), gelatin (Jongjareonrak et al., 2006a,b,c; Ou et al., 2002), and egg white (Gennadios et al., 1996). Protein-based films show the excellent barrier property to gases but not to water vapor. Nevertheless, such films provide the nutritional value for coated food (Tharanathan, 2003). Films prepared from plasticized myofibrillar proteins are flexible and semitransparent (Thongnuanchan et al., 2011). Soy protein films have received considerable attention due to their excellent film-forming abilities, low cost, and barrier properties against oxygen permeation, but they are brittle films (Rhim et al., 2006). The packaging films made from an alcohol-soluble protein-like corn zein have relatively high water vapor barrier properties, compared to films from other proteins, due to its relatively high hydrophobicity (Shukla and Cheryan, 2001). Wheat gluten films are poor water vapor barriers because of the inherent hydrophilicity of the proteins (Krochta, 2002). Whey protein films are transparent and flexible. Mechanical properties of films and barrier properties of films/coatings mainly depend on the molecular weight distribution and amino acid composition of proteins (Tharanathan, 2003). Additionally, other components, especially plasticizer, and technology used in conjunction with film and coating formation are of prime factors affecting the properties of film or coating.

6.3 Collagen and Gelatin

6.3.1 Sources

Most of the collagen and gelatin are derived from bovine hides and bones as well as pigskins (Nagarajan et al., 2013a). Apart from mammalian source, the discards from fish processing including the skin, scales, fins, and bones can be used as raw materials for collagen and gelatin extraction (Karim and Bhat, 2009). Recently, swim bladder has been reported as the excellent source of collagen and gelatin (Sinthusamran et al., 2014, 2013). Gelatin is not a naturally occurring protein, and it can be obtained by thermal denaturation and partial hydrolysis of fibrous collagen (United States Pharmacopeia, 1990). During the collagen-to-gelatin transition, many noncovalent bonds are broken along with some covalent inter- and intramolecular bonds (Schiff's base and aldol condensation bonds) (Foegeding et al., 1996). This results in the conversion of helical collagen structure to an amorphous form, known as gelatin. Gelatin has served as an important hydrocolloid with numerous applications in food products because of its wide range of functional properties (Karim and Bhat, 2009).

6.3.2 Molecular Characteristics

Characteristics and properties of collagen and gelatin are dependent on the source, animal age, quality of raw material, pretreatment and extraction methods, etc. (Karim and Bhat, 2009). Collagen is abundant in animal connective tissues. It has a triple helix structure with three long polypeptide chains.

Each polypeptide is a left-handed triple helix but the three helices are wrapped around each other toward the right. Each polypeptide is made up of roughly 1000 amino acid residues with a repeated Glycine-X-Y sequence (Benjakul et al., 2012a; Mathew, 2002). Glycine-X-Y repeats with the frequent occurrence of proline and hydroxyproline in the X and Y position, respectively. Hydroxyproline is found only in position Y, as is hydroxylysine, while proline can be found in either the X- or Y-position (Fratzl, 2008). Collagen can be classified into 27 types based on the composition of α-chain. Collagen from the skin and bone is typically belonging to type I, in which it consists of 2 α_1-chains and 1 α_2-chain (Benjakul et al., 2012a; Gomez-Guillen et al., 2011; Karim and Bhat, 2009).

Gelatin can be classified into two types, depending on denaturation processes or pretreatment conditions of native collagen. Type A (pI 6–9) and type B (pI 5) gelatins are derived from collagenous starting material with acid and alkaline pretreatment, respectively (Cole and Roberts, 1997). The primary structure of gelatin closely resembles the parent collagen (Benjakul et al., 2012b). This similarity has been substantiated for several tissues and species. Differences are associated with varying raw material together with different pretreatment and extraction procedures (Karim and Bhat, 2009). It is mostly composed of glycine (Gly, 34%), proline and hydroxyproline (Pro + Hyp-imino acids, 16%), and alanine residues (Ala, 10%) (Gomez-Guillen et al., 2002).

6.4 Collagen/Gelatin Films and Coatings

Films and coatings are thin material, in which inter- and intramolecular associations or cross-linking of polymer chains take place in network/matrix (Tharanathan, 2003). Formation of collagen and gelatin films or coatings requires three steps (Cuq et al., 1998):

1. The rupture of low-energy intermolecular bonds that stabilize polymers in the native state
2. The arrangement and orientation of polymer chains
3. The formation of a three-dimensional network stabilized by new interactions and bonds

6.4.1 Film Formation Processes

There are two types of film formation processes: dry and wet (Guerrero et al., 2010; Guilbert et al., 1997) (Figure 6.1). The dry process or thermal processing of film production does not use solvent or medium, such as water or alcohol. Molten casting, extrusion, and heat pressing are good examples of dry process. For the dry process, heat is applied to film-forming materials or resins to increase the temperature to above the glass transition temperature (T_g) or the melting point of the film-forming materials, thereby causing them to flow. Film formation by thermal processing can be carried out by several methods, for example, compression molding, blow molding, and extrusion (Figure 6.1). The solvents for the dispersion of film-forming materials are used in the wet process or solution casting, followed by drying to remove the solvent and form a film structure (Krochta, 2002). In the wet process, the selection of solvents is one of the most important factors. Since the FFS should be edible and biodegradable, only water, ethanol, and their mixtures are appropriate as solvents (Krochta, 2002). All the ingredients of film-forming materials should be dissolved or homogeneously dispersed in the solvents to produce FFSs (Cuq et al., 1995).

Films from collagen and gelatin of different sources, including bovine (Gomez-Estaca, et al., 2009a, 2011; Martucci and Ruseckaite, 2010a), channel catfish (Liu et al., 2014), silver carp (Safandowska and Pietrucha, 2013), bigeye snapper and brownstripe red snapper (Jongjareonrak et al., 2006a,b,c), Baltic cod (Kolodziejska et al., 2006), tilapia (Pranoto et al., 2007), and tuna (Gomez-Guillen et al., 2007) skins, have been produced by solution casting methods. Recently, films from pigskin gelatin resins developed using single-screw extruder (Park et al., 2008); beef, pork, and fish gelatin films using twin screw extruder (Hanani et al., 2012); and fish gelatin films using extrusion and compression molding method (Krishna et al., 2012) have been prepared.

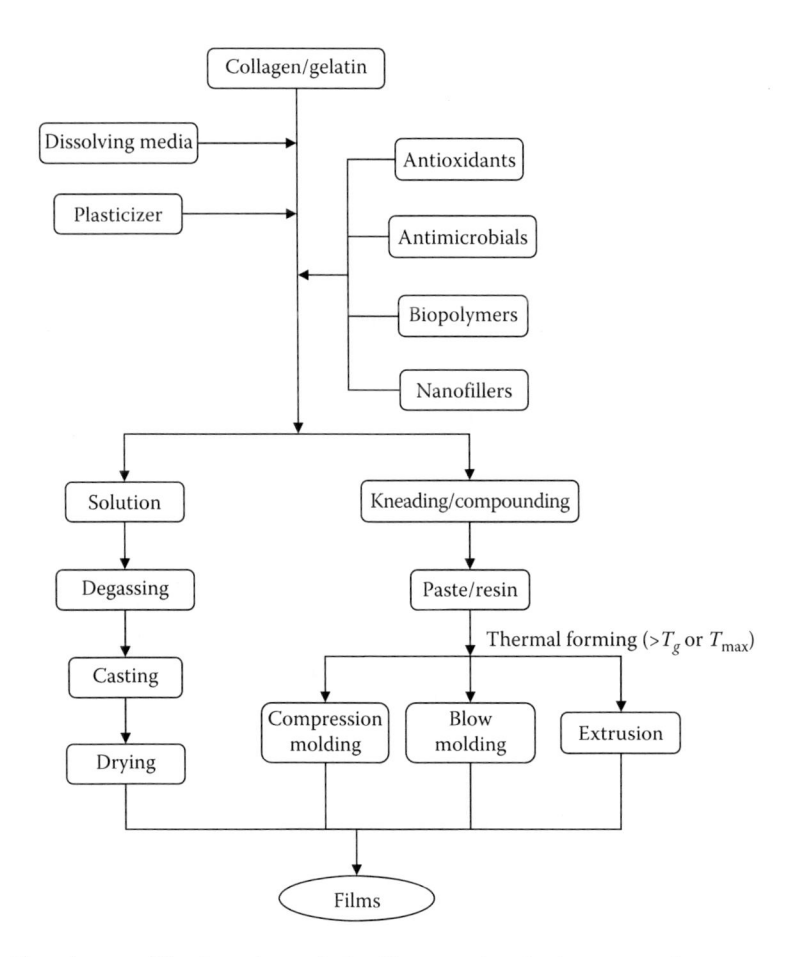

FIGURE 6.1 Flow diagram of film formation method and incorporation of active compounds.

6.4.2 Parameters Affecting Properties of Film/Coating

6.4.2.1 Intrinsic Factors

Amino acid composition, molecular weight distribution, and sources of film-forming material are key factors governing the properties of films or coatings from collagen and gelatin. Imino acid (proline and hydroxyproline) content is found to be an important intrinsic factor affecting the properties of collagen and gelatin films. The pyrrolidine rings of imino acids may impose conformational constraints, imparting a certain degree of molecular rigidity that can affect film deformability (Gomez-Guillen et al., 2009). Gomez-Estaca et al. (2009a) reported that breaking deformation value of tuna skin gelatin films was higher than that of bovine gelatin films. This was plausibly due to the higher imino acid content of bovine gelatin (210 residues/1000 residues) than tuna skin gelatin (185 residues/1000 residues). Moreover, fish gelatin film showed lower water vapor permeability (WVP) than bovine gelatin (Karim and Bhat, 2009). Gomez-Guillen et al. (2007) also reported that films from tuna skin gelatin had lower WVP than pigskin gelatin. The lower WVP of films based on fish gelatins from several species, compared to those from bovine or porcine, can be explained in terms of the amino acid composition. As compared to mammalian gelatins, fish gelatins are known to have much higher hydrophobicity because of their lower proline and hydroxyproline contents. Hydroxyl group of hydroxyproline is normally available to form hydrogen bonds with water (Avena-Bustillos et al., 2006). Gelatin films prepared from cold-water fish and warmwater fish also exhibited different WVP. WVP of cold-water fish gelatin films was significantly lower than that of warmwater fish gelatin films. This was attributed to the lower amounts of proline and hydroxyproline in cold-water fish gelatins (Avena-Bustillos et al., 2006). There were significant differences in molecular distribution and amino acid composition between gelatins obtained from skins of tuna and halibut and tunics of squid (Aleman et al., 2011). Stronger protein–protein interactions occur with higher imino acid content (Gomez-Guillen et al., 2009). Hydrophobic interactions between cellulose and aromatic rings of amino acids of collagen are the primary driving forces of binding in collagen–cellulose blend films (Pei et al., 2013). The imino acid content of gelatin affects polyphenol–protein interactions.

Properties and characteristics of films or coatings from collagen/gelatin depend on the sources, age of animal, quality of raw material, and preprocess/extraction conditions (Karim and Baht, 2009). Apart from amino acid composition, molecular chain length plays a crucial role in film formation and properties, particularly the mechanical performance. Longer molecular chains could form film matrix effectively by introducing more number of junction zones. Increasing content of low-molecular-weight fragments may impair the formation of junction or chain interaction. Renaturation of gelatin chains into helix coil structure possibly takes place during the conditioning of the gelatin films, leading to a decrease in the puncture force of the films (Gimenez et al., 2009a). On the other hand, the increase in components of lower molecular weight favors the plasticizer effect of glycerol and sorbitol in the films (Thomazine et al., 2005). Gomez-Guillen et al. (2009) also reported that gelatin containing higher amount of lower-molecular-weight fractions yielded the film with lower tensile strength (TS) but higher percent elongation. Molecular weight of gelatin is governed in part by the extraction condition. Films from splendid squid skin gelatin extracted at higher temperatures (70°C and 80°C) rendered the films with poor mechanical and thermal properties (Nagarajan et al., 2012), more likely due to the shorter molecular chains. Gelatin extracted from Nile perch bones by severe heating resulted in more low-molecular-weight protein fractions (Muyonga et al., 2004). However, combined alkaline and acid pretreatments rendered the gelatin with high-molecular-weight protein fractions, and films from this gelatin exhibited the excellent properties (Gimenez et al., 2009b).

6.4.2.2 Processing Parameters

Molecular properties and characteristics of collagen or gelatin are affected by processing conditions, for example, heat treatment and pH adjustment of FFS. As a consequence, properties of films or coatings can be influenced by processes used for preparing FFS of gelatin. Unfolding of proteins by heat treatment is considered as promising approach to improve the film-forming ability, and it could extend or unfold the protein molecule to favor the interaction among molecules, in which the junction zones could be formed to a higher extent (Krochta, 2002). When the heating temperature of FFS of cuttlefish skin gelatin was higher than 70°C, the degradation of gelatin could occur, and the shorter chains of gelatin molecules could not form the strong film network (Hoque et al., 2010). Gelatin molecules with the shorter chain most likely formed the weaker chain-to-chain interaction or less junction zones via hydrogen bond. In addition, the increasing number of short chains of gelatin with thermal degradation directly enhanced the mobility of chains (Hoque et al., 2011b). As a result, the weaker film network was formed. Small peptides could be easily inserted in the protein network and establish hydrogen bondings with the gelatin chains, thereby decreasing the density of intermolecular interactions and increasing free volume between gelatin chains (Gimenez et al., 2009a).

The changes in net charge and number of ionized groups as pH varies would be expected to exert a significant effect on forces between protein molecules, resulting in varying film properties. Gelatin films became very elastic with pH adjustment toward extreme alkaline conditions. It is generally accepted that at an alkaline pH above the isoelectric point of gelatin, protein unfolding and solubilization are enhanced (Artharn et al., 2009). During solubilization, the cohesive forces between the protein macromolecules are neutralized by complexing with the solvent molecules. The same charged protein groups also repelled each other and stretched the polymer chain during dissolution. This phenomenon facilitates molecular orientation and the formation of a fine-stranded network. As a consequence, stronger films are formed.

6.4.2.3 Film Drying Condition

Drying rate and environmental conditions (temperature and relative humidity) determine structural characteristics of the films. Very-well-controlled drying process should be performed (Campos et al., 2011). Gontard et al. (1996) stated that the barrier properties of edible films are affected greatly by film composition and environmental conditions. During drying, interconnection of protein molecules leads to the formation of film matrix (Hoque et al., 2010). Gelatin films dried at different temperatures exhibited different molecular arrangements and properties, depending on whether they were dried above or below their gelation temperature (Bradbury and Martin, 1952; Kozlov and Burdygina, 1983). Films can be classified into two groups based on their drying condition including cold-cast and hot-cast films (Chiou et al., 2009). Gelatin chains in a solution dried below gelation temperature can form triple helical structures before the complete evaporation of water. This dried film, generally considered as cold-cast film, can retain these helical structures, depending on moisture content (Chiou et al., 2009; Gomez-Guillen et al., 2002). On the contrary, gelatin chains in a solution dried above gelation temperature remain as random coils during the drying process, and it can retain this amorphous structure. This dried film is termed as hot-cast film (Chiou et al., 2009). A partial renaturation of collagen takes place when the gelatin film is prepared at room temperature (Ghoshal et al., 2010). Chiou et al. (2009) reported that gelatin films dried at different temperatures showed varying properties. Films dried at temperature lower than gelation temperature had the better mechanical properties, compared to those dried at temperature higher than gelation temperature. Cold-cast films showed poor barrier property against water and had the higher weight loss. This might be due to the triple helical structures being able to form more hydrogen bonds with water than amorphous gelatin chains. Consequently, film had higher moisture content (Tanioka et al., 1974). Collagen films dehydrated thermally (dehydrothermal [DHT]-collagen films) at 105°C for 24 h showed higher cross-linking than the control collagen films as determined by TNBS assay (Safandowska and Pietrucha, 2013). Tharanathan (2003) stated that infrared drying chambers are advantageous to accelerate the drying process for coatings. Thus, drying conditions directly affect properties of collagen or gelatin films.

6.4.2.4 Some Additives and Protein Modifiers

Different types of plasticizer, cross-linker, and protein modifier have been incorporated with collagen/gelatin films and coatings (Table 6.1).

6.4.2.4.1 Plasticizers

Plasticizers play a vital role in the preparation of edible films and coatings, especially from proteins. Those films are often brittle and stiff due to the extensive interactions between polymer molecules (Krochta, 2002). Plasticizers are low-molecular-weight agents incorporated into the polymeric film-forming materials to decrease inherent brittleness of films by reducing intermolecular forces and decrease the glass transition temperature of polymers (Cuq et al., 1997). They are able to position themselves between polymer molecules and interfere with the polymer–polymer interaction, resulting in an increase of the mobility of polymeric chains. As a consequence, flexibility and processability can be enhanced (Guilbert and Gontard, 1995; Krochta, 2002). Most plasticizers used for protein-based films are very hydrophilic and hygroscopic. Water molecules in the films also function as plasticizers. Water is actually a very good plasticizer, but it can easily be lost through dehydration at a low relative humidity (Guilbert and Gontard, 1995). Therefore, the addition of hydrophilic plasticizers to films can reduce water loss through dehydration, increase the amount of bound water, and maintain a high water activity (Guilbert and Gontard, 1995).

Hydroxyl compounds and polyols are often used as good plasticizers for protein-based materials. Glycerol is the most widely used because of its better stability and compatibility with hydrophilic biopolymeric chains in comparison with sorbitol, polyethylene glycol (PEG), and sugars (Audic and Chauffeur, 2005). In general, plasticizers such as glycerol and sorbitol are required for making flexible gelatin films (Gomez-Guillen et al., 2009; Sobral et al., 2001). They could easily fit into gelatin networks and form hydrogen bonds with the reactive groups on amino acid residues, thereby reducing protein–protein interactions (Gomez-Guillen et al., 2009).

Different plasticizers with varying amounts directly determine the properties of collagen and gelatin films. Lipids and waxes have been used as hydrophobic plasticizers and can lower WVP of films. Bertan et al. (2005) used triacetin as plasticizer at 15% (w/w on dried gelatin) in gelatin-elemi composite film. Gelatin films had glycerol as plasticizer at low and middle concentrations, about 0%–50% (Cuq et al., 1997; Jongjareonrak et al., 2006a; Nagarajan et al., 2012). However, higher plasticizer concentration

TABLE 6.1

Plasticizers, Cross-Linkers, and Other Protein Modifiers Incorporated with Collagen/Gelatin Films and Coatings

Additives	Sources of Collagen/Gelatin	References
Plasticizers		
Glycerol, sorbitol, and sugar	Pigskin gelatin	Arvanitoyannis et al. (1998)
Triacetin	Bovine hide gelatin	Bertan et al. (2005)
Glycerol	Atlantic sardine fish myofibrillar protein	Cuq et al. (1997)
Glycerol	Bigeye snapper and brownstripe red snapper skin gelatins	Jongjareonrak et al. (2006a, 2008)
Glycerol, ethylene glycerol, sorbitol, and PEG 200 and 400	Bigeye snapper and brownstripe red snapper skin gelatins	Jongjareonrak et al. (2006b)
Glycerol	Tilapia fish skin gelatin	Nagarajan et al. (2014a,b)
Glycerol	Splendid squid skin gelatin	Nagarajan et al. (2012, 2013b)
Sorbitol	Bovine hide gelatin and pigskin gelatin	Sobral et al. (2001)
Blend of glycerol and sorbitol	Pigskin gelatin	Thomazine et al. (2005)
Glycerol	Tilapia fish skin gelatin	Tongnuanchan et al. (2012)
Glycerol, propylene glycol, diethylene glycol, and ethylene glycol	Gelatin	Vanin et al. (2005)
Cross-linkers		
MTGase	Fish gelatin	Bae et al. (2009b)
Alkyl diols	Collagen	Boni (1988)
Glycerol, glutaraldehyde, and PVA	Channel catfish skin collagen	Liu et al. (2014)
Genipin	Collagen hydrolysate	Pei et al. (2013)
EDC, NHS	Silver carp collagen (type I)	Safandowska and Pietrucha (2013)
Glutaraldehyde	Collagen	Weadock et al. (1984)
Other protein modifiers		
Fenton's reagent	Cuttlefish skin gelatin	Hoque et al. (2011c)

generally yields the collagen or gelatin films with lower stiffness and rigidness and higher stretchability (Arvanitoyannis, 2002). Vanin et al. (2005) studied plasticization of gelatin film by using four polyols, that is, glycerol, propylene glycol, diethylene glycol, and ethylene glycol. Diethylene glycol showed the highest plasticizing effect and ethylene glycol increased the thermal properties of gelatin films. Jongjareonrak et al. (2006b) compared the effects of glycerol, ethylene glycerol, sorbitol, and PEG 200 and 400 on the properties of gelatin films and found that different plasticizers showed varying effects on the properties of resulting films. The combination of sorbitol and glycerol was also used as plasticizer for gelatin film (Thomazine et al., 2005).

6.4.2.4.2 Cross-Linkers

Properties of collagen or gelatin films can be improved by cross-linking through physical, chemical, or enzymatic treatments. Properties of collagen films were improved by reducing internal hydrogen bonding while increasing intermolecular spacing with the addition of lower alkyl diols with 4–8 carbon atoms (Boni, 1988). Chemical and enzymatic cross-linkings were induced by different cross-linkers such as aldehydes, microbial transglutaminase (MTGase), genipin, 1-ethyl-3-(3-dimethylaminopropyl) carbodiimide (EDC), and N-hydroxysulfosuccinimide (NHS) (Gomez-Guillen et al., 2009; Pei et al., 2013; Safandowska and Pietrucha, 2013). Aldehydes such as formaldehyde, glutaraldehyde, and glyoxal promote inter- and intramolecular cross-linking of proteins (Feeney et al., 1975; Habeeb and Hiramoto, 1968). Formaldehyde, glutaraldehyde, or glyoxal was used to cross-link films from collagen (Lieberman and Gilbert, 1973; Tomihata et al., 1992; Weadock et al., 1984) and gelatin (Tomihata et al., 1992). The ε-amino groups of lysine, the guanidine group of arginine, the imidazole ring of histidine, and

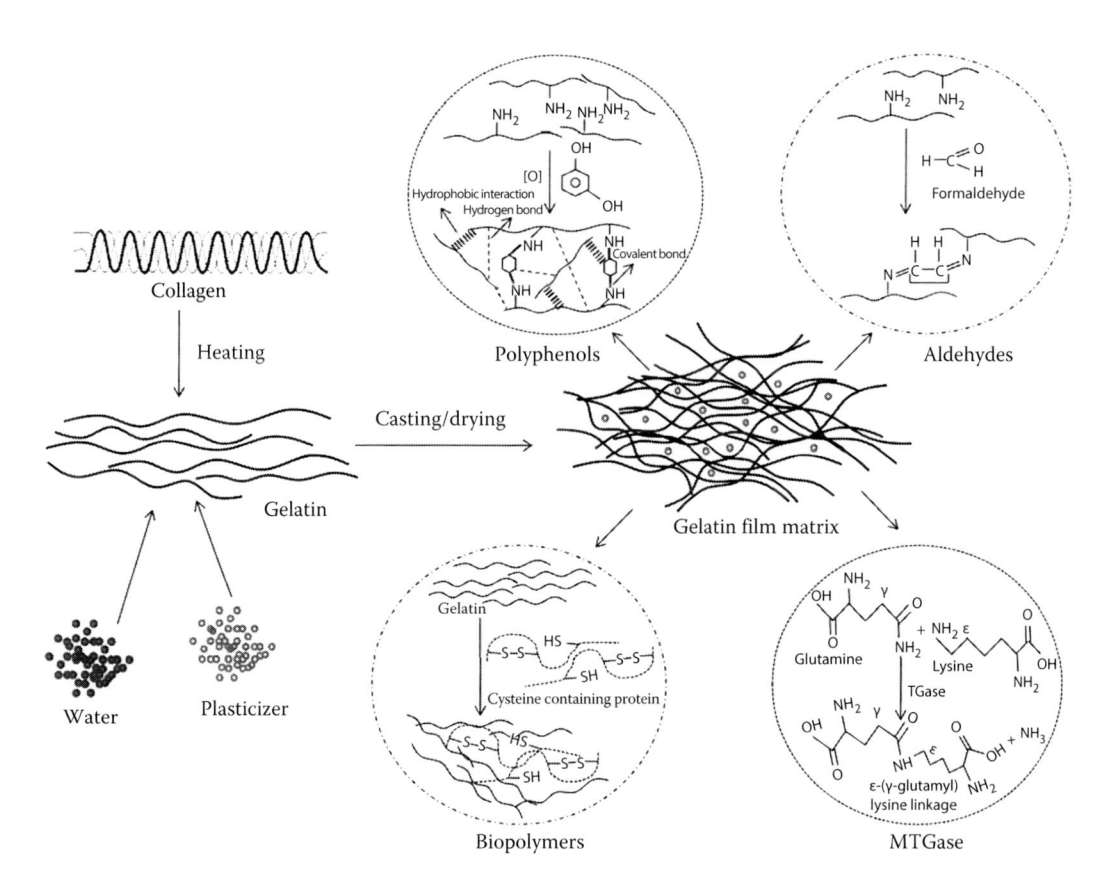

FIGURE 6.2 Schematic diagram of film-forming mechanism and their interactions.

the phenolic ring of tyrosine have the ability to react with aldehydes (Habeeb and Hiramoto, 1968). In particular, the ε-amino group of lysine was considered as the primary reactive site between collagen and aldehydes (Nayudamma et al., 1961) (Figure 6.2). Nevertheless, the inherent toxicity of the afore-mentioned aldehydes (Speer et al., 1980) limits their uses in improving properties of collagen or gelatin films and coatings. Genipin was used as a cross-linker for collagen–cellulose blend film but the films had porous structure (Pei et al., 2013). However, further increases in protein concentration reduced the pores in film matrix. TS, solubility, and brittleness of films normally increased and elongation decreased by chemical or enzymatic cross-linking. Safandowska and Pietrucha (2013) reported that collagen films treated with EDC/NHS had higher cross-linking than the control collagen films. This might be due to the newly formed cross-links between collagen polypeptide chains by cross-linkers (Safandowska and Pietrucha, 2013). Bae et al. (2009b) reported that fish gelatin films cross-linked by MTGase had the decreased TS but increased elongation at break (EAB). MTGase can induce the formation of isopeptides via acyl transfer from glutamine to acyl acceptor, especially lysine (Figure 6.2).

Polyphenol–protein interaction via hydrophobic bonds and hydrogen bonds can affect the bioactivity of phenols. First, polyphenol, which contained hydrophobic groups such as galloyl group, entered into the hydrophobic district of protein by hydrophobic reaction. Then, phenolic hydroxyl group of polyphenol combined with polar group of protein by hydrogen bonds (Shi and Di, 2000) (Figure 6.2). Wu et al. (2013) stated that polyphenolic compounds contain many hydrophobic groups, which can form hydrophobic interaction with hydrophobic region of gelatin molecule. Hydroxyl groups of polyphenolic compounds were able to combine with hydrogen acceptors of gelatin molecule by hydrogen bonds (Gomez-Guillen et al., 2009; Hoque et al., 2011a). Consequently, films with improved mechanical and thermal properties and lowered solubility, WVP, and transparency were obtained (Gomez-Estaca et al., 2009b; Gomez-Guillen et al., 2009; Hoque et al., 2011a; Rattaya et al., 2009; Wu et al., 2013). Polyphenols could form the hydrogen and hydrophobic bonds with the polar groups of polypeptide in gelatin, and these bonds ultimately limited the availability of hydrogen groups to form hydrophilic bonding with water and led to a decrease in the affinity of gelatin films with water. This results in the lower WVP. Oxidized phenolic compounds could also improve the properties of films (Hoque et al., 2011a; Rattaya et al., 2009). After the oxidation process, phenolic compounds were converted to quinone under alkaline pH in the presence of oxygen. The greatest extent of quinone formation occurred with star anise extract (SAE) when compared to clove and cinnamon extracts (Hoque et al., 2011a). Quinones react with amino group of gelatin to form C–N covalent bonds (Strauss and Gibson, 2004) (Figure 6.2). TS and WVP of cuttlefish skin gelatin films were more increased when oxidized herb extracts were incorporated, in comparison with those found with the addition of extracts (without oxidation). Oxidized phenolic compounds in different plant extracts might contribute to the formation of nondisulfide covalent bond. Thus, the incorporation of oxidized phenolic compounds effectively improved the properties of resulting film.

6.4.2.4.3 Other Protein Modifiers

Fenton's reagent is another protein modifier, which can generate the active radical, hydroxyl radicals (OH·), from H_2O_2 in the presence of Fe^{2+} by Fenton reaction (Krochta et al., 1997). Hydroxyl radicals are reactive species, which can alter protein composition and configuration (Liu and Xiong, 2000). Stadtman (2001) stated that the OH· radical involves abstraction of the α-hydrogen atom from amino acid residues to form a carbon-centered radical derivative. Two different carbon-centered amino acid radicals can react with one another to form –C–C– protein cross-linked products. Recently, properties of cuttlefish skin gelatin films were improved by adding Fenton's reagent into the FFS (Hoque et al., 2011c). TS of resulting films increased, while EAB and solubility were decreased. However, higher level of Fenton's reagent resulted in the poor film properties. Higher amounts of OH· radical could induce protein fragmentation. The fragmentation of proteins is a consequence of a direct attack by OH· radicals on the polypeptide backbone or on the side chains of glutamyl or prolyl residues (Stadtman and Berlett, 1997). Peptide bond cleavage can occur by reactive oxygen substance–mediated oxidation of glutamyl side chains (Stadtman, 2001). Lysine, arginine, proline, and threonine residues of proteins are particularly sensitive to metal-catalyzed oxidation.

6.4.2.5 Ultraviolet, γ-Irradiation, and Other Physical Treatments

Physical cross-linking for film improvement can be conducted using several methods including ultraviolet (UV), γ-irradiation, and DHT treatment (Bigi et al., 1998). UV and γ-irradiation treatments have been applied in pharmaceutical and medical studies to cross-link gelatin and collagen films (Bessho et al., 2007). Ionization of a polymer material induced by the radiation gives rise to radicals, which further induce the subsequent alteration of the structure of polymer material (Inamura et al., 2013). High-energy ionizing radiation not only causes the structural modifications of polymer through cross-linking, chain scission, oxidation, and change in the number and nature of double bonds but also contributes to the trapped charge within the material (Kacarevic-Popovic et al., 2004). Mechanical and barrier properties of protein-based films were improved by applying UV and γ-irradiation treatments. This was plausibly due to the enhanced cross-linking of protein molecules (Jo et al., 2005; Sung and Chen, 2014). However, higher dose treatments generally result in poor film properties, plausibly due to the degradation of protein molecules instead of cross-linkings (Jo et al., 2005). Cataldo et al. (2008) reported that ionizing radiation is able to induce cross-linking of collagen, gelatin, and their films. However, collagen/PEG blend films are less stable under UV irradiation than pure collagen films. The surface characteristics of collagen and collagen/PEG blends were not drastically altered after UV irradiation (Sionkowska, 2006). DHT treatment is another physical treatment, which was applied to cross-link the protein molecules and modified strength and solubility of collagen-based products (Wess and Orgel, 2000). With DHT treatment, water is driven off from the collagen molecules, and there are changes in amino acids of the collagen chain associated with oxidative damage or cross-linking (Gorham et al., 1992). Collagen films with DHT treatment at 105°C for 24 h showed higher cross-linking than control collagen films (Safandowska and Pietrucha, 2013).

6.5 Blend and Composite Film/Coating Based on Collagen/Gelatin

Blend and composite films and coatings could be prepared by combining two different polymers with varying advantages for improved property. Normally, these blend components are also biodegradable polymers to ensure that the final blend/composite films and coatings can still be regarded as "green" (Rhim et al., 1998). Blend films are defined as two different biopolymers combined or mixed together to obtain uniform film matrix. Nevertheless, the constituent parts are still indistinguishable from one another. Films/coatings made with a blend of biopolymers generally exhibit improved properties, compared with film prepared from single polymer. The compatibility of blend components is an important issue when dealing with mixtures of biopolymers as this might drastically alter the performance of these materials (Diab et al., 2001). Those blends can interact each other by ionic interaction, hydrogen bonds, or other bondings. Some hydrocolloids have been used to blend with gelatin for film making. Cuttlefish skin gelatin–mung bean protein isolate blend films showed the improved properties due to strong interaction in film matrix via hydrogen bonds and hydrophobic interaction (Hoque et al., 2011d). Mung bean proteins are rich in cysteine, in which disulfide bonds can be formed during film formation (Figure 6.2). Gelatin–konjac glucomannan blend films were successfully prepared as an edible inner packaging material with different blending ratios by using a solvent casting technique (Li et al., 2006). Gum arabic with or without gelatin has been used to produce protective films for chocolates, nuts, cheese, and pharmaceutical tablets (Colloides Naturels Inc., 1988). Ciesla et al. (2006) reported that the addition of polysaccharides had a significant effect on the properties of protein-based edible films. Gelatin and chitosan are hydrophilic biopolymers with good affinity and compatibility, and they are expected to form blend films with good mechanical and optical properties (Arvanitoyannis et al., 1998). Nevertheless, they are highly sensitive to moisture and exhibit poor barrier property against water (Guilbert et al., 1996). Gelatin–polysaccharide blend films could be developed when the control of WVP is not the goal. These blend films are effective barrier to O_2, CO_2, and lipids (Tharanathan, 2003).

The miscibility and compatibility of collagen or gelatin with other polymers depend on their ability to form specific interaction (Alizadeh and Behfar, 2013). Collagen hydrolysate and cellulose blend films cross-linked with genipin had both cellulose framework and protein cross-linked network in film matrix (Pei et al., 2013). Moreover, blend films from collagen hydrolysate/cellulose were rich in glycine, proline, hydroxyproline, glutamic acid, and alanine, revealing that there were many intermolecular hydrogen bonds between the –OH of the cellulose and the $-NH_2$, –COOH, and –CONH– groups in the collagen hydrolysate (Pei et al., 2013). Blends of collagen and synthetic polymer had different hydrogen bonds in their film matrix (Sionkowska et al., 2009). Hydrogen bonds can be formed between two hydroxyl groups (OH–OH) and/or between the amide groups. Furthermore, oxygen of hydroxyl group (NH–OH) also can interact with the hydrogen of heteroatoms and the carbonyl group of the amide (C=O–HO, C=O–HN) in blends of collagen-synthetic polymer films (Sionkowska et al., 2009).

Edible coatings from biopolymers are normally sensitive to temperature and relative humidity. Blend solution of gelatin and polyvinyl alcohol (PVA) increased the resistance to humidity of resulting films (Bergo et al., 2006). Coating made from megrim skin gelatin–chitosan blends maintained the sensory properties with delayed spoilage of cold patties (Lopez-Caballero et al., 2005). This was plausibly due to the strong gel formed over the product by blend coating, which acted as a thin barrier (Gomez-Guillen et al., 2009). This protective barrier could be molten and removed during cooking process. Gelatin–starch coatings were applied to delay the ripening process and respiratory climacteric pattern of avocados, as indicated by better pulp firmness and retention of skin color, and lower weight loss of coated fruits in comparison with the control (Aguilar-Mendez et al., 2008).

The composite system is formulated from two or more other different biopolymers and fillers, in which individual phases of components remain distinct from each other after being combined. Composite film/coating based on collagen/gelatin could be prepared by incorporating lipids or oils. Basically, biopolymers, such as proteins and polysaccharides, serve as the supporting matrix for films/coatings, and lipids provide a good barrier to water vapor (Baldwin et al., 1997). Emulsification of oil or lipid in FFSs (Jongjareonrak et al., 2006c) to ensure uniform distribution of oil droplet affected the vapor barrier property of gelatin films. However, the phase separation during casting might bring about bilayer gelatin films with improved water vapor barrier property (Tongnuanchan et al., 2012).

6.6 Nanocomposite Film/Coating Based on Collagen/Gelatin

Polymer nanocomposites have received great interest since nanosized material fillers significantly improve properties of polymers in comparison with polymer alone or microscale composites (Bae et al., 2009a). Nanocomposite films developed from biopolymers known as "bio-nanocomposites" showed the improved barrier and mechanical properties and thermal stability, due to the enhanced polymer–filler interfacial interaction (Martucci and Ruseckaite, 2010b; Ray and Okamoto, 2003). Several nanofillers such as clay, silica, talc, ZnO, and TiO_2 have been used (Rhim, 2007). The layered silicates, naturally occurring smectite clays, such as hectorite and montmorillonite (MMT), have been used to improve the property of films based on biopolymers including gelatin (Ray and Okamoto, 2003; Rhim et al., 2009; Sothornvit et al., 2009). The improved water and gas barrier properties of nanoclay incorporated nanocomposite films are believed to be due to the presence of ordering dispersed silicate layers with large aspect ratios in the polymer matrix. This forces water/gas travelling through the film via an increased "tortuous path" of the polymer matrix surrounding the nanofillers, thereby increasing the effective path length for diffusion (Figure 6.3) (Ray and Okamoto, 2003; Rhim, 2007). Apart from their improved material properties, biodegradability makes them eco-friendly and alternative to traditional packaging (Sozer and Kokini, 2009).

Depending on the compatibility of polymer and clay, three polymer hybrids are possibly formed, namely, conventionally phase separated, intercalated, and delaminated/exfoliated, as schematically shown in Figure 6.4 (Sozer and Kokini, 2009).

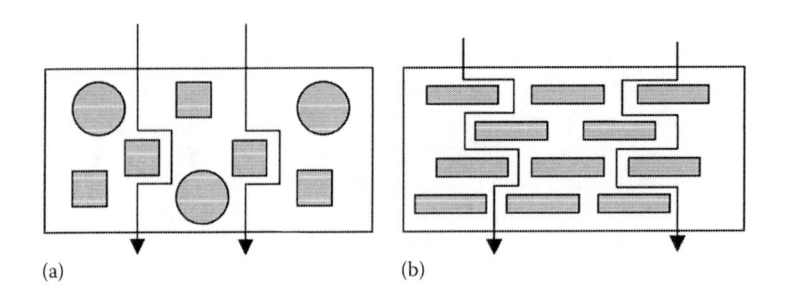

(a)   (b)

FIGURE 6.3 Schematic illustration of formation of "tortuous path" in nanocomposite. Conventional filler reinforced composites (a), polymer/layered silicate nanocomposites (b). (From Rhim, J. and Ng, P.K.W., *Crit. Rev. Food Sci.*, 47, 411, 2007.)

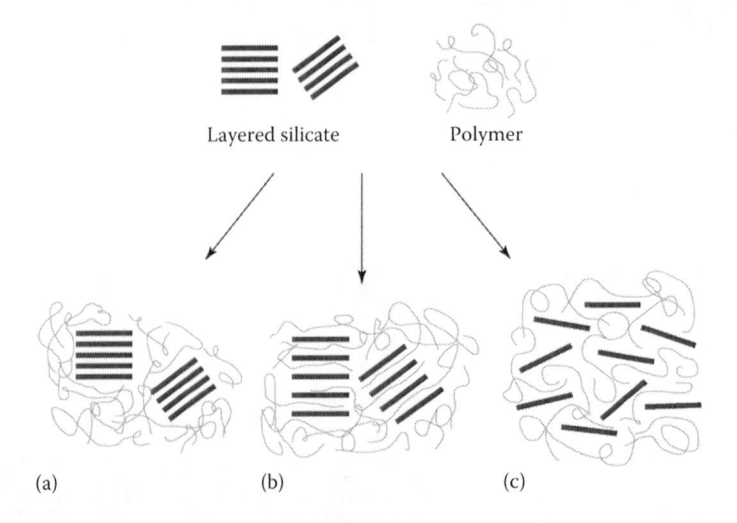

FIGURE 6.4 Types of composite derived from the interaction between clays and polymers. (a) Phase separated (microcomposite). (b) Intercalated (nanocomposite). (c) Exfoliated (nanocomposite). (From Sozer, N. and Kokini, J.L., *Trend Biotechnol.*, 27, 82, 2009.)

Those three hybrids are as follows:

1. *The phase-separated hybrids*: They are corresponding to the conventionally filled polymers, in which the clay particles agglomerate, generally leading to the poor mechanical performance (LeBaron et al., 1999).
2. *The intercalated hybrids*: The insertion of extended polymer chains into the clay layers occurs in regular multilayers with a repeat distance of a few nanometers (Burnside and Giannelis, 1995).
3. *The delaminated/exfoliated hybrids*: The clay platelets (1 nm thick) are expected to be individually dispersed in the polymer matrix, resulting in high aspect ratio of about 1000 (fully dispersed). The value of 10 is found in phase-separated systems. The interlayer expansion of these delaminated/exfoliated hybrids is comparable to the radius of gyration of the polymer molecules (Burnside and Giannelis, 1995).

Two particular characteristics of layered silicates are generally considered for polymer/layered silicate nanocomposites. The first is the ability of the silicate particles to disperse into individual layers. The second characteristic is the ability to fine-tune their surface chemistry through ion exchange reactions with organic and inorganic cations (Ray and Okamoto, 2003).

Nanofillers/nanoclays have been incorporated with collagen/gelatin films and coatings to improve their properties (Table 6.2). Nanoclay, particularly the MMT Cloisite® Na$^+$, which is hydrophilic in nature, has been used to improve mechanical and barrier properties of fish gelatin–based biopolymers (Bae et al., 2009a; Shakila et al., 2012).

Li et al. (2003) studied swelling property of gelatin/MMT nanocomposites and reported that the swelling rate and solvent uptake decreased when MMT were incorporated. This might be due to the structural change of the original gelatin network in the presence of MMT nanoclays. TS of MTGase-treated fish gelatin nanofilms incorporated with MMT Na$^+$ decreased but EAB increased. This was coincidental with the increases in molecular weight and viscosity of gelatin solutions (Bae et al., 2009b). Types of clays could affect their application when incorporated into films (Sothornvit et al., 2009). Gelatin film added with hydrophobic nanoclay, MMT 20A, had poorer mechanical properties, compared to those added with hydrophilic nanoclay, MMT Na$^+$ (Nagarajan et al., 2014a). It might be due to the poor dispersion of nanoclay and lesser interaction between hydrophilic polymer and hydrophobic nanoclay. However, the film had the improved water vapor barrier property, plausibly due to the hydrophobicity of nanoclay (Rhim, 2011). Hydrophobic nanoclays are hardly dispersed in the FFS, and further treatments such as high-pressure homogenization and ultrasonication could be implemented (Alamsi et al., 2010; Rhim, 2011). Homogenizing at various pressure ranges and passes could affect the film matrix and resulted in different barrier and mechanical properties and thermal stability of films (Nagarajan et al., 2014b).

TABLE 6.2

Nanofillers/Nanoclays Incorporated with Collagen/Gelatin Films and Coatings

Nanofillers/Nanoclays	Sources of Nanofilms/Coating	References
Cloisite Na$^+$	Fish gelatin	Bae et al. (2009a)
Graphene	Type I collagen	Ebrahimi et al. (2013)
Cloisite® 20A	Bovine hide gelatin	Farahnaky et al. (2014)
Hydroxyapatite	Bovine tendon and calf skin collagens	Gelinsky et al. (2008)
Cloisite Na$^+$	Bovine skin gelatin	Li et al. (2003)
Cloisite Na$^+$	Bovine hide gelatin	Martucci and Ruseckaite (2010a,b)
Cloisite Na$^+$	Bovine hide gelatin	Martucci et al. (2007)
Cloisite Na$^+$, Cloisite® 15A, Cloisite 20A, and Cloisite® 30B	Tilapia skin gelatin	Nagarajan et al. (2014a)
Laponite® RDS, Cloisite Na$^+$	Gelatin	Rao (2007)
Cloisite Na$^+$	Red snapper and grouper bone gelatin	Shakila et al. (2012)
Indium oxide	Collagen (type I and III)	Shan et al. (2004)
Cloisite Na$^+$	Bovine skin gelatin	Zheng et al. (2002)

6.7 Active Film/Coating Based on Collagen/Gelatin

Films and coatings from collagen or gelatin can be used as smart/active packaging, in which anti-oxidants/antimicrobials can be incorporated. Active packaging is an innovative packaging, which can be used to delay oxidation, inhibit microbial growth, and control respiration rate (Ahvenainen, 2003). Antioxidant/antimicrobial packaging technology is important and very promising for extending the shelf life of foods (Lopez-De-Dicastillo et al., 2012). The direct addition of antioxidants/antimicrobials to food results in some loss of activity due to leaching into the bulk of foods (Vojdani and Torres, 1989), or interaction with other food components such as lipids, carbohydrates, and proteins (Henning et al., 1986; Jung et al., 1992). When antioxidant/antimicrobial agents are gradually released from the films/coatings for a prolonged period, its activity could be extended during transportation and storage. The incorporation of antioxidants/antimicrobials with the protein matrix may not only affect the release of active components to the food but also alter the physico-chemical properties of the films, thereby altering the film solubility and their barrier properties (Gomez-Guillen et al., 2007). Active films have been developed by the addition of different kinds of antioxidant/antimicrobial in gelatin films and coatings (Table 6.3).

6.7.1 Antioxidant Film/Coating

Incorporation of antioxidants into packaging films/coatings becomes very popular since oxidation is one of the main causes of loss in food quality (Byun et al., 2010). Owing to the superior oxygen barrier property, fish gelatin–based film could prevent the lipid oxidation in food systems

TABLE 6.3

Antioxidants/Antimicrobials Incorporated with Gelatin Films and Coatings

Antioxidants/Antimicrobials	Sources of Gelatin	References
Brazilian elemi	Bovine hide gelatin	Bertan et al. (2005)
Lysozyme	Cold-water fish skin gelatin	Bower et al. (2006)
Squid skin gelatin hydrolysate	Squid skin gelatin	Gimenez et al. (2009a)
Essential oils of clove, fennel, cypress, lavender, thyme, herb of the cross, pine, and rosemary	Bovine hide gelatin	Gomez-Estaca et al. (2010)
Oregano or rosemary extracts	Tuna skin gelatin	Gomez-Estaca et al. (2009b)
Clove essential oil	Catfish skin gelatin or blend with chitosan	Gomez-Estaca et al. (2009c)
Borage seed extract	Sole skin and catfish skin gelatins	Gomez-Estaca et al. (2009d)
Oregano or rosemary extracts or chitosan	Porcine skin gelatin	Gomez-Estaca et al. (2007)
Murta leaf extracts	Tuna skin gelatin	Gomez-Guillen et al. (2007)
Cinnamon, clove, and SAEs	Cuttlefish skin gelatin	Hoque et al. (2011a)
α-Tocopherol and BHT	Bigeye snapper and brownstripe red snapper skin gelatins	Jongjareonrak et al. (2008)
Green tea extract, grape seed extract (proanthocyanidins), grape seed extract (polyphenols), ginger extract, gingko leaf extract	Silver carp skin gelatin	Li et al. (2014)
Sodium magnesium and sodium copper chlorophyllins	Gelatin	Lopez-Carballo et al. (2008)
Lignosulfunate from eucalyptus wood	Deepwater fish skin gelatin	Nunez-Flores et al. (2012)
Benzoic acid	Gelatin	Ou et al. (2002)
Chitosan	Bovine hide gelatin	Pereda et al. (2011)
Seaweed extract	Bigeye snapper skin gelatin	Rattaya et al. (2009)
Citrus essential oils	Tilapia skin gelatin	Tongnuanchan et al. (2012)
Green tea extract	Silver carp skin gelatin	Wu et al. (2013)

(Jongjareonrak et al., 2006a). Several antioxidants including synthetic and natural antioxidants have been added into gelatin film to produce active film with antioxidative activity (Jongjareonrak et al., 2008). Essential oils (Gomez-Estaca et al., 2010; Tongnuanchan et al., 2013, 2012), fatty acids (Bertan et al., 2005), plant extracts (Gomez-Estaca et al., 2009b; Gomez-Guillen et al., 2007; Nunez-Flores et al., 2013; Rattaya et al., 2009; Rubilar et al., 2006), natural extracts (Jongjareonrak et al., 2008), and spices (Hoque et al., 2011a) were added as antioxidants into gelatin films. Antioxidative activity and properties of gelatin films incorporated with butylated hydroxytoluene (BHT) and α-tocopherol were investigated by Jongjareonrak et al. (2008). WVP of films decreased when both antioxidants were incorporated. However, films incorporated with α-tocopherol became more transparent. Hoque et al. (2011a) reported that incorporation of herb extracts enhanced the antioxidative property of cuttlefish skin gelatin films. Films incorporated with plai and turmeric essential oils showed the higher DPPH and ABTS radical scavenging activity, respectively, compared with the control film and ginger essential oil added film (Tongnuanchan et al., 2013). Gelatin-based films incorporated with oregano or rosemary extracts (1.25% and 20%, respectively) had increased phenol content and antioxidant activity (Gomez-Estaca et al., 2007). The extracts served as antioxidants that terminate chain reactions by removing free radicals and inhibit other oxidation reactions. The extracts contained various reducing agents (i.e., thiols, polyphenols). The edible films containing plant extracts lowered lipid oxidation in cold-smoked sardine (Gomez-Estaca et al., 2007).

However, the higher amount of antioxidants incorporated into films may induce the development of a heterogeneous structure with the presence of discontinuous areas. This resulted in the lower TS than control gelatin film (Li et al., 2014). A slight decrease in mechanical properties was observed, depending on antioxidants added. This was attributed to the differences in composition of antioxidants used, which might have the influence on the degree of protein–antioxidant interactions (Gomez-Guillen et al., 2009). The interactions between the antioxidants and gelatin film matrix might be formed via hydrogen bonding, particularly during film formation (Jongjareonrak et al., 2008). However, the release of antioxidants from films can be impeded since some active compounds can be tightly associated with the matrix. Gimenez et al. (2009a) reported that control gelatin film (without any kind of antioxidants) also possessed antioxidant activity, plausibly due to their inherent amino acid composition, which could act as electron donors.

6.7.2 Antimicrobial Film/Coating

Antimicrobial films and coatings have been paid great interest to extend the shelf life of food products by preventing microbiological contamination and termed as "active packaging" (Lopez-Carballo et al., 2008). Antimicrobials (Campos et al., 2011; Ou et al., 2002), plant extracts (Gomez-Estaca et al., 2007; Lopez-Carballo et al., 2008), essential oils (Gomez-Estaca et al., 2009c, 2010), lysozyme (Bower et al., 2006), etc., were added into films to gain the antimicrobial activity. Composite (gelatin–chitosan) and bilayer (gelatin–chitosan) antimicrobial films incorporated with essential oil were prepared (Gomez-Estaca et al., 2010; Pereda et al., 2011). Photoactivated chlorophyllins immobilized in gelatin film had a strong bactericidal effect on the viability of Gram-positive bacteria (Lopez-Carballo et al., 2008). Pereda et al. (2011) reported that antimicrobial activity of the gelatin solution was more likely attributed to the presence of oligopeptide chains from the hydrolysis of gelatin, which might have antimicrobial activity due to the presence of side-chain amino groups. Shelf life of tilapia fillet was extended under refrigeration after being coated with gelatin containing benzoic acid as an antimicrobial agent (Ou et al., 2002). Gelatin–chitosan film incorporated with clove extract was less effective to inhibit microbial growth for raw sliced salmon than the gelatin one. This difference might be due to the different degrees of gelatin interaction with polyphenols and chitosan, which may affect the capacity of the antimicrobial compounds (Gomez-Estaca et al., 2007, 2009c). Additionally, gelatin–chitosan films exhibited an excellent antimicrobial and oxygen barrier properties (Gomez-Estaca et al., 2007). Chitosan appears to involve in altering the membrane of Gram-negative bacteria and increasing its permeability, thereby leading to cell death (Helander et al., 2001). The film also acted as a barrier to oxygen, thus inhibiting the growth of aerobic microorganisms (Gomez-Estaca et al., 2007; Jeon et al., 2002).

6.8 Conclusions

Collagen and gelatin can be used as the bio-based materials for film or coating preparation. Nevertheless, properties of films/coatings can be determined by sources, starting raw material, species, age, processing conditions, intrinsic parameters, cross-linkers, etc. To maximize the utility, the improvement of film or coating can be achieved by the incorporation of different substances such as polysaccharides, lipids, and nanoclays, or physical treatment. Nanotechnology is another approach for improvement of film or coating from collagen and gelatin. Furthermore, films or coatings with antimicrobial and/or antioxidant activity known as "active" can be developed for shelf life extension of foods or other products.

REFERENCES

Aguilar-Mendez, M.A., Martin-Martinez, E.S., Tomas, S.A., Cruz-Orea, A., and Jaime-Fonseca, M.R. 2008. Gelatine starch films: Physicochemical properties and their application in extending the post-harvest shelf life of avocado (*Persea americana*). *J. Sci. Food Agr.* 88: 185–193.

Ahvenainen, R. 2003. *Active and Intelligent Packaging: Novel Food Packaging Techniques.* New York: CRC Press.

Alamsi, H., Ghanbarzadeh, B., and Entezami, A.A. 2010. Physicochemical properties of starch-CMC-nanoclay biodegradable films. *Int. J. Biol. Macromol.* 46: 1–5.

Aleman, A., Gimenez, B., Montero, P., and Gomez-Guillen, M.C. 2011. Antioxidant activity of several marine skin gelatins. *LWT Food Sci. Technol.* 44: 407–413.

Alizadeh, A. and Behfar, S. 2013. Properties of collagen based edible films in food packaging: A review. *Ann. Biol. Res.* 4: 253–256.

Arcan, I. and Yemenicioglu, A. 2011. Incorporating phenolic compounds opens a new perspective to use zein films as flexible bioactive packaging materials. *Food Res. Int.* 44: 550–556.

Artharn, A., Prodpran, T., and Benjakul, S. 2009. Round scad protein-based film: Storage stability and its effectiveness for shelf-life extension of dried fish powder. *LWT Food Sci. Technol.* 42: 1238–1244.

Arvanitoyannis, I.S. 2002. Formation and properties of collagen and gelatin films and coatings. In A. Gennadios (ed.), *Protein-Based Films and Coatings*, pp. 275–304. Boca Raton, FL: CRC Press.

Arvanitoyannis, I.S., Nakayama, A., and Aiba, S. 1998. Chitosan and gelatin based edible films; state diagrams, mechanical and permeation properties. *Carbohydr. Polym.* 37: 371–382.

ASTM. 1999, D 6400. Standard specification for compostable plastics. *Annual Book of ASTM Standards.* Philadelphia, PA: ASTM, pp. 1–3.

Audic, J.L. and Chauffeur, B. 2005. Influence of plasticisers and crosslinking on the properties of biodegradable films made from sodium caseinate. *Eur. Polym. J.* 41: 1934–1942.

Avena-Bustillos, R.J., Olsen, C.W., Chiou, B., Yee, E., Bechtel, P.J., and McHugh, T.H. 2006. Water vapor permeability of mammalian and fish gelatin films. *J. Food Sci.* 71: E202–E207.

Bae, H.J., Darby, D.O., Kimmel, R.M., Park, H.J., and Whiteside, W.S. 2009b. Effects of transglutaminase-induced cross-linking on properties of fish gelatin-nanoclay composite film. *Food Chem.* 114: 180–189.

Bae, H.J., Park, H.J., Hong, S.I., Byun, Y.J., Darby, D.O., Kimmel, R.M., and Whiteside, W.S. 2009a. Effect of clay content, homogenisation RPM, pH, and ultrasonication on mechanical and barrier properties of fish gelatin/montmorillonite nanocomposite films. *LWT Food Sci. Technol.* 42: 1179–1186.

Baldwin, E.A., Nisperos-Carriedo, M.O., Hagenmaier, R.D., and Baker, R.A. 1997. Use of lipids in coatings for food products. *Food Technol.* 51: 56–64.

Benjakul, S., Kittiphattanabawon, P., and Regenstein, J.M. 2012b. Fish gelatin. In B.K. Simpson (ed.), *Food Biochemistry and Food Processing*, pp. 388–405. Ames, IA: John Wiley & Sons, Inc.

Benjakul, S., Nalinanon, S., and Shahidi, F. 2012a. Fish collagen. In B.K. Simpson (ed.), *Food Biochemistry and Food Processing*, pp. 365–387. Ames, IA: John Wiley & Sons, Inc.

Bergo, P., Carvalho, R.A., Sorbal, P.J.A., and Bevilacqua, F.R.S. 2006. Microwave transmittance in gelatin-based films. *Meas. Sci. Technol.* 17: 3261–3264.

Bertan, L.C., Tanada-Palmu, P.S., Siani, A.C., and Grosso, C.R.F. 2005. Effect of fatty acids and "Brazilian elemi" on composite films based on gelatin. *Food Hydrocolloids* 19: 73–82.

Bessho, M., Kojima, T., Okuda, S., and Hara, M. 2007. The radiation-induced cross-linking of gelatin by using γ-rays: Insoluble gelatin hydrogel formation. *Bull. Chem. Soc. Jpn.* 80: 979–985.

Bigi, A., Bracci, B., Cojazzi, G., Panzavolta, S., and Roveri, N. 1998. Drawn gelatin films with improved mechanical properties. *Biomaterials* 19: 2335–2340.

Boni, K.A. 1988. Strengthened edible collagen casing and method of preparing same. US Patent 4,794,006.

Bower, C.K., Avena-bustillos, R.J., Olsen, C.W., McHugh, T.H., and Bechtel, P.J. 2006. Characterization of fish-skin gelatin gels and films containing the antimicrobial enzyme lysozyme. *J. Food Sci.* 71: 141–145.

Bradbury, E. and Martin, C. 1952. The effect of the temperature of preparation on the mechanical properties and structure of gelatin films. *Proc. R. Soc. Lond. A* 214: 183–192.

Burnside, S.D. and Giannelis, E.P. 1995. Synthesis and properties of new poly(dimethylsiloxane) nanocomposites. *Chem. Mater.* 7: 1597–1600.

Byun, Y., Kim, Y.T., and Scott, W. 2010. Characterisation of an antioxidant polylactic acid (PLA) film prepared with α-tocopherol, BHT and polyethylene glycol using film cast extruder. *J. Food Eng.* 100: 239–244.

Campos, C.A., Gerschenson, L.N., and Flores, S.K. 2011. Development of edible films and coatings with antimicrobial activity. *Food Bioprocess Technol.* 4: 849–875.

Cataldo, F., Ursini, O., Lilla, E., and Angelini, G. 2008. Radiation-induced crosslinking of collagen gelatin into a stable hydrogel. *J. Radioanal. Nucl. Chem.* 275: 125–131.

Chiou, B., Avena-Bustillos, R.J., Bechtel, P.J., Imam, S.H., Glenn, G.M., and Orts, W.J. 2009. Effects of drying temperature on barrier and mechanical properties of cold-water fish gelatin films. *J. Food Eng.* 95: 327–331.

Ciesla, K., Salmieri, S., and Lacroix, M. 2006. Modification of the properties of milk protein films by gamma radiation and polysaccharide addition. *J. Sci. Food Agr.* 86: 908–914.

Cole, C.G.B. and Roberts, J.J. 1997. Gelatine colour measurement. *Meat Sci.* 45: 23–31.

Colloides Naturels Inc. 1988. Sealgum: Something new in films. *Product Bulletin.* Colloides Naturels Inc., Bridgewater, NJ.

Cuq, B., Aymad, C., Cuq, J., and Quilbert, S. 1995. Edible packaging films based on fish myofibrillar proteins: Formulation and functional properties. *J. Food Sci.* 60: 1369–1373.

Cuq, B., Gontard, N., Cuq, J.L., and Guilbert, S. 1997. Selected functional properties of fish myofibrillar protein-based films as affected by hydrophilic plasticisers. *J. Agr. Food Chem.* 45: 622–626.

Cuq, B., Gontard, N., and Guilbert, S. 1998. Proteins as agricultural polymers for packaging production. *Cereal Chem.* 75: 1–9.

Diab, T., Biliaderis, C.G., Gerasopoulos, D., and Sfakiotakis, E. 2001. Physico-chemical properties and application of pullulan edible films and coatings in fruit preservation. *J. Sci. Food Agr.* 81: 988–1000.

Ebrahimi, S., Montazeri, A., and Rafii-Tabar, H. 2013. Molecular dynamics study of the interfacial mechanical properties of the graphene–collagen biological nanocomposite. *Comp. Mater. Sci.* 69: 29–39.

Farahnaky, A., Mohammad, S., Dadfar, M., and Shahbazi, M. 2014. Physical and mechanical properties of gelatin–clay nanocomposite. *J. Food Eng.* 122: 78–83.

Feeney, R.E., Blankenhorn, G., and Dixon, H.B.F. 1975. Carbonylamine reactions in protein chemistry. *Adv. Protein Chem.* 29: 135–203.

Foegeding, E.A., Lanier, T.C., and Hultin, H.O. 1996. Characteristics of edible muscle tissue. In O.R. Fennema (ed.), *Food Chemistry*, pp. 879–942. New York: Marcel Dekker.

Fratzl, P. 2008. Structure and mechanics: An introduction. In P. Fratzl (ed.), *Collagen, Structure and Mechanics*, pp. 1–12. New York: Springer.

Gelinsky, M., Welzel, P.B., Simon, P., Bernhardt, A., and Konig, U. 2008. Porous three-dimensional scaffolds made of mineralised collagen: Preparation and properties of a biomimetic nanocomposite material for tissue engineering of bone. *Chem. Eng. J.* 137: 84–96.

Gennadios, A., Brandenburg, A.H., Park, J.W., Weller, C.L., and Testin, R.F. 1994. Water vapor permeability of wheat gluten and soy protein isolate films. *Ind. Crop. Prod.* 2: 189–195.

Gennadios, A., Weller, C.L., Hanna, M.A., and Froning, G.W. 1996. Mechanical and barrier properties of egg albumen films. *J. Food Sci.* 61: 585–589.

Ghoshal, S., Mattea, C., and Stapf, S. 2010. Inhomogeneity in the drying process of gelatin film formation: NMR microscopy and relaxation study. *Chem. Phys. Lett.* 485: 343–347.

Gimenez, B., Gomez-Estaca, J., Aleman, A., Gomez-Guillen, M.C., and Montero, M.P. 2009a. Improvement of the antioxidant properties of squid skin gelatin films by the addition of hydrolysates from squid gelatin. *Food Hydrocolloids* 23: 1322–1327.

Gimenez, B., Gomez-Estaca, J., Aleman, A., Gomez-Guillen, M.C., and Montero, M.P. 2009b. Physico-chemical and film forming properties of giant squid (*Dosidicus gigas*) gelatin. *Food Hydrocolloids* 23: 585–592.

Gomez-Estaca, J., Bravo, L., Gomez-Guillen, M.C., Aleman, A., and Montero, P. 2009b. Antioxidant properties of tuna-skin and bovine-hide gelatin films induced by the addition of oregano and rosemary extracts. *Food Chem.* 112: 18–25.

Gomez-Estaca, J., Gimenez, B., Montero, P., and Gomez-Guillen, M.C. 2009d. Incorporation of antioxidant borage extract into edible films based on sole skin gelatin or a commercial fish gelatin. *J. Food Eng.* 92: 78–85.

Gomez-Estaca, J., Gomez-Guillen, M.C., Fernandez-Martin, F., and Montero, P. 2011. Effects of gelatin origin, bovine-hide and tuna-skin, on the properties of compound gelatin-chitosan films. *Food Hydrocolloids* 25: 1461–1469.

Gomez-Estaca, J., Lopez de Lacey, A., Gomez-Guillen, M.C., Lopez-Caballero, M.E., and Montero, P. 2009c. Antimicrobial activity of composite edible films based on fish gelatin and chitosan incorporated with clove essential oil. *J. Aquat. Food Prod. Technol.* 18: 46–52.

Gomez-Estaca, J., Lopez de Lacey, A., Lopez-Caballero, M.E., Gomez-Guillen, M.C., and Montero, P. 2010. Biodegradable gelatin-chitosan films incorporated with essential oils as antimicrobial agents for fish preservation. *Food Microbiol.* 27: 889–896.

Gomez-Estaca, J., Montero, P., Fernandez-Martin, F., and Gomez-Guillen, M.C. 2009a. Physico-chemical and film forming properties of bovine-hide and tuna-skin gelatin: A comparative study. *J. Food Eng.* 90: 480–486.

Gomez-Estaca, J., Montero, P., Gimenez, B., and Gomez-Guillen, M.C. 2007. Effect of functional edible films and high pressure processing on microbial and oxidative spoilage in cold-smoked sardine (*Sardina pilchardus*). *Food Chem.* 105: 511–520.

Gomez-Guillen, M.C., Gimenez, B., Lopez-Caballero, M.E., and Montero, M.P. 2011. Functional and bioactive properties of collagen and gelatin from alternative sources: A review. *Food Hydrocolloids* 25: 1813–1827.

Gomez-Guillen, M.C., Ihl, M., Bifani, V., Silva, A., and Montero, P. 2007. Edible films made from tuna-fish gelatin with antioxidant extracts of two different murta ecotypes leaves (*Ugni molinae Turcz*). *Food Hydrocolloids* 21: 1133–1143.

Gomez-Guillen, M.C., Perez-Mateos, M., Gomez-Estaca, J., Lopez-Caballero, E., Gimenez, B., and Montero, P. 2009. Fish gelatin: A renewable material for developing active biodegradable films. *Trends Food Sci. Technol.* 20: 3–16.

Gomez-Guillen, M.C., Turnay, J., Fernandez-Diaz, M.D., Ulmo, N., Lizarbe, M.A., and Montero, P. 2002. Structural and physical properties of gelatin extracted from different marine species: A comparative study. *Food Hydrocolloids* 16: 25–34.

Gontard, N., Thibault, R., Cuq, B., and Guilbert, S. 1996. Influence of relative humidity and film composition on oxygen and carbon dioxide permeabilities of edible films. *J. Agr. Food Chem.* 44: 1064–1069.

Gorham, S.D., Light, N.D., Diamond, A.M., Willins, M.J., Bailey, A.J., Wess, T.J., and Leslie, N.J. 1992. Effect of chemical modifications on the susceptibility of collagen to proteolysis. II. Dehydrothermal crosslinking. *Int. J. Biol. Macromol.* 14: 129.

Guerrero, P., Retegi, A., Gabilondo, N., and de-la-Caba, K. 2010. Mechanical and thermal properties of soy protein films processed by casting and compression. *J. Food Eng.* 100: 145–151.

Guilbert, S., Cuq, B., and Gontard, N. 1997. Recent innovations in edible and/or biodegradable packaging materials. *Food Addit. Contam.* 14: 741–751.

Guilbert, S. and Gontard, N. 1995. Edible and biodegradable food packaging. In Ackermann, P. et al. (eds.), *Foods and Packaging Materials: Chemical Interactions*, pp. 159–168. Cambridge, U.K.: The Royal Society of Chemistry.

Guilbert, S., Gontard, N., and Gorris, L.G.M. 1996. Prolongation of the shelf-life of perishable food products using biodegradable films and coatings. *LWT Food Sci. Technol.* 29: 10–17.

Habeeb, A.F.S.A. and Hiramoto, R. 1968. Reaction of proteins with glutaraldehyde. *Arch. Biochem. Biophys.* 126: 16–26.

Han, J.H., Aristippos, G., and Jung, H.H. 2005. Edible films and coatings: A review. In J.H. Han (ed.), *Innovations in Food Packaging*, pp. 239–262. London, U.K.: Academic Press.

Hanani, Z.A.N., Beatty, E., Roos, Y.H., Morris, M.A., and Kerry, J.P. 2012. Manufacture and characterization of gelatin films derived from beef, pork and fish sources using twin screw extrusion. *J. Food Eng.* 113: 606–614.

Helander, I.M., Nurmiaho-Lassila, E.-L., Ahvenainen, R., Rhoades, J., and Roller, S. 2001. Chitosan disrupts the barrier properties of the outer membrane of Gram-negative bacteria. *Int. J. Food Microbiol.* 71: 235–244.

Henning, S., Metz, R., and Hammes, W. 1986. New aspects for the application of nisin to food products based on its mode of action. *Int. J. Food Microbiol.* 3: 135–141.

Hoque, M.S., Benjakul, S., and Prodpran, T. 2011a. Properties of film from cuttlefish (*Sepia pharaonis*) skin gelatin incorporated with cinnamon, clove and star anise extracts. *Food Hydrocolloids* 25: 1085–1097.

Hoque, M.S., Benjakul, S., and Prodpran, T. 2011b. Effects of partial hydrolysis and plasticiser content on the properties of film from cuttlefish (*Sepia pharaonis*) skin gelatin. *Food Hydrocolloids* 25: 82–90.

Hoque, M.S., Benjakul, S., and Prodpran, T. 2011c. Effects of hydrogen peroxide and Fenton's reagent on the properties of film from cuttlefish (*Sepia pharaonis*) skin gelatin. *Food Chem.* 128: 878–888.

Hoque, M.S., Benjakul, S., and Prodpran, T. 2010. Effect of heat treatment of film forming solution on the properties of film from cuttlefish (*Sepia pharaonis*) skin gelatin. *J. Food Eng.* 96: 66–73.

Hoque, M.S., Benjakul, S., Prodpran, T., and Songtipya, P. 2011d. Properties of blend film based on cuttlefish (*Sepia pharaonis*) skin gelatin and mung bean protein isolate. *Int. J. Biol. Macromol.* 49: 663–673.

Inamura, P.Y., Kraide, F.H., Drumond, W.S., de Lima, N.B., Moura, E.A.B., and del Mastro, N.L. 2013. Ionizing radiation influence on the morphological and thermal characteristics of a biocomposite prepared with gelatin and Brazil nut wastes as fiber source. *Radiat. Phys. Chem.* 84: 66–69.

Jeon, Y., Kamil, J.Y.V.A., and Shahidi, F. 2002. Chitosan as an edible invisible film for quality preservation of herring and Atlantic cod. *J. Agr. Food Chem.* 50: 5167–5178.

Jo, C., Kang, H., Lee, N.Y., Joong Ho Kwon, J.H., and Byun, M.W. 2005. Pectin- and gelatin-based film: Effect of gamma irradiation on the mechanical properties and biodegradation. *Radiat. Phys. Chem.* 72: 745–750.

Jongjareonrak, A., Benjakul, S., Visessanguan, W., Prodpran, T., and Tanaka, M. 2006a. Characterisation of edible films from skin gelatin of brownstripe red snapper and bigeye snapper. *Food Hydrocolloids* 20: 492–501.

Jongjareonrak, A., Benjakul, S., Visessanguan, W., and Tanaka, M. 2006b. Effects of plasticisers on the properties of edible films from skin gelatin of bigeye snapper and brownstripe red snapper. *Eur. Food Res. Technol.* 222: 229–235.

Jongjareonrak, A., Benjakul, S., Visessanguan, W., and Tanaka, M. 2006c. Fatty acids and their sucrose esters affect the properties of fish skin gelatin-based film. *Eur. Food Res. Technol.* 222: 650–657.

Jongjareonrak, A., Benjakul, S., Visessanguan, W., and Tanaka, M. 2008. Antioxidant activity and properties of skin gelatin films incorporated with BHT and α-tocopherol. *Food Hydrocolloids* 22: 449–458.

Jung, D., Bodyfelt, F., and Daeschel, M. 1992. Influence of fat emulsifiers on the efficacy of nisin in inhibiting *Listeria monocytogenes* in fluid milk. *J. Dairy Sci.* 75: 387–393.

Kacarevic-Popovic, Z., Kostoski, D., Novakovic, L., Miljevic, N., and Cecerov, B. 2004. Influence of the irradiation conditions on the effect of radiation on poly-ethylene. *J. Serb. Chem. Soc.* 69: 1029–1041.

Karim, A.A. and Bhat, R. 2009. Fish gelatin: Properties, challenges, and prospects as an alternative to mammalian gelatins. *Food Hydrocolloids* 23: 563–576.

Kolodziejska, I., Piotrowska, B., Bulge, M., and Tylingo, R. 2006. Effect of transglutaminase and 1-ethyl-3-(3-dimethylaminopropyl) carbodiimide on the solubility of fish gelatin–chitosan films. *Carbohydr. Polym.* 65: 404–409.

Kozlov, P.V. and Burdygina, G.I. 1983. The structure and properties of solid gelatin and the principles of their modification. *Polymer* 24: 651–666.

Krishna, M., Nindo, C.I., and Min, S.C. 2012. Development of fish gelatin edible films using extrusion and compression molding. *J. Food Eng.* 108: 337–344.

Krochta, J.M. 2002. Proteins as raw materials for films and coatings: Definitions, current status, and opportunities. In A. Gennadios (ed.), *Protein-Based Films and Coatings.* Boca Raton, FL: CRC Press.

Krochta, T., Yamaguchi, M., Ohtaki, H., Fukuda, T., and Aoyagi, T. 1997. Hydrogen peroxide-mediated degradation of protein: Different oxidation modes of copper- and iron-dependent hydroxyl radicals on the degradation of albumin. *Biochim. Biophys. Acta* 1337: 319–326.

LeBaron, P.C., Wang, Z., and Pinnavaia, T.J. 1999. Polymer-layered silicate nanocomposites: An overview. *Appl. Clay Sci.* 15: 11–29.

Li, B., Kennedy, J.F., Jiang, Q.G., and Xie, B.J. 2006. Quick dissolvable, edible and heatsealable blend films based on konjac glucomannan-gelatin. *Food Res. Int.* 39: 544–549.

Li, J., Miao, J., Wu, J., Chen, S., and Zhang, Q. 2014. Preparation and characterisation of active gelatin-based films incorporated with natural antioxidants. *Food Hydrocolloids* 37: 166–173.

Li, P., Zheng, J.P., Ma, Y.L., and Yao, K.D. 2003. Gelatin/montmorillonite hybrid nanocomposite. II. Swelling behavior. *J. Appl. Polym. Sci.* 88: 322–326.

Lieberman, E.R. and Gilbert, S.G. 1973. Gas permeation of collagen films as affected by cross-linkage, moisture, and plasticiser content. *J. Polym. Sci.* 41: 33–43.

Limpisophon, K., Tanaka, M., and Osako, K. 2010. Characterisation of gelatin–fatty acid emulsion films based on blue shark (*Prionace glauca*) skin gelatin. *Food Chem.* 122: 1095–1101.

Liu, G. and Xiong, Y.L. 2000. Electrophoretic pattern, thermal denaturation, and *in vitro* digestibility of oxidised myosin. *J. Agr. Food Chem.* 48: 624–630.

Liu, H., Zhao, L., Guo, S., Xia, Y., and Zhou, P. 2014. Modification of fish skin collagen film and absorption property of tannic acid. *J. Food Sci. Technol.* 51(6): 1102–1109.

Lopez-Caballero, M.E., Gomez-Guillen, M.C., Perez-Mateos, M., and Montero, P. 2005. A chitosan-gelatin blend as a coating for fish patties. *Food Hydrocolloids* 19: 303–311.

López-Carballo, G., Hernández-Muñoz, P., Gavara, R., and Ocio, M.J. 2008. Photoactivated chlorophyllin-based gelatin films and coatings to prevent microbial contamination of food products. *Int. J. Food Microbiol.* 126: 65–70.

Lopez-De-Dicastillo, C., Gomez-Estaca, J., Catala, R., Gavara, R., and Hernandez-Munoz, P. 2012. Active antioxidant packaging films: Development and effect on lipid stability of brined sardines. *Food Chem.* 131: 1376–1384.

Martucci, J.F. and Ruseckaite, R.A. 2010a. Biodegradable three-layer film derived from bovine gelatin. *J. Food Eng.* 99: 377–383.

Martucci, J.F. and Ruseckaite, R.A. 2010b. Biodegradable bovine gelatin/Na+-montmorillonite nanocomposite films: Structure, barrier and dynamic mechanical properties. *Polym. Plast. Technol.* 49: 581–588.

Martucci, J.F., Vazquez, A., and Ruseckaite, R.A. 2007. Nanocomposites based on gelatin and montmorillonite: Morphological and thermal studies. *J. Therm. Anal. Calorim.* 89: 117–122.

Mathew, S. 2002. Fish Collagens. CIFT Technology Advisory Series: 8. Agricultural Technology Information Centre. CIFT, Kochi, India.

McHugh, T.H. and Krochta, J.M. 1994. Milk protein based edible film and coatings. *Food Technol.* 48: 97–103.

Muyonga, J.H., Cole, C.G.B., and Duodu, K.G. 2004. Characterisation of acid soluble collagen from skins of young and adult Nile perch (*Lates niloticus*). *Food Chem.* 85: 81–89.

Nagarajan, M., Benjakul, S., Prodpran, T., and Songtipya, P. 2012. Properties of film from splendid squid (*Loligo formosana*) skin gelatin with various extraction temperatures. *Int. J. Biol. Macromol.* 51: 489–496.

Nagarajan, M., Benjakul, S., Prodpran, T., and Songtipya, P. 2014a. Characteristics of bio-nanocomposite films from tilapia skin gelatin incorporated with hydrophilic and hydrophobic nanoclays. *J. Food Eng.* 143: 195–204.

Nagarajan, M., Benjakul, S., Prodpran, T., and Songtipya, P. 2014b. Properties of bio-nanocomposite films from tilapia skin gelatin as affected by different nanoclays and homogenising conditions. *Food Bioprocess Tech.* 7: 3269–3281.

Nagarajan, M., Benjakul, S., Prodpran, T., Songtipya, P., and Nuthong, P. 2013b. Film forming ability of gelatins from splendid squid (*Loligo formosana*) skin bleached with hydrogen peroxide. *Food Chem.* 138: 1101–1108.

Nagarajan, M., Robinson, J.S., Durairaj, S., and Geevaretinam, J. 2013a. Skin, bone and muscle collagen extraction from the trash fish, leather jacket (*Odonus niger*) and their characterisation. *J. Food Sci. Technol.* 50: 1106–1113.

Nayudamma, Y., Joseph, K.T., and Bose, S.M. 1961. Studies on the interaction of collagen with dialdehyde starch. *Am. Leather Chem. Assoc. J.* 56: 548–567.

Nunez-Flores, R., Gimenez, B., Fernandez-Martin, F., Lopez-Caballero, M.E., Montero, M.P., and Gomez-Guillen, M.C. 2012. Role of lignosulphonate in properties of fish gelatin films. *Food Hydrocolloids* 27: 60–71.

Nunez-Flores, R., Gimenez, B., Fernandez-Martin, F., Lopez-Caballero, M.E., Montero, M.P., and Gomez-Guillen, M.C. 2013. Physical and functional characterisation of active fish gelatin films incorporated with lignin. *Food Hydrocolloids* 30: 163–172.

Ou, C., Tsay, S., Lai, C., and Weng, Y. 2002. Using gelatin-based antimicrobial edible coating to prolong shelf-life of tilapia fillets. *J. Food Quality* 25: 213–222.

Ou, S.Y., Kwok, K.C., and Kang, Y.J. 2004. Changes in *in vitro* digestibility and available lysine of soy protein isolate after formation of film. *J. Food Eng.* 64: 301–305.

Park, J.W., Whiteside, W.S., and Cho, S.Y. 2008. Mechanical and water vapor barrier properties of extruded and heat-pressed gelatin films. *Food Sci. Technol.* 41: 692–700.

Pei, Y., Yang, J., Liu, P., Xu, M., Zhang, X., and Zhang, L. 2013. Fabrication, properties and bioapplications of cellulose/collagen hydrolysate composite films. *Carbohydr. Polym.* 92: 1752–1760.

Pereda, M., Ponce, A.G., Marcovich, N.E., Ruseckaite, R.A., and Martucci, J.F. 2011. Chitosan-gelatin composites and bi-layer films with potential antimicrobial activity. *Food Hydrocolloids* 25: 1372–1381.

Pranoto, Y., Lee, C.M., and Park, H.J. 2007. Characterisations of fish gelatin films added with gellan and κ-carrageenan. *LWT Food Sci. Technol.* 40: 766–774.

Rao, Y. 2007. Gelatin-clay nanocomposites of improved properties. *Polymer* 48: 5369–5375.

Rattaya, S., Benjakul, S., and Prodpran, T. 2009. Properties of fish skin gelatin film incorporated with seaweed extract. *J. Food Eng.* 95: 151–157.

Ray, S.S. and Okamoto, M. 2003. Polymer/layered silicate nanocomposites: A review from preparation to processing. *Prog. Polym. Sci.* 28: 1539–1641.

Rhim, J. 2007. Potential use of biopolymer-based nanocomposite films in food packaging applications. *Food Sci. Biotech.* 16: 691–709.

Rhim, J. 2011. Effect of clay contents on mechanical and water vapor barrier properties of agar-based nanocomposite films. *Carbohydr. Polym.* 86: 691–699.

Rhim, J., Hong, S., and Ha, C. 2009. Tensile, water vapor barrier and antimicrobial properties of PLA/nanoclay composite films. *LWT Food Sci. Technol.* 42: 612–617.

Rhim, J., Mohanty, K.A., Singh, S.P., and Ng, P.K.W. 2006. Preparation and properties of biodegradable multilayer films based on soy protein isolate and poly(lactide). *Ind. Eng. Chem. Res.* 45: 3059–3066.

Rhim, J. and Ng, P.K.W. 2007. Natural biopolymer-based nanocomposite films for packaging applications. *Crit. Rev. Food Sci.* 47: 411–433.

Rhim, J.W., Gennadios, A., Weller, C.L., Cezeirat, C., and Hanna, M.A. 1998. Soy protein isolate dialdehyde starch films. *Ind. Crop. Prod.* 8: 195–203.

Rubilar, M., Pinelo, M., Ihl, M., Scheuermann, E., Sineiro, J., and Nunez, M.J. 2006. Murta leaves (*Ugni molinae Turcz*) as a source of antioxidant polyphenols. *J. Agr. Food Chem.* 54: 59–64.

Safandowska, S. and Pietrucha, K. 2013. Effect of fish collagen modification on its thermal and rheological properties. *Int. J. Biol. Macromol.* 53: 32–37.

Senaratne, L.S, Park, P., and Kim, S. 2006. Isolation and characterisation of collagen from brown backed toadfish (*Lagocephalus gloveri*) skin. *Biores. Technol.* 97: 191–197.

Shakila, R.J., Jeevithan, E., Varatharajakumar, A., Jeyasekaran, G., and Sukumar, D. 2012. Comparison of the properties of multi-composite fish gelatin films with that of mammalian gelatin films. *Food Chem.* 135: 2260–2267.

Shan, Y., Zhou, Y., Cao, Y, Xu, Q., Ju, H., and Wu, Z. 2004. Preparation and infrared emissivity study of collagen-g-PMMA/In$_2$O$_3$ nanocomposite. *Mater. Lett.* 58: 1655–1660.

Shi, B. and Di, Y. 2000. *Plant Polyphenols*, 1st ed. Beijing, China: Science Press, pp. 6, 73–91.

Shukla, R. and Cheryan, M. 2001. Zein: The industrial protein from corn. *Ind. Crop. Prod.* 13: 171–192.

Sinthusamran, S., Benjakul, S., and Kishimura, H. 2013. Comparative study on molecular characteristics of acid soluble collagens from skin and swim bladder of sea bass (*Lates calcarifer*). *Food Chem.* 138: 2435–2441.

Sinthusamran, S., Benjakul, S., and Kishimura, H. 2014. Characteristics and gel properties of gelatin from skin of sea bass (*Lates calcarifer*) as influenced by extraction conditions. *Food Chem.* 152: 276–284.

Sionkowska, A. 2006. The influence of UV light on collagen/poly(ethylene glycol) blends. *Polym. Degrad. Stabil.* 91: 305–312.

Sionkowska, A., Skopinska-Wisniewska, J., and Wisniewski, M. 2009. Collagen–synthetic polymer interactions in solution and in thin films. *J. Mol. Liq.* 145: 135–138.

Sobral, P.J.A., Menegalli, F.C., Hubinger, M.D., and Roques, M.A. 2001. Mechanical, water vapor barrier and thermal properties of gelatin based edible films. *Food Hydrocolloids* 15: 423–432.

Sothornvit, R., Rhim, J., and Hong, S. 2009. Effect of nano-clay type on the physical and antimicrobial properties of whey protein isolate/clay composite films. *J. Food Eng.* 91: 468–473.

Sozer, N. and Kokini, J.L. 2009. Nanotechnology and its applications in the food sector. *Trends Biotechnol.* 27: 82–89.

Speer, D.P., Chvapil, M., Eskelson, C.D., and Ulreich, J. 1980. Biological effects of residual glutaraldehyde in glutaraldehyde-tanned collagen biomaterials. *J. Biol. Mater. Res.* 14, 753–764.

Stadtman, E.R. 2001. Protein oxidation in aging and age-related diseases. *Ann. NY Acad. Sci.* 928: 22–38.

Stadtman, E.R. and Berlett, B.S. 1997. Reactive oxygen-mediated protein oxidation in aging and disease. *Chem. Res. Toxicol.* 10: 485–494.

Strauss, G. and Gibson, S.M. 2004. Plant phenolics as cross-linkers of gelatin gels and gelatin-based coacervates for use as food ingredients. *Food Hydrocolloids* 18: 81–89.

Sung, W. and Chen, Z. 2014. UV treatment and γ-irradiation processing on improving porcine and fish gelatin and qualities of their premix mousse. *Radiat. Phys. Chem.* 97: 208–211.

Tanioka, A., Tazawa, T., Miyasaka, K., and Ishikawa, K. 1974. Effects of water on the mechanical properties of gelatin films. *Biopolymers* 13: 753–764.

Tharanathan, R.N. 2003. Biodegradable films and composite coatings: Past, present and future. *Trends Food Sci. Technol.* 14: 71–78.

Thomazine, M., Carvalho, R.A., and Sobral, P.J.A. 2005. Physical properties of gelatin films plasticised by blends of glycerol and sorbitol. *J. Food Sci.* 71: 172–176.

Tomihata, K., Burczak, K., Shiraki, K., and Ikada, Y. 1992. Crosslinking and biodegradation of native and denatured collagen. *Polym. Preprints* 33: 534–535.

Tongnuanchan, P., Benjakul, S., and Prodpran, T. 2012. Properties and antioxidant activity of fish skin gelatin film incorporated with citrus essential oils. *Food Chem.* 134: 1571–1579.

Tongnuanchan, P., Benjakul, S., and Prodpran, T. 2013. Physico-chemical properties, morphology and antioxidant activity of film from fish skin gelatin incorporated with root essential oils. *J. Food Eng.* 117: 350–360.

Tongnuanchan, P., Benjakul, S., Prodpran, T., and Songtipya, P. 2011. Characteristics of film based on protein isolate from red tilapia muscle with negligible yellow discoloration. *Int. J. Biol. Macromol.* 48: 758–767.

United States Pharmacopeia. 1990. Gelatin. Official monographs for USP XXII/NF.

Vanin, F.M., Sobral, P.J.A., Menegalli, F.C., Carvalho, R.A., and Habitante, A.M.Q.B. 2005. Effects of plasticisers and their concentrations on thermal and functional properties of gelatin-based films. *Food Hydrocolloids* 19: 899–907.

Vartiainen, J., Tammelin, T., Pere, J., Tapper, U., and Harlin, A. 2010. Biohybrid barrier films from fluidized pectin and nanoclay. *Carbohydr. Polym.* 82: 989–996.

Vojdani, F. and Torres, J.A. 1989. Potassium sorbate permeability of methylcellulose and hydroxypropyl methylcellulose multi-layer films. *J. Food Process. Preserv.* 13: 417–430.

Weadock, K. Olson, R.M., and Silver, F.H. 1984. Evaluation of collagen crosslinking techniques. *Biomater. Med. Dev. Art. Org.* 11: 293–318.

Wess, T.J. and Orgel, J.P. 2000. Changes in collagen structure: Drying, dehydrothermal treatment and relation to long term deterioration. *Thermochim. Acta* 365: 119–128.

Wu, J., Chen, S., Ge, S., Miao, J., Li, J., and Zhang, Q. 2013. Preparation, properties and antioxidant activity of an active film from silver carp (*Hypophthalmichthys molitrix*) skin gelatin incorporated with green tea extract. *Food Hydrocolloids* 32: 42–51.

Zheng, J.P., Li, P., Ma, Y.L., and Yao, K.D. 2002. Gelatin/montmorillonite hybrid nanocomposite. 1. Preparation and properties. *J. Appl. Polym. Sci.* 86: 1189–1194.

7

Films and Coatings from Starch and Gums

Florencia Cecilia Menegalli

CONTENTS

7.1 Starch

7.1.1 Chemical and Physical Characteristics

Starches commonly occur in seeds, roots, and tubers. The main kinds of starches originate from several sources: corn, cassava, wheat, sorghum, rice, sago, arrowroot, barley, potato, and pea. Banana fruit and some pseudocereals such as amaranth and quinoa grains have recently arisen as new sources of starches.

On the basis of the amylose content, starches can be classified as high-amylose, regular, and waxy starches. Other than natural, starches can be achieved by chemical treatment or genetic mutation. In its native form, starch has quite a variable structure and composition, depending on the conditions in which the raw material grows. Moreover, genetic mutants with specific characteristics can be obtained.

Starch granules consist primarily of two polymers, amylose and amylopectin. Amylose is a linear polymer bearing 1→4 linked α-D-glucopyronasyl units with an average chain length ranging from 100 to 10,000 glucosyl units. Although this polymer is almost linear, a variable grade of branching may exist in the molecule. The number of branches can vary—from 2.9 for cornstarch to 9.8 for sweet potato starch (Jay-Lin, 2009). Interestingly, cereal amyloses have lower molecular weight than the amyloses of other starches.

As for amylopectin, it is a branched polymer composed of α-D-glucopyronasyl units 1→4. Its branches result from 1→6 linkages with an average chain length of 20–30 units (Shannon et al., 2009). Three types of branch chains exist, namely, A, B, and C. The amylopectin molecule bears only one C-chain, which carries the reducing end of the polymer. A-chains link with B-chains, and B-chains link with B- and C-chains, to afford a complex cluster.

In natural starches, the amylose content ranges from 11% to 37%, but in commercial genetic mutants, this value can rise up to 70% (Schwartz and Whistler, 2009). Waxy starches present high amylopectin content, up to 100%. Also, the shape and size of the starch granules are extremely diverse and tightly depend on the raw material. Small granules occur in starches from pseudocereals like amaranth and quinoa (≈ 1 μm). In contrast, potato starch displays granules with diameters larger than 120 μm. In addition, starch granules can acquire spherical, polygonal, or ellipsoidal geometry, platelet form, or any other irregular morphology. The granule structure is very complex: it consists of amorphous and crystalline rings, with radial organization of the amylopectin molecules (Pérez et al., 2009). The crystallinity of native starches (NS) lies between 15% and 45%, and the crystalline structure varies. According to the diffractometric spectra, starches can be classified into types A (cereals), B (tubers), and C (generally present in legumes). These different crystalline structures somehow depend on the average chain length of amylopectin. In general, A-type starches bear long and short chains that are smaller than the long and short chains of B-type starches. Also, A-type starches have larger ratio of short chains as compared with the B-type. As for C-type starches, their x-ray pattern reveals a mixture of the features of A- and B-type starches (Jay-Lin, 2009). The crystal of A-type starches consists of a left-handed double helix packed in a monoclinical space group with eight water molecules inside. The crystal of B-type starches has the shape of a double helix in a hexagonal unit that traps 36 water molecules (Myllärinen et al., 2002a).

The molecular structure of amylose and amylopectin also influences the properties of the starch granules and of the starch suspension that emerges after gelatinization of the granules. These properties will determine the application of starches. Therefore, the functional properties of starches rely on a series of issues such as composition, molecular structure, interchain organization, and the presence of minor constituents like lipids, phosphate ester groups, and proteins.

Before being used, starches have to undergo heat treatment in the presence of water. This process swells and disrupts the structure of the granule, causing loss of crystallinity and amylose lixiviation to the solution. Depending on the structure of the granule and gelatinization conditions, amylopectin may remain in the empty granule (ghost). More severe conditions can destroy the granule. Gelatinization depends on water diffusion into the granule (water content), on the temperature, and obviously on the characteristics of the starch. The temperature range in which gelatinization proceeds will be a function of these factors that will also affect the characteristics of the gelatinized starch, such as the rheological behavior of the starch suspension, the retrogradation of starches during storage, and the mechanical properties of the film produced from this raw material.

7.1.2 Starch Films

7.1.2.1 Casting and Extrusion Process

Casting and extrusion are the two main processes through which it is possible to prepare starch films. The casting process generally finds application at the laboratory scale. In this process, starches mixed with other biopolymers and additives like plasticizers, lipids, nanoparticles, or fibers undergo heat treatment. The mixture gelatinizes, to give an aqueous suspension with low content of solid components. The starch can also be gelatinized prior to being mixed with the plasticizer and the other components. Gelatinization occurs at larger water content and at temperatures higher than the final temperature of gelatinization. In these conditions, disruption of the granular structure is complete and irreversible: the H-bonds are cleaved, water is taken up, and swelling is maximized. Consequently, the crystallites melt, the apparent viscosity increases, and the suspension probably adopts a gel-like rheological behavior, depending on the lixiviated amylose content. In fresh suspension, the amylose molecule assumes a random coil structure. After cooling in the water suspension, amylose forms a double helix that underlies the viscoelastic behavior of the suspension. The empty, partially collapsed granules of the remaining amylopectin may act to reinforce the network. In this situation, the gel could be considered as a composite. Additionally, in the presence of lipids or other complexing agents, the amylose molecules can complex molecules, fatty acids, and diverse alcohols in the inner portion of a single helix. Both single and double helices are thermodynamically stable, although their formation may be slow.

During the casting process, the drying conditions can impact the crystallinity of the films. Rindlav et al. (1997) tailored the crystallinity of unplasticized potato starch films by controlling the relative humidity and temperature of the air used during drying. The higher the relative humidity and the lower the temperature of the air, the larger the degree of film crystallinity achieved. Lower drying rates allow a longer time for the chains to arrange into more favorable conformations, that is, crystals. In other words, films produced under higher air humidity display improved crystallinity (Rindlav et al., 1997). Films generated at 20°C clearly exhibit a B-type crystalline structure. The water content increases as a function of the degree of crystallinity. This is attributed to a larger channel between the six double helices, which can accommodate about 36 water molecules per three glucose units in the B-type crystalline structure.

Rindlav-Westling et al. (1998) have studied how film formation conditions affect the structure as well as the mechanical and barrier properties of pure amylose and amylopectin films. These authors have obtained such films by solution–gel casting of amylose and amylopectin from potato, with or without the addition of glycerol as a plasticizer. They found that amylose films without glycerol plasticization exhibit a relatively high degree of B-type crystallinity (35%), almost irrespective of the relative humidity of the air used during drying. In contrast, the amylopectin films were completely amorphous. Plasticization of the films with glycerol elicits a different behavior: amylose crystallinity increases slightly with the relative humidity, the amylopectin films also present B-type crystallinity, and the degree of crystallinity depends on the air humidity during film formation for relative humidity higher than 50%. In both films, development of crystallinity reduces elongation and enhances tensile stress. In relation to barrier properties, oxygen and water permeability does not depend on the relative humidity, and the amylopectin film presents higher permeability values than the amylose film.

Starch is a thermoplastic material that can be processed by extrusion when large concentrations, temperatures, and shear forces are applied on its granules. During extrusion, the material is processed in continuous equipment, generally in the presence of a plasticizer, like water, glycerol, or sorbitol (Moscicki et al., 2012). The plasticizer penetrates into the starch granules and partially or totally disrupts the initial crystallographic structure. The temperature and shear forces may promote melting, so that the material forms a continuous, amorphous, or semicrystalline mass depending on the amount of plasticizer and on the composition of starch.

7.1.3 Starch Film Properties

During film development, the main properties to consider are the mechanical properties (tensile and puncture test), the barrier properties (water, fat, oxygen, and carbon dioxide permeability), and the solubility in water. These properties are a function of the production process (casting, extrusion, or blowing); the origin, structure, and composition of starch; the plasticizer content; and the process variables. Table 7.1 compiles the properties of starch films selected from the extensive data available in the literature.

Taking only mechanical properties into account, Dieulot and Skurtys (2013) have recently classified starch-based films using literature data on a series of starches from tubers (potato, cassava, and yam) and cereals (maize and wheat). The authors considered up to 316 different data. Their classification considered how much the origin, concentration, and amylose content of starch, the glycerol content, the relative humidity during storage, and the aging time influence the tensile properties, providing five groups of mechanical behavior. Class one (C1) starch-based films present low shear rate and high shear stress, which indicate their stiffness and brittleness. Examples of this class are potato and cassava starch with a starch concentration lower than 20%, amylose content lower than 25%, glycerol plasticizer content lower than 10%, low relative humidity during storage (≤50%), and aging times shorter than 120 h. The second type (C2) corresponds to strong and tough films that can stretch before breaking. Corn with high amylose (>55%) and starch (>43%) content, up to 20% glycerol content, and storage time longer than 500 h at 60% relative humidity affords this type of film. Unlike films belonging to C2, C3 films are soft and ductile; that is, they are neither strong nor tough. These films have high

TABLE 7.1

Properties of Films Prepared with Different Types of Starches

Film Matrix	Film Thickness (µm)	Plasticizer	Plasticizer Content (g/100 g Polymer)	Tensile Strength (MPa)	Elongation (%)	Solubility	Water Permeability ($\times 10^{-13}$ kg/ms Pa)	Oxygen Permeability ($cm^3\ µm/m^3$ dkPa)	References
Achira S.	83	Glycerol	17	18.6	1.4	28.4	3.2	—	Andrade-Mahecha (2009)
Cassava S./tape casting	118	Glycerol/30% fibers	20	30.71	7	—	—	—	Moraes et al. (2013)
Banana S./extruded	1500	Glycerol	26.1	2.8–6.5	58–17	12	19–5	—	Garcia-Tejeda et al. (2013)
Oxidized banana S./extruded	1500	Glycerol	26.1	3.5–8.5	65–22	12	15–5	—	Garcia-Tejeda et al. (2013)
Plantain banana S.	80	Glycerol							Pelissari et al. (2013a)
Wheat S./casting	30	Glycerol	20/50	12/1.8	7/19	7/28	0.008/0.04	26.3	Farahnak et al. (2013)
Potato S./extrusion/blowing	330	Glycerol	20	6.49	38.34				Mościcki et al. (2012)
Sago/casting	130–140	Glycerol	30	2.9	65	60	0.0047	—	Rajeev et al. (2013)
Potato S. Native/oxidized/HMT/casting	102	Glycerol	30	3.5/5.2/6.1	85/57/85	17.3/14.2/17.5	8.3/6.4/10.6	22.03	Zavareze et al. (2012)

(Continued)

TABLE 7.1 (Continued)

Properties of Films Prepared with Different Types of Starches

Film Matrix	Film Thickness (μm)	Plasticizer	Plasticizer Content (g/100 g Polymer)	Tensile Strength (MPa)	Elongation (%)	Solubility	Water Permeability ($\times 10^{-13}$ kg/ms Pa)	Oxygen Permeability (cm³ μm/m³ dkPa)	References
Low/high amylose corn S./casting	56/140	Glycerol	25	27.6/30.6	13.2/4.6	—	1.2/0.73(0%–53%)	25/80	Muscat et al. (2013)
Waxy corn/ACLS/OS/H-OSA/casting	41	Sorbitol	$80/6.6 \times 10^{-3}/6.6 \times 10^{-2}$	ND	ND	44/41/35/42	1.4–1.7	ND	Pérez-Gallardo et al. (2012)
Native/oxidized cassava S./casting	80	Glycerol	25	20/82	40/13	–/41	–/4.4	ND	Pauli et al. (2011)
Native/oxidized banana S./extrusion	1500	Glycerol	80	13.6/17.5	9/14.1	22.5/27.9	0.15/.27	32.2/7.5	Alanis-Lopez et al. (2011)
Potato S./casting	60	Glycerol	0/30	46/2.5	5/37	—	—	—	Osés et al. (2009)
Yam S./casting	110/110/70	Glycerol	60/39.4/32	2.8/4.7/7.84	34/47/10.3			30	Mali et al. (2005)
Potato amylose/casting	70–100	Glycerol	40	20	31		11.9	7	Rindlav-Westling et al. (1998)
Potato amylopectin	70–100	Glycerol	40	6	29		14	14	Rindlav-Westling et al. (1998)
Rice S./casting	100	Glycerol	20/30	10.9/1.6	2.8/59.8	—	0.46/0.86	—	Dias et al. (2010)
Quinoa S. (wet/dry grinding)/casting	80	Glycerol	21.2	7.6	58	15.9	0.56	4.34	Araujo-Farro (2010)
Amaranth S. (caudatus)	83	Glycerol	22.5	6.5	2.3	62.5	—	—	Tapia-Blácido (2003)

shear rate but low shear stress and are obtained from starches of various origins with lower amylose content ($\approx 25\%$). Also, these films have glycerol content ranging from 20% up to 30% and aging time up to 250 h at 60% relative humidity. Finally, the authors mentioned two small classes for very low tensile stress (<0.35): C4, with elongation at break higher than 80%, and C5, with medium elongation (50%). Wheat starch at high concentration (70%), 28% glycerol content, and high relative humidity (90%) furnishes C4 films. C5 films can be obtained from wheat and yam starch in conditions similar to those of C4, but at lower relative humidity (50%). This classification casts light on the main factors that influence film quality.

The point to discuss is how the process and process conditions, raw materials, amylose contents, and storage conditions determine the structure of starch films.

7.1.3.1 Influence of Amylose Content

The amylose content is a key point when it comes to mechanical properties. Myllärinen et al. (2002a,b) have studied the characteristics of pure amylopectin and amylose films. Although the equilibrium moisture content and glass transition temperatures are similar in both materials, tensile properties are completely different: the amylopectin films are brittle, whereas the amylose films display higher tensile stress and elongation.

Cassava films enriched with amylose (6.3%–25%) are stronger but more permeable films, although the equilibrium moisture content decreases upon addition of amylose to the casting suspension (Alves et al., 2007). Overall, the tensile stress increases with the amylose content. However, Mali et al. (2004) have shown that it is crucial to consider not only the amylose content but also the molecular mass and the degree of branching of biopolymers. These authors have studied three types of starch: cassava, corn, and yam, with amylose contents of 19%, 25%, and 30%, respectively. Cassava amylose has the highest molecular weight, while yam amylose has the lowest; corn has an intermediate value. In relation to the amylopectin molecular weights, the order is reversed—cassava and yam have the lowest and highest values, respectively. On the basis of the debranching of the amylopectin fractions, the ratio between short and long branches is 5.32, 2.7, and 2.3 for corn, cassava, and yam, respectively. Film opacity and retrogradation of the starch gels are a direct function of the amylose content. However, according to the values of tensile strength and elongation at break, the cassava film is weaker and flexible, in agreement with its lower amylose content. As for yam and corn films, they present similar values of tensile strength and elongation at break, even though yam contains less amylose. The authors (Mali et al., 2004) suggested that the higher degree of short chains in corn amylopectin and its lower molecular weight favor the interactions and contribute to the formation of the film matrix.

The glass transition temperature (T_g) of an amorphous material is a measure of the state of the polymeric matrix (vitreous or rubbery) and directly affects the properties of the material. Liu et al. (2010) have measured the glass transition temperature of different samples of cornstarch film without any plasticizer, with an amylose content ranging from 0% to 80%, molecular weight ranging from 6×10^7 to 6×10^5, and type A and type B crystallinity. T_g increases with the amylose content. B-type starches (50% and 80% amylose content) have higher T_g than the A-type (waxy and normal corn). Hence, starches with high amylopectin content have larger molecular weight and improved crystallinity, but lower T_g. In a previous study, Yu and Christie (2005) proposed that the double-helical, crystalline structures consisting of short, branched chains of amylopectin do not undergo complete destruction during gelatinization. Instead, these chains form gel balls composed mainly of chains from the same submain chain. The amylopectin molecule itself may generate a relatively separate super globe. The molecular entanglements between gel balls and super globes may move more easily than long linear chains. This might underlie the smaller modulus, improved elongation, and lower T_g (Liu et al., 2010). Garcia et al. (2009) found similar results when they used the casting technique to prepare films from waxy corn (99% amylopectin) and cassava starch (28% amylose) aged for 2 weeks at 43% relative humidity and 25°C. Analysis of the crystalline structure of both films revealed type B crystallinity as well as incipient A-structure in the case of the waxy film. The latter event could be due to a local order generated by the association of amylopectin chains. Morphological characterization of the waxy film attested to the presence of these cluster zones (Garcia et al., 2009).

The same trends emerge upon application of these materials in an extrusion process. Young's modulus and tensile strength increase with rising amylose content. The same happens to elongation, but to a lesser extent, due to orientation of the long linear amylose chains during the haul-off process. The orientation of a highly branched amylopectin is less achievable during the extrusion process (Li et al., 2011). It has also been described that the crystalline structure of the film with high amylose content is not completely destroyed during the extrusion process.

7.1.3.2 Influence of Plasticizer

That plasticizers affect the characteristics of films such as water sorption, mechanical and barrier properties, and glass transition temperature is well known and has been extensively studied. Glycerol is one of the plasticizers that are most frequently employed (Table 7.1); other polyols have also been tested; for example, sorbitol and xylitol as well as sugars like sucrose (da Silva et al., 2012), mannose, fructose, and glucose (Zhang and Hang, 2006). The addition of a plasticizer usually improves the flexibility and ductility of films; thus the tensile stress and the rupture force diminish, while the tensile elongation, the strain at break, and the permeability and solubility of the film increase. Interestingly, very low glycerol (2.5%) and water ($aw \leq 0.22$) content can elicit an antiplasticizer effect, attributed to higher affinity of glycerol for water removed from the starch matrix, thereby inhibiting water–matrix plasticization (Chang et al., 2006).

The optimal concentration of a plasticizer such as glycerol usually ranges from 20% to 30% of the weight of starch. At lower concentration, the film is too brittle or even breaks; at higher concentration, the film becomes sticky due to plasticizer phase separation. Comparison between the effects of glycerol and sorbitol on cassava starch film has shown that the glycerol molecules can interact with the starch chains, increasing the molecular mobility of the film matrix. Moreover, glycerol has a more hydrophilic character than sorbitol (Mali et al., 2005). Therefore, glycerol exerts a more important plasticizing effect than sorbitol, stressing that plasticizers with lower molecular weight penetrate the starch matrix more easily.

Plasticization with polyols also depends on water sorption. At water activities lower than 0.7, potato starch films with different contents of glycerol, sorbitol, and xylitol have lower equilibrium moisture content than the films without plasticizer, as demonstrated by the water sorption isotherms (Talja et al., 2007). At higher water activities, the films containing plasticizer present a substantially larger water sorption. The film plasticized with glycerol has the highest water sorption value, followed by xylitol- and sorbitol-plasticized films. The two latter films require a larger content of plasticizer to afford more flexible films. Indeed, fresh potato starch-based films plasticized with 20% and 30% glycerol are flexible, but fresh potato starch-based films plasticized with xylitol and sorbitol become flexible at polyol contents above 30% and 40%, respectively.

The barrier properties also rely on the plasticizer type and content. Due to the hydrophilic character of biopolymers and plasticizers, the water permeability of the film increases with the relative humidity. Considering glycerol, sorbitol, and xylitol (Talja et al., 2007), glycerol has the greatest effect on the amount of water permeated through the film, whereas sorbitol affects the water vapor permeability the least. Water permeability decreases with the molecular mass of the plasticizer. In addition, at constant starch/plasticizer ratios, glycerol and sorbitol impact the mechanical properties of the films the most and the least, respectively. For the same content of plasticizer, xylitol and sorbitol afford films with lower tensile stress and higher elongation. The use of a binary mixture of plasticizers does not promote crystallization of the polyols during storage (Talja et al., 2008). The application of a xylitol–sorbitol mixture as plasticizer gives the best potato starch properties.

Zhang and Han (2006) have compared the behavior of monosaccharides (mannose, glucose, and fructose) as plasticizer with the behavior of glycerol and sorbitol in a pea starch film. Pea starch has almost 40% amylose content and C-type crystallinity. In that study, the authors used a high plasticizer content, from 4.3 to 10.87 mmol/g of starch, which is equivalent to 40–100 g of glycerol/100 g of starch. The authors found that glucose segregates and crystallizes at concentrations higher than 8 mmol/g; the films become brittle. Sorbitol also elicits the same phenomenon, but the films continue to be flexible. Therefore, only a certain proportion of the plasticizer is compatible with the starch network, and phase separation occurs. The authors verified that water also acts as plasticizer with increasing concentration of polyols; the exception is sorbitol, for which the water equilibrium content remains practically constant. The films plasticized with monosaccharides exhibit higher tensile strength and elongation and lower permeability than those treated with polyols.

7.1.4 Chemically Modified Starch Films

Starch modification by chemical methods involves introduction of functional groups into the starch molecule by means of derivatization (etherification, esterification, cross-linking, and grafting) or decomposition (acid or enzymatic hydrolysis and oxidation) reactions. In this sense, Lopez et al. (2010) have verified the film-forming capacity of acetylated (AS), acetylated cross-linked, hydroxypropylated cross-linked, acid-modified, and native (NS, used as control) corn starch. Chemical treatment reduces the crystallinity of all the modified starches and reduces the amylose content. According to rheological characterization, AS is the only starch that develops a rigid gel structure upon chemical treatment ($G' \gg G''$). Hence, only NS and AS can form films.

Several authors have investigated whether starch oxidation can improve film properties. During such oxidation, amylose and amylopectin can depolymerize; the hydroxyl groups of the starch molecules are first oxidized to carbonyl groups and then to carboxyl groups. The oxidation of potato starch (Table 7.1) culminates in increased tensile stress, without significant changes in elongation (Zavareze et al., 2012). The use of 30% glycerol as plasticizer does not alter solubility or permeability. However, the oxidation of waxy cornstarch reduces the solubility to some extent (Pérez-Gallardo et al., 2012). Upon oxidation, the tensile strength and the elongation of banana starch increases and decreases, respectively; unfortunately, water permeability rises as a function of the level of oxidation (Zamudio-Flores et al., 2006, 2011). Other authors (Pauli et al., 2011) have also verified these trends for cassava starch treated with 0.8% active chlorine, but the mechanical properties of the film decay at higher concentration of the oxidation agent (2%).

Oxidation impacts the extrusion of banana starch (Alanis-Lopez et al., 2011) as noted from increased tensile stress, elongation, and permeability (Table 7.2). In this case, the water content and crystallinity augment, whereas the amylose content diminishes. On the basis of microscopy analysis, intact starch granules exist in the nonoxidized starch sheets. Such granules are not detected in the sample obtained after starch oxidation. Thus, during the extrusion process, chemical treatment may affect the gelatinization occurring in the extrusion equipment.

Cross-linking with agents like citric acid (Reddy and Yang, 2010), epichlorohydrin (Jimenez-Elizondo, 2007), and glutaraldehyde (Parra et al., 2004) can also enhance film properties.

7.1.5 Flour Films

The use of natural mixtures (polysaccharides, proteins, fat, or fibers) obtained from the same raw material is a current trend in the development of films and coatings. These materials basically embrace flours

TABLE 7.2

Composition of Starches and Flours of Pseudo Cereals

Specie	Amylose	Protein	Lipid	Fiber	Ash
F. *Amaranthus caudatus*[a]	11.9	14.2	8.9	—	2.1
F. *Amaranthus cruentus*[a]	—	19.9	6.9	4.2	3.2
F. Quinoa v. Real[b]	—	12.8	6.3	2.5	2.4
A. Quinoa v. Chupacapa[c]	9.3	1.09	1.94	0.24	1.46
A. Canihua K-081[c]	10.7	1.55	2.4	0.96	0.97
F. Canihua v. Illpa[d]	12.6	16.7	11.9	ND	5.3
F. Canihua v. Cupi[d]	12.1	15.5	11.8	ND	4.9

[a] Tapia-Blácido (2003) and Tapia-Blácido et al. (2007).
[b] Araujo-Farro (2008).
[c] Steffolani et al. (2013).
[d] Sales-Valero et al. (2015).

of diverse sources like grains, pseudocereals, tubers, and fruit. These flours display variable proportions of starch, protein, and fibers (Table 7.2). This chapter will only consider fruit with high starch content. Table 7.3 summarizes the properties of the films.

Flour film properties heavily rely on the interactions taking place between the various polymers (starch, proteins, pectin, and cellulose from fibers), lipids, and other saccharides (fructose or sucrose) present in fruit and plasticizers. It is crucial to consider how the distribution of these interactions along the matrix, the balance between hydrophobic and hydrophilic links, and the content of each component will determine the mechanical and barrier properties of the final film (Tapia-Blácido et al., 2007; Andrade-Mahecha et al., 2012).

Flour films obtained from pseudocereals have good barrier properties as compared with other starch and tuber films (Table 7.2). However, mechanical properties like the tensile strength assume low values, while elongation is commonly high. Flour films are flexible because they display high protein and lipid content, especially natural fatty acids that may act as plasticizers (Tapia-Blácido et al., 2007).

Tapia-Blácido et al. (2011) have formulated and optimized films of amaranth flour and starch extracted from *Amaranthus caudatus* and *A. cruentus* using glycerol and sorbitol as plasticizer. The films present a yellowish color, moderate opacity, high flexibility, low mechanical resistance, and good barrier to water and oxygen transfer (Tapia-Blácido et al., 2005a,b).

Tapia-Blácido et al. (2007) have analyzed the role that the constituents of amaranth film play by comparing the properties of films made with each amaranth component and their mixtures: protein, lipid-protein, and flour. In the flour films, a strong association exists between the lipids and the proteins, which are homogenously distributed throughout the starch phase. Stabilization of these phases culminates in the absence of polymorphism transition of the fatty acids (oleic acid) present in the lipid phase, in contrast to protein–lipid films. Moreover, the lipid phase in the flour film matrix does not separate, contributing to the good plasticization and the excellent barrier properties of amaranth flour films. In other words, the properties of the flour film result from interactions between the components (starch, protein, and lipid) and from the natural state in which they exist in the flour (Tapia-Blácido et al., 2007).

Araujo-Farro (2008) has optimized the formulation and the casting process conditions to obtain films from two kinds of flours (wet and dry grinding) and the starch extracted from quinoa grains *Real* variety. The starch films present moderate tensile strength, higher elongation, and low water and oxygen permeability in relation to starch films from other sources (Table 7.2). In the case of the films obtained from the flours, wet grinding provides materials with lower fiber content, so the film has improved properties as compared with those made with integral flour (dry milling) (Table 7.3).

Cereal flours have furnished different results. Dias et al. (2010) have developed films from rice flour and starch using glycerol and sorbitol as plasticizers. Both types of films present compact and homogeneous structure. The starch and flour films have similar mechanical properties too, although the flour has 7% protein content. The starch film displays lower water permeability as compared with the flour films (Table 2.3). The mechanical properties of the films are very sensitive to the plasticizer content between 20% and 30% plasticizer. Wheat flour also seems be to an adequate raw material: it is possible to obtain 300 μm sheets by extrusion and pressing using 20%–25% glycerol as plasticizer (Sreekumar et al., 2013). In contrast to pseudocereals, films obtained from corn flours have poorer properties as a consequence of the lipid and protein contents (Ayadi et al., 2011).

Andrade-Mahecha et al. (2012) have used achira rhizomes to obtain flour films with the same features discussed earlier: they are more flexible and more permeable to water vapor, and they have lower mechanical tensile stress. The flour films are more hydrophilic, and their high-fiber content may introduce some imperfection in the starch matrix, accounting for the observed facts (Table 2.3).

Several authors have also used banana, especially green banana, due to its high starch content (Table 7.3). Sothornvit and Pitak (2007) have worked with banana flour enriched with pectin and with banana starch. Pelissari et al. (2013b) have optimized the formulation and process conditions to obtain plantain flour films. Because the banana flour is a natural blend consisting of starch, protein, lipids, and fiber, the film originating from this raw material has mechanical, barrier, and optical properties comparable to those of other films made from the flour of other plant species: achira (Andrade-Mahecha et al., 2012), amaranth (Tapia-Blácido et al., 2005a,b), quinoa (Araujo-Farro, 2008), and rice (Dias et al., 2010).

TABLE 7.3

Properties of Films Prepared with Different Types of Flour

Film Matrix	Film Thickness (μm)	Plasticizer	Plasticizer Content (g/100 g Polymer)	Tensile Strength (MPa)	Elongation (%)	Solubility	Water Permeability (×10^{-13} kg/ms Pa)	Oxygen Permeability (cm^3 μm/m^3 dkPa)	References
Achira flour	84	Glycerol	17	7	14.6	38.3	5.3	—	Andrade-Mahecha et al. (2012)
Amaranth flour (*Caudatus v.*)	83	Glycerol	22	1.5	83.7	42.3	0.7	5.6	Tapia-Blácido et al. (2005a)
Amaranth flour (*Cruentus v.*)	80	Glycerol	20	5.4	16.9	40.6	3.7	22	Tapia-Blácido et al. (2011)
Amaranth flour (*Cruentus v.*)	80	Sorbitol	29	7.4	12.9	50.8	3.1	13.5	Tapia-Blácido et al. (2011)
Banana flour	87	Glycerol	19	9.2	24.2	27.9	2.1	—	Pelissari et al. (2013b)
Banana flour	190	Glycerol	30	12	8	100	—	40	Sothornvit and Pitak (2007)
Kañiwa flour v. Cupi[a]	81	Glycerol	20	1.5	73	40.3	1.0	—	Sales-Valero et al. (2015)
Kañiwa flour v. Ilpa[a]	82	Glycerol	20	2.0	46.5	37.8	0.4	—	Sales-Valero et al. (2015)
Quinoa flour (wet/dry grinding)	80	Glycerol	21/20	4.1/1.58	88.4/93	18.7/24.4	0.6/0.8	5.36/7.52	Araujo-Farro (2008)
Rice flour	100	Glycerol	20/30	10.3/1.3	2.7/66.4	—	1.1/1.7	—	Dias et al. (2010)
Wheat flour	300	Glycerol	20/23	4.9/1.9	14/34	—	—	—	Sreekumar et al. (2013)

[a] Determined RH gradient: 33–65.

7.2 Gums

Polysaccharides consist of monosaccharides joined by O-glycosidic links. The food industry employs a wide diversity of polysaccharides with different structures and molecular weights for various purposes, such as thickening, gelling, encapsulating, and bulking agents as well as water binder. The classification of polysaccharides depends on their origin; that is, plant, seaweed, microbial, and animal polysaccharides exist. Cellulose and its derivatives, hemicelluloses, pectins, exudates, mucilage gums, and fructans are the main plant polysaccharides (Izydorczyk and Wang, 2005).

7.2.1 Cellulose and Derivatives

The polymer cellulose is widely distributed in nature. It is one of the constituents of the cell walls of higher plants and algae; it is also a component of fungal membranes. It is a polymer with a high molecular weight, and it comprises linear chains of $(1\rightarrow4)$-β-D-glucopyranosyl units. Because cellulose is insoluble, it is necessary to modify this polymer chemically and physically before its use in foodstuff. The main cellulose derivatives are microcrystalline cellulose, carboxymethyl cellulose (CMC), methylcellulose (MC), hydroxypropylmethylcellulose (HPMC), and hydroxypropylcellulose (HPC).

Aggressive chemical treatment of cellulose with carbon dioxide, sodium hydroxide, and sulfuric acid generates cellophane films. Although these derivatives are relatively costly, they are potentially applicable in coating and film production.

7.2.2 Hemicelluloses

Hemicellulose refers to a series of polymers present in the cell wall of higher plants; they generally bind to cellulose and lignins. Overall, hemicelluloses consist of heteropolysaccharides substituted with other carbohydrates and noncarbohydrates. The main classes of hemicelluloses are mannans and galactomannans (locust bean gum [LBG], senna gum, guar gam, tara gum, and fenugreek gum) and glucomannans (konjac gum) (Izydorczyk and Wang, 2005).

7.2.3 Pectins

Pectins are a family of very complex polysaccharides; they are one of the principal components of the cell wall of fruits and vegetables. They are very useful in the food industry: several commercial products originate from apple pulp and citrus peel. The pectin backbone contains $(1\rightarrow4)$-α-D-galacturonic acid units linked to single $(1\rightarrow2)$-α-L-rhamnose residues (Ridley et al., 2001). The carboxyl groups of the galacturonic acid units are partly esterified with methyl groups. The classification of pectins as a low methyl esterification degree and high methyl esterification degree depends on the degree of esterification (Solava-Fortuny et al., 2012).

7.2.4 Seaweed Gums

Alginates, carrageenans, and agar are the main gums obtained from seaweed extracts. Alginates, which occur in brown algae, are salts of alginic acid, a polysaccharide with high molecular weight that consists of varying proportions of D-mannuronic and L-guluronic acids. Carrageenans, obtained from red algae, bear repeating dimers of an $\alpha(1\rightarrow4)$–linked D-galactopyranose or 3,6-anydro-D-galactopyranose residue, and a $\beta(1\rightarrow3)$–linked D-galactopyranose residue (Soliva-Fortuny et al., 2012). Three main types of carrageenans exist; they differ in terms of the position and number of sulfate groups, which gives ι-, κ-, and λ-carrageenan. Agar is a galactan polysaccharide, a linear polymer made up of disaccharides composed of $\beta(1,3)$- and $\alpha(1,4)$-linked galactose residues; it is a fine gelling agent and is compatible with other gums.

7.2.5 Microbial and Exudative Gums

The main gums of microbial origin are xanthan, pullulan, and gellan gums. Gums produced by plant exudate include arabic gum, tragacanth gum, karaya gum, and ghatti gum. All of these gums contain high proportion of glucuronic and galacturonic acid residues (Izydorczyk and Wang, 2005).

7.2.6 Gum Films and Coating

Among gums, the nonstarch polysaccharides that exhibit the best filmogenic capacity are CMC, MC, HPC, HPMC, alginate, carrageenan, pectin, chitosan (CH), and gellan (Han and Gennadios, 2005). Table 7.4 lists some recent research articles that have dealt with gum films.

It is possible to obtain HPMC films without any plasticizer. These films present high tensile strength, but they display low elongation and high water permeability (Bilbao-Sainz et al., 2013). One can improve the properties of these films by incorporating crystalline nanomicrocellulose. In general, gums help to enhance protein or starch film properties (Ghanbarzadeh et al., 2010; Yoo and Krocha, 2012).

Mannans and galactomannans have found wide application in the food industry. Cerqueira et al. (2013) have published a recent review on the use of galactomannans to develop edible films/coatings for food applications. Martins et al. (2012) have used LBG as raw material for films, mixed with k-carrageenan—addition of k-carrageenan (k-car) to LBG improves the barrier properties of the films, leading to reduced water vapor permeability. A k-car/LBG ratio of 80/20 or 40/60 (% w/w) gives better values of elongation at break (Table 7.4). LBG can also be employed as coating of sausages and mandarins (Rojas-Argudo et al., 2009; Dilek et al., 2011) to abate water loss in both cases. Preparation of a series of films with different ratios of CH and guar gum (GG) by the casting method has shown that the film produced with 85% CH and 15% GG (v/v) affords lower oxygen permeability and better mechanical properties, while retaining antibacterial properties similar to those of a CH film without GG (Rao et al., 2010).

To be used as polymeric matrix, alginate gums have to be cross-linked with divalent or trivalent cations, especially calcium ions (Ca^{2+}). Sodium alginate is soluble in water, but the substitution of Na^+ with Ca^{2+} culminates in a structure named eggbox, to furnish a rather strong and rigid film. Several works have employed alginate as a biopolymer to obtain films and coatings containing alginate and pure fruits plasticized with corn syrup for chopped food (Azeredo et al., 2012b). Azeredo et al. (2012a) have investigated the properties of films elaborated with mixtures of sodium alginate and cashew tree gum cross-linked with calcium. Juck et al. (2010) have analyzed the efficacy of coatings for deli turkey products based on pectin, carrageenan, and xanthan gums with respect to the growth of *Listeria monocytogenes*. Nussitovitch and Hershko (1996) have coated garlic bulbs with gellan and alginate gums. Indeed, tough pectin and alginate gum coats can be shaped over the food by action of the cross-linking agent, so they can function as a protective agent prior to fruit drying (Lago-Vanzela, 2013) or osmotic dehydration (Ferrari et al., 2013).

Additionally, xanthan has been incorporated into pea starch to elaborate films by casting (da Matta et al., 2011). Veiga-Santos (2005a,b) have evaluated the effect of deacetylated xanthan gum, additives (sucrose, soybean oil, sodium phosphate, and propylene glycol), and pH modifications on the mechanical properties, hydrophilicity, and water activity of cassava starch–xanthan gum films. These authors examined and optimized the effects of pullulan (Pul), glycerin (Gly), xanthan gum (Xa), and locust bean (Lb) concentrations on pullulan film properties. Optimal ingredient combination has allowed for determination of the antimicrobial activity of films combined with sakacin A against *L. monocytogenes* (Trinetta et al., 2011). The authors prepared edible films using various ratios of pullulan and rice wax. Shih et al. (2011) have achieved freestanding composite films with up to 46.4% rice wax. Leon and Rojas (2007) have evaluated edible gellan films as carriers to stabilize L-(+)-ascorbic acid (AA) for nutritional purposes and antioxidant effect in foods.

It is clear that the most interesting use of coatings is their ability to serve as carriers of additives with functional properties (Sitonio-Eça et al., 2014). Table 7.5 lists some examples of films produced with different biopolymers and additives.

TABLE 7.4

Properties of Films Prepared with Different Types of Gums

Film Matrix	Film Thickness (μm)	Plasticizer	Plasticizer Content (g/100 g Polymer)	Tensile Strength (MPa)	Elongation (%)	Solubility	Water Permeability ($\times 10^{-13}$ kg/ ms Pa)	References
Sodium alginate/casting	—	Glycerol	25	18	11.5	100	2	Abdollahi et al. (2013)
Galactomannan/casting	58/65	Glycerol	33/100	12/2.7	10/41	65/57.8	7.2/8.3	Cerqueira et al. (2013)
Pullulan–alginate–carboxymethyl cellulose	—	Glycerol/fructose	20	25/30	45/30	—	3/12	Tong et al. (2013)
Kefiran	74	Glycerol	42	5	57	28	0.3	Motedayen et al. (2013)
Xylan/2-OSA (DS 0/0.064/0.17)[a]	—	Glycerol	25	20.8/44/23	3.4/7.9/18	—	—	Zhong et al. (2013)
Methylcellulose	—	Glycerol/PEG	50	22/28	82/110	—	—	Arık Kibar and Us (2013)
Carboxymethyl cellulose	—	Glycerol/PEG	50	7/24	70/35	—	—	Arık Kibar and Us (2013)
Hydroxymethylcellulose/MC (0/2.6%)[b]	20	—	0	35.6/53	4.9/6.5	—	1300	Bilbao-Sainz et al. (2010)
k-Carrageenan/locust bean gum (%K-C:100/40/0)	52/42/61	Glycerol	30	20/23/8	16/15/28	—	0.55/0.51/0.8	Martins et al. (2012)

[a] Octenylsuccinic anhydride.
[b] Mycrocrystalline cellulose nanoparticle.

TABLE 7.5

Coating and Films Incorporated with Pure Compounds

Film Composition	Additive	Food	Results	References
Low methoxyl pectin	Ascorbic acid	—	Ascorbic acid degradation was less sensitive than film browning upon increasing storage relative humidity.	De'Nobili et al. (2013)
High methoxyl pectin (HMP)	Ascorbic acid (AA)	—	AA was the least stable in 80% HMP, with higher retention in 50% and 70% HMP networks.	Pérez et al. (2012)
Methylcellulose	Nanoparticles of poly-ε-caprolactone and β-carotene	—	The films developed with 70% β-carotene nanoparticles showed low antioxidant capacity.	Lino (2012)
Carboxymethyl cellulose	α-Tocopherol	—	The films had low values of α-tocopherol release in ethanol solution. The film matrix was stable.	Motta (2012)
Methylcellulose	Nanocapsules of poly-ε-caprolactone (NCs)	—	The film control had no radical scavenging activity that significantly increased with NCs concentration.	Noronha (2012)
Calcium alginate-Capsul	Ascorbic acid	—	AA was stable during 5 months stored at refrigeration in the dark and at room temperature, for 3 months.	Bastos et al. (2009)
Gellan gum	AA	—	AA retention that varied between 103% and 99% after film casting.	León and Rojas (2007)
Alginate	Ascorbic acid Citric acid	Mango	The addition of these antioxidants contributed not only to color retention but also to the antioxidant potential of fresh-cut mangoes.	Robles-Sanchez et al. (2013)
Cassava starch	Citric acid	Fresh-cut mango	This combination delayed the quality deterioration of fresh-cut mangoes, decreasing the fruit respiration rate and inhibiting the metabolic.	Chiumarelli et al. (2010)
Alginate, gellan or pectin	N-acetylcysteine glutathione	Pears	Significantly reduced vitamin C loss occurred for fresh-cut pears during more than 1 week.	Oms-Oliu et al. (2008)

REFERENCES

Abdollahi, M., Alboofetileh, M., Rezaei, M. et al. Comparing physico-mechanical and thermal properties of alginate nanocomposite films reinforced with organic and/or inorganic nanofillers. *Food Hydrocolloids* 32(2) (2013): 416–424.

Alanis-Lopez, P., Perez-Gonzalez, J., Rendon-Villalobos, R. et al. Extrusion and characterization of thermoplastic starch sheets from "macho" banana. *Journal of Food Science* 76(6) (2011): E465–E471.

Alves, V.D., Mali, S., Beléia, A., and Grossmann, M.V.E. Effect of glycerol and amylose enrichment on cassava starch film properties. *Journal of Food Engineering* 78 (2007): 941–946.

Andrade-Mahecha, M.M. Development and characterization of film from Achira flour. Master Degree. Food Engineering Program. University of Campinas, Campinas, Brazil, 2009.

Andrade-Mahecha, M.M., Tapia-Blácido, D., and Menegalli, F.C. Development and optimization of biodegradable films based on achira flour. *Carbohydrate Polymers* 88 (2012): 449–458.

Araujo-Farro, P.C. Development of biodegradable film from quinoa (*Chenopodium quinoa, Willdenow*) v. real. 2008. PhD thesis. Food Engineering Program. University of Campinas, Campinas, Brazil, 2008.

Araujo-Farro, P.C., Podadera, G., Sobral, P.J.A., and Menegalli, F.C. Development of films based on quinoa (*Chenopodium quinoa*, Willdenow) starch. *Carbohydrate Polymers* 81 (2010): 839–848.

Arik Kibar, E.A. and Us, F. Thermal, mechanical and water adsorption properties of corn starch carboxymethylcellulose/methylcellulose biodegradable films. *Journal of Food Engineering* 114(1) (2013): 123–131.

Ayadi, F., Bliard, C., and Dole, P. Materials based on maize biopolymers: Effect of flour components on mechanical and thermal behavior. *Starch–Stärke* 63(10) (2011): 604–615.

Azeredo, H.M.C., Magalhaes, U.S., Oliveira, S.A. et al. Tensile and water vapour properties of calcium-crosslinked alginate-cashew tree gum films. *International Journal of Food Science & Technology* 47(4) (2012a): 710–715.

Azeredo, H.M.C., Miranda, K.W.E., Rosa, M.F. et al. Edible films from alginate-acerola puree reinforced with cellulose whiskers. *LWT: Food Science and Technology* 46(1) (2012b): 294–297.

Bastos, D.D.S., Araújo, K.G.D.L., and Leão, M.H.M.D.R. Ascorbic acid retaining using a new calcium alginate-Capsul based edible film. *Journal of Microencapsulation* 26(2) (2009): 97–103.

Bilbao-Sainz, C., Wood, R.J., Avena-Bustillos, D.F. et al. Composite edible films based on hydroxypropyl methylcellulose reinforced with microcrystalline cellulose nanoparticles. *Journal of Agricultural and Food Chemistry* 58(6) (2010): 3753–3760.

Cerqueira, M.A., Souza, B.W.S., Teixeira, J.A. et al. Utilization of galactomannan from gleditsia triacanthos in polysaccharide-based films: Effects of interactions between film constituents on film properties. *Food and Bioprocess Technology* 6(6) (2013): 1600–1608.

Chang, Y.P., Abd, K.A., and Seow, C.C. Interactive plasticizing–antiplasticizing effects of water and glycerol on the tensile properties of tapioca starch films. *Food Hydrocolloids* 20 (2006): 1–8.

Chiumarelli, M., Pereira, L.M., Ferrari, C.C., Sarantópoulos, C.I.G.L., and Hubinger, M.D. Cassava starch coating and citric acid to preserve quality parameters of fresh-cut "Tommy Atkins" mango. *Journal of Food Science* 75(5) (2010): E297–E304.

da Matta, M.D., Sarmento, S., Bruder, S. et al. Mechanical properties of pea starch films associated with xanthan gum and glycerol. *Starch–Stärke* 63(5) (2011): 274–282.

da Silva, J.B.A., Pereira, F.V., and Druzian, J.I. Cassava starch-based films plasticized with sucrose and inverted sugar and reinforced with cellulose nanocrystals. *Journal of Food Science* 77(6) (2012): N14–N19.

De'nobili, M., Pérez, C., Navarro, D., Stortz, C., and Rojas, A. Hydrolytic stability of L-(+)-ascorbic acid in low methoxyl pectin films with potential antioxidant activity at food interfaces. *Food and Bioprocess Technology* 6(1) (2013): 186–197.

Dias, A.B., Müller, C.M.O., Larotonda, F., and Laurindo, J.B. Biodegradable films based on rice starch and rice flour. *Journal of Cereal Science* 51 (2010): 213–219.

Dieulot, J.-Y. and Skurtys, O. Classification, modeling and prediction of the mechanical behavior of starch-based films. *Journal of Food Engineering* 119(2) (2013): 188–195.

Dilek, M., Polat, H., Kezer, F. et al. Application of locust bean gum edible coating to extend shelf life of sausages and garlic-flavored sausage. *Journal of Food Processing and Preservation* 35(4) (2011): 410–416.

Farahnaky, A., Saberi, B., and Majzoobi, M. Effect of glycerol on physical and mechanical properties of wheat starch edible films. *Journal of Texture Studies* 44(3) (2013): 176–186.

Ferrari, C.C., Sarantópoulos, C., Carmello-Guerreiros, S. et al. Effect of osmotic dehydration and pectin edible coatings on quality and shelf life of fresh-cut melon. *Food and Bioprocess Technology* 6 (2013): 80–91.

Garcia, N.L., Fama, L., Dufresne, A. et al. A comparison between the physico-chemical properties of tuber and cereal starches. *Food Research International* 42(8) (2009): 976–982.

Garcia-Tejeda, V., Lopez-Gonzalez, Y., Perez-Orozco, C.P. et al. Physicochemical and mechanical properties of extruded laminates from native and oxidized banana starch during storage. *LWT: Food Science and Technology* 54(2) (2013): 447–455.

Ghanbarzadeh, B., Almasi, H., and Entezami, A.A. Physical properties of edible modified starch/carboxymethyl cellulose films. *Innovative Food Science & Emerging Technologies* 11(4) (2010): 697–702.

Han, J.H. and Gennadios, A. Edible films and coating: A review. In J.H. Han (ed.), *Innovation in Food Packaging*. New York: Elsevier Science, 2005, pp. 239–262.

Izydorczyk, S.C. and Wang, Q. Chapter 6: Polysaccharide gums: Structure, functional properties, and application. In S.W. Cui (ed.), *Food Carbohydrates: Chemistry, Physical Properties, and Applications*. Boca Raton, FL: CRC Press, 2005.

Jay-Lin, J. Chapter 6: Structural features of starch granules II. In J. BeMiller and R. Whistler (eds.), *Starch Chemistry and Technology*, 3rd ed. London, U.K.: Elsevier, 2009, pp. 193–236.

Jimenez-Elizondo, N. Mechanical and barrier properties, solubility and microstructure of film of amaranth flour cross-linked with epichlorohydrin and mixture of poly(vinyl)alcohol. Master Degree. Food Engineering Program. University of Campinas, Campinas, Brazil, 2007.

Juck, G., Neetoo, H., and Chen, H. Application of an active alginate coating to control the growth of Listeria monocytogenes on poached and deli turkey products. *International Journal of Food Microbiology* 142 (2010): 302–308.

Lago-Vanzela, E.S., Nascimento, P., Fontes, E.A.F., Mauro, M.A., and Kimura, M. Edible coatings from native and modified starches retain carotenoids in pumpkin during drying. *LWT: Food Science and Technology* 50 (2013): 420–425.

León, P.G. and Rojas, A.M. Gellan gum films as carriers of L-(+)-ascorbic acid. *Food Research International* 40(5) (2007): 565–575.

Li, M., Liu, P., Zou, W. et al. Extrusion processing and characterization of edible starch films with different amylose contents. *Journal of Food Engineering* 106(1) (2011): 95–101.

Lino, R.C. Desenvolvimento de filmes de metilcelulose incorporados por nanopartículas de poli-ε-caprolactona/β-caroteno. 2012. Dissertação (Mestrado em Ciência de Alimentos). Centro de Ciências Agrárias, Universidade Federal de Santa Catarina, Florianópolis, Brazil.

Liu, P., Yu, L., Wang, X. et al. Glass transition temperature of starches with different amylose/amylopectin ratios. *Journal of Cereal Science* 51(3) (2010): 388–391.

Lopez, O.V., Zaritzky, N.E., and Garcia, M.A. Physicochemical characterization of chemically modified corn starches related to rheological behavior, retrogradation and film forming capacity. *Journal of Food Engineering* 100(1) (2010): 160–168.

Mali, S., Grossmann, M.V.E., García, M.A., Martino, M.N., and Zaritzky, N.E. Mechanical and thermal properties of yam starch films. *Food Hydrocolloids* 19(1) (2005): 157–164.

Mali, S., Karam, L.B., Grossmann, M.V.E., and Ramos, L.P. Relationship among the composition and physicochemical properties of starches with the characteristics of their films. *Journal of Agricultural and Food Chemistry* 52(25) (2004): 720–725.

Martins, J.T., Cerqueira, M.A., Bourbon, A.I. et al. Synergistic effects between kappa-carrageenan and locust bean gum on physicochemical properties of edible films made thereof. *Food Hydrocolloids* 29(2) (2012): 280–289.

Moraes, J.O., Scheibe, A.S., Sereno, A. et al. Scale-up of the production of cassava starch based films using tape-casting. *Journal of Food Engineering* 119(4) (2013): 800–808.

Móscicki, L., Mitrus, M., Wojtowicz, A. et al. Application of extrusion-cooking for processing of thermoplastic starch (TPS). *Food Research International* 47(2) (2012): 291–299.

Motedayen, A.A., Khodaiyan, F., and Salehi, E.A. Development and characterisation of composite films made of kefiran and starch. *Food Chemistry* 136(3–4) (2013): 1231–1238.

Motta, C. Incorporação do antioxidante natural α-tocoferol em filmes de carboximetilcelulose. 2012. Dissertação (Mestrado em Química). Centro de Ciências Físicas e Matemáticas, Universidade Federal de Santa Catarina, Florianópolis, Brazil.

Muscat, D., Adhikari, R., McKnight, S. et al. The physicochemical characteristics and hydrophobicity of high amylose starch-glycerol films in the presence of three natural waxes. *Journal of Food Engineering* 119(2) (2013): 205–219.

Myllärinen, P., Buleon, A., Lahtinen, R., and Forssell, P. The crystallinity of amylose and amylopectin films. *Carbohydrate Polymers* 41(1) (2002a): 41–48.

Myllärinen, P., Partanen, R., Seppala, J., and Forssell, P. Effect of glycerol on behaviour of amylose and amylopectin films. *Carbohydrate Polymers* 50(4) (2002b): 355–361.

Noronha, C.M. Incorporação de nanocápsulas de poli(ε-caprolactona) contendo α-tocoferol em biofilmes de metilcelulose. 2012. Dissertação (Mestrado em Ciência de Alimentos). Centro de Ciências Agrarias, Universidade Federal de Santa Catarina, Florianópolis, Brazil.

Nussinovitch, A. and Hershko, V. Gellan and alginate vegetable coatings. *Carbohydrate Polymers* 30(2–3) (1996): 185–192.

Oms-Oliu, G., Soliva-Fortuny, R., and Martín-Belloso, O. Edible coatings with antibrowning agents to maintain sensory quality and antioxidant properties of fresh-cut pears. *Postharvest Biology and Technology* 50(1) (2008): 87–94.

Osés, J., Niza, S., Ziani, K., and Maté, J.I. Potato starch edible films to control oxidative rancidity of polyunsaturated lipids: Effects of film composition, thickness and water activity. *International Journal of Food Science & Technology* 44 (2009): 1360–1366.

Parra, D., Tadini, C., Ponce, P., and Lugão, A. Mechanical properties and water vapor transmission in some blends of cassava starch edible films. *Carbohydrate Polymers* 58(4) (2004): 475–481.

Pauli, R.B., Quast, L., Demiate, I. et al. Production and characterization of oxidized cassava starch (*Manihot esculenta* Crantz) biodegradable films. *Starch–Stärke* 63(10) (2011): 595–603.

Pelissari, F.M., Andrade-Mahecha, M.M., Sobral, P.J., and Menegalli, F.C. Comparative study on the properties of flour and starch films of plantain bananas (*Musa paradisiaca*). *Food Hydrocolloids* 30 (2013a): 681–690.

Pelissari, F.M., Andrade-Mahecha, M.M., Sobral, P.J., and Menegalli, F.C. Optimization of process conditions for the production of films based on the flour from plantain bananas (*Musa paradisiaca*). *LWT: Food Science and Technology* 52 (2013b): 1–11.

Pérez, C.D., Fissore, E.N., Gerschenson, L.N., Cameron, R.G., and Rojas, A.M. Hydrolytic and oxidative stability of L-(+)-ascorbic acid supported in pectin films: Influence of the macromolecular structure and calcium presence. *Journal of Agricultural and Food Chemistry* 60(21) (2012): 5414–5422.

Pérez-Gallardo, A., Bello-Perez, L.A., Garcia-Almendarez, B. et al. Effect of structural characteristics of modified waxy corn starches on rheological properties, film-forming solutions, and on water vapor permeability, solubility, and opacity of films. *Starch–Stärke* 64(1) (2012): 27–36.

Pérez, S., Baldwin, P.M., and Gallan, D.J. Chapter 5: Structural features of starch granules I. In J. BeMiller and R. Whistler (eds.), *Starch Chemistry and Technology*, 3rd ed. London, U.K.: Elsevier, 2009, pp. 149–192.

Rajeev, B., Abdullah, A., Rozman, D.H. et al. Producing novel sago starch based food packaging films by incorporating lignin isolated from oil palm black liquor waste. *Journal of Food Engineering* 119(4) (2013): 707–713.

Rao, M.S., Kanatt, S.R., Chawla, S.P. et al. Chitosan and guar gum composite films: Preparation, physical, mechanical and antimicrobial properties. *Carbohydrate Polymers* 82(4) (2010): 1243–1247.

Reddy, N. and Yang, Y. Citric acid cross-linking of starch films. *Food Chemistry* 118 (2010): 702–711.

Ridley, B.L., O'Neill, M.A., and Mohnen, D. Pectins: structure, biosynthesis, and oligogalacturonide- related signaling. *Phytochemistry* 57 (2001): 929–967.

Rindlav, A., Hulleman, S.H.D., and Gatenholma, P. Formation of starch films with varying crystallinity. *Carbohydrate Polymers* 34(1–2) (1997): 25–30.

Rindlav-Westling, A., Stading, M., Hermansson, A.M., and Gatenholm, P. Structure, mechanical and barrier properties of amylose and amylopectin films. *Carbohydrate Polymers* 36 (1998): 217–224.

Robles-Sánchez, R.M., Rojas-Graü, M.A., Odriozola-Serrano, I., González-Aguilar, G., and Martin-Belloso, O. Influence of alginate-based edible coating as carrier of antibrowning agents on bioactive compounds and antioxidant activity in fresh-cut Kent mangoes. *LWT: Food Science and Technology* 50(1) (2013): 240–246.

Rojas-Argudo, C., del Rio, M.A., and Perez-Gago, M.B. Development and optimization of locust bean gum (LBG)-based edible coatings for postharvest storage of "Fortune" mandarins. *Postharvest Biology and Technology* 52(2) (2009): 227–234.

Sales-Valero, L.M., Tapia-Blácido, D.R., and Mengalli, F.C. Biofilms based on canihua flour (*Chenopodium Pallidicaule*): Design and characterization. *Quimica Nova* 38(1) (2015): 14–21.

Schwartz, D. and Whistler, R.L. Chapter 1: History and future of starch. In J. BeMiller and R. Whistler (eds.), *Starch Chemistry and Technology*, 3rd ed. London, U.K.: Elsevier, 2009, pp. 1–10.

Shannon, J.C., Garwood, D.L., and Boyer, C.D. Chapter 3: Genetic and physiology of starch development. In J. BeMiller and R. Whistler (eds.), *Starch Chemistry and Technology*, 3rd ed. London, U.K.: Elsevier 2009, pp. 193–236.

Shih F.F., Daigle, K.W., and Champagne, E.T. Effect of rice wax on water vapour permeability and sorption properties of edible pullulan films. *Food Chemistry* 127 (2011) 118–121.

Sitonio-Eça, K., Tanara Sartori, T., and Menegalli, F.C. Films and edible coatings containing antioxidants—A review. *Brazilian Journal of Food Technology* 17 (2014): 98–112.

Soliva-Fortuny, R., Rojas-Graü, M.A., and Martín-Belloso, O. Polysaccharide coatings. In E.A. Baldwin, R.D. Hagenmaier, and J. Bai (eds.), *Edible Coatings and Films to Improve Food Quality*. Boca Raton, FL: Taylor & Francis Group, 2012, pp. 103–115.

Sothornvit, R. and Pitak, K. Oxygen permeability and mechanical properties of banana films. *Food Research International* 40(3) (2007): 365–370.

Sreekumar, P.A., Leblanc, N., and Saiter, J.M. Effect of glycerol on the properties of 100% biodegradable thermoplastic based on wheat flour. *Journal of Polymers and the Environment* 21(2) (2013): 388–394.

Steffolani, M.E., Leon, A.E., and Perez, G.T. Study of the physicochemical and functional characterization of quinoa and kañiwa starches. *Starch–Stärke* 65 (2013): 976–983.

Talja, R.A., Helén, H., Ross, Y.H., and Jouppilla, K. Effect of various polyols and polyol contents on physical and mechanical properties of potato starch-based films. *Carbohydrate Polymers* 67(3) (2007): 288–295.

Talja, R.A., Helén, H., Ross, Y.H., and Jouppilla, K. Effect of type and content of binary polyol mixtures on physical and mechanical properties of starch-based edible films. *Carbohydrate Polymers* 71(2) (2008): 269–276.

Tapia-Blácido, D. Development and characterization of film from Amaranth flour. Master Degree. Food Engineering Program. University of Campinas, Campinas, Brazil, 2003.

Tapia-Blácido, D., Mauri, A., Menegalli, F.C., Sobral, P.J., and Añon, M.C. Contribution of the starch, protein, and lipid fractions to the physical, thermal, and structural properties of amaranth (*Amaranthus caudatus*) flour films. *Journal of Food Science* 71 (2007): 293–300.

Tapia-Blácido, D., Sobral, P.J.A., and Menegalli, F.C. Development and characterization of biofilms based on Amaranth flour (*Amaranthus caudatus*). *Journal of Food Engineering* 67(1–2) (2005a): 215–223.

Tapia-Blácido, D., Sobral, P.J., and Menegalli, F.C. Effects of drying temperature and relative humidity on the mechanical properties of amaranth flour films plasticized with glycerol. *Brazilian Journal of Chemical Engineering* 22 (2005b): 249–256.

Tapia-Blácido, D., Sobral, P.J., and Menegalli, F.C. Optimization of amaranth flour films plasticized with glycerol and sorbitol by multi-response analysis. *LWT: Food Science and Technology* 44 (2011): 1731–1738.

Tong, Q., Xiao, Q., and Lim, L.-T. Effects of glycerol, sorbitol, xylitol and fructose plasticisers on mechanical and moisture barrier properties of pullulan-alginate-carboxymethylcellulose blend films. *International Journal of Food Science & Technology* 48(4) (2013): 870–878.

Trinetta, V., Cutter, C.N., and Floros, J.D. Effects of ingredient composition on optical and mechanical properties of pullulan film for food-packaging applications. *LWT: Food Science and Technology* 44(10) (2011): 2296–2301.

Veiga-Santos, P., Oliveira, L.M., Cereda, M.P. et al. Mechanical properties, hydrophilicity and water activity of starch-gum film: Effect of additives and deacetylated xanthan gum. *Food Hydrocolloids* 19(2) (2005a): 341–349.

Veiga-Santos, P., Suzuki, C.K., Cereda, M.P. et al. Microstructure and color of starch-gum films: Effect of gum deacetylation and additives: Part 2. *Food Hydrocolloids* 19(6) (2005b): 1064–1073.

Yoo, S.R. and Krochta, J.M. Starch-methylcellulose-whey protein film properties. *International Journal of Food Science & Technology* 47(2) (2012): 255–261.

Yu, L. and Christie, G. Microstructure and mechanical properties of orientated thermoplastic starches. *Journal of Materials Science* 40 (2005): 111–116.

Zamudio-Flores, P.B., Gutierrez-Meraz, F., and Bello-Perez, L.A. Effect of dual modification of banana starch and storage time on thermal and crystallinity characteristics of its films. *Starch–Stärke* 63(9) (2011): 550–557.

Zamudio-Flores, P.B., Vargas-Torres, A., Pérez-González, J., Bosquez-Molina, E., and Bello-Pérez, L.A. Films prepared with oxidized banana starch: Mechanical and barrier properties. *Starch–Stärke* 58 (2006): 274–282.

Zavareze, E.R., Pinto, V.Z., Klein, B. et al. Development of oxidised and heat-moisture treated potato starch film. *Food Chemistry* 132(1) (2012): 344–350.

Zhang, Y. and Han, J.H. Mechanical and thermal characteristics of pea starch films plasticized with monosaccharides and polyols. *Journal of Food Science* 71(2E) (2006): 109–118.

Zhong, L.-X., Peng, X.-W., Yang, D. et al. Long-chain anhydride modification: A new strategy for preparing xylan films. *Journal of Agricultural and Food Chemistry* 61(3) (2013): 655–661.

8

Films and Coatings from Lignocellulosic Polymers

Véronique Aguié-Béghin, Gabriel Paës, Michael Molinari, and Brigitte Chabbert

CONTENTS

Abstract

Lignocellulosic polymers represent abundant and renewable resources in the world and have attracted great interest as novel organic feedstocks in nanocomposites. They belong to three main categories: cellulose, hemicelluloses, and lignins. These components are interconnected through a variety of covalent and noncovalent interactions in plant cell walls (PCWs) and form a highly complex organized network. This organization plays a key role in plants for cell growth, mechanical properties, and biodegradation resistance. It thus provides inspiration to nanostructured composite designing. Relying on the long experience for extracting and preparing different types of polymers, the more recent development of nanoscale techniques provides new means to build and characterize bioinspired assemblies (thin films and coatings). These are generally based on cellulose (cellulose nanocrystals, cellulose nanofibers, cellulose derivatives), which offer a great technical and scientific potential for optical, biological, medical, packaging, and electronical applications. Moreover, they are well suited for fundamental researches aiming at understanding the physicochemical properties of the PCW polymers and their reactivity to biological agents.

8.1 Introduction

Over the past 20 years, interest in the exploration of the potential of lignocellulosic biomass, which gathers resources such as trees and plants, increased significantly for the preparation of bionanocomposites used in buildings, automotives, biomedical applications, and food packaging. The interest in polymers from lignocellulose (LC), more particularly cellulose crystalline and their derivatives (with chemical substitutions), is in nanocomposites due to their reinforcement capacity, contribution to low density, and biodegradability (Dufresne 2013; Fernandes et al. 2013). In addition, cellulose nanocrystals (CNCs) and nanofibrils (CNFs) attract much attention for their use in nanotechnology, microelectronics, optical and biomimetic systems, functional surfaces, self-organization assemblies, etc. (Eichhorn et al. 2010; Moon et al. 2011), whereas the two other LC polymers (hemicelluloses and lignins) have still an unexplored potential. Emergence of LC-based nanocomposites relies on the recent development of two main industrial and research fields. The first one is the development of biorefineries, which are LC-based industries aiming at substituting conventional fossil products with renewable sources, which support economic growth, national energy security, and environmental goals. The second one is the high interest for nanotechnology, in particular for designing nanomaterials that consist in nanostructured biopolymers under confinement in thin films or coatings ranging from nanometer (monolayer) to several micrometers in thickness. Nanodroplets, nanotubes, and nanoporous assemblies from biopolymers are also explored to create nanostructured materials with new functionalities (e.g., nanodevices delivering biomolecules) (Dosch and Van de Voorde 2009). The increasing knowledge in molecular structure of LC polymers (cellulose, hemicelluloses, and lignins) in combination with the control of their molecular self-organization by means of chemical and physical methods can enhance the properties of nanomaterials (Lucia and Rojas 2009). Natural nanostructured assemblies found in plant cell walls (PCWs) may be used as templates on the road to develop new bioinspired materials with advanced structures and functions (Dujardin and Mann 2002).

In this context, the first section of this review presents a brief description of LC polymers before and after their extraction from PCW. Then, the most recent and advanced techniques and conditions to design LC-based nanomaterials from elementary bricks (CNCs or CNFs, hemicelluloses, and lignins) are discussed. Most of these nanomaterials are bioinspired from the PCW; they consist in thin film and coating systems with specific interfacial interactions required to connect native polymers together without harsh or complex chemical substitution or functionalization. The last section is an overview of the current and potential applications of LC nanomaterials.

8.2 LC Polymers in Plant Cell Wall

PCWs are complex structures composed of two main layers, the primary and secondary walls (Cosgrove and Jarvis 2012). LCs are essentially secondary cell walls that are made of three main categories of polymers: cellulose, hemicelluloses, and lignins (Figure 8.1).

In PCWs, these polymers are interconnected by numerous covalent and noncovalent interactions to form cohesive networks that can match cell functions *in planta* (Cosgrove and Jarvis 2012). Cellulose is one of the main PCW components (40%–70%) in secondary walls and is composed of aligned (1→4)-linked β-D-glucose chains. A distinct number of glucan linear chains are tightly linked by hydrogen bonds during biosynthesis to form parallel nanofibrils with large crystalline zones and some amorphous junctions (Kondo 2005). The weak contrast observed in neutron small-angle scattering experiments and the small weight loss during acid hydrolysis suggest that the amorphous regions are extremely small compared to the crystalline zones (Nishiyama 2009). Nanofibrils are self-assembled with various polysaccharidic or aromatic structures (pectins in primary walls, hemicelluloses in primary and secondary walls, lignins in secondary walls) to form microfibrils (Gorshkova et al. 2012). In secondary walls, these microfibrils form successive, concentric, and cohesive layers, within which they are aligned (Roland et al. 1992). Noncellulosic polysaccharides in PCWs are classified as pectins and hemicelluloses according to the chemicals used to extract these components. However, this distinction is poorly reliable owing to their large structural heterogeneity depending on botanical origin, cell type, and cell wall layer. Noncellulosic polysaccharides have a lower degree of polymerization and do not lead to crystalline organization as compared to cellulose. Pectins are typically polysaccharides of the primary cell walls. They consist of galacturonic acid–rich polysaccharides, which can be classified in different groups based on polysaccharide

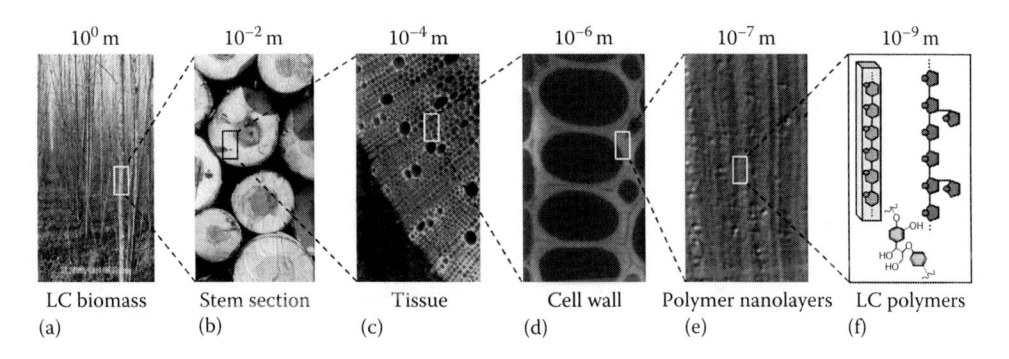

FIGURE 8.1 Multiscale architecture of plant cell wall. Macroscopic biomass (a) contains different organs, among them is the stem (b) which accounts for the largest proportion in weight. Stem is formed by tissues (c) composed by cells whose walls (d) consist of a matrix of lignocellulosic nanolayers (e) made of cellulose, hemicellulose, and lignin (f). (a,b: Courtesy of INRA, Paris, France.)

composition and ramification (homogalacturonans, xylogalacturonans, rhamnogalacturonans I and II). Different glycosidic linkages, calcium cross-linking, borate ester cross-linking, and covalent linkages to phenolic are involved in cross-linking the pectin network (Caffall and Mohnen 2009). Hemicelluloses are another group of polysaccharides of the primary and secondary PCWs. In contrast to cellulose, they are nonlinear heteropolymers with significant different chemical properties. For example, xyloglucan (XG) is the major hemicellulose in the primary cell walls of dicots, made of a linear β-(1,4) backbone of glucose branched in a repetitive pattern with xylose or a short chain of xylose, fucose, and arabinose. The type and site of substitution on the XG main chain show great variation according to taxa group and development (Zhou et al. 2007). Xylans and mannans are the main hemicelluloses of secondary cell walls (Scheller and Ulvskov 2010). Mannans consist of a β-(1,4)-linked mannose backbone, or an alternating mannose, and glucose chains, which can be acetylated. Mannose units are branched with α-D-galactose in the galactoglucomannans, which are the major hemicelluloses in softwood (Willfor et al. 2005). Xylans are made of β-D-xylopyranose residues linked in β-(1,4) displaying various side chains such as 4-O-methyl glucuronic acid, α-L-arabinose, and ester-linked acetyl groups (Scheller and Ulvskov 2010). The branching pattern of xylans shows large variation within taxonomic groups. Hardwood xylans are rich in glucuronic and acetyl groups, therefore, called glucuronoxylans, while arabinose is the main substituent in grass species, so xylans are named arabinoxylans (AXs). In addition, grass xylans are characterized by feruloyl arabinose decoration that play an essential role in interconnecting xylan chains through ester–ether linkages like ferulate dimers, or multimers (Bunzel 2010). Grass cell walls are also characterized by the presence of mixed β-(1,3) and α-(1,4) glucans (Vogel 2008).

Lignin is an amorphous polymer of three main hydroxyphenylpropanoid units: syringyl, guaiacyl, and *p*-hydroxyphenyl, whose proportions correlate with taxonomy (Ralph 2007). Gymnosperm lignins contain mainly guaiacyl units (G) and a weak proportion of hydroxyphenyl units, whereas angiosperm lignins are a mixed syringyl–guaiacyl (S-G) polymer. Lignin structural heterogeneity is strengthened by a large panel of interunit-bonding pattern owing to the monolignol composition and the polymerization mode of monolignols involving oxidase-generated phenoxyl radicals. The so-called noncondensed bonds (alkyl aryl ether β-O-4 bond) are the most abundant interunit linkages in angiosperms (Ralph 2007). In contrast, gymnosperm lignin contains higher proportion of C–C and diaryl ether linkages ether including 5-5, 4-O-5, β-1, β-5, and β-β also referred to as condensed bonds. Additional complexity holds in the possible acylation of lignin by phenolic acids: p-hydroxybenzoic acid in some hardwood species (poplar or willow) and *p*-coumaric acid or ferulic acid in grass species (maize, wheat). Given these diversity of monomer and bonding pattern, G-type lignins are more ramified than S–G lignins (Ralph 2007).

Overall, PCWs are a highly complex mix of polymers made of polysaccharides and phenolic compounds that can be considered as a large resource of chemicals that first need to be isolated and characterized.

8.3 Isolated LC Polymers: Structure and Physicochemical Properties according to Their Isolation Methods

8.3.1 Cellulose

Many products are derived from cellulose, having a large range of size and aspect ratio depending on their natural source and the chemical–mechanical processes used: cellulose microfibrils (CMFs), CNFs, and CNCs (Eichhorn et al. 2010; Klemm et al. 2005; Miao and Hamad 2013; Moon et al. 2011) (Table 8.1).

CMFs or CNFs were first produced several years ago by fibrillation process from pulp sheets (Kentaro et al. 2007). After microgrinding, homogenization, and microfluidization, they can be obtained with a length below 100 nm (Chinga-Carrasco et al. 2013), notably with a homogeneous width of 15 nm, identical to that of natural CMFs found in the PCW and obtained by a grinding treatment in an undried state after the removal of the matrix substance (Kentaro et al. 2007). Fibrillation processes can be optimized by adding chemical, enzymatic, or physical treatment steps (high-speed blending, TEMPO (2,2,6,6-tetramethylpiperidine-1-oxyl radical)-mediated oxidation) (Jiang and Hsieh 2013) or ultrasonication (Li and Renneckar 2011). The first CNCs were obtained 20 years ago (Revol et al. 1994; Samir et al. 2005). Typically, they are extracted by mild acid hydrolysis of cellulose after removing hemicelluloses and lignins by bleaching treatments of wood and fibers and also cleaning off noncellulosic polymers from nonplant biomass (some bacteria and animals produce cellulose). Main characteristics of CNCs are presented in Table 8.1 and shown in Figure 8.2.

CNC size and production yield decrease with increasing reaction time (Jiang and Hsieh 2013). Because of the complicated multilayered structure of plant fibers, all these cellulose preparations can be contaminated by hemicelluloses and lignins due to the fractionation process severity when extracted from wood and higher plants (Ferrer et al. 2012). Small contents (limited to a few percent) of lignin and hemicelluloses associated with cellulose crystallites and the presence of intrinsic disorder of cellobiose chains at crystallite interfaces bring amorphous monolayers surrounding native crystal phase (cellulose I) with different hydration properties (Driemeier and Bragatto 2013). CNFs and CNCs have the advantage to be the most pure native cellulose, namely, cellulose I (constituted by two crystalline forms Iβ and Iα with a different ratio according to the origin of nanocrystal). Nevertheless, their crystallinity and thermostability can be influenced by the drying method (Peng et al. 2013; Ramanen et al. 2012). Their low density ($1.566 \, \text{g} \cdot \text{cm}^{-3}$) makes them sensitive to the pressure and temperature at different water contents (Jallabert et al. 2013). The use of sulfuric acid during their preparation contributes to produce charged CNCs. The coverage of sulfate groups on the surface of CNCs is different depending on their origin

TABLE 8.1

Characteristics of Micro- and Nanocellulose Materials

Cellulose Materials	Sources	Extraction Processes	Characteristic Sizes	References
Cellulose microfibrils (CMFs)	Soft- and hardwood	Fibrillation process with or without bleaching	$d = 0.1–1 \, \mu m$ $L > 10,000$ $L/d > 1,000$	Chinga-Carrasco et al. (2013), Miao and Hamad (2013)
Cellulose nanofibrils (CNFs)	Wood	Fibrillation process with bleaching; microfluidization	$d = 10–20 \, \text{nm}$ $L > 10,000$ $L/d > 1,000$	Kentaro et al. (2007), Ferrer et al. (2012)
Cellulose nanocrystals (CNCs)	Wood, cotton, flax, hemp, ramie, rice straw, tunicates, bacteria, and alga (*Valonia*)	Acid hydrolysis with bleaching	$d = 5–20 \, \text{nm}$ $L = 100–400 \, \text{nm}$ (from LCs) L/d: 25–30 $L = 100 \, \text{nm}$ to $1–3 \, \mu m$ (from tunicates, bacteria, alga) $L/d = 70–160$	Miao and Hamad (2013), Klemm et al. (2005), Moon et al. (2011), Lahiji et al. (2012), Beck-Candanedo et al. (2005), Jiang and Hsieh (2013), Ramanen et al. (2012), Fernandes et al. (2013)

Note: *L*, length; *d*, cross section; *L/d*, aspect ratio.

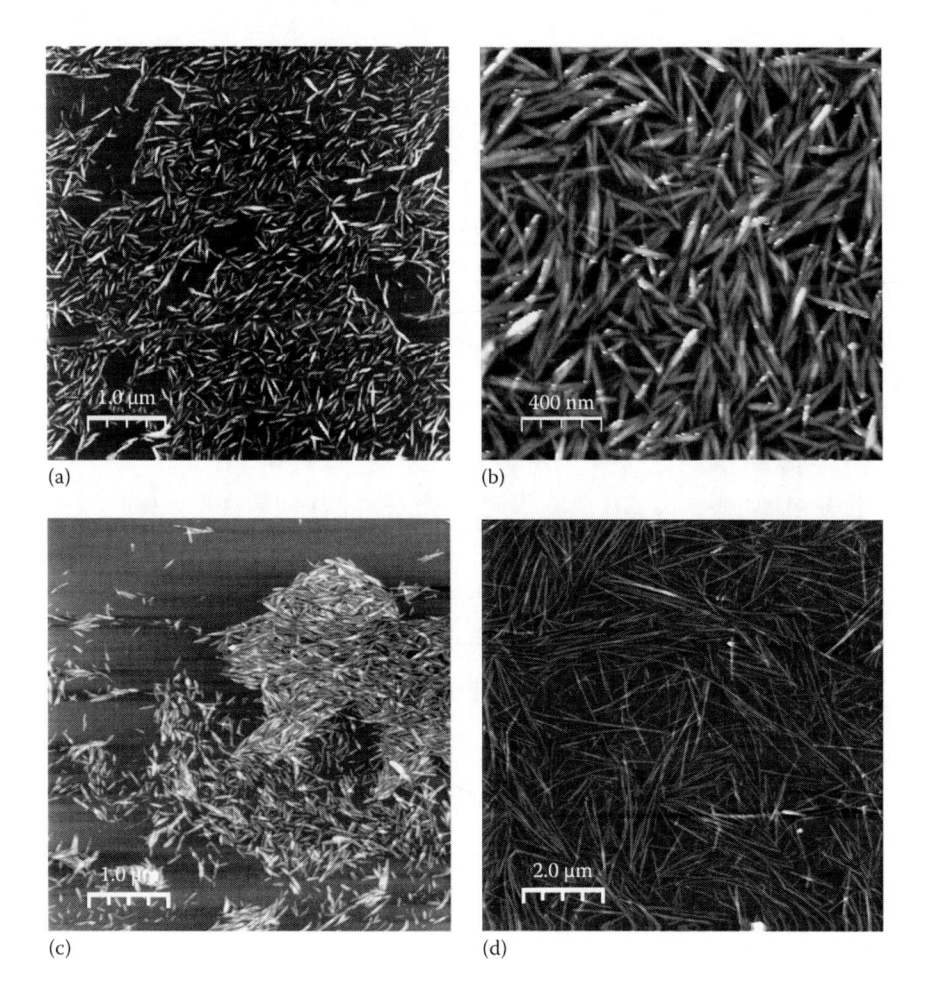

FIGURE 8.2 Height AFM images of isolated cellulose nanocrystals from different sources: (a) ramie, (b) flax, (c) hemp, and (d) tunicin.

(Lahiji et al. 2012) and hydrolysis conditions (Voronova et al. 2013), and can vary from 0 to 0.333 ester sulfate group · nm^{-2} for CNC prepared from softwood sulfite pulp (Jiang et al. 2010). Partial desulfation catalyzed by HCl leads to aggregation of the particles as observed for HCl-hydrolyzed CNCs (Jiang et al. 2010). The surface chemistry of CNCs (sulfate group density) has a large impact on adhesion properties (Lahiji et al. 2012), self-assembling behavior (Jiang and Hsieh 2013), enzymatical hydrolysis by cellulases, and interactions with carbohydrate-binding modules (Jiang et al. 2013).

8.3.2 Noncellulosic Polysaccharides

In addition to cellulose, several noncellulosic polysaccharides can be extracted from plant material with salt solutions, various chaotropic reagents, alkali, or enzymatic treatments (Fry 1988). Hemicelluloses can be also isolated from plant materials by various selective fractionation methods (acid hydrolysis, alkaline treatments, ionic liquids, organic solvent, supercritical fluids, enzymatic hydrolysis, as well as steam or microwave treatments) (Girio et al. 2010; Hansen and Plackett 2008). A delignification step is generally required before isolation of pectins and hemicelluloses. The noncellulosic fractions vary in their degree of polymerization and branching according to the procedure used in the fractionation, isolation, and purification from cell wall materials. Some are commercially exploited for many years (e.g., konjac glucomannan derived from the tuber *Amorphophallus konjac*, XG from tamarind seed) and used in numerous areas such as in medicine (drug delivery, cellular therapy) and food (encapsulation, gelification).

8.3.3 Lignins

Lignins are hardly isolated in native form. Partial disruption of the cross-linkages between lignin and noncellulosic polysaccharides is required prior to isolation of lignin fractions. This procedure includes a combination of ultramilling and enzymatic pretreatments to solubilize polysaccharides before recovering lignin using an appropriate solvent (Lange et al. 2013; Lundquist 1992; Sipponen et al. 2013). Sulfite, Kraft, and Soda lignins are the most common methods to extract lignin fractions from black liquor of pulping industries (Doherty et al. 2011). They involve sulfite or alkaline solutions at medium temperature to solubilize lignin, the harsh step being the recovery of inorganic chemicals employed in the process (Joffres et al. 2014; Notley and Norgren 2012). In these conditions, lignin structure is more or less transformed (weight average molecular mass [MW] is in order to 3000 g · mol^{-1} with mainly β-O-4 linkages between phenyl-propane monomers) (Norgren et al. 2006). In the organosolv process, sulfur-free lignin fractions are soluble in organic solvent/water mixture with preserved native structure in MW of 1500–4000 g · mol^{-1} and a polydispersity of 1.4–1.7 according to its plant origin (Aguié-Béghin et al. 2002; Argyropoulos and Bolker 1987; Hambardzumyan et al. 2011; Lange et al. 2013; Pereira et al. 2007). Isolating enzymatic mild acidolysis lignin makes lignin samples highly representative of the lignin in wood cell wall by increasing the average MW according to wood species (MW from 7,500 to 83,200 g · mol^{-1}) with a polydispersity range from 2.7 to 8.4 (Crestini et al. 2011; Guerra et al. 2007). More recently, in new biorefineries, ionic liquids were used to extract lignin directly from a raw material with high purity (>90%) (Prado et al. 2013). Alternatively, lignin analog dehydrogenation polymers (DHPs) can be prepared using peroxidase/hydrogen peroxide and different lignin monomers (Barakat et al. 2007; Tobimatsu et al. 2012). These synthetic lignins are thus not contaminated by sugars like biomass-extracted lignins and have MWs similar to that of organosolv lignins (Méchin et al. 2007). In addition, their chemical composition is more controlled since only one or two different monolignols are incorporated simultaneously.

8.4 LC Bioinspired Films and Coatings

8.4.1 LC Films and Coatings Fabrication Techniques

Many designing processes have been developed for several years to prepare model biopolymer-based films or coatings at a nanometric scale, with thickness ranging from a few nanometers to several micrometers (Table 8.2). The layer-by-layer (LbL) technique forms thin films by depositing alternating layers of oppositely charged materials with wash steps between. They were used to obtain polyelectrolyte multilayer buildup essentially by noncovalent interactions between polymers (electrostatic type, hydrogen bonds, and hydrophobic interactions). The LbL technique has the advantage to control the thickness during the linear growth of the film, with a 1 nm resolution. The Langmuir–Blodgett (LB) technique consists in transferring a monolayer of polymer from the surface of a liquid onto a solid (silicon, mica) by vertical immersion (or emersion) of the substrate into (or from) the liquid at controlled speed. The density and the thickness of the monolayer, usually composed of amphiphilic molecules, are controlled by the surface pressure measured at the air–liquid interface before and during the transfer onto the solid. Alternatively, the solid can be applied horizontally onto the surface of the liquid. Spin coating is a procedure to spread a coating material on a solid by centrifugal force. A small and fixed volume of the solution is applied on the center of the flat solid substrate, which is then accelerated to a rotational speed of a few thousand revolutions per minute. As a result, the solution is more or less spread out, becoming a thin film after evaporation of the solvent. The thickness, homogeneity, and smoothness of the film depend on several factors including size and structure of the polymer, concentration and viscosity of the solution, solvent nature, and adhesion properties of the solid. Multilayer assemblies can be built up by spin coating from two distinct coating materials with high affinity between them. Solvent casting is a method in which a solution of the coating polymer is deposited onto a flat solid substrate or onto a mold followed by a controlled evaporation of the solvent. Other evaporation processes are used under electric, magnetic fields or shear flow to favor the spatial arrangement of polymer nanoparticles in the film. More recently, capillary flow–based approaches

TABLE 8.2

Processing Methods of LCs Films and Coatings

Processing Methods	LC Materials—Factors to Control Deposition	References
Layer by layer Polymer 1 Water Polymer 2 Water	Multilayer of LC polymers with opposite charges or polyelectrolytes Thickness of ~10 nm/bilayer Dipping time in order of s to min	Podsiadlo et al. (2005), Podsiadlo et al. (2007), Cranston and Gray (2006), Jean et al. (2008), Aulin et al. (2010), Olszewska et al. (2013), Cranston et al. (2010), Guyomard-Lack et al. (2012)
Langmuir–Blodgett	Monolayer of CNCs: from 5 to 15 nm Lignins ca. 0.5–1 nm Controlled density by moving barriers, vertical or horizontal position of solid substrate	Pereira et al. (2007), Habibi et al. (2010), Aguié-Béghin et al. (2009), Aguié-Béghin et al. (2002), Zeder-Lutz et al. (2013)
Spin coating	Multilayer of CNCs, hemicelluloses, lignins, proteins, and polyelectrolytes Thickness from 10 nm to several hundreds nm Room temperature, centrifugal force	Notley and Norgren (2012), Cranston and Gray (2008), Eronen et al. (2011), Cerclier et al. (2010), Gustafsson et al. (2012), Saxena et al. (2011), Hambardzumyan et al. (2011), Notley and Norgren (2010), Hoeger et al. (2012), Suchy et al. (2011)
Casting	Film of CNFs, CNCs, hemicelluloses, and lignins (pure or mixture) Thickness: several μm Controlled temperature, under vacuum, shear flow, and electric or magnetic fields	Yoshiharu et al. (1997), Stevanic et al. (2012), Mikkonen et al. (2010), Peng et al. (2011), Saxena et al. (2011), Aguié-Béghin et al. (2009), Hambardzumyan et al. (2012, 2015), Mulder et al. (2011), Muraille et al. (2015), Stepan et al. (2013), Ying et al. (2013), Tang et al. (2013), Fukuzumi et al. (2013), Zhang et al. (2013), Paes et al. (2013)
Others Microfluidics, electrospinning, drying of hydrogels, and drying on template	CNCs, lignins or hemicelluloses (nanobeads, nanofillers, porous nanofilm)	Ago et al. (2013), Sehaqui et al. (2011), Ten et al. (2014)

like microfluidics and electrospinning were used to generate LC-based Janus microbeads and dry-spinning fillers, respectively.

The primary load-bearing assembly of LC cell walls consists in CMF, which can behave as a gel or colloidal suspension in a hydrated medium with appropriate salts. The amphiphilic molecular character of crystalline cellulose favors interactions (hydrogen bonds, hydrophobic interactions even in polar environments, van der Waals interactions) (Nishiyama 2009). CNCs are thus the most used polymers in bioinspired nanocomposites films to create smooth cellulosic films with various thicknesses ranging from 5 nm to a few μm. These cellulose-based assemblies have the advantage to be used under controlled-hydrated conditions (from 5% to 60% of water content) in a concentrated regime of cellulose, analogous to the conditions found in PCWs. Moreover, they make it possible for the development of new molecular approaches to study their properties (affinity with polymers, morphology of nanoparticles, self-assembly, spectroscopic properties) at the local scale with appropriate techniques. They can be complexified with other LC polymers by noncovalent and covalent interactions to develop new nanomaterials with advanced structures and properties (Figure 8.3).

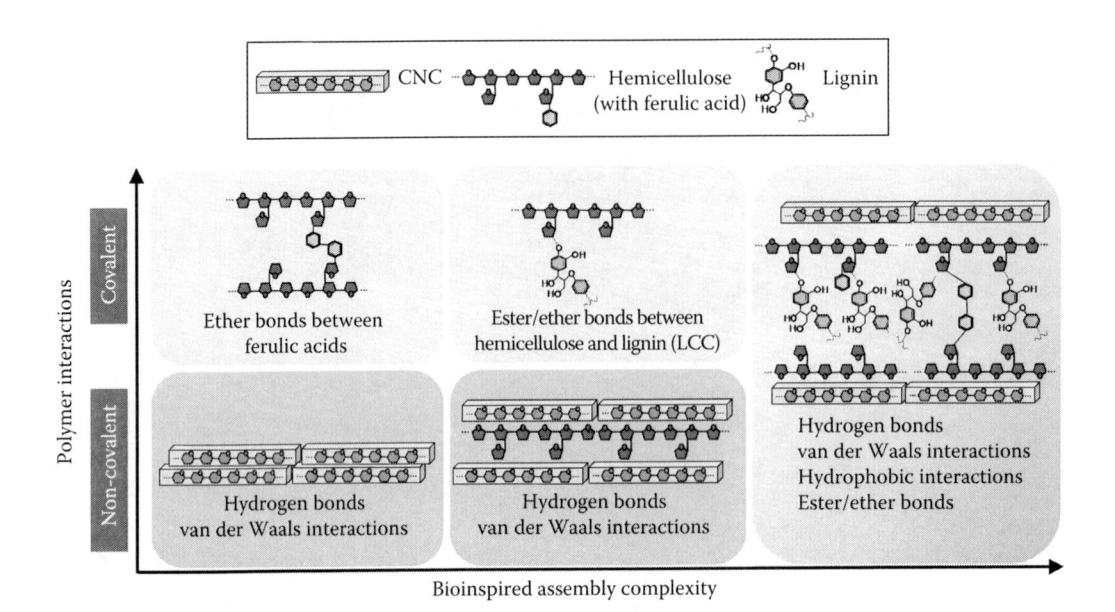

FIGURE 8.3 Bioinspired lignocellulosic assemblies with increasing complexity of cross-linking between polymers (non-covalent and covalent interactions).

8.4.2 Noncovalent LC Assemblies and Their Properties

8.4.2.1 Cellulosic Film Assemblies

Many designing processes are in development since several years to prepare model cellulose-based films or coatings at a nanometric scale (Roman 2009). These models offer the possibility to study surface properties and interactions between polymers in a simplified and controlled environment at a molecular scale. Most of them have been prepared from CNCs or CNFs to mimic native multilayering network of cellulosic microfibrils in PCW by spin coating or electrostatic LbL processing (Aulin et al. 2010; Cranston and Gray 2006; Edgar and Gray 2003; Jean et al. 2008; Kontturi et al. 2007; Notley et al. 2006), more recently pure cellulose multilayer with long nanofibrils and short nanocrystals (Olszewska et al. 2013), or with anionically and cationically modified microfibrillated cellulose (Aulin et al. 2010). Another experiments have been performed to control the orientation of CNCs in films by drying a gel layer under shear flow conditions (Yoshiharu et al. 1997), in magnetic and electric fields (Aguié-Béghin et al. 2009; Bordel et al. 2006; Habibi et al. 2008), and in monolayers using the LB technique (Aguié-Béghin et al. 2009; Habibi et al. 2007, 2010). Using the latter technique, large and uniform coverage of nanocrystals can be obtained. Moreover, the nanorods can be oriented in large domains when the transfer of cellulosic gel layer is performed by positioning vertically the silicon support at weak-controlled speed (Figure 8.4). The high occurrence of ordered domains is larger for tunicin CNCs with an aspect ratio of 100 (Figure 8.4c) compared to ramie or flax CNC with an aspect ratio of 26 to 50 (Figure 8.4a and b). Oriented CNCs in ultrathin films provided anisotropy in the system with enhanced optical properties as birefringence (Cranston and Gray 2008), higher surface mechanical strength, and wear resistance (Hoeger et al. 2011).

These pure cellulose films are often useful templates for measuring enzyme activity by surface plasmon resonance (SPR) (Allen et al. 2012) and flow ellipsometry (Maurer et al. 2012), molecular affinity interactions of cellulose with lignins by spectroscopic ellipsometry (Hambardzumyan et al. 2011), with hemicelluloses (Eronen et al. 2011; Kohnke et al. 2011), with proteins as carbohydrate-binding module (Mengmeng et al. 2012) by combining AFM and single-molecule dynamic force spectroscopy using functionalized AFM tip or bead (Gustafsson et al. 2012).

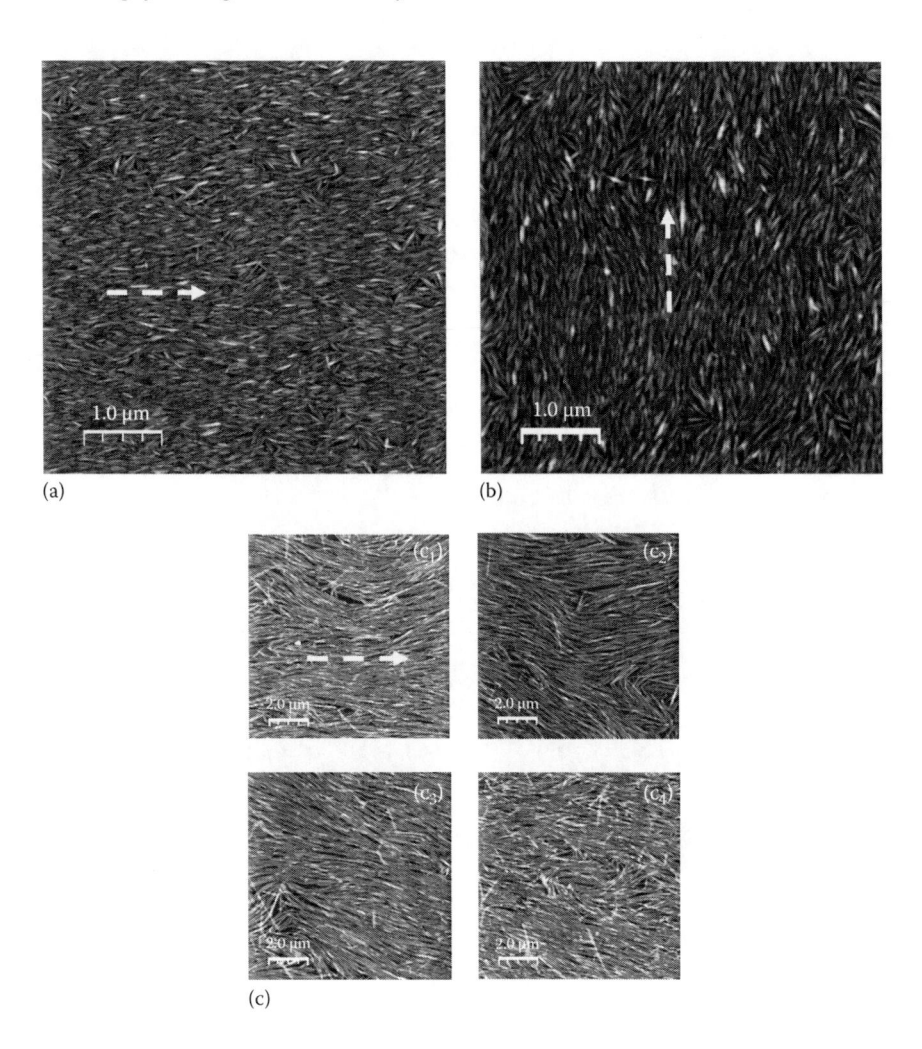

FIGURE 8.4 Topographic AFM images of oriented domains of nanocrystals in cellulosic Langmuir–Blodgett (LB) films: (a) ramie, (b) flax, and (c) tunicin. Height AFM images of the middle surface area of ramie and flax cellulose nanocrystal LB films transferred on 1 cm^2 silicon wafer surface area (Aguié-Béghin et al. 2009), except for tunicin where c_1–c_4 are respectively upper, down, left, and right of the surface area. The arrow indicates the principal orientation of CNCs on each image.

8.4.2.2 Cellulosic Film Assemblies Including Another Cell Wall Polymer

Among all cellulose nanoassemblies, only a few are prepared from pure cell wall biopolymer fractions without functionalization (i.e., silylation, oxidation of cellulose), compatibilizers, and polyelectrolytes (Aulin et al. 2010). Coupling agents are often used to improve the adhesion between the nanorods themselves or between the nanorods and the matrix (Berglund and Peijs 2010). Other pure LC polymers can be used as coupling agents. Among them, pure hemicelluloses like konjac glucomannan, galactoglucomannan, and xylans, in combination with an external plasticizer like glycerol or sorbitol, have been known in exhibiting good barrier properties against gas, making them valuable materials for food packaging applications (Cheng et al. 2008; Hartman et al. 2006; Mikkonen et al. 2007, 2010; Sedlmeyer 2011; Ying et al. 2013). However, the commercial potential of hemicellulose-based films is still limited by their poor resistance to high relative humidity and their low resistance properties. The reinforcement of hemicellulose-based films with CNFs or CNCs allows to improve their physical and mechanical properties (flexibility, stiffness, and strength) and their moisture sorption properties even in the absence of glycerol as a plasticizer and without the chemical modification of the polymer (Mikkonen et al. 2010; Peng et al. 2011; Stevanic et al. 2012). Because of their common hydrophilic nature, cellulose–hemicellulose films (such as xylan with sulfuric CNCs or glucuronoxylan with bacterial cellulose) are homogeneous at micrometric scale and display nodular structure when observed by SEM and AFM (Peng et al. 2011) with strong hydrogen-bonding interactions between the surface of CNCs and the matrix (Dammstrom et al. 2009; Saxena et al. 2011). Structural variation in the hemicellulose polymers leads to different cross-linking in nanocomposites, observed from centimeter to nanometric scales (Whitney et al. 2006). For example, high-MW XG promotes strong noncovalent binding to cellulose, and the ultrastructure of the nanocomposite is modulated by the nature of the side chains (Chambat et al. 2005). Other model assemblies showed that AX chains adsorbed on bacterial cellulose through both linear regions of the AX chains by hydrogen bonds and hydrophobic interactions due to the remaining lignin residues (Linder et al. 2003). This adsorption on coated microfibrillated cellulose is apparently irreversible and dependent on the degree of substitution pattern (Kohnke et al. 2011). Cross-linked feruloylated AX gels are more organized after the introduction of CNCs, probably revealing some noncovalent interactions between feruloylated AXs and CNCs (Paës and Chabbert 2012) (Figure 8.5a and b).

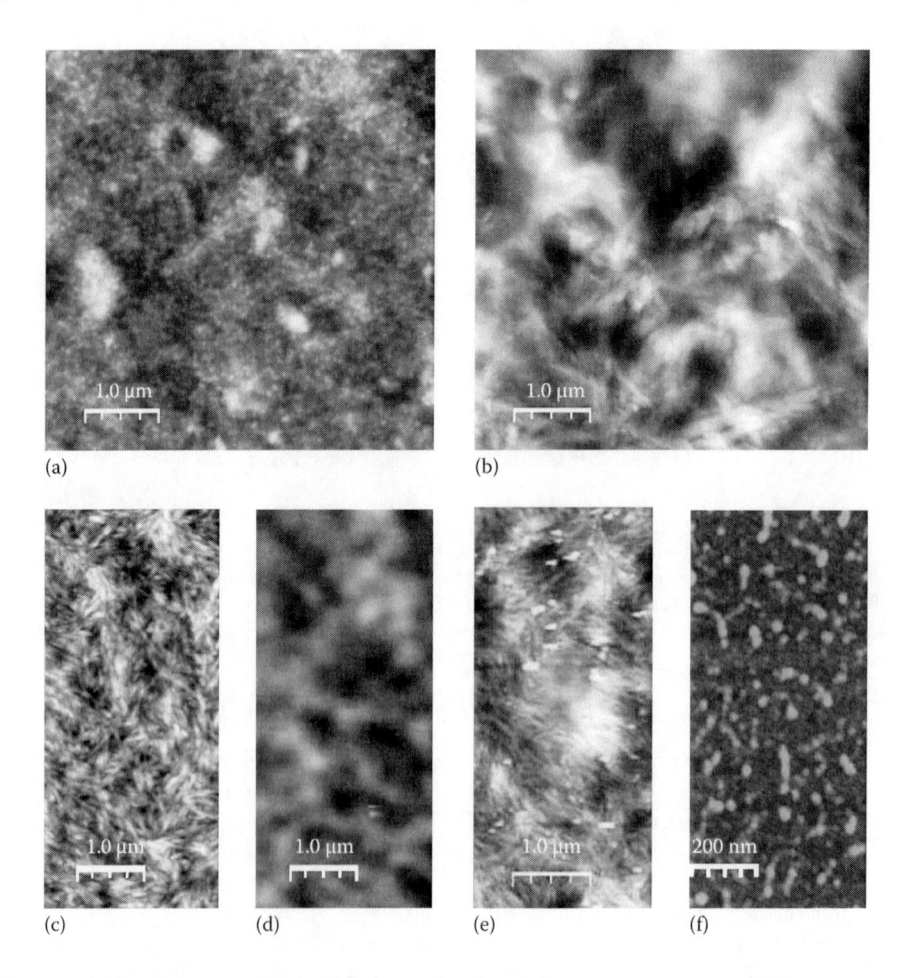

FIGURE 8.5   Height AFM images of lignocellulosic complex films and coatings. Feruolylated AX film (a) without and (b) with Ramie CNC (1/1, w/w); (c) Ramie cellulose nanocrystal (CNC) and (d) dehydrogenation polymer (DHP) layers in multilayer film; (e) Ramie CNC–DHP in mixed casting film; (f) DHP Langmuir–Blodgett layer on self-assembled monolayer (SAM) on gold sensor. (a,b: From Paës, G. et al., *Biomacromolecules*, 14(7), 2196, 2013; c,d: From Hambardzumyan, A. et al., *Comptes rendus Biologies*, 334(11), 839, 2011; e: From Hambardzumyan, A. et al., *Biomacromolecules*, 13(12), 4081, 2012; f: From Zeder-Lutz, G. et al., *Analyst*, 138(22), 6889, 2013.)

Furthermore, the enzymatic surface esterification of AX films with lipase or cutinase can improve its water resistance (Stepan et al. 2013). Then, the incorporation of lignin in cellulose-based films leads to an increase on the thermomechanical stability, hydrophobicity (Notley and Norgren 2010), and resistance to cellulolytic reactions (Hoeger et al. 2012). Lignin on solid substrate or in contact with CNCs without a linking agent has a tendency to form nodules observed from topography analysis by AFM and from fluorescence by SNOM in reflection mode (Hambardzumyan et al. 2011, 2012; Keplinger et al. 2014) (Figure 8.5c through e). When a model or extracted lignins are immobilized on self-assembled monolayer (SAM)-gold sensor by the LB technique (Zeder-Lutz et al. 2013) (Figure 8.5f), spin coating (Notley and Norgren 2010), or controlled polymerization of monolignols after immobilization of peroxidase on gold sensor (Wang et al. 2013), they form monolayers with different arrangements between nodular structures according to the composition in syringyl, guaiacyl, and coumaryl monomers. Increasing concentration of lignin nodules to cover cellulose layer in spin-coating multilayer film leads to adhesive cohesion between cellulose layers by noncovalent interactions and brings antireflective properties to the film (Hambardzumyan et al. 2011) (Figure 8.5d). On the contrary, when lignin nodules are dispersed in a cellulose film by mixing these two polymers in colloidal suspension before casting process on quartz support, the transparent film has higher UV-absorbent properties (Hambardzumyan et al. 2012) (Figure 8.5e). These results clearly show that noncovalent intermolecular interactions (electrostatic, hydrogen bonds, and hydrophobic interactions) between cellulose, hemicellulose, and lignin and the final nanostructure of the network have a strong impact on functional properties of the LC-based films.

8.4.3 LC Complex Films or Coatings with Covalent Interactions

In PCW, cohesiveness relies on the formation of interpolymer covalent and noncovalent linkages during wall deposition and lignification, thereby inducing major chemical and organizational changes of the preexisting polysaccharides. Noteworthy, phenol components (lignin, phenolic acid, and ferulic acid in particular) are important chemical functions in the organization of PCW (Koshijima and Watanabe 2003; Ralph 2007). *In vitro* cross-linking between polysaccharides and lignins can be achieved enzymatically or chemically in suspension before coating processes (Barakat et al. 2007; Boukari et al. 2009). Ferulic acid decorations on AX are critical features acting on DHP/AX supramolecular architecture, either by cross-linking the AX chains (Paës and Chabbert 2012) or as a nucleation point for lignin monomers with oxidative enzyme (Ralph 2007). As a result, ferulic or model lignins enzymatically complexed to noncellulosic polysaccharides improve water sorption capacity of corresponding films (thickness 15–30 µm) (Muraille 2015; Paës et al. 2013). The same behavior is observed in films prepared from synthetic or extracted lignin fractions grafted to CNC by using Fenton reagent. Thin films (thickness from 1 to 5 µm) coated on quartz slide with a CNC/lignin weight ratio from 5 to 10 display higher water resistance with high UV-absorbant properties (Hambardzumyan 2015). Covalent interactions such as xylan–DHP associations limit the hydrolysis of glucuronoarabinoxylan (GAX) by enzymes (Boukari et al. 2009). Irradiation (UV, electron beam) of polysaccharide and aromatic blends is another method to induce interchain covalent linkages between components, increasing the MW of the mix (Akrman and Prikryl 2008; Khandal et al. 2013), which thus have a potential for the development of renewable thermoplastic materials and UV absorbers textiles.

8.4.3.1 Relevant Applications of Nanostructured LC-Based Assemblies

Knowledge increase in the LC polymers preparation and characterization has established the possibility to build LC–polymer assemblies like films and coatings used in several domains (Figure 8.6).

In the biorefinery area, which aims at optimizing the use of biomass for energy, materials and chemicals, LC-based assemblies are used to investigate interfacial phenomena relevant to LC biocatalysis by QCM-D, often in association with another physicochemical technique for surface characterization such as atomic force microscopy, X-ray photoelectron spectroscopy, and water contact angle. The information obtained brings new relevant hypothesis regarding PCW dynamics and deconstruction that should be assessed *in planta* and in a processed LC for understanding the complex mechanism that

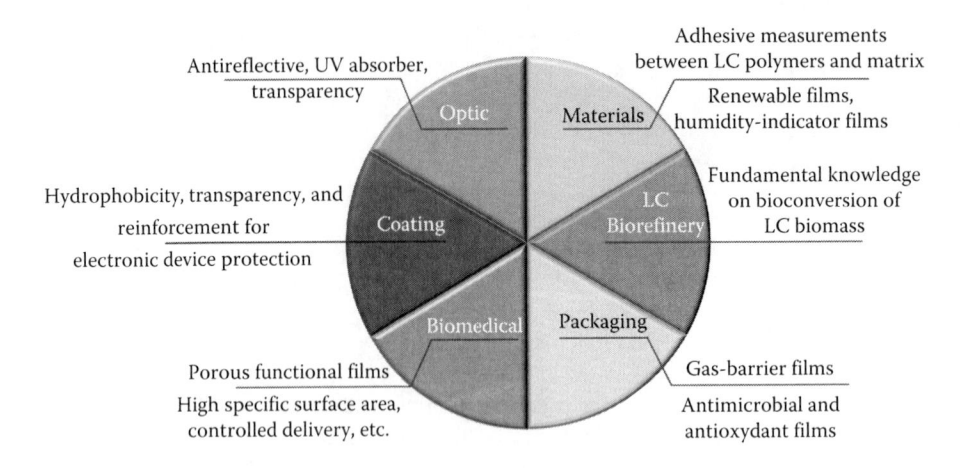

FIGURE 8.6 Synthetic illustration of applications and use of lignocellulosic films and coatings.

occurs during the bioconversion of cellulosic biomass (Cerclier et al. 2011; Guyomard-Lack et al. 2012; Hoeger et al. 2012; Kumagai et al. 2013; Martin-Sampedro et al. 2013; Suchy et al. 2011). In combination with biotechnological approaches on plants and enzymes and with molecular dynamics simulation (Chundawat et al. 2011), increasing knowledge in LC assemblies design should allow the optimization of the PCW properties for subsequent application in biorefinery.

In material science, physical approaches like adhesive measurements between LC polymers model surfaces are investigated to gain fundamental understanding of interactions both within the fiber wall and in fiber/fiber joints in paper or the fiber/matrix in material composites. At first, this study was carried out in water using surface force apparatus (Holmberg et al. 1997), then using colloidal probe in AFM (Cranston et al. 2010; Notley et al. 2006; Notley and Norgren 2012; Rutland et al. 1997). More recently, similar measurements were investigated in ambient conditions at controlled relative humidity of 25%–35% (Estephan et al. 2011; Gustafsson et al. 2012). These methods require highly smooth and chemically homogeneous surfaces. New generation of membranes based on derived TEMPO-oxidized cellulose nanofibrils (TOCN) are studied as biobased H_2 gas separation in fuel cell electric power generation systems (Fukuzumi et al. 2013). CNCs can be casted in thick iridescent films (40 μm thickness) to become a humidity indicator whose color varies from blue green in dry state to red orange in wet state (Zhang et al. 2013) under controlled-vacuum condition (Tang et al. 2013). This reversible shift in the film iridescence is attributed to the enlargement of the pitch tuning along the helicoidal axis in chiral-nematic phase solid film created by self-assembly of the rigid crystallites.

In packaging, green renewable and degradable polymers can be of a great interest that is why hemicelluloses and CNF films, which present good oxygen barrier properties, are considered (Hansen and Plackett 2008; Li et al. 2013; Ryberg et al. 2011; Stevanic et al. 2012). To overcome their instability at high relative humidity, enzymatic surface modifications are performed (Hoije et al. 2008).

In coating domain, CNCs are added to various finished products like varnish, acrylic lacquer after UV polymerization to increase mechanical properties of the coating (adhesion, resistance to the impact, to the scratches and abrasion). Their low refractive index (*n* ca. 1.5) contrary to several inorganic particles used (aluminum oxide) allows to maintain a better transparency to the coating even with a high level of reinforcement (from 1% to 5%) (www.fpinnovations.ca). If CNCs are associated to lignin, antireflective, anti-UV and antioxydant properties are obtained according to the designing process used (Hambardzumyan et al. 2012; Aguié-Béghin et al. 2015). Functional cellulose paper obtained after deposition of transparent resin on the surface of cellulose nanofibers sheet offers a large potential in an electronic device, thanks to their high-transparency, high-modulus, high-strength, and minimal thermal expansion (Nogi et al. 2009; Nogi and Yano 2009). Besides, lignin shows high potential as hydrophobic coating material to control the solubility of fertilizer as urea beads (Mulder et al. 2011).

In the biomedical field, flexible lignin nanotubes synthesized from a sacrificial alumina template have been recently developed as vehicles for gene delivery into human cells (Ten et al. 2014). They have been also produced by electrospinning with CNCs as reinforcement for the production of porous films with high specific surface area, thus offering great potential for tailoring functional materials in chemical, medical, physical, and electronic domains (Ago et al. 2012).

Finally, nanoreinforcement of hydrogel by CNCs can offer nanostructured fiber networks with high specific surface area (up to 480 m^2 g^{-1}), porosity, low density, and interesting mechanical properties comparable to common thermoplastic films (Sehaqui et al. 2011).

8.5 Conclusion

LC biomass offers significant potential for new PCW-biobased products free of additive and bring large opportunities in the context of biorefineries. LC films and coatings are an environmental-friendly alternative to petroleum-based polymers due to their low cost, availability, biodegradability, low density, physical–mechanical performance, and antimicrobial properties in particular when lignin is incorporated inside the nanomaterials. With the emergence of the production CNC on a large scale on the one hand and the by-products production (hemicelluloses, lignins) from second-generation biorefineries on the other hand, several projects intensified in the world to develop a new platform of research and innovations in a green nanostructured material domain with high added value and unexpected properties.

ACKNOWLEDGMENTS

Funding of UMR FARE work was supported by the French National Research Agency under projects ANR-08-BIO2-004 Hemili and 08-0009-02 Analogs, by the Conseil Régional Champagne-Ardenne, Fonds Européen de Développement Régional and INRA. AFM measurements were provided through Nano'Mat Platform, University of Reims-Champagne-Ardenne. L. Foulon, A. Habrant, and D. Crônier are gratefully acknowledged for their technical support.

REFERENCES

Ago M., Jakes J. E., Johansson L. S. et al. Interfacial properties of lignin-based electrospun nanofibers and films reinforced with cellulose nanocrystals. *ACS Applied Materials & Interfaces* 4(12) (2012):6848–6855.

Ago M., Jakes J. E., and Rojas O. J. Thermomechanical properties of lignin-based electrospun nanofibers and films reinforced with cellulose nanocrystals: A dynamic mechanical and nanoindentation study. *ACS Applied Materials & Interfaces* 5(22) (2013):11768–11776.

Aguié-Béghin V., Baumberger S., Monties B. et al. Formation and characterization of spread ligin layers at the air water interface. *Langmuir* 18 (2002):5190–5196.

Aguié-Béghin V., Molinari M., Hambardzumyan A. et al. Preparation of ordered films from cellulose nanocrystals. In *Model Cellulosic Surfaces*, ed. M. Roman, pp. 115–136. Washington, DC: ACS, 2009.

Aguié-Béghin V., Foulon L., Soto P. et al. Use of food and packaging model matrices to investigate the antioxydant properties of biorefinery grass lignins. *Journal of Agricultural and Food Chemistry* 63(45) (2015):10022–10031.

Akrman J. and Prikryl J. Application of benzotriazole reactive UV absorbers to cellulose and determining sun protection of treated fabric spectrophotometrically. *Journal of Applied Polymer Science* 108(1) (2008):334–341.

Allen S. G., Tanchak O. M., Quirk A. et al. Surface plasmon resonance imaging of the enzymatic degradation of cellulose microfibrils. *Analytical Methods* 4(10) (2012):3238–3245.

Argyropoulos D. S. and Bolker H. I. Condensation of lignin in dioxane-water-HCl. *Journal of Wood Chemistry and Technology* 7(1) (1987):1–23.

Aulin C., Johansson E., Wagberg L. et al. Self-organized films from cellulose I nanofibrils using the layer-by-layer technique. *Biomacromolecules* 11(4) (2010):872–882.

Barakat A., Winter H., Rondeau-Mouro C. et al. Studies of xylan interactions and cross-linking to synthetic lignins formed by bulk and end-wise polymerization: A model study of lignin carbohydrate complex formation. *Planta* 226(1) (2007):267–281.

Beck-Candanedo S., Roman M., and Gray D. G. Effect of reaction conditions on the properties and behavior of wood cellulose nanocrystal suspensions. *Biomacromolecules* 6(2) (2005):1048–1054.

Berglund L. A. and Peijs T. Cellulose biocomposites from bulk moldings to nanostructured systems. *MRS Bulletin* 35 (2010):201–207.

Bordel D., Putaux J. L., and Heux L. Orientation of native cellulose in an electric field. *Langmuir* 22(11) (2006):4899–4901.

Boukari I., Putaux J. L., Cathala B. et al. In vitro model assemblies to study the impact of lignin-carbohydrate interactions on the enzymatic conversion of xylan. *Biomacromolecules* 10(9) (2009):2489–2498.

Bunzel M. Chemistry and occurrence of hydroxycinnamate oligomers. *Phytochemistry Reviews* 9(1) (2010):47–64.

Caffall K. H. and Mohnen D. The structure, function, and biosynthesis of plant cell wall pectic polysaccharides. *Carbohydrate Research* 344(14) (2009):1879–1900.

Cerclier C., Cousin F., Bizot H. et al. Elaboration of spin-coated cellulose-xyloglucan multilayered thin films. *Langmuir* 26(22) (2010):17248–17255.

Cerclier C., Guyomard-Lack A., Moreau C. et al. Coloured semi-reflective thin films for biomass-hydrolyzing enzyme detection. *Advanced Materials* 23(33) (2011):3791–3795.

Chambat G., Karmous M., Costes M. et al. Variation of xyloglucan substitution pattern affects the sorption on celluloses with different degrees of crystallinity. *Cellulose* 12(2) (2005):117–125.

Cheng L. H., Abd K. A., and Seow C. C. Characterisation of composite films made of konjac glucomannan (KGM), carboxymethyl cellulose (CMC) and lipid. *Food Chemistry* 107(1) (2008):411–418.

Chinga-Carrasco G., Averianova N., Gibadullin M. et al. Micro-structural characterisation of homogeneous and layered MFC nano-composites. *Micron* 44 (2013):331–338.

Chundawat S. P. S., Bellesia G., Uppugundla N. et al. Restructuring the crystalline cellulose hydrogen bond network enhances its depolymerization rate. *Journal of the American Chemical Society* 133(29) (2011):11163–11174.

Cosgrove D. J. and Jarvis M. C. Comparative structure and biomechanics of plant primary and secondary cell walls. *Frontiers in Plant Science* 3 (2012):204.

Cranston E. D. and Gray D. G. Morphological and optical characterization of polyelectrolyte multilayers incorporating nanocrystalline cellulose. *Biomacromolecules* 7(9) (2006):2522–2530.

Cranston E. D. and Gray D. G. Birefringence in spin-coated films containing cellulose nanocrystals. *Colloids and Surfaces A—Physicochemical and Engineering Aspects* 325(1–2) (2008):44–51.

Cranston E. D., Gray D. G., and Rutland M. W. Direct surface force measurements of polyelectrolyte multi layer films containing nanocrystalline cellulose. *Langmuir* 26(22) (2010):17190–17197.

Crestini C., Melone F., Sette M. et al. Milled wood lignin: A linear oligomer. *Biomacromolecules* 12(11) (2011):3928–3935.

Dammstrom S., Salmen L., and Gatenholm P. On the interactions between cellulose and xylan, a biomimetic simulation of the hardwood cell wall. *Bioresources* 4(1) (2009):3–14.

Doherty W. O. S., Mousavioun P., and Fellows C. M. Value-adding to cellulosic ethanol: Lignin polymers. *Industrial Crops and Products* 33(2) (2011):259–276.

Dosch H. and Van de Voorde M. H. Gennesys White Paper: A new european partnership between nanomaterials science & nanotechnology and Synchrotron radiation and Neutron facilities, eds. H. Dosch and M. H. Van de Voorde. Stuttgart, Germany: Max-Planck-Institut für Metallforschung, 2009.

Driemeier C. and Bragatto J. Crystallite width determines monolayer hydration across a wide spectrum of celluloses isolated from plants. *Journal of Physical Chemistry B* 117(1) (2013):415–421.

Dufresne A. Nanocellulose: A new ageless bionanomaterial. *Materials Today* 16(6) (2013):220–227.

Dujardin E. and Mann S. Bio-inspired materials chemistry. *Advanced Engineering Materials* 4(7) (2002):461–474.

Edgar C. D. and Gray D. G. Smooth model cellulose I surfaces from nanocrystal suspensions. *Cellulose* 10 (2003):299–306.

Eichhorn S. J., Dufresne A., Aranguren M. et al. Review: Current international research into cellulose nanofibres and nanocomposites. *Journal of Materials Science* 45(1) (2010):1–33.

Eronen P., Osterberg M., Heikkinen S. et al. Interactions of structurally different hemicelluloses with nanofibrillar cellulose. *Carbohydrate Polymers* 86(3) (2011):1281–1290.

Estephan E., Aguié-Béghin V., Muraille L. et al. Substrate and film structure impacts on adhesion properties between lignocellulosic polymers. In *MRS Symposium Proceedings/MRS Online Proceedings MRS2011*, Boston, MA, 2011.

Fernandes E. M., Pires R. A., Mano J. F. et al. Bionanocomposites from lignocellulosic resources: Properties, applications and future trends for their use in the biomedical field. *Progress in Polymer Science* 38(10–11) (2013):1415–1441.

Ferrer A., Quintana E., Filpponen I. et al. Effect of residual lignin and heteropolysaccharides in nanofibrillar cellulose and nanopaper from wood fibers. *Cellulose* 19(6) (2012):2179–2193.

Fry S. C. *The Growing Plant Cell Wall: Chemical and Metabolic Analysis*. Essex, U.K.: Longman Scientific and Technical, 1988.

Fukuzumi H., Fujisawa S., Saito T. et al. Selective permeation of hydrogen gas using cellulose nanofibril film. *Biomacromolecules* 14(5) (2013):1705–1709.

Girio F. M., Fonseca C., Carvalheiro F. et al. Hemicelluloses for fuel ethanol: A review. *Bioresource Technology* 101(13) (2010):4775–4800.

Gorshkova T., Brutch N., Chabbert B. et al. Plant fiber formation: State of the art, recent and expected progress, and open questions. *Critical Reviews in Plant Sciences* 31(3) (2012):201–228.

Guerra A., Gaspar A. R., Contreras S. et al. On the propensity of lignin to associate: A size exclusion chromatography study with lignin derivatives isolated from different plant species. *Phytochemistry* 68(20) (2007):2570–2583.

Gustafsson E., Johansson E., Wagberg L. et al. Direct adhesive measurements between wood biopolymer model surfaces. *Biomacromolecules* 13(10) (2012):3046–3053.

Guyomard-Lack A., Cerclier C., Beury N. et al. Nano-structured cellulose nanocrystals-xyloglucan multilayered films for the detection of cellulase activity. *European Physical Journal* 213(1) (2012):291–294.

Habibi Y., Foulon L., Aguié-Béghin V. et al. Langmuir-Blodgett films of cellulose nanocrystals: Preparation and characterization. *Journal of Colloid and Interface Science* 316(2) (2007):388–397.

Habibi Y., Heim T., and Douillard R. AC electric field-assisted assembly and alignment of cellulose nanocrystals. *Journal of Polymer Science* 46 (2008):1430–1436.

Habibi Y., Hoeger I., Kelley S. S. et al. Development of Langmuir-Schaeffer cellulose nanocrystal monolayers and their interfacial behaviors. *Langmuir* 26(2) (2010):990–1001.

Hambardzumyan A., Foulon L., Bercu N.B. et al. Organosolv lignin as natural grafting additive to improve the water resistance of films using cellulose nanocrystals. *Chemical Engineering Journal* 264 (2015):780–788.

Hambardzumyan A., Foulon L., Chabbert B. et al. Natural organic UV-absorbent coatings based on cellulose and lignin: Designed effects on spectroscopic properties. *Biomacromolecules* 13(12) (2012):4081–4088.

Hambardzumyan A., Molinari M., Dumelie N. et al. Structure and optical properties of plant cell wall bio-inspired materials: Cellulose-lignin multilayer nanocomposites. *Comptes rendus Biologies* 334(11) (2011):839–850.

Hansen N. M. L. and Plackett D. Sustainable films and coatings from hemicelluloses: A review. *Biomacromolecules* 9(6) (2008):1493–1501.

Hartman J., Albertsson A. C., and Sjoberg J. Surface- and bulk-modified galactoglucomannan hemicellulose films and film laminates for versatile oxygen barriers. *Biomacromolecules* 7(6) (2006):1983–1989.

Hoeger I., Rojas O. J., Efimenko K. et al. Ultrathin film coatings of aligned cellulose nanocrystals from a convective-shear assembly system and their surface mechanical properties. *Soft Matter* 7(5) (2011):1957–1967.

Hoeger I. C., Filpponen I., Martin-Sampedro R. et al. Bicomponent lignocellulose thin films to study the role of surface lignin in cellulolytic reactions. *Biomacromolecules* 13(10) (2012):3228–3240.

Hoije A., Sternemalm E., Heikkinen S. et al. Material properties of films from enzymatically tailored arabinoxylans. *Biomacromolecules* 9(7) (2008):2042–2047.

Holmberg M., Berg J., Stemme S. et al. Surface force studies of Langmuir-Blodgett cellulose films. *Journal of Colloid and Interface Science* 186 (1997):369–381.

Jallabert B., Vaca-Medina G., Cazalbou S. et al. The pressure-volume-temperature relationship of cellulose. *Cellulose* 20(5) (2013):2279–2289.

Jean B., Dubreuil F., Heux L. et al. Structural details of cellulose nanocrystals/polyelectrolytes multilayers probed by neutron reflectivity and AFM. *Langmuir* 24(7) (2008):3452–3458.

Jiang F., Esker A. R., and Roman M. Acid-catalyzed and solvolytic desulfation of H_2SO_4-hydrolyzed cellulose nanocrystals. *Langmuir* 26(23) (2010):17919–17925.

Jiang F. and Hsieh Y. L. Chemically and mechanically isolated nanocellulose and their self-assembled structures. *Carbohydrate Polymers* 95(1) (2013):32–40.

Jiang F., Kittle J. D., Tan X. Y. et al. Effects of sulfate groups on the adsorption and activity of cellulases on cellulose substrates. *Langmuir* 29(10) (2013):3280–3291.

Joffres B., Lorentz C., Vidalie M. et al. Catalytic hydroconversion of a wheat straw soda lignin: Characterization of the products and the lignin residue. *Applied Catalysis B—Environmental* 145 (2014):167–176.

Kentaro A., Shinichiro I., and Hiroyuki Y. Obtaining cellulose nanofibers with a uniform width of 15 nm from wood. *Biomacromolecules* 8 (2007):3276–3278.

Keplinger T., Konnerth J., Aguié-Béghin V. et al. A zoom into the nanoscale texture of secondary cell walls. *Plant Methods* 10 (2014):1.

Khandal D., Mikus P. Y., Dole P. et al. Radiation processing of thermoplastic starch by blending aromatic additives: Effect of blend composition and radiation parameters. *Radiation Physics and Chemistry* 84 (2013):218–222.

Klemm D., Heublein B., Fink H. P. et al. Cellulose: Fascinating biopolymer and sustainable raw material. *Angewandte Chemie, International Edition* 44(22) (2005):3358–3393.

Kohnke T., Ostlund A., and Brelid H. Adsorption of arabinoxylan on cellulosic surfaces: Influence of degree of substitution and substitution pattern on adsorption characteristics. *Biomacromolecules* 12(7) (2011):2633–2641.

Kondo T. Hydrogen bonds in cellulose and cellulose derivatives. In *Polysaccharides*, ed. A. Steinbüchel, 2nd ed., pp. 69–98. Weinheim, Germany: Wiley, 2005.

Kontturi E., Johansson L. S., Kontturi K. S. et al. Cellulose nanocrystal submonolayers by spin coating. *Langmuir* 23(19) (2007):9674–9680.

Koshijima T. and Watanabe T. *Association between Lignin and Carbohydrates in Wood and Other Plant Tissues.* Berlin, Germany: Springer-Verlag, 2003.

Kumagai A., Lee S. H., and Endo T. Thin film of lignocellulosic nanofibrils with different chemical composition for QCM-D study. *Biomacromolecules* 14(7) (2013):2420–2426.

Lahiji R. R., Boluk Y., and McDermott M. Adhesive surface interactions of cellulose nanocrystals from different sources. *Journal of Materials Science* 47(9) (2012):3961–3970.

Lange H., Decina S., and Crestini C. Oxidative upgrade of lignin—Recent routes reviewed. *European Polymer Journal* 49(6) (2013):1151–1173.

Li F., Biagioni P., Bollani M. et al. Multi-functional coating of cellulose nanocrystals for flexible packaging applications. *Cellulose* 20(5) (2013):2491–2504.

Li Q. and Renneckar S. Supramolecular structure characterization of molecularly thin cellulose I nanoparticles. *Biomacromolecules* 12(3) (2011):650–659.

Linder A., Bergman R., Bodin A. et al. Mechanism of assembly of xylan onto cellulose surfaces. *Langmuir* 19 (2003):5072–5077.

Lucia L. A. and Rojas O. J. *The Nanoscience and Technology of Renewable Biomaterials.* Chichester, U.K.: John Wiley & Sons, 2009.

Lundquist K. Wood. In *Methods in Lignin Chemistry*, eds. C. W. Dence and S. Y. Lin, pp. 65–69. Berlin, Germany: Springer-Verlag, 1992.

Martin-Sampedro R., Rahikainen J. L., Johansson L. S. et al. Preferential adsorption and activity of monocomponent cellulases on lignocellulose thin films with varying lignin content. *Biomacromolecules* 14 (2013):1231–1239.

Maurer S. A., Bedbrook C. N., and Radke C. J. Cellulase adsorption and reactivity on a cellulose surface from flow ellipsometry. *Industrial & Engineering Chemistry Research* 51(35) (2012):11389–11400.

Méchin V., Baumberger S., Pollet B. et al. Peroxidase activity can dictate the in vitro lignin dehydrogenative polymer structure. *Phytochemistry* 68(4) (2007):571–579.

Mengmeng Z., Sheng-Cheng W., Wen Z. et al. Imaging and measuring single-molecule interaction between a carbohydrate-binding module and natural plant cell wall cellulose. *The Journal of Physical Chemistry B* 116 (2012):9949–9956.

Miao C. W. and Hamad W. Y. Cellulose reinforced polymer composites and nanocomposites: A critical review. *Cellulose* 20(5) (2013):2221–2262.

Mikkonen K. S., Heikkila M. I., Helen H. et al. Spruce galactoglucomannan films show promising barrier properties. *Carbohydrate Polymers* 79(4) (2010):1107–1112.

Mikkonen K. S., Rita H., Helen H. et al. Effect of polysaccharide structure on mechanical and thermal properties of galactomannan-based films. *Biomacromolecules* 8(10) (2007):3198–3205.

Moon R. J., Martini A., Nairn J. et al. Cellulose nanomaterials review: Structure, properties and nanocomposites. *Chemical Society Reviews* 40(7) (2011):3941–3994.

Mulder W. J., Gosselink R. J. A., Vingerhoeds M. H. et al. Lignin based controlled release coatings. *Industrial Crops and Products* 34(1) (2011):915–920.

Muraille L., Pernes M., Habrant A. et al. Impact of lignin on water sorption of bioinspired self-assemblies of lignocellulosic polymers. *European Polymer Journal* 64 (2015):21–35.

Nishiyama Y. Structure and properties of the cellulose microfibril. *Journal of Wood Science* 55(4) (2009):241–249.

Nogi M., Iwamoto S., Nakagaito A. N. et al. Optically transparent nanofiber paper. *Advanced Materials* 21(16) (2009):1595.

Nogi M. and Yano H. Optically transparent nanofiber sheets by deposition of transparent materials: A concept for a roll-to-roll processing. *Applied Physics Letters* 94(23) (2009):233117.

Norgren M., Notley S. M., Majtnerova A. et al. Smooth model surfaces from lignin derivatives. I. Preparation and characterization. *Langmuir* 22(3) (2006):1209–1214.

Notley S. M., Eriksson M., Wgberg L. et al. Surface forces measurements of spin-coated cellulose thin films with different crystallinity. *Langmuir* 22(7) (2006):3154–3160.

Notley S. M. and Norgren M. Surface energy and wettability of spin-coated thin films of lignin isolated from wood. *Langmuir* 26(8) (2010):5484–5490.

Notley S. M. and Norgren M. Study of thin films of kraft lignin and two DHPs by means of single-molecule force spectroscopy (SMFS). *Holzforschung* 66(50) (2012):615–622.

Olszewska A. M., Kontturi E., Laine J. et al. All-cellulose multilayers: Long nanofibrils assembled with short nanocrystals. *Cellulose* 20(4) (2013):1777–1789.

Paës G., Burr S., Saab M. B. et al. Modeling progression of fluorescent probes in bioinspired lignocellulosic assemblies. *Biomacromolecules* 14(7) (2013):2196–2205.

Paës G. and Chabbert B. Characterization of arabinoxylan/cellulose nanocrystals gels to investigate fluorescent probes mobility in bioinspired models of plant secondary cell wall. *Biomacromolecules* 13(1) (2012):206–214.

Peng X. W., Ren J. L., Zhong L. X. et al. Nanocomposite films based on xylan-rich hemicelluloses and cellulose nanofibers with enhanced mechanical properties. *Biomacromolecules* 12(9) (2011):3321–3329.

Peng Y. C., Gardner D. J., Han Y. et al. Influence of drying method on the material properties of nanocellulose I: Thermostability and crystallinity. *Cellulose* 20(5) (2013):2379–2392.

Pereira A. A., Martins G. F., Antunes P. A. et al. Lignin from sugar cane bagasse: Extraction, fabrication of nanostructured films, and application. *Langmuir* 23(12) (2007):6652–6659.

Podsiadlo P., Choi S. Y., Shim B. et al. Molecularly engineered nanocomposites: Layer-by-layer assembly of cellulose nanocrystals. *Biomacromolecules* 6(6) (2005):2914–2918.

Podsiadlo P., Sui L., Elkasabi Y. et al. Layer-by-layer assembled films of cellulose nanowires with antireflective properties. *Langmuir* 23(15) (2007):7901–7906.

Prado R., Erdocia X., and Labidi J. Lignin extraction and purification with ionic liquids. *Journal of Chemical Technology and Biotechnology* 88(7) (2013):1248–1257.

Ralph J. 2007. Lignins. In *Encyclopedia of Life Science*, eds. F. Rose and K. Osborne. Chichester, U.K.: John Wiley & Sons.

Ramanen P., Penttila P. A., Svedstrom K. et al. The effect of drying method on the properties and nanoscale structure of cellulose whiskers. *Cellulose* 19(3) (2012):901–912.

Revol J. F., Godbout L., Dong X. M. et al. Chiral nematic suspensions of cellulose crystallites—Phase separation and magnetic-field orientation. *Liquid Crystals* 16(1) (1994):127–134.

Roland J. C., Reis D., and Vian B. Liquid crystal order and turbulence in the planar twist of the growing plant cell walls. *Tissue and Cell* 24(3) (1992):335–345.

Roman M. Model cellulosic surfaces: History and recent advances. In *Model Cellulosic Surfaces*, ed. M. Roman, pp. 3–53. Washington, DC: ACS, 2009.

Rutland M. W., Carambassis A., Willing G. A. et al. Surface force measurements between cellulose surfaces using scanning probe microscopy. *Colloids and Surfaces, A: Physicochemical and Engineering Aspects* 123/124 (1997):369–374.

Ryberg Y. Z. Z., Edlund U., and Albertsson A. C. Conceptual approach to renewable barrier film design based on wood hydrolysate. *Biomacromolecules* 12(4) (2011):1355–1362.

Samir A., Alloin F., and Dufresne A. Review of recent research into cellulosic whiskers, their properties and their application in nanocomposite field. *Biomacromolecules* 6(2) (2005):612–626.

Saxena A., Elder T. J., and Ragauskas A. J. Moisture barrier properties of xylan composite films. *Carbohydrate Polymers* 84(4) (2011):1371–1377.

Scheller H. V. and Ulvskov P. Hemicelluloses. *Annual Review of Plant Biology* 61 (2010):263–289.

Sedlmeyer F. B. Xylan as by-product of biorefineries: Characteristics and potential use for food applications. *Food Hydrocolloids* 25(8) (2011):1891–1898.

Sehaqui H., Zhou Q., Ikkala O. et al. Strong and tough cellulose nanopaper with high specific surface area and porosity. *Biomacromolecules* 12(10) (2011):3638–3644.

Sipponen M. H., Lapierre C., Méchin V. et al. Isolation of structurally distinct lignin-carbohydrate fractions from maize stem by sequential alkaline extractions and endoglucanase treatment. *Bioresource Technology* 133 (2013):522–528.

Stepan A. M., Anasontzis G. E., Matama T. et al. Lipases efficiently stearate and cutinases acetylate the surface of arabinoxylan films. *Journal of Biotechnology* 167(1) (2013):16–23.

Stevanic J. S., Bergstrom E. M., Gatenholm P. et al. Arabinoxylan/nanofibrillated cellulose composite films. *Journal of Materials Science* 47(18) (2012):6724–6732.

Suchy M., Linder M. B., Tammelin T. et al. Quantitative assessment of the enzymatic degradation of amorphous cellulose by using a quartz crystal microbalance with dissipation monitoring. *Langmuir* 27(14) (2011):8819–8828.

Tang H., Guo B., Jiang H. T. et al. Fabrication and characterization of nanocrystalline cellulose films prepared under vacuum conditions. *Cellulose* 20(6) (2013):2667–2674.

Ten E., Ling C., and Wang Y. Lignin nanotubes as vehicles for gene delivery into human cells. *Biomacromolecules* 15(1) (2014):327–338.

Tobimatsu Y., Elumalai S., Grabber J. H. et al. Hydroxycinnamate conjugates as potential monolignol replacements: In vitro lignification and cell wall studies with rosmarinic acid. *Chemsuschem* 5(4) (2012):676–686.

Vogel J. Unique aspects of the grass cell wall. *Current Opinion in Plant Biology* 11(3) (2008):301–307.

Voronova M. I., Surov O. V., and Zakharov A. G. Nanocrystalline cellulose with various contents of sulfate groups. *Carbohydrate Polymers* 98(1) (2013):465–469.

Wang C., Qian C., Roman M. et al. Surface-initiated dehydrogenative polymerization of monolignols: A quartz crystal microbalance with dissipation monitoring and atomic force microscopy study. *Biomacromolecules* 14(11) (2013):3964–3972.

Whitney S. E. C., Wilson E., Webster J. et al. Effects of structural variation in xyloglucan polymers on interactions with bacterial cellulose. *American Journal of Botany* 93(10) (2006):1402–1414.

Willfor S., Sundberg A., Pranovich A. et al. Polysaccharides in some industrially important hardwood species. *Wood Science and Technology* 39(8) (2005):601–617.

Ying R. F., Rondeau-Mouro C., Barron C. et al. Hydration and mechanical properties of arabinoxylans and beta-D-glucans films. *Carbohydrate Polymers* 96(1) (2013):31–38.

Yoshiharu N., Shigenori K., Masahisa W. et al. Cellulose microcrystal film of high uniaxial orientation. *Macromolecules* 30 (1997):6395–6397.

Zeder-Lutz G., Renau-Ferrer S., Aguié-Béghin V. et al. Novel surface-based methodologies for investigating GH11 xylanase-lignin derivative interactions. *Analyst* 138(22) (2013):6889–6899.

Zhang Y. P., Chodavarapu V. P., Kirk A. G. et al. Structured color humidity indicator from reversible pitch tuning in self-assembled nanocrystalline cellulose films. *Sensors and Actuators B* 176 (2013):692–697.

Zhou Q., Rutland M. W., Teeri Tuula T. et al. Xyloglucan in cellulose modification. *Cellulose* 14(6) (2007):625–641.

9

Films and Coatings from Chitosan

Gloria Sánchez, María José Fabra, José María Lagarón, and María José Ocio

CONTENTS

Antimicrobial packaging is a very promising system for the future improvement of food quality and preservation during processing and storage. In fact, antimicrobial packaging can be helpful in extending the shelf life of food products. Chitosan is a versatile and promising biodegradable polymer for food packaging applications since it presents an inherent antimicrobial character against the growth of pathogen and spoilage microorganisms. In this chapter, state-of-the-art information regarding the mechanism of action and the main factors affecting the antimicrobial properties of chitosan films have been summarized. Besides, changes on the biocide properties of the compound when incorporated into other polymer or when tested in food products have been also reviewed.

9.1 Introduction

Since society is experiencing a trend toward "green consumerism," with a desire of fewer synthetic food additives and products with a smaller impact on the environment, the use of naturally derived antimicrobials has substantially increased in the last decade.

In this sense, chitosan, which is a linear copolymer of β-1,4-linked D-glucosamine and N-acetyl-D-glucosamine, has attracted much attention. The term chitosan describes a heterogeneous group of polymers differing in molecular weight (Mw), viscosity, degree of deacetylation, pK_a, etc. Chitosan is obtained by partial alkaline N, the second most abundant natural polymer in nature after cellulose is widely distributed in nature (e.g., exoskeleton of crustaceans, insects, and certain fungi). It is mostly available from waste bioproducts in the shellfish industry, and therefore, abundant commercial suppliers are offered currently. It can also be obtained from the chitin component of fungal cell walls.

Chitosan has excellent film- and coating-forming properties when coming from organic acidic water solutions. Due to its unique biological characteristics, including biodegradability and nontoxicity, many applications have been found either alone or blended with other natural polymers (starch, gelatin, alginates) in the food, pharmaceutical, textile, agriculture, water treatment, and cosmetics industries (reviewed by Kong et al., 2010).

Several studies have already demonstrated the antibacterial and antifungal action of this compound for both bioactive preservative and bioactive packaging applications (Fernandes et al., 2008; Li et al., 2007; No et al., 2002, 2007; Qin et al., 2006; Rhoades and Roller, 2000). The virucide activity of chitosan able to be incorporated into food packaging has only been evaluated on virus suspensions (Davis et al., 2012; Su et al., 2009), and no one has investigated their virucide activity when incorporated in food packaging.

Although the exact mechanism by which chitosan exerts its antimicrobial activity is currently unknown, it has been suggested that the polycationic nature of this polymer that forms from acidic solutions below pH 6.5 is a crucial factor. Thus, it has been proposed that the positively charged amino groups of the glucosamine units interact with negatively charged components in microbial cell membranes altering their barrier properties, thereby preventing the entry of nutrients or causing the leakage of intracellular contents (Fernandez-Saiz et al., 2006, 2008). Another reported mechanism involves the penetration of low-Mw chitosan in the cell, the binding to DNA, and the subsequent inhibition of RNA and protein synthesis (Hadwiger et al., 1985). Chitosan has also been shown to activate several defense processes in plant tissues, and it inhibits the production of toxins and microbial growth due to its ability to chelate metal ions (Cuero et al., 1991; El Ghaouth et al., 1992).

9.2 Mechanism of Antimicrobial Action of Chitosan

While there is no doubt that chitosan has shown a tremendous potential as antimicrobial component in a number of cases, a clear understanding between the phenomenology of the chitosan biopolymer as a biocide agent and its molecular structure, functional chemistry, and physical state for optimum performance is of paramount importance from an applied view point. Attenuated total reflectance infrared spectroscopy (ATR-FTIR) has become one of the most versatile techniques for the quick identification and characterization of substances because it gathers information from virtually any type of sample form and shape, provided that adequate accessories are employed, and, furthermore, it can be used to, for instance, follow film-forming processes and other in situ processes. Fernandez-Saiz et al. (2006) and Lagaron et al. (2007) studied the mechanism of antimicrobial action of chitosan using ATR-FTIR spectroscopy. The intensity of the carboxylate biocide groups ($-NH_3^+ -OOCH$) was studied at 1546 and at 1405 cm^{-1} in various chitosan samples. They demonstrated that high-Mw chitosan appears to act only in its acetate (salt) form and as carrier of the activated protonated species, in which bypassing from the film to the microbial solution leads to the inactivation of certain microorganisms. Moreover, a novel methodology based on the use of a normalized infrared band centered at 1405 cm^{-1} was proposed, which can be correlated with the antimicrobial character of the biopolymer.

In relation to the feasible mechanisms by which chitosonium acetate decreases the number of "active" protonated groups, Lagaron et al. (2007) argued two feasible mechanisms. First, the biocide carboxylate groups that form when chitosan is cast from acetic acid solutions are being continuously evaporated from the formed film in the form of acetic acid (mechanism I) in the presence of environmental humidity, rendering weak biocide film systems. Second, upon direct contact of the cast chitosonium acetate film with liquid water, water solutions, or the high moisture TSA hydrogel, a positive rapid migration, with diffusion coefficient faster than 3.7×10^{-12} m^2 s^{-1}, of protonated glucosamine water-soluble molecular fractions (mechanisms II) takes place from the film into the liquid phase, yielding strong antimicrobial performance and leaving in the remaining cast film only the non-water-soluble chitosan fractions.

In a more recent work, Fernandez-Saiz et al. (2008) showed again that only dissolved polysaccharide molecules from the gliadin–chitosonium acetate blends were able to act as the antimicrobial agent. These observations are in accordance with those obtained by Zhai et al. (2004), who evaluated the antimicrobial capacity of starch/chitosan blends. In this particular case, films presented significant biocide properties even with a low content of chitosan, but uniquely when formed under the action of irradiation. Degradation of chitosan by the irradiation treatment improved its dissolution into the nutrient broth, enabling a detectable bactericidal effect. These results appear to contradict, however, a previous study performed by Vartiainen et al. (2005), on the antimicrobial capacity of chitosan immobilized in a BOPP matrix, where chitosan was reported to preserve a strong bactericidal activity without any notable leaching. This finding was probably obtained because bacteria were directly exposed over the film, and therefore, no migration was required for the microbes to be exposed. Tanabe et al. (2002) reported that the antimicrobial capacity of chitosan was probably due to the irreversible adsorption of bacteria into the film. On the contrary, several studies have reported that chitosan acts more efficiently from its solution form by precipitation onto, or even by penetration through, the microbial membrane (Chung et al., 2004; Fernandez-Saiz et al., 2006; Liu et al., 2004; Qin et al., 2006).

However, it seems essential for preservation of the chitosonium acetate full biocide capacity to maintain during storage a minimum number of losses in the activated ionic species present in the as-cast film. Previous studies have also suggested that the type of acid used for the solubilization of chitosan does also have an influence on this and other structural processes (Demarger-Andre and Domard, 1994). Thus, by controlling the film-forming process and storage conditions, one should be capable of optimizing biocide capacity.

9.3 Optimization of the Biocide Properties of Chitosan

9.3.1 Effects of Chitosan Formulation

Researchers used chitosan from various sources and with varying physicochemical properties. Thus, the question arises as to how to globally produce chitosans with consistent properties. Each batch of chitosan produced from the same manufacturer may differ in its quality. For proper quality control in the chitosan production, there is a critical need to establish less expensive and reliable analytical methods, especially for the evaluation of Mw and degree of deacetylation (No et al., 2007). Moreover, the traditional chitosan production involves deproteinization, demineralization, decolorization, and deacetylation. The physicochemical characteristics of chitosan can be variously affected by production methods and crustacean species. Any modification of these methodologies could affect chitosan physicochemical and functional properties (i.e., Mw, viscosity, degree of deacetylation, hydrophilicity, water and fat absorption capacities) and therefore influence the effectiveness of chitosan as antimicrobial agent (No et al., 2007).

9.3.1.1 Effect of Molecular Weight of Chitosan

Different studies evaluating the influence of Mw of chitosan have demonstrated that COS, which are soluble in water, were the least effective in terms of biocide properties (No et al., 2002; Uchida et al., 1989; Zivanovic et al., 2004). Qin et al. (2006) have evaluated the antimicrobial capacity of chitosan solutions against the growth of *Candida albicans*, *Escherichia coli*, and *Staphylococcus aureus* showing that only water-insoluble chitosan in organic acidic solutions, that is, chitosonium salts, exhibit efficient antibacterial activity. Moreover, these authors reported that of the chitosan materials tested, that is, 310–375 and 50–190 kDa, they did not lead to significant variations in biocide properties. Therefore, in spite of the great number of reports on this subject, the reported conclusions are divergent probably since the methodologies used for the susceptibility tests vary from work to work. Besides, as mentioned earlier, published works are not all comparable as individual researchers used chitosans with varying physicochemical properties (affected by production methods) and perhaps from various sources (No et al., 2007).

9.3.1.2 Effect of the Acid Concentration to Obtain Chitosan Dispersions

Although several studies report the influence of the organic solvent employed to obtain chitosan films on their physical or antimicrobial properties (Caner, 2005; Kim et al., 2006; Park et al., 2002), the effect of the acetic acid concentration to obtain chitosan dispersions has been scarcely studied in the literature. Only Fernandez-Saiz et al. (2009a) compared the biocide properties of films of chitosonium acetate obtained from 0.5 and 1 wt.% of acetic acid in the solution. Although no statistical differences were obtained between samples, a certain trend could also be observed in which chitosan dispersions prepared from an acid–water solution with 1% of acetic acid have demonstrated to lead to matrices with superior antimicrobial properties. A feasible explanation may be the fact that solubilization of chitosan is influenced by acetic acid concentration in the solution. In this sense, a previous work performed by Rinaudo et al. (1999) demonstrated through potentiometric assays that complete solubilization of chitosan in acidic water caused by a maximum degree of protonation occurs with a stoichiometry [AcOH]/[Chit-NH$_2$] = 0.6. Such ratio corresponds to the one used in the mentioned work, that is, for chitosan dispersions of 1.5 wt.% of polymer and 1 wt.% of acetic acid.

9.3.1.3 Effect of Film Neutralization

Generally, chitosan is dissolved in acidic solution to form chitosan solution. Thereafter, the samples are neutralized in either sodium hydroxide aqueous solution or ethanol solution. The neutralization is a necessary step before the rehydration of the chitosan matrices. However, there is still very limited information about the effects of the different neutralization methods on the antimicrobial activity.

When the biocide properties of a neutralized chitosan film were studied, no antimicrobial activity was observed, indicating that amine groups are no longer activated by titration with an alkaline component (Fernandez-Saiz et al., 2006; Ouattara et al., 2000). Neutralization results in chitosan inactivation of the active groups and in the loss of biopolymer solubility in aqueous media. In apparent opposition to these results, Tang et al. (2003) obtained a bactericidal effect toward *S. aureus* and *E. coli* when measured biocide properties for suspensions of both neutralized chitosan films and neutralized cross-linked chitosan films. This may be due to either the incomplete neutralization process of the tested films or a protonation reaction during the preparation of the studied film suspensions since the solvent composition to obtain the suspensions was not specified in this study.

9.3.1.4 Effect of Film Sterilization

For chitosan matrices designed to be applied as food contact materials, in many cases a humid heat sterilization process becomes necessary. Even though the sterilization of chitosan solutions showed no significant impact on the biocide properties of the cast material, the antimicrobial activity of just formed autoclaved films (with and without steam contact) demonstrated that the sterilized films lost their biocide properties by this treatment (Fernandez-Saiz et al., 2009). Both dry and humid heat sterilization processes of a cast film did cause to the resultant biopolymer an increase in yellowness or even an important browning effect, respectively, probably due to the formation of insaturations or colored complexes (Marreco et al., 2005; Zotkin et al., 2004). These chemical changes are thought to be responsible for the insoluble character of the annealed samples.

9.3.1.5 Effect of the Thickness of the Film

Fernandez-Saiz et al. (2009) analyzed this variable and showed no significant differences in chitosan films ranged from 25 to 100 μm. When the ATR-FTIR spectra were obtained for these particular samples, only minor differences were obtained between the different samples in terms of the intensity of the active carboxylate active bands. These results suggest that the amount of protonated groups in the final film does not seem to depend on thickness for the range evaluated in this particular work. Thus, even though the films were thicker (and that entails among other things that the corresponding castings take longer to dry out), given the fact that chitosan presents a very strong hydrophilic character, water diffuses very rapidly into the film and an immediate release of a significant fraction of the carboxylate groups is produced. This event leads to a maximal antimicrobial effect regardless of the thickness of the biopolymer in the range studied.

9.3.2 Effects of the Methods Used for Determining the Biocide Activity

Another important issue is the lack of proper standardized methods to determine the effectiveness of antimicrobial polymers. According to the standard method described by the National Committee for Clinical Laboratory Standards (NCCLS) (NCCLS, 1999), minimal inhibitory concentration (MIC) is the lowest concentration of an antimicrobial agent that inhibits visible growth, after a predetermined incubation period (usually 18–24 h). The minimal bactericide concentration corresponds to the lowest concentration of an antimicrobial agent that causes at least a 3 log reduction in the number of surviving cells after incubation and is quantified by subculture of all nonturbid tubes. When Fernandez-Saiz et al. (2009) studied the biocide properties of chitosonium acetate solutions or films, some turbidity was detected in the test tubes containing the nutrient broth even before bacterial inoculation. This additional turbidity causes an overestimation of bacterial final concentrations when calculated by optical density. Furthermore, important differences on turbidity do not necessarily correspond to significant differences in bacterial counts. For that reason, some previous published works on the antimicrobial capacity of chitosan films showed alterations or even a lack of performance when evaluated by optical density (Coma et al., 2002; Fernandes et al., 2008; Jeon et al., 2001; Liu et al., 2006). Devlieghere et al. (2004) also used the turbidimetric method, but in this particular work, the culture media with chitosan solutions were previously centrifuged to remove formed complexes between the polysaccharide and components in the media so that visual detection of growth was possible. Nevertheless, and due to this "solution clearing" procedure, the antibacterial properties of chitosan could have been underestimated since migrated chitosan fractions could have been eliminated from the solution.

The biocide properties of chitosan films (Pranoto et al., 2005) and chitosan solutions (Coma et al., 2002) have also been investigated by the agar diffusion method with no satisfactory results. Zivanovic et al. (2005) evaluated the antimicrobial capacity of chitosan films enriched with essential oils (EOs) by the agar diffusion test. These authors concluded that chitosan films without EOs did not exert antimicrobial activity when they were tested on an agar surface. The authors concluded that possibly the initial counts of inoculum were too high or that no dissolution of chitosan took place. On the other hand, other studies have showed considerable antimicrobial activity when tested by means of other techniques, such as macrodilution method and agar plate counts or microcalorimetry (Fernandez-Saiz et al., 2006; No et al., 2002; Qin et al., 2006).

9.3.2.1 Effect of Initial Inoculum Size

A recent work performed by Fernandez-Saiz et al. (2009) indicated that the lower initial bacterial populations led to a higher antimicrobial performance of chitosan. The result agrees well with the work performed by Tsai and Su (1999), who studied the effect of NaCl on the antibacterial activity of chitosan against *E. coli* at different initial cell densities. Another work reported the action of chitosan films on bologna slices against the growth of *Listeria monocytogenes*. This work showed that there is a relation between the initial bacterial number and the biocide properties of the film (Zivanovic et al., 2005). These results are in agreement with the premises reported by the NCCLS in which the inoculum size is considered to be the single most important factor in susceptibility testing, and changes in this variable may alter results by one or more dilutions (NCCLS, 1999).

9.3.2.2 Effect of the Physiological State of the Bacteria

Fernandez-Saiz et al. (2009) showed that the sensitivity of *S. aureus* seems to be higher when bacteria were inoculated in mid-log phase. A similar effect was found by Tabak et al. (2007), who studied the influence on the growth phase of *Salmonella typhimurium* using triclosan as the antimicrobial agent. Liu et al. (2006) obtained comparable results when testing chitosonium acetate solutions at different growth stages of *E. coli*. However, not all of the literature is coincident on this issue. For instance, the aforementioned results do not agree with those obtained by Chen and Chou (2005) who studied the effect of water-soluble lactose chitosan derivative against *S. aureus* growth in different physiological states, obtaining a greater susceptibility when late-log phase cells were tested. According to Tsai and Su (1999), the susceptibility of *E. coli* cells to chitosan may depend on the cells' surface electronegativity, which progressively increases from the early exponential phase to the late exponential phase. At the beginning of the stationary phase, this parameter starts to decrease. Nevertheless, Tsai et al. (2006) showed that *Bacillus cereus* cells in stationary phase were less sensitive to the action of low-Mw chitosan. In any case, a reduction of susceptibility with cell age was generally observed, which suggests that cells in stationary phase are probably more able to implement adaptive stress responses. In addition, Fux et al. (2005) suggested that the observed tolerance seems a reversible phenomenon, which proves that it must be caused by phenotypic changes rather than by the product of genetic alterations.

9.3.2.3 Influence of Microbial Species

Although owning a broad spectrum of antimicrobial activity, chitosan exhibits differing inhibitory efficiency against different microorganisms. Overall, chitosan generally showed stronger antibacterial activity against gram-positive bacteria than gram-negative bacteria (Fernandez-Saiz et al., 2009; Jeon et al., 2001; No et al., 2002, 2006). It is worth noting that whereas some antimicrobial agents (for instance, bacteriocins) appear not to be able to act against gram-negative bacteria due to the resistance conferred by the outer membrane (Rodriguez et al., 2005), chitosan has demonstrated to have a significant antimicrobial effect against the growth of gram-negative bacteria such as *Salmonella* spp. These results are in accordance with the hypothesis of an electrostatic interaction between chitosan and the cell wall. In this sense, gram-positive microorganisms should be more susceptible than gram-negative since their wall is composed by a thick peptidoglycan layer and by polymers called teichoic acids. This teichoic acid backbone is highly charged by phosphate groups with negative charge, which could establish electrostatic interactions with cationic antimicrobial compounds such as the chitosan salts. On the other hand, a study performed by Devlieghere et al. (2004) showed that gram-negative bacteria seemed to be generally more sensitive to chitosan solutions (MIC $\leq 0.006\%$ w/v) than gram-positive bacteria. In addition, Chung et al. (2004) supported further in their work the major susceptibility of the gram-negative bacteria to chitosan on the reasoning that gram-negative bacteria possess higher negative charge values on the cell surface. Nevertheless, because of the variability of the methods used, further investigations are required to fully unravel the mechanisms of interaction between chitosan and microbial cell wall.

9.3.2.4 Influence of the pH

Chitosan is normally insoluble in aqueous solutions with pH value above 7; however, the molecules become soluble in dilute acids (pH < 6). pH has demonstrated to present a large effect on the antimicrobial effectiveness of chitosan, showing that at lower pHs stronger antibacterial activity was reported (Fernandez-Saiz et al., 2009; No et al., 2002). These findings are related to the particular pK_a of chitosan (i.e., 6.4), which is near to the pH 6.2 value. At this pH, the amount of positively charged amino groups (active groups) is close to 75%, while at pH of 7.4, this quantity drops until approximately 10% (Igarashi and Nakano, 2003). Additionally, it has been published that at lower pH values, the physiology of the cell is more prone to suffer damage. Thus, at lower pH conditions, the carboxyl and phosphate groups of the bacterial surface are anionic and offer potential sites for electrostatic binding of chitosan (Helander et al., 2001).

9.3.2.5 Effect of Temperature

In a recent research on the antimicrobial properties of medium-Mw chitosan against *Salmonella enterica*, chitosan presented strong antimicrobial activity at 10°C, while the inhibition was almost inexistent at

20°C (Marques et al., 2008). Likewise, the addition of low-Mw chitosan in cooked rice samples inhibited or retarded increases in total aerobic counts and *B. cereus*, being this effect more significant at lower temperatures (Tsai et al., 2006). Thus, at low temperatures, even if bacterial cells presented an optimal growth, they could be somehow injured, and these damages hampered their repair mechanisms in the presence of chitosan, showing a higher susceptibility to the polymer at this condition. Contrarily, Chen and Chou (2005) observed that as the temperature was increased from 5°C to 37°C, the population of *S. aureus* in the chitosan derivative–containing deionized water decreased. Furthermore, Fernandez-Saiz et al. (2009) demonstrated that *L. monocytogenes* showed to be more resistant to the biocide action of chitosan at 12°C than at other studied temperatures (4°C and 37°C). The reason of such behavior could be since this microorganism is known to adapt to lower temperatures by producing phospholipids with shorter and more branched fatty acids. Also, the high susceptibility at 37°C could be attributed to exhaustion caused by efforts to multiply at a growth promoting temperature (Simpson et al., 2008). Similarly, Briandet et al. (1999) found that *L. monocytogenes* grown at 8°C had significantly less surface electronegativity than cells grown at 15°C, 20°C, and 37°C. They speculated that the reduced charge at 8°C was the result of synthesis of acclimation proteins and fewer carboxylic groups in the cell wall composition.

9.4 Effect of Film-Forming Conditions and Storage on Physicochemical Properties of Chitosan

From an application point of view, the effectiveness of an antimicrobial agent (for instance, chitosan) when applied in or coated over foods or packaging materials may deteriorate during film forming, distribution, and storage. Hence, the chemical stability of an incorporated substance is likely to be affected by the forming process or the storage conditions until its use. Consequently, in order to obtain highly efficient biocides, all these parameters should be assessed and optimized for the final application.

The effects of film-forming conditions and storage conditions on mechanical and barrier properties of chitosan-based films have been investigated to some extent. Thus, Butler et al. (1996) demonstrated that water vapor permeability and elongation at break of low-Mw chitosan-based films decreased with storage time at environment conditions (23°C and 50%RH). Also, it has been shown that a thermal treatment of 120°C during 3 h of chitosan films led to a significant strengthening of the films and reduced their solubility in aqueous media (Zotkin et al., 2004). This last effect was also observed after storage of chitosan films at various temperature and relative humidity conditions (Kam et al., 1999; Ritthidej et al., 2002).

Concerning to film-forming conditions, Srinivasa et al. (2004) demonstrated that infrared illumination drying showed to be faster and superior in preserving water or oxygen barrier properties than those prepared by oven-drying or room temperature drying. Furthermore, chemical changes of chitosan salts during film formation and the effect of storage conditions have been extensively evaluated through infrared spectroscopy (Demarger and Domard, 1994; Fernandez-Saiz et al., 2006; Kam et al., 1999; Lagaron et al., 2007; Osman and Arof, 2003; Zotkin et al., 2004).

Fernandez-Saiz et al. (2009) have extensively evaluated the optimal procedure to preserve the full biocide capacity of chitosan-based films with minimum losses of the activated antimicrobial species in the materials. Thus, it has been demonstrated that just formed low-Mw chitosonium acetate films present significant biocide properties against *S. aureus* and *Salmonella* when cast at 37°C or 80°C, while this capacity is reduced when cast at 120°C. In addition, the films preserved significant biocide properties when maintained at low temperatures and dry conditions (4°C, 23°C, 0%RH) during the storage time. On the other hand, physical and antimicrobial capacity of these films were greatly affected when they were maintained at high relative humidity conditions (i.e., 75%) or higher temperatures (i.e., 37°C), showing a progressive yellow coloration and a gradual loss of their antimicrobial capacity, due to water resistance induced by chemical and/or physical alterations in the films. These color changes can be ascribed to nonenzymatic browning reactions, which led to the presence of conjugated double bonds in the structure of the polymer. Similarly, Larena and Caceres (2004) also studied the color changes of chitosan films after a storage period and obtained a decrease in the membrane transmittance with temperature. Furthermore, Srinivasa et al. (2004) also observed an increase in yellowness in oven-dried chitosan films. The same findings were reported by Kam et al. (1999) when chitonium acetate films for 2 h at 120°C are heated. In this latter work, color changes were more intense with higher amounts of acetic acid in the films, which suggests that color formation is an indicator visible factor underlying chemical alterations in the biomaterial. These alteration processes result in a substantial loss in the release of antimicrobial protonated species as a result of cross-linking and/or other molecular entanglement processes and/or by chemical alterations of a more hydrophobic nature in the biomaterial. In line with these findings, Fernandez-Saiz et al. (2009) observed by ATR-FTIR spectroscopy that chitosonium acetate films cast at 120°C or stored at 75%RH preserved a significant fraction of the biocide carboxylate chemistry after contact with water due to a strong reduction in cast film solubility. Since biopolymer active species migration from the film to the culture media is needed for the biomaterial to exhibit measurable antimicrobial effect, a proper control of temperature and humidity during film formation and storage is necessary to design the optimum performance of chitosan as a biocide.

9.5 Water-Resistant Chitosan-Based Films

As known, chitosan forms films with excellent oxygen barrier properties, good mechanical properties, and good antimicrobial activity. However, a major drawback of chitosan is its high water vapor permeability. Biopolymer films made from mixtures of different biopolymers may advantageously use the distinct functional characteristics of each film-forming ingredient. Thus, in the design of a chitosan-based packaging system for food packaging applications, the presence of the high levels of humidity typically found in many foods has to be taken into account. In this sense, it has been found that the development of chitosan active coatings leads to both a very rapid release of the biocide groups onto the food surface and the film partial or total dissolution. In order to solve this problem, physical and chemical treatments must be taken into account. These actions could prevent film disintegration in water, and, at the same time, dissolution of the protonated glucosamine groups in water would be blocked or deactivated, hence reducing the biocide character of the formulation. In this sense, the incorporation of cross-linking agents such as glutaraldehyde, glyoxal, or epichlorohydrin forms covalent bonds between the chains of chitosan preventing polysaccharide–water interactions that could be a means to hold the film structure (Suto and Ui, 1996; Tual et al., 2000; Zheng et al., 2000). However, Tang et al. (2003) confirmed that the antimicrobial capacity of chitosan films diminishes with an increase in the cross-linking agent content. On the other hand, Mi et al. (2006) showed that cross-linking agents like geniposidic acid improve mechanical properties of chitosan films and do not alter their antibacterial capability when the cross-linking degree is relatively low. Of additional concern is, of course, the inherent toxicity of these cross-linking agents that can restrict their use. Another approach to overcome this aspect is to blend chitosan with other more moisture-resistant polymer matrices. In this sense, some researchers have included chitosan into other biopolymer matrices such as cellulose, starch, or konjac glucomannan in order to improve mechanical and water-swelling properties of antimicrobial chitosan films (Li et al., 2006; Sebti et al., 2007; Zhai et al., 2004). The similarity of cellulose and chitosan in primary structures hypothesized that they could form homogeneous composite films. Consequently, many works have been conducted by blending both components with the aim of improving the mechanical properties of chitosan after wetting (Wu et al., 2004), or reducing the cost of materials, as cellulose and derivatives are inexpensive biopolymers to produce (Moller et al., 2004). Nevertheless, the incorporation of chitosan within any of the aforementioned matrices does not necessary imply an improvement in water resistance. In fact, many of the published studies on this matter generated biocomposite materials with optimum antimicrobial activity but poor water barrier properties (Park et al., 2004). Thus, Moller et al. (2004) showed that composite films made of chitosan and hydroxypropyl methylcellulose (HPMC), with and without stearic acid, preserved its antilisterial activity compared to a chitosan control but the composite was completely soluble in distilled water, despite the fact that the incorporation of chitosan reduced the water vapor transmission rate of HPMC by 20%. On the other hand, the same films with chemical cross-linking showed a decrease in their solubility in water but simultaneously lost their antimicrobial activity. Wu et al. (2004) studied the mechanical and antimicrobial properties of chitosan/cellulose blends and obtained a slight reduction in water vapor transmission rate while a substantial bactericidal effect was maintained. Li et al. (2006) observed that the addition of chitosan to konjac glucomannan edible films incorporating nisin improved their antimicrobial activity and their water barrier properties.

In our research group, Fernandez-Saiz et al. (2008) reported that composite films based on chitosonium acetate and gliadins did not lose their physical integrity after immersion in the nutrient broth. Nevertheless, although film disintegration was prevented, some turbidity in the nutrient medium was detected. Indeed, the gliadins protein network must exert a blocking effect against the dissolution of biocide chitosonium acetate chains, albeit some of the latter chains do still migrate to exert their required antimicrobial role. Similar findings were observed by Tanabe et al. (2002), when chitosan was blended with keratin. Fernandez-Saiz et al. (2009) studied the water barrier and the antimicrobial activity of high and low-Mw chitosonium acetate–based solvent-cast blends with ethylene vinyl alcohol copolymers against *S. aureus* and *Salmonella* spp. These samples showed excellent antimicrobial activity as well as enhanced water barrier when low-Mw chitosan was used as the dispersed phase in the blend.

9.6 Application of Chitosan-Based Films

During the last decade, leading researchers in the field have studied the benefits of the application of chitosan-based films for food products. Edible chitosan-based films and coatings have been used to extend the shelf life of fruits, vegetables, seafood, meats, and egg products by preventing dehydration, oxidation rancidity, and surface browning among others. Furthermore, the effectiveness of chitosan against food spoilage microorganisms has been tested in recent years in order to assess the potential of using chitosan as a natural preservative when applied as an edible coating. For instance, Tsai et al. (2002) studied its application on fish preservation obtaining an increase of the shelf life of the product from 5 to 9 days. Other researches performed by Sagoo et al. (2002) evaluated the effect of treatments with chitosan solutions and reached an increase of the shelf life of raw sausages stored at chill temperatures from 7 to 15 days. Fernandez-Saiz et al. (2009) demonstrated the biocide properties of chitosonium acetate film in fish soup. In this study, the effects of the microorganism, the temperature of incubation, and the food substrate were analyzed. The results showed an increase in the lag phase of *L. monocytogenes* of ca. 7 days when 20 mg of chitosan film was added into 10 mL of fish soup at 4°C.

Lopez-Caballero et al. (2005) studied the effect of chitosan–gelatin blends as a coating for fish patties, showing a difference of around 2 log cycles between the control and the coated batches for total bacterial counts, *Pseudomonas*, and enterobacteria at 8–11 days of storage. In this study, the addition of powered chitosan to the mixture of patty ingredients was also investigated, but no antimicrobial effect was observed due to its poor solubility at neutral pH. The effects of the addition of low-Mw chitosan to raw rice before steam cooking were also analyzed by Tsai et al. (2006) who found an effective inhibition of the total anaerobic bacteria and *B. cereus* in cooked rice stored at 37°C and 18°C for 48 and 72 h, respectively. Another work performed by Juneja et al. (2006) demonstrated that the incorporation of chitosan glutamate into beef or turkey may reduce the potential risk of *Clostridium perfringens* spore germination and development during abusive cooling from 54.4°C to 7.2°C in 12, 15, or 18 h. In a more recent work, Guo et al. (2014) also compared the effectiveness of chitosan-based films containing lauric arginate ester or nisin against *Listeria innocua* in ready-to-eat deli turkey meat. Nevertheless, in all aforementioned works, chitosan was always applied as a food coating. Ye et al. (2008) studied the control of *L. monocytogenes* on ham steaks by the use of chitosan-coated plastic films. In this research, the antimicrobial properties of chitosan became negligible when this polysaccharide was tested in the form of insoluble films since they seem to be incapable of diffusing through a solid medium such as agar. In contrast, Ouattara et al. (2000)

obtained satisfactory results when studying the inhibition of surface spoilage bacteria in processed meats by the application of chitosan-based films. Moreover, in order to examine potential effects on egg quality properties, Caner and Cansiz (2007) coated eggshells with three chitosan-based coatings (produced with organic acids: acetic [C–AA], lactic [C–LA], and propionic [C–PA]). All chitosan-coated eggs showed greater inner food quality than the noncoated eggs. When compared with the controlled, noncoated egg specimens, the coated eggs significantly kept better from weight loss and better maintained the nutritional value amounts of minerals (especially calcium, iron, and magnesium concentrations). In a more recent work, Yuceer and Caner (2014) evaluated the interior quality, shell impact strength, and functional characteristics of eggs coated with chitosan and lysozyme–chitosan combinations for enhancing egg freshness during storage. They concluded that the 10%, 20%, and 60% lysozyme–chitosan coatings, considered active packaging, showed promising attributes. They could be a viable alternative to existing techniques for maintaining the internal quality of fresh eggs during long-term storage. Chitosan coatings also improved shell strength. This study also confirms that measurements of albumen quality (pH, dry matter, viscosity, and RWC) are excellent indicators of egg freshness.

The application of water-soluble polysaccharides on fruits and vegetables has become popular and extensive due to their ability to reduce O_2 and increase CO_2 levels in internal atmospheres. This effect modifies the internal atmospheres and reduces respiration rates, thereby prolonging the shelf life of fruits and vegetables, in a manner analogous to controlled atmospheres (Shiekh et al., 2013). The effects of edible chitosan coating on the quality and shelf life of sliced mango fruits were studied by Chien et al. (2007). The chitosan coating retarded the deterioration of sensory qualities and delayed water loss while increasing the soluble solid content, titratable acidity, and ascorbic acid content. It also inhibited the growth of microorganisms. The data gathered reveal that applying a chitosan coating effectively prolongs the quality attributes and extends the shelf life of sliced mango fruit. Casariego et al. (2008) demonstrated that chitosan films inhibited fungi growth, reduced ethylene production, increased internal CO_2, and decreased O_2 levels when applied to tomatoes. Adriano et al. (2009) proposed a combined application of a chitosan-based coating and the use of modified atmosphere packaging as a postharvest treatment process to maintain quality and prolong shelf life of carrots. The study showed that the use of a chitosan-containing edible coating preserved the overall visual quality of carrots and reduced surface whiteness during storage. The combined application of an edible coating containing chitosan, along with moderate O_2 and CO_2 levels, maintained quality and enhanced phenolic content in carrot sticks.

When chitosan or chitosan films are tested on real food systems, their biocide character is generally lower than on lab conditions, probably due to their irreversibly bound by some of the negatively charged compounds in the food. This finding was observed by Rhoades and Roller (2000) when evaluating chitosan effects against the natural microflora in an apple–elderflower juice and in hummus. Fernandez-Saiz et al. (2009) also observed a significant reduction of the biocide properties of chitosan films in tests performed in the fish soup compared to the laboratory nutrient medium (i.e., tryptone soy broth). These results also agree with a work performed by Devlieghere et al. (2004) who studied the influence of food components on the biocide properties of chitosan against *Candida lambica* and found a protective effect for bacteria when starch, wheat proteins, and NaCl were added to the nutrient medium. Oil, conversely, had no influence in this particular work.

ACKNOWLEDGMENTS

G. Sánchez and M.J. Fabra are recipients of a "Ramon y Cajal" and "Juan de la Cierva" contract from the Spanish Ministry of Economy and Competitiveness.

REFERENCES

Adriano D. N., Simões, Tudela J. A., Allende A., Puschmann R., and Gil M. I. Edible coatings containing chitosan and moderate modified atmospheres maintain quality and enhance phytochemicals of carrot sticks. *Postharvest Biology and Technology*, 51(3) (2009): 364–370.

Briandet R. T., Meylheuc T., Maher C., and Bellon-Fontaine, M. N. *Listeria monocytogenes* Scott A: Cell surface charge, hydrophobicity, and electron donor and acceptor characteristics under different environmental growth conditions. *Applied Environmental Microbiology*, 65 (1999): 5328–5333.

Butler B. L., Vergano P. J., Testin R. F., Bunn J. M., and Wiles J. L. Mechanical and barrier properties of edible chitosan films as affected by composition and storage. *Journal Food Science*, 61 (1996): 953–955.

Caner C. The effect of edible eggshell coatings on egg quality and consumer perception. *Journal of the Science of Food and Agriculture*, 85 (2005): 1897–1902.

Caner C. and Cansiz O. Effectiveness of chitosan-based coating in improving shelf-life of eggs. *Journal of the Science of Food and Agriculture*, 87(2) (2007): 227–232.

Casariego A., Souza B. W. S., Vicente A. A., Teixeira J. A., Cruz L., and Díaz R. Chitosan coating surface properties as affected by plasticizer, surfactant and polymer concentrations in relation to the surface properties of tomato and carrot. *Food Hydrocolloids*, 22(8) (2008): 1452–1459.

Chen Y. L. and Chou C. C. Factors affecting the susceptibility of *Staphylococcus aureus* CCRC 12657 to water soluble lactose chitosan derivative. *Food Microbiology*, 22 (2005): 29–35.

Chien P. J., Sheu F., and Yang F. H. Effects of edible chitosan coating on quality and shelf life of sliced mango fruit. *Journal of Food Engineering*, 78(1) (2007): 225–229.

Chung Y., Su Y., Chen C., Jia G., Wang H., Wu J. C. G., and Lin J. Relationship between antibacterial activity of chitosan and surface characteristics of cell wall. *Acta Pharmacology Sinica*, 25 (2004): 932–936.

Coma V., Martial-Gros A., Garreau S., Copinet A., Salin F., and Deschamps A. Edible antimicrobial films based on chitosan matrix. *Journal of Food Science*, 67 (2002): 1162–1169.

Cuero R. G., Osuji G., and Washington A. N-carboxymethyl chitosan inhibition of aflatoxin production: Role of zinc. *Biotechnology Letters*, 13 (1991): 441–444.

Davis R., Zivanovic S., D'Souza D. H., and Davidson P. M. Effectiveness of chitosan on the inactivation of enteric viral surrogates *Food Microbiology*, 32(1) (2012): 57–62.

Demarger-Andre S. and Domard A. Chitosan carboxylic acid salts in solution and in the solid state. *Carbohydrate Polymers*, 23 (1994): 211–219.

Devlieghere F., Vermeulen A., and Debevere J. Chitosan: Antimicrobial activity, interactions with food components and applicability as a coating on fruits and vegetables. *Food Microbiology*, 21 (2004): 703–714.

El Ghaouth A., Arul J., Asselin A., and Benhamou N. Antifungal activity of chitosan on post-harvest pathogens: Induction of morphological and cytological alterations and rhizopus stolonifer. *Mycology Research*, 96 (1992): 769–779.

Fernandes J. C., Tavaria F. K., Soares J. C., Ramos O. S., Monteiro M. J., Pintado M. E., and Malcata F. X. Antimicrobial effects of chitosan and chitooligosaccharides upon *Staphylococcus aureus* and *Escherichia coli* in food model systems. *Food Microbiology*, 25 (2008): 922–928.

Fernandez-Saiz P., Lagaron J. M., Hernandez-Muñoz P., and Ocio M. J. Characterization of the antimicrobial properties against *S. aureus* of novel renewable blends chitosan and gliadins of interest in active food packaging and coating applications. *International Journal of Food Microbiology*, 124 (2008): 13–20.

Fernandez-Saiz P., Lagaron J. M., and Ocio M. J. Optimization of the biocide properties of chitosan for its application in the design of active films of interest in the food area. *Food Hydrocolloids*, 23 (2009a): 913–921.

Fernandez-Saiz P., Lagarón J. M., and Ocio M. J. Optimization of the film-forming and storage conditions of chitosan as an antimicrobial agent. *Journal Agricultural and Food Chemistry*, 57 (2009b): 3298–3307.

Fernandez-Saiz P., Ocio M. J., and Lagarón J. M. Film forming and biocide assessment of chitosan as determined by combined ATR FTIR spectroscopy and antimicrobial assays. *Biopolymers*, 83 (2006): 577–583.

Fux C. A., Costerton J. W., and Stewart P. S. Survival strategies of infectious biofilms. *Trends in Microbiology*, 13 (2005): 34–40.

Guo M., Jin T. Z., Wang L., Scullen O. J., and Sommers C. H. Antimicrobial films and coatings for inactivation of *Listeria innocua* on ready-to-eat deli turkey meat. *Food Control*, 40(1) (2014): 64–70.

Hadwiger L. A., Kendra D. F., Fristensky B. W., and Wagoner W. Chitosan both activates genes in plants and inhibits RNA synthesis in fungi. In: Muzzarelli R. A. A., Jeuniaux C., and Gooday G. W. (eds.), *Chitin in Nature and Technology*. Plenum Press, New York, pp. 209–222, 1985.

Helander I. M., Nurmiaho-Lassila E. L., Ahvenainen R., Rhoades J., and Roller S. 2001. Chitosan disrupts the barrier properties of the outer membrane of Gram-negative bacteria. *International Journal of Food Microbiology*, 71 (2001): 235–244.

Igarashi K. and Nakano Y. Dissociation properties of amino groups in the chitosan gel particles prepared by the suspension evaporation method. *Journal of Chemical Engineering of Japan*, 36 (2003) 716–719.

Jeon Y. J., Park P. J., and Kim S. K. Antimicrobial effect of chitooligosaccharides produced by bioreactor. *Carbohydrate Polymers*, 44 (2001): 71–76.

Juneja V. K., Thippareddi H., Bari L., Inatsu Y., Kawamoto S., and Friedman M. Chitosan protects cooked ground beef and turkey against *Clostridium perfringens* spores during chilling. *Journal of Food Science*, 71 (2006): 236–240.

Kam H. M., Khor E., and Lim L. Y. Storage of partially deacetylated chitosan films. *Journal of Biomedical Materials Research*, 48(6) (1999): 881–888.

Kim K. M., Son J. H., Kim S. K., Weller C., and Hanna M. A. Properties of chitosan films as function of pH and solvent type. *Journal of Food Science*, 71 (2006): 119–124.

Kong M., Chen X. G., Xing K., and Park H. J. Antimicrobial properties of chitosan and mode of action: A state of the art review. *International Journal of Food Microbiology*, 144(1) (2010): 51–63.

Lagaron J. M., Fernandez-Saiz P., and Ocio M. J. Using ATR-FTIR spectroscopy to design active antimicrobial food packaging structures based on high molecular weight chitosan polysaccharide. *Journal of Agricultural and Food Chemistry*, 55 (2007): 2554–2562.

Larena A. and Caceres D. A. Variability between chitosan membrane surface characteristics as function of its composition and environmental conditions. *Applied Surface Science*, 238 (2004): 273–277.

Li B., Peng J., Yie X., and Xie B. Enhancing physical properties and antimicrobial activity of konjac glucomannan edible films by incorporating chitosan and nisin. *Journal of Food Science*, 71 (2006): 174–178.

Li Y., Chen X. G., Liu N. et al. Physicochemical characterization and antibacterial property of chitosan acetates. *Carbohydrate Polymers*, 67 (2007): 227–232.

Liu H., Du Y., Xiaohui Wang X., and Sun L. Chitosan kills bacteria through cell membrane damage. *International Journal Food Microbiology*, 95 (2004): 147–155.

Liu N., Chen X. G., Park H. J., Liu C. G., Liu C.-S., Meng X. H., and Yu L. J. Effect of MW and concentration of chitosan on antibacterial activity of *Escherichia coli*. *Carbohydrate Polymers*, 64 (2006): 60–65.

Lopez-Caballero M. E., Gómez-Guillén M. C., Pérez-Mateos M., and Montero P. A chitosan-gelatin blend as a coating for fish patties. *Food Hydrocolloids*, 19 (2005): 303–311.

Marques A., Encarnaçao S., Pedro S., and Nunes M. L. In vitro antimicrobial activity of garlic, oregano and chitosan against *Salmonella enterica*. *World Journal Microbiology Biotechnology*, 24 (2008): 2375–2360.

Marreco P. R., Moreira P. L., Genari S. C., and Moraes A. Effects of different sterilization methods on the morphology, mechanical properties, and cytotoxicity of chitosan membranes used as wound dressings. *Journal of Biomedical Materials Research, Part B*, 71 (2005): 268–277.

Mi F. L., Huang C. T., and Liang H. F. Physicochemical, antimicrobial, and cytotoxic characteristics of a chitosan film cross-linked by a naturally occurring cross-linking agent, aglycone geniposidic acid. *Journal of Agriculture and Food Chemistry*, 54 (2006): 3290–3296.

Moller H., Grelier S., Pardon P., and Coma V. Antimicrobial and physicochemical properties of chitosan-HPMC-based films. *Journal of Agriculture and Food Chemistry*, 52 (2004): 6585–6591.

NCCLS. *Methods for Determining Bactericidal Activity of Antimicrobial Agents: Approved Guideline*. M26-A, Vol. 19(18), Wayne, PA, 1999.

No H. K., Kim S. H., Lee S. H., Park N. Y., and Prinyawiwatkul W. Stability and antibacterial activity of chitosan solutions affected by storage temperature and time. *Carbohydrate Polymers*, 65 (2006): 174–178.

No H. K., Meyers S. P., Prinyawiwatkul W., and Xu Z. Applications of chitosan for improvement of quality and shelf life of foods: A review. *Journal of Food Science*, 72 (2007): 87–100.

No H. K., Park N. Y., Lee S. H., and Meyers S. P. Antibacterial activity of chitosan and chitosan oligomers with different molecular weights. *International Journal of Food Microbiology*, 74 (2002): 65–72.

Osman Z. and Arof A. K. FTIR studies of chitosan acetate based polymer electrolytes. *Electrochimica Acta*, 48 (2003): 993–999.

Ouattara B., Simard R. E., Piette G., Bégin A., and Holley R. A. Inhibition of surface spoilage bacteria in processed meats by application of antimicrobial films prepared with chitosan. *International Journal Food Microbiology*, 62 (2000): 139–148.

Park S. I., Daeschel M. A., and Zhao Y. Functional properties of antimicrobial lysozyme-chitosan composite films. *Journal of Food Science*, 69(8) (2004): M215–M221.

Park S. Y., Marsh K. S., and Rhim J. W. Characteristics of different molecular weight chitosan films affected by the type of organic solvents. *Journal of Food Science,* 67 (2002): 194–197.

Pranoto Y., Rakshit S. K., and Salokhe V. M. Enhancing antimicrobial activity of chitosan films by incorporating garlic oil, potassium sorbate and nisin. *LWT: Food Science and Technology,* 38 (2005): 859–865.

Qin C., Li H., Xiao Q., Liu Y., Zhu J., and Du Y. Water-solubility of chitosan and its antimicrobial activity. *Carbohydrate Polymers,* 63 (2006): 367–374.

Rhoades J. and Roller S. Antimicrobial actions of degraded and native chitosan against spoilage organisms in laboratory media and foods. *Applied and Environmental Microbiology,* 66 (2000): 80–86.

Rinaudo M., Pavlov G., and Desbrieres J. Influence of acetic acid concentration on the solubilization of chitosan. *Polymer,* 40 (1999): 7029–7032.

Ritthidej G. C., Phaechamud T., and Koizumi T. Moist heat treatment on physicochemical change of chitosan salt films. *International Journal Pharmaceutics,* 232 (2002): 11–22.

Rodriguez E., Calzada J., Arqués J. L., Rodriguez J. M., Nuñez M., and Medina M. Antimicrobial activity of pediocin-producing *Lactococcus lactis* on *Listeria monocytogenes, Staphylococcus aureus* and *Escherichia coli* O157:H7 in cheese. *International Dairy Journal,* 15 (2005): 51–57.

Sagoo S., Board R., and Roller S. Chitosan inhibits growth of spoilage micro-organisms in chilled pork products. *Food Microbiology,* 19 (2002): 175–182.

Sebti I., Chollet E., Degraeve C. N., and Peyrol E. Water sensitivity, antimicrobial, and physicochemical analyses of edible films based on HPMC and/or chitosan. *Journal of Agriculture and Food Chemistry,* 55 (2007): 693–699.

Shiekh R. A., Malik M. A., Al-Thabaiti S. A., and Shiekh M. A. Chitosan as a novel edible coating for fresh fruits. *Food Science and Technology Research,* 19(2) (2013): 139–155.

Simpson C. A., Geornaras I., Yoon Y., Scanga J. A., Kendall P. A., and Sofos J. N. Effect of inoculum preparation procedure and storage time and temperature on the fate of *Listeria monocytogenes* on inoculated salami. *Journal Food Protection,* 71 (2008): 494–501.

Srinivasa P. C., Ramesh M. N., Kumar K. R., and Tharanathan R. N. 2004. Properties of chitosan films prepared under different drying conditions. *Journal Food Engineering,* 63 (2004): 79–85.

Su X., Zivanovic S., and D'Souza D. H. Effect of chitosan on the infectivity of murine norovirus, feline calicivirus, and bacteriophage MS2. *Journal of Food Protection,* 72(12) (2009): 2623–2628.

Suto S. and Ui N. Chemical crosslinking of hydroxypropyl cellulose and chitosan blends. *Journal of Applied Polymer Science,* 61 (1996): 2273–2278.

Tabak M., Scher K., Hartog E., Romling U., Matthews K. R., Chikindas M. L., and Yaron S. Effect of triclosan on *Salmonella* typhimurium at different growth stages and biofilms. *FEMS Microbiology Letters,* 267 (2007): 200–2006.

Tanabe T., Okitsu N., Tachinaba A., and Yamauchi K. Preparation and characterization of keratin-chitosan composite film. *Biomaterials,* 23 (2002): 817–825.

Tang R., Du Y., and Fan L. Dialdehyde starch-crosslinked chitosan films and their antimicrobial effects. *Journal of Polymer Science,* 41 (2003): 993–997.

Tsai G. J. and Su W. H. Antibacterial activity of shrimp chitosan against *Escherichia coli. Journal of Food Protection,* 62(3) (1999): 239–243.

Tsai G. J., Su W. H., Chen H. C., and Pan C. L. Antimicrobial activity of shrimp chitin and chitosan from different treatments and applications of fish preservation. *Fisheries Science,* 68 (2002): 170–177.

Tsai G. J., Tsai M. T., Lee J. M., and Zhong M. Z. Effects of chitosan and low-molecular-weight chitosan on *Bacillus cereus* and application in the preservation of cooked rice. *Journal of Food Protection,* 69 (2006): 2168–2175.

Tual C., Espuche E., Escoubes M., and Domard A. Transport properties of chitosan membranes: Influence of crosslinking. *Journal of Polymer Science,* 38 (2000): 1521–1529.

Uchida Y., Izume M., and Ohtakara A. Preparation of chitosan oligomers with purified chitosanase and its application. In: Skjåk-Bræk G., Anthonsen T., and Sandford, P. (eds.), *Chitin and Chitosan: Sources, Chemistry, Biochemistry, Physical Properties and Applications.* Elsevier, London, U.K., pp. 373–382, 1989.

Vartiainen J., Rättöa M., Tapperb U., Paulussenc S., and Hurmea E. Surface modification of atmospheric plasma activated BOPP by immobilizing chitosan. *Polymer Bulletin,* 54 (2005): 343–352.

Wu Y.-B., Yu S.-H., Mi F.-L., Wu C.-W., Shyu S.-S., Peng C.-K., and Chao A.-C. Preparation and characterization on mechanical and antibacterial properties of chitosan/cellulose blends. *Carbohydrate Polymers*, 57 (2004): 435–440.

Ye M., Neetoo H., and Chen H. 2008. Control of *Listeria monocytogenes* on ham steaks by antimicrobials incorporated into chitosan-coated plastic films. *Food Microbiology*, 25 (2008): 260–268.

Yuceer M. and Caner C. Antimicrobial lysozyme-chitosan coatings affect functional properties and shelf life of chicken eggs during storage. *Journal of the Science of Food and Agriculture*, 94(1) (2014): 153–162.

Zhai M., Zhao L., Yoshii F., and Kume T. Study on antibacterial starch/chitosan blend film formed under the action of irradiation. *Carbohydrate Polymers*, 57 (2004): 83–88.

Zheng H., Du Y.-M., Yu J.-H., and Xiao L. The properties and preparation of crosslinked chitosan films. *Chemical Journal of Chinese Universities*, 21 (2000): 809–812.

Zivanovic S., Basurto C. C., Chi S., Davidson P. M., and Weiss J. Molecular weight of chitosan influences antimicrobial activity in oil-in-water emulsions. *Journal of Food Protection*, 67 (2004): 952–959.

Zivanovic S., Chi S., and Draughon A. F. Antimicrobial activity of chitosan films enriched with essential oils. *Journal of Food Science*, 70 (2005): 45–51.

Zotkin M. A., Vikhoreva G. A., Smotrina T. V., and Derbenev M. A. Thermal modification and study of the structure of chitosan films. *Fibre Chemistry*, 36 (2004): 16–20.

10

Films and Coatings from Lipids and Wax

Amparo Chiralt and Alberto Jiménez

CONTENTS

Abstract

The use of films obtained from hydrophobic materials, mainly waxes such as carnauba wax or candelilla wax, is well known since many fruits and vegetables, such as citrics, have been traditionally coated with these kinds of compounds. These are able to reduce the transpiration/respiration rate of fruits and limit the moisture loss while at the same time increasing gloss, thus improving their appearance. Nevertheless, lipid-based films and coatings exhibit a poor mechanical performance, mainly due to their fragility and low transparency. One method of taking advantage of the beneficial properties of lipids while reducing their drawbacks is the formulation of lipid–polymer (protein or polysaccharide) blends to obtain composite matrices, where lipids are embedded in the polymer network. In this way, composite films may exhibit improved properties in comparison with pure lipid- or pure biopolymer–based films. In the last few years, the properties of these kinds of films have been studied by many authors as a function of the lipid characteristics, polymer affinity, component dispersion method, and film-drying conditions. The obtained results show great differences in the films' behavior, depending on the blend considered. The physical state of the lipids, their particle size in the film-forming dispersion and film, and the specific lipid–polymer interactions play an important role in the final properties and functionality of the film. Indeed, some properties of the polymeric matrix may be improved by lipid incorporation, whereas others can be negatively affected. For instance, biopolymer films containing lipids show lower water vapor permeability (WVP), as compared with the corresponding lipid-free film, although oxygen permeability can be enhanced and the film transparency may be reduced. Likewise, the mechanical properties of the biopolymer matrices containing lipids are positively or negatively affected, depending on several factors, such as the component compatibility, lipid–polymer ratio, and the final nano- and microstructure of the blend. In this chapter, the effect of lipids on the properties of biopolymer films has been discussed as a function of the nature of the lipid, the lipid–polymer ratio, and the polymer's characteristics, through the data reported from different studies. The most recent achievements and trends in the formulation of lipids–biopolymers are also discussed.

10.1 Films Obtained from Lipids and Waxes

Edible films and coatings are defined as thin layers obtained from those components suitable for food consumption, which are able to protect a food product by acting as a barrier against external agents, thus increasing its shelf life (Guilbert et al., 1996). An edible film may also avoid the mass transfer between different phases of a food product. Whereas the term *film* refers to a stand-alone structure, coating is directly formed on the food product surface by applying a forming solution (Guilbert and Biquet, 1986; Gennadios and Weller, 1990).

The use of films and coatings that are compatible with food products is not new. Indeed, wax coatings have been used to prevent moisture loss in citrus fruits in China since the twelfth century (Guilbert and Biquet, 1986). Another way coatings are employed to protect foodstuffs is the use of fats to wrap meat cuts, which dates back to the sixteenth century (Kester and Fennema, 1986). The first reports on edible films correspond to the nineteenth century, whereas the first patents only date back to the 1950s. From this date on, research into this field has steadily increased, from the application of simple wax coatings to protect citrus fruits to the incorporation of nanoparticles to obtain highly functional smart materials.

Over the last few decades, many natural, but also synthetic, substances have been tested as hydrophobic materials to be used as coatings for food preservation. Nowadays, all food ingredients and additives allowed by Codex Alimentarius, U.S. Food and Drug Administration, or European Union regulations can be potentially used in the formulation of edible films and coatings (Debeaufort and Voilley, 2009). A classification of these components is proposed in Table 10.1.

Oils and fats from vegetal or animal origin, such as waxes, were the first components used to protect foodstuffs from external agents. Nowadays, these materials are put to very few uses, although some of them, such as cocoa butter and its derivatives, are extensively used in the confectionery and biscuit industries to avoid mass transfer processes: for instance, nut oxidation in chocolate or the moisturizing of different layers in chocolate bars. Other oils and fats are also used in film formulation, but predominantly in the form of emulsions, in combination with structural polymers, such as polysaccharides or proteins.

The main group of hydrophobic materials used to form coatings is waxes, probably due to the fact that they are more efficient barriers than lacs, fatty acids, or acetylated glycerides (Debeaufort and Voilley, 2009). Currently, their main application is the protection of fresh fruits. The main functions of wax-based coatings when applied on fruits are the reduction of both moisture loss and respiration rate, as well as the improvement in their appearance brought about by the increase in gloss. The word *wax* is sometimes used to denote any coating, whether or not it includes a lipid substance such as beeswax (BW), carnauba wax, polyethylene wax, or candelilla wax (Hall, 2011).

Carnauba wax, which is extracted from the leaves of the Brazilian palm (*Copernicia cerifera*), has been commonly used for fruit coating to prevent moisture loss and extend shelf life (Barman et al., 2011). Different studies analyzed the ability of this wax to maintain the postharvest quality of different fruits, such as avocado (Feygenberg et al., 2005) or mango (Khuyen et al., 2008).

Candelilla wax is also widely used for food coating. This substance, soluble in organic solvents, is composed of hydrocarbons (approximately 50%) and low quantities of volatile esters (Saucedo-Pompa et al., 2009). Candelilla wax is extracted from the candelilla plant (*Euphorbia cerifera, Euphorbia antisyphilitica, Pedilanthus pavonis, Pedilanthus aphyllus*) by immersion in boiling water. The wax

TABLE 10.1

Hydrophobic Materials Used to Obtain Films and Coatings

Family of Compounds	Examples
Oils and fats from vegetal origin	Sunflower oil, peanut oil, cocoa butter, palm oil, soybean oil, olive oil
Oils and fats from animal origin	Milk butter, tallow, lard
Waxes	Carnauba wax, candelilla wax, beeswax, paraffin wax, oxidized polyethylene
Fatty acids	Lauric acid, myristic acid, palmitic acid, stearic acid, oleic acid, linoleic acid
Emulsifiers	Lecithins, mono- and diglycerides, mono- and diglycerides esters
Essential oils	Oregano, mint, citrus, basil, cinnamon, ginger

is skimmed off the surface, refined, and bleached (Hall, 2011). In the food industry, candelilla wax is mainly used to maintain the quality of fresh fruits and in the production of some confectionery products, such as chewing gum or hard candy.

BW, produced by honeybees, is another important wax used in the food industry. It has been reported to be one of the most effective materials that decrease the water vapor permeability (WVP) of edible composite films due to its high hydrophobicity and solid state at room temperature (Yang and Paulson, 2000; Morillon et al., 2002). Nevertheless, it has also been used as fruit coating (alone or in combination with other waxes) or in confectionery (Hall, 2011).

Apart from waxes obtained from natural resources, such as those previously described, there are others from nonrenewable resources, called synthetic waxes. The most relevant are paraffin wax, polyethylene wax, and oxidized polyethylene, which are obtained from petroleum. These materials have been used in the coating of raw fruits; cheese; root crops such as rutabaga, turnip, and yucca; and the stems of sugarcane (Hall, 2011). Nevertheless, there is a tendency to substitute synthetic, nonrenewable waxes by other naturally occurring ones in food applications. In fact, most of the recent studies into paraffin wax concern its nonfood, industrial applications (Zhang et al., 2012; Trigui et al., 2014).

Resins are another important group of hydrophobic substances that have been used over the last few decades in the food industry to form coatings. Of the resins, the best-known compounds are shellac, wood rosin, and coumarone indene (Rhim and Shellhammer, 2005), the former being the most widely used. Shellac is secreted by the insect *Laccifer lacca*, and it has been applied in pharmaceuticals, in confectionery, and on fruits and vegetables (Rhim and Shellhammer, 2005). However, shellac presents serious stability problems, as a result of the interaction of its hydroxyl and carboxyl groups (Limmatvapirat et al., 2007), leading to significant changes in its physicochemical properties (Soradech et al., 2013).

Other lipids, such as fatty acids, have been extensively used in film formation, including in polymer matrices, as discussed in the following section (Tanaka et al., 2001; Fabra et al., 2009a; Jiménez et al., 2012). Despite their ability to limit the water vapor exchanges in food products, they are less hydrophobic than resins and lacs, high-melting-point fats, and waxes (Debeaufort and Voilley, 2009).

Different hydrophobic substances, other than fatty acids, such as surfactants or essential oils have also been incorporated into composite films in order to modulate the properties of the polymer matrices. The addition of these compounds normally leads to an increase in the film's resistance to moisture action, in some cases to an improvement of the mechanical performance and, when essential oils are used, to antimicrobial and/or antioxidant activity.

10.2 Films Based on Biopolymers Containing Lipids

The most recent papers concerning the use of lipid compounds for food packaging applications focus on composite films. As previously commented on, these types of materials have better properties than those of purely lipid-based films. In fact, lipid-based films are very good barriers against moisture transfer and significantly improve the fruit appearance by increasing the gloss, but other properties, such as the mechanical behavior, are poorer in comparison with those of polysaccharide or protein-based films and coatings. These biopolymers allow us to obtain more resistant and flexible structures with, generally, better gas barrier properties. Indeed, due to the high oxygen solubility in the lipid phases, oxygen permeability increases in polymeric films containing lipids, especially if they are liquid (Fabra et al., 2012; Jiménez et al., 2013a; Valenzuela et al., 2013).

Three different component arrangements may be found in films containing lipids (Debeaufort and Voilley, 2009). In one of them, the biopolymers act as a support on whose surface a lipid layer is deposited, leading to a bilayer film. The barrier properties in this case are highly improved due to the parallel specific resistance to mass transfer of each layer. In another configuration, the hydrophobic component may form a continuous phase in which sugar crystals or biopolymer chains are dispersed. Finally, the biopolymers (proteins and/or polysaccharides) can constitute a continuous phase, where the hydrophobic components are dispersed. This type of film or coating structure, which is the most common, corresponds with a solid emulsion. Very different components have been considered for the purposes of obtaining emulsion-based films. Table 10.1 shows some of the lipids used in the formulation of these kinds of films.

A significant number of the studies have been carried out for emulsion-based films where lipids are dispersed as droplets or particles of different sizes, depending on the interactions among components and the homogenization method used for blending. Although there are some studies (García et al., 2000; Salgado et al., 2013) where lipids and polymers are dry blended (by using roll mills or extruders), most of them used the casting method to prepare the films, with the subsequent step of the evaporation of a great amount of solvent (usually water or ethanol) under given drying conditions. This implies a previous step of component homogenization where the order and conditions in which compounds are blended may have a great influence on the film's microstructure and properties. One important aspect in the formation of the emulsion is the physical state of the hydrophobic compounds. If they are solid, heating is required to obtain stable film-forming emulsions with adequate lipid particle size (Soazo et al., 2013; Ortega-Toro et al., 2014). Some lipid components, which are liquid at room temperature and have antioxidant properties, may be thermosensitive. So, the emulsification process should be carried out at low temperatures to avoid oxidation phenomena.

The blend of the components in film-forming emulsions requires a high shear stress to promote the reduction of the lipid droplet size and the emulsion stability. The equipment used to obtain the polymer–lipid blends varies in the different studies, although the most commonly used is the rotor–stator homogenizer and/or high-pressure homogenizers (Vargas et al., 2011; Ma et al., 2012). The film's properties are greatly affected by the structure of the film-forming emulsion (lipid volume fraction and droplet size and stability), component interactions, and drying conditions. During the film-drying step, the emulsion can partially destabilize giving rise to flocculation, coalescence, and creaming phenomena, which will affect the film's microstructure and properties. The incorporation of lipids into hydrocolloid matrices generally improves their resistance to moisture action, in comparison with lipid-free films, but other properties may be negatively affected, as commented in the following text. Table 10.2 shows some studies where lipids are incorporated into different polymer matrices and the main changes induced by the lipids in the properties of the film, as compared with the lipid-free one.

Essential oils and their components are especially relevant lipids, widely used as fillers of hydrocolloid films, since they can act as antioxidants and antimicrobials, thus imparting bioactivity to the film. Essential oils have been used to obtain bioactive films with both protein (Aguirre et al., 2013; Salgado et al., 2013) and polysaccharide (Sánchez-Gonzalez et al., 2010, 2011a; Jouki et al., 2014a,b; Shojaee-Aliabadi et al., 2014) or blend (Montero, 2010) matrices. Some essential oils, such as those obtained from oregano, clove, thyme, or pennyroyal, are effective at inhibiting the growth of both Gram-negative (*Escherichia coli, Pseudomonas aeruginosa, Shewanella putrefaciens*, or *Salmonella typhimurium*) and, especially, Gram-positive (*Staphylococcus aureus* or *Listeria monocytogenes*) bacteria. Essential oil compounds are also effective at providing the films with antioxidant activity.

10.3 Microstructural Features

The heterogeneous structure of the film-forming emulsion (different phases) is also present in the dried film, which shows a continuous matrix of polymer and miscible compounds, such as plasticizer, and dispersed lipid particles of different sizes and shapes, depending on the structural properties of the initial dispersion and the film-drying conditions (Villalobos et al., 2005). Figure 10.1 shows the microstructure of some films obtained with different hydrocolloids: sodium caseinate, hydroxypropyl methylcellulose (HPMC), and corn starch, without lipids and those containing stearic or oleic acid (OA). In all cases, film-forming dispersions were obtained under similar homogenization conditions and film-drying conditions were also similar. The effect of lipid dispersion on the film's microstructure can be appreciated through the cross-sectional scanning electron microscopy (SEM) images, showing clear differences, depending on the lipid and polymer composition. OA droplets are only clearly visible in the starch matrix, whereas it is very finely dispersed in the protein matrix and HPMC. Stearic acid forms laminar structures in sodium caseinate films, but this was not observed in HPMC and starch films. The greater viscosity of the polysaccharide dispersions could inhibit the progressive lipid shelf association during film drying, which occurs in the protein films (Fabra et al., 2010a; Jiménez et al., 2010). Mechanisms contributing to the stabilization of the film-forming emulsion, such as protein interactions or viscous

TABLE 10.2

Films Obtained from Biopolymers and Lipids

Polymer Matrix	Type of Lipid	Main Features	References
Triticale protein	Oregano essential oil	WVP did not change after oil addition, but this component had a plasticizing effect on the films. Oil conferred antimicrobial activity on the films, mainly against Gram-positive bacteria.	Aguirre et al. (2013)
Wheat starch–chitosan blend	Thyme essential oil	Oil addition reduced the OP but no changes in WVP were found. Elongation, gloss, and transparency of the films were significantly reduced by oil addition, whereas their antioxidant activity was significantly improved.	Bonilla et al. (2013)
	α-Tocopherol	α-Tocopherol promoted phase separation in the hydrocolloid matrix due to the different interaction with the polymers. OP was reduced by lipid addition due to a chemical blocking, concordant with the great antioxidant capacity of α-tocopherol.	
Sodium caseinate	Oleic acid–BW blend[a]	Oleic acid, pure or mixed with beeswax, had a plasticizing effect on the films, increasing their elasticity, flexibility, and stretchability. WVP was also reduced by oleic acid addition in comparison with pure NaCas films.	Fabra et al. (2008)
Lipophilic maize starch–gelatin blend	Fatty acids[b]	Although the addition of fatty acids to the biopolymer films increased their opacity and elongation, their incorporation reduced the TS and WVP. In addition, the fatty acid concentration found to be effective at reducing the biopolymer film's permeability that varied between 15% and 25%.	Fakhouri et al. (2009)
Whey protein isolate	Fatty acids[c]	The effect of fatty acids was different depending on the degree of unsaturation. In this way, stearic acid was more effective at reducing the WVP than oleic and linoleic acids. On the contrary, the mechanical properties were less affected when unsaturated fatty acids were incorporated.	Fernández et al. (2006)
Corn starch or amylomaize starch	Sunflower oil	Water vapor barrier ability was improved by lipid addition without affecting the gas barrier properties. On the contrary, the incorporation of very high concentrations of oil increased the WVP due to lipid migration.	García et al. (2000)
Pea starch	Beeswax	The addition of BW affected the mechanical properties, reducing TS and increasing EM. WVP was reduced and OP was increased as a result of BW incorporation. More significant changes in the physical properties took place as the BW content in films increased.	Han et al. (2006)
HPMC	Fatty acids	The addition of saturated fatty acids gave rise to the formation of crystallized lipid layers in the films, which significantly improved the moisture barrier properties. On the contrary, these structures gave rise to more brittle, less stretchable, more opaque, and less glossy films in comparison with lipid-free films.	Jiménez et al. (2010)
Starch–sodium caseinate blend	Lecithin nanoliposomes	Nanoliposomes containing essential oils were properly incorporated into the hydrocolloid matrix. However, while their addition led to a decrease in the mechanical resistance and extensibility of the films, the natural color of lecithin conferred a loss of lightness, a gain in chrome, and a redder hue to the films, which were also less transparent than the lipid-free films.	Jiménez et al. (2014)

(Continued)

TABLE 10.2 (*Continued*)

Films Obtained from Biopolymers and Lipids

Polymer Matrix	Type of Lipid	Main Features	References
Cod gelatin	Sunflower oil	Oil addition did not improve the films' water vapor barrier properties. In addition, it led them to become less transparent and whiter. Authors related these changes with some lipid–protein interactions (hydrogen bonds and ester formation) as observed in the FTIR analysis.	Pérez-Mateos et al. (2009)
HPMC	Tea tree essential oil	TTO contributed to the decrease in the TS and EM of the polysaccharide films. In addition, the degree of transparency and gloss fell as the TTO content in the film rose, in agreement with the film's microstructure, as observed by SEM.	Sánchez-González et al. (2010)
Chitosan	Oleic acid	Interactions between CH and OA in the FFD and their development during film drying affected the final structure of the composite films and their physical properties. WVP and stiffness of composite films were reduced by OA addition when the CH–OA ratio was lower than 1:1. Gloss and transparency of the films increased as the OA content in the films rose.	Vargas et al. (2009)
Whey protein isolate	Stearic acid	Stearic acid addition contributed to the decrease in protein solubility and WVP, whereas the mechanical properties were negatively affected by fatty acid addition.	Yoshida and Antunes (2004)

Note: Effects of lipid incorporation.

Abbreviations: BW, beeswax; CH, chitosan; EM, elastic modulus; FFD, film-forming dispersion; FTIR, Fourier transform infrared spectroscopy; NaCas, sodium caseinate; OA, oleic acid; OP, oxygen permeability; SEM, scanning electron microscopy; TS, tensile strength; TTO, tea tree essential oil; WVP, water vapor permeability.

[a] OA incorporated in the films was both pure and in combination with beeswax.

[b] Fatty acids incorporated were palmitic (C_{16}), myristic (C_{14}), lauric (C_{12}), capric (C_{10}), caprylic (C_8), and caproic (C_6).

[c] Fatty acids incorporated were stearic ($C_{18:0}$), oleic ($C_{18:1}$), and linoleic ($C_{18:2}$).

effects, play an important role in the final film microstructure: the size and shape of the lipid particles in the polymer matrix.

In polar lipids, such as fatty acids, the lipid self-association phenomenon plays an important role in the film's microstructure. This phenomenon is determined by molecular interactions in the film-forming dispersions and their changes in line with the water evaporation during the film formation. Several authors observed a multilayered structure for saturated fatty acids (lauric, palmitic, and stearic) in sodium caseinate (Fabra et al., 2010a) and HPMC (Jiménez et al., 2010) films. This was associated with the growth of the molecular aggregations in the bilayer self-association of the lipid molecules during the film drying and the final crystallization of fatty acids in the polymer matrix. The length of lipid layers in the matrix was greater for lauric acid and decreased as the fatty acid chain length increased. This indicates that the growth of the molecule aggregates during the film drying was more effective when the fatty acid chain was shorter, probably due to its greater molecular mobility. These successive lipid layers in the film matrix greatly contribute to the reduction in WVP; the longer the lipid layers, the lower the WVP values.

The heterogeneity of the film structure is also reflected on the film surface, as different authors have observed by means of atomic force microscopy (AFM) (Villalobos et al., 2005; Fabra et al., 2009b). The different surface topography is the result of the initial lipid particle size in the film-forming dispersion and the destabilization phenomena, which occur during film drying, associated with flocculation, coalescence, and creaming of lipid particles. The more unstable the dispersions, the rougher the film surface due to the greater progression of destabilization phenomena, which implies that a greater quantity of lipids migrate to the film surface. Surface roughness has a significant impact on the film gloss and the films with very rough surfaces usually exhibit low gloss levels (Villalobos et al., 2005; Fabra et al., 2009b).

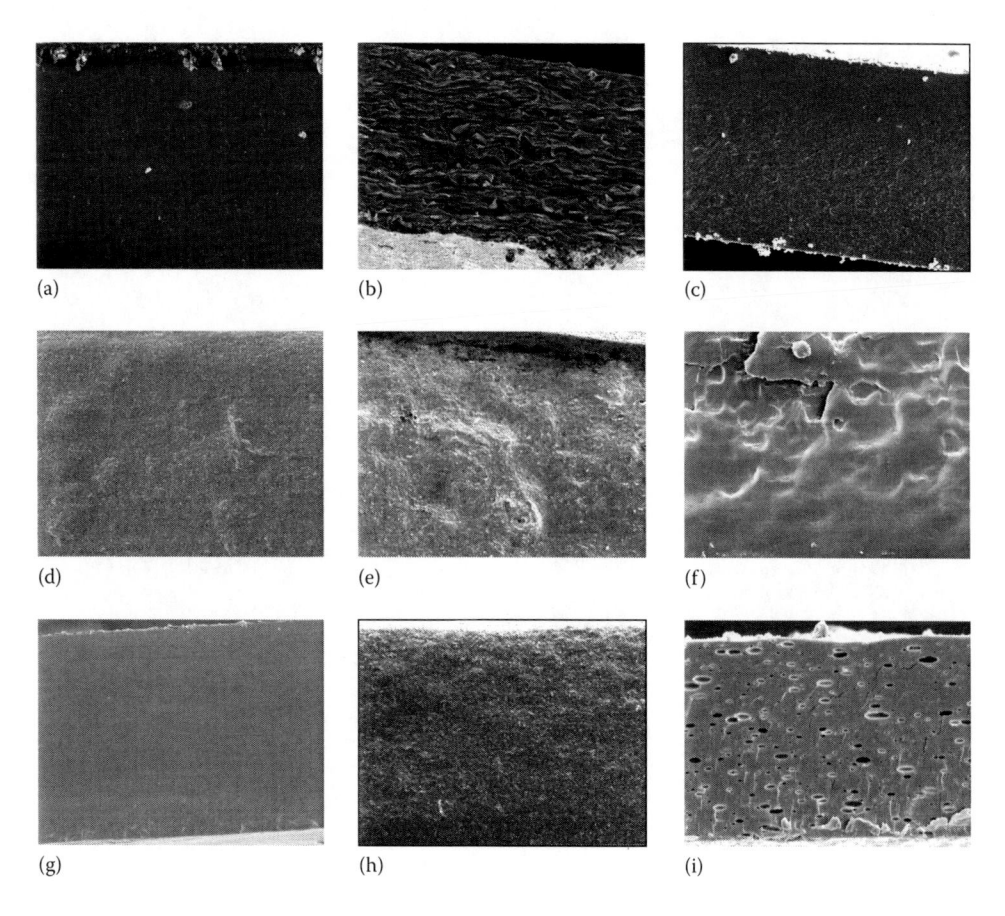

FIGURE 10.1 Scanning electron microscopy micrographs of biopolymer-based films containing fatty acids. (a) (Sodium caseinate [NaCas]); (b) (NaCas–stearic acid); (c) (NaCas–oleic acid); (d) (hydroxypropylmethylcellulose [HPMC]); (e) (HPMC–stearic acid); (f) (HPMC–oleic acid); (g) (corn starch); (h) (corn starch–stearic acid); (i) (corn starch–oleic acid).

When the incorporated lipid interacts with the polymer chains, structural changes can also be related with molecular-level modifications in the matrix. These interactions are also reflected in the mechanical behavior of the films, since these produce changes in the phase transition temperatures of the polymers. When they are near the film handling temperature, sharp changes in flexibility or mechanical resistance can be produced. Likewise, the difference between the product temperature and T_g determines the rate at which diffusion-dependent processes, such as mass transport, occur in the system, and so it has a great incidence on film barrier properties.

Table 10.3 shows the values of the glass transition temperature (T_g) of different polymer matrices as affected by the incorporation of several lipids. The incorporation of OA into sodium caseinate matrices barely affects T_g values when the films are equilibrated at a very low relative humidity, while it acts as a plasticizer, decreasing T_g, when the films are equilibrated at a relative humidity of over 53% (Fabra et al., 2010b). This makes the films more extensible when they contain OA. Palmitic acid also provokes a T_g decrease in corn starch–glycerol films, regardless of the equilibrium relative humidity. Similar effects were observed for essential oils in other polymer matrices (Salgado et al., 2013; Jouki et al., 2014a).

The protein–lipid interactions may also affect lipid phase transitions: the melting temperature range and enthalpy. Fabra et al. (2010b) analyzed lipid phase transition in the sodium caseinate films containing OA and BW, equilibrated at different a_w values, in order to obtain complementary information about their interactions in the film matrix. No significant differences were observed between pure BW melting characteristics and those obtained for BW in films equilibrated at different a_w, which seems to indicate that no notable interactions were developed between protein and wax. Nevertheless, significant differences were

TABLE 10.3

Glass Transition Temperatures in Several Biopolymer Films

Film Composition	Film Conditioning before Test	Glass Transition Temperature (°C)	References
NaCas[a]	25°C; 0% RH	59.7±0.3	Fabra et al. (2010b)
NaCas[a]/OA		62.1±0.7	
NaCas[a]	25°C; 23% RH	57.2±0.7	
NaCas[a]/OA		58.2±0.1	
NaCas[a]	25°C; 53% RH	55.1±1.5	
NaCas[a]/OA		51.6±0.2	
NaCas[a]	25°C; 75% RH	51.1±0.6	
NaCas[a]/OA		34.5±0.4	
Starch/glycerol	25°C; 0% RH	45±4	Jiménez et al. (2013b)
Starch/glycerol/PA 1/0.25/0.15		22.1±0.9	
Starch/glycerol	25°C; 53% RH	37.4±1.9	
Starch/glycerol/PA 1/0.25/0.15		22.5±1.3	
Starch/glycerol	25°C; 68% RH	11.2±1.2	
Starch/glycerol/PA 1/0.25/0.15		15.6±1.1	
Starch/glycerol	25°C; 75% RH	5±6; −49±7	
Starch/glycerol/PA 1/0.25/0.15		−5.3±1.2; −41±3	
Mucilage[b]	25°C; 50% RH	−35.2±1.3	Jouki et al. (2014a)
Mucilage[b] + 1% OEO (v/v)		−43.3±1.2	
Mucilage[b] + 1.5% OEO (v/v)		−45.8±1.9	
Mucilage[b] + 2% OEO (v/v)		−46.7±1.6	
SPC/glycerol/CEO 1/0.3/0	20°C; 58% RH	−29.0±0.3	Salgado et al. (2013)
SPC/glycerol/CEO 1/0.3/0.75[c]		−33.0±0.7	
PGP[d]/SA 1/0	25°C; 55% RH	127.19	Zahedi et al. (2010)
PGP[d]/SA 1/0.06		125.01	

Note: Effect of lipid incorporation.

Abbreviations: CEO, clove essential oil; NaCas, sodium caseinate; OA, oleic acid; OEO, oregano essential oil; PA, palmitic acid; PGP, pistachio globulin protein; SA, stearic acid; SPC, sunflower protein concentrate.

[a]   Films contained 0.3 g glycerol/g protein.

[b]   Mucilage refers to a solution of 1 wt.% quince seed mucilage containing 0.35 g glycerol/g mucilage.

[c]   CEO content was 0.75 mL/g SPC.

[d]   Films contained 1 g glycerol/g protein. Stearic acid containing film included 0.1 g Tween 80/g fatty acid.

observed between OA melting endotherms of the pure compound and OA films equilibrated at different a_w values. A significant enthalpy reduction was observed in line with the a_w increase, which indicates that OA crystallization was inhibited to quite an extent in the films, in line with the increase in sample a_w, despite the fact that the associated increase in molecular mobility usually promotes crystallization rates. These results agree with the binding of OA molecules to the caseinate chains promoted when a_w increases. These bonds weaken the polymeric cross-linking in the matrix, making it softer (lower T_g values). This has a great impact on the film's mechanical and barrier properties due to the influence of molecular mobility on both rheological properties and diffusional processes (Fabra et al., 2010b).

10.4   Mechanical Behavior

One of the most widely studied features of edible and biodegradable films is their mechanical behavior. Usually, the elastic modulus, the tensile strength, and the elongation at break are measured in order to determine the mechanical performance of these materials (Monedero et al., 2009; Aguirre et al., 2013;

Pires et al., 2013; Shojaee-Aliabadi et al., 2014). Universal test machines for mechanical assays are used to obtain the force–distance curves, which are transformed into stress–Hencky strain relationships, using film dimensions, to calculate the mentioned parameters.

As commented on earlier, pure lipid films present poorer mechanical properties than composite ones, where the continuous polysaccharide or protein matrices predominantly govern the mechanical performance. Nevertheless, the type of lipid and its content in films, as well as its interaction with the continuous matrix, also affect the films' mechanical behavior. Other factors, such as the equilibrium moisture content, also have a great influence. Jiménez et al. (2013b) found that the mechanical resistance of starch–fatty acid films decreased as the equilibrium moisture content increased, due to the increase in water–polymer interactions. In films equilibrated at a water activity of up to 0.75, they observed glycerol separation from the matrix since water–polymer interactions predominated over glycerol–polymer ones. This was revealed from the appearance of two different glass transitions in DSC analyses: one from the starch-rich phase and other from the glycerol-rich phase.

Table 10.4 shows the values of the aforementioned parameters, where the effect of lipid incorporation and the equilibrium relative humidity on some biopolymer matrices can be observed. No clear tendencies may be seen in these effects, since factors, such as the chemical structure of lipids, responsible for the established interactions with the other film components, are relevant in determining the lipid effects. In some cases, the incorporation of liquid lipids (essential oils, OA, or α-tocopherol, etc.) into biopolymer matrices gives rise to less stretchable, less resistant films, as observed by different authors (Fabra et al., 2008; Aguirre et al., 2013; Jiménez et al., 2013a). These results have been related with the heterogeneous structure of the film matrix where lipids interrupt the continuity of the polymer network, thus implying a reduction of the matrix cohesion forces. When liquid lipids are dispersed, the oil droplets may be easily deformed during the extension of the films, which exert a positive impact on their stretchability. Nevertheless, other studies revealed that the addition of essential oils did not notably affect the mechanical properties of films or that lipid-containing films were stiffer and less stretchable. Bonilla et al. (2013) found that the addition of basil essential oil significantly increased the elastic modulus and the tensile strength of wheat starch–chitosan films, whereas thyme essential oil did not produce any significant effects on the final mechanical behavior. Pires et al. (2013) found that, depending on the essential oil incorporated into the matrix, the mechanical resistance of hake protein films increased or decreased. The addition of citronella essential oil to hake protein films increased the tensile strength and decreased the elongation at break. However, thyme essential oil produced the opposite effect. Authors related this result with the complex composition of the essential oils. In particular, the increase in the mechanical resistance of films containing citronella essential oil was related with the presence of β-citronellal, which may promote the protein cross-linking.

In general, for lipids that do not interact with the polymer chains, the introduced discontinuities provoke a decrease in the cohesion forces of the network, which leads both to films that are less resistant to fracture and ones that are less extensible, depending on the lipid ratio. Solid lipids reduce film resistance to a lesser extent than liquid lipids, but enhance the brittleness of the film. Nevertheless, when lipid molecules significantly interact with the biopolymers in the matrices, the mechanical behavior cannot be predicted.

Of the solid lipids, waxes and saturated fatty acids have been included in both protein (Tanaka et al., 2001; Monedero et al., 2009) and polysaccharide matrices (Jiménez et al., 2010; Velickova et al., 2013). These compounds led to an increase in the mechanical resistance and the tensile strength of the films, while elongation at break was significantly reduced. Han et al. (2006) obtained films based on pea starch and BW and observed that the elastic modulus increased with the BW content. On the contrary, tensile strength did not change significantly and elongation at break decreased to 53% when wax content was 30%. They explained that, although starch was responsible for the strength of the matrix, the presence of homogeneously distributed solid BW particles in the matrix enhances the film elastic modulus despite the interruption of the matrix continuity. Monedero et al. (2009) studied the effect of lipid mixtures (OA–BW) on the tensile properties of soy protein isolate (SPI) films. They found that the BW ratio in the lipid mixture did not affect the mechanical behavior to a great extent, except when this component was pure in the lipid phase, where an increase in the film's stiffness, resistance to break, and brittleness

TABLE 10.4

Effect of Lipids Incorporation on Mechanical Properties of Some Biopolymer-Based Films

Film Composition	Storage Conditions before Tests	EM (MPa)	TS (MPa)	E (%)	References
Triticale protein	25°C; 32% RH	147.6±0.7	2.90±0.03	9.7±1.3	Aguirre et al.
Triticale protein + 1% OEO (w/v)[a]		83±11	2.08±0.19	12.1±0.3	(2013)
Triticale protein + 2% OEO (w/v)[a]		22±11	0.85±0.16	85±5	
SPI	21°C; 33% RH	1.30±0.06	2.18±0.15	73±20	Bilbao-Sáinz
SPI/Acetem/T60 10/1/0.5		1.12±0.09	2.3±0.3	74±17	et al. (2010)
SPI/Acetem/T60/OEO 10/1/0.5/0.67		1.22±0.09	3.2±0.5	87±14	
SPI/Acetem/T60/CiEO 10/1/0.5/0.67		1.31±0.09	3.3±0.3	84±19	
Wheat starch–chitosan[b]	25°C; 53% RH	1460±150	44±3	4.3±0.8	Bonilla et al.
Wheat starch–chitosan + BEO[c]		1710±50	54.6±1.2	4.3±0.6	(2013)
Wheat starch–chitosan + TEO[c]		1450±200	43±7	3.6±0.6	
Wheat starch–chitosan + Toc[c]		1400±600	34±12	3.0±1.0	
Pea starch/glycerol/BW 3/2/0	Room temperature; 50% RH	12.1	2.30	51.44	Han et al. (2006)
Pea starch/glycerol/BW 3/2/0.3		24.7	2.20	45.38	
Pea starch/glycerol/BW 3/2/0.6		33.0	2.39	50.29	
Pea starch/glycerol/BW 3/2/0.9		41.8	1.92	21.94	
Corn starch/glycerol/PA 1/0.25/0.15	25°C; 0% RH	3000±300	27.4±1.6	0.8±0.3	Jiménez et al.
	25°C; 53% RH	640±110	6.8±0.3	4±3	(2013b)
	25°C; 75% RH	370±20	4.1±0.8	4.2±1.2	
	25°C; 75% RH	170±60	3.2±0.7	8.3±1.4	
Corn starch/glycerol/OA 1/0.25/0.15	25°C; 0% RH	1760±160	6.1±1.7	0.45±0.04	
	25°C; 53% RH	170±30	3.3±0.3	13±7	
	25°C; 75% RH	233±10	3.72±0.13	5.1±1.4	
	25°C; 75% RH	230±10	4.1±0.2	9.4±1.4	
Hake protein	Room temperature; 57% RH	—	6.2±1.7	148±30	Pires et al.
Hake protein + CiEO[d]		—	7.1±1.6	182±30	(2013)
Hake protein + TEO[d]		—	4.3±0.9	111±40	
κ-Carrageenan[e]	25°C; 53% RH	—	26±3	36±1	Shojaee-Aliabadi
κ-Carrageenan[e] + 1% MEO (v/v)		—	26±5	36±3	et al. (2014)
κ-Carrageenan[e] + 2% MEO (v/v)		—	22.2±1.6	37±3	

Abbreviations: BEO, basil essential oil; BW, beeswax; CiEO, cinnamon essential oil; E, elongation at break; EM, elastic modulus; MEO, *Mentha pulegium* essential oil; OA, oleic acid; OEO, oregano essential oil; PA, palmitic acid; SPI, soy protein isolate; T60, Tween 60; TEO, thyme essential oil; Toc, tocopherol; TS, tensile strength.

[a] Essential oils were added in the film-forming solutions containing 7.5 g of protein/100 g solution.
[b] Ratio starch/CH was 4/1.
[c] Lipids concentration was 0.1 g/g starch.
[d] Essential oils concentration was 0.25 mL/g protein.
[e] κ-Carrageenan refers to a blend 1:0.5 κ-carrageenan–glycerol dissolved in water at 1.5% (w/v).

were observed. This was related with the combined effect of OA plasticization and the discontinuities introduced in the matrix.

The effect of lipids on the film mechanical behavior is directly related with the distribution of the film's components and their interactions. Indeed, lipid components may either appear in a separate phase with scarcely any interactions with the biopolymers or interact with their chains, even giving rise to an almost homogeneous structure as observed in Figure 10.1. When interactions take place, the developed nanostructure must be analyzed in order to understand the final properties of the matrix.

In starch matrices, lipid incorporation greatly affects the film's nanostructure since complexes of amylose helical forms and lipids are formed, affecting the crystalline structure of the films and their mechanical behavior (Jiménez et al., 2012; Ortega-Toro et al., 2014). Amylose–fatty acid complexes associate in V-type structures, as observed in the x-ray diffraction patterns, which make the films very brittle, negatively affecting mechanical behavior. Likewise, starch films become stiffer and less stretchable after storage time due to progressive crystallization and chain aggregation (Forssell et al., 1999).

10.5 Optical Properties

Optical properties of the films, such as gloss and transparency, are relevant functional properties, since they have an impact on the product appearance, while they are greatly affected by the degree of the film's structural heterogeneity (Villalobos et al., 2005; Ozdemir and Floros, 2008; Fabra et al., 2009b). Therefore, lipid incorporation in biopolymer matrices has an impact on the film's optical properties since it implies the presence of a dispersed phase that greatly contributes to the light–film interactions, responsible for gloss, color, and transparency. The particle size distribution in the matrix, as well as the volume fraction of the dispersed phase and the differences between the refractive indexes of the phases, affects the changes in the film's optical properties.

The gloss of the edible coatings, as affected by lipids, relative humidity, and storage time, has been studied by Trezza and Krochta (2000). They found that for lipid/surfactant dispersion coatings, the increase in the surfactant level greatly decreased gloss with respect to the pure hydrocolloid film, depending on the lipid type and the particle size in the dispersed phase. An increase in the size or number of particles implied a loss of glossiness. The gloss of the films containing surfactants has been well correlated with the surface roughness, affected by the particle size (Villalobos et al., 2005). Nevertheless, other factors such as the angle of incident light or the intrinsic properties (refractive index) of the materials also affect the film gloss (Trezza and Krochta, 2001).

Fabra et al. (2009b) analyzed the effect of internal and surface microstructure of the films on the optical properties (transparency and gloss) of sodium caseinate films containing lipids (OA–BW mixtures). Both transparency and gloss were greatly affected by the arrangement of film components in the matrix, which, in turn, depended on the structure (droplet size distribution) of the film-forming dispersion. Lipid-free films exhibited the greatest transparency and gloss, whereas lipids introduce opacity and lead to a loss in gloss. The greater the content of BW in the lipid mixture, the less transparent and glossy the film. Significant correlations were established between the film surface roughness, analyzed by AFM, and the optical (transparency and gloss) parameters.

Table 10.5 shows the effect of the incorporation of some lipids into protein or polysaccharide films on optical properties such as the opacity, whiteness index, or gloss. All the parameters were modified to differing extents, depending on the type of lipid and polymer and their ratio in the film.

10.6 Film Barrier Properties

Lipids are also usually added to the film formulation in order to improve water barrier properties. The incorporation of these compounds also changes the water sorption capacity of the film, which greatly affects the barrier properties. The effect of lipids can be very different, depending on the hydrocolloid–lipid interactions in the matrix and the film microstructure reached after drying. In this sense, different authors (McHugh and Krochta, 1994; Perez-Gagó and Krochta, 2001) report how lipid particle size affects the water vapor barrier properties of the films. The reduction of the particle size leads to an increase in the tortuosity factor for mass transport in the film matrix, which leads to an improvement of barrier properties.

The incorporation of lipids usually reduces the film's water sorption capacity, which also contributes to the effect they have on the film's barrier properties. This is, in part, due to the fact that lipids correspond to a fraction of solids with a small water uptake capacity. Nevertheless, lipid interactions with the polymer chains can also modify the number of water sorption active points. This has been observed for

TABLE 10.5

Effect of Lipids Addition on Optical Properties of Some Biopolymer-Based Films

Film Composition	Storage Conditions before Tests	Optical Property	Value	References
Hake protein	Room temperature; 57% RH	Opacity[b] 600 nm	1.84±0.08	Pires et al. (2013)
Hake protein + CitEO[a]			3.3±0.5	
Hake protein + TEO[a]			2.8±0.3	
SPC/glycerol/CEO 1/0.3/0	20°C; 58% RH	Opacity 500 nm	12.9±0.7	Salgado et al. (2013)
SPC/glycerol/CEO 1/0.3/0.75[c]			13.1±0.7	
κ-Carrageenan[d]	25°C; 53% HR	Opacity 600 nm	0.81±0.02	Shojaee-Aliabadi et al. (2014)
κ-Carrageenan[d] + 1% MEO (v/v)			1.04±0.16	
κ-Carrageenan[d] + 2% MEO (v/v)			1.32±0.07	
SPI/G/Gly 1/0.3/0.15	25°C; 50% RH	Whiteness index	67.0±0.4	Guerrero et al. (2011)
SPI/G/Gly/OO 1/0.3/0.15/0.05			68.6±0.8	
SPI/G/Gly/OO 1/0.3/0.15/0.10			68.8±0.3	
SPI/G/Gly/OO 1/0.3/0.15/0.15			71.7±0.6	
HPMC	20°C; 54% RH	Whiteness index	93±1	Jiménez et al. (2010)
HPMC + LA			93.8±0.3	
HPMC + MA			94.9±0.2	
HPMC + PA			93±2	
HPMC + SA			85±1	
HPMC + OA			86±1	
κ-Carrageenan[d]	25°C; 53% HR	Whiteness index	88.3±0.5	Shojaee-Aliabadi et al. (2014)
κ-Carrageenan[d] + 1% MEO (v/v)			88.21±0.14	
κ-Carrageenan[d] + 2% MEO (v/v)			85.6±1.7	
Wheat starch–CH[e]	25°C; 53% HR	Gloss 60°	51±16	Bonilla et al. (2013)
Wheat starch–CH + BEO[f]			19±14	
Wheat starch–CH + TEO[f]			13±2	
Wheat starch–CH + α-tocopherol[f]			13±3	
CS–NaCas[g]	25°C; 53% HR	Gloss 60°	76±8	Jiménez et al. (2013a)
CS–NaCas[g]/OA 1/0.15			90±3	
CS-NaCas[g]/OA/Toc 1/0.15/0.1			78±2	
CS–NaCas[g]/Toc 1/0.1			92±1	
κ-Carrageenan[d]	25°C; 53% HR	Gloss 60°	46±2	Shojaee-Aliabadi et al. (2014)
κ-Carrageenan[d] + 1% MEO (v/v)			43±3	
κ-Carrageenan[d] + 2% MEO (v/v)			42±3	

Parameters: Opacity = Absorbance$_{wavelength}$/film thickness (mm).

Abbreviations: BEO, basil essential oil; CitEO, citronella essential oil; CEO, clove essential oil; G, gelatin; Gly, glycerol; HPMC, hydroxypropylmethylcellulose; LA, lauric acid; MA, myristic acid; MEO, *Mentha pulegium* essential oil; NaCas, sodium caseinate; OA, oleic acid; OO, virgin extra olive oil; PA, palmitic acid; SA, stearic acid; SPC, sunflower protein concentrate; SPI, soy protein isolate; TEO, thyme essential oil; Toc, α-tocopherol.

[a]   Essential oils concentration was 0.25 mL/g protein.

[b]   Authors talked about transparency and indicated that the higher the transparency value, the lower the transparency.

[c]   CEO content was 0.75 mL/g SPC.

[d]   κ-Carrageenan refers to a blend 1:0.5 κ-carrageenan–glycerol dissolved in water at 1.5% (w/v).

[e]   Ratio starch/CH was 4/1.

[f]   Lipids concentration was 0.1 g/g starch.

[g]   Ratio corn starch/NaCas in films was 1/1. Films contained 0.25 g glycerol/g hydrocolloid.

different kinds of films, such as sodium caseinate containing OA and BW mixtures (Fabra et al., 2010b) or corn starch containing fatty acids (Jimenez et al., 2013b) and surfactants (Ortega-Toro et al., 2014).

The addition of lipids to biopolymer films usually provokes a reduction of their WVP, while promoting their gas permeability. This opposite effect is related to the different chemical affinity of lipids with water or gas molecules. Whereas the solubility of water molecules is very low in the lipid fraction, gas molecules show greater solubility in the lipid fraction than in the biopolymer matrix. When liquid lipids are used, the greater the gas solubility in the lipid fraction, the higher the gas permeability in the film (Fabra et al., 2012). The solid state of lipids contributes to limit the mass transport of gas molecules due to the low molecular mobility in this phase. This has been observed by Fabra et al. (2012) for sodium caseinate films containing OA–BW mixtures. They analyze the effect of lipid addition as well as the influence of the relative humidity, or the equilibrium water content of the films, on the permeability to water vapor, oxygen, and carbon dioxide. The effect of lipid addition was dependent both on the composition of OA–BW mixtures and the film's moisture content. The addition of lipid mixtures reduced water vapor transfer as compared to the lipid-free films, whereas pure OA or BW was less effective. Both lipid-free films and films with pure BW showed the lowest O_2 and CO_2 permeability, whereas the incorporation of OA exponentially increased this parameter. Predictive equations for water vapor and gas permeability values were established as a function of water content and lipid composition.

The incorporation of essential oils into polysaccharide matrices leads to a slight promotion of the barrier properties to water vapor and, in some cases, improves the oxygen barrier properties. This has been observed for HPMC and chitosan films containing different ratios of lemon, bergamot, and tree essential oils in the film; the higher the oil ratio, the lower the WVP (Sánchez-González et al., 2011b). Nevertheless, when basil and thyme essential oils were incorporated into chitosan matrices at different ratios, the effect was only notable on WVP when the homogenization of the film-forming dispersion was carried out at high pressure and there was a low ratio of essential oils (Bonilla et al., 2012). However, these essential oils reduced the oxygen permeability of wheat starch–chitosan films to half, probably due to their antioxidant effect (Bonilla et al., 2013). This promotion of the oxygen barrier has also been observed for other antioxidant lipids, such as α-tocopherol, in corn starch–sodium caseinate films (Jiménez et al., 2013a).

Whereas in starch films, lipids, such as fatty acids or sorbitan esters of fatty acids, did not notably enhance the water barrier properties, they did promote oxygen permeability (Jiménez et al., 2012; Ortega-Toro et al., 2014). Table 10.6 shows the values of water vapor and oxygen permeability of several biopolymer films as affected by the incorporation of different lipids, where the general effects commented earlier can be observed, to a different extent, depending on the type and ratio of the film components.

10.7 Trends in the Use of Biopolymer–Lipid Films

The impact of incorporating lipids into biopolymer films for the purpose of improving their functionality as water vapor barriers leads in some cases to undesirable changes in other properties, such as mechanical or optical parameters. The components' immiscibility gives rise to a lipid-dispersed phase, which greatly affects the film microstructure. Likewise, lipid–polymer interactions, which mainly occur in protein matrices, induced better or poorer properties in the continuous matrices depending on the established lipid–polymer bonds. In this sense, in films obtained through the emulsification of components, two kinds of lipids could be defined: inactive or active. Whereas the former only gives rise to discontinuities in the polymer's continuous matrix, the latter greatly modifies the properties of this matrix, by changing the cross-linking behavior of the polymer chains, which leads to more marked differences in the film behavior. The physical properties of these systems are much more affected by the characteristics of the continuous phase than those of the dispersed phase. So, when active lipids are incorporated into the matrix, the product can behave very differently than the lipid-free films. Phase transition analysis as a function of water activity is a good means of identifying the degree of the polymer–lipid interactions.

TABLE 10.6

Effect of the Incorporation of Lipids on the Barrier Properties of Biopolymer-Based Films

Film Composition	Test Conditions WVP	WVP × 10^{10} (g·m^{-1}·s^{-1}·Pa^{-1})	Test Conditions OP	OP × 10^{12} (cm^3·m^{-1}·s^{-1}·Pa^{-1})	References
Corn starch[a]	20°C; 33.3%–98.2% RH	2.6 ± 1.0	20°C; 63.8% RH	460 ± 50	García et al. (2000)
Corn starch[a] + SO[b]		1.9 ± 0.5		Nd	
Amylomaize starch[a]		2.1 ± 0.8		321 ± 19	
Amylomaize starch[a] + SO[b]		1.8 ± 0.4		Nd	
Pea starch/glycerol/BW 3/2/0	25°C; 0%–100% RH	21.61	23°C; 50% RH	0.887	Han et al. (2006)
Pea starch/glycerol/BW 3/2/0.3		20.81		0.826	
Pea starch/glycerol/BW 3/2/0.6		20.14		0.898	
Pea starch/glycerol/BW 3/2/0.9		18.64		0.884	
CS–NaCas[c]	25°C; 53%–100% RH	25.1 ± 0.5	25°C; 53% RH	0.120 ± 0.004	Jiménez et al. (2013a)
CS–NaCas[c]/OA 1/0.15		19.4 ± 0.6		0.99 ± 0.30	
CS–NaCas[c]/OA/ Toc 1/0.15/0.1		22 ± 3		0.39 ± 0.09	
CS–NaCas[c]/Toc 1/0.1		20 ± 2		0.1653 ± 0.0003	
Mucilage[d]	25°C; 0%–75% RH	0.76 ± 0.10	23°C; 50% RH	0.43 ± 0.03	Jouki et al. (2014a,b)
Mucilage[d] + 1% OEO (v/v)		0.92 ± 0.09		0.46 ± 0.03	
Mucilage[d] + 1.5% OEO (v/v)		1.11 ± 0.13		0.52 ± 0.03	
Mucilage[d] + 2% OEO (v/v)		1.76 ± 0.09		0.59 ± 0.03	
Mucilage[d] + 1% TEO (v/v)		0.88 ± 0.11		0.47 ± 0.03	
Mucilage[d] + 1.5% TEO (v/v)		0.99 ± 0.08		0.54 ± 0.04	
Mucilage[d] + 2% TEO (v/v)		1.49 ± 0.11		0.65 ± 0.05	
QP–CH[e]		—	23°C; 0% RH	0.28 ± 0.05	Valenzuela et al. (2013)
QP–CH[e] + 27.82% SO (d.b.)		—		1.07 ± 0.12	
QP–CH[e] + 36.46% SO (d.b.)		—		1.48 ± 0.13	
QP–CH[e] + 47.02% SO (d.b.)		—		1.73 ± 0.19	

Abbreviations: BW, beeswax; CH, chitosan; CS, corn starch; NaCas, sodium caseinate; Nd, not determined by authors; OA, oleic acid; OEO, oregano essential oil; OP, oxygen permeability; QP, quinoa protein; SO, sunflower oil; TEO, thyme essential oil; Toc, α-tocopherol; WVP, water vapor permeability.

[a] Films contained 34.4% glycerol (d.b.).

[b] Oil concentration was 2 g/L.

[c] Ratio corn starch/NaCas in films was 1/1. Films contained 0.25 g glycerol/g hydrocolloid.

[d] Mucilage refers to a solution of 1% quince seed mucilage (w/w) containing 0.35 g glycerol/g mucilage.

[e] OP/CH ratio was 0.1. In SO containing films, the oil concentration (d.b.) has been calculated considering the information included in the paper.

The incorporation of lipids with antimicrobial or antioxidant activity, such as essential oils, is an interesting alternative method via which active films for food preservation can be obtained, while modulating as their physical properties. The incorporation of this kind of volatile compounds causes problems in the different methods used to obtain films; by casting, due to the partial evaporation of volatile compounds during the drying step and, additionally, in thermoformation (extrusion or compression molding) processes, the thermosensitive nature of the lipid compounds makes them nonfeasible. An alternative means of introducing active thermosensitive lipids into biopolymer films could be their previous encapsulation by an adequate technique in order to protect them and to ensure their bioactive role in the film, while modulating their release to the coated product surface.

ACKNOWLEDGMENT

The authors acknowledge the financial support from the Spanish Ministerio de Educación y Ciencia throughout the project AGL2010-20694.

REFERENCES

Aguirre, A., Borneo, R., and León, A.E. (2013). Antimicrobial, mechanical and barrier properties of triticale protein films incorporated with oregano essential oil. *Food Bioscience*, 1, 2–9.

Barman, K., Asrey, R., and Pal, R.K. (2011). Putrescine and carnauba wax pretreatments alleviate chilling injury, enhance shelf life and preserve pomegranate fruit quality during cold storage. *Scientia Horticulturae*, 130(4), 795–800.

Bilbao-Sáinz, C., Avena-Bustillos, R.J., Wood, D.F., Williams, T.G., and McHugh, T.H. (2010). Nanoemulsions prepared by a low-energy emulsification method applied to edible films. *Journal of Agriculture and Food Chemistry*, 58(22), 11932–11938.

Bonilla, J., Atarés, L., Vargas, M., and Chiralt, A. (2012). Effect of essential oils and homogenization conditions on properties of chitosan-based films. *Food Hydrocolloids*, 26(1), 9–16.

Bonilla, J., Talón, E., Atarés, L., Vargas, M., and Chiralt, A. (2013). Effect of the incorporation of antioxidants on physicochemical and antioxidant properties of wheat starch-chitosan films. *Journal of Food Engineering*, 118, 271–278.

Debeaufort, F. and Voilley, A. (2009). Lipid-based edible films and coatings. In M.E. Embuscado and K.C. Huber (eds.), *Edible Films and Coatings for Food Applications* (Springer: New York), pp. 135–168.

Fabra, M.J., Jiménez, A., Atarés, L., Talens, P., and Chiralt, A. (2009a). Effect of fatty acids and beeswax addition on properties of sodium caseinate dispersions and films. *Biomacromolecules*, 10(6), 1500–1507.

Fabra, M.J., Pérez-Masià, R., Talens, P., and Chiralt, A. (2010a). Influence of the homogenization conditions and lipid self-association on properties of sodium caseinate based films containing oleic and stearic acids. *Food Hydrocolloids*, 25(5), 1112–1121.

Fabra, M.J., Talens, P., and Chiralt, A. (2008). Tensile properties and water vapour permeability of sodium caseinate films containing oleic acid-beeswax mixtures. *Journal of Food Engineering*, 85(3), 393–400.

Fabra, M.J., Talens, P., and Chiralt, A. (2009b). Microstructure and optical properties of sodium caseinate films containing oleic acid–beeswax mixtures. *Food Hydrocolloids*, 23, 676–683.

Fabra, M.J., Talens, P., and Chiralt, A. (2010b). Water sorption isotherms and phase transitions of sodium caseinate–lipid films as affected by lipid interactions. *Food Hydrocolloid*, 24, 384–391.

Fabra, M.J., Talens, P., Gavara, R., and Chiralt, A. (2012). Barrier properties of sodium caseinate films as affected by lipid composition and moisture content. *Journal of Food Engineering*, 109(3), 372–379.

Fakhouri, F.M., Fontes, L.C.B., Innocentini-Mei, L.H., and Collares-Queiroz, F.P. (2009). Effect of fatty acid addition on the properties of biopolymer films based on lipophilic maize starch and gelatin. *Starch/Stärke*, 61, 528–536.

Fernández, L., Díaz de Apodaca, E., Cebrián, M., Villarán, M.C., and Maté, J.I. (2006). Effect of the unsaturation degree and concentration of fatty acids on the properties of WPI based edible films. *European Food Research and Technology*, 224(4), 415–420.

Feygenberg, O., Hershkovitz, V., Ben-Arie, R., Nikitenko, T., Jacob, S., and Pesis, E. (2005). Postharvest use of organic coating for maintaining bio-organic avocado and mango quality. *Acta Horticulturae*, 682(3), 507–512.

Forssell, P.M., Helleman, S.H.D., Myllärinen, P.J., Moates, G.K., and Parker, R. (1999). Ageing of rubbery thermoplastic barley and oat starches. *Carbohydrate Polymers*, 39, 43–51.

García, M.A., Martino, M.N., and Zaritzky, N.E. (2000). Lipid addition to improve barrier properties of edible starch-based films and coatings. *Journal of Food Science*, 65(6), 941–944.

Gennadios, A. and Weller, C.L. (1990). Edible films and coatings from wheat and corn proteins. *Food Technology*, 44(10), 63–69.

Guerrero, P., Nur Hanani, Z.A., Kerry, J.P., and de la Caba, K. (2011). Characterization of soy protein-based films prepared with acids and oils by compression. *Journal of Food Engineering*, 107(1), 41–49.

Guilbert, S. and Biquet, B. (1986). Technology and application of edible protective films. In M. Mathlouthi (ed.), *Food Packaging and Preservation: Theory and Practice* (Elsevier Applied Science: London, U.K.), pp. 371–394.

Guilbert, S., Gontard, N., and Gorris, L.G.M. (1996). Prolongation of the shelf-life of perishable food products using biodegradable films and coatings. *LWT: Food Science and Technology*, 29(1–2), 10–17.

Hall, D.J. (2011). Edible coatings from lipids, waxes, and resins. In E.A. Baldwin (ed.), *Edible Coatings and Films to Improve Food Quality* (CRC Press: Boca Raton, FL), pp. 79–101.

Han, J.H., Seo, G.H., Park, I.M., Kim, G.N., and Lee, D.S. (2006). Physical and mechanical properties of pea starch edible films containing beeswax emulsions. *Journal of Food Science*, 71(6), 290–296.

Jiménez, A., Fabra, M.J., Talens, P., and Chiralt, A. (2010). Effect of lipid self-association on the micro-structure and physical properties of hydroxypropyl-methylcellulose edible films containing fatty acids. *Carbohydrate Polymers*, 82(3), 585–593.

Jiménez, A., Fabra, M.J., Talens, P., and Chiralt, A. (2012). Effect of re-crystallization on tensile, optical and water vapour barrier properties of cornstarch films containing fatty acids. *Food Hydrocolloids*, 26(1), 302–310.

Jiménez, A., Fabra, M.J., Talens, P., and Chiralt, A. (2013a). Physical properties and antioxidant capacity of starch-sodium caseinate films containing lipids. *Journal of Food Engineering*, 116(3), 695–702.

Jiménez, A., Fabra, M.J., Talens, P., and Chiralt, A. (2013b). Phase transitions in starch based films con-taining fatty acids: Effect on water sorption and mechanical behaviour. *Food Hydrocolloids*, 30(1), 408–418.

Jiménez, A., Sánchez-González, L., Desobry, S., Chiralt, A., and Arab Tehrany, E. (2014). Influence of nano-liposomes incorporation on properties of film forming dispersions and films based on cornstarch and sodium caseinate. *Food Hydrocolloids*, 35, 159–169.

Jouki, M., Mortazavi, S.A., Yazdi, F.T., and Koocheki, A. (2014b). Characterization of antioxidant-antibacterial quince seed mucilage films containing thyme essential oil. *Carbohydrate Polymers*, 99(2), 537–546.

Jouki, M., Yazdi, F.T., Mortazavi, S.A., and Koocheki, A. (2014a). Quince seed mucilage films incorporated with oregano essential oil: Physical, thermal, barrier, antioxidant and antibacterial properties. *Food Hydrocolloids*, 36, 9–19.

Kester, J.J. and Fennema, O.R. (1986). Edible films and coatings: A review. *Food Technology*, 54, 47–58.

Khuyen, T.H.D., Singh, Z., and Swinny, E.E. (2008). Edible coatings influence fruit ripening, quality, and aroma biosynthesis in mango fruit. *Journal of Agricultural and Food Chemistry*, 56(4), 1361–1370.

Limmatvapirat, S., Limmatvapirat, C., Puttipipatkhachorn, S., Nuntanid, J., and Luangtana-Anan, M. (2007). Enhanced enteric properties and stability of shellac films through composite salts formation. *European Journal of Pharmaceutics and Biopharmaceutics*, 67(3), 690–698.

Ma, W., Tang, C.H., Yin, S.W., Yang, X.Q., Wang, Q., Liu, F., and Wei, Z.H. (2012). Characterization of gelatin-based edible films incorporated with olive oil. *Food Research International*, 49(1), 572–579.

McHugh, T.H. and Krochta, J.M. (1994). Dispersed phase particle size effects on water vapour permeability of whey protein–beeswax emulsion films. *Journal of Food Processing and Preservation*, 18, 173–188.

Monedero, M., Fabra, M.J., Talens, P., and Chiralt, A. (2009). Effect of oleic acid-beeswax mixtures on mechanical, optical and water barrier properties of soy protein isolate based films. *Journal of Food Engineering*, 91(4), 509–515.

Montero, P. (2010). Biodegradable gelatin–chitosan films incorporated with essential oils as antimicrobial agents for fish preservation. *Food Microbiology*, 27(7), 889–896.

Morillon, V., Debeaufort, F., Blond, G., Capelle, M., and Voilley, A. (2002). Factors affecting the moisture permeability of lipid-based edible films: A review. *Critical Reviews in Food Science and Nutrition*, 42(1), 67–89.

Ortega-Toro, R., Jiménez, A., Talens, P., and Chiralt, A. (2014). Effect of the incorporation of surfactants on the physical properties of cornstarch films. *Food Hydrocolloids*, 38, 66–75.

Ozdemir, M. and Floros, J.D. (2008). Optimization of edible whey protein films containing preservatives for mechanical and optical properties. *Journal of Food Engineering*, 84(1), 116–123.

Perez-Gagó, M.B. and Krochta, J.M. (2001). Lipid particle size effect on water vapour permeability and mechanical properties of whey protein/beeswax emulsion films. *Journal of Agricultural and Food Chemistry*, 49(2), 996–1002.

Pérez-Mateos, M., Montero, P., and Gómez-Guillén, M.C. (2009). Formulation and stability of biodegradable films made from cod gelatin and sunflower oil blends. *Food Hydrocolloids*, 23(1), 53–61.

Pires, C., Ramos, C., Teixeira, B., Batista, I., Nunes, M.L., and Marques, A. (2013). Hake proteins edible films incorporated with essential oils: Physical, mechanical, antioxidant and antibacterial properties. *Food Hydrocolloids*, 30(1), 224–231.

Rhim, J.W. and Shellhammer, T.H. (2005). Lipid-based edible films and coatings. In J.H. Han (ed.), *Innovations in Food Packaging* (Elsevier Academic Press: London, U.K.), pp. 362–383.

Salgado, P.R., López-Caballero, M.E., Gómez-Guillén, M.C., Mauri, A.N., and Montero, M.P. (2013). Sunflower protein films incorporated with clove essential oil have potential application for the preservation of fish patties. *Food Hydrocolloids*, 33(1), 74–84.

Sánchez-González, L., Cháfer, M., Hernández, M., Chiralt, A., and González-Martínez, C. (2011a). Antimicrobial activity of polysaccharide films containing essential oils. *Food Control*, 22(8), 1302–1310.

Sánchez-González, L., González-Martínez, C., Chiralt, A., and Cháfer, M. (2010). Physical and antimicrobial properties of chitosan-tea tree essential oil composite films. *Journal of Food Engineering*, 98(4), 443–452.

Sánchez-González, L., González-Martínez, C., Chiralt, A., and Chafer, M. (2011b). Effect of essential oils on properties of film forming emulsions and films based on hydroxypropylmethylcellulose and chitosan. *Journal of Food Engineering*, 105(2), 246–253.

Saucedo-Pompa, S., Rojas-Molina, R., Aguilera-Carbó, A.F., Saenz-Galindo, A., de la Garza, H., Jasso-Cantú, D., and Aguilar, C.N. (2009). Edible film based on candelilla wax to improve the shelf life and quality of avocado. *Food Research International*, 42(4), 511–515.

Shojaee-Aliabadi, S., Hosseini, H., Mohammadifar, M.A., Mohammadi, A., Ghasemlou, M., Hosseini, S.M., and Khaksar, R. (2014). Characterization of κ-carrageenan films incorporated plant essential oils with improved antimicrobial activity. *Carbohydrate Polymers*, 101(30), 582–591.

Soazo, M., Pérez, L.M., Rubiolo, A.C., and Verdini, R.A. (2013). Effect of freezing on physical properties of whey protein emulsion films. *Food Hydrocolloids*, 31(2), 256–263.

Soradech, S., Limatvapirat, S., and Luangtana-Anan, M. (2013). Stability enhancement of shellac by formation of composite film: Effect of gelatin and plasticizers. *Journal of Food Engineering*, 116(2), 572–580.

Tanaka, M., Ishizaki, S., Suzuki, T., and Takai, R. (2001). Water vapour permeability of edible films prepared from fish water soluble proteins as affected by lipid type. *Journal of Tokyo University of Fisheries*, 87, 31–37.

Trezza, T.A. and Krochta, J.M. (2000). The gloss of edible coatings as affected by surfactants, lipids, relative humidity and time. *Journal of Food Science*, 65(4), 658–662.

Trezza, T.A. and Krochta, J.M. (2001). Specular reflection, gloss, roughness and surface heterogeneity of biopolymer coatings. *Journal of Applied Polymer Science*, 79(12), 2221.

Trigui, A., Karkri, M., and Krupa, I. (2014). Thermal conductivity and latent heat thermal energy storage properties of LDPE/wax as a shape-stabilized composite phase change material. *Energy Conversion and Management*, 77, 586–596.

Valenzuela, C., Abugoch, L., and Tapia, C. (2013). Quinoa protein-chitosan-sunflower oil edible film: Mechanical, barrier and structural properties. *LWT: Food Science and Technology*, 50(2), 531–537.

Vargas, M., Albors, A., Chiralt, A., and González-Martínez, C. (2009). Characterization of chitosan-oleic acid composite films. *Food Hydrocolloids*, 23, 536–547.

Vargas, M., Perdones, A., Chiralt, A., Cháfer, M., and González-Martínez, C. (2011). Effect of homogenization conditions on physicochemical properties of chitosan-based film forming dispersions and films. *Food Hydrocolloids*, 25(5), 1158–1164.

Velickova, E., Winkelhausen, E., Kuzmanova, S., Alves, V.D., and Moldão-Martins, M. (2013). Impact of chitosan-beeswax edible coatings on the quality of fresh strawberries (*Fragaria ananassa* cv Camarosa) under commercial storage conditions. *LWT: Food Science and Technology*, 52(2), 80–92.

Villalobos, R., Hernández, P., Chanona, J., Gutiérrez-López, G., and Chiralt, A. (2005). Gloss and transparency of hydroxypropylmethylcellulose films containing surfactants as affected by their microstructure. *Food Hydrocolloids*, 19, 53–61.

Yang, L. and Paulson, A.T. (2000). Effect of lipids on mechanical and moisture barrier properties of edible gellan film. *Food Research International*, 33(7), 571–578.

Yoshida, C.M.P. and Antunes, A.J. (2004). Characterization of whey protein emulsion films. *Brazilian Journal of Chemical Engineering*, 21(2), 247–252.

Zahedi, Y., Ghanbarzadeh, B., and Sedaghat, N. (2010). Physical properties of edible emulsified films based on pistachio globulin protein and fatty acids. *Journal of Food Engineering*, 100, 102–108.

Zhang, K., Han, B., and Yu, X. (2012). Electrically conductive carbon nanofiber/paraffin wax composites for electric thermal storage. *Energy Conversion and Management*, 64, 62–67.

11

Films and Coatings from Agro-Industrial Residues

Mariana S.L. Ferreira, Renata Linhares, and Milena Martelli

CONTENTS

11.1 Introduction

Disposal of wastes from nondegradable packaging represents a worldwide huge problem; this fact has challenged the scientific community to develop biodegradable packaging to reduce the environmental impact (Krochta and Johnson, 1997; Tharanathan, 2003). Hence, the development of polymeric materials, based on renewable sources, has become increasingly important over the last two decades, due to the inevitable rising prices of petroleum-based materials and the relevant environmental concerns.

Proteins and polysaccharides have historically been the most tested renewable and biodegradable materials for film or coating processing with the potential to replace many of the currently used hydrocarbon-derived plastics. Indeed, a wide range of naturally occurring polymers derived from renewable resources are nowadays available for biomaterials applications, such as polysaccharides from plants (starch, cellulose, pectin, alginate, carrageenan, gums), polysaccharides from animals (hyaluronic acid, chitin, chitosan), polysaccharides from fungi (pullulan, elsinan, scleroglucan), polysaccharides from bacteria (xanthan, polygalactosamine, curdlan, gellan, dextran, chitin), proteins (soy, zein, wheat gluten, casein, serum, albumin, collagen/gelatin), lipids/surfactants (acetoglycerides, waxes, surfactants), and other polymers (lignin, natural rubber) (Yu, 2009). However, even if these polymers present a renewable feature, they have been extracted from sources that could be further used for human consumption. Therefore, the utilization of alternative sources from agro-industrial residues would be a promising way of recycling such wastes into useful products in an eco-friendly manner.

Bio-derived polymers can be alternatively obtained from agro-industrial residues, which represent a nutritional, inexpensive, and eco-friendly raw material. In food agro-industry, about one-third of the total production has been annually discarded. Fruits and vegetables are extensively processed generating a very large amount of residues, mainly composed of peels, seed, stalks, and pomace, which are frequently discarded. In the same way, agricultural and forestry wastes (sugarcane bagasse, cereal straw, hulls and bran, oil cake, fruit pomace, etc.) constitute very profitable biomass residues for biopolymer production. Chitin and chitosan are also important bio-derived polymers generated by the seafood processing industry. Seafood by-products (head, skin, fin, scales, bones, cartilages, crustacean shell) represent novel sources that can be processed enzymatically for the production of these polymers. In this way, this chapter presents an overview of bio-derived polymers processed from agro-industrial residues in order to produce coating and film-forming matrices.

In brief, biopolymers, such as polysaccharides and proteins obtained from by-products of agricultural origin, have been more and more proposed for the formulation of biodegradable materials, since they are harmless, biocompatible, and susceptible to biodegradation, except when severe chemical modifications are applied, derived from renewable sources, and nontoxic to the soil and the environment. Additionally, in this chapter, some critical issues and strategies for future applications of alternative sources are discussed, highlighting some efforts to overcome the poor barrier properties, as well as some nutritional or biodegradability aspects, in order to boost their use as packaging material. Depending on specific applications of the films, targeted film functionality can be achieved by incorporating proper matrices and by improving biopolymeric extraction. More recently, nano-based materials from agro-industry by-products have been developed and applied into biopolymeric films, showing significant improvement in both mechanical and barrier properties.

11.2　Films and Coatings from Plant Industry Residues

11.2.1　Fruit and Vegetable Industry Residues

Plant residues are extensively processed for the beverage manufacture generating a large amount of residue, which is frequently discarded, causing disposal problems. This intense process entails the production of large amount of waste, estimated between 30% and 40% of agro-industrial waste. Classically, the outer layers and extremities of fruits and vegetables are removed during processing, mainly by peeling and pressing; they comprise essentially stalks, peels, seeds, and crashed pulps, which still contain large amounts of bioactive molecules and biopolymers that can be used for the preparation of biodegradable films and coatings.

11.2.1.1 Pectin Films

Pectin is one of the major cell wall structural polysaccharides of higher plants and thus widely available from underutilized agricultural waste. Since it is readily modified, through demethylation, considerable attention has been given to pectin in the preparation of biodegradable films and coatings. Pectin is composed of water-soluble pectinic acids (colloidal polygalacturonic acids) of different methyl ester contents and degrees of neutralization, able to form gels under appropriate conditions, such as with the presence of sugars and acids (Gitco, 1999; Ranganna, 1986).

Due to the gel-forming properties, pectin presents the potential to form a cohesive structural matrix prior to the preparation of biodegradable films (Arevalo et al., 2009; Batista, 2004; Nascimento et al., 2012). However, its operating efficiency depends on the nature of the ingredients added to the formulation; when glycerol and starch were added to the pectin filmogenic solution, the films showed good mechanical and flexibility properties (Fishman et al., 2000; Nascimento et al., 2012). In some cases, the use of glycerol or other suitable plasticizer is required to make starch-based films sufficiently flexible and nonbrittle. Indeed, the increase in glycerol concentration was directly related with the increase of strength and flexibility of pectin-based films (Coffin and Fishman, 1993, 1994). Pectin has been demonstrating high potential for the development of biodegradable films such as single or reinforcement material together with other matrices in composite films (Espitia et al., 2014; Galus and Lenart, 2013; Penhasi and Meidan, 2014).

TABLE 11.1

Variability in Anhydrouronic Acid, Methoxyl Content, and Gel Grade

Fruit Wastes Used for Pectin Extraction	Anhydrouronic Acid (%)	Methoxyl Content (%)	Gel Grade (%)
Mango peel	56.7	7.3	199
Jackfruit ring	66.0	7.7	159
Banana peel	53.0	7.0	99
Nutmeg ring	59.5	7.5	167
Pumello peel	64.2	8.6	202
Passion fruit ring	46.2	5.0	73
Cocoa pod husk	52.8	7.0	129
Lime peel	72.5	9.9	213
Mangosteen ring	73.2	10.5	171

Madhav and Pushpalatha (2002) showed that very pure pectin, presenting high levels of anhydrouronic acid, can be obtained from different fruit by-products, such as mangosteen rinds and lime peel (Table 11.1). Currently, the main sources of industrial by-products for the extraction of pectin are apple pomace and citrus peels (Videcoq et al., 2011). Previous works have shown that films made from elastic methoxyl citrus pectin and high amylose starch have very good mechanical properties (Coffin and Fishman, 1993, 1994). The films had tensile strength of the order of 3×10^8 dyn/cm^2, approaching the values found in commercial plastics (Coffin and Fishman, 1993). Further blend films based on sugar beet and almond pectin were developed, presenting similar mechanical properties comparable to citrus pectin films, but very low oxygen permeability (OP) (Coffin and Fishman, 1994).

Overall, fruit pomaces are mainly obtained from the fruit juice industry and represented a very rich source of pectin. The passion fruit shell (mesocarp) is a by-product of the industrial juice production and represents about 55%–90% of the fresh fruit (Arvanitoyannis and Varzakas, 2008; Kulkarni and Vijayanand, 2010). About 15% (db) of pectin can be extracted from the dried mesocarp of passion fruit (Kulkarni and Vijayanand, 2010). Previous study showed that films prepared with passion fruit mesocarp flour, using glycerol as plasticizer, presented higher viscosity, endurance, and strength when compared to starch films. On the other hand, it showed that they had low flexibility and high hydrophilic character (Nascimento et al., 2012).

Hence, water extracts of fruit pomace have also been appointed as a new film-forming material, able to form natural colors and fruit flavors of edible films. Cranberry pomace extracts were used to form films added with low or high methoxyl pectin, sorbitol, or glycerol (Park and Zhao, 2006). Films incorporated with low methoxyl pectin and sorbitol showed higher tensile strength, lower elongation at break, and lower permeability to water vapor when compared to other films obtained. Hence, targeted film functionality can be achieved by incorporating proper pectin type and concentration and plasticizer into pomace extracts (Park and Zhao, 2006).

Fruit pomace represents also good candidates for thermoforming applications, since some of these components aside from pectin, such as proteins, organic acids, and sugars, have thermoplastic properties. For this reason, fruit pomaces can also be processed to create biocomposites through the incorporation of other biopolymers. Park et al. (2010) demonstrated the feasibility of creating biocomposite boards from berry fruit pomaces (blueberry, cranberry, and wine grape pomaces) combined with soy flour, which can be applied in the packaging industry.

11.2.1.2 Starch Films

Globally, starch has been considered as one of the most promising resources on the development of biomaterials. The starch can be also extracted from fruit by-products, and according to its origin, this biopolymer will have different physicochemical properties and functionalities. Therefore, starch-based films can present different mechanical and barrier properties (Liu et al., 2005). Starch-based films provide effective barriers against oils and fats, but moisture barriers are ineffective and thus considered as limiting attribute (Durango et al., 2006; Larotonda et al., 2004; Oliveira and Cereda, 2003). Similar to pectin films, polyols are commonly used as plasticizers in starchy films to improve their flexibility (Gontard et al., 1993; Laohakunjit and Noomhorm, 2004). However, the hygroscopic characteristics of glycerol-plasticized starch films contributed to increase even more film hydrophilicity (García et al., 2009).

Remarkably, the unexplored parts of fruits have also been used to obtain starch-based films. Barbosa et al. (2011) produced films based on starch extracted from jackfruit pits added with glycerol. The plasticized starch films prepared from jackfruit pits showed high hydrophilicity and water vapor permeability (WVP) and low stability. Typically, the hydrophilic behavior depended directly on the amount of glycerol and water activity of the starch films. In the same context, Ooi et al. (2012) prepared biodegradable films with tropical fruit waste flours blended with polyvinyl alcohol (PVA). Rambutan skin waste flour and PVA, in the presence of glycerol as plasticizer or cross-linker agents, were used to prepare biodegradable films. In the preparation of plasticized biodegradable PVA/tropical fruit waste flour blends, these authors showed an improvement of the tensile strength and Young's modulus, lower elongation at break, and lower absorption and WVP throughout the cross-linking reaction.

11.2.1.3 Films from Fruit and Vegetable Purees

Since 1996, the fruit and vegetable purees obtained from industrial by-products or destined to waste, especially from seasonal fruits, have been suggested as an alternative source for the production of flexible films (McHugh et al., 1996). These films are mainly composed of cellulose and pectic compounds. Banana, for example, is a very fragile and highly perishable fruit, which can be rapidly rejected by the consumers (Martelli et al., 2013); from its harvest to the market, losses can reach up to 50% of the total volume produced (Sebrae, 2008). Therefore, overripe fruits represent an interesting raw material for plastic processing (Martelli et al., 2013).

The banana (Martelli et al., 2013; Sothornvit and Pitak, 2007), tomato (Du et al., 2008, 2009), mango (Azeredo et al., 2009; Sothornvit and Rodsamran, 2008, 2010), and carrot (Wang et al., 2011) purees are being processed and evaluated with respect to film production, characterization, and bactericide and fungicide properties, resulting in materials with good oxygen barrier, moisture, carbon dioxide, lipids, flavors, and acceptable mechanical properties. In Table 11.2, some results of mechanical properties, WVP, and OP of the films produced from mashed pulp of fruits and vegetables are summarized.

Films based on fruit purees offer sensitive mechanical properties and little flexibility, requiring the addition of plasticizers or other polysaccharides to improve resistance and processability. The incorporation of pectin, cellulose nanofibers, and chitosan nanoparticles has been evaluated such as reinforcing materials within fruit puree matrix for film production. The formation of nanocomposites, either fiber or particle forms, also proved very useful in improving the mechanical and barrier properties (Azeredo et al., 2009; Lorevice et al., 2012; McHugh and Olsen, 2004; McHugh and Senesi, 2000). Taken both aspects together, Martelli et al. (2013) demonstrated that overripe bananas with chitosan nanoparticles and small concentrations of pectin and glycerol proved satisfactory results when employed as a raw material for processing edible films. The obtained films showed good mechanical properties, although lacking of antimicrobial activity, due to the small amount of nanoparticles added to the chitosan films.

Multicomposite plant residues obtained from the juice processing of whole fruits and vegetables (Ferreira et al., 2015), which means including all edible and nonedible parts, such as peels, seeds, and stalks, represent a rich source of biopolymers, especially dietary fibers (Andrade et al., 2014). Therefore, it has been recently applied to develop biodegradable films and coatings (Andrade et al., 2016; Fai et al., 2016; Ferreira et al., 2016). Although the rheological behavior of the filmogenic solutions revealed the predominantly liquid-like character of the samples, the use of fruit and vegetable residue flour resulted in stand-alone and very flexible films without the addition of plasticizers (Ferreira et al., 2016). Hence, fruit and vegetable residues flour has been applied in the film packaging and coating of fruits and minimally fresh-cut carrots (Fai et al., 2016; Ferreira et al., 2016).

The formulated films exhibited promising characteristics as homogeneous aspect and high water solubility, which can be used for specific purposes in the food industry. Incorporation of potato skins

TABLE 11.2

Mechanical Properties (Maximum Strength at Break [σ_{max}] and Maximum Elongation [ε_{max}] and Elastic Modulus or Young's Modulus [E]) and Barrier (Water Vapor Permeability and Oxygen [O_2P]) of Biodegradable Films Based on Mashed Fruits and Vegetables

Film	σ_{max} (MPa)	ε_{max} (%)	E (MPa)	WVP (g·mm/ kPa·h·m²)	O_2P (cm³·μm/ m²d·kPa)	Test Conditions (WVP, O_2P)
Peach (McHugh et al., 1996)	—	—	—	4.2	69.6	RH = 80% T = 25°C
Damascus (McHugh et al., 1996)	—	—	—	4.3	—	RH = 80% T = 25°C
Apple (McHugh et al., 1996)	—	—	—	5.8	—	RH = 76% T = 25°C
Pear (McHugh et al., 1996)	—	—	—	7.8	—	RH = 74% T = 25°C
Apple, alginate (~1.4%) (Rojas-Grau et al., 2007)	2.9	51.1	7.1	4.95	0.43[c]	RH = 83% T = 25°C
Banana flour (4%), glycerol (30%) (Sothornvit and Pitak, 2007)	6.0	8.2	1.8	—	23	RH = 50% T = 25°C
Tomato, pectin (~2.1%) (Du et al., 2008)	11.4[a], 13.7[b]	11.2[a], 9.6[b]	248[a], 317[b]	2.4[a], 2.2[b]	—	RH = 81% T = 25°C
Mango (Sothornvit and Rodsamran, 2008)	1.2	18.5	8.3	8.9	41.2	RH = 50% T = 27°C
Mango (Azeredo et al., 2009)	4.1	44.1	19.9	2.7	—	RH = 83% T = 25°C
Overripe banana (4.5%) (Martelli et al., 2013)	1.1 ± 0.1	15 ± 2	11 ± 1	3.03	—	RH = 54% T = 25°C
Overripe banana (4.5%), pectin (0.5%) (Martelli et al., 2013)	3.2 ± 0.5	23 ± 3	21 ± 3	2.95	—	RH = 54% T = 25°C
Overripe banana (4.5%), pectin (0.5%), chitosan nanoparticles (0.2%) (Martelli et al., 2013)	4.5 ± 0.7	18 ± 2	43 ± 3	2.33	—	RH = 54% T = 25°C
Carrot (Wang et al., 2011)	7.5–21.9	4.4–46.2	—	0.26–0.99	10.8–17.5[c]	RH = 81% T = 23°C

[a] Batch process.
[b] Continuous process.
[c] Relative humidity (RH); RH, 50% and T, 25°C.

improved the tensile strength of films (Andrade et al., 2016; Ferreira et al., 2016). However, the high content of soluble compounds, such as sugars and globular proteins, has been appointed as a crucial factor in providing more flexibility but also less strength and stretchability of films than fruit starch–based films (Andrade et al., 2016; Sothornvit and Rodsamran, 2008).

11.2.2 Sugarcane Bagasse

Sugarcane bagasse (or "bagasse," as it is commonly called) is one of the largest cellulosic agro-industrial by-products. It represents a fibrous residue of cane stalks leftover after the crushing and extraction of sugarcane juice for ethanol and sugar production. The sugar and growing ethanol production makes the sugarcane industry one of the main economic segments in Brazil, the country leader in the sugarcane ethanol production. Generally, sugarcane mills generate approximately 225 kg of bagasse (db) per ton of cane; especially 70–80 million tons of dried bagasse is annually available in China, while in Brazil considering jointly sugar and ethanol production, it reaches 140 million tons (McKendry, 2002; Porto et al., 2013).

The bagasse contains up to 75% of cellulose and hemicellulose. It represents a complex polymer that chemically is mainly composed of, in dry-weight basis, about 40%–50% of cellulose forming a crystalline structure; 25%–35% of hemicelluloses, which are mainly composed of xylose, arabinose, galactose, and mannose monomers; and about 19% lignin and minor amounts of minerals, waxes, and other compounds (Adsul et al., 2004; Pandey et al., 2000; Jacobsen and Wyman, 2002).

Especially, in the biotechnology industry, bagasse has been considered a low-cost source for the extraction of lignocelluloses, which are being used as raw material for the development of various products, offering significant economic, environmental, and scientific benefits (Adsul et al., 2004; McKendry, 2002; Sabiha-Hanim and Siti-Norsafurah, 2012). The use of lignocellulosic feedstock for the development of biodegradable films or coatings can be considered an important and promising feature due to the potential to form an excellent barrier against WVP, which is generally increased in films based on biopolymers having very hydrophilic characteristics (Doherty et al., 2007; Driemeier et al., 2011).

Hemicelluloses extracted from crop residues present a good potential for the development of biodegradable films. Recently, Sabiha-Hanim and Siti-Norsafurah (2012) developed biodegradable films with alkaline hemicellulose extracted from crushed sugarcane. The different extracts of hemicelluloses resulted in films with thickness between 0.13 and 0.15 mm and are presented with wide-ranging properties, for instance, 36.9%–67.1% water solubility, 250.4–483.3 g·m²/day WVP, and 0.31–1.72 MPa tensile strength.

Previous work has shown that biodegradable films developed from hemicelluloses of oil palm fronds extracted by using different concentrations of alkali solution presented good barrier properties such as tensile strength between 11 and 15 MPa, 64%–93% water solubility, and 180–210 g·m²/day WVP (Noor Haliza et al., 2006). All hemicelluloses alkaline extracted produced self-supporting films with different properties, according to the difference in hemicelluloses composition, particularly the lignin content (Noor Haliza et al., 2006; Sabiha-Hanim and Siti-Norsafurah, 2012).

The bagasse contains biopolymers and antioxidant compounds mainly represented by the lignin–hemicellulosic complex, which have shown film-forming potential with hydrophobic and good mechanical characteristics. In addition to this film-forming potential, bagasse has also been tested as reducing free radicals (Adsul et al., 2004; Dizhbite et al., 2004; Fengel and Wegener, 1989). Indeed, the lignin extracted from crushed sugarcane is a natural phenolic polymer, which has a potential antioxidant. Li and Ge (2012) showed that lignin alkali extracted from bagasse has a significant free radical scavenging activity, due to their large amounts of phenolic hydroxyl (OH) and methoxyl groups (OCH_3) obtained during alkaline processing treatment. These functional groups played more important roles in the antioxidant activity of lignin than the molecular weight and polydispersity. Even if no relationship between film-forming and antioxidant capacities of bagasse is available in the literature, the alkali treatment applied for hemicelluloses extraction can add an antioxidant attribute to the bagasse-based films.

11.2.3 Residues from Cereal Straw, Bran, and Other Fiber Sources

The global agricultural sector produces very large amounts of biomass from several other crops in addition to sugarcane, such as soybeans, maize, rice, and wheat. These agricultural crop residues, mainly represented by the dry plant stalks, such as cereal straw and also bran, are produced in billions of tons around the world representing an abundant, inexpensive, and readily available source of lignocellulosic biomass.

By-products from the cultivation of corn, wheat, rice, sorghum, barley, sugarcane, pineapple, banana, and coconut are the major sources of agro-based biofibers (Reddy and Yang, 2005). These lignocellulosic fibers include advantageous characteristics when added as reinforcement to traditional biopolymeric fillers such as gluten and starch. They have low density, nonabrasive nature, and high levels of fillers, availability, and renewability (Pervaiz and Sain, 2004; Rouison et al., 2004; Woodhams et al., 1984). Moreover, these fibers provide high stiffness and tensile strength to the films (Satyanarayana et al., 2009).

Therefore, agricultural activities generate a substantial volume of waste that provides renewable fiber sources that can be readily incorporated in the production of biodegradable films. Several types of natural fibers from plant by-products have been explored as fillers in starch- and gluten-based biocomposite films, for example, pea hulls (Chen et al., 2009), coconut, sisal and jute fibers (Corradini et al., 2009), wheat straw (Montaño-Leyva et al., 2013), and lignin extracted from wheat straw (El-Wakil, 2009).

Several blends of composite films based on fibrous lignocellulosic components derived from agro-industrial wastes resulting from sugarcane, citrus fruits, corn, wheat, and wood processing, in conjunction with gelatin, starch, and polyvinyl alcohol (PVA), have been proposed (Chiellini et al., 2001, 2004). Notwithstanding the hydrophilic character of the biopolymers used as fillers, these agro-industrial residues showed to be suitable for blending in higher amounts in the production of cast films in the presence of cross-linking agents.

The wheat straw is an important agricultural waste applied in the biodegradable films production. Over 500 million of tons of wheat straw are annually produced in the world (Zhang et al., 2012). In Canada, the 6th in the world rank of wheat production, tons of unused wheat straws are produced each year and only a very small percentage has been applied as biomaterial or energy production (Tampier and Probe, 2002). Similar to other fibrous materials, the wheat straw has a suitable chemical composition for a film-forming material, consisting mainly of cellulose, hemicelluloses, and lignin, as summarized by Carvalheiro et al. (2009) (Table 11.3).

More recently, studies have also added natural lignocellulosic fiber from wheat straw (Montano-Leyva et al., 2013) by using a thermomechanical process and also other natural sources of lignocellulosic fiber as hemp and wood (Kunanopparat et al., 2008a,b) to form wheat gluten–based composite materials. Addition of natural fiber, in compression-molded films, has significantly improved the mechanical properties of these matrices, by increasing tensile strength and elastic Young's modulus, despite the decrease in the elongation at the break.

Hemicelluloses from wheat straw were also applied as reinforcement material in gum bases, such as κ-carrageenan/locust bean gum polymeric blend films (Ruiz et al., 2013). The incorporation of hemicelluloses from wheat straw, under certain proportion and in the presence of glycerol, caused decrease in WVP and increased in tensile strength, presenting a good potential as reinforcement for biodegradable films.

Rice bran is the most important rice by-product available worldwide. It is produced in large quantities and is considered as a low-cost underutilized by-product of rice-milling industry, which presents a good protein quality. Defatted rice bran contains about 12%–20% of total protein (Hamada, 2000), among which are different types of albumin, globulin, prolamin, and glutelin (Adebiyi et al., 2008b). In the last decades, its film-forming potential has been investigated in the literature (Adebiyi et al., 2008a; Gnanasambandam et al., 1997; Shih, 1996; Shin et al., 2011). Globally, the rice bran protein films presented functional properties comparable to those of the soy protein–based ones (Adebiyi et al., 2008a).

The corn stalk and corn cob are also widely produced and also represent interesting agro-industrial residues. Kayserilioglu et al. (2003) showed that the extracted xylan corn stalk added to a wheat gluten matrix has the potential to form suitable biodegradable films. Moreover, corn cobs have been proposed as filler in chitosan-based films in the presence of a cross-linking agent (Yeng et al., 2013).

TABLE 11.3

Average Macromolecular Chemical Composition of Wheat Straw (% Dry Weight)

Component	Learmonth (1971)	Carvalheiro et al. (2009)	Kabel et al. (2007)	Nabarlatz et al. (2007)
Cellulose	56.7	38.9	31	28.4
Lignin	16.6	—	—	—
Pentosans	28.4	—	—	—
Hemicelluloses	—	23.5	24.2	22.5
Xylan	—	18.1	20	17.4
Arabinoxylan	—	3.0	2.5	2.5
Acetyl groups	—	2.5	1.7	2.6
Klason lignin	—	18.0	25	15.9
Proteins	—	4.5	—	—

The interesting composition of soy hulls gives to this soy by-product the film-forming potential. Soy hulls present an average of 56.4% alpha cellulose, 12.5% hemicelluloses, 18% lignin (Alemdar and Sain, 2008b), and 7%–16% pectin (Monsoor and Proctor, 2001). Even if not widespread in the literature, Park et al. (2010) showed a feasibility of a soy–pomace system to provide appropriate mechanical properties for a variety of biomaterial applications where biodegradability was the key factor.

Cotton stalk consists of the left biomass available in the field after the harvest of seed cotton. About 25 million tons of cotton stalks are generated annually in India (Shaikh et al., 2009), and cotton stalk is the major agricultural waste in Turkey (Akpinar et al., 2007). Cotton stalk contains about 69% holocellulose (36% cellulose and 21% hemicelluloses) and 27% lignin (Akpinar et al., 2007; Shaikh et al., 2009). Goksu et al. (2007) produced cast films based on xylan extracted from cotton stalk. However, self-supporting continuous films could not be produced using pure cotton stalk xylan; film formation was achieved by using 8%–14% (w/w) xylan and about 1% (w/w) lignin. The WVP decreased when xylan concentration increased, rising the films thickness. The glycerol addition resulted in more stretchable films presenting higher water permeability and lower water solubility values.

It is interesting to note that the use of all these biobased and renewable materials can be envisaged not only to offer an alternative to petroleum-based materials but also to offer lower environmental global impact. In this way, singular green composite materials have been studied and proposed as a source of bio-derived polymers, such as fibrous residues of *Posidonia oceanica*, an endemic Mediterranean alga, which are deposited in large quantities on the beache coasts (Ferrero et al., 2013). These residues reach the coast in a continuous movement forming very big dried balls. Recently, these authors proposed films by using a fibrous material derived from *P. oceanica* wastes with high cellulose content (90%, db) on a wheat gluten matrix by hot-press molding (Ferrero et al., 2013). The formed films showed cohesive matrices, and the water sensitivity depended on the cellulose content present in the dry waste used.

11.2.4 Edible Oil and Beverage Industry Residues

The use of inexpensive agricultural and food processing by-products also includes oil cakes as feedstock and has been highly favored in view of facilitating better utilization of edible oil cakes as sources of protein, for instance (Nigam et al., 2009). As mentioned earlier, rice bran protein can be used as a film-forming material. However, due to its poor solubility and tendency for aggregation (Adebiyi et al., 2008b), the protein isolated from deoiled rice bran residue has been considered as a more suitable material for film production (Shin et al., 2011). Indeed, rice bran protein is not able to form film with good physical properties, requiring different plasticizers, such as sucrose, fructose, glycerol, polypropylene glycol, and sorbitol, resulting in poor mechanical properties (Shin et al., 2011). In contrast, protein extracts from rice bran oil residues added by gelatin and red algae matrix were efficient for the production of multicomposite biodegradable films (Shin et al., 2011).

The wine grape pomace is a by-product of the alcoholic beverage industry, generally composed of 30% seeds and 70% skins and small stems (Guendez et al., 2005; Mattick and Rice, 1976). Compared to the stems, wine grape pomace seeds and skins have more oil, protein, pectin, and sugar (Llobera and Cañellas, 2007), besides being a rich source of dietary fibers and polyphenols (Katalinić et al., 2010). Overall, the grape skins include 39 types of anthocyanins, hydroxycinnamic acids, catechins, and flavonoids (Kammerer et al., 2004). Wine grape pomace extracts may be utilized as a film-forming material, since these extracts contain pectin, celluloses, and sugars (Deng et al., 2011; Deng and Zhao, 2011). The natural pigments, flavors, and polyphenols from wine grape pomace extracts would provide additional benefits to its applications, such as antimicrobial activity and antioxidant properties (Deng and Zhao, 2011; Tseng and Zhao, 2012).

In pomace-based edible films, the addition of small amounts of film-forming materials (protein or polysaccharide) is generally required to obtain films with adequate mechanical and barrier properties (Cerruti et al., 2011; Corrales et al., 2009; Deng and Zhao, 2011; Mayachiew and Devahastin, 2010). Cerruti et al. (2011) used the polyphenol-containing extract from winery waste as an additive to starch-based films (Mater-Bi®). As a result, an improved productivity in the film processing was achieved, together with an increase in elongation at break and delayed thermal aging of the films. The extract also assigned antimicrobial activity, reducing the rate of disintegration of Mater-Bi films.

Noticeably, food industry residues represent a rich source of biopolymers and functional compounds. Taken all these studies together, plant residues have demonstrated a good potential for application in the preparation of biodegradable films and may be a means to promote the use of these residues largely discarded.

11.3 Films and Coatings from Animal Industry Residues

During the manufacturing of animal products for human consumption (meat and dairy products) or for other human needs (leather), a large production of residues is generated. In the slaughter process, the main available waste products are the blood, liver, hair, bones, feathers, fat, and wastewater. In dairy plants, by-products obtained depend on the type of product produced (e.g., milk, cheese, butter, milk powder, condensate). In cheese production, the wastewater could have a considerable amount of whey. In addition, a large quantity of processing by-products are accumulated from marine bioprocessing plants, including fins, frames, heads, skin, viscera, and shells of crustaceans and shellfish. These residues or by-products present some additional applications, including the biopolymers extraction for developing biodegradable films and coatings.

11.3.1 Animal Bones, Skins, and Feathers

The main animal by-products obtained from the animal processing industry consist of parts of carcasses, catering waste (including used cooking oil), butcher and slaughterhouse waste, blood, feathers, wool, hides and skins, and fallen stock. In the animal industry, the quantity of animal by-products often exceeds 50% of the live weight, and the dressing percentages of carcasses range from 57% (standard cattle grades) to 70% (chicken), depending on the animal (FAOSTAT, 2014). The main polymers produced from animal bones, skins, and feathers are collagen, gelatin, and keratin, respectively.

Collagen is a fibrous structural protein, with a particular amino acid composition, rich in glycine (33%), proline (12%), alanine (11%), hydroxyproline (10%), and hydroxylysine (1%), and occurs as a significant component of the skin, bone, tendon, and connective tissues (Damodaran et al., 2008). For industrial production, first, the insoluble native collagen is converted into a suitable form for extraction by water heating at temperatures higher than 45°C. The production of collagen involves subsequently chemical and enzymatic treatments. One of the most important applications of collagen as packaging material is in the sausage production and also in other meat products, which, due to their preserved fibrous structure, display excellent mechanical and oxygen barrier characteristics (Langmaler et al., 2008).

Remarkable, collagen biomaterials have also wide applications in biomedicine, as reviewed by Lee et al. (2001): drug delivery systems as collagen shields in ophthalmology; sponges for burns/wounds; minipellets and tablets for protein delivery; gel formulation in combination with liposomes for sustained drug delivery, as controlling material for transdermal delivery; and nanoparticles for gene delivery and basic matrices for cell culture systems. It has been also studied and applied for tissue engineering including skin replacement, bone substitutes, and artificial blood vessels and valves.

Alternatively, gelatin is produced by partial hydrolysis of collagen under moist heating. The amino acid glycine is present in the most concentrated form with 20.6 g per 100 g gelatin, followed by proline with 11.7 g and lysine (3.4 g per 100 g gelatin) (GME, 2014). The degree of collagen conversion into gelatin is related to the severity of both the pretreatment and the warm-water extraction process, as a function of pH, temperature, and extraction time (Gällstedt et al., 2011). Hence, two types of gelatin can be obtained, the type A gelatin (isoelectric point at pH $\sim$8–9) and type B gelatin (isoelectric point at pH $\sim$ 4–5) obtained under acid and alkaline pretreatment conditions, respectively. As a result, the gel-forming properties are different between the two types of gelatin.

Industrial applications claim for one or the other gelatin type, depending on the degree of collagen cross-linking in the raw material. A recent review was published with wide information about collagen and gelatin extraction from new sources (Gomez-Guillen et al., 2011). Moreover, new processing conditions and potential novel or improved applications were also related. Indeed, many of which are largely based on induced cross-linking, blending with other biopolymers, or enzymatic hydrolysis (Gomez-Guillen et al., 2011).

Gelatin films have been widely studied as an application in food preservation, either as coatings or as films (Bergo et al., 2013; Gomez-Estaca et al., 2009; Gomez-Guillen et al., 2009; Tongnuanchan et al., 2013; Vanin et al., 2005). Although gelatin is considered highly hydrophilic, globally gelatin films present very good processability and acceptable barrier and mechanical properties. Edible coating was also studied as an oil barrier on deep fat frying of chicken nuggets (Martelli et al., 2006). Interestingly, the gelatin origin and film-processing parameters have significant influence on the functional properties of the resulting gelatin-based films (Gomez-Estaca et al., 2009). Polysaccharide–gelatin interactions have been applied for microencapsulation (Bruschi et al., 2003; Chilvers and Morris, 1987; Nakagawa, 2013), drug release (Kumar, 2000; Tabata and Ikada, 1998; Young et al., 2005), and tissue adhesion (Ohya et al., 2005).

In the same line, keratin comprehends a family of another fibrous protein, rich in cysteine, glycine, proline, and serine. Keratin can be mainly found in hair, wool, feathers, and other epithelial coverings (Perez-Gago, 2012). The amino acid contents vary between different keratins, depending on their nature and conformations: α-keratin is the principal protein of cytoskeleton intermediate filaments in mammalian epithelia, while α-keratin is mainly found in feathers, claws, beaks of birds, and reptilian skins (Kessel and Ben-Tal, 2011). For instance, chicken feathers are agricultural residues rich in keratin contents, probably the most abundant keratinous material (85%–99%) (Shi and Dumont, 2014). An estimated 5 million tons is produced annually as a waste stream from the production of chicken meat, of which over 65 million tons was produced worldwide in 2007 (Poole et al., 2009).

Keratin-based films can be processed after extraction procedures, and it was found that films were mechanically strong (Katoh et al., 2004; Yamauchi et al., 1996). Appropriate selection of plasticizer type and concentration can be helpful in controlling film properties, like WVP (Martelli et al., 2006). According to Shi and Dumont (2014), keratin-based films could have a promising use in the tissue engineering and medical fields, due to their suitable medium for the attachment and proliferation of mouse fibroblast and good mammalian cell adhesion and proliferation.

11.3.2 Milk By-Products

In 2011, the total milk and dairies production in the European Union reaches 156 million tons and 142 million tons, respectively, 98% of which was cows' milk (EUROSTAT, 2014). Bovine milk proteins consist of about 80% casein and 20% whey proteins (Damodaran et al., 2008). Whey proteins are beta-lactoglobulin, alpha-lactalbumin, bovine serum albumin, and immunoglobulin. The composition and functional properties of whey are highly variable as the main by-product of cheese production (Morr and Ha, 1993). About 9 kg of whey are formed when 1 kg of cheese is produced. Because this product is highly perishable, new processes are needed to make use of the possible residues.

Whey proteins have been extracted in order to produce coatings and films. Whey protein films have low tensile strength and high WVP due to the high proportion of hydrophilic amino acid in their structures (McHugh et al., 1994). Their mechanical properties can be improved by blending the protein with other biopolymers, like natural latex and egg white albumin (Sharma and Luzinov, 2013). Due to their good oxygen barrier, whey protein coatings substantially reduced oxygen uptake and rancidity of roasted peanuts (Maté et al., 1996) and improved microbial quality of poultry products (Fernandez-Pan et al., 2013).

11.3.3 Marine Food Processing

During marine food processing, a considerable amount of by-products is produced and that includes trimmings, fins, frames, heads, skin, viscera, and residues from crab, shrimp, and crawfish. Studies revealed that current discards from the worldwide fisheries would exceed 20 million tons, which is equivalent to 25% of the total production of marine capture fisheries (Kim and Mendis, 2006). However, some compounds could be isolated from these by-products, such as fish skin collagen and gelatin, fish oils, omega-3 fatty acids, fish bone as potential calcium or mineral sources, and chitin, chitosan, and their oligomers (Kim and Mendis, 2006). Apart from collagen and gelatin, chitin and chitosan are important bioactive compounds extracted from marine food processing.

Chitin is the second most abundant polysaccharide in nature after cellulose and is found in the exoskeleton of crustaceans, in fungal cell walls, and in other biological materials. Chitin is represented as a linear polysaccharide composed of β-(1→4) linked units of N-acetyl-2-amino-2-deoxy-D-glucose (Soares, 2009). Chitin can be processed in the form of films and fibers, as binders in the paper maker (as revised by Rinaudo, 2006). However, the most important chitin derivatives, in terms of application in biodegradable films and coatings, are the chitosan.

Chitosan (β-(1,4)-2-amino-2-deoxy-D-glucose) is produced by extensive deacetylation of chitin, as revised by Kumar et al. (2004), when the degree of deacetylation reaches values of about 50% and then becomes soluble in aqueous acidic media (Soares, 2009). Chitosan is a biodegradable, biocompatible, nontoxic amino polysaccharide that can easily form gels. Moreover, chitosan presents the potential interest as an inherent antimicrobial film-forming material (No et al., 2002). Due to these characteristics, several works have been studied using chitosan-based films/coatings of food products, as revised by No et al. (2007) mainly for improvement of quality and shelf life of fruits and vegetables (Assis and Britto, 2011; Chien et al., 2007; Devlieghere et al., 2004; Dong et al., 2004; Goy and Assis, 2014; Pilon et al., 2013).

Most of these studies have shown the antimicrobial activity of chitosan and the improvement of shelf life of the fresh food products. The typical mean of chitosan-processing films has been the casting method based on organic acidic water solutions. It is noteworthy that chitosan is already widely used internationally for different applications, but its potential for film packaging and coating applications is not completely explored, especially with regard to active packaging concerns (Fernandez-Saiz and Lagaron, 2011).

11.4 Nanoparticles Obtained from Alternative Sources and Their Application on Biopolymeric Films and Coatings

Among the main polymers that can be extracted from agro-industrial residues, especially chitosan and fibers have been produced in nanoscale in order to improve film/coating properties. In the main polymers, the reduced particle size has higher superficial surface and is more reactive (Durán et al., 2006). The higher reactivity is desired according to the chemical processes involved during film formation.

The study of cellulosic nanofibers as a reinforcing phase in nanocomposites started 19 years ago (Favier et al., 1995). Since then, a huge amount of literature has been devoted to cellulose nanofibers, and it is becoming an increasingly topical subject. Table 11.4 presented the main agro-industrial residues used to produce cellulose nanofibers. Different descriptions of these nanofibers are often referred to in the literature. These include "nanowhiskers" (or just simply "whiskers"), "nanocrystals," or even "monocrystals." These crystallites have also often been referred to in the literature as "microfibrils," "microcrystals," or "microcrystallites," despite their nanoscale dimensions. The term "whiskers" is used to designate elongated crystalline rodlike nanoparticles, whereas the designation "nanofibrils" should be used to designate long flexible nanoparticles consisting of alternating crystalline and amorphous strings (Eichhorn, 2011; Silva et al., 2009). The shape and size distribution of fiber crystalline and amorphous strings depends on the lignocellulosic biomass (Elazzouzi-Hafraoui et al., 2008).

Cellulose nanofibers are essentially obtained according to the following steps: partial hydrolysis, using acid (HCl or H_2SO_4) or enzymatic treatments (xylanases and cellulases) to break fiber structure into crystals, and fragmentation using mechanical treatments (high-pressure or ultrasonic treatments). Combined processes can be used, and different nanofiber characteristics are obtained. In general, the addition of cellulose nanofiber improves mechanical and barrier properties of films (Alemdar and Sain, 2008a; Chauve et al., 2005; Samir et al., 2004).

TABLE 11.4

Main Agro-Industrial Residues Used to Produce Cellulose Nanofibers

Agro-Industrial Residue	Procedure	Nanoparticle Dimension	Applications	References
Wheat straw and soy hulls	Alkaline, acid, and mechanical treatments	WS, diameter 30–40 nm SB, diameter 20–120 nm Length, >100 nm	Reinforcement of starch-based thermoplastic polymer (improvement of tensile strength and modulus and glass transition shifted to higher values)	Alemdar and Sain (2008a,b)
Wheat straw	Alkaline and steam explosion, followed by acid treatment (HCl)	Diameter, 10–50 nm	Reinforcements of thermoplastic corn starch composites (improvement of tensile strength and modulus and a reduction in water sorption)	Kaushik et al. (2010), Kaushik and Singh (2011)
Curauá (C) and sugarcane bagasse	Alkaline and bleaching, followed by enzymatic preparation (hemicell/pectinase and endoglucanase) and sonification	C diameter, 55–109 nm Length, 1.3–4.1 µm SB diameter, 20–40 nm Length, 0.25–0.82 µm	Potential for reinforcing polymer composites	Campos et al. (2013)
Rice straw pulp	Grinding and high-pressure homogenization	Diameter, 4–13 nm	Reinforcement of chitosan films (improvement of mechanical and thermal properties)	Hassan et al. (2012)
Sugar beet	Chemical (NaOH, $NaClO_2$) and mechanical treatments (ultrahigh-pressure homogenizer)	Diameter, 30–100 nm Length, >1 µm	• Cellulose microfibrils: stable suspensions • Reinforcement of PVA and phenol formaldehyde resin	Dinand et al. (1996) Leitner et al. (2007)
Potato tuber cells	Alkaline, bleaching, and mechanical treatment (ultrahigh-pressure homogenizer)	Diameter, 2–4 nm Length, >1 µm	Reinforcement of potato starch nanocomposite films (improvement of mechanical properties and reduction in water uptake and water-diffusion coefficient)	Dufresne et al. (2000)
Alfa, eucalyptus, and pine fibers	Catalytic oxidation and mechanical treatment (ultrahigh-pressure homogenizer)	Diameter, 2–4 nm	Reinforcement of unbleached eucalyptus fiber matrix (enhancement of the mechanical properties and reduction in porosity)	Besbes et al. (2011), Alcala et al. (2013)
Sisal fibers	Acid hydrolysis, chlorination, alkaline extraction, and bleaching	Diameter, 2–11 nm Length, 360–1700 nm		Moran et al. (2008)

(Continued)

TABLE 11.4 (*Continued*)

Main Agro-Industrial Residues Used to Produce Cellulose Nanofibers

Agro-Industrial Residue	Procedure	Nanoparticle Dimension	Applications	References
Pea hull fiber	Acid hydrolysis, bleaching, and dialysis	Diameter, 7–12 nm Length, 240–400 nm	Reinforcement of pea starch films (improvement of mechanical properties, higher ultraviolet absorption, transparency, and water resistance)	Chen et al. (2009)
Hemp (H) and flax (F) fibers	Acid hydrolysis, bleaching, and dialysis or sonification	H diameter, 20–40 nm F diameter, 10–30 nm Length, 100–500 nm	Reinforcement of pea starch films (improvement of mechanical properties and water resistance)	Cao et al. (2008a,b)
Cassava bagasse	Acid hydrolysis, dialysis, and sonification	Diameter, 2–11 nm Length, 360–1700 nm	Reinforcement of cassava starch films (decrease of hydrophilic character and capacity of water uptake)	Teixeira et al. (2009)

The formation of nanocomposites by nanofiber addition proved very useful in enhancing both mechanical and barrier properties of puree films; the Young's modulus increased 16 times when 36 g of nanofibers/100 g of mango puree (db) was added (Azeredo et al., 2009). For the same concentration used, the tensile ultimate stress increased from 4.0 to 8.8 MPa and WVP decreased by 37%. The possible formation of an entangled network could be responsible for the strong increase in thermomechanical stability of films (Dalmas et al., 2007).

Chitosan nanoparticles have been also studied to improve mechanical properties of films/coatings. Despite their ability to form films, chitosan is known to possess good antifungal and antibacterial properties (Devlieghere et al., 2004; Dutta et al., 2009; No et al., 2007; Vasconez et al., 2009), both desirable for food applications. Recently, chitosan nanoparticles were successfully used as vitamin (B9, B12, C) carrier, with potential applications in foodstuffs (Britto et al., 2012).

The literature reports several attempts to produce chitosan particles with different particle sizes, by ion tropic gelation with sodium tripolyphosphate (Janes and Alonso, 2003) or methacrylic acid (MAA) polymerization (Moura et al., 2008). In the second case, the particle size is dependent on the chitosan concentration used in the nanoparticle preparation and is greatly influenced by the solution pH (pH sensitive). Nanoparticles obtained from both methodologies were incorporated on hydroxypropyl methylcellulose (HPMC) edible films (Moura et al., 2009). Nanoparticles obtained from MAA polymerization with the concentration of 0.2% (w/v) of chitosan presented the most important results for improving HPMC film properties: tensile ultimate stress increased from 30 to 67 MPa and WVP decreased by 40%.

Improvements on mechanical properties and WVP were also observed when the same concentration of MAA chitosan nanoparticles was used as reinforcement of banana puree films (Martelli et al., 2013). In low concentrations, chitosan nanoparticles have been confirmed to be nontoxic according to the analysis performed by Lima et al. (2010).

The approach of nanotechnology as reinforcements in composites offers a way for improving the agro-industrial residue uses, due to nanoparticle ability to chemically modify film/coating surface. A number of methods have been reviewed that enable nanoparticles to be extracted from either plant or animal sources. It has to be remembered that in order to do this, some efforts are needed in order to reduce the large amounts of energy used and effluent generation.

11.5 Concluding Remarks

The generation of immense quantity of agro-industrial residues has long been recognized as wastes, and huge efforts have been made to use these materials in different applications, especially as source of biologically active compounds, such as bio-derived polymers. Both plant and animal residues or by-products are rich sources of some biopolymers that have been studied as an alternative to produce films and coatings, such as starch, pectin, collagen, gelatin, chitin, chitosan, and fats.

Noticeably, the main drawback of most biopolymer-based films/coatings is the mechanical and moisture barrier properties. It can be concluded that depending on specific applications of the films, targeted film functionality can be achieved by incorporating proper matrices and by improving biopolymeric extraction. To straight this point, several extraction processes have been described in the literature, as well as film and coating production based on blends or multicomposite formulations.

Hence, by producing blend composites, such as combining proteins (e.g., milk proteins, soy protein, collagen, and gelatin) with polysaccharides (e.g., starches, alginates, cellulose, and chitosan) or other polymers, it is possible to improve the barrier and physical properties of films. In other cases, cross-linking techniques could be an interesting process that accounts with chemical, enzymatic, and physical processes to attain materials with better properties. On the other hand, innovative applications need also to be explored in order to further enable uses of biopolymer-based eco-friendly packaging materials.

Recent progress has also been done in the area of nanocomposites with the purpose of improving biopolymeric film and coating properties. The main nanoparticles obtained from agro-industrial residues are cellulose nanofibers based on lignocellulosic biomass (wheat straw, soy hulls, rice straw, sugarcane bagasse, cassava bagasse, sisal, pea, hemp, flax, and others) and also chitosan nanoparticles from marine food processing or bacterial origin. It is important to emphasize that concerns about the toxicity and environmental impact of nanocomposites remain not fully understood, thus requiring more studies.

REFERENCES

Adebiyi, A.P., Adebiyi, A.O., Jin, D.H., Ogawa, T., and Muramoto, K. Rice bran protein-based edible films. *International Journal of Food Science & Technology* 43(3) (2008a): 476–483.

Adebiyi, A.P., Adebiyi, A.O., Ogawa, T., and Muramoto, K. Purification and characterisation of antioxidative peptides from unfractionated rice bran protein hydrolysates. *International Journal of Food Science & Technology* 43(1) (2008b): 35–43.

Adsul, M.G., Ghule, J.E., Singh, R., Shaikh, H., Bastawde, K.B., Gokhale, D.V., and Varma, A.J. Polysaccharides from bagasse: Applications in cellulase and xylanase production. *Carbohydrate Polymers* 57(1) (2004): 67–72.

Akpinar, O. et al. Enzymatic production of xylooligosaccharides from cotton stalks. *Journal of Agricultural and Food Chemistry* 55(14) (2007): 5544–5551.

Alcala, M., Gonzalez, I., Boufi, S., Vilaseca, F., and Mutje, P. All-cellulose composites from unbleached hardwood kraft pulp reinforced with nanofibrillated cellulose. *Cellulose* 20(6) (2013): 2909–2921.

Alemdar, A. and Sain, M. Biocomposites from wheat straw nanofibers: Morphology, thermal and mechanical properties. *Composites Science and Technology* 68(2) (2008a): 557–565.

Alemdar, A. and Sain, M. Isolation and characterization of nanofibers from agricultural residues—Wheat straw and soy hulls. *Bioresource Technology* 99(6) (2008b): 1664–1671.

Andrade, R.M.S., Ferreira, M.S.L., and Gonçalves, E.C.B.A. Evaluation of the functional capacity of fruit and vegetable residue flour. *International Food Research Journal* 21(4) (2014): 1675–1681.

Andrade, R.M.S., Ferreira, M.S.L., Gonçalves, E.C.B.A. Development and characterization of edible films based on fruit and vegetable residues. *Journal of Food Science* 81(2): (2016): 412–418.

Arevalo, K., Aleman, E., Rojas, G., Morales, L., and Galan, L.J. Properties and biodegradability of cast films based on agroindustrial residues, pectin and polivinilic alcohol (PVA). *New Biotechnology* 25 (2009): 287–288.

Arvanitoyannis, L.S. and Varzakas, T.H. Fruit/fruit juice waste management: Treatment methods and potential uses of treated waste. *Waste Management for the Food Industries* 2 (2008): 569–628.

Assis, O.B.G. and Britto, D. Evaluation of the antifungal properties of chitosan coating on cut apples using a non-invasive image analysis technique. *Polymer International* 60 (2011): 932–936.

Azeredo, H.M.C., Mattoso, L.H.C., Wood, D., Williams, T.G., Avena-Bustillos, R.J., and McHugh, T.H. Nanocomposite edible films from mango puree reinforced with cellulose nanofibers. *Journal of Food Science* 74(5) (2009): N31–N35.

Barbosa, H.R., Ascheri, D.P.R., Ascheri, J.L.R., and Carvalho, C.W.P. Permeabilidade, estabilidade e funcionalidade de filmes biodegradáveis de amido de caroço de jaca (*Artocarpus heterophyllus*). *Revista Agrotecnologia* 2(1) (2011): 73–88.

Batista, J.A. Desenvolvimento, Caracterização e Aplicações de Biofilmes à Base de Pectina, Gelatina e Ácidos Graxos em Bananas e Sementes de Brócolos. MSc dissertation (in Portuguese). Campinas, Brazil: Universidade Estadual de Campinas, 2004.

Bergo, P., Moraes, I.C.F., and Sobral, P.J.A. Effects of plasticizer concentration and type on moisture content in gelatin films. *Food Hydrocolloids* 32(2) (2013): 412–415.

Besbes, I., Vilar, M.R., and Boufi, S. Nanofibrillated cellulose from Alfa, Eucalyptus and Pine fibres: Preparation, characteristics and reinforcing potential. *Carbohydrate Polymers* 86(3) (2011): 1198–1206.

Britto, D., Moura, M.R., Aouada, F.A., Mattoso, L.H.C., and Assis, O.B.G. N,N,N-Trimethyl chitosan nanoparticles as a vitamin carrier system. *Food Hydrocolloids* 27(2) (2012): 487–493.

Bruschi, M.L., Cardoso, M.L.C., Lucchesi, M.B., and Gremiao, M.P.D. Gelatin microparticles containing propolis obtained by spray-drying technique: Preparation and characterization. *International Journal of Pharmaceutics* 264(1–2) (2003): 45–55.

Campos, A., Correa, A.C., Cannella, D., Teixeira, E.D.M., Marconcini, J.M., Dufresne, A., Mattoso, L.H.C., Cassland, P., and Sanadi, A.R. Obtaining nanofibers from curaua and sugarcane bagasse fibers using enzymatic hydrolysis followed by sonication. *Cellulose* 20(3) (2013): 1491–1500.

Cao, X., Chen, Y., Chang, P.R., Muir, A.D. and Falk, G. Starch-based nanocomposites reinforced with flax cellulose nanocrystals. *Express Polymer Letters* 2(7) (2008a): 502–510.

Cao, X., Chen, Y., Chang, P.R., Stumborg, M., and Huneault, M.A. Green composites reinforced with hemp nanocrystals in plasticized starch. *Journal of Applied Polymer Science* 109(6) (2008b): 3804–3810.

Carvalheiro, F., Silva-Fernandes, T., Duarte, L.C., and Gírio, F.M. Wheat straw autohydrolysis: Process optimization and products characterization. *Applied Biochemistry and Biotechnology* 153(1–3) (2009): 84–93.

Cerruti, P., Santagata, G., Gomez d'Ayala, G., Ambrogi, V., Carfagna, C., Malinconico, M., and Persico, P. Effect of a natural polyphenolic extract on the properties of a biodegradable starch-based polymer. *Polymer Degradation and Stability* 96(5) (2011): 839–846.

Chauve, G., Heux, L., Arouini, R., and Mazeau, K. Cellulose poly(ethylene-co-vinyl acetate) nanocomposites studied by molecular modeling and mechanical spectroscopy. *Biomacromolecules* 6 (2005): 2025–2031.

Chen, Y., Liu, C., Chang, P.R., Cao, X., and Anderson, D.P. Bionanocomposites based on pea starch and cellulose nanowhiskers hydrolyzed from pea hull fibre: Effect of hydrolysis time. *Carbohydrate Polymers* 76(4) (2009): 607–615.

Chiellini, E., Cinelli, P., Chiellini, F., and Imam, S.H. Environmentally degradable bio-based polymeric blends and composites. *Macromolecular Bioscience* 4(3) (2004): 218–231.

Chiellini, E., Cinelli, P., Imam, S.H., and Mao, L. Composite films based on biorelated agro-industrial waste and poly(vinyl alcohol): Preparation and mechanical properties characterization. *Biomacromolecules* 2(3) (2001): 1029–1037.

Chien, P.-J., Sheu, F., and Yang, F.-H. Effects of edible chitosan coating on quality and shelf life of sliced mango fruit. *Journal of Food Engineering* 78 (2007): 225–229.

Chilvers, G.R. and Morris, V.J. Coacervation of gelatin gellan gum mixtures and their use in microencapsulation. *Carbohydrate Polymers* 7(2) (1987): 111–120.

Coffin, D.R. and Fishman, M.L. Viscoelastic properties of pectin/starch blends. *Journal of Agricultural and Food Chemistry* 41(8) (1993): 1192–1197.

Coffin, D.R. and Fishman, M.L. Physical and mechanical properties of highly plasticized pectin/starch films. *Journal of Applied Polymer Science* 54(9) (1994): 1311–1320.

Corradini, E., Imam, S.H., Agnelli, J.A.M., and Mattoso, L.H.C. Effect of coconut, sisal and jute fibers on the properties of starch/gluten/glycerol matrix. *Journal of Polymers and the Environment* 17 (2009): 1–9.

Corrales, M., Han, J.H., and Tauscher, B. Antimicrobial properties of grape seed extracts and their effectiveness after incorporation into pea starch films. *International Journal of Food Science & Technology* 44(2) (2009): 425–433.

Dalmas, F., Cavaillíe, J.Y., Gauthier, C., Chazeau, L., and Dendievel, R. Viscoelastic behavior and electrical properties of flexible nanofiber filled polymer nanocomposites. Influence of processing conditions. *Composites Science and Technology* 67 (2007): 829–839.

Damodaran, S., Parkin, K.L., and Fennema, O.R. *Fennemás Food Chemistry*. New York: CRC Press Taylor & Francis Group, 2008.

Deng, Q., Penner, M.H., and Zhao, Y. Chemical composition of dietary fiber and polyphenols of five different varieties of wine grape pomace skins. *Food Research International* 44(9) (2011): 2712–2720.

Deng, Q. and Zhao, Y. Physicochemical, nutritional, and antimicrobial properties of wine grape (cv. Merlot) pomace extract-based films. *Journal of Food Science* 76(3) (2011): E309–E317.

Devlieghere, F., Vermeulen, A., and Debevere, J. Chitosan: Antimicrobial activity, interactions with food components and applicability as a coating on fruit and vegetables. *Food Microbiology* 21(6) (2004): 703–714.

Dinand, E., Chanzy, H., and Vignon, M.R. Parenchymal cell cellulose from sugar beet pulp: Preparation and properties. *Cellulose* 3(3) (1996): 183–188.

Dizhbite, T., Telysheva, G., Jurkjane, V., and Viesturs, U. Characterization of the radical scavenging activity of lignins-natural antioxidants. *Bioresource Technology* 95(3) (2004): 309–317.

Doherty, W., Halley, P., Edye, L., Rogers, D., Cardona, F., Park, Y., and Woo, T. Studies on polymers and composites from lignin and fiber derived from sugar cane. *Polymers for Advanced Technologies* 18(8) (2007): 673–678.

Dong, H., Cheng, L., Tan, J., Zheng, K., and Jiang, Y. Effects of chitosan coating on quality and shelf life of peeled litchi fruit. *Journal of Food Engineering* 64 (2004): 355–358.

Driemeier, C., Oliveira, M.M., Mendes, F.M., and Gómez, E.O. Characterization of sugarcane bagasse powders. *Powder Technology* 214(1) (2011): 111–116.

Du, W.X., Olsen, C.W., Avena-Bustillos, R.J., McHugh, T.H., Levin, C.E., and Friedman, M. Antibacterial activity against E-*coli* O157: H7, physical properties, and storage stability of novel carvacrol-containing edible tomato films. *Journal of Food Science* 73(7) (2008): M378–M383.

Du, W.X., Olsen, C.W., Avena-Bustillos, R.J., McHugh, T.H., Levin, C.E., and Friedman, M. Effects of allspice, cinnamon, and clove bud essential oils in edible apple films on physical properties and antimicrobial activities. *Journal of Food Science* 74(7) (2009): M372–M378.

Dufresne, A., Dupeyre, D., and Vignon, M.R. Cellulose microfibrils from potato tuber cells: Processing and characterization of starch-cellulose microfibril composites. *Journal of Applied Polymer Science* 76(14) (2000): 2080–2092.

Durán, N., Mattoso, L.H.C., and Moraes, P.C. *Nanotecnologia—Introdução, preparação e caracterização de nanomateriais e exemplos de aplicação*. São Carlos, Brazil: Artiliber Editora Ltda, 2006.

Durango, A.M., Soares, N.F.F., Benevides, S., Teixeira, J., Carvalho, M., Wobeto, C., and Andrade, N.J. Development and evaluation of an edible antimicrobial film based on yam starch and chitosan. *Packaging Technology and Science* 19(1) (2006): 55–59.

Dutta, P.K., Tripathi, S., Mehrotra, G.K., and Dutta, J. Perspectives for chitosan based antimicrobial films in food applications. *Food Chemistry* 114(4) (2009): 1173–1182.

Eichhorn, S.J. Cellulose nanowhiskers: Promising materials for advanced applications. *Soft Matter* 7 (2011): 303–315.

Elazzouzi-Hafraoui, S., Nishiyama, Y., Putaux, J.L., Heux, L., Dubreuil, F., and Rochas, C. The shape and size distribution of crystalline nanoparticles prepared by acid hydrolysis of native cellulose. *Biomacromolecules* 9(1) (2008): 57–65.

Elsabee, M.Z. and Abdou, E.S. Chitosan based edible films and coatings: A review. *Materials Science and Engineering C: Materials for Biological Applications* 33(4) (2013): 1819–1841.

Espitia, P.J.P., Du, W.X., Avena-Bustillos, R.D.J., Soares, N.D.F.F., and McHugh, T. Edible films from pectin: Physical-mechanical and antimicrobial properties-a review. *Food Hydrocolloids* 35 (2014): 287–296.

El-Wakil, N.A. Use of lignin strengthened with modified wheat gluten in biodegradable composites. *Journal of Applied Polymer Science* 113 (2009): 793–801.

EUROSTAT. Available from: http://epp.eurostat.ec.europa.eu/statistics_explained/index.php/Milk_and_dairy_production_statistics. Accessed on January 13, 2014.

Fai, A.E.C., Souza, M.R.A, Barros, S.T., Bruno, N.V., Ferreira, M.S.L., Gonçalves, E.C.B.A. Development and evaluation of biodegradable films and coatings obtained from fruit and vegetable residues applied to fresh-cut carrot (*Daucus carota* L.). *Postharvest Biology and Technology* 112 (2016): 194–204.

FAOSTAT. Available from: http://www.fao.org/wairdocs/lead/x6114e/x6114e04.htm#b6-2.1.2.%20Quantities%20of%20byproducts. Accessed on January 13, 2014.

Favier, V., Chanzy, H., and Cavaille, J.Y. Polymer nanocomposites reinforced by cellulose whiskers. *Macromolecules* 28(18) (1995): 6365–6367.

Fengel, D. and Wegener, G. *Wood: Chemistry, Ultrastructure, Reactions.* Berlin, Germany: Walter de Gruyter, 1989.

Fernandez-Pan, I., Mendoza, M., and Mate, J.I. Whey protein isolate edible films with essential oils incorporated to improve the microbial quality of poultry. *Journal of the Science of Food and Agriculture* 93(12) (2013): 2986–2994.

Fernandez-Saiz, P. and Lagaron, J.M. Chitosan for film and coating applications. In D.V. Plackett (ed.), *Biopolymers—New Materials for Sustainable Films and Coatings*, pp. 87–106. Chichester, U.K.: Wiley, 2011.

Ferreira, M.S., Fai, A.E., Andrade, C.T., Picciani, P.H., Azero, E.G., Gonçalves, É.C. Edible films and coatings based on biodegradable residues applied to acerolas (*Malpighia punicifolia* L.). *Journal of the Science of Food and Agriculture* 96 (2016): 1634–1642.

Ferreira, M.S.L., Santos, M.C.P., Moro, T.M.A., Basto, G.J., Andrade, R.M.S., and Gonçalves, E.C.B.A. Formulation and characterization of functional foods based on fruit and vegetable residue flour. *Journal of Food Science and Technology* 52(2) (2015): 822–830.

Ferrero, B., Boronat, T., Moriana, R., Fenollar, O., and Balart, R. Green composites based on wheat gluten matrix and posidonia oceanica waste fibers as reinforcements. *Polymer Composites* 34(10) (2013): 1663–1669.

Fishman, M.L., Coffin, D.R., Konstance, R.P., and Onwulata, C.I. Extrusion of pectin/starch blends plasticized with glycerol. *Carbohydrate Polymers* 41(4) (2000): 317–325.

Gällstedt, M., Hedenqvist, M.S., and Ture, H. Production, chemistry and properties of proteins. In D.V. Plackett (ed.), *Biopolymers—New Materials for Sustainable Films and Coatings*, pp. 107–129. Chichester, U.K.: Wiley, 2011.

Galus, S. and Lenart, A. Development and characterization of composite edible films based on sodium alginate and pectin. *Journal of Food Engineering* 115(4) (2013): 459–465.

García, M.A., Pinotti, A., Martino, M.N., and Zaritzky, N.E. Characterization of starch and composite edible films and coatings. In M.E. Embuscado and K.C. Huber (eds.), *Edible Films and Coatings for Food Applications*, pp. 169–210. New York: Springer, 2009.

Gitco, H. *Twenty-Five Prospective Food Processing Projects.* Ahmadabad, India: Gujarat Industrial and Technical Consultancy Organization Limited, Vol. 2, 1999.

GME Market Data. Official website of GME e Gelatin manufacturers of Europe. Brussels, Belgium: GME Market Data, 2014. Available from: http://www.gelatine.org.

Gnanasambandam, R., Hettiarachchy, N.S., and Coleman M. Mechanical and barrier properties of rice bran films. *Journal of Food Science* 62(2) (1997): 395–398.

Goksu, E.I., Karamanlioglu, M., Bakir, U., Yilmaz, L., and Yilmazer, U. Production and characterization of films from cotton stalk xylan. *Journal of Agricultural and Food Chemistry* 55(26) (2007): 10685–10691.

Gomez-Guillen, M.C., Gimenez, B., Lopez-Caballero, M.E., and Montero, M.P. Functional and bioactive properties of collagen and gelatin from alternative sources: A review. *Food Hydrocolloids* 25(8) (2011): 1813–1827.

Gomez-Estaca, J., Gomez-Guillen, M.C., Fernandez-Martin, F., and Montero, P. Effects of gelatin origin, bovine-hide and tuna-skin, on the properties of compound gelatin-chitosan films. *Food Hydrocolloids* 25(6) (2011): 1461–1469.

Gomez-Estaca, J., Montero, P., Fernandez-Martin, F., and Gomez-Guillen, M.C. Physico-chemical and film-forming properties of bovine-hide and tuna-skin gelatin: A comparative study. *Journal of Food Engineering* 90(4) (2009): 480–486.

Gomez-Guillen, M.C., Perez-Mateos, M., Gomez-Estaca, J., Lopez-Caballero, E., Gimenez, B., and Montero, P. Fish gelatin: A renewable material for developing active biodegradable films. *Trends in Food Science & Technology* 20(1) (2009): 3–16.

Gontard, N., Guilbert, S., and Cuq, J.L. Water and glycerol as plasticizers affect mechanical and water vapor barrier properties of an edible wheat gluten film. *Journal of Food Science* 58(1) (1993): 206–211.

Goy, R.C. and Assis, O.B.G. Antimicrobial analysis of films processed from chitosan and n,n,n-trimethylchitosan. *Brazilian Journal of Chemical Engineering* 31(3) (2014): 643–648.

Guendez, R., Kallithraka, S., Makris, D.P., and Kefalas, P. Determination of low molecular weight polyphenolic constituents in grape (*Vitis vinifera* sp.) seed extracts: Correlation with antiradical activity. *Food Chemistry* 89 (2005): 1–9.

Hamada, J.S. Characterization and functional properties of rice bran proteins modified by commercial exoproteases and endoproteases. *Journal of Food Science* 65(2) (2000): 305–310.

Hassan, M.L., Fadel, S.M., El-Wakil, N.A., and Oksman, K. Chitosan/rice straw nanofibers nanocomposites: Preparation, mechanical, and dynamic thermomechanical properties. *Journal of Applied Polymer Science* 125 (2012): E216–E222.

Jacobsen, S.E. and Wyman, C.E. Xylose monomer and oligomer yields for uncatalyzed hydrolysis of sugarcane bagasse hemicellulose at varying solids concentration. *Industrial & Engineering Chemistry Research* 41(6) (2002): 1454–1461.

Janes, K.A. and Alonso, M.J. Depolymerized chitosan nanoparticles for protein delivery: Preparation and characterization. *Journal of Applied Polymer Science* 88(12) (2003): 2769–2776.

Kabel, M.A., Bos, G., Zeevalking, J., Voragen, A.G., and Schols, H.A. Effect of pretreatment severity on xylan solubility and enzymatic breakdown of the remaining cellulose from wheat straw. *Bioresource Technology* 98(10) (2007): 2034–2042.

Katalinić, V., Možina, S.S., Skroza, D., Generalić, I., Abramovič, H., Miloš, M., and Boban, M. Polyphenolic profile, antioxidant properties and antimicrobial activity of grape skin extracts of 14 Vitis vinifera varieties grown in Dalmatia (Croatia). *Food Chemistry* 119 (2010): 715–723.

Kammerer, D., Claus, A., Carle, R., and Schieber, A. Polyphenol screening of pomace from red and white grape varieties (*Vitis vinifera* L.) by HPLC-DAD-MS/MS. *Journal of Agricultural and Food Chemistry* 52 (2004): 4360–4367.

Katoh, K., Shibayama, M., Tanabe, T., and Yamauchi, K. Preparation and physicochemical properties of compression-molded keratin films. *Biomaterials* 25(12) (2004): 2265–2272.

Kaushik, A. and Singh, M. Isolation and characterization of cellulose nanofibrils from wheat straw using steam explosion coupled with high shear homogenization. *Carbohydrate Research* 346(1) (2011): 76–85.

Kaushik, A., Singh, M., and Verma, G. Green nanocomposites based on thermoplastic starch and steam exploded cellulose nanofibrils from wheat straw. *Carbohydrate Polymers* 82(2) (2010): 337–345.

Kayserilioglu, B.Ş., Bakir, U., Yilmaz, L., and Akkaş, N. Use of xylan, an agricultural by-product, in wheat gluten based biodegradable films: Mechanical, solubility and water vapor transfer rate properties. *Bioresource Technology* 87(3) (2003): 239–246.

Kessel, A. and Ben-Tal, N. *Introduction to Proteins: Structure, Function, and Motion.* New York: CRC Press Taylor & Francis Group, 2011.

Kim, S.-K. and Mendis, E. Bioactive compounds from marine processing byproducts—A review. *Food Research International* 39 (2006): 383–393.

Krochta, J.M. and Johnson, C.D.M. Edible and biodegradable polymer films: Challenges and opportunities. *Food Technology* 51 (1997): 61–73.

Kulkarni, S.G. and Vijayanand, P. Effect of extraction conditions on the quality characteristics of pectin from passion fruit peel (*Passiflora edulis* f. *flavicarpa* L.). *Food Science and Technology* 43 (2010): 1026–1031.

Kumar, M. Nano and microparticles as controlled drug delivery devices. *Journal of Pharmacy and Pharmaceutical Sciences* 3(2) (2000): 234–258.

Kumar, M., Muzzarelli, R.A.A., Muzzarelli, C., Sashiwa, H., and Domb, A.J. Chitosan chemistry and pharmaceutical perspectives. *Chemical Reviews* 104(12) (2004): 6017–6084.

Kunanopparat, T., Menut, P., Morel, M.H., and Guilbert, S. Plasticized wheat gluten reinforcement with natural fibers: From mechanical improvement to deplasticizing effect. *Composites Part A: Applied Science and Manufacturing* 39(5) (2008a): 777–785.

Kunanopparat, T., Menut, P., Morel, M.H., and Guilbert, S. Plasticized wheat gluten reinforcement with natural fibers: Effect of thermal treatment on the fiber/matrix adhesion. *Composites Part A: Applied Science and Manufacturing* 39(12) (2008b): 1787–1792.

Langmaler, F., Mokrejs, P., Kolomamik, K., and Mladek, M. Biodegradable packing materials from hydrolysates of collagen waste proteins. *Waste Management* 28(3) (2008): 549–556.

Laohakunjit, N. and Noomhorm, A. Effect of plasticizers on mechanical properties of rice starch film. *Starch/ Stärke* 56(8) (2004): 348–356.

Larotonda, F.D.S., Matsui, K.N., Soldi, V., and Laurindo, J.B. Biodegradable films made from raw and acetylated cassava starch. *Brazilian Archive of Biological and Technology* 47(3) (2004): 477–484.

Learmonth, G.S. *Fillers for Plastics*. London, U.K.: Iliffe, 1971.

Lee, C.H., Singla, A., and Lee, Y. Biomedical applications of collagen. *International Journal of Pharmaceutics* 221(1–2) (2001): 1–22.

Leitner, J., Hinterstoisser, B., Wastyn, M., Keckes, J., and Gindl, W. Sugar beet cellulose nanofibril-reinforced composites. *Cellulose* 14(5) (2007): 419–425.

Li, Z. and Ge, Y. Antioxidant activities of lignin extracted from sugarcane bagasse via different chemical procedures. *International Journal of Biological Macromolecules* 51(5) (2012): 1116–1120.

Lima, R., Feitosa, L., Pereira, A.D.S., de Moura, M.R., Aouada, F.A., Mattoso, L.H.C., and Fraceto, L.F. Evaluation of the genotoxicity of chitosan nanoparticles for use in food packaging films. *Journal of Food Science* 75(6) (2010): N89–N96.

Liu, L., Kerry, J.F., and Kerry, J.P. Selection of optimum extrusion technology parameters in the manufacture of edible/biodegradable packaging films derived from food-based polymers. *Journal of Food, Agriculture & Environment* 3(3/4) (2005): 51–58.

Llobera, A. and Cañellas, J. Dietary fibre content and antioxidant activity of Manto Negro red grape (*Vitis vinifera*): Pomace and stem. *Food Chemistry* 101 (2007): 659–666.

Lorevice, M.V., De Moura, M.R., Aouada, F.A., and Mattoso, L.H.C. Development of novel guava puree films containing chitosan nanoparticles. *Journal of Nanoscience and Nanotechnology* 12 (2012): 2711–2717.

Madhav, A. and Pushpalatha, P.B. Characterization of pectin extracted from different fruit wastes. *Journal of Tropical Agriculture* 40 (2002): 53–55.

Martelli, M.R., Barros, T.T., de Moura, M.R., Mattoso, L.H.C., and Assis, O.B.G. Effect of chitosan nanoparticles and pectin content on mechanical properties and water vapor permeability of banana puree films. *Journal of Food Science* 78(1) (2013): N98–N104.

Martelli, M.R., Santos, J.S., Sobral, P.J.A., and Carvalho, R.A. Efeito de cobertura comestível a base de gelatina na fritura de nuggets. In *CIBIA V—Libro de articulos en extenso del 5° Congresso Ibero-Americano de Engenharia de Alimentos*, Puerto Vallarta, México, 2006, pp. 1–5.

Martelli, M.S., Moore, G., Silva Paes, S., Gandolfo, C., and Laurindo, J.B. Influence of plasticizers on the water sorption isotherms and water vapor permeability of chicken feather keratin films. *LWT: Food Science and Technology* 39(3) (2006): 292–301.

Maté, J.I., Frankel, E.N., and Krochta, J.M. Whey protein isolate edible coatings: Effect on the rancidity process of dry roasted peanuts. *Journal of Agricultural and Food Chemistry* 44 (1996): 1736–1740.

Mattick, L.R. and Rice, A.C. Fatty acid composition of grape seed oil from native American and hybrid grape varieties. *Journal of American Enology and Viticulture* 27 (1976): 88–90.

Mayachiew, P. and Devahastin, S. Effects of drying methods and conditions on release characteristics of edible chitosan films enriched with Indian gooseberry extract. *Food Chemistry* 118 (2010): 594–601.

McHugh, T.H., Aujard, J.F., and Krochta, J.M. Plasticized whey protein edible films: Water vapor permeability properties. *Journal of Food Science* 59(2) (1994): 416–419.

McHugh, T.H., Huxsoll, C.C., and Krochta, J.M. Permeability properties of fruit puree edible films. *Journal of Food Science* 61(1) (1996): 88–91.

McHugh, T.H. and Olsen, C.W. Tensile properties of fruit and vegetable edible films. United States–Japan Cooperative Program in Natural Resources, 2004, pp. 104–108.

McHugh, T.H. and Senesi, E. Apple wraps: A novel method to improve the quality and extend the shelf-life of fresh-cut apples. *Journal of Food Science* 65(3) (2000): 480–485.

Mckendry, P. Energy production from biomass (part 2): Conversion technologies. *Bioresource Technology* 83(1) (2002): 47–54.

Monsoor, M.A. and Proctor, A. Preparation and functional properties of soy hull pectin. *Journal of the American Oil Chemists' Society* 78(7) (2001): 709–713.

Montaño-Leyva, B., da Silva, G.D., Gastaldi, E., Torres-Chávez, P., Gontard, N., and Angellier-Coussy, H. Biocomposites from wheat proteins and fibers: Structure/mechanical properties relationships. *Industrial Crops and Products* 43 (2013): 545–555.

Moran, J.I., Alvarez, V.A., Cyras, V.P., and Vazquez, A. Extraction of cellulose and preparation of nanocellulose from sisal fibers. *Cellulose* 15(1) (2008): 149–159.

Morr, C.V. and Ha, E.Y.W. Whey protein concentrates and isolates: Processing and functional properties. *Critical Reviews in Food Science and Nutrition* 33(6) (1993): 431–476.

Moura, M.R., Aouada, F.A., Avena-Bustillos, R.J., McHugh, T.H., Krochta, J.M., and Mattoso, L.H.C. Improved barrier and mechanical properties of novel hydroxypropyl methylcellulose edible films with chitosan/tripolyphosphate nanoparticles. *Journal of Food Engineering* 92(4) (2009): 448–453.

Moura, M.R., Aouada, F.A., and Mattoso, L.H.C. Preparation of chitosan nanoparticles using methacrylic acid. *Journal of Colloid and Interface Science* 321(2) (2008): 477–483.

Nabarlatz, D., Ebringerová, A., and Montané, D. Autohydrolysis of agricultural by-products for the production of xylo-oligosaccharides. *Carbohydrate Polymers* 69(1) (2007): 20–28.

Nakagawa, K. Characterization of freeze-dried core-shell nanoparticles prepared via gelatin-acacia complex coacervation: A study on particle formation upon freezing. *Drying Technology* 31(13–14) (2013): 1466–1476.

Nascimento, T.A., Calado, V., and Carvalho, C.W.P. Development and characterization of flexible film based on starch and passion fruit mesocarp flour with nanoparticles. *Food Research International* 49(1) (2012): 588–595.

Nigam, P., Pandey, A., Sivaramakrishnan, S., and Gangadharan, D. Edible oil cakes. In P.S. Nigam and A. Pandey (eds.), *Biotechnology for Agro-Industrial Residues Utilisation*, pp. 253–271. Dordrecht, the Netherlands: Springer, 2009.

No, H.K., Meyers, S.P., Prinyawiwatkul, W., and Xu, Z. Applications of chitosan for improvement of quality and shelf life of foods: A review. *Journal of Food Science* 72(5) (2007): R87–R100.

No, H.K., Park, N.Y., Lee, S.H., and Meyers, S.P. Applications of chitosan oligomers with different molecular weights. *International Journal of Food Microbiology* 74(1–2) (2002): 65–72.

Noor Haliza, A.H., Fazilah, A., and Mohd Azemi, M.N. Development of hemicelluloses biodegradable films from oil palm frond (*Elaeis guineensis*). In *International Conference on Green and Sustainable Innovation*, Energy Management and Conservation Center (EMAC), Chiang Mai, Thailand, 2006, pp. 260–265.

Ohya, S., Sonoda, H., Nakayama, Y., and Matsuda, T. The potential of poly(N-isopropylacrylamide) (PNIPAM)-grafted hyaluronan and PNIPAM-grafted gelatin in the control of post-surgical tissue adhesions. *Biomaterials* 26(6) (2005): 655–659.

Oliveira, M.A. and Cereda, M.P. Pós-colheita de pêssegos (*Prunus pérsica* L. Bastsch) revestidos com filmes a base de amido como alternativa à cera comercial. *Ciência e Tecnologia de Alimentos* 23 (2003): 28–33.

Ooi, Z.X., Ismail, H., Bakar, A.A., and Aziz, N.A.A. Properties of the crosslinked plasticized biodegradable poly (vinyl alcohol)/rambutan skin waste flour blends. *Journal of Applied Polymer Science* 125(2) (2012): 1127–1135.

Pandey, A., Soccol, C.R., Nigam, P., and Soccol, V.T. Biotechnological potential of agro-industrial residues. I: Sugarcane bagasse. *Bioresource Technology* 74(1) (2000): 69–80.

Park, S., Jiang, Y., Simonsen, J., and Zhao, Y. Feasibility of creating compression-molded biocomposite boards from berry fruit pomaces. *Journal of Applied Polymer Science* 115 (2010): 127–136.

Park, S. and Zhao, Y. Development and characterization of edible films from cranberry pomace extracts. *Journal of Food Science* 71(2) (2006): E95–E101.

Penhasi, A. and Meidan, V.M. Preparation and characterization of *in situ* ionic cross-linked pectin films: Unique biodegradable polymers. *Carbohydrate Polymers* 102 (2014): 254–260.

Perez-Gago, M. Protein-based films and coatings. In E.A. Baldwin, R.D. Hagenmaier, and J. Bai (eds.), *Edible Coatings and Films to Improve Food Quality*, pp. 13–77. New York: CRC Press Taylor & Francis Group, 2012.

Pervaiz, M. and Sain, M. High performance natural fiber thermoplastics for automotive interior parts. *Training* 2013 (2004): 11–11.

Pilon, L., Britto, D., Assis, O.B.G., Calbo, A.G., and Ferreira, M.D. Effects of antibrowning solution and chitosan-based edible coating on the quality of fresh-cut apple. *International Journal of Postharvest Technology and Innovation* 3 (2013): 151.

Poole, A.J., Church, J.S., and Huson, M.G. Environmentally sustainable fibers from regenerated protein. *Biomacromolecules* 10 (2009): 1–8.

Porto, S.I., Silva, A.C.P., and Oliveira, E.P. Relatório da produção brasileira de cana-de-açúcar. CONAB— Companhia Nacional de Abastecimento. Disponível em: http://www.conab.gov.br/OlalaCMS/uploads/arquivos/13_08_08_09_39_29_boletim_cana_portugues_-_abril_2013_1o_lev.pdf, acesso em January 19, 2014.

Ranganna, S. *Handbook of Analysis and Quality Control for Fruit and Vegetable Products.* New Delhi, India: Tata Me Graw-Hill Publishing Company, 1986, p. 1112.

Rinaudo, M. Chitin and chitosan: Properties and applications. *Progress in Polymer Science* 31(7) (2006): 603–632.

Reddy, N. and Yang, Y. Biofibers from agricultural byproducts for industrial applications. *Trends in Biotechnology* 23(1) (2005): 22–27.

Rojas-Grau, M.A., Raybaudi-Massilia, R.M., Soliva-Fortuny, R.C., Avena-Bustillos, R.J., McHugh, T.H., and Martin-Belloso, O. Apple puree-alginate edible coating as carrier of antimicrobial agents to prolong shelf-life of fresh-cut apples. *Postharvest Biology and Technology* 45(2) (2007): 254–264.

Rouison, D., Couturier, M., and Sain, M. The effect of surface modification on the mechanical properties of hemp fiber/polyester composites. *Training* 2013 (2004): 11–11.

Ruiz, H.A., Cerqueira, M.A., Silva, H.D., Rodríguez-Jasso, R.M., Vicente, A.A., and Teixeira, J.A. Biorefinery valorization of autohydrolysis wheat straw hemicellulose to be applied in a polymer-blend film. *Carbohydrate Polymers* 92(2) (2013): 2154–2162.

Sabiha-Hanim, S. and Siti-Norsafurah, A.M. Physical properties of hemicellulose films from sugarcane bagasse. *Procedia Engineering* 42 (2012): 1518–1523.

Samir, M., Alloin, F., Sanchez, J.Y., El Kissi, N., and Dufresne, A. Preparation of cellulose whiskers reinforced nanocomposites from an organic medium suspension. *Macromolecules* 37 (2004): 1386–1393.

Satyanarayana, K.G., Arizaga, G.G.C., and Wypych, F. Biodegradable composites based on lignocellulosic fibers-an overview. *Progress in Polymer Science* 34(9) (2009): 982–1021.

Sebrae. Banana: Estudos de Mercado. Série Mercado: Report Sebrae/ESPM, 86. Brasília, Brazil, 2008.

Shaikh, A.J., Gurjar, R.M., Patil, P.G., Paralikar, K.M., Varadarajan, P.V., and Balasubramanya, R.H. Particle boards from cotton stalk. Central Institute for Research on Cotton Technology, Mumbai, India, 2009.

Sharma, S. and Luzinov, I. Whey based binary bioplastics. *Journal of Food Engineering* 119 (2013): 404–410.

Shi, W. and Dumont, M.-J. Review: Bio-based films from zein, keratin, pea, and rapeseed protein feedstocks. *Journal of Materials Science* 49 (2014): 1915–1930.

Shih, F.F. Edible films from rice protein concentrate and pullulan. *Cereal Chemistry* 73(3) (1996): 406–409.

Shin, Y.J., Jang, S.A., and Song, K.B. Preparation and mechanical properties of rice bran protein composite films containing gelatin or red algae. *Food Science and Biotechnology* 20(3) (2011): 703–707.

Silva, R., Haraguchi, S.K., Muniz, E.C., and Rubira, A.F. Aplicações de fibras lignocelulósicas na química de polímeros e em compósitos. *Química Nova* 32(3) (2009): 661–671.

Soares, N.F.F. Chitosan—Properties and application. In L. Yu (ed.), *Biodegradable Polymer Blends and Composites from Renewable Resources*, pp. 107–128. Hoboken, NJ: John Wiley & Sons, 2009.

Sothornvit, R. and Pitak, N. Oxygen permeability and mechanical properties of banana films. *Food Research International* 40(3) (2007): 365–370.

Sothornvit, R. and Rodsamran, P. Effect of a mango film on quality of whole and minimally processed mangoes. *Postharvest Biology and Technology* 47(3) (2008): 407–415.

Sothornvit, R. and Rodsamran, P. Mango film coated for fresh-cut mango in modified atmosphere packaging. *International Journal of Food Science and Technology* 45(8) (2010): 1689–1695.

Tabata, Y. and Ikada, Y. Protein release from gelatin matrices. *Advanced Drug Delivery Reviews* 31(3) (1998): 287–301.

Tampier, M. and Probe, P. *Promoting Green Power in Canada.* Pollution Probe Organization, Toronto, Ontario, Canada, 2002.

Teixeira, E.D.M., Pasquini, D., Curvelo, A.A.S., Corradini, E., Belgacem, M.N., and Dufresne, A. Cassava bagasse cellulose nanofibrils reinforced thermoplastic cassava starch. *Carbohydrate Polymers* 78(3) (2009): 422–431.

Tharanathan, R.N. Biodegradable films and composite coatings: Past, present and future. *Trends in Food Science and Technology* 14(3) (2003): 71–78.

Tongnuanchan, P., Benjakul, S., and Prodpran, T. Characteristics and antioxidant activity of leaf essential oil-incorporated fish gelatin films as affected by surfactants. *International Journal of Food Science and Technology* 48(10) (2013): 2143–2149.

Tseng, A. and Zhao, Y. Effect of different drying methods and storage time on the retention of bioactive compounds and antibacterial activity of wine grape pomace (Pinot noir and Merlot). *Journal of Food Science* 77(9) (2012): H192–H201.

Vanin, F.M., Sobral, P.J.A., Menegalli, F.C., Carvalho, R.A., and Habitante, A. Effects of plasticizers and their concentrations on thermal and functional properties of gelatin-based films. *Food Hydrocolloids* 19(5) (2005): 899–907.

Vasconez, M.B., Flores, S.K., Campos, C.A., Alvarado, J., and Gerschenson, L.N. Antimicrobial activity and physical properties of chitosan-tapioca starch based edible films and coatings. *Food Research International* 42(7) (2009): 762–769.

Videcoq, P., Garnier, C., Robert, P., and Bonnin, E. Influence of calcium on pectin methylesterase behaviour in the presence of medium methylated pectins. *Carbohydrate Polymers* 86(4) (2011): 1657–1664.

Wang, X.W., Sun, X.X., Liu, H., Li, M., and Ma, Z.S. Barrier and mechanical properties of carrot puree films. *Food and Bioproducts Processing* 89(C2) (2011): 149–156.

Woodhams, R.T., Thomas, G., and Rodgers, D.K. Wood fibers as reinforcing fillers for polyolefins. *Polymer Engineering & Science* 24(15) (1984): 1166–1171.

Yamauchi, K., Yamauchi, A., Kusunoki, T., Kohda, A., and Konishi, Y. Preparation of stable aqueous solution of keratins, and physiochemical and biodegradational properties of films. *Journal of Biomedical Materials Research* 31(4) (1996): 439–444.

Yeng, C.M., Husseinsyah, S., and Ting, S.S. Chitosan/corn cob biocomposite films by cross-linking with glutaraldehyde. *Bio Resources* 8(2) (2013): 2910–2923.

Young, S., Wong, M., Tabata, Y., and Mikos, A.G. Gelatin as a delivery vehicle for the controlled release of bioactive molecules. *Journal of Controlled Release* 109(1–3) (2005): 256–274.

Yu, L. *Biodegradable Polymer Blends and Composites from Renewable Resources*. Hoboken, NJ: John Wiley & Sons, Inc., 2009.

Zhang, Y., Ghaly, A.E., and Li, B. Physical properties of wheat straw varieties cultivated under different climatic and soil conditions in three continents. *American Journal of Engineering and Applied Sciences* 5(2) (2012): 98–106.

Section III

Strategies to Optimize Coating and Film Functionality

12

Conventional and Alternative Plasticizers and Cross-Linkers

Delia R. Tapia-Blácido and Bianca C. Maniglia

CONTENTS

12.1 Introduction

The plastic industry has used plasticizers since the 1800s. Some plasticizers, like di(2-ethylhexyl) phthalate (DEHP), have raised concern, because they can migrate and cause toxicity. Legislation and health safety issues have prompted the development of commercial plasticizers. Nowadays, interest in biopolymer-based films and coatings has increased, and nonconventional plasticizers can serve as additive in these materials. Polyols, fatty acids (FA), monosaccharides, ethanolamine (EA), urea, triethanolamine (TEA), vegetable oils, lecithin, waxes, amino acids, surfactants, and water can function as plasticizers in polysaccharides, proteins, and polysaccharides/protein mixed films. The plasticizer renders biopolymer-based films more flexible, processable, and extensible. The plasticizer type and amount affect the biofilm mechanical properties, barrier properties to oxygen and water vapor, optical clarity, degree of crystallinity, and glass transition temperature (T_g).

Cross-linkers can also function as additives that improve polysaccharides and protein films' performance. Indeed, cross-linking via physical, chemical, or enzymatic treatments constitutes a viable method to enhance the mechanical strength and barrier properties of starch, protein, and blended films. Protein films are effective lipid, oxygen, and aroma barriers at low relative humidity (RH) (Bamdad et al., 2006), but they constitute unsatisfactory barriers to water vapor because of the presence of hydrophilic groups in their molecular structure (Mokrejs et al., 2009). On the other hand, starch films exhibit good oxygen barrier properties and moderate tensile strength, not to mention that they become markedly brittle at low moisture (Forssell et al., 2002; Talja et al., 2007). The hydrophilic character of starch films makes them very sensitive to moisture, which limits their use as food packaging material.

Cross-linkers like glutaraldehyde, glyceraldehyde, formaldehyde, and glyoxal have found wide application in protein films. Other cross-linking agents such as Cymel 303, trisodium trimetaphosphate (TSTP), sodium trimetaphosphate (STMP), sodium tripolyphosphate (STPP), and epichlorohydrin help to achieve cross-linking in starch films. Most of these agents are not edible, which restricts their use in edible packaging. The use of enzymes together with ultraviolet (UV) and ionizing radiation to obtain biopolymer film cross-linking is a possible way to overcome this limitation. This chapter presents the advances in plasticizers and cross-linkers and discusses how they affect the properties of biopolymer-based films.

12.2 Plasticizers

In the late nineteenth century, the concept of plasticizers was first introduced by applying natural camphor and castor oil to plasticize celluloid or celluloid lacquers (Gachter and Muller, 1990). A plasticizer is an additive that can be incorporated in a material (usually plastic or elastomer) to increase its flexibility, elongation, processing (workability), and distensibility. The addition of a plasticizer generally reduces the cohesive intermolecular forces along the polymer chains, which can then move more freely relative to one another, decreasing polymer stiffness (Vieira et al., 2011).

The mechanism of plasticization may refer to one of the three concepts: lubricity, gel theories, and free-volume theory (Marcilla and Beltran, 2012). According to the lubricity theory, the plasticizer acts to diminish the intermolecular friction just as oil lubricates a bearing. Some solvent action has also been suggested. This theory is not complete, because it cannot explain why a plasticizer is more efficient than others. The gel theory suggests that what makes and unplasticizes resinous mass rigid is an internal three-dimensional honeycomb structure or a gel formed by loose attachments between the resin macromolecules at intervals along the molecular chains. Plasticization diminishes the relative number of polymer–polymer unions, relieving rigidity of the truss-like three-dimensional structure and allowing the film to deform without breaking. The plasticizer acts by masking these attachment centers from each other, to prevent them from reattaching. Although some neighboring molecules have to be displaced for thick sections to be deformed, the main resistance to deformation is not viewed as internal friction between molecules, but rather as somewhat elastic resistance of the three-dimensional gel resulting from interlocking segments of the resin macromolecules. Therefore, instead of lubricating the internal "glide planes," the plasticizer makes the plastic less rigid by opposing more complete aggregation of the resin macromolecules.

The free-volume theory is a more expanded theory that allows for quantitative analysis of polymer–plasticizer interaction. It is possible to describe the free volume of a polymer as the "empty internal space" available for the polymer chains to move.

$$V_f = V_t - V_o \qquad (12.1)$$

where
V_f is the free volume of the resin
V_t is the specific volume at a temperature t
V_o is the specific volume of an arbitrary reference point, usually taken as zero degrees Kelvin

Plasticizers act to increase the free volume of the polymer and to ensure this free volume remains constant as the polymer–plasticizer mixture cools after melting. The free volume of a polymer increases markedly when it reaches the glass transition temperature, the point where significant molecular motion begins to occur. This motion, which corresponds to the desired increase in the free volume of the polymer, can result from the motion of the chain itself, of the chain ends, or of the side chains attached to the polymer. The addition of the plasticizer facilitates this process because it has lower molecular weight, thus favoring a larger free volume per volume of material.

12.2.1 Conventional Plasticizers

Conventional plasticizers can be categorized in several ways. A more comprehensive ranking classifies plasticizers into three large families: monomeric plasticizers, hybrid plasticizers, and polymeric plasticizers (Marcilla and Beltran, 2012).

Monomeric plasticizers, such as dioctyl adipate, dibutyl phthalate (DBP), and tri-isononyl trimellitate, display a relatively simple molecular structure. Their molecular weight generally ranges between 300 and 500 g/mol rarely approaching 600 g/mol. These plasticizers are usually liquids with high boiling point, low viscosity, and good solvency in most cases. As a rule, they exhibit very good thermal and chemical stability.

Hybrid plasticizers comprise plasticizers with molecular weight between 600 and 1000 g/mol. They represent a good combination between the isolated monomeric polymeric plasticizers, to afford satisfactory and exhibit good solvency and viscosity as well as relatively low volatility and good performance at temperature.

Polymeric plasticizers, such as polymers, consist of repeated units of the same molecule. Like polymers they do not possess a single molecular weight but are characterized by an average molecular weight and molecular weight distribution. The molecular weight lies between 1000 and 10,000. Plasticizers with higher molecular weight provide better retention in vulcanizates, because they are less volatile. Nevertheless, as adverse factors, they are more viscous and expensive. Examples of this class are polyesters (adipates, phthalates, and sebacates), polyethers, aromatic polyester ethers, liquid polybutadiene, and polyisobutylene.

Another way to classify plasticizers is to use the name of the chemical family they belong to, which is perhaps the most usual classification (Table 12.1).

The most commonly used plasticizers worldwide are esters of phthalic acid (e.g., DEHP, diisodecyl phthalate [DIDP], diisotridecyl phthalate [DITDP], and diisononyl phthalate [DINP]). Among the more than 30 different phthalates in the market, EHP is by far the most widely employed.

Phthalate esters have been extensively applied in thermoplastic cellulose ester molding compounds, PVC, and other vinyl chloride copolymers for over 60 years. Phthalates account for 92% of the plasticizers produced worldwide, whereas DEHP represents 51% of the phthalates (Murphy, 2001). In general, phthalates combine most of the desirable properties of a plasticizer, such as minimal interaction with resins at room temperature, good fusion properties, satisfactory insulation for cables, production of highly elastic compounds with reasonable cold strength, relatively low volatility at ambient conditions, and low cost.

In the case of phthalates, the polarizable benzene nucleus is highly compatible with PVC, which makes the polymer chains flexible, but compatibility decreases with the longer disubstituted alkyl esters. Shorter-chain phthalates are easier to formulate because they diffuse faster; however, they are more volatile (Gachter and Muller, 1990).

Terephthalates, oligoesters of o-phthalic acids, and solid phthalate esters (e.g., dicyclohexyl phthalate and diphenyl phthalate) have some plasticizing properties. Nevertheless, they are seldom employed because they are expensive. At the first phthalate, esters were thought to be benign to human beings, so they were used in various products such as children's toys and medical plastics, situations during which they may come in close contact with the human body (Rahman and Brazel, 2004). Unfortunately, more recent reports have criticized these petroleum-derived products. They are suspected to display endocrine-disruptive activity in laboratory rats, especially after phthalate plasticizers have been found to leach from medical plastics such as intravenous (IV) bags and dialysis tubing (Tickner et al., 2001).

Nowadays, there is increasing interest in the use of "green" plasticizers that are characterized by low toxicity, low migration, and biodegradability. In addition, this search for "green" plasticizers is also related to the increased interest of material researchers and industries in the development of new bio-based materials, made from renewable and biodegradable resources with the potential to reduce the use of conventional plastic goods (Vieira et al., 2011).

TABLE 12.1

Properties and Applications of Some Commercial Plasticizers

Plasticizer Type	Main Characteristics	Common Examples	End Uses
Phthalates	Have excellent compatibility, high gelling capacity, low volatility, good water resistance, and low cost (terephthalates having high migration resistance)	DEHP, DIDP, DINP, DITDP, DBP	Medical plastics (IV bags and tubing), kitchen floor, vinyl wall coverings, carpet backing, wires and cables, toys, horses, shower curtains, food packaging, automobile parts
Phosphates	Have good flame retardance and heat resistance, are highly solvating for vinyl resin, have lower fusion temperature than DEHP, but accelerate thermal degradation of PVC, not suitable for low-temperature and food-contact applications	Triphenyl phosphate, tris(2-ethylhexyl) phosphate, tricresyl phosphate, Kronitex®	Flame-retardant plasticizer in calendered goods, extrusions, plastisol-derived products with nylon, sulfonamides and other highly polar compounds, PVC, polyacrylates, cellulose derivatives, synthetic rubbers
Adipates	Have low viscosity and higher gelation and fusion temperature than DEHP, cause less brittleness than phthalates, and are relatively volatile and extractable, but give superior low-temperature flexibility	Dibutyl adipate, bis(2-ethylhexyl) adipate or DEHA, diisodecyl adipate	In combination with phthalates, with improved low temperature (even arctic) and flexibility for automobile parts and aircraft interiors
Azelates	Improve low-temperature flexibility and less water sensitive than adipates	Bis(2-ethylhexyl) azelate	With cellulosic resins and elastomers, and food-contact applications with PET and polyester
Sebacates	Have excellent low-temperature performance	Dibutyl sebacate, dioctyl sebacate	Dibutyl sebacate used specially for polyisoprene in food-contact, medical, and pharmaceutical plastics
Epoxidized FA esters	Impart cold strength, have very low volatility, are pigment dispersing agents in plasticized PVC, establish synergistic thermostabilizing effect with Ca-Zn stabilizers, and stabilize other plasticizers by offering migration resistance	Butyl epoxystearate, cyclohexyl epoxystearate	Low-temperature applications of PVC and its copolymers
Benzoates	Are highly solvating and have low moisture sensitivity; excellent resistance to organic extraction; excellent stain and UV resistance; good gelation properties; desirable environmental, health, and safety profiles; and high viscosity (limits application)	Benzoplast®, Benzoflex®	Vinyl flooring, PVA adhesives, PU castle and sealant, latex caulks, coatings, plastisol formation, processing aids, inks, hot-melt adhesives
Polyesters/ polymeric plasticizers (MW between 850 and 3500)	Have very low volatility, are highly resistant to extraction and migration, extend lifetime of flexible articles, improve weathering resistance, are highly viscous, and are usually blended with lower-viscosity plasticizers	Poly(1,3-butyleneglycl adipate), PEG, Admex®, Paraplex®	Compatible with PVC, cellulose acetate butyrate, and cellulose nitrate, applicable in vinyl dispersions, films, sheets, floor coverings, cable insulation, and sheathing resins only where oil and fat resistance is required PVC tubes, blood storage bags, hemodialysis tubing, catheters

(*Continued*)

TABLE 12.1 (*Continued*)

Properties and Applications of Some Commercial Plasticizers

Plasticizer Type	Main Characteristics	Common Examples	End Uses
Trimellitates	Have low volatility, good water resistance, and high-temperature stability, are similar to phthalates in compatibility and plasticizing effectiveness, have less migration tendency and high extraction by oils and hydrocarbons as phthalates, and are of high prices	Trioctyl trimellitate, octyl dibenzyl trimellitate	PVC tubes, blood storage bags, hemodialysis tubing, catheters
Sulfonic acid esters and sulfonamides	Are less volatile than phthalates, tend to discolor, are weather resistant, have slightly better gelation effectiveness than DEHP, have good hydrolytic resistance in alkaline solution, and are not compatible with PVC	n-Butylbenzenesul-fonamide, toluenesulfonamide	Sulfonic acid esters used with PVC, sulfonamides used especially with polyamide, and cellulose-based molding resins
Monocarboxylic acid esters	Are too volatile and sensitive to water (especially esters with low alkanols) and show poor gelation properties; longer chain esters of FAs being not too compatible with polymers	n-Butyl formate, ethyl lactate	Important as low-temperature secondary plasticizers and as lubricants in processing rigid and plasticized PVC
Epoxidized vegetable oils	Have good heat and light stability and good resistance to extraction; epoxidized soybean oil having high MW (~1000) and bulky structures, which provide resistance to migration	ESO, epoxidized linseed oil, tallates	Primarily used as heat stabilizers
Chlorinated hydrocarbons	Are somewhat flame retardant, have limited compatibility, are often colored, and have odor, are aromatic (high viscosity, fairly good compatibility, poor heat and light stability), are aliphatic (have low viscosity and may exude from aged fused products)	Polychlorinated biphenyls, polychlorinated 1-dodecene, 1-tetradecene, 1-hexadecene	Mainly used as secondary plasticizers, for cost reduction
Citrates	Have good solvating power for PVC and cellulose acetate and high efficiency and are nontoxic; nonacetyl citrates being relatively volatile and water sensitive compared to DEHP, which are good for medical plastics if not exposed to high lipid media	(Acetyltri-n-hexyl citrate) Citroflez® A-6, (n-butyryltri-n-hexyl citrate) Citroflex® B-6	With cellulosics, PVC, PVA, and other polymer films and flexible tubings used in medical plastics and food-contact plastics
Oligomers (low-MW polymers)	Offer better extraction and migration resistance; have lower volatility, weathering resistance, and reduced odor; and may exude particularly at high temperature and humidity	(Resorcinol bidiphenyl phosphate) Fyrolflex® RDP-B, poly(butadiene dimethacrylate)	Automotive, marine, and aeronautical applications
Polymerizable plasticizers	Polymerize at elevated temperature during the gelling of PVC	Allyl phthalate, acrylic esters, monochlorostyrene	Toys, shoe heels, and certain industrial articles that must have high stiffness

(*Continued*)

TABLE 12.1 (*Continued*)

Properties and Applications of Some Commercial Plasticizers

Plasticizer Type	Main Characteristics	Common Examples	End Uses
Elastomers	Are used instead of or together with low-MW plasticizers for PVC to improve resistance to migration and saponification and have low-temperature flexibility and mechanical and oxidative stability at high temperatures	Nitrile rubber, ethylene/vinyl acetate copolymers, acrylonitrile/butadiene/styrene terpolymers	Coverings of instrument panels, cushion covers, shoe soles, insulations for cables, tubes, and articles for aircraft interior

The following paragraphs contain examples of green plasticizers:

Isosorbide diesters are biobased plasticizers produced from the reactions between FAs of plant origin (e.g., n-octanoate) and isosorbide produced by dehydration of sorbitol, a glucose derivative.

Epoxidized vegetable oils are a biobased class of plasticizers obtained via esterification of vegetable oil with polyols.

Diisononyl cyclohexane-1,2-dicarboxylate (DINCH) with the trade name Hexamoll® DINCH is originated from the catalytic hydrogenation of DINP.

Citrates, marketed under the trade name of Citroflex®, are based on biobased citric acid feedstock and arise mainly from the fermentation of corn.

Dibenzoates, commercially known as Benzoflex® plasticizers and developed by Genovique Specialties (formerly Velsicol Chemical), have been commercially available for over 40 years.

12.2.2 Nonconventional Plasticizer

Edible films produced from polysaccharides, proteins, and lipids, among others, bear a structure that produces a strong cohesive film upon dehydration, which usually requires a plasticizer. In this case, it is necessary to add a nonconventional plasticizer, to decrease intermolecular forces along the polymer chains and consequently improve flexibility and chain mobility. Water is the most powerful "natural" plasticizer for hydrocolloid-based films. In addition, water, polyols, and mono-, di-, and oligosaccharides are also common plasticizers. Polyols are particularly effective plasticizers for hydrophilic polymers (Zhang and Han, 2006), and glycerol (GLY) is almost always systematically incorporated in most hydrocolloid films (Cuq et al., 1997). Indeed, GLY is a highly hygroscopic molecule; its addition to film-forming solutions prevents film brittleness (Karbowiak et al., 2006).

Many studies have focused on the use of polyols such as GLY, ethylene glycol (EG), diethylene glycol (DEG), triethylene glycol (TEG), tetraethylene glycol, polyethylene glycol (PEG), propylene glycol (PG), sorbitol, mannitol, and xylitol, FAs, monosaccharides (glucose, mannose, fructose, sucrose), EA, urea, TEA, vegetable oils, lecithin, waxes, amino acids, surfactants, and water. In these cases, the plasticizers increase the free-volume or intermolecular spacing, molecular mobility, and the flexibility and stretchability of biopolymer films produced from polysaccharide and protein sources. Without a plasticizing agent, biodegradable films tend to become brittle due to extensive intermolecular forces (forces that act between functional groups within the macromolecules) involving polymer chain-to-chain interactions.

12.3 Plasticization of Protein Films

Amino acid composition, distribution, and polarity, ionic cross-links between the amino and carboxyl groups, hydrogen bonds, and intra- and intermolecular disulfide bonds can affect the film-forming ability of a protein (Gennadios and Weller, 1991). The protein molecules interconnect during the drying process, to form the film matrix. Therefore, extension or unfolding of the protein molecules could favor interaction between them, to give rise to more junction zones (Hoque et al., 2010). Proteins have a unique structure (based on 20 different monomers), which provides them with a wider range of functional properties, especially high intermolecular binding potential (Cuq et al., 1997). The molecular weight and the number and positions of hydroxyl groups in a plasticizer are all variables that impact its ability to plasticize a protein-based polymer (Bourtoom, 2009). Table 12.2 lists natural plasticizers used in biodegradable protein films.

In zein protein, the addition of oleic and linoleic acids as plasticizers results in flexible, elongated, tough, and clear sheets with low modulus and tensile strength. FA separation causes zein to aggregate, resulting in loss of flexibility and increased water absorption. Linoleic acid is more effective than oleic acid at reducing water absorption by the sheets. However, zein plasticization with oleic acid affords relatively tough and water-resistant sheets that may find application in thermoformed packaging trays (Santosa and Padua, 1999). The literature contains numerous reports on how the type of

TABLE 12.2

Natural Plasticizers Used in Biodegradable Films from Proteins and Lipids

System of Application	Plasticizer	References
Zein	Oleic and linoleic acids	Santosa et al. (1999)
Caseinate–pullulan	Water and sorbitol	Kristo and Biliaderis (2006)
Whey protein	GLY and sorbitol	Kim and Ustunol (2001)
Whey protein/beeswax (BW) emulsion	GLY	Galietta et al. (1998)
β-Lactoglobulin	Sorbitol, GLY, and PEG	Bourtoom (2009)
	Sorbitol, GLY, EG, PEG 200, and PEG 400	Jongjareonrak et al. (2006)
	PG, GLY, sorbitol, PEG 200, PEG 400, and sucrose	Sothornvit and Krochta (2000)
	GLY, EG, DEG, TEG, and PG	Orliac et al. (2003)
Sunflower protein	Saturated FAs	Pommet et al. (2003)
Peanut protein	Glycerin, sorbitol, PEG, and PG	Jangchud and Chinnan (1999)
Wheat gluten	Glycerin	Sobral et al. (2005)
Feather keratin	GLY	Moore et al. (2006)
Fish mince from Atlantic sardines (*Sardina pilchardus*)	Sorbitol, GLY, and sucrose	Cuq et al. (1997)
Fish skin protein	FAs and sucrose esters	Jongjareonrak et al. (2006)
Water-soluble fish proteins	GLY and PEG	Suyatma et al. (2005)
	GLY, PEG, EG, sucrose, and sorbitol	Tanaka et al. (2001)
Fish muscle proteins	GLY, PG, DEG, and EG	Vanin et al. (2005)
Fish myofibrillar protein	Glycerin and water	Sobral et al. (2002)
Gelatin	GLY and sorbitol	Thomazine et al. (2005)
	Sucrose, oleic acid, citric acid, tartaric acid, malic acid, PEG of different molecular weights (300, 400, 600, 800, 1500, 4,000, 10,000, 20,000), sorbitol, mannitol, EG, DEG, TEG, EA, DEA, and TEA	Cao et al. (2009)
Pigskin gelatin	GLY	Bergo and Sobral (2007)
	Sorbitol	
Bovine gelatin	FAs	Bertan et al. (2005)
	Sorbitol	Sobral et al. (2001)
	Glycerol	Carvalho et al. (2008)
Fish gelatin	GLY and sorbitol	Al-Hassan and Norziah (2012)
Skin of unicorn leatherjacket	GLY	Ahmad et al. (2012)

natural plasticizers such as polyols (GLY, PEG, EG, DEG, TEG, and PG), sorbitol, and sucrose and the concentration of plasticizer affect the properties of protein- and lipid-based films (Audic and Chaufer, 2005; Cuq et al., 1995, 1997; Galietta et al., 1998; Jangchud and Chinnan, 1999; Jongjareonrak et al., 2006; Kristo and Biliaderis, 2006; Orliac et al., 2003; Pérez-Gago and Krochta, 2001b; Sobral et al., 2002, 2005; Sothornvit and Krochta, 2000; Tanaka et al., 2001; Van de Velde and Kiekens, 2002; Vanin et al., 2005). Larger GLY content increases film solubility in water and reduces the mechanical resistance of whey protein–based films (Galietta et al., 1998). On the other hand, the addition of sorbitol, GLY, or sucrose, at the same molecular concentration, to myofibrillar protein–based films prepared with fish mince from Atlantic sardines (*Sardina pilchardus*) does not elicit significant differences in film properties, because these plasticizers are structurally similar (Lai et al., 1997). The use of GLY as plasticizer for fish protein films reduces opacity, color, and T_g (Sobral et al., 2005). Similarly, increased plasticizer (GLY and PEG) concentration decreases the tensile strength, with concomitant enhancement in elongation at break and water vapor permeability of water-soluble fish protein-edible films (Bourtoom, 2009). Other studies corroborate with these results (Jongjareonrak et al., 2006; Tanaka et al., 2001).

Fish protein films plasticized with EG, sucrose, or sorbitol are too brittle and fragile to handle, making their preparation unfeasible. PEG concentration influences the tensile strength of these films, whereas GLY affects elongation at break the most. Results have clearly demonstrated the plasticizing effect of highly hydrophilic GLY, which acts by reducing internal hydrogen bonding within the protein, thereby diminishing the internal forces and augmenting the intermolecular spacing. The addition of combined plasticizers (GLY and PEG) can modify the mechanical properties and water vapor permeability of fish protein films (Tanaka et al., 2001).

β-Lactoglobulin films have been plasticized with PG, GLY, sorbitol, PEG 200, PEG 400, or sucrose, aiming to improve mechanical properties. GLY and PEG 200 are the plasticizers that most efficiently afford films with the desirable mechanical properties (Sothornvit and Krochta, 2000).

The use of GLY, EG, DEG, TEG, and PG as plasticizers for films of sunflower protein isolate produces soft, brown, and smooth films with good mechanical properties and high level of impermeability to water vapor (Orliac et al., 2003). No marked loss of GLY or TEG occurs over the 3-month aging period, being both substances the most suitable plasticizers for sunflower proteins. Because GLY is a completely nontoxic plasticizer, it is the compound of choice in the food industry. In other research, the addition of a series of saturated FAs with different carbon chain lengths (from 6 to 10 carbons) to wheat gluten film (Pommet et al., 2003) provided promising and opened up new horizons for plasticization and improvement in the properties of gluten-based plastics.

The addition of polyols (GLY, PG, DEG, and EG) as plasticizers improves the thermal and functional properties of pigskin gelatin–based films. After being tested at five different concentrations, the plasticizers proved to be compatible with gelatin, to give flexible and easy-to-handle films in the studied concentration range. No typical phase separation occurs during thermal analyses. In terms of functional properties, GLY furnishes higher plasticizing effect and better efficiency. Application of other plasticizers such as sucrose, oleic acid, citric acid, tartaric acid, malic acid, PEG, sorbitol, mannitol, EG, DEG, TEG, EA, diethanolamine (DEA), and TEA in gelatin films modifies its mechanical and barrier properties. With regard to mechanical and visual properties, malic acid, PEG 300, sorbitol, EG, DEG, TEG, EA, DEA, and TEA give rise to the most promising plasticizing effect. Gelatin films plasticized with EG, DEG, or TEG have the highest water vapor permeability and water content values, in contrast to the gelatin plasticized with malic acid or sorbitol (Cao et al., 2009).

12.4 Plasticization of Polysaccharide Films

Polysaccharides have good film-forming properties and constitute efficient barriers against oils and lipids, but they do not bar moisture effectively. Generally, polysaccharide films consisting of starch, alginate, cellulose ethers, chitosan, carrageenan, or pectins exhibit good gas barrier properties. The linear structure of some of these polysaccharides, for example, cellulose (1,4-b-D-glucan), amylose (a component of starch, 1,4-a-D-glucan), and chitosan (1,4-b-D-glucosamine polymer), renders their films tough, flexible, transparent, and resistant to fats and oils (Tharanathan, 2003). However, their hydrophilic nature makes them poor water vapor barriers. Among polysaccharides and biopolymers in general, starch is potentially applicable in biodegradable plastics (Ma et al., 2009).

Starch films commonly contain incorporated hydrophilic compounds, such as polyols (GLY and sorbitol) that are commonly used in starch films (Bergo et al., 2008; Cheng et al., 2006; Galdeano et al., 2009; Garcia et al., 2000; Honary and Orafai, 2002; Mali et al., 2005; Müller et al., 2008; Navarro-Tarazaga et al., 2008; Talja et al., 2007; Zhang and Han, 2006), but some sugars (Galdeano et al., 2009; Veiga-Santos et al., 2007), surfactants (Ghebremeskel et al., 2007; Rodriguez et al., 2006; Van Soest and Knooren, 1997), amino acids, and FAs (Rotta et al., 2009) could help to improve their mechanical and barrier properties (Table 12.3).

An ideal plasticizer for starch-based materials should impart flexibility and suppress retrogradation of thermoplastic starch (TPS) during aging (Ma et al., 2009). The changes in crystallinity of GLY-containing potato starch plastic sheets are clearly a function of the initial amount of plasticizer and moisture migration during aging. The differences in material properties could originate from formation of an entangled starch matrix and from starch chain-to-chain associations related to plasticizer content (Van Soest and Knooren, 1997).

TABLE 12.3

Natural Plasticizers Used in Biodegradable Films from Polysaccharides

System of Application	Plasticizer	References
Citric acid–modified pea starch and citric acid–modified rice starch	GLY	Ma et al. (2009)
γ-Carrageenan edible films	GLY and water	Karbowiak et al. (2006)
Rice starch	GLY	Dias et al. (2010)
Potato starch	GLY and EG	Smits et al. (2003)
Potato starch	GLY, xylitol, and sorbitol	Talja et al. (2007)
Waxy maize starch, maize starch, and amylomaize starch	GLY	Rodríguez et al., (2006), VanSoest and Knooren (1997)
	GLY, sorbitol, and water	Funke et al. (1998)
Soluble starch/gelatin	GLY, sorbitol, and sucrose	Arvanitoyannis et al. (1997)
Cornstarch	EA	Huang et al. (2005)
	Caproic acid, lauric acid, and glycerol triacetate (triacetin)	Fringant et al. (1998)
	GLY, acetamide, formamide, anhydrous glucose, and urea	Ma et al. (2009)
	Sorbitol and GLY	Mali et al. (2005), Garcia et al. (2000)
	GLY and amino acids	Stein et al. (1999)
Cassava starch	GLY	Bergo et al. (2008)
	GLY and sorbitol	Müller et al. (2008)
Oat starch	GLY, sorbitol, urea, sucrose, and GLY–sorbitol mixture	Galdeano et al. (2009)
Pea starch	Mannose, glucose, fructose, GLY, and sorbitol	Zhang and Han (2006)
Chitosan films	GLY, EG, PEG, and PG	Suyatma et al. (2005)
Hydroxypropyl methylcellulose–BW	GLY and mannitol	Navarro-Tarazaga et al. (2008)
Cellulose from sugarcane bagasse and cellulose acetates	Residual xylan acetate	Shaikh et al. (2009)
Konjac glucomannan	Sorbitol and GLY	Cheng et al. (2006)
Alginate/pectin	GLY	Silva et al. (2009)
Sago starch	GLY and sorbitol	Al-Hassan and Norziah (2012)

EA is a novel plasticizer with application in TPS processing. It destroys the native starch granules and gives rise to a uniform continuous phase. EA-plasticized TPS could restrain the recrystallization of traditional GLY-plasticized TPS, to improve their mechanical properties and thermal stability (Huang et al., 2005).

The plasticizer and starch interact very specifically. In plasticized crystalline amylose and crystalline and amorphous amylopectin systems, the plasticizers (GLY or EG) interact via hydrogen bonding with the polysaccharide when the temperature rises and during film storage at room temperature. Crystalline amylopectin and amylose display similar behavior—the plasticizer and the polymer interact at a slower rate as plasticizer/polymer interaction, compared with amorphous amylopectin. A marked interaction occurs upon increasing temperature, probably due to H-bond formation. Consequently, matrix mobility increases, viscosity reduces, and the material behaves like a rubber (Smits et al., 2003).

Plasticizers containing amide groups (urea, formamide, and acetamide) have been tested for TPS plasticization, using GLY as reference. Amide groups seem to affect TPS retrogradation suppression and mechanical properties by establishing mainly hydrogen bonds with the starch molecules. Hydrogen bonding formation decreases in the following order: urea > formamide > acetamide > polyol (Ma et al., 2009).

In TPS, which is hydrophilic, the plasticizer interferes in how the chains associate, thereby increasing flexibility and reducing mechanical strength (Chen and Lai, 2008). Another effect is the rise in hydrophilicity and water vapor permeability of the laminated films, since most of the plasticizers employed in starch films are hydrophilic (Mali et al., 2004). Plasticizers are generally added in proportions ranging from 10 to 60 g/100 g of dry matter, depending on the degree of material rigidity (Gontard et al., 1993). However, the concentration of the plasticizer may cause an effect called antiplasticizing, that is, instead of increasing polymer flexibility and hydrophilicity, it may prompt the opposite effect (Gaudin et al., 1999, 2000). Typically, this occurs at low plasticizer concentrations (below 20 g/100 of starch), when the plasticizer interacts with the polymer matrix but this interaction is not sufficient to enhance the molecular mobility, a phenomenon that also depends on the storage conditions (Lourdin et al., 1997). Compounds with low molecular mass or diluents can act as external plasticizers, and be an integral part of polymeric systems, to improve the flexibility and workability of otherwise rigid neat polymers. Nevertheless, at low concentrations, they may serve as mechanical antiplasticizers, making the polymer–diluent blends become stiffer than the neat polymer (Sears and Darby, 1992).

12.5 Plasticization of Other Biodegradable Films

Natural mixtures of starch, protein, and lipids can be obtained in the form of flour from raw materials of plant origin such as cereals and legumes. The characteristics of the films based on flour are a result of the natural interactions occurring between the starch, protein, and lipids during drying of the filmogenic suspension (Tapia-Blácido et al., 2007), which can also be influenced by the different variables employed during the production process such as heating temperature, pH, drying conditions, and plasticizer type and concentration (Tapia-Blácido et al., 2005, 2011). Flours obtained from whole materials such as amaranth, achira, soy, wheat, turmeric residue, and plantain bananas have been used to film production (Andrade-Mahecha et al., 2012; Dias et al., 2010; Maniglia et al., 2014, 2015; Mariniello et al., 2003; Pelissari et al., 2013; Rayas et al., 1997; Tapia-Blácido et al., 2005, 2011). GLY and sorbitol were used as plasticizers. Sorbitol was the most suitable plasticizer for amaranth and turmeric flour. It furnished films that were more resistant to break and less permeable to oxygen, due to its greater miscibility with the biopolymers present in the flour and its lower affinity for water. The achira and banana flour film display low solubility in water, good WVP and flexibility, and excellent mechanical strength at low GLY concentration, 17 and 19 g/100 g flour, respectively (Andrade-Mahecha et al., 2012; Pelissari et al., 2013).

Biodegradable polymer blends that enhance the degradation of the final product constitute another type of edible films. In recent years, poly-β-hydroxyalkanoates have attracted attention as potential biocompatible and biodegradable thermoplastics. Poly(3-hydroxybutyrate) (PHB) is a natural thermoplastic polyester with many mechanical properties that are comparable to those of synthetically produced degradable polyesters (Bucci et al., 2005). Addition of biodegradable plasticizers such as soybean oil (SO), epoxidized SO (ESO), DBP, and triethyl citrate (TEC) to poly(3-hydroxybutyrate-co-3-hydroxyvalerate) (PHBV) films enhances the film's thermal and mechanical properties. TEC and DBP are better plasticizers for PHBV than SO and ESO (Choi and Park, 2004).

The use of additives (dodecanol, lauric acid, tributyrin, and trilaurin) modifies the structure of PHB films (Yoshie et al., 2000), decreasing their T_g and T_{cc} (cold crystallization temperature). These additives are miscible with PHB and improve the mobility of the molecules in the amorphous phase. Besides acting as plasticizers, at a fairly small amount (1 wt%), these additives could accelerate enzymatic degradation of the polymer chains. The addition of a biodegradable plasticizer, di-n-butyl phthalate, to PHB films promoted the same effect (Ceccorulli et al., 1992).

12.6 Cross-Linking

Cross-linking is one of the oldest methods to chemically modify a polymer. This method makes a polymer more resistant to heat, light, and other physical agents, providing it with a high degree of dimensional stability, mechanical strength, and chemical and solvent resistance. The degree of cross-linking, the regularity of the resulting network, and the presence and absence of crystallinity in the polymer primarily affect the physical properties of the polymer (Bhattacharya and Ray, 2009). In polysaccharides, introduction of various degrees of cross-linking into molecules can generate larger molecular aggregates with enhanced viscosity or furnish insoluble products with a wide range of swelling characteristics (Dumitriu et al., 1996). Protein cross-linking refers to the formation of covalent bonds between polypeptide chains within a protein (intramolecular cross-links) or between proteins (intermolecular cross-links). Thus, protein networks can establish stronger intermolecular covalent bonds, to achieve closer molecular packing and reduced polymer mobility (Wihodo and Moraru, 2013). The cross-linking reaction can follow different pathways (chemical, thermal, radiation, and enzymatic pathways).

12.7 Thermal Cross-Linking

Heat curing of protein-based films promotes intra- and intermolecular cross-linking, for example, between lysine and cysteine and polar groups of the protein chains, to furnish films with increased tensile strength and reduced elongation at break, as well as enhanced hydrophobicity (Pérez-Gago, 2012). Most proteins denature at high temperatures: heat disrupts hydrogen bonds and nonpolar hydrophobic groups in the protein molecule, exposing the amino acid groups to the solvent and producing a more open structure (Damodaran, 2008; Wihodo and Moraru, 2013). Some authors have studied how the temperature and time of heating affect the properties of protein films. Concerning soy protein films subjected to heat curing (90°C for 24 h), the tensile strength increases from 8.2 to 14.7 MPa, the elongation at break decreases from 30% to 6% and the moisture content in the film diminishes (Rhim et al., 2000). Another research described that heating of a soy protein isolate (SPI) solution at 85°C reduces the water vapor permeability, augments the tensile strength, and enhances the elongation at break (%) of soy protein film (Stuchell and Krochta, 1994). Heating at 90°C for 30 min of whey protein film-forming solution produces films that are less permeable to oxygen than films made from unheated solution (Pérez-Gago and Krochta, 2001). The ability of heat-treated whey protein isolate (WPI) films to withstand higher deformation may stem from unfolding of the globular structure of the whey protein, which exposes the previously inward-oriented S-H groups and allows strong, covalent disulfide intermolecular bonds to form (Pérez-Gago et al., 1999; Pérez-Gago and Krochta, 2001; Wihodo and Moraru, 2013). Another study reported that heating at 70°C–100°C for 5–20 min produces stronger and more extensible WPI films (Pérez-Gago et al., 1999).

12.8 Chemical Cross-Linking

Chemical cross-linking can also improve the mechanical and barrier properties of polysaccharide, protein, and polysaccharide–protein blended films. Such cross-linking of starch polymers stabilizes the granules and ensures that they do not overswell or rupture, especially in cases where food processing demands stability to excessive heat, shear, and reduced pH. Cross-linking reinforces the hydrogen bonds that already exist in the granules. Chemical cross-links remain intact upon hydrogen bonds cleavage during heating or shearing.

Polysaccharide cross-linking agents include di- and trifunctional reagents, such as epichlorohydrin, bisepoxides, dihalogenated reagents, glutaraldehyde, acetaldehyde, formaldehyde, maleic, boric and oxalic acid, dimethylurea, polyacrolein, diisocyanates, divinyl sulfate, ceric redox systems, borax, and s-triazine (Dumitriu et al., 1996). Other studied natural agents include genipin, hydroxycinnamic, citric, ferulic and tannic acid, and proanthocyanidin (Rivero et al., 2010).

Incorporation of citric acid into films based on renewable material improves the mechanical and barrier properties of such materials (Ning et al., 2010; Olivato et al., 2012; Olsson et al., 2013; Reddy and Yang, 2010; Wang et al., 2007). The reaction mechanism for the cross-linking, that is, intermolecular diester formation, consists in the well-known Fischer esterification between the carboxylic acid groups of citric acid and the hydroxyl groups in starch, which happens twice within the same citric acid molecule. The addition of Lewis acids or pH reduction catalyzes formation of an ester bond (Olsson et al., 2013). The addition of tannic acid to the chitosan matrix yields a more rigid structure, increasing the tensile strength and decreasing the elongation at break and WVP (Rivero et al., 2010). Cross-linking potato starch with malonic acid makes the resulting starch films brittle and less hydrophilic, but raises Young's modulus (Dastidar and Netravali, 2012).

Glutaraldehyde, epichlorohydrin, citric acid, boric acid, borax, STMP, Cymel 303, sodium hexametaphosphate, and TSTP are the cross-linking reagents that react with the hydroxyl groups in starch or PVA, to modify TPS/poly(vinyl alcohol) (PVA) blend films (Li et al., 2009; Mao et al., 2006; Reddy and Yang, 2010; Sreedhar et al., 2005, 2006; Wang and Hsieh, 2010; Yin et al., 2005). Liu et al. (2012) studied cross-linking of TPS/PVA blend films by sequentially soaking the films in sodium carbonate aqueous solution and sodium hexametaphosphate aqueous solution, followed by heating in an oven. These authors investigated how the concentrations of the sodium carbonate aqueous solution and the sodium hexametaphosphate aqueous solution, soaking time, heating temperature, and time affect the properties of the TPS/PVA blend films. These authors observed that cross-linking increases the tensile strength and Young's modulus but reduces the elongation at break of the TPS/PVA blend films. The hydrophilic characteristic also diminishes.

In the case of proteins, cross-linking procedures generally employ bifunctional reagents that modify amino groups or sulfhydryl groups, that is, lysine or cysteine residues. The reactive group in a cross-linker can be identical or different, providing a diversity of reagents that can bring about covalent bonding between any chemical species, either intra- or intermolecularly (Wong, 1993). Cross-linking of gelatin, casein, albumin, and ovalbumin by addition of organic acids improves their barrier properties (Guilbert, 1986). In zein films, cross-linking agents such as polymeric dialdehyde starch, 1,2-epoxy-3-chloropropane, 1-(3-dimethylaminopropyl)-3-ethylcarbodiimide hydrochloride, and N-hydroxysuccinimide (NHS) enhances the moisture barrier (Kim et al., 2004; Parris and Coffin, 1997; Parris et al., 2001; Yamada et al., 1995). As cross-linking agents, formaldehyde, glutaraldehyde, and citric acid increase the tensile strength of zein films up to twofold (Parris and Coffin, 1997; Yang et al., 1996). In wheat gluten films, cross-linking agents such as formaldehyde, calcium chloride, SDS, transglutaminase (TGase), 1-ethyl-3-3-dimethylaminopropyl carbodiimide, and NHS improve the mechanical properties of these films (Larre et al., 2000; Micard and Guilbert, 2000; Tropini et al., 2004). Glutaraldehyde, formaldehyde, dialdehyde starch, and carbonyldiimidazole enhance the tensile strength and the insolubility behavior of cross-linked whey protein films, whereas elongation at break remains unaffected. The cross-linked WPI films also display higher water vapor permeability and lower oxygen permeability (Galietta et al., 1998; Ustunol and Mert, 2004). In collagen film, cross-linking with formaldehyde or chromium ions raises gas permeability, due to increase moisture uptake by the cross-linked films via the greater spacing between the collagen chains (Pérez-Gago, 2012). On the other hand, cross-linking with glutaraldehyde furnishes stronger and more coherent films of collagen fibrils with improved longitudinal and transverse strength, regardless of the casing being rewetted or dry (Ustunol, 2009). Glyceraldehyde is another aldehyde that functions as cross-linker. This agent also cross-links of collagen fibrils and increases film strength and temperature resistance (Gennadios et al., 1994).

12.9 Enzymatic Cross-Linking

TGase is an enzyme that can cross-link proteins. Its complete name is R-glutaminyl-peptide: amine γ-glutamyltransferase (E.C. 2.3.2.13). TGase catalyzes acyl-transfer reactions between λ-carboxamide groups of glutamine residues (acyl donor) and ε-amino groups of lysine residues (acyl acceptor), to form ε-(λ-glutaminyl) lysine intra- and intermolecular cross-linked proteins (De Jong and Koppelman, 2002). This cross-linking does not reduce the nutritional quality of the food, because the lysine residue remains available for digestion (Chambi and Grosso, 2006; Seguro et al., 1996; Yokoyama et al., 2004).

Industrial TGase production relies on extraction and purification from tissues (bovine, fish, pork), bovine blood, genetic manipulation of microorganisms (*E. coli*, *Bacillus*, and *Aspergillus*), or microorganisms using traditional fermentation technologies (Carvalho and Grosso, 2004). In 1989, isolation and characterization of a microbial TGase (mTGase) from *Streptoverticillium* sp. indicated that this isoform could be extremely useful as a biotechnological tool (Ando et al., 1989). The protein structure of the bacterial enzyme is quite different from that of the mammalian enzymes: mTGase has a smaller molecular mass (about 40 kDa as compared with 80–100 kDa of mammalian TGases) and a very low-sequence identity with other TGases (Kanaji et al., 1993; Micanovic et al., 1994; Porta et al., 2011). However, the hydrophobic environment of the catalytic site, including a single cysteine residue, is similar to those of other isoforms, while no sequence identity with the calcium-binding domain exists (Kanaji et al., 1993; Porta et al., 2011). In fact, mTGase possesses calcium-independent activity and exhibits wide substrate specificity, being also active over a wide range of temperature and pH values (Porta et al., 2011; Yokoyama et al., 2004). Proteins such as SPIs, sodium caseinate, porcine plasma, deamidated gluten, egg white, and gelatin have been cross-linked with this enzyme (Carvalho and Grosso, 2004; Jiang and Tang, 2014; Juvonen et al., 2011; Larre et al., 2000; Lim et al., 1998; Nuthong et al., 2009; Oh et al., 2004; Patzsch et al., 2010; Piotrowska et al., 2008; Su et al., 2007; Sztuka and Kolodziejska, 2009; Tang et al., 2005; Yi et al., 2006). Taylor et al. (2002) observed that increasing concentration of the cross-linking agent improves the tensile strength and water absorption properties, making the gelatin films less soluble in water. However, Carvalho and Grosso (2004) observed that enzymatic treatment with TGase does not alter the tensile strength of the modified film, but it lowers the WVP of gelatin films. In other research, fish gelatin films treated with TGase possess better mechanical resistance and oxygen barrier property, but lower elongation at break; the WVP does not change significantly (Yi et al., 2006). Cross-linking of fish skin gelatin with TGase decreases film solubility in aqueous medium of different pH, but it does not improve the water barrier properties (Piotrowska et al., 2008). Jiang and Tang (2014) observed that treatment with TGase increases the tensile strength and the mean contact angle by 8.4%–25.6% and 2.1°–2.3°, respectively; the moisture content falls by 2.6%–9% and the moisture adsorption rate of gelatin films decreases.

Cross-linking with TGase reduces the water vapor permeability and solubility of whey protein films, while the tensile strength rises by twofold as compared with the control (Yildirim and Hettiarachchy, 1998). SPI films cross-linked with TGase have improved functional properties as a function of mTGase concentration and the duration of the cross-linking reaction (Su et al., 2007). Tang et al. (2005) observed that treatment with four units per SPI (U g-1) of mTGase raises the tensile strength and surface hydrophobicity by 10%–20% and 17%–56%, respectively, and diminishes the elongation at break, moisture content, and transparency. Cross-linking of gluten films with TGase, with or without addition of external diamines, elevates the tensile strength and elongation at break but decreases the contact angle of films. Sodium caseinate films cross-linked with mTGase exhibit 66% and 40% higher tensile strength and elongation at break, respectively. Cross-linking of protein–protein and polysaccharide–protein blends with TGase has also been reported in the literature (Chambi and Grosso, 2006;

Di Pierro et al., 2006, 2007, 2013; Kolodziejska and Piotrowska, 2007; Kolodziejska et al., 2006; Marquez et al., 2014; Oh et al., 2004; Sheng and Zhao, 2013). Films produced from WPI, soybean 11S globulin, and a mixture of the two proteins (1:1; wt/wt) cross-linked with TGase have different tensile strengths (Yildirim and Hettiarachchy, 1998). The soybean 11S globulin and the mixture are more resistant than the TGase-cross-linked whey protein. Chambi and Grosso (2006) produced casein–gelatin films (100:0, 75:25, 50:50, 25:75, and 0:100) by cross-linking with TGase. The gelatin and casein mixture produces a synergistic effect on the film properties, which is the most evident, the (75:25) casein–gelatin formulation, with elongation values of 27.2% and 56.8% for the samples without and with enzymatic modification, respectively. This formulation also presents the lowest WVP value (5.06 ± 0.31 g mm/m^2 d kPa). In some cases, fish gelatin–casein modification with TGase reduces the solubility but increases the fragility of the films, which calls for plasticization. Indeed, plasticization of enzymatically modified films does not increase the solubility or WVP of the film (Kolodziejska and Piotrowska, 2007). Chitosan–whey protein films cross-linked with TGase have enhanced mechanical resistance, reduced deformability, lower WVP, and better oxygen and carbon dioxide barrier properties (Di Pierro et al., 2006). Whey protein/pectin films prepared at pH 5.1 in the presence of TGase display increased tensile strength (twofold) and elongation at break (tenfold) (Di Pierro et al., 2013). Marquez et al. (2014) demonstrated that the coating with an edible whey protein/pectin film cross-linked with TGase is a useful way of reducing the oil content in widely consumed deep-fat fried foods and of preventing water absorption by baked foods over time. Whey protein/zein hydrolysate and chitosan–ovalbumin treated with TGase also have lower solubility (Di Pierro et al., 2007; Oh et al., 2004). The whey protein/zein film becomes more flexible after enzymatic treatment with TGase but the tensile strength decreases (Oh et al., 2004). Meanwhile, the mechanical resistance of the chitosan–ovalbumin films increases from 24 to 35 MPa (Di Pierro et al., 2007).

12.10 Cross-Linking by Radiation

Cross-linking is the most important effect originating from polymer irradiation. The degree of cross-linking is proportional to the radiation dose. The cross-linking mechanism depends on the polymer. The universally accepted mechanism involves cleavage of a C–H bond on a polymer chain, to form a hydrogen atom, followed by abstraction of a second hydrogen atom from a neighboring chain, to produce molecular hydrogen. Then, the two adjacent polymeric radicals combine, to form a cross-link (Bhattacharya, 2000). Generally, materials cross-linked by radiation are stronger and more resistant to impact and stress cracking. They have improved creep resistance and, in many cases, enhanced chemical resistance. Some of the advantages of radiation cross-linking as compared with conventional cross-linking with chemical additives are cost, speed, ability to cross-link a preformed part at or near room temperature, reduction of chemical ingredients and chemical residues for environmental or toxicological reasons, and superior material properties in the final product (Clough, 2001).

It is possible to mix additives, typically multifunctional monomers, with the basic polymer, to enhance the cross-linking effect and reduce the dose requirement. Adding antioxidants UV stabilizers and flame retardants is another possibility, to meet industrial performance specifications, but such additives may reduce the cross-linking effect (Cleland et al., 2003).

UV and ionizing radiation can modify functional properties of protein films. Ionizing radiation, like γ-irradiation, affects proteins because it elicits conformational changes, oxidation of amino acids, rupture of covalent bonds, formation of protein free radicals, and recombination and polymerization reactions (Gennadios et al., 1998). Depending on the nature of a protein and the irradiation dosage, the net result of irradiating proteins in the solid state could be cross-linking (aggregate formation) or molecular degradation (Urbain, 1977). In zein films, the effect of γ-rays of various doses, namely, 10, 20, 30, and 40 kGy, at a dose rate of 10.5 kGy/h, indicated that treating the film-forming solution with γ-irradiation can improve the water barrier properties, the color, and the appearance of films (Soliman et al., 2009). Cross-linking by γ-irradiation also improves the mechanical properties, water vapor barrier ability, and resistance to attack by proteolytic enzymes of protein films. In wheat gluten isolate film, γ-irradiation at 10 kGy dose increased tensile strength and decreased elongation at break (32% reduction). This effect stems from the formation of dityrosine cross-links. Radiation doses greater than 10 kGy reduce this effect, because the amount of insoluble glutenin polymer diminishes (Micard et al., 2000). Calcium caseinate and whey protein cross-linked by γ-irradiation have altered band intensities associated with larger β-sheet structure and smaller α-helix and unordered fractions of cross-linked proteins as compared with the nonirradiated control (Vu et al., 2012).

In protein films, double bonds and aromatic rings absorb UV radiation, to generate free radicals in amino acids (such as tyrosine and phenylalanine), which in turn can form intermolecular covalent bonds (Gennadios et al., 1998; Rhim et al., 1999). The UV radiation–induced cross-linking of egg albumin film involves aromatic amino acids, like phenylalanine and tyrosine, and reduces film solubility (Rhim et al., 1999). UV radiation at 51.8 J/m^2 over 24 h significantly augments the tensile strength of wheat gluten, corn zein, and egg albumin films, but it does not affect sodium caseinate films (Rhim et al., 1999). The tensile strength and elongations at break of SPI films treated with UV radiation increase and decrease linearly with UV dosage, respectively. However, UV irradiation does not affect the WVP of these films (Gennadios et al., 1998). Micard et al. (2000) also reported increased tensile strength, decreased film elongation, and no effect on the WVP of wheat gluten films. UV radiation of WPI at a dosage of 324 J/cm^2 for 3 h produces WPI films with increased WVP, larger tensile strength, and smaller elongation at break (Ustunol and Mert, 2004).

The use of UV-curable resins to generate cross-linked polymers can improve the mechanical properties of materials subjected to UV irradiation. Unfortunately, controversies related to the formation of toxic compounds or other health-related issues have limited the use of photo polymerization to modify the surface of food packaging polymers (Ozdemir et al., 1999; Wihodu and Moraru, 2013). Wihodu and Moraru (2013) applied UV-curable resins in wheat gluten, zein, and casein films, and more specifically, they applied epoxy acrylate and urethane acrylate to the surface of wheat gluten films and treated the coated films with UV light using benzophenone as photoinitiator. The authors found that the tensile strength of the treated films increases by up to 20%, depending on the % RH conditions, while the WVP decreases by more than 50% (Irissin-Mangata et al., 2000). Shi et al. (2009) were able to control the surface properties of zein films by alternating two solvents (ethanol and acetic acid) to spin-cast zein films in combination with an UV/ozone treatment. The different surface morphologies of the zein films prepared in distinct solvents originate from the diverse interactions between zein molecules and the Si wafers used to cast the films. Subsequent oxidation of these films for up to 180 s using UV/ozone treatment further helps to control their surface hydrophilicity. For casein films, the authors treated the films with pulsed light (PL), to improve the mechanical and barrier properties (Wihodo, 2009). PL consists of intense and short duration pulses of broad spectrum light, ranging from UV to near infrared. At 53% RH, the tensile strength of casein films rises by 31.6% and elongation increases by 90% after PL treatment of 15 pulses (23.5 J/cm^2) on each side. An increase in tensile strength also takes place in casein films when polyethylene glycol (400) diacrylate is incorporated as a photoinitiator and the film is subsequently treated with PL (Wihodo, 2009).

The radiation of microcrystalline cellulose–starch composites with UV light using sodium benzoate as photosensitizer reduces water absorption and swelling degree along the photoirradiation time and upon increasing cellulose content. Meanwhile, the tensile strength improves with increasing photoirradiation and cellulose content (Kumar and Singh, 2008).

12.11 Conclusions

The plasticization and cross-linking techniques can modify the mechanical and functional properties of biopolymer-based films. The degree of plasticity of biopolymers is largely dependent on the chemical structure of the plasticizer, including chemical composition, molecular weight, and functional groups. The plasticizer renders a film more flexible, processable, and extensible, but plasticizer type and quantity can make the film less mechanically resistant and more permeable to water vapor. Plasticizers also affect the optical properties, degree of crystallinity, and glass transition temperature (T_g) of biofilms. Cross-linking via physical, chemical, or enzymatic treatment constitutes a viable method to improve the mechanical strength and moisture barrier properties of polysaccharides, protein, and blend films.

ACKNOWLEDGMENTS

The authors wish to thank Fundação de Amparo à Pesquisa do Estado de São Paulo (São Paulo Research Support Foundation–FAPESP), CAPES (Coordenacão de Aperfeicionamento de Pessoal de Nível Superior) and CNPq (Conselho Nacional de Desenvolvimento Científico e Tecnológico) for their financial support.

REFERENCES

Ahmad, M., Benjakul, S., Prodpran, T. et al. Physico-mechanical and antimicrobial properties of gelatin film from the skin of unicorn leather jacket incorporated with essential oils. _Food Hydrocolloids_ 28 (2012): 189–199.

Al-Hassan, A. A. and Norziah, M. H. Starch—Gelatin edible films: Water vapor permeability and mechanical properties as affected by plasticizers. _Food Hydrocolloids_ 26 (2012): 108–117.

Ando, H., Adachi, M., Umeda, K. et al. Purification and characteristics of a novel transglutaminase derived from microorganisms. _Agricultural and Biological Chemistry_ 53 (1989): 2613–2617.

Andrade-Mahecha, M. M., Tapia-Blácido, D. R., and Menegalli, F. C. Development and optimization of biodegradable films based on achira flour. _Carbohydrate Polymers_ 88(2) (2012): 449–58.

Arvanitoyannis, I., Psomiadou, E., Nakayama, A. et al. Edible films made from gelatin, soluble starch and polyols. 3. _Food Chemistry_ 60(4) (1997): 593–604.

Audic, J. and Chaufer, B. Influence of plasticizers and crosslinking on the properties of biodegradable films made from sodium caseinate. _European Polymer Journal_ 41(8) (2005): 1934–1942.

Bamdad, F., Goli, A. H., and Kadivar, M. Preparation and characterization of proteinous film from lentil (_Lens culinaris_)—Edible film from lentil (_Lens culinaris_). _Food Research International_ 39(1) (2006): 106–111.

Bhattacharya, A. Radiation and industrial polymers. _Progress in Polymer Science_ 25 (2000): 371–401.

Bhattacharya, A. and Ray, P. Basic features and techniques in polymer grafting and crosslinking. In _Polymer Grafting and Crosslinking_, Bhattacharya, A., Rawlins, J. W., and Ray, P. (eds.) (New York: John Wiley & Sons, 2009), pp. 7–34.

Bergo, P. and Sobral, P. J. A. Effects of plasticizer on physical properties of pigskin gelatin films. _Food Hydrocolloids_ 21(8) (2007): 1285–1289.

Bergo, P. V. A., Carvalho, R. A., Sobral, P. J. A. et al. Physical properties of edible films based on cassava starch as affected by the plasticizer concentration. _Packaging Technology and Science_ 21(2) (2008): 85–89.

Bertan, L. C., Tanada-Palmu, P. S., Siani, A. C. et al. Effect of fatty acids and 'Brazilian elemi' on composite films based on gelatin. _Food Hydrocolloids_ 19(1) (2005): 73–82.

Bourtoom, T. Edible protein films: Properties enhancement. _International Food Research Journal_ 16(1) (2009): 1–9.

Bucci, D. Z., Tavares, L. B. B., and Sell, I. PHB packaging for the storage of food products. _Polymer Testing_ 24(5) (2005): 564–71.

Cao, N., Yang, X., and Fu, Y. Effects of various plasticizers on mechanical and water vapor barrier properties of gelatin films. _Food Hydrocolloids_ 23(3) (2009): 729–735.

Carvalho, R. A. and Grosso, C. R. F. Characterization of gelatin based films modified with transglutaminase, glyoxal and formaldehyde. _Food Hydrocolloids_ 18 (2004): 717–726.

Carvalho, R. A., Grosso, C. R. F., and Sobral, P. J. A. Effect of chemical treatment on the mechanical properties, water vapor permeability and sorption isotherms of gelatin-based films. _Packaging Technology and Science_ 21(3) (2008): 165–169.

Ceccorulli, G., Pizzoli, M., and Scandola, M. Plasticization of bacterial poly(3-hydroxybutyrate). _Macromolecules_ 25(12) (1992): 3304–3306.

Chambi, H. and Grosso, C. Edible films produced with gelatin and casein cross-linked with transglutaminase. _Food Research International_ 39 (2006): 458–466.

Chen, C. H. and Lai, L. S. Mechanical and water vapor barrier properties of tapioca starch/decolorized hsiantsao leaf gum films in the presence of plasticizer. _Food Hydrocolloids_ 22 (2008): 1584–1595.

Cheng, L. H., Karim, A. A., and Seow, C. C. Effects of water-glycerol and water-sorbitol interactions on the physical properties of Konjac. _Journal of Food Science_ 71(2) (2006): 62–67.

Choi, J. S. and Park, W. H. Effect of biodegradable plasticizers on thermal and mechanical properties of poly(3-hydroxybutyrate). *Polymer Testing* 23(4) (2004): 455–460.

Cleland, M. R., Parks, L. A., and Cheng, S. Applications for radiation processing of materials. *Nuclear Instruments & Methods in Physics Research Section B—Beam Interactions with Materials and Atoms* 208 (2003): 66–73.

Clough, R. L. High-energy radiation and polymers: A review of commercial processes and emerging applications. *Nuclear Instruments & Methods in Physics Research Section B—Beam Interactions with Materials and Atoms* 185 (2001): 8–33.

Cuq, B., Aymard, C., Cuq, J. L. et al. Edible packaging films based on fish myofibrillar proteins: Formulation and functional properties. *Journal of Food Science* 60(6) (1995): 1369–1374.

Cuq, B., Gontard, N., Cuq, J. L. et al. Selected functional properties of fish myofibrillar protein-based films as affected by hydrophilic plasticizers. *Journal of Agricultural and Food Chemistry* 45(3) (1997): 622–626.

Damodaran, S. Amino acids, peptides, and proteins. In *Fennema's Food Chemistry*, Damodaran, S., Parkin, K. L., and Fennema, O. R. (eds.) (Boca Raton, FL: CRC Press, 2008), pp. 217–329.

Dastidar, T. G. and Netravali, A. N. 'Green' crosslinking of native starches with malonic acid and their properties. *Carbohydrate Polymers* 90 (2012): 1620–1628.

De Jong, G. A. H. and Koppelman, S. J. Transglutaminase catalyzed reactions: Impact on food applications. *Journal of Food Science* 67 (2002): 2798–2806.

Dias, A. B., Muller, C. M. O., Larotonda, F. D. S. et al. Biodegradable films based on rice starch and rice flour. *Journal of Cereal Science* 51 (2010): 213–219.

Di Pierro, P., Chico, B., Villalonga, R. et al. Chitosan-Whey protein edible films produced in the absence or presence of transglutaminase: Analysis of their mechanical and barrier properties. *Biomacromolecules* 7(3) (2006): 744–749.

Di Pierro, P., Chico, B., Villalonga, R. et al. Transglutaminase-catalyzed preparation of chitosan-ovalbumin films. *Enzyme and Microbial Technology* 40 (2007): 437–441.

Di Pierro, P., Marquez, G. R., Mariniello, L. et al. Effect of transglutaminase on the mechanical and barrier properties of whey protein/pectin films prepared at complexation pH. *Journal of Agricultural and Food Chemistry* 61 (2013): 4593–4598.

Dumitriu, S., Vidal, P. F., and Chornet, E. Hydrogels based on polysaccharides. In *Polysaccharides in Medicinal Applications*, Dumitriu, S. (ed.) (New York: Marcel Dekker, 1996), pp. 125–223.

Forssell, P., Lahtinen, R., Lahelin, M. et al. Oxygen permeability of amylose and amylopectin films. *Carbohydrate Polymers* 47(2) (2002): 125–129.

Fringant, C., Rinaudo, M., Foray, M. F. et al. Preparation of mixed esters of starch or use of an external plasticizer: Two different ways to change the properties of starch acetate films. *Carbohydrate Polymers* 59(1–3) (1998): 97–106.

Funke, U., Bergthaller, W., and Lindhauer, M. G. Processing and characterization of biodegradable products based on starch. *Polymer Degradation and Stability* 59(1–3) (1998): 293–298.

Gachter, R. and Muller, H. (eds.). *Plastics Additives Handbook*, 3rd ed. New York: Carl Hanser, 1990.

Galdeano, M. C., Grossmann, M. V. E., Mali, S. et al. Effects of production process and plasticizers on stability of films and sheets of oat starch. *Materials Science & Engineering C—Biomimetic and Supramolecular Systems* 29(2) (2009): 492–98.

Galdeano, M. C., Mali, S., Grossmann, M. V. E. et al. Effects of plasticizers on the properties of oat starch films. *Materials Science & Engineering C—Biomimetic and Supramolecular Systems* 29(2) (2009): 532–538.

Galietta, G., Di Gioia, L., Guilbert, S. et al. Mechanical and thermomechanical properties of films based on whey proteins as affected by plasticizer and crosslinking agents. *Journal Dairy Science* 81(12) (1998): 3123–3130.

Garcia, M. A., Martino, M. N., and Zaritzky, N. E. Barrier properties of edible starch-based films and coatings. *Journal of Food Science* 65(6) (2000): 941–947.

Gaudin, S., Lourdin, D., Forssell, P. M. et al. Antiplasticisation and oxygen permeability of starch-sorbitol films. *Carbohydrate Polymers* 43 (2000): 33–37.

Gaudin, S., Lourdin, D., Le Botlan, D. et al. Plasticisation and mobility in starch-sorbitol films. *Journal of Cereal Science* 29 (1999): 273–284.

Gennadios, A., McHugh, T., Weller, C. L. et al. Edible coatings and films based on proteins. In *Edible Coatings and Films to Improve Food Quality*, Krochta, J. M., Baldwin, E. A., Nisperos-Carriedo, M. O. (eds.) (Lancaster, PA: Technomic Publishing, 1994), pp. 201–277.

Gennadios, A., Rhim, J. W., Handa, A. et al. Ultraviolet radiation affects physical and molecular properties of soy protein films. *Journal of Food Science* 63 (1998): 225–228.

Gennadios, A. and Weller, C. L. Edible films and coatings from soy milk and soy protein. *Cereal Foods World* 36 (1991): 1004–1009.

Ghebremeskel, N. A., Vemavarapu, C., and Lodaya, M. Use of surfactants as plasticizers in preparing solid dispersions of poorly soluble API: Selection of polymer–surfactant combinations using solubility parameters and testing the processability. *International Journal of Pharmacy* 328(2) (2007): 119–129.

Gontard, N., Guilbert, S., and Cuq, J. L. Water and glycerol as plasticizers affect mechanical and water vapor barrier properties of an edible wheat gluten film. *Journal of Food Science* 58(1) (1993): 206–211.

Guilbert, S. 1986. Technology and application of edible protective films. In *Food Packaging and Preservation: Theory and Practice*, Mathlouthi, M. (ed.) (London, U.K.: Elsevier Applied Science, 1986), pp. 371–394.

Honary, S. and Orafai, H. The effect of different plasticizer molecular weights and concentrations on mechanical and thermomechanical properties of free films. *Drug Development Industrial Pharmaceutical* 28(6) (2002): 711–715.

Hoque, M. S., Benjakul, S., and Prodpran, T. Effect of heat treatment of film forming solution on the properties of film from cuttlefish (*Sepia pharaonis*) skin gelatin. *Journal of Food Engineering* 96(1) (2010): 66–73.

Huang, M., Yu, J., and Ma, X. Ethanolamine as a novel plasticizer. *Polymer Degradation and Stability* 90(3) (2005): 501–507.

Irissin-Mangata, J., Bauduin, G., and Boutevin, B. Bilayer films composed of wheat gluten film and UV-cured coating: Water vapor permeability and other functional properties. *Polymer Bulletin* 44 (2000): 409–416.

Jangchud, A. and Chinnan, M. S. Properties of peanut protein film: Sorption isotherm and plasticizer effect. *LWT: Food Science and Technology* 32(2) (1999): 79–84.

Jiang, Y. and Tang, C. H. Effects of transglutaminase on sorption, mechanical and moisture-related properties of gelatin films. *Food Science and Technology International* 19(2) (2014): 99–108.

Jongjareonrak, A., Benjakul, S., Visessanguan, W. et al. Fatty acids and their sucrose esters affect the properties of fish skin gelatin based film. *European Food Research and Technology* 222(5–6) (2006): 650–657.

Juvonen, H., Smolander, M., Boer, H. et al. Film formation and surface properties of enzymatically cross-linked casein films. *Journal Applied Polymer Science* 119 (2011): 2205–2213.

Kanaji, T., Ozaki, H., Takao, T. et al. Primary structure of microbial transglutaminase from *Streptoverticillium* sp. Strain S-8112. *Journal of Biological Chemistry* 268 (1993): 11565–11572.

Karbowiak, T., Hervet, H., Leger, L. et al. Effect of plasticizers (water and glycerol) on the diffusion of a small molecule in iota-carrageenan biopolymer films for edible coating application. *Biomacromolecules* 7(6) (2006): 2011–2019.

Kim, S., Sessa, D. J., and Lawton, J. W. Characterization of zein modified with a mild cross-linking agent. *Industrial Crops Products* 20 (2004): 291–300.

Kim, S. J. and Ustunol, Z. Solubility and moisture sorption isotherms of whey protein-based edible films as influenced by lipid and plasticizer incorporation. *Journal of Agricultural and Food Chemistry* 49(9) (2001): 4388–4391.

Kolodziejska, I. and Piotrowska, B. The water vapour permeability, mechanical properties and solubility of fish gelatin-chitosan films modified with transglutaminase or 1-ethyl-3-(3-dimethylaminopropyl) carbodiimide (EDC) and plasticized with glycerol. *Food Chemistry* 103 (2007): 295–300.

Kolodziejska, I., Piotrowska, B., Bulge, M. et al. Effect of transglutaminase and 1-ethyl-3-(3-dimethylamino-propyl) carbodiimide on the solubility of fish gelatin-chitosan films. *Carbohydrate Polymers* 65 (2006): 404–409.

Krauskopf, L. G. Plasticizers: Types, properties and performance. In *Encyclopedia of PVC*, Nass, L. I. and Heiberger, C. A. (eds.), Vol. 2 (New York: Marcel Dekker, 1988), p. 214.

Krauskopf, L. G. Monomers for polyvinyl chloride (phthalates, adipates and trimelliates). In *Plastics Additives and Modifiers Handbook*, Edenbaum, J. (ed.) (London, U.K.: Chapman & Hall, 1996), pp. 359–378.

Kristo, E. and Biliaderis, C. G. Water sorption and thermo-mechanical properties or water/sorbitol-plasticized composite biopolymer films: Caseinato pullulan bilayers and blends. *Food Hydrocolloids* 20(7) (2006): 1057–1071.

Kumar, A. P. and Singh, R. P. Biocomposites of cellulose reinforced starch: Improvement of properties by photo-induced crosslinking. *Bioresource Technology* 99 (2008): 8803–8809.

Lai, H. M., Padua, G. W., and Wei, L. S. Properties and micro-structure of zein sheets plasticized with palmitic and stearic acids. *Cereal Chemistry* 74 (1997): 49–59.

Larre, C., Desserme, J., Barbot, J. et al. Properties of deamidated gluten films enzymatically cross-linked. *Journal of Agricultural and Food Chemistry* 48 (2000): 5444–5449.

Li, B. Z., Wang, L. J., Li, D. et al. Physical properties and loading capacity of starch-based microparticles crosslinked with trisodium trimetaphosphate. *Journal of Food Engineering* 92 (2009): 255–260.

Lim, L. T., Mine, Y., and Tung, M. A. Transglutaminase cross-linked egg white protein films: Tensile properties and oxygen permeability. *Journal of Agricultural and Food Chemistry* 46(10) (1998): 4022–4029.

Liu, Z., Dong, Y., Men, H. et al. Post-crosslinking modification of thermoplastic starch/PVA blend films by using sodium hexametaphosphate. *Carbohydrate Polymers* 89 (2012): 473–477.

Lourdin, D., Coignard, L., Bizot, H. et al. Influence of equilibrium relative humidity and plasticizer concentration on the water content and glass transition of starch materials. *Polymer* 38(21) (1997): 5401–5406.

Ma, X., Chang, P. R., Yu, J. et al. Properties of biodegradable citric acid-modified granular starch/thermoplastic pea starch composites. *Carbohydrate Polymer* 75(1) (2009): 1–8.

Mali, S., Karam, L. B., Ramos, L. P. et al. Relationships among the composition and physicochemical properties of starches with the characteristics of their films. *Journal of Agricultural and Food Chemistry* 52 (2004): 7720–7725.

Mali, S., Sakanaka, L. S., Yamashita, F. et al. Water sorption and mechanical properties of cassava starch films and their relation to plasticizing effect. *Carbohydrate Polymers* 60(3) (2005): 283–289.

Maniglia, B. C., Domingos, J. R., de Paula, R. L. et al. Development of bioactive edible film from turmeric dye solvent extraction residue. *LWT: Food Science and Technology* 56 (2014): 269–277.

Maniglia, B. C., de Paula, R. L., Domingos, J. R., et al. Turmeric dye extraction residue for use in bioactive film production: Optimization of turmeric film plasticized with glycerol. *LWT Food Science and Technology* 64 (2015): 1187–1195.

Mao, G. J., Wang, P., Meng, X. S. et al. Crosslinking of cornstarch with sodium trimetaphosphate in solid state by microwave irradiation. *Journal of Applied Polymer Science* 102 (2006): 5854–5860.

Marcilla, A. and Beltran, M. Mechanism of plasticizer action. In *Handbook of Plasticizers*, 2nd ed., Wypych, G. (ed.) (Toronto, Ontario, Canada: ChemTec Publishing, 2012).

Mariniello, L., Di Pierro, P., Esposito, C. et al. Preparation and mechanical properties of edible pectin-soy flour films obtained in the absence or presence of transglutaminase. *Journal of Biotechnology* 102(2) (2003): 191–198.

Marquez, G. R., Di Pierro, P., Esposito, M. et al. Application of transglutaminase-crosslinked whey protein/pectin films as water barrier coatings in fried and baked foods. *Food and Bioprocess Technology* 7 (2014): 447–455.

Micanovic, R., Procyk, R., Lin, W. et al. Role of histidine 373 in catalytic activity of coagulation factor XIII. *Journal of Biological Chemistry* 269 (1994): 9190–9194.

Micard, V., Belamri, R., Morel, M. H. et al. Properties of chemically and physically treated wheat gluten films. *Journal of Agricultural and Food Chemistry* 48(7) (2000): 2948–2953.

Micard, V. and Guilbert, S. Thermal behavior of native and hydrophobized wheat gluten, gliadin and glutenin-rich fractions by modulated DSC. *International Journal of Biological Macromolecules* 27 (2000): 229–236.

Mokrejs, P., Langmaier, F., Janacova, D. et al. Thermal study and solubility tests of films based on amaranth flour starch-protein hydrolysate. *Journal of Thermal Analysis and Calorimetry* 98(1) (2009): 299–297.

Moore, G. R. P., Martelli, S. M., Gandolfo, C. et al. Influence of glycerol concentration on some physical properties of feather keratin films. *Food Hydrocolloids* 20(7) (2006): 975–982.

Müller, C. M. O., Yamashita, F., and Borges-Laurindo, J. Evaluation of the effects of glycerol and sorbitol concentration and water activity on the water barrier properties of cassava starch films through a solubility approach. *Carbohydrate Polymers* 72(1) (2008): 82–87.

Murphy, J. *Additives for Plastics Handbook*, 2nd ed. New York: Elsevier, 2001.

Navarro-Tarazaga, M. L., Sothornvit, R., and Pérez-Gago, M. B. Effect of plasticizer type and amount on hydroxypropyl methylcellulose beeswax edible film properties and postharvest quality of coated plums (*Cv. Angeleno*). *Journal of Agricultural and Food Chemistry* 56(20) (2008): 9502–9509.

Ning, W., Xingxiang, Z., Na, H. et al. Effects of water on the properties of thermoplastic starch poly(lactic acid) blend containing citric acid. *Journal of Thermoplastic Composite Materials* 23 (2010): 19–34.

Nuthong, P., Benjakul, S., and Prodpran, T. Characterization of porcine plasma protein-based films as affected by pretreatment and cross-linking agents. *International Journal of Biological Macromolecules* 44 (2009): 143–148.

Oh, J. H., Wang, B., Field, P. D. et al. Characteristics of edible films made from dairy proteins and zein hydrolisates cross-linked with transglutaminase. *International Journal of Food Science and Technology* 39 (2004): 287–294.

Olivato, J. B., Grossmann, M. V. E., Bilck, A. P. et al. Effect of organic acids as additives on the performance of thermoplastic starch/polyester blown films. *Carbohydrate Polymers* 90 (2012): 159–164.

Olsson, E., Menzel, C., Johansson, C. et al. The effect of pH on hydrolysis, cross-linking and barrier properties of starch barriers containing citric acid. *Carbohydrate Polymers* 98 (2013): 1505–1513.

Orliac, O., Rouilly, A., Silvestre, F. et al. Effects of various plasticizers on the mechanical properties, water resistance and aging of thermo-moulded films made from sunflower proteins. *Industrial Crops and Products* 18(2) (2003): 91–100.

Ozdemir, M., Yurteri, C. U., and Sadikoglu, H. Physical polymer surface modification methods and applications in food packaging polymers. *Critical Reviews in Food Science and Nutrition* 39(5) (1999): 457–477.

Parris, N. and Coffin, D. R. Composition factors affecting the water vapor permeability and tensile properties of hydrophilic zein films. *Journal Agricultural and Food Chemistry* 45 (1997): 1596–1599.

Parris, N., Dickey, L. C., Tomasula, P. M. et al. Films and coatings from commodity agroproteins. *ACS Symposium Series* 786 (2001): 118–131.

Patzsch, K., Riedel, K., and Pietzsch, M. Parameter optimization of protein film production using microbial transglutaminase. *Biomacromolecules* 11 (2010): 896–903.

Pelissari, F. M., Andrade-Mahecha, M. M., Sobral, P. J. A. et al. Optimization of process conditions for the production of films based on the flour from plantain bananas (*Musa paradisiaca*). *LWT: Food Science and Technology* 52 (2013): 1–11.

Pérez-Gago, M. B. Protein-based films and coatings. In *Edible Coatings and Films to Improve Food Quality*, Baldwin, E. A., Hagenmaier, R. D., and Bai, J. (eds.) (Boca Raton, FL: CRC Press, 2012), pp. 12–76.

Pérez-Gago, M. B. and Krochta, J. M. Lipid particle size effect on water vapor permeability and mechanical properties of whey protein/beeswax emulsion films. *Journal of Agricultural and Food Chemistry* 49(2) (2001a): 996–1002.

Pérez-Gago, M. B. and Krochta, J. M. Denaturation time and temperature effects on solubility, tensile properties, and oxygen permeability of whey protein edible films. *Journal of Food Science* 66(5) (2001b): 705–710.

Pérez-Gago, M. B., Nadaud, P., and Krochta, J. M. Water vapor permeability, solubility and tensile properties of heat-denatured versus native whey protein films. *Journal of Food Science* 64(6) (1999): 1034–1037.

Piotrowska, B., Sztuka, K., Kolodziejska, I. et al. Influence of transglutaminase or 1-ethyl-3-(3-dimethylaminopropyl) carbodiimide (EDC) on the properties of fish-skin gelatin films. *Food Hydrocolloids* 22 (2008): 1362–1371.

Pommet, M., Redl, A., Morel, M. H. et al. Study of wheat gluten plasticization with fatty acids. *Polymer* 44(1) (2003): 115–122.

Porta, R., Mariniello, L., Di Pierro, P. et al. Transglutaminase crosslinked pectin and chitosan-based edible films: A review. *Critical Reviews in Food Science and Nutrition* 51 (2011): 223–238.

Rahman, M. and Brazel, C. S. The plasticizer market: An assessment of traditional plasticizers and research trends to meet new challenges. *Progress in Polymer Science* 29 (2004): 1223–1248.

Rayas, L. M., Hernandez, R. J., and Ng, P. K. W. Development and characterization of biodegradable/edible wheat protein films. *Journal of Food Science* 62(1) (1997): 160–162.

Reddy, N. and Yang, Y. Citric acid cross-linking of starch films. *Food Chemistry* 118 (2010): 702–711.

Rhim, J. W., Gennadios, A., Fu, D. et al. Properties of ultraviolet irradiated protein films. *LWT: Food Science and Technology* 32 (1999): 129–133.

Rhim, J. W., Gennadios, A., Handa, A. et al. Solubility, tensile, and color properties of modified soy protein isolate films. *Journal of Agricultural and Food Chemistry* 48(10) (2000): 4937–4941.

Rivero, S., García, M. A., and Pinotti, A. Crosslinking capacity of tannic acid in plasticized chitosan films. *Carbohydrate Polymers* 82 (2010): 270–276.

Rodriguez, M., Oses, J., Ziani, K. et al. Combined effect of plasticizers and surfactants on the physical properties of starch based edible films. *Food Research International* 39(8) (2006): 840–846.

Rotta, J., Ozório, R. A., Kehrwald, A. M. et al. Parameters of color, transparency, water solubility, wettability and surface free energy of chitosan/hydroxypropylmethylcellulose (HPMC) films plasticized with sorbitol. *Materials Science & Engineering C—Biomimetic and Supramolecular Systems* 29(2) (2009): 619–623.

Santosa, F. X. B. and Padua, G. W. Tensile properties and water absorption of zein sheets plasticized with oleic and linoleic acids. *Journal of Agricultural and Food Chemistry* 47(5) (1999): 2070–2074.

Sears, J. K. and Darby, J. R. Thermodynamic parameters of the junction zones in thermoreversible maltodextrin gels. *Carbohydrate Polymers* 12 (1992): 245–253.

Seguro, K., Kumazawa, Y., Kuraishi, C. et al. The épsilon-(gamma-glutamyl) lysine moiety in cross-linked casein is an available source of lysine for rats. *Journal of Nutrition* 126 (1996): 2557–1562.

Shaikh, H. M., Pandare, K. V., Nair, G. et al. Utilization of sugarcane bagasse cellulose for producing cellulose acetates: Novel use of residual hemicellulose as plasticizer. *Carbohydrate Polymers* 76(1–2) (2009): 23–29.

Sheng, W. and Zhao, X. Functional properties of a cross-linked soy protein-gelatin composite towards limited tryptic digestion of two extents. *Journal of the Science of Food and Agriculture* 93(15) (2013): 3785–3791.

Shi, K., Kokini, J. L., and Huang, Q. Engineering zein films with controlled surface morphology and hydrophilicity. *Journal of Agricultural and Food Chemistry* 57 (2009): 2186–2192.

Silva, M. A., Bierhalz, A. C. K., and Kieckbusch, T. G. Alginate and pectin composite films crosslinked with Ca^{2+} ions: Effect of the plasticizer concentration. *Carbohydrate Polymers* 77(4) (2009): 736–742.

Smits, A. L. M., Kruiskamp, P. H., Van Soest, J. J. G. et al. Interaction between dry starch and glycerol or EG, measured by differential scanning calorimetry and solid state NMR spectroscopy. *Carbohydrate Polymers* 53(4) (2003): 409–416.

Sobral, P. J. A., Menegalli, F. C., Hubinger, M. D. et al. Mechanical, water vapor barrier and thermal properties of gelatin based edible films. *Food Hydrocolloids* 15 (2001): 423–432.

Sobral, P. J. A., Monterrey-Quintero, E. S., and Habitante, A. M. Q. B. Glass transition of Nile Tilapia myofibrillar protein films plasticized by glycerin and water. *Journal of Thermal Analysis and Calorimetry* 67(2) (2002): 499–494.

Sobral, P. J. A, Santos, J. S., and García, F. T. Effect of protein and plasticizer concentrations in film forming solutions on physical properties of edible films based on muscle proteins of a Thai Tilapia. *Journal of Food Engineering* 70(1) (2005): 93–100.

Soliman, E. A., Eldin, M. S. M., and Furuta, M. Biodegradable Zein-based films: Influence of gamma-irradiation on structural and functional properties. *Journal of Agricultural and Food Chemistry* 57(6) (2009): 2529–2535.

Sothornvit, R. and Krochta, J. M. Water vapor permeability and solubility of films from hydrolyzed whey protein. *Journal of Food Science* 65(4) (2000): 700–703.

Sreedhar, B., Chattopadhyay, D. K., Karunakar, M. S. H. et al. Thermal and surface characterization of plasticized starch polyvinyl alcohol blends crosslinked with epichlorohydrin. *Journal of Applied Polymer Science* 101 (2006): 25–34.

Sreedhar, B., Sairam, M., Chattopadhyay, D. K. et al. Thermal, mechanical, and surface characterization of starch-poly(vinyl alcohol) blends and borax-crosslinked films. *Journal of Applied Polymer Science* 96 (2005): 1313–1322.

Stein, T. M., Gordon, S. H., and Greene, R. V. Amino acids as plasticizers—II. Use of quantitative structure-property relationships to predict the behavior of mono ammonium monocarboxylate plasticizers in starch-glycerol blends. *Carbohydrate Polymers* 39(1) (1999): 7–16.

Stuchell, Y. M. and Krochta, J. M. Enzymatic treatments and thermal effects on edible soy protein films. *Journal of Food Science* 59(6) (1994): 1332–1337.

Su, G., Cai, H., Zhou, C. et al. Formation of edible soybean and soybean-complex protein films by a cross-linking treatment with a new *Streptomyces* transglutaminase. *Food Technology and Biotechnology* 45(4) (2007): 381–388.

Suyatma, N. E., Tighzert, L., and Copinet, A. Effects of hydrophilic plasticizers on mechanical, thermal, and surface properties of chitosan films. *Journal of Agricultural and Food Chemistry* 53(10) (2005): 3950–3957.

Sztuka, K. and Kolodziejska, I. The influence of hydrophobic substances on water vapor permeability of fish gelatin films modified with transglutaminase or 1-ethyl-3-(3-dimethylaminopropyl) carbodiimide (EDC). *Food Hydrocolloids* 23 (2009): 1062–1064.

Talja, R. A., Helen, H., Roos, Y. H. et al. Effect of various polyols and polyol contents on physical and mechanical properties of potato starch-based films. *Carbohydrate Polymers* 67(3) (2007): 288–295.

Tanaka, M., Iwata, K., Sanguandeekul, R. et al. Influence of plasticizers on the properties of edible films prepared from fish water-soluble proteins. *Fish Science* 67(2) (2001): 346–351.

Tang, C.H, Jiang, Y., Wen, Q. B. et al. Effect of transglutaminase treatment on the properties of cast films of soy protein isolates. *Journal of Biotechnology* 120 (2005): 296–307.

Tapia-Blácido, D., Mauri, A. N., Menegalli, F. C. et al. Contribution of the starch, protein, and lipid fractions to the physical, thermal, and structural properties of Amaranth (*Amaranthus caudatus*) flour films. *Journal of Food Science* 72 (2007): 293–300.

Tapia-Blácido, D., Sobral, P. J., and Menegalli, F. C. Development and characterization of biofilms based on Amaranth flour (*Amaranthus caudatus*). *Journal of Food Engineering* 67 (2005): 215–223.

Tapia-Blácido, D., Sobral, P. J. A., and Menegalli, F. C. Optimization of amaranth flour films plasticized with glycerol and sorbitol by multi-response analysis. *LWT: Food Science and Technology* 44(8) (2011): 1731–1738.

Taylor, M. M., Liu, C. K., Latona, N. et al. Enzymatic modification of hydrolysis products from collagen using microbial transglutaminase. II. Preparation of films. *Journal of the American Leather Chemists Association* 97(6) (2002): 225–234.

Tharanathan, R. N. Biodegradable films and composite coatings: Past, present and future. *Trends in Food Science and Technology* 14(3) (2003): 71–78.

Thomazine, M., Carvalho, R. A., and Sobral, P. J. A. Physical properties of gelatin films plasticized by blends of glycerol and sorbitol. *Journal of Food Science* 70(3) (2005): 172–176.

Tickner, J. A., Schettler, T., Guidotti, T. et al. Health risks posed by use of di-2-ethylhexyl phthalate (DEHP) in PVC medical devices: A critical review. *American Journal of Industrial Medicine* 39(1) (2001): 100–111.

Tropini, V., Lens, L. P., Mulder, W. J. et al. Wheat gluten films cross-linked with 1-ethyl-3-(3-dimethylaminopropyl) carbodiimide and N-hydroxysuccinimide. *Industrial Crops and Products* 20 (2004): 281–289.

Urbain, W. M. Radiation chemistry of proteins. In *Radiation Chemistry of Major Food Components*, Elias, P. S. and Cohen, A. J. (eds.), Chapter 4 (Amsterdam, the Netherlands: Elsevier Scientific Publishing Company, 1977), pp. 63–130.

Ustunol, Z. Edible films and coatings for meat and poultry. In *Edible Films and Coatings for Food Applications*, Embuscado, M. E. and Huber, K. C. (eds.) (New York: Springer, 2009), pp. 245–268.

Ustunol, Z. and Mert, B. Water solubility, mechanical, barrier, and thermal properties of cross-linked whey protein isolate-based films. *Journal of Food Science* 69 (2004): 129–133.

Van de Velde, K. and Kiekens, P. Biopolymers: Overview of several properties and consequences on their applications. *Polymeric Test* 21(4) (2002): 433–442.

Vanin, F. M., Sobral, P. J. A., Menegalli, F. C. et al. Effects of plasticizers and their concentrations on thermal and functional properties of gelatin-based films. *Food Hydrocolloids* 19(5) (2005): 899–907.

Van Soest, J. J. G. and Knooren, N. Influence of glycerol and water content on the structure and properties of extruded starch plastic sheets during aging. *Journal of Applied Polymer Science* 64(7) (1997): 1411–1422.

Veiga-Santos, P., Oliveira, L. M., Cereda, M. P. et al. Sucrose and inverted sugar as plasticizer. Effect on cassava starch–gelatin film mechanical properties, hydrophilicity and water activity. *Food Chemistry* 103(2) (2007): 255–262.

Vieira, M. G. A., Da Silva, M. A., Dos Santos, L. O. et al. Natural based plasticizers and biopolymer films: A review. *European Polymer Journal* 47 (2011): 254–263.

Vu, K. D., Hollingsworth, R. G., Salmieri, S. et al. Development of bioactive coatings based on γ-irradiated proteins to preserve strawberries. *Radiation Physics and Chemistry* 81 (2012): 1211–1214.

Wang, N., Yu, J., Chang, P. R. et al. Influence of citric acid on the properties of glycerol-plasticized dry starch (DTPS) and DTPS/poly(lactic acid) blends. *Starch/Stärke* 59 (2007): 409–417.

Wang, Y. H. and Hsieh, Y. L. Crosslinking of polyvinyl alcohol (PVA) fibrous membranes with glutaraldehyde and PEG diacylchloride. *Journal of Applied Polymer Science* 116 (2010): 3249–3255.

Wihodo, M. Effect of pulsed light treatment on the functional properties of protein films. MS thesis, Cornell University, Ithaca, NY, 2009.

Wihodo, M. and Moraru, C. I. Physical and chemical methods used to enhance the structure and mechanical and properties of protein films: A review. *Journal of Food Engineering* 114 (2013): 292–302.

Wong, S. S. *Chemistry of Protein Conjugation and Cross-Linking.* Boca Raton, FL: CRC Press, 1993, pp. 30–45.

Wypych, G. Plasticizers types. In *Handbook of Plasticizers*, Wypych, G. (ed.) (Toronto, Ontario, Canada: Chem Tec Laboratories, 2004), pp. 7–10.

Yamada, K., Takahashi, H., and Noguchi, A. Improved water resistance in edible zein films and composites for biodegradable food packaging. *International Journal of Food Science and Technology* 30 (1995): 599–608.

Yang, Y., Wang, L., and Li, S. Formaldehyde-free zein fiber preparation and investigation. *Journal of Applied Polymer Science* 59 (1996): 433–441.

Yi, J. B., Kim, Y. T., Bae, H. J. et al. Influence of transglutaminase-induced cross-linking on properties of fish gelatin films. *Journal of Food Science* 71(9) (2006): 376–383.

Yildirim, M. and Hettiarachchy, N. S. Properties of films produced by cross-linking whey proteins and 11S globulin using transglutaminase. *Journal of Food Science* 63 (1998): 248–252.

Yin, Y. P., Li, J. F., Liu, Y. C. et al. Starch crosslinked with poly(vinyl alcohol) by boric acid. *Journal of Applied Polymer Science* 96 (2005): 1394–1397.

Yokoyama, K., Nio, N., and Kikuchi, Y. Properties and applications of microbial transglutaminase. *Applied Microbiology and Biotechnology* 64 (2004): 447–454.

Yoshie, N., Nakasato, K., Fujiwara, M. et al. Effect of low molecular weight additives on enzymatic degradation of poly(3-hydroxybutyrate). *Polymer* 41(9) (2000): 3227–3234.

Zhang, Y. and Han, J. H. Mechanical and thermal characteristics of pea. *Journal of Food of Science* 71(2) (2006): 109–118.

13

Nanocompounds as Formulating Aids

María Cecilia Condés, Ignacio Echeverría,
María Cristina Añón, and Adriana Noemí Mauri

CONTENTS

Nanotechnology involves the characterization, fabrication, and/or manipulation of structures, devices, or materials that, or at least, contain components with at least one dimension that is approximately 1–100 nm long. When particle size is reduced below this threshold, the resulting material exhibits physical and chemical properties that are significantly different from the properties of macroscale materials composed of the same substances.

Undeniably, the most active area of food nanoscience research and development is packaging, where significant advances in the nanoreinforcement of biobased materials provide a more solid ground toward increasing the technical and economic competitiveness of renewable polymers for different applications. Because of their size, the nanoreinforcements have a high aspect ratio that involves a relative larger surface area per mass of filler than the microreinforcements, which causes a strong interaction with the polymer matrix. As a result, it is expected that the resulting materials could exhibit better mechanical and barrier properties, thermal stability, chemical resistance, and surface appearance. Interestingly, these improvements can be achieved with low loading levels ($\leq 5\%$–10%), while traditional compounds need higher filler contents (40%–50%) to achieve the same reinforcement effect. This phenomenon leads to reductions in the weight for the same behavior—which is important for various applications—or increased resistance and better barrier properties to similar structural dimensions. These facts make them very interesting to be used in packaging materials (Rhim et al., 2007; Zhao et al., 2008).

This chapter resumes the recent advances in bionanocomposites and their preparation techniques, properties, and applications.

13.1 Bionanocomposites

Over the last decades, there has been an increase in the need to develop new biodegradable materials from renewable sources to replace synthetic polymers, at least in some applications. This increase is due to the fact that plastic materials present environmental problems associated with their production and accumulation (mainly because of the high resistance to degradation of plastic) and they are made from petroleum, a nonrenewable resource.

Bioplastics possess at least any of these required characteristics: they are biodegradable and/or are obtained from renewable sources. The "American Society of Testing and Materials" defines a *biodegradable* material as "that can be decomposed into carbon dioxide, methane, inorganic components or biomass, by the enzymatic activity of microorganisms and can be measured by standard assays over a determined period of time" (ASTM, 2002).

Based on their origin, *bioplastics* can be classified into three groups (Reddy et al., 2013) (Figure 13.1):

1. *Renewable resource–based bioplastics*: Those that are either synthesized naturally from plants and animals or entirely synthesized from renewable resources by chemical or biological methods (Sudesh and Iwata, 2008).

2. *Petroleum-based bioplastics*: These polymers are synthesized from petroleum resources but are biodegradable at the end of their functionality. Poly(caprolactone) (PCL) and poly(butylene adipate-co-terephthalate) are included in this category.

3. *Bioplastics from mixed sources*: These are made from combinations of biobased and petroleum monomers; they include polymers such as poly(trimethylene terephthalate), biothermosets, and biobased blends.

The fact that a material is obtained from renewable resources does not necessarily imply that it is biodegradable and vice versa. Therefore, the combination of both characteristics is what makes the great interest in this area. The *renewable resource–based bioplastics* mentioned earlier are both *biodegradable* and *renewable*. They can be classified into three categories according to the method of production (Figure 13.1) (van Tuil et al., 2000):

- *Polymers produced by classical chemical synthesis from natural monomers*: Among these polymers, the most studied are the polylactic acid (PLA), prepared from the polymerization of lactic acid that is produced by fermentation of carbohydrates, and polyglycolic acid synthesized

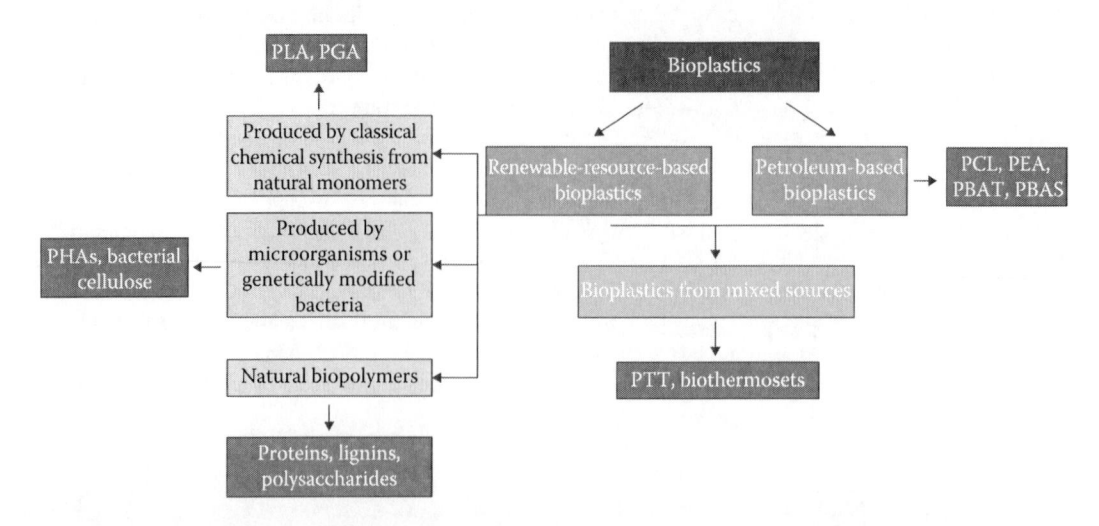

FIGURE 13.1 Classification of bioplastics based on their origin and on their method of production.

from glycolic acid. From a commercial standpoint, PLA is an interesting material since it has good mechanical properties and it is transparent and biodegradable; however, the level of industrial production is costly compared to the most used thermoplastic materials (Bohlmann, 2005; van Tuil et al., 2000).

- *Polymers produced by wild-type microorganisms or genetically modified bacteria*: This group is constituted by polyesters produced by a wide variety of microorganisms as a source of carbon and energy storage (Suriyamongkol et al., 2007). Among them, the best known polymers are the polyhydroxyalkanoates, among which the most studied ones are the polyhydroxybutyrates (PHB). There are other polymers produced by microorganisms that are currently being studied, such as bacterial cellulose (BC), among others. It is noteworthy that the production costs of these polymers are still high.

- *Natural biopolymers*: This class includes polysaccharides (cellulose derivatives, alginate, pectin, starch, chitosan, carrageenan, agar, and gums), lignins, and proteins (soy protein, wheat gluten, corn zein, gelatin, whey, casein, and keratin, among others) that are of plant or animal origin. All these compounds contain hydrolyzable bonds, a property that makes them very susceptible to biodegradation by hydrolytic enzymes of microorganisms. This feature has great impact on the performance and durability of these materials when stored under high moisture conditions.

Compared to petroleum-based synthetic plastics, biopolymer films normally exhibit relatively poor mechanical and barrier properties, brittleness, low heat distortion temperature, low melt viscosity for further processing, etc., which limit their industrial use (Bharadwaj, 2001; Koh et al., 2008; Sorrentino et al., 2006). Furthermore, biopolymers present other problems such as those associated with performance, processing, and cost that are common to all biodegradable polymers regardless their origin (Pandey et al., 2005; Scott, 2000; Trznadel, 1995).

As with synthetic polymers, the new generation of materials in the area of biopolymers is being focused toward obtaining *nanocomposites* because it has already proved to be an effective means to improve their properties. These compounds can be defined as composite materials where at least one of the phases has any of their dimensions in the nanometer range (1–100 nm). As mentioned earlier, these nanocomposites are able to improve their performance with loading levels that are lower than those required in traditional composites to obtain the same behavior (Giannelis, 1996).

13.2 Building Strategies Employed in Nanotechnology: "Top-Down" and "Bottom-Up"

In the area of nanotechnology, there are two strategies for building structures: bottom-up and top-down.

Bottom-up is a process of self-assembly where a nanostructure is assembled to create a larger structure. This self-assembly is commonly defined as molecular nanotechnology. Self-assembly relies on balancing attraction and repulsion forces that are generated between a pair of molecules as building blocks to form more functional supramolecular structures (Sanguansri and Augustin, 2006).

Nature is a good example of bottom-up, where each living organism has been created by the self-assembly of nanostructures, such as atoms and different vital biomolecules. This occurs through optimized processes that tend to minimize the free energy. This process is used in the production of nanostructures for use in food science and technology, so using directed self-assembly processes thermodynamically. Areas of research that could prove useful in the near future include molecular design of protective surface systems (Charpentier, 2005), surface engineering (Krajewska, 2004), and various methods of manufacturing, such as electrospinning (Min et al., 2004) and nanofiltration (van der Graaf et al., 2005).

Top-down is a process by which nanostructures are obtained from larger structures. This is the process used to obtain nanoparticles or nanostructured materials in industry. There are many methods to prepare these nanomaterials such as ball milling, ultrasonication, reverse micelling, chemical reduction, chemical and physical vapor depositions, solid-state reaction, hydrothermal, nanolithography, and microwaves. Their applications can be as wide as refractories, textiles, energy, biomedicals, functional barriers, and environmental fields (Guodong, 2005; Rajendran, 2009).

It has been reported that at the nanoscale (below about 100 nm), the properties of a material can change dramatically. With only a reduction in size and no change in the substance itself, materials can exhibit new properties such as electrical conductivity, insulating behavior, elasticity, greater strength, different color, and greater reactivity characteristics that the very same substances do not exhibit at the micro- or macroscale.

13.3 Nanoreinforcements

Nanoreinforcements are fillers that have at least one dimension on the nanometer scale (<100 nm) that when dispersed in a polymer matrix, they offer tremendous improvement in the performance properties of the resulting nanocomposite.

One of the most widespread classifications of nanoreinforcements is that taking into consideration the number of dimensions of the dispersed fillers in the nanometer range:

1. *Isodimensional nanoparticles*: When three dimensions are in the order of nanometers, such as spherical silica nanoparticles, semiconductor nanoclusters, and metallic nanoparticles (Herron and Thorn, 1998; Vladimirov et al., 2006). These nanoparticles generally show moderate reinforcement effect due to their low aspect ratio, but they are used to enhance resistance to flammability and to decrease permeability or costs.
2. *Elongated particles*: When two dimensions are present in the nanometer scale and the third one is larger, such as carbon nanotubes (CNTs) or cellulose whiskers that are extensively studied as reinforcing nanofillers yielding materials with exceptional properties (Calvert, 1997; Siqueira et al., 2010).
3. *Layered particles*: When only one dimension is in the nanometer range, in this case, the filler is present in the form of sheets of one to a few nanometers thick to hundreds to thousands nanometers long, such as layered crystals or clays (Ojijo and Sinha Ray, 2013).

As dimensions reach the nanometer level, interactions at the interfaces become largely improved, provoking an important improvement in the properties of materials. In this context, the surface area/volume ratio of reinforcement materials employed in the preparation of nanocomposites is crucial to understand their structure–property relationships (McCrum et al., 1996). Figure 13.2 shows the ratio superficial area (A)/volume (V) vs. the aspect ratio a: length (l)/diameter (d) of reinforcements. This figure shows that the ratio A/V is higher for sheets (a ≪ 1) and rods (a ≫ 1), explaining how they can produce a higher reinforcement effect than isodimensional nanoparticles. This figure also shows that the A/V ratio increases much more abruptly for sheets than for rods. This phenomenon suggests that the fibers are much easier to incorporate into a material because they have less surface contact than sheets, generally leading to an improvement of the mechanical potential (reinforcing), as recently described theoretically by Gusev (2001).

Nanoreinforcements added on the polymer matrix can improve the polymer properties and also provide value-added properties not present in the neat matrix, without sacrificing the matrix's inherent processability (Schwartz, 2005). Because of this, over the last decade, the number of studies related to the use of nanoreinforcements for manufacturing and optimization of bionanocomposite materials has increased considerably. Several types of nanofillers, such as layered silicate clay minerals, nanoparticles derived from polysaccharides (starch nanoparticles, cellulosic, and chitosan whiskers), metal nanoparticles, CNTs, graphene, silica, hydroxyapatite, and organic nanofillers, have been incorporated in biodegradable polymer formulations (Famá et al., 2011; Hong et al., 2005; Ojijo and Sinha Ray, 2013; Pan et al., 2011).

It should be noted that smaller sizes combined with a homogeneous dispersion decrease the likelihood of finding large stress concentration sites within the material, thus improving the mechanical properties.

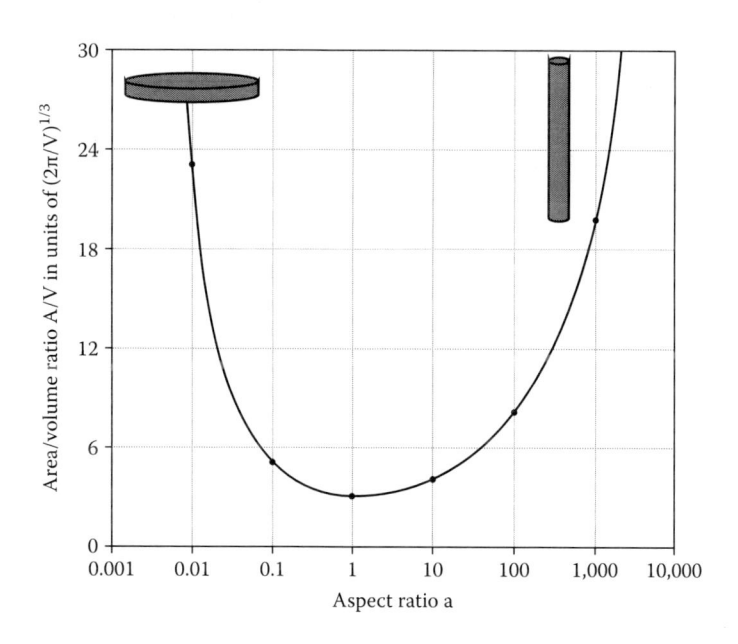

FIGURE 13.2 Surface area to volume ratio A/V of a cylindrical particle of given volume versus aspect ratio a = length (l)/ diameter (d). (From McCrum, N.G. et al., *Principles of Polymer Engineering*, Oxford Science, New York, 1996.)

As nanofillers can provide exceptionally large interfacial area in composites, the matrix material properties related to local chemistry, polymer cross-linking, polymer chain mobility, and conformation and the degree of polymer chain ordering or crystallinity are significantly affected in the vicinity of the reinforcement and can vary significantly and continuously from the interface with the reinforcement into the bulk of the matrix (Ajayan et al., 2003).

13.4 Promising Nanoreinforcements in Biodegradable Films and Coatings

Some organic and inorganic materials, such as clay minerals, polysaccharides, and some metals, have been tested as potential sources of nanoparticles and nanofibers. These materials have the advantage of being renewable, which means they have a much smaller environmental impact than their synthetic equivalents. Therefore, when they are incorporated into bioplastics formulations, they are useful to prepare green or eco-friendly materials. Their main features are described in the following texts.

13.4.1 Nanoclays and Silicates

Layered silicate clay minerals are among the most studied nanoreinforcements, not only because of their easy availability, low cost, and being environmentally safe but also because of their relatively simple processability and the significant improvements they generate when being incorporated into polymeric materials (Giannelis, 1996; Sinha Ray and Bousmina, 2005; Sinha Ray and Okamoto, 2003a,b). Among the layered clay minerals, the most common are the silicates with 2:1 layered or phyllosilicate, being the montmorillonite (MMT) the most studied one. MMT crystal lattice consists of 1 nm thin layers formed by an octahedral alumina sheet sandwiched between two tetrahedral silica sheets. It has a high surface area (aspect ratio of about 100) and is negatively charged (Sinha Ray and Okamoto, 2003a). The stacking of these layers leads to a van der Waals gap or gallery, in which alkaline cations, such as Na^+, Li^+, or Ca^{2+}, are located and neutralize the charge. The major problem in preparing these composites is to separate the initially agglomerated clay layers, which is a necessary step because polymer properties are improved when the clay layers are well dispersed in the polymer matrix (Alexandre and Dubois, 2000; Giannelis, 1996; Sinha Ray and Bousmina, 2005).

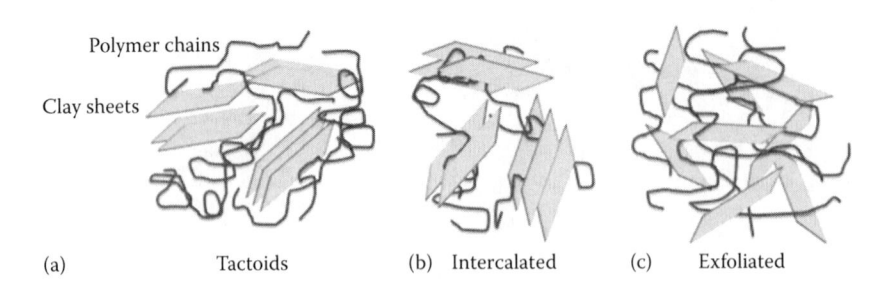

Polymer chains

Clay sheets

(a)   Tactoids   (b)   Intercalated   (c)   Exfoliated

FIGURE 13.3   Scheme of different types of composite based on layered silicates and polymers: (a) phase separated micro-composite, (b) intercalated nanocomposite, and (c) exfoliated nanocomposite.

Three types of thermodynamically possible composites can be obtained when a polymer and the clay are associated (Figure 13.3). The result depends on the preparation conditions and the polymer matrix–clay nanolayer affinity. When the polymer is unable to intercalate between the silicate sheets, a phase separated composite is obtained, also called "pellets" or "tactoids," whose properties are similar to those of a microcomposite (Figure 13.3a). But when the polymer intercalation is possible, two types of nano-composites can be recovered: (1) intercalated, where the insertion of polymer chains in the structure of the silicates occurs at regular crystallographic phases regardless of the proportion of fillers and with a repeated distance of a few nanometers (Figure 13.3b), and (2) exfoliated, wherein the platelets of silicate are separated and dispersed into the polymer matrix and the average distance between them depends on the level of charge (Figure 13.3c) (Sinha Ray and Bousmina, 2005).

It is known that exfoliated clays yield better mechanical properties than intercalated ones. The com-plete dispersion of clay layers in a polymer must optimize the number of strengthening elements avail-able to support a load and avoid cracks in the material, improving the mechanical properties. In addition, the clay layers generate a tortuous pathway through which the permeable elements have much greater difficulty in penetrating the nanocomposite.

The addition of nanoclay has improved the performance of synthetic polymer matrices—both thermo-plastic and thermoset (Aranda and Ruis-Hitzky, 1992; Kim and White, 2005; Kojima et al., 1993)—and biodegradable polymer matrix materials based on proteins (soy protein, gluten, whey protein, and gelatin), polysaccharides (starch, carboxymethylcellulose, chitosan), PLA, poly(butylene succinate) (PBS), PCL, and PVA, obtained by different processing techniques. Generally, this improvement includes any of the following characteristics: higher elastic modulus, tension at break, flexibility and thermal stability, lower permeability to water and gases, and lower density, as well as better dimensional stability and higher clar-ity (Cho et al., 2010; Echeverría et al., 2014; Giannelis, 1998; Guilherme et al., 2010; Hedenqvist et al., 2006; Kumar et al., 2010; Quilaqueo Gutiérrez et al., 2012; Rao, 2007; Sinha Ray and Bousmina, 2005; Sinha Ray and Okamoto, 2003a; Thellen et al., 2005; Tunc et al., 2007; Wang and Zhang, 2005).

When the incompatibility between the clay particles and the polymer hinders exfoliation of MMT at the nanoscale level, a possible alternative is to use organically modified clays. The nature of MMT can be modified by substituting the cations in the layer galleries with cationic organic surfactants such as alkylammonium or alkylphosphonium. This surface modification, apart from increasing the interlayer spacing of clay sheets that could facilitate the entry of the polymer chains into the clay galleries, could improve the compatibility between the hydrophilic surface chemistry of the clay and the hydrophobic polymer matrix, descending the surface energy of the inorganic host and improving its wetting charac-teristics with the polymer.

There are some studies reported in the literature on the addition of organically modified MMTs to biopolymer-based films (Guilherme et al., 2010; Kumar et al., 2010; Sothornvit et al., 2009, 2010).

13.4.2 Nanoparticles Derived from Polysaccharides

Polysaccharides are considered a good option to be used as raw material for producing renewable nano-fillers, and since as they are partially crystalline, they can provide interesting properties apart from their biocompatibility and biodegradability, upgradability, multiple reacting groups, and low cost. In addition, the functional groups of the polysaccharide backbone allow facile chemical modification to develop nanoparticles with diverse structures that are compatible with different polymer systems. The most studied polysaccharide that is used to produce nanoparticles is cellulose, followed by starch and chitosan, the latter used mainly as drug carrier (Azizi Samir et al., 2005; Dubief et al., 1999; LeCorre et al., 2010; Nagpal et al., 2010).

The main method used to obtain nanoparticles from polysaccharides is the acid hydrolysis, which consists in the remotion of the amorphous regions present in the polysaccharides while leaving the crystalline regions intact (Gardner et al., 2008); however, the process may vary slightly depending on the nature of the polysaccharide used.

Cellulose nanowhiskers or nanofibers: Cellulose is naturally organized as microfibril chains of poly-β-(1→4)-D-glucosyl residues formed during the biosynthesis, linked together through intermolecular hydrogen bonds to form cellulose fibers. Individual cellulose microfibrils have diameters ranging from 2 to 20 nm and can be considered as a string of cellulose crystals linked along the microfibril axis by disordered amorphous domains, for example, twists and kinks (Siqueira et al., 2010). The amorphous regions are susceptible to acid attack, and under strictly controlled conditions of time and temperature, they may be removed leaving the crystalline regions intact. A series of preliminary steps to hydrolysis (i.e., an alkaline treatment and bleaching) are necessary to remove lignin and hemicelluloses, which could affect the properties of nanocomposites. Cellulose nanowhiskers are nanofibers that have been grown under controlled conditions that lead to the formation of highly pure single crystals (Azizi Samir et al., 2005).

Cellulose whiskers can be prepared from microcrystalline cellulose, BC, algal cellulose (valonia), hemp, tunicin, cotton, ramie, sisal, cassava bagasse, sugar beet, and wood (Siqueira et al., 2010). Their geometrical characteristics such as size, dimensions, and shape depend on the nature of the cellulose source as well as the hydrolysis conditions such as time, temperature, ultrasound treatment, and purity of materials. Typical dimensions of whiskers range from 5 to 10 nm in diameter and from 100 to 500 nm in length. As they contain only a small number of defects, their Young's modulus ranges from 130 to 250 GPa and the experimental strength is near 10 GPa (Azizi Samir et al., 2005; Zimmermann et al., 2004).

Another method used to obtain cellulose nanoparticles is a mechanical disintegration process, which allows obtaining microfibrillated cellulose (MFC) (Siqueira et al., 2010). Contrary to straight cellulose whiskers, cellulose microfibrils are long and flexible nanoparticles consisting of alternating crystalline and amorphous domains, presenting a weblike structure (Andresen et al., 2006). Some authors have demonstrated that it is necessary to repeat the procedure of homogenization several times, in order to increase the disintegration process, which requires a larger amount of energy. A progress in MFC production has been made by combining a mechanical process with enzymatic treatments or acid hydrolysis, as well as through a new process based on a TEMPO reaction with a vigorous mixing (Siqueira et al., 2010). The high strength, flexibility, and aspect ratio of MFC make them interesting as reinforcements in bionanocomposites.

Finally, BC is secreted to the extracellular space as synthesized cellulose nanofibers (2–4 nm in diameter and several 100 μm in length) by some bacterial species such as those belonging to the *Acetobacter* genus. These nanofibers aggregate on the top of the culture medium; they incorporate water and form a 3D coherent network (Grande et al., 2009). BC has higher purity, crystallinity (above 60%), degree of polymerization, and tensile strength than plant cellulose. Some authors have prepared BC nanocomposites by impregnation of the cellulose nanofiber network with other polymers. The disintegration of the cellulose network in order to blend it as standard nanofiller (Iguchi et al., 2000; Nakagaito et al., 2005) has also been tried.

Starch nanoparticles: Starch is a polysaccharide of natural origin that is renewable, biodegradable, and semicrystalline and known to be formed by amylose and amylopectin. Starch nanoparticles can be obtained directly by acid hydrolysis—normally with sulfuric acid—of native starch granules by strictly controlling the temperature, acid and starch concentrations, and time and stirring speed (Siqueira et al., 2010). The study of the kinetics of starch hydrolysis has enabled, in part, to understand the evolution of the reaction and its dependence on certain parameters (Jayakody and Hoover, 2002): the first hydrolysis step involves mainly the amorphous region of the granule and is influenced by the size and the pores in the granule surface, the amylose content, and the amylose chains that are complexed with lipids, and the second stage of hydrolysis involves the crystalline region and is influenced by the amylopectin, the distribution of the $\alpha(1{\to}6)$ bonds between the amorphous region and the crystalline region, and the degree of packaging of the double helices.

Starch nanocrystals are crystalline square-like platelets about 10 nm thick and 50–100 nm equivalent diameters. Depending on the botanic origin of starch, these platelets show different features (Le Corre et al., 2010).

These nanocrystals can gelatinize in hot water, which could be a drawback when preparing bionanocomposites. In order to avoid this problem, a second method has been developed in which starch nanocrystals are prepared by ethanol precipitation into a gelatinized starch solution with constant stirring and also further modified by citric acid in a dry preparation technique (Ma et al., 2008).

Chitin nanowhiskers/chitosan nanoparticles: Chitosan is a naturally occurring nontoxic, biocompatible, biodegradable, and cationic polysaccharide (Shahidi et al., 1999). Chitin nanowhiskers can be made by hydrolysis with a boiling HCl solution with vigorous stirring, with a previous deproteinization step in a boiling alkaline (KOH) solution (Gopalan et al., 2003). On the other hand, chitosan nanoparticles can be obtained by physical cross-linking by electrostatic interactions between tripolyphosphate and protonated chitosan with vigorous stirring and sonication (Chang et al., 2010). The origin of chitin, namely, the type of crystallinity, determines the structure and morphology of the chitin nanowhiskers. The nanoparticles occur as rodlike or spindle-like nanowhiskers with properties comparable to perfect crystals (Mincea et al., 2012).

Generally, nanowhiskers or nanoparticles obtained from polysaccharides tend to aggregate due to association by strong hydrogen bonding, especially cellulose nanowhiskers and starch nanoparticles. This explains why high level of nanofiller addition is not necessarily good; on the contrary, it can adversely affect the properties of the final material (Chen et al., 2009). Besides, the nanofiller structure may be destroyed at high temperature or be highly hydrolyzed, thus hindering its reinforcing ability. Because of this, the solution casting method has been the most widely used one when nanofillers are incorporated into the polymer matrix, as demonstrated by Xie et al. (2013).

These nanoreinforcements have been incorporated into biodegradable matrices like protein, starch, and PLA and have been studied (Arora and Padua, 2010). For example, the addition of cellulose nanofibers obtained from microcrystalline cellulose improved the mechanical properties of casein films. The same effect was observed by adding starch nanocrystals in soy protein formulations (Pereda et al., 2011; Zheng et al., 2009). These nanocrystals have also been used as reinforcement in other environmentally safe polymer matrices such as organic solvent-free polyurethane (Chen et al., 2008b), PLA (Yu et al., 2008), polyvinylalcohol (Chen et al., 2008a), polyhydroxyalkanoates (Bordes et al., 2010), pullulan (Kristo and Biliaderis, 2007), waxy maize starch (Angellier et al., 2006; Viguié et al., 2007), and manioc starch (García et al., 2009). Although in most cases, an improvement in mechanical properties was achieved; in some cases, they were also able to improve the WVP and water uptake by the nanocomposites (Cyras et al., 2008; García et al., 2009; Kristo and Biliaderis, 2007).

13.4.3 Metal Nanoparticles

The metal nanoparticles have started to be studied over the last years due to the great contribution they have made in the field of clinical diagnosis and as therapeutic agents. In food science and technology, the most important are the metal ions (Ag°, Cu°, Au°, Pt°) and the metal oxides (TiO_2, ZnO, MgO) mainly because when added to polymers networks, they can provide antimicrobial properties to the final material (Rhim et al., 2013).

Silver nanoparticles (Ag-NPs) are the most widely used ones for the development of innovative packaging materials. These nanoparticles, with diameters 45–50 nm and minimum inhibitory concentrations, are potent broad-spectrum antimicrobials against many bacterial species (*Escherichia coli*, *Enterococcus faecalis*, *Staphylococcus* spp., *Vibrio cholerae*, *Pseudomonas* spp., *Shigella flexneri*, *Bacillus* spp., *Proteus mirabilis*, *Salmonella enterica* Typhimurium, *Micrococcus luteus*, *Klebsiella pneumoniae*, *Listeria monocytogenes*), fungi (*Candida albicans*, *Aspergillus niger*, *Trichophyton mentagrophytes*), yeasts isolated from bovine mastitis, algae, and phytoplankton, and at least two viruses (HIV and monkeypox) (Duncan, 2011). The most conservative viewpoint of bacterial effect of Ag-NPs is that silver atoms detach from the surfaces nanoparticles and cause cellular damage by the same mechanisms observed for conventional silver antimicrobials; basically they are known to interfere with the respiratory chain and cell division, the disruption of DNA replication, and the induction of oxidative stress (Duncan, 2011; Rai et al., 2009). The toxicity of Ag-NPs depends on their diameter because smaller nanoparticles have larger relative surface areas for Ag^+ release; they have higher protein-binding efficiencies and pass through pores in bacterial membranes more easily (Baker et al., 2005). The toxicity also depends on the nanoparticle shape (triangular particles are thought to have a better bactericidal activity than spherical- or rod-shaped particles), surface charge, solubility, and degree of agglomeration as well as on surface coating (Kvítek et al., 2008; Lok et al., 2007; Morones et al., 2005; Sondi and Salopek-Sondi, 2004). The antimicrobial activity also depends on factors affecting the Ag^+ release rate, such as the degree of polymer crystallinity, the hydrophobicity of the matrix, and the presence of any type of filler (Duncan, 2011).

The most common method for preparing Ag-NPs involves the reduction of a silver salt solution with a reducing agent such as sodium borohydride, citrate, or ascorbate. The trend of synthesis methods currently focuses on the use of compounds that do not harm the environment. In this regard, enzymes/proteins, amino acids, polysaccharides, and vitamins have been used, which are environmentally benign but chemically complex (Sharma et al., 2009). The addition of a surface stabilizer agent to maintain a uniform particle size and a stable dispersion in the course of time is often essential.

Several applications in food packaging have been studied showing the antimicrobial effect of Ag-NPs added at different bioplastic films and coatings based on different polymers such as cellulose, sodium alginate, chitosan, and starch (de Moura et al., 2012; Fayaz et al., 2009) and also with the presence of another nanofiller such as cellulose nanofibers (Fortunati et al., 2013). In addition, since silver particles could catalyze the destruction of ethylene gas, fruits stored in the presence of Ag-NPs have shown slower ripening times and thus extended shelf lives (Fernández et al., 2010).

13.5 Nanocomposite Processing Techniques

The preparation of nanocomposites involves the incorporation of nanoparticles into polymer matrices in order to improve the material functionality provided by the proper interaction between the nanoparticles and the polymer. This process should be aimed at achieving a good distribution of the nanoparticles in the polymer matrix, thus overcoming the tendency of agglomeration of these particles due to their high superficial area. Two primary factors determine the level of dispersion of the nanoparticles within the polymer matrix during processing: (1) the chemical affinity of the polymer and the nanoparticles and (2) the processing conditions.

Although other preparation techniques, such as electrospinning and processing under supercritical conditions, have recently gained interest, there are three main techniques for preparing nanocomposites (Ojijo and Sinha Ray, 2013):

1. *In situ polymerization*: In which, the nanoparticles are incorporated within the liquid monomer or monomer solution followed by the polymerization step that is initiated by heat, radiation, or another initiator.

2. *Solution casting*: Based on a solvent system in which the polymer and the nanoparticles are dispersed. Typically, both components are separately swollen and dispersed in the same or different solvents before being mixed. Sometimes, special dispersion technique steps are incorporated to the process, such as sonication to favor the nanoparticles dispersion. Finally, when the solvent is evaporated, interactions between the nanofiller and the polymers increase resulting in a nanocomposite. Although a better dispersion of nanoparticles can be achieved using the solvent-casting method, this process uses many chemicals and solvents, most of which could be hazardous and hence impractical in industrial-scale production (Beyer, 2002).

3. *Melt intercalation*: The nanoparticles are mixed with the polymer matrix in the molten state. The process involves heating the mixture above the softening point of the polymer, statically or under shear. This method is environmentally safe because it does not require the use of solvents and is compatible with current industrial processes, such as extrusion and injection molding. It also allows the use of biopolymers that were not suitable for *in situ* polymerization or could not be dissolved in similar solvents to those required to disperse the nanoparticles. However, in the case of bionanocomposites, the mechanical shearing force or the temperature applied during processing can degrade certain biopolymers and organic nanofillers. For example, it has been reported that if the conditions are not well controlled, PLA can undergo thermal, oxidative, and hydrolytic degradation during processing, leading to the cleavage of polymer chains and consequently decreasing the molecular weight (Ojijo and Sinha Ray, 2013).

Figure 13.4 schematizes the three processes where clays are used as nanofillers.

In all methods, the polymer needs to be sufficiently compatible with the nanoparticle surface to ensure proper dispersion. To enhance the dispersion of nanoparticles in some biopolymers, strategies such as surface modifications are deemed necessary and should be further studied. The use of compatibilizer compounds could also be a good alternative to improve the interactions in the interphases. Polymers can also be grown from nanoparticles using the surface reactive groups as initiating sites, or the surface can be modified to introduce different initiator sites needed for controlled polymerization techniques such as atom transfer radical polymerization or reverse addition–fragmentation radical polymerization, such as using the surface hydroxyl groups of cellulose nanowhiskers. For example, the grafting of poly(ε-caprolactone) from the surface of ramie cellulose nanowhiskers has been performed and these PCL-grafted cellulose nanowhiskers were subsequently shown to increase the mechanical properties of PCL–cellulose nanowhisker composites to a higher extent than the unmodified cellulose ones (Eichhorn et al., 2010).

The selection of any of the techniques depends on the type of biopolymer involved and, to a large extent, on the nanoparticle in question. Optimization of the processing conditions to have well-dispersed

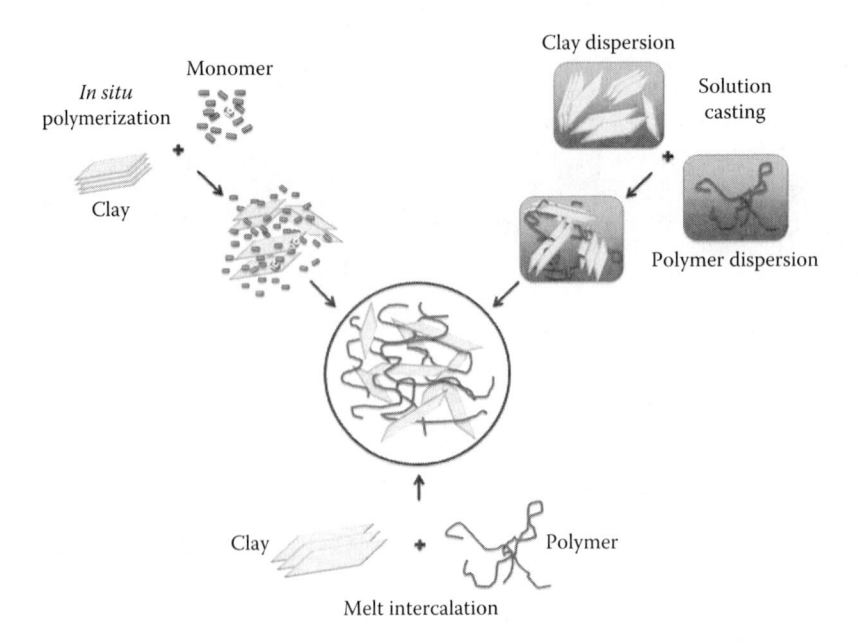

FIGURE 13.4 Scheme of different processes for preparing nanocomposites: *in situ* polymerization, solution casting, and melt intercalation.

nanoparticles while simultaneously ensuring structural integrity of the nanoparticles and minimal adverse effects on the polymer (i.e., degradation) is essential. The use of the most environmentally safe processing routes and those that can easily be aligned with the currently available industrial processes are the most attractive to prepare bionanocomposites.

13.6 Nanostructure Characterization

Recent advances in characterization techniques, especially for structure elucidation, have allowed the advancement of nanotechnology. When going from a nanocrystal to a bulk material, the structure is affected, thus leading to changes in lattice parameters and consequently the possibility of having new phases associated with the residual strain on the particles arises. This phenomenon has been observed for many materials for different types of particles. Important aspects of nanocomposite structure characterization include the study of particle dispersion, the changes in the bulk matrix, and the nature of the particle–polymer interface (Souza Filho and Fagan, 2011). The most common techniques used to probe nanocomposite structures are transmission electron microscopy (TEM); scanning electron microscopy (SEM); atomic-force microscopy (AFM); x-ray diffraction (XRD), both wide angle and small angle; and infrared spectroscopy (Karger-Kocsis and Zhang, 2005; Sinha Ray and Okamoto, 2003a).

TEM is the preferred method to examine nanoparticle dispersions. Polymer structure, void size and shape, filler size, shape, distribution, local crystallinity, and crystal size can be determined by TEM through direct visualization. In the case of a clay nanocomposite, TEM allows a qualitative assessment of the degree of dispersion of clay in polymer matrices and the observation of the structure of silicates (Morgan and Gilman, 2003). However, to obtain accurate information about the dispersion of silicate layers in the 3D biopolymer matrix, the electron tomography has been recently applied (Sinha Ray, 2009). Changes in the polymer matrix may also be assessed by polarized light microscopy with assistance from TEM and/or AFM.

SEM fracture analysis continues to be the best method for assessing structure–property relations, especially for toughness (Karger-Kocsis and Zhang, 2005). Recently, SEM combined with energy dispersive x-ray spectroscopy has been used to study the degree of dispersion of nanoparticles on the fractured surface of bionanocomposites.

Optical microscopy (OM), with less resolution, has also been used to study the degree of dispersion of nanoparticles in bionanocomposites. If the nanoparticles are dispersed at the nanoscale, the nanocomposite sample should exhibit no turbidity because the fundamental particle size is less than $\lambda/4$. Therefore, aggregates significantly less than 1 µm in size will not be resolvable by OM.

The XRD technique is one of the most used for evaluating structural properties of nanomaterials. In general, the diffraction pattern of crystalline nanoscale materials exhibits broadened and shifted peaks as compared to bulk, and these changes are associated with both size and strain (Souza Filho and Fagan, 2011). The degree of intercalation, exfoliation, and dispersion has been traditionally characterized by XRD (Alexandre and Dubois, 2000). When the basal spacing of a mixture is the same as that of the clay cluster, the structure is considered a tactoid with no polymer chains inside the clay gallery (interlayer space). In intercalated structures, the d-space is increased as the interlayer space is expanded, thus decreasing the 2θ position in the x-ray spectra (smaller value). Exfoliated structures show no peaks in XRD, indicating that polymer chains have penetrated the gallery and widened the interlayer space until the regular stacks of clay layers become disordered so that x-ray cannot detect any regular structure. Exfoliation is achieved when clay stacks no longer show an XRD peak. Protein–clay nanocomposite studies carried out by Chen and Zhang (2006) have demonstrated that MMT tactoids were delaminated into thin lamellas in soy protein. The d-spacing values increased from 1.4 nm for the MMT tactoid to a value ranging from 2 to 3 nm (Arora and Padua, 2010). Nevertheless, conclusions concerning the mechanism for the formation of bionanocomposites and their structures based solely on XRD patterns are only tentative because this technique does not inform about the spatial distribution of the clay layers or any structural inhomogeneities in bionanocomposites. Therefore, XRD and TEM analysis complement each other.

13.7 Physicochemical Properties of Nanocomposite Films

As mentioned earlier, bioplastics have several disadvantages such as poor gas and water barrier properties, unbalanced mechanical properties, low softening temperature, and weak resistivity, as well as their performance, processing, and cost compared to the petroleum-based synthetic plastics that limit their use in a wide range of applications, such as food packaging. Nanotechnology could help in overcoming these problems since the nanofillers generally improve the earlier properties of the bioplastics.

Mechanical properties of biopolymers have been improved by the addition of nanofillers, and this improvement depends strongly on the high rigidity and aspect ratio of nanoparticles and on the amount added along with the good affinity through interfacial interaction between polymer matrix and dispersed nanofiller, leading to achieve a good dispersion of the fillers in the polymer matrix.

Several authors have reported improvements in mechanical properties such as increases in the elastic modulus and stress at break and decrease in the elongation at break in tensile test in nanocomposites prepared by adding different nanoparticles to almost all types of bioplastics, for example, nanocomposites based on MMT and starch or soy proteins, cellulose nanowhiskers and PLA, CNTs and PCL, or starch nanoparticles and natural rubber (Angellier et al., 2005a,b, 2006; Chrissafis et al., 2007; Echeverría et al., 2014; Huang and Yu, 2006; Petersson et al., 2007).

Barrier properties: In this regard, gas and water vapor permeabilities have been found to decrease, in some cases to a large extent, in the nanocomposites. It is well known that the incorporation of nanofillers especially nanoclays and nanostarches into the polymeric matrix can lead to significant enhancement in the *barrier properties* (LeCorre et al., 2010; Sinha Ray and Okamoto, 2003a,b). These improved barrier properties in nanocomposites are explained on the basis of the generation of an increased path length due to the presence of nanofillers mainly affecting the passage of vapors and gas molecules, which they need to traverse while diffusing through the matrix. The scheme shown in Figure 13.5 illustrates this effect. Therefore, the barrier properties of bionanocomposites depend on the aspect ratio of nanofillers and their orientation and dispersion in the polymer matrix (Bharadwaj, 2001). However, the cellulose nanofibers are not as effective as that of nanoclay, probably due to their shape, which limits the increment in the tortuous path.

Generally, different morphologies coexist within the nanocomposites leading to different permeabilities that cause complex transport phenomena. Moreover, it is well known that the semicrystalline polymers have both a crystalline and an amorphous region leading to different permeabilities as crystalline regions are impermeable to penetrant molecules. However, these fillers also have an effect on the crystallinity and chain mobility of the polymer matrix leading to the reduction in permeation. Hence, both crystallinity changes and the presence of a tortuous path have to be taken into consideration when analyzing the effect of nanofillers on the permeability of nanocomposites (Reddy et al., 2013).

The improvement of gas permeability of nanocomposites with clays strongly depends on the type of clay, aspect ratio of clay platelets, and structure of the nanocomposites (Reddy et al., 2013). The best improvements are associated to exfoliate clays nanocomposites. These improvements in water barrier properties have been reported, for example, for nanocomposites prepared from thermoplastic starch and

Water vapor, oxygen

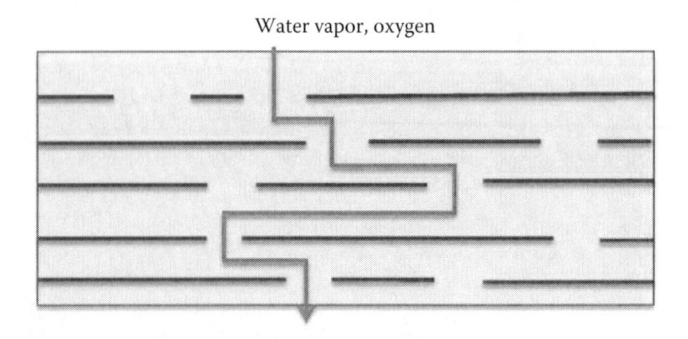

FIGURE 13.5   Scheme that illustrates the tortuous path length in a polymer nanocomposite when used as a gas or vapor barrier.

clay, cellulose acetate and clay, or starch and cellulose whiskers (Dufresne et al., 2000; Park et al., 2003, 2004). Improvements for oxygen barrier properties due to the addition of nanofillers have been reported for different nanocomposite systems based on bioplastics, for example, for nanoclays added to PLA or to PHB (Rhim and Ng, 2007; Sanchez-Garcia and Lagaron, 2010; Thellen et al., 2005).

Optical properties are important characteristics of polymers. Amorphous polymers are transparent to the visible range (Trotignon et al., 1989). The incorporation of reinforcements at nanolevels should have significant effects on the transparency and haze characteristics of films. The addition of small quantities of nanofillers could maintain the transparency of the polymer in spite of being sufficient to generate a significant improvement in the mechanical properties and heat resistance of the nanocomposite (Bharadwaj et al., 2002).

The influence on optical properties of the addition of nanoclays to different materials depends on the size and shape of the particles (Bharadwaj et al., 2002). Nanoclay incorporation has been shown to induce a significant enhancement in transparency and a reduction of haze in different polymer systems. When single layers are dispersed in a polymer matrix, the resulting nanocomposite is optically clear in the visible region while there is a loss of intensity in the UV region (for $\lambda < 300$ nm), mostly due to scattering by the MMT particles. A plausible reason for the aforementioned observations could be that the size of nanoclay particles is less than the wavelength of visible light; hence, visible light rays are not appreciably scattered by nanoclay particles. However, visible light rays could be appreciably scattered by regions where clay particles form agglomerates (Anandhan and Bandyopadhyay, 2011). Petersson and Oksman (2006) reported that no reduction in the amount of light being transmitted through the nanocomposite films is an indication that the nanoreinforcements are fully exfoliated. Echeverría et al. (2014) have observed that higher concentrations of MMT in soy protein–based films resulted in a lower level of film opacity and suggested that they could be due to three possible effects: (1) the high degree of exfoliation achieved; (2) the possible better dispersion of protein in water in the presence of clay and plasticizer, preventing the formation of protein microaggregates that might scatter light, reducing transparency and increasing opacity; and (3) the different types of interactions that stabilized the protein network in the presence or absence of MMT, inducing different levels of molecular aggregation, which finally allows higher light transparency in the nanocomposite films.

Biodegradability of biomaterials is one of the most interesting characteristics to retain after the formation of nanocomposites. The increase or conservation of the biodegradability by addition of nanofillers could be due to different causes depending on the nature of the material and the nanofiller. For example, Tetto et al. (1999) tested the biodegradability of nanocomposites based on PCL and clay and observed an improved biodegradability that could be attributed to the catalytic role of the organoclay during the biodegradation process. However, Sinha Ray et al. (2003a,b) have performed different methods for evaluation of biodegradability of nanocomposites of PLA and clays and concluded that the biodegradability was significantly enhanced compared to neat PLA, due to the presence of terminal hydroxylated edge groups in the clay layers.

On the other hand, several authors have shown a decrease of biodegradability for addition of nanofillers. Nanocomposites prepared with PBS and organoclay exhibited a decrease of biodegradability, which was explained by the inability of microorganism to go through in the films by tortuous paths (Lee et al., 2002). Rhim et al. (2006) concluded that the decrease of biodegradability of nanocomposites of PBS and organoclay is due to a strong antimicrobial activity of the quaternary ammonium group in the modified clay.

It is of interest to recognize that the nanoparticles can have two opposite effects on polymer nanocomposites, for example, degradation or stabilization depending on processing and environmental conditions. Even though the proper stability and durability of the bionanocomposite packaging materials should be maintained during their useful shelf life to perform their packaging functions, a rapid biodegradation is desirable once they are discarded. Such innovative properties of nanoparticles can be exploited in the packaging industry depending on the final use.

Nanoscale dispersion of filler or controlled nanostructures in the composite can also introduce new physical properties and novel behaviors (e.g., fire resistance or flame retardancy and accelerated biodegradability) that are absent in the unfilled matrices (Morgan and Wilkie, 2007). These composite materials are sometimes recognized as *genuine nanocomposites* or *hybrids* (Manías, 2007).

The improvements in these properties through the addition of nanofillers into bioplastics formulations make these materials especially interesting for food packaging applications.

13.8 Food Packaging Applications

The use of nanocomposites in food packaging is expected to enhance the shelf life of foods considerably. Due to the improved properties of nanocomposites, such as mechanical and barrier properties, clarity, and biodegradability, but also thermal, chemical, and dimensional stability and heat resistance, besides the possible development of active antimicrobial and antifungal surfaces and sensing and signaling microbiological and biochemical changes, food packaging has been one of the most concentrated nanocomposite technology developments. This improvement can lead to lower weight packages because less material is needed to obtain the same or even better barrier properties, leading to reduced package cost with less packaging waste.

Nanoclay-based bionanocomposites have gained more importance in the packaging industry, due to the ease of availability, processing, and low costs compared to other nanofillers such as nanocellulose or CNTs. They have been proved in different food packaging applications, such as processed meats, cheese, confectionery, cereals, and boil-in-bag foods, as well as in extrusion coating applications for fruit juices and dairy products or coextrusion processes for the manufacture of bottles for beer and carbonated beverages (Smolander and Chaudhry, 2010). But most applications have been proved with nanocomposites made from synthetic thermoset and thermoplastic polymers. Those based on bionanocomposites are mainly being studied and developed.

13.9 Nanocomposite Systems as Modulators of the Release of Active Compounds

Nanotechnology allows the development of carriers forming submicron particles such as nanospheres, nanocapsules, and micellar systems, which can be strategically used to protect drugs or bioactive compounds against the chemical and enzymatic degradation (especially proteins, peptides, and nucleic acids), to mask flavor and odor, to increase the solubility in water, and to decrease the rate of dissolution and/or the controlled release of these compounds. The knowledge needed for the development of these products can be obtained from the pharmaceutical and biomedical field, where the development of matrixes for controlled release of drugs is already a fact and it is an active research area in constant improvement (Lopez-Rubio, 2011).

Compared to the micronsized particles, nanoparticles offer increased surface area and solubility and reduced sedimentation rate. By reducing particle size, the properties of bioactive compounds are improved, such as delivery properties, solubility, prolonged residence time in the gastrointestinal tract, and efficient absorption by cells (Fang and Bhandari, 2010; Gibbs et al., 1999; Lakkis, 2007; Madene et al., 2006; Shahidi and Han, 1993; Zuidam and Nedovic, 2010). For example, the formation of food-grade nanoemulsions and nanoparticles may be particularly useful to increase the bioactivity of lipophilic compounds whose absorption is usually poor. Compounds such as omega 3 and omega 6 fatty acids, probiotics, prebiotics, vitamins, and minerals have been used as bioactive compounds using these methodologies to protect them during their processing and storage.

The established biocompatibility of cellulose supports the use of nanocellulose for a similar purpose. The very large surface area and negative charge of crystalline nanocellulose suggest that large amounts of drugs might be bound to the surface of this material with the potential for high payloads and optimal control of dosing. The abundant surface hydroxyl groups on crystalline nanocellulose also provide a site for the surface modification of the material with a range of chemical groups by a variety of methods. Surface modification may be used to modulate the loading and release of drugs that would not normally bind to nanocellulose, such as nonionized and hydrophobic drugs. For example, Lönnberg et al. (2008) suggested that PCL chains might be conjugated onto nanocrystalline cellulose for such a purpose. Other nanocrystalline materials, such as nanocrystalline clays, have been shown to bind and subsequently release drugs in a controlled manner via ion exchange mechanisms and are being investigated for use in pharmaceutical formulations (Shaikh et al., 2007).

Bioactive packaging materials would be capable of withholding and protecting desired bioactive principles or functional ingredients within the packaging walls in optimum conditions until their eventual release into the food product through either controlled or fast release during storage, or just before consumption, taking into account the specific product/functional substance characteristics or requirements (Lopez-Rubio et al., 2006). The possible use of nanocomposites as carriers of active agents and controlled release systems in the field of food packaging materials is newly started to be evaluated (Mascheroni et al., 2010). Transfer properties of these compounds can be modulated by the addition of varying proportions of fillers or nanofillers (Tunc et al., 2007), which can be attributed to differences in polymer network structure due to the presence of these reinforcements and the increased tortuosity of the road ahead, modifying the diffusion rate of active compounds.

Apart from exhibiting improvements in the mechanical and rheological properties and reducing the water absorption, polymer–clay nanocomposites can also control the release of active compounds. Based on properties such as swelling, film-forming ability, bioadhesion, and cell capture properties, these materials are being targeted to the conception of new forms of drug release with highly specific dosage and in the improvement of the technological and biopharmaceutical properties. For example, Costa et al. (2011, 2012) showed that by the addition of Ag-MMT nanoparticles into the bottom of the polypropylene boxes of fresh fruit salad, or to fresh-cut carrots, the shelf life can be expanded. Furthermore, the addition of Ag-MMT nanoparticles on fior di latte cheese packaged using modified atmosphere packaging allowed the extension of the shelf life from 3 to 5 days (Gammariello et al., 2011).

13.10 Legislation and Future Trends

The rapid proliferation of nanotechnologies in a wide range of consumer products has also raised a number of questions related to their safety, environmental, ethical, policy, and regulatory issues (Maynard et al., 2006; Royal Society and Royal Academy of Engineering, 2004). The main concerns stem from the lack of knowledge with regard to the interactions of nanosized materials at the molecular or physiological levels and their potential effects and impacts on consumer's health and the environment.

However, the nanotechnology applicability to the food industry is increasing steadily. Nanotechnologies offer a variety of possibilities for application in the food and feed area and in production/processing technology to improve food contact materials, to monitor food quality and freshness, to improve traceability and product security, to modify taste, texture, sensation, consistency, and fat content, and to enhance nutrient absorption. Food packaging, including bionanocomposites, makes up the largest share of current and short-term predicted markets.

The successes of these advancements will be strictly dependent on exploration of worldwide regulatory issues for nanotechnology. But there are some aspects to be studied in order to demonstrate product safety. They include the nanomaterial migration through polymer films; the interaction of nanomaterial biomolecules and cellular components; the value of mass-based definitions of dosage in the context of nanomaterials; the interrelationships between nanoparticle characteristics (size, shape, surface charge, etc.) and toxicity or pharmacokinetic properties; the appropriate and consistent methods to identify, characterize, and quantify nanomaterials in complex food matrices; the chronic toxicity of nanomaterials or toxicity following oral routes of exposure; and the biodegradability of nanomaterials or the toxicity of nanomaterials to ecologically important organisms (Duncan, 2011).

REFERENCES

Ajayan, P.M., Schadler, L.S., and Braun, P.V. *Nanocomposite Science and Technology.* Wiley, Weinheim, Germany, 2003.

Alexandre, M. and Dubois, P. Polymer-layered silicate nanocomposites: Preparation, properties and uses of a new class of materials. *Material Science and Engineering*, 28 (2000): 1–63.

Andresen, M., Johansson, L.S., Tanem, B.S., and Stenius, P. Properties and characterization of hydrophobized microfibrillated cellulose. *Cellulose*, 13 (2006): 665–677.

Anandhan, S. and Bandyopadhyay, S. Polymer nanocomposites: From synthesis to applications. In *Nanocomposites and Polymers with Analytical Methods*, J. Cuppoletti (ed.). InTech, Rijeka, Croatia, 2011.

Angellier, H., Molina-Boisseau, S., Dole, P., and Dufresne, A. Thermoplastic starch- waxy maize starch nanocrystals nanocomposites. *Biomacromolecules*, 7(2) (2006): 531–539.

Angellier, H., Molina-Boisseau, S., and Dufresne, A. Mechanical properties of waxy maize starch nanocrystal reinforced natural rubber. *Macromolecules*, 38(22) (2005a): 9161–9170.

Angellier, H., Putaux, J.-L., Molina-Boisseau, S., Dupeyre, D., and Dufresne, A. Starch nanocrystal fillers in an acrylic polymer matrix. *Macromolecular Symposium*, 221 (2005b): 95–104.

Aranda, P. and Ruis-Hitzky, E. Poly(ethylene oxide)-silicate intercalation materials. *Chemistry of Materials*, 4 (1992): 1395–1403.

Arora, A. and Padua, G.W. Review: Nanocomposites in food packaging. *Journal of Food Science*, 75(1) (2010): 43–49.

ASTM. *Standard Specification for Compostable Plastics*. ASTM International, West Conshohocken, PA, 2002.

Azizi Samir, M.A.S., Alloin, F., and Dufresne, A. Review of recent research into cellulosic whiskers, their properties and their application in nanocomposite field. *Biomacromolecules*, 6 (2005): 612–626.

Baker, C., Pradhan, A., Pakstis, L., Pochan, D.J., and Shah, S.I. Synthesis and antibacterial properties of silver nanoparticles. *Journal of Nanoscience and Nanotechnology*, 5 (2005): 244–249.

Beyer, G. Nanocomposites: A new class of flame retardants for polymers. *Plastics, Additives & Compounding*, 4 (2002): 22–28.

Bharadwaj, R., Mehrabi, A., Hamilton, C., Trujillo, C., Murga, M., Fan, R., Chavira, A., and Thompson, A. Structure-property relationships in cross-linked polyester clay nanocomposites. *Polymer*, 43 (2002): 3699–3705.

Bharadwaj, R.K. Modeling the barrier properties of polymer-layered silicate nanocomposites. *Macromolecules*, 34 (2001): 9189–9192.

Bohlmann, G.M. General characteristics, processability, industrial applications and market evolution of biodegradable polymers. In *Handbook of Biodegradable Polymers*, C. Bastioli (ed.), pp. 183–217. Rapra Technology, Shropshire, U.K., 2005.

Bordes, P., Pollet, E., and Avérous, L. Potential use of polyhydroxyalkanoate (PHA) for biocomposite development. In *Nano and Biocomposites*, A.K. Lau, F. Hussain, and K. Lafdi (eds.). CRC Press, Boca Raton, FL, 2010.

Calvert, P. Potential applications of nanotubes. In *Carbon Nanotubes*, T.W. Ebbesen (ed.), pp. 277–292. CRC Press, Boca Raton, FL, 1997.

Chang, P.R., Jian, R., Yu, J., and Ma, X. Fabrication and characterisation of chitosan nanoparticles/plasticised-starch composites. *Food Chemistry*, 120 (2010): 736–740.

Charpentier, J.C. Four main objectives for the future of chemical and process engineering mainly concerned by the science and technologies of new materials production. *Chemical Engineering Journal*, 107(1–3) (2005): 3–17.

Chen, G., Wei, M., Chen, J., Huang, J., Dufresne, A., and Chang, P.R. Simultaneous reinforcing and toughening: New nanocomposites of waterborne polyurethane filled with low loading level of starch nanocrystals. *Polymer*, 49 (2008a): 1860–1870.

Chen, P. and Zhang, L. Interaction and properties of highly exfoliated soy protein/montmorillonite nanocomposites. *Biomacromolecules*, 7(6) (2006): 1700–1706.

Chen, Y., Cao, X., Chang, P.R., and Huneault, M.A. Comparative study on the films of poly(vinyl alcohol)/pea starch nanocrystals and poly(vinyl alcohol)/native pea starch. *Carbohydrate Polymers*, 73(1) (2008b): 8–17.

Chen, Y., Liu, C., Chang, P.R., Anderson, D.P., and Huneault, M.A. Pea starch-based composite films with pea hull fibers and pea hull fiber-derived nanowhiskers. *Polymer Engineering & Science*, 49 (2009): 369–378.

Cho, S.-W., Gällstedt, M., Johansson, E., and Hedenqvist, M.S. Injection-molded nanocomposites and materials based on wheat gluten. *International Journal of Biological Macromolecules*, 48(1) (2010): 146–152.

Chrissafis, K., Antoniadis, G., Paraskevopoulos, K.M., Vassiliou, A., and Bikiaris, D.N. Comparative study of the effect of different nanoparticles on the mechanical properties and thermal degradation mechanism of in situ prepared poly(ε-caprolactone) nanocomposites. *Composites Science and Technology*, 67 (2007): 2165–2174.

Costa, C., Conte, A., Buonocore, G.G., and Del Nobile, M.A. Antimicrobial silver-montmorillonite nanoparticles to prolong the shelf life of fresh fruit salad. *International Journal of Food Microbiology*, 148 (2011): 164–167.

Costa, C., Conte, A., Buonocore, G.G., Lavorgna, M., and Del Nobile, M.A. Calcium-alginate coating loaded with silver-montmorillonite nanoparticles to prolong the shelf-life of fresh-cut carrots. *Food Research International*, 48 (2012): 164–169.

Cyras, V.P., Manfredi, L.B., Ton-That, M.T., and Vazquez, A. Physical and mechanical properties of thermoplastic starch/montmorillonite nanocomposite films. *Carbohydrate Polymers*, 73 (2008): 55–63.

de Moura, M.R., Mattoso, L.H.C., and Zucolotto, V. Development of cellulose-based bactericidal nanocomposites containing silver nanoparticles and their use as active food packaging. *Journal of Food Engineering*, 109(3) (2012): 520–524.

Dubief, D., Samain, E., and Dufresne, A. Polysaccharide microcrystals reinforced amorphous poly(α-hydroxyoctanoate) nanocomposite materials. *Macromolecules*, 32(18) (1999): 5765–5771.

Dufresne, A., Dupeyre, D., and Vignon, M.R. Cellulose microfibrils from potato tuber cells: Processing and characterization of starch–cellulose microfibril composites. *Journal of Applied Polymer Science*, 76 (2000): 2080–2092.

Duncan, T.V. Applications of nanotechnology in food packaging and food safety: Barrier materials, antimicrobials and sensors. *Journal of Colloid and Interface Science*, 363 (2011): 1–24.

Echeverría, I., Eisenberg, P., and Mauri, A.N. Nanocomposites films based on soy proteins and montmorillonite processed by casting. *Journal of Membrane Science*, 449 (2014): 15–26.

Eichhorn, S.J., Dufresne, A., Aranguren, M., Marcovich, N.E., Capadona, J.R., Rowan, S.J., Weder, C. et al. Review: Current international research into cellulose nanofibers and nanocomposites. *Journal of Material Science*, 45 (2010): 1–33.

Famá, L.M., Pettarin, V., Goyanes, S.N., and Bernal, C.R. Starch/multi-walled carbon nanotubes composites with improved mechanical properties. *Carbohydrate Polymers*, 83 (2011): 1226–1231.

Fang, Z. and Bhandari, B. Encapsulation of polyphenols—A review. *Trends in Food Science & Technology*, 21 (2010): 510–523.

Fayaz, A.M., Balaji, K., Girilal, M., Kalaichelvan, P.T., and Venkatesan, R. Mycobased synthesis of silver nanoparticles and their incorporation into sodium alginate films for vegetable and fruit preservation. *Journal of Agricultural and Food Chemistry*, 57(14) (2009): 6246–6252.

Fernández, A., Picouet, P., and Lloret, E. Cellulose-silver nanoparticle hybrid materials to control spoilage-related microflora in absorbent pads located in trays of fresh-cut melon. *International Journal of Food Microbiology*, 142 (2010): 222–228.

Fortunati, E., Peltzer, M., Armentano, I., Jiménez, A., and Kenny, J.M. Combined effects of cellulose nanocrystals and silver nanoparticles on the barrier and migration properties of PLA nano-biocomposites. *Journal of Food Engineering*, 118 (2013): 117–124.

Gammariello, D., Conte, A., Buonocore, G.G., and Del Nobile, M.A. Bio-based nanocomposite coating to preserve quality of Fior di latte cheese. *Journal of Dairy Science*, 94 (2011): 5298–5304.

García, N.L., Ribba, L., Dufresne, A., Aranguren, M., and Goyanes, S. Physico-mechanical properties of biodegradable starch nanocomposites. *Macromolecular Materials and Engineering*, 294 (2009): 169–177.

Gardner, D.J., Oporto, G.S., Mills, R., and Azizi Samir, M.A.S. Adhesion and surface issues in cellulose and nanocellulose. *Journal of Adhesion Science and Technology*, 22 (2008): 545–567.

Giannelis, E.P. Polymer-layered silicate nanocomposites. *Advanced Material*, 8 (1996): 29–35.

Giannelis, E.P. Polymer-layered silicate nanocomposites: Synthesis, properties and applications. *Applied Organometallic Chemistry*, 12 (1998): 675–680.

Gibbs, B.F., Kermasha, S., Alli, I., and Mulligan, C.N. Encapsulation in the food industry. *International Journal of Food Sciences and Nutrition*, 50 (1999): 213–224.

Gopalan, N.K., Dufresne, A., Gandini, A., and Belgacem, M.N. Crab shell chitin whiskers reinforced natural rubber nanocomposites. 3. Effect of chemical modification of chitin whiskers. *Biomacromolecules*, 4 (2003), 1835–1842.

Grande, C.J., Torres, F.G., Gomez, C.M., Troncoso, O.P., Canet-Ferrer, J., and Martínez-Pastor, J. Development of self-assembled bacterial cellulose-starch nanocomposites. *Materials Science and Engineering C*, 29 (2009): 1098–1104.

Guilherme, M.R., Mattoso, L.H.C., Gontard, N., Guilbert, S., and Gastaldi, E. Synthesis of nanocomposites films from wheat gluten matrix and MMT intercalated with different quaternary ammonium salts by way of hydroalcoholic solvent casting. *Composites Part A: Applied Science and Manufacturing*, 41(3) (2010): 375–382.

Guodong, Y. Natural and modified nanomaterials as sorbents of environmental contaminants. *Journal of Environmental Science and Health Part A*, 39 (2005): 2661–2670.

Gusev, A.A. Numerical identification of the potential of whisker- and platelet-filled polymers. *Macromolecules*, 34 (2001): 3081–3093.

Hedenqvist, M.S., Backman, A., Gällstedt, M., Boyd, R.H., and Gedde, U.W. Morphology and diffusion properties of whey/montmorillonite nanocomposites. *Composites Science and Technology*, 66(13) (2006): 2350–2359.

Herron, N. and Thorn, D.L. Nanoparticles. Uses and relationships to molecular clusters. *Advances Materials*, 10 (1998): 1173–1184.

Hong, Z., Zhang, P., He, C., Qiu, X., Liu, A., and Chen, L. Nano-composite of poly(1-lactide) and surface grafted hydroxyapatite: Mechanical properties and biocompatibility. *Biomaterials*, 26 (2005): 6296–6304.

Huang, M. and Yu, J. Structure and properties of thermoplastic corn starch/montmorillonite biodegradable composites. *Journal of Applied Polymer Science*, 99 (2006): 170–176.

Iguchi, M., Yamanaka, S., and Budhiono, A. Bacterial cellulose—A masterpiece of nature's arts. *Journal of Materials Science*, 35 (2000): 261–270.

Jayakody, L. and Hoover, R. The effect of lintnerization on cereal starch granules. *Food Research International*, 35(7) (2002): 665–680.

Karger-Kocsis, J. and Zhang, Z. Structure-property relationships in nanoparticle/semicrystalline thermoplastic composites. In *Mechanical Properties of Polymers Based on Nanostructures and Morphology*, J.F. Balta Calleja and G. Michler (eds.), pp. 547–596. CRC Press, New York, 2005.

Kim, Y. and White, J.L. Formation of polymer nanocomposites with various organoclays. *Journal of Applied Polymer Science*, 96 (2005): 1888–1896.

Koh, H.C., Parka, J.S., Jeong, M.A., Hwang, H.Y., Hong, Y.T., Ha, S.Y., and Nam, S.Y. Preparation and gas permeation properties of biodegradable polymer/layered silicate nanocomposite membranes. *Desalination*, 233 (2008): 201–209.

Kojima, Y., Usuki, A., Kawasumi, M., Okada, A., Karauchi, T., and Kamigaito, O. Synthesis of nylon 6-clay hybrid by montmorillonite intercalated with ε-caprolactam. *Journal of Polymer Science Part A*, 31 (1993): 983–986.

Krajewska, B. Application of chitin- and chitosan-based materials for enzyme immobilizations: A review. *Enzyme and Microbial Technology*, 35(2–3) (2004): 126–139.

Kristo, E. and Biliaderis, C.G. Physical properties of starch nanocrystal-reinforced pullulan films. *Carbohydrate Polymers*, 68 (2007): 146–158.

Kumar, P., Sandeep, K.P., Alavi, S., Truong, V.D., and Gorga, R.E. Preparation and characterization of bio-nanocomposite films based on soy protein isolate and montmorillonite using melt extrusion. *Journal of Food Engineering*, 100(3) (2010): 480–489.

Kvítek, L., Panáček, A., Soukupová, J., Kolář, M., Večeřová, R., Prucek, R., Holecová, M., and Zbořil, R. Effect of surfactants and polymers on stability and antibacterial activity of silver nanoparticles (NPs). *Journal of Physical Chemistry C*, 112 (2008): 5825–5834.

Lakkis, J.M. *Encapsulation and Controlled Release Technologies in Food Systems*. Blackwell Publishing Limited, Hoboken, NJ, 2007.

LeCorre, D., Bras, J., and Dufresne, A. Starch nanoparticles: A review. *Biomacromolecules*, 11 (2010): 1139–1153.

Lee, S.R., Park, H.M., Lim, H.T., Kang, K.Y., Li, L., Cho, W.J., and Ha, C.S. Microstructure, tensile properties, and biodegradability of aliphatic polyester/clay nanocomposites. *Polymer*, 43 (2002): 2495–2500.

Lok, C.-N., Ho, C.-M., Chen, R., He, Q.-Y., Yu, W.-Y., Sun, H., Tam, P.K.-H., Chiu, J.-F., and Che, C.-M. Silver nanoparticles: Partial oxidation and antibacterial activities. *Journal of Biological Inorganic Chemistry*, 12 (2007): 527–534.

Lönnberg, H., Fogelström, L., Samir, M.A.S.A., Berglund, L., Malmström, E., and Hult, A. Surface grafting of microfibrillated cellulose with poly(ε-caprolactone)-synthesis and characterization. *European Polymer Journal*, 44(9) (2008): 2991–2997.

Lopez-Rubio, A. Bioactive food packaging strategies. In *Multifunctional and Reinforced Polymers for Food Packaging*, J.M. Lagaron (ed.). Woodhead, Cambridge, U.K., 2011.

Lopez-Rubio, A., Gavara, R., and Lagaron, J.M. Bioactive packaging: Turning foods into healthier foods through biomaterials. *Trends in Food Science & Technology*, 17 (2006): 567–575.

Ma, X., Jian, R., Chang, P.R., and Yu, J. Fabrication and characterization of citric acid-modified starch nanoparticles/plasticized-starch composites. *Biomacromolecules*, 9 (2008): 3314–3320.

Madene, A., Jacquot, M., Scher, J., and Desobry, S. Flavour encapsulation and controlled release—A review. *International Journal of Food Science and Technology*, 41(1) (2006): 1–21.

Manías, E. Nanocomposites: Stiffer by design. *Nature Materials*, 6(1) (2007): 9–11.

Mascheroni, E., Chalier, P., Gontard, N., and Gastaldi, E. Designing of a wheat gluten/montmorillonite based system as carvacrol carrier: Rheological and structural properties. *Food Hydrocolloids*, 24 (2010): 406–413.

Maynard, A.D., Aitken, R.J., Butz, T., Colvin, V., Donaldson, K., Oberdorster, G., Philbert, M.A. et al. Safe handling of nanotechnologies. *Nature*, 444 (2006): 267–269.

McCrum, N.G., Buckley, C.P., and Bucknall, C.B. *Principles of Polymer Engineering*. Oxford Science, New York, 1996.

Min, B.M., Lee, S.W., Lim, J.N., You, Y., Lee, T.S., Kang, P.H., and Park, W.H. Chitin and chitosan nanofibers: Electrospinning of chitin and deacetylation of chitin nanofibers. *Polymer*, 45(21) (2004): 7137–7142.

Mincea, M., Negrulescu, A., and Ostafe, V. Preparation, modification, and applications of chitin nanowhiskers: A review. *Reviews on Advanced Materials Science*, 30 (2012): 225–242.

Morgan, A.B. and Gilman, J.W. Characterization of polymer-layered silicate (clay) nanocomposites by transmission electron microscopy and X-ray diffraction: A comparative study. *Journal of Applied Polymer Science*, 87 (2003): 1329–1338.

Morgan, A.B. and Wilkie, C.A. (eds.). *Flame Retardant Polymer Nanocomposites*. Wiley, Hoboken, NJ, 2007.

Morones, J.R., Elechiguerra, J.L., Camacho, A., Holt, K., Kouri, J.B., Ramírez, J.T., and Yacaman, M.J. The bactericidal effect of silver nanoparticles. *Nanotechnology*, 16 (2005): 2346–2353.

Nagpal, K., Singh, S.K., and Mishra, D.N. Chitosan nanoparticles: A promising system in novel drug delivery. *Chemical and Pharmaceutical Bulletin*, 58(11) (2010): 1423–1430.

Nakagaito, A.N., Iwamoto, S., and Yano, H. Bacterial cellulose: The ultimate nano-scalar cellulose morphology for the production of high-strength composites. *Applied Physics A*, 80 (2005): 93–97.

Ojijo, V. and Sinha Ray, S. Processing strategies in bionanocomposites. *Progress in Polymer Science*, 38 (2013): 1543–1589.

Pan, Y., Wu, T., Bao, H., and Li, L. Green fabrication of chitosan films reinforced with parallel aligned graphene oxide. *Carbohydrate Polymers*, 83 (2011): 1908–1915.

Pandey, J.K., Kumar, A.P., Misra, M., Mohanty, A.K., Drzal, L.T., and Singh, R.P. Recent advances in biodegradable nanocomposites. *Journal for Nanoscience and Nanotechnology*, 5 (2005): 497–526.

Park, H.M., Lee, W.K., Park, C.Y., and Ha, C.S. Environmentally friendly polymer hybrids. Part I mechanical, thermal, and barrier properties of thermoplastic starch/clay nanocomposites. *Journal of Materials Science*, 38 (2003): 909–915.

Park, H.M., Mohanty, A., Misra, M., and Drzal, L.T. Green nanocomposites from cellulose acetate bioplastic and clay: Effect of eco-friendly triethyl citrate plasticizer. *Biomacromolecules*, 5 (2004): 2281–2288.

Pereda, M., Amica, G., Rácz, I., and Marcovich, N.E. Structure and properties of nanocomposite films based on sodium caseinate and nanocellulose fibers. *Journal of Food Engineering*, 103(1) (2011): 76–83.

Petersson, L., Kvien, I., and Oksman, K. Structure and thermal properties of poly(lactic acid)/cellulose whiskers nanocomposites materials. *Composites Science and Technology*, 67(11–12) (2007): 2535–2544.

Petersson, L. and Oksman, K. Preparation and properties of biopolymer-based nanocomposite films using microcrystalline cellulose. *American Chemical Society Symposium Series*, 938 (2006): 132–150.

Quilaqueo Gutiérrez, M., Echeverría, I., Ihl, M., Bifani, V., and Mauri, A.N. Carboxymethylcellulose–montmorillonite nanocomposite films activated with murta (*Ugni molinae* Turcz) leaves extract. *Carbohydrate Polymers*, 87 (2012): 1495–1502.

Rai, M., Yadav, A., and Gade, A. Silver nanoparticles as a new generation of antimicrobials. *Biotechnology Advances*, 27 (2009): 76–83.

Rajendran, V. Development of nanomaterials from natural resources for various industrial applications. *Advanced Materials Research*, 67 (2009): 71–76.

Rao, Y.Q. Gelatine-clay nanocomposites of improved properties. *Polymer*, 48(18) (2007): 5369–5375.

Reddy, M.M., Vivekanandhan, S., Misra, M., Bhatia, S.K., and Mohanty, A.K. Biobased plastics and bionano-composites: Current status and future opportunities. *Progress in Polymer Science*, 38 (2013): 1653–1689.

Rhim, J.W., Hong, S.I., Park, H.M., and Ng, P.K.W. Preparation and characterization of chitosan-based nano-composite films with antimicrobial activity. *Journal of Agricultural and Food Chemistry*, 54 (2006): 5814–5822.

Rhim, J.W., Lee, J.H., and Ng, P.K.W. Mechanical and barrier properties of biodegradable soy protein isolate-based films coated with polylactic acid. *LWT: Food Science and Technology*, 40(2) (2007): 232–238.

Rhim, J.W. and Ng, P.K. Natural biopolymer-based nanocomposite films for packaging applications. *Critical Reviews in Food Science and Nutrition*, 47 (2007): 411–433.

Rhim, J.W., Park, H.M., and Ha, C.S. Bio-nanocomposites for food packaging applications. *Progress in Polymer Science*, 38 (2013): 1629–1652.

Royal Society and Royal Academy of Engineering. Nanoscience nanotechnologies: Opportunities uncertainties. Available: www.nanotec.org.uk/finalReport.htm, 2004. Accessed: November 20, 2005.

Sanchez-Garcia, M. and Lagaron, J. Novel clay-based nanobiocomposites of biopolyesters with synergistic barrier to UV light, gas, and vapour. *Journal of Applied Polymer Science*, 118 (2010): 188–199.

Sanguansri, P. and Augustin, M.A. Nanoscale materials development—A food industry perspective. *Trends in Food Science & Technology*, 17(10) (2006): 547–556.

Schwartz, M. *New Materials, Processes, and Methods Technology*. CRC Press, Boca Raton, FL, 2005.

Scott, G. 'Green' polymer. *Polymer Degradation and Stability*, 68 (2000): 1–7.

Shahidi, F., Arachchi, J.K.V., and Jeon, Y.J. Food applications of chitin and chitosan. *Trends in Food Science & Technology*, 10 (1999): 37–51.

Shahidi, F. and Han, X.Q. Encapsulation of food ingredients. *Critical Reviews in Food Science and Nutrition*, 33 (1993): 501–547.

Shaikh, S., Birdi, A., Qutubuddin, S., Lakatosh, E., and Baskaran, H. Controlled release in transdermal pressure sensitive adhesives using organosilicate nanocomposites. *Annals of Biomedical Engineering*, 35(12) (2007): 2130–2137.

Sharma, V.K., Yngard, R.A., and Lin, Y. Silver nanoparticles: Green synthesis and their antimicrobial activities. *Advances in Colloid and Interface Science*, 145(1–2) (2009): 83–96.

Sinha Ray, S. Visualisation of nanoclay dispersion in polymer matrix by high-resolution electron microscopy combined with electron tomography. *Macromolecular Materials and Engineering*, 294 (2009): 281–286.

Sinha Ray, S. and Bousmina, M. Biodegradable polymers and their layered silicate nanocomposites: In greening the 21st century materials world. *Progress in Materials Science*, 50 (2005): 962–1080.

Sinha Ray, S. and Okamoto, M. Polymer/layered silicate nanocomposites: A review from preparation to processing. *Progress in Polymer Science*, 28 (2003a): 1539–1641.

Sinha Ray, S. and Okamoto, M. New polylactide/layered silicate nanocomposites: Open a new dimension for plastics and composites. *Macromolecular Rapid Communications*, 24 (2003b): 815–840.

Sinha Ray, S., Yamada, K., Okamoto, M., and Ueda, K. New polylactide- layered silicate nanocomposites. 2. Concurrent improvements of materials properties, biodegradability and melt rheology. *Polymer*, 44 (2003a): 857–866.

Sinha Ray, S., Yamada, K., Okamoto, M., and Ueda, K. Biodegradable poly- lactide/montmorillonite nano-composites. *Journal for Nanoscience and Nanotechnology*, 3 (2003b): 503–510.

Siqueira, G., Bras, J., and Dufresne, A. Cellulosic bionanocomposites: A review of preparation, properties and applications. *Polymers*, 2 (2010), 728–765.

Smolander, M. and Chaudhry, Q. Nanotechnologies in food packaging. In *Nanotechnologies in Food*, Q. Chaudhry, L. Castle, and R. Watkins (eds.), pp. 86–101. RSC Publishing, Cambridge, U.K., 2010.

Sondi, I. and Salopek-Sondi, B. Silver nanoparticles as antimicrobial agent: A case study on *E. coli* as a model for Gram-negative bacteria. *Journal of Colloid and Interface Science*, 275 (2004): 177–182.

Sorrentino, A., Tortora, M., and Vittoria, V. Diffusion behavior in polymer–clay nanocomposites. *Journal of Polymer Science Part B: Polymer Physics*, 44 (2006): 265–274.

Sothornvit, R., Hong, S.I., An, D.J., and Rhim, J.W. Effect of clay content on the physical and antimicrobial properties of whey protein isolate/organo-clay composite films. *LWT: Food Science and Technology*, 43 (2010): 279–284.

Sothornvit, R., Rhim, J.W., and Hong, S.I. Effect of nano-clay type on the physical and antimicrobial properties of whey protein isolate/clay composite films. *Journal of Food Engineering*, 91 (2009): 468–473.

Souza Filho, A.G. and Fagan, S.B. Nanomaterials properties. In *Nanostructured Materials for Engineering Applications*, C.P. Bergmann and M.J. de Andrade (eds.). Springer, Berlin, Germany, 2011.

Sudesh, K. and Iwata, T. Sustainability of biobased and biodegradable plastics. *Clean*, 36(5–6) (2008): 433–442.

Suriyamongkol, P., Weselake, R., Narine, S., Moloney, M., and Shah, S. Biotechnological approaches for the production of polyhydroxyalkanoates in microorganisms and plants: A review. *Biotechnology Advances*, 25 (2007): 148–175.

Tetto, J.A., Steeves, D.M., Welsh, E.A., and Powell, B.E. Biodegradable poly(caprolactone)/clay nanocomposites. *ANTEC*, Lancaster, PA, 1999, pp. 1628–1632.

Thellen, C., Orroth, C., Froio, D., Ziegler, D., Lucciarini, J., Farrell, R., D'Souza, N.A., and Ratto, J.A. Influence of montmorillonite layered silicate on plasticized poly (L-lactide) blown films. *Polymer*, 46 (2005): 11716–11727.

Trotignon, J., Verdu, J., Piperaud, M., and Dobraczinski, A. *Précis de matières plastiques: Structures, Propriétés, Mise en oeuvre et Normalisation*. AFNOR, Paris, France, 1989.

Trznadel, M. Biodegradable polymer materials. *International Polymer Science and Technology*, 22 (1995): 58–65.

Tunc, S., Angellier, H., Cahyana, Y., Chalier, P., Gontard, N., and Gastaldi, E. Functional properties of wheat gluten/montmorillonite nanocomposite films processed by casting. *Journal of Membrane Science*, 289(1–2) (2007): 159–168.

van der Graaf, S., Schroen, C.G.P.H., and Boom, R.M. Preparation of double emulsions by membrane emulsification—A review. *Journal Membrane Science*, 251(1–2) (2005): 7–15.

van Tuil, R., Fowler, P., Lawther, M., and Weber, C.J. Properties of biobased packaging materials. In *Biobased Packaging Materials for the Food Industry: Status and Perspectives*, C.J. Weber (ed.), pp. 13–44. KVL, Frederiksberg, Denmark, 2000.

Viguié, J., Molina-Boisseau, S., and Dufresne, A. Processing and characterisation of waxy maize starch films plasticized by sorbitol and reinforced with starch nanocrystals. *Macromolecular Bioscience*, 7(11) (2007): 1206–1216.

Vladimirov, V., Betchev, C., Vassiliou, A., Papageorgiou, G., and Bikiaris, D. Dynamic mechanical and morphological studies of isotactic polypropylene/fumed silica nanocomposites with enhanced gas barrier properties. *Composites Science and Technology*, 66(15) (2006): 2935–2944.

Wang, N.G. and Zhang, L.N. Preparation and characterization of soy protein plastics plasticized with waterborne polyurethane. *Polymer International*, 54 (2005): 233–239.

Xie, F., Pollet, E., Halley, P.J., and Avérous, L. Starch-based nano-biocomposites. *Progress in Polymer Science*, 38 (2013): 1590–1628.

Yu, J., Ai, F., Dufresne, A., Gao, S., Huang, J., and Chang, P.R. Structure and mechanical properties of poly(lactic acid) filled with (starch nanocrystal)-graft-poly(ε-caprolactone). *Macromolecular Material Engineering*, 293 (2008): 763–770.

Zhao, R., Torley, P., and Halley, P.J. Emerging biodegradable materials: Starch and protein-based bionanocomposites. *Journal of Materials Science*, 43(9) (2008): 3058–3071.

Zheng, H., Ai, F., Chang, P.R., Huang, J., and Dufresne, A. Structure and properties of starch nanocrystal-reinforced soy protein plastics. *Polymer Composites*, 30(4) (2009): 474–480.

Zimmermann, T., Pohler, E., and Geiger, T. Cellulose fibrils for polymer reinforcement. *Advanced Engineering Materials*, 6 (2004): 754–761.

Zuidam, N.J. and Nedovic V.A. *Encapsulation Technologies for Active Food Ingredients and Food Processing*. Springer Science, New York, 2010.

14

Antioxidant Films and Coatings

Mónica Ihl, Andrea Silva-Weiss, and Valerio Bifani

CONTENTS

Abstract

Bioactive films and coatings show the benefits of using natural antioxidants incorporated to improve the antioxidant/antibrowning activities and the barrier properties to O_2, CO_2, water vapor, and light. Natural extracts, polyphenol compounds, and vitamins having antioxidant capacity, incorporated in edible films and coatings, represent a new approach to solve the detrimental impact of oxygen on films and fresh food. In good correlation with food and health studies, it is interesting to keep in mind that the current trend is the use of the natural antioxidant products, whose activities take place not only in the packaged food but also as nutraceuticals, once the food has been ingested. Outcomes of these studies can help develop new food products with better antioxidant activity, sensorial quality, and beneficial health aspects.

14.1 Introduction

Edible antioxidant films and coatings is an environmentally friendly technology that can be used as an active packaging, extending the shelf life in a wide variety of food applications by preventing oxidative rancidity, surface browning, oil diffusion, and dehydration. Furthermore, the inherent biodegradability and edibility of these films and coatings must be maintained by the addition of food-grade ingredients in facilities that are acceptable for food processing, in which the solvents used are restricted to water and ethanol (Zaritzky, 2011).

The ultimate functionality of edible antioxidant films and coatings is related to the following:

1. Their bioactivity is due to the antioxidant additives, which means
   a. Proton donors that counteract the toxic-reactive oxygen and nitrogen species formed during the metabolism. This occurs in living aerobic cells either of coated fresh food

(Li et al., 2012; Wang and Gao, 2013) or of the human who will eat this coated food, and therefore will have a better chance to strengthen his or her own health prevention (Lasagni Vitar et al., 2014; Yousaf et al., 2014).

b. Antibrowning properties related to inhibition of the polyphenol oxidase enzyme system or restriction to the oxygen availability of this same enzyme (Amiot et al., 1992; Kubo et al., 2003; Caillet et al., 2007). Phenolic acids, such as gallic acid, exhibit antioxidant and antibrowning activity that has been attributed to their capacity to reduce ortho-quinones (colored) to the corresponding colorless phenolic compounds (Kubo et al., 2003) and flavonoids such as quercetin, rutin, and morin, which have significant anti-oxidant activities (Caillet et al., 2007; Boots et al., 2008). Because antioxidants control the diffusion of oxygen through membranes or films by diminishing their permeability (Ayranci and Tunc, 2004; Pokorný, 2007), these antioxidant flavonoids, when incorporated in edible films or coatings, might act as antibrowning compounds by delaying the contact of oxygen with the enzyme polyphenoloxidase (PPO). Onion extract, which possesses flavonols such as quercetin and quercetin derivates, can be used as an additive to prevent browning (Kim et al., 2005; Roldán et al., 2008). In addition, some flavonoids have also been found to protect ascorbic acid from degradation (Sarma et al., 1997). Another possibility to prevent browning in fresh products is to use packaging, films, or coatings that prevent the oxygen diffusion produced by mechanical damage during transport, storage, and distribution. Browning is inhibited in atmospheres containing less than 1 kPa of oxygen (Gorny, 1997), so that the incorporation of plant extracts on films or coatings is expected to enhance the oxygen and light barrier, preventing oxygen from coming in contact with polyphenol oxidase and maintaining the stability of ascorbic acid in the food.

2. Functional properties, such as their ability to serve as a barrier to water vapor, oxygen, carbon dioxide, and UV–visible light (Bifani et al., 2007; Tongnuanchan et al., 2012).

3. Mechanical properties, such as tensile strength (TS), elongation at break (EB), and physical properties, such as opacity and color (Gómez-Guillén et al., 2007; Tongnuanchan et al., 2012).

Research has shown that incorporation of antioxidant extracts and components from plants, spices, herbs, and seaweeds on coatings and films improves their active properties as antioxidants and/or anti-brownings (Gómez-Guillén et al., 2007; Siripatrawan and Harte, 2010; Jouki et al., 2013; Blanco-Pascual et al., 2014a,b; Cian et al., 2014) by adding value increasing food product's shelf life, for example, in beef burgers (Georgantelis et al., 2007), Özvural et al., 2016), large yellow croakers (Li et al., 2012), mangos (Wang et al., 2007), strawberries (Wang and Gao, 2013), meat, fish, nuts, fruits, and vegetables (Camo et al., 2008; Giménez et al., 2011; Bonilla et al., 2012; Siripatrawan and Noipha, 2012).

Moreover, in foods, edible antioxidant coatings or films can preserve or enhance the sensorial properties and may have the ability to modify the internal atmosphere. Therefore, manufacturers may choose the most appropriate type of packaging or edible films, depending on the nature and requirements of the product, the degree and nature of the protection need, the method of distribution, the shelf life, and the environmental impact (Silva-Weiss et al., 2013b).

14.2 Current Trends for Edible Antioxidant Films/Coatings: Concept, Formulation, Requirements, and Some Health Promotion Effects

In general, ingredients in food packaging that may be added to food are considered food additives and must meet food additive standards (FAO/WHO, 2005). Additives are incorporated into foods to contribute to the overall quality, safety, nutritive value, organoleptic characteristics (color, smell, and taste), convenience, and economy (IFT, 2010). Currently, edible films and coatings with natural additives are applied in a variety of food applications (Kerch, 2015; Salgado et al., 2015; Silva-Weiss et al., 2013b).

There are edible antioxidant films and coatings that can transport food additives that act as vehicles to fortify foods with vitamins, such as ascorbic acid, omega-3 fatty acids, and other nutrients such as polyphenols, which are the major plant compounds with antioxidant activity, and chiefly flavonoids, which have potent antioxidant activities and positive effects on human health. When incorporated to edible films, they can also control the oxidation of foods that causes browning (León et al., 2008).

To maintain the quality, the films and coatings should create a semipermeable barrier against moisture migration, carbon dioxide (CO_2), oxygen (O_2), aromas, and lipid oxidation from the foodstuff. To increase the shelf life of fruits and vegetables, ethylene levels must be decreased through a reduced O_2 availability, which contributes to decrease or stop ethylene biosynthesis. Also, an increment of CO_2 reduces ripening, respiration rate, and water losses. For meat products, edible films and coatings can serve as a carrier not only for herbs and spices but also for antioxidants, antimicrobials (Özvural et al., 2016), nutrients, and colors (Cardoso et al., 2016).

Thus, the new generation of edible antioxidant films and coatings or active packaging has been specifically designed to increase functionality by incorporating bioactive/functional natural additives; for example, green tea extracts in alginate (Zhang et al., 2016) or chitosan (Özvural et al., 2016; Sabaghi et al., 2015) based edible coatings.

A potential storage improvement has been observed when antioxidant and/or antibrowning compounds are introduced into edible films and coatings based on polysaccharides (such as chitosan, starch, carrageenan, agar, gellan, alginate and pectin) and proteins (such as whey protein concentrate and fish skin, pig skin and bovine skin gelatin) (Silva-Weiss et al., 2013b). In fish as an example, an extension of 8–10 days of refrigerated shelf life has been obtained with tea and rosemary extracts in chitosan coatings (Li et al., 2012).

The selection of the natural additives and their application depends on their availability, cost, effectiveness, and effect on the sensory attributes of the final product, as well as on consumer awareness (Perumalla and Hettiarachchy, 2011). Natural antioxidants found in plants are now very good alternatives to synthetic antioxidants. Plant extracts, such as basil, laurel, and rosemary, and tea leaf extracts slow down the oxidation of several oils more effectively than the synthetic antioxidants BHA and BHT (Frankel et al., 1996; Hinneburg et al., 2006; Wambura et al., 2011).

Besides incorporating dietary antioxidant phytochemicals also in the coatings, a major possibility may exist to promote their important roles as chemopreventive or chemotherapeutic agents in the prevention of many diseases:

1. Green tea has many biologically active compounds, like flavanols and polyphenols. Catechins are flavanols that constitute the majority of soluble solids of green tea, having epigallocatechin gallate as its major component, contributing with more than 50% of the polyphenols. Green tea has many health-related characteristics, like hypoglycemic, hypocholesterolemic, anticancer, antiviral, and antihypertensive activities. In studies done with normal, hyperglycemic, and hypercholesterolemic rats (Yousaf et al., 2014), using three types of diets described as control (normal), functional (with green tea powder), and nutraceutical (with green tea extracts), showed that green tea can be effective against hypercholesterolemia and hyperglycemia.

2. Another example, the aqueous herb extract of the South American native plant *Aloysia triphylla*, known as "cedron," "Yerba Luisa," or "Verbena de Indias" (with a popular favorable response to better night sleep), is rich in antioxidant biological activities. A recent publication (Lasagni et al., 2014) evaluates the protective effect of aqueous extracts (infusion and decoction) of *A. triphylla* against lipid peroxidation of brain homogenates to determine changes in the prooxidant/antioxidant balance when the plant material is added. The protective effect of this plant extract against brain lipid peroxidation shows that the antioxidants present in the extracts of *A. triphylla* could act as strong scavengers of reactive oxygen species not only at the initiation of the lipid-peroxidation chain reaction but also at the propagation steps. Therefore, it could be used as a prophylactic and therapeutic agent for those diseases where the occurrence of oxidative stress and lipid peroxidation contributes to the progression of damage.

It is interesting to keep in mind that these health studies, in good correlation with the food studies, can be potentiated when, besides the amount of the antioxidant plant extracts in the food, they are also included in the edible films around the food. Therefore, now, the current trend should be to use natural antioxidant products, whose activities take place not only in packaged food but also as a nutraceutical once the food has been ingested.

14.3 Interactions among Antioxidant Additives and Films/Coatings

When a coating gets in contact with a food, the evaporation flux at the interfaces coating/environment (liquid/air) should be evaluated as well as the coating absorption flux due to the interface between the food and the coating (solid/liquid) (Karbowiak et al., 2009). Therefore, the interaction between the film and the foodstuff (solid/solid interface) depends on the molecular size of the biopolymer (weight and volume), the chemical nature of the compounds (polarity, planarity, etc.), the temperature and process conditions, the film structure (crystallinity, plasticizing, morphology, glass transition temperature, etc.), the thickness of the film, and the nature of the material (Silva-Weiss et al., 2013b).

Studies should specifically address the compatibility of the coating agent and active compounds, because the properties of the resulting film depend on their binary combination and the ratio between them, which, in turn, depends on the required dosage of active compounds in the material (Silva-Weiss et al., 2013b).

Some properties of coatings, films, and food that could be affected by interactions with active agents or natural additives are as follows:

1. For natural additive–coating interaction, compatibility and stability with respect to temperature and pH; and steady and dynamic rheology properties with respect to temperature, pH, active agent concentration, and shear rate
2. For natural additive–film interaction, physical–mechanical properties such as resistance, elasticity, color, gloss, and lightness; barrier properties to O_2, CO_2, water vapor, and UV light; and thermal properties such as glass transition temperature, crystallization, melting and decomposition point, and morphology and structure (crystallinity, bond type, and shape)
3. For active film/coating and food, surface properties such as adhesion and wettability (W_S); physiological properties such as respiration rate, bioactive additive control delivery, and bioavailability with respect to temperature and pH; sensorial properties such as texture, color, odor, and flavor; and biological properties such as antioxidant, antibrowning, and also antimicrobial activities

When plant extracts from green tea, murta leaves, oregano, rosemary, etc. rich in polyphenols are incorporated in polymeric matrixes based on chitosan, gelatin, and chitosan–starch mixture, the antioxidant and light-barrier properties increase (Silva-Weiss et al., 2013b), even more than with the antioxidants BHT and tocopherol (Gómez-Estaca et al., 2009b). However, it has been reported that the polyphenol compounds interact on the polymeric matrix (Siripatrawan and Harte, 2010; Silva-Weiss et al., 2013a) and modify the physical and functional properties of the formed film (Silva-Weiss et al., 2013b), which sometimes makes difficult the liberation of the active agents incorporated in it (Mayachiew and Devahastin, 2010) and therefore, consequently, the benefit over the food and/or consumer. Recent studies propose encapsulating active compounds before they are incorporated into films, with the aim of reducing its interaction with the polymer matrix and improving their release (Bustos et al., 2016; Navarro et al., 2016).

The chemical or physical interactions between biopolymers and antioxidant bioactive compounds depend on the nature, chemical characteristics, concentration, and pH and can affect the structure and functionality. The nature of the interaction between biopolymers and additives depends on the nature, chemical characteristics, concentration, and pH of both the biopolymer and additives, as well as on the structural parameters of the active compounds (stereochemistry, conformational flexibility, and molecular weight). Interactions between natural/active compounds and the film material may be identified by the presence of new absorption bands in the UV–visible regions of the electronic absorption spectra, as reported between the tetrahydrocurcuminoid and chitosan film (Portes et al., 2009).

14.4 Functionality of Edible Antioxidant Films and Coatings

The semipermeable barrier of films and coatings regulates the atmosphere around foods, preventing their desiccation and controlling the migration of ingredients and additives in the food systems. The functionality depends on the bioactive and physical properties. The bioactive properties include the antioxidant and antibrowning activities. Among the physical properties that can be mentioned are the barrier to oxygen, carbon dioxide, and UV–visible light, water vapor permeability (WVP), tensile strength (TS), elongation at break (EB), opacity, and color. The incorporation of the natural additives to active packaging systems or biopolymer-based edible films can modify the film structure and, as a result, modify their functionality and consequently the application to foods (Silva-Weiss et al., 2013a).

14.4.1 Mechanical Properties

TS and EB are parameters that are related to the film's mechanical properties and chemical structure (McHugh and Krochta, 1994). The TS (MPa) is the maximum strength measuring the resistance of the film, whereas the EB (%) is a measure of the stretching capacity or flexibility of the film prior to breaking (Krochta and De Mulder-Johnston, 1997).

As an example, the addition of vitamin E to films based on fish skin gelatin from *Priacanthus macracanthus* and *Lutjanus vitta* modifies film mechanical properties over time, showing a reduction in the film's mechanical properties because of the interaction between vitamin E and fish gelatin, which decreases the movement of macromolecules in the biopolymer film (Jongjareonrak et al., 2008). However, as shown in Table 14.1 on chitosan and methylcellulose films with resveratrol added, the reduction of mechanical properties occurred due to nonmiscible compounds that provoke structural discontinuities in the polymer network and a reduction in the overall cohesion forces of the matrix (Pastor et al., 2013).

As can also be seen in Table 14.1, changes in the mechanical properties of biopolymer-based films occurred when green tea water extract (GTE) was added: slightly increases the mechanical resistance (or TS) and extensibility (or EB) of chitosan-based films (Siripatrawan and Harte, 2010) but decreases the TS and EB of fish skin gelatin and agar and agar–gelatin films (Giménez et al., 2013; Li et al., 2014). Polyphenolic compounds in these antioxidant extracts could form hydrogen and covalent bonds with amino and hydroxyl groups of the gelatin polypeptides, which could probably weaken the protein–protein interactions and stabilize the protein network (Li et al., 2014). The mechanical properties of chitosan film did not significantly change when the GTE concentration increased from 0% to 5% but significantly increased when the GTE concentration changed from 5% to 20% (Siripatrawan and Harte, 2010). The improvement in the mechanical properties of films containing GTE may be attributed to the interaction between the chitosan matrix and the polyphenolic compounds from the GTE.

In general, as can be appreciated in Table 14.2, the incorporation of polyphenol-rich aqueous extracts reduces the mechanical properties (TS and EB) of the films (Moradi et al., 2012). The addition of phenolic acid to the film has been shown to increase slowly these properties (Mathew and Abraham, 2008; Rivero et al., 2010). However, antioxidant leaf extracts of murta (*Ugni molinae* Turcz), of ecotype SC containing a higher concentration of phenolic compounds, strongly reduced the TS and EB of gelatin-based films, while films containing extract of ecotype SG behaved similarly to the control film. Therefore, a negative relationship between the concentration of phenolic compounds in the plant extract and the film's mechanical properties can also be seen with the SC murta leaf extract (Gómez-Guillén et al., 2007; Silva-Weiss et al., 2013a).

14.4.2 Oxygen-, Carbon Dioxide–, and Water Vapor–Barrier Properties

The permeability of the biomaterials to oxygen, carbon dioxide, and water vapor should be considered for regulating the atmosphere around foods. WVP should be as low as possible, since the main

TABLE 14.1

Antioxidant Plant Extracts or Vitamins in Edible Films and Coatings

Antioxidant	Film Based On	Effects on Film Properties	Reference
Trolox and catechin[a]	Fish muscle protein	Leads to the retardation of lipid oxidation and yellow discoloration in films.	Tongnuanchan et al. (2012)
Resveratrol[b]	Chitosan and methylcellulose	Are less stretchable and resistant to fracture, more opaque, and less glossy, reduces the WVP, and reduces oxygen permeability only in chitosan film.	Pastor et al. (2013)
Green tea[a] (GTE)	Agar and agar–fish gelatin	Increases film-water solubility and decreases TS and extensibility (EB) in both films. WVP and water resistance was not affected.	Giménez et al. (2013)
GTE, quercetin, ferulic acid, or ascorbic acid	EVOH	Presence of ascorbic and ferulic acids resulted in a significant decrease in WVP.	Lopez-Dicastillo et al. (2012)
GTE[a]	Chitosan	Increases AOX activity and polyphenolic content, increases mechanical properties, and reduces the WVP.	Siripatrawan and Harte (2010)
GTE[a] and epigallocatechin gallate	Fish gelatin	Antioxidants do not enhance or alter the physical and barrier properties (WVP) of the films to an extent that the application of films to food systems is affected.	Tammineni et al. (2012)
GTE, grape seed (GSP), ginger (GE), and gingko leaf (GBE) extracts[a]	Fish skin gelatin	The incorporation of GTE, GSP, or GBE into film caused a significant decrease of TS compared to that of control. GSP also reduces EB. UV light barrier was greatly improved with GBE (value of light transmission was 23.242% at 400 nm). Also, the addition of GTE causes a significant decrease WVP.	Li et al. (2014)
Betalains and anthocyanins of red color[b]	HPMC	Color became darker and redder as NRC increased, while an increase effect of light exposure was noticed on color stability.	Akhtar et al. (2012)
Betacyanin	HPMC	Enhances barrier to oxygen and WVP.	Akhtar et al. (2013)
Tea polyphenols[a]	Chitosan	Increases AOX activity and water solubility, decreases the WVP, and increases the opacity and color of chitosan films.	Wang et al. (2013)

Note: Physical and chemical properties.

[a] Aqueous extract or water-soluble compound.

[b] Ethanol: water extract or ethanol/water-soluble compound; AOX, antioxidant; TS, tensile strength (MPa); EB, elongation at break (%); WVP, water vapor permeability; CMC, sodium carboxymethylcellulose; HPMC, hydroxypropyl methylcellulose; EVOH, ethylene vinyl alcohol copolymer with a 29% ethylene molar content.

function of a food packaging is to avoid or at least to decrease moisture transfer between the food and the surrounding atmosphere or between two components of heterogeneous food products (Gontard et al., 1992, 1993).

The barrier properties of hydrophilic films and coatings are dependent on the surrounding relative humidity conditions (López-de-Dicastillo et al., 2012). For this reason, tests should be carried out at constant storage conditions of relative humidity and temperature. Increasing polarity, hydrogen-bond forces, and crystallinity decreases the free volume of the film, enhances the cohesion, and reduces the flexibility, thereby increasing the barrier properties of the polymer (Mali et al., 2006).

Oxygen has a detrimental effect on the quality of a wide variety of food products. The application of edible films and coatings to food products represents a new approach to solve this problem. Oxygen permeability through the film can be measured directly by oxygen permeation tests and involves the measurement of oxygen gas transmission through films, the flowing of an oxygen gas stream on one side

TABLE 14.2

Antioxidant Plant Extracts or Vitamins in Edible Films and Coatings

Antioxidant	Film Based On	Effects on Film Properties	Reference
Red seaweeds *Mastocarpus stellatus*	Red seaweeds *M. stellatus*	Decreases WVP, with suitable mechanical properties, in both tensile and puncture tests	Blanco-Pascual et al. (2014a)
Red seaweeds *Porphyra columbina*	Red seaweeds *P. columbina*	For films composed of 75 phycobiliproteins and 25 phycocolloids, decreases WVP, tensile strength, and elastic modulus and increases elongation at break	Cian et al. (2014)
Kiam wood[a]	HPMC	Increases WVP and film solubility, decreases TS and EB, darkens film color, and decreases transparency	Chana-Thaworn et al. (2011)
Oregano and rosemary[a]	Fish skin and bovine gelatin	Increases AOX activity and UV light barrier	Gómez-Estaca et al. (2009a)
Murta leaves SC and SG[a]	Fish skin gelatin	SC increasing AOX activity and UV light barrier and cohesiveness, but reducing TS, EB, and WVP; SG increasing AOX activity and UV light barrier	Gómez-Guillén et al. (2007)
Murta leaves SG[a]	CMC	Decreases WVP and forms a selective barrier to gases (CO_2/O_2)	Bifani et al. (2007)
Murta leaves SC[a]	CMC	Increases O_2 barrier	Bifani et al. (2007)
Borage[b]	Sole skin gelatin and fish gelatin	Increases AOX activity, irrespective of the type of gelatin employed, to a level higher than films with tocopherol and BHT and produces minor changes in physicochemical properties	Gómez-Estaca et al. (2009b)
Grape seed[a]	Chitosan	Have strong scavenging activity	Moradi et al. (2012)
Acacia seed[a]	Galactomannan	Increases radical scavenging activity and the phenolic content of the film	Cerqueira et al. (2010)

Note: Physical and chemical properties.

[a] Aqueous extract or water-soluble compound.

[b] Ethanol/water extract or ethanol/water-soluble compound; AOX, antioxidant; TS, tensile strength (MPa); EB, elongation at break (%); WVP, water vapor permeability; CMC, sodium carboxymethyl cellulose; HPMC, hydroxypropyl methylcellulose; EVOH, ethylene vinyl alcohol copolymer with a 29% ethylene molar content.

of the film and a nitrogen stream on the other side that carries the transmitted oxygen gas to the analyzer. A colorimetric sensor, an infrared sensor, a gas chromatograph, or a dedicated oxygen analyzer may be used for monitoring (Ayranci and Tunc, 2003). The differences in the oxygen-barrier properties depend mainly on the chemical composition and the molecular structure of the polymers (Mokwena and Tang, 2012), as well as the relative humidity and temperature (Hong and Krochta, 2006; Bonilla et al., 2012). Films made from proteins and carbohydrates are excellent barriers to oxygen, because of their tightly packed, ordered hydrogen-bonded network structure (Yang and Paulson, 2000). The films providing a semipermeable barrier to oxygen/carbon dioxide could also prevent the leaching of vitamins during food washing (Shrestha et al., 2003). High temperature promotes gas transference across the film in an exponential way (Maté and Krochta, 1998).

Edible films and coatings can include antioxidant agents in their formulation, and at the same time, they represent a barrier to oxygen, which results in a better preservation and quality (Bonilla et al., 2012). Depending on the relative humidity, some active compounds can reduce the interactions between the hydroxyl groups of the polymer chains, thereby the hydrogen bonds, by attaching themselves to chains, consequently decreasing the oxygen permeability, due to a tortuous pathway for the pass of oxygen molecules through the amorphous zones in the matrix as can also be appreciated in Table 14.1 (Pastor et al., 2013) and Table 14.2 (Bifani et al., 2007). As relative humidity increases, more water molecules interact with the material, and the film becomes more plasticized. In these conditions, the mobility and the extensive mass transfer across the film are favored. Therefore, the antioxidant activity of edible films should always be tested under controlled relative humidity conditions. The increase in the film water content reduces the oxygen-barrier effect, but can enhance the chemical action of the antioxidants. For this reason, the moisture content of the foodstuff and the relative humidity in the ambient should be taken into account in order to develop effective films and coatings with antioxidant activity (Bonilla et al., 2012; Silva-Weiss et al., 2013b).

Therefore, the WVP of films mainly depends on its chemical structure and morphology (Li et al., 2014). Hydrophilic compounds could increase the WVP of films and water loss from food products (Rojas-Graü et al., 2007, 2009; López-de-Dicastillo, 2012). However, aqueous extracts of green tea or tea polyphenols (Table 14.1) (Li et al., 2014) of murta leaf ecotype SG (Table 14.2) (Bifani et al., 2007) have shown decreased film WVP. In the case of gelatin films with water (Table 14.1), the polyphenols of green tea aqueous extracts are responsible for forming hydrogen and covalent bonds with the polar groups of the biopolymer chains, limiting the availability to form hydrophilic bonding with water and leading to a decrease in the affinity to water (Curcio et al., 2009; Ubonrat and Bruce, 2010; Li et al., 2014). On the other hand, the hydrophobic molecule α-tocopherol (vitamin E) can reduce the WVP of calcium caseinate film (Mei and Zhao, 2003) and of fish skin gelatin (Jongjareonrak et al., 2008). In the last case, incorporating vitamin E into fish skin gelatin films, initially reduces the WVP by 38%, achieving a further reduction of approximately 11%–16% in 6 weeks. This reduction in the film's WVP could be caused by an increase in the cross-linking between protein chains and/or the protein and vitamin E during storage and the recrystallization of the gelatin in the matrix of the film (Arvanitoyannis et al., 1997), which produces a matrix, that is, less permeable and has a higher density.

14.4.3 Light-Barrier Properties and Appearance

Transparent packaging induces the oxidation and degradation of nutritional compounds, because light acts as a catalyst for these processes. Therefore, opaque packaging and packaging containing specific compounds that absorbs light in the UV–visible spectrum have been developed to prevent these reactions. The light-barrier property is related to the color and opacity of the edible films.

For films, good barrier to UV and visible light is indicated by low transmission in the range of 200–280 nm and 350–800 nm, respectively. Lipid oxidation is the main factor inducing yellow discoloration of fish muscle protein film exposed to oxygen, and the incorporation of antioxidants (Trolox and catechin) in films prepared at acidic pH was able to prevent yellow discoloration of the resulting film (Tongnuanchan et al., 2012) (Table 14.1).

For fish skin films with murta leaf extracts of ecotypes SC and SG (Table 14.2), the spectroscopic scanning (between 200 and 690 nm) of the films at the beginning of the visible spectrum, at about 400 nm, presented an absorbance level of approximately sixfold higher than the control gelatin film (with water instead of extract). The difference in the ultraviolet spectrum was even higher. This could mean that the tuna-fish skin gelatin films enriched with murta extracts can be an excellent barrier to prevent UV light–induced lipid oxidation, when applied in food systems (Gómez-Guillén et al., 2007). Similar results (Table 14.2) are found for fish skin and bovine gelatin films enriched with oregano and rosemary extracts, with improved light barrier properties and antioxidant activity, irrespective of the type of gelatin employed (Gómez-Estaca et al., 2009a).

Also, for fruits and vegetables, gloss is an expected sensorial characteristic of edible coatings, and the relation between the opacity and gloss factors must be evaluated *in vivo* on the product to obtain equilibrium between the sensory and nutritional qualities of the food.

Liposoluble vitamins such as vitamin E increase the transparency of films during storage, allowing the normal passage of light, which catalyzes the rancidity of fats. However, film gloss is not affected by the incorporation of vitamin E. In hydrocolloid matrices of fish gelatin, vitamin E interacts with proteins through bridging hydrogen links, causing an increase in the absorption of UV–visible light (Jongjareonrak et al., 2008).

Plant extracts are commonly used to provide color and opacity to polymers (Hong et al., 2000; Wang et al., 2012a,b), and as a result, films containing plant extracts are less transparent than films without plant extracts, as seen in Table 14.1 (Wang et al., 2013) and Table 14.2 (Gómez-Guillén et al., 2007).

Thus, the incorporation of plant extracts in films provides an adequate barrier to light, which is also important for preventing the degradation of ascorbic acid and, consequently, browning. These results confirm the beneficial effects of adding an antioxidant to the applied packaging.

14.4.4 Wettability and Contact Angle

The addition of natural plant extracts as functional compounds to the film-forming solution can change some surface properties of these films that affect an adequate adherence and thickness of the coating on the food surface (Ramírez et al., 2012). Thicker films can produce anaerobic conditions that would cause the deterioration of food, which is possible to avoid, with an adequate spread of the film-forming solution on the food. Wettability of the surface (Ws) by coating solutions is the main factor that influences the effective spreading, and the main parameter used to characterize the wetting of a surface is the equilibrium contact angle (Choi et al., 2002; Skurtys et al., 2011).

The W_S effect of a liquid on a solid is the function of the adhesive forces that cause the liquid to spread on the solid flat surface and the cohesive forces of the liquid that cause the shrink of the droplet. The balance of these forces can be negative or zero, and when the value is closer to 0, the surface is more wetting (Choi et al., 2002; Ramírez et al., 2012).

The method of contact angle measurement and determination of interfacial forces among vapor, liquid, and solid are well described (Zisman, 1964; de Gennes, 1985; Lee et al., 2008, 2009; Ramírez et al., 2012).

Ramírez et al. (2012) studied the contact angle and the W_S in carboxymethyl cellulose (CMC)-based, film-forming solution with the addition of murta (*U. molinae* Turcz) leaf extract applied on apple and quince skin. The addition of the extract reduces the contact angle formed between the solution and the surface measured at t = 0 s, both in apple and quince skin, but there are no differences at t = 10 s in the measured contact angles for the different droplets of film-forming solution deposited on the apples and quince skin. The critical surface tension obtained for apple (18.56 mN m^{-1}) and for quince (18.71 mN m^{-1}) is similar to those reported for Fuji apple skin (18.70 mN m^{-1}) (Choi et al., 2002) and for strawberry (18.80 mN m^{-1}) (Ribeiro et al., 2007). An interesting result was the increase in surface tension of the film-forming solution with the addition of the murta leaf extract, which implies an interaction between the CMC and the active extract (Ramírez et al., 2012).

14.4.5 Antioxidant Capacity of Films and Coatings

As pointed out by Huang et al. (2005), the antioxidant assays can roughly be classified into two types: assays based on hydrogen atom transfer (HAT) reactions and assays based on electron transfer (ET). The majority of HAT-based assays apply a competitive reaction scheme, in which antioxidant and substrate compete for thermally generated peroxyl radicals through the decomposition of azo compounds. These assays include inhibition of induced low-density lipoprotein autoxidation, oxygen radical absorbance capacity (ORAC), total radical trapping antioxidant parameter, and crocin bleaching assays. ET-based assays measure the capacity of an antioxidant in the reduction of an oxidant, which changes color when reduced. The degree of color change is correlated with the sample's antioxidant concentrations. ET-based assays include the total phenols assay by Folin–Ciocalteu reagent (FCR), Trolox equivalent antioxidant capacity (TEAC), ferric-reducing antioxidant power (FRAP), *total antioxidant potential* assay using a Cu(II) complex as an oxidant, and the synthetic radical 2,2-diphenyl-1-picryhydrazyl (DPPH). In addition, other assays intended to measure a sample's scavenging capacity of biologically relevant oxidants such as singlet oxygen, superoxide anion, peroxynitrite, and hydroxyl radical. On the basis of this analysis, it was suggested that the total phenols assay by FCR be used to quantify an antioxidant's reducing capacity and the ORAC assay to quantify peroxyl radical scavenging capacity. To comprehensively study different aspects of antioxidants, validated and specific assays are needed in addition to these two commonly accepted assays.

Antioxidant capacity in filmogenic solutions and dry films is now also reported. In films, it can be evaluated after dissolving a known part of the film in a known volume of solvent and knowing the time of solubilization of the film. Bonilla et al. (2012) refers to the antioxidant efficiency tests for edible films using different approaches. For instance, chitosan films can be dissolved in distilled water (Siripatrawan and Harte, 2010), whereas a more elaborated procedure (freezing, grinding, and extraction with methanol) was necessary for alginate films (Norajit et al., 2010). The methods used to measure antioxidant capacity in films are assays based on ET; radical scavenging, which uses DPPH (Siripatrawan and Harte, 2010); and ABTS 2,2′-azinobis (3-ethylbenzothiazoline-6-sulphonate), Folin-reactive substances (total phenolic compounds), and ferric-reducing antioxidant power (FRAP) (Gómez-Guillén et al., 2007; Gómez-Estaca et al., 2009a; Blanco-Pascual et al., 2014b; Cian et al., 2014). Another method used is thiobarbituric acid reactive substances (TBARSs) (Tongnuanchan et al., 2012).

It has been shown that proteins significantly decrease the antioxidant capacity (Ozdal et al., 2013). For example, tuna-fish gelatin shows some antioxidant capacity through the FRAP method, attributed to the amino acids glycine and proline from gelatin (Mendis et al., 2005).

As can be seen from Tables 14.3 through 14.5, there are controversies about the results obtained through different antioxidant capacity methods, which are mainly due to the differences in the principles or fundamentals of these methods. Furthermore, it becomes difficult to compare the antioxidant capacity, even measured by the same method, because many times the units of measurement or standards used in the calibration curves are different. When antioxidant capacity is measured in edible films, the time required for the solubility of films is very determining in the results obtained. Being conscientious of all these problems, the antioxidant capacity results of different authors were put together in Tables 14.3 through 14.5 as to have a general vision of the problem.

Unquestionable is the fact that films with antioxidant extracts incorporated show an increased antioxidant capacity. In Table 14.4, results on tuna-fish gelatin films with murta (*U. molinae* Turcz) leaf extract of two different ecotypes show that the antioxidant capacity of the film with the extract of the ecotype with higher concentration of polyphenols is higher than that of the film with the extract lower in polyphenols, being even much lower in the films without extract (Gómez-Guillén et al., 2007). In another example in Table 14.5, films of silver carp fish skin with different antioxidant tea extracts (green tea, grape seed, ginger, gingko) show an increase in antioxidant capacity of these films, but the addition of gingko to the gelatin-based film shows the highest DPPH radical scavenging activity (Li et al., 2014). In films made from red (Blanco-Pascual et al., 2014a; Cian et al., 2014) and brown (Blanco-Pascual et al., 2014b) seaweeds, with antioxidant components from the same seaweeds added, can be appreciated a considerable antioxidant capacity (Table 14.3).

The antioxidant activity of edible films and coatings is greatly influenced by the water availability, which in turn is affected by both the moisture of the product and the ambient relative humidity (Bonilla et al., 2012). According to Tongnuanchan et al. (2012), antioxidant activity of films after casting and drying without conditioned is less than films conditioned in an environmental chamber at 25°C and 50% ± 5% RH for 24 h. Furthermore, the amount of phenolics released from films differed considerably according to testing temperature. In active films prepared from chitosan and polyvinyl alcohol containing aqueous mint extract/pomegranate peel extract, at a higher temperature (37°C), the phenolics released from the film was maximum and very little phenolics were released from films kept at 15°C (Kanatt et al., 2012).

About the effects of antioxidant films on retardation of oil oxidation, Camo et al. (2008) hypothesized that the mechanism of action by which the antioxidant-active packaging lowers the number of molecules reactive to 2-thiobarbituric acid reactive substances (TBARS) could be (1) inactivation of free radicals by migration of antioxidant molecules from the active film to the food product or (2) scavenging of those oxidant molecules from the food product onto the active film. According to Tammineni et al. (2012), the rate of oil oxidation in an oiled model food was more dependent upon the inherent oxygen-barrier property of the films (based on gelatin) than the presence of antioxidants (epigallocatechin gallate and green tea).

In addition, the pH also affects strongly the antioxidant capacity of the films on lipid oxidation. Tongnuanchan et al. (2012) studied lipid peroxidation of films based on red tilapia mine at different pH values, with or without the addition of antioxidants (Trolox or catechin). The results show that in the film without an antioxidant, the TBARS are 90% reduced at pH 11, compared to the results at pH 3, while the incorporation of Trolox or catechin, in the range of 100–400 mg/L into film-forming solutions, was able to prevent oxidation of lipid film prepared at acidic pH.

TABLE 14.3

Antioxidant Capacity of the Water-Soluble Fraction of Films

Antioxidant	Film Based On	pH	Folin-Reactive Substances	ABTS	DPPH	FRAP	Reference
Red seaweed *Porphyra columbina*	Red seaweed *P. columbina*	7.0	6.99 ± 0.13[6]	256.3 ± 0.00[2]			Cian et al. (2014)
Red seaweed *Mastocarpus stellatus*	Red seaweeds *M. stellatus*	6.5	41.32 ± 0.73[1]	70.60 ± 0.47[1]		1.16 ± 0.04[1]	Blanco-Pascual et al. (2014a)
Brown seaweeds *Ascophyllum nodosum*	Brown seaweeds *A. nodosum*	10	44.02 ± 1.26[1]	20.36 ± 0.90[1]		4.11 ± 0.51[1]	Blanco-Pascual et al. (2014b)
Brown seaweeds *Laminaria digitata* extract	Brown seaweeds *L. digitata* extract	10	35.76 ± 0.67[1]	4.20 ± 0.51[1]		1.74 ± 0.21[1]	Blanco-Pascual et al. (2014b)
α-Tocopherol	Wheat, starch, and chitosan			96[2]			Bonilla et al. (2013)
Citric acid	Wheat, starch, and chitosan			177[2]			Bonilla et al. (2013)
Resveratrol	Chitosan	4.5			0.71–0.73[1]		Pastor et al. (2013)
Resveratrol	MC	6.5			0.87–0.95[1]		Pastor et al. (2013)
Phenolic compounds from sunflower protein concentrates(SFCs)	SFC	9.0	98.1 ± 2.1[2]	65.26 ± 2.16[2]		545.80 ± 18.60[2]	Salgado et al. (2012)
Quince seed mucilage	Quince seed mucilage				18.39[2]		Jouki et al. (2013)

Notes: Folin-reactive substances: [1] mg gallic acid equivalent/g film, [2] mg chlorogenic acid/g film, [3] g caffeic acid eq./mL, [4] μg eq. gallic acid/mL, [5] mg catechin equivalent/g film, [6] mg gallic acid equivalent/100 g F.W.; ABTS: [1] mg Vit C equivalent/g, [2] TEAC mg/g dry film, [3] TEAC mg/mL; DPPH: [1] EC_{50} (moles R/mole DPPH), [2] % DPPH scavenging, [3] g Vit C equivalent/100 g film, [4] mmol/g F.W., [5] mg Trolox/g film ó F.W.; FRAP: [1] μmol Fe^{2+}/g film, [2] mmol Fe^{2+}/mL or g film; EC_{50}: the amount of antioxidant needed to reduce the initial DPPH concentration to 50%, TEAC: Trolox equivalent antioxidant capacity; EVOH: ethylene vinyl alcohol copolymer with a 29% ethylene molar content; PVA: polyvinyl alcohol, HPMC: hydroxypropyl methylcellulose; MC: methylcellulose.

TABLE 14.4

Antioxidant Capacity of the Water-Soluble Fraction of Films

Antioxidant	Film Based On	pH	Folin-Reactive Substances	ABTS	DPPH	FRAP	Reference
Mint extract	Chitosan–PVA		~5–14[5]		70–75[2]		Kanatt et al. (2012)
Pomegranate peel extract	Chitosan–PVA		~3–10[5]		10–15[2]		Kanatt et al. (2012)
Ascorbic acid	EVOH				4.06 ± 0.52[3]		Lopez-de-Dicastillo et al. (2012)
Ferulic acid	EVOH				2.18 ± 0.06[3]		Lopez-de-Dicastillo et al. (2012)
Quercetin	EVOH				4.58 ± 0.10[3]		Lopez-de-Dicastillo et al. (2012)
Green tea extract	EVOH				4.76 ± 0.11[3]		Lopez-de-Dicastillo et al. (2012)
Borage extract	Sole skin and fish gelatin		1542[3]				Gómez-Estaca et al. (2009b)
Murta leaf SC extract	Fish skin gelatin		283.2[4]		22.14 ± 1.14[5]	943.50 ± 44.71[1]	Gómez-Guillén et al. (2007)
Murta leaf SG extract	Fish skin gelatin		224.2[4]		10.35 ± 0.40[5]	667.387 ± 70.42[1]	Gómez-Guillén et al. (2007)
Tea polyphenols	Chitosan		~170[1]		~60[2]		Wang et al. (2013)

Notes: Folin-reactive substances: [1] mg gallic acid equivalent/g film, [2] mg chlorogenic acid/g film, [3] g caffeic acid eq./mL, [4] µg eq. gallic acid/mL, [5] mg catechin equivalent/g film, [6] mg gallic acid equivalent/100 g F.W.; ABTS: [1] mg Vit C equivalent/g, [2] TEAC mg/g dry film, [3] TEAC mg/mL; DPPH: [1] EC_{50} (moles R/mole DPPH), [2] % DPPH scavenging, [3] g Vit C equivalent/100 g film, [4] mmol/g F.W., [5] mg Trolox/g film ó F.W.; FRAP: [1] µmol Fe^{2+}/g film, [2] mmol Fe^{2+}/mL or g film; EC_{50}: the amount of antioxidant needed to reduce the initial DPPH concentration to 50%; TEAC: Trolox equivalent antioxidant capacity; EVOH: ethylene vinyl alcohol copolymer with a 29% ethylene molar content; PVA: polyvinyl alcohol; HPMC: hydroxypropyl methylcellulose; MC: methylcellulose.

TABLE 14.5

Antioxidant Capacity of the Water-Soluble Fraction of Films

Antioxidant	Film Based On	pH	Folin-Reactive Substances	ABTS	DPPH	FRAP	Reference
Green tea extract	Agar–gelatin		~60[1]	~90[1]		~1100[1]	Giménez et al. (2013)
Green tea extract	Agar		~80[1]	~110[1]		~1600[1]	Giménez et al. (2013)
Green tea extract	Fish skin gelatin				~95[2]		Li et al. (2014)
Gingko leaf extract	Fish skin gelatin				~95[2]		Li et al. (2014)
Ginger extract	Fish skin gelatin				~18[2]		Li et al. (2014)
Grape seed extract	Fish skin gelatin				~90[2]		Li et al. (2014)
None	Chitosan		70.5[6]		20.0[4]		Wang and Gao (2013)
Carvacrol and methyl cinnamate	Strawberry		~150[6]		~5[5]		Peretto et al. (2014)

Notes: Folin-reactive substances: [1] mg gallic acid equivalent/g film, [2] mg chlorogenic acid/g film, [3] g caffeic acid eq./mL, [4] µg eq. gallic acid/mL, [5] mg catechin equivalent/g film, [6] mg gallic acid equivalent/100 g F.W.; ABTS: [1] mg Vit C equivalent/g, [2] TEAC mg/g dry film, [3] TEAC mg/mL; DPPH: [1] EC_{50} (moles R/mole DPPH), [2] % DPPH scavenging, [3] g Vit C equivalent/100 g film, [4] mmol/g F.W., [5] mg Trolox/g film ó F.W.; FRAP: [1] µmol Fe^{2+}/g film, [2] mmol Fe^{2+}/mL or g film; EC_{50}: the amount of antioxidant needed to reduce the initial DPPH concentration to 50%; TEAC: Trolox equivalent antioxidant capacity; EVOH: ethylene vinyl alcohol copolymer with a 29% ethylene molar content; PVA: polyvinyl alcohol; HPMC: hydroxypropyl methylcellulose; MC: methylcellulose.

REFERENCES

Akhtar, M.J., Jacquot, M., Jasniewski, J., Jacquot, C., Imran, M., Jamshidian, M., Paris, C., and Desobry, S. (2012). Antioxidant capacity and light-aging study of HPMC films functionalized with natural plant extract. *Carbohydrate Polymers*, 89, 1150–1158.

Alkan, D. and Yemenicioğlu, A. (2016). Potential application of natural phenolic antimicrobials and edible film technology against bacterial plant pathogens. *Food Hydrocolloids*, 55, 1–10.

Amiot, M.J., Tacchini, M., Aubert, S., and Nicolas, J. (1992). Phenolic composition and browning susceptibility of various apple cultivars at maturity. *Journal of Food Science*, 57(4), 958–962.

Arvanitoyannis, I., Psomiadou, E., Nakayama, A., Aiba, S., and Yamamoto, N. (1997). Edible film made from gelatine, soluble starch and polios. *Food Chemistry*, 60, 593–604.

Ayranci, E. and Tunc, S. (2003). A method for the measurement of the oxygen permeability and the development of edible films to reduce the rate of oxidative reactions in fresh foods. *Food Chemistry*, 80(3), 423–431.

Ayranci, E. and Tunc, S. (2004). The effect of edible coatings on water and vitamin C loss of apricots (*Armenia vulgarism* Lam.) and green peppers (*Capsicum annul L.*). *Food Chemistry*, 87(3), 339–349.

Bifani, V., Ramírez, C., Ihl, M., Rubilar, M., García, A., and Zaritzky, M. (2007). Effects of murta (*Ugni molinae* Turcz) extract on gas and water vapor permeability of carboxymethylcellulose-based edible films. *LWT: Food Science and Technology*, 40(8), 1473–1481.

Blanco-Pascual, N., Gómez-Guillén, M.C., and Montero, M.P. (2014a). Integral *Mastocarpus stellatus* use for antioxidant edible development. *Food Hydrocolloids*, 40, 128–137.

Blanco-Pascual, N., Montero, M.P., and Gómez-Guillén, M.C. (2014b). Antioxidant film development from unrefined extracts of brown seaweeds *Laminaria digitata* and *Ascophyllum nodosum*. *Food Hydrocolloids*, 37, 100–110.

Bonilla, J., Atarés, L., Vargas, M., and Chiralt, A. (2012). Edible films and coatings to prevent the detrimental effect of oxygen on food quality: Possibilities and limitations. *Journal of Food Engineering*, 110(2), 208–213.

Boots, A.W., Haenen, G.R.M.M., and Bast, A. (2008). Review health effects of quercetin: From antioxidant to nutraceutical. *European Journal of Pharmacology*, 585(2–3), 325–337.

Bustos, C.R.O., Albert, R.F.V., and Matiasevich, S.B. (2016). Edible antimicrobial films based on microencapsulated lemograss oil. *Journal of Food Science and Technology*, 53(1), 832–839.

Caillet, S., Yu, H., Lessard, S., Lamoureux, G., Ajdukovic, D., and Lacroix, M. (2007). Fenton reaction applied for screening natural antioxidants. *Food Chemistry*, 100(2), 542–552.

Camo, J., Beltrán, J.A., and Roncalés, P. (2008). Extension of the display life of lamb with an antioxidant active packaging. *Meat Science*, 80, 1086–1091.

Cardoso, G.P., Dutra, M.P., Fontes, P.R., Ramos Ade, L., Gomide, L.A., and Ramos, E.M. (2016). Selection of chitosan-gelatine-based edible coating for color preservation of beef in retail display. *Meat Science*, 111, 85–94.

Choi, W.Y., Park, H.J., Ahn, D.J., Lee, J., and Lee, C.Y. (2002). Wettability of chitosan coating solution on 'Fuji' apple skin. *Journal of Food Science*, 67(7), 2668–2672.

Cian, R.E., Salgado, P.R., Drago, S.R., González, R.J., and Mauri, A.N. (2014). Development of naturally activated edible films with antioxidant properties from red seaweed *Porphyra columbina* biopolymers. *Food Chemistry*, 146, 6–14.

Curcio, M., Puoci, F., Iemma, F., Parisi, O.I., Cirillo, G., and Spizzirri, U.G. (2009). Covalent insertion of antioxidant molecules on chitosan by a free radical grafting procedure. *Journal of Agricultural and Food Chemistry*, 57, 5933–5938.

de Gennes, P.G. (1985). Wetting: Statics and dynamics. *Reviews of Modern Physics*, 57(3), 827–863.

FAO/WHO. (2005). Food Standards. Codex general standard for food additives. Available online at: http://www.codexalimentarius.net/download/standards/4/CXS192e.pdf. (Accessed May 3, 2008).

Frankel, E.N., Huang, S.-H., Aeschbach, R., and Prior, E. (1996). Antioxidant Activity of a rosemary extract and its constituents, carnosic acid, carnosol, and rosmarinic acid, in bulk oil and oil-in-water emulsion. *Journal of Agricultural and Food Chemistry*, 44(1), 131–135.

Georgantelis, D., Blekas, G., Katikou, P., Ambrosiadis, I., and Fletouris, D.J. (2007). Effect of rosemary extract, chitosan and α-tocopherol on lipid oxidation and colour stability during frozen storage of beef burgers. *Meat Science*, 75(2), 256–264.

Giménez, B., Gómez-Guillén, M.C., Pérez-Mateos, M., Montero, P., and Márquez-Ruiz, G. (2011). Evaluation of lipid oxidation in horse mackerel patties covered with borage-containing film during frozen storage. *Food Chemistry*, 124(4), 1393–1403.

Giménez, B., López de Lacey, A., Pérez-Santín, E., López-Caballero, M.E., and Montero, P. (2013). Release of active compounds from agar and agar–gelatin films with green tea extract. *Food Hydrocolloids*, 30(1), 264–271.

Gómez-Estaca, J., Bravo, L., Gómez-Guillén, M.C., Alemán, A., and Montero, P. (2009a). Antioxidant properties of tuna-skin and bovine-hide gelatin films induced by the addition of origanum and rosemary extracts. *Food Chemistry*, 112(1), 18–25.

Gómez-Estaca, J., Giménez, B., Montero, P., and Gómez-Guillén, M.C. (2009b). Incorporation of antioxidant borage extract into edible films based on sole skin gelatin or a commercial fish gelatin. *Journal of Food Engineering*, 92(1), 78–85.

Gómez-Guillén, C., Ihl, M., Bifani, V., Silva, A., and Montero, P. (2007). Edible films made from Tuna-fish gelatin with antioxidant extracts of two different murta ecotypes leaves (*Ugni molinae* Turcz). *Food Hydrocolloids*, 21(7), 1130–1143.

Gontard, N., Guilbert, S., and Cuq, J.L. (1992). Edible wheat gluten films: Influence of the main process variables on film properties using response surface methodology. *Journal of Food Science*, 57,190–199.

Gontard, N., Guilbert, S., and Cuq, J.L. (1993). Water and glycerol as plasticizer affect mechanical and water vapor barrier properties of an edible wheat gluten film. *Journal of Food Science*, 58(1), 206–211.

Gorny, J.R. (1997). Summary of CA and MA requirements and recommendations for fresh-cut fruit and vegetables. In: J. Gorny (ed.), *Proceedings: Fresh Cut Fruit and Vegetables and MAP*, Vol. 5, pp. 30–33. Davis, CA: University of California.

Hinneburg, I., Dorman, D., and Hiltunen, R. (2006). Antioxidant activities of extracts from selected culinary herbs and spices. *Food Chemistry*, 97(1), 122–129.

Hong, S., Park, J., and Kim, D. (2000). Antimicrobial and physical properties of food packaging films incorporated with some natural compounds. *Food Science and Biotechnology*, 9(1), 38–42.

Hong, S.I. and Krochta, J.M. (2006). Oxygen barrier performance of whey-protein-coated plastic films as affected by temperature, relative humidity, base film and protein type. *Journal of Food Engineering*, 77, 739–745.

Huang, D., Ou, B., and Prior, R.L. (2005). The Chemistry behind antioxidant capacity assays. *Journal of Agricultural Food Chemistry*, 53(6), 1841–1856.

Institute of Food Technologists (IFT). (2010). Feeding the world today and tomorrow: The importance of food science and technology. *Comprehensive Reviews in Food Science and Food Safety*, 9(5), 572–599.

Jongjareonrak, A., Benjakul, S., Visessanguan, W., and Tanaka, M. (2008). Antioxidative activity and properties of fish skin gelatin films incorporated with BHT and α-tocopherol. *Food Hydrocolloids*, 22(3), 449–458.

Jouki, M., Yazdi, F.T., Mortazavi, S.A., and Koocheki, A. (2013). Physical, barrier and antioxidant properties of a novel plasticized edible film from quince seed mucilage. *International Journal of Biological Molecules*, 62, 500–507.

Kanatt, S., Rao, M.S., Chawla, S.P., and Sharma, A. (2012). Active chitosan-polyvinyl alcohol films with natural extracts. *Food Hydrocolloids*, 29, 290–297.

Karbowiak, T., Debeaufort, F., Voilley, A., and Trystram, G. (2009). From macroscopic to molecular scale investigations of mass transfer of small molecules through edible packaging applied at interfaces of multiphase food products. *Innovative Food Science and Emerging Technologies*, 10(1), 116–127.

Kerch, G. (2015). Chitosan films and coatings prevent losses of fresh fruits nutritional quality. A review. *Trend in Foos Science and Technology*, 46(2), 159–166.

Kim, M., Kim, C.Y., and Park, I. (2005). Prevention of enzymatic browning of pear by onion extract. *Food Chemistry*, 89(2), 181–184.

Krochta, J.M. and De Mulder-Johnston, C. (1997). Edible and biodegradable polymer films challenges and opportunities. *Food Technology*, 51(2), 61–74.

Kubo, I., Chen, Q.-X., and Nihei, K.-I. (2003). Molecular design of antibrowning agents: Antioxidative tyrosinase inhibitors. *Food Chemistry*, 81(2), 241–247.

Lasagni Vitar, R.M., Reides, C.G., Ferreira, S.M., and Llesuy, S.F. (2014). The protective effect of *A. triphylla* extracts against brain lipid—Peroxidation. *Food & Function*, 5(3), 557–563.

Lee, B.B., Pogaku, R., and Chan, E.S. (2008). A critical review: Surface and interfacial tension measurement by the drop weight method. *Chemical Engineering Communications*, 195(8), 889–924.

Lee, B.B., Ravindra, P., and Chan, E.S. (2009). New drop weight analysis for surface tension determination of liquids. *Colloids and Surfaces A: Physicochemical and Engineering Aspects*, 332(2–3), 112–120.

León, P.G., Lamanna, M.E., Gerschenson, L.N., and Rojas, A.M. (2008). Influence of composition of edible films based on gellan polymers on L-(+)-ascorbic acid stability. *Food Research International*, 41(6), 667–675.

Li, J.-H., Miao, J., Wu, J.-L., Chen, S.-F., and Zhang, Q.-Q. (2014). Preparation and characterization of active gelatin-based films incorporated with natural antioxidants. *Food Hydrocolloids*, 37, 166–173.

Li, T., Hu, W., Li, J., Zhang, X., Zhu, J., and Li, X. (2012). Coating effects of tea polyphenols and rosemary extract combined with chitosan on the storage quality of large yellow croaker (*Pseudosciaena crocea*). *Food Control*, 25(1), 101–106.

López-de-Dicastillo, C., Gómez-Estaca, J., Catalá, R., Gavara, R., and Hernández-Muñoz, P. (2012). Active antioxidant packaging films: Development and effect on lipid stability of brined sardines. *Food Chemistry*, 131(4), 1376–1384.

Mali, S., Grossmann, M.V., García, M., Martino, M.N., and Zaritzky, N.E. (2006). Effects of controlled storage on thermal, mechanical and barrier properties of plasticized films from different starch sources. *Journal of Food Engineering*, 75(4), 453–460.

Maté, J.I. and Krochta, J.M. (1998). Oxygen uptake model for uncoated and coated peanuts. *Journal of Food Engineering*, 35, 299–312.

Mathew, S. and Abraham, T.E. (2008). Characterisation of ferulic acid incorporated starch-chitosan blend films. *Food Hydrocolloids*, 22(5), 826–835.

Mayachiew, P. and Devahastin, S. (2010). Effects of drying methods and conditions on release characteristics of edible chitosan films enriched with Indian gooseberry extract. *Food Chemistry*, 118(1), 594–601.

McHugh, T.H. and Krochta, J.M. (1994). Sorbitol vs glycerol-plasticized whey protein edible films. Integrated oxygen permeability and tensile property evaluation. *Journal of Agricultural and Food Chemistry*, 42(4), 841–845.

Mei, Y. and Zhao, Y. (2003). Barrier and mechanical properties of milk protein-based edible films containing nutraceuticals. *Journal of Agricultural and Food Chemistry*, 51(7), 1914–1918.

Mendis, E., Rajapakse, N., and Kim, S.K. (2005). Antioxidant properties of a radical-scavenging peptide purified from enzymatically prepared fish skin gelatin hydrolysate. *Journal of Agricultural and Food Chemistry*, 53(3), 581–587.

Mokwena, K. and Tang, J. (2012). Ethylene vinyl alcohol: A review of barrier properties for packaging shelf stable foods. *Critical Reviews in Food Science and Nutrition*, 52, 640–650.

Moradi, M., Tajik, H., Razavi Rohani, S.M., Oromiehie, A.R., Malekinejad, H., Aliakbarlu, J., and Hadian, M. (2012). Characterization of antioxidant chitosan film incorporated with *Zataria multiflora* Boiss essential oil and grape seed extract. *LWT: Food Science and Technology*, 46(2), 477–484.

Navarro, R., Arancibia, C., Herrera, M.L., and Matiacevich, S. (2016). Effect of type of encapsulating agent on physical properties of edible films based on alginate and thyme oil. *Food and Bioproducts Processing*, 97, 63–75.

Norajit, K., Kim, K.M., and Ryu, G.H. (2010). Comparative studies on the characterization and antioxidant properties of biodegradable alginate films containing ginseng extract. *Journal of Food Engineering*, 98(3), 377–384.

Ozdal, T., Capanoglu, E., and Altay, F. (2013). A review on protein–phenolic interactions and associated changes. *Food Research International*, 51, 954–970.

Özvural E.V., Huang, Q., and Chikindas, M.L. (2016). The comparison of quality and microbiological characteristic of hambuerger patties enrichedwith green tea extract using three techniques: Direct addition, and encapsulation. *LWT: Food Science and Technology*, 68, 385–390.

Pastor, C., Sánchez-González, L., Chiralt, A., Cháfer, M., and González-Martínez, C. (2013). Physical and antioxidant properties of chitosan and methylcellulose based films containing resveratrol. *Food Hydrocolloids*, 30, 272–280.

Peretto, G., Dub, W.-X., Avena-Bustillos, R.J., Sarreal, S.B., Huab, S.S., Sambo, P., and McHugh, T.H. (2014). Increasing strawberry shelf-life with carvacrol and methyl cinnamate antimicrobial vapors released from edible films. *Postharvest Biology and Technology*, 89, 11–18.

Perumalla, A.V.S. and Hettiarachchy, N.S. (2011). Review: Green tea and grape seed extracts—Potential applications in food safety and quality. *Food Research International*, 44(4), 827–839.

Pokorný, J. (2007). Antioxidant in food preservation. In: S. Rahman (ed.), *Handbook of Food Preservation*, Chapter 10, pp. 309–335. Boca Raton, FL: CRC Press.

Portes, E., Gardrat, C., Castellan, A., and Coma, V. (2009). Environmentally friendly films based on chitosan and tetrahydrocurcuminoid. *Carbohydrate Polymers*, 76(4), 578–584.

Ramírez, C., Gallegos, I., Ihl, M., and Bifani, V. (2012). Study of contact angle, wettability and water vapor permeability in carboxymethylcellulose (CMC) based film with murta leaves (*Ugni molinae* Turcz) extract. *Journal of Food Engineering*, 109, 424–429.

Ribeiro, C., Vicente, A.A., Teixeira, J.A., and Miranda, C. (2007). Optimization of edible coating composition to retard strawberry fruit senescence. *Postharvest Biology and Technology*, 44(1), 63–70.

Rivero, S., García, M.A., and Pinotti, A. (2010). Crosslinking capacity of tannic acid in plasticized chitosan films. *Carbohydrate Polymers*, 82(2), 270–276.

Rojas-Graü, M.A., Soliva-Fortuny, R., and Martín-Belloso, O. (2009). Edible coatings to incorporate active ingredients to fresh-cut fruits: A review. *Trends in Food Science & Technology*, 20(10), 438–447.

Rojas-Graü, M.A., Tapia, M.S., Rodríguez, F.J., Carmona, A.J., and Martin-Belloso, O. (2007). Alginate and gellan-based edible coating as carriers of antibrowning agents applied on fresh-cut Fuji apples. *Food Hydrocolloids*, 21(1), 118–127.

Roldán, E., Sánchez-Moreno, C., De Ancos, B., and Cano, M.P. (2008). Characterization of onion (*Allium cepa* L.) by-products as food ingredients with antioxidant and antibrowning properties. *Food Chemistry*, 108(3), 907–916.

Sabaghi, M., Maghsoudlou, Y., Khomeiri M., and Ziaiifar, A.M. (2015). Active edible coating from chitosan incorporating green tea extract as an antioxidant and antifungal on fresh walnut kernel. *Postharvest Biology and Technology*, 110: 224–228.

Salgado, P.R., Ortiz, C.M., Mussi, Y.S., Di Giorgio, L., and Mauri, A.N. (2015). Edible films and coatings containing bioactives. *Current Opinion in Food Science*, 51, 86–92.

Sarma, A., Sreelakshmi, Y., and Sharma, R. (1997). Antioxidant ability of anthocyanins against ascorbic acid oxidation. *Phytochemistry*, 45(4), 671–674.

Shrestha, A.K., Arcot, J., and Paterson, J.L. (2003). Edible coating materials-their properties and use in the fortification of rice with folic acid. *Food Research International*, 36 (9–10), 921–928.

Silva-Weiss, A., Bifani, V., Ihl, M., Sobral, P.J.A., and Gómez-Guillén, M.C. (2013a). Structural properties of films and rheology of film-forming solutions based on chitosan and chitosan-starch blend enriched with murta leaf extract. *Food Hydrocolloids*, 31(2), 458–466.

Silva-Weiss, A., Ihl, M., Sobral, P.J.A., Gómez-Guillén, M.C., and Bifani, V. (2013b). Natural additives in bioactive edible films and coatings: Functionality and applications in foods. *Food Engineering Reviews*, 5, 200–216.

Siripatrawan, U. and Harte, B.R. (2010). Physical properties and antioxidant activity of an active film from chitosan incorporated with green tea extract. *Food Hydrocolloids*, 24(8), 770–775.

Siripatrawan, U. and Noipha, S. (2012). Active film from chitosan incorporating green tea extract for shelf life extension of pork sausages. *Food Hydrocolloids*, 27(1), 102–108.

Skurtys, O., Velásquez, P., Henriquez, O., Matiacevich, S., Enrione, J., and Osorio, F. (2011). Wetting behavior of chitosan solutions on blueberry epicarp with or without epicuticular waxes. *LWT: Food Science and Technology*, 44(6), 1449–1457.

Tammineni, N., Unlu, G., Rasco, B., Powers, J., Sablani, S., and Nind, C. (2012). Trout-skin gelatin-based edible films containing phenolic antioxidants: Effect on physical properties and oxidative stability of cod-liver oil model food. *Journal of Food Science*, 77(11), E342–E347.

Tongnuanchan, P., Benjakul, S., and Prodpran, T. (2012). Effects of oxygen and antioxidants on the lipid oxidation and yellow discolouration of film from red tilapia mince. *Journal of the Science of Food and Agriculture*, 92, 2507–2517.

Ubonrat, S. and Bruce, R. (2010). Physical properties and antioxidant activity of an active film from chitosan incorporated with green tea extract. *Food Hydrocolloids*, 24, 770–775.

Wambura, P., Yang, W., and Mwakatage, N.R. (2011). Effects of sonication and edible coating containing rosemary and tea extracts on reduction of peanut lipid oxidative rancidity. *Food and Bioprocess Technology*, 4, 107–115.

Wang, J., Wang, B., Jiang, W., and Zhao, Y. (2007). Quality and shelf life of mango (*Mangifera indica* L. cv. 'Tainong') coated by using chitosan and polyphenols. *Food Science and Technology International*, 13(4), 317–322.

Wang, L., Dong, Y., Men, H., Tong, J., and Zhou, J. (2012a). Preparation and characterization of active films based on chitosan incorporated tea polyphenols. *Food Hydrocolloids*, 32(1), 35–41.

Wang, L., Dong, Y., Men, H., Tong, J., and Zhou, J. (2013). Preparation and characterization of active films based on chitosan incorporated tea polyphenols. *Food Hydrocolloids*, 32, 35–41.

Wang, S., Marcone, M.F., Barbut, S., and Lim, L.-T. (2012b). Review: Fortification of dietary biopolymers-based packaging material with bioactive plant extracts. *Food Research International*, 49(1), 80–91.

Wang, S.Y. and Gao, H. (2013). Effect of chitosan-based edible coating on antioxidants, antioxidant enzyme system, and postharvest fruit quality of strawberries (*Fragaria x ananassa* Duch.). *LWT: Food Science and Technology*, 52(2), 71–79.

Yang, L. and Paulson, A.T. (2000). Effects of lipids on mechanical and moisture barrier properties of edible gellan film. *Food Research International*, 33, 571–578.

Yousaf, S., Butt, M.S., Suleria, H.A.R., and Iqbal, M.J. (2014). The role of green tea extract and powder in mitigating metabolic syndrome with special reference to hypoglycemia and hypercholesterolemia. *Food & Function*, 5(3), 545–556.

Zhang, L., Li, S., Dong, Y., Zhi, H., and Zong, W. (2016). Tea polyphenolsincorporated to alginate-based edible coating for quality maintenances of Chinese winter jujube under ambient temperature. *LWT: Food Science and Technology*, 70, 155–161.

Zaritzky, N. (2011). Edible coating to improve food quality and safety. In: J.M. Aguilera, G.V. Barbosa-Cánovas, R. Simpson, J. Welti-Chanes, and D. Bermúdez-Aguirre (eds.), *Food Engineering Interfaces*, Chapter 27. Food Engineering Series. New York: Springer.

Zisman, W.A. (1964). Contact angle wettability and adhesion. In: F. Fowkes (ed.), *Advances in Chemistry*, pp. 1–51. Washington, DC: American Chemical Society.

15

Antimicrobial Edible Films and Coatings

Ximena Carrión-Granda, Idoya Fernández-Pan, and Juan Ignacio Maté

CONTENTS

15.1 Introduction

Edible films and coatings (EFCs) based on polysaccharides, proteins, or lipids can be applied to most foodstuffs in order to increase the shelf life and to enhance food quality, stability, and safety. Thus, EFCs are used on fresh and processed food products to delay moisture loss, to reduce lipid oxidation and decoloration, and to improve the appearance of the product. In addition, these matrices can be used as carriers of antimicrobials to improve the safety and extend the shelf life of food systems (Alvarez et al., 2013; Gennadios et al., 1997; Gómez-Estaca et al., 2009; Quintavalla and Vicini, 2002).

Foods are ecosystems susceptible to the colonization and the development of a large number and variety of altering microorganisms and pathogens that are then transmitted through the food to consumers. Important ones are *Listeria monocytogenes*, *Salmonella enteritidis*, *Aeromonas hydrophila*, and *Escherichia coli* O157:H7 among others. Their presence can cause major food safety problems affecting consumers, industries, and the entire economy. Thus, it is essential to implement appropriate preservation technologies since the prevalence of foodborne pathogens and the numbers of outbreaks annually detected worldwide are high. Besides these important safety problems, the growth of altering microorganisms on food products is the responsible for food spoilage. Food spoilage is understood as the process by which food deteriorates to the point it is considered nonedible or its quality is reduced, making it undesirable or unsuitable for sale or consumption. This is usually the result of the metabolic activity of a variety of microorganisms (bacteria, yeasts, molds, and fungi) belonging to the genera *Pseudomonas*, *Acinetobacter*, *Micrococcus*, *Moraxella*, *Shewanella*, lactic acid bacteria (LAB), Enterobacteriaceae, *Alternaria*, *Fusarium*, *Aspergillus*, among others, causing changes in taste, odor, and appearance of such products, which limit their quality and shelf life leading to substantial economic losses in the industry (Arvanitoyannis and Stratakos, 2012; Ray, 2005; Sun and Holley, 2012). This fact has made the food industry search for innovative preservation technologies to extend the shelf life while maintaining nutritional quality and ensuring food safety. Thus, prevention of spoilage has become an important challenge for food industry, and over the last few years, there has been a strong movement toward the development of EFCs that incorporate antimicrobial agents to improve the preservation of food products.

The microbiota that spoils food depends largely on the intrinsic properties of the product (free moisture, pH, protein content), contamination during first handling (slaughter, harvesting, washing), number of bacteria initially present, and their ability to grow, as well as extrinsic factors including packaging and the environment in which food products are stored (T, water activity, atmosphere composition). As a consequence, different preservatives are currently used (in combination with other technologies) in the food industry in order to delay its development. Antimicrobials, defined as those substances that affect the development of bacteria or fungi (by delaying their growth, known as bacteriostatics, or by killing them, known as bactericides), are commonly employed both in food preservation to control natural spoilage processes and in food safety to prevent/control the growth of pathogens. One of the traditional ways employed to apply these antimicrobial agents on the food surfaces is by immersion or by pulverization. However, in these particular applications, the effectiveness of antimicrobials is limited during storage periods because of two main factors: uncontrolled migration into the food and partial inactivation because of interaction with other food components.

As explained in previous chapters, EFCs have a high potential to carry active and functional compounds such as antimicrobials. As such, antimicrobial EFCs are promising emerging technologies because their effectiveness is based on the controlled release of the antimicrobials retained in the structural matrix and have been proposed as a novel way to improve the safety and extend the shelf life of food systems by optimizing and localizing additive doses. The antimicrobials incorporated into the EF or coating migrate selectively and gradually from the matrix to the food surface, with one main objective throughout storage: maintaining effective concentrations of the agents on the food surface when they are needed and over a prolonged exposure.

15.2 Basic Formulation of Antimicrobial Edible Films and Coatings

There are two different ways of developing antimicrobial EFCs: to use a biopolymer with inherent antimicrobial activity, for example, chitosan, or, as it is the case in this chapter, to employ the structural matrix as carrier of antimicrobials. The basic formulation of antimicrobial EFCs is composed of at least one component able to form a structural matrix, one antimicrobial compound to confer or enhance its functionality, and, if necessary, other technological additives. The main materials used to develop EFCs are hydrocolloids (proteins or carbohydrates) and lipids (waxes, triglycerides, fatty acids, and resins). Owing to their nonpolymeric nature, lipids do not generally form cohesive films. As a consequence, hydrocolloids are the main materials preferred for developing antimicrobial EFCs, as it can be seen in Table 15.1. The physicochemical characteristics of the hydrocolloid selected for making films and coatings definitely determine their final properties, and both proteins (i.e., gelatin, corn gluten, wheat gluten, soy protein, casein, and whey proteins) and carbohydrates (i.e., cellulose derivatives, starches, pectins, and gums) can be considered good film formers, with very good barrier properties for oxygen, odors, and lipids (at low RH). However, they are generally poor barriers for moisture (Fernández-Pan and Maté, 2011; Han and Gennadios, 2005). With the objective of improving the overall characteristics of an EFC, composite films based on the combination of hydrocolloids and lipids have been developed. In this sense, lipids provide water vapor resistance and the hydrocolloids provide the structural stability and cohesion (Avena-Bustillos et al., 1994; Park et al., 1994; Pérez-Gago and Krochta, 2005).

In addition, to improve the film's technological properties, additives such as plasticizers and surfactants are included in the formulation of antimicrobial EFCs. Plasticizers, which lend flexibility and manageability, are currently used to modify mechanical properties. In order to form adequate coatings over extremely different types of surface areas of food systems, surfactants are used to improve the stability of emulsions and to improve the wettability of solutions. In this way, it is important to emphasize that all film components, including any functional additives, should be food grade or generally recognized as safe (GRAS) and be used within any limitations specified by the particular legal system.

The functional effectiveness of the formulated antimicrobial EFCs depends largely on (1) the nature and type of hydrocolloid used in the structural matrix, (2) the selected antimicrobial agent, (3) the physicochemical characteristics of the food product to be protected, and (4) all their interactions. Many types of hydrocolloids with specific physicochemical properties, such as whey protein isolate (WPI)

TABLE 15.1

Examples of Effective Antimicrobial Edible Films Containing Essential Oils, Organic Acids, Nisin, and Other Minor Antimicrobials

Antimicrobial	Matrix	Target	Method	Effect	Reference
Citronella, tarragon, thyme, coriander (0.25 mL/g protein)	Hake protein	*L. monocytogenes, L. innocua, B. thermosphacta, E. coli, P. putida, S. typhimurium, Shewanella putrefaciens*	Agar diffusion	Films incorporated with citronella and thyme inhibited *L. monocytogenes.* Films with tarragon and thyme oil also inhibited *L. innocua.*	Pires et al. (2013)
Oregano, clove, tea tree, coriander, mastic thyme, laurel, rosemary, sage (0.5–9 w/w)	WPI	*L. innocua, S. aureus, S. enteritidis, P. fragi*	Agar diffusion	Films with oregano or clove EO resulted effective against all four strains tested.	Fernández-Pan et al. (2012)
Tea tree (0.5%–2% w/w)	Chitosan	*L. monocytogenes*	Plate count	Up to 5 days of total inhibition.	Sánchez-González et al. (2010)
Oregano (1%, 2%, 4% w/w)	WPI	*L. innocua, S. aureus, S. enteritidis*	Agar diffusion	Effectiveness against all three microorganisms. The highest inhibition zones were against *L. innocua.*	Royo et al. (2010)
Oregano or thyme (1%–5%)	SPI	*E. coli, E. coli* O157:H7, *S. aureus, Pseudomonas aeruginosa, Lactobacillus plantarum*	Agar diffusion	Effectiveness against the five strains. *L. plantarum* and *P. aeruginosa* resulted more resistant bacteria.	Emiroglu et al. (2010)
Clove (0.75 mL/g biopolymer)	Gelatin, chitosan	*L. acidophilus, P. fluorescens, L. innocua, E. coli*	Agar diffusion	Formulations with clove EO resulted the most effective.	Gómez-Estaca et al. (2009)
Oregano (0.1%–1%)	Starch–chitosan	*B. cereus, E. coli, S. enteritidis, S. aureus*	Agar diffusion	Effectiveness against the four bacteria. Films were more effective against *B. cereus.*	Pelissari et al. (2009)
Potassium sorbate (15% w/w)	Potato starch	*E. coli, S. aureus*	Agar diffusion	Just *E. coli* inhibition.	Shen et al. (2010)
Lactic, malic, or citric acids (3% w/v)	WPI	*L. monocytogenes*	Agar diffusion	Effectiveness was in the following order: lactic < citric < malic acid.	Pintado et al. (2009)
Potassium sorbate (0.3% w/w)	Tapioca starch	*Zygosaccharomyces bailii*	Plate count	External *Z. bailii* contamination prevention. Yeast growth control.	Flores et al. (2007)
Citric, lactic, malic, and tartaric acids (0.9/1.8/2.6% w/w)	SPI	*L. monocytogenes, E. coli* O157:H7, *S. gaminara*	Agar diffusion and plate count	Total inhibition of *L. monocytogenes.* Malic and tartaric were more effective than citric against *S. gaminara.* Citric and tartaric were more effective than malic and lactic against *E. coli* O157:H7.	Eswaranandam et al. (2004)

(Continued)

TABLE 15.1 (*Continued*)

Examples of Effective Antimicrobial Edible Films Containing Essential Oils, Organic Acids, Nisin, and Other Minor Antimicrobials

Antimicrobial	Matrix	Target	Method	Effect	Reference
Nisin (3,000 IU/mL) and potassium sorbate (0.3% w/w)	Tapioca starch, HPMC	*L. innocua, Z. bailii*	Agar diffusion	Combination was more effective than individual incorporation.	Basch et al. (2012)
Nisin (10,000 IU/g)	HPMC	*Listeria, Enterococcus, Bacillus* spp.	Agar diffusion	Film bioactivity demonstrated efficacy against the three spp.	Imran et al. (2010)
Nisin (50 IU/mL) with lactic, malic, or citric acids (3% w/v)	WPI	*L. monocytogenes*	Agar diffusion	The largest mean zone of inhibition was for malic acid with nisin.	Pintado et al. (2009)
Nisin (10,000 IU/g), grape seed extract (1% w/w) (EDTA, 0.16% w/w)	SPI	*L. monocytogenes, S. typhimurium, E. coli* O157:H7	Plate count	Films were able to reduce *Listeria* by 2.9 log CFU/mL, while the population of *E. coli* O157:H7 and *S. typhimurium* were reduced by 1.8 and 0.6 log CFU/mL, respectively.	Sivarooban et al. (2008)
Kiam wood extract (300–1,500 mg/L)	HPMC	*E. coli* O175:H7, *S. aureus*, *L. monocytogenes*	Agar diffusion	More effective against *Listeria*.	Chana-Thaworn et al. (2011)
Lysozyme (50 mg/100 mL)	Sodium caseinate	*Micrococcus lysodeikticus, S. aureus*	Agar diffusion	Effectiveness against the strains.	Mendes de Souza et al. (2010)
Aloe leaf gel powder (4:1, 3:2, 2:3, 1:4)	Gelatin	*Citrobacter freundii, E. coli, Enterobacter aerogenes, Serratia marcescens, S. aureus, B. cereus*	Agar diffusion	Antimicrobial activity increased as the amount of aloe gel powder used in the composite films increased.	Chen et al. (2010)
LPS (0.2–0.6 mg/g)	Alginate	*L. innocua, P. fluorescens, E. coli*	Plate count	The growth of all tested bacteria was prevented for at least a 6 h period.	Yener et al. (2009)
Lysozyme (1,400–2,800 IU/cm^2)	Zein	*E. coli, B. subtilis*	Agar diffusion	Antimicrobial effectiveness against *E. coli* and *B. subtilis*.	Güçbilmez et al. (2007)

(Fernández-Pan et al., 2014), soy protein isolate (SPI) (Atarés et al., 2010), alginate (Rojas-Graü et al., 2007), starch (Maizura et al., 2008), and chitosan (Ziani et al., 2009), have been employed on antimicrobial EFC's development (Table 15.1) with different and particular levels of effectiveness. It is important to note that the hydrocolloid that forms the structural matrix is not only useful as a carrier of antimicrobials but also provides complementary benefits, for example, humidity loss prevention during storage, lipid oxidation rate reduction, and limited aroma transfer (Quintavalla and Vicini, 2002).

The release rate of the antimicrobial from the EFC is a key element when talking about their effectiveness. There are several recent works focused on the release characterization and control for different EFC formulations, test conditions, and food products. It depends on different factors that could include (1) antimicrobial–hydrocolloid electrostatic interactions of the matrix structure, (2) ion-induced osmosis and structural changes caused by the antimicrobial presence, (3) product characteristics (such as water activity, pH, and fat content), and (4) its environmental storage conditions (T, RH, time) (Del Nobile et al., 2008; Durango et al., 2006; Kurek et al., 2012, 2014; Mastromatteo et al., 2009).

15.3 Antimicrobials Currently Used for Edible Film and Coating Formulations

There is a great variety of antimicrobials that can be incorporated to EFCs to minimize the risk of foodborne contamination by pathogens and inhibit the development of spoiler microbes. When selecting the antimicrobial agent to be incorporated in an EFC, one should consider first its effectiveness against the target microorganisms, but it is equally important to note its potential interaction with the hydrocolloid film former and with the food components over which it will act. The most common antimicrobial substances used in EFCs are plant extracts, organic acids (OAs), bacteriocins, and their combinations.

15.3.1 Plant Extracts and Essential Oils

Essential oils (EOs) are defined as a mixture of volatile water-insoluble substances that are distilled or extracted from plant material (flowers, sprouts, seeds, leaves, branches, barks, herbs, wood, fruits, and vegetable roots). Although the antioxidant and/or antimicrobial properties of EOs have been recognized for a long time, their use in the food industry as natural antimicrobials is relatively recent (Burt, 2004) and now they are among the most demanded natural antimicrobials.

There is abundant scientific evidence regarding the efficacy of different fractions of EOs from many plant species as active antimicrobials, antifungals, and antivirals (Burt, 2004; Tiwari et al., 2009). As examples, Fernández-Pan et al. (2012) reported the antibacterial effectiveness of EOs from oregano, clove, tea tree, mastic thyme, laurel, cilantro, rosemary, and sage against different potential spoilers or pathogenic bacteria of interest in meat and poultry industry such as *Listeria innocua*, *Staphylococcus aureus*, *S. enteritidis*, and *Pseudomonas fragi*. Among them, in general, the group formed by the oregano, clove, and tea tree EOs has been presented as the most active. Also against similar strains, Chorianopoulos et al. (2004) evaluated the antimicrobial effectiveness of various EOs from different species of oregano and thyme and found that they were active against the pathogens *S. enteritidis*, *S. aureus*, and *L. monocytogenes*. Otoni et al. (2014) found that clove bud EO was more effective than oregano EO against *Aspergillus niger* and *Penicillium* sp. common fungi spoilage microorganisms in bread and bakery products.

EOs and their constituents have a wide spectrum of antimicrobial activity. Factors such as the specific composition, structure, and functional groups play a fundamental role in their efficacy. The chemical composition of EOs is extremely complex and highly depends on the extraction procedure, the plant selected for extraction, the harvest season, and its geographic origin (Burt, 2004). The specific composition of each EO and the structure and functional groups of its components play a key role in their degree of reactivity. The major EO components with antimicrobial effects include phenolic compounds, terpenes, aliphatic alcohols, aldehydes, ketones, acids, and isoflavonoids (Bakkali et al., 2008; Burt, 2004; Tiwari et al., 2009). Examples of principal constituents are carvacrol, thymol, linalool, 1,8-cineol, eugenol, cinnamaldehyde, and their precursors (Bakkali et al., 2008; Tajkarimi et al., 2010). EOs with high levels of carvacrol (oregano, thyme) or eugenol (cinnamon, clove, allspice) are usually characterized as having strong antimicrobial activity with a wide spectrum of action (Lambert et al., 2001; Tiwari et al., 2009). However, it has been reported that other minor ingredients also have a critical influence in the antimicrobial and antioxidant activity, acting in synergy with other components (Nakatsu et al., 2000; van Vuuren and Viljoen, 2007).

Many studies that have investigated the activity of EOs against spoilers and food pathogens agree that, in general, the Gram-positive bacteria are more sensitive to the antimicrobial effect of the EOs than the Gram-negative ones (Pelissari et al., 2009; Zivanovic et al., 2005). This difference in the sensitivity is related to differences in the structure and composition of the cell wall between Gram-negative and Gram-positive bacteria (Burt, 2004; Pranoto et al., 2005). Despite the well-known biological activity of EOs, they show some disadvantages such as their relative biological and chemical instability, their insolubility in water, and their potentially poor distribution over a food matrix. Besides, one of the major problems confronted when using directly EOs over food is that to achieve the desired inhibitory effect of these natural antimicrobials, high concentrations of EOs must be used, which might alter the sensory characteristics of the products (Petrou et al., 2012; Sánchez-González et al., 2011). This limitation can be potentially overcome by lowering the EO concentration and controlling its release from a biopolymeric matrix that is adequately designed as an edible coating. There is a large list of EOs tested for antimicrobial activity when incorporated into EFCs. Table 15.1 shows some *in vitro* studies of antimicrobial EFs with EOs against different target microorganisms.

Owing to its demonstrated high level of antimicrobial activity and broad spectrum of action, the oregano EO has been widely studied in different matrices and in many cases has become the reference EO against different food pathogens and spoilers. As such, Fernández-Pan et al. (2012, 2013) and Royo et al. (2010) developed EFs based on WPI with oregano EO incorporated that were capable of inhibiting the development of *L. innocua*, *S. enteritidis*, and *S. aureus* among others. Pelissari et al. (2009) developed an edible film based on starch with oregano EO that proved to be effective against *Bacillus cereus*, *S. aureus*, *E. coli*, and *S. enteritidis*. Moreover, they were more effective against strains of Gram-positive bacteria than the Gram-negatives *E. coli* and *S. enteritidis*.

Maqbool et al. (2011) developed gum arabic films and incorporated at the formulation cinnamon and lemongrass EO. The tested EFs showed antifungal activity against *Colletotrichum musae* and *Colletotrichum gloeosporioides*, causal organisms of banana and papaya anthracnose. Cinnamon EO showed higher mycelial and spore germination inhibition than lemongrass. Both formulations were more effective against *C. musae*. Avila-Sosa et al. (2010) reported mold inhibition caused by Mexican oregano incorporated in starch, chitosan, and amaranth films. Starch films containing 0.5% of EO showed the highest inhibition against *A. niger* and *Penicillium* sp.

Among the most effective EOs controlling microbial growth on food products are those from different types of thyme and oregano, which can be attributed to their high content of the phenolic compounds carvacrol and/or thymol (Burt, 2004; Oussalah et al., 2004, 2006).

Based on all these studies, it can be deduced that the antimicrobial activity of films that incorporate EOs in their formulation depends on different factors: (1) type of structural matrix, (2) type and concentration of EO, (3) target bacteria species, and (4) their interactions. But when developing EFCs, another factor must be addressed: the type of product over which it is applied and, of course, the interaction with it. Thus, after checking the antimicrobial activity of the EOs by *in vitro* tests, many of them have been applied in food systems. Fernández-Pan et al. (2014), based on their previous studies, developed WPI coating for skinned chicken breast as carrier of oregano and clove EOs and demonstrated its effectiveness as compared with EO direct additions. They reported up to a 50% shelf-life extension for the coated chicken breasts as compared with the uncoated ones. Emiroglu et al. (2010) evaluated the effects of the SPI-based edible films that contained oregano or thyme at 5% and a mixture of both in vacuum-packed minced-beef hamburgers during 12 days of refrigerated storage (4°C). The films applied on the hamburgers were effective against coliform and *Pseudomonas* spp. groups, although they were not significantly effective against the total viable microorganisms, LAB, or *Staphylococcus* spp.

Cinnamon EO has also shown high effectiveness in the reduction of spoiler microorganisms. Its efficacy when incorporated in chitosan EF was demonstrated by Ojagh et al. (2010). The films applied over rainbow trout fillets stored at 4°C for 16 days reduced the total viable counts in 5 log units and the psychrotrophic bacteria in 4 log units, extending the shelf life of fish fillets in approximate 4 days. Another polysaccharide used for EF applications is pullulan, which was tested by Gniewosz et al. (2013), concluding that coatings incorporated with caraway EO were effective in extending the shelf life of baby carrots stored at room temperature for 7 days. Such coatings reduced the growth of inoculated *S. aureus*, *A. niger*, and *Saccharomyces cerevisiae* by 3, 5, and 4 log CFU/g, respectively. There was no reduction in counts of *S. enteritidis*.

As exposed, differences in the inhibitory effects of EFCs can be attributed to the susceptibility of each target bacteria, to the active biological components of the oils, to its concentrations, and to the interactions with the food matrix affecting the diffusivity of the active compounds and therefore the final antimicrobial activity. Finally, according to the literature, effective antimicrobial EFCs can be developed by using relatively low concentrations of EOs. Nevertheless, it would be necessary to take into account that their application can exceed the standard organoleptic acceptable levels and requires an exhaustive sensory study. In order to reduce the sensory impact of EOs, the use of the isolated active compounds mainly responsible of the antimicrobial activity should be considered. Thus, formulations with a similar concentration of an effective antimicrobial should be tested with the objective of decreasing the final concentration of EO in the EFC. However, the use of a single compound could lead to possible loss of effectiveness due to the lack of a synergistic effect with other minor compounds present in the EO (e.g., 1,8-cineol with camphor).

15.3.2 Organic Acids and Salts

OAs are naturally obtained from plants and fermented products, although they can also be obtained through chemical synthesis. OAs such as lactic (E-270), acetic (E-260), tartaric (E-334), malic (E-296), and citric (E-330) acids among others are widely used for food preservation. For example, sorbic acid and its salts can be cited as common preservatives in the food industry. The good solubility, stability, and the manufacturing process make potassium sorbate the most widely used form of sorbic salt in food systems. It is an effective antifungal agent.

The antimicrobial activity of OAs is based on the fact that protonated acids are membrane soluble and can enter the bacterial cytoplasm by simple diffusion. As a result, the depression of the internal pH of the microbial cell due to ionization of the undissociated acid molecules, the disruption of substrate transport by alteration of the permeability of the cellular membrane, and the reduction of the motor force of the protons are achieved (Eswaranandam et al., 2004; Gadang et al., 2008). From the point of view of antimicrobial effectiveness, OAs have different effects as function of the type of acid, its concentration, environmental conditions, and the target microorganism. Moreover, as OAs have specific sensitivities to microorganisms, mixtures of OAs present a wider antimicrobial spectrum and stronger activity than a single OA (Han and Gennadios, 2005).

A large number of studies have shown that the shelf life of fresh products can also be extended by the use of OAs and their salts as antimicrobial sprays or dips. The classic role of OAs in EFC formulations is their function as acidulants. In several cases and in order to form the film or coating, a modified hydrocolloid interaction is necessary to obtain their complete dispersion in the solvent, so the film-forming solution must be pH adjusted with the corresponding acidulant. In addition, based on the chemical composition and structure, researches on the effect of the acidulant type and concentration may provide opportunities for achieving the desired mechanical and barrier properties (Caner et al., 1998). Thus, when developing EFC formulations, it should be considered that the OA addition will definitively alter the mechanical and barrier properties of the final films and coatings. This trend is partially related to the fact that acids may exert a plasticizing effect on films since they possess hydroxyl groups that can take part in polymer–polymer interactions by developing hydrogen bonds. Additionally, OAs are also frequently used in combination with other antimicrobial agents such as chitosan or nisin that are more effective under acidic conditions (Gadang et al., 2008; Pintado et al., 2009).

Focusing on the function of EFCs as carriers of OAs as antimicrobials, different scientific works have reported both the incorporation and the effectiveness of different OA-based formulations in order to obtain an improvement on food safety and shelf-life extension (Table 15.1). Regarding food safety, the use of OAs in SPI-based EFs resulted in the effective growth control of pathogens such as *L. monocytogenes*, *E. coli*, and *Salmonella gaminara*. The incorporation of lactic and malic acids in the EFs caused a reduction in the population of *S. gaminara* by 3 log cycles. Similarly, citric acid was also effective, but to a lesser extent, achieving a reduction of 1 log unit. The differences in efficacy were attributed to the greater ease of penetration of the malic and lactic acids into the cellular cytoplasm because of their lower molecular weight as compared to the citric acid (Eswaranandam et al., 2004). Pintado et al. (2009) developed WPI-based EFs that incorporated citric and malic acids at 3% (w/w), and they obtained greater activity against *L. monocytogenes* than they did with lactic acid films. This was primarily due to the lower pKa of citric and malic acids that achieved a greater acidification of the medium.

When applying the EFC over food products, Jiang et al. (2011) prepared catfish skin gelatin films incorporating potassium sorbate, sodium tripolyphosphate, and a combination of both. Fresh white shrimps were coated and stored in ice (0°C) under aerobic conditions for a 30-day period. They found the coatings extended the shelf life of shrimps up to 10 days, retarding the lag phase of mesophilic aerobic and psychrotrophic microorganisms. Cagri et al. (2002) developed WPI-based antimicrobial films that contained sorbic acid or *p*-aminobenzoic acid that were tested on slices of two types of emulsified meats, bologna, and sausages inoculated with *L. monocytogenes*, *E. coli* O157:H7, and *Salmonella Typhimurium*. After 21 days of refrigerated storage and aerobic conditions, the films achieved a significant reduction in the populations of both microorganisms. In the test of the films on beef, the EFs were effective in the inhibition of the development of mesophilic aerobic bacteria, LAB, molds, and yeasts. Sayanjali et al. (2011) coated pistachios using carboxymethyl cellulose (CMC) solutions incorporated with potassium sorbate at 1, 0.5 and 0.25 g/100 mL of coating solution. After coating, pistachios were analyzed for mold growth. All treatments showed no growth of molds. Besides they observed that in samples treated just with CMC coatings, the growth of molds was partially inhibited, due to the reduction in oxygen level.

As it can be observed before, numerous studies concerning the antimicrobial activity of EFC formulations with OA have been reported and different results have been obtained. It is necessary to highlight the high variation on the formulation efficacy obtained by the compounds included in the matrix, the own matrix, test microorganisms, and products. Thus, all factors and interactions explained earlier for those formulations containing EOs have direct application when using OAs. However, in order to develop new formulations, three main advantages compared to EOs must be highlighted:

1. Water solubility: The high water solubility of OAs should result in greater stability for matrices and a better distribution of the antimicrobials when released to the product. In this sense, with OA formulations, the release rate must be greatly influenced by the product type, on the assumption that when applying the EFC onto dried surfaces, the release would be slower.
2. Nonvolatility: The release and migration of OA always occurs toward the product surface. There will not be efficiency losses due to the migration of volatile compounds to the headspace of the package as occurs with EOs.
3. Less or even no sensory impact of OAs as compared with EOs.

15.3.3 Lactic Acid Bacteria

There is an increasing interest in biopreservation in the food industry. It makes use of the antimicrobial potential of naturally occurring organisms in food and/or their metabolites with the main purpose of extending the shelf life and improving the hygienic quality, minimizing the impact on the nutritional and organoleptic properties of perishable food products (García et al., 2010).

LAB are organisms that could be used for this purpose. They are normally part of the microbiota of fresh and packed food and are able to produce different metabolic products, with a proved antibacterial effect, such as OAs, diacetyl, acetoin, hydrogen peroxide, reuterin, reutericyclin, antifungal peptides, and bacteriocins (Ghanbari et al., 2013; Muñoz-Atienza et al., 2013).

The mode of action of LAB as antimicrobials is related to one of the synergistic effects of several mechanisms: competition for nutrients with spoilage and pathogen microbiota, reduction of food pH (with the production of OAs), gas composition of atmosphere, and the production of the aforementioned metabolites (Calo-Mata et al., 2008; Stiles, 1996). Most LAB have a GRAS status given by the FDA, which makes them an appropriate natural preservative alternative. Murillo et al. (2013) tested three different strains of LAB named *Pediococcus acidilactici*, *Lactobacillus animalis*, and *Lactobacillus amylovorus* against *L. monocytogenes*. Results showed that both the whole cell culture and the cell-free supernatant of *P. acidilactici* inhibited the growth of the target bacteria. The other two LAB did not show any inhibition effect.

Bacteriocins are antibacterial peptides produced by LAB. These antimicrobial agents are generally heat stable, apparently hypoallergenic, and readily degraded by proteolytic enzymes in the human intestinal tract (Coma, 2008). Bacteriocins are produced by various bacteria, so numerous bacteriocins have been characterized displaying different mechanisms of action, antimicrobial spectra, and chemical properties. Although some bacteriocins, such as pediocin PA-1 and lacticin 3147, have been developed for approval and use, the most well-known bacteriocin is nisin (E-234), which is recognized as GRAS by the U.S. Food and Drug Administration, and it is commonly used as an additive in the conservation of cheese.

In a quite complete study, Cizeikine et al. (2013) tested five different LAB, *Lactobacillus sakei*, *P. acidilactici*, and three different strains of *Pediococcus pentosaceus*, against several pathogen bacteria, yeast, and molds isolated from different food products like vegetables, meat, fats, and water. The results showed that the strains *Bacillus subtilis, Bacillus thuringiensis, Pseudomonas fluorescens, Penicillium gladioli,* and *Pseudomonas facilis* were the most sensitive bacteria to the action of the five LAB. Different inhibition zones varied from 6 to 23 mm diameter were measured. In the case of molds and yeast, delay of spore formation with a clear inhibition zone around the cut well of *Fusarium culmorum* was produced by the five LAB strains tested. *Candida parapsilosis* was the only yeast showing small sensitivity to the action of LAB. Besides they also studied the bacteriocin-like inhibitory substances (BLIS) produced by the five LAB tested. Results showed that *B. subtilis* was the only strain sensitive to the action of all BLIS: sakacin 05-6, pediocin Ac05-07, pediocin 05-8, pediocin 05-9, and pediocin 05-10. *P. fluorescens, P. gladioli,* and *P. facilis* were inhibited only by sakacin 05-6. The authors concluded that the antibacterial and antifungal activities depend on the LAB strain and on the target microorganism species and strain.

LAB can be incorporated also in EFC to use them as a preservation technology. In such line, Sánchez-González et al. (2014) developed films of sodium caseinate and methylcellulose and incorporated *Lactobacillus acidophilus* and *Lactobacillus reuteri,* two well-known bacteriocin-producing LAB, and tested such films against *L. innocua.* Films were stored at 5°C for 12 days. Both LAB completely inhibited the growth of the target bacteria after 1 week of storage. At the end of the storage period, *L. innocua* was reduced in approximate 1.5 log CFU/g. The two biopolymers are presented as a good alternative to improve food safety.

Giamalas et al. (2010) tested sodium caseinate films in food laboratory model and in fresh beef inoculated with *L. monocytogenes. L. sakei* was either incorporated in the film or sprayed onto its surface. After 12 days of storage at 4°C, films that incorporated *L. sakei* in its formulation reduced in about 3 log cycles the counts of *L. monocytogenes.* In the case of sprayed films, the reduction was about 3.6 log cycles. When working with fresh beef, the counts of the target bacteria was reduced in 2 log cycles. Concha-Meyer et al. (2001) put in contact pieces of cold smoked salmon, inoculated with *L. monocytogenes,* with alginate films containing a mixture of LAB-A and LAB-B (bacteria isolated by the authors from cold smoked salmon). After a storage period of 28 days at 4°C, 3 log cycles reduction on the counts of *L. monocytogenes* was found.

As it can be seen, LAB can be a very good alternative when talking about new preservation technologies, either using them in direct application to the products or incorporating them in edible films. The effectiveness of LAB in extending the shelf life of different products has been proven.

15.3.4 Nisin

Nisin is a natural amphiphilic polypeptide that is produced by *Lactococcus lactis* subsp. *lactis* group and remains the most commercially important bacteriocin because of its relatively long history of safe use and documented effectiveness against important Gram-positive foodborne pathogens and spoilage agents. It is capable of inhibiting a wide spectrum of Gram-positive bacteria such as *L. monocytogenes* and *S. aureus* (Coma, 2008). Nisin acts by destroying the integrity of the cytoplasmic membrane through the formation of pores, which provoke the loss of small compounds and change the motor force of the protons, which are necessary for the production of energy and the synthesis of nucleic acid and proteins. Similar as other antimicrobial agents, nisin is more effective in acidic conditions due to its greater stability, solubility, and lower pH activity (Pintado et al., 2009). The inhibitory effects of nisin can be expanded to the group of Gram-negative bacteria, thanks to its combination with chelating agents such as EDTA or lysozyme, which are capable of altering the permeability of the outer membrane of the bacteria (Gadang et al., 2008). The effects of nisin have been tested on meat and meat products, dairy products, and different vegetables. In their study, Turgis et al. (2012) determined the minimum inhibitory concentration of nisin against several pathogens and food spoilers. In order to inhibit the growth of *B. cereus,* *L. monocytogenes,* and *Lactobacillus sakei,* 172 ppm of nisin was needed. On the other hand, nisin was not effective against *E. coli, S. aureus, S. typhimurium,* and *Pseudomonas putida.*

As it is shown in Table 15.1, nisin has been incorporated as an antimicrobial agent in EFCs based on hydrocolloids such as tapioca starch (Sanjurjo et al., 2006), whey protein (Ko et al., 2001; Pintado et al., 2009), sodium caseinate (Kristo et al., 2008), soy protein (Eswaranandam et al., 2004), hydroxypropylmethylcellulose (HPMC) (Coma et al., 2001), corn zein (Ku and Song, 2007), or glucomannan (Li et al., 2006).

The antimicrobial activity of the nisin incorporated in EFCs is very sensitive to the final formulation of the system. As such, edible films based on HPMC that contain nisin are more effective against the pathogens *S. aureus* and *L. monocytogenes* (Coma et al., 2001). In these films, the use of stearic acid intended to improve their barrier properties (specifically water vapor permeability) affected the inhibitory activity of the nisin against the two pathogens.

Murillo-Martínez et al. (2013) formulated films based on WPI incorporating 3 and 12 mg/mL of nisin and tested against some pathogens and spoilage bacteria. The results confirmed that films with the higher nisin content were more effective and bacteria showed different sensitivity (expressed as inhibition zones): *Enterococcus faecalis* > *L. innocua* > *E. coli*. Films did not inhibit the growth of *Brochothrix thermosphacta*. Ko et al. (2001) reported that the nisin incorporated in WPI films was more effective against *Listeria* than when it was incorporated in wheat gluten films, suggesting that the antimicrobial activity of nisin against the bacteria was better in hydrophobic films. Additionally, they detected greater activity against *Listeria* under acidic conditions. The results confirmed strong dependence on the efficacy of the nisin, the characteristics of the system, and the environmental conditions. Basch et al. (2012) evaluated the viability of *L. innocua* after 48 h of exposure to HPMC plus tapioca films incorporated with nisin. The growth of *L. innocua* was 4 log CFU/g lower when bacteria was in contact with the developed films than the control films. McCormick et al. (2005) developed wheat gluten films with nisin that were effective in the reduction of *L. monocytogenes*, even though they were not effective against the Gram-negative *S. Typhimurium*.

The effectiveness of nisin-incorporated EFCs has also been studied in food products. Martinez et al. (2010) coated ricotta cheese using galactomannan films that contained nisin. They found 1 log reduction of *L. monocytogenes* after 2 days of storage. After 7 days of storage, 2.2 log CFU/g reduction was found. They concluded that the activity of nisin against this pathogen could be due to its easy adsorption when in contact with the hydrophilic surface of the microorganism. Natrajan and Sheldon (2000) developed different formulations, based on agar and calcium alginate with nisin, and applied these to the surface of fresh chicken meat. The mean log reduction in the population of *S. Typhimurium* was more than 3 and 4 log units after 72 and 96 h, respectively, when maintained at 4°C.

Millette et al. (2007) developed alginate films containing nisin and were able to reduce the growth of *S. aureus* in beef fillets. Ollé et al. (2014) evaluated the effectiveness of tapioca starch films containing nisin over Port Salut cheese surface inoculated with a mix of *L. innocua* and *S. cerevisiae* and incubated at 25°C for 8 days. The films were able to retard the growth of lag phase of microorganisms. Besides, they found that films reduced in 2 and 4 log CFU/g final counts of *S. cerevisiae* and *L. innocua*, respectively.

The efficacy of nisin can be increased by combining it with other antimicrobial agents. Pintado et al. (2009) developed WPI-based antimicrobial films with OAs (malic, lactic, and citric) and nisin to control the development of *L. monocytogenes*. The OAs improved the antimicrobial effectiveness of the nisin, and its combination with malic acid was the most effective against the different strains of *L. monocytogenes* isolated from cheese. Sivarooban et al. (2008) developed SPI-based edible films that incorporated combinations of grape seed, nisin, and EDTA as antimicrobial agents. They demonstrated their antimicrobial efficacy *in vitro* against three pathogens of great interest to the food industry: *S. Typhimurium*, *L. monocytogenes*, and *E. coli* O157:H7.

Similarly, Gadang et al. (2008) demonstrated the effectiveness of different combinations of vegetable extracts, OAs, nisin, and EDTA incorporated in WPI-coatings that could be used to improve the safety of ready-to-eat food products stored at refrigerated temperatures. As such, they evaluated the inhibition caused by films on the growth of *S. Typhimurium*, *L. monocytogenes*, and *E. coli* O157:H7 inoculated into turkey sausage and stored at 4°C for 28 days. The synergistic effect of the antimicrobial agents was confirmed against *L. monocytogenes* and *E. coli* O157:H7, but they were not effective against *S. Typhimurium*. The most effective formulation against *S. Typhimurium* was the one containing only malic acid, with a reduction of 3.3 log units. Moreover, the most effective combination against *L. monocytogenes* was the one containing grape seed extract, malic acid, and nisin. It was able to achieve a reduction of 4.8 log units. By incorporating nisin, malic acid, and EDTA, they achieved reductions of 4.6 log units in the *E. coli* O157:H7 population.

Lu et al. (2010) formulated alginate-calcium coatings incorporating cinnamon EO, nisin, and EDTA and coated snakehead fish fillets, which were stored at 4°C for 15 days. They found that the combination of cinnamon, nisin, and EDTA was the most effective in inhibiting the growth of mesophilic and psychrotrophic microorganisms and *Pseudomonas* spp. compared with just nisin or cinnamon EO. Reductions of 2 log units in the case of mesophilic and *Pseudomonas* and 4 log units in the case of psychrotrophic microorganisms after 6 days of storage were registered. After 15 days of storage, and even the samples already reported counts higher than 6 log CFU/g (legislation limit for fresh refrigerated fish), the mix treatment kept the counts lower than the control samples.

Most of the studies agreed that the most important factors influencing the antimicrobial effectiveness of nisin are those related to its concentration and the microbial target, but it is important to highlight the decisive influence that additives have on the EFC matrix, especially acidulants and EDTA, to produce the final success of formulations.

15.3.5 Other Minor Antimicrobials

Table 15.1 shows some different formulations based on antimicrobial agents, other than those aforementioned, employed to a lesser extent in EFCs to increase safety and to maintain the quality of food. They are vegetable extracts (Corrales et al., 2009), lysozyme (Güçbilmez et al., 2007), lactoferrin (Brown et al., 2008), ovotransferrin (Seol et al., 2009), and lactoperoxidase system (LPS) (Yener et al., 2009) among others. As examples, works reported by Min and Krochta (2005) and Min et al. (2005, 2009) incorporating different concentrations of the LPS in WPI films that showed strong antimicrobial activity against *Salmonella enterica*, *E. coli*, *L. monocytogenes*, and *Penicillium commune* could be cited.

Applied to food products, pea starch-based edible films enriched with grape seed extracts were developed by Corrales et al. (2009). The effect of these edible films was evaluated against the growth of *B. thermosphacta* inoculated in vacuum-packed pork loins. They achieved a reduction of 1.3 log CFU/mL during the first 4 days of storage. Seol et al. (2009) developed edible films based on κ-carrageenan, which incorporated ovotransferrin as an antimicrobial agent combined with EDTA, to increase the shelf life of fresh chicken breast. After 7 days of storage at 5°C, the chicken breasts wrapped in the films showed a reduction in their number of total viable microorganisms and in *E. coli* with respect to the untreated chicken breasts. Ünalan et al. (2013) formulated films using zein and wax and incorporated them with lysozyme and a combination of the latest one with catechin and gallic acid. Pieces of fresh Kashar cheese were coated, inoculated with *L. monocytogenes*, and stored for 8 weeks at 4°C. At the end of the storage period, 2.5 log cycle reductions in the counts of *L. monocytogenes* were reached.

15.4 Conclusions

The high nutrient composition of food products makes them an ideal environment for the growth and development of foodborne pathogens and spoilers. Based on the multiple-hurdle barrier theory, antimicrobial EFCs have been introduced as an effective, innovative, and emerging technology that can enhance food safety and increase shelf life of food products. A great variety of hydrocolloids can be used as effective carriers of antimicrobials, that is, cellulose derivatives, alginate, casein, soy protein, etc. The antimicrobials most commonly used as part of the aforementioned procedure are EOs, OAs, and bacteriocins. The effectiveness of antimicrobial EFCs depends on a gradual, selective, and controlled release of their active compounds by maintaining the precise dose of additive on the surface of the food product. Although the research and investigation up to date has been quiet successful, more research is needed in order to have a better understanding on the controlled release of such compounds and interactions between matrices and food systems.

REFERENCES

Alvarez M, Ponce A, Moreira M. Antimicrobial efficiency of chitosan coating enriched with bioactive compounds to improve the safety of fresh cut broccoli. *LWT—Food Sci Technol* 50 (2013):78–87.
Arvanitoyannis IS, Stratakos AC. Application of modified atmosphere packaging and active/smart technologies to red meat and poultry: A review. *Food Bioprocess Technol* 5 (2012):1423–1446.

Atarés L, De Jesús C, Talens P, Chiralt A. Characterization of SPI-based edible films incorporated with cinnamon or ginger essential oils. *J Food Eng* 99(3) (2010):384–391.

Avena-Bustillos RJ, Cisneros-Zevallos LA, Krochta JM, Saltveit ME. Application of casein-lipid edible film emulsions to reduce white blush on minimally processed carrots. *Postharvest Biol Technol* 4(4) (1994):319–329.

Avila-Sosa R, Hernánadez-Zamoran E, López-Mendoza I, Palou E, Jiménez M et al. Fungal inactivation by Mexican oregano (*Lippia berlandier* Schauer) essential oil added to amaranth, chitosan or starch edible films. *J Food Sci* 75(3) (2010):M127–M133.

Bakkali F, Averbeck S, Averbeck D, Idaomar M. Biological effects of essential oils—A review. *Food Chem Toxicol* 46(2) (2008):446–475.

Basch C, Jagus R, Flores S. Physical and antimicrobial properties of tapioca starch-HPMC edible films incorporated with nisin and/or potassium sorbate. *Food Bioprocess Technol* 6 (2012):2416–2428.

Brown CA, Wang B, Oh JH. Antimicrobial activity of lactoferrin against foodborne pathogenic bacteria incorporated into edible chitosan film. *J Food Prot* 71(2) (2008):319–324.

Burt S. Essential oils: Their antibacterial properties and potential applications in foods—A review. *Int J Food Microbiol* 94(3) (2004):223–253.

Cagri A, Ustunol Z, Ryser ET. Inhibition of three pathogens on bologna and summer sausage using antimicrobial edible films. *J Food Sci* 67(6) (2002):2317–2324.

Calo-Mata P, Arlindo S, Boechme K, de Miguel T, Pascoal A et al. Current applications and future trends of lactic acid bacteria and their bacteriocins for the biopreservation of aquatic food products. *Food Bioprocess Technol* 1 (2008):43–63.

Caner C, Vergano PJ, Wiles JL. Chitosan film mechanical and permeation properties as affected by acid, plasticizer and storage. *J Food Sci* 63(6) (1998):1049–1053.

Chana-Thaworn J, Chanthachum S, Wittaya T. Properties and antimicrobial activity of edible films incorporated with kiam wood (*Cotyleobium lanceotatum*) extract. *LWT—Food Sci Technol* 44(1) (2011):284–292.

Chen CP, Wang BJ, Weng YM. Physicochemical and antimicrobial properties of edible aloe/gelatin composite films. *Int J Food Sci Technol* 45(5) (2010):1050–1055.

Chorianopoulos N, Kalpoutzakis E, Aligiannis N, Mitaku S, Nychas GJ et al. Essential oils of *Satureja*, *Origanum*, and *Thymus* species: Chemical composition and antibacterial activities against foodborne pathogens. *J Agric Food Chem* 52 (2004):8261–8267.

Cizeikiene D, Juodeikiene G, Paskevicius A, Bartkiene E. Antimicrobial activity of lactic acid bacteria against pathogenic and spoilage microorganism isolated from food and their control in wheat bread. *Food Control* 31 (2013):539–545.

Coma V. Bioactive packaging technologies for extended shelf-life of meat-based products. *Meat Sci* 78(1–2) (2008):90–103.

Coma V, Sebti I, Pardon P, Deschamps A, Pichavant FH. Antimicrobial edible packaging based on cellulosic ethers, fatty acids, and nisin incorporation to inhibit *Listeria innocua* and *Staphylococcus aureus*. *J Food Prot* 64(4) (2001):470–475.

Concha-Meyer A, Schöbitz R, Brito C, Fuentes R. Lactic acid bacteria in an alginate film inhibit *Listeria monocytogenes* growth on smoked salmon. *Food Control* 22 (2001):485–489.

Corrales M, Han JH, Tauscher B. Antimicrobial properties of grape seed extracts and their effectiveness after incorporation into pea starch films. *Int J Food Sci Technol* 44(2) (2009):425–433.

Del Nobile MA, Conte A, Incoronato AL, Panza O. Antimicrobial efficacy and release kinetics of thymol from zein films. *J Food Eng* 89(1) (2008):57–63.

Durango AM, Soares NFF, Andrade NJ. Microbiological evaluation of an edible antimicrobial coating on minimally processed carrots. *Food Control* 17(5) (2006):336–341.

Emiroglu ZK, Yemiş GP, Coşkun KB, Candogan K. Antimicrobial activity of soy edible films incorporated with thyme and oregano essential oils on fresh ground beef patties. *Meat Sci* 86(2) (2010):283–288.

Eswaranandam S, Hettiarachchy NS, Johnson MG. Antimicrobial activity of citric, lactic, malic, or tartaric acids and nisin-incorporated soy protein film against *Listeria monocytogenes*, *Escherichia coli* O157:H7, and *Salmonella gaminara*. *J Food Sci* 69(3) (2004):FMS79–FMS84.

Fernández-Pan I, Carrión-Granda X, Maté JI. Antimicrobial efficiency of edible coatings on the preservation of chicken breast fillets. *Food Control* 36(1) (2014):69–75.

Fernández-Pan I, Maté JI. 2011. Biopolymers for edible films and coatings in food applications. In *Films and Coatings from Renewable Resources: An Applications Perspective*, ed. D. Plackett, pp. 233–254. New York: John Wiley & Sons, Inc.

Fernández-Pan I, Mendoza M, Maté JI. Whey protein isolate edible films with essential oils incorporated to improve the microbial quality of poultry. *J Sci Food Agric* 93(12) (2013):2986–2994.

Fernández-Pan I, Royo M, Maté JI. Antimicrobial activity of whey protein isolate edible films with essential oils against food spoilers and foodborne pathogens. *J Food Sci* 77(7) (2012):M383–M390.

Flores S, Haedo AS, Campos C, Gerschenson L. Antimicrobial performance of potassium sorbate supported in tapioca starch edible films. *Eur Food Res Technol* 225(3–4) (2007):375–384.

Gadang VP, Hettiarachchy NS, Johnson MG, Owens C. Evaluation of antibacterial activity of whey protein isolate coating incorporated with nisin, grape seed extract, malic acid, and EDTA on a turkey frankfurter system. *J Food Sci* 73(8) (2008):M389–M394.

García P, Rodríguez L, Rodríguez A, Martínez B. Food biopreservation: Promising strategies using bacteriocins, bacteriophages and endolysins. *Trends Food Sci Technol* 21 (2010):373–382.

Gennadios A, Hanna MA, Kurth LB. Application of edible coatings on meats, poultry and seafoods: A review. *LWT—Food Sci Technol* 30(4) (1997):337–350.

Ghanbari M, Jami M, Domig K, Kneifel W. Seafood biopreservation by lactic acid bacteria—A review. *LWT—Food Sci Technol* 24 (2013):315–324.

Giamalas H, Zinoviadou K, Biliaderis C, Koutsoumanis K. Development of a novel bioactive packaging based on the incorporation of *Lactobacillus sakei* into sodium-caseinate films for controlling *Listeria monocytogenes* in foods. *Food Res Int* 43 (2010):2402–2408.

Gniewosz M, Krasniewska K, Woreta M, Kosakowska O. Antimicrobial activity of a pullulan-caraway essential oil coating on reduction of food microorganisms and quality in fresh baby carrot. *J Food Sci* 78(8) (2013):M1242–M1248.

Gómez-Estaca J, López de Lacey A, Gómez Guillén M, López-Caballero E, Montero P. Antimicrobial activity of composite edible films based on fish gelatin and chitosan incorporated with clove essential oil. *J Aquat Food Prod Technol* 18(1–2) (2009):46–52.

Güçbilmez CM, Yemenicioglu A, Arslanoglu A. Antimicrobial and antioxidant activity of edible zein films incorporated with lysozyme, albumin proteins and disodium EDTA. *Food Res Int* 40(1) (2007):80–91.

Han JH, Gennadios A. 2005. Edible films and coatings: A review. In *Innovations in Food Packaging*, ed. J.H. Han, pp. 239–262. San Diego, CA: Elsevier Academic.

Imran M, El-Fahmy S, Revol-Junelles AM, Desobry S. Cellulose derivative based active coatings: Effects of nisin and plasticizer on physico-chemical and antimicrobial properties of hydroxypropyl methylcellulose films. *Carbohydr Polym* 81(2) (2010):219–225.

Jiang M, Liu S, Wang Y. Effects of antimicrobial coating from catfish skin gelatin on quality and shelf life of fresh white shrimp (*Penaeus vannamei*). *J Food Sci* 76(3) (2011):M204–M209.

Ko S, Janes ME, Hettiarachchy NS, Johnson MG. Physical and chemical properties of edible films containing nisin and their action against *Listeria monocytogenes*. *J Food Sci* 66(7) (2001):1006–1011.

Kristo E, Koutsoumanis KP, Biliaderis CG. Thermal, mechanical and water vapor barrier properties of sodium caseinate films containing antimicrobials and their inhibitory action on *Listeria monocytogenes*. *Food Hydrocoll* 22(3) (2008):373–386.

Ku K, Song KB. Physical properties of nisin-incorporated gelatin and corn zein films and antimicrobial activity against *Listeria monocytogenes*. *J Microbiol Biotechnol* 17(3) (2007):520–523.

Kurek M, Descours E, Galic K, Voilley A, Debeaufort F. How composition and process parameters affect volatile active compounds in biopolymer films. *Carbohydr Polym* 88 (2012):646–656.

Kurek M, Guinault A, Voilley A, Galic K, Debeaufort F. Effect of relative humidity on carvacrol release and permeation properties of chitosan based films and coatings. *Food Chem* 144 (2014):9–17.

Lambert RJW, Skandamis PN, Coote PJ, Nychas GJE. A study of the minimum inhibitory concentration and mode of action of oregano essential oil, thymol and carvacrol. *J Appl Microbiol* 91(3) (2001):453–462.

Li B, Peng J, Yie X, Xie B. Enhancing physical properties and antimicrobial activity of konjac glucomannan edible films by incorporating chitosan and nisin. *J Food Sci* 71(3) (2006):C174–C178.

Lu F, Ding Y, Ye X, Liu D. Cinnamon and nisin in alginate-calcium coating quality of fresh northern snakehead fish fillets. *LWT—Food Sci Technol* 43 (2010):1331–1335.

Maizura M, Fazilah A, Norziah MH, Karim AA. Antibacterial activity of modified sago starch-alginate-based edible film incorporated with lemongrass (*Cymbopogon citratus*) oil. *Int Food Res J* 15(2) (2008):233–236.

Maqbool M, Ali A, Alderson P, Mohamed M, Siddiqui Y, Zahid N. Postharvest application of gum Arabic and essential oils for controlling anthracnose and quality of banana and papaya during cold storage. *Postharvest Biol Technol* 62 (2011):71–76.

Martinez J, Cerqueira M, Souza B, Avides M, Vicente A. Shelf life extension of ricotta cheese using coatings of galactomannans from nonconventional sources incorporating nisin against *Listeria monocytogenes*. *J Agric Food Chem* 58 (2010):1884–1891.

Mastromatteo M, Barbuzzi G, Conte A, Del Nobile MA. Controlled release of thymol from zein-based film. *Innov Food Sci Emerg Technol* 10(2) (2009):222–227.

McCormick KE, Han IY, Acton JC, Sheldon BW, Dawson PL. In-package pasteurization combined with biocide-impregnated films to inhibit *Listeria monocytogenes* and *Salmonella typhimurium* in turkey bologna. *J Food Sci* 70(1) (2005):M52–M57.

Mendes de Souza P, Fernández A, López-Carballo G, Gavara R, Hernández-Muñoz P. Modified sodium caseinate films as releasing carriers of lysozyme. *Food Hydrocoll* 24(4) (2010):300–306.

Millette M, Le Tien C, Smoragiewicz W, Lacroix M. Inhibition of *Staphylococcus aureus* on beef by nisin-containing modified alginate films and beads. *Food Control* 18(7) (2007):878–884.

Min BJ, Oh JH. Antimicrobial activity of catfish gelatin coating containing origanum (*Thymus capitatus*) oil against gram-negative pathogenic bacteria. *J Food Sci* 74(4) (2009):M143–M148.

Min S, Harris LJ, Krochta JM. Antimicrobial effects of lactoferrin, lysozyme, and the lactoperoxidase system and edible whey protein films incorporating the lactoperoxidase system against *Salmonella enterica* and *Escherichia coli* O157:H7. *J Food Sci* 70(7) (2005):M332–M338.

Min S, Krochta JM. Inhibition of *Penicillium commune* by edible whey protein films incorporating lactoferrin, lactoferrin hydrolysate, and lactoperoxidase systems. *J Food Sci* 70(2) (2005):M87–M94.

Muñoz-Atienza E, Gómez-Sala B, Araujo C, Campanero C, del Campo R et al. Antimicrobial activity, antibiotic susceptibility and virulence factors of Lactic Acid Bacteria of aquatic origin intended for use as probiotics in aquaculture. *BMC Microbiol* 13 (2013):15.

Murillo S, Story R, Pak D, O'Bryan C, Crandall P et al. Antimicrobial properties of three lactic acid bacterial cultures and their free supernatants against *L. monocytogenes*. *J Environ Sci Health B: Pesticides Food Contam Agric Wastes* 48(1) (2013):63–68.

Murillo-Martínez M, Tello-Solís S, García-Sánchez M, Ponce-Alquicira E. Antimicrobial activity and hydrophobicity of edible whey protein isolate films formulated with nisin and/or glucose oxidase. *J Food Sci* 78(4) (2013):M560–M566.

Nakatsu T, Lupo AT, Chinn JW, Kang RKL. Biological activity of essential oils and their constituents. *Studies Nat Prod Chem* 21 (2000):571–631.

Natrajan N, Sheldon BW. Inhibition of *Salmonella* on poultry skin using protein- and polysaccharide-based films containing a nisin formulation. *J Food Prot* 63(9) (2000):1268–1272.

Ojagh S, Rezaei M, Razavi S, Hosseini S. Effect of chitosan coatings enriched with cinnamon oil on the quality of refrigerated rainbow trout. *Food Chem* 120 (2010):193–198.

Ollé C, Gerschenson L, Jagus R. Natamycin and nisin supported on starch edible films for controlling mixed culture growth on model systems and Port Salut cheese. *Food Control* 44 (2014):146–151.

Otoni C, Pontes S, Medeiros E, Soares N. Edible films from methylcellulose and nanoemulsions of clove bud (*Syzygium aromaticum*) and oregano (*Origanum vulgare*) essential oils as shelf life extenders for sliced bread. *J Agric Food Chem* 62 (2014):5214–5219.

Oussalah M, Caillet S, Salmiéri S, Saucier L, Lacroix M. Antimicrobial and antioxidant effects of milk protein-based film containing essential oils for the preservation of whole beef muscle. *J Agric Food Chem* 52(18) (2004):5598–5605.

Oussalah M, Caillet S, Salmiéri S, Saucier L, Lacroix M. Antimicrobial effects of alginate-based film containing essential oils for the preservation of whole beef muscle. *J Food Prot* 69(10) (2006):2364–2369.

Park JW, Testin RF, Park HJ, Vergano PJ, Weller CL. Fatty acid concentration effect on tensile strength, elongation, and water vapor permeability of laminated edible films. *J Food Sci* 59(4) (1994):916–919.

Pelissari FM, Grossmann MVE, Yamashita F, Pined EAG. Antimicrobial, mechanical, and barrier properties of cassava starch-chitosan films incorporated with oregano essential oil. *J Agric Food Chem* 57(16) (2009):7499–7504.

Pérez-Gago MB, Krochta JM. 2005. Emulsions and bi-layer edible films. In *Innovations in Food Packaging*, ed. J.H. Han, pp. 384–402. San Diego, CA: Elsevier Academic.

Petrou S, Tsiraki M, Giatrakou V, Savvaidis IN. Chitosan dipping or oregano oil treatments, singly or combined on modified atmosphere-packaged chicken breast meat. *Int J Food Microbiol* 156 (2012):264–271.

Pintado CMBS, Ferreira MASS, Sousa I. Properties of whey protein-based films containing organic acids and nisin to control *Listeria monocytogenes*. *J Food Prot* 72(9) (2009):1891–1896.

Pires C, Ramos C, Teixeira B, Batista I, Nunes ML, Marques A. Hake proteins edible films incorporated with essential oils: Physical, mechanical, antioxidant and antibacterial properties. *Food Hydrocoll* 30(1) (2013):224–231.

Pranoto Y, Rakshit SK, Salokhe VM. Enhancing antimicrobial activity of chitosan films by incorporating garlic oil, potassium sorbate and nisin. *LWT—Food Sci Technol* 38(8) (2005):859–865.

Quintavalla S, Vicini L. Antimicrobial food packaging in meat industry. *Meat Sci* 62(3) (2002):373–380.

Ray B. 2005. *Fundamental Food Microbiology*. Boca Raton, FL: CRC Press.

Rojas-Graü MA, Avena-Bustillos RJ, Olsen C, Friedman M, Henika PR et al. Effects of plant essential oils and oil compounds on mechanical, barrier and antimicrobial properties of alginate-apple puree edible films. *J Food Eng* 81(3) (2007):634–641.

Royo M, Fernández-Pan I, Maté JI. Antimicrobial effectiveness of oregano and sage essential oils incorporated into whey protein films or cellulose-based filter paper. *J Sci Food Agric* 90(9) (2010):1513–1519.

Sánchez-González L, González-Martínez C, Chiralt A, Cháfer M. Physical and antimicrobial properties of chitosan-tea tree essential oil composite films. *J Food Eng* 98(4) (2010):443–452.

Sánchez-González L, Quintero I, Chiralt A. Antilisterial and physical properties of biopolymer films containing lactic acid bacteria. *Food Control* 35 (2014):200–206.

Sánchez-González L, Vargas M, González-Martínez Ch, Chiralt A, Cháfer M. Use of essential oils in bioactive edible coatings. *Food Eng Rev* 3 (2011):1–16.

Sanjurjo K, Flores S, Gerschenson L, Jagus R. Study of the performance of nisin supported in edible films. *Food Res Int* 39(6) (2006):749–754.

Sayanjali S, Ghanbarzadeh B, Ghiassifar S. Evaluation of antimicrobial and physical properties of edible films based on carboxymethyl cellulose containing potassium sorbate on some mycotoxigenic *Aspergillus* species in fresh pistachios. *LWT—Food Sci Technol* 44 (2011):1133–1138.

Seol KH, Lim DG, Jang A, Jo C, Lee M. Antimicrobial effect of κ-carrageenan-based edible film containing ovotransferrin in fresh chicken breast stored at 5°C. *Meat Sci* 83(3) (2009):479–483.

Shen XL, Wu JM, Chen Y, Zhao G. Antimicrobial and physical properties of sweet potato starch films incorporated with potassium sorbate or chitosan. *Food Hydrocoll* 24(4) (2010):285–290.

Sivarooban T, Hettiarachchy NS, Johnson MG. Physical and antimicrobial properties of grape seed extract, nisin, and EDTA incorporated in soy protein edible films. *Food Res Int* 41(8) (2008):781–785.

Stiles M. Biopreservation by lactic acid bacteria. *Antonie van Leeuwenhoek* 70(2–4) (1996):331–345.

Sun XD, Holley RA. Antimicrobial and antioxidative strategies to reduce pathogens and extend the shelf-life of fresh red meats. *Compr Rev Food Sci Food Saf* 11 (2012):340–354.

Tajkarimi MM, Ibrahim SA, Cliver DO. Antimicrobial herb and spice compounds in food. *Food Control* 21(9) (2010):1199–1218.

Tiwari BK, Valdramidis VP, O'Donnell CP, Muthukumarappan K, Bourke P et al. Application of natural antimicrobials for food preservation. *J Agric Food Chem* 57(14) (2009):5987–6000.

Turgis M, Vu K, Dupont C, Lacroix M. Combined antimicrobial effect of essential oils and bacteriocins against foodborne pathogens and food spoilage bacteria. *Food Res Int* 48 (2012):696–702.

Ünalan I, Arcan I, Korel F, Yemenicioglu A. Application of active zein-based films with controlled reléase properties to control *Listeria monocytogenes* growth and lipid oxidation in fresh Kashar cheese. *Innov Food Sci Emerg Technol* 20 (2013):208–214.

van Vuuren SF, Viljoen AM. Antimicrobial activity of limonene enantiomers and 1,8-cineole alone and in combination. *Flavour Frag J* 22(6) (2007):540–544.

Yener FYG, Korel F, Yemenicioglu A. Antimicrobial activity of lactoperoxidase system incorporated into cross-linked alginate films. *J Food Sci* 74(2) (2009):M73–M79.

Ziani K, Fernandez-Pan I, Royo M, Maté J. Antifungal activity of films and solutions based on chitosan against typical seed fungi. *Food Hydrocoll* 23 (2009):2309–2314.

Zivanovic S, Chi S, Draughon AF. Antimicrobial activity of chitosan films enriched with essential oils. *J Food Sci* 70(1) (2005):M45–M51.

Section IV

Encapsulation and Controlled Release in Films and Coatings

16

Methods of Encapsulation

Izabela D. Alvim, Ana S. Prata, and Carlos R.F. Grosso

CONTENTS

16.1 Introduction

Microencapsulation is a technology that permits the formation of structures whose main functions are the protection and controlled release of substances. Enzymes, cells, bioactive substances, dyes, flavors, and adjuvants of technology, among other named core or active substances, can be microencapsulated and applied to various products, having preserved their properties and activities against processing conditions, storage, or end uses. Several industrial areas, including the food products, cosmetics, pharmaceutics, veterinary, agricultural, chemical, automotive paints, and printing industries, are using microencapsulation to produce differentiated ingredients that are used to obtain innovative products (Shahidi and Han 1993). In addition to the classic applications in the food industry to mask undesirable flavors and odors and to facilitate handling of volatile liquid compounds by converting them into solid powders (Byun et al. 2010), encapsulation also aims to protect the encapsulated material from the adverse conditions of the medium, including light, oxygen, temperature, humidity, pH (Gibbs et al. 1999), and biological conditions, present during product ingestion and absorption in the gastrointestinal tract (Jones and McClements 2010). Moreover, encapsulation ensures the controlled release of the encapsulated active material at the desired site and time for maximum effectiveness of the active agent (Desai and Park 2005). These effects can be obtained by coating the active material or by inserting it into a film or wall material matrix, resulting in the development of encapsulated materials with different shapes and sizes, including nano-, micro-, and macrosized (Augustin and Hemar 2009; Thies 1995). Many different methods for preparing encapsulated materials are available, some of which have been widely adopted and are known with respect to their chemical, physical, or physicochemical properties. These methods include spray drying, spray chilling, and cooling, fluid bed coating, complex coacervation, liposome production, extrusion processes using carbohydrates or ionic gelation, spinning disc/centrifugal extrusion, and molecular inclusion, among others (Shahidi and Han 1993). Recently, we have observed an exponential growth of new methods that are still little explored regarding the principles that allow the encapsulation of active agents and the properties of encapsulated materials. These methods include supercritical systems and the use of combined techniques to produce encapsulated materials with novel functionalities (Gouin 2004). Usually, the encapsulation methods are classified as physical, physicochemical, or chemical processes, and the selection of each method depends on the physicochemical properties of the active material, the wall material used, the particle size required, the release mechanism and kinetics desired, the difficulty in moving production to the industrial scale, and the costs involved (Ré 1998). Other important considerations include the desired amount of active ingredient; the method of transfer and incorporation into food ingredients; the conditions associated with processing, handling, and storage; and the preparation conditions of food products containing the encapsulated material (Shahidi and Han 1993).

The types of materials that can be encapsulated are varied and exhibit distinct chemical, physical, and physicochemical characteristics, including size, polarity, and solubility. For materials with biological activity, their susceptibility to the conditions observed during passage through the mouth, stomach, and intestinal tract, as well as specific target sites and colonization sites, should be considered (Jones and McClements 2010). The materials to be encapsulated may be liquid, solid, or gas, and can form solutions, emulsions, and suspensions (Shahidi and Han 1993). Concerning their application in the food industry, compounds or materials to be encapsulated can be classified into technological adjuvants as dyes, fermentation agents, flavoring agents, acidulants, and microbiological preservatives, among others. More recently, a second class of substances with specific physiological functionalities, known as bioactive substances, is being widely used for nutritional and health purposes. These bioactive substances include vitamins, minerals, probiotic microorganisms, compounds with prebiotic and symbiotic activity, omega-3 polyunsaturated fatty acids, and natural antioxidants, among others (Augustin and Hemar 2009). Consumers who are better informed and interested in food products with novel properties associated with well-being and health generate an increased demand for innovative products, technologies, ingredients, and manufacturing processes in the food industry (Onwulata 2013).

Both classes of products need to meet strict safety standards regarding handling, processing, and storage. The technological adjuvants must adequately achieve their technological function, which means that these substances need to be active, used in adequate amounts, and effectively incorporated into food products during preparation and consumption. With respect to bioactive materials, in addition to being active and used in adequate amounts, the vehicle (particle) must provide protection to the active material after its passage through the mouth and stomach and ensure its release in the intestines (De Vos et al. 2010). The incorporation of encapsulated active materials into food products, regardless of their type or function, should not alter the physical, chemical, or sensory characteristics of the food product, including its texture, appearance, stability, flavor, and aroma (Jones and McClements 2010).

When used in food products, wall materials need to be generally regarded as safe (GRAS) and of food grade. The number of compounds with these characteristics observed in nature is limited, and novel materials or modifications of existing ones are both difficult to find and costly (Jones and MacClements 2010). These compounds include proteins, carbohydrates, lipids, and, in some cases, mixtures thereof, and each compound has specific physical and chemical properties, which determine which ones should be used depending on the purpose of the application (Augustin and Hemar 2009).

The release of the active material can be controlled by the trigger release, according to the action it should perform. The trigger release can be started by environmental conditions, such as pH and temperature, mechanical stimulation such as chewing or friction, mouth and gastrointestinal conditions, or properties of the matrix particle such as its solubility and swelling. When the particle matrix is inert in the environment in which it is inserted, the core can still be released by diffusion (Shahidi and Han 1993).

As an artifact used to control the release of active substances, the particles may be associated with other structures, such as edible/biodegradable films or coatings, usually produced with the same materials used in the construction of the particles. The inclusion of particles in films is a challenge that requires intensive research. The insertion of the particles homogeneously into the films is a major challenge, as is the release behavior for the combination of particles/films (Cerqueira et al. 2013; Silvestre et al. 2013).

This chapter presents various microencapsulation techniques with a description of the particle characteristics and the core and wall materials best suited to each method and trigger release, thus trying to give the reader many particle options to help in the selection of a delivery system/protection most suitable for the desired application.

Several encapsulation methods have been discussed in detail (spray drying, spray cooling, fluid bed coating, coacervation, ionic gelation, combined methods such as ionic gelation and electrostatic interaction) because these methods are widely used and relatively simple. Furthermore, these methods can easily be used for the insertion/adsorption/combination of materials into edible and biodegradable films. Many other methods, including supercritical technology, extrusion technology, spinning disc, liposome, and inclusion complex/complexation, are also used; however, these methods will not be discussed in this chapter. Their characteristics, advantages, disadvantages, and challenges can be found in the review articles consulted during the preparation of this study.

16.2 Wall Materials: Properties and Physicochemical Characterization

The wall materials used in microencapsulation in the food industry are in general biomaterials and include proteins, carbohydrates, and lipids (Shahidi and Han 1993). Wall materials are selected based on their characteristics, type of protection, desired particle type and size, method of release of the active ingredient (trigger release or controlled release), encapsulation process available and/or desired, cost, and regulatory aspects (Desai and Park 2005; Onwulata 2013). Other important considerations include physical and chemical characteristics, such as the composition, size, emulsifying capacity, solubility, viscosity in solution, and electric charge of the components (Jones and McClements 2010). These characteristics determine the microparticle wall structure, encapsulation efficiency, degree of protection of the active material, and the method of release of the latter (Augustin and Hemar 2009).

Proteins exhibit amphiphilic properties and are formed by the sequence and number of amino acids that are linked together by peptide bonds. The exposed amino acid residues along the polypeptide chain confer specific characteristics to each type of protein, including its structural conformation, solubility, hydrophilic nature, hydrophobic nature, interactions, electric charge, and chemical reactivity. These characteristics are affected by the conditions of the medium, including the pH, temperature, and ionic strength (Damodaran 1996).

Proteins can form matrices during microencapsulation using drying, electrostatic interactions with other polymers, or gelation/thermal denaturation or can be triggered by a change in pH. Furthermore, proteins can aggregate during denaturation, which can be employed as encapsulating agents. These effects can be produced using heat, ionic strength, pH, solvent type, dehydration, or chemical treatment (Jones and McClements 2010). Their emulsifying properties are very useful in the stabilization of emulsions at the stage before the formation of microparticles. In some cases, the high viscosity or poor solubility in water can limit the methods available for particle formation, and the association of proteins with carbohydrates is common for the generation of more efficient mixed matrices (Augustin and Hemar 2009; Desai and Park 2005). Proteins can be modified using various reactions. Most proteins are available in nature in a globular conformation. Proteins from milk (caseins and whey proteins), soy, corn zein, egg albumin, gelatin, and protein hydrolysates are the most frequently cited sources/types of wall material used for the encapsulation of different types of active material using distinct encapsulation methods (Augustin and Hemar 2009; Desai and Park 2005). Gelatin is water soluble, thermoreversible, nontoxic, and inexpensive and is the protein most widely used, together with gum arabic, to produce active particles using complex coacervation, whereas whey proteins are used in the production of polysaccharide-containing particles using electrostatic interactions (Jones and McClements 2010; Shahidi and Han 1993).

Carbohydrates are biopolymers that, depending on their chemical composition and molecular size, can be classified as polysaccharides, gums, starches, dextrins, maltodextrins, and sugars. Carbohydrates can form dried matrices using processes such as spray drying and lyophilization or gels using ionic interaction with salts or electrostatic interactions (e.g., alginate + Ca^{2+}, chitosan + TPP, gelatin + gum arabic, low-methoxyl pectin + chitosan). Carbohydrates can also form amorphous or crystalline structures but exhibit low emulsifying power, and therefore, the addition of surfactants is required for emulsion stabilization. Despite their low emulsifying power, carbohydrates increase viscosity and can consequently indirectly help stabilize emulsions (Augustin and Hemar 2009). Gum arabic, a polysaccharide derived from plant exudates, contains a small protein content (<5%, according to Shahidi and Han 1993) and therefore constitutes an exception, as it exhibits not only emulsifying power but also low viscosity and high solubility in water; it is used as a major wall material in the production of active particles using spray drying for the encapsulation of flavoring agents. Its disadvantages include its cost and limited availability in large quantities (Gibbs et al. 1999). Alternatives to gum arabic include starch hydrolysates—known as maltodextrins—which are obtained via chemical or enzymatic hydrolysis and can be produced with varying degrees of hydrolysis and in various sizes. Gum arabic and maltodextrins are highly soluble in water and have low viscosity, which enable working at high solid concentrations during spray drying for the encapsulation of flavoring agents. However, gum arabic has a higher encapsulation efficiency compared with maltodextrins. In contrast, active particles produced with maltodextrins are more efficient protectors of encapsulated flavoring agents against lipid oxidation (Reineccius 1989). Starches are highly viscous and weakly soluble in water, which limits their use. However, chemical modifications of starches by succinylation can improve their solubility, consequently making them amphiphilic emulsifiers and also increasing their viscosity in solution, thus allowing the production of good encapsulating materials. Cyclodextrins (α, β, and γ) are produced by microbial enzymes (*Bacillus macerans*) using starch as a substrate. Cyclodextrins comprise cyclic molecules with a hydrophobic interior, where hydrophobic compounds can be attached, including flavoring agents, and a hydrophilic surface, which ensures their solubility in water. Cyclodextrins are exclusively used as wall materials for the production of encapsulated materials by molecular inclusion; however, this method is not significant compared with other production methods (Shahidi and Han 1993).

Various types of lipids, including oils, fats, phospholipids, waxes, and mono-, di-, and triglycerides (e.g., milk fat and cocoa butter) with different compositions, can be used in microencapsulation. This particle type requires a temperature increase above its melting temperature to allow the release of the encapsulated material. Lipids have little space available for encapsulation compared with other methods (Müller et al. 2002). Moreover, lipids can be inserted as core materials into other particle types using ionic gelation or fluid bed coating. For this purpose, solid core materials, hydrophilic and hydrophobic liquid core materials, cells, and microorganisms can be used. When a lipid or lipid mixture is used as a wall material in the production of particles using spray cooling or spray chilling, the melting temperature of the lipid or lipid mixture is extremely important for the establishment of the production conditions and the structural properties of the particles formed. The melting point of these matrices can be adjusted by mixing more rigid lipids with less rigid ones, allowing temperature-driven release triggers, which can be very useful in food applications. Solid lipids containing fat crystals can take different polymorphic forms (α, β, and β'), which are determined by the lipid composition and the cooling rate used to solidify the lipid solution, suspension, or emulsion that will produce the particles. Depending on the polymorphic state, a higher degree of structural organization results in a greater potential of expulsion of the encapsulated core material from the lipid. Core material expulsion due to crystallization is a limitation that can be prevented by mixing lipids of different compositions, and these mixtures help stabilize the active material in the matrix (Müller et al. 2002).

Many lipids are hydrophobic and are thus excellent barriers to water permeation. Therefore, these lipids can be used for the direct coating of fruits and vegetables to prevent weight loss during ripening and storage or can be applied as a secondary coating material over particles that are produced with other encapsulation techniques. Polar lipids (e.g., monoglycerides, phospholipids, and glycolipids) have an amphiphilic nature and active surfaces and can be used as emulsion stabilizers (Augustin and Hemar 2009). Soy lecithin and egg phospholipids are used in the production of liposomes (Shahidi and Han 1993) but have limitations in terms of their low physical stability when inserted into very complex systems such as food products.

16.3 Core and Bioactive Materials

The physical, chemical, and physicochemical characteristics, as well as the desired and/or expected action of the active material, should initially be considered to define the best wall material and the most appropriate microencapsulation process (Desai and Park 2005). Many wall materials and processes are known and available, and several active, and/or functional substances with applications in the food industry can be microencapsulated to guarantee their protection, transport, and controlled release. Nonetheless, each active substance has specific characteristics, indicating that no universal wall material and/or microencapsulation process can be adopted during the development of active materials (Wilson and Shan 2007). An active material can be encapsulated using more than one process and coating material such that, in the final application, the desired functions of the active material and the type of release dictate the most suitable combination.

Active materials can be classified as solid, liquid, or gas. Nonetheless, the number of studies on the encapsulation of gaseous substances is small compared with those on the other two groups. For liquid active materials, hydrophilicity and hydrophobicity are important parameters for the selection of the encapsulation process. Some microencapsulation methods, such as spray drying and spray chilling, allow the use of materials with both properties (Augustin and Hemar 2009; Zuidan and Shimoni 2010). Other methods, conducted in aqueous solutions, are suitable for the production of hydrophobic active materials, with the formation of emulsions during the initial stage of the microencapsulation process, such as ionic gelation and simple/complex coacervation (Zuidan and Shimoni 2010). The incorporation of active hydrophilic materials in these processes can be performed by immobilizing active materials in a hydrophobic matrix in lipid microparticles or simple/multiple emulsions and subsequently including these into other particles using a second method.

In addition to their polarity, the volatility of the active materials is an important characteristic to be considered in the development of microparticles because, in this case, the wall material should be efficient at forming films and should rapidly form barriers and walls/matrices without porosity to maximize the retention of the volatile compounds in these structures. These active materials are usually unsaturated compounds and therefore require antioxidant protection of the wall material (Reineccius 1989).

Solid active materials, including cells and microorganisms, must have low or no solubility in the wall material, where the microparticles will be formed. In addition, the maintenance of the active material in the solid form as salt crystals and minerals may be necessary. In this case, the electrostatic charges on the particle surface and particle size are some of the characteristics of the active material that must be considered. Microorganisms are "particles" ranging from 1 to 5 μm in size and require processes capable of producing larger particles and in sufficient amount to adequately accommodate these microorganisms. In this case, the use of particles or processes at the nanoscale level becomes infeasible. Furthermore, crystals may have inadequate sizes and may therefore need to be ground or micronized for size adjustment to become suitable for microencapsulation (De Vos et al. 2010).

The susceptibility of the active material to the application conditions (food products), processing conditions, and product use is an essential aspect to be considered when defining a wall material/process. If the active material is susceptible to oxidation, the particle produced must become a barrier to oxygen; if the material is susceptible to temperature, the wall must protect the active material, and the microencapsulation process must be less aggressive. Probiotic microorganisms must reach the gastrointestinal tract not only alive but also maintaining enough of their bioactive properties to exert the expected beneficial effects. Therefore, the microencapsulation process should not significantly affect the survival of these microorganisms, should protect them from the harsh conditions of the stomach, and should deliver them alive and active to the target sites in the intestines (De Vos et al. 2010).

In the area of food products, an exponential growth in the number of studies and publications on bioactive compounds has been observed recently. These studies have focused on the novel sources, the identification and characterization of active fractions, and the assessment of the antioxidant capacity using *in vitro* and *in vivo* activity assays. Bioactive materials can be classified into three main types according to their biological activity. One large group of materials is associated with antioxidant activity, a second group relates to nutritional essentiality, and a third group involves microorganisms with probiotic activity. These materials exhibit specific physical, chemical, and physicochemical characteristics and susceptibilities and may require protection and specific release mechanisms to exert their bioactive effects at the appropriate site and time, which can be performed with the aid of encapsulation (Onwulata 2013).

16.4 Triggers and Release Mechanisms

The active material can be released immediately via solubilization, disintegration, or disorganization of the wall material or instead gradually through the wall. In the first case, the mechanisms that rupture the particle wall are used when a rapid release under a particular stimulus is needed. These specific stimuli are called release triggers and may occur by mechanical rupture of the wall and manipulation of the medium conditions where the particle is placed, such as temperature, pH, and enzymatic activity, or through the selection of solvents used for solubilization of the microparticle wall material (Reineccius 1989). The distribution of the material encapsulated in the particle (mononucleated, multinucleated, matrix type, or reservoir type) also affects the release (Thies 1995).

The chemical characteristics of the wall material, its interactions with the encapsulated active material, the resulting vitreous transition temperature, and the particle morphology, including the presence of pores or cracks in the particle wall, determine the release/loss of active material as well as the volatility of the encapsulated material (Augustin and Hemar 2009). Particles in the form of powders may suffer loss of the active material by evaporation or oxidation, and the latter occurs by the permeation of oxygen through the particle matrix and/or wall (Reineccius 1989). Moreover, water absorption due to relative humidity, temperature, and the packaging used affect the release. In this sense, a higher water content increases the release rate of the active material (Shahidi and Han 1993). In general, food powders containing active particles are designed for immediate particle solubilization when in contact with water during food preparation, and the immediate release of the active material is often associated with an increase in temperature to accelerate solubilization (Soottitantawat et al. 2005; Vega and Roos 2006). The primary effect of wet particles, or particles that rehydrate without loss of wall integrity, may be the release of the encapsulated active agent to the external medium through diffusion.

Microcapsules that rupture upon application of a force (shearing or external pressure) were developed and generated an important patent in the history of encapsulation, known as carbonless copy paper (Versic 1988). Other microparticles were developed along those lines, including those used in chewing gums with synthetic flavors and sweeteners (Meyers 2014; Rømer Rassing 1994), deodorants, fungicides applied to footwear (Abderrahmen et al. 2011; Meirowitz 2010), and polishing pastes (Perfetti et al. 2012).

The medium conditions may favor particle solubilization by selecting solvents capable of dissolving the wall-forming material—including particles designed by spray drying—or by changing their physical state by setting the temperature above the melting point of the coating wall, as is the case with lipid particles (Binks and Rocher 2009; Jannin and Cuppok 2013) used in the encapsulation of chemical leaveners for bakery products (Al-Widyan and Small 2004; Toublan 2014). Other applications of this trigger mechanism include temperature indicators for frozen or baked foods (Lee and Rahman 2014), cosmetic components that are released as a function of the body temperature (Cheng et al. 2008; Kim et al. 2012), and flavoring agents for teas and cakes (Milanovic et al. 2010).

The wall can also be ruptured through the adjustment of the medium's ionic strength, which is associated with the presence of enzymes, and by changing the temperature (biodegradation). Polymers that are sensitive to these changes degrade rapidly and release the active component, as occurs for enteric microcapsules used to transport active components that need to pass intact through the stomach and target the intestines, where the particle wall solubilizes and releases the active compound (De Vos et al. 2010). Several studies and commercial products associated with the encapsulation of probiotic microorganisms are available (Doherty et al. 2012; Gbassi and Vandamme 2012; Picot and Lacroix 2004). In the cases described earlier, the release is not sustained. However, the release of encapsulated compounds at specific sites and times can be achieved.

When the wall structure remains unchanged, the release occurs slowly. These mechanisms are associated with diffusion-controlled release through solvent penetration or chemically controlled release, with or without particle swelling, and may involve the slow degradation of the wall material. These mechanisms, despite being accompanied by diffusion after solvent penetration, exhibit different solubilization rates of the encapsulated active ingredient, which modifies the temporal profile of particle release.

The release of active components by diffusion is due to the wall permeability and to the solubility of these components in the wall material and release medium. The difference in concentration between the particle interior and the medium is the driving force of release (Higuchi 1963). In addition to the wall permeability and solubility of the active material, the particle size and wall thickness affect the kinetics of diffusion to the release medium (Vandenberg et al. 2001; Zhang et al. 2012). In studies on product release by diffusion, limitations were reported concerning the selection of release media that represent real situations, such as the passage of particles through the mouth and gastrointestinal tract or the product release into food matrices (Brannon-Peppas and Peppas 1989; Ortakci et al. 2012; Prata et al. 2008).

Microcapsules that contain weak or permeable walls can be modified with cross-linking agents, which strengthen the particle wall and can decrease its porosity, and these effects modulate the release of the active component and prolong its effect. Although microparticles obtained by complex coacervation can encapsulate large amounts of hydrophobic materials, these particles are fragile and require a cross-linking agent after production (Thies 1995).

16.5 Spray Drying

Drying by atomization is a well-established process in the food industry and is based on the conversion of liquids into powders by spraying the liquid into a heated chamber, followed by evaporation of the solvent, usually water, and collecting the dried material in the form of a powder (Augustin and Hemar 2009; Gharsallaoui et al. 2007; Ré 1998).

In microencapsulation, spray drying can be used in the formation of structures for the protection, retention, and transport of active materials. Initially, during atomization, a series of microdroplets of a solution/emulsion/dispersion of the wall material is formed, followed by rapid solvent evaporation and fixation of the dry microparticle, resulting in the entrapment of the compound of interest (Augustin and Hemar 2009; Gharsallaoui et al. 2007).

Among the microencapsulation techniques commercially available in the food industry, spray drying is the most widely used. Some of its advantages include economic feasibility in addition to being a well-established process. Furthermore, equipment for large-scale production is available and can yield particles of good quality (Ré 1998). Its disadvantages include the small particle size (poor reconstitution, high surface area) and the limited number of wall materials (Desai and Park 2005; Gharsallaoui et al. 2007; Gouin 2004).

Microencapsulation by spray drying is recommended for the protection/transport of active materials with technological functions (e.g., dyes, preservatives, salts, flavoring agents, enzymes) and bioactive functions (e.g., vitamins, minerals, microorganisms, polyunsaturated oils, phytonutrients such as antioxidants, polyphenols). These materials can have hydrophilic properties (solutions and dispersions) or hydrophobic properties (emulsions). Despite the high temperatures used during drying (over 100°C for aqueous solutions), spray drying is very efficient and suitable for the microencapsulation of flavoring agents and of volatile and heat-sensitive compounds (Reineccius 1989). For flavoring agents and other volatile compounds, despite the high volatility of some of their constituents, their retention in the microparticle matrix is high because of the phenomenon of selective diffusion, which is based on the principle that the diffusion coefficients of volatile compounds decrease by several orders of magnitude relative to the coefficient of water in concentrated solutions. During drying, the diffusion coefficients of the volatile compounds in the dry surface region become much smaller than those of water, thus drastically reducing the possibility of loss of the active compounds, which are retained in the matrix (Ré 1998).

For thermosensitive substances, the short contact time of droplets/particles with the hot air in the drying chamber, together with the rapid evaporation of water, keeps the particle temperature relatively low, consequently avoiding or minimizing the damaging effects of temperature on the active material (Augustin and Hemar 2009).

The wall materials used in microencapsulation by spray drying for application in food products should be GRAS and exhibit high solubility and low viscosity at high concentrations, good film formation, low cost, and increased availability as their main characteristics. Among the available materials, the most commonly used include gums (essentially acacia gum), modified starches, maltodextrins and their mixtures, proteins (whey, caseinates, and soy), and high-viscosity polysaccharides (alginates, carboxymethylcellulose, and guar gum).

The preparation of the mixture containing the wall and active materials is a very important step in the process of microencapsulation by spray drying. Furthermore, for hydrophobic active materials, emulsion stability throughout the entire process, including stability inside the micro droplets, is essential for the proper retention of the active material within the matrix. If retention does not occur, a substantial amount of the active material may be deposited on the microparticle surface, which thus loses its protective effect.

The microparticles obtained using this process can have varying sizes and exhibit high polydispersity. These particles are predominantly microspheres; however, special atomizers, such as three-fluid atomizers, can produce true (reservoir type) capsules. The release is mainly associated with the wall material composition—in addition to the wall material application—and can be triggered primarily by solubilization or erosion. The partial cross-linking of certain coating materials can decrease the solubility and enable the formation of particles released by diffusion.

16.6 Spray Chilling/Cooling

Spray chilling/cooling is a technique of microencapsulation of lipid-coated active materials. As in spray drying, this technique is based on the atomization of a liquid solution/dispersion/emulsion, comprising the wall and active materials. In contrast to the drying process, in which the hardening of microdroplets and the formation of microparticles occur by solvent evaporation in a heated chamber, during the chilling/cooling process, droplet structuring and formation occur through liquid cooling in a cooling chamber. The particles exhibit varied and polydisperse sizes and a matrix-type spherical structure and are therefore classified as microspheres. The release trigger is based on the melting temperature of the lipid matrix. Microencapsulation by chilling/cooling spray is considered economically viable, and equipment is commercially available on an industrial scale. The type of equipment used is very similar to that used in spray drying, offers high process output, and can be run in batch or continuously (Nedovic et al. 2011). Particle formation by spray cooling and spray chilling is similar. In spray chilling, the melting point of the wall material is in the range of $34°C–42°C$, whereas in spray cooling, the melting point is higher (Nedovic et al. 2011).

Various methods can be used to produce lipid microparticles (e.g., cold emulsification and high-pressure homogenization). However, spray chilling has the advantage of the direct obtainment of particles in a single step, without the need for subsequent particle separation from water. The disadvantages associated with spray chilling/cooling include the limited loading capacity for active materials (generally 30%) compared with reservoir-type microparticles as well as the limited expulsion of the active material from inside the matrix to the matrix surface during storage (Müller et al. 2002).

The expulsion of the active material from inside the lipid structure may occur through crystallization of the molecules that form this matrix. Depending on the lipid composition, lipid molecules can rearrange themselves during cooling or storage in a highly ordered and compact manner, consequently expelling the active material from the structure. To avoid this problem, imperfections can be created in the matrix to prevent crystallization, such that the lipid structure remains solid but not crystalline. The use of mixtures of solid and liquid lipids in the formulation of the lipid wall material can produce more efficient matrices regarding the retention of the active material because the active material can more easily fit into the imperfections created by the mixture containing molecules of different sizes and spatial conformations (Muller et al. 2002). Parameters such as the maintenance temperature of the lipid in the liquid state and the cooling temperature of the chamber affect the crystalline arrangement of the lipid matrix, whereas the stability of the emulsion/dispersion is associated with the encapsulation efficiency and increased protection of the active material.

The active material to be microencapsulated by spray chilling/cooling can be either hydrophilic or hydrophobic. In the first case, a simple or multiple emulsions are prepared, in which the final continuous phase involves wall lipids, and in the second case, the active material can be solubilized in the lipids.

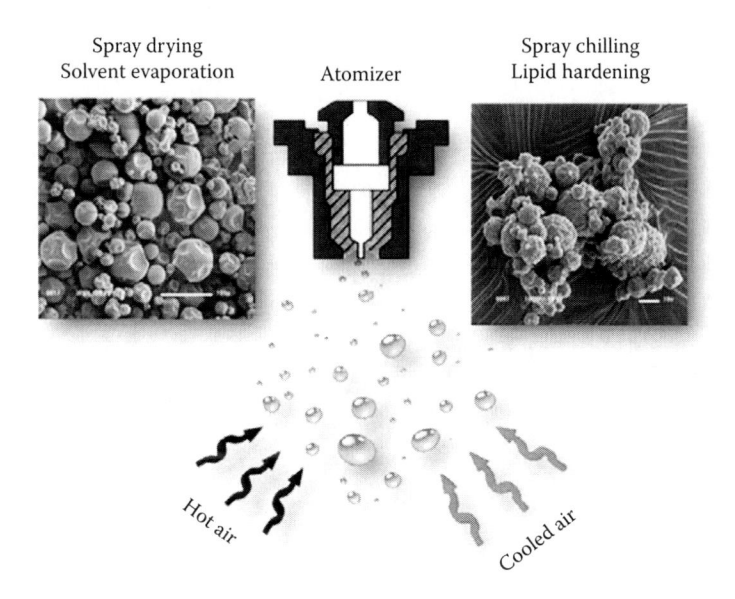

Spray drying
Solvent evaporation

Atomizer

Spray chilling
Lipid hardening

Hot air

Cooled air

FIGURE 16.1 Spray drying/spray chilling scheme: principle of particle formation and morphology of microparticles.

Particles containing insoluble materials (microorganisms, salt crystals, and sugars) may also be coated with lipids using this technique, as long as size adjustments are made and the homogeneity of the dispersion is maintained. Figure 16.1 illustrates the processes of spray drying and spray chilling/cooling.

16.7 Fluid Bed Coating

The principle of fluidization, that is, the suspension of particles in an upward or downward airflow, can be used to produce coated particles. For this purpose, a spray nozzle similar to that used in spray drying is used to atomize an aqueous solution or a liquid lipid over the suspended particles, which will be deposited by solvent evaporation or by the solidification of a molten compound (Sauer et al. 2013). The particles may be coated with a single layer or multiple layers of the same material or of different materials by passing these particles through the spray zone several times (Prata et al. 2012).

In contrast to encapsulation by spray drying, which in most cases yields particles with matrices that are mixed homogeneously with the active material (Dziezak 1988), the layers created using fluid bed coating generate particles with a defined core and particles that are much larger than those produced by spray drying.

This technique was first used as a drying technique due to its efficient heat transfer, which was ensured by particle separation and by the convection generated by the passage of hot air. Moreover, this approach has other advantages, including the ability to coat irregular shapes, the high-coating-rate yields, the relatively low cost, and the ability to operate continuously (Ivanova et al. 2005). A limitation involves the particle size obtained, which ranges from 100 μm to 2 mm (Leeke et al. 2014). The thickness and roughness of the coating layer and the presence of cracks and pores in the coating are factors associated with product release.

The same parameters as those of spray drying affect the particle coating using fluid bed coating. Additional process variables include the parameters adopted for obtaining a good fluidization regime (the relative particle density as a function of the fluidizing air and airflow rate), the contact angle between the coating solution and the particle, the number of cycles, and the equipment settings (Bacelos et al. 2007; Guignon et al. 2003).

Two different positions of the atomizer nozzle relative to airflow can be used. However, the bottom position, in which the atomized liquid is applied in the same direction as the fluidizing air, provides a coating that is more uniform and adequate to the controlled release compared with the top spray position (atomized liquid against the fluidizing airflow). The quality of the obtained coating using the top spray is not excellent but is sufficient to mask the taste or modify the appearance of the powder (color and texture) (Ronsse et al. 2008).

The bottom spray setting has the advantage of reducing the risk of drying before droplet deposition, and this effect is further improved by inserting a tube that concentrates the particles in the nozzle region. This tube, called a Wurster tube, is not fixed on the distribution plate, and its spacing can be adjusted to allow the flow of particles that sediment nearly in a dry state on the annular region, allowing for a new coating cycle (Arsenijevic et al. 2002; Su et al. 2014).

The coating may be formed by evaporation of the solvent, which corresponds to not less than 70% of the composition of the solution. In this case, a large amount of energy is consumed during this operation to obtain a very low percentage of coating. Coatings with hot-melt materials have been increasingly studied because of their ability to solidify on the surface of particles in contact with air at room temperature in addition to requiring no solvent and less processing time. One of the limitations of the hot-melt process is the need to operate with temperature-sensitive products. Alternatively, dry coating can be employed through the combined use of micronized powders, a suitable vitreous transition temperature, and plasticizers.

16.8 Coacervation

The phases of a complex mixture can be separated using segregative or associative processes (Polyakov et al. 1997). Associative processes, which occur in many mixtures of two oppositely charged polyelectrolytes, are known as complex coacervation (Doublier et al. 2000). The word coacervation derives from the Latin co (together) and acerv (amorphous mass) to indicate the aggregation of colloidal particles, and this process can be simple or complex. Aggregation in simple coacervation is obtained by adding a highly hydrophilic agent to a colloidal solution or by decreasing the temperature. The addition of a non-solvent, as well as the change in pH or ionic strength, can also be regarded as inducing factors (Bachtsi and Kiparissides 1996; Magdassi and Vinetsky 1996).

Complex coacervation involves more than one polymer in an aqueous medium, where polymers interact by electrostatic attraction (Burgess and Carless 1985). The insoluble complexes formed between the polymers are concentrated in the liquid droplets that sediment and coalesce to form a coacervate phase (De Kruif et al. 2004). Because of this mixture, the separation occurs in two liquid phases in colloidal systems. In the colloid, the coacervate is the most concentrated phase, and the other phase is the equilibrium solution (Weinbreck et al. 2004).

The associative phase separation between proteins and polysaccharides is important in many biological systems (De Kruif et al. 2004) and can be used to microencapsulate hydrophobic compounds. Coacervation is based on complex physical and chemical characteristics and involves variables such as the stirring rate, the core/wall ratio, the polymer concentration and charge, the characteristics of the active material, and the addition, stirring, and cooling rates (Schmitt et al. 1998; Thimma and Tammishetti 2003). Data on the formation mechanism of coacervates and their structure are limited, and this lack of information restricts the prediction of technofunctional properties of encapsulated materials and their processing (Schmitt 2000).

The formation of biopolymer complexes is mainly due to electrostatic interactions and depends on the degree of ionization of the polymers and therefore on the system pH (Weinbreck et al. 2004). Coacervation depends primarily on the net charge of the system and is affected by several factors, including the stoichiometry, parameters of the structural polymers (proper conformation and chain length), and medium conditions, including the pH, ionic strength, temperature, and nature of reagents (De Kruif et al. 2004). Microencapsulation, in turn, occurs in the presence of an immiscible core and in suitable proportions (Rabiskova and Valaskova 1998).

In general, complex coacervation comprises three stages: (1) formation of a system comprising three immiscible chemical phases (one solvent, one coating material, and one core material), (2) deposition of the liquid polymeric material loaded with a charge opposite to that of the colloidal species, which will be responsible for coating formation, and (3) coating solidification.

The microcapsules produced by coacervation may have a small diameter (4 µm) and high encapsulation efficiency (Liu et al. 2010). The main disadvantage is the need to maintain strict control of the concentration of colloidal material and the coacervation initiator because coacervation will occur only within a limited range of pH and concentration of colloid and/or electrolyte (Burgess and Carless 1985).

Not every oppositely charged polyelectrolyte mixture will form coacervates, and not every coacervate will form a microcapsule. For microencapsulation by complex coacervation to occur, the complexes need to be formed around the active material to be encapsulated.

The local variations in the concentration of the biopolymer within the mixture volume may originate from inefficient component mixing, temperature fluctuations, or specific and nonspecific interactions. Thus, phase separation is a kinetic process that may not occur immediately, depending on the conformation of the polymers in solution (Tolstoguzov 2003).

16.9 Ionic Gelation

Ionic gelation is formed by the interaction between the carboxylic groups present in anionic polysaccharides, such as alginates, carrageenan, and low-methoxyl pectins, and divalent ions, such as calcium ions, which are extensively used in food applications. During gelation, gels may be molded into different shapes, including microcapsules (Krasaekoopt et al. 2003). Ionic gelation is used for the encapsulation of emulsions containing hydrophilic or hydrophobic compounds (McClements 2005), and many different types of active materials can be used, including antioxidants, essential fatty acids, and compounds with biological activity, such as probiotic microorganisms, enzymes, and cells (Patil et al. 2010).

The concentration of the cations and polysaccharides, pH, and ionic strength determines the kinetics, stability, and volume of the gel formed. The construction of phase diagrams (polysaccharide concentration vs. calcium ion concentration) is useful for identifying regions where the mixtures remain as solutions or as gels or gel very rapidly, with expulsion of water through a process known as syneresis, which is an undesirable condition during production (Mestdagh and Axelos 1998).

The technique does not require high temperatures, organic solvents, or extreme pH and uses materials naturally present in food products that are safe and nontoxic for human consumption in addition to being low cost (Patil et al. 2010). The aforementioned polysaccharides resist the conditions in the stomach or upper intestine. However, in the colon, these polysaccharides are hydrolyzed through the activity of bacterial enzymes normally present in this region (Liu et al. 2003; Vandamme et al. 2002). Alginate gels are also heat resistant. This technique has high encapsulation efficiency for hydrophobic compounds. Its disadvantages include the limited production of small-sized particles as well as its susceptibility to the presence of ions such as citrates and phosphates, which act as calcium ion scavengers and can solubilize the gel formed (Poncelet et al. 1992). Although ionic gelation is a simple and mild technique, the gel formed is porous, which can accelerate the permeation of oxygen into the matrix or the release of both hydrophilic and low-molecular-weight encapsulated active compounds (Sezer and Akbuga 1999).

Both external and internal ionic gelation can be used. In external ionic gelation, the anionic polysaccharide is dripped onto a cationic solution, both at appropriate concentrations, and this process enables the formation of gels of different shapes and sizes, producing a 3D structure containing the encapsulated active material and a large amount of water. The interactions of ions with the carboxylate groups of the polysaccharides result in the formation of insoluble gels (Gombotz and Wee 1998).

The process of internal ionic gelation involves the release of calcium ions from an initially insoluble calcium source. The calcium salt is added to the polysaccharide solution/emulsion, and the pH is adjusted to avoid salt solubilization. The mixture is inserted into a vegetable oil bath, after which an organic acid is added to the oil and diffuses into the polysaccharide solution/emulsion. This diffusion acidifies the system and enables salt dissociation and the release of calcium ions, which in turn form complexes with the carboxylate groups present in the polysaccharide solution/emulsion, thus enabling gelation in the form of particles due to the controlled stirring in the oil bath (Draget et al. 1990; Poncelet et al. 1992).

To control the solution–gel transition during external gelation, the most important factors to be adjusted are the calcium concentration, polymer composition, and polymer concentration. During internal ionic gelation, the most important factors to be considered are the calcium solubility and calcium concentration—which are modified by adjusting the pH of the system—as well as the organic acid concentration (Draget et al. 2000).

Arguments associated with industrial scale-up found in the literature are controversial but indicate increased difficulty in the use of external ionic gelation in contrast to internal gelation, with the latter allowing industrial-scale production (Poncelet et al. 1992). However, because of intensive research on novel production alternatives, scale-up currently appears feasible even with the use of external ionic gelation (De Vos et al. 2010).

16.10 Combination of Ionic Gelation and Electrostatic Interaction

Polysaccharides are abundant in nature and are available from renewable sources, including plant exudates, microorganisms, seeds, fruits, algae, and microalgae. Moreover, polysaccharides have structures and properties that cannot be easily mimicked by synthetic materials (Coviello et al. 2007; Krasaekoopt et al. 2003). Ionic gelation is a mild process that does not require the use of organic solvents or high temperature. This process occurs by the ionic association between the carboxylic groups present in anionic polysaccharides and divalent ions, such as Ca^{2+}, and is considered a low-cost technique (Patil et al. 2010).

However, the particles produced by ionic gelation are porous and may allow oxygen migration and lipid oxidation of the encapsulated active material; migration of H^+ ions from the medium to the particle core, which results in decreased viability of the encapsulated probiotic microorganisms; or the release of low-molecular-weight hydrophilic compounds (Gouin 2004; Sezer and Akbuğa 1999). Previous studies examining calcium alginate microspheres by electron microscopy revealed that the pore diameters varied from 5 to 200 nm (Smidsrød 1974; Stewart and Swaigood 1993).

To overcome this limitation and improve particle functionality, some authors propose mixing polysaccharides with other biopolymers (Devi and Kakati 2013; Krasaekoopt et al. 2003; Yu et al. 2009) or coating them with a layer of oppositely charged polyelectrolytes via electrostatic interaction (Burgain et al. 2011; Gbassi et al. 2011; Hébrard et al. 2013).

During ionic gelation, not all carboxyl groups interact with the calcium ions, and consequently, many particles acquire a negative charge, which favors associations between the particles and positively charged polyelectrolytes, such as proteins. This strategy may ensure greater protection of the encapsulated compounds (De Vos et al. 2007). The formation of complexes between pectins and whey proteins during the production of nanoparticles was investigated to determine the best conditions of nanocomplexation between these biopolymers using zeta potential measurements to identify the best stoichiometry as well as the effect of pH and the hydrocolloid concentration used (Santipanichwong et al. 2008). Recently, microparticles using whey protein coated with alginate were used for the encapsulation of probiotic microorganisms (Doherty et al. 2012). The strategy used, mentioned earlier, involved the initial denaturation of β-lactoglobulin and subsequent electrostatic coating with polysaccharides.

Alginate particles obtained by ionic gelation using alginates and subsequently coated with undenatured whey proteins have been used and exhibited excellent performance in the encapsulation of probiotics with the specific aim of protecting microorganisms from gastric activity (Gbassi et al. 2011). This strategy for the production of alginate particles may be used to improve the stability of encapsulated functional lipids, such as fish oil and olive oil (Sun-waterhouse et al. 2011).

The electrostatic interaction between particles (solid support and solution) and polyelectrolyte solutions can be used for successive coating using the sequential adsorption of oppositely charged polyelectrolytes. This strategy is known as the layer-by-layer technique (Decher and Hong 1991). Although conceptually possible, few studies have investigated its use in food applications, and many experimental details need to be investigated. Figure 16.2 shows the morphology of particles produced by the methods described earlier.

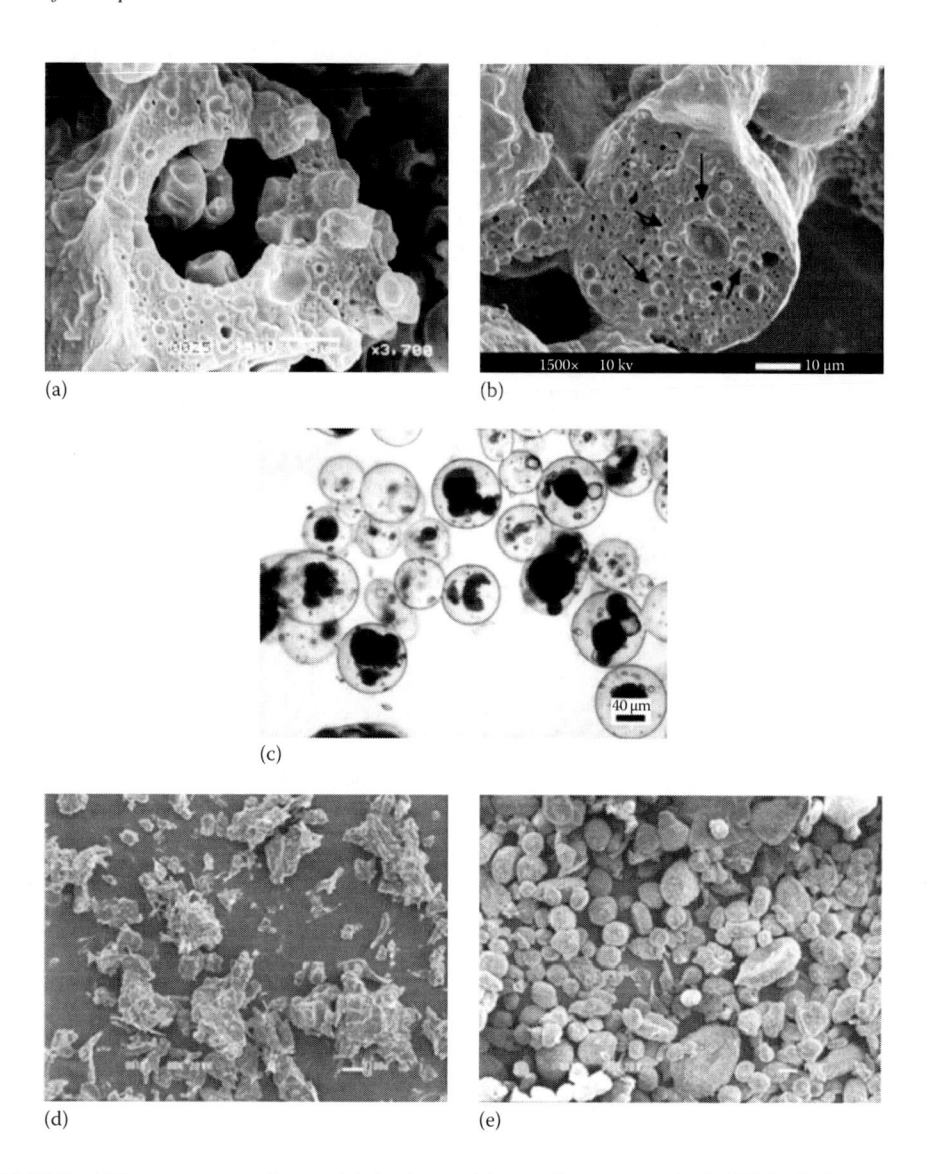

FIGURE 16.2 (a) Inner structure of spray dried microparticle containing essential oil (SEM). (b) Inner structure of coacervated microparticle containing paprika oilresin (SEM). (c) Combined method: lipid particle containing protein and afterward covered throughout with complex coacervation (optical microscopy). (d) Freeze-dried ionic gelled pectin particle containing omega-3 fatty acid covered with whey protein using electrostatic interaction. (e) Freeze-dried ionic gelled alginate particle containing omega-3 fatty acid covered with whey protein using electrostatic interaction.

REFERENCES

Abderrahmen, R., Gavory, C., Chaussy, D., Briançon, S., Fessi, H., and Belgacem, M. N. Industrial pressure sensitive adhesives suitable for physicochemical microencapsulation. *International Journal of Adhesion and Adhesives* 31(7) (2011): 629–633.

Al-Widyan, O. and Small, D. M. Microencapsulation of bakery ingredients and the impact on bread characteristics: Effect of tartaric acid encapsulated with carnauba wax. In: Cauvain, S. P., Young, L. S., and Salmon, S. (eds.), *Using Cereal Science and Technology for the Benefit of Consumers*, pp. 158–162. Woodhead Publishing, Harrogate, U.K., 2004.

Arsenijevic, Z., Grbavcic, Ž., and Garic-Grulovic, R. Drying of solutions and suspensions in the modified spouted bed with draft tube. *Thermal Science* 6(2) (2002): 47–70.

Augustin, M. A. and Hemar, Y. Nano- and micro-structured assemblies for encapsulation of food ingredients. *Chemical Society Reviews* 38(4) (2009): 902–912.

Bacelos, M. S., Passos, M. L., and Freire, J. T. Effect of interparticle forces on the conical spouted bed behavior of wet particles with size distribution. *Powder Technology* 174(3) (2007): 114–126.

Bachtsi, A. R. and Kiparissides, C. Synthesis and release studies of oil-containing poly(vinyl alcohol) microcapsules prepared by coacervation. *Journal of Controlled Release* 38(1) (1996): 49–58.

Binks, B. P. and Rocher, A. Effects of temperature on water-in-oil emulsions stabilised solely by wax microparticles. *Journal of Colloid and Interface Science* 335(1) (2009): 94–104.

Brannon-Peppas, L. and Peppas, N. A. Solute and penetrant diffusion in swellable polymers. IX. The mechanisms of drug release from pH-sensitive swelling-controlled systems. *Journal of Controlled Release* 8(3) (1989): 267–274.

Burgain, J., Gaiani, C., Linder, M., and Scher, J. Encapsulation of probiotic living cells: From laboratory scale to industrial applications. *Journal of Food Engineering* 104(4) (2011): 467–483.

Burgess, D. J. and Carless, J. E. Manufacture of gelatin/gelatin coacervate microcapsules. *International Journal of Pharmaceutics* 27(1) (1985): 61–70.

Byun, Y., Kim, Y. T., Desai, K. G. H., and Park, H. J. Microencapsulation techniques for food flavour. In: Herrmann, A. (ed.), *The Chemistry and Biology of Volatiles*, pp. 307–332. John Wiley & Sons, Inc., Hoboken, NJ, 2010.

Cerqueira, M. A., Bourbon, A. I., Pinheiro, A. C., Silva, H. D., Quintas, M. A. C., and Vicente, A. A. Edible nano-laminate coatings for food applications. In: Silvestre, C. and Cimmino, S. (eds.), *Ecosustainable Polymer Nanomaterials for Food Packaging: Innovative Solutions, Characterization Needs, Safety and Environmental Issues*, pp. 221–252. CRC Press, Abingdon, UK, 2013.

Cheng, S., Yuen, C., Kan, C., and Cheuk, K. Development of cosmetic textiles using microencapsulation technology. *Research Journal of Textile and Apparel* 12(4) (2008): 41–51.

Coviello, T., Alhaique, F., Dorigo, A., Matricardi, P., and Grassi, M. Two galactomannans and scleroglucan as matrices for drug delivery: Preparation and release studies. *European Journal of Pharmaceutics and Biopharmaceutics* 66(2) (2007): 200–209.

Damodaran, S. Amino acids, peptides, and proteins. In: Fennema, O. (ed.), *Food Chemistry*, pp. 321–430. Marcel Dekker, Inc., New York, 1996.

De Kruif, C. G., Weinbreck, F., and de Vries, R. Complex coacervation of proteins and anionic polysaccharides. *Current Opinion in Colloid & Interface Science* 9(5) (2004): 340–349.

De Vos, P., De Haan, B. J., Kamps, J. A., Faas, M. M., and Kitano, T. Zeta-potentials of alginate-PLL capsules: A predictive measure for biocompatibility? *Journal of Biomedical Materials Research, Part A* 80A (2007): 813–819.

De Vos, P., Faas, M. M., Spasojevic, M., and Sikkema, J. Encapsulation for preservation of functionality and targeted delivery of bioactive food components. *International Dairy Journal* 20(4) (2010): 292–302.

Decher, G. and Hong, J.-D. Buildup of ultrathin multilayer films by a self-assembly process: I. Consecutive adsorption of anionic and cationic bipolar amphiphiles. *Makromolekulare Chemie. Macromolecular Symposia* 46(1) (1991): 321–327.

Desai, K. G. H. and Park, H. J. Recent developments in microencapsulation of food ingredients. *Drying Technology: An International Journal* 23(7) (2005): 1361–1394.

Devi, N. and Kakati, D. K. Smart porous microparticles based on gelatin/sodium alginate polyelectrolyte complex. *Journal of Food Engineering* 117(2) (2013): 193–204.

Doherty, S. B., Auty, M. A., Stanton, C., Ross, R. P., Fitzgerald, G. F., and Brodkorb, A. Application of whey protein micro-bead coatings for enhanced strength and probiotic protection during fruit juice storage and gastric incubation. *Journal of Microencapsulation* 29(8) (2012): 713–728.

Doublier, J., Garnier, C., Renard, D., and Sanchez, C. Protein–polysaccharide interactions. *Current Opinion in Colloid & Interface Science* 5(3–4) (2000): 202–214.

Draget, K. I., Ostgaard, K., and Smidsrod, O. Homogeneous alginate gels: A technical approach. *Carbohydrate Polymers* 14(2) (1990): 159–178.

Draget, K. I., Strand, B., Hartmann, M., Valla, S., Smidsrød, O., and Skjåk-Bræk, G. Ionic and acid gel formation of epimerised alginates; the effect of AlgE4. *International Journal of Biological Macromolecules* 27(2) (2000): 117–122.

Dziezak, J. D. Microencapsulation and encapsulated ingredients. *Food Technology* 42(April) (1988): 136–159.

Gbassi, G. K. and Vandamme, T. Probiotic encapsulation technology: From microencapsulation to release into the gut. *Pharmaceutics* 4(1) (2012): 149–163.

Gbassi, G. K., Vandamme, T., Yolou, F. S., and Marchioni, E. In vitro effects of pH, bile salts and enzymes on the release and viability of encapsulated *Lactobacillus plantarum* strains in a gastrointestinal tract model. *International Dairy Journal* 21(2) (2011): 97–102.

Gharsallaoui, A., Roudaut, G., Chambin, O., Voilley, A., and Saurel, R. Applications of spray drying in micro-encapsulation of food ingredients: An overview. *Food Research International* 40(9) (2007): 1107–1121.

Gibbs, B. F., Kermasha, S., Ali, I., and Mulligan, C. N. Encapsulation in the food industry: A review. *International Journal of Food Science and Nutrition* 50(3) (1999): 213–234.

Gombotz, W. R. and Wee, S. F. Protein release from alginate matrices. *Advanced Drug Delivery Reviews* 31(3) (1998): 267–285.

Gouin, S. Microencapsulation: Industrial appraisal of existing technologies and trends. *Trends in Food Science & Technology* 15(7–8) (2004): 330–347.

Guignon, B., Regalado, E., Duquenoy, A., and Dumoulin, E. Helping to choose operating parameters for a coating fluid bed process. *Powder Technology* 130(1–3) (2003): 193–198.

Hébrard, G., Hoffart, V., Cardot, J. M., Subirade, M., and Beyssac, E. Development and characterization of coated-microparticles based on whey protein/alginate using the Encapsulator device. *Drug Development and Industrial Pharmacy* 39(1) (2013): 128–137.

Higuchi, T. Mechanism of sustained-action medication. Theoretical analysis of rate of release of solid drugs dispersed in solid matrices. *Journal of Pharmaceutical Sciences* 52(12) (1963): 1145–1149.

Ivanova, E., Teunou, E., and Poncelet, D. Encapsulation of water sensitive products: Effectiveness and assessment of fluid bed dry coating. *Journal of Food Engineering* 71(2) (2005): 223–230.

Jannin, V. and Cuppok, Y. Hot-melt coating with lipid excipients. *International Journal of Pharmaceutics* 457(2) (2013): 480–487.

Jones, O. G. and McClements, D. J. Functional biopolymer particles: Design, fabrication, and applications. *Comprehensive Reviews in Food Science and Food Safety* 9(4) (2010): 374–397.

Kim, D.-H., Park, W. R., Kim, J. H. et al. Fabrication of pseudo-ceramide-based lipid microparticles for recovery of skin barrier function. *Colloids and Surfaces B: Biointerfaces* 94(1) (2012): 236–241.

Krasaekoopt, W., Bhandari, B., and Deeth, H. Evaluation of encapsulation techniques of probiotics for yoghurt. *International Dairy Journal* 13(1) (2003): 3–13.

Lee, S. J. and Rahman, A. T. M. Intelligent packaging for food products. In: Han, J. H. (ed.), *Innovations in Food Packaging*, pp. 171–209. Elsevier, London, UK, 2014.

Leeke, G. A., Lu, T., Bridson, R. H., and Seville, J. P. K. Application of nano-particle coatings to carrier particles using an integrated fluidized bed supercritical fluid precipitation process. *The Journal of Supercritical Fluids* 91(July) (2014): 7–14.

Liu, L., Fishman, M. L., Kost, J., and Hicks, K. B. Pectin-based systems for colon-specific drug delivery via oral route. *Biomaterials* 24(19) (2003): 3333–3343.

Liu, S., Elmer, C., Low, N. H., and Nickerson, M. T. Effect of pH on the functional behavior of pea protein isolate–gum Arabic complexes. *Food Research International* 43(2) (2010): 489–495.

Magdassi. S. and Vinetsky, Y. Microencapsulation of O/W Emulsions by Proteins. In: S. Benita (ed.), *Microencapsulation Methods and Industrial Applications*, Marcel Dekker, 1996.

McClements, D. J. Theoretical analysis of factors affecting the formation and stability of multilayered colloidal dispersions. *Langmuir* 21(21) (2005): 9777–9785.

Meirowitz, R. Microencapsulation technology for coating and lamination of textiles. In: Smith, W. C. (ed.), *Smart Textile Coatings and Laminates*, pp. 125–154. Elsevier, Boca Raton, FL, 2010.

Mestdagh, M. M. and Axelos, M. A. V. Physico-chemical properties of polycarboxylate gel phase and their incidence on the retention/release of solutes. In: Colonna, P. and Guilbert, S. (eds.), *Biopolymer Science: Food and Non-Food Applications*, pp. 303–314. INRA Editions, Montpellier, France, 1998.

Meyers, M. A. Flavor release and application in chewing gum and confections. In: Gaonkar, A. G., Vasisht, N., Khare, A. R., and Sobel, R. (eds.), *Microencapsulation in the Food Industry*, pp. 443–453. Academic Press, Cambridge, UK, 2014.

Milanovic, J., Manojlovic, V., Levic, S., Rajic, N., Nedovic, V., and Bugarski, B. Microencapsulation of flavors in carnauba wax. *Sensors* 10(1) (2010): 901–912.

Müller, R. H., Radtke, M., and Wissing, S. A. Nanostructured lipid matrices for improved microencapsulation of drugs. *International Journal of Pharmaceutics* 242(1–2) (2002): 121–128.

Nedovic, V., Kalusevic, A., Manojlovic, V., Levic, S., and Bugarski, B. An overview of encapsulation technologies for food applications. *Procedia Food Science* 1 (2011): 1806–1815.

Onwulata, C. Microencapsulation and functional bioactive foods. *Journal of Food Processing and Preservation* 37(5) (2013): 510–532.

Ortakci, F., Broadbent, J. R., McManus, W. R., and McMahon, D. J. Survival of microencapsulated probiotic *Lactobacillus paracasei* LBC-1e during manufacture of Mozzarella cheese and simulated gastric digestion. *Journal of Dairy Science* 95(11) (2012): 6274–6281.

Patil, J. S., Kamalapur, M. V., Marapur, S. C., and Kadam, D. V. Ionotropic gelation and polyelectrolyte complexation: The novel techniques to design hydrogel particulate sustained, modulated drug delivery system: A review. *Digest Journal of Nanomaterials and Biostructures* 5(1) (2010): 241–248.

Perfetti, G., Depypere, F., Zafari, S., van Hee, P., Wildeboer, W. J., and Meesters, G. M. H. Attrition and abrasion resistance of particles coated with pre-mixed polymer coating systems. *Powder Technology* 230(1) (2012): 1–13.

Picot, A. and Lacroix, C. Encapsulation of bifidobacteria in whey protein-based microcapsules and survival in simulated gastrointestinal conditions and in yoghurt. *International Dairy Journal* 14(6) (2004): 505–515.

Polyakov, V. I., Grinberg, V. Y., and Tolstoguzov, V. B. Thermodynamic incompatibility of proteins. *Food Hydrocolloids* 11(2) (1997): 171–180.

Poncelet, D., Lencki, R., Beaulieu, C., Halle, J. P., Neufeld, R. J., and Fournier, A. Production of alginate beads by emulsification/internal gelation. I. Methodology. *Applied Microbiology and Biotechnology* 38(1) (1992): 39–45.

Prata, A. S., Maudhuit, A., Boillereaux, L., and Poncelet, D. Development of a control system to anticipate agglomeration in fluidised bed coating. *Powder Technology* 224(July) (2012): 168–174.

Prata, A. S., Zanin, M. H. A., Ré, M. I., and Grosso, C. R. F. Release properties of chemical and enzymatic crosslinked gelatin-gum Arabic microparticles containing a fluorescent probe plus vetiver essential oil. *Colloids and Surfaces B: Biointerfaces* 67(2) (2008): 171–178.

Rabiskova, M. and Valaskova, J. The influence of HLB on the encapsulation of oils by complex coacervation. *Journal of Microencapsulation* 15(6) (1998): 747–751.

Ré, M. I. Microencapsulation by spray drying. *Drying Technology* 16(6) (1998): 1195–1236.

Reineccius, G. Flavor encapsulation. *Food Reviews International* 5(2) (1989): 147–176.

Rømer Rassing, M. Chewing gum as a drug delivery system. *Advanced Drug Delivery Reviews* 13(1–2) (1994): 89–121.

Ronsse, F., Pieters, J., and Dewettinck, K. Modelling side-effect spray drying in top-spray fluidised bed coating processes. *Journal of Food Engineering* 86(4) (2008): 529–541.

Santipanichwong, R., Suphantharika, M., Weiss, J., and McClements, D. J. Core-shell biopolymer nanoparticles produced by electrostatic deposition of beet pectin onto heat-denatured beta-lactoglobulin aggregates. *Journal of Food Science* 73(6) (2008): 23–30.

Sauer, D., Cerea, M., DiNunzio, J., and McGinity, J. Dry powder coating of pharmaceuticals: A review. *International Journal of Pharmaceutics* 457(2) (2013): 488–502.

Schmitt, C. Effect of protein aggregates on the complex coacervation between β-lactoglobulin and acacia gum at pH 4.2. *Food Hydrocolloids* 14(4) (2000): 403–413.

Schmitt, C., Sanchez, C., Desobry-Banon, S., and Hardy, J. Structure and technofunctional properties of protein-polissacarídeo complexes: A review. *Critical Reviews in Food Science and Nutrition* 38(8) (1998): 689–753.

Sezer, A. D. and Akbuğa, J. Release characteristics of chitosan treated alginate beads: II. Sustained release of a low molecular drug from chitosan treated alginate beads. *Journal of Microencapsulation* 16(6) (1999): 687–696.

Shahidi, F. and Han, X. Q. Encapsulation of food ingredients. *Critical Reviews in Food Science and Nutrition* 33(6) (1993): 501–547.

Silvestre, C., Peezzuto, M., Cimmino, S., and Duraccio, D. Polymer nanomaterials for food packaging. In: Silvestre, C. and Cimmino, S. (eds.), *Ecosustainable Polymer Nanomaterials for Food Packaging: Innovative Solutions, Characterization Needs, Safety and Environmental Issues*, pp. 221–252. CRC Press, Boca Raton, FL, 2013.

Smidsrød, O. Molecular basis for some physical properties of alginates in the gel state. *Faraday Discussions of the Chemical Society* 57(January) (1974): 263–274.

Soottitantawat, A., Takayama, K., Okamura, K. et al. Microencapsulation of l-menthol by spray drying and its release characteristics. *Innovative Food Science & Emerging Technologies* 6(2) (2005): 163–170.

Stewart, W. W. and Swaisgood, H. E. Characterization of calcium alginate pore diameter by size-exclusion chromatography using protein standards. *Enzyme and Microbial Technology* 15(11) (1993): 922–927.

Su, G., Huang, G., Li, M., and Liu, C. Study on the flow behavior in spout-fluid bed with a draft tube of sub-millimeter grade silicon particles. *Chemical Engineering Journal* 237(February) (2014): 277–285.

Sun-Waterhouse, D., Zhou, J., Miskelly, G. M., Wibisono, R., and Wadhwa, S. S. Stability of encapsulated olive oil in the presence of caffeic acid. *Food Chemistry* 126(3) (2011): 1049–1056.

Thies, C. Chapter 2: Characterization. In: Thies, C. (ed.), *How to Make Microcapsules: Lecture and Laboratory Manual*. Thies Technology, St. Louis, MO, 1995.

Thimma, R. T. and Tammishetti, S. Study of complex coacervation of gelatin with sodium carboxymethyl guar gum: Microencapsulation of clove oil and sulphamethoxazole. *Journal of Microencapsulation* 20(2) (2003): 203–210.

Tolstoguzov, V. Some thermodynamic considerations in food formulation. *Food Hydrocolloids* 17(1) (2003): 1–23.

Toublan, F. J. J. Fats and waxes in microencapsulation of food ingredients. In: Gaonkar, A. G., Vasisht, N., Khare, A. R., and Sobel, R. (eds.), *Microencapsulation in the Food Industry*, pp. 253–266. Academic Press, Cambridge, UK, 2014.

Vandamme, Th. F., Lenourry, A., Charrueau, C., and Chaumeil, J.-C. The use of polysaccharides to target drugs to the colon. *Carbohydrate Polymers* 48(3) (2002): 219–231.

Vandenberg, G. W., Drolet, C., Scott, S. L., and de la Noüe, J. Factors affecting protein release from alginate-chitosan coacervate microcapsules during production and gastric/intestinal simulation. *Journal of Controlled Release* 77(3) (2001): 297–307.

Vega, C. and Roos, Y. H. Invited review: Spray-dried dairy and dairy-like emulsions—Compositional considerations. *Journal of Dairy Science* 89(2) (2006): 383–401.

Versic, R. Flavour encapsulation: An overview. In: Reineccius, G. A. and Risch, S. J. (eds.), *Flavour Encapsulation*, pp. 1–6. American Chemical Society, Washington, DC, 1988.

Weinbreck, F., Minor, M., and de Kruif, C. G. Microencapsulation of oils using whey protein/gum Arabic coacervates. *Journal of Microencapsulation* 21(6) (2004): 667–679.

Wilson, N. and Shah, N. P. Microencapsulation of vitamins. *ASEAN Food Journal* 14(1) (2007): 1–14.

Yu, C. Y., Yi, B. C., Zhang, W., Cheng, S. X., Zhang, X. Z., and Zhuo, R. X. Composite microparticle drug delivery systems based on chitosan, alginate and pectin with improved pH-sensitive drug release property. *Colloids Surfaces B: Biointerfaces* 68(2) (2009): 245–249.

Zhang, K., Zhang, H., Hu, X., Bao, S., and Huang, H. Synthesis and release studies of microalgal oil-containing microcapsules prepared by complex coacervation. *Colloids and Surfaces B: Biointerfaces* 89(1) (2012): 61–66.

Zuidan, N. J. and Shimoni, E. Overview of microencapsulates for use in food products or processes and methods to make them. In: Zuidam, N. J. and Nedovic, V. (eds.), *Encapsulation Technologies for Active Food Ingredients and Food Processing*, pp. 3–29. Springer, New York, 2010.

17

Encapsulation of Flavors and Aromas: Controlled Release

Bojana Isailović, Verica Djordjević, Steva Lević, Jelena Milanović, Branko Bugarski, and Viktor Nedović

CONTENTS

17.1 Introduction

Flavors are one of the most important ingredients in the food industry since they have a huge influence on consumer satisfaction and make product taste more attractive to the user (Madene et al. 2006). There are varieties of natural and synthetic flavors, and all of them are defined as combinations of taste, smell, and trigeminal stimuli (Zuidam and Heinrich 2010). In general, volatile flavors that interact with receptors in the mouth and nose cavity are usually called aroma. Aromas contain many volatile and fragrance organic molecules, which can be classified as esters, linear terpenes, cyclic terpenes, aromatic, amines, etc. At room temperature, most of aromas are liquids (usually oils) but they can be also in gas or even in solid state (e.g., vanillin, camphor, and menthol). It is known that aroma molecules are very sensitive to light, oxygen, humidity, and high temperatures. Moreover, some of the chemically unstable flavors (e.g., citral) when degrading over time start to create off-flavors (Maswal and Dar 2014).

Flavors are often characterized as reactive components with varying solubility and partition coefficients in the oil and water phases (Zasypkin and Porzio 2004). The vapor pressure values are an important characteristic of aroma since they provide a method for ranking the relative volatilities of compounds (Rusu 2006). The partition coefficient (log P [o/w]) is another important parameter as the underlying principle that governs the release of flavors from the food matrix into the gas phase. Table 17.1

TABLE 17.1

List of Natural Aroma Compounds Used in Food Product in Encapsulated Form

Aroma Compound	Molecular Formula	Fragrance	Natural Source	Description	Melting Point (°C)	Boiling Point (°C)	Vapor Pressure (mm Hg at 25°C)	log P (o/w)	Application
Furaneol	$C_6H_8O_3$	Strawberry	Strawberries, raspberries, pineapple	Light-yellow powder	73–77	259	0.0320	1.4 (Relkin et al. 2004)	Ice cream, gelatin, and puddings, candy, bakery products, beverages
L-Menthol	$C_{10}H_{20}O$	Menthol	Mentha	White or colorless crystalline solid	31–33	212	0.0637	3.40 (Griffin et al. 1999)	Chewing gum and candy
Isoamyl acetate	$C_7H_{14}O_2$	Banana	Banana plant	Colorless liquid	−78	142	5.6000	2.26 (Žnidaršić-Plazl and Plazl 2009)	Honey, butterscotch, artificial coffee, beverages
Geraniol	$C_{10}H_{18}O$	Rose, flowery	Lavender, jasmine, coriander, and others	Clear, colorless liquid	−15	230	0.0210	3.47 (Griffin et al. 1999)	Alcoholic and nonalcoholic beverages, ice cream, candies, baked goods
Limonene	$C_{10}H_{16}$	Lemon, orange	Orange, lemon	Colorless liquid	−74.3	176	0.1980	4.83 (Lim et al. 2008)	Foods, beverages, and chewing gum
Camphor	$C_{10}H_{16}O$	Camphor	Camphor, laurel	White, translucent crystals	177	204	4.000	2.38 (Ran and Yalkowsk 2001)	Cooking, mainly for dessert dishes in Asia
Carvone	$C_{10}H_{14}O$	Caraway or spear mint	Carum carvi, dill, mentha spicata	Clear, colorless liquid	25.2	231	0.1600	2.71 (Griffin et al. 1999)	Chewing gum and candy
Cinnamaldehyde	C_9H_8O	Cinnamon	Cinnamon	Yellow oil	−7.5	248	0.0265	1.90 (Hansch and Hoekman 1995)	Chewing gum, ice cream, candy, and beverages
Vanillin	$C_8H_8O_3$	Vanilla	Vanilla	White crystals	82	285	0.0020	1.21 (Hansch and Hoekman 1995)	Usually in sweet foods
Anethole	$C_{10}H_{12}O$	Anise	Anise, sweet basil	Colorless to pale-yellow liquid	21	234	0.0690	3.45 (Brauss et al. 1999)	Alcoholic drinks ouzo and pernod
Methyl salicylate	$C_8H_8O_3$	Mint	Gaultheria	Colorless to pink clear liquid	−9	220–224	0.0343	2.55 (Hansch and Hoekman 1995)	Foods and beverages

lists the most abundant natural food aromas together with some of the physicochemical characteristics important from the viewpoint of encapsulation.

Various changes in typical aroma of foodstuffs may also occur during manufacturing, packaging, or storage due to interactions of aroma with the other food components (even other flavors), due to migration in the food, or due to high volatility of aroma compounds. As most of aroma substances are actually very expensive, their degradation or loss is unacceptable for the food industry. A good example for industrial processing that is often accompanied by considerable loss in flavoring is extrusion. Extrusion has gained immense popularity in the snack food industry. During extrusion, ingredients go through mixing, heating, and shearing at a high temperature, with shear and pressure resulting in changes in molecular properties and sensory attributes (Wen et al. 1990). In the snack food industry, flavoring the extrudates is still a major problem because of the high volatility of flavors and their instability under extrusion conditions. The extrusion process can cause loss of preextrusion-added flavor compounds because of thermal degradation, oxidation, polymerization, and reaction with other ingredients (Yuliani et al. 2004). In addition, the flavor compounds are flashed off with steam during expansion of the extrudate due to a sudden pressure drop at the die (Bonnet et al. 2005; Thepkunya et al. 2006; Decker and McClements 2007; McClements et al. 2007). Because of these inherent problems, flavors are mostly added after a product has been extruded. Although this postaddition of flavors is used extensively, it suffers from numerous inherent problems, such as the need for high dosage levels, uneven distribution (large portion of a flavor stay on the surface, which then causes sticking to the inner side of its packaging and consumer's fingers), oxidation of flavors applied to the food surface, insolubility of some flavors in the chosen base medium, the need for an addition of fat to bind the flavors to the product, and an increase in product weight by up to 25% (Maga and Sizer 1979).

Therefore, preserving aroma and making it as stable as possible is usually a major concern of food manufacturers. Furthermore, for some food products (like cookies), it is essential that the aroma is released when appropriate stage of processing is reached (baking or cooking) or when the food is placed in the mouth. In order to retain the characteristics of aroma, to protect it from evaporation, undesirable interactions, light- or oxygen-induced oxidation, to prolong its shelf life, or to achieve a controlled release, it is important to encapsulate aroma prior to application.

Overall, encapsulation can be defined as a technology for storing solid, liquid, or gas actives in small, sealed capsules that can release the insides at optimized and controlled rates under specific conditions (Fang and Bhandari 2010). The retention of actives, especially volatile aroma compounds, depends on the nature of the compound (molecular weight, physicochemical characteristics, polarity, and relative volatility). The type of encapsulation process and the composition of the flavor carrier also determine the extent of protection achieved. The capsules can be made from pure materials or mixtures. The capsule material is also called matrix, carrier, coating material, membrane, or shell. Nowadays, it is common to encapsulate aroma ingredients within the membrane of different materials that are food approved. The widely used materials are natural polymers, gums, waxes, polysaccharides, proteins, lipids, or some combinations of those (Jeon et al. 2003).

The selection of the optimal encapsulation method and matrix varies in relation to the final application of the product and to the manufacturing processing conditions. The encapsulation process should be efficient, scalable to production, and without significant flavor loss; it should ensure preservation of flavor fidelity and minimize reactions between flavor components, carrier, and oxygen. This chapter focuses on technologies currently used for aroma encapsulation and highlights benefits and drawbacks of each. Special focus is on release properties of different encapsulation systems during storage and thermal treatments, in food or during eating.

17.2 General Aspects of Controlled Release in Food Flavoring

One of the most difficult tasks for food technologists is to achieve control release of compounds such as flavors. In food applications, control release often refers to retarded flavor release during food processing in order to minimize flavor losses. Sometimes, control release means the opposite—acceleration of release of flavors, which, otherwise, will not be ever released (due to strong interactions with food matrix constituents). Then, in some situations, control release is attributed to a release triggered by one specific factor prior to consumption (moisture, temperature, pH) or in the mouth (chewing, dissolution, or saliva enzyme activity). This chapter reviews delivery systems aimed at control release in various situations.

Possible controlled-release benefits during heat-involved food preparation are improved aroma retention, aroma burst, or aroma differentiation. Improved aroma retention can be achieved by delayed release from encapsulates, but then encapsulates should not dissolve quickly during heat treatment, and aroma should diffuse slowly with time out of encapsulates. Because of these reasons, not all encapsulates can be considered as controlled-release delivery systems. Coacervation encapsulation and inclusion complexation are typical technological solutions for controlled release, while glass encapsulation usually is not. Glass encapsulates are made of water-soluble materials, and as such, they dissolve easily in the food products that contain a lot of water. However, in some situations, for example, frying of high-fat food in oil, glass encapsulates can also provide retarded release.

Since the last decade, there is an increasing interest for developing different methods for following flavor release to the air phase. The most popular are based on static headspace, followed by gas chromatography (GC) with flame ionization detection or GC combined with mass spectrometry. These methods are used for separation, quantification, and/or identification of the components, sampling at a fixed time during flavor release. They have been able to give accurate data on volatile compounds when applied on model flavors of relatively simple composition (Roberts et al. 1996; Gunning et al. 1999). However, real flavors are complex mixtures typically of more than 20 compounds with different properties. Separation, identification, and quantification of each aroma compound in the mixture, and even more, following the changes in the flavor release pattern in real-time are nearly impossible to accomplish. However, there have been some progresses in development of analytical tools worth to be mentioned. Thus, the portable electronic nose has been designed; it is composed of an array of gas sensors, with nonspecific responses that have pattern recognition ability from multivariate data analysis (Gardner and Bartlett 1999; Young et al. 1999; Gilsenan et al. 2000; O'Connell et al. 2001; Branca et al. 2003). The information is obtained from the sensors' array responses by pattern recognition techniques, such as principal component analysis or artificial neural networks. The inventers describe this device as promising for real-time, in-line, in situ determinations and nondestructive sensing, especially useful in the food industry (Monge et al. 2004). In another study, pulsed field gradient magic angle spinning nuclear magnetic resonance was used to obtain in situ the microscopic restricted pore diffusion coefficient of the aroma molecules entrapped in sol-gel-made silica particles (Veith et al. 2004a). Recently, a linearly ramped humidity system coupled with automatic sampling GC has been developed to measure the dynamic release flux of D-limonene encapsulated in spray-dried β-cyclodextrin powders under linearly ramped humidity (0.375%/min) at constant temperature (50°C) (Yamamoto et al. 2012).

17.3 Glass Encapsulation Systems for Delivery of Flavors

The term "glass encapsulation" refers to the encapsulation in the glassy state of amorphous carbohydrates. At the glassy state, water-soluble carbohydrates have high physical and chemical stability, providing very high barrier properties for oxygen and organic molecules. The glassy state controls or has an effect on most of the key properties necessary for commercialized encapsulated flavors, including caking, milling response, cracking, and specific gravity, among others (Zasypkin and Porzio 2004). Spray drying, extrusion encapsulation, vacuum drying, freeze-drying, and fluidized bed drying are some of processing techniques that give carbohydrate microcapsules containing flavors (Yuliani et al. 2004). Spray drying and extrusion encapsulation are two main techniques used commercially to produce encapsulated flavor (Reineccius 1989). The stability of glassy carbohydrate encapsulates depends on storage conditions, but the typical expiration date is between 6 and 12 months. Glass encapsulates dissolve quickly in water-containing food products; thus, they cannot provide a controlled-release benefit during thermal treatments and eating.

17.3.1 Carrier Materials for Glass Encapsulation of Flavors

Usually, when selecting a carrier material for glass encapsulation, a compromise is made between the physical stability of the encapsulation matrix (which is favored by increasing the molecular weight of the encapsulation matrix) and the oxygen permeability (which is usually minimized by reducing the molecular weight of the encapsulation matrix). Therefore, in most commercial products, the encapsulation matrix is composed of a mixture of intermediate- or high-molecular-weight carbohydrates (e.g., starches and maltodextrins of low dextrose equivalent [DE]) and low-molecular-weight carbohydrates (usually disaccharides, e.g., sucrose). In blends subjected to spray drying, the portion of low-molecular-weight carbohydrates (such as sugars, corn syrup, polyols) amounts between 10% and 35% of the total mass of carrier. Crystallinity of the matrix material may change when changing portions of matrix compounds. Thus, Tackenberg et al. (2014) found that increasing the sucrose content combined with increasing water content resulted in increased crystalline fraction within the matrix.

Maltodextrins are a good compromise between cost and effectiveness. They have high ability to absorb liquid flavors (Morris 1981) and provide oxidative stability to encapsulated oil but exhibit poor emulsifying capacity, emulsion stability, and low oil retention (Buffo and Reineccius 2000). Maltodextrins are starch hydrolysates with DE value between 3 and 20. The higher the DE value implicates, the shorter the glucose chains as well as the higher sweetness, higher solubility, and lower heat resistance. Retention of volatile flavor compounds increases with an increase of the maltodextrin DE.

Starch is a carbohydrate consisting of two types of molecules: the linear and helical amylose and the branched amylopectin. The helical structure of the amylose fraction allows physical entrapment of flavor molecules. Depending on the plant, starch forms more or less porous granules, and granules of some varieties have naturally occurring 1–3 μm diameter surface porous. The porosity of starch granules can be modified by different strategies. One is to treat starch granules by amylase enzymes in order to create a highly porous structure (Yamada et al. 1995; Zeller et al. 1999). Another way is to agglomerate granules with an aid of some protein or polysaccharide-binding agent. Then, a flavor can be efficiently entrapped in the interior of the interconnecting cavities between these agglomerates (Zhao and Whistler 1994; Zaho et al. 1996). Octenylsuccinic acid anhydride-modified starches (OSA or OSAN starches) are extensively used as carriers for both spray drying and melt extrusion in the food industry (Zasypkin and Porzio 2004). The first generation of OSAN starches was made by concomitant OSA anhydride derivatization and their maltodextrinization. This process, however, resulted in a carrier with an undesirable off-flavor character (Zasypkin and Porzio 2004). A second generation of OSAN starches have been recently commercialized based on either acid or enzymatic hydrolyses and OSA anhydride derivatization. These starches are virtually bland and still retain excellent emulsification and film-forming properties. Carriers based on this modified starches (Buffo and Reineccius 2000) and their blends with other components (Boskovic et al. 1992) have been successfully optimized for flavor encapsulation by spray drying. The carrier obtained by combining two OSAN starches with low-molecular-weight carbohydrates in the presence of water as a plasticizer exposed better responses to melt extrusion than compositions of individual OSAN starches with sugars or polyols (Zasypkin and Porzio 2004). The most critical parameter controlling the kinetics of OSAN starches melting, melt viscosity, and setting into the glassy state is the amount of plasticizer added to the composition (Zasypkin and Porzio 2004). Processing of OSAN starches involves low pH values (about 3), which may cause degradation of acid-sensitive flavors, such as citral and acetal (Porzio 2004, 2007). Modified starches are not considered as *natural* for labelling purposes; they have an unattractive off-taste and give poor protection to oxidizable aromas (Reineccius 1989).

Gum arabic (also known as acacia gum, chaar gund, char goond, or meska) is a complex mixture of glycoproteins and polysaccharides. In encapsulation technologies, gum arabic has been used as a protective wall material for spray drying of liposoluble and sensitive compounds as it fulfills the roles of being both a surface-active agent and a drying matrix. The big advantage over the other carriers lies in the fact that it is noncarcinogenic, low in caloric value, and a good source of soluble fibers. Also, gum arabic is a prominent natural emulsifier.

However, in recent years, the use of gum arabic decreases because of its limited availability, fluctuations in supply, expensiveness, and impurities. Moreover, aqueous solutions of solely gum arabic may be difficult to process due to high viscosity. Therefore, many authors investigated the use of the blend of gum arabic with other wall materials or even tried to replace it completely (McNamee et al. 2001; Krishnan et al. 2005). Krishnan et al. (2005) found that the gum arabic/maltodextrin/modified starch (4/6:1/6:1/6) blend proved to be more efficient than the other blends, even better than 100% gum arabic. On the other hand, blends of gum arabic and maltodextrin could not provide the protection of cumin oleoresin offered by gum arabic alone (Kanakdande et al. 2007). Along with other trials to substitute gum arabic is the usage of Angum gum (a natural extrudate from mountain almond tree) as an emulsifier to create more stable emulsions of D-limonene in maltodextrin compared to emulsions stabilized with gum arabic (Jafari et al. 2013a,b).

17.3.2 Drying Processes: Spray Drying and Freeze-Drying

At first, spray drying was used for transformation of a liquid form (solution, dispersion, emulsion) into dried particulates. Then in the 1930s spray drying was employed for encapsulation of flavors into gum arabic, and since then, it became one of the most widely used encapsulation methods (Jafari et al. 2008b). Numerous authors reported that this method is effective for protection of flavors and aromas, and nowadays more than 90% of encapsulated aromas are prepared by spray drying (Gibbs et al. 1999; Madene 2006; Vaidya et al. 2006). The reason lies in the fact that spray drying is a flexible low-cost process that uses readily available equipment and enables the production of encapsulated forms on a large-scale and in a continuous manner. Besides, spray drying allows the choice of different carriers depending on the end-product application and gives the end product that is stable and of good quality.

The selection of an appropriate carrier (or matrix) is the first step in spray drying of an aroma. The suitable matrix should have further features (Zuidam and Heinrich 2010; Carneiro et al. 2013; Turchiuli et al. 2013):

- To be highly soluble in water
- To form stabile emulsion
- To have good spraying ability (i.e., an aqueous solution should not be too viscous to be pumped and sprayed)
- Not to be too sticky or hygroscopic (for rapid and efficient drying)
- To be nonreactive
- To release the aroma when it is required in a finished food product
- To be low cost and available
- To be tasteless and stable
- To have good film-forming and drying properties
- To provide good protection to the encapsulated aroma compound

In addition to carbohydrate materials described earlier, milk or soy proteins, hydrolyzed gelatin, and new biopolymers are also core materials suitable and generally available for encapsulation of different food flavors and aromas by this method. Easy of drying, good retention during processing, and flow properties of protein powders are among the most important attributes of proteins as carriers for flavors. Proteins may provide better protection of some aromas (e.g., limonene) from oxidation during storage than carbohydrates (gum arabic and modified starch), which is attributed to limited oxygen permeability (>70% retained upon storage at 40°C for 28 days) (Charve and Reineccius 2009). However, protein powders (especially of whey protein isolate) with encapsulated limonene or aldehydes suffered from nonenzymatic browning upon 1 week storage. In fact, a lot of flavor compounds contain carbonyl groups (e.g., ketones and aldehydes) that can react with the amino groups of a protein (Schiff base formation) initiating the Maillard reaction and resulting in brown pigments as well as flavor loss.

The first step in spray drying is preparation of a dispersion of aroma (optimum 20–25 wt.%) in the aqueous solution of a chosen matrix (35–45 wt.%) followed by homogenization. Homogenization is usually done by Ultra Turax, but some studies showed that the more effective retention of aroma (e.g., cheese aroma) was achieved when the ultrasound was applied for emulsification instead of Ultra Turax (Mongenot et al. 2000). In case of hydrophilic aromas, the first step is preparation of water-in-oil-in-water double emulsions, with aroma trapped within the inner water droplets (Brückner et al. 2007).

The next step is to feed the mixture into a spray dryer wherein the mixture is atomized and then sprayed into a hot chamber. The atomization can be carried out using the high-pressure nozzle or the spinning wheel. The use of these two types of atomizers is equally present in the food industry, although each has its own advantages and disadvantages (Jafari et al. 2008b). In the hot chamber, the atomized droplets fall through the hot air (160°C–220°C) and form spherical microcapsules, which are further transported to a cyclone separator for recovery. The final product consists of dry microparticles typically between 10 and 100 µm in diameter, containing a dispersion of fine aroma droplets. In general, the microparticles with a diameter above 50 µm are noticeable for costumers during eating, and when used in beverages, they are even visible by the costumer's eye. A schematic presentation of a spray drier connected to a cyclone for a product recovery is presented in Figure 17.1.

Many studies showed that each step in the drying process affects the properties of the end product (Reineccius 2004; Vaidya et al. 2006). As already mentioned, the composition of matrix and its properties have big influence on retention of aroma and formation of a stabile final product. The stability increases with an increase in the portion of a carrier material of spray-dried encapsulates, which is attributed to less porous structure of, for example, gum arabic or modified starch (Rosenberg et al. 1990; Charve and Reineccius 2009), but not necessarily (with soy or way proteins this was not confirmed). The type of flavor compound also has a large influence. The aroma's retention upon spray drying will be better if the aroma is less volatile and less polar and has a lower molecular weight (Jafari et al. 2008b; Zuidam and Heinrich 2010). The properties of the initial emulsion (such as the composition of both hydrophilic and lipophilic phases, their weight ratio, the dry matter content, oil droplet size, size distribution, and viscosity) were found to be key parameters for aroma encapsulation (Soottitantawat et al. 2005a,b; Jafari et al. 2008a,b; Frascareli et al. 2012). Hence, it has been shown that a decrease of the emulsion oil droplets size (down from 2 µm) leads to an increase in encapsulation efficiency, a better retention of flavors during spray drying, a lower portion of a flavor compound on the surface of particles, and a better protection of the oil against oxidation (Risch and Reineccius 1988; Soottitantawat et al. 2003; Jafari et al. 2008a).

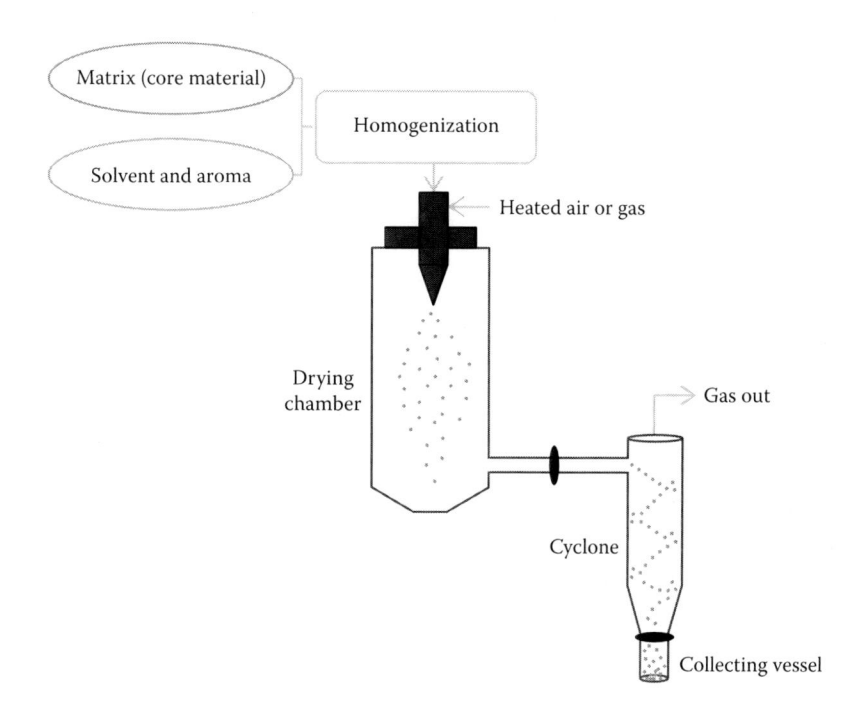

FIGURE 17.1   Schema of a spray drying process. The final product is collected in the collection vessel.

This is important from the economical aspect of the process, and it is convenient for the manufacturer as well as for the user. For water-soluble aromas, as stated by Soottitantawat et al. (2003), the optimal emulsion size is 2–3 µm, depending on the type of aroma. It is also found that high exit air temperatures are beneficial to the retention of hydrophilic aroma compounds (Reineccius 1989). Further, moisture content of the final microcapsules and the humidity of the exhaust air have a strong impact on the degree of aroma retention (Madene 2006). It is interesting to mention that some authors found the relationship between the size of final microparticles and aroma retention, while the others negated the existence of any correlation. Reineccius (1989) found an explanation about this controversy. As he reported, the size of microparticles is irrelevant if high infeed solid levels are used. However, one should be careful with increasing the solid level; namely, at some point, the carrier material becomes nondissolved and, thus, too viscous to be efficiently atomized. Then, inlet and outlet air temperatures affect aroma retention during processing and encapsulation efficiency. Bringas-Lantigua et al. (2012) via response surface methodology determined that an inlet air temperature of 220°C and an outlet air temperature of 85°C provided the maximum evaporative rate (7.7 kg/h), volatile oil retention (95.7%), and microencapsulation efficiency (99.9%), and the values agreed well with experimental data. According to some authors (Jafari et al. 2008b), process yield increases with increasing inlet air temperature and decreasing feed flow rate, but studies of some others diverged from these conclusions (Wang et al. 2009).

The flexibility of spray drying is one of the main advantages over the other encapsulation methods. It is applicable for heat label and volatile materials as aromas, and the obtained microparticles are stable, rapidly soluble, and of small size. However, spray drying method has also some disadvantages. Some aromas with low boiling pint (e.g., acetaldehyde, diacetyl, dimethylsulfide) can be partially lost during this process since processing temperatures are high (minimum of 100°C if all precautions are undertaken to avoid overheating, Porzio 2007). Also, the emulsification process (prior to spray drying) involving high-pressure or high-speed homogenization may lead to a loss of aroma. For example, Penbunditkul et al. (2012) investigated encapsulation of multiflavor bergamot oil via spray drying by using octenyl succinylated waxy maize starch as a wall material; they discovered that some chemical functional groups were lost during the high-pressure homogenization.

Spray drying transforms a carrier into the glassy state by the rapid drying and evaporative cooling process. The low-moisture content of the carrier at the end of processing assures desirable glassy state of the carrier. The end product of spray drying is a powdered emulsion where oil droplets containing the flavor molecules are dispersed within the solid polymer matrix of powdered particles. Encapsulates produced by spray drying have hollow-sphere morphology. An example is given in Figure 17.2 showing powder morphology of flaxseed oil microencapsulated by spray drying using maltodextrin-modified starch (Hi-Cap 100™) mixture as a wall material (Carneiro et al. 2013). The powder form is stable, allows easier dosage, and mixes with other food constituents. For easy handling, the powder should have good flow ability and mixing ability and allow reconstitution of the initial emulsion upon rehydration

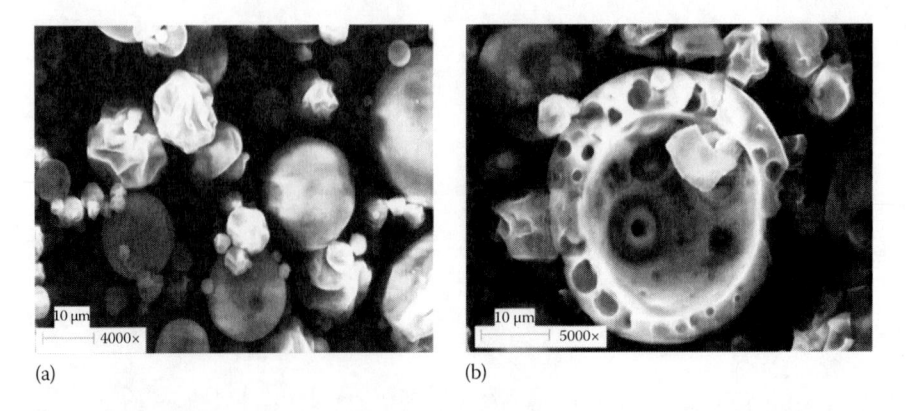

(a)   (b)

FIGURE 17.2   SEM micrographs of external (a) and internal (b) microstructures of powders produced from maltodextrin/ Hi-Cap 100™ (1:3) wall material mixture. (From Carneiro, H.C.F. et al., *J. Food Eng.*, 115, 443, 2013. With permission.)

in water (Christensen et al. 2001). However, there is a strong possibility that in the final product a large portion of aroma stays on the surface of microparticles. In this way, it becomes directly exposed to the environment and consequently, easily prone to oxidation, which directly affects the taste of the final product. The nonuniformity of the microparticles in the end powder product can also be pointed out as a drawback of this method (Reineccius 1989).

Spray drying sets restrictions on concentration of both agent and core material and requires high temperature for drying and low viscosity of feed solution; all the earlier limitations can be complemented by freeze-drying. Few studies on freeze-drying of flavor oil emulsions have been published (Tobitsuka et al. 2006; Kaushik and Roos 2007; Lee et al. 2009; Kaasgaard and Keller 2010; Chranioti and Tzia 2013). The main difference between freeze-drying and spray drying is that emulsions are exposed to different stresses in the two techniques. Spray drying involves atomization and heating, whereas freeze-drying involves freezing. Despite these obvious differences, the encapsulation materials and emulsion characteristics used in spray drying generally work well for freeze-drying too. However, freeze-drying is a more expensive technology than spray drying and, currently, far less understood. Kaushik and Roos (2007) investigated the effect of wall-material composition and high-pressure homogenization pressure on retention levels of freeze-dried limonene emulsions containing different blends of gelatin, sucrose, and gum arabic as encapsulation materials. Gum arabic was found to be the best of the tested encapsulation materials as these emulsions resulted in high retention levels (up to 75%) as well as good drying properties, whereas emulsions with high gelatin contents retained high levels of limonene but collapsed during drying and were difficult for conversion into a powder. Sucrose alone resulted in very low retention levels. Tobitsuka et al. (2006) have also reported good retention levels in freeze-dried flavor oil emulsions when gum arabic was used as an encapsulation material, whereas hydrophobically modified starch resulted in good retention of limonene in another recent study by Lee et al. (2009). Furthermore, results with freeze-dried nonvolatile oil emulsions have shown improved encapsulation efficiency when using stable emulsions with smaller particle sizes (Kaushik and Roos 2007). It is shown that a combination of a charged small-molecule emulsifier and an oppositely charged polysaccharide improves retention and redispersibility of freeze-dried flavor oil emulsions (Kaasgaard and Keller 2010). Figure 17.3 shows the interior structure of freeze-dried chitosan-coated emulsions of R-carvone oil in maltodextrin obtained under optimal conditions (0.4% chitosan, 10% maltodextrin), which provided flavor retention of about 95%.

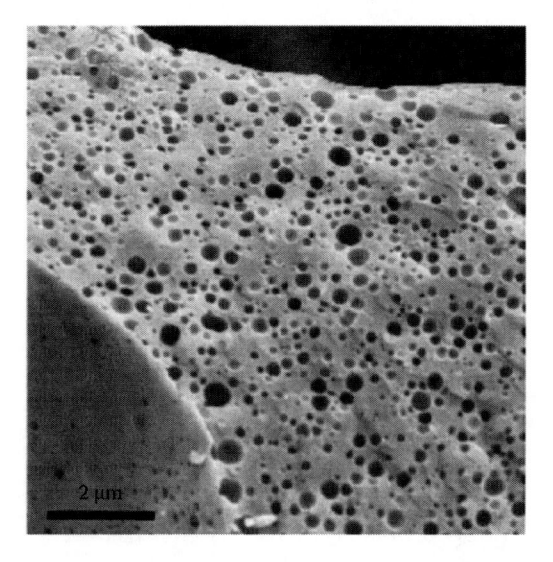

FIGURE 17.3 SEM micrographs showing the interior structure of freeze-dried chitosan-coated carvone emulsions containing 0.4% (w/w) chitosan and 10% (w/w) maltodextrin. (From Kaasgaard, T. and Keller, D., *J. Agric. Food Chem.*, 58, 2446, 2010. With permission.)

17.3.3 Melt Extrusion and Melt Injection

For production of larger particles that create visual appearance and textural effects and/or have controlled-release properties, the melt extrusion technique has an advantage over spray drying. It has been commercialized since 1995 (Popplewell et al. 1995; Porzio and Popplewell 1997). The melt extrusion is a cost-effective and environmental-friendly process, which can be continuous, and therefore, it is utilized for industrial production of aroma encapsulates. The procedure for this method is relatively simple. The first step is mixing and melting of the chosen carriers (one or a mixture of few). It is usual to add an emulsifier (1%–2.5%) and a small amount of water (5%–10%) to the mixture in order to form stabile emulsion of a proper viscosity (Baines and Knights 2005). The water addition affects the glass transition (aroma melt extrudates have it between 30°C and 70°C) and robustness of the extrudate. Further, the aroma is added as oil, as oil-in-water emulsion, or as a dried form (5%–10%) to the melted mixture just before extrusion or sometimes at about halfway the extrusion (Figure 17.4). By this way, the extended contact of aroma with the high temperatures is avoided. The extrusion is generally performed at temperatures from 105°C to 120°C (Zuidam and Heinrich 2010). The mixture is then expelled under high pressure through die by a single- or twin-screw extruder or in some cases by a costume-made extruder. The obtained extrudate is further cooled by air, cold bath, or cold solution of an aroma solvent to form the amorphous or solid glass as a final product. The use of an aroma solvent is suggested since any residual or free aroma is washed from the extrudate surfaces during the process, which provides a rather odor-free particulate resistant to oxidation during storage (Shahidi and Han 1993). The size of the extrudate can be reduced using standard processes such as chopping or cutting by a rotating knife immediately after the die or by grinding after the cooling step (Zuidam and Heinrich 2010). This step can be followed by size classification to eliminate the smallest or oversized particles.

The shape of the extrudate depends on die geometry and it can be a sheet, rope, or thread of different dimensions. Other processing parameters such as screw speed and configuration, throughput, and die geometry have also an important impact on the product quality parameters (Yilmaz et al. 2001). As reported by Yuliani et al. (2006a), the temperature increase provides an extensive reduction in the hardness of the extrudate and also an increase in aroma retention. The same authors also found that the aroma retention was reduced with the increase in screw speed. Emin et al. (2013) investigated the influence of processing parameters such as screw configuration, feed rate, and screw speed on the dynamics

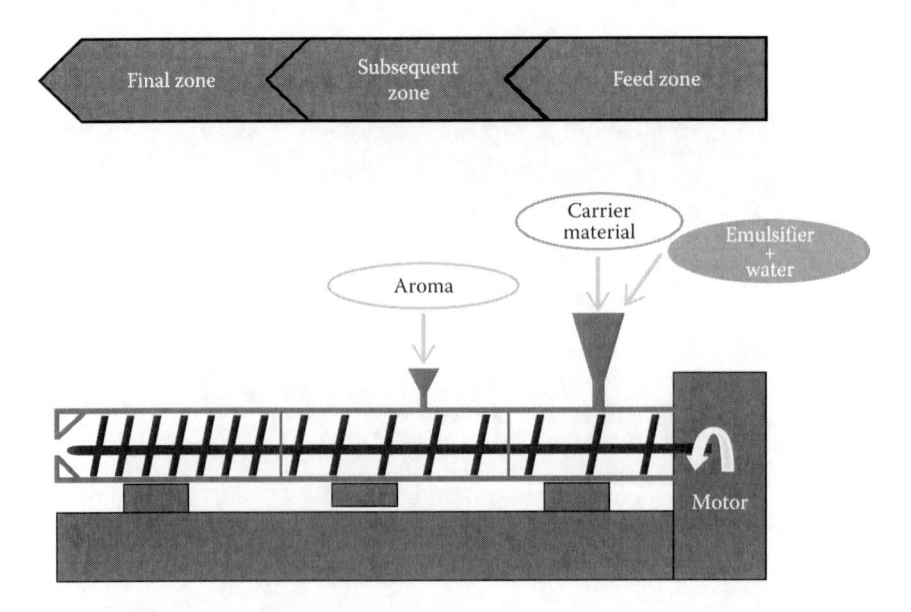

FIGURE 17.4 Setup of a melt extruder. Emulsifier and water can be added simultaneously with matrix or separately: feed zone, low pressure generated to homogenize the feed; subsequent zone, gradual increase in pressure to melt, homogenize, and compress the extrudate; and final zone, constant high pressure to transport melted material out of the extruder.

of morphological changes during extrusion of a starch-based matrix. The changes were monitored by analyzing droplet breakup and coalescence separately. It appeared that the increase in screw speed did not necessarily result in droplets of smaller sizes and that the increasing screw speed not only expanded droplet breakup but also increased the degree of coalescence. Smaller droplet sizes were obtained by increasing the feed rate, that is, when higher blend viscosities were processed as a consequence of a reduced degree of coalescence. In the same study, it is shown that increasing oil content enlarged the coalescence rate and provided bigger droplets.

Melt injection is a very similar process to melt extrusion. The main difference is that melt injection is performed in a vertical position without screws. In brief, the process starts with mixing of the chosen matrix, aroma, and emulsifier when they are molten (at 110°C–140°C). Then, the mixture is pushed through an orifice at 1–7 bars and at the end quenched by cooling to form a glass. It is usual to use isopropanol for cooling because the matrix hardens on contact with the isopropanol and thus encapsulate aroma (Fulger and Popplewell 1999). It is known that the use of isopropanol requires a specialized equipment for handling and storage, but still it is the widest used cooling solvent. Subramaniam et al. (2006) showed that liquid nitrogen might be used as a substitute for isopropanol.

If the carrier is adequately chosen, it is slowly dissolved in an aqueous surrounding in food product providing the release of the aroma. Therefore, the choice of carrier depends on the specific application and physical properties of the amorphous matrix and encapsulant (Fulger and Popplewell 1999). These include the temperature stability, solubility, pH resistance, particle hardness, and release properties of the encapsulated aroma applied in the food product. The overall flexibility of the process allows the use of a variety of carriers such as glucose, lactose, fructose, sucrose, maltose, sorbitol, mannitol, maltitol, glycerol, and propylene glycol. The saccharides are the first choice because of their glass transition state, flow and melt properties as well as plasticizing ability, and flavor-release properties (Baines and Knights 2005). Shahidi and Han (1993) suggested that the use of emulsifying starches for melt extrusion provide a *sugar-free* product. They also reported that the use of modified starches instead of sucrose permits longer cooking times at higher temperatures. Hydroxypropyl cellulose is usually used for the encapsulation of aroma intended for confectionary products because it protects aroma at elevated temperatures during confectionary production (Lou and Popplewell 2003). Maltodextrins with the low DE are in use because of their good protection of aroma against oxidation, while gum arabic is preferable when the increased encapsulation strength and aroma release are required (Baines and Knights 2005). The use of other materials such as gluten, sodium alginate, and gelatin in melt extrudates is also typical because they exhibit good properties in cross-linking; they easily form films and provide good aroma binding.

The most critical parameters controlling the properties of extruded encapsulates are the kinetics of material melting, melt viscosity, and setting into the glassy state. And these properties are tightly connected with processing conditions. To illustrate, Zasypkin and Porzio (2004) investigated glassy states of identical spray-dried and extruded compositions based on OSAN starches with respect to glass transition temperature, heat capacity change, and enthalpy relaxation. They revealed two glass transitions in the case of extruded encapsulates, while the spray-dried compositions showed only one broad glass transition. Two structural phases are attributed rather to incomplete mixing of matrix constituents (OSAN starches and lactose/dextrose) in the highly viscous melt during the short process of extrusion than to incompatibility of polymers.

Upon encapsulation by extrusion, a flavor is entrapped within an impermeable glass. The melt extrusion/melt injection process provides long shelf life of the product, which usually reaches 2 years but in some cases even 4 or 5 years (Shahidi and Han 1993; Gouin 2004). However, some structural defects in the form of cracks, thinned walls, or pores occurred during or after processing may enhance diffusion of flavors out of carbohydrate matrices. These structural imperfections may limit shelf life of extruded flavor encapsulates. In comparison to the spray-dried process, the melt extrusion and melt injection enable better protection of the encapsulated aroma against oxidation since aroma is totally encapsulated inside the carrier. The extended shelf life of oxidation-sensitive aroma (e.g., citrus oils) is attributed to better barrier properties to oxygen transport (Ubbink and Schoonman 2003). The melt extrudates are typically applied in the production of food subjected to elevated temperatures where they can remain intact. As well as the others, this method has its disadvantages. The aroma loading is low (up to 10%), process temperatures are too high for some extremely volatile aromas, and the end extrudates are not stable in water-containing food products, but only in dry ones (Zuidam and Heinrich 2010). Formation of quite large encapsulates (500–1000 μm in the case of melt extrusion and 200–2000 μm in melt injection) limits the use of extruded flavors in application where mouthfeel is a crucial factor (Gouin 2004). This technology is commercially used for flavoring dry beverages, mostly tea bags, to prevent aroma losses and oxidation during storage (Schleifenbaum et al. 2000). On the contrary, spray-dried powder particles are so small that they can pass through filter bags. The production of boiled sweets and candies is also improved by using the extruded encapsulates of aromas (Baines and Knights 2005).

17.3.4 Fluidized Bed Coating

The basic principle of fluid bed spray coating is placing a coating onto powdered particles suspended in air. The powdered particles are aroma encapsulates obtained from some another encapsulation process (mainly it is spray drying). In this way, aroma will be protected with more than one material after several processing steps so that the final encapsulates will be ready to meet many challenges. Therefore, fluidized bed coating is maybe the most suitable technology for aroma encapsulation. At the beginning, particles to be coated enter through the bottom of the temperature and humidity-controlled chamber into the hot atmosphere where they become fluidized by an air current. Then, the coating material (e.g., fat, wax) is pumped through a nozzle and sprayed onto the particles. This step contributes to the formation of the film or layering on the particles surface. After a series of wetting and drying stages, a uniform film is readily obtained. If the particles are between 0.1 and 10 mm, the layer is usually about 10 mm thick (Guignon et al. 2002). Air-suspension particle coating is a complex process involving as many as 20 different variables (Werner et al. 2007). Some of the processing parameters affecting the product quality are flow rate and pressure of the spraying material, composition and physical properties of the coating matrix, and flow rate and temperature of the fluidizing air (Guignon et al. 2002). Configuration characteristics of fluidized bed towers are important too. For example, the choice of a binary or pneumatic nozzle instead of a hydraulic type enables better control of the droplet size and size distribution; also, the nozzle should be positioned so that the droplet travel distance is minimal (Dewettinck and Huyghebaert 1999).

Various coating materials can be used such as hydrogenated vegetable oil, gums, fatty starches, maltodextrins, and waxes. The water-soluble coatings release the encapsulate aroma when found in a water surrounding, while the hot-melt coatings release the core material at increased temperatures or by physical breakage during chewing. The choice of coating material strongly depends on the application which encapsulates are designed for. For example, when cooking in water, encapsulates should have a coating of water-insoluble material in order to achieve delayed aroma release. On the contrary, when they are ought to be frying in oil, it is better if the coating is composed of oil-insoluble compounds. Coated encapsulates can be designed so that the permeability of the coating material changes is triggered by pH change to induce or reduce leakage of the flavor. In addition, the high glass transition temperature of the carrier limits caking and enhances aroma retention upon storage. Unwanted interparticle agglomeration is the most serious problem encountered in the spray-coating process particularly for particles smaller than 100 μm. Physical and chemical strategies have been used to prevent agglomeration. The implementation of fluidized bed coating involves expensive laboratory and pilot-scale tastings since each application is specific and requires optimization of processing conditions (Werner et al. 2007). Additionally, large amounts of coating materials are needed to ensure complete coverage. The fluid bed coated encapsulates are claimed as resistant on confectionary processing conditions. Encapsulates prepared by fluid bed coating (as those produced via melt extrusion) can be used in dry food in order to prevent aroma migration.

17.4 Spray-Cooling Technology for Production of Flavor-Delivery Encapsulates

Spray cooling and spray chilling (also referred as spray congealing or melt spraying) are low-priced encapsulation processes. They are routinely employed for encapsulation of aroma compounds in order to improve their heat stability, provide delay release in wet environments, and/or convert liquid aromas into powders (Gouin 2004). The equipment utilized in spray cooling is basically the same as in spray drying with the main difference in heating of the feed line and using of ambient or chilled air for solidification of molten droplets (Gavory et al. 2014).

Spray cooling/chilling starts with the emulsification of the aroma compound and the molten matrix, and the molten feed is further atomized into fine droplets followed by their solidification. The atomization is performed at the expense of centrifugal, kinetic, pressure, or ultrasonic energy. Pneumatic spraying device includes a system of air nozzles, airless nozzles, or two-fluid nozzles. In a rotary (centrifugal) atomizer, a melt is dispersed onto a disk rotating at high speeds. With a rotary atomizer, even viscous molten mixtures are possible to process. However, the spray pattern is wide and short, so the cooling chambers, wherein solidification of microdroplets occurs, have to be wide. The cooling is performed with an aid of air (cold or at room temperature) flowing cocurrently, mix-currently, or countercurrently. The cooling rate is largely determined by construction characteristics. Thus, small particles (50–150 μm) are usually produced by high-speed rotary atomization in a device having cocurrent spray cooling chamber. Bigger particles (150–500 μm) that solidify slowly are the end product of a process conducted in a spray-cooling chamber with a fountain nozzle. For extra-large particles (500–2000 μm) (also called "prills"), a low-speed rotary atomizer with a specially designed prilling wheel is generally in use. The size of the particles is important for further food processing applications during which encapsulates can be destroyed and aroma released (especially large particles are risky). The cooling chambers may have integrated filters for direct separation of products in one unit (Albertini et al. 2008). Otherwise, the final separation of the product can be achieved in a separate cyclone unit.

The only difference between the spray cooling and spray chilling is in the melting point of the matrix used in process. In spray chilling, the matrix is usually a fractionated or hydrogenated vegetable oil (melting point in the range of 32°C–42°C), while in spray cooling, the matrix is typically a vegetable oil that can be combined with other materials such as waxes, hydrocolloids, gums, and mono- and diglyceride with the melting point in the range of 45°C–122°C (Risch 1995).

The process parameters known to have an impact on microparticles characteristics are heating temperature, feed flow rate, compressed air pressure (or wheel speed), and air flow rate, but also the relationship between all these variables. There are only limited literature data about the specific impact of each of the process variables on the product features, and the results are often nonconsistent between different reports. The smaller particles will be obtained if the nozzle orifice is smaller in diameter (Ilic et al. 2009; Gavory et al. 2014), if the energy consumption is higher (pressure or kinetic energy) (Maschke et al. 2007; Ilic et al. 2009), and when the feed rate is lower (Ilic et al. 2009). Ilic et al. (2009) showed that increased feed rate led to the decrease of the global product yields due to an inadequate cooling before the molten droplets reached the device walls. Some authors reported about the influence of the molten feed viscosity on the particle size, but their conclusions diverge (Cusimano and Becker 1968; Maschke et al. 2007).

The spray-cooled products are insoluble in water-based food, but cooking or baking of the food product with encapsulated aroma leads to the release of aroma. By choosing a matrix with the right melting point, it is possible to control release of aroma. Therefore, the main application of encapsulated aroma by spray cooling/chilling is in bakery products, dry soup mixes, and food containing a high level of fat such as margarines. On the other hand, this kind of encapsulation leaves a high level of the aroma on the surface of particles; thus, the prolonged release of water-soluble aroma by spray cooling/chilling is problematic for application in food of relatively high water content (Gouin 2004).

17.5 Emulsions as Flavor-Delivery Systems

Conventional oil-in-water emulsions have been largely used to encapsulate flavors. Oil-in-water emulsions consist of small oil droplets dispersed in aqueous phase where the interface is stabilized by emulsifiers. This type of emulsion is prepared conventionally by homogenizing oil, water, and emulsifier together using a mechanical device like colloid mill, sonicator, microfluidizer, or high-shear mixer. Lipophilic components (such are most of the flavors) could be incorporated into the oil phase, while hydrophilic components could be integrated into the aqueous phase either before or after homogenization. Typical emulsifiers form an interfacial film on oil droplets. There are some others (e.g., monoglyceride) that can form different structures, such as micelles or other self-assembly structures, which may accommodate flavor molecules.

The major disadvantage of oil-in-water emulsions is that they are only kinetically stable. Namely, emulsions have weak physical stability when exposed to environmental stresses such as heating, refrigeration, freezing, drying, pH, and ionic strength changes and high mineral concentrations during transport, storage, or utilization. Moreover, they have poor capacity to protect the encapsulated flavor from release, due to the low thickness of the interfacial membranes and, consequently, extremely high rate of molecular diffusion of the encapsulated compounds. Nanoemulsions containing smaller oil droplets (r < 100 nm) are more resistant to oxidation than microemulsions (Maswal and Dar 2014), but, at the same time, they are prone to growth in particle size (by Oswald ripening) (Sagalowicz and Leser 2010).

Engineering the interface of oil-in-water emulsion droplets with biopolymers that modify its permeability is a novel strategy in improvement of flavor retention. Multilayer oil-in-water emulsions are prepared by coating the emulsion droplets with nanolaminated interfacial layers formed by using layer-by-layer deposition method (Caruso 2001; Moreau et al. 2003; Ogawa et al. 2004). The resulting encapsulates have multilayered membranes. In multilayered emulsions, each oil droplet is surrounded by a multilayer interfacial membrane composed of alternating layers of an emulsifier and oppositely charged biopolymer (Grigoriev et al. 2008). Electrostatic interactions between proteins and polysaccharides are responsible for creation of stronger interfacial complexes than with a single biopolymer. It has been shown that multilayer emulsions have better stability than conventional emulsions toward particle aggregation at high salt concentrations, freeze–thaw cycling, thermal processing, and high calcium concentrations, the conditions commonly found in foods (Ogawa et al. 2003a,b, 2004; Gu et al. 2004, 2005a,b; Mun et al. 2005).

Most importantly, multilayered emulsions are able to better retain flavors than conventional ones. For example, when a pea protein isolate and pea protein isolate/pectin complex have been used to stabilize dry emulsions, an increased retention of three ethyl esters (belonging to the strawberry flavor note) during spray drying was achieved (Gharsallaoui et al. 2012). By tuning the properties of interfacial layers such as thickness, it is possible to control the release rate of flavors. In general, flavor release from emulsions occurs through partitioning and diffusion through the oil phase, the interface, and the water space toward the headspace (McClements et al. 2007). The release is determined by thermodynamic and kinetic factors. Thermodynamics of aroma release involve the partitioning of aroma compounds under equilibrium conditions, while the rate at which equilibrium is achieved is defined by kinetic factors. The driving force of aroma release under dynamic conditions is the difference in aroma concentration in the food and the air above the food. The release rate depends on compositional and structural properties of oil-in-water emulsions. Emulsion characteristics such as size of oil droplets, lipid fraction, and viscosity may influence to some extent flavor release (Landy et al. 2007; Benjamin et al. 2012) but not necessarily (Carey et al. 2002; Rabe et al. 2003). Furthermore, not only properties of the flavor molecules such as solubility, polarity, and molecular size but also the interactions between flavor molecules and other emulsion constituents such as oils, emulsifiers, and thickening agents influence flavor release from emulsions (Guyot et al. 1996; Van Ruth et al. 2002; Roberts et al. 2003; Karaiskou et al. 2008; Mao et al. 2012). Van Ruth et al. (2002) investigated the release of 20 aroma compounds by in-mouth conditions. They determined that the decrease in lipid fraction and emulsifier fraction as well as the increase in particle diameter led to an increased aroma release.

The physicochemical properties of flavor molecules have an impact on their release behavior as confirmed by static headspace analysis at thermodynamic equilibrium (Philippe et al. 2003). Thus, flavor partitioning was governed by hydrophobicity of the flavor molecules and the concentration of the oil phase (Carey et al. 2002; Philippe et al. 2003). Good retention at elevated temperatures has been obtained when using carnauba wax as an oil phase. Milanovic et al. (2010, 2011) showed that the release of ethyl vanillin from wax particles occurred under heating proceeded at different temperatures; vanillin evaporation occurred close to 200°C, while matrix degradation started at 250°C.

Not only the choice of emulsifier is an important issue from the aspect of emulsion stability, but it also reflects on flavor release. If the flavor molecules tend to bind to the emulsifier molecules, then there will be a less number of flavor molecules, which can freely diffuse from the oil phase to the water phase, and consequently, the mass transfer rate at the emulsion–gas interface is reduced (Harrison et al. 1997; Guichard and Langourieux 2000). In some cases, an interfacial film formed by the emulsifier acts as a barrier to mass transfer. Emulsifiers that form self-assembly structures can retain aroma compounds in structured emulsions. Thus, Phan et al. (2008) discovered that, under dynamic conditions, the structured emulsion (where unsaturated monoglycerides were added to the oil) had exhibited delayed release compared to the oil-in-water emulsion containing the same lipid content of 5%.

Flavors have been widely encapsulated in emulsions for applications in beverages. These emulsions are specific as very dilute containing as little as 20 mg/L of oil phase. Moreover, they must remain physically stable for a long time period, which means the absence of creaming, sedimentation, slicking, or droplet aggregation over the product shelf life (Given 2009). The oil phase in beverage emulsions consists of oily flavors in which a weighting agent is dissolved to increase the density close to the value of the final beverage product. The weighting agents are glycerol esters of wood, gum, oil resins (ester gum), sucrose acetate, isobutyrate, and brominated vegetable oil. Gum acacia, modified corn starch, whey proteins, arabinogalactan, and modified pectin are typically used as emulsifiers. The acid or enzyme hydrolysis has been employed to modify a wide range of hydrocolloids in order to improve their emulsifying properties. Various combinations of protein (net positive) and polysaccharides (net negative) or low-molecular-weight, anionic emulsifiers (like lecithin) have been utilized in order to form robust interfacial complexes, which should stabilize oil-in-water emulsions under acidic conditions for beverages (Bonnet et al. 2005; Thepkunya et al. 2006; Decker and McClements 2007; McClements et al. 2007). The first step of the preparation protocol is high-shear mixing of an emulsifier solution with an oil phase. The primary emulsion is then homogenized in a two-stage homogenizer in one or two passes employing pressures between 2000 and 5000 psi (Dluzewska and Leszczynski 2005). The obtained beverage macroemulsions have a particle size between 0.2 and 2 μm.

17.6 Hydrogel-Based Systems for Flavor Delivery

Protein or polysaccharide gels as flavor carriers have been frequently designed in the form of edible films or coatings. Edible barrier films provide good mechanical properties, and they are effective barriers to gases and aroma compounds (Miller et al. 1997). The transfer rate of aroma compounds from films to packaging is an important characteristic of edible films since it affects organoleptic profile of food during storage. The transfer rate depends on the physicochemical properties of the aroma compound (such as molecular weight, structure, hydrophobicity, polarity, solubility, volatility, and partial vapor pressure) and specific aroma–hydrocolloid interactions (Dury-Brun et al. 2007). Marcuzzo et al. (2010) investigated the release of 10 aroma compounds (methyl ketones, ethyl esters, and alcohols) from the iota-carrageenan emulsion–based edible film and compared those to lipid-based edible films. They found that carrageenan films can be used to control flavor release, since the release with time was retarded (a second order kinetics) in comparison to release from a lipid matrix (a first order release).

However, hydrocolloid films give poor protection against water diffusion (Miller et al. 1997). One way to improve barrier properties is to include lipidic materials in the hydrocolloid-based formulation, such as fatty acids or waxes (Fabra et al. 2009). There are two ways to associate lipids with a hydrophilic material: (1) by forming an emulsion of the lipid in which the outcome is a homogenous hydrocolloid network in which lipids are dispersed (Hambleton et al. 2008) and (2) by laminating the hydrocolloid film with a lipid layer (Kristo et al. 2007). Lipid content and specific aroma–lipid interactions affect film permeability of the emulsified films. It has been shown that at the lipid portion about 30% of the water permeability is reduced (Anker et al. 2002; Karbowiak et al. 2007). Waxes are especially effective in reducing films moisture permeability because they are highly hydrophobic (Morillon et al. 2002). In this type of films, the presence of a flavor changes some properties of the films. Thus, Hambleton et al. (2008) showed that the presence of a flavor compound (n-hexanal) in iota-carrageenan-fats-wax films (fats used are mono- and diglycerides blended with 20.2 wt.% bees wax) increased significantly the water vapor permeability, but had no effect on the oxygen permeability. Indeed, physicochemical interactions between aroma compounds and film components may induce structural changes such as plasticization, for example, in methylcellulose films containing 2-heptanon as a model compound (Quezada Gallo et al. 1999). When it happens, films become opaque and swell, while glass transition temperature (T_g) of the polymer decreases as well as the barrier properties. The heat treatments corresponding to those classically encountered in baking processes have not significantly changed fat globule size and water permeability of iota-carrageenan-fats-wax films (Karbowiak et al. 2007).

Apart from films, microspheres are another form of hydrogels used for aroma encapsulation. In general, gel microspheres are made by applying one of the extrusion (dropping) methods or by the emulsion method. Alginate, chitosan, agarose, gelating, gellan gum, and κ-carrageenan have been used to create gel matrix. Ca-alginate microspheres (~450 μm) have been able to prolong the release of ethyl vanillin under heating conditions, which mimicked usual food processing (Manojlovic et al. 2008). Namely, up to 230°C, most of the vanillin remained *intacta*, while prolonged heating at elevated temperatures led to the entire loss of the aroma compound. Ethyl–vanillin encapsulates made from Ca-alginate and Ca-alginate-poly(vinyl alcohol) blend were stable for a period of 1 month after being dried (Levic et al. 2013). In these systems, aroma (as a hydrophobic compound) is inhomogenously dispersed within the gel matrix, which is a hydrophilic material. However, in most applications, aroma-containing microspheres are prepared in the presence of aroma in (vegetable) oil droplets. The retention of aroma is controlled by both their initial lipophilic medium and the porosity of the gel matrix surrounding the oil droplets. With increasing polymer concentration and/or cross-linking time, it is possible to retard flavor release. Increased aroma retention has been shown for Ca-alginate microspheres when backing crackers (De Roos 2003). Microgel emulsion particles have been used for the controlled release of lipophilic volatiles during eating. They are complex systems consisting of oil-in-water emulsions entrapped within a gel matrix. By encapsulating oil droplets within biopolymer-gelled particles (70–5000 μm), the initial flavor release maxima were reduced by kinetically inhibiting the mass transfer of flavor through the particle (Malone and Appelqvist 2003). Factors such as particle size, oil phase volume, and the partition coefficient of the volatile all affect the rate of volatile release. Lian et al. (2004) developed a mathematical model describing in-mouth volatile release of aroma compounds from Ca-alginate-gelled emulsion particles. The model was tested on several aroma compounds (butanone, heptanone, octanone, nonanone, ethyl hexanoate). From this model, three regimes of in-mouth release of aroma from the dispersion of gelled emulsion particles were identified including the emulsion regime, the transition regime, and the gel particle regime. In the emulsion regime, changes in the size of gelled emulsion particles had negligible impact on the overall release. In the transition regime, the release was controlled by the interaction of flavor transfer from the particles with that across the air–water interface. In the gel particle regime, aroma release at long times was governed by the particles and that at short times was governed by the air–water interface.

17.7 Flavor Encapsulation via Physical Adsorption

Physical adsorption is the least developed encapsulation technology. Only microporous carriers often called molecular sieves have adequate surfaces for effective physical adsorption of flavors. Some microporous materials have a significant portion of mesopores (with openings 20–500 Å in diameter) in which capillary condensation of volatile flavors occurs apart from physical adsorption. The examples are activated carbons (surface 500–1400 m²/g) and silicas (surface are of 100–1000 m²/g), but their use is severely restricted in food. On the other hand, only physical adsorption can provide dynamic, equilibrium-controlled flavor release and consistent headspace aroma throughout the product use. Therefore, there have been some trials in the past to create highly microporous adsorbent sugars (sucrose, glucose, fructose, corn starch, calcium alginates, sugar salts). One of the successful treatments proposed by Zeller et al. (1987) and Zeller and Saleeb (1996) is based on spray atomization of a carbohydrate solution into cryogenic liquids (liquid nitrogen or cold ethanol) followed by drying below their collapse temperature (freeze-drying or vacuum drying). The obtained carriers with surface areas of several hundred m²/g have been tasted for immobilization of flavors. The manufacture of high-porosity carbohydrates is relatively complex compared to the production of conventional flavor carriers and can be justified if some unique functionalities are ought to be achieved, for example, adsorption of some other compounds, apart from flavor (e.g., unpleasant compounds, impurities, carbon dioxide). Veith et al. (2004a,b) investigated retention performance of silica particles made by hydrolysis of tetraethyl orthosilicate. Since particle morphology, porosity, and pore size distribution can be controlled by the sol-gel preparation method, the preparation conditions seem to be important from the aspect of aroma retention as well. The retention performance decreased from alcohols > aldehydes ≥ esters > terpenes. The release of molecules that just physically interact with the silica surface were associated to a simple desorption process (e.g., reversible hydrogen bonds). In the case of chemical adsorption, these bonds are believed to be reversible as well. Open sol-gel-made silica particles showed an increased retention with increasing aroma load, while denser silica matrices show a maximum retention with increasing load.

17.8 Cyclodextrin Inclusion Complexes for Flavor Delivery

Cyclodextrin inclusion complexes are widely used in the food industry to stabilize volatile and sensitive flavors and other food additives. Cyclodextrins are cyclic oligosaccharides having a cone-shaped molecular structure with a hydrophobic internal cavity and a hydrophilic part faced outside (Del Valle 2004). The three most common cyclodextrins used are α-cyclodextrins, β-cyclodextrins, and γ-cyclodextrins having 6, 7, and 8 glucopyranose units in the cyclic structure, respectively. The cavity diameter varies between 0.5 and 0.8 nm, which is relatively small allowing the solubilization of only small molecules (such as flavors), but the cavity depth is the same for all cyclodextrins, which is ~8 Å (Szejtli 1998). From the viewpoint of aroma encapsulation, β-cyclodextrins are of the most interest as they usually form complexes with aromatics and heterocyclics (Del Valle 2004). About 1% of all aroma encapsulates are based on the use of β-cyclodextrins (Porzio 2004). Cyclodextrins have a unique ability to form host–guest inclusion complexes with a variety of molecules. The interactions responsible for complexation are noncovalent, such as van der Waals forces, hydrophobic interactions, and other forces. The main driving force of complex formation is the release of the solvent molecules from the cavity. Water molecules are replaced by more hydrophobic guest molecules present in the solution to attain an apolar–apolar association and decrease of cyclodextrin ring strain resulting in a more stable lower-energy state (Szejtli 1998). Complexes can be formed either in a solution or in the crystalline state. In the crystalline form, only the surface molecules of the cyclodextrin crystals are available for complexation unlike in the solution where the probability for complexation is higher. It is also better if the guest compound is in a solution, or at least dispersed in fine particles. In addition, heating promotes complexation, but up from some point (usually above 60°C), the initially formed complexes start to decompose. Thermostable complexes are those formed via strong bonds between the host and guest molecules or when they are highly insoluble. Water is typically the solvent of choice for complexation in general, but it can be accomplished in the presence of any nonaqueous cosolvent (Del Valle 2004). As most of the flavors are nonsoluble in water, an organic solvent (which should not complex well with cyclodextrin) must be used to dissolve the flavor, such as ethanol and diethyl ether. In order to avoid aroma loss during processing, especially in case of highly volatile compounds, a sealed reactor has to be used or even operating with a refluxing the solution of flavor.

Some beneficial changes induced by complexation include alteration of guest's solubility, stabilization against effects of light and heat, oxidation and reactions, guest's physiological responses, and inhibition of its volatility. Generally, complexation greatly enhances the solubility of poorly water-soluble compounds. Once the cavity is occupied by molecule, other reactive molecules are excluded from occupying the cavity at the same time, preventing interaction and reaction. In addition, steric hindrance prevents the interaction with exposed portions of the guest molecule (Maswal and Dar 2014). Cyclodextrin-flavor inclusion complexes provide prolonged shelf life and high-temperature stability of the flavor (up to 200°C). The release of the complexed flavors starts at above 160°C, whereas noncomplexed flavors readily evaporate at about 100°C. The inclusion of volatiles in β-cyclodextrin molecules reduces the activity coefficient of the volatiles; thereby, their relative volatility to water is depressed. Flavor complexes in dry conditions are very stable, with oxygen uptake less than 10% of that of noncomplexed compounds (Yuliani et al. 2004). In addition, flavor complexes only lose 25%–30% of their aroma content after 24 h at 150°C in vacuum, whereas pure spice aromatics evaporate completely under similar conditions (Szejtli et al. 1979). The crystalline formed from inclusion complexation are very stable with a strongly reduced water solubility (Lindner et al. 1981). Furthermore, they are effective for controlled/sustained release of the flavors. For example, cyclodextrins are used in chewing gums to retain flavor for longer duration (Mabuchi and Ngoa 2001). In order to produce microcapsules, which have controlled-release properties, Yamamoto et al. (2012) made D-limonene-β-cyclodextrin inclusion complex, which was coated with a polymer-based agent (polyethylene resin or ethylene-vinyl acetate copolymer resins) during spray drying. The release of the flavor was almost negligible (at relative humidity less than 90% and in the temperature range 40°C–60°C) when polyethylene resin was used as a coating agent at a concentration of 12 wt.%. Such behavior is linked to a particle's surface structure composed of the fine complex particles bound by the coating agents, whereas the inside of the coated complex-powder particles consists of fine complex particles packed in a random manner.

However, superior retention of β-cyclodextrin-based encapsulates at some point is not so welcome. Namely, during eating or even prior to eating, pleasant aroma compounds (such as those used in confectionery products) have to be released in order to enable sensory perception. Encapsulation of flavors in cyclodextrin inclusion complexes is not the best solution for such applications. However, the load of cyclodextrins with aromas is low, between 8% and 10% (Szente and Szejtli 2004). Furthermore, cyclodextrin encapsulates are not convenient for beverages with fat content (e.g., milk, yogurt) since the aroma will migrate quickly into the fat. β-Cyclodextrins are metabolized in the cecum and colon, and only small amounts are (1%–2%) absorbed in the upper intestinal tract after oral administration. High doses of β-cyclodextrins may cause harmful effects during degradation in the colon. Therefore, the use of β-cyclodextrin for food application is very limited, possibly due to regulatory requirements in a number of countries. Namely, the use of β-cyclodextrin for food application has been approved by the FDA since 1998 at a level of 2% (Szente and Szejtli 2004), but it has not been permitted in Australia yet (Yuliani et al. 2006b).

17.9 Coacervation Complexes for Flavor Delivery

Coacervation, often called "phase separation," is considered as a true microencapsulation technique, because the core material is completely entrapped by the matrix. Protein–polysaccharide complex coacervation has been used as a unique and promising method for flavor encapsulation. Two oppositely charged biopolymer molecules are mixed at a pH below the protein isoelectric point (pI) leading to separation of the polymer-rich phase and the formation of a complex, that is, coacervate, which precipitates. At the same time the aroma material is retained at the coacervate phase. Only mixtures with quite low-polymer concentrations are able to undergo this phase separation transition. For example, with the polymer system gelatin/gum arabic, the concentration of both polymers has to be below 8.0% (w/w) for coacervation to occur (Lemetter et al. 2009). The process is governed by pH manipulation. For the same polymer system, the starting pH value is above 7 (to create a stable solution of hydrated gelatin and gum arabic), and then it should be set to a pH of 4.3 (an acidification step) to have enough opposite electrical charges, which should lead to the phase separation. Also, the temperature is an important factor. At first, the temperature must be above the gelling temperature of the polymer, and then it should be lowered below the gelation temperature of the protein during coacervation and formation of the shell. The agitation speed has influence on distribution of the coacervate phase over the oil surface. Simple coacervation has been less explored in flavor encapsulation (Hsieh et al. 2006). The process is based on separation of an oil-in-water emulsion into (only one) polymer-rich phase (coacervate) and a polymer-poor phase.

Flavor encapsulation by complex coacervation of proteins has been the subject of some recently published research papers (Yeo et al. 2005; Prata et al. 2008; Leclercq et al. 2009; Jun-xia et al. 2011; Koupantsis et al. 2014). In these studies, various combinations including gelatin or soybean protein isolate with gum arabic, xanthan, pectin, and carboxymethyl cellulose have been used for the encapsulation of essential oils or individual flavor compounds. Optionally, these polymers can be replaced by some other negatively charged molecules, such as alginate, alginate derivates, or phosphate (Meyer 1992). Fish gelatin can replace mammalian gelatin for halal food (Reilly and Subramaniam 2004).

The liquid coacervate layer can be transformed to a rigid membrane by gelatin cross-linking. For this purpose, a cross-linking agent is needed, such as glutaraldehyde and transglutaminase (Thies 2007). However, aldehyde agents are not considered as safe for the food industry in many countries. Tannins, a widely distributed plant polyphenols, could be employed because of their ability to complex with conformationally open proteins, such as gelatin, primarily through hydrogen bonding and hydrophobic interactions (Hagerman and Butler 1981; Xing et al. 2004). Complex coacervates might be postloaded with an aroma. The process starts with dissolving aroma in ethanol; this solution is then mixed with water-swollen complex coacervates under agitation for several hours, followed by ethanol removal (Thies 2007).

The difficulties mentioned here can be avoided with a right choice of protein and other parameters (such as pH, temperature, shear, residence time, flavor/protein ratio).

Polymer concentrations and homogenization rate affect particle morphology, size distribution, and flavor release from coacervation complexes. For example, when complex coacervate microcapsules encapsulating flavor oil made from gelatin and gum arabic were exposed to heating at 100°C or higher, univesicular microcapsules (prepared with a lower-homogenization rate) released almost all of the encapsulated oil, while multivesicular microcapsules (produced by high-homogenization rates) had lesser degrees of release (Yeo et al. 2005). Lemetter et al. (2009) were able to determine the effect of the impeller rotation speed (and, thus, the effect of the level of shear) on mononucleated and/or polynucleated capsule formation and on the corresponding capsule size distribution. Moreover, they determined what is the turbulence level (the Reynold number) above which the agitation should be maintained to avoid agglomeration. Taking the turbulence level as the main scale-up criteria, the authors performed a successful scale up from a bench (2 L) to a pilot-plant scale (50 L).

Microcapsules produced by the complex coacervation method possess excellent controlled-release characteristics, heat-resistant properties, and high encapsulation efficiency (aroma load between 20% and 90%) (Reineccius 1989; Yang et al. 2014). To illustrate, in one of very recent studies, vanilla oil was encapsulated using the complex coacervation approach; microcapsules produced under optimized conditions (regarding viscosity of chitosan and vanilla oil/chitosan ratio) showed high encapsulation efficiency (94%), superior thermostability of vanilla oil, and a retention of about 60% after release for 1 month (Yang et al. 2014). In the same study, it was determined that the chitosan viscosity had a critical influence on the microcapsule fabrication since only with moderate viscosity chitosan (vs low viscosity chitosan and high viscosity chitosan) it was possible to obtain microcapsules, which were spherical in shape and smooth in surface with a good dispersibility, shown in Figure 17.5.

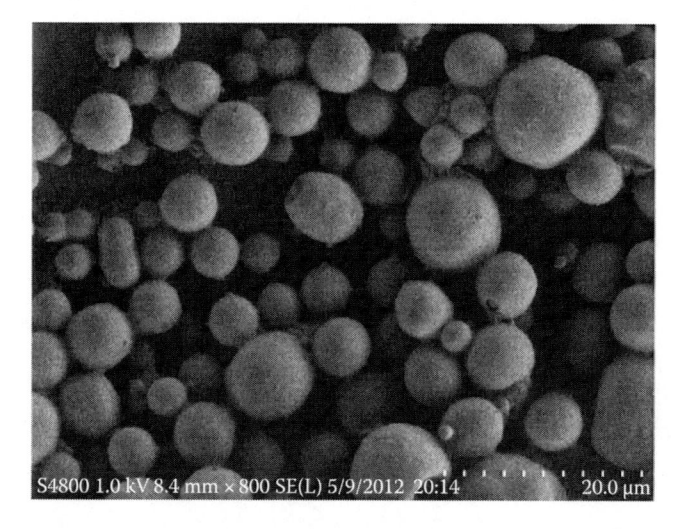

FIGURE 17.5 SEM micrographs of freeze-dried microcapsules containing vanilla oil and produced by complex coacervation: vanilla-oil-in-gum-arabic emulsion was coacervated with medium-viscosity chitosan. (From Yang, Z. et al., *Food Chem.*, 145, 272, 2014. With permission.)

However, a challenging issue with complex coacervation is the burst release effect, aggregation, and conglutination problems, which is not desirable in most applications. Thus, it is known that processing with gelatin (even if it is a commonly used wall material for complex coacervation) may lead to aggregation and/or formation of polynucleated capsules (because it is highly viscous even at low concentrations) resulting in a hardly controllable capsule size.

17.10 Yeast Encapsulates for Flavor Delivery

The use of microorganisms such as baker's yeast as microcontainers for the encapsulation of actives has been widely investigated since the 1970s (Shank 1977; Pannell 1994; Bishop et al. 1998; Normand et al. 2005; Shi et al. 2007, 2008; Paramera et al. 2011). Encapsulation of flavors by using yeast cells (*Saccharomyces cerevisiae*) as a wall material is considered as a low-cost and safe process (Bishop et al. 1998). Apart from yeast, only water and the flavor to be encapsulated are needed; no other additives are used. According to Ciamponi et al. (2012), who investigated kinetics and mechanistic aspects of encapsulation in yeast (terpenes were used as model hydrophobic compounds), the process is essentially a phenomenon of passive diffusion with negligible relevance of active transport. Further, they showed that the major determinant of the encapsulation kinetics is the solubility of the hydrophobes in the cell wall, which is inversely related to partition coefficient (log P). Therefore, only hydrophobic flavors (log P between 0.5 and 4) can penetrate inside the yeast and thus be encapsulated, and the loading process is driven by the affinity between the flavor molecules and the hydrophobic bag formed by the phospholipid bilayer (plasma membrane) (Dardelle et al. 2007). Then, a flavor ought to be encapsulated should be small enough (molecular weight below 1000) (Zuidam and Heinrich 2010). If a water-soluble flavor compound is ought to be absorbed by yeast, then a pretreatment with a plasmolyser or by chemicals inducing autolysis precedes the encapsulation.

The most important benefit of yeast-based encapsulates is that they are thermo-stable delivery systems even at very high temperatures. Namely, yeasts are stable to temperatures as high as 250°C (Bishop et al. 1998), and flavors encapsulated in yeast cells leaked out only at higher temperatures than 243°C. Furthermore, during drying of the cells, cell wall permeability reduces with temperature increasing and water losing. This explains superior flavor protection during thermal treatment characteristic for cooking process. Thus, yeast capsules behave better in dry food products (Zuidam and Heinrich 2010). Heinrich (2006) determined that the rate of ginger aroma diffusion through the yeast cell bilayer membrane remains constant above a critical temperature of about 60°C. Interestingly, the release of flavor from loaded yeasts does not occur in pure fat. Actually, water is needed to trigger release starting with swelling of the external layer (identified as mannoproteins) and formation of the holes estimated to several tenths of nanometers in diameter (Normand et al. 2005). When limonene was used as a model flavor, commercial yeast-based delivery systems showed higher organoleptic scores compared to conventional spray-dried powders (Dardelle et al. 2007). Yeast-cell capsules have found applications in flavored pasta and rice, while aroma stability becomes jeopardized in food containing triglyceride fats (Zuidam and Heinrich 2010).

17.11 Outlook on Applications of Encapsulates as Flavor-Delivery Systems

The flavoring industry is a pioneer in implementation of encapsulation technology in the food sector. Still, it utilizes about three times more free flavors compared to encapsulated forms. The main reason behind this fact is the high cost of encapsulation, so that it, as a minimum, doubles the cost of any product. Some decade ago, it has been estimated that the maximum cost acceptable for the food industry was €0.1/kg (Gouin 2004). Precise information about the costs of encapsulation technologies is not available because a lot of developed applications are keeping as secret. Even so, we believe in increased production and consumption of encapsulated flavors having in mind current consumers' demands. Namely, people's dietary habits are changing and are becoming more and more oriented to food that is healthy, safe, and, at the same time, tasty, even if it is more expensive. Then, there is a trend toward easier food preparation procedures, which implies a need for protection and controlled release of often complex and sensitive flavors. Therefore, the processes and carriers for encapsulation are continually being improved on both laboratory scale and industrial level in order to further promote aroma preservation and delivery. With improved encapsulation techniques and by increasing capacities, manufactures can reduce costs of microencapsulation, especially, if they subtract savings derived from benefits, which can be achieved by encapsulation (increased productivity, improved yields, extended shelf life, and more consistent product quality).

Apart from price, the lack of knowledge on release of encapsulated aroma when incorporated in real food products is another limiting factor. There have not been enough studies discussing this matter, while literature that is dealing with the release of the aroma upon application and with stability of encapsulated aroma (not incorporated in the food) is far more available.

It is likely to expect that in the future, glass encapsulation will again dominate the food industry. Variations in the supply of gum acacia and increasing prices of starch-based materials lead the flavor industry to seek alternative materials for glass encapsulation. Legal status, price, availability, and, most of all, functionality of the material are crucial for their selection. Proteins (such as milk proteins) will be getting more attention of scientists in the next future. Some operating conditions of fluidized bed coating (such as dilute particle–air volume ratio, low spray flow rates) are not economically viable for most food-particle-coating applications. Moreover, because the role of many processing variables is still not completely resolved, it is difficult to predict coating morphology. Despite all these, fluidized bed coating is experiencing a rapid growth because some additional benefits (particle size, shape, color) related to this technology are paying off.

Currently, the main controlled-release encapsulates on the market are spray-chilled encapsulates and coacervate capsules. There are many more at the R&D level, but it seems unreasonable to expect their commercial application (too expensive and/or have unwanted side effects). Achieving a delicate balance between shelf life stability and release is what makes aroma encapsulation successful. Encapsulation in cyclodextrins means processing at a molecular level, and this makes this technology so attractive. Recent biotechnological advancements have resulted in dramatic improvements in the manufacture of cyclodextrins. This resulted in lowering the cost of these materials making highly purified cyclodextrins and cyclodextrins derivates available. Combining the emulsion formulations with the naturally occurring antioxidants will have a future for enhancing chemical stability of some sensitive flavors and also from the viewpoint of customer acceptance (antioxidant-rich products).

ACKNOWLEDGMENT

This work was supported by the Ministry of Education, Science and Technological Development, Republic of Serbia (Projects No. III46010 and No. III46001).

REFERENCES

Albertini, B., Passerini, N., Pattarino, F., and Lorenzo, R. New spray congealing atomizer for the microencapsulation of highly concentrated solid and liquid substances. *Eur. J. Pharm. Biopharm.* 69 (2008), 348–357.

Anker, M., Bensten, J., Hermansson, A., and Stading, M. Improved water vapor barrier of whey protein films by addition of an acetylated monoglyceride. *Innov. Food Sci. Emerg. Technol.* 3 (2002), 81–92.

Baines, D. and Knights, J. Applications I: Flavors. Chapter 12. In: D.J. Rowe (ed.), *Chemistry and Technology of Flavors and Fragrances.* De Monchy Aromatics Ltd Poole, Poole, U.K., 2005, pp. 274–304.

Benjamin, O., Silcock, P., Leus, M., and Everett, D.W. Multilayer emulsions as delivery systems for controlled release of volatile compounds using pH and salt triggers. *Food Hydrocoll.* 27 (2012), 109–118.

Bishop, J.R.P., Nelson, G., and Lamb, J. Microencapsulation in yeast cells. *J. Microencapsul.* 15 (1998), 761–773.

Bonnet, C., Corredig, M., and Alexander, M. Stabilization of caseinate-covered oil droplets during acidification with high methoxyl pectin. *J. Agric. Food Chem.* 53 (2005), 8600–8606.

Boskovic, M.A., Vidal, S.M., and Saleeb, F.Z. Stabilization of flavorants without the use of antioxidants by spray-drying in a carbohydrate substrate. U.S. Patent 5,124,162 (1992).

Branca, A., Simonian, P., Ferrante, M., Novas, E., and Negri, R.M. Electronic nose based discrimination of a perfumery compound in a fragrance. *Sens. Actuators B* 92 (2003), 222–227.

Brauss, M.S., Balders, B., Linforth, R.S.T., Avison, S., and Taylor, A.J. Fat content, baking time, hydration and temperature affect flavour release from biscuits in model-mouth and real systems. *Flavour Frag. J.* 14 (1999), 351–357.

Bringas-Lantigua, M., Valdes, D., and Pino, J.A. Influence of spray-dryer air temperatures on encapsulated lime essential oil. *Int. J. Food Sci. Technol.* 47 (2012), 1511–1517.

Brückner, M., Bade, M., and Kunz, B. Investigations into stabilization of a volatile aroma compound using a combined emulsification and spray drying process. *Eur. Food Res. Technol.* 226 (2007), 137–146.

Buffo, R. and Reineccius, G.A. Optimization of Gum Acacia/modified starch/maltodextrin blends for the spray drying of flavors. *Perfumer Flavor.* 25 (2000), 45–54.

Carey, M.E., Asquith, T., Linforth, R.S.T., and Taylor, A.J. Modeling the partition of volatile aroma compounds from a cloud emulsion. *J. Agric. Food Chem.* 50 (2002), 1985–1990.

Carneiro, H.C.F., Tonon, R.V., Grosso, C.R.F., and Hubinger, M.D. Encapsulation efficiency and oxidative stability of flax seed oil microencapsulated by spray drying using different combinations of wall materials. *J. Food Eng.* 115 (2013), 443–451.

Caruso, F. Generation of complex colloids by polyelectrolyte-assisted electrostatic self-assembly. *Aust. J. Chem.* 54 (2001), 349–353.

Charve, J. and Reineccius, G. Encapsulation performance of proteins and traditional materials for spray dried flavors. *J. Agric. Food Chem.* 57 (2009), 2486–2492.

Chranioti, C. and Tzia, C. Binary mixtures of modified starch, maltodextrin and chitosan as efficient encapsulating agents of fennel oleoresin. *Food Bioprocess Technol.* 6 (2013), 3238–3246.

Christensen, K.L., Pedersen, G.P., and Kistensen, H.G. Preparation of redispersible dry emulsions by spray drying. *Int. J. Pharm.* 212 (2001), 187–194.

Ciamponi, F., Duckham, C., and Tirelli, N. Yeast cells as microcapsules. Analytical tools and process variables in the encapsulation of hydrophobes in *S. cerevisiae. Appl. Microbiol. Biotechnol.* 95 (2012), 1445–1456.

Cusimano, A.G. and Becker, C.H. Spray-congealed formulations of sulfaethylthiadiazole (SETD) and waxes for prolonged-release medication—Effect of wax. *J. Pharm. Sci.* 57 (1968), 1104–1112.

Dardelle, G., Normand, V., Steenhoudt, M., Bouquerand, P.-E., Chevalier, M., and Baumgartner, P. Flavour-encapsulation and flavour-release performances of a commercial yeast-based delivery system. *Food Hydrocoll.* 21 (2007), 953–960.

De Roos, K.B. Effect of texture and microstructure on flavor release. *Int. Dairy J.* 13 (2003), 593–605.

Decker, E.A. and McClements, D.J. Stable acidic beverage emulsions and methods of preparation. WO 2007038624 A2 (2007).

Del Valle, E.M.M. Cyclodextrins and their uses: A review. *Process Biochem.* 39 (2004), 1033–1046.

Dewettinck, K. and Huyghebaert, A. Fluidized bed coating in food technology. *Trends Food Sci. Technol.* 10 (1999), 163–168.

Dluzewska, E. and Leszczynski, K. Effect of carrier on the quality of encapsulated flavours. *Pol. J. Food Nutr. Sci.* 14 (2005), 293–298.

Dury-Brun, C., Chalier, P., Desobry, S., and Voilley, A. Multiple mass transfers of small volatile molecules through flexible food packaging. *Food Rev. Int.* 23 (2007), 199–255.

Emin, M.A. and Schuchmann, H.P. Droplet breakup and coalescence in a twin-screw extrusion processing of starch based matrix. *J. Food Eng.* 116 (2013), 118–129.

Fabra, M.J., Hambleton, A., Talens, P., Debeaufort, F., Chiral, A., and Voilley, A. Influence of interactions on water and aroma permeabilities of iota-carrageenan oleic acid-beeswax films used for flavor encapsulation. *Carbohydr. Polym.* 76 (2009), 325–332.

Fang, Z. and Bhandari, B. Encapsulation of polyphenols—A review. *Trends Food Sci. Technol.* 21 (2010), 510–523.

Frascareli, E.C., Silva, V.M., Tonon, R.V., and Hubinger, M.D. Effect of process conditions on the microencapsulation of coffee oil by spray drying. *Food Bioprod. Process.* 90 (2012), 413–424.

Fulger, C.V. and Popplewell, L.M. Flavor encapsulation. U.S. Patent 5,958,502 (1999).

Gardner, J.W. and Bartlett, P.N. *Electronic Noses: Principles and Applications*. Oxford University Press, New York, 1999.

Gavory, C., Abderrahmen, R., Bordes, C., Chaussy, D., Belgacem, M.N., Fessi, H., and Briançon, S. Encapsulation of a pressure sensitive adhesive by spray-cooling: Optimum formulation and processing conditions. *Adv. Powder Technol.* 25 (2014), 292–300.

Gharsallaoui, A., Roudaut, G., Beney, L., Chambin, O., Voilley, A., and Saurel, R. Properties of spray-dried food flavours microencapsulated with two-layered membranes: Roles of interfacial interactions and water. *Food Chem.* 132 (2012), 1713–1720.

Gibbs, B.F., Kermasha, S., Alli, I., and Mulligan, C.N. Encapsulation in the food industry: A review. *Int. J. Food Sci. Nutr.* 50 (1999), 213–224.

Gilsenan, P.M., Richardson, R.K., and Morris, E.R. Thermally reversible acid-induced gelation of low-methoxy pectin. *Carbohydr. Polym.* 41 (2000), 339–349.

Given, Jr., P.S. Encapsulation of flavors in emulsions for beverages. *Curr. Opin. Colloid Interface Sci.* 14 (2009), 43–47.

Gouin, S. Micro-encapsulation: Industrial appraisal of existing technologies and trends. *Trends Food Sci. Technol.* 15 (2004), 330–347.

Griffin, S., Grant Wyllie, S., and Markham, J. Determination of octanol–water partition coefficient for terpenoids using reversed-phase high-performance liquid chromatography. *J. Chromatogr. A* 864 (1999), 221–228.

Grigoriev, D.O., Bukreeva, T., Möhwald, H., and Shchukin, D.G. New method for fabrication of loaded micro- and nanocontainers: Emulsion encapsulation by polyelectrolyte layer-by-layer deposition on the liquid core. *Langmuir* 24 (2008), 999–1004.

Gu, Y.S., Decker, A.E., and McClements, D.J. Influence of pH and iota-carrageenan concentration on physicochemical properties and stability of beta-lactoglobulin-stabilized oil-in-water emulsions. *J. Agric. Food Chem.* 52 (2004), 3626–3632.

Gu, Y.S., Decker, A.E., and McClements, D.J. Production and characterization of oil-in-water emulsions containing droplets stabilized by multilayer membranes consisting of beta-lactoglobulin, iota-carrageenan and gelatin. *Langmuir* 21 (2005a), 5752–5760.

Gu, Y.S., Regnier, L., and McClements, D.J. Influence of environmental stresses on stability of oil-in-water emulsions containing droplets stabilized by beta-lactoglobulin-iota-carrageenan membranes. *J. Colloid Interface Sci.* 286 (2005b), 551–558.

Guichard, E. and Langourieux, S. Interactions between β-lactoglobulin and flavour compounds. *Food Chem.* 71 (2000), 301–308.

Guignon, B., Duquenoy, A., and Dumoulin, E.D. Fluid bed encapsulation of particles: Principles and practice. *Dry. Technol.* 20 (2002), 419–447.

Gunning, Y.M., Gunning, P.A., Kemsley, E.K., Parker, R., Ring, S.G., Wilson, R.H., and Blake, A. Factors affecting the release of flavour encapsulated in carbohydrate matrixes. *J. Agric. Food Chem.* 47 (1999), 5198–5205.

Guyot, C., Bonnafont, C., Lesschaeve, I., Issanchou, S., Voilley, A., and Spinnler, H.E. Effect of fat content on odor intensity of three aroma compounds in model emulsions: δ-decalactone, diacetyl, and butyric acid. *J. Agric. Food Chem.* 44 (1996), 2341–2348.

Hagerman, A.E. and Butler, L.G. The specificity of proanthocyanidin–protein interaction. *J. Biol. Chem.* 256 (1981), 4494–4497.

Hambleton, A., Debeaufort, F., Beney, L., Karbowiak, T., and Voilley, A. Protection of active aroma compound against moisture and oxygen by encapsulation in biopolymeric emulsion-based edible films. *Biomacromolecules* 9 (2008), 1058–1063.

Hansch, L. and Hoekman, D. *Exploring QSAR: Hydrophobic, Electronic and Steric Constants*. American Chemical Society, Washington, DC, 1995.

Harrison, M., Hills, B.P., Bakker, J., and Clothier, T. Mathematical models of flavor release from liquid emulsions. *J. Food Sci.* 62 (1997), 653–664.

Heinrich, E. Aroma release properties from yeast cell encapsulates in watery applications. In: *Proceeding of the XIVth International Workshop on Bioencapsulation*, Lausanne, CH, October 6–7, 2006.

Hsieh, W.C., Chang, C.P., and Gao, Y.L. Controlled release properties of chitosan encapsulated volatile citronella oil microcapsules by thermal treatments. *Colloids Surf. B* 53 (2006), 209–214.

Ilic, I., Dreua, R., Burjak, M., Homarb, M., Kerc, J., and Srcic, S. Microparticle size control and glimepiride microencapsulation using spray congealing technology. *Int. J. Pharm.* 381 (2009), 176–183.

Jafari, S.M., Assadpoor, E., Bhandari, B., and He, Y. Nano-particle encapsulation of fish oil by spray drying. *Food Res. Int.* 41 (2008a), 172–183.

Jafari, S.M., Assadpoor, E., He, Y., and Bhandari, B. Encapsulation efficiency of food flavours and oils during spray drying. *Dry. Technol.* 26 (2008b), 816–835.

Jafari, S.M., Beheshti, P., and Assadpoor, E. Emulsification properties of a novel hydrocolloid (Angum gum) for D-limonene droplets compared with Arabic gum. *Int. J. Biol. Macromol.* 61 (2013a), 182–188.

Jafari, S.M., Beheshti, P., and Assadpoor, E. Rheological behavior and stability of D-limonene emulsions made by a novel hydrocolloid (Angum gum) compared with Arabic gum. *J. Food Eng.* 109 (2013b), 1–8.

Jeon, Y.J., Vasanthan, T., Temelli, F., and Song, B.K. The suitability of barley and corn starches in their native and chemically modified forms for volatile meat flavor encapsulation. *Food Res. Int.* 36 (2003), 349–355.

Jun-xia, X., Hai-yan, Y., and Jian, Y. Microencapsulation of sweet orange oil by complex coacervation with soybean protein isolate/gum arabic. *Food Chem.* 125 (2011), 1267–1272.

Kaasgaard, T. and Keller, D. Chitosan coating improves retention and redispersibility of freeze-dried flavour oil emulsions. *J. Agric. Food Chem.* 58 (2010), 2446–2454.

Kanakdande, D., Bhosale, R., and Singhal, R.S. Stability of cumin oleoresin microencapsulated in different combinations of gum arabic, maltodextrin, and modified starch. *Carbohydr. Polym.* 67 (2007), 536–541.

Karaiskou, S., Blekas, G., and Paraskevopoulou, A. Aroma release from gum arabic or egg yolk/xanthan-stabilized oil-in-water emulsions. *Food Res. Int.* 41 (2008), 637–645.

Karbowiak, T., Debeaufort, F., and Voilley, A. Influence of thermal process on structure and functional properties of emulsion-based edible films. *Food Hydrocoll.* 21 (2007), 879–888.

Kaushik, V. and Roos, Y.H. Limonene encapsulation in freeze-drying of gum Arabic–sucrose–gelatin systems. *LWT Food Sci. Technol.* 40 (2007), 1381–1391.

Koupantsis, T., Pavlidou, E., and Paraskevopoulou, A. Flavour encapsulation in milk proteins—CMC coacervate-type complexes. *Food Hydrocoll.* 37 (2014), 134–142.

Krishnan, S., Bhosale, R., and Singhal, R.S. Microencapsulation of cardamom oleoresin: Evaluation of blends of gum arabic, maltodextrin and a modified starch as wall materials. *Carbohydr. Polym.* 61 (2005), 95–102.

Kristo, E., Biliaderis, C.G., and Zampraka, A. Water vapour barrier and tensile properties of composite caseinate-pullulan films: Biopolymer composition effects and impact of bees wax lamination. *Food Chem.* 101 (2007), 753–764.

Landy, P., Pollien, P., Pytz, A., Leser, M.E., Sagalowicz, L., Blank, I., and Spadone, J.-C. Model studies on the release of aroma compounds from structured and nonstructured oil systems using proton-transfer reaction mass spectrometry. *J. Agric. Food Chem.* 55 (2007), 1915–1922.

Leclercq, S., Harlander, K.R., and Reineccius, G.A. Formation and characterization of microcapsules by complex coacervation with liquid or solid aroma cores. *Flavour Fragr. J.* 24 (2009), 17–24.

Lee, S.W., Kang, S.Y., Han, S.H., and Rhee, C. Influence of modification method and starch concentration on the stability and physical properties of modified potato starch as wall materials. *Eur. Food Res. Technol.* 228 (2009), 449–455.

Lemetter, C.Y.G., Meeuse, F.M., and Zuidam, N.J. Control of the morphology and the size of complex coacervate microcapsules during scale-up. *AIChE J.* 55 (2009), 1487–1496.

Levic, S., Djordjevic, V., Rajic, N., Milivojevic, M., Bugarski, B., and Nedovic, V. Entrapment of ethyl vanillin in calcium alginate and calcium alginate/poly(vinyl alcohol) beads. *Chem. Pap.* 67 (2013), 221–228.

Lian, G., Malone, M.E., Homan, J.E., and Norton, I.T. A mathematical model of volatile release in mouth from the dispersion of gelled emulsion particles. *J. Control. Release* 98 (2004), 139–155.

Lim, P.F.C., Liu, X.Y., Kang, L., Ho, P.C.L., and Chan, S.Y. Physicochemical effects of terpenes on organogel for transdermal drug delivery. *Int. J. Pharm.* 358 (2008), 102–107.

Lindner, K., Szente, L., and Szejtli, L. Food flavouring with b-cyclodextrin complexed flavour substances. *Acta Alimentaria* 10 (1981), 175–186.

Lou, W.C. and Popplewell, L.M. Hydroxypropyl cellulose encapsulation material. U.S. Patent 20,030,077,378 (2003).

Mabuchi, N. and Ngoa, M. Controlled release powdered flavour preparations and confectioneries containing preparations. Japanese Patent 128:638 (2001).

Madene, A., Jacquot, M., Scher, J., and Desobry, S. Flavour encapsulation and controlled release—A review. *Int. J. Food Sci. Technol.* 41 (2006), 1–21.

Maga, J.A. and Sizer, C.E. Pyrazine formation during the extrusion of potato flakes. *Lebensm. Wiss. Technol.* 12 (1979), 15–16.

Malone, M.E. and Appelqvist, I.A.M. Gelled emulsion particles for the controlled release of lipophilic volatiles during eating. *J. Control. Release* 90 (2003), 227–241.

Manojlovic, V., Rajic, N., Djonlagic, J., Obradovic, B., Nedovic, V., and Bugarski, B. Application of electrostatic extrusion—Flavour encapsulation and controlled release. *Sensors* 8 (2008), 1488–1496.

Mao, L., O'Kennedy, B.T., Roos, Y.H., Hannon, J.A., and Miao, S. Effect of monoglyceride self-assembled structure on emulsion properties and subsequent flavor release. *Food Res. Int.* 48 (2012), 233–240.

Marcuzzo, E., Sensidoni, A., Debeaufort, F., and Voilley, A. Encapsulation of aroma compounds in biopolymeric emulsion based edible films to control flavour release. *Carbohydr. Polym.* 80 (2010), 984–988.

Maschke, A., Becker, C., Eyrich, D., Kiermaier, J., Blunk, T., and Göpferich, A. Development of a spray congealing process for the preparation of insulin-loaded lipid microparticles and characterization thereof. *Eur. J. Pharm. Biopharm.* 65 (2007), 175–187.

Maswal, M. and Dar, A.A. Formulation challenges in encapsulation and delivery of citral for improved food quality. *Food Hydrocoll.* 37 (2014), 182–195.

McClements, D.J., Decker, E.A., and Weiss, J. Emulsion-based delivery systems for lipophilic bioactive components. *J. Food Sci.* 72 (2007), 109–124.

McNamee, B.F., O'Riordan, E.D., and O'Sullivan, M. Effect of partial replacement of gum arabic with carbohydrates on its microencapsulation properties. *J. Agric. Food Chem.* 49 (2001), 3385–3388.

Meyer, A. Perfume microencapsulation by complex coacervation. *Chimia* 46 (1992), 101–102.

Milanovic, J., Levic, S., Manojlovic, V., Nedovic, V., and Bugarski, B. Carnauba wax microparticles produced by melt dispersion technique. *Chem. Pap.* 65 (2011), 213–220.

Milanovic, J., Manojlovic, V., Levic, S., Rajic, N., Nedovic, V., and Bugarski, B. Microencapsulation of flavors in carnauba wax. *Sensors* 10 (2010), 901–912.

Miller, K.S. and Krochta, J.M. Oxygen and aroma barrier properties of edible films: A review. *Trends Food Sci. Technol.* 8 (1997), 228–237.

Monge, M.E., Bulone, D., Giacomazza, D., Bernik, D.L., and Negri, R.M. Detection of flavour release from pectin gels using electronic noses. *Sens. Actuators B* 101 (2004), 28–38.

Mongenot, N., Charrier, S., and Chalier, P. Effect of ultrasound emulsification on cheese aroma encapsulation by carbohydrates. *J. Agric. Food Chem.* 48 (2000), 861–867.

Moreau, L., Kim, H.J., Decker, E.A., and McClements, D.J. Production and characterization of oil-in-water emulsions containing droplets stabilized by beta-lactoglobulin-pectin membranes. *J. Agric. Food Chem.* 51 (2003), 6612–6617.

Morillon, V., Debeaufort, F., Blond, G., Capelle, M., and Voilley, A. Factors affecting the moisture permeability of lipid-based edible films: A review. *Crit. Rev. Food Sci. Nutr.* 42 (2002), 67–89.

Morris, C.E. New form of maltodextrin has unique properties. *Food Eng.* 53 (1981), 94–95.

Mun, S., Decker, E.A., and McClements, D.J. Influence of droplet characteristics on the formation of oil-in-water emulsions stabilized by surfactant chitosan layers. *Langmuir* 21 (2005), 6228–6234.

Normand, V., Dardelle, G., Bouquerand, P.-E., Nicolas, L., and Johnston, D. Flavor encapsulation in yeasts: Limonene used as a model system for characterization of the release mechanism. *J. Agric. Food Chem.* 53 (2005), 7532–7543.

O'Connell, M., Valdora, G., Peltzer, G., and Negri, R.M. A practical approach for fish freshness determinations using a portable electronic nose. *Sens. Actuators B* 80 (2001), 149–154.

Ogawa, S., Decker, E.A., and McClements, D.J. Influence of environmental conditions on the stability of oil in water emulsions containing droplets stabilized by lecithin-chitosan membranes. *J. Agric. Food Chem.* 51 (2003a), 5522–5527.

Ogawa, S., Decker, E.A., and McClements, D.J. Production and characterization of O/W emulsions containing cationic droplets stabilized by lecithin-chitosan membranes. *J. Agric. Food Chem.* 51 (2003b), 2806–2812.

Ogawa, S., Decker, E.A., and McClements, D.J. Production and characterization of O/W emulsions containing droplets stabilized by lecithin-chitosan-pectin multilayered membranes. *J. Agric. Food Chem.* 52 (2004), 3595–3600.

Pannell, N.A. Encapsulating materials in microbial cells—By passive diffusion in absence of solvent or plasmolyser. Patent EP242135-A2 (1994).

Paramera, E.I., Konteles, S.J., and Karathanos, V.T. Microencapsulation of curcumin in cells of *Saccharomyces cerevisiae*. *Food Chem.* 125 (2011), 892–902.

Penbunditkul, P., Yoshii, H., Ruktanonchai, U., Charinpanitkul, T., Assabumrungrat, S., and Soottitantawat, A. The loss of OSA-modified starch emulsifier property during the high-pressure homogeniser and encapsulation of multi-flavour bergamot oil by spray drying. *Int. J. Food Sci. Technol.* 47 (2012), 2325–2333.

Phan, V.A., Liao, Y.C., Antille, N., Sagalowicz, L., Robert, F., and Godinot, N. Delayed volatile compound release properties of self-assembly structures in emulsions. *J. Agric. Food Chem.* 56 (2008), 1072–1077.

Philippe, E., Seuvre, A.-M., Colas, B., Langendorff, V., Shippa, C., and Voilley, A. Behavior of flavor compounds in model food systems: A thermodynamic study. *J. Agric. Food Chem.* 51 (2003), 1393–1398.

Popplewell, L.M., Black, J.M., Norris, L.M., and Porzio, M. Encapsulation system for flavors and colors. *Food Technol.* 49 (1995), 76–82.

Porzio, M. Flavour encapsulation: A convergence of science and art. *Food Technol.* 58 (2004), 40–47.

Porzio, M. Spray drying: An in-depth look at the steps in spray drying and the different options available to flavorists. *Perfum. Flavor.* 32 (2007), 34–39.

Porzio, M.A. and Popplewell, L.M. Encapsulation compositions. U.S. Patent 5,603,971 (1997).

Prata, A.S., Zanin, M.H.A., Ré, M.I., and Grosso, C.R.F. Release properties of chemical and enzymatic crosslinked gelatin-gum Arabic microparticles containing a fluorescent probe plus vetiver essential oil. *Colloids Surf. B* 67 (2008), 171–178.

Quezada Gallo, J.A., Debeaufort, F., and Voilley, A. Interactions between aroma and edible films. 1. Permeability of methylcellulose and low-density polyethylene films to methyl ketones. *J. Agric. Food Chem.* 47 (1999), 108–113.

Rabe, S., Krings, U., and Berger, R.G. Influence of oil-in-water emulsion characteristics on initial dynamic flavour release. *J. Sci. Food Agric.* 83 (2003), 1124–1133.

Ran, Y. and Yalkowsky, S.H. Prediction of drug solubility by the general solubility equation (GSE). *J. Chem. Inf. Comput. Sci.* 41 (2001), 354–357.

Reilly, A. and Subramaniam, A. Preparation of microcapsules. WO2004022221 (2004).

Reineccius, G.A. Flavour encapsulation. *Food Rev. Int.* 5 (1989), 147–176.

Reineccius, G.A. The spray drying of food flavours. *Dry. Technol.* 22 (2004), 1289–1324.

Relkin, P., Fabre, M., and Guichard, E. Effect of fat nature and aroma compound hydrophobicity on flavor release from complex food emulsions. *J. Agric. Food Chem.* 52 (2004), 6257–6263.

Risch, S.J. Encapsulation: Overview of uses and techniques. In: S. Risch and G.A. Reineccius (eds.), *Encapsulation and Controlled Release of Food Ingredients*. American Chemical Society, Washington, DC, 1995, pp. 2–7.

Risch, S.J. and Reineccius, G.A. Spray dried orange oil: Effect of emulsion size on flavor retention and shelf life stability. In: G.A. Reineccius and S.J. Risch (eds.), *Flavor Encapsulation*. ACS Symposium Series, Washington, DC, 1988, pp. 66–77.

Roberts, D.D., Elmore, J.S., Langley, K.R., and Bakker, J. Effects of sucrose Guar Gum and carboxymethylcellulose on the release of volatile flavour compounds under dynamic conditions. *J. Agric. Food Chem.* 44 (1996), 1321–1326.

Roberts, D.D., Pollien, P., and Watzke, B. Experimental and modeling studies showing the effect of lipid type and level on flavor release from milk-based liquid emulsions. *J. Agric. Food Chem.* 51 (2003), 189–195.

Rosenberg, M., Kopelman, I.J., and Talmon, Y. Factors affecting retention in spray-drying microencapsulation of volatile materials. *J. Agric. Food Chem.* 38 (1990), 1288–1294.

Rusu, M. Food matrices—Impact on odorant partition coefficients and flavour perception. PhD thesis, Faculty of Mathematics and Natural Sciences Department of Food Chemistry, Bergische Universität Wuppertal, Wuppertal, Germany (2006).

Sagalowicz, L. and Leser, M.E. Delivery systems for liquid food products. *Curr. Opin. Colloid Interface Sci.* 15 (2010), 61–72.

Schleifenbaum, B., Uhlemann, J., and Renz, K.-H. Encapsulated flavorings. U.S. Patent 6,902,751 (2000).

Shahidi, F. and Han, X.-Q. Encapsulation of food ingredients. *Crit. Rev. Food Sci. Nutr.* 33 (1993), 501–547.

Shank, J.L. Encapsulation process utilizing microorganisms and products produced thereby. Patent US 4001480 A (1977).

Shi, G., Rao, L., Yu, H., Xiang, H., Yang, H., and Ji, R. Stabilization of photosensitive resveratrol within yeast cell. *Int. J. Pharm.* 349 (2008), 83–93.

Shi, G.R., Rao, L.Q., Yu, H.Z., Xiang, H., Pen, G.P., Long, S., and Yang, C. Yeast-cell-based microencapsulation of chlorogenic acid as a water-soluble antioxidant. *J. Food Eng.* 80 (2007), 1060–1067.

Soottitantawat, A., Bigeard, F., Yoshii, H., Furuta, T., Ohkawara, M., and Linko, P. Influence of emulsion and powder size on the stability of encapsulated D-limonene by spray drying. *Innov. Food Sci. Emerg. Technol.* 6 (2005a), 107–114.

Soottitantawat, A., Takayama, K., Okamura, K., Muranaka, D., Yoshii, H., Furuta, T., Ohkawara, M., and Linko, P. Microencapsulation of l-menthol by spray drying and its release characteristics. *Innov. Food Sci. Emerg. Technol.* 6 (2005b), 163–170.

Soottitantawat, A., Yoshii, H., Furuta, T., Ohkawara, M., and Linko, P. Microencapsulation by spray drying: Influence of emulsion size on the retention of volatile compounds. *J. Food Sci.* 68 (2003), 2256–2262.

Subramaniam, A., McIver, R.C., and Van Sleeuwen, R.M.T. Process for the incorporation of a flavor or fragrance ingredient or composition into a carbohydrate matrix. WO2006038067 (2006).

Szejtli, J. Introduction and general overview of cyclodextrin chemistry. *Chem. Rev.* 98 (1998), 1743–1754.

Szejtli, J., Szente, L., and Banky-Elod, E. Molecular encapsulation of volatile, easily oxidizable labile flavour substances by cyclodextrins. *Acta Chim. Sci. Hung.* 101 (1979), 27–46.

Szente, L. and Szejtli, J. Cyclodextrins as food ingredients. *Trends Food Sci. Technol.* 15 (2004), 137–142.

Tackenberg, M.W., Thommes, M., Schuchmann, H.P., and Kleinebudde, P. Solid state of processed carbohydrate matrices from maltodextrin and sucrose. *J. Food Eng.* 129 (2014), 30–37.

Thepkunya, H., Rungnaphar, P., and McClements, D.J. Stabilization of model beverage cloud emulsions using protein–polysaccharide electrostatic complexes formed at the oil–water interface. *J. Agric. Food Chem.* 54 (2006), 5540–5547.

Thies, C. Microencapsulation of flavours by complex coacervation. In: J.M. Lakkis (ed.), *Encapsulation and Controlled Release Technologies in Food Systems*. Blackwell, Ames, IA, 2007, pp. 149–170.

Tobitsuka, K., Miura, M., and Kobayashi, S. Retention of a European pear aroma model mixture using different types of saccharides. *J. Agric. Food Chem.* 54 (2006), 5069–5076.

Turchiuli, C., Lemarié, N., Cuvelier, M.E., and Dumoulin, E. Production of fine emulsions at pilot scale for oil compounds encapsulation. *J. Food Eng.* 115 (2013), 452–458.

Ubbink, J. and Schoonman, A. Flavor delivery systems. In: A. Seidel (ed.), *Kirk-Othmer Encyclopedia of Chemical Technology*, on-line edition, Vol. 11. Wiley-Interscience, New York, 2003.

Vaidya, S., Bhosale, R., and Singhal, R.S. Microencapsulation of cinnamon oleoresin by spray drying using different wall materials. *Dry. Technol.* 24 (2006), 983–992.

Van Ruth, S.M., King, C., and Giannouli, P. Influence of lipid fraction, emulsifier fraction, and mean particle diameter of oil-in-water emulsions on the release of 20 aroma compounds. *J. Agric. Food Chem.* 50 (2002), 2365–2371.

Veith, S.R., Hughes, E., and Pratsinis, S.E. Restricted diffusion and release of aroma molecules from sol-gel-made porous silica particles. *J. Control. Release* 99 (2004a), 315–327.

Veith, S.R., Pratsinis, S.E., and Perren, M. Aroma retention in sol-gel-made silica particles. *J. Agric. Food Chem.* 52 (2004b), 5964–5971.

Wang, Y., Lu, Z., Lv, F., and Bie, X. Study on microencapsulation of curcumin pigments by spray drying. *Eur. Food Res. Technol.* 229 (2009), 391–396.

Wen, L.F., Rodis, P., and Wasserman, B.P. Starch fragmentation and protein insolubilisation during twin-screw extrusion of corn meal. *Cereal Chem.* 67 (1990), 268–275.

Werner, S.R.L., Jones, J.R., Paterson, A.H.J., Archer, R.H., and Pearce, D.L. Air-suspension particle coating in the food industry: Part I—State of the art. *Powder Technol.* 171 (2007), 25–33.

Xing, F., Cheng, G., Yang, B., and Ma, L. Microencapsulation of capsaicin by the complex coacervation of gelatin, acacia and tannins. *J. Appl. Polym. Sci.* 91 (2004), 2669–2675.

Yamada, T., Hisamatsu, M., Teranishi, K., Katsuro, K., Hasegawa, N., and Hayashi, T. Components of the porous maize starch granule prepared by amylase treatment. *Starch—Stärke* 46 (1995), 358–361.

Yamamoto, C., Neoh, T.L., Honbou, H., Yoshii, H., and Furuta, T. Kinetic analysis and evaluation of controlled release of D-limonene encapsulated in spray-dried cyclodextrin powder under linearly ramped humidity. *Dry. Technol.* 30 (2012), 1283–1291.

Yang, Z., Peng, Z., Li, J., Li, S., Kong, S., Li, P., and Wang, Q. Development and evaluation of novel flavour microcapsules containing vanilla oil using complex coacervation approach. *Food Chem.* 145 (2014), 272–277.

Yeo, Y., Bellas, E., Firestone, W., Langer, R., and Kohane, D.S. Complex coacervates for thermally sensitive controlled release of flavour compounds. *J. Agric. Food Chem.* 53 (2005), 7518–7525.

Yılmaz, G., Jongboom, R.O.J., Feil, H., and Hennink, W.E. Encapsulation of sunflower oil in starch matrices via extrusion: Effect of the interfacial properties and processing conditions on the formation of dispersed phase morphologies. *Carbohydr. Polym.* 45 (2001), 403–410.

Young, H., Rossiter, K., Wang, M., and Miller, M. Characterization of royal gala apple aroma using electronic nose technology-potential maturity indicator. *J. Agric. Food Chem.* 47 (1999), 5173–5177.

Yuliani, S., Bhandari, B., Rutgers, R., and D'Arcy, B. Application of microencapsulated flavor to extrusion product. *Food Rev. Int.* 20 (2004), 163–185.

Yuliani, S., Torley, P.J., D'Arcy, B., Nicholson, T., and Bhandari, B. Effect of extrusion parameters on flavour retention, functional and physical properties of mixtures of starch and D-limonene encapsulated in milk protein. *Int. J. Food Sci. Technol.* 41 (2006a), 83–94.

Yuliani, S., Torley, P.J., D'Arcy, B., Nicholson, T., and Bhandari, B. Extrusion of mixtures of starch and D-limonene encapsulated with β-cyclodextrin: Flavour retention and physical properties. *Food Res. Int.* 39 (2006b), 318–331.

Zasypkin, D. and Porzio, M. Glass encapsulation of flavours with chemically modified starch blends. *J. Microencapsul.* 21 (2004), 385–397.

Zeller, B.L., McKay, R.P., and Saleeb, F.Z. Method and manufacture for moisture-stable, inorganic, microporous saccharide salts. U.S. Patent 4,659,390 (1987).

Zeller, B.L. and Saleeb, F.Z. Production of microporous sugars for adsorption of volatile flavors. *J. Food Sci.* 61 (1996), 749–752.

Zeller, B.L., Saleeb, F.Z., and Ludescher, R.D. Trends in development of porous carbohydrate food ingredients for use in flavor encapsulation. *Trends Food Sci. Technol.* 9 (1999), 389–394.

Zhao, J., Madson, M.A., and Whistler, R.L. Cavities in porous corn starch provide a large storage space. *Cereal Chem.* 73 (1996), 379–380.

Zhao, J. and Whistler, R.L. Spherical aggregates of starch granules as flavor carriers. *Food Technol.* 48 (1994), 104–105.

Žnidaršić-Plazl, P. and Plazl, I. Modelling and experimental studies on lipase-catalyzed isoamyl acetate synthesis in a microreactor. *Process Biochem.* 44 (2009), 1115–1121.

Zuidam, N.J. and Heinrich, J. Encapsulation of aroma. In: N.J. Zuidam and V.A. Nedovic (eds.), *Encapsulation Technologies for Food Active Ingredients and Food Processing.* Springer, Dordrecht, the Netherlands, 2010, pp. 127–160.

18

Encapsulation of Active/Bioactive/Probiotic Agents

Carmen S. Favaro-Trindade, Talita A. Comunian, Volnei B. Souza, Milla G. dos Santos, and Mariana S. de Oliveira

CONTENTS

Abstract

The food added with active/bioactive molecules and bioactive living cells (probiotics) has increased all over the world. Despite the incontestable importance and consumer interest for this kind of food products, there are lots of challenges in handling and using some bioactives because many of them are highly sensitive or reactive and some can reduce the sensory acceptance of the food product. In this context, microencapsulation has been studied and applied to overcome such drawbacks; besides, this technique can promote a controlled release of the bioactive in their site of action. More than this, microencapsulation can produce a complete change in the properties of bioactive materials, providing new applications for them.

18.1 Introduction

Food bioactive compounds are food constituents (essential and nonessential) and bioactive living cells that have some biological activities, such as antioxidant, anti-inflammatory, antithrombotic, and anticarcinogenic. Due to their functional properties, bioactive compounds are usually associated with positive effects on human health.

Food bioactive compounds are spread in many foodstuffs of both animal and plant origin, but they occur mainly in vegetables. Probiotics and food bioactive compounds such as prebiotics, dietary fibers, vitamins, minerals, phytosterols, polyphenols, peptides, omega-3, lutein, and lycopene are being intensively studied all over the world.

Knowledge of the role of food bioactive compounds in specific pathologies has advanced fast and also has increased the development of functional foods in the food industry, particularly concerning probiotic products. The global functional food market is huge, despite the fact that the development of this kind of product is not an easy task. In fact, normally, the development of functional foods represents a big challenge because most of the bioactive compounds are very sensitive to many factors, such as oxygen, light, high temperature, and low pH. In addition, some compounds have poor water solubility and off-flavors and are highly hygroscopic and reactive. However, the microencapsulation technology has been used as a successful alternative to solve many problems related to limitations in food bioactive compounds and probiotic applications.

Microcapsules or microspheres can protect food bioactive compounds and probiotic from degradation processes under adverse environmental conditions as well as promote delivery in their sites of action, facilitate manipulation and application; alter solubility and functionality; reduce volatility, reactivity, and hygroscopicity; and mask undesirable tastes or aromas. It explains why the microencapsulation technology became so crucial for the food industry.

One frequent concern about the utilization of bioactive compounds is that they should not affect the sensory properties of functional products, such as color, flavor, and texture, so the food industry usually aims to prevent this (Champagne and Fustier, 2007). This prevention is largely achieved by the encapsulation of bioactive compounds. For instance, the control of microcapsule size must be precise due to the fact that the addition of large particles in food is undesirable in most cases, since large particle size (normally bigger than 50–70 μm) affects texture. On the other hand, the addition of particles visible to the naked eye (larger than 60 μm) has been used by the food industry to create a nice and desirable appearance in some foods and beverages, mainly food for children, and the food industry take advantage of this aspect for promoting the commercialization of this kind of foods.

Unlike the pharmaceutical and cosmetic industries, another concern of the food industry is the cost of microencapsulated ingredients. According to Gouin (2004), considering that bioactive compounds generally are used at low concentrations in foods (0.5%–5%), a maximum cost for a microencapsulation process in the food industry can be roughly estimated at €0.1/kg.

Besides low cost, encapsulation processes for food ingredients have some other requisites; the wall material, carrier, and encapsulating agent must be of food grade, and organic solvents and the use of high temperature for a long time are prohibitive. In this way, despite the fact that some encapsulation technologies are very effective, they might not be suitable for encapsulating food bioactive compounds or any other food ingredient.

Many microencapsulation technologies have been developed for use in the bioactive compounds and show promise for the production of functional foods. However, according to De Vos et al. (2010), many encapsulation procedures have been proposed but none of them can be considered as a universally applicable procedure for all bioactive food components.

This chapter focuses on the uses of encapsulation processes that allow the addition of bioactive compounds into food products.

18.2 Microencapsulation Technologies Used for Bioactive Food Compounds

Desai and Park (2005) gave a simplified definition and reported some utilities of the technology:

> Microencapsulation involves the incorporation of food ingredients, enzymes, cells, or other materials in small capsules. Microcapsules offer food processors a means with which to protect sensitive food components, ensure against nutritional loss, utilize otherwise sensitive ingredients, incorporate unusual or time-release mechanisms into the formulation, mask or preserve flavors and aromas, and transform liquids into easily handled solid ingredients.

Nowadays, microencapsulation has the same definition and still the same utilities cited by Desai, besides many others. In fact, this technology is limited just for human imagination.

The most common methods of encapsulation are spray drying, spray chilling/cooling/congealing, liposome encapsulations, simple and complex coacervation, micelle encapsulations, processing of microfluidic devices, extrusion processes, supercritical fluids encapsulation processes, processing of nanostructured lipid matrices, solvent evaporation, spray coating, emulsions, and molecular inclusion.

The particles produced by encapsulation methods are known as microcapsules and microspheres (Figure 18.1). Microcapsules are reservoir-type structures, where the bioactive compound is surrounded by a shell. Liposome encapsulations, complex coacervation, processing of microfluidic devices, spray coating, and molecular inclusion are examples of methodologies that produce microcapsules. Microspheres are matrix-type structures, that is, particles where the bioactive compound is distributed all over the volume of the particle, including its surface. Spray drying and spray chilling produce microspheres. Structures from both methods are very important for the food industry and can have similar and different utilities.

The size of particles can vary from a few nanometers to many micrometers (Favaro-Trindade et al., 2008), depending on the method applied for encapsulation as well as, for some methods of encapsulation, on the conditions of the operations such as airflow and feed flow (for spraying process), the nozzle configurations (for extrusion process), and fluid flow and the diameter size of capillaries (for processing of microfluidic devices). The appearance and shape are variable, but most microparticles have a round shape, as we can see in Figure 18.2, for different methods of encapsulation.

The cost of encapsulated materials depends on the level of complexity, capacity, and operating cost of the method of encapsulation applied, besides the cost of the core and the wall material (Figure 18.3).

The choice of the encapsulating agent/carrier/wall material/shell depends on many factors, such as the process of encapsulation (lipids cannot be used to spray drying while polysaccharides cannot be used to spray chilling process, for instance), the expected functionality for the microparticle and the expected release mechanisms (if the release of the core is into the gut, an enteric polymer is always a good choice), the composition of the food where the microparticle will be added (carrier cannot react with food compounds and also with the core), and last but not the least the cost of the material. The materials most widely used for encapsulation of food ingredients are polysaccharides, proteins, and lipids.

The release mechanisms vary with the wall material and process of encapsulation applied. Diffusion, dissolution, digestion, mechanical disruption (by chewing, for instance), and triggered release with change of temperature, pH, pressure, and ionic force are some possibilities of release processes.

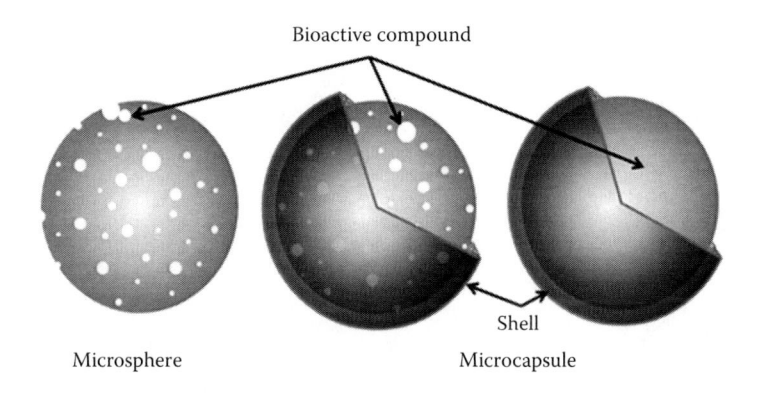

Microsphere   Microcapsule

FIGURE 18.1 Microparticles structures.

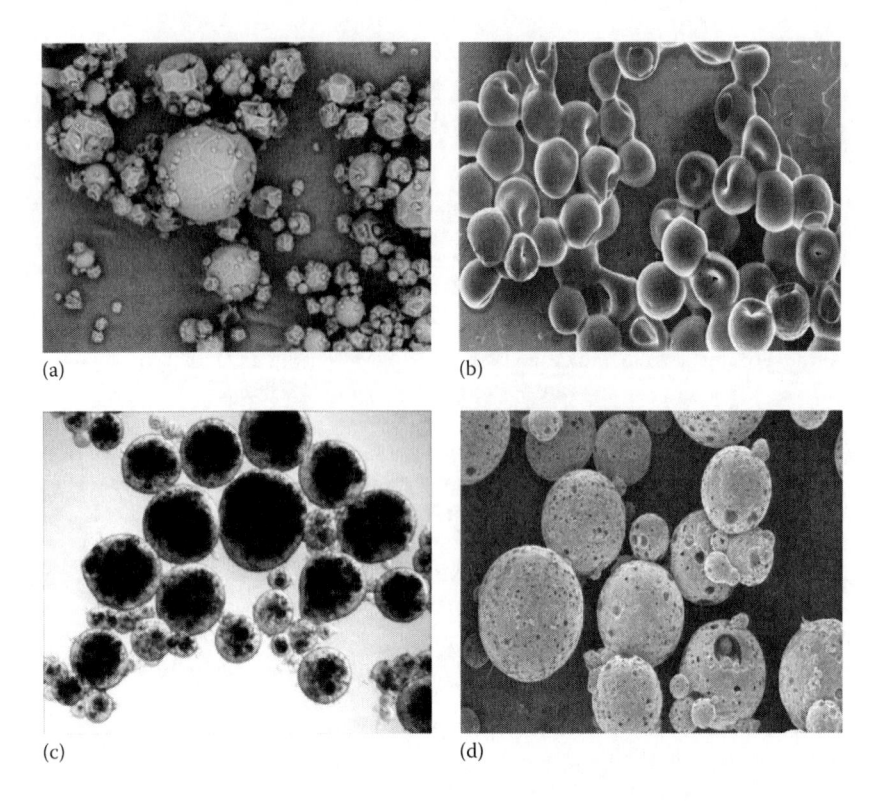

FIGURE 18.2 Microparticles produced by (a) spray drying, (b) microfluidic devices, (c) complex coacervation, and (d) spray chilling.

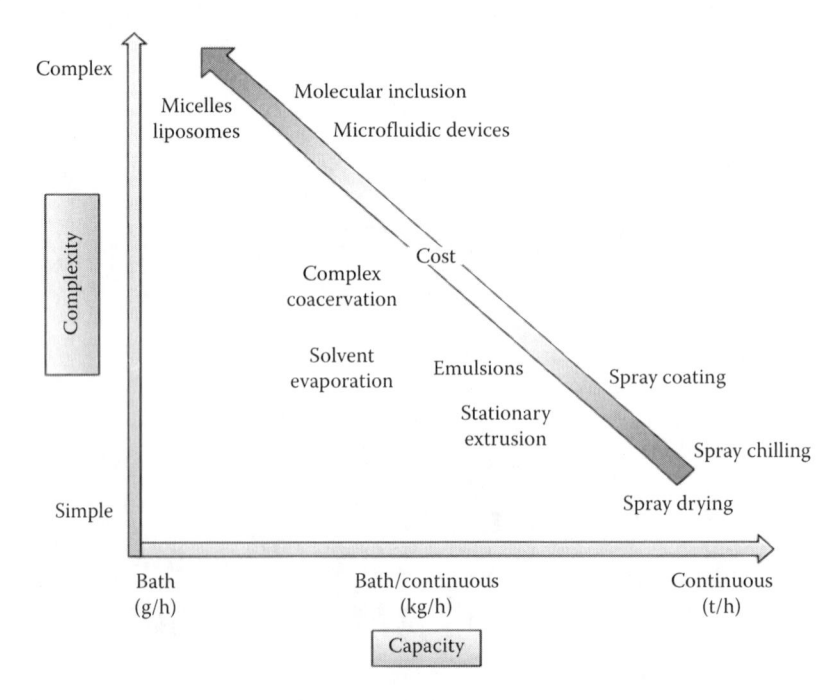

FIGURE 18.3 Levels of complexity, capacity, and operating cost of some encapsulation processes. (Adapted from Southwest Research Institute, Catalog of Capability Brochures for: SwRI Chemistry & Chemical Engineering brochure, 2012, available at: http://www.swri.org/3pubs/brochure/d01/mne/MicroNanoEncap.pdf. Accessed February, 2014.)

18.3 Encapsulated Food Bioactive Compounds

18.3.1 Bioactive Living Cells

Probiotics are considered bioactive living cells, that is, microorganisms that can confer human health benefits and well-being when it is consumed alive. The most studied and used probiotic genera are *Lactobacillus* and *Bifidobacterium*, although there are many different species and strains within these genera. The research results of one strain generally do not apply to others. Many probiotic properties are well established and many others are under investigation.

Microencapsulation of probiotics is an industrial practice and the subject of studies of many industrial and academic researchers.

The aims of probiotic encapsulation are (1) protecting probiotics during the application of unit operations in food process, such as thermal process, freezing, and homogenization; (2) protecting probiotics against adverse environmental conditions during food shelf life, such as oxygen, low pH, and the presence of other cultures; (3) protecting cells against gastrointestinal secretions during ingestion; (4) avoiding multiplication of probiotics in some foods that can cause food sensory quality reduction; (5) releasing probiotics on their site of action, normally in the gut; (6) converting the inoculums into a powder form for convenient use; and (7) cell concentrations.

Many methods were studied to encapsulate probiotics, such as complex coacervation, alginate extrusion (ionotropic gelation), emulsion, spray drying, spray chilling, and polyelectrolyte complexation. However, all of them produced micro- or macroparticles, because processes for the elaboration of nanoparticles are prohibitive, since probiotic cells have size varying from 1 to 7 µm.

The method more studied to promote probiotic encapsulation is extrusion by ionotropic gelation using alginate as the carrier (Favaro-Trindade et al., 2011). This method is simple, low cost, and mild (there is no need of using high temperatures, organic solvents, and high pressures). Although the particles produced, known as beads, are porous ones, they can compromise the efficiency of the method and are susceptible to the action of chelating agents, which can cause disruption or dissolution. Besides, the beads need an additional process of dehydration because the wet environment also is detrimental for living cells. So some advances in this method include the coating of the beads (Table 18.1).

Spray chilling is a promising technology for probiotic encapsulation because it is also a mild method and lipid microparticles can release the probiotic in the intestine due to the digestion of the fat (carrier).

Complex coacervation can generate low-cost enteric polymer, but some points in this method must be clarified, such as the load capacity for probiotic encapsulation. Some remarkable and recent developments in probiotic encapsulation are shown in Table 18.1.

18.3.2 Phenolic Compounds

Phenolic compounds, such as phenolic acids, flavonoids, and proanthocyanidins, are secondary metabolites of plants found in large quantities in tea, coffee, nuts, cocoa, fruits, wines, and juices, among others. Its bioactivity is linked to antioxidant, antimicrobial, anti-inflammatory, antiviral, and antidiabetic activities and reduces the risk of cardiovascular disease and some cancers (Fang and Bhandari, 2010; Ho et al., 2010).

The encapsulation of phenolics is used to maintain their stability, bioactivity, and bioavailability, providing protection to the compounds against adverse processing and environmental conditions. In addition to masking unwanted sensory aspects such as bitterness and astringency, the encapsulation also allows the controlled release of the active ingredient. Several techniques are used to encapsulate phenolic compounds targeting various applications (Fang and Bhandari, 2010).

The encapsulation of green tea catechins and cocoa polyphenols can be performed using the techniques of complex coacervation and emulsification/internal gelation in alginate microspheres, respectively. This makes it possible to achieve high retention and encapsulation efficiency of the compounds.

TABLE 18.1

Some Studies about Probiotic Encapsulation

Microencapsulation Technology	Shell/Carrier	Remark	References
Complex coacervation followed by spray drying	Casein/pectin complex	Encapsulated probiotics were more resistant to acid conditions than free ones. Microencapsulated *Lactobacillus acidophilus* maintained its viability for a longer storage period, including at temperature of 37°C.	Oliveira et al. (2007a)
Spray chilling	Fat obtained upon interesterification of fully hydrogenated palm and palm-kernel oil	Mild process. Protection under simulated gastrointestinal secretions. Promising results for viability were obtained when refrigerated and frozen storage was applied.	Pedroso et al. (2012)
	Fat obtained upon interesterification of fully hydrogenated palm and palm-kernel oil	Symbiotic particles were produced, which made probiotics resistant to simulated gastrointestinal secretions. Polydextrose were the most effective symbiotic studied. Encapsulated cells were shown to have high viability when refrigerated and frozen with a controlled relative humidity (11%).	Okuro et al. (2013)
Spray drying	Cellulose acetate phthalate (CAP)	Cells resisted to high temperature. CAP is an enteric polymer that protected the probiotic under acid solutions and bile solutions.	Favaro-Trindade and Grosso (2002)
	Skim milk	Resistance to simulated gastrointestinal digestion was enhanced by spray drying. Mild heat treatment before spray drying enhances cell survival during storage and the resistance to gastrointestinal digestion.	Paez et al. (2012)
	Whey protein	Exploiting whey as a bacterial substrate and encapsulation matrix within a coupled fermentation and spray drying.	Jantzen et al. (2013)
Ionotropic gelation	Alginate	Porous structure: low viability and resistance to acid and bile solutions.	Favaro-Trindade and Grosso (2000)
Ionotropic gelation and coating	Alginate Coating: chitosan, sureteric, or acryl-eze	Ionic gelation followed by air-drying. Enteric polymers improved survival at low pH.	Liserre et al. (2007)
	Alginate Coating: whey protein	Ionic gelation followed by freeze-drying. Whey proteins are efficient materials for coating alginate beads loaded with bacteria.	Gbassi et al. (2009)
	Alginate Coating: chitosan	Microencapsulation effectively protects the probiotic from heating at 65°C and refrigerating at 4°C.	Trabelsi et al. (2013)
	Pectin Coating: whey protein	High encapsulation yield positively affected the resistance of the microorganism when exposed to conditions simulating the transit through the gastrointestinal tract.	Gebara et al. (2013)

(Continued)

TABLE 18.1 (*Continued*)

Some Studies about Probiotic Encapsulation

Microencapsulation Technology	Shell/Carrier	Remark	References
Polyelectrolyte complexation	Alginate, xanthan gum (XG), and CAP	Incorporation of the 0.5% (w/v) of XG or the 1% (w/v) of CAP within the 3% (w/v) of alginate solution increased the survival of the probiotic bacteria in acid conditions. High viability.	Albertini et al. (2010b)
Supercritical carbon dioxide	Poly(vinyl pyrrolidone)-poly(vinyl acetate-co-crotonic acid) interpolymer complex	Encapsulation matrix is stable at low pH, but disintegrates at higher pH. Need more studies.	Moolman et al. (2006)

In addition, there is an increase in protection, allowing controlled release in simulated gastric system (Wang et al., 2007; Lupo et al., 2014).

Another interesting possibility in the encapsulation of phenolic compounds is the simultaneous use of two or more compounds as active materials in the same capsule. In the coencapsulation of the pairs gallic acid/curcumin and ascorbic acid (AA)/quercetin in types of niosomal systems, the combined use of antioxidants allows to obtain systems with a greater amount of encapsulated bioactive and increased possibility of regulation and release when compared with systems containing only one of the active compounds. There is also the possibility of increasing the solubility of quercetin and curcumin. Furthermore, the synergistic effect of combined antioxidants displays a clear superiority of this biological activity (Tavano et al., 2014).

Some challenges are encountered in the encapsulation of phenolics, for example, the encapsulation efficiency values that are typically lower than that of the hydrophobic compounds. However, there are enormous possibilities for encapsulation of these bioactive, in view of the many techniques available and the possibility of combining different molecules in the same system. In addition, tests of bioactivity and bioavailability *in vivo* are poorly explored, which open up a range of opportunities for various studies.

18.3.3 Natural Pigments

The natural pigments or colorants are substances found in plant and animal tissues and are responsible for their characteristic colors. Among the pigments found in nature, including the heme compounds, chlorophylls, carotenoids, anthocyanins, and betalains, two groups stand out because, besides having the color power, they have proven biological activity. They are the carotenoids and anthocyanins.

Among the carotenoids that have biological activity are beta-carotene, lycopene, zeaxanthin, and lutein, and the main benefits are the known activity of provitamin A of the beta-carotene and some other carotenoids, plus the antioxidant capacity and prevention of some cancers. Anthocyanin in turn has its bioactivity mainly linked to antioxidant capacity, but also has antimicrobial activity and reduces the risk of cardiovascular disease and cancer, among other benefits.

The coating of natural pigments using encapsulation techniques can provide greater durability of color power and greater stability against mainly oxidative reactions, which guarantees the bioactivity of the compounds.

Beta-carotene, lycopene, and lutein can be encapsulated using nanoparticles of polylactic acid, complex coacervation, and chitosan nanocapsules, respectively. These techniques promote increased stability of the beta-carotene and lycopene against oxidation and increase the bioavailability of the lutein *in vitro* and *in vivo* (Cao-Hoang et al., 2011; Arunkumar et al., 2013; Rocha-Selmi et al., 2013).

Anthocyanins from grape pomace, jabuticaba peels, and blueberry can be encapsulated using the spray drying technique with a carrier agent. The powdered pigment obtained from grape exhibits stable color during storage, and in the case of jabuticaba, biological activities such as antioxidant, antimicrobial, and inhibition of the *Leishmania amazonensis* arginase can be observed (Souza et al., 2015; Silva et al., 2014). Besides, the pigment obtained from blueberry exhibits property of controlled release of the phenolic compounds in simulated gastric system (Flores et al., 2014).

The encapsulation techniques allow obtaining natural pigments with high coloring strength while protecting these molecules against degradation reactions during storage and processing of products. Encapsulation makes the challenges of low stability of natural colorants are overcome, allowing these compounds, which besides the color provide many health benefits, can replace more and more the traditional synthetic pigments.

18.3.4 Essential Oils

The essential oils are one of the most ancient and known classes of substances. In recent years, a considerable interest in researching and utilizing of the beneficial properties of essential oils has been increased to develop natural health products for humans and animals, besides general fragrances and flavor products (Sutaphanit and Chitprasert, 2014). They are natural plant products represented by multicomponent systems and consist of mixtures of several chemical compounds. Their compounds are derived from hydrocarbons including alcohols, aldehydes, esters, ketones, and phenols (Răileanu et al., 2013).

Essential oils are also known as volatile oils due to their evaporation at room temperature. Rightfully so, their flavor is realized almost immediately (Zitune and Januário, 2005). These substances are volatile and chemically unstable in the presence of oxygen, moisture, light, and heat. Thus, these compounds are microencapsulated to improve their stability.

Essential oils are encapsulated for several reasons: (1) to provide protection against light, heat, moisture, or air-induced reactions or oxidation; (2) to retain them in a food product during storage, preventing premature evaporation; (3) to convert liquid food flavorings into a dry and free-flowing powder form, which is easy to handle and is incorporated into a dry food system; (4) to protect the flavor from undesirable interactions with the food; (5) to minimize flavor/flavor interactions within a mixture; and (6) to mask undesirable flavors.

Besides protecting the flavors, the encapsulation provides controlled release. The controlled release of ingredients at the right place and at the right time can improve the effectiveness of food additives, broaden the application range, and ensure optimal dosage (Baranauskiene et al., 2007).

According to Dronen and Reneccius (2006), it is necessary to attempt to separate losses due to chemical reactions from losses due to diffusion. The authors concluded that chemical reactions may contribute substantially to the loss of desirable flavor components and thereby may be a limiting factor in the shelf life of dry foods.

Numerous studies have been conducted in order to understand the interaction of volatile compounds with the wall material. Studies of interactions of proteins with volatile components have shown that proteins usually possess a high binding capacity for the flavor compounds (Baranauskienė et al., 2006). On the other hand, the use of carbohydrates with a high capacity for emulsification (i.e., gum arabic and modified starch) as wall materials also exhibit good retention of volatiles (Błaszczak et al., 2013).

Furthermore, studies have shown that the junction of two polymers (proteins and carbohydrates) can increase the efficiency of encapsulation. Bylaitë et al. (2001) microencapsulated caraway oil and concluded that the encapsulation efficiency can be increased by the selection of wall components that exhibit different functional properties: partial replacement of whey protein concentrate by surface active carbohydrates increases retention of volatiles during spray drying and enhances protective properties of solidified capsules to oxidation and release of volatiles during the storage.

In order to protect the bioactive essential and retain its functional properties, the researchers have studied different microencapsulation methods and wall materials. Some of microencapsulated bioactive compounds are shown in Table 18.2.

18.3.5 Vitamins

Vitamins are sensitive molecules when in contact with environmental adverse conditions and are classified into two groups: (1) water-soluble vitamins (vitamin C and B complex), whose replacement is

TABLE 18.2

Studies about Essential Oil Encapsulation

Encapsulation Technology	Essential Oil	Shell/Carrier	References
Spray drying	Peppermint (*Mentha piperita* L.) oil	Modified starches	Baranauskiene et al. (2007)
	Oregano essential oil	Modified starch, arabic gum, and maltodextrin	Botrel et al. (2012)
	Orange essential oil	Modified starch, arabic gum, and maltodextrin	Aburto et al. (1998)
	Rosemary essential oil	Gum arabic, starch, maltodextrin, inulin	Fernandes et al. (2014)
Complex coacervation	Peppermint oil	Gelatin and gum arabic	Dong et al. (2011)
Simple coacervation	Holy basil essential oil	Gelatin	Sutaphanit and Chitprasert (2014)

TABLE 18.3

Some Studies on Ascorbic Acid Encapsulation

Technology	Encapsulating Agents	Encapsulation Efficiency (%)	References
Microfluidic device	Palm oil	96	Comunian et al. (2014)
Double emulsion followed by complex coacervation	Gelatin/gum arabic	97–99	Comunian et al. (2013)
Double emulsion	Paraffin oil	—	Akhtar et al. (2010)
Extrusion	Maltodextrin	—	Chang et al. (2010)
Spray drying	Rice starch/gum arabic	97.6–99.7	Trindade and Grosso (2000)
Ionic gelation	Chitosan/sodium tripolyphosphate	—	Alishahi et al. (2011)

TABLE 18.4

Studies on Fat-Soluble Vitamins Encapsulation

Technology	Vitamin	Encapsulating Agents	Encapsulation Efficiency (%)	References
Solid lipid nanoparticles	Vitamin A	Glyceryl behenate	—	Jenning et al. (2000)
Ionotropic gelation/ complex coacervation	Vitamin A	Chitosan/chitosan and alginate	94	Albertini et al. (2010)
Membrane emulsification	Vitamin E	—	99.7	Laouini et al. (2012)
Solvent evaporation	Vitamin E	Polylactic acid	—	Chaiyasat et al. (2013)
Ionic gelation	Vitamin D_3	Carboxymethyl chitosan and soy isolate protein	96.8	Teng et al. (2013)
Micelles	Vitamin D_3	Chitosan	—	Li et al. (2014)

required daily, and (2) fat-soluble vitamins (vitamin A, E, D, and K), which require the assistance of fat to be absorbed. One of the most used alternatives for protection and controlled release of these compounds is the microencapsulation method (Tables 18.3 and 18.4).

Vitamin C, also known as ascorbic acid (AA), is frequently found in fruits and vegetables (Damodaran et al., 2010). Besides its function as a vitamin, AA is often used as an antioxidant; however, its application in foods is difficult because of its instability. Because of this instability, many studies on the encapsulation of AA are found in the literature (Table 18.3). Unlike vitamin C, vitamin B complex presents higher stability. For this reason, studies involving the encapsulation of these types of vitamins are not often found in the literature.

Vitamin A refers to a group of polyunsaturated hydrocarbons and is essential for the growth, vision, and embryonic development. Regarding vitamin E, a fat-soluble antioxidant, this compound is widely used in pharmaceuticals and cosmetic products and can be found in eight different molecular forms, classified as tocopherols and tocotrienols. From the same group of vitamin, vitamin D can prevent bone disease, cancer, diabetes, and multiple sclerosis (Fuleihan and Vieth, 2007; Pettifor and Prentice, 2011). The enrichment of foods with these vitamins is a challenge for the food industry due to the sensitivity of these compounds when in contact with air, light, or ultraviolet process, and their addition into food becomes more difficult due to their hydrophobic nature.

Vitamin K, an antibleeding compound and with an importance in the early skeletal development and maintenance of mature bone, is more stable than other vitamins mentioned earlier, so studies on encapsulation of this kind of vitamin are not found in the literature.

18.3.6 Fatty Acids

Fatty acids are organic compounds formed by one carbon chain and one carboxylic group that are usually esterified to glycerol-forming acylglycerides. Some examples of important fatty acids to human health are those that belong to the omega-3 (Ω-3) and omega-6 (Ω-6) families, obtained from the diet or produced by the body from linoleic acid and alpha-linolenic acid (Rubio-Rodriguez et al., 2010).

Omega-3 fatty acids have been studied to decrease triglycerides, promoting the reduction of risk of cardiovascular disease (Poole et al., 2013), in addition to its anti-inflammatory, antiarrhythmic, and hypotriglyceridemic effects and its ability to reduce blood pressure functionality (Klaypradit and Huang, 2008). Omega-6 fatty acids are sources of compounds with anti-inflammatory and anticancer action, improve the conduction velocity in diabetic patients, increase blood flow, and help reduce tingling of the extremities (Kapoor and Huang, 2006). However, these fatty acids are easily oxidized and have a strong odor (omega-3 from fish oil) (Klaypradit and Huang, 2008). For this reason, many studies on the encapsulation of these compounds are found in the literature.

Xu et al. (2013) studied the stability and controlled release of Ω-3/Ω-6 when they were encapsulated by a complex of dextrin. The obtained complexes helped in improving the stability of α-linolenic acid and linoleic acid. Chen et al. (2013) studied the coencapsulation of fish oil with phytosterol and limonene by spray drying technique using whey isolate protein and sodium caseinate as wall materials. The values of encapsulation efficiency range from 21.9% to 95.1%. The encapsulation of fish oil was also studied by a method consisting of three steps: emulsification, ultrasonic atomization, and freeze-drying, using chitosan, maltodextrin, and whey isolate protein as encapsulating materials. The encapsulation efficiency was in the range of 79%–83% (Klaypradit and Huang, 2008).

18.3.7 Proteins, Hydrolysates, and Peptides

Proteins are natural polymers formed by a large chain of amino acids linked by peptide bonds. The possible combinations in the sequence of amino acid and the different conformations in proteins allow that them to exercise many functions. A peptide is a short chain of amino acid. Protein hydrolysates are obtained by acid or enzymatic hydrolysis of the proteins; they are composed by a mixture of peptides with different chain sizes and free amino.

Proteins, peptides, hydrolysates, and amino acids may also be incorporated as bioactive compounds in food due to their biological functions, such as immunomodulatory, antioxidant, and antimicrobial activity, increase of muscle mass, and control of blood glucose levels and blood pressure, and their effects on satiety and capability to serve as precursors of serotonin synthesis.

The proteins most frequently used in the process of obtaining hydrolysates are derived from milk (casein and whey) due to their high availability, low cost, and capacity of production of bioactive hydrolysates. Some protein hydrolysates, particularly from milk protein, are associated with bitter taste, and this is attributed to the exposure of hydrophobic amino acid residues during the hydrolysis process.

Despite the wide range of biological benefits offered on intake of certain proteins, peptides, and amino acids, the bioavailability and incorporation of these compounds in food products still have challenges to overcome. According to McClements et al. (2009), these are the main challenges: (1) addition of these compounds without matrix interactions or food sensory changes (some proteins and protein hydrolysates present bitter taste, which makes their incorporation in food more complicated); (2) the low stability of these compounds during production, storage, transportation, and consumption of food (certain proteins can lose their bioactivity during the extraction, purification, or thermal processing steps); and (3) assurance that these compounds remain intact within the human body before being absorbed into the required site of action (certain proteins or peptides have to arrive in the site of absorption in the small intestine with an intact conformation to exercise a beneficial effect to the health).

Microencapsulation has been studied as an alternative to minimize the limitations in the use of these compounds with biological benefits to the consumer. The choice of the microencapsulation method should take into account the possible physical or chemical changes of the proteins, which can be occasioned by the use of high temperature and speed of agitation, organic solvents, and ultrasound. The microencapsulation process becomes difficult due to the complex and three-dimensional structure of the protein when compared with the peptides structure. It can be explained due to its native state, which should remain intact during the process to keep its biological functions.

Microencapsulation of protein hydrolysates has been studied by different methods as shown in Table 18.5. It was possible to reduce the bitter taste and improve the stability of the encapsulated material due to the reduced exposure of hydrophobic groups and hygroscopicity. Although each food matrix allowed the incorporation of different particle sizes, there is no consensus about the perception of the

TABLE 18.5
Studies on Microencapsulation of Hydrolysates

Technology	Carrier	Particle Size (μm)	Encapsulation Efficiency (%)	Remark	References
Complex coacervation	Soy protein isolate (SPI) and pectin	16.24–24.12	78.8–91	Bitterness reduction	Mendanha et al. (2009)
Spray drying	Gelatin and maltodextrin	10.26–17.72	—	Bitterness reduction	Favaro-Trindade et al. (2010)
	SPI	9.18–11.32	—	Bitterness reduction	Molina Ortiz et al. (2009)
	Maltodextrin	13.06–15.57	—	Bitterness reduction	Rocha et al. (2009)
	Maltodextrin and gum arabic	—	—	High stability, low hygroscopicity of hydrolyzed chicken meat	Kurozawa et al. (2009)
Ionic gelation	Alginate	—	—	Beads were insoluble in water with good buoyancy	Mukai-Correa et al. (2005)
Liposome	Soy phosphatidylcholine Phosphatidylglycerol Cholesterol	—	56–62	Bitterness reduction	Morais et al. (2003)
	Nonpurified soy lecithin Trehalose Sucrose	0.53–5.00	30.8–46	Bitterness reduction	Yokota et al. (2012)
Liposphere	Soy phosphatidylcholine Stearic acid	3.8	50–83	Bitterness reduction	Barbosa et al. (2002)
Lipid microparticles	Cupuacu butter Stearic acid	—	74	Not measured	Silva and Pinho (2013)
Spray chilling	Stearic acid + lauric acid	21.7	100*	Not measured	Chambi et al. (2008)
	Stearic acid + oleic acid				

* Approximate value.

consumers. The variation in the average size of the microparticles is an important factor to be studied for future applications into foods. Small particle sizes cannot be perceived by the consumer and do not alter the product. However, larger particles can cause undesirable grittiness aspect in the product.

18.4 Conclusions

Future research must explore the potential of the coencapsulation technologies, where two or more bioactive compounds are combined to have a synergistic effect or double functionality for the particle.

Despite the several advances in the area of encapsulation, there are still many challenges in this area to be overcome, such as determining the release mechanisms of the bioactive compounds of the microcapsules and the development of low-cost enteric polymers.

REFERENCES

Aburto, L.C., D. de Queiroz Tavares, and E.T. Martucci. Microencapsulção de Óleo Essencial de Laranja. *Food Science and Technology* 18(1) (1998): 45–48.

Akhtar, N., N. Zulfiqar, G. Fishan, M. Ahmed, H.M.S. Khan, and T. Saeed. Effect of L-ascorbic acid on the formulation and characterization of a multiple emulsion from paraffin oil. *Journal of Chemical Society of Pakistan* 32(6) (2010): 724–730.

Albertini, B., M.D. Sabatino, G. Calogerà, N. Passerini, and L. Rodriguez. Encapsulation of vitamin A palmitate for animal supplementation: Formulation, manufacturing and stability implications. *Journal of Microencapsulation* 27(2) (2010a): 150–161.

Albertini, B., B. Vitali, N. Passerini et al. Development of microparticulate systems for intestinal delivery of *Lactobacillus acidophilus* and *Bifidobacterium lactis*. *European Journal of Pharmaceutical Sciences* 40(4) (2010b): 359–366.

Alishahi, A., A. Mirvaghefi, M.R. Tehrani et al. Shelf life and delivery enhancement of vitamin C using chitosan nanoparticles. *Food Chemistry* 126 (2011): 935–940.

Arunkumar, R., K.V.H. Prashanth, and V. Baskaran. Promising interaction between nanoencapsulated lutein with low molecular weight chitosan: Characterization and bioavailability of lutein *in vitro* and *in vivo*. *Food Chemistry* 141 (2013): 327–337.

Baranauskiene, R., E. Bylaite, J. Zukauskaite, and R.P. Venskutonis. Flavor retention of peppermint (*Mentha piperita* L.) essential oil spray-dried in modified starches during encapsulation and storage. *Journal of Agricultural and Food Chemistry* 55 (2007): 3027–3036.

Baranauskienė, R., P.R. Venskutonis, K. Dewettinck, and R. Verhé. Properties of oregano (*Origanum vulgare* L.), citronella (*Cymbopogon nardus* G.) and Marjoram (*Majorana hortensis* L.) flavors encapsulated into milk protein-based matrices. *Food Research International* 39 (2006): 413–425.

Barbosa, C.M., H.A. Morais, D.C.F. Lopes, H.S. Mansur, M.C. Oliveira, and M.P.C. Silvestre. Microencapsulation of casein hydrolysates in liposomes for debittering flavor: Physicalchemistry and sensorial evaluation. *Brazilian Journal of Pharmaceutical Sciences* 38(3) (2002): 361–370.

Błaszczak, W., T.A. Misharina, D. Fessas, M. Signorelli, and A.R. Górecki. Retention of aroma compounds by corn, sorghum and amaranth starches. *Food Research International* 54 (2013): 338–344.

Botrel, D.A., S.V. Borges, R.V. de Barros Fernandes et al. Evaluation of spray drying conditions on properties of microencapsulated oregano essential oil. *International Journal of Food Science & Technology* 47 (2012): 2289–2296.

Bylaitë, E., P.R. Venskutonis, and R. Maþdþierienë. Properties of caraway (*Carum carvi* L.) essential oil encapsulated into milk protein-based matrices. *European Food Research and Technology* 212 (2001): 661–670.

Cao-Hoang, L., R. Fougère, and Y. Waché. Increase in stability and change in supramolecular structure of β-carotene through encapsulation into polylactic acid nanoparticles. *Food Chemistry* 124 (2011): 42–49.

Chaiyasat, P., A. Chaiyasat, P. Teeka, S. Noppalit, and U. Srinorachun. Preparation of poly (L-lactic acid) microencapsulated vitamin E. In *10th Eco-Energy and Materials Science and Engineering, Energy Procedia*, Vol. 34, 2013, Muang, Ubon-Ratchathani, Thailand, pp. 656–663.

Chambi, H.N.M., I.D. Alvim, D. Barrera-arellano, and C.R.F. Grosso. Solid lipid microparticles containing water-soluble compounds of different molecular mass: Production, characterisation and release profiles. *Food Research International* 41(1) (2008): 229–236.

Champagne, C.P. and P. Fustier. Microencapsulation for the improved delivery of bioactive compounds into foods. *Food Biotechnology* 18 (2007): 184–190.

Chang, D., S. Abbas, K. Hayat et al. Encapsulation of ascorbic acid in amorphous maltodextrin employing extrusion as affected by matrix/core ratio and water content. *International Journal of Food Science & Technology* 45(9) (2010): 1895–1901.

Chen, Q., D. McGillivray, J. Wen, F. Zhong, and S.Y. Quek. Co-encapsulation of fish oil with phytosterol esters and limonene by milk proteins. *Journal of Food Engineering* 117 (2013): 505–512.

Comunian, T.A., A. Abbaspourrad, C.S. Favaro-Trindade, and D.A. Weitz. Fabrication of solid lipid microcapsules containing ascorbic acid using a microfluidic technique. *Food Chemistry* 152 (2014): 271–275.

Comunian, T.A., M. Thomazini, A.J.G. Alves, F.E.M. Junior, J.C.C. Balieiro, and C.S. Favaro-Trindade. Microencapsulation of ascorbic acid by complex coacervation: Protection and controlled release. *Food Research International* 52 (2013): 373–379.

Damodaran, S., K.L. Parkin, and O.R. Fennema. *Química de Alimentos de Fennema*, pp. 366–374. Porto Alegre, Brazil: Artmed, 2010.

Desai, K.G.H. and H.J. Park. Recent developments in microencapsulation of food ingredients. *Drying Technology* 23 (2005): 1361–1394.

de Vos, P., M.M. Faas, M. Spasojevic, and J. Sikkema. Encapsulation for preservation of functionality and targeted delivery of bioactive food components. *International Dairy Journal* 20(4) (2010): 292–302.

Dong, Z., Y. Ma, K. Hayat, C. Jia, S. Xia, and X. Zhang. Morphology and release profile of microcapsules encapsulating peppermint oil by complex coacervation. *Journal of Food Engineering* 104 (2011): 455–460.

Dronen, D.M. and G.A. Reineccius. Volatile loss from dry food polymer systems resulting from chemical reactions. *Flavour Science: Recent Advances and Trends* 43 (2006): 469–472.

Fang, Z. and B. Bhandari. Encapsulation of polyphenols—A review. *Trends in Food Science and Technology* 21 (2010): 510–523.

Favaro-Trindade, C.S. and C.R.F. Grosso. The effect of the immobilisation of *Lactobacillus acidophilus* and *Bifidobacterium lactis* in alginate on their tolerance to gastrointestinal secretions. *Milchwissenschaft* 55 (2000): 496–499.

Favaro-Trindade, C.S. and C.R.F. Grosso. Microencapsulation of *L. acidophilus* (La-05) and *B. lactis* (Bb-12) and evaluation of their survival at the values of stomach and in bile. *Journal of Microencapsulation* 19(4) (2002): 485–494.

Favaro-Trindade, C.S., R.J.B. Heinemann, and D.L. Pedroso. Developments in probiotic encapsulation. *CAB Reviews* 6(4) (2011): 1–8.

Favaro-Trindade, C.S., S.C. Pinho, and G.A. Rocha. Revisão: Microencapsulação de ingredientes alimentícios. *Brazilian Journal of Food Technology* 11(2) (2008): 103–112.

Favaro-Trindade, C.S., A.S. Santana, E.S. Monterrey-Quintero, M.A. Trindade, and F.M. Netto. The use of spray drying technology to reduce bitter taste of casein hydrolysate. *Food Hydrocolloids* 24(4) (2010): 336–340.

Fernandes, R.V.D.B., S.V. Borges, and D.A. Botrel. Gum arabic/starch/maltodextrin/inulin as wall materials on the microencapsulation of rosemary essential oil. *Carbohydrate Polymers* 101 (2014): 524–532.

Flores, F.P., R.K. Singh, W.L. Kerr, R.B. Pegg, and F. Kong. Total phenolics content and antioxidant capacities of microencapsulated blueberry anthocyanins during *in vitro* digestion. *Food Chemistry* 153 (2014): 272–278.

Fuleihan, G.E. and R. Vieth. Vitamin D insufficiency and musculoskeletal health in children and adolescents. *International Congress Series* 1297 (2007): 91–108.

Gbassi, G.K., T. Vandamme, S. Ennahar, and E. Marchioni. Microencapsulation of *Lactobacillus plantarum* spp. in an alginate matrix coated with whey proteins. *International Journal of Food Microbiology* 129 (2009): 103–105.

Gebara, C., K.S. Chaves, M.C.E. Ribeiro, C.R.F. Grosso, and M.L. Gigante. Viability of *Lactobacillus acidophilus* La5 in pectin-whey protein microparticles during exposure to simulated gastrointestinal conditions. *Food Research International* 51(2) (2013): 872–878.

Gouin, S. Microencapsulation: Industrial appraisal of existing technologies and trends. *Trends in Food Science and Technology* 15 (2004): 330–347.

Ho, C.-T., M.M. Rafi, and G. Ghai. Substâncias Bioativas: Nutracêuticas E Tóxicas. In: S. Damodaran., K.L. Parkin and O.R. Fennema (Eds.), *Química de Alimentos de Fennema*, pp. 585–609. Porto Alegre, Brazil: Artmed, 2010.

Jantzen, M., A. Göpel, and C. Beermann. Direct spray drying and microencapsulation of probiotic *Lactobacillus reuteri* from slurry fermentation with whey. *Journal of Applied Microbiology* 115 (2013): 1029–1036.

Jenning, V., A. Gysler, M. Schäfer-Korting, and S.H. Gohla. Vitamin A loaded solid lipid nanoparticles for topical use: Occlusive properties and drug targeting to the upper skin. *European Journal of Pharmaceutics and Biopharmaceutics* 49 (2000): 211–218.

Kapoor, R. and Y.-S. Huang. Gamma linolenic acid: An anti-inflammatory omega-6 fatty acid. *Current Pharmaceutical Biotechnology* 7 (2006): 531–534.

Klaypradit, W. and Y.-W. Huang. Fish oil encapsulation with chitosan using ultrasonic atomizer. *LWT—Food Science and Technology* 41 (2008): 1133–1139.

Kurozawa, L.E., K.J. Park, and M.D. Hubinger. Effect of carrier agents on the physicochemical properties of a spray dried chicken meat protein hydrolysate. *Journal of Food Engineering* 94(3–4) (2009): 326–333.

Laouini, A., H. Fessi, and C. Charcosset. Membrane emulsification: A promising alternative for vitamin E encapsulation within nano-emulsion. *Journal of Membrane Science* 423–424 (2012): 85–96.

Li, W., H. Peng, F. Ning et al. Amphiphilic chitosan derivative-based core-shell micelles: Synthesis, characterization and properties for sustained release of vitamin D_3. *Food Chemistry* 152 (2014): 307–315.

Liserre, A.M., M.I. Re, and B.D.G.M. Franco. Microencapsulation of *Bifidobacterium animalis* subsp. *lactis* in modified alginate–chitosan beads and evaluation of survival in simulated gastrointestinal conditions. *Food Biotechnology* 21 (2007): 1–16.

Lupo, B., A. Maestro, M. Porras, J.M. Gutiérrez, and C. González. Preparation of alginate microspheres by emulsification/internal gelation to encapsulate cocoa polyphenols. *Food Hydrocolloids* 38 (2014): 56–65.

McClements, D.J., E.A. Decker, Y. Park, and J. Weiss. Structural design principles for delivery of bioactive components in nutraceuticals and functional foods. *Critical Reviews in Food Science and Nutrition* 49(6) (2009): 577–606.

Mendanha, D.V., S.E.M. Ortiz, C.S. Favaro-Trindade et al. Microencapsulation of casein hydrolysate by complex coacervation with SPI/pectin. *Food Research International* 42(8) (2009): 1099–1104.

Molina Ortiz, S.E., A. Mauri, E.S. Monterrey-Quintero et al. Production and properties of casein hydrolysate microencapsulated by spray drying with soybean protein isolate. *LWT—Food Science and Technology* 42(5) (2009): 919–923.

Moolman, F.S., P.W. Labuschagne, M.S. Thantsa et al. Encapsulating probiotics with an interpolymer complex in supercritical carbon dioxide. *South African Journal of Science* 102(7–8) (2006): 349–354.

Morais, H.A., C.M.S. Barbosa, F.M. Delvivo et al. Estabilidade e Avaliação Sensorial de Lipossomas Contendo Hidrolisados de Caseína. *Brazilian Journal of Food Technology* 6(2) (2003): 213–220.

Mukai-Correa, R., A.S. Prata, I.D. Alvim, and C. Grosso. Caracterização de Microcápsulas Contendo Caseína e Gordura Vegetal Hidrogenada Obtidas por Gelificação Iônica. *Brazilian Journal of Food Technology* 8(1) (2005): 73–80.

Okuro, P.K., M. Thomazini, J.C.C. Balieiro, R.D.C.O. Liberal, and C.S. Fávaro-Trindade. Co-encapsulation of *Lactobacillus acidophilus* with inulin or polydextrose in solid lipid microparticles provides protection and improves stability. *Food Research International* 53 (2013): 96–103.

Oliveira, A.C., T.S. Moretti, C. Boschini, J.C.C. Baliero, O. de Freitas, and C.S. Favaro-Trindade. Stability of microencapsulated *B. lactis* (BI 01) and *L. acidophilus* (LAC 4) by complex coacervation followed by spray drying. *Journal of Microencapsulation* 24 (2007): 685–693.

Paez, R., L. Lavari, G. Vinderola, G. Audero, A. Cuatrin, N. Zaritzky, and J. Reinheimer. Effect of heat treatment and spray drying on lactobacilli viability and resistance to simulated gastrointestinal digestion. *Food Research International* 48 (2012): 748–754.

Pedroso, D.L., M. Thomazini, R.J.B. Heinemann, and C.S. Favaro-Trindade. Protection of *Bifidobacterium lactis* and *Lactobacillus acidophilus* by microencapsulation using spray-chilling. *International Dairy Journal* 26 (2012): 127–132.

Pettifor, J.M. and A. Prentice. The role of vitamin D in paediatric bone health. *Best Practice & Research Clinical Endocrinology & Metabolism* 25 (2011): 573–584.

Poole, C.D., J.P. Halcox, S. Jenkins-Jones et al. Omega-3 fatty acids and mortality outcome in patients with and without type 2 diabetes after myocardial infarction: A retrospective, matched-cohort study. *Clinical Therapeutics* 35(1) (2013): 40–51.

Răileanu, M., L. Todan, M. Voicescu, C. Ciuculescu, and M. Maganu. A way for improving the stability of the essential oils in an environmental friendly formulation. *Materials Science and Engineering C* 33 (2013): 3281–3288.

Rocha, G.A., M.A. Trindade, F.M. Netto, and C.S. Favaro-Trindade. Microcapsules of a casein hydrolysate: Production, characterization, and application in protein bars. *Food Science and Technology International* 15(4) (2009): 407–413.

Rocha-Selmi, G.A., C.S. Favaro-Trindade, and C.R.F. Grosso. Morphology, stability, and application of lycopene microcapsules produced by complex coacervation. *Journal of Chemistry* 2013 (2013): 1–7.

Rubio-Rodriguez, N., S. Beltran, I. Jaime, S.M. Diego, M.T. Sanz, and J. Carballido. Production of omega-3 polyunsaturated fatty acid concentrates: A review. *Innovative Food Science and Emerging Technologies* 11 (2010): 1–12.

Silva, J.C. and S.C. Pinho. Viability of the microencapsulation of a casein hydrolysate in lipid microparticles of cupuacu butter and stearic acid. *International Journal of Food Studies* 2(April) (2013): 48–59.

Silva, M.C., V.B. Souza, M. Thomazini et al. Use of the Jabuticaba (*Myrciaria cauliflora*) depulping residue to produce a natural pigment powder with functional properties. *LWT—Food Science and Technology* 55 (2014): 203–209.

Southwest Research Institute. Catalog of Capability Brochures for: SwRI Chemistry & Chemical Engineering brochure, 2012. Available at: http://www.swri.org/3pubs/brochure/d01/mne/MicroNanoEncap.pdf. Accessed February, 2014.

Souza, V.B., M. Thomazini, J.C.C. Balieiro, and C.S. Favaro-Trindade. Effect of spray drying on the physicochemical properties and color stability of the powdered pigment obtained from vinification byproducts of the bordo grape (*Vitis labrusca*). *Food and Bioproducts Processing* 93 (2015): 39–50.

Sutaphanit, P. and P. Chitprasert. Optimisation of microencapsulation of Holy basil essential oil in gelatin by response surface methodology. *Food Chemistry* 150 (2014): 313–320.

Tavano, L., R. Muzzalupo, N. Piccia, and B. de Cindio. Co-encapsulation of antioxidants into niosomal carriers: Gastrointestinal release studies for nutraceutical applications. *Colloids and Surfaces B: Biointerfaces* 114 (2014): 82–88.

Teng, Z., Y. Luo, and Q. Wang. Carboxymethyl chitosan-soy protein complex nanoparticles for the encapsulation and controlled release of vitamin D_3. *Food Chemistry* 141 (2013): 524–532.

Trabelsi, I., W. Bejar, D. Ayadi et al. Encapsulation in alginate and alginate coated-chitosan improved the survival of newly probiotic in oxgall and gastric juice. *International Journal of Biological Macromolecules* 61 (2013): 36–42.

Trindade, M.A. and C.R.F. Grosso. The stability of ascorbic acid microencapsulated in granules of rice starch and in gum arabic. *Journal of Microencapsulation* 17(2) (2000): 169–176.

Wang, X., Y. Jiang, and Q. Huang. Encapsulation technologies for preserving and controlling the release of enzymes and phytochemicals. In: J.M. Lakkis (Ed.), *Encapsulation and Controlled Release Technologies in Food Systems*, pp. 135–147. Oxford, U.K.: Blackwell Publishing, 2007.

Xu, J., W. Zhao, Y. Ning et al. Improved stability and controlled release of omega-3/omega-6 polyunsaturated fatty acids by spring dextrin encapsulation. *Carbohydrate Polymers* 92 (2013): 1633–1640.

Yokota, D., M. Moraes, and S.C. Pinho. Characterization of lyophilized liposomes produced with non-purified soy lecithin: A case study of casein hydrolysate microencapsulation. *Brazilian Journal of Chemical Engineering* 29(2) (2012): 325–335.

Zitune, G. and S. Januário. *Óleos Essenciais na Culinária, Cosmética e Saúde*. São Paulo, Brazil: Optionline Ltda., 2005.

Section V

Applications of Films and Coatings in Foodstuffs

19

Application of Edible Films and Coatings on Fruits and Vegetables

Rajinder Kumar Dhall

CONTENTS

19.1 Introduction

Fresh fruits and vegetables contain considerable percentage of water, the amount of which is maintained during production in their natural environment. Respiration is maintained naturally at an appropriate equilibrium between oxygen, carbon dioxide and water by skin, which control transmission to and from surrounding environment. As soon as the fruit or vegetable is harvested, there is an increase in the respiration rate of these living tissues, resulting in metabolic loss that ultimately takes the fruit or vegetable to a gradual maturation and eventual senescence. During transit and marketing, the fruits and vegetables get bruises, in spite of the care, lose water, shine appeal and nutrients and thus require protective treatment. In fresh-cut fruits and vegetables, the wounding of tissues leads to texture breakdown, development of off-flavor, and brown discoloration on the cut surface. Controlled respiration of these living tissues would improve storability and extend the shelf life of fresh and fresh-cut produce. Several techniques like controlled atmosphere (CA) storage and modified atmosphere packaging (MAP) have been used in fruits and vegetables to minimize quality changes and quantity losses during storage. Edible coatings can provide an alternative to CA storage and MAP as these coatings modify and control the internal atmosphere of the individual fruit or vegetable. Edible coatings may extend the shelf life of fresh and fresh-cut produce by regulating the transfer of moisture, oxygen, and carbon dioxide, by retaining the aroma and taste compounds in a food system, and by improving the mechanical handling property of foodstuff (Baldwin et al. 1996; Park 1999). The ideal edible coating should create a barrier that can retard loss of desirable flavor volatile and water vapor, while restricting the CO_2 and O_2, thus creating a modified atmosphere (MA) (relatively elevated CO_2 and reduced O_2). Such a MA would slow down respiration and ethylene production and inhibit ethylene action (elevated CO_2). The MA created by the coatings should not, however, create anaerobic conditions that could cause anaerobic respiration, undesirable flavor changes and growth of anaerobic microbes. Edible coatings may also be used to improve structural integrity of frozen fruits and vegetables and prevent moisture absorption and oxidation of freeze-dried fruits or vegetables (Baker et al. 1994; Baldwin and Baker 2002; Olivas and Barbosa-Vanovas 2005; Park 2005). In addition, edible coatings can carry active ingredients such as antibrowning agents, colorants, antioxidants, antimicrobials, nutrients, flavors, and spices to further enhance food stability, quality, functionality, and safety (Krochta et al. 1994; Krochta and De Mulder-Johnson 1997; Debeaufort et al. 1998; Min and Krochta 2005). Besides this, edible coatings may replace plastic packaging to some extent by natural and biodegradable substances, and their use could lead to reduction in the overall packaging requirements and waste disposal problems.

Fruits or vegetables are usually coated by dipping in or by spraying with a range of edible materials, so that a semipermeable membrane is formed on fruit surface (Ukai et al. 1976; Thompson 2003b). The use of edible coating on fruits and vegetables is conditioned by the fulfillment of various characteristics such as the cost, availability, functional attributes, mechanical properties (flexibility, tension), optical properties (brightness and opacity), barrier effect against flow of gases, structural resistance to water and microorganisms, and sensory acceptability. These characteristics are influenced by parameters such as the kind of material implemented as structural matrix (composition, molecular weight), the conditions under which coatings are preformed (type of solvent, pH, components concentration, and temperature), and the type and concentration of additives (plasticizers, antimicrobials, antioxidants, or emulsifiers) (Guilbert et al. 1996; Rojas-Grau et al. 2009).

19.2 Coatings versus Films

An edible coating is a thin layer of edible material formed as a coating on the surface of food product, while an edible film is a preformed stand-alone thin layer of edible material, which once formed can be placed on or between food components (McHugh and Senesi 2000). The main difference between them is that the edible coating is applied in liquid form on the food, usually by immersing the product in a solution-generating substance formed by the structural matrix (carbohydrate, protein, lipid, or multicomponent mixture), and edible films are first molded as solid sheets, which are then applied as a wrapping on the food product. The main use of edible films is to test the structures of the material used for its barrier, mechanical, solubility, and other properties. In this chapter, the author uses the terminology edible coatings or edible films; this should be treated as the same.

19.3 Structural Matrix: Hydrocolloids and Lipids

There is a very wide range of compounds that can be used in the formulation of edible coatings or films and the choice depends mainly on the target application. According to their components, edible films and coatings can be divided into three categories: hydrocolloids, lipids, and composites. Hydrocolloids include proteins and polysaccharides. Lipids include waxes, resins, triglycerides (natural lipids), and fatty acids. Composites contain both hydrocolloid and lipids (Greener-Donhowe and Fennema 1994; Nisperos-Carriedo 1994; Baldwin et al. 1999). Several other compounds such as plasticizers and emulsifiers can be added to edible coatings when lipids and hydrocolloids are combined. The purpose of adding plasticizers in film-forming material is to decrease the intermolecular forces between polymer chains, which results in greater film flexibility, elongation, toughness, and permeability. Plasticizers used for edible coatings include glycerol, sorbitol, sucrose, propylene glycol, polyethylene glycol, fatty acids, and monoglycerides. Emulsifiers or surfactants are surface-active agents that are used to improve the stability of the emulsion and ensure good surface wetting, spreading, and adhesion of the coating to the food surface. Common emulsifiers used for coatings are fatty acids, ethylene glycol monostearate, glycerol monostearate, esters of fatty acids, lecithin, sucrose ester, and sorbitan monostearate or polysorbates

(Tweens). In addition, food additives include antioxidants, colorants, flavoring agents, and antimicrobial compounds that can also be added to edible coatings (Han 2002; Cha and Chinnan 2004).

19.4 Application of Hydrocolloids Coatings

Hydrocolloids include proteins and polysaccharides. Proteins and polysaccharides generally have a good barrier to oxygen at low relative humidity (RH) due to their tightly packed hydrogen-bonded network structure but have a poor moisture barrier due to their hydrophilic nature (McHugh and Krochta 1994). Therefore, hydrophobic compounds are generally added to make films or coatings. Proteins used to make edible coatings include corn zein, wheat gluten, soy protein, whey protein, casein, collagen/gelatin, pea protein, rice bran protein, cotton seed protein, peanut protein, and keratin (Baldwin and Baker 2002; Han and Gennadios 2005). Polysaccharides used in edible film formation include starch and starch derivatives, cellulose derivatives, alginate, carrageenan, pectin, pullulan, chitosan, and various gums (Han and Gennadios 2005).

Zein-based coatings have been applied to fresh and dried fruits, often as a substitute for shellac coatings. Compared to shellac coatings, zein coatings are favorable for gloss and other quality characteristics on apples (Bai et al. 2002, 2003a,b). Zein coatings were able to maintain the firmness and color of broccoli head (Rakotonirainy et al. 2001). Park et al. (1994a,b) reported that the application of zein coating delayed color change and weight loss and maintained the firmness of tomato during storage. Carlin et al. (2001) reported that zein coatings provide a continuous and stable coating with satisfactory sensory properties and also reduce the growth of *Listeria monocytogenes* on cooked sweet corn. Caseinate- and whey protein–based coatings have been applied on frozen peas and peanuts to provide a barrier to oxygen and moisture transfer for extending their shelf life (Chen 1995; Mate and Krochta 1995). Caseinate and whey protein isolate coatings were reported to delay the browning of apple and potato slices (Le Tien et al. 2001). Lafortune et al. (2005) reported that caseinate and whey protein coatings along with MAP protected carrots against dehydration and retained their firmness during storage. Soy protein coatings generally exhibit poor moisture resistance and water vapor barrier properties but were able to preserve freshness of apple slices (Kinzel 1992) and to retard the senescence process of kiwifruit (Xu et al. 2001). Eswaranandam et al. (2006) applied soy protein coating containing malic or lactic acid on whole apple with the purpose to study its effect on the sensory quality of the fruit and observed that incorporation of organic acids to films did not adversely affect the sensory properties of coated apples after cold storage.

Carboxymethylcellulose (CMC) is by far the most important cellulose derivative for food applications (Sanderson 1981). Edible coatings made of CMC, methylcellulose (MC), hydroxypropyl cellulose, and hydroxypropyl methylcellulose (HPMC) have been applied to some fruits and vegetables for providing barriers to oxygen, oil, or moisture transfer (Morgan 1971; Sacharow 1972; Krumel and Lindsay 1976; Maftoonazad and Ramaswamy 2005) and for improving batter adhesion (Meyers 1990; Dziezak 1991). CMC is the most used coating that helped to maintain the original firmness and crispness of apples, berries, peaches, celery, lettuce, and carrots (Mason 1969), preserved important flavor components of some fresh fruits and vegetables (Nisperos-Carriedo and Baldwin 1990), and reduced O_2 uptake without increasing CO_2 level in the internal environment of coated apples and pears by simulating a CA environment (Lowings and Cutts 1982; Banks 1985; Meheriuk and Lau 1988; Santerre et al. 1989). Ribeiro et al. (2007) studied the effect of polysaccharide-based edible coatings (starch, carrageenan, and chitosan) on the shelf life of strawberry and observed minimum weight loss in strawberry by using carrageenan and chitosan-based edible coatings with added calcium chloride and lower microbial growth rate in strawberries coated with chitosan and calcium chloride.

Chitosan, which is mainly obtained from deacetylation of crustacean chitin, has been one of the most promising coating materials for fresh produce because of its excellent film-forming property, broad antimicrobial activity, and compatibility with other substances, such as vitamins, minerals, and antimicrobial agents (Li et al. 1992; Shahidi et al. 1999; Park and Zhao 2004; Durango et al. 2006; Chien et al. 2007; Ribeiro et al. 2007). Application of chitosan induces the production of plant defense enzymes such as chitinase. Chitosan-based coatings were reported to protect highly perishable fruits like strawberries, raspberries, and grapes from fungal decay (El Ghaouth et al. 1991a; Zhang and Quantick 1998;

Romanazzi et al. 2002; Devlieghere et al. 2004; Park et al. 2005; Vargas et al. 2006). Reuck et al. (2009) found that the combination of chitosan (1.0 g L⁻¹) + MAP was effective in preventing decay, browning, and retaining the pericarp color in the cultivar McLean's Red compared with single MAP. Shao et al. (2012) revealed that apple (cv. Gala) fruits treated with heat (38°C) + chitosan showed the lowest respiration rate, ethylene evolution, malondialdehyde and membrane leakage, and the highest firmness and consumer acceptance among the treatments. At the same time, this combined treatment could inhibit the lost of green color, titratable acidity, and weight loss compared with heat treatment alone. Chien et al. (2007) reported the effectiveness of chitosan coating for extending shelf life and prolonging quality of sliced mango fruit. The application of chitosan coating delayed changes in the contents of anthocyanin, flavonoid, and total phenolics, reduced weight loss and browning of litchi fruit, and improved its storability (Zhang and Quantick 1997). Chitosan also reduced decay and improved appearance of carrots (Cheah et al. 1997; Li and Barth 1998).

Carrageenan, extracted from several red seaweeds, is a complex mixture of several polysaccharides and is another potential coating material for fruits and vegetables. Carrageenan-based coatings have been applied to fresh apples for reducing moisture loss, oxidation, or disintegration of the apples (Lee et al. 2003). Vina et al. (2007) study the effect of maize starch edible coating with glycerol (as a plasticizer) on brussels sprouts. The buds were treated with the solution, stored in polystyrene trays, and covered with polyvinyl chloride film, preserving the quality parameters regarding different factors such as weight loss, firmness, surface color of the food, commercial acceptability, and nutritional quality.

19.5 Application of Lipid Coatings

Lipids due to their hydrophobic nature are used in edible films to provide a barrier to moisture. In addition, they are often used to provide gloss to food surfaces (Greener and Fennema 1994). However, lipids are not polymers; they form films or coatings with poor mechanical properties. Lipids used for the preparation of lipid-based edible films or coatings include neutral lipids, fatty acids, waxes (beeswax, candelilla wax, carnauba wax, rice bran wax), and resins (shellac, wood rosin). Most fatty acids derived from vegetable oils are considered generally recognized as safe (GRAS) and have been suggested as substitutes for the petroleum-based mineral oils used in the preparation of edible coatings (Hernandez 1994; Baldwin et al. 1997).

Waxes (carnauba wax, beeswax, paraffin wax, and others) have been commercially used for coating fresh fruits and vegetables since the 1930s with the purpose of reducing moisture loss and surface abrasion during fruit handling (Lawrence and Iyengar 1983) and controlling internal gas composition of the fruits (Kester and Fennema 1986). In general, wax coatings are substantially more resistant to moisture transport than other lipid or nonlipid coatings (Schultz et al. 1949; Landmann et al. 1960; Watters and Brekke 1961; Kaplan 1986). Wax coatings are commercially applied to citrus, apples, mature green tomatoes, rutabagas, cucumbers, and other vegetables such as asparagus, carrot, radish, turnip, brinjal, knol khol, okra, parsnip, pepper, potato, sweet potato, and squash, where high gloss and shiny surface are desired (Alleyne and Hagenmaier 2000; Hagenmaier 2000; Bai et al. 2002, 2003b; Da-Mota et al. 2003; Fallika et al. 2005; Porat et al. 2005). McGuire and Hagenmaier (2001) applied shellac coating to grapefruit and oranges over a CMC layer and observed that a shellac formulation at pH 9.0 with 5.2% ethanol was more toxic to the coliform bacteria *Enterobacter aerogenes* and *Escherichia coli* than a formulation at pH 7.25 with 12% ethanol.

19.6 Application of Composite Coatings

A composite coating may consist of lipids and hydrocolloids combined to form a bilayer or a stable emulsion (Krochta et al. 1994). In bilayer composite coatings, the lipid forms a second layer over the polysaccharide or protein layer. In emulsion composite coatings, the lipid is dispersed and entrapped in the supporting matrix of protein or polysaccharide (Shellhammer and Krochta 1997; Perez-Gago and Krochta 2005). In some recent studies, the production of edible and biodegradable coatings by combining various polysaccharides, proteins, and lipids is considered with the aim of taking advantage of the properties of each compound and the synergy between them. The mechanical and barrier properties of these films not only depend on the material used in the polymer matrix, but also on their compatibility (Altenhofen et al. 2009). The optimization of composition of edible coatings is in one of the most important steps to consider while film formation, and they must be formulated according to the properties of the fruits and vegetables to which they have to be applied (Rojas-Grau et al. 2009). Thus, it is very important to characterize and test different coating solutions on fresh and fresh-cut produce, since both have different quality attributes to be maintained during the storage period (Oms-Oliu et al. 2008a). Selecting a coating material with appropriate permeability for various gases is critical in modifying the internal environment of fresh produce for its preservation. In addition, storage temperature and RH are also critical in modifying the internal environment of fresh produce since coating permeability and produce respiration are affected by these parameters.

Composite coating formulation can be tailor-made to suit to the needs of a specific commodity or farm produce. Recently, many researchers have extensively explored the development of composite films using lipid and HPMC (Hagenmaier and Shaw 1990), MC and lipid (Greener and Fennema 1989), MC and fatty acid (Sapru and Labuza 1994), corn zein, MC and fatty acid (Park et al. 1996), whey isolate and lipids (McHugh et al. 1994), casein and lipids (Avena-Bustillos and Krochta 1993), gelatin and soluble starch (Arvanitoyannis et al. 1997), hydroxypropyl starch and gelatin (Arvanitoyannis et al. 1998), corn zein and corn starch (Ryu et al. 2002), gelatin and fatty acid (Bertana et al. 2005), soy protein isolate and gelatin (Cao et al. 2002), and soy protein isolate and polylactic acid (Rhim et al. 2007).

Composite coatings with soy or zein protein with amylose ester or fatty acids provided an effective moisture barrier on carrot and fresh-cut apple slices (Williams 1968). Wheat gluten with lipid (beeswax, stearic acid, and palmitic acid)-based bilayer coatings significantly retained firmness and reduced weight loss of fresh strawberries (Tanada-Palmu and Grosso 2005). Chitosan–lauric acid composite coatings prevented fresh-cut apple slices from browning and water loss (Pennisi 1992). A casein–lipid emulsion coating formed a tight matrix that binds to the cut apple surfaces and protects apple slices from moisture loss and oxidative browning (Krochta et al. 1988, 1990). A sodium caseinate/stearic acid emulsion coating reduced white blush and respiration rate of peeled carrots, and a calcium caseinate acetylated monoglyceride emulsion coating reduced water loss of apples, celery sticks, and zucchini as a result of increased water vapor resistance of the emulsion coatings (Avena-Bustillos et al. 1994a,b, 1997). HPMC–lipid composite coatings consisting of beeswax or shellac significantly reduced texture loss and internal breakdown of plums (Perez-Gago et al. 2003). Composite coatings prepared from whey protein concentrate as the hydrophilic phase and beeswax or carnauba wax as the lipid phase exerted an antibrowning effect on fresh-cut apples (Perez-Gago et al. 2003, 2005, 2006). Locust bean gum, shellac, and beeswax coatings prolonged the storability of the cherries by reducing moisture loss (Rojas-Argudo et al. 2005). An emulsion coating with CMC as the hydrophilic phase and paraffin wax, beeswax, or soybean oil as the hydrophobic phase also extended shelf life and reduced weight loss of apples, peaches, and pears (Togrul and Arslan 2004, 2005).

Chitosan-based composite coatings significantly retarded color development, reduced weight loss and respiration rate, maintained firmness, and reduced titratable acidity of coated fruits (mango, banana, capsicum) compared to uncoated controls (Kittur et al. 2001; Srinivasa et al. 2002). Maqbool et al. (2010) applied gum arabic + chitosan edible coating on fresh banana fruits and observed that composite edible coating formed by 10% gum arabic and 1% chitosan coating showed a synergistic behavior that allowed maintaining sensory quality and microbiological parameters, without phytotoxic effects on bananas stored for 33 days. Munoz et al. (2008) reported that strawberry fruits coated with 1% chitosan + 0.5% calcium gluconate maintain shelf life for 1 week at 10°C and 70% ± 5% RH without any fungal decay and observed that addition of calcium to the chitosan solution increased the firmness and nutrients of the fruit. Yu et al. (2012) reported that composite coating with 1% chitosan and 1% phytic acid could decrease the weight loss rate and malondialdehyde content of fresh-cut lotus root; postpone browning; restrain the activities of peroxidase (POD), polyphenol oxidase (PPO), and phenylalanine ammonia-lyase; and maintain the content of vitamin C and polyphenol at relatively high level. The combination of 1-MCP and chitosan coating could effectively extend the storage life and maintain the quality of Indian jujube fruit at room temperature storage (Zhong and Xia 2007). Zhang et al. (2004) reported that ozone water treatment (concentration of 4.2 mg m^{-3}) and a composite coating made of polyvinyl alcohol (1%), chitosan (1%), lithium chloride (0.5%), glacial acetic acid (2.5%), and sodium benzoate (0.05%) inhibit respiration, chlorophyll breakdown, and PPO activity.

19.7 Addition of Active Compounds

Nowadays, edible coatings have different applications, and their use is expected to be expanded with the development of active coating. These second generations of coating materials include chemicals, enzymes, or microorganisms that prevent, for example, microbial growth or lipids oxidation in coated food products. In this sense essential oils, in combination with structural polymers, can be a promising source since different scientists showed the evidence of their effectiveness as antimicrobial and antioxidant compounds (Atares et al. 2010; Sanchez-Gonzalez et al. 2010). Coatings of second generation may contain nutrients or other bioactive compounds that have a positive effect on health, especially due to the application of new microencapsulation or nanoencapsulation techniques. In this way, coating materials would act as carriers of these bioactive compounds to be transported to target sites such as the intestine without losing its activity, being within a matrix during its passage through the gastrointestinal tract (Korhonen 2005).

19.7.1 Antimicrobial Coatings

Edible films and coatings with antimicrobial properties have evolved the concept of active packaging, being developed to reduce, inhibit, or stop the growth of microorganisms on food surface (Appendini and Hotchkiss 2002) as antimicrobial agents comprise natural or synthetic compounds with known and minimal toxicological effects on mammals and the environment. Natural antimicrobial agents include chitosan, polypeptides, and essential oils from fruits, vegetable, and spices. It has been proved that essential oils of angelica, anise, carrot, cardamom, coriander, mint, vanillin, cinnamon, cloves, coriander, dill weed, fennel, garlic, nutmeg, oregano, parsley, rosemary, sage, or thymol have antimicrobial properties (Cagri et al. 2004). Synthetic antimicrobial agents are mostly organic acids that include acetic, benzoic, citric, fumaric, lactic, malic, propionic, sorbic, succinic, and tartaric acid. These acids typically inhibit the outgrowth of bacterial and fungal cells. Potassium sorbate, sodium benzoate, and benzoic acid are organic acid salts more widely used as antimicrobial food additives. Additionally, hydrophobic compounds such as tea tree essential oil in HPMC-based films have also been used (Sanchez-Gonzalez et al. 2009). Antimicrobials used for the formulation of edible films and coatings must be classified as food-grade additives or compounds GRAS by the relevant regulations. In most fresh or fresh-cut products, microbial contamination is found maximum on their surface. Therefore, in such cases, the incorporation of antimicrobial additives in edible films or coatings is very important as these additives can migrate selectively and gradually from the wrapping compounds to the surface of the food (Ouattara et al. 2000).

Starch-based coatings containing potassium sorbate were applied on the surface of fresh strawberries for reducing microbial growth and extending its storage life (Garcia et al. 1998). In addition, chitosan coatings containing potassium sorbate increased antifungal activity against the growth of *Cladosporium* and *Rhizopus* on fresh strawberries (Park et al. 2005). Lysozyme was incorporated into chitosan coatings for enhancing the antimicrobial activity of chitosan against *E. coli* and *Streptococcus faecalis* (Park et al. 2004). A new patented edible film comprising organic acids, protein, and glycerol (e.g., 0.9% glycerol, 10% soy protein, and 2.6% malic acid) can inhibit pathogen growth including *L. monocytogenes*, *S. gaminara*, and *E. coli* 0157:H7. Such film also provides a method for coating comestible products with edible films without masking the color but increasing the shelf life (Hettiarachchy and Satchithanandam 2007).

Chitosan has the ability to slow the growth of certain microorganisms that are deleterious in fruit post-harvest such as *Fusarium* spp., *Colletotrichum musae*, and *Lasiodiplodia theobromae* in banana (Kyu Kyu et al. 2007; Maqbool et al. 2010). Functional properties and antimicrobial effects of chitosan are related to its deacetylation degree and molecular weight. Chitosan-treated longan fruit had reduced firmness loss, titratable acidity, total soluble solids, decay, respiration rates, and PPO activity associated with peel discoloration compared to control fruits (Jiang and Li 2001). Chitosan increased vitamin C content, reduced ethylene production, and delayed ripening in peach fruit as indicated by the high content of titratable acidity (Li and Yu 2000). Moreover, El Ghaouth et al. (1992) reported that the chitosan coatings

(1%–2%) reduced the deterioration in tomato caused by *Botrytis cinerea*. The application of chitosan significantly reduces postharvest penicillium decay and delay of fruit senescence during cold storage of different citrus fruits (Chien and Chou 2006). The application of chitosan to apples at 0.25%–2.0% concentration significantly induced resistance to postharvest blue mold, caused by the fungus *Penicillium expansum* (de Capdeville et al. 2002). The application of 2.0% chitosan to Gala apples considerably reduced the severity of blue and grey mold (Wu et al. 2005). Li and Yu (2000) reported that the application of chitosan to peaches artificially inoculated with *Monilinia fructicola* significantly reduced the incidence of brown rot and delayed disease development as compared to control fruit. Moreover, chitosan-treated peaches were firmer and had higher titratable acidity and vitamin C content than control peaches. In tomatoes, chitosan coatings greatly reduced postharvest black rot caused by the pathogen *Alternaria alternata* (Reddy et al. 2000). According to Liu et al. (2007), the application of chitosan significantly controlled tomato gray and blue molds, caused by *B. cinerea* and *P. expansum*, respectively. Han et al. (2004a,b) reported that chitosan coatings on fresh strawberries and red raspberries extended their shelf life by decreasing weight loss and delaying changes in color, titratable acidity, and pH during cold storage. Park et al. (2005) demonstrated the antifungal function of chitosan coatings on fresh strawberries and observed the excellent compatibility of chitosan with other antifungal agents. Vargas et al. (2006) reported that the addition of oleic acid to chitosan coating not only enhanced the chitosan antimicrobial activity but also improved water vapor resistance of coated produce. The chitosan-based coatings reduced the incidence of anthracnose (caused by *Colletotrichum gloeosporioides*) by about 40% in papaya (Bautista-Banos et al. 2003). Chitosan coatings containing natamycin were also effective in controlling the decay of "Hami" melons caused by natural infections of *A. alternata* and *Fusarium semitectum*. Sangsuwan et al. (2008) evaluated the inhibitory effect of chitosan/MC stand-alone films, with and without vanillin, against *E. coli* and *Saccharomyces cerevisiae* on wrapped fresh-cut cantaloupe and pineapple and reported that films inhibited the growth of *E. coli* and *S. cerevisiae* on fresh-cut cantaloupe, the film with vanillin being more effective. However, the application of the antimicrobial film reduced the ascorbic acid content in pineapple, remaining only 10% of its original content after storage. In general, the quality attributes of coated fresh-cut cantaloupe and pineapple were reported as acceptable in this study. This coating also improved the quality properties of coated fruit (Cong et al. 2007).

HPMC-based coatings containing 0.4% sorbic acid enhanced the inactivation of *Salmonella Montevideo* on tomato fruit surface. However, these coatings caused a chalky appearance of the fruit surface, limiting its potential for commercial application (Zhuang et al. 1996). Raybaudi-Massilia et al. (2008) indicated that the addition of cinnamon, clove, or lemongrass oils at 0.7% (v/v) or their active compounds (citral, cinnamaldehyde, and eugenol) at 0.5% (v/v) into an alginate-based coating increased their antimicrobial effect, reduced the population of *E. coli* O157:H7 by more than 4 log CFU/g, and extended the microbiological shelf life of Fuji apples for at least 30 days. However, they observed that lemongrass and citral acted faster against *E. coli* O157:H7 at day 0 than the other compounds, suggesting that both essential oils entered into the bacteria more easily (higher rate of diffusion), causing irreversible damage and cell death. Seydim and Sarikus (2006) developed whey protein–based antimicrobial coatings by adding oregano, rosemary, and garlic essential oils; Rojas-Grau et al. (2006) used apple puree and high methoxyl pectin combined with oregano, lemon grass, or cinnamon oil at different concentrations. However, these coatings showed their efficacy in vitro against a wide spectrum of microorganisms but they were not tested in real food systems, and, as a result, there is a lack of available information about their possible impact on the aroma and flavor of the coated products (Han 2002).

Ozone is a strong antimicrobial agent with high reactivity, penetrability, and spontaneous decomposition to a nontoxic product (Kim et al. 1999; Grass et al. 2003). Several researchers have shown that treatment with ozone appears to have a beneficial effect in extending the storage life of fresh noncut commodities such as broccoli, cucumber, apples, grapes, oranges, pears, raspberries, and strawberries by reducing microbial populations and by oxidation of ethylene (Beuchat et al. 1998; Kim et al. 1999; Skog and Chu 2001). The use of ozonated water has been applied to fresh-cut vegetables for sanitation purposes reducing microbial populations and extending the shelf life of some of these products (Beltran et al. 2005). However, scarce information is currently available about inactivation of food borne pathogens. Ozone has been declared in many countries to have potential use for food processing and declared in the United States as GRAS (FDA 1997; Smilanick et al. 1999). When compared to chlorine, ozone has a greater effect against certain microorganisms leaving no residues (White 1992). However, a higher corrosiveness and initial capital cost for generator are the main disadvantages compared to the use of chlorine (Smilanick et al. 1999).

19.7.2 Antioxidant Coatings

Antioxidants are added to edible coatings to protect fruits and vegetables against oxidative rancidity, degradation, and discoloration (Baldwin et al. 1995). The sensory aspects are very important to ensure the success of new edible coatings and color is one of the most important parameter that must be maintained during different postharvest operations. The color change is usually caused by the enzyme PPO, which in the presence of oxygen converts phenolic compounds into dark-colored pigments (Zawistowski et al. 1991). Some researchers have proved the effectiveness of edible coatings on the control of browning and PPO activity. Application of antioxidant treatments as dipping after peeling and/or cutting is the most common way to control browning of fresh-cut fruits. Ascorbic acid (L-ascorbic acid) and its various neutral salts and other derivatives have been GRAS (Dorantes-Alvarez et al. 1998; Rocha et al. 1998; Agar et al. 1999; Buta et al. 1999; Gorny et al. 1999; Senesi et al. 1999; Soliva-Fortuny et al. 2001, 2002). The carrageenan-based coatings in combination with ascorbic acid resulted in positive sensory results and reduction of microbial levels on minimally processed apple slices (Lee et al. 2003).

Lettuce can undergo changes in color due to different biochemical processes, mainly chlorophyll degradation and browning appearance, which is the limiting factor for the marketability of fresh-cut lettuce (Bolin and Huxsoll 1991; Couture et al. 1993; Lopez-Galvez et al. 1996). Carrots effected by environmental and postharvest stresses can synthesize phenolic compounds along with wound barriers such as lignin (Babic et al. 1993; Howard et al. 1994; Talcott and Howard 1999), and lignin formation increases whiteness in carrots and the activation of certain enzymes after the processing can lead to degradation of antioxidants (Howard and Griffin 1993; Howard and Dewi 1996).

Perez-Gago et al. (2006) reported a substantial reduction in browning of fresh-cut apples when whey protein concentrate beeswax coating containing ascorbic acid, cysteine, or 4-hexylresorcinol was used. Brancoli and Barbosa-Canovas (2000) decreased surface discoloration of apple slices with maltodextrin and MC coatings including ascorbic acid. Baldwin et al. (1996) found that CMC-based coating with antioxidants, including ascorbic acid, reduced browning and retarded water loss of cut apple. McHugh and Senesi (2000) reported that the coated with a mixture of apple puree, pectin, and vegetable oils containing ascorbic acid and citric acid delayed browning of fresh-cut apples. Olivas et al. (2003) preserved surface browning of fresh-cut pear wedges by using an MC-based coating containing ascorbic acid and citric acid. Similar results were obtained by Lee et al. (2003), who studied the effect of carrageenan and whey protein concentrate edible coatings in combination with antibrowning agents on fresh-cut apple slices and observed that the incorporation of ascorbic, citric, and oxalic acids was advantageous in maintaining color during 2 weeks. Oms-Oliu et al. (2008a) observed that the incorporation of *N*-acetylcysteine and glutathione into gellan, alginate, or pectin formulations helped to control enzymatic browning of fresh-cut pears during 2 weeks of storage. Ponce et al. (2008) applied chitosan films enriched with olive and rosemary oleoresins on pumpkin slices, which showed a clear antioxidant effect by slowing the action of PPO and POD within 5 days of storage. Chitosan coatings were also used by Eissa (2008), who found that they delayed discoloration associated with reduced enzyme activity of PPO and other enzymes and had a good effect on color characteristics of fresh-cut mushroom during storage at 4°C. Hui-Min et al. (2009) studied the effects of three different edible coatings (carrageenan, CMC, and sodium alginate) and their combinations on browning of fresh-cut peach during storage at 5°C and observed that sodium alginate coating and the various composite combinations increased peach color and reduced browning due to inhibition of PPO activity.

19.7.3 Texture Enhances

Fruits and vegetables that remain firm, crunchy texture are highly desirable because consumers associate these characters with freshness. Textural changes in fruits and vegetables are related to certain enzymatic and nonenzymatic processes. The most common way of controlling softening phenomena in fruits and vegetables is by treatments with calcium salts (Garcia et al. 1996), and mostly calcium chloride ($CaCl_2$)

is used as dip treatment in the range of 0.1%–1.0% (Rosen and Kader 1989; Sapers and Miller 1998; Bett et al. 2001; Soliva-Fortuny et al. 2002, 2003). Calcium ions interact with pectin polymers to form a cross-linked network that increases mechanical strength, thus delaying senescence and controlling physiological disorders in fruits and vegetables (Poovaiah 1986). Calcium lactate has been used as a firming agent for fruit such as cantaloupes and strawberry (Morris et al. 1985; Main et al. 1986). It has been reported to be a good alternative to $CaCl_2$ because it avoids bitterness or off-flavors associated with this salt (Luna-Guzman and Barret 2000). Hernandez-Munoz et al. (2008) observed that the addition of calcium gluconate to chitosan (1%) coating formulation increased the firmness of strawberries during refrigerated storage. Han et al. (2004) observed that a chitosan-based coating containing 5% calcium (Gluconal Cal) increased the firmness of frozen-thawed raspberries by approximately 25% in comparison to uncoated fruits.

Rojas-Grau et al. (2008) observed that apple wedges coated with alginate and gellan edible coatings cross-linked with calcium salts outstandingly maintained their initial firmness during refrigerated storage. Similar results were obtained by Oms-Oliu et al. (2008b) on fresh-cut melon coated with different calcium-cross-linked polysaccharide-based edible coatings (alginate, gellan, and pectin). Lee et al. (2003) indicated that incorporating 1% of calcium chloride within the whey protein concentrate coating formulation helped to maintain firmness of fresh-cut apple pieces. Some workers have reported that dip treatment of apple slices with $CaCl_2$ solution have no or slight effect on the preservation of texture. Olivas et al. (2007) maintained firmness of fresh-cut "Gala" apples practically constant using alginate edible coatings. Optimal concentration of $CaCl_2$ treatments in minimally processed melon was found to be 2.5% (Luna-Guzman et al. 1999). Agar et al. (1999) could not observe significant differences between 1% and 2% $CaCl_2$ treatments in kiwifruit slices.

19.7.4 Nutraceutical Coatings

Several researchers have incorporated minerals, vitamins, and fatty acids into edible film and coating formulations to enhance the nutritional value of some fruits and vegetables. Tapia et al. (2008) reported that the addition of ascorbic acid (1% w/v) to the alginate and gellan based edible coatings preserve the natural ascorbic acid content in fresh-cut papaya throughout its storage. Han (2002) reported that chitosan-based coatings have the capability to hold high concentrations of calcium or vitamin E and therefore significantly increased their content in fresh and frozen strawberries and red raspberries. Similarly, Hernandez-Munoz et al. (2006) observed that chitosan-coated strawberries retained more calcium gluconate than strawberries dipped into calcium solutions. Han et al. (2004b) improved the nutritional and physicochemical quality of strawberries and raspberries by means of chitosan-based coatings enriched with calcium and vitamin E.

Xanthan gum coating was utilized to contain a high concentration of calcium and vitamin E for not only preventing moisture loss and surface whitening, but also significantly increasing the calcium and vitamin E contents of the carrots (Mei et al. 2002). The development of chitosan coatings containing high concentrations of calcium, zinc, or vitamin E also provided alternative ways to fortify fresh fruits and vegetables that otherwise could not be accomplished with common processing approaches (Park and Zhao 2004).

The addition of probiotics to obtain functional edible coatings and films has been less studied. Tapia et al. (2007) developed first edible probiotic film for coating on fresh-cut apple and papaya and observed that both fruits were successfully coated with alginate or gellan film-forming solutions containing viable 10^6 cfu/g bifidobacteria.

19.8 Nanotechnology

Nowadays nanotechnology is applied in many research areas. A nanoparticle is an ultrafine particle of nanometer size, which is able to form nanobiocomposite films when it is combined with natural polymers. Some of the applications associated with nanotechnology include improved taste, color, flavor, texture, and consistency of foodstuffs and increased absorption and bioavailability of food or food ingredients (nutrients) and the development of new food packaging materials with improved mechanical, barrier, and antimicrobial properties (Restuccia et al. 2010). Traditionally, mineral fillers such as clay, silica, and talc have been incorporated in film preparation in order to reduce its cost or to improve its performance to some extent (Rhim and Ng 2007). Both proteins (Shotornvit et al. 2009) and polysaccharides (Tang et al. 2008; Casariego et al. 2009) have given rise to films in combination with clay nanoparticles. Rhim et al. (2006) incorporated different types of nanoparticles (nano-silver, siver zeolite, montmorillonites) into a chitosan matrix, obtaining composites with better mechanical, water vapor barrier and antimicrobial properties than the traditional chitosan coating. Cellulose nanofibers have also shown good possibilities as reinforcements in composite coatings for food packaging. However, other nanoparticles such as tripolyphosphate–chitosan (De Moura et al. 2009), microcrystalline cellulose (Bilbao-Sainz et al. 2010), and silicon dioxide (Tang et al. 2009) have also been added to biopolymers to obtain films.

19.9 Commercial Application of Edible Coatings

Nowadays, many commercial edible coatings for use on fresh and fresh-cut fruits and vegetables are available in the market to reduce weight loss and physiological disorders and maintain produce quality. Most of them are assigned to maintain the quality of citrus and apples and, to a lesser extent, mangoes, papayas, pomegranates, avocados, or tomatoes (Olivas et al. 2007). Several commercial products available in the market for postharvest use, such as Biosave (*Pseudomonas syringae*) registered in the United States and "Shemer" (*Metschnikowia fructicola*) registered in Israel (Droby et al. 2009). Prolong, TAL-Prolong (Courtaulds Group, London, United Kingdom), Semperfresh (AgriCoat Industries Ltd., Berkshire, United Kingdom), and Natural Shine 9000 (Pace International, Seattle, WA) are commercially available composite coating formulations based on CMC (Nisperos-Carriedo et al. 1992). They contain sucrose fatty acid ester, sodium salt of CMC, and emulsifier. Available in powder or granular form, their aqueous solution (0.5%–2% concentration) is used to extend shelf life of many fruits. Coating of TAL-Prolong (1%) on mangoes showed delayed ripening with an extended shelf life. Nature Seal (EcoScience Product System Division, Orlando, FL) is another cellulose-based edible coating formulation used for delayed ripening of tomatoes and mangoes. Nature-Seal has also been used (in combination with antimicrobials, plasticizers, antioxidants, etc.) to coat fresh-cut apples and potatoes. The coating significantly reduced weight loss of apples and potatoes more than those treated with water solutions and was not objectionable in taste during several weeks of storage (Baldwin et al. 1996). The coating has also been used to effectively reduce the discoloration of mini-peeled carrots without affecting microbial and chemical quality (Howard and Dewi 1995). A commercial fruit coating, Nutri-Save (Nova Chem, Halifax, Nova Scotia, Canada), was developed to serve as both film former and natural preservative and to create a modified atmosphere for whole apples and pears to reduce respiration rate and desiccation of these commodities (Elson et al. 1985). Nutri-save reduces the respiration rate, weight loss, and microbial decay in fresh-cut pears and apples (Baldwin et al. 1995). Table 19.1 shows some of the formulations that have already been applied to fresh fruits and vegetables (Dhall 2013). Nowadays, there is a new generation of edible coatings that is being applied to fresh-cut fruits and vegetables (Perez 2003). Some examples of the application of these coatings on fresh-cut fruits and vegetables are shown in Table 19.2.

19.10 Problems Associated with Edible Coatings

Success of edible coatings on fruits and vegetables mainly depends on selecting films or coatings that can give a desirable internal gas composition that is appropriate for a specific product. Shewfelt et al. (1987) stated that color change, firmness loss, ethanol fermentation, decay ratio, and weight loss of edible film coated fruits are important quality parameters for some products and these parameters must be monitored throughout the storage period.

  Thick coating on fruit and vegetable surface becomes an undesirable barrier between the external and internal atmosphere and restricts exchange of respiratory gases (CO_2 and O_2) causing accumulation of high levels of ethanol and development of off-flavors (Miller et al. 1983; El Ghaouth et al. 1992; Howard and Dewi 1995; Cisneros-Zevallos and Krochta 2003). Therefore, it is necessary to adjust the thickness of the coating according to the variety, storage conditions, and marketing temperature. Park et al. (1994)

TABLE 19.1

Application of Edible Coatings to Fresh Fruits and Vegetables

Coating Material	Composition	Fruit/Vegetable	Effect of Coating	References
Semperfresh™	Sucrose esters with high proportion of short chain unsaturated fatty acid esters, sodium salts of CMC, and mixed mono- and diglycerides	Zucchini	Reduced water loss and internal CO_2 of zucchini fruit	Avena-Bustillos et al. (1994b)
		Cucumber, summer squash	$O_2/H_2O/CO_2$ barrier	Kaynas and Ozelkok (1999)
		Apple	Reduced color changes, retained acid, increased shelf life, and maintained quality	Drake et al. (1987)
		Tomato	Reduced color changes, retained acid, increased shelf life, and maintained quality	Tasdelen and Bayindirli (1998)
		Banana	Decreased ethylene production and delayed chlorophyll loss	Nisperos-Carriedo et al. (1992)
		Quince	$O_2/H_2O/CO_2$ barrier	Yurdugul (2005)
Prolong™	Sucrose polyester of fatty acid esters, sodium CMC, and mono- and diglycerides	Mango	Retard ripening, reduce weight loss and chlorophyll loss	Dhalla and Hanson (1988) and Motlagh and Quantick (1998)
		Pear	Retention of firmness, green skin color, and titratable acidity	Farber et al. (2003)
Tal-Prolong™	Sucrose fatty acid esters, sodium salt of CMC, and mono- and diglycerides	Mango	Delayed ripening with extended shelf life	Nisperos-Carriedo et al. (1992)
NatureSeal™	Cellulose-based edible coating	Mango and tomato Pome fruit	Delayed ripening —	Nisperos-Carriedo et al. (1992) Farber et al. (2003)
		Carrot	Retard discoloration and carotene loss; a barrier for O_2 diffusion	Chen et al. (1996)
Chitosan	Chitosan	Citrus	$O_2/H_2O/CO_2$ barrier	Fornes et al. (2005)
		Raspberry	H_2O barrier; Ca^{2+}, vit. E carrier	Han et al. (2004a,b)
	Chitosan; HPMC	Raspberry	H_2O barrier; carrier (antimicrobial)	Park et al. (2005)
	Chitosan and Tween 80	Strawberry, cucumber, and bell pepper	Antimicrobial coatings	El Ghaouth et al. (1991a,b)
		Apple, pear plum, and peach	Act as gas barrier	Elson and Hayes (1985) and Davies et al. (1989)
		Litchi	Reduce weight loss and browning, improve storability	Zhang and Quantick (1997)

(Continued)

TABLE 19.1 (*Continued*)

Application of Edible Coatings to Fresh Fruits and Vegetables

Coating Material	Composition	Fruit/Vegetable	Effect of Coating	References
		Carrot	Reduce decay and improve appearance	Li and Barth (1998) and Cheah et al. (1997)
		Tomato	Reduce respiration rate and ethylene production and increase titratable acidity	El Ghaouth et al. (1992)
		Pear	Reduce ethylene production and delay ripening as indicated by high content of titratable acidity	Li and Yu (2000)
Polysaccharide–lipid	Maltodextrin, CMC, propylene glycol, fatty acid esters, sodium benzoate	Mango		Diaz-Sobac et al. (1996)
	MC, PEG, stearic acid, citric acid, ascorbic acid	Apricot		Ayranci and Tunc (2004)
	Paraffin wax, beeswax, soybean oil, CMC from sugar beet pulp, eumulgin PE, oleic acid, and sodium oleate	Peach, pear mandarin		Togrul and Arslan (2004)
	HPMC, beeswax, shellac, stearic acid, and glycerol	Plum		Togrul and Arslan (2004b)
Carrageenan	Carrageenan, glycerol, and Tween 80	Strawberry	—	Ribeiro et al. (2007)
Protein–Polysaccharide	CMC, WPI, caseinates and glycerol	Strawberry	—	Vachon et al. (2003)
Starch	Starch and glycerol	Strawberry	—	Garcia et al. (1998)
Soy protein	Soy protein, glycerol, malic acid, lactic acid	Apple		Eswaranandam et al. (2006)
Zein	Corn zein protein	Tomato	Delayed color change, lost firmness and weight and extended shelf life	Park et al. (1994)
	Corn zein protein and propylene glycol	Apple	—	Bai et al. (2003)
	Corn zein protein	Mango	$O_2/H_2O/CO_2$ barrier	Hoa et al. (2002)

(Continued)

TABLE 19.1 (*Continued*)

Application of Edible Coatings to Fresh Fruits and Vegetables

Coating Material	Composition	Fruit/Vegetable	Effect of Coating	References
Nutri-Save™	*N,O*-Carboxymethyl chitosan	Apple, pear, pomegranate, bread fruit, cherry	Increases its shelf life	Kittur et al. (2001), Srinivasa et al. (2002), Farber et al. (2003), Worrell et al. (2002), and Lau and Meheriuk (1994)
Brillo shine	Sucrose esters and wax (shine)	Apple, avocado, melons, citrus	Protect, increase shine, and extend shelf life	Baldwin (1994)
Nu-coat Flo, Ban-seel	Sucrose esters of fatty acids and sodium salt of CMC	Apple, banana, cucumber, guava, melon, pear, plum	Protect, increase shine, and extend shelf life	Baldwin (1994)
Crisp Coat 868	Starch	Strawberry	—	Ribeiro et al. (2007)
Food Coat	Composite polysaccharides	Strawberry	—	Farber et al. (2003)
AgriCoat	Composite polysaccharides	Avocado, pear, pome fruit	—	Farber et al. (2003)
Citrashine	Sucrose ester and wax	Mandarins	Increase shine and extend shelf life	Ahmad and Khan (1987)
Casein	Calcium caseinate	Sweet pepper	Effective gas barrier to internal O_2 and CO_2; inhibit color change and reduce decay	Lerdthanangkul and Krochta (1996)
	Sodium caseinate and stearic acid	Carrot	Reduce moisture loss	Krochta et al. (1993)
Cellulose	MC and glycerol	Carrot	Extend storage life, result more carotene than control	Li and Barth (1998)
		Strawberry, avocado	—	Maftoonazad and Ramaswamy (2005)
Fresh Seal™	Polyvinyl alcohol, starch, and surfactant	Fruits	Extend shelf life	Farber et al. (2003) and Posey et al. (2005)

TABLE 19.2

Application of Edible Coatings to Fresh-Cut Fruits and Vegetables

Coating Material	Functional Ingredient	Fruit/Vegetable	Effect	References
Starch	Potassium sorbate, citric acid	Strawberry	Antimicrobial effect, inhibited the growth of mesophilic aerobes, mold, and yeast counts.	Garcia et al. (2001)
Alginate	Cinnamon, clove, lemongrass, cinnamaldehyde, eugenol, citral	Apple	Antimicrobial effect, inhibited native microbiota during 30 days and reduced >4 log CFU/g of *Escherichia coli* O157:H7 in the first week of storage.	Raybaudi-Massilia et al. (2008b)
	Malic acid, palmarosa, cinnamon, lemongrass	Melon	Antimicrobial effect, inhibited the microbial growth and reduced up to 3.1 log CFU/g after 30 days of storage.	Raybaudi-Massilia et al. (2008a)
Alginate, apple puree	Lemongrass, oregano, vanillin	Apple	Antimicrobial effect, reduced native psychrophilic aerobes, molds, and yeast. Lemongrass (1.0%–1.5%) and oregano (0.5%) reduced >4 log CFU/g of inoculated *Listeria innocua*.	Rojas-Grau et al. (2007)
Methylcellulose	Chitosan	Melon	Antimicrobial effect, reduced the growth of mesophilic aerobes, psychrotrophs, yeast, and molds, and maintained the growth of *E. coli*, *Staphylococcus aureus*, *Salmonella* sp. <10 CFU/g.	Krasaekoopt and Mabumrung (2008)
Chitosan	Chitosan	Water chestnut	O_2 barrier	Pen and Jiang (2003)
Chitosan	Chitosan	Litchi (peeled)	O_2/H_2O barrier	Tay and Perena (2004)
Whey protein, Carrageenan	Ascorbic acid (1.0%), citric acid (1.0%), oxalic acid (0.05%)	Apple	Antioxidant effect, maintained the original color during storage without changes in sensory properties.	Lee et al. (2003)
Whey protein, beeswax	Ascorbic acid (0.5%–1.0%), cysteine (0.1%–0.5%), 4-hexylresorcinol (0.005%–0.02%)	Apple	Antioxidant effect, reduced surface browning (4-hexylresorcinol showing the least effectiveness at reducing browning).	Perez-Gago et al. (2006)
Corn zein protein	—	Apple	$O_2/H_2O/CO_2$ barrier, gloss.	Bai et al. (2003a)
Methylcellulose maltodextrin	Ascorbic acid	Apple	Antioxidant effect, reduced surface discoloration.	Brancoli and Barbosa-Canovas (2000)
Methylcellulose	Ascorbic acid	Pear	Antioxidant effect, prolonged shelf life by retarding browning.	Olivas et al. (2003)
CMC, soy protein	Ascorbic acid	Apple	Antioxidant effect, delayed browning more effectively when applied in an edible coating than in an aqueous solution.	Baldwin et al. (1996)

(*Continued*)

TABLE 19.2 (*Continued*)

Application of Edible Coatings to Fresh-Cut Fruits and Vegetables

Coating Material	Functional Ingredient	Fruit/Vegetable	Effect	References
Pectin, apple puree	Ascorbic acid, citric acid	Apple	Antioxidant effect, preserved color for 12 days at 5°C.	McHugh and Senesi (2000)
Alginate, gellan	*N*-Acetylcysteine, glutathione	Apple	Antioxidant effect, prevented surface discoloration.	Rojas-Grau et al. (2007)
Alginate, gellan	*N*-Acetylcysteine	Apple	Antioxidant effect, maintained the original color by 2 weeks of storage.	Rojas-Grau et al. (2008)
Alginate, gellan	*N*-Acetylcysteine, glutathione	Pear	Antioxidant effect, prevented browning for 2 weeks.	Oms-Oliu et al. (2008a)
Alginate, gellan	Calcium chloride	Apple	Texture enhancement, maintained firmness by 2 weeks.	Rojas-Grau et al. (2008)
Alginate, gellan, pectin	Calcium chloride	Melon	Texture enhancement, helped to maintain fruit firmness during 15 days.	Oms-Oliu et al. (2008b)
Alginate	Calcium chloride	Apple	Texture enhancement, maintained firmness during storage.	Olivas et al. (2007)
Alginate	Calcium lactate	Melon	Texture enhancement, maintained firmness in coated samples.	Raybaudi-Massilia et al. (2008)
Alginate	Calcium lactate	Apple	Texture enhancement, maintain the texture for more than 30 days.	Raybaudi-Massilia et al. (2008)
Alginate	Calcium chloride	Pineapple	Texture enhancement, helped to retain internal liquids.	Montero-Calderon et al. (2008)
Whey protein	Calcium chloride	Apple	Texture enhancement, inhibited the loss of firmness.	Lee et al. (2003)
Chitosan	Calcium gluconate	Strawberry	Texture enhancement, did not improve the firmness retention.	Hernandez-Munoz et al. (2008)
Chitosan	Calcium gluconate	Raspberry	Texture enhancement, helped to maintain textural quality.	Han et al. (2004)
Chitosan	Calcium gluconate, vitamin E	Strawberry, raspberry	Nutraceutical effect, increased the content of these nutrients in both fruits.	Han et al. (2004)
Chitosan	Calcium gluconate	Strawberry	Increased nutritional value of the strawberries.	Hernandez-Munoz et al. (2008)
Alginate	Ascorbic acid	Papaya	Preserved the natural ascorbic acid content of the fruit.	Tapia et al. (2008)
Alginate, gellan	*Bifidobacterium lactis*	Apple, papaya	Maintained values of *B. lactis* >106 CFU/g by 10 days during storage.	Tapia et al. (2007)

reported that tomatoes coated with 2.6 mm zein film produced alcohol and off-flavors internally, which is due to low oxygen and high carbon dioxide concentration. Smith et al. (1987) summarized that the use of coatings resulted in physiological disorders like core flush, flesh breakdown, accumulation of ethanol, and alcoholic off-flavor, which is due to modification of internal atmosphere. Waxes and shellac tend to restrict the gas exchange of O_2 and CO_2 between atmosphere and fruit to the extent that the internal O_2 level becomes too low to support aerobic respiration, resulting in high levels of internal ethanol, acetaldehyde, and internal CO_2 (Petracek et al. 1998; Alleyne and Hagenmaier 2000). Smith and Stow (1984) reported that the application of sucrose fatty acid ester coating on apples (cv. Cox's Orange Pippin) resulted in reduction of fruit firmness, yellowing, and weight loss, but also increased core flush incidence. Ben-Yehoshua (1985) reported that the coatings having poor water vapor barrier properties could result in weight or moisture loss of the product, but it could prevent water vapor condensation, which could be a potential source of microbial spoilage in fruit and vegetable packaging. Films that have good gas barrier properties could cause anaerobic respiration and interfere with normal ripening (Meheriuk and Lau 1988). Therefore, it is clear that the control of gas permeability should be a priority in the development of edible coatings (Parra et al. 2004).

Edible coatings are usually consumed with the coated products. Therefore, the incorporation of compounds such as antimicrobials, antioxidants, and nutraceuticals should not affect consumer acceptance. Some authors have indicated that the incorporation of antimicrobial agents into edible coatings could impart undesirable sensorial modifications in foods, especially when essential oils are used (Burt 2004). Oms-Oliu et al. (2008a) summarized that the coating of fresh-cut melon with gellan gum increased the phenolic compounds during storage, which further affect the sensory properties such as odor, color, and flavor. Rojas-Grau et al. (2008) observed the synthesis of ethanol and acetaldehyde in apple slices coated with alginate and gellan gum after 2 weeks of storage, which decreased its sensory quality especially fruit flavors. Sometimes the incorporation of certain antibrowning agents into edible coatings can yield an unpleasant odor, particularly when high concentrations of sulfur-containing compounds such as *N*-acetylcysteine and glutathione are used as dipping agents (Richard et al. 1992; Iyidogan and Bayindirli 2004; Rojas-Grau et al. 2006). The addition of nutraceutical compounds to edible coatings may impart bitter taste, astringent, or off-flavor (Drewnowski and Gomez-Carneros 2000) that can lead to rejection of the product by consumers (LeClair 2000).

Many edible coatings are made from ingredients that could cause allergic reactions. Within these allergens, milk, soybeans, fish, peanuts, nuts, and wheat are the most important. Therefore, the presence of a coating with a known allergen on a food must be also clearly labelled (Franssen and Krochta 2003).

19.11 Regulatory Status and Food Safety Issues

According to the European Directive (ED 1995, 1998) and U.S. regulations (FDA 2012), edible coatings are those coatings that are formed from food products, food ingredients, food additives, food contact substances, or food packaging materials. To maintain edibility, all film-forming components as well as any functional additives in the film-forming materials should be food grade and nontoxic, and all process facilities should meet high standards of hygiene (Guilbert and Gontard 1995; Guilbert et al. 1996; Han 2002; Nussinovitch 2003). According to the U.S. Code of Federal Regulations, the amount of edible coating ingredients used must be only that which is necessary to accomplish the intended effect, and the ingredients have to be GRAS and be listed in the aforementioned code.

In the regulation of most countries, chemical substances added as antimicrobials are regarded as food additives if the primary purpose of the substances is shelf life extension. However, each country has its own regulations defining a list of approved additives (ED 1995; USDA 2006). For instance, according to U.S. regulations, organic acids including acetic, lactic, citric, malic, propionic, tartaric, and their salts are GRAS for miscellaneous and general purpose usage (Doores 1993). On the other hand, many essential oils are used widely in the food, health, and personal care industries and are also classified as GRAS substances or permitted as food additives. According to European Directive (1995), the ingredients that can be incorporated into the formulations of edible coating include arabic and karaya gum, pectins, shellac, beeswax, candelilla wax, and carnauba wax. This directive was modified in 1998 by introducing new ingredients such as lecithin, polysorbates, fatty acids, and fatty acid salts. On the other hand, additives used by U.S. Food and Drug Administration (FDA) for coating fresh fruits and vegetables include morpholine, polydextrose, sorbitan monostearate, sucrose fatty acid esters, cocoa butter, and castor oil (FDA 2012). In India, the PFA section ZZZ (23) of Rule 42 (2006) states that "The fresh fruits and vegetables may be coated with bees wax or carnauba wax under proper label declaration." However, some consumers have concerns about their use. Vegetarians and others who avoid animal products may worry that fruits and vegetables contain animal-based waxes, such as oleic acid. Some people fear that the edible coatings and films trap pesticides, making the fruit or vegetable unsafe to eat. If consumers want to avoid edible coated fruits and vegetables, FDA regulations that took effect in 1994 may help them to identify the appropriate products for them. These regulations require produce packers or grocers to provide information about the presence of coatings on fresh fruits and vegetables. The information will say that the product is (a) coated with food-grade animal-based wax or (b) coated with food-grade vegetable-, petroleum-, beeswax-, and/or shellac-based wax or resin to maintain freshness. FDA will also allow the statement "no wax or resin coating" on fresh fruits and vegetables that do not contain any coating.

There are many safety concerns about nanomaterials, as their size may allow them to penetrate into cells and eventually remain in the human organism. So the need for accurate information on the effects of nanomaterials on human health following chronic exposure is imperative before any nanostructured food packaging is available for commercialization.

19.12 Conclusion

The application of edible films and coatings to fresh, minimally processed, and processed fruits and vegetables extend their commercial shelf life, maintain their microbiological, sensory, and nutritional quality and have a similar effect on the storage under controlled or modified atmospheres. Some formulations have been specifically tested on their ability to inhibit PPO activity and delay browning reactions. In addition, edible films and coatings are able to transport substances that bring some benefits not only for the food itself but also for the consumer, through the encapsulation of bioactive compounds, developing new products with nutraceutical, or functional effect. Their use in highly perishable products such as horticultural ones is based on some particular properties such as the cost, availability, functional attributes, mechanical properties (flexibility, tension), optical properties (brightness and opacity), barrier effect against gas flow, structural resistance to water and microorganisms, and sensory acceptability. In some cases, edible coatings were not successful. The success of edible coatings for fresh products totally depends on the control of internal gas composition. Edible coating technology seems to be very promising as long as consumer accept this technology as safe and friendly. Basic information on film-coating formulation, properties, methods of application to fruit or vegetable surface, and demonstration of effectiveness is lacking. The interaction of proteins, polysaccharides and lipids in coatings and the interaction between coatings and plastic packaging need to be explored further for successful commercialization of this technology. Tremendous research is required in the area of applications of edible coatings and films on fresh and fresh-cut fruits and vegetables.

REFERENCES

Agar IT, Massantini R, Hess-Pierce B, and Kader AA (1999). Postharvest CO_2 and ethylene production and quality maintenance of fresh-cut kiwifruit slices. *J. Food Sci.* 64:433–440.

Ahmad M and Khan I (1987). Effects of waxing and cellophane lining on chemical quality indices of citrus fruits. *Plant Foods Hum. Nutr.* 37:47–57.

Alleyne V and Hagenmaier R (2000). Candelilla-shellac: An alternative formulation for coating apples. *Hort. Sci.* 35:691–693.

Altenhofen M, Krause AC, and Guenter T (2009). Alginate and pectin composite films crosslinked with Ca^{+2} ions: Effect of the plasticizer concentration. *Carbohydr. Polym.* 77:736–742.

Appendini P and Hotchkiss JH (2002). Review of antimicrobial food packaging. *Innovative Food Sci. Emerg. Technol.* 3:113–126.

Arvanitoyannis I, Nakayama A, and Aiba S (1998). Edible films made from hydroxypropyl starch and gelatin and plasticized by polyols and water. *Carbohydr. Polym.* 36:105–119.

Arvanitoyannis I, Psomiadou E, Nakayama A, Aiba S, and Yamamoto N (1997). Edible films made from gelatin, soluble starch and polyols, Part 3. *Food Chem.* 60:593–604.

Atares L, Bonilla J, and Chiralt A (2010). Characterization of sodium caseinate-based edible films incorporated with cinnamon or ginger essential oils. *J. Food Eng.* 100:678–687.

Avena-Bustillos RJ, Cisneros-Zevallos LA, Krochta JM, and Saltveit ME (1994a). Application of casein-lipid edible film emulsions to reduce white blush on minimally processed carrots. *Postharvest Biol. Technol.* 4:319–329.

Avena-Bustillos RJ and Krochta JM (1993). Water vapor permeability of caseinate-based films as affected by pH, calcium crosslinking, and lipid content. *J. Food Sci.* 58:904–907.

Avena-Bustillos RJ, Krochta JM, and Saltveit ME (1997). Water vapor resistance of Red Delicious apples and celery sticks coated with edible caseinate-acetylated monoglyceride films. *J. Food Sci.* 62:351–354.

Avena-Bustillos RJ, Krochta JM, Saltveit ME, Rojas-Villegas RJ, and Sauceda-Perez JA (1994b). Optimization of edible coating formulations on zucchini to reduce water loss. *J. Food Eng.* 21:197–214.

Ayranci E and Tunc S (2004). The effect of edible coatings on water and vitamin C loss of apricots (*Armeniaca vulgaris* Lam.) and green peppers (*Capsicum annuum* L.). *Food Chem.* 87:339–342.

Babic I, Amiot MJ, Nguyen-the C, and Aubert S (1993). Changes in phenolic content in fresh ready-to-use shredded carrots during storage. *J. Food Sci.* 58:351–356.

Bai J, Alleyne V, Hagenmaier RD, Mattheis JP, and Baldwin EA (2003a). Formulation of zein coatings for apples (*Malus domestica* Borkh). *Postharvest Biol. Technol.* 28:259–268.

Bai J, Baldwin EA, and Hagenmaier RH (2002). Alternatives to shellac coatings provide comparable gloss, internal gas modification, and quality for 'Delicious' apple fruit. *Hort. Sci.* 37:559–563.

Bai J, Hagenmaier RD, and Baldwin EA (2003b). Coating selection for 'Delicious' and other apples. *Postharvest Biol. Technol.* 28:381–390.

Baker RA, Baldwin EA, and Nisperos-Carriedo MO (1994). Edible coatings and films for processed foods. In: Krochta JM, Baldwin EA, and Nisperos-Carriedo MO (eds.), *Edible Coatings and Films to Improve Food Quality*. Lancaster, PA: Technomic Publishing Co., Inc., pp. 89–104.

Baldwin EA (1994). Edible coatings for fresh fruits and vegetables: Past, present, and future. In: Krochta JM, Baldwin EA, and Nisperos-Carriedo MO (eds.), *Edible Coatings and Films to Improve Food Quality*. Lancaster, PA: Technomic Publishing Company, Inc., pp. 25–64.

Baldwin EA and Baker RA (2002). Use of proteins in edible coatings for whole and minimally processed fruits and vegetables. In: Gennadios A (ed.), *Protein Based Films and Coatings*. Boca Raton, FL: CRC Press, pp. 501–515.

Baldwin EA, Burns JK, Kazokas W, Brecht JK, Hagenmaier RD, Bender RJ, and Pesis E (1999). Effect of two edible coatings with different permeability characteristics on mango *Mangifera indica* L. ripening during storage. *Postharvest Biol. Technol.* 17:215–226.

Baldwin EA, Nisperos-Carriedo MO, and Baker RA (1995a). Edible coatings for lightly processed fruits and vegetables. *Hort. Sci.* 30:35–38.

Baldwin EA, Nisperos-Carriedo MO, Chen X, and Hagenmaier RD (1996). Improving storage life of cut apple and potato with edible coating. *Postharvest Biol. Technol.* 9:151–163.

Baldwin EA, Nisperos-Carriedo MO, Hagenmaier RD, and Baker RA (1997). Use of lipids in coatings for food products. *Food Technol.* 51:56–62.

Baldwin EA, Nisperos-Carriedo MO, Show PE, and Burns JK (1995b). Effect of coatings and prolonged storage conditions on fresh orange flavor volatiles, degrees brix, and ascorbic acid levels. *J. Agric. Food Chem.* 43:1321–1331.

Banks NH (1985). Internal atmosphere modification in Pro-long coated apples. *Acta Hort.* 157:105.

Bautista-Banos S, Hernandez-Lopez M, Bosquez-Molina E, and Wilson CL (2003). Effect of chitosan and plant extracts on growth of *Colletotrichum gloeosporioides*, anthracnose levels and quality of papaya fruit. *Crop Prot.* 22:1087–1092.

Beltran D, Selma MV, Tudela JA, and Gil MI (2005). Effect of different sanitizers on microbial and sensory quality of fresh-cut potato strips stored under modified atmosphere or vacuum packaging. *Postharvest Biol. Technol.* 37:37–46.

Ben-Yehoshua S (1969). Gas exchange, transportation, and the commercial deterioration in storage of orange fruit. *J. Am. Soc. Hortic. Sci.* 94:524–528.

Ben-Yehoshua S (1985). Individual seal-packaging of fruit and vegetables in plastic film—A new postharvest technique. *Hort. Sci.* 20:32–37.

Bertana LC, Tanada-Palmua PS, Sianib AC, and Grosso CRF (2005). Effect of fatty acids and 'Brazilian elemi' on composite films based on gelatin. *Food Hydrocoll.* 19:73–82.

Bett KL, Ingram DA, Grimm CC, Lloyd SW, Spanier AM, Miller JM, Gross KC, Baldwin EA, and Vinyard BT (2001). Flavor of fresh-cut gala apples in barrier film packaging as affected by storage time. *J. Food Quality* 24:141–156.

Beuchat LR, Nail BV, Adler BB, and Clavero MRS (1998). Efficacy of spray application of chlorine in killing pathogenic bacteria on raw apples, tomatoes, and lettuce. *J. Food Prot.* 61:1305–1311.

Bilbao-Sainz C, Avena-Bustillos RJ, Wood DF, Williams TG, and McHugh TH (2010). Composite edible films based on hydroxypropyl methylcellulose reinforced with microcrystalline cellulose nanoparticles. *J. Agric. Food Chem.* 58:3753–3760.

Bolin HR and Huxsoll CC (1991). Effect of preparation procedures and storage parameters on quality retention of salad-cut lettuce. *J. Food Sci.* 56:60–67.

Brancoli N and Barbosa-Canovas GV (2000). Quality changes during refrigerated storage of packaged apple slices treated with polysaccharide films. In: Barbosa-Canovas GV and Gould GW (eds.), *Innovations in Food Processing.* Pennsylvania, PA: Technomic Publishing Co., pp. 243–254.

Burt S (2004). Essential oils: Their antibacterial properties and potential applications in foods: A review. *Int. J. Food Microbiol.* 94:223–253.

Buta JG, Moline HE, Spaulding DW, and Wang CY (1999). Extending storage life of fresh-cut apples using natural products and their derivatives. *J. Agric. Food Chem.* 47:1–6.

Cagri A, Ustunol Z, and Ryser ET (2004). Antimicrobial edible films and coatings. *J. Food Prot.* 67:833–848.

Cao N, Fua Y, and He Y (2002). Preparation and physical properties of soy protein isolate and gelatin composite films. *Food Sci. Technol.* 35:680–686.

Carlin F, Gontard N, Reich M, and Nguyen-The C (2001). Utilization of zein coating and sorbic acid to reduce *Listeria monocytogenes* growth on cooked sweet corn. *J. Food Sci.* 66:1385–1389.

Casariego A, Souza BWS, Cerqueira MA, Teixeira JA, Cruz L, Diaz R et al. (2009). Chitosan/clay films' properties as affected by biopolymer and clay micro/nanoparticles' concentrations. *Food Hydrocoll.* 23:1895–1902.

Cha DS and Chinnan MS (2004). Biopolymer-based antimicrobial packaging: A review. *Crit. Rev. Food Sci. Nutr.* 44:223–237.

Cheah LH, Page BBC, and Shepherd R (1997). Chitosan coating for inhibition of sclerotinia rot of carrots. *New Zeal. J. Crop Hort. Sci.* 25:89–92.

Chen H (1995). Functional properties and applications of edible films made of milk proteins. *J. Dairy Sci.* 78:2563–2583.

Chen XH, Campbell CA, Grant LA, Li P, and Barth M (1996). Effect of nature seal® on maintaining carotene in fresh-cut carrots. *Proc. Florida State Hortic. Soc.* 109:258–259.

Chien PJ and Chou CC (2006). Antifungal activity of chitosan and its application to control post-harvest quality and fungal rotting of Tankan citrus fruit (*Citrus tankan* Hayata). *J. Sci. Food Agric.* 86:1964–1969.

Chien PJ, Sheu F, and Yang FH (2007). Effects of edible chitosan coating on quality and shelf life of sliced mango fruit. *J. Food Eng.* 78:225–229.

Cisneros-Zevallos L and Krochta JM (2003). Dependence of coating thickness on viscosity of coating solution applied to fruits and vegetables by dipping method. *J. Food Sci.* 68:503–510.

Cong F, Zhang Y, and Dong W (2007). Use of surface coatings with natamycin to improve the storability of Hami melon at ambient temperature. *Postharvest Biol. Technol.* 46:71–75.

Couture R, Cantwell MI, Ke D, and Saltveit ME (1993). Physiological attributes and storage life of minimally processed lettuce. *Hort. Sci.* 28:723–725.

Da-Mota WF, Salomao LCC, Cecon PR, and Finger FL (2003). Waxes and plastic film in relation to the shelf life of yellow passion fruit. *Sci. Agric.* 60:51–57.

Davies DH, Elson CM, and Hayes ER (1989). N,O-carboxymethyl chitosan, a new water soluble chitin derivative. In: Skjak-Braek G, Anthosen T, and Sandford P (eds.), *Chitin and Chitosan: Source, Chemistry, Biochemistry, Physical Properties, and Application.* New York: Elsevier Applied Science, pp. 467–472.

Debeaufort F, Quezada-Gallo JA, and Voilley A (1998). Edible films and coatings: Tomorrow packaging: A review. *Crit. Rev. Food Sci. Nutr.* 38:299–313.

De Capdeville G, Wilson CL, Beer SV, and Aist JR (2002). Alternative disease control agents induce resistance to blue mold in harvested "Red Delicious" apple fruit. *Phytopathology* 92:900–908.

De Moura MR, Aouada FA, Avena-Bustillos RJ, McHugh TH, Krochta JM, and Mattoso LHC (2009). Improved barrier and mechanical properties of novel hydroxypropyl methylcellulose edible films with chitosan/tripolyphosphate nanoparticles. *J. Food Eng.* 92:448–453.

Devlieghere F, Vermeulen A, and Debevere J (2004). Chitosan: Antimicrobial activity, interactions with food components and applicability as a coating on fruit and vegetables. *Food Microbiol.* 21:703–714.

Dhall RK (2013). Advances in edible coatings for fresh fruits and vegetables: A review. *Crit. Rev. Food Sci. Nutr.* 53:435–450.

Dhalla R and Hanson W (1988). Effect of permeable coating on the storage life of fruits. II. Pro-Long treatments of mangoes (*Mangifera indica* L. cv. Julie). *Int. J. Food Sci. Technol.* 23:107–112.

Diaz-Sobac R, Luna AV, Beristain CI, de la Cruz J, and Garcia HS (1996). Emulsion coating to extend postharvest life of mango (*Mangifera indica* cv. Manila). *J. Food Process. Preserv.* 20:191–202.

Doores S (1993). Organic acids. In: Davidson PM and Branen AL (eds.), *Antimicrobials in Food*. New York: Marcel Dekker, Inc., pp. 95–136.

Dorantes-Alvarez L, Parada-Dorantes L, Ortiz-Moreno A, Santiago-Pineda T, Chiralt-Boix A, and Barbosa-Canovas G (1998). Effect of anti-browning compounds on the quality of minimally processed avocados. *Food Sci. Technol. Int.* 4:107–113.

Drake SR, Fellman JK, and Nelson JW (1987). Postharvest use of sucrose polyesters for extending the shelf-life of stored 'Golden Delicious' apples. *J. Food Sci.* 52:685–690.

Drewnowski A and Gomez-Carneros C (2000). Bitter taste, phytonutrients and the consumer: A review. *Am. J. Clin. Nutr.* 72:1424–1435.

Droby S, Wisniewski M, Macarisin D, and Wilson C (2009). Twenty years of postharvest biocontrol research: Is it time for a new paradigm? *Postharvest Biol. Technol.* 52:137–145.

Durango AM, Soares NF, and Andrade NJ (2006). Microbiological evaluation of an edible antimicrobial coating on minimally processed carrots. *Food Control* 17:336–341.

Dziezak JD (1991). Special report: A focus on gums. *Food Technol.* 45:116–132.

ED-European Parliament and Council Directive N 95/2/EC (1995). On food additive other than colors and sweeteners. Available from http://ec.europa.eu/food/fs/sfp/addit flavor/flav11en.pdf.

ED-European Parliament and Council Directive N 98/72/EC (1998). On food additive other than colors and sweeteners. Available from http://ec.europa.eu/food/fs/sfp/addit flavor/flav11en.pdf.

Eissa HAA (2008). Effect of chitosan coating on shelf-life and quality of fresh-cut mushroom. *Polish J. Food Nutr. Sci.* 58:95–105.

El Ghaouth A, Arul J, and Ponnampalam R (1991a). Use of chitosan coating to reduce water loss and maintain quality of cucumber and bell pepper fruits. *J. Food Process. Preserv.* 15:359–368.

El Ghaouth A, Arul J, Ponnampalam R, and Boulet M (1991b). Chitosan coating effect on storability and quality of fresh strawberries. *J. Food Sci.* 12:1618–1632.

El Ghaouth A, Ponnampalam R, Castaigne F, and Arul J (1992). Chitosan coating to extend the storage life of tomatoes. *Hort. Sci.* 27:1016–1018.

Elson CM and Hayes ER (1985). Development of the differentially permeable fruit coating Nutri-Save® for the modified atmosphere storage of fruit. In: *Proceedings of the Fourth National Controlled Atmosphere Research Conference: Controlled Atmosphere for Storage and Transport of Perishable Agricultural Commodities*, Raleigh, NC, pp. 248–262.

Elson CM, Hayes ER, and Lidster PK (1985). Development of the differentially permeable fruit coating 'Nutri-Save' for the modified atmosphere storage of fruit. In: Blankenship M (ed.), *Controlled Atmosphere for Storage and Transport of Perishable Agricultural Commodities*. Raleigh, NC: North Carolina State University, 248pp.

Eswaranandam S, Hettiarachchy NS, and Meullenet JF (2006). Effect of malic and lactic acid incorporated soy protein coatings on the sensory attributes of whole apple and fresh-cut cantaloupe. *J. Food Sci.* 71:307–313.

Fallika E, Shaloma Y, Alkalai-Tuviaa S, Larkovb O, Brandeisb E, and Ravidb U (2005). External, internal and sensory traits in Galia-type melon treated with different waxes. *Postharvest Biol. Technol.* 36:69–75.

Farber JN, Harris LJ, Parish ME, Beuchat LR, Suslow TV, Gorney JR, Garrett EH, and Busta FF (2003). Microbiological safety of controlled and modified atmosphere packaging of fresh and fresh-cut produce. *Compr. Rev. Food Sci. Food Saf.* 2:142–160.

Food and Drug Administration (FDA) (1997). Substances generally recognized as safe, proposed rule. *Fed. Reg.* 62:18937–18964.

Food and Drug Administration (FDA) (1998). Department of Health and Human Services. Secondary Direct Food Additive for Human Consumption. 21 CFR. Part 173.300 Chlorine dioxides.

Food and Drug Administration (FDA) (2012). CFR Title 21: Foods and Drugs, CFR Part 172: Food additives permitted for direct addition to food for human consumption, CFR Subpart C: Coatings, films and related substances. Silver Spring, MD. http://www.accessdata.fda.gov/scripts/cdrh/cfdocs/cfCFR/CFRSearch.cfm.

Fornes F, Almela V, Abad M, and Agusti M (2005). Low concentration of chitosan coating reduce water spot incidence and delay peel pigmentation of clementine mandarin fruit. *J. Sci. Food Agric.* 85:1105–1112.

Franssen LR and Krochta JM (2003). Edible coatings containing natural antimicrobials for processed foods. In: Roller S (ed.), *Natural Antimicrobials for Minimal Processing of Foods.* Boca Raton, FL: CRC Press, pp. 250–262.

Garcia JM, Herrera S, and Morilla A (1996). Effects of postharvest dips in calcium chloride on strawberry. *J. Agric. Food Chem.* 44:30–33.

Garcia MA, Martino MN, and Zaritzky NE (1998a). Plasticized starch-based coatings to improve strawberry (*Fragaria ananassa*) quality and stability. *J. Agric. Food Chem.* 46:3758–3767.

Garcia MA, Martino MN, and Zaritzky NE (1998b). Starch-based coatings: Effect on refrigerated strawberry (*Fragaria ananassa*) quality. *J. Sci. Food Agric.* 76:411–420.

Garcia MA, Martino MN, and Zaritzky NE (2001). Composite starch-based coatings applied to strawberries (*Fragaria ananassa*). *Nahrung Food* 45:267–272.

Gorny JR, Hess-Pierce B, and Kader AA (1999). Quality changes in fresh-cut peach and nectarine slices as affected by cultivar storage atmosphere and chemical treatments. *J. Food Sci.* 64:429–432.

Grass ML, Vidal D, Betoret N, Chiralt A, and Fito P (2003). Calcium fortification of vegetables by vacuum impregnation. *J. Food Eng.* 56:279–284.

Greener I and Fennema O (1994). Edible films and coatings: Characteristics, formation, definitions, and testing methods. In: Krochta JM, Baldwin EA, and Nisperos-Carriedo MO (eds.), *Edible Coatings and Films to Improve Food Quality.* Lancaster, PA: Technomic Publishing Company, Inc., pp. 1–24.

Greener IK and Fennema O (1989). Barrier properties and surface characteristics of edible, bilayer films. *J. Food Sci.* 54:1393–1399.

Guilbert S and Gontard N (1995). Edible and biodegradable food packaging. In: Ackermann P, Jagerstad M, and Ohlsson T (eds.), *Foods and Packaging Materials—Chemical Interactions.* Cambridge, England: The Royal Society of Chemistry, pp. 159–168.

Guilbert S, Gontard N, and Gorris LGM (1996). Prolongation of the shelf-life of perishable food products using biodegradable films and coatings. *LWT—Food Sci. Technol.* 29:10–17.

Hagenmaier RD (2000). Evaluation of a polyethylene–candelilla coating for 'Valencia' oranges. *Postharvest Biol. Technol.* 19:147–154.

Hagenmaier RD and Shaw PE (1990). Moisture permeability of edible films made with fatty acid and (hydroxy-propyl) methyl cellulose. *J. Agric. Food Chem.* 38:1799–1803.

Han C, Lederer C, McDaniel M, and Zhao Y (2004a). Sensory evaluation of fresh strawberries (*Fragaria ananassa*) coated with chitosan-based edible coatings. *J. Food Sci.* 70:172–178.

Han C, Zhao Y, Leonard SW, and Traber MG (2004b). Edible coatings to improve storability and enhance nutritional value of fresh and frozen strawberries (*Fragaria ananassa*) and raspberries (*Rubus idaeus*). *Postharvest Biol. Technol.* 33:67–78.

Han JH (2002). Protein-based edible films and coatings carrying antimicrobial agents. In: Gennadios A (ed.), *Protein-Based Films and Coatings.* Boca Raton, FL: CRC Press, pp. 485–499.

Han JH and Gennadios A (2005). Edible films and coatings: A review, Chapter 15. In: Han JH (ed.), *Innovations in Food Packaging.* Amsterdam, the Netherlands: Elsevier Academic Press, pp. 239–262.

Hernandez E (1994). Edible coatings from lipids and resins. In: Krochta JM, Baldwin EA, and Nisperos-Carriedo MO (eds.), *Edible Coatings and Films to Improve Food Quality.* Lancaster, PA: Technomic Publishing Co., Inc., pp. 279–303.

Hernandez-Munoz P, Almenar E, Ocio MJ, and Gavara R (2006). Effect of calcium dips and chitosan coatings on postharvest life of strawberries (*Fragaria ananassa*). *Postharvest Biol. Technol.* 39:247–253.

Hernandez-Munoz P, Almenar E, Valle VD, Velez D, and Gavara R (2008). Effect of chitosan coating combined with postharvest calcium treatment on strawberry quality during refrigerated storage. *Food Chem.* 110:428–435.

Hettiarachchy NS and Satchithanandam E (2007). Inventors; The Board of Trustees for the University of Arkansas, assignee. January 1, 2007. Organic acids incorporated edible antimicrobial films. U.S. Patent 7,160,580.

Hoa TT, Ducamp MN, Lebrun M, and Baldwin EA (2002). Effect of different coating treatments on the quality of mango fruit. *J. Food Quality* 25:471–468.

Howard LR and Dewi T (1995). Sensory, microbiological and chemical quality of mini-peeled carrots as affected by edible coating treatment. *J. Food Sci.* 60:142–144.

Howard LR and Dewi T (1996). Minimal processing and edible coating effects on composition and sensory quality of mini-peeled carrots. *J. Food Sci.* 61:643–645.

Howard LR and Griffin LE (1993). Lignin formation and surface discoloration of minimally processed carrot sticks. *J. Food Sci.* 58:1065–1067.

Howard LR, Griffin LE, and Lee Y (1994). Steam treatment of minimally processed carrot sticks to control surface discoloration. *J. Food Sci.* 59:356–358.

Hui-Min J, To H, Li-Ping L, and Hai-Ying Z (2009). Effects of edible coatings on browning of fresh-cut peach fruits. *Trans. Chin. Soc. Agric. Eng.* 25:282–286.

Iyidogan NF and Bayindirli A (2004). Effect of L-cysteine, kojic acid and 4-hexylresorcinol combination on inhibition of enzymatic browning in Amasya apple juice. *J. Food Eng.* 62:299–304.

Jiang Y and Li Y (2001). Effect of chitosan coating on postharvest life and quality and longan fruit. *Food Chem.* 73:39–143.

Kaplan HJ (1986). Washing, waxing, and color adding. In: Wardowdki WF, Nagy S, and Grierson W (eds.), *Fresh Citrus Fruit*. Westport, CT: AVI Publishing Co., 379pp.

Kaynas K and Ozelkok IS (1999). Effect of Semperfresh on postharvest behavior of cucumber (*Cucumis sativus* L.) and summer squash (*Cucurbita pepo* L.) fruits. *Acta Hort.* 492:213–220.

Kester JJ and Fennema OR (1986). Edible films and coatings: A review. *Food Technol.* 40:47–59.

Kim JG, Yousef AE, and Chism GW (1999). Applications of ozone for enhancing the microbiological safety and quality of foods: A review. *J. Food Prot.* 62:1071–1087.

Kinzel B (1992). Protein-rich edible coatings for food. *Agric. Res.* 40:20–21.

Kittur FS, Saroja N, Habibunnisa S, and Tharanathan RN (2001). Polysaccharide based composite coating formulations for shelf life extension of fresh banana and mango. *Eur. Food Res. Technol.* 213:306–311.

Korhonen H (2005). Technology options for new nutritional concepts. *Int. J. Dairy Technol.* 55:79–88.

Krasaekoopt W and Mabumrung J (2008). Microbiological evaluation of edible coated fresh-cut cantaloupe. *Kasetsart J. Nat. Sci.* 42:552–557.

Krochta JM, Avena-Bustillos RJ, Cisneros-Zevallos LA, and Saltveit ME (1993). Optimization of edible coatings on minimally processed carrots using response surface methodology. *Trans. Am. Soc. Agric. Eng.* 36:801–805.

Krochta JM, Baldwin EA, and Nisperos-Carriedo MO (1994). *Edible Coatings and Films to Improve Food Quality*. Lancaster, PA: Technomic Publishing.

Krochta JM and DeMulder-Johnston C (1997). Edible and biodegradable polymer films: Challenges and opportunities. *Food Technol.* 51:61–74.

Krochta JM, Hudson JS, Camirand WM, and Pavlath AE (1988). Edible films for lightly processed fruits and vegetables. *Am. Soc. Agric. Eng.* 88:6523.

Krochta JM, Pavlath AE, and Goodman N (1990). Edible films from casein-lipid emulsions for lightly-processed fruits and vegetables. In: Spiess WE and Schubert H (eds.), *Engineering and Food*, Vol. 2: *Preservation Processes and Related Techniques*. New York: Elsevier Science Publishers, pp. 329–340.

Krumel KL and Lindsay TA (1976). Nonionic cellulose ethers. *Food Technol.* 30:36–38, 40, 43.

Kyu Kyu WN, Jitareerat P, Kanlayanarat S, and Sangchote S (2007). Effects of cinnamon extract, chitosan coating, hot water treatment and their combinations on crown rot disease and quality of banana fruit. *Postharvest Biol. Technol.* 45:333–340.

Lafortune R, Caillet S, and Lacroix M (2005). Combined effects of coating, modified atmosphere packaging, and gamma irradiation on quality maintenance of ready-to-use carrots (*Daucus carota*). *J. Food Prot.* 68:353–359.

Landmann W, Lovegren NV, and Feuge RO (1960). Permeability of some fat products to moisture. *J. Am. Oil Chem. Soc.* 37:1–4.

Lau OL and Meheriuk M (1994). The effect of edible coatings on storage quality of Mcintosh, Delicious and Spartan apples. *Can. J. Plant Sci.* 74:847–852.

Lawrence JF and Iyengar JR (1983). Determination of paraffin wax and mineral oil on fresh fruits and vegetables by high temperature gas chromatography. *J. Food Sci.* 5:119–129.

LeClair K (2000). Breaking the sensory barrier for functional foods. *Food Prod. Des.* 7:59–63.

Lee JY, Park HJ, Lee CY, and Choi WY (2003). Extending shelf-life of minimally processed apples with edible coatings and antibrowning agents. *LWT—Food Sci. Technol.* 36:323–329.

Lerdthanangkul S and Krochta JM (1996). Edible coating effects on post harvest quality of green bell peppers. *J. Food Sci.* 61:176–179.

Le Tien C, Vachon C, Mateescu MA, and Lacroix M (2001). Milk protein coatings prevent oxidative browning of apples and potatoes. *J. Food Sci.* 66:512–516.

Li H and Yu T (2000). Effect of chitosan on incidence of brown rot, quality and physiological attributes of postharvest peach fruit. *J. Sci. Food Agric.* 81:269–274.

Li P and Barth MM (1998). Impact of edible coatings on nutritional and physiological changes in lightly processed carrots. *Postharvest Biol. Technol.* 14:51–60.

Li Q, Dunn ET, Grandmaison EW, and Goosen MFA (1992). Applications and properties of chitosan. *J. Bioact. Compatible Polym.* 7:370–397.

Liu J, Tian S, Meng X, and Xu Y (2007). Effects of chitosan on control of postharvest diseases and physiological responses of tomato fruit. *Postharvest Biol. Technol.* 44:300–306.

Lopez-Galvez G, Saltveit ME, and Cantwell MI (1996). The visual quality of minimally processed lettuce stored in air or controlled atmospheres with emphasis on romaine and iceberg types. *Postharvest Biol. Technol.* 8:179–190.

Lowings PH and Cutts DF (1982). The preservation of fresh fruits and vegetables. In: *Proceedings of the Institute of Food Science and Technology Annual Symposium*, Nottingham, U.K., July 1981, 52pp.

Luna-Guzman I and Barrett DM (2000). Comparison of calcium chloride and calcium lactate effectiveness in maintaining shelf stability and quality of fresh-cut cantaloupe. *Postharvest Biol. Technol.* 19:61–72.

Luna-Guzman I, Cantwell M, and Barrett DM (1999). Fresh-cut cantaloupe: Effects of $CaCl_2$ dips and heat treatments on firmness and metabolic activity. *Postharvest Biol. Technol.* 17:201–213.

Maftoonazad N and Ramaswamy HS (2005). Postharvest shelf-life extension of avocados using methylcellulose edible coating. *LWT—Food Sci. Technol.* 38:617–624.

Main GL, Morris JR, and Wehunt EJ (1986). Effect of preprocessing treatment on the firmness and quality characteristics of whole and sliced strawberries after freezing and thermal processing. *J. Food Sci.* 51:391–394.

Maqbool M, Ali A, Ramachandran S, Smith DR, and Alderson PG (2010). Control of postharvest anthracnose of banana using a new edible composite coating. *Crop Prot.* 29:1136–1141.

Mason DF (1969). Fruit preservation process. U.S. Patent 3,472,662.

Mate JI and Krochta JM (1995). Effect of WPI coatings on the oxygen uptake of dry roasted peanuts. In: *Annual Meeting of the Institute of Food Technologists*, Anaheim, CA, June 3–7, 1995.

McGuire R and Hagenmaier R (2001). Shellac formulation to reduce epiphytic survival of coliform bacteria on citrus fruit postharvest. *J. Food Prot.* 64:1756–1760.

McHugh TH, Aujard JF, and Krochta JM (1994). Plasticized whey protein edible films: Water vapor permeability properties. *J. Food Sci.* 59:416–419.

McHugh TH and Krochta JM (1994). Water-vapor permeability properties of edible whey protein-lipid emulsion films. *J. Am. Oil Chem. Soc.* 71:307–312.

McHugh TH and Senesi E (2000). Apple wraps: A novel method to improve the quality and extend the shelf life of fresh-cut apples. *J. Food Sci.* 65:480–485.

Meheriuk M and Lau LO (1988). Effect of two polymeric coatings on fruit quality of *Bartlett* and *Anjou* pears. *J. Am. Soc. Hortic. Sci.* 113:226–227.

Mei Y, Zhao Y, Yang J, and Furr HC (2002). Using edible coating to enhance nutritional and sensory qualities of baby carrots. *J. Food Sci.* 67:1964–1968.

Meyers MA (1990). Functionality of hydrocolloids in batter coating system. In: Kulp K and Loewe R (eds.), *Batters and Breadings in Food Processing*. St. Paul, MN: American Association of Cereal Chemists, pp. 117–141.

Miller WR, Spalding DL, and Risse LA (1983). Decay firmness and color development of Florida bell peppers dipped in chlorine and imazalil and film wrapped. *Proc. Florida State Hortic. Soc.* 96:347–350.

Min S and Krochta JM (2005). Antimicrobial films and coatings for fresh fruit and vegetables. In: Jongen W (ed.), *Improving the Safety of Fresh Fruit and Vegetables*. New York: CRC Press, pp. 455–492.

Montero-Calderon M, Rojas-Grau MA, and Martin-Belloso O (2008). Effect of packaging conditions on quality and shelf-life of fresh-cut pineapple (*Ananas comosus*). *Postharvest Biol. Technol.* 50:182–189.

Morgan BH (1971). Edible packaging update. *Food Prod. Dev.* 5:75–77, 108.

Morris JR, Sistrunk WA, Sims CA, Main GL, and Wehunt EJ (1985). Effects of cultivar, postharvest storage, pre-processing dip treatments and style of pack on the processing quality of strawberries. *J. Am. Soc. Hortic. Sci.* 110:172–177.

Motlagh HF and Quantick PC (1998). Effect of permeable coatings on the storage life of fruits. I. Pro-long treatment of limes (*Citrus aurantifolia* cv. Persian). *Int. J. Food Sci. Technol.* 23:99–105.

Munoz PH, Almenar E, Valle VD, Velez D, and Gavar R (2008). Effect of chitosan coating combined with postharvest calcium treatment on strawberry (*Fragaria ananassa*) quality during refrigerated storage. *J. Food Chem.* 110:428–435.

Nisperos-Carriedo MO (1994). Edible films and coatings based on polysaccharides. In: Krochta JM, Baldwin EA, and Nisperos-Carriedo MO (eds.), *Edible Coatings and Films to Improve Food Quality*. Lancaster, PA: Technomic Publishing Co., Inc., pp. 305–335.

Nisperos-Carriedo MO and Baldwin EA (1990). Edible coatings for fresh fruits and vegetables. In: *Subtropical Technology Conference Proceedings*, Lake Alfred, FL, October 18, 1990.

Nisperos-Carriedo MO, Baldwin EA, and Shaw PE (1992). Development of an edible coating for extending postharvest life of selected fruits and vegetables. *Florida State Hort. Soc.* 104:122–125.

Nussinovitch A (2003). *Water Soluble Polymer Applications in Foods*. Oxford, U.K.: Blackwell Science.

Olivas GI and Barbosa-Canovas GV (2005). Edible coatings for fresh-cut fruits. *Crit. Rev. Food Sci. Nutr.* 45:657–670.

Olivas GI, Mattinson DS, and Barbosa-Canovas GV (2007). Alginate coatings for preservation of minimally processed 'Gala' apples. *Postharvest Biol. Technol.* 45:89–96.

Olivas GI, Rodriguez JJ, and Barbosa-Canovas GV (2003). Edible coatings composed of methylcellulose, stearic acid, and additives to preserve quality of pear wedges. *J. Food Process. Preserv.* 27:299–320.

Oms-Oliu G, Soliva-Fortuny R, and Martin-Belloso O (2008a). Using polysaccharide-based coatings to enhance quality and antioxidant properties of fresh-cut melon. *LWT—Food Sci. Technol.* 41:1862–1870.

Oms-Oliu G, Soliva-Fortuny R, and Martin-Belloso O (2008b). Edible coatings with antibrowning agents to maintain sensory quality and antioxidant properties of fresh-cut pears. *Postharvest Biol. Technol.* 50:87–94.

Ouattara B, Simard R, Piette G, Begin A, and Holley RA (2000). Inhibition of surface spoilage bacteria in processed meats by application of antimicrobial films prepared with chitosan. *Int. J. Food Microbiol.* 62:139–148.

Park HJ (1999). Development of advanced edible coatings for fruits. *Trends Food Sci. Technol.* 10:254–260.

Park HJ (2005). Edible coatings for fruit. In: Jongen W (ed.), *Fruit and Vegetable Processing*. Boca Raton, FL: CRC Press LLC.

Park HJ, Chinnan MS, and Shewfelt RL (1994a). Edible coating effects on storage life and quality of tomatoes. *J. Food Sci.* 59:568–570.

Park HJ, Chinnan MS, and Shewfelt RL (1994b). Edible corn-zein film coating to extend storage life of tomatoes. *J. Food Process. Preserv.* 18:317–331.

Park HJ, Rhim JW, and Lee HY (1996). Edible coating effects on respiration and storage life of "Fuji" apples and 'Shingo' pears. *Food Biotechnol.* 5:59–63.

Park S, Stan SD, Daeschel MA, and Zhao Y (2005). Antifungal coatings on fresh strawberries (*Fragaria ananassa*) to control mold growth during cold storage. *J. Food Sci.* 70:202–207.

Park SI, Daeschel MA, and Zhao Y (2004). Functional properties of antimicrobial lysozyme–chitosan composite films. *J. Food Sci.* 69:215–221.

Park SI and Zhao Y (2004). Incorporation of a high concentration of mineral or vitamin into chitosan-based films. *J. Agric. Food Chem.* 52:1933–1939.

Parra DF, Tadini CC, Ponce P, and Lugao A (2004). Mechanical properties and water vapor transmission in some blends of cassava starch edible films. *Carbohydr. Polym.* 58:475–481.

Pen LT and Jiang YM (2003). Effects of chitosan coating on shelf life and quality of fresh-cut Chinese water chestnut. *Lebens. Wissen. Technol.* 36:359–364.

Pennisi E (1992). Sealed in (plastic) edible film. *Sci. News* 141:12.

Perez L (2003). Aplicacion de metodos combinados para el control del desarrollo del pardeamiento enzimatico en pera (variedad *Blanquilla*) mınimamente procesada, Doctoral thesis, Universidad Politecnica de Valencia, Valencia, Spain.

Perez-Gago MB, Rojas C, and del Rio MA (2003). Effect of hydroxypropyl methylcellulose-lipid edible composite coatings on plum (cv. Autumn giant) quality during storage. *J. Food Sci.* 68:879–883.

Perez-Gago MB, Serra M, Alonso M, Mateos M, and del Rıo MA (2005). Effect of whey protein- and hydroxypropyl methylcellulose-based edible composite coatings on color change of fresh-cut apples. *Postharvest Biol. Technol.* 36:77–85.

Perez-Gago MB, Serra M, and del Rio MA (2006). Color change of fresh-cut apples coated with whey protein concentrate-based edible coatings. *Postharvest Biol. Technol.* 39:84–92.

Petracek PD, Dou H, and Pao S (1998). Influence of applied waxes on postharvest physiological behavior and pitting of white grapefruit. *Postharvest Biol. Technol.* 14:99–106.

Ponce AG, Roura SI, del Valle CE, and Moreira MR (2008). Antimicrobial and antioxidant activities of edible coatings enriched with natural plant extracts: In vitro and in vivo studies. *Postharvest Biol. Technol.* 49:294–300.

Poovaiah BW (1986). Role of calcium in prolonging storage life of fruits and vegetables. *Food Technol.* 40:86–89.

Porat R, Weiss B, Cohena L, Dausa A, and Biton A (2005). Effects of polyethylene wax content and composition on taste, quality, and emission of off-flavor volatiles in 'Mor' mandarins. *Postharvest Biol. Technol.* 38:262–268.

Posey R, Culbertson EC, and Westermeier JC (2005). Clear barrier coating and coated film. U.S. Patent, 6,911,255 B2.

Rakotonirainy AM, Wang Q, and Padua GW (2001). Evaluation of zein films as modified atmosphere packaging for fresh broccoli. *J. Food Sci.* 66:1108–1111.

Raybaudi-Massilia RM, Mosqueda-Melgar J, and Martin-Belloso O (2008a). Edible alginate-based coating as carrier of antimicrobials to improve shelf-life and safety of fresh-cut melon. *Int. J. Food Microbiol.* 121:313–327.

Raybaudi-Massilia RM, Rojas-Grau MA, Mosqueda-Melgar J, and Martin-Belloso O (2008b). Comparative study on essential oils incorporated into an alginate-based edible coating to assure the safety and quality of fresh-cut Fuji apples. *J. Food Prot.* 71:1150–1161.

Reddy MVB, Angers P, Castaigne F, and Arul J (2000). Chitosan effects on blackmold rot and pathogenic factors produced by *Alternaria alternata* in postharvest tomatoes. *J. Am. Soc. Hortic. Sci.* 125:742–747.

Restuccia D, Spizzirri UG, Parisi OI, Cirillo G, Curcio M, Iemma F et al. (2010). New EU regulation aspects and global market of active and intelligent packaging for food industry applications. *Food Control* 21:1425–1435.

Reuck KD, Sivakumar D, and Korsten L (2009). Effect of integrated application of chitosan coating and modified atmosphere packaging on overall quality retention in litchi cultivars. *J. Sci. Food Agric.* 89:915–920.

Rhim JW, Hong SI, Park HM, and Perry KW (2006). Preparation and characterization of chitosan based nanocomposite films with antimicrobial activity. *J. Agric. Food Chem.* 54:5814–5822.

Rhim JW, Lee JH, and Perry KW (2007). Mechanical and barrier properties of biodegradable soy protein isolate-based films coated with polylactic acid. *LWT—Food Sci. Technol.* 40:232–238.

Rhim JW and Ng PKW (2007). Natural biopolymer-based nanocomposite films for packaging applications. *Crit. Rev. Food Sci. Nutr.* 47:411–433.

Ribeiro C, Vicente AA, Teixeira JA, and Miranda C. 2007. Optimization of edible coating composition to retard strawberry fruit senescence. *Postharvest Biol. Technol.* 44:63–70.

Richard FC, Goupy PM, and Nicolas JJ (1992). Cysteine as an inhibitor of enzymatic browning. 2. Kinetic studies. *J. Agric. Food Chem.* 40:2108–2114.

Rocha AMCN, Brochado CM, and Morais AMMB (1998). Influence of chemical treatment on quality of cut apple (cv. Jonagored). *J. Food Quality* 21:13–28.

Rojas-Argudo C, Perez-Gago MB, and del Rio MA (2005). Postharvest quality of coated cherries cv. 'Burlat' as affected by coating composition and solids content. *Food Sci. Technol. Int.* 11:417–424.

Rojas-Grau MA, Avena-Bustillos RJ, Friedman M, Henika PR, Martin-Belloso O, and McHugh TH (2006). Mechanical, barrier, and antimicrobial properties of apple puree edible films containing plant essential oils. *J. Agric. Food Chem.* 54:9262–9267.

Rojas-Grau MA, Raybaudi-Massilia RM, Soliva-Fortuny RC, Avena-Bustillos RJ, McHugh TH, and Martin-Belloso O (2007a). Apple puree alginate edible coating as carrier of antimicrobial agents to prolong shelf-life of fresh-cut apples. *Postharvest Biol. Technol.* 45:254–264.

Rojas-Grau MA, Soliva-Fortuny R, and Martın-Belloso O (2009). Edible coatings to incorporate active ingredients to fresh-cut fruits: A review. *Trends Food Sci. Technol.* 20:438–447.

Rojas-Grau MA, Tapia MS, and Martin-Belloso O (2008). Using polysaccharide-based edible coatings to maintain quality of fresh cut Fuji apples. *LWT—Food Sci. Technol.* 41:139–147.

Rojas-Grau MA, Tapia MS, Rodriguez FJ, Carmona AJ, and Martin-Belloso O (2007b). Alginate and gellan-based edible coatings as carriers of antibrowning agents applied on fresh-cut Fuji apples. *Food Hydrocoll.* 21:118–127.

Romanazzi G, Nigro F, Ippolito A, Venere DD, and Salerno M (2002). Effects of pre- and postharvest chitosan treatments to control storage grey mold of table grapes. *J. Food Sci.* 67:1862–1867.

Rosen JC and Kader AA (1989). Postharvest physiology and quality maintenance of sliced pear and strawberry fruits. *J. Food Sci.* 54:656–659.

Ryu SY, Rhim JW, Roh HJ, and Kim SS (2002). Preparation and physical properties of zein-coated high-amylose corn starch film. *LWT—Food Sci. Technol.* 35:680–686.

Sacharow S (1972). Edible films. *Packaging* 43:6–9.

Sanchez-Gonzalez L, Gonzalez-Martınez C, Chiralt A, and Chafer M (2010). Physical and antimicrobial properties of chitosan-tea tree essential oil composite films. *J. Food Eng.* 98:443–452.

Sanchez-Gonzalez L, Vargas M, Gonzalez-Martinez C, Chiralt A, and Chafer M (2009). Characterization of edible films based on hydroxypropylmethylcellulose and tea tree Essentials oil. *Food Hydrocoll.* 23:2102–2109.

Sanderson GR (1981). Polysaccharides in foods. *Food Technol.* 35:50–57, 83.

Sangsuwan J, Rattanapanone N, and Rachtanapun P (2008). Effect of chitosan/methyl cellulose films on microbial and quality characteristics of fresh-cut cantaloupe and pineapple. *Postharvest Biol. Technol.* 49:403–410.

Santerre CR, Leach TF, and Cash JN (1989). The influence of the sucrose polyester, Semperfresh on the storage of Michigan grown 'McIntosh' and 'Golden Delicious' apples. *J. Food Process. Preserv.* 13:293–305.

Sapers GM and Miller RL (1998). Browning inhibition in fresh-cut pears. *J. Food Sci.* 63:342–346.

Sapru V and Labuza TP (1994). Dispersed phase concentration effect on water vapor permeability in composite methyl cellulose-stearic acid edible films. *J. Food Process. Preserv.* 18:359–368.

Schultz TH, Miers JC, Owens HS, and Maclay WD (1949). Permeability of pectinate films to water vapor. *J. Phys. Colloid Chem.* 53:1320–1330.

Senesi E, Galvis A, and Fumagalli G (1999). Quality indexes and internal atmosphere of packaged fresh-cut pears (Abate fetel and Kaiser varieties). *Italian J. Food Sci.* 2:111–120.

Seydim AC and Sarikus G (2006). Antimicrobial activity of whey protein based edible films incorporated with oregano, rosemary and garlic essential oils. *Food Res. Int.* 39:639–644.

Shahidi F, Arachchi JKV, and Jeon Y (1999). Food applications of chitin and chitosans. *Trends Food Sci. Technol.* 10:37–51.

Shao XF, Tu K, Tu S, and Tu J (2012). A combination of heat treatment and chitosan coating delays ripening and reduces decay in "Gala" apple fruit. *J. Food Quality* 35:83–92.

Shellhammer TH and Krochta JM (1997). Whey protein emulsion film performance as affected by lipid type and amount. *J. Food Sci.* 62:390–394.

Shewfelt RL, Prussia SE, Resurreccion AVA, Hurst WC, and Campbell DT (1987). Quality changes of vine-ripened tomatoes within the postharvest handling system. *J. Food Sci.* 52:661–672.

Shotornvit R, Rhim J, and Hong S (2009). Effect of nano-clay type on the physical and antimicrobial properties of whey protein isolate/clay composite films. *J. Food Eng.* 91:468–473.

Smilanick JL, Crisosto C, and Mlikota F (1999). Postharvest use of ozone on fresh fruit. *Perishables Handling* 99:10–14.

Smith SM, Geeson J, and Stow J (1987). Production of modified atmospheres in deciduous fruits by the use of films and coatings. *Hort. Sci.* 22:772–776.

Smith SM and Stow JR (1984). The potential of a sucrose ester coating material for improving the storage and shelf-life qualities of cox's orange pippin apples. *Ann. Appl. Biol.* 104:383–391.

Soliva-Fortuny RC, Biosca-Biosca M, Grigelmo-Miguel N, and Martin-Belloso O (2002a). Browning, polyphenol oxidase activity and headspace gas composition during storage of minimally processed pears using modified atmosphere packaging. *J. Agric. Food Chem.* 82:1490–1496.

Soliva-Fortuny RC, Grigelmo-Miguel N, Hernando I, Lluch MA, and Martin-Belloso O (2002b). Effect of minimal processing on the textural and structural properties of fresh-cut pears. *J. Agric. Food Chem.* 82:1682–1688.

Soliva-Fortuny RC, Grigelmo-Miguel N, Odriozola-Serrano I, Gorinstein S, and Martin-Belloso O (2001). Browning evaluation of ready-to-eat apples as affected by modified atmosphere packaging. *J. Agric. Food Chem.* 49:3685–3690.

Soliva-Fortuny RC, Lluch MA, Quiles A, Grigelmo-Miguel N, and Martin-Belloso O (2003). Evaluation of textural properties and microstructure during storage of minimally processed apples. *J. Food Sci.* 68:312–317.

Srinivasa PC, Revathy B, Ramesh MN, Harish Prashanth KV, and Tharanathan RN (2002). Storage studies of mango packed using biodegradable chitosan films. *Storage Technol.* 215:504–508.

Talcott ST and Howard LR (1999). Chemical and sensory quality of processed carrot puree as influenced by stress-induced phenolic compounds. *J. Agric. Food Chem.* 47:1362–1366.

Tanada-Palmu PS and Grosso CRF (2005). Effect of edible wheat gluten-based films and coatings on refrigerated strawberry (*Fragaria ananassa*) quality. *Postharvest Biol. Technol.* 36:199–208.

Tang H, Xiong H, Tang S, and Zou P (2009). A starch-based biodegradable film modified by nano silicon dioxide. *J. Appl. Polym. Sci.* 113:34–40.

Tang X, Alavi S, and Herald TJ (2008). Effect of plasticizers on the structure and properties of starch-clay nanocomposite films. *Carbohydr. Polym.* 74:552–558.

Tapia MS, Rojas-Grau MA, Carmona A, Rodriguez FJ, Soliva-Fortuny R, and Martin-Belloso O (2008). Use of alginate and gellan-based coatings for improving barrier, texture and nutritional properties of fresh-cut papaya. *Food Hydrocoll.* 22:1493–1503.

Tapia MS, Rojas-Grau MA, Rodriguez FJ, Ramirez J, Carmona A, and Martin-Belloso O (2007). Alginate- and gellan-based edible films for probiotic coatings on fresh-cut fruits. *J. Food Sci.* 72:190–196.

Tasdelen OE and Bayindirli L (1998). Controlled atmosphere storage and edible coating effects on storage life and quality of tomatoes. *J. Food Process. Preserv.* 22:303–320.

Tay SL and Perera CO (2004). Effect of 1-methylcyclopropene treatment and edible coatings on the quality of minimally processed lettuce. *J. Food Sci.* 69:131–135.

Thompson AK (2003). Postharvest treatments. In: *Fruit and Vegetables.* Ames, IA: Blackwell Publishing Ltd., pp. 47–52.

Togrul H and Arslan N (2004). Extending shelf-life of peach and pear by using CMC from sugar beet pulp cellulose as a hydrophilic polymer in emulsions. *Food Hydrocoll.* 18:215–226.

Togrul H and Arslan N (2005). Carboxymethyl cellulose from sugar beet pulp cellulose as a hydrophilic polymer in coating of apples. *J. Food Sci. Technol.* 42:139–144.

Ukai YN, Tsutsumi T, and Marakami K (1976). Preservation of agricultural products. U.S. Patent 3,997,674.

USDA—U.S. Food and Drug Administration (2006). Food additive status list. Available from http://www.cfsan.fda.gov/dms/opa-appa.html. Accessed September 25, 2008.

Vachon C, D'Aprano G, Lacroix M, and Letendre M (2003). Effect of edible coating process and irradiation treatment of strawberry *fragaria* spp. on storage-keeping quality. *J. Food Sci.* 68:608–612.

Vargas M, Albors A, Chiralt A, and Gonzalez-Martinez C (2006a). Application of chitosan-methylcellulose edible coatings to strawberry fruit. In: *Proceedings of the 13th World Congress of Food Science and Technology, IUFoST-2006 Food is Life*, pp. 389–390.

Vargas M, Albors A, Chiralt A, and Gonzalez-Martinez C (2006b). Quality of cold-stored strawberries as affected by chitosan-oleic acid edible coatings. *Postharvest Biol. Technol.* 41:164–171.

Vina SZ, Mudridge A, Garcia MA, Ferreyra RM, Martino MN, Chaves AR et al. (2007). Effects of polyvinylchloride and edible starch coatings on quality aspects of refrigerated Brussels sprouts. *Food Chem.* 103:701–709.

Watters GG and Brekke JE (1961). Stabilized raisins for dry cereal products. *Food Technol.* 15:236–238.

White GC (1992). Ozone. In: *Handbook of Chlorination and Alternative Disinfectants*, 3rd edn. New York: Van Nostrand Reinhold, pp. 1046–1110.

Williams LG (1968). Process for coating low moisture fruits. U.S. Patent 3,406,078.

Worrell DB, Carrington CMS, and Huber DJ (2002). The use of low temperature and coatings to maintain storage quality of breadfruit. *Artocarpus altilis* (Parks.) Fosb. *Postharvest Biol. Technol.* 25:33–40.

Wu T, Zivanovic S, Draughon FA, Conway WS, and Sams CE (2005). Physicochemical properties and bioactivity of fungal chitin and chitosan. *J. Agric. Food Chem.* 53:3888–3894.

Xu S, Chen X, and Sun DW (2001). Preservation of kiwifruit coated with an edible film at ambient temperature. *J. Food Eng.* 50:211–216.

Yu YW, Li H, Jinhua D, and Ren YZ (2012). Study of natural film with chitosan combining phytic acids on preservation of fresh-cutting lotus root. *J. Chin. Inst. Food Sci. Technol.* 12:131–136.

Yurdugul S (2005). Preservation of quinces by the combination of an edible coating material, Semperfresh, ascorbic acid and cold storage. *Eur. Food Res. Technol.* 220:579–586.

Zawistowski J, Biliaderis CG, and Eskin NAM (1991). Polyphenol oxidases. In: Robinson DS and Eskin NAM (eds.), *Oxidative Enzymes in Foods.* London, U.K.: Elsevier, pp. 217–273.

Zhang D and Quantick PC (1997). Effects of chitosan coating on enzymatic browning and decay during postharvest storage of litchi (*Litchi chinensis* Sonn.) fruit. *Postharvest Biol. Technol.* 12:195–202.

Zhang D and Quantick PC (1998). Antifungal effects of chitosan coating on fresh strawberries and raspberries during storage. *J. Hort. Sci. Biotechnol.* 73:763–767.

Zhang M, Xiao G, Luo G, Peng J, and Salokhe VM (2004). Effect of coating treatments on the extension of the shelf-life of minimally processed cucumber. *Int. Agrophys.* 18:97–102.

Zhong QP and Xia WS (2007). Effect of 1-methylcyclopropene and/or chitosan coating treatments on storage life and quality maintenance of Indian jujube fruit. *Food Sci. Technol.* 40:404–411.

Zhuang R, Beuchat LR, Chinnan MS, Shewfelt RL, and Huang YW (1996). Inactivation of *Salmonella montevideo* on tomatoes by applying cellulose-based films. *J. Food Prot.* 59:808–812.

20

Edible Film and Coating Applications for Fresh-Cut and Minimally Processed Fruits and Vegetables

Adriana Izquier, Maria S. Tapia, Robert Soliva-Fortuny, and Olga Martín-Belloso

CONTENTS

20.1 Introduction

Consumers' demands toward healthier, higher-quality, and more convenient plant-based foods have driven research and innovation regarding minimal processing technologies. As a result, the fresh-cut fruit and vegetable industry has rapidly grown during the last two decades and is expected to continue to grow for the next years (Toivonen, 2006).

Fruit and vegetable quality is greatly determined by preharvest, harvest, and postharvest factors. Selection of appropriate cultivars, as well as climatic and cultural conditions, highly affects the quality of raw materials (Wang et al., 2007b). Among the aspects related to harvest factors that may limit the shelf life of produce, the handling practices, the state of maturity, and the water content at harvest date outstand as the most relevant. Finally, postharvest treatments and storage conditions may substantially extend the shelf life of fruit and vegetable commodities without compromising their quality (Beaulieu et al., 2004). Keeping the fresh-like characteristics of fruits and vegetables is thus a thrilling endeavor due to the high number of involved factors. Nevertheless, fresh-cut fruits and vegetables are even more difficult to handle. Minimal processing operations start a cascade of metabolic reactions leading to accelerated ripening and senescence, off-flavors production, discoloration, texture changes, and other undesirable deleterious events that can render the product unmarketable (Baldwin and Bai, 2011). Increased respiration rates and ethylene production, together with microbial proliferation, underlie those processes profoundly impacting the stability of the plant tissues.

Packaging strategies designed for fresh-cut products aim at reducing the extent of such changes. During the last years, significant contributions have led to the development of novel packaging materials specifically adapted to the physiological features of fresh-cut commodities. In this regard, oriented polypropylene films and microperforated layers have stood out as a way for successfully keeping low oxygen concentrations inside the packages while maintaining relatively high carbon dioxide levels (Toivonen et al., 2009). Semipermeable plastic packaging systems are, though, frequently insufficient to prevent quality decay as a result of inappropriate conditions. Edible coatings offer excellent prospects for reducing quality loss of minimally processed fruits and vegetables. Countless examples of their ability to reduce water migration phenomena, prevent gas diffusion, and control microbial growth are presented in literature. In addition, more unconventional uses, for example, when acting as carriers of active ingredients, offer new perspectives for their application to lightly processed products. This chapter presents an update of the main advantages, opportunities, requirements, and limitations of edible coating applications for an improved preservation of fresh-cut fruits and vegetables.

20.2 Quality Decay in Fresh-Like Fruit and Vegetable Commodities

Fruit and vegetable tissues exhibit particular physiological responses after harvest, handling, and storage or even minimal processing operations. Physicochemical transformations underlying quality deterioration and senescence are tightly linked to respiration, transpiration, and, in some cases, ethylene production. These issues are specially increased in minimally processed fruits and vegetables. On the one hand, it is because the natural cuticle that protects the fruit is often removed. On the other hand, cutting operations may cause severe injury to plant cell tissues, first, promoting the development of a stress response, which, second, leads to deleterious changes caused by an increase in the metabolic activity of the product (Salveit, 1997). A summary of the main deteriorative changes related to quality decay is provided next.

20.2.1 Moisture Loss

Moisture losses caused by transpiration during postharvest storage of fruits and vegetables are one of the most significant phenomena with economical implications. Water stress is associated with turgidity changes and firmness decay in plant tissues, as well as with metabolic alterations triggering senescence and spoilage, flavor loss, and nutritional decay. In fresh-cut fruits and vegetables, moisture loss can be substantially increased due to the removal of the cuticle and the increment of the surface/volume ratio (Toivonen and DeEll, 2002). The use of edible formulations to control moisture transfer phenomena is one of the most relevant applications in both the fresh and fresh-cut produce fields.

20.2.2 Browning

Oxidative reactions catalyzed by enzymes and leading to color changes are especially remarkable in fruits with a high phenolic content. Polyphenol oxidase (PPO) contributes to the o-hydroxylation of monophenols to o-diphenols and their subsequent conversion to o-quinones, which in turn polymerize to form colored compounds. Some of the phenolic compounds, for example, chlorogenic acid, gallic acid, caffeic acid, catechin, and epicatechin, typically present in fruits and vegetables such as apples, pears, bananas, potatoes, eggplant, artichoke, and lettuce are oxidized through these enzymatically mediated reactions (Macheix et al., 1990). Edible coatings successfully contribute to minimize color changes in fresh-cut tissues by successfully compartmentalizing the released cellular fluids containing enzymes and substrates of browning reactions.

20.2.3 Changes in Aroma

Flavor modifications in fresh and fresh-cut fruits are frequently associated to undesirable atmosphere composition, which alters the metabolic behavior of the fruit throughout storage. Many aromatic compounds are produced and noticed only once the cellular structure is altered and enzymes are put in contact with their substrates. For instance, lipoxygenase enzymes account for the generation of volatile compounds that are determinant to the characteristic flavor of many fruits. Edible coatings may be used to control these deteriorative phenomena by generating atmosphere conditions yielding appropriate internal oxygen and carbon dioxide concentrations in the plant tissues. The evolvement of aromatic volatiles by controlling enzymatic processes is also possible with this strategy (Rojas-Graü et al., 2011a).

20.2.4 Microbial Decay

Microbial spoilage of fruits and vegetables is a major cause of postharvest losses. Microbial contamination can occur in any step of the production chain and may be associated to preharvest, harvest, and postharvest factors (Beuchat, 1996). Fresh-cut fruits and vegetables are especially susceptible to microbial decay because of the release of nutrients, as a consequence of cell injuring, and the exposure to conditions that favor the proliferation of microorganisms. Molds and yeasts are the naturally occurring microorganisms in fruits because of their relatively low pH. However, both fruits and vegetables can be contaminated with bacterial species such as *Pseudomonas* spp., *Enterobacter agglomerans*, *Leuconostoc mesenteroides*, and *Lactobacillus* spp., among others (Zagory, 1999). These microorganisms, if found in high counts, are responsible for the production of deteriorative enzymes that promote cell wall modifications and thus texture decay. Pathogenic microorganisms can grow on low acidic fruits such as melon and watermelon and on several vegetables because of their relative high pH values. Edible films may be used to control microbial growth by limiting exposure of the cut tissues to deleterious conditions. Nevertheless, the reduction of microbial counts through the design of systems that are capable of carrying and successfully delivering antimicrobial compounds comes to be a promising alternative for the development of safe and high-quality fresh-cut products.

20.3 Applications of Edible Films and Coatings for Fresh-Like Fruits and Vegetables

The use of edible films and coatings to extend microbial, nutritional, and sensory quality of fresh-cut or minimally processed fruits and vegetables has been studied extensively and is reviewed recurrently in order to keep up with new approaches on the traditional uses of biopolymer materials: active food packaging, as well as composite films with enhanced antimicrobial, antioxidant, and barrier properties (Lin and Zhao, 2007; Vargas et al., 2008; Embuscado and Huber, 2009; Rojas-Grau et al., 2009; González-Aguilar et al., 2010; Janjarasskul and Krochta, 2010; Oms-Oliu et al., 2010; Skurtys et al., 2010; Falguera et al., 2011; Fernández-Pan and Maté Caballero, 2011; Valencia-Chamorro et al., 2011; Dhanapal et al., 2012; Dhall, 2013; Moncayo et al., 2013; Pascall and Lin, 2013). Diverse natural antimicrobial agents as well as enhanced methodologies have been proposed. Among these, one of the most outstanding approaches is the use of nanocomposite materials to improve mechanical and transport properties, thus allowing the optimization of the concentration of functional ingredients and a reduction of their impact on the sensory attributes of the treated commodities. The ultimate goal of this approach is to broaden the application and efficiency of films and coatings for the development of lightly processed fruits and vegetables.

Hydrocolloid materials have been studied extensively for the formation of edible films and coatings in foods. They are hydrophilic polymers of vegetable, animal, microbial, or synthetic origin, containing many hydroxyl groups. These substances are usually polyelectrolytes, capable of forming a cohesive structure, comprising animal- or plant-based proteins, polysaccharides, and composite combinations of two or more of these components. Their selection depends on their capability to act as a barrier against microorganisms and as a semipermeable barrier to gases and water vapor, providing modified atmosphere conditions that decrease respiration, and thus protect the flavor, texture, and color of the product. Diverse polymeric materials have been applied over a variety of fresh-like fruits and vegetables, which have been discussed elsewhere (Wong et al., 1994; Baldwin et al., 1995a,b; Nisperos-Carriedo and Baldwin, 1996; Olivas and Barbosa-Cánovas, 2005, 2009; Lin and Zhao, 2007; Valencia-Chamorro et al., 2011; de Azevedo, 2012).

Edible films and coatings carrying antimicrobial and antioxidant compounds represent a novel fashion to improve the safety and shelf life of food systems. Antimicrobials and antioxidants suitable for incorporation into edible films and coatings include organic acids, fatty acid esters, polypeptides, plant essential oils (EOs), antioxidants, colorants, flavors, and texture enhancers (Petersen et al., 1999; Franssen and Krochta, 2003; Olivas and Barbosa-Cánovas, 2005; Pranoto et al., 2005; Conte et al., 2009; González-Aguilar et al., 2010; Ahmed et al., 2012; Fagundes et al., 2013). Nowadays, the list has increased significantly.

Table 20.1 compiles significant developments involving the use of various hydrocolloids as coatings for fruits and vegetables, incorporating diverse compounds and different approaches to improve characteristics of films and keeping quality of coated produces.

20.3.1 Polysaccharide Layers

Polysaccharides are a group of polymers frequently utilized as a base component of biodegradable films and coatings that include materials such as cellulose and starch (and their derivatives), chitosan (CH), seaweed extracts (carrageenan and alginates), plant exudates (arabic gum), seed extracts (guar gum), compounds obtained by microbial fermentation (xanthan and gellan gum), and pectin (Han and Gennadios, 2005). Polysaccharide-based films are excellent oxygen, aroma, and oil barriers and provide good strength and structural integrity, but are not good moisture barrier due to their hydrophilic nature (Kester and Fennema, 1986; Krochta, 2002).

20.3.1.1 Cellulose Derivatives

The inexpensive and abundant cellulose can be used as a coating matrix. This natural polymer can yield by etherification of four soluble polysaccharides, methylcellulose (MC), hydroxypropyl cellulose (HPC), hydroxypropyl methylcellulose (HPMC), and carboxymethylcellulose (CMC), which overcome limitations associated with native cellulose. CMC forms flexible and transparent films, with moderate strength, resistant to oil and fat migration that act as semipermeable barriers to moisture and oxygen (Hagenmaier and Shaw, 1990). These derivatives tend to produce films that have moderate strength, are resistant to oils and fats, and are flexible, transparent, flavorless, colorless, tasteless, and water soluble, with moderate moisture and oxygen transmission properties (Krochta and Mulder-Johnson, 1997; Bourtoom, 2008) but have poor water vapor barrier and mechanical properties, which could be overcome by incorporation of hydrophobic compounds such as fatty acids (Morillon et al., 2002). Edible coatings made of CMC, MC, HPC, and HPMC have been applied to some fruits and vegetables for providing barriers to oxygen, oil, or moisture transfer. Cellulose nanoreinforcements have been proposed to improve mechanical and barrier properties of biopolymers, whose performance is usually poor when compared to those of synthetic polymers. HPMC-based films have promising applications in the food industry because of their environmental appeal, low cost, flexibility, and transparency. Nevertheless, their mechanical and moisture barrier properties should be improved, and this has been investigated with good results, reinforcing the films with microcrystalline cellulose at the nanoscale level (Bilbao-Sáinz et al., 2010).

Nowadays there are cellulose products on the market for specific application of fruit and vegetable produce. Nature Seal™ is a coating formulation composed of cellulose derivatives and blends of vitamins and minerals with a claimed shelf life extension of 21 days for fresh-cut fruits and vegetables. The product has been the subject of multiple studies; for instance, a combination of Nature Seal and soy protein

TABLE 20.1

Application of Different Biopolymer-Based Coatings on Minimally Processed Fruits and Vegetables

Fruit/Vegetable	Coating Material	Additives or Condition	Main Results	References
Fruits				
Fresh-cut melon	Alginate	Citral, CH	Quality and safety of fresh-cut melon preserved.	Poverenov et al. (2014a)
Pomegranate arils	*Aloe vera* gel	Ascorbic acid, citric acid	Microbial spoilage reduced and quality parameters such as firmness, color, and bioactive compounds maintained through storage.	Martínez-Romero et al. (2013)
"Hayward" kiwifruit slices	*Aloe vera* gel		Reduction in respiration rates and microbial spoilage and improved texture and color quality.	Benítez et al. (2013)
Minimally processed prickly pear	Guar gum or xanthan gum		Weight loss and microbial growth reduction; less softening; increase in carotenoids content.	Mohamed et al. (2013)
Fresh-cut "Fuji" apple	Pullulan	Glutathione, chitooligosaccharides	Delay of enzymatic browning, firmness loss, and weight loss. Microbial growth and respiration rate were slowed down during storage.	Wu and Chen (2013)
Fresh-cut tomatoes	Delactosed whey permeate		Reduction of overall microbial counts, and texture changes through 10 days; higher levels of vitamin C, total phenols, and antioxidant activity.	Ahmed et al. (2012)
Slices apples (cv. Gala)	CH or zein		Increased firmness preservation in zein-coated samples. Both coatings accelerated browning.	Assis et al. (2012)
Fresh-cut pineapples	Alginate, gellan	Sunflower oil	Optimal formulations reduced weight loss and respiration rate and maintained firmness of coated samples for 10 days.	Azarakhsh et al. (2012)
Strawberries and apricots	Beeswax–coconut oil, sunflower oil	Coconut oil, monolaurin	Coatings reduced vitamin C loss and prevented dehydration.	Mladenoska (2012)
Fresh-cut pineapple	Cassava starch	Pretreatment with ascorbic acid, citric acid, calcium lactate	Reduction in respiration rate, weight loss, and juice leakage. Firmness and sensory acceptance were maintained. Shelf life was not substantially extended due to increased browning and vitamin C loss.	Bierhals et al. (2011)
Apple slices (var. Fuji)	Shellac and AG (separately and in combination)	Pretreatment with ascorbic acid, citric acid, and sodium benzoate for 10 min	Reduced respiration and ethylene synthesis rates as well as electrolyte leakage. AG coatings reduced PPO and POD activities followed by composite (shellac + AG) and shellac coatings. Overall quality was preserved.	Chauhan et al. (2011)

(Continued)

TABLE 20.1 (*Continued*)

Application of Different Biopolymer-Based Coatings on Minimally Processed Fruits and Vegetables

Fruit/Vegetable	Coating Material	Additives or Condition	Main Results	References
Fresh-cut "Tommy Atkins" mango	Cassava starch or sodium alginate	Glycerol, citric acid (pretreatment)	Weight loss and respiration rate were reduced and texture and color parameters were better preserved. Sodium alginate coatings did not maintain quality characteristics.	Chiumarelli et al. (2011)
Fresh-cut melon	Pectin	Osmotic dehydration with 40°Bx sucrose solution containing calcium lactate	Improved fruit sensory acceptance; reduced respiration rates; maintenance of quality parameters through 14 days.	Ferrari et al. (2011)
Minimally processed "Hayward" kiwifruit	Sodium alginate under passive and active modified atmosphere packaging (MAP)	Dipping into an hydroalcoholic solution or including in the coating formulation grape fruit extract solution	Reduced dehydration and respiratory activity under both passive and active MAP; increase of the sensorial acceptability limit.	Mastromatteo et al. (2011)
Fresh-cut "Fuji" apples	CH	Ascorbic acid, calcium chloride	Delay of enzymatic browning; reduced initial respiration rate and tissue softening; mediocre water vapor barrier properties.	Qi et al. (2011)
Fresh-cut pears	CH or carboxymethyl CH (CMCH) coating	Sodium chloride dipping previous to coating	Tissue firmness and weight were kept with both coatings. CH coatings accelerated discoloration of cut surfaces and increased PPO activity; CMCH coatings prevented the browning reaction and inhibited PPO activity.	Xiao et al. (2011)
Fresh-cut mango	Cassava starch	Citric acid dipping	Decreased respiration rate and improved color and texture; however, incorporation of glycerol promoted weight loss, impaired fruit texture, increased carotenogenesis, and favored microbial growth during storage.	Chiumarelli et al. (2010)
Minimally processed strawberry	Cassava starch	Presence or not of potassium sorbate	Coatings without potassium sorbate did not cause changes in mechanical properties and sensory attributes; respiration was reduced and water vapor resistance increased with increased concentrations of starch in the formulation.	García et al. (2010)
Fresh-cut "Tommy Atkins," "Kent," and "Keitt" mangoes	CMC, CH, or carrageenan	Previous dipping calcium ascorbate, citric acid and N-acetyl-L-cysteine	CMC maintained the visual quality; carrageenan or CH caused modification of the color attributes.	Plotto et al. (2010)

(Continued)

TABLE 20.1 (Continued)

Application of Different Biopolymer-Based Coatings on Minimally Processed Fruits and Vegetables

Fruit/Vegetable	Coating Material	Additives or Condition	Main Results	References
Fresh-cut mango	Mango puree	MAP	Shelf life extension similar to that obtained under MAP conditions; the coating enhanced quality and consumer acceptance.	Sothornvit and Rodsamran (2010)
Fresh-cut banana (cv. Cavendish)	Carrageenan	Dipping (calcium chloride, ascorbic acid, and cysteine) and controlled atmosphere	Overall quality attributes maintained for 5 days at 5°C.	Bico et al. (2009)
Fresh-cut papaya "Maradol"	CH of low (LMWC), medium (MMWC), and high molecular weights (HMWC)		MMWC maintained an attractive fruit color, firmness, and low microbial load.	Gonzalez-Aguilar et al. (2009)
Minimally processed strawberries	CH	MAP	The growth of microorganisms was controlled without affecting the visual appearance and the overall sensory quality.	Campaniello et al. (2008)
Minimally processed Royal Gala apples	Sodium alginate or cassava starch	Ascorbic acid, citric acid, sodium chloride, calcium chloride	Alginate coatings were the most effective. Respiration rate was decreased by 38% and ethylene production by 50%.	Fontes et al. (2008)
Fresh-cut cantaloupe	MC and CH		Incorporation of CH in the MC coating improved the microbiological quality.	Krasaekoopt and Mabumrung (2008)
Fresh-cut pears	Alginate, pectin, or gellan	N-Acetylcysteine and glutathione	Water vapor resistance was increased and ethylene production inhibited; pectin and alginate coatings better maintained the sensory attributes.	Oms-Oliu et al. (2008a)
Fresh-cut "Fuji" apples	Alginate or gellan	Calcium chloride as a cross-linked, N-acetylcysteine	Reduced ethylene production; increased firmness and color; decreased microbial deterioration.	Rojas-Grau et al. (2008)
Fresh-cut "Piel de Sapo" melón	Alginate, pectin, or gellan		Increased water vapor resistance; reduced ethylene production; better preservation of sensory attributes with pectin or alginate coatings.	Oms-Oliu et al. (2008b)
Fresh-cut "Piel de Sapo" melon	Alginate	Malic acid, EOs and their main active compounds	Shelf life improved in view of microbiological and physicochemical aspects.	Raybaudi-Massilia et al. (2008)
Minimally processed mangoes	Mango film		Extended shelf life; however, its highly hydrophilic character limited its application.	Sothornvit and Rodsamran (2008)
Vegetables				
Fresh-cut broccoli	CH	BC (pomegranate, resveratrol), EO (tea tree, rosemary, pollen, propolis)	Reduced microbial populations without affecting sensory attributes.	Alvarez et al. (2013)

(Continued)

TABLE 20.1 (*Continued*)

Application of Different Biopolymer-Based Coatings on Minimally Processed Fruits and Vegetables

Fruit/Vegetable	Coating Material	Additives or Condition	Main Results	References
Fresh-cut carrots	Tapioca starch/decolorized hsian-tsao leaf gum (dHG)	Cinnamon EO or grape seed extract (GSE)	Visual appearance was maintained and white blush formation retarded.	Lai et al. (2013)
Minimally processed carrots	Sodium alginate	Treatment of dipping into hydroalcoholic solution before and after coating	Coatings preserved dehydration and delayed respiration and microbial proliferation, thus extending its shelf life.	Mastromatteo et al. (2012)
Minimally processed broccoli	CH or CMC, with or without previous application of a mild heat shock		Reduced weight loss; inhibition of florets opening and yellowing; better retention of green color; lower degradation of ascorbic acid and chlorophyll retention with CH, which also reduced microbial growth and preserved firmness.	Ansorena et al. (2011)
Fresh-cut broccoli	CH or CMC combined or not with previous application of mild heat shock		CH coating inhibited the florets opening and controlled microbiological counts, except for lactic acid bacteria; CMC coating did not exert any antibacterial effect.	Moreira et al. (2011)
Carrot sticks	CH	MAP	Overall visual quality was preserved. Respiration rate was increased; vitamin C was reduced and carotenoids content of carrots decreased through storage. Microbial loads not affected by the coating.	Simões et al. (2009)
Fresh-cut carrot slices	HPMC	Surfactant mixtures of sorbitan monostearate and sucrose palmitate	Color mainly affected by the solvent dispersion media; the changes induced for aqueous dispersion were more significant.	Villalobos-Carvajal et al. (2009)
Fresh-cut carrot	CH (HMWC)-based edible coating applied by immersion and by vacuum impregnation	MC or oleic acid	Control of the occurrence of white blush during storage; coating application with a vacuum pulse improved water vapor resistance and better preserved color and mechanical properties.	Vargas et al. (2009)
White asparagus	CMC, whey protein isolate, and pullulan	Sucrose fatty acid esters, polyethylenoglycol, sorbitol, and stearic acid	Beneficial impact on quality by retarding weight loss; reduced hardening and purple color development.	Tzoumaki et al. (2009)
Minimally processed lampascioni	Sodium alginate	Dipping in a solution of citric acid and calcium chloride	Reduced respiratory activity, browning, and microbial growth.	Conte et al. (2009)
Minimally processed garlic quality	Agar-agar	CH and acetic acid	Increased color control; reduced moisture loss and respiration rate; the incorporation of CH reduced the water vapor transmission.	Geraldine et al. (2008)

coatings carrying antibrowning agents and preservatives prolonged the shelf life of cut apples by 1 week when stored at 4°C (Baldwin et al., 1996). Some further investigations on cellulose-based edible coatings for fruit and vegetables are presented in Table 20.1.

20.3.1.2 Pectin

Pectin is a water-soluble anionic heteropolysaccharide obtained from fruit and vegetables, readily modified through demethylation (Marudova et al., 2005). In general, pectin forms films that are poor moisture barrier (Baldwin, 2007). Due to its biodegradability, biocompatibility, edibility, and versatile chemical and physical properties (such as gelation, selective gas permeability), pectin is a suitable polymeric matrix for the elaboration of edible films intended as active food packaging. Pérez-Espitia et al. (2014) published an extensive review on edible films from pectin. In 1996, McHugh et al. developed the first edible films made from fruit purees (rich in pectin) and characterized their water vapor and oxygen permeability properties. Apple-based edible films were excellent oxygen barriers, particularly at low to moderate relative humidities (RHs) but were not very good moisture barriers. Later, McHugh and Senesi (2000) added various concentrations of fatty acids, fatty alcohols, beeswax, and vegetable oil to apple pieces coated with apple puree or wrapped in preformed films. Lipid addition significantly reduced the water vapor permeability of apple films. Apple-based wraps significantly reduced moisture loss and browning in fresh-cut apples. This work has evolved in the development of antimicrobial films adding various plant EOs as natural antibacterial agents (Rojas-Graü et al., 2006). Besides apple puree, antimicrobial films can also be obtained from broccoli, tomato, carrot, mango, peach, pear, and a variety of other produce items. Nonantimicrobial versions of these food wraps are now being made commercially by California-based Origami Foods® in cooperation with the USDA for use in a small but growing number of food applications, including sushi wraps. Nanocomposite edible films from fruit purees have also been developed by adding cellulose nanofibers in different concentrations as nanoreinforcement (Azeredo et al., 2009). Some recent studies with pectin are presented in Table 20.1.

20.3.1.3 Alginate and Gellan

Alginate, isolated from brown seaweed, possesses excellent colloidal properties and, in association with divalent cations, yields strong, uniform, translucent, glossy films effective for coating minimally processed fruits and vegetables (Olivas et al., 2007). These films are characterized by a reduced water resistance, low oxygen permeability, and high tensile strength (Wang et al., 2007a). As presented in Table 20.1, sodium alginate (SA) coatings have been reported to preserve the quality parameters of fresh-cut apple (Lee et al., 2003; Olivas et al., 2007; Fontes et al., 2008; Rojas-Graü et al., 2008), peach (Maftoonazad et al., 2008), papaya (Tapia et al., 2008), pear (Oms-Oliu et al., 2008a), melon (Oms-Oliu et al., 2008b; Raybaudi-Massilia et al., 2008; Ferrari et al., 2011), mango (Chiumarelli et al., 2011), kiwifruits (Mastromatteo et al., 2011), and pineapple (Azarakhsh et al., 2012).

Gellan gum is a high-molecular-weight polysaccharide gum produced by fermentation by *Sphingomonas elodea*. Gellan, just as alginate, have a high sensitivity to moisture due to their hydrophilic nature. In a classical work, Yang and Paulson (2000a) investigated mechanical and water vapor barrier properties of edible gellan films adding glycerol and evaluating its effect on these properties. Next, the authors (Yang and Paulson, 2000b) incorporated beeswax or a blend of stearic–palmitic acids into gellan films through emulsification, to form gellan/lipid composite films, improving their water sensitivity. Since then, various active agents such as probiotic microorganisms, natural antioxidants, and antimicrobials, for example, ascorbic acid, acetylcysteine, glutathione, and plant EOs, have been incorporated into SA, pectin, and gellan gum films applied onto various fresh-cut products (Table 20.1) (Tapia et al., 2007; Olivas and Barbosa-Canovas, 2008; Oms-Oliu et al., 2008b).

20.3.1.4 Chitosan

Chitosan is a biodegradable cationic copolymer obtained by *N*-deacetylation of chitin, produced from shellfish waste, adequate for application in fresh-cut commodities. CH produces without the incorporation of additives films that are transparent, slightly bright, homogenous, and generally cohesive, with good mechanical properties that yield a semipermeable barrier with control of gas exchange and, what is most important, with excellent antimicrobial properties against a wide spectrum of microorganisms (Coma et al., 2002; Romanazzi et al., 2002; Kim et al., 2003; Devlieghere et al., 2004; Dong et al., 2004; Vartiainen et al., 2004; Abdou et al., 2007; Thommohaway et al., 2007; Campaniello et al., 2008; Elsabee and Abdou, 2013). These properties altogether make CH one of the most extensively investigated polymers with many recent examples of application (Youwei and Yinzhe, 2013; Kaur and Singh, 2014). However, CH films have relatively high water vapor permeability (Vargas et al., 2009), which is a major drawback since adequate water vapor barrier properties are desirable in order to avoid dehydration processes during fruit/vegetable storage. The addition of lipid materials to CH-based films can improve their moisture barrier properties as has been proven over whole fruits such as strawberries (Vargas et al., 2006) and apple slices (Assis et al., 2012). The addition of lipid materials to CH-based films may improve their moisture barrier properties (Wong et al., 1992; Vargas et al., 2006, 2009).

The antimicrobial action of CH with the positive repercussion in shelf life extension has been reported by various authors working with minimally processed produce, and some examples are provided in Table 20.1.

20.3.1.5 Starch

Starch appears to be an interesting alternative for edible layers due to it is abundant, inexpensive, biodegradable, and easy-to-use condition (Durango et al., 2006). Starches can be obtained from cereals, legumes, roots, and tubers, which are granules composed of two macromolecules, linear amylose and highly branched amylopectin in a ratio that depends on the source. The amylose fraction is responsible for the film-forming capacity of starch. Commercial starch extraction is carried out in limited number of conventional sources; however, diverse botanical commodities are being investigated, as is the case of tropical roots and tubers that represent unexploited sources of starch with interesting characteristics in terms of functional properties and potential uses as starch-based plastics (Tapia et al., 2012; Gutiérrez et al., 2014). As discussed by Embuscado and Huber (2009), in general, films and coatings developed from starch are isotropic, odorless, tasteless, colorless and transparent, nontoxic, and semipermeable to CO_2 but highly resistant to O_2 and are often very brittle and with poor mechanical properties. Higher proportion of amylose (Han et al., 2006) or chemically modified starch may create starch with unique properties and improve mechanical properties of starch-based films. Additionally, the presence of plasticizers to improve the flexibility of starch films is required (Mali et al., 2002). The addition of lipids can reduce the water vapor permeability of the coatings or films but may affect their transparency, permeability, and mechanical properties (Rhim and Shellhammer, 2005; Chiumarelli et al., 2010). Since starch films are hydrophilic, their barrier properties decrease with increasing RH. Starch is therefore not the best option when working with minimally processed high water activity commodities (Olivas and Barbosa-Cánovas, 2009). Table 20.1 compiles some studies on the use of starch-based coating on minimally processed fruits and vegetables.

20.3.1.6 Carrageenan

Carrageenans are water-soluble and anionic linear sulfated polysaccharides extracted from red edible seaweeds considered for numerous industrial food applications, including coating formulations in the presence of monovalent or divalent cations for minimally processed fruits (Olivas and Barbosa-Cánovas, 2005; Bico et al., 2009). Table 20.1 presents studies conducted with carrageenan-based films on some commodities.

20.3.1.7 Aloe vera

Aloe gel (AG) consists mainly of polysaccharides and has been successfully investigated as potential edible film material. As revised by several authors, *Aloe vera* gel reduces ethylene production and acts as a barrier to O_2 and CO_2 thus generating modified atmosphere conditions. In addition, *Aloe vera* gels act as a moisture barrier preventing weight loss, browning, texture decay, and growth of yeast and molds (Jasso de Rodríguez et al., 2005; Valderde et al., 2005; Baldwin, 2007; Chauhan et al., 2011; Martínez-Romero et al., 2013). *Aloe vera* gel has been evaluated as an edible coating for various whole and minimally processed fruits, as seen in Table 20.1.

20.3.1.8 Pullulan

Pullulan is a polysaccharide produced by a yeastlike fungus *Aureobasidium pullulans*. It is a linear glucan consisting mainly of 1,6-linked maltotriose and some interspersed maltotetraose units. Pullulan is a thickener capable of forming edible films with several advantages over other polysaccharides (Qi et al., 2011). Films based on pullulan are colorless, tasteless, resistant to oil, heat sealable, and semipermeable to oxygen (Diab et al., 2001). Table 20.1 presents some studies conducted on pullulan as base for edible films and coatings of various commodities.

20.3.2 Protein Layers

Proteins are macromolecules generally used in film and coating formulation that can be obtained either from animal (milk and whey proteins, collagen, egg albumin, casein) or plant sources (wheat gluten, corn zein, soy protein, whey protein). As discussed by Baldwin et al. (1995a), protein films exhibit better gas barrier and mechanical properties than polysaccharide films due to their ordered hydrogen-bonded network structure. However, because of their hydrophilic nature, they generally are a poor water vapor barrier. The protein structure can be modified by heat, pressure, or other agents with the intention of improving physical and chemical properties of films and coatings.

Water-insoluble proteins, such as corn zein and wheat gluten, produce insoluble coatings, whereas proteins that are soluble in water produce coatings of varying solubility, depending on the protein and the conditions of coating formation and treatment. In soybeans, most of the protein is insoluble in water but soluble in dilute neutral salt solutions. Because of this, soy proteins are not considered as a hydrocolloid (Skurtys et al., 2010).

20.3.2.1 Whey Protein

Whey proteins constitute the soluble part of milk protein system. Whey protein–based edible films normally involve heat denaturation of proteins in aqueous solutions (Perez-Gago and Krochta, 2002) to ensure network integrity. They form transparent films and have demonstrated good mechanical and barrier properties and even better than other protein-based films as corn zein, wheat gluten, and soy isolate. However, its formulation requires addition of plasticizer to improve moisture barrier and to avoid brittleness while enhance flexibility and extensibility (Ramos et al., 2012). Good results have been obtained combining whey proteins with other substances to form films. Table 20.1 presents some uses of whey proteins as coating of fruits and vegetables.

20.3.2.2 Corn Zein

Zein is a storage protein of maize. Corn contains 9%–12% protein, half of it represented by an industrially useful protein called zein, constantly revised in the literature (Anderson and Lamsal, 2011). It is a prolamine that dissolves in ethanol and when blended with plasticizers has excellent film-forming properties, producing flexible, heat-sealable, and transparent films, suitable for applications as gas and moisture barrier. Zein coatings may have a light yellow color due to the presence of carotenoids β-carotene, zeaxanthin, and lutein (Sessa et al., 2003) yielding relatively opaque materials with less shiny appearance. Technically, the films made from an alcohol soluble protein like zein have relatively high barrier properties but still need improvement. The resulting films are brittle requiring the addition of plasticizers. Zein coatings have shown the ability to reduce moisture and loss of firmness and delay color change (the reduction of oxygen and carbon dioxide transmission) in fresh fruit.

In an attempt to improve the flexibility of zein films, the addition of oleic acid as a plasticizer has been proposed for biodegradable packaging applications. Conversion treatments including lamination and coating films with tung oil were reported to improve water vapor and gas barrier properties of films. A recent research (Assis et al., 2012) that studied the zein-based coating on slice apples is presented in Table 20.1.

20.3.2.3 Wheat Gluten

Wheat gluten is formed by globular water-insoluble proteins that yield wheat gluten–based films homogenous, transparent, mechanically strong, water insoluble, and semipermeable to oxygen and carbon dioxide, and its properties have been thoroughly investigated (Gontard and Ring, 1996; Mojumdar et al., 2011). Their barrier properties highly depend on the RH, T, and plasticizers utilized to avoid brittleness of films (Fernández-Pan and Maté Caballero, 2011). It has been tested on strawberries with promising results (Tanada-Palmu and Grosso, 2005; Amal et al., 2010). However, the role of gluten in the celiac disease has to be taken into consideration.

20.3.3 Lipid Layers

Lipids and hydrophobic substances are used as moisture barrier in edible films and coatings technology. Since they are not polymers, they do not generally form cohesive matrixes, and emulsions should be made within the polymeric base. They are used either as coatings or incorporated into biopolymers to form composite films. The emulsion stability, structure, degree of saturation, chain length, physical state, shape, and dimension of crystals and distribution of lipids into the film influence the functional properties of the film, like barrier properties of the emulsified coatings, which make the study of physical stability important (Callegarin et al., 1997; Morillon et al., 2002). Lipids commonly used in formulation of edible films and coatings are stearic acid, palmitic acid, and some vegetable oils, such as soybean and sunflower (Garcia et al., 2000; Martín-Belloso et al., 2005; Colla et al., 2006; Tapia et al., 2008). Natural and synthetic waxes are also used (beeswax, resins, carnauba, etc.) with enhanced gas and moisture barrier properties than coatings containing only fatty acids (Rhim and Shellhammer, 2005; Talens and Krochta, 2005; Rojas-Argudo et al., 2009). Chiumarelli and Hubinger (2012) added carnauba wax to cassava starch coatings to preserve fresh-cut apples.

20.3.4 Composite Layers and Emulsions

Edible films are classified into three categories taking into account the nature of their components: hydrocolloids (containing proteins, polysaccharides or alginates), lipids (constituted by fatty acids, acylglycerols, or waxes), and composites (made by combining substances from the two categories). None of the basic components of edible films (polysaccharides, proteins, or lipids) can provide by themselves all the properties desired for specific food applications, so they can be used in combination looking for enhanced properties of final films. Composite films and coatings are expected to combine the advantages of both lipid and hydrocolloid components, specifically the good water barrier of lipids and good gas barrier and film integrity of proteins or polysaccharides. There are though many challenges since added antimicrobial, antioxidants, plasticizers, and other constituents need to be compatible for a good performance of the composite film (Lai et al., 2013; Poverenov et al., 2014b). For instance, the interaction between the product surface and the coating can modify the barrier properties of the film (Chiumarelli and Hubinger, 2012), and natural antibrowning or antibacterial agents in the coating can affect the sensory quality of coated minimally processed fruits and vegetables (Rojas-Graü et al., 2007b; Raybaudi-Massilia et al., 2008; Lai et al., 2013; Pan et al., 2013). Composite layers have a heterogeneous structure that can be obtained by way of an emulsion (a continuous matrix with inclusion of lipid globules) and by way of multilayer coating. The importance of the emulsion stability on the coating's final structure and properties, as well as methods of emulsification, stability evaluation of emulsions, viscoelastic characteristics of lipids, lipid content, lipid particle size, film-drying conditions, etc., has been discussed elsewhere (Baldwin et al., 1997; Perez-Gago and Krochta, 1999; Becher, 2001; Pinnamaneni et al., 2003; Perez-Gago and Krochta, 2005; Urban et al., 2006; Vargas et al., 2011; Perdones et al., 2012).

Composite films can also be obtained by means of multilayer coating, composed of several layers of distinct materials. A multilayered film has better mechanical and barrier efficiencies than emulsion-based films but requires additional steps during its manufacture (Dutta et al., 2009). Bilayer films can potentially provide better moisture barriers; however, they tend to delaminate and exhibit poor mechanical properties compared to emulsions (Fernández-Pan and Maté Caballero, 2011). Thicker coatings can be produced either by increasing polymer concentration or by using layer-by-layer (LBL) assembly (Skurtys et al., 2010; Assis et al., 2012). As discussed by Campos et al. (2011) and Brasil et al. (2012), the application of edible coating to minimally processed fruits has some technical problems including the difficult adhesion of materials to the hydrophilic surface of the cut fruit, degradation of the antimicrobial compound, or inadequate diffusion rates. Multilayered coatings obtained by a serial LBL electrodeposition of oppositely charged natural polysaccharides, which if mixed do not form a homogeneous solution, could solve these challenges (Soliva-Fortuny and Martín-Belloso, 2003; Soliva-Fortuny, 2010). The use of composite antimicrobial edible coatings to increase the shelf life of fresh-cut fruit has been extensively investigated and reviewed with promising results as shown in Table 20.2.

Nanotechnology developments offer great potential to be implemented for the preservation of fresh-cut fruits and vegetables. Many advantages are expected when using a nanocomposite films and coatings in terms, for instance, of the much lower amounts of functional ingredients needed with less impact in the sensory attributes of foods. Some of these works are presented and summarized in Table 20.2.

20.4 Considerations for Industrial Implementation

Different methods can be used for the application of edible coatings on minimally processed fruits and vegetables. The coating technique is usually conditioned by the geometry and surface characteristics of the food product. Although spraying applications provide even distribution of the coating material, and thus uniform thickness, immersion of the product into a coating solution is by far the simplest method in terms of operation. Coatings formed by immersion may be less uniform than those obtained by other methods and usually a subsequent step is required to drain the excess of coating solution.

The structural integrity of the coating in terms of cohesiveness and adhesiveness to the food surface is affected by the nature of its composition. Hydrocolloids used as a base for a coating formulation are usually charged molecules. The sign and extent of these charges modulate the affinity for a food surface and also some physical properties of the coating material, such as surface tension and plasticity, which condition the mechanical and barrier properties of the coating. In this sense, surfactants can be added to the coating in order to reduce surface tension, whereas plasticizers such as glycerol, mannitol, and sorbitol are used to prevent brittleness of the cast material (Pavlath and Orts, 2009). In most cases, after draining the excess of coating solution, the adsorbed material needs to be allowed to dry on the product surface. In specific cases, for example, SA and pectins, a subsequent step is required in order to cross-link the hydrocolloid molecules with divalent ions such as calcium (Ca^{2+}). Gelation occurs as soon as the bivalent ions replace monovalent ions contained in the polymeric structure and form bonds that facilitate molecular entanglement.

As well, edible films can be cast from hydrocolloid solutions and used in a further stage as a wrap that can be cut according to the dimension and geometry of the food to be coated. Some published works suggest the application of wraps containing fruit and vegetable purees to extend the shelf life of fresh-cut products (McHugh and Senesi, 2000). These authors patented wraps containing apple puree plus various concentrations of lipidic compounds used to confer improved water barrier properties (McHugh and Senesi, 1999). According to their results, wraps offered better performance than other coating applications. However, dipping remains as the most preferred technique for edible layer formation because of its inexpensive way of operation.

TABLE 20.2

Application of Composite Coatings on Minimally Processed Fruits and Vegetables

Fruit/Vegetable	Edible Coating	Additives or Condition	Main Results	References
Mandarin, orange, and grapefruit	CMC/CH (bilayer)		Gloss enhancement and increased fruit firmness. In mandarins, bilayer coating slightly impaired flavor quality.	Arnon et al. (2014)
Fresh-cut cantaloupe	CH and pectin (multilayered)	Encapsulated *trans*-cinnamaldehyde	Multilayered coating extended the shelf life of fresh-cut cantaloupe up to 9 days at 4°C.	Martiñon et al. (2014)
Cherry tomato fruit	HPMC/lipid (emulsion)	PS, SA, SMP, SPP, SEP, SP,SB, PC, Aph, AC, Psi, SF, SBC, PBC, ABC	HPMC/lipid coatings more effectively controlled black spot caused by *Alternaria alternata* than gray mold caused by *Botrytis cinerea*.	Fagundes et al. (2013)
Sweet orange	CMC/corn starch (CS)	*Moringa oleifera* extract	Better firmness retention and shelf life extension of orange with CS than with CMC.	Adetunji et al. (2013)
Fresh-cut pineapple	SA, pectin, and calcium chloride as a multilayered coating	*Trans*-cinnamaldehyde microencapsulated in beta-cyclodextrin	Shelf life extension to 15 days at 4°C by inhibiting microbial growth. Also better preservation of color, texture, and pH.	Mantilla et al. (2013)
Fresh-cut "Fuji" apple	Tapioca starch/dHG	Ascorbic acid, calcium chloride, and cinnamon oil	Better quality retention and shelf life extension for up to 5–7 days in comparison with control samples (less than 2 days).	Pan et al. (2013)
Fresh-cut melon	Alginate–CH (bilayer)		Antimicrobial protection and maintained firmness for more than 14 days of storage.	Poverenov et al. (2014b)
Fresh-cut watermelon	Multilayered alginate, pectin	Beta-cyclodextrin, microencapsulated *trans*-cinnamaldehyde, calcium lactate	Shelf life extension from 7 to 12–15 days. During this period, the product maintained the quality and sensory acceptance.	Sipahi et al. (2013)
Fresh-cut "Maradol" papaya	Multilayered (CH and pectin)	Microencapsulated beta-cyclodextrin and *trans*-cinnamaldehyde complex	Better quality preservation and improved retention of vitamin and total carotenoids, with no negative impact on flavor.	Brasil et al. (2012)
Fresh-cut "Gala" apples	Emulsion of cassava starch–carnauba wax	Glycerol, stearic acid	Emulsion stability, respiration rate of apple slices coated, and water vapor resistance strongly influenced by carnauba-wax-to-stearic-acid ratio. Mechanical properties and solubility of films mainly influenced by glycerol content and carnauba-wax-to-stearic-acid ratio.	Chiumarelli and Hubinger (2012)
Fresh-cut "Rocha" pears	κ-Carrageenan–lysozyme (nanolayered)		Positive effect on fruit quality and shelf life extension.	Medeiros et al. (2012)

(*Continued*)

TABLE 20.2 (Continued)

Application of Composite Coatings on Minimally Processed Fruits and Vegetables

Fruit/Vegetable	Edible Coating	Additives or Condition	Main Results	References
Apple "Fuji" slices	Shellac and AG (separately and in combination)	Pretreatment with ascorbic acid, citric acid, and sodium benzoate for 10 min	Reduction of respiration and ethylene synthesis rates as well as electrolyte leakage. AG coatings reduced PPO and POD activity followed by composite coatings (shellac + AG) and shellac alone. Overall quality was also preserved.	Chauhan et al. (2011)
Fruit-based salads and romaine hearts	Tapioca starch/dHG	Green tea (GT) extracts	Microbial growth was substantially reduced.	Chiu and Lai (2010)
Fresh baby carrot	Pullulan–caraway EO		Antimicrobial activity against both Gram-negative and Gram-positive bacteria and fungi. Visual quality was also preserved.	Gniewosz et al. (2013)
Fresh-cut carrots	Tapioca starch/dHG	Cinnamon EO or GSE	Visual appearance was maintained. Formation of "white blush" on the surface of carrots during refrigeration. Coated samples exhibited higher microbial counts in some cases.	Lai et al. (2013)
Potato round slices	SA and CMC	GT, AA	Color of potato slices better maintained during storage. The mixture of SA and CMC with AA or with GT exhibited good moisture barrier properties.	Spanou and Giannouli (2013)
Fresh-cut cantaloupe and pineapple	CH/MC	Vanillin	Inhibitory effect against *Escherichia coli* on fresh-cut cantaloupe. Counts of *Saccharomyces cerevisiae* inoculated on cantaloupe and pineapple were also reduced.	Sangsuwan et al. (2008)

CMC, carboxymethylcellulose; PS, potassium sorbate; SA, sodium acetate; SMP, sodium methylparaben; SPP, sodium propylparaben; SEP, sodium ethylparaben; SP, sodium propionate; SB, sodium benzoate; PC, potassium carbonate; Aph, ammonium phosphate; AC, ammonium carbonate; Psi, potassium silicate; SF, sodium formate; SBC, sodium bicarbonate; PBC, potassium bicarbonate; ABC, ammonium bicarbonate.

20.5 Regulatory Issues

As an integral part of the edible portion of a fruit or vegetable product, edible coatings need to be regarded as a food ingredient. Therefore, they should be food-grade nontoxic materials and any hygiene standard required for a food ingredient should be obeyed. In this regard, the Food and Drug Administration requires that any substance included in the formulation has to be considered as generally recognized as safe or regulated as a food additive and used within specified limitations (FDA, 2006). In the European Union, ingredients incorporated into edible coating formulations are usually regarded as food additives. Only pectins, acacia and karaya gums, beeswax polysorbates, fatty acids, and lecithin are specifically mentioned for coating uses (Rojas-Graü et al., 2011b).

20.6 Final Considerations

Edible films and coatings offer promising opportunities for quality improvement, shelf life extension, and safety enhancement of fresh and fresh-cut fruits and vegetables. Edible formulations offer the possibility of valorizing extracts obtained from by-products of the fruit industry. However, although many studies have focused on the development of edible coatings with beneficial effects on fruits and vegetables, their mechanisms of action are yet to be elucidated. The source of the extracts is important in order to identify potential issues linked to possible allergenicity of the used compounds. On the other hand, the great diversity of fruits and vegetables poses thrilling challenges to food technologists. Tailor-made applications need to be developed for each product in order to minimize quality losses without affecting its intrinsic properties.

REFERENCES

Abdou, E.S., Nagy, K.S.A., and Elsabee, M.Z. Extraction and characterization of chitin and chitosan from local sources. *Bioresource Technology* 99(5) (2007): 1359–1367.

Adetunji, C.O., Fawole, O.B., Arowa, K.A. et al. Performance of edible coatings from carboxymethylcellulose (CMC) and corn starch (CS) incorporated with *Moringa oleifera* extract on *Citrus sinensis*. *Agrosearch* 13(1) (2013): 77–85.

Ahmed, L., Martin-Diana, A.B., Rico, D., and Barry-Ryan, C. Quality and nutritional status of fresh-cut tomato as affected by spraying of delactosed whey permeate compared to industrial washing treatment. *Food and Bioprocess Technology* 5 (2012): 3103–3114.

Alvarez, M.V., Ponce, A.G., and Moreira, M.R. Antimicrobial efficiency of chitosan coating enriched with bioactive compounds to improve the safety of fresh cut broccoli. *LWT—Food Science and Technology* 50 (2013): 78–87.

Amal, S.H.A., El-Mogy, M.M., Aboul-Anean, H.E., and Alsanius, B.W. Improving strawberry fruit storability by edible coating as a carrier of thymol or calcium chloride. *Journal of Horticultural Science & Ornamental Plants* 2(3) (2010): 88–97.

Anderson, T.J. and Lamsal, B.P. Zein extraction from corn, corn products, and coproducts and modifications for various applications: A review. *Cereal Chemistry* 88(2) (2011): 159–173.

Ansorena, M.R., Marcovich, N.E., and Roura, S.I. Impact of edible coatings and mild heat shocks on quality of minimally processed broccoli (*Brassica oleracea* L.) during refrigerated storage. *Postharvest Biology and Technology* 59(1) (2011): 53–63.

Arnon, H., Zaitsev, Y., Porat, R., and Poverenov, E. Effects of carboxymethyl cellulose and chitosan bilayer edible coating on postharvest quality of citrus fruit. *Postharvest Biology and Technology* 87 (2014): 21–26.

Assis, O.B.G., Scramin, J.A., Correa, T.A., de Brito, D., and Forato, L.A. A comparative evaluation of integrity and colour preservation of slice apples protected by chitosan and zein edible coatings. *Revista Iberoamericana de Tecnología Postcosecha* 13(1) (2012): 76–85.

Azarakhsh, N., Osman, A., Ghazali, H.M., Tan, C.P., and Adzahan, N.M. Optimization of alginate and gellan-based edible coatings formulations for fresh-cut pineapples. *International Food Research Journal* 19(1) (2012): 279–285.

Azeredo, H.M., Mattoso, L.H., Wood, D., Williams, T.G., Avena-Bustillos, R.J., and McHugh, T.H. Nanocomposite edible films from mango puree reinforced with cellulose nanofibers. *Journal of Food Science* 74(5) (2009): N31–N35.

Baldwin, E. Chapter 21. Surface treatments and edible coating in food preservation. In M. Shafiur Rahman (ed.), *Handbook of Food Preservation*, pp. 477–578. Boca Raton, FL: CRC Press, 2007.

Baldwin, E.A. and Bai, J. Physiology of fresh-cut fruits and vegetables. In O. Martín-Bellloso and R. Soliva Fortuny (eds.), *Advances in Fresh-Cut Fruits and Vegetables Processing*, pp. 1–11. Boca Raton, FL: CRC Press, 2011.

Baldwin, E.A., Nisperos, M.O., Chen, X., and Hagenmaier, R.D. Improving storage life of cut apple and potato with edible coating. *Postharvest Biology and Technology* 9 (1996): 151–163.

Baldwin, E.A., Nisperos, M.O., Hagenmaier, R.D., and Baker, R.A. Use of lipids in coatings for food products. *Food Technology* 51(6) (1997): 56–64.

Baldwin, E.A., Nisperos-Carriedo, M.O., and Baker, R.A. Edible coatings for lightly processed fruits and vegetables. *The Journal of Horticultural Science and Biotechnology* 30 (1995a): 35–38.

Baldwin, E.A., Nisperos-Carriedo, M., Shaw, P.E., and Burns, J.K. Effect of coatings and prolonged storage conditions on fresh orange flavor volatiles, degrees brix, and ascorbic acid levels. *Journal of Agricultural Food Chemistry* 43(8) (1995b): 1321–1331.

Beaulieu, J.C., Ingram, D.A., Lea, J.M., and Bett-Garber, K.L. (2004). Effect of harvest maturity on the sensory characteristics of fresh-cut cantaloupe. *Journal of Food Science* 69(7): S250–S258.

Becher, P. *Emulsions: Theory and Practice*. Oxford, U.K.: Oxford University Press, 2001.

Benítez, S., Achaerandio, I., Sepulcre, F., and Pujola, M. *Aloe vera* based edible coatings improve the quality of minimally processed 'Hayward' Kiwifruit. *Postharvest Biology and Technology* 81 (2013): 29–36.

Beuchat, L.R. Pathogenic microorganisms associated with fresh produce. *Journal of Food Protection* 59 (1996): 204–216.

Bico, S.L.S., Raposo, M.F.J., Morais, R.M.S.C., and Morais, A.M.M.B. Combined effects of chemical dip and/ or carrageenan coating and/or controlled atmosphere on quality of fresh-cut banana. *Food Control* 20 (2009): 508–514.

Bierhals, V.S., Chiumarelli, M., and Hubinger, M.D. Effect of cassava starch coating on quality and shelf life of fresh-cut pineapple (*Ananas Comosus* L. Merril cv "Pérola"). *Journal of Food Science* 76(1) (2011): E62–E72.

Bilbao-Sáinz, C., Avena-Bustillos, R.J., Wood, D.F., Williams, T.G., and McHugh, TH. Composite edible films based on hydroxypropyl methylcellulose reinforced with microcrystalline cellulose nanoparticles. *Journal of Agricultural and Food Chemistry* 58(6) (2010): 3753–3760.

Bourtoom, T. Edible films and coatings: Characteristics and properties. *International Food Research Journal* 15(3) (2008): 1–12.

Brasil, I.M., Gomes, C., Puerta-Gomez, A., Castell-Perez, M.E., and Moreira, R.G. Polysaccharide-based multilayered antimicrobial edible coating enhances quality of fresh-cut papaya. *LWT—Food Science and Technology* 47(1) (2012): 39–45.

Callegarin, F., Quezada-Gallo, J.A., Debeaufort, F., and Voilley, A. Lipids and biopackaging. *Journal of the American Oil Chemists' Society* 74(10) (1997): 1183–1192.

Campaniello, D., Bevilacqua, A., Sinigaglia, M., and Corbo, M.R. Chitosan: Antimicrobial activity and potential applications for preserving minimally processed strawberries. *Food Microbiology* 25 (2008): 992–1000.

Campos, C., Gerschenson, L.N., and Flores, S.K. Development of edible films and coatings with antimicrobial activity. *Food and Bioprocess Technology* 4(6) (2011): 849–875.

Chauhan, O.P., Raju, P.S., Singh, A., and Bawa, A.S. Shellac and aloe-gel-based surface coatings for maintaining keeping quality of apples slices. *Food Chemistry* 126(3) (2011): 961–966.

Chiu, P.-E. and Lai, L.-S. Antimicrobial activities of tapioca starch/decolorized Hsian-Tsao leaf gum coatings containing green tea extracts in fruit-based salads, Romaine hearts and pork slices. *International Journal of Food Microbiology* 139(1–2) (2010): 23–30.

Chiumarelli, M., Ferrari, C.C., Sarantópoulos, C.I.G.L., and Hubinger, M.D. Fresh cut "Tommy Atkins" mango pre-treated with citric acid and coated with Cassava (*Manihot esculenta* crantz) starch or sodium alginate. *Innovative Food Science and Emerging Technologies* 12(3) (2011): 381–387.

Chiumarelli, M. and Hubinger, M.D. Stability, solubility, mechanical and barrier properties of Cassava Starch-Carnauba wax edible coatings to preserve fresh-cut apples. *Food Hydrocolloids* 28 (2012): 59–67.

Chiumarelli, M., Pereira, L.M., Ferrari, C.C., Sarantópoulos, C.I., and Hubinger, M.D. Cassava starch coating and citric acid to preserve quality parameters of fresh-cut "Tommy Atkins" Mango. *Journal of Food Science* 75(5) (2010): E297–E304.

Colla, E., Sobral, P.J.A., and Menegalli, F.C. *Amaranthus cruentus* flour edible films: Influence of stearic acid addition, plasticizer concentration, and emulsion stirring speed on water vapor permeability and mechanical properties. *Journal of Agricultural and Food Chemistry* 54 (2006): 6645–6653.

Coma, V., Martial-Gros, A., Garreau, S., Copinet, A., and Deschamps, A. Edible antimicrobial film based on chitosan matrix. *Journal of Food Science* 67(3) (2002): 1162–1169.

Conte, A., Scrocco, C., Brescia, I., and Del Nobile, M.A. Packaging strategies to prolong the shelf life of minimally processed lampascioni (*Muscari comosum*). *Journal of Food Engineering* 90 (2009): 199–206.

de Azevedo, H.M.C. Edible coatings. In S. Rodrigues, F. A. N. Fernandes (Eds.), *Advances in Fruit Processing Technologies*, pp. 345–361. Boca Raton, FL: CRC Press, 2012.

Devlieghere, F., Vermeulen, A., and Debevere, J. Chitosan: Antimicrobial activity, interactions with food components and applicability as a coating on fruit and vegetables. *Food Microbiology* 21 (2004): 703–714.

Dhall, R.K. Advances in edible coatings for fresh fruits and vegetables: A review. *Critical Reviews in Food Science and Nutrition* 53(5) (2013): 435–450.

Dhanapal, A., Sasikala, P., Lavanya, R., Kavitha, V., Yazhini, G., and Shakila-Banu, M. Edible films from polysaccharides. *Food Science and Quality Management* 3 (2012): 9–18.

Diab, T., Biliaderis, C.G., Gerasopoulos, D., and Sfekiotakis, E. Physicochemical properties and application of pullulan edible films and coatings in fruit preservation. *Journal of the Science of Food and Agriculture* 81 (2001): 988–1000.

Dong, H., Cheng, L., Tan, J., Zheng, K., and Jiang, Y. Effects of chitosan coating on quality and shelf life of peeled litchi fruit. *Journal of Food Engineering* 64 (2004): 355–358.

Durango, A., Soares, N., and Andrade, N. Microbiological evaluation of an edible antimicrobial coating on minimally processed carrots. *Food Control* 17 (2006): 336–341.

Dutta, P.K., Tripathi, S., Mehrotra, G.K., and Dutta, J. Review: Perspectives for chitosan based antimicrobial films in food applications. *Food Chemistry* 114(4) (2009): 1173–1182.

Elsabee, M.Z. and Abdou, E.S. Chitosan based edible films and coatings: A review. *Materials Science and Engineering C* 33 (2013): 1819–1841.

Embuscado, M.E. and Huber, K.C. *Edible Films and Coatings for Food Applications*. New York: Springer, 2009.

Fagundes, C., Pérez-Gago, M.B., Monteiro, A.R., and Palou, L. Antifungal activity of food additives in vitro and as ingredients of hydroxypropyl methylcellulose-lipid edible coatings against *Botrytis cinerea* and *Alternaria alternata* on cherry tomato fruit. *International Journal of Food Microbiology* 166 (2013): 391–398.

Falguera, V., Quintero, J.P., Jimenez, A., Muñoz, J.A., and Ibarz, A. Edible films and coatings: Structures, active functions and trends in their use. *Trends in Food Science & Technology* 22 (2011): 292–303.

FDA, U.S. Food and Drug Administration. 2006. Food additives permitted for direct addition to food for human consumption 21CFR172, subpart C. Coatings, Films and Related Substances.

Fernández-Pan, I. and Maté Caballero, J.I. Biopolymers for edible films and coatings in food applications. In David Plackett (Ed.), *Biopolymers—New Materials for Sustainable Films and Coatings*. Chichester, UK: John Wiley & Sons, Ltd., 2011.

Ferrari, C.C., Sarantopoulos, C.I.G.L., Carmello-Guerreiro, S.M., and Hubinger, M.D. Effect of osmotic dehydration and pectin edible coatings on quality and shelf life of fresh-cut melon. *Food and Bioprocess Technology* 6(1) (2011): 80–91.

Fontes, L.C.B., Sarmento, S.B.S., Spoto, M.H.F., and Dias, C.T.dosS. Preservation of minimally processed apple using edible coatings. *Ciência e Tecnologia de Alimentos* 28 (2008): 872–880.

Franssen, L.R. and Krochta, J.M. Edible coatings containing natural antimicrobials for processed foods. In *Natural Antimicrobials for Minimal Processing of Foods*, pp. 250–262. Boca Raton, FL: CRC Press, 2003.

García, L.C., Pereira, L.M., de Luca Sarantópoulos, C.I.G., and Hubinger, M.D. Selection of edible starch coating for minimally processed strawberry. *Food and Bioprocess Technology* 3 (2010): 834–842.

Garcia, M.A., Martino, M.N., and Zaritzky, N.E. Lipid addition to improve barrier properties of edible starch-based films and coatings. *Journal of Chemistry and Toxicology* 65(6) (2000): 941–947.

Geraldine, R.M., Soares, N.F.F., Botrel, D.A., and Gonçalves, L.deA. Characterization and effect of edible coatings on minimally processed garlic quality. *Carbohydrate Polymers* 72 (2008): 403–409.

Gniewosz, M., Kraśniewska, K., Woretam, M., and Kosakowska, O. Antimicrobial activity of a Pullulan-Caraway essential oil coating on reduction of food microorganisms and quality in fresh baby carrot. *Journal of Food Science* 78(8) (2013): M1242–M1248.

Gontard, N. and Ring, N.S. Edible wheat gluten film: Influence of water content on glass transition temperature. *Journal of Agricultural and Food Chemistry* 44(11) (1996): 3474–3478.

González-Aguilar, G.A., Ayala-Zavala, J.F., Olivas, G.I., de la Rosa, L.A., and Álvarez-Parrilla, E. Preserving quality of fresh-cut products using safe technologies. *Journal fur Verbraucherschutz und Lebensmittelsicherheit* 5 (2010): 65–72.

Gonzalez-Aguilar, G.A., Valenzuela-Soto, E., Lizardi-Mendoza, J. et al. Effect of chitosan coating in preventing deterioration and preserving the quality of fresh-cut papaya 'Maradol'. *Journal of the Science of Food and Agriculture* 89 (2009): 15–23.

Gutiérrez, T.J., Morales, N.J., Tapia, M.S., Pérez, E., and Famá, L. Corn Starch 80:20 "Waxy": Regular, "Native" and phosphated, as bio-matrixes for edible films. *Procedia Materials Science* 8 (2014): 304–310.

Hagenmaier, R.D. and Shaw, P.E. Moisture permeability of edible films made with fatty acid and hydroxypropyl methyl cellulose. *Journal of Agricultural and Food Chemistry* 38(9) (1990): 1799–1803.

Han, J.H. and Gennadios, A. Edible films and coatings: A review. In J. H. Han (Ed.), *Innovations in Food Packaging*, pp. 239–259. San Diego, CA: Elsevier Academic Press, 2005.

Han, J.H., Seo, G.H., Park, I.M., Kim, G.N., and Lee, D.S. Physical and mechanical properties of pea starch edible films containing beeswax emulsions. *Journal of Food Science* 71(6) (2006): E290–E296.

Janjarasskul, T. and Krochta, J.M. Edible packaging materials. *Annual Review of Food Science and Technology* 1 (2010): 415–448.

Jasso de Rodríguez, D., Hernández-Castillo, D., Rodríguez-García, R., and Angulo-Sámchez, J.L. Antifungal activity in vitro of *Aloe vera* pulp and liquid fraction against plant pathogenic fungi. *Industrial Crops and Products* 21 (2005): 81–87.

Kaur, S. and Singh, D.G. The versatile biopolymer chitosan: Potential sources, evaluation of extraction methods and applications. *Critical Reviews in Microbiology* 40(2) (2014): 155–175.

Kester, J.J. and Fennema, O.R. Edible films and coatings: A review. *Food Technology* 40(12) (1986): 47–59.

Kim, K.W., Thomas, R.L., Lee, C., and Park, H.J. Antimicrobial activity of native, degraded and O-carboxymethylated chitosan. *Journal of Food Protection* 66 (2003): 1495–1498.

Krasaekoopt, W. and Mabumrung, J. Microbiological evaluation of edible coated fresh-cut cantaloupe. *Kasetsart Journal: Natural Sciences* 42(3) (2008): 552–557.

Krochta, J.M. Proteins as raw materials for films and coatings: Definitions, current status and opportunities. In A. Gennadios (Ed.), *Protein-Based Films and Coatings*. Boca Raton, FL: CRC Press, 2002.

Krochta, J.M., Baldwin, E.A., and Nisperos-Carriedo, M. *Edible Coatings and Films to Improve Food Quality*. Lancaster, PA: Technomic Publishing Co., 1994.

Krochta, J.M. and Mulder-Johnston, C.D. Edible and biodegradable polymer films: Challenges and opportunities. *Food Technology* 51 (1997): 61–74.

Lai, T.-Y., Chen, C.-H., and Lai, L.-S. Effects of tapioca starch/decolorized Hsian-Tsao leaf gum based active coatings on the quality of minimally processed carrots. *Food and Bioprocess Technology* 6 (2013): 249–258.

Lee, J.Y., Park, H.J., Lee, C.Y., and Choi, W.Y. Extending shelf-life of minimally processed apples with edible coatings and antibrowning agents. *Food Science and Technology* 36 (2003): 323–329.

Lin, D. and Zhao, Y. Coatings for fresh and minimally processed fruits and vegetables. *Comprehensive Reviews in Food Science and Food Safety* 6(3) (2007): 60–75.

Macheix, J., Fleuriet, A., and Billot, J. Phenolic compounds in fruit processing. In J. Macheix, A. Fleuriet, J. Billot (Eds.), *Fruit Phenolics*, pp. 295–357. Boca Raton, FL: CRC Press, 1990.

Maftoonazad, N., Ramaswamy, H.S., and Marcotte, M. Shelf-life extension of peaches through sodium alginate and methyl cellulose edible coatings. *International Journal of Food Science & Technology* 43 (2008): 951–957.

Mali, S., Grossmann, M.V.E., Garcia, M.A., Martino, M.N., and Zaritzky, N.E. Microstructural characterization of Yam starch films. *Carbohydrate Polymers* 50 (2002): 379–386.

Mantilla, N., Castell-Perez, M.E., Gomes, C., and Moreira, R.G. Multilayered antimicrobial edible coating and its effect on quality and shelf-life of fresh-cut pineapple (*Ananas comosus*). *LWT—Food Science and Technology* 51 (2013): 37–43.

Martín-Belloso, O., Soliva-Fortuny, R., and Baldwin, E.A. Conservación mediante recubrimientos comestibles. In G. A. González-Aguilar, A. A. Gardea, F. Cuamea-Navarro (Eds.), *Nuevas Tecnologías de Conservación de Productos Vegetales Frescos Cortados*, pp. 61–74. México, Mexico: CIAD, A. C., Hermosillo, Sonora, 2005.

Martínez-Romero, D., Castillo, S., Guillén, F. et al. *Aloe vera* gel coating maintains quality and safety of ready-to-eat pomegranate arils. *Postharvest Biology and Technology* 86 (2013): 107–112.

Martiñon, M.E., Moreira, R.G., Catel-Pérez, M.E., and Gomes, C. Development of a multilayered antimicrobial edible coating for shelf-life extension of fresh-cut cantaloupe (*Cucumis melo* L.) stored at 4°C. *LWT—Food Science and Technology* 56(2) (2014): 341–350.

Marudova, M., Lang, S., Brownsey, G.J., and Ring, S.G. Pectin-chitosan multilayer formation. *Carbohydrate Research* 340 (2005): 2144–2149.

Mastromatteo, M., Conte, A., and Del Nobile, M.A. Packing strategies to prolong the shelf life of fresh carrots (*Daucus carota* L.). *Innovative Food Science and Emerging Technologies* 13 (2012): 215–220.

Mastromatteo, M., Mastromatteo, M., Conte, A., and Del Nobile, M.A. Combined effect of active coating and MAP to prolong the shelf life of minimally processed Kiwifruit (*Actinidia deliciosa* cv. Hayward). *Food Research International* 44 (2011): 1224–1230.

McHugh, T.H., Huxsoll, C.C., and Krochta, J.M. Permeability properties of fruit puree edible films. *Journal of Food Science* 61(1) (1996): 88–91.

McHugh, T.H. and Senesi, E. Apple wraps: A novel method to improve the quality and extend the shelf life of fresh-cut apples. *Journal of Food Science* 65(3) (2000): 480–485.

McHugh, T.H. and Senesi, E. (Inventors). USDA-ARS-WRRC assignee. Filed 1999 June 11. Fruit and vegetable edible film wraps and methods to improve and extend the shelf life of foods. U.S. Patent Application Serial No. 09/330,358, 1999.

Medeiros, B.G., de Pinheiro, S.A.C., Carneiro-da-Cunha, M.G., and Vicente, A.A. Development and characterization of a nanomultilayer coating of pectin and chitosan. Evaluation of its gas barrier properties and application on 'Tommy Atkins' mangoes. *Journal of Food Engineering* 110(3) (2012): 457–464.

Mladenoska, I. The potential application of novel beeswax edible coatings containing coconut oil in the minimal processing of fruits. *Advanced Technologies* 1(2) (2012): 26–34.

Mohamed, A.Y.I., Aboul-Anean, A., and Hassan, A.M. Utilization of edible coating in extending the shelf life of minimally processed prickly pear. *Journal of Applied Sciences Research* 9(2) (2013): 1202–1208.

Mojumdar, S.C., Moresoli, C., Simon, L.C., and Legge, R.L. Edible wheat gluten (WG) protein films. *Journal of Thermal Analysis and Calorimetry* 104(3) (2011): 929–936.

Moncayo, D., Buitrago, G., and Algecira, N. The surface properties of biopolymer-coated fruit: A review. *Ingeniería e Investigación* 33(3) (2013): 11–16.

Moreira, M.R., Ponce, A., Ansorrena, R., and Roura, S.I. Effectiveness of edible coatings combined with mild heat shocks on microbial spoilage and sensory quality of fresh cut broccoli (*Brassica oleracea* L.). *Journal of Food Science* 76(6) (2011): M367–M374.

Morillon, V., Debeaufort, F., Blond, G., Capelle, M., and Voilley, A. Factors affecting the moisture permeability of lipid-based edible films: A review. *Critical Reviews in Food Science and Nutrition* 42(1) (2002): 67–89.

Nisperos-Carriedo, M.O. and Baldwin, E.A. Edible coatings for whole and minimally processed fruits and vegetables. *Food Australia* 48 (1996): 27–31.

Olivas, G.I. and Barbosa-Cánovas, G.V. Edible coatings for fresh-cut fruits. *Critical Reviews in Food Science and Nutrition* 45(7–8) (2005): 657–670.

Olivas, G.I. and Barbosa-Canovas, G.V. Alginate-calcium films: Water vapor permeability and mechanical properties as affected by plasticizer and relative humidity. *LWT—Food Science and Technology* 41 (2008): 359–336.

Olivas, G.I. and Barbosa-Canovas, G.V. Chapter 7. Edible films & coatings for fruits and vegetables. In K. Huber, M. E. Embuscado (Eds.), *Edible Films and Coatings for Food Applications*. New York: Springer, 2009.

Olivas, G.I., Mattinson, D.S., and Barbosa-Cánovas, G.V. Alginate coatings for preservation of minimally processed 'Gala' apples. *Postharvest Biology and Technology* 45 (2007): 89–96.

Oms-Oliu, G., Rojas-Grau, M.A., Alandes-Gonzalez, L. et al. Recent approaches using chemical treatments to preserve quality of fresh-cut fruit: A review. *Postharvest Biology and Technology* 57 (2010): 139–148.

Oms-Oliu, G., Soliva-Fortuny, R., and Martin-Belloso, O. Edible coatings with antibrowning agents to maintain sensory quality and antioxidant properties of fresh-cut pears. *Postharvest Biology and Technology* 50 (2008a): 87–94.

Oms-Oliu, G., Soliva-Fortuny, R., and Martin-Belloso, O. Using polysaccharide-based edible coatings to enhance quality and antioxidant properties of fresh-cut melon. *LWT—Food Science and Technology* 41 (2008b): 1862–1870.

Pan, S.-Y., Chen, C.-H., and Lai, L.-S. Effect of Tapioca starch/decolorized Hsian-Tsao leaf gum-based active coatings on the qualities of fresh-cut apples. *Food and Bioprocess Technology* 6(8) (2013): 2059–2069.

Pascall, M.A. and Lin, S.-J. The application of edible polymeric films and coatings in the food industry. *Journal of Food Processing & Technology* 4(2) (2013): e116.

Pavlath, A.E. and Orts, W. Edible films and coatings: Why, what, and how? In M.E. Embuscado and K.C. Huber (eds.), *Edible Films and Coatings for Food Applications*, pp. 1–23. New York: Springer, 2009.

Perdones, A., Sánchez-González, L., Chiralt, A., and Vargas, M. Effect of chitosan-lemon essential oil coatings on storage-keeping quality of strawberry. *Postharvest Biology and Technology* 70 (2012): 32–41.

Pérez-Espitia, P.J., Du, W.-X., Avena-Bustillos, R.deJ., and Ferreira-Soares, N.deF. Edible films from pectin: Physical–mechanical and antimicrobial properties—A review. *Food Hydrocolloids* 35 (2014): 287–296.

Perez-Gago, M.B. and Krochta, J.M. Water vapor permeability of whey protein emulsion films as affected by pH. *Journal of Food Science* 64(4) (1999): 695–698.

Perez-Gago, M.B. and Krochta, J.M. Formation and properties of whey protein films and coatings. In A. Gennadios (Ed.), *Protein-Based Films and Coatings*, pp. 159–180. Boca Raton, FL: CRC Press, 2002.

Perez-Gago, M.B. and Krochta, J.M. Emulsions and bi-layer edible films. In J. H. Han (Ed.), *Innovations in Food Packaging*, pp. 384–402. San Diego, CA: Elsevier Academic Press, 2005.

Petersen, K., Nielsen, P.V., Lawther, M., Olsen, M.B., Nilsson, N.H., and Mortensen, G. Potential of biobased materials for food packaging. *Trends in Food Science & Technology* 10 (1999): 52–68.

Pinnamaneni, S., Das, N.G., and Das, S.K. Comparison of oil-in-water emulsions manufactured by microfluidization and homogenization. *Pharmazie* 58 (2003): 554–558.

Plotto, A., Narciso, J.A., Rattanapanone, N. et al. Surface treatments and coatings to maintain fresh-cut mango quality in storage. *Journal of the Science of Food and Agriculture* 90 (2010): 2333–2341.

Poverenov, E., Cohen, R., Yefremov, T., Vinokur, Y., and Rodov, V. Effects of polysaccharide-based edible coatings on fresh-cut melon quality. *Acta Horticulturae (ISHS)* 1015 (2014a): 145–151.

Poverenov, E., Danino, S., Horev, B., Granit, R., Yakov Vinokur, Y., and Rodov, V. Layer-by-layer electrostatic deposition of edible coating on fresh cut melon model: Anticipated and unexpected effects of alginate–chitosan combination. *Food and Bioprocess Technology* 7 (2014b): 1424–1432.

Pranoto, Y., Salokhe, V., and Rakshit, K.S. Physical and antibacterial properties of alginate-based edible film incorporated with garlic oil. *Food Research International* 38 (2005): 267–272.

Qi, H., Hu, W., Jiang, A., Tian, M., and Li, Y. Extending shelf-life of fresh-cut 'Fuji' apples with chitosan-coatings. *Innovative Food Science and Emerging Technologies* 12 (2011): 62–66.

Ramos, O.L., Fernandez, J.C., Silva, S.I., Pintado, M.E., and Malcata, X. Edible films and coating from whey proteins: A review on formulation, and on mechanical and bioactive properties. *Critical Review in Food Science and Nutrition* 52 (2012): 533–552.

Raybaudi-Massilia, R.M., Rojas-Graü, M.A., Mosqueda-Melgar, J., and Martin-Belloso, O. Comparative study on essential oils incorporated into an alginate-based edible coating to assure the safety and quality of fresh-cut Fuji apples. *Journal of Food Protection* 71 (2008): 1150–1161.

Rhim, J.W. and Shellhammer, T.H. Lipid-based edible films and coatings. In J. H. Han (ed.), *Innovations in Food Packaging*. San Diego, CA: Elsevier Academic Press, 2005.

Rojas-Argudo, C., del Río, M.A., and Pérez-Gago, M.B. Development and optimization of locust bean gum (LBG) based edible coatings for postharvest storage of 'Fortune' mandarins. *Postharvest Biology and Technology* 52 (2009): 227–234.

Rojas-Graü, M.A., Avena-Bustillos, R.J., Friedman, M., Henika, P.R., Martín-Belloso, O., and McHugh, T.H. Mechanical, barrier, and antimicrobial properties of apple puree edible films containing plant essential oils. *Journal of Agricultural and Food Chemistry* 54(24) (2006): 9262–9267.

Rojas-Graü, M.A., Garner, E., and Martín-Belloso, O. The fresh-cut fruit and vegetables industry: Current situation and market trends. In O. Martín-Belloso and R. Soliva Fortuny (eds.), *Advances in Fresh-Cut Fruits and Vegetables Processing*, pp. 1–11. Boca Raton, FL: CRC Press, 2011a.

Rojas-Graü, M.A., Raybaudi-Massilia, R.M., Soliva-Fortuny, R., Avena-Bustillos, R.J., McHugh, T.H., and Martín-Belloso, O. Apple puree-alginate coating as carrier of antimicrobial agents to prolong shelf life of fresh-cut apples. *Postharvest Biology and Technology* 45 (2007a): 254–264.

Rojas-Grau, M.A., Soliva-Fortuny, R., and Martin-Belloso, O. Edible coatings to incorporate active ingredients to fresh-cut fruits: A review. *Trends in Food Science & Technology* 20 (2009): 438–447.

Rojas-Graü, M.A., Soliva-Fortuny, R.C., and Martín-Belloso, O. Use of edible coatings for fresh-cut fruits and vegetables. In O. Martín-Belloso and R. Soliva Fortuny (eds.), *Advances in Fresh-Cut Fruits and Vegetables Processing*, pp. 285–311. Boca Raton, FL: CRC Press, 2011b.

Rojas-Grau, M.A., Tapia, M.S., and Martin-Belloso, O. Using polysaccharide-based edible coatings to maintain quality of fresh-cut Fuji apples. *LWT—Food Science and Technology* 41 (2008): 139–147.

Rojas-Graü, M.A., Tapia, M.S., Rodríguez, F.J., Carmona, A.J., and Martín-Belloso, O. Alginate and gellan-based edible coatings as carriers of antibrowning agents applied on fresh-cut Fuji apples. *Food Hydrocolloids* 21 (2007b): 118–127.

Romanazzi, G., Nigro, F., Ippolito, A., Venere, D.D., and Salerno, M. Effects of pre- and postharvest chitosan treatments to control storage grey mold of table grapes. *Journal of Food Science* 67 (2002): 1862–1867.

Salveit, M.E. Physical and physiological changes in minimally processed fruits and vegetables. In F.A. Tomás-Barberán and R.J. Robins (eds.), *Phytochemistry of Fruit and Vegetables*, pp. 205–220. Oxford, U.K.: New York Press, 1997.

Sangsuwan, J., Rattanapanone, N., and Rachtanapun, P. Effect of chitosan/methyl cellulose films on microbial and quality characteristics of fresh-cut cantaloupe and pineapple. *Postharvest Biology and Technology* 49(3) (2008): 403–410.

Sessa, D.J., Eller, F.J., Palmquist, D.E., and Lawton, J.W. Improved methods for decolorizing corn Zein. *Industrial Crop and Products* 18 (2003): 55–65.

Simões, A.D.N., Tudela, J.A., Allende, A., Puschmann, R., and Gil, M.I. Edible coatings containing chitosan and moderate modified atmospheres maintain quality and enhance phytochemicals of carrot sticks. *Postharvest Biology and Technology* 51 (2009): 364–370.

Sipahi, R.E., Castell-Perez, M.E., Moreira, R.G., Gomes, C., and Castillo, A. Improved multilayered antimicrobial alginate-based edible coating extends the shelf-life of fresh cut melon (*Citrullus lanatus*). *Food Science and Technology* 51 (2013): 9–15.

Skurtys, O., Acevedo, C., Pedreschi, F., Enrione, J., Osorio, F., and Aguilera, J.M. *Food Hydrocolloid Edible Films & Coatings*, In C.S. Hollingworth, Food Hydrocolloids: Characteristics, Properties, and Structures. Hauppauge NY: Nova Science Pub Inc., 2010.

Soliva-Fortuny, R. Polysaccharide coatings extend fresh-cut fruit shelf life. *Emerging Food Research and Development Report* 21 (2010): 1–2.

Soliva-Fortuny, R.C. and Martín-Belloso, O. New advances in extending the shelf-life of fresh-cut fruits: A review. *Trends in Food Science and Technology* 14(9) (2003): 341–353.

Sothornvit, R. and Rodsamran, P. Effect of a mango film on quality of whole and minimally processed mangoes. *Postharvest Biology and Technology* 47 (2008): 407–415.

Sothornvit, R. and Rodsamran, P. Mango film coated for fresh-cut mango in modified atmosphere packaging. *International Journal of Food Science & Technology* 45 (2010): 1689–1695.

Spanou, A. and Giannouli, P. Extend of shelf-life of potato round slices with edible coating, green tea and ascorbic acid. *World Academy of Science, Engineering and Technology* 79 (2013): 464–468.

Talens-Oliag, P. and Krochta, J.M. Plasticizing effects of beeswax and carnauba wax on tensile and water vapor permeability properties of whey protein films. *Journal of Food Science* 70(3) (2005): E239–E243.

Tanada-Palmu, P.S. and Grosso, C.R.F. Effect of edible wheat gluten-based films and coatings on refrigerated strawberry (*Fragaria ananassa*) quality. *Postharvest Biology and Technology* 36 (2005): 199–208.

Tapia, M.S., Pérez, E., Rodríguez, P.E. et al. Some properties of starch and starch edible films from under-utilized roots and tubers from the Venezuelan Amazons. *Journal of Cellular Plastics* 48 (2012): 526–544.

Tapia, M.S., Rojas-Grau, M.A., Carmona, A., Rodriguez, F.J., Soliva-Fortuny, R., and Martin-Belloso, O. Use of alginate- and gellan-based coatings for improving barrier, texture and nutritional properties of fresh-cut papaya. *Food Hydrocolloids* 22 (2008): 1493–1503.

Tapia, M.S., Rojas-Graü, M.A., Rodríguez, F.J., Ramírez, J., Carmona, A., and Martin-Belloso, O. Alginate- and gellan-based edible films for probiotic coatings on fresh-cut fruits. *Journal of Food Science* 72(4) (2007): E190–E196.

Thommohaway, C., Kanlayanarat, S., Uthairatanakij, A., and Jitareerat, P. Quality of fresh-cut guava (*Psidium guajava* L.) as affected by chitosan treatment. *Acta Horticulturae (ISHS)* 746 (2007): 449–454.

Toivonen, P.M.A. Fresh-cut apples: Challenges and opportunities for multi-disciplinary research. *Canadian Journal of Plant Science* 86 (2006): 1361–1368.

Toivonen, P.M.A., Brandenburg, A.J.S., and Yuo, L. (2009). Modified atmosphere packaging for fresh-cut produce. In E. Yahia (ed.), *Modified and Controlled Atmospheres for Storage, Transportation and Packaging of Horticultural Commodities*, pp. 463–489. Boca Raton, FL: CRC Press.

Toivonen, P.M.A. and DeEll, J.R. Physiology of fresh-cut fruits and vegetables. In O. Lamikanra (ed.), *Fresh-Cut Fruits and Vegetables*, pp. 463–489. Boca Raton, FL: CRC Press, 2002.

Tzoumaki, M.V., Biliaderis, C.G., and Vasilakakis, M. Impact of edible coatings and packaging on quality of white Asparagus (*Asparagus officinalis*, L.) during cold storage. *Food Chemistry* 117 (2009): 55–63.

Urban, K., Wagner, G., Schaffner, D., Roglin, D., and Ulrich, J. Rotor–stator and disc systems for emulsification processes. *Chemical Engineering & Technology* 29 (2006): 24–31.

Valderde, J.M., Valero, D., Martínez-Romero, D., Guillen, F., Castillo, S., and Serrano, M. Novel edible coating based on *Aloe vera* gel to maintain table grape quality and safety. *Journal of Agricultural and Food Chemistry* 53 (2005): 7807–7813.

Valencia-Chamorro, S.A., Palou, L., Del Río, M.A., and Pérez-Gago, M.B. Antimicrobial edible films and coatings for fresh and minimally processed fruits and vegetables: A review. *Critical Reviews in Food Science and Nutrition* 51(9) (2011): 872–900.

Vargas, M., Albors, A., Chiralt, A., and Martínez-González, C. Quality of cold-stored strawberries as affected by chitosan-oleic acid edible coatings. *Postharvest Biology and Technology* 41 (2006): 164–171.

Vargas, M., Chiralt, A., Albors, A., and González-Martínez, C. Effect of chitosan-based edible coatings applied by vacuum impregnation on quality preservation of fresh-cut carrot. *Postharvest Biology and Technology* 51(2) (2009): 263–271.

Vargas, M., Pastor, C., Chiralt, A., McClements, D.J., and González-Martínez, C. Recent advances in edible coatings for fresh and minimally processed fruits. *Critical Reviews in Food Science and Nutrition* 48(6) (2008): 496–511.

Vargas, M., Perdones, Á., Chiralt, A., Cháfer, M., and González-Martínez, C. Effect of homogenization conditions on physicochemical properties of chitosan-based film-forming dispersions and films. *Food Hydrocolloids* 25 (2011): 1158–1164.

Vartiainen, J., Motion, R., Kulonen, H., Ratto, M., Skytta, E., and Ahvenainen, R. Chitosan-coated paper: Effects of nisin and different acids on the antimicrobial activity. *Journal of Applied Polymer Science* 94(3) (2004): 986–993.

Villalobos-Carvajal, R., Hernández-Muñoz, P., Albors, A., and Chiralt, A. Barrier and optical properties of edible hydroxypropyl methylcellulose coatings containing surfactants applied to fresh cut carrot slices. *Food Hydrocolloids* 23 (2009): 526–535.

Wang, C.Y., Wang, S.Y., Yin, J.J., Parry, J., and Yu, L. Enhancing antioxidant, antiproliferation, and free radical scavenging activities in strawberries with essential oils. *Journal of Agricultural and Food Chemistry* 55 (2007a): 6527–6532.

Wang, Q.L., Khanizadeh, S., and Vigneault, C. Preharvest ways of enhancing the phytochemical content of fruits and vegetables. *Stewart Postharvest Review* 3(3) (2007b): Article #3.

Wong, D., Gastineau, F., Gregorski, K.S., Tillin, S.J., and Pavlath, A.E. Chitosan–lipid films microstructure and surface energy. *Journal of Agricultural and Food Chemistry* 40 (1992): 540–544.

Wong, D.W.S., Tillin, S.J., Hudson, J.S., and Pavlath, A.E. Gas exchange in cut apples with bilayer coatings. *Journal of Agricultural and Food Chemistry* 42 (1994): 2278–2285.

Wu, S. and Chen, J. Using pullulan-based edible coatings to extend shelf-life of fresh-cut 'Fuji' apples. *International Journal of Biological Macromolecules* 55 (2013): 254–257.

Xiao, Z., Luo, Y., Luo, Y., and Wang, Q. Combined effects of sodium chlorite dip treatment and chitosan coatings on the quality of Fresh-Cut d'Anjou pears. *Postharvest Biology and Technology* 62 (2011): 319–326.

Yang, L. and Paulson, A.T. Effects of lipids on mechanical and moisture barrier properties of edible gellan film. *Food Research International* 33(7) (2000a): 571–578.

Yang, L. and Paulson, A.T. Mechanical and water vapour barrier properties of edible gellan films. *Food Research International* 33(7) (2000b): 563–570.

Youwei, Y. and Yinzhe, R. Effect of chitosan coating on preserving character of post-harvest fruit and vegetable: A review. *Journal of Food Processing & Technology* 4(8) (2013): 4–8.

Zagory, D. Effects of post-processing handling and packaging on microbial populations. *Postharvest Biology and Technology* 15 (1999): 313–321.

21

Edible Packaging in Muscle Food

**M. Elvira López-Caballero, M. Carmen Gómez-Guillén,
Begoña Giménez, and María Pilar Montero García**

CONTENTS

Abstract

Fresh meat and fish muscle food are highly perishable foods, although some differences in the rate of spoilage can be found among species. Meat from animals for slaughter is more resistant to spoilage due to their intrinsic characteristics and postmortem phenomena. However, spoilage is usually faster in the case of fish since they are poikilotherms and their metabolic rate is determined by environmental temperatures. Oxidation processes and microbial contamination are the main causes for spoilage of muscle food, with predominance of one or another depending on the type of product and the preservation system (chilled or frozen). Therefore, one of the primary functions of both traditional plastic packages and biodegradable edible packages is to prevent food spoilage. Edible packages can be of different types, but gelatin-based coatings, combined or not with other biopolymers, are especially relevant since gelatin sets to a gel on cooling, avoiding unnecessary thermal treatments to muscle food. Recent research trends are focused on the development of bioactive coatings and films, with bioactive compounds that may migrate into the muscle food to prevent spoilage processes and that may exert a beneficial effect in human health when they are consumed. These edible coatings and films may be applied in chilled or frozen meat and fish products. The choice between coating and film will depend on the type of food product as well as the preservation conditions to which they are subjected.

21.1 Introduction

Active packaging has been applied to maintain or enhance the quality and safety of food for years. With the increasing awareness of environmental hazards that are associated with the use of synthetic packaging materials, an urgent need for biodegradable and edible packages, whether as films or coatings, has been realized. Edible packages are thin layers of edible material applied to the product surface to provide a barrier to moisture, oxygen, and solute movement for the food. This type of packaging has been long used for food packaging, for example, cheese coatings, sausage casings, or even caramelized apples.

Research into edible packages has progressed considerably in recent years, achieving a breakthrough in the potential and diversification of this type of packages, as well as in the range of applications. This has led to the successful application of edible packages to preserve certain foods, such as meat and seafood products, allowing the obtaining of safe, stable, and quality foods that are able to face economic and marketing challenges. Furthermore, the coatings can go completely unnoticed or even make food more visually appealing by choosing an appropriate development strategy.

Edible coatings and films can be applied to fresh and processed food products to extend the shelf life and improve the overall eating quality, both in chilling and freezing conditions, by acting as barriers to water, gases, and vapor (Coma, 2006; Sánchez-Ortega et al., 2014). The mechanisms involved in food deterioration during storage are enzymatic autolysis, lipid oxidation, and microbial spoilage, which are usually accompanied by a marked desiccation and discoloration. The contribution of each mechanism to the overall food deterioration and loss of sensory quality will depend on the preservation method and the type of food product. As a result, the properties that edible packages must fulfill will be different in each case. Carbohydrates, proteins, and lipids have been used as materials in the development of edible packages. Various types of proteins have been successfully used as edible films, such as gelatin, casein, whey protein, corn zein, wheat gluten, soy protein, bean protein, and peanut protein (Gennadios et al., 1997; Bourtoom, 2008). Among carbohydrate sources, alginate, carrageenan, starch, and cellulose have been shown to be suitable to produce edible packages (Gennadios et al., 1997; Phadke et al., 2011). Several waxes, fats, oils, glycerides, and acetylated glycerides have been used as protective lipid-based coatings in foods (Stuchell and Krochta, 1995; Gennadios et al., 1997). Furthermore, antioxidant and antimicrobial compounds are often incorporated to the formulation of edible packages (Gennadios et al., 1997; Cutter and Sumner, 2002; López de Lacey et al., 2014). In this way, edible coatings and films can extend the shelf life of muscle food by maintaining high concentrations of preservatives on the surface of food for a longer time (active packaging), and they may even exert a beneficial effect on human health when consumed (bioactive packaging).

This chapter is dealing with the recent trends in the application of edible packages to muscle food (meat, poultry, and seafood) and their role in preservation.

21.2 Application of Edible Packaging to Muscle Food

Shelf life of fresh muscle food is very limited, although there are some differences in the rate of spoilage among species. Thus the rate of spoilage in land animals and tropical warmwater fish species (due to their high body temperature) is greater in cold-water fish species, although chilling is equally crucial for their preservation.

The use of edible packaging on muscle food offers several potential benefits such as reduction of moisture loss and volatile flavor loss, color changes, lipid oxidation, and microbial growth, improving their quality, safety, and shelf life (Gennadios et al., 1997; Cutter and Sumner, 2002; López-Caballero et al., 2005). In recent years, there have been many studies on edible films and coatings, where numerous biopolymers have been assayed: waxes, oils, carbohydrates, and proteins from animal and vegetable sources. However, most of the studies dealing with the application of edible packaging in muscle food are focused on the use of coatings, and only a few are dealing with the application of edible films (Coma, 2006; Gómez-Guillén et al., 2009). This is partly because the industrial implementation of edible coatings is easier and they can be produced in the same food processing company, whereas edible films require an entire industrial development that just stands out as a future possibility. Nevertheless, some studies are focused on the development of edible films with interesting sensory and bioactive properties, as will be shown throughout this chapter.

In spite of the potential benefits related to edible coatings and films, some negative aspects of their use have been also described. The application of protein-based coatings on meats, poultry, and seafood has limitations due to the susceptibility of proteins to the activity of muscle enzymes, microbial spoilage, or potential allergenicity of the protein material. Furthermore, muscle water content can migrate to the edible package leading to film swelling or even disintegration, together with eventual food moisture loss, which may affect adversely both product appearance and texture. On the other hand, edible packaging could also be useful to apply as absorbent material to avoid the excess of drip loss of certain muscles (meats, poultry, and fish) during storage.

21.2.1 Coatings

The biopolymers and matrices that can be used for muscle food coating development are highly diverse, and the choice of one or the other will depend on the distinctive features and positive aspects they confer in that food. Besides the type of food, other aspects should be taken into account, such as postslaughter handling of muscle food and storage temperatures as well as the type of processing applied. Several studies were performed in the late 1990s dealing with coating development for muscle food using several types of matrices such as waxes, oils, proteins (collagen, gelatin, cereal proteins, oilseed proteins, milk proteins), and polysaccharides (starch, alginate, carrageenan, dextran, cellulose) (Cutter and Sumner, 2002; Cha and Chinnan, 2004; Khwaldia et al., 2004; Cutter, 2006). However, lipid-based coatings have shown application (thickness and homogeneity, greasy surface, cracking) and organoleptic (rancidity, waxy taste) issues that rendered their use unpractical (Gennadios et al., 1997; Guilbert, 2000).

Current research trends are focused on using complex matrices made up of two or more biopolymers and on the application of these composite films for meat, poultry, and seafood preservation in combination with other preservation technologies (vacuum packaging, modified atmosphere packaging, high pressure, flash pasteurization, and irradiation) (Cutter, 2006; Gómez-Estaca et al., 2007; Duan et al., 2010; Baranenko et al., 2013; Guo et al., 2014).

A number of methods have been employed for the application of edible coatings to muscle food such as dipping, spraying, brushing, wrapping, or rolling (Cutter, 2006). Furthermore, coatings may be applied as a film-forming solution with varying viscosity rates, but also as an emulsion or foam. A high viscosity hampers the ability to extend the solution to form a very thin, barely perceptible film, so as to be sensory acceptable. When a thinner, more uniform coating is required for certain surfaces, coatings may be best applied by spraying (Cutter and Sumner, 2002). However, spraying does not always lead to a uniform distribution of the coating on the surface of muscle food, leaving uncoated areas. After the coating has been applied, this is allowed to solidify on the muscle food. The drying process often involves a heating treatment that may be unsuitable for certain food products, especially fresh muscle food.

21.2.2 Collagen/Gelatin Coatings

Collagen and gelatin, which is derived by partial hydrolysis of collagen, became biopolymers of particular interest in the development of edible coatings for muscle food since these biopolymers are able to gel upon cooling. This feature allows collagen and gelatin edible coatings to be applied on fresh/raw and processed muscle food without leading to loss of quality or changes in their organoleptic properties (Cutter and Miller, 2004; López-Caballero et al., 2005).

In the case of mammalian or fish gelatin–based coatings, the polymeric matrix is composed only of gelatin or gelatin combined with other biopolymers such as chitosan (Baranenko et al., 2013; Poverenov et al., 2014) or carboxymethylcellulose (Silva-Weiss et al., 2014). In any case, the percentage of gelatin in the coating should be enough to maintain the ability to gel on cooling. The plasticizers usually used in gelatin-based, film-forming solutions are glycerol or a glycerol–sorbitol mixture at the same percentage. Gelatin coatings show good barrier characteristics against oxygen and aroma transfers at low and intermediate relative humidity. However, they have poor barrier properties against water vapor transfer

 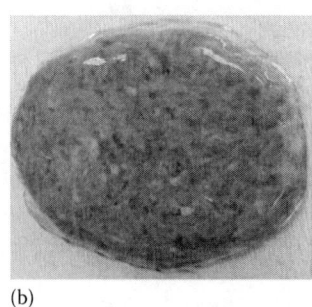 

(a)   (b)   (c)

FIGURE 21.1   (a) Hamburger covered with a gelatin–chitosan-based coating very fine as refrigerated product, (b) hamburger covered with a gelatin–chitosan-based coating very thick as glazed in frozen product, (c) fish sausages covered with a gelatin–chitosan–carotenoid concentrate–based coating. (Photo courtesy of the Research Group: Development, Valorisation and Innovation of Fish Products (INNOVAPESCA), of ICTAN-CSIC.)

due to their hydrophilic nature (Jongjareonrak et al., 2006; Carvalho et al., 2008; Limpisophon et al., 2009). Many attempts have been done to reduce the moisture affinity of gelatin-based coatings and films, including the incorporation of cellulose nanocrystals (George and Siddaramaiah, 2012).

The application of gelatin-based coatings on refrigerated or frozen muscle food leads to the prevention of moisture loss, reduction of the presence of oxygen at the surface, and off-flavor adsorption or volatile flavor loss. Furthermore, gelatin-based coatings have shown a good potential as carriers of bioactive compounds. Thus, several antimicrobial and antioxidant compounds have been added to the gelatin-based coating formulations and applied on muscle food, both meat products and seafood (Cutter, 2006; Andevari and Rezaei, 2011; Giménez et al., 2011; Liang et al., 2011; Matos de Oliveira et al., 2013). A special benefit of gelatin-based coatings is that their removal before consuming the food is not necessary, since they are completely edible and furthermore they melt and disappear during cooking.

Regarding collagen, when heated, intact collagen films can form an edible skin that becomes an integral part of the meat product (Cutter and Miller, 2004). Collagen films prevent shrink loss, increase juiciness, and adsorb fluid exudate. Furthermore, they may increase permeability of smoke to the meat product and allow for easy removal of nets after cooking or smoking (Cutter and Sumner, 2002).

Figure 21.1 shows some products coated with different edible coatings. The coating on the burger to be kept refrigerated is hardly noticeable, while the coating that will protect the burger during freezing, as a glaze, is thicker. In addition, coatings can be a way to protect other product such as sausages; depending on the composition and properties of the coatings, they also confer surface coloring and bioactive properties with protective or nutraceutical function.

21.2.3 Casings

An interesting application, although not new, is the use of collagen or gelatin for casing manufacture. Collagen casings are used in sausage manufacture to offer sausages with edible casings obtained from a food ingredient, and it is estimated that 80% of all edible casings are made from manufactured collagen (Cutter, 2006). From an industrial point of view, the manufacture of edible sausage casings is one of the most successful applications of edible packaging in meat products. The collagen used for manufactured collagen casings is extracted from the corium layer of bovine hides. Once the corium layer has been removed, it is decalcified and ground. Acid is then added to induce swelling of the collagen and the swollen collagen dough is extruded (Harper et al., 2012). However, while production and application of collagen sausage casings are well-established technologies, little research has been focused on the manufacture of sausage casings using other food ingredients although it has been resumed in recent years.

According to Liu et al. (2005), casings formed from pectin and blends of alginate and gelatin showed good properties and produced intact and stable sausage products. Quality and stability of these casings manufactured by extrusion were greatly enhanced on the incorporation of vegetable oils to the formulations (Liu et al., 2006). Afterward, pectin and alginate/gelatin casings were assessed for the potential manufacture of pork sausages (Liu et al., 2007). Weight loss was lower in the case of sausages stuffed with alginate/gelatin casings, although shrinkage of the product was observed after some days of storage. Furthermore, sausages manufactured using casings containing emulsified corn oil were more stable to lipid oxidation rates than those containing olive oil. These are promising results, since, in spite of the successful marketing of collagen casings, it is possible to further improve their properties throughout the incorporation of other ingredients that enhance their stability or even give a more attractive product.

21.2.4 Use of Coatings on Deep-Fat Fried Products

Deep-fat frying is a dry cooking process that involves the immersion of food pieces in hot vegetable oils (Moyano et al., 2002). A soft and moist core and a crispy crust are desirable characteristics of most fried foods. Coating the food with different types of biopolymers can be an effective method to reduce the oil uptake and moisture loss during deep-fat frying as well as to reduce lipid oxidation. This application has become increasingly important in recent years, as oil uptake in fried products has become a health concern, related to obesity and coronary diseases (García et al., 2008). Among the different coating materials, hydrocolloids such as cellulose derivatives, alginates, or proteins are the most attractive because they are good barriers to fat (Garmakhani et al., 2014). Besides forming a coating, hydrocolloids may be added to the batter among its other ingredients, where they are able to avoid oil absorption but also they can act as viscosity control agents; improve adhesion, pick-up control, and freeze–thaw stability; or help to retain the crispness of the battered/breaded fried foods (Varela and Fiszman, 2011).

Although there are many studies dealing with the use of coatings on fat-fried potatoes, a few studies have been focused on the effect that certain coatings may have on breaded or battered meat, poultry, or fish products during frying (Holownia et al., 2000, 2001; Kilincceker et al., 2009; Dilek et al., 2011; Kilincceker and Hepsag, 2011; Kurt and Kilincceker, 2011). These products are usually marketed as prefried and frozen products, and the coatings used may consist of one biopolymer or mixtures of them, in a single layer or multiple layers that give different properties to the product (Garmakhani et al., 2014). Optimizing the coating materials is necessary to increase the performance of the coating. Some studies have been focused on the evaluation of the effectiveness of multilayer coatings, both the type of biopolymer used and the order of the layers. It was more advantageous to use a protein-based coating as the first coating, a gum as the second coating, and flours as the last coating, when protein-, gum-, and flour-based coatings were applied on prefried and frozen fish fillets. As for fillet quality, the lowest thiobarbituric acid (TBA) index and volatile basic nitrogen levels were found in the samples with the protein-based coating, whereas the highest levels were reported in the those samples with gum-based coatings (Kilincceker et al., 2009). In another study, these authors used yellow lentic flour and chickpea flour as coating materials on fried fish balls. Both flours and their mixtures increased efficiently the quality of the battered fish balls, being a suitable alternative to produce battered fish balls with good sensory acceptability even at low frying temperatures (Kilincceker and Hepsag, 2011).

The application of coating materials on deep-fried chicken strips has been reported to give noticeable fat-uptake reductions. The use of hydroxypropyl methylcellulose (HPMC) as an ingredient of the breading mixture in prefried marinated chicken strips resulted in a significant lower fat absorption (Holownia et al., 2000). Furthermore, the use of HPMC in prefried chicken strips marinated in pickle juice, either as an edible coating or as an ingredient in the breading mixture, reduced tocopherol losses in the peanut oil used for frying. This was probably because the coating material acted as a hydrophilic barrier to migration of the prooxidant acetic acid from the product to the oil (Holownia et al., 2001). Methylcellulose-coated nuggets showed lower fat uptake than control nuggets both in the crust (26%) and in the core (14%) when frying was performed at 175°C and 190°C (Lalam et al., 2013). Protein-based coatings have been also successfully used to reduce fat uptake in deep-fried chicken (Kurt and Kilincceker, 2011). Thus, Dragich and Krochta (2010) reported that whey protein–based coatings reduced the fat uptake up to 37% in deep-fried chicken breast strips when compared with the uncoated control.

21.3 Active Packaging

Edible coatings and films may act as carriers for active compounds with different functionalities. This is an advantage compared to the direct application of active compounds (antioxidants, antimicrobials, among others), which usually is less effective due to rapid migration into the food bulk or reaction with the food components (Siragusa and Dickinson, 1992). The objects of applying these active compounds as part of the packaging are a controlled and progressive diffusion of the active agent to the surface of the food as well as the use of lower doses. The activities and properties of the packaging will depend on the type of polymeric matrix, the active compounds added, and the interactions that are established between them. Based on the relation between the polymeric structure and the transport of the active molecules through the network (Papadokostaki et al., 1997), coatings and films containing active compounds enable controlled release of active molecules into the food (Lacroix et al., 2004). On the other hand, film application is usually less effective than coating to fulfill its intended purpose, since edible films may have difficulties in attaching to the food surface, limiting migration and the efficacy of the active compound (Chiu and Lai, 2010).

The most commonly used active compounds are plant extracts, from leaves, seeds, or even roots. The addition of plant extracts to edible coatings and films has been shown to protect against spoilage and pathogenic microorganisms as well as lipid oxidation while also enhancing sensory properties of foods. The activity of these extracts is mainly attributed to their content of polyphenolic compounds. In general, it may be considered that their activity is proportional to the content of these compounds, although their effectiveness will also depend on the type of polyphenols they contain. Some authors consider that the presence of other compounds in the plant extracts such as minerals or vitamins may enhance their activity (Chiu and Lai, 2010). Essential oils or oily extracts usually show higher effectiveness than the corresponding water-soluble extracts. This differences would probably be due to a different composition profile, as essential oils are mainly composed of nonpolar or low-polarity volatile compounds, whereas water-soluble extracts are composed of polar compounds (Iturriaga et al., 2012). However, both color and smell of essential oils are very intense, and their incorporation into edible coatings and films is limited by the organoleptic changes they produce when used for muscle food preservation. In spite of this, essential oils have been added to edible films and coatings for the preservation of both meat and seafood products (Gómez-Estaca et al., 2010; Moradi et al., 2011; Salgado et al., 2013; Alparslan et al., 2014; Bonilla et al., 2014). However, a sensory evaluation to assess the sensory repercussions derived from the incorporation of essential oils is not performed in most of these studies. In connection with this, bilayer films based on agar and sodium alginate were developed by the incorporation of cinnamon essential oil in the upper layer in order to reduce negatively the sensory repercussions derived from the essential oil (Arancibia et al., 2014). Both agar and alginate bilayer films allowed to reduce significantly the microbial growth, including *Listeria monocytogenes*, in peeled shrimps during the chilled storage without a negative impact on the organoleptic properties.

As a consequence of the foregoing considerations, numerous studies that apply edible coatings and films on meat, poultry, or seafood products are based on the incorporation of aqueous or ethanol/water extracts, so they do not confer smell or coloring to the muscle food (Gómez-Estaca et al., 2007; Hong et al., 2009; Giménez et al., 2011; Song et al., 2011; Shin et al., 2012; Siripatrawan and Noipha, 2012; López de Lacey et al., 2014). For example, Chiu and Lai (2010) have described that a green tea–water extract as an active compound of a polysaccharide-based coating does not affect the sensorial characteristics of ready-to-eat salads and freshly cut foods, and they even enhance the consumer's decision to purchase probably due to the residual tea flavor or the slightly shiny appearance of the food. The type of muscle food to be preserved must be also taken into account when considering the limitations of certain types of ingredients in packaging. Thus, packaging for white-colored mild-tasting hake fillet is not the same as the one for beef steak, with its intense color and strong flavor, or packaging for sardines or salmon. Thus, Lacroix et al. (2004) reported that the combination of ascorbic acid and coatings with powdered spices does not affect the sensorial quality of minces beef. With regard to essential oils, whose incorporation into coatings is limited by the organoleptic changes they originate, Chi et al. (2006) propose that the addition of oregano essential oil (45 ppm or less) to bologna is acceptable to the consumer. The high lipid and moisture content of bologna facilitate diffusion of essential oils from the chitosan matrix to the product. Thus, the addition of an emulsifier is important to slow down the liberation of volatile compounds in the oil and by this, to control their liberation.

Other active compounds used in edible films and coatings for muscle food preservation include chitosan (Jeon et al., 2002; López-Caballero et al., 2005; Jiang et al., 2011; Guo et al., 2014), organic acids and their salts such as acetic acid or lactic acid (Beverlya et al., 2008; Jiang et al., 2011; Baranenko et al., 2013), antioxidant peptides such as nisin or antioxidant protein hydrolyzates (Giménez et al., 2009; Gómez-Guillén et al., 2010; Liang et al., 2011; Guo et al., 2014), enzymes such as lysozyme (Unalan et al., 2011), and nanoparticles (Morsy et al., 2014), among others. Chitosan deserves special attention because it is a biopolymer with good film-forming ability (Jeon et al., 2002; Gómez-Estaca et al., 2010; Ojagh et al., 2010; Huang et al., 2012; Asik and Candogan, 2014; Bonilla et al., 2014) that can be also used in combination with other biopolymers such as gelatin, sodium caseinate, and starch (López-Caballero et al., 2005; Gómez-Estaca et al., 2010; Moreira et al., 2011; Baranenko et al., 2013), omega-3 fatty acids (Duan et al., 2010), or essential oils (Gómez-Estaca et al., 2010; Ojagh et al., 2010; Bonilla et al., 2014). Chitosan has good antimicrobial activity against many pathogenic and spoilage microorganisms, including gram-positive and gram-negative bacteria, molds, and yeasts, because of its polycationic property (Agulló et al., 2003). Chitosan also exhibits antioxidant activity when used as a food additive because of its ability to chelate metal ions (Kamil et al., 2002; Agulló et al., 2003; López-Caballero et al., 2005).

21.3.1 Packaging with Antioxidant Properties

The addition of antioxidant compounds is the most widely used strategy to confer this property to edible films and coatings. Given the multiple options of compounds that have these properties, the array of possibilities is extensive. Several aspects of the coatings and films with antioxidant properties must be considered because they may act as visible and ultraviolet light filters, they prevent lipid oxidation, and they may maintain their bioactive potential, giving the packaging a possible beneficial effect in human health when consumed.

21.3.1.1 Light Barrier

Some films and coatings are highly transparent and uncolored, such as those elaborated with gelatin, cellulose derivatives, pullulan, chitosan, starch, and sodium caseinate (Akhtar et al., 2010; Moreira et al., 2011; Baranenko et al., 2013; Morsy et al., 2014). Other films have a high degree of transparency, but with some yellow hues, such as isolated soy protein films. Sometimes, by incorporating plant extracts into gelatin films, transparency is maintained and a pleasant tone is achieved from a sensory point of view, with color variations depending on the tone of the extract. Such is the case of the addition of aqueous extracts of oregano, rosemary, green tea leaves, ethanol/water extracts of borage seeds, and essential oils, among others (Gómez-Estaca et al., 2009a,b,c). It is also possible to achieve high opacity, for example, by emulsifying the biopolymer with oil (Pérez-Mateos et al., 2009). Figure 21.2 shows the appearance of films with different ingredients.

Although they show different hues, all of them may be suitable for product packaging (Figure 21.3). Consumers usually demand translucent food packages to be able to see the food packed in them. However, this exposes the food to visible and ultraviolet light during marketing, thus accelerating oxidation reactions. Films enriched with plant extracts frequently absorb ultraviolet and visible light, preventing oxidation due to light (Gómez-Guillén et al., 2007; Gómez-Estaca et al., 2009a,b). The light absorption level in gelatin–lignin composite films is also very high (Nuñez-Flores et al., 2013) and could even be higher than that obtained with vegetable extracts in some parts of the spectrum. The chromophoric nature of lignin is known to be highly capable of protecting against UV radiation (Ban et al., 2007; Pereira et al., 2007).

Another mechanism to prevent photooxidation is the addition of edible colors (blue, green, yellow, red, and white) to HPMC-based films (Akhtar et al., 2010). As mentioned earlier, films of these polysaccharides are highly transparent, but transparency is lost to some degree by adding these colors. According to the authors, the decreasing order of transparency was yellow, green, blue, red, and white (yellow with values that were very close to those of uncolored HPMC film). These types of packaging were adequate

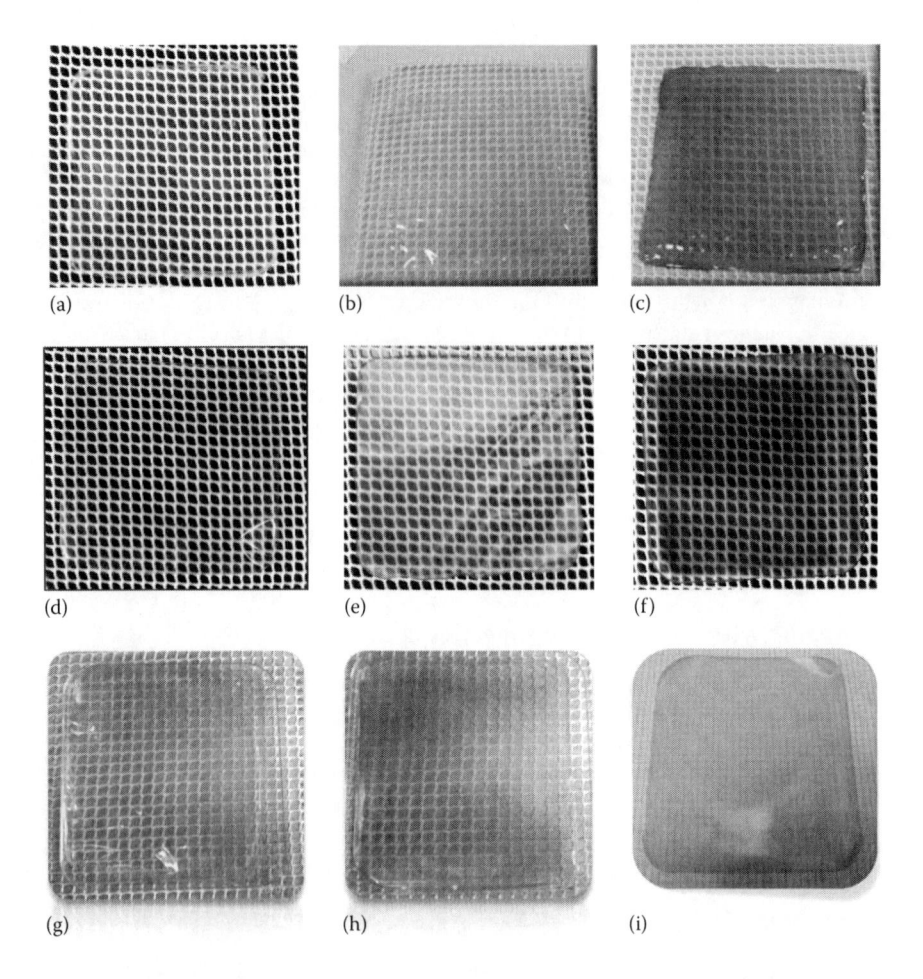

FIGURE 21.2 Appearance of the films developed with (a) myofibrillar squid protein, (b) fish gelatin, (c) gelatin with lignosulfonate solution, (d) carrageenan extracts from mastocarpus seaweed, (e) carrageenan with bioactive extracts from mastocarpus seaweed, (f) carrageenan with bioactive and hydrolyzate extracts from mastocarpus seaweed, (g) gelatin and chitosan, (h) gelatin and chitosan with carotenoids extract, (i) ε-polylysine-coated liposomes entrapping bioactive peptides incorporated in an edible film from cooked shrimp muscle. (Photo courtesy of the Research Group: Development, Valorisation and Innovation of Fish Products (INNOVAPESCA), of ICTAN-CSIC.)

light barriers to prevent photooxidation of salmon oil stored under fluorescent light at 20°C for 8 days. Furthermore, white-, red-, and yellow-colored films prevented oil oxidation in the same extent as darkness. This fact indicates that not only transparency is the only factor involved in oxidation but also there are other factors that are also involved such as color of edible films and the wavelength of maximum absorption of colors. Packaging with lower light transmission rates (white, red, and yellow) reduced lipid oxidation in oil salmon, evaluated by both conjugated dienes and fatty acid composition. This fact was related to lower oxygen consumption in the packaging. Considering these results, yellow is a good choice if a transparent packaging is required, whereas red and white are also interesting if transparency is not required. Other authors reported that when horsemeat was packaged with red plastic film that was permeable to oxygen, the color of the film delayed the formation of cholesterol oxidation products after 8 h of exposure to neon light, compared to transparent film (Boselli et al., 2010).

21.3.1.2 Bioactive Potential

Covering the muscle food with edible coatings and films enriched with plant extracts may cause deposition of phenols in the muscle, conferring antioxidant power to the muscle (Chi et al., 2006;

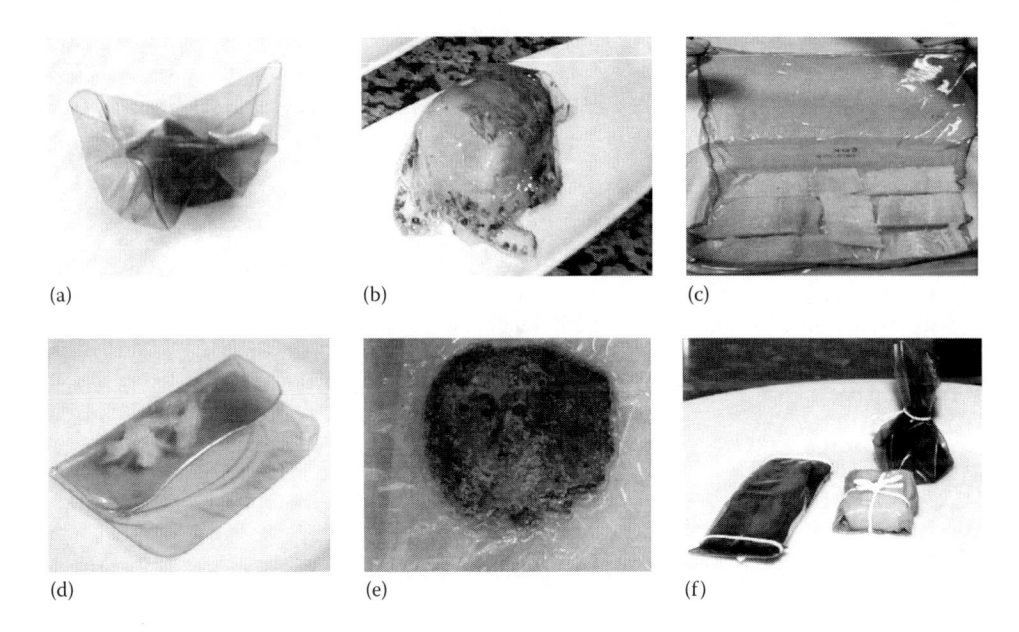

(a)　　　　　　　　(b)　　　　　　　　(c)

(d)　　　　　　　　(e)　　　　　　　　(f)

FIGURE 21.3 Muscle products coated with edible films highly transparent and colored due to bioactive compounds. (a) Pork meat coated with agar–tea extract film, (b) monkfish muscle coated with gelatin-spiced film, (c) hake fillets coated with gelatin–ruda extract films, (d) poultry pieces coated with *Ascophyllum* seaweed extract, (e) frozen beef meat hamburger coated with gelatin–borage seed extract film, and (f) different kinds of muscles packed for preparation in *novel cuisine*. (Photo courtesy of the Research Group: Development, Valorisation and Innovation of Fish Products (INNOVAPESCA), of ICTAN-CSIC.)

Gómez-Estaca et al., 2007), as has been observed in smoked sardine covered with film enriched with aqueous extracts of oregano (*Origanum vulgare*) or rosemary (*Rosmarinus officinalis*) leaves. Antioxidant power (as ferric reducing ability of plasma [FRAP]) of the samples coated with the oregano and rosemary extract–enriched films was similar, regardless of the film applied, despite those coated with rosemary-enriched films that had lower phenol content. This suggests a quantitative and qualitative difference in the phenolic compounds of each extract, as well as the different degree of interaction with the matrix protein. Throughout the storage, different behaviors were observed among the two coated samples with regard to the phenol content. Although phenols were released mainly at the beginning of the storage in both coated samples, a more gradual release of antioxidant compounds were observed in samples coated with rosemary-enriched films. Oussalah et al. (2004) also evaluated the antioxidant capacity in beef muscle stored in refrigeration (7 days), appreciating an increase of phenol contents at the beginning, stabilization later, and finally a slight decrease, depending on the source of antioxidant compounds used (oregano or pepper). The antioxidant capacity of mackerel patties, measured as ferric reducing ability, increased when patties were coated with gelatin-based films containing a borage-seed extract (Giménez et al., 2011). FRAP values noticeably increased until day 30 of frozen storage most likely as a consequence of the diffusion of phenolic compounds from the borage films to the patties, increasing the antioxidant capacity of the muscle. However, the FRAP values sharply decreased from day 30 to 120 and remained stable until the end of the frozen storage, probably due to the role of phenolic compounds in lipid oxidation, which consequently involves a reduction of the availability of the antioxidant compounds as well as the antioxidant capacity of muscle in spite of the migration from the films. However, around a twofold increase of reducing activity was shown in samples after thawing and subsequent chilled storage, probably due to an increase of the migration rate of phenolic compounds from the film to the muscle.

As commented earlier, chitosan provides antioxidant power. Kamil et al. (2002) indicated that the antioxidant power of chitosan is due not to its reducing power, but rather to its ability to prevent the formation of reactive oxygen species (ROS) via metal chelation. Heme iron plays an important role, since it is one of the main factors causing lipid oxidation in fish and meat products due to its catalytic effect. Some researchers have also reported that chitosan is capable of scavenging ROS directly (Xue et al., 1998;

Xing et al., 2004). In another study, chitosan films with oregano essential oil were applied on bologna slices (Chi et al., 2006). Carvacrol was not detected in the films after application on bologna slices for 5 days at 4°C, mainly due to its diffusion into bologna. Moisture and high lipid content of bologna seemed to help the diffusion of the oregano essential oil from the chitosan film matrix into the product.

21.3.1.3 Rancidity Prevention

In most studies, the effect of antioxidant edible coatings and films on muscle lipid oxidation is evaluated through the decrease of peroxide value and the TBA index compared to the uncoated muscle. Edible coatings and films can be used to prevent lipid oxidation in muscle food by the incorporation of antioxidant compounds in the formulation, but at the same time, they also represent a barrier to oxygen, which results in a better preservation of quality (Bonilla et al., 2012). Therefore, oxidation could be reduced by selecting materials of limited oxygen permeability. According to Kester and Fennema (1986), hydrophilic films and coatings (polysaccharide or protein based) generally provide a good barrier to oxygen transference. This property is in turn greatly affected by the water availability and temperature (Bonilla et al., 2012). Whey protein edible coatings were effective in preventing lipid oxidation in gutted kilka (*Clupeonellia delitula*) during frozen storage. Peroxide and TBA index values were significantly lower in samples coated by dipping in whey protein solution at 12%–13% than in uncoated samples, probably due to the oxygen barrier properties of whey protein coatings (Rostami et al., 2010; Motalebi and Seyfzadeh, 2011).

An antioxidant composite edible film was developed from xanthan gum and defatted mustard meal without incorporation of external antioxidants and applied to cold-smoked salmon to retard lipid oxidation (Kim et al., 2012). This composite film significantly reduced lipid oxidation, and the TBARS values of coated samples showed were significantly lower than those of uncoated samples irrespective of the storage temperature (4°C, 10°C, and 20°C).

A number of studies are dealing with the prevention of lipid oxidation in muscle food by applying edible coatings and films incorporated with plant extracts, essential oils, or water-soluble extracts. Thus, gelatin films enriched with oregano or rosemary applied to smoked sardine significantly decreased both indices compared to the uncoated samples (Gómez-Estaca et al., 2007). Free fatty acid content also decreased in smoked sardine coated with oregano enriched films. However, no protecting effect was observed in beef coated with whey protein films enriched with oregano or pepper. This may be attributed to the different nature of meat and fish lipids, because the latter are more unsaturated and more susceptible to oxidation (Oussalah et al., 2004). Lipid oxidation, measured by peroxide and TBARS values, was retarded by gelatin-based films enriched with laurel essential oil in rainbow trout fillets stored at 4°C, due to the antioxidant properties of laurel essential oil together with the low oxygen permeability of the gelatin film (Alparslan et al., 2014). In another study, gelatin films enriched with an ethanol/water extract of borage seeds prevented lipid oxidation in mackerel patties (*Trachurus trachurus*) during frozen storage and subsequent refrigeration. Compared to the effect of vacuum packaging, the gelatin films with borage extract showed a similar effect up to advanced stages of frozen storage, although higher protection was observed once the hamburgers were thawed and exposed to oxygen during refrigeration, with the additional advantage of increasing the antioxidant capacity of muscle (expressed as ferric reducing ability) (Giménez et al., 2011). Sunflower protein films incorporated with clove essential oil have been applied for the preservation of sardine patties during chilled storage (Salgado et al., 2013). These films significantly reduced lipid oxidation, measured as TBARS throughout the storage period when compared to uncoated samples.

Chitosan has been successfully incorporated to edible coatings and films to prevent lipid oxidation in muscle food. Artharn et al. (2009) applied round scad (*Decapterus maruadsi*) protein-based films with and without the addition of palm oil and chitosan during the preservation of dried mackerel powder. The results showed that edible films with chitosan and palm oil noticeably reduced the TBA values and might therefore function as an oxygen barrier, while hydrophobic polymer films such as high-density polyethylene films that were relatively poor oxygen barriers but good barrier to water vapor showed the highest TBA values. Chitosan films performed well as oxygen barriers (Butler et al., 1996), but provide limited barrier to water vapor due to their hydrophilic nature, although this disadvantage may be counteracted to some extent by the addition of oil. Jeon et al. (2002) reported that chitosan coatings were effective to reduce lipid oxidation in fresh cod and herring. Furthermore, the inhibitory effect of chitosan against oxidation was viscosity dependent, and the higher the apparent viscosity of chitosan, the higher the efficiency of chitosan coatings to reduce the lipid oxidation, probably due to the presence of a large number of ionic functional groups, which create strong polymer interactions that restrict the chain motion in high-viscosity chitosans, consequently providing good oxygen barrier properties. According to Ojagh et al. (2010), the application of chitosan coatings with and without cinnamon essential oil was effective in retarding the production of hydroperoxides in rainbow trout fillets during chilled storage. In another study, chitosan films efficiently protected minced pork meat from lipid oxidation, and the highest protection was achieved when chitosan films were enriched with basil and thyme essential oils (Bonilla et al., 2014). However, no preventive effect of rancidity was reported in cod hamburgers with gelatin–chitosan coatings (López-Caballero et al., 2005), although it was attributed to the low values of oxidation observed. Gelatin–chitosan films did not prevent lipid oxidation in smoked sardine (Gómez-Estaca et al., 2007).

When the applications of edible films enriched with oregano extract were combined with high-pressure treatment, it was observed that they prevented pressure-induced lipid oxidation (Gómez-Estaca et al., 2007). Pressure facilitates phenol diffusion from the film matrix to the muscle, promoting interaction between phenols and lipids, retarding rancidity. However, a beneficial effect was not observed regarding the decrease in free fatty acids, indicating that the pressure level applied did not inactivate lipases. The combined use of high-pressure and gelatin–lignin films was proposed as an alternative to a more aggressive conventional heat treatment (Ojagh et al., 2011). The objective was to improve the appearance and quality of salmon fillets in ready-to-eat or semi-prepared products. The film reduced the levels of carbonyl groups formed immediately after treatment, preventing high-pressure-induced oxidation in advanced stages of preservation. The combined use of pectin-based coatings enriched with green tea extracts (ethanol/water) with irradiation has also been evaluated in pork patties. Addition of green tea extract to the coating materials for irradiated pork patties reduced lipid oxidation, and the tea extract was deposited in the muscle during storage (Kang et al., 2007). Lipid oxidation is attributed to the combination of free radicals with O_2 to form hydroperoxides. The combination of ascorbic acid and edible coatings with powdered spices reduces lipid oxidation and the production of SH radicals without affecting the sensory quality of minced beef (Ouattara et al., 2002a; Lacroix et al., 2004). According to the authors, ascorbic acid and other compounds in the spices act as oxygen and free radical scavengers, thus preventing oxidation processes.

21.3.2 Packaging with Antimicrobial Properties

When a packaging acquires antimicrobial properties, the created material or system prevents microbial growth, increasing the lag phase and reducing the growth rate or decreasing the live microorganism counts (Han, 2000). In the last years, a large number of studies dealing with antimicrobial properties of edible coatings and films have been reported. Packaging from polysaccharide matrices achieve the functionality through the addition of green tea in the starch of tapioca/rubber from hsian-tsao leaves (*Mesona procumbens*) (Chiu and Lai, 2010), organic acids immobilized in calcium alginate gels (Siragusa and Dickinson, 1992), EDTA, citric acid, polyoxyethylene monolaureate and nisin in agar coating (Natrajan and Sheldon, 1995), and trisodium phosphate and sodium chlorite in pea starch and calcium alginate, respectively (Mehyar et al., 2007). In the case of films with protein matrices, the functionality is achieved by adding thyme, rosemary, and sage in caseinate and whey protein films (Lacroix et al., 2004); oregano and pepper essentials oils in soy and whey protein matrices (Ouattara et al., 2002b); thyme, sage, and rosemary in calcium caseinate and whey protein isolate coatings (Ouattara et al., 2002a); sorbic acid in wheat gluten matrices (Guillard et al., 2009); chitosan in fish gelatin coatings (López-Caballero et al., 2005); clove essential oil in fish gelatin (Gómez-Estaca et al., 2009c) and bovine gelatin (Gómez-Estaca et al., 2010); and aqueous extracts of rosemary and oregano in fish gelatin films (Gómez-Estaca et al., 2007).

The antimicrobial activity in lipid coatings has also been described, for example, with sorbic acid in bees wax matrices (Guillard et al., 2009). At present, the application of coatings to confer stability to foods is a broad field of study. Many of these works are related to the use of chitosan, which, as mentioned earlier, has a great potential as a packaging material due to its antimicrobial properties and the lack of toxicity. This, in addition to its filmogenic capacity, makes it an ideal biodegradable material, contributing to food preservation and shelf life increase (Dutta et al., 2009). Furthermore, the functional properties of chitosan films may be improved when combined with other polymers.

Chitosan films may have an antimicrobial effect on food due to the properties of chitosan, as well as the acid in which it is solubilized. Low pH is usually used to solubilize chitosan, resulting in an acid pH on the surface of the muscle, creating a dysgenesic medium therein and thus demonstrating its inhibitory capacity. Some authors dissolve chitosan in acetic acid (at 1%) with a final pH of about 4.5–5.5 (Duan et al., 2010; Wu, 2014). Others dissolve it in 0.5 M acetic acid (López-Caballero et al., 2005) or the combination of acetic and lactic acid at 0.5 M (Beverlya et al., 2008), later adjusting the pH at ≥5.5 in order to appreciate the antimicrobial properties of the chitosan, although not all the authors state the conditions. The application of chitosan-based films with chitooligosaccharides and glutathione effectively inhibited bacterial growth and reduced total volatile basic nitrogen in white shrimp during partially frozen storage (Wu, 2014). In fish, the application of bovine gelatin and chitosan films with clove essential oil contributed to the stability of fresh cod (Gómez-Estaca et al., 2010). A decrease in the total flora, pseudomonas, and enterobacteriaceae was observed during storage, showing a bactericide effect on H_2S-producing microorganisms. This fact may be important in fish preservation because *Shewanella putrefaciens* belongs to this group, a specific microorganism of fish deterioration in warmwater fish preserved in ice (Gram and Huss, 1996). The antimicrobial activity of clove films may be attributed to its hydrophobic nature, which enables the essential oils to accumulate in cell and mitochondrial membranes, distorting structures and making them more permeable (Sikkema et al., 1994). Fish gelatin– and/or chitosan-based films with the addition of clove essential oil contributed to the stability of fresh salmon, because a reduction in total flora was achieved (>2 logarithmic cycles) after 11 days of storage at 2°C (Gómez-Estaca et al., 2009c). The antimicrobial activity of the clove essential oil was maintained once it was incorporated into the edible films, although differences were observed related to the matrix used. The different degrees of interaction based on the gelatin origin with the polyphenols (Gómez-Estaca et al., 2010) and chitosan (Gómez-Estaca et al., 2007) may affect the antimicrobial capacity of the compounds. When edible films are applied on the surface of a food, their solubility largely determines the release of antimicrobial compounds (Papadokostaki et al., 1997). When clove essential oil is incorporated into gelatin films, solubility increases. This fact may be attributed to the protein–polyphenol interactions, weakening the interactions that stabilize the protein network. On the other hand, when the oil is incorporated into the gelatin–chitosan matrix, film solubility is maintained. This may be attributed to the specific interactions between the gelatin and the chitosan, which stabilize the structure regardless of the clove oil. These films released less amounts of antimicrobial compounds, but maintained their integrity to a greater extent (Gómez-Estaca et al., 2010).

Chitosan-based edible coatings and films incorporated with other essential oils, such as cinnamon or garlic essential oils, have also been applied for seafood preservation. Chitosan-based coatings with and without cinnamon essential oil efficiently reduced total viable and psychrotrophic counts in rainbow trout fillets during refrigerated storage (Ojagh et al., 2010). Shrimps coated with chitosan solutions enriched with garlic oil showed significant lower aerobic plate counts than those samples coated with chitosan without oil at the end of the chilled storage (days 9 and 11). However, the incorporation of garlic essential oil in chitosan coatings did not have any significant effect on the total volatile base nitrogen and trimethylamine nitrogen contents (Asik and Candogan, 2014). Chitosan-based coatings and films enriched with salts of organic acids have been applied in cold-smoked salmon preservation, showing high efficacy against *L. monocytogenes* (Jiang et al., 2011). Chitosan coatings, with or without the antimicrobials, consistently showed higher efficacy against *L. monocytogenes* than chitosan films having the same compositions.

Gelatin and chitosan coatings were effective for cod hamburger preservation (López-Caballero et al., 2005), and reductions of total basic nitrogen and microbial counts (total flora, luminescent colonies, *Enterobacteriaceae*, *Pseudomonas*, and *Staphylococcus aureus*), and especially gram-negative bacteria, were achieved. These coatings also demonstrated the advantage of their cold application, preventing unnecessary heat treatments and preserving their quality. At the same time, the presence of chitosan stimulated the growth of lactic flora to some extent, probably because the slightly lower pH of the coated hamburgers due to the acid solution in which chitosan was dissolved (López-Caballero et al., 2005). In this regard, Lee et al. (2002) described that chitosan oligosaccharides had a stimulating effect on bifidobacterium in concentrations between 0.1% and 0.5% and on *Lactobacillus casei* and *Lactobacillus brevis* at 0.1%. Jeon et al. (2002) described that the microorganisms (total PCA at 20°C) reached the stationary phase in all the samples of cod and herring with chitosan coatings after 6 days of storage, although a difference was observed of up to 3 logarithmic cycles between coated and uncoated samples within 12 days. Several factors affect the antimicrobial activity of chitosan (Cuero, 1999), and its mechanism of action seems to be related to the alteration of the lipopolysaccharide external membrane of gram-negative bacteria (Nikaido, 1996; Helander et al., 2001) as well as to oxygen transfer hindering (Jeon et al., 2002).

In recent years, the antimicrobial effect of chitosan films has also been widely studied in meat and processed meat products such as cuts of veal and rabbit meat, bologna, boiled ham, pastrami, roast beef, pork meat hamburgers, chicken breast, sliced turkey deli meat, pork sausages, minced pork meat, or mortadella-type sausages (Ouattara et al., 2000; Beverlya et al., 2008; Moradi et al., 2011; Petrou et al., 2012; Siripatrawan and Noipha, 2012; Baranenko et al., 2013; Higueras et al., 2013; Guo et al., 2014). Ouattara et al. (2000) applied chitosan-based films for the preservation of bologna, boiled ham, and pastrami, to which several organic acids were added (acetic and propionic acids, alone or in combination with lauric or cinnamaldehyde acids). The release of propionic acid into the product was completed at 48 h of storage, while that of acetic acid was more limited and most of it remained in the film matrix after 163 h. The presence of lauric acid reduced the acetic acid release from the film, but the cinnamaldehyde did not show this effect. On the other hand, the film on the products behaved differently. When it was applied to bologna, the release of acid was lower than with boiled ham and pastrami. Growth of *Enterobacteriaceae* and *Serratia liquefaciens* added to the surface of the products was completely inhibited, while lactic acid bacteria were not affected. Inhibition of *Enterobacteriaceae* was more effective in bologna than in pastrami, possibly due to their different formulation and the lower exudates of bologna.

The presence of organic acids as constituents of the packaging plays an important role. In slightly cooked meats (roast beef) that were inoculated with *L. monocytogenes*, the addition of lactic acid to chitosan films was less effective than the addition of acetic acid in inhibiting this microorganism (Beverlya et al., 2008). Gelatin–chitosan-based coatings with organic acids showed the strongest bacteriostatic effect for meat and meat products such as retail cuts of veal and rabbit meat, boiled sausages, smoked sausages, and smoked–boiled pork brisket (Baranenko et al., 2013). The incorporation of antimicrobial proteins may be another option to confer active properties to the edible coatings and films. This is the case of the addition of ovotransferrin or conalbumin to k-carrageenan films applied on fresh chicken breasts (Seol et al., 2009). They inhibited the growth of *Escherichia coli* during the storage at 5°C, and although it was not very evident, the inhibition rate increased by incorporating EDTA 5 mM into the films. However, the action against *Salmonella typhimurium*, *Bacillus cereus*, and *Candida albicans* was either very slight or nil. The incorporation of bacteriocins (nisin, pediocin, etc.), by their own or by a starter culture that produce them, to edible films and coatings has been reported previously (Han, 2005) in order to inhibit the growth of pathogenic and spoiler microorganisms. Factors such as processing techniques, properties of the antimicrobial compound, and compatibility with the polymer could modify the activity of these compounds. The antimicrobial peptide nisin has also been successfully added to edible coatings and films, such as gelatin-based or chitosan-based coatings and films, for muscle food preservation (Nguyen et al., 2008; Liang et al., 2011; Guo et al., 2014). Nisin has been reported to be a very effective antimicrobial against gram-positive microorganisms, including *Listeria*. Addition of nisin to chitosan-coating solutions inhibited the growth of *Listeria* to a greater extent than the chitosan-coating solutions without nisin, when these coatings were applied on ready-to-eat deli turkey meat (Guo et al., 2014). Nguyen et al. (2008) coated Frankfurt sausages with cellulose films containing nisin before they were vacuum packed. The inhibition of *L. monocytogenes* as well as total bacteria was significant and produces higher content of nisin in the film. On the other hand, whey protein films enriched with oregano essential oil demonstrated a clear antimicrobial effect in meat coating, reducing total flora, and pseudomonas by 1 or 2 logarithmic cycles, while lactic bacterial growth was completely inhibited (Zinoviadou et al., 2009). *L. monocytogenes* was inhibited using pediocin or nisin fixed on a cellulose casing and applied on turkey breast meat, beef, and ham (Ming et al., 1997).

Studies in the model system based on muscle food have also been an object of interest. One experiment compared the application of edible coatings (calcium alginate and $CaCl_2$) incorporated with organic acids (lactic and acetic acids) versus treatment with active compounds (lactic and acetic acids, without coating) on beef (lean and fat) contaminated with *L. monocytogenes* (Siragusa and Dickenson, 1992). The results showed an important decrease in the number of *Listeria* adhered to the surface of the lean sample with the coating, compared to the acid treatment without coating. However, in the fat sample, the inhibition produced by treatment with acid did not increase with the immobilization in alginate coatings. In the case of agar coatings enriched with EDTA, citric acid, polyoxyethylene monolaureate, and nisin applied to chicken skin contaminated with *S. typhimurium*, reductions of over 2 logarithmic cycles were observed after 4 days at 4°C (Natrajan and Sheldon, 1995). The antimicrobial effect of pea starch coating with trisodium phosphate and calcium alginate with sodium chloride applied to chicken skin contaminated with different strains of *Salmonella* largely depends on factors such as pH, adhesion of the coating, and its absorption (Mehyar et al., 2007). The proliferation of microorganisms, especially of *L. monocytogenes*, was reduced by applying a coating of hydrocolloids, organic acids (lactic and acetic), and antimicrobial compounds to the fish (Sensidoni and Peressini, 1997).

Application of edible coatings and films on food to increase its shelf life through microbial control is one of the ultimate goals of edible packaging. However, not all the studies found in literature have been satisfactory. For example, acetylated monoglyceride (Dermatex®) coatings did not affect the microbiological characteristics of fresh beef steaks or that of grilled samples during vacuum preservation (Leu et al., 1987). Application of alginate Flavor-Tex® coating to lamb carcasses stored at 4°C and in cow parts stored at 5°C did not affect the total aerobic microbial counts either (Lazarus et al., 1976; Williams et al., 1978). Casein–whey protein films did not affect the microbial counts of meat *Carpaccio* inoculated with *S. aureus*, possibly because the high moisture content of the product reduced the oxygen barrier properties of the coating (Nortje et al., 2006).

As described earlier, edible coatings and films are often applied together with other technologies, resulting in combined treatments, based on the principle of barrier technologies. Combined treatments of irradiation and ascorbic acid, with or without coatings incorporating spices, did not significantly improve the effect of irradiation (Lacroix et al., 2004). These authors attributed the results to different factors such as the gram-negative/gram-positive ratio of the indigenous microflora of meat samples and the low concentration of active compounds that powdered spices have in comparison with, for example, essential oils. Combined treatments of irradiation and coating, with the addition of spices, significantly reduced microbial growth, especially *S. aureus* and gram-negative bacteria (coliform bacteria, *Enterobacteriaceae*, and *Pseudomonas* spp.) and stabilized the biochemical parameters of ground beef (Ouattara et al., 2002a). Irradiation of restructured pork meat, together with the application of pectin coatings with extract of green tea leaves, resulted in the decrease of total aerobic flora (Kang et al., 2007). In chilled shrimp and refrigerated pizza, Ouattara et al. (2002b) described a synergistic effect between irradiation and coating with essential oils to reduce microbial growth (total microorganisms and *Pseudomonas putida*), in terms of extension of the lag phase, lower growth rates, and therefore, shelf life extension, because the microorganisms surviving irradiation are probably more sensitive to environmental conditions (pH, nutrients, among others).

The combined application of vacuum and protective gelatin–chitosan coatings enriched with organic acids provided the strongest suppressing effect on microflora in meat and meat products such as retail cuts of veal and rabbit meat, boiled sausages, smoked sausages, and smoked–boiled pork brisket (Baranenko et al., 2013). In another study, the combination of chitosan-based antimicrobial coatings or films and flash pasteurization, which uses short burst of steam under pressure, provided over a 5 log CFU/cm^2 reductions of *Listeria innocua*, whereas chitosan antimicrobial coating reduced *L. innocua* by 4.5 log CFU/cm^2 (Guo et al., 2014).

High-pressure treatment (300 MPa/10 min/5°C) combined with coating with a gelatin–lignin film maintained the quality (in terms of protein oxidation and degradation) of a precooked product based on salmon (Ojagh et al., 2011). However, the effect of the combined treatment was not very distinct. The antimicrobial properties of lignin have been described in scientific literature, but unlike the mentioned article, other studies do not work with the complete polymer. Since lignin is a very heterogeneous natural product, its biological activity may be related to other compounds to which it is attached. In-depth studies are necessary to determine the antimicrobial activity of lignin on strains of microorganisms *in vitro*, as well as the study of this polymer once it is incorporated into the film.

21.4 Other Contributions to Film Functionality

Edible coatings and films may be enriched with compounds that, besides producing a benefit to the food, may have a positive impact on the consumers' health. There are recent studies in which chitosan films are enriched with a blend of omega-3 fatty acids to coat cod fillets (*Ophiodon elongatus*), in order to offer a supplement of this compound and to protect the fish during chilled and frozen storage (Duan et al., 2010). Besides reducing the TBA index values and inhibiting the growth of total bacteria and psychotropic bacteria, the films promoted an increase of the total lipid and omega-3 content in the fish, did not interfere with the color, and reduced loss of exudates during preservation in frozen state.

Gelatin-based films enriched with hydrolyzed and bioactive peptides have been elaborated recently with similar objectives. Although the antioxidant power of compounds such as polyphenols is much higher, their application is limited by the taste, color, and smell they give food. The incorporation of compounds obtained from gelatin, such as gelatin hydrolyzates or peptides, gave transparent, colorless, and odorless edible packaging and therefore is very suitable for numerous applications. Furthermore, these protein hydrolyzates confer a protective effect against rancidity and a potential bioactive effect (antioxidant, antihypertensive, and anticancer) (Giménez et al., 2009; Alemán et al., 2011a,b,c).

21.5 Conclusions

Active/bioactive edible packaging is a system with attributes beyond mere barrier properties, and this is achieved by the addition of active/bioactive compounds into the packaging system and/or by the use of active functional polymers. The functionality depends on the nature of the components and the composition and structure of the packaging. The choice of a substance with filmogenic capacity and/or of an active compound depends on the purpose, nature of the food, application method, and the sensorial characteristics they confer.

Immobilization of active/bioactive compounds in filmogenic solutions has great advantages for food preservation. The resulting edible coatings or films contribute to the stability of food by controlling the diffusion process and by maintaining high concentration of active molecules on the surface of the food for a longer time (Ouattara et al., 2000). The effects of these compounds are influenced by the molecular interactions that occur in the film and coating matrices, which lead to higher or lower release and diffusion. It is therefore necessary to gain insight into the mechanisms of controlled release of active compounds in order to obtain greater diffusion with lower concentration. This is one of the key aspects to be considered to maintain the quality of foods. The packaging design as an edible part of the food has gained interest in the last few years not only because it contributes to food preservation but also due to its attractive characteristics from the sensory point of view and its possible positive impact on consumers' health.

REFERENCES

Agulló, E., Rodríguez, M.S., Ramos, V., and Albertengo, L. (2003). Present and future role of chitin and chitosan in food. *Macromolecular Bioscience*, 3(10), 521–530.

Akhtar, M.J., Jacquot, M., Arab-Tehrany, E., and Gaïani, C. (2010). Control of salmon oil photo-oxidation during storage in HPMC packaging film: Influence of film colour. *Food Chemistry*, 120, 395–401.

Alemán, A., Giménez, B., Montero, P., and Gómez-Guillén, M.C. (2011a). Antioxidant activity of several marine skin gelatins. *LWT—Food Science and Technology*, 44, 407–413.

Alemán, A., Giménez, B., Pérez-Santín, E., Gómez-Guillén, M.C., and Montero, P. (2011b). Contribution of Leu and Hyp residues to antioxidant and ACE-inhibitory activities of peptides sequences isolated from squid gelatin hydrolysate. *Food Chemistry*, 125, 334–341.

Alemán, A., Pérez-Santín, E., Bordenave-Juchereau, S., Arnaudin, I., Gómez-Guillén, M.C., and Montero, P. (2011c). Squid gelatin hydrolysates with antihypertensive, anticancer and antioxidant activity. *Food Research International*, 44(4), 1044–1051.

Alparslan, Y., Baygar, T., Baygar, T., Hasanhocaoglu, H., and Metin, C. (2014). Effects of gelatin-based edible films enriched with laurel essential oil on the quality of rainbow trout (*Oncorhynchus mykiss*) fillets during refrigerated storage. *Food Technology Biotechnology*, 52(3), 325–333.

Andevari, G.T. and Rezaei, M. (2011). Effect of gelatin coating incorporated with cinnamon oil on the quality of fresh rainbow trout in cold storage. *International Journal of Food Science and Technology*, 46(11), 2305–2311.

Arancibia, M., Giménez, B., López-Caballero, M.E., Gómez-Guillén, M.C., and Montero, P. (2014). Release of cinnamon essential oil from polysaccharide bilayer films and its use for microbial growth inhibition in chilled shrimps. *LWT—Food Science and Technology*, 59, 989–995.

Artharn, A., Prodpran, T., and Bejakul, S. (2009). Round scad protein-based film: Storage stability and its effectiveness for shelf-life extension of dried fish powder. *LWT—Food Science and Technology*, 42, 1238–1244.

Asik, E. and Candogan, K. (2014). Effects of chitosan coatings incorporated with garlic oil on quality characteristics of shrimp. *Journal of Food Quality*, 37, 237–246.

Ban, W., Song, J., and Lucia, L.A. (2007). Influence of natural biomaterials on the absorbency and transparency of starch-derived films: An optimization study. *Industrial and Engineering Chemistry Research*, 46(20), 6480–6485.

Baranenko, D.A., Kolodyaznaya, V.S., and Zabelina, N.A. (2013). Effect of composition and properties of chitosan-based edible coatings on microflora of meat and meat products. *Acta Scientiarum Polonorum, Technologia Alimentaria*, 12(2), 149–157.

Beverlya, R., Janes, M., Prinyawiwatkula, W., and No, H. (2008). Edible chitosan films on ready-to-eat roast beef for the control of *Listeria monocytogenes*. *Food Microbiology*, 25, 534–537.

Bonilla, J., Vargas, M., Atarés, L., and Chiralt, A. (2012). Edible films and coatings to prevent the detrimental effect of oxygen on food quality: Possibilities and limitations. *Journal of Food Engineering*, 110, 208–213.

Bonilla, J., Vargas, M., Atarés, L., and Chiralt, A. (2014). Effect of chitosan essential oil films on the storage-keeping quality of pork meat products. *Food and Bioprocess Technology*, 7(8), 2443–2450.

Boselli, E., Rodriguez-Estrada, M.T., Ferioloi, F., Caboni, M.F., and Lercker, G. (2010). Cholesterol photo-sensitised oxidation of horse meat slices stored under different packaging films. *Meat Science*, 85(3), 500–505.

Bourtoom, T. (2008). Edible films and coatings: Characteristics and properties. *International Food Research Journal*, 15(3), 237–248.

Butler, B.L., Vergano, P.J., Testin, R.F., Bunn, J.N., and Wiles, J.L. (1996). Mechanical and barrier properties of edible chitosan films as affected by composition and storage. *Journal of Food Science*, 61, 953–955, 961.

Carvalho, R.A., Sobral, P.J.A., Thomazine, M., Habitante, A.M.Q.B., Giménez, B., Gómez-Guillén, M.C. et al. (2008). Development of edible films based on differently processed Atlantic halibut (*Hippoglossus hippoglossus*) skin gelatin. *Food Hydrocolloids*, 22(6), 1117–1123.

Cha, D.S. and Chinnan, M.S. (2004). Biopolymer-based antimicrobial packaging a review. *Critical Reviews in Food Science & Nutrition*, 44, 223–227.

Chi, S., Zivanovic, S., and Penfield, M.P. (2006). Antibacterial of chitosan films enriched with oregano essential oil on bologna-active compounds and sensory attributes. *Food Science and Technology International*, 12, 111–117.

Chiu, P.E. and Lai, L.S. (2010). Antimicrobial activities of tapioca starch/decolorized Hsian-tsao leaf gum coatings containing green tea extracts in fruit-based salads, romaine hearts and pork slices. *International Journal of Food Microbiology*, 139, 23–30.

Coma, V. (2006). Bioactive packaging technologies for extended shelf life of meat-based products. *Meat Science*, 78, 90–103.

Cuero, R.G. (1999). Antimicrobial action of exogenous chitosan. In: Jolles, P. and Muzarelli, R.A.A. (eds.), *Chitin and Chitinases*. Basel, Switzerland: Birkhauser Verlag, pp. 315–333.

Cutter, C.N. (2006). Opportunities for bio-based packaging technologies to improve the quality and safety of fresh and further processed muscle food. *Meat Science*, 74, 131–142.

Cutter, C.N. and Miller, B.J. (2004). Incorporation of nisin into a collagen film retains antimicrobial activity against *Listeria monocytogenes* and *Brochothrix themosphacta* associated with a ready-to-eat meat product. *Journal of the Association of Food and Drug Officials*, 68(4), 64–77.

Cutter, C.N. and Sumner, S.S. (2002). Application of edible coatings on muscle food. In: Gennadios, A. (ed.), *Protein-Based Films and Coating*. Boca Raton, FL: CRC, pp. 467–484.

Dilek, M., Polat, H., Kezer, F., and Korcan, E. (2011). Application of locust bean gum edible coating to extend shelf life of sausages and garlic-flavored sausage. *Journal of Food Processing and Preservation*, 35(4), 410–416.

Dragich, A.M. and Krochta, J.M. (2010). Whey protein solution coating for fat-uptake reduction in deep-fried chicken breast strips. *Journal of Food Science*, 75(1), S43–S47.

Duan, J., Cherian, G., and Zhao, Y. (2010). Quality enhancement in fresh and frozen lingcod (*Ophiodon elongates*) fillets by employment of fish oil incorporated chitosan coatings. *Food Chemistry*, 119, 524–532.

Dutta, P.K., Tripathi, S., Mehrotra, G.K., and Dutta, J. (2009). Perspectives for chitosan based antimicrobial films in food applications. *Food Chemistry*, 114, 1173–1182.

García, M., Bifani, V., Campos, C., Martino, M.N., Sobral, P., Flores, S. et al. (2008). Edible coating as an oil barrier or active system. In: Gutiérrez-López, G.F., Barbosa-Cánovas G.V., Welti-Chanes, J., and Parada-Arias, E. (eds.), *Food Engineering: Integrated Approaches*. Food Engineering Series. New York: Springer, pp. 225–241.

Garmakhani, A.D., Mirzaei, H.O., Maghsudlo, Y., Kashaninejad, M., and Jarafi, S.M. (2014). Production of low fat French-fries with single and multi-layer hydrocolloids coatings. *Journal of Food Science and Technology*, 51(7), 1334–1341.

Gennadios, A., Hanna, M.A., and Kurth, L.B. (1997). Application of edible coatings on meats, poultry and seafoods: A review. *LWT—Food Science and Technology*, 30, 337–350.

George, J. and Siddaramaiah. (2012). High performance edible nanocomposite films containing bacterial cellulose nanocrystals. *Carbohydrate Polymers*, 87(3), 2031–2037.

Giménez, B., Gómez-Estaca, J., Alemán, A., Gómez-Guillén, M.C., and Montero, M.P. (2009). Improvement of the antioxidant properties of squid skin gelatin films by the addition of hydrolysates from squid gelatin. *Food Hydrocolloids*, 23, 1322–1327.

Giménez, B., Gómez-Guillén, M.C., Pérez-Mateos, M., Montero, P., and Márquez-Ruiz, G. (2011). A throughout evaluation of lipid oxidation in horse mackerel patties coated with borage-containing film during frozen storage. *Food Chemistry*, 124(4), 1393–1403.

Gómez-Estaca, J., Bravo, L., Gómez-Guillén, M.C., Alemán, A., and Montero, P. (2009a). Antioxidant properties of tuna skin and bovine hide gelatin films induced by addition of oregano and rosemary extract. *Food Chemistry*, 112, 18–25.

Gómez-Estaca, J., Giménez, B., Montero, P., and Gómez-Guillén, M.C. (2009b). Incorporation of antioxidant borage extract into edible films based on sole skin gelatin. *Journal of Food Engineering*, 92(1), 78–85.

Gómez-Estaca, J., López de Lacey, A., Gómez-Guillén, M.C., López-Caballero, M.E., and Montero, P. (2009c). Antimicrobial activity of composite edible films based on fish gelatin and chitosan incorporated with clove essential oil. *Journal of Aquatic Food Products Technology*, 18, 46–52.

Gómez-Estaca, J., López de Lacey, A., López-Caballero, M.E., Gómez-Guillén, M.C., and Montero, P. (2010). Biodegradable gelatin–chitosan films incorporated with essential oils as antimicrobial agents for fish preservation. *Food Microbiology*, 27(7), 889–896.

Gómez-Estaca, J., Montero, P., Giménez, B., and Gómez-Guillén, M.C. (2007). Effect of functional edible films and high pressure-processing on microbial and oxidative spoilage in cold-smoked sardine (*Sardina pilchardus*). *Food Chemistry*, 105, 511–520.

Gómez-Guillén, M.C., Ihl, M., Bifani, V., Silva, A., and Montero, P. (2007). Edible films made from tuna fish gelatin with antioxidant extracts of two different murta ecotypes leaves (*Ugni molinae* Turcz). *Food Hydrocolloids*, 21, 1133–1143.

Gómez-Guillén, M.C., López-Caballero, M.E., Alemán, A., López de Lacey, A., Giménez, B., and Montero, P. (2010). *Antioxidant Peptide Fractions from Squid and Tuna Skin Gelatin*, Valorisation of Marine by Products. Kerala, India: Research Signpost.

Gómez-Guillén, M.C., Pérez-Mateos, M., Gómez-Estaca, J., López-Caballero, M.E., Giménez, B., and Montero, P. (2009). Fish gelatin: A renewable material for developing active biodegradable films. *Trends in Food Science and Technology*, 20(1), 3–16.

Gram, L. and Huss, H.H. (1996). Microbiological spoilage of fish and fish products. *International Journal of Food Microbiology*, 33, 121–137.

Guilbert, S. (2000). Technology and application of edible protective films. In: Mathlouthi, M. (ed.), *Food Packaging and Preservation*. London, U.K.: Elsevier Applied Science Publishers, pp. 371–394.

Guillard, V., Issoupov, V., Redl, A., and Gontard, N. (2009). Food preservative content reduction by controlling sorbic acid release from a superficial coating. *Innovative Food Science and Emerging Technology*, 10, 108–115.

Guo, M., Jin, T.Z., Wang, L., Scullen, O.J., and Sommers, C.H. (2014). Antimicrobial films and coatings for inactivation of *Listeria innocua* on ready-to-eat deli turkey meat. *Food Control*, 40, 64–70.

Han, J.H. (2000). Antimicrobial food packaging. *Food Technology*, 54(3), 56–65.

Han, J.H. (2005). Antimicrobial packaging systems, Chapter 6. In: Han, J.H. (ed.), *Innovations in Food Packaging*. Oxford, England: Elsevier Academic, pp. 80–107.

Harper, B.A., Barbut, S., Lim, L.T., and Marcone, M.F. (2012). Microstructural and textural investigation of various manufactured collagen sausage casings. *Food Research International*, 49(1), 494–500.

Helander, I.M., Nurmiaho-Lassila, E.L., Ahvenainen, R., Rhoades, J., and Roller, S. (2001). Chitosan disrupt the barrier properties of the outer membrane of Gram-negative bacteria. *International Journal of Food Microbiology*, 71, 235–244.

Higueras, L., López-Carballo, G., Hernández-Muñoz, P., Gavara, R., and Rollini, M. (2013). Development of a novel antimicrobial film based on chitosan with LAE (ethyl-Nα-dodecanoyl-L-arginate) and its application to fresh chicken. *International Journal of Food Microbiology*, 165(3), 339–345.

Holownia, K.I., Chinnan, M.S., Erickson, M.C., and Mallikarjunan, P. (2000). Quality evaluation of edible film-coated chicken strips and frying oils. *Journal of Food Science*, 65(6), 1087–1090.

Holownia, K.I., Ericsson, M.C., Chinnan, M.S., and Eitenmiller, R.R. (2001). Tocopherol losses in peanut oil during pressure frying of marinated chicken strips coated with edible films. *Food Research International*, 34, 77–80.

Hong, Y.-H., Lim, G.-O., and Song, K.B. (2009). Physical properties of Gelidium corneum–gelatin blend films containing grapefruit seed extract or green tea extract and its application in the packaging of pork loins. *Journal of Food Science*, 74(1), C6–C10.

Huang, J., Chen, Q., Qiu, M., and Li, S. (2012). Chitosan-based edible coatings for quality preservation of postharvest whiteleg shrimp (*Litopenaeus vannamei*). *Journal of Food Science*, 77(4), C491–C496.

Iturriaga, L., Olabarrieta, I., and Martínez de Marañón, I. (2012). Antimicrobial assay of natural extracts and their inhibitory effect against *Listeria innocua* and fish spoilage bacteria, after incorporation into biopolymer edible films. *International Journal of Food Microbiology*, 158, 58–64.

Jeon, Y.J., Kamil, J.Y.V.A., and Shahidi, F. (2002). Chitosan as an edible invisible film for quality preservation of herring and Atlantic cod. *Journal of Agricultural and Food Chemistry*, 20, 5167–5178.

Jiang, Z., Neetoo, H., and Chen, H. (2011). Control of *Listeria monocytogenes* on cold-smoked salmon using chitosan-based antimicrobial coatings and films. *Journal of Food Science*, 76(1), M22–M26.

Jongjareonrak, A., Benjakul, S., Visessanguan, W., Prodpran, T., and Tanaka, M. (2006). Characterization of edible films from skin gelatin of brownstripe red snapper and bigeye snapper. *Food Hydrocolloids*, 20(4), 492–501.

Kamil, J., Jeon, J., and Shahidi, F. (2002). Antioxidative activity of chitosan of different viscosity in cooked comminuted flesh of herring (*Clupea harengus*). *Food Chemistry*, 79, 69–77.

Kang, H.J., Jo, C., Kwon, J.H., Kim, J.H., Chung, H.L., and Byun, M.W. (2007). Effect of a pectin-based edible coating containing green tea powder on the quality of irradiated pork patty. *Food Control*, 18, 430–435.

Kester, J.J. and Fennema, O. (1986). Edible films and coatings: A review. *Food Technology*, 40, 47–59.

Khwaldia, K., Pérez, C., Banon, S., Desobry, S., and Hardy, J. (2004). Milk proteins for edible film and coatings. *Critical Reviews in Foods Science and Nutrition*, 44, 239–251.

Kilincceker, O., Dogan, I., and Kucukoner, E. (2009). Effect of edible coatings on the quality of frozen fish fillets. *Food Science and Technology*, 42, 868–873.

Kilincceker, O. and Hepsag, F. (2011). Performance of different coating batters and frying temperatures for fried fish balls. *Journal of Animal and Veterinary Advances*, 10, 2256–2262.

Kim, I.H., Yang, H.J., Noh, B.S., Chung, S.J., and Min, S.C. (2012). *Food Chemistry*, 133, 1501–1509.

Kurt, S. and Kilincceker, O. (2011). Performance optimization of soy and whey protein isolates as coating materials on chicken meat. *Poultry Science*, 90(1), 195–200.

Lacroix, M., Ouattara, B., Saucier, L., Giroux, M., and Smoragiewicz, W. (2004). Effect of gamma irradiation and presence of ascorbic acid on microbial composition and TBARS concentration of ground beef coated with an edible active coating. *Radiation Physics and Chemistry*, 71, 71–75.

Lalam, S., Sandhu, J.S., Takhar, P.S., Thompson, L.D., and Alvarado, C. (2013). Experimental study on transport mechanisms during deep fat frying of chicken nuggets. *LWT—Food Science and Technology*, 50, 110–119.

Lazarus, C.R., West, R.L., Oblinger, J.L., and Palmer, A.Z. (1976). Evaluation of calcium alginate coating and a protective plastic wrapping for the control of lamb carcass shrinkage. *Journal of Food Science*, 41, 639–641.

Lee, H.W., Park, Y.S. Jung, J.S., and Shin, W.S. (2002). Chitosan oligosaccharides, dp 2–8, have prebiotic effect on the *Bifidobacterium bifidum* and *Lactobacillus* sp. *Anaerobe*, 8(6), 319–324.

Leu, R., Keeton, J.T., Griffin, D.B., Savell, J.W., and Vanderzant, C. (1987). Microflora of vacuum packaged beef steaks and roast treated with an edible acetylated monoglyceride. *Journal of Food Protection*, 50, 554–556.

Liang, R.R., Zhang, X.B., Wang, X.J., Wang, R.H., Mao, Y.W., Zhang, Y.M., and Luo, X. (2011). Application of gelatin-based antimicrobial edible coatings on the preservation of chicken meat and prepared products. *Advanced Material Research*, 236–238, 2255–2258.

Limpisophon, K., Tanaka, M., Weng, W., Abe, S., and Osako, K. (2009). Characterization of gelatin films prepared from under-utilized blue shark (*Prionace glauca*) skin. *Food Hydrocolloids*, 23(7), 1993–2000.

Liu, L., Kerry, J.F., and Kerry, J.P. (2005). Selection of optimum extrusion technologies parameters in the manufacture of edible biodegradable packaging films derived from food-based polymers. *Journal of Food Agriculture and Environment*, 3, 51–58.

Liu, L., Kerry, J.F., and Kerry, J.P. (2006). Effect of food ingredients and selected lipids on the physical properties of extruded edible films/casings. *International Journal of Foods Science and Technology*, 41, 295–302.

Liu, L., Kerry, J.F., and Kerry, J.P. (2007). Application and assessments of extruded edible casing manufactured from pectin and gelatin/sodium alginate blends for use with breakfast pork sausage. *Meat Science*, 73, 196–202.

López-Caballero, M.E., Gómez-Guillén, M.C., Pérez-Mateos, M., and Montero, P. (2005). A chitosan gelatin blend as a coating for fish patties. *Food Hydrocolloids*, 19, 303–311.

López de Lacey, A., López-Caballero, M.E., and Montero, P. (2014). Agar films containing green tea extract and probiotic bacteria for extending fish shelf-life. *LWT—Food Science and Technology*, 55, 559–564.

Matos de Oliveira, M.M., Brugnera, D.F., and Piccoli, R.H. (2013). Essential oils of thyme and rosemary in the control of *Listeria monocytogenes* in raw beef. *Brazilian Journal of Microbiology*, 44(4), 1181–1188.

Mehyar, G.F., Han, J.H., Holley, R.A., Blank, G., and Hydamaka, A. (2007). Suitability of pea starch and calcium alginate as antimicrobial coatings of chicken skin. *Poultry Science*, 86, 386–393.

Ming, X., Weber, G.H., Ayres, J.W., and Sandine, W.E. (1997). Bacteriocins applied to food packaging materials to inhibit *Listeria monocytogenes* on meats. *Journal of Food Science*, 62(2), 413–415.

Moradi, M., Tajik, H., Rohani, S.M.R., and Oromiehie, A.R. (2011). Effect of *Zataria multiflora* Boiss essential oil and grape seed extract impregnated chitosan film on ready-to-eat mortadella-type sausages during refrigerated storage. *Journal of the Science of Food and Agriculture*, 91, 2850–2857.

Moreira, M.R., Pereda, M., Marcovich, N.E., and Roura, S.I. (2011). Antimicrobial effectiveness of bioactive packaging materials from edible chitosan and casein polymers: Assessment of carrot, cheese and salami. *Journal of Food Science*, 76(1), M54–M63.

Morsy, M.K., Khalaf, H.H., Sharoba, A.M., El-Tanahi, H.H., and Cutter, C.N. (2014). Incorporation of essential oils and nanoparticles in pullulan films to control foodborne pathogens on meat and poultry products. *Journal of Food Science*, 79(4), M675–M684.

Motalebi, A.A. and Seyfzadeh, M. (2011). Effects of whey protein edible coating on bacterial, chemical and sensory characteristics of frozen common Kilka. *Iranian Journal of Fisheries Science*, 11(1), 132–144.

Moyano, P.C., Rioseco, V.K., and González, P.A. (2002). Kinetics of crust color changes during deep-fat frying of impregnated French fries. *Journal of Food Engineering*, 54(3), 249–255.

Natrajan, N. and Sheldon, B. (1995). Evaluation of bacteriocin-based packaging and edible film delivery system to reduce *Salmonella* in fresh poultry. *Poultry Science*, 74(1), 31–37.

Nguyen, V.T., Gidley, M.J., and Dykes, G.A. (2008). Potential of a nisin-containing bacterial cellulose film to inhibit *Listeria monocytogenes* on processed meats. *Food Microbiology*, 25, 471–478.

Nikaido, H. (1996). Outer membrane. In: Neidhardt, F.C. (ed.), *Escherichia coli and Salmonella: Cellular and Molecular Biology*, Vol. 1. Washington, DC: American Society for Microbiology, pp. 23–37.

Nortje, K., Buys, E.M., and Minaar, A. (2006). Use of gamma irradiation to reduce high levels of *Staphylococcus aureus* on casein-whey protein coated moist beef biltong. *Food Microbiology*, 23, 723–737.

Nuñez-Flores, R., Giménez, B., Fernández-Martín, F., López-Caballero, M.E., Montero, M.P., and Gómez-Guillén, M.C. (2013). Physical and functional characterization of active fish gelatin films incorporated with lignin. *Food Hydrocolloids*, 30, 163–172.

Ojagh, M., Núñez, R., López-Caballero, M.E., Gómez-Guillén, M.C., and Montero, P. (2011). Lessening of high-pressure-induced changes in Atlantic salmon muscle by the combined use of a gelatin-lignin film. *Food Chemistry*, 125(2), 595–606.

Ojagh, M., Rezaei, M., Razavi, S.H., and Hosseini, S.M.H. (2010). Effect of chitosan coatings enriched with cinnamon oil on the quality of refrigerated rainbow trout. *Food Chemistry*, 120, 193–198.

Ouattara, B., Giroux, M., Yefsah, R., Smoragiewicz, W., Saucier, L., Borsa, J., and Lacroix, M. (2002a). Microbiological and biochemical characteristics of ground beef as affected by gamma irradiation, food additives and edible coating film. *Radiation Physics and Chemistry*, 63, 299–304.

Ouattara, B., Sabato, S.F., and Lacroix, M. (2002b). Use of gamma-irradiation technology in combination with edible coating to produce shelf-stable foods. *Radiation Physics and Chemistry*, 63, 305–310.

Ouattara, B., Simard, R.E., Begin, A., and Holley, R. (2000). Inhibition of surface spoilage bacteria in processed meats by application of antimicrobial films prepared with chitosan. *International Journal of Food Microbiology*, 62, 139–148.

Oussalah, M., Caillet, S., Salmiéri, S., Saucier, L., and Lacroix, M. (2004). Antimicrobial and antioxidant effects of milk protein-based film containing essential oils for the preservation of whole beef muscle. *Journal of Agricultural and Food Chemistry*, 52, 5598–5605.

Papadokostaki, K., Amanratos, S.G., and Petropoulus, J.H. (1997). Kinetics of release of particles solutes incorporated in cellulosic polymer matrices as a function of solute solubility and polymer swellability. I. Sparingly soluble solutes. *Journal of Applied Polymer Science*, 67, 277–287.

Pereira, A.A., Martins, G.F., Antunes, P.A., Conrrado, R., Pasquini, D., Job, A.E. et al. (2007). Lignin from sugar cane bagasse: Extraction, fabrication of nanostructured films, and application. *Langmuir*, 23(12), 6652–6659.

Pérez-Mateos, M., Montero, P., and Gómez-Guillén, M.C. (2009). Formulation and stability of biodegradable films made from cod gelatin and sunflower oil blends. *Food Hydrocolloids*, 23(1), 53–61.

Petrou, S., Tsiraki, M., Giatrakou, V., and Savvaidis, I.N. (2012). Chitosan dipping or oregano oil treatments, singly or combined on modified atmosphere packaged chicken breast meat. *International Journal of Food Microbiology*, 156(3), 264–271.

Phadke, G.G., Pagarkar, A.U., Sehgal, K., and Mohanta, K.N. (2011). Application of edible and biodegradable coatings in enhancing seafood quality and storage life: A review. *Ecology, Environment and Conservation*, 17(3), 619–623.

Poverenov, E., Rutenberg, R., Danino, S., Horev, B., and Rodov, V. (2014). Gelatin–chitosan composite films and edible coatings to enhance the quality of food products: Layer-by-layer vs. blended formulations. *Food and Bioprocess Technology*, 7(11), 3319–3327.

Rostami, H.R., Motalebi, A.A., Khanipour, A.A., Soltani, M., and Khanedan, N. (2010). Effect of whey protein coating on physico-chemical properties of gutted Kilka during frozen storage. *Iranian Journal of Fisheries Science*, 9(3), 412–421.

Salgado, P.R., López-Caballero, M.E., Gómez-Guillén, M.C., Mauri, A.N., and Montero, M.P. (2013). Sunflower protein films incorporated with clove essential oil have potential application for the preservation of fish patties. *Food Hydrocolloids*, 33, 74–84.

Sánchez-Ortega, I., García-Almendárez, B.E., Santos-López, E.M., Amaro-Reyes, A., Barboza-Corona, J.E., and Regalado, C. (2014). *Scientific World Journal*, Volume 2014, Article ID 248935, pp. 18. Hindawi Publishing Corporation. http://dx.doi.org/10.1155/2014/248935.

Sensidoni, A. and Peressini, D. (1997). Edible films: Potential innovation for fish products. *Industrie Alimentari*, 36(356), 129–133.

Seol, K., Lim, D., Jang, A., Jo, C., and Lee, M. (2009). Antimicrobial effect of k-carrageenan-based edible film containing ovotransferrin in fresh chicken breast stored at 5°C. *Meat Science*, 83, 479–483.

Shin, Y.J., Song, H.-Y., Seo, Y.B., and Song, K.B. (2012). Preparation of red algae film containing grapefruit seed extract and application for the packaging of cheese and bacon. *Food Science and Biotechnology*, 21(1), 225–231.

Sikkema, J., de Bont, J.A.M., and Poolman, B. (1994). Interactions of cyclic hydrocarbons with biological membranes. *Journal of Biological Chemistry*, 269(11), 8022–8028.

Silva-Weiss, A., Bifani, V., Ihl, M., Sobral, P.J.A., and Gómez-Guillén, M.C. (2014). Polyphenol-rich extract from murta leaves on rheological properties of film-forming solutions based on different hydrocolloid blends. *Journal of Food Engineering*, 140, 28–38.

Siragusa, G.A. and Dickinson, J.S. (1992). Inhibition of *Listeria monocytogenes* on beef tissue by application of organic acids immobilized in a calcium alginate gel. *Journal of Food Science*, 57, 293–296.

Siripatrawan, U. and Noipha, S. (2012). Active film from chitosan incorporating green tea extract for shelf life extension of pork sausages. *Food Hydrocolloids*, 27(1), 102–108.

Song, Y., Liu, L., Shen, H., You, J., and Luo, Y. (2011). Effect of sodium alginate-based edible coating containing different anti-oxidants on quality and shelf life of refrigerated bream (*Megalobrama amblycephala*). *Food Control*, 22, 608–615.

Stuchell, M. and Krochta, J.M. (1995). Edible coatings on frozen king salmon: Effect of whey protein isolate and acetylated monoglyceride on moisture loss and lipid oxidation. *Journal of Food Science*, 60(1), 28–31.

Unalan, I.U., Korel, F., and Yemenicioglu, A. (2011). Active packaging of ground beef patties by edible zein films incorporated with partially purified lysozyme and Na₂EDTA. *International Journal of Food Science and Technology*, 46, 1289–1295.

Varela, P. and Fiszman, S.M. (2011). Hydrocolloids in fried foods. A review. *Food Hydrocolloids*, 25(8), 1801–1812.

Williams, S.K., Oblinger, J.L., and West, R.L. (1978). Evaluation of a calcium alginate film for use on beef cuts. *Journal of Food Science*, 43, 292–296.

Wu, S. (2014). Effect of chitosan-based edible coating on preservation of white shrimp during partially frozen storage. *International Journal of Biological Macromolecules*, 65, 325–328.

Xing, R., Liu, S., Yu, H., Zang, Q., Li, Z., and Li, P. (2004). Preparation of low-molecular-weight and high sulfate-content chitosan under microwave radiation and their potential antioxidant activity *in vitro*. *Carbohydrate Research*, 339, 2515–2519.

Xue, C., Yu, G., Hirat, T., Terao, J., and Lin, H. (1998). Antioxidative activities of several marine polysaccharides evaluated in a phosphatidylcholine-liposomal suspension and organic solvents. *Bioscience, Biotechnology and Biochemistry*, 62, 206–209.

Zinoviadou, K., Koutsoumanis, P., and Biliaderis, G. (2009). Physico-chemical properties of whey protein isolate films containing oregano oil and their antimicrobial action against spoilage flora of fresh beef. *Meat Science*, 82, 338–345.

22

Applications of Films and Coatings in Intermediate Moisture and Thermally Processed Food

Aurora Valdez-Fragoso, Vito Verardo, and Hugo Mújica-Paz

CONTENTS

22.1 Introduction

Intermediate moisture foods (IMF) are partially dehydrated foods (10%–40%) that have been added of glycerol, sorbitol, salt, sugars, or organic acids, having water activity between 0.5 and 0.9 (Li et al. 2009). Under these conditions, some of the water in the semimoist foods is bound and unavailable, which limits growth of bacteria, molds, and yeasts and controls undesirable enzymatic activity (Rao 1997).

The application of heat is another effective preservation method, which can also permit to address variations or changes in food products. Pasteurization, blanching, cooking, baking, and frying are among the heat processing methods to obtain safe and stable foods and to lengthen their shelf life (Wu et al. 2013). However, one of the main challenges regarding the thermally processed foods (TPF) is to maintain their quality characteristics over their handling, distribution, and storage.

Although IMF and TPF are generally regarded as stable products, they are very sensitive to deleterious factors after processing. One of the important inexpensive strategies for maintaining quality, safety, and stability of these products can be the use of edible films and coatings. For instance, hydrophobic films can prevent moisture gain of IMF, and oxygen barrier films can protect oxidizable compounds of TPF from oxidation.

Films are defined as a thin layer of edible material that can wrap or contain a food, while coating is a thin layer that is directly formed on the surface of a food (Tharanathan 2003). Films and coatings are prepared with biopolymers obtained from marine and agricultural raw material sources or produced by microorganisms.

Films and coatings act as an interface between the food and its surroundings. Besides facilitating the handling of food products, they can improve the appearance and structure of the product; intensify flavor; enhance color; present a barrier to moisture, oxygen, and undesirable aroma; limit oil uptake; and carry antioxidants, antimicrobial, or some other functional additives (Tharanathan 2003; Falguera et al. 2011).

The purpose of this chapter is to present current information on edible films and coatings used to preserve the safety, quality, and functionality of IMF and TPF.

22.2 Application of Edible Films and Coatings in IMF

Edible films and coatings have a wide range of applications in a large variety of IMF. Applications involve the food or food systems and their processing or postprocessing operations.

22.2.1 Films and Coatings Use in IMF

Many ready-to-cook and ready-to-eat products belong to the category of the IMF, which are very susceptible to deteriorative phenomena. Therefore, studies using edible films and coatings added with active agents have been undertaken to maintain or improve the IMF properties, safety, and stability.

The capacity of corn zein coating formulations containing antimicrobials and antioxidants agents (0.1% potassium sorbate and 1% ascorbic acid) has been evaluated to maintain the quality of intermediate moisture (IM) apricots ($a_w = 0.80$), obtained by osmodehydration in a hypertonic sucrose solution. It was found that coatings effectively retarded color changes during a storage period of 10 months at 5°C and 10°C, as well as the microbial growth, around 2 log between control and coated samples (Baysal et al. 2010).

Single-baked Mustafakemalpasa cheese, a Turkish dessert, is preferred for its palatability and flavor; however, the high moisture content limits its shelf life to no more than 3 days. This shelf life was extended to 10 days, without significant loss of organoleptic properties, when the dessert was coated either by whey protein concentrate or by corn zein (Guldas et al. 2010).

A hurdle technology approach, using reduced a_w, chitosan (CH) coating, and gamma irradiation, was applied to obtain shelf stable IM meat products (Rao et al. 2005). It was found that the inherent antioxidant activity of CH was considerably improved by the irradiation treatment. The combined effect of these hurdles delayed lipid oxidation in diverse IM meat products.

In other applications, a firm cell structure was observed in apple cubes during osmotic dehydration (OD) treatments, when samples were coated with a maltodextrin coating before osmodehydration. Coating pumpkin slices with native and modified maize and cassava starch, previously to air-drying, resulted in a good provitamin A retention during air-drying at 70°C (Lago-Vanzela et al. 2013).

To counteract the high susceptibility of nut products to oxidative reactions, the antioxidant capacity of hydroxypropyl methylcellulose and carboxymethylcellulose (CMC), with α-tocopherol, BHA, and BTH, was evaluated in toasted pecans (Baldwin and Wood 2006). Pecans, coated with CMC containing α-tocopherol, underwent less oxidation of lipids and were less rancid during storage.

A coating formulation including hydroxypropyl methylcellulose, ascorbic acid, citric acid, and ginger oil had a protective effect against lipid oxidation of toasted almonds (Atarés et al. 2011). More recently, addition of montmorillonite nanoparticles to starch and cashew tree gum coatings reduced the oxygen permeability of coating and the lipid oxidation of cashew (Pinto et al. 2015).

22.2.2 Films and Coatings Use in IMF Processing

A number of edible coatings have been evaluated for improving the OD of foods, with respect to the mass transfer phenomena. During OD extensive solute gain and water loss can occur, affecting the product characteristics and efficiency of the process. Coatings of maltodextrin, CH, low methoxyl pectinate (LMP), CMC, and corn starch have shown efficiency to reduce solutes uptake and water loss of several vegetal products, which were coated before OD treatments (Table 22.1) (Khin et al. 2006, 2007; García et al. 2010; Jalaee et al. 2011). Additionally, compared to noncoated samples, a higher dehydration efficiency or performance ratio (water loss/solute gain) was observed in the OD of strawberries covered with a double coating of sodium alginate (SA) (Matuska et al. 2006). A high water loss/solute gain ratio has been observed too, in the OD of apple rings coated with CMC and osmodehydrated in 60% sucrose solution (Jalaee et al. 2011).

Edible coatings have potential applications to control moisture exchange between internal components of a multicomponent food, having different water activity. This effect can be illustrated with caseinate-based coatings that considerably reduced moisture migration from osmodehydrated, impregnated, and air-dried pineapple cylinders (IMF of $a_w = 0.750$) to a breakfast cereal ($a_w = 0.3$–0.4) (Talens et al. 2012).

TABLE 22.1

Edible Films and Coatings for Intermediate Moisture Fruits and Vegetables

Product	Coating/Film Material	Additive	Effect	References
Pineapple	Casein		Increased shelf life of the pineapple–cereal system through application of coatings by vacuum impregnation and air-drying after coating	Talens et al. (2012)
Apple	Maltodextrin		Reduced solute gain during osmotic dehydration	Khin et al. (2007)
Apple	LMP, CMC, corn starch		Reduced water content and solute gain	Jalaee et al. (2011)
Papaya	Chitosan	Tween 80, oleic acid	Reduced water content and solute gain	García et al. (2010)
Potato	SA, LMP	$CaCl_2$	Effective control of solute uptake at high temperatures	Khin et al. (2006)
Strawberry	SA, carrageenan, Mixture of carrageenan and guar gum (C + GG)		High dehydration efficiency (high water loss/solute gain ratio)	Matuska et al. (2006)

In recent studies, edible films or coatings have been shown to be an effective way for maintaining the probiotic viability, during the processing and storage of diverse probiotic products. Functional apple snacks, coated with methylcellulose films containing fructooligosaccharides and *Lactobacillus plantarum*, were developed; apple snacks were organoleptically accepted after dehydration (60°C) and storage (90 days) and presented a cultivable lactobacilli load higher than the minimum considered necessary (10^6–10^7 CFU/g) (Tavera-Quiroz et al. 2015). Similarly, probiotic baked bread was obtained by covering the surface with an SA or whey protein concentrate coatings containing *Lactobacillus rhamnosus* guar gum (GG). Surface coated by films effectively maintained the *L. rhamnosus* GG viability on pan bread during processing and storage; besides, SA films protected probiotic microorganism during in vitro digestion tests (Soukoulis et al. 2014).

22.3 Application of Edible Films and Coatings in TPF

22.3.1 Films and Coatings in TPF

22.3.1.1 Antimicrobial Films and Coatings

TPF such as cooked jam, chicken, meat, and salmon can be exposed to the environment during peeling, slicing, and repackaging operations, leading to a potential postprocessing contamination by microorganisms. The main microorganisms related to health and safety include *Salmonella*, *Campylobacter jejuni*, *Escherichia coli*, *Staphylococcus aureus*, *Clostridium botulinum*, and *Listeria monocytogenes*. Growth of aerobic bacteria on surfaces of TPF might be inhibited by incorporating antimicrobials agents, like benzoic, acetic, citric and lactic acids, benzoates, propionates, sorbates, nisin, and natural preservatives, into edible films and coatings (Tauxe 2002; Joerger 2007; Velusamy et al. 2010).

Antimicrobial films and coatings, presented in Table 22.2, were chiefly prepared with polysaccharides (pectin, starch, CH, pullulan, κ-carrageenan) and proteins (zein, whey protein, gelatin). These materials impart an oxygen barrier property to films and coatings, because of their ordered network structure at low relative humidity (Lin and Zhao 2007).

The retarding or killing effect of antimicrobials on foodborne pathogens could be complemented by the low oxygen permeability of the films and coatings or by the inherent antimicrobial activity of the biomaterial. CH is one of the few biomaterials that exhibit antibacterial and antifungal activity by itself, which is attributed to a change in microbial cell permeability caused by interactions between the

TABLE 22.2

Antimicrobial Edible Films and Coatings for Thermally Processed Meat, Poultry, and Sea Products

Product	Coating/Film Material	Additive	Antimicrobial Effect	References
Bologna, regular cooked ham, and beef pastrami	CH	Lauric acid, trans-cinnamaldehyde	*Enterobacteriaceae* and *S. liquefaciens* (delayed or completely inhibited); lactic acid bacteria (not affected)	Ouattara et al. (2000)
Ready-to-eat chicken	Zein	Nisin, calcium propionate	*L. monocytogenes* (inhibited)	Janes et al. (2002)
Sliced bologna and summer sausage	Whey protein isolate	p-Aminobenzoic acid, sorbic acid	*L. monocytogenes, E. coli,* and *S. typhimurium* (reduced)	Cagri et al. (2002)
Hot dogs	Whey protein isolate	p-Aminobenzoic acid	*L. monocytogenes* (reduced)	Cagri et al. (2003)
Roast beef	CH	Acetic and lactic acid	*L. monocytogenes* (high reduction by acetic acid and small reduction by lactic acid)	Beverlya et al. (2008)
Frankfurters	Bacterial cellulose	Nisin	*L. monocytogenes* (effectively inhibited)	Nguyen et al. (2008)
Turkey frankfurter	Whey protein isolate	Grape seed extract, nisin, malic acid, EDTA	*L. monocytogenes, E. coli* O157:H7, *S. typhimurium* (effectively inhibited)	Gadang et al. (2008)
Sliced ham	Cellulose acetate	Pediocin	*L. innocua* (high reduction), *Salmonella* sp. (small reduction)	Santiago-Silva et al. (2009)
Poached and deli turkey products	Alginate, κ-carrageenan, pectin, xanthan gum, and starch	Nisin, NovaGARD CB1, Guardian NR100, sodium lactate, sodium diacetate, potassium sorbate	*L. innocua* (high reduction)	Juck et al. (2010)
Ham	Gelidium corneum (agarose)	Carvacrol	*L. monocytogenes* (reduced)	Lim et al. (2010)
Turkey bologna	Gelatin	Nisaplin and Guardian	*L. monocytogenes* (reduced)	Min et al. (2010)
Salami	CH, SC, and SC/CH applied as coatings or wrappers		Native flora (reduced)	Moreira et al. (2011)
Roasted turkey	Starch, CH, alginate, pectin	Sodium lactate, sodium diacetate	*L. monocytogenes* (reduced)	Jiang et al. (2011)
Mortadella sausages	CH	*Zataria multiflora* Boiss essential oil, grape seed extract	*L. monocytogenes* (reduced)	Moradi et al. (2011)
Ham and bologna sausages	Pectin-based apple, carrot, and hibiscus	Carvacrol, cinnamaldehyde	*L. monocytogenes* (reduced)	Ravishankar et al. (2012)
Pork sausages	CH	Green tea extract	Yeasts, molds, lactic acid bacteria (reduced)	Siripatrawan and Noipha (2012)
Cooked sliced ham	Lauric arginate	Polylactic acid	*L. monocytogenes* and *S. typhimurium* (reduced)	Theinsathid et al. (2012)

(Continued)

TABLE 22.2 (*Continued*)

Antimicrobial Edible Films and Coatings for Thermally Processed Meat, Poultry, and Sea Products

Product	Coating/Film Material	Additive	Antimicrobial Effect	References
Ready-to-eat turkey breast	Pullulan	Essential oils of oregano and rosemary, silver and ZnO nanoparticles	*S. aureus*, *L. monocytogenes, E. coli* O157:H7, and *S. typhimurium* (reduced)	Morsy et al. (2014)
Ready-to-eat deli turkey meat	CH, lauric arginate ester	Nisin	*L. innocua* (reduced)	Guo et al. (2014)
Cooked pork patties	Pectin	Green tea extract polyethylene glycol	Total aerobic bacteria (reduced)	Kang et al. (2007)

positively charged CH molecules and the negatively charged microbial cell membranes (Wang 1992; Pranoto et al. 2005; No et al. 2007).

Smoking and frying are thermal process methods to preserve fish products. In order to keep them protected from oxidation and microbial spoilage, they can be coated of wrapped with edible coatings or films. Thus, it has been demonstrated that calcium alginate coatings containing oyster lysozyme and nisin controlled the growth of *L. monocytogenes* and *Salmonella anatum* on the surface of ready-to-eat smoked salmon, at refrigerated temperatures (Datta et al. 2008). The antimicrobial activity of whey protein isolate films and coatings with lysozyme was also studied on smoked salmon; films retarded the growth of *L. monocytogenes*, total aerobes, yeasts, and molds in the samples, at 4°C and 10°C (Min et al. 2005). As for meat and poultry foods, *L. monocytogenes* is of primary importance for fish products.

22.3.1.2 Antioxidant Films and Coatings

Lipid oxidation is an important reaction affecting the quality of meat and fish products. It leads to undesirable flavors and taste, nutrient and quality losses, and production of toxic compounds. These undesirable changes can be minimized using films or coating with high oxygen barrier, which is expected to reduce the food–oxygen interactions. For instance, lipid oxidation was effectively reduced in cooked turkey meat slices wrapped with sodium caseinate films (Caprioli et al. 2009), which can be explained by the low oxygen permeability of the wrapping film ($\sim18.23 \times 10^{-14}$ cm^3 m^{-1} s^{-1} Pa^{-1}) (Jiménez et al. 2012).

The incorporation of antioxidants into edible films and coatings is another approach to control oxidation in foods like TPF (Bonilla et al. 2012; Coskun et al. 2014).

In spite of the low cost, high stability, and effectiveness of the synthetic antioxidants, there is an increasing tendency for using natural antioxidants. Besides α-tocopherol and ascorbic and citric acids, plant extracts from butterfly pea (*Clitoria ternatea* Linn.), clove (*Syzygium aromaticum* Linn.), ginger (*Zingiber officinale* Rosc.), cinnamon (*Cinnamomum iners*), green tea (*Camellia sinensis*), rosemary (*Rosmarinus officinalis*), thyme (*Thymus vulgaris* Linn.), oregano (*Origanum heracleoticum* L.), garlic (*Allium sativum*), and laurel (*Laurus nobilis*) have shown antioxidant activity (Bonilla et al. 2012; Erkan et al. 2015).

Several studies on antioxidant films and coatings applied to TPF are summarized in Table 22.3. CH film and pectin coatings containing green tea extracts inhibited lipid oxidation in pork sausages and cooked pork patties, respectively (Kang et al. 2007; Siripatrawan and Noipha 2012). Turkey breast, wrapped in corn zein film with butylated hydroxyanisole, was better protected against lipid oxidation than turkey samples in polyvinylidene chloride films (Herald et al. 1996). The antioxidant efficacy of the antioxidant films and coatings is determined by a number of factors like type of the film and its oxygen

TABLE 22.3

Antioxidant Edible Films and Coatings for Thermally Processed Meat and Poultry Products

Product	Coating/Film Material	Additive	Effect	References
Cooked turkey	Corn zein	BHA, bacterial enzyme, emulsifier	Low lipid oxidation	Herald et al. (1996)
Ham	*Gelidium corneum* (agarose)	Carvacrol	Low lipid oxidation	Lim et al. (2010)
Mortadella sausages	Chitosan	*Zataria multiflora* Boiss essential oil, grape seed extract	Lipid oxidation reduced	Moradi et al. (2011)
Pork sausages	Chitosan	Green tea extract	Lipid oxidation inhibited	Siripatrawan and Noipha (2012)
Cooked pork patties	Pectin	Green tea extract polyethylene glycol	Lipid oxidation decreased	Kang et al. (2007)

barrier properties, specific activity of the incorporated antioxidant and its releasing kinetics, moisture content of the foodstuff, and relative humidity in the ambient, among others.

22.3.1.3 Other Applications of Films and Coatings in TPF

Deep-fat frying is a widely used thermal process for preparing foods. The soft and moist interior along with the porous crispy crust of fried foods provides their characteristic palatability. Because one of the main drawbacks of fried foods is their high oil content, different strategies have been proposed to reduce the oil absorbed levels. The use of edible films and coatings is among the tested strategies. Hence, it was reported that coating potato French fries with pectin or SA and a $CaCl_2$, as a first coating, followed by a second coating of CMC, reduced the oil content of the final product to less than 50% of its initial value. Additionally, double coating prevented water evaporation during deep-fat frying and enhanced the sensory characteristics of the potato French fries (Kahlil 1999).

On the other hand, soy protein/gellan gum coatings were developed to reduce fat transfer in doughnut mix discs and in potato discs during deep-fat frying (Rayner et al. 2000). The application of these coatings allowed a significant fat reduction content in coated discs of doughnut mix compared to uncoated fried samples. Sensory testing indicated that some panelists observed a slight difference between coated and uncoated potato fries, but the majority of the panelists preferred the coated fries over the uncoated ones. Penetration test on potato fries showed no significant difference between the texture of the coated and uncoated fried samples.

Mallikarjunan et al. (1997) demonstrated the potential of zein, hydroxypropyl methylcellulose, or methylcellulose coatings for favoring moisture retention and reducing fat absorption during frying of mashed potato balls. Compared to the uncoated control samples, coated mashed potato balls had reduced oil absorption and reduced moisture loss.

Crispy batter coatings are a critical part of fried foods. However, batter coatings normally absorb significant amounts of water during microwave reheating. To overcome this problem, it has been reported that an additional coating (hydroxypropyl methylcellulose) between mackerel fish meat and batter provided a good water barrier. This led to a lower diffusion of water molecules from fish meat to the crust during microwave reheating (Chen et al. 2008) (Table 22.4).

22.3.1.4 Trends in Films and Coatings Applications

Edible films and coatings can accomplish different functions such as carriers of antioxidants, antimicrobials, aromatic compounds, and probiotics or modulators of mass transfer phenomena between the product and its surroundings during or after processing. At present, there is an increasing research trend in the development of new films and coatings formulations with nanoparticles to improve the film properties and satisfy the food industry requirements.

TABLE 22.4

Effect of Edible Films and Coatings in Several Thermally Processed Foods

Product	Coating/Film Material	Other Constituents	Effect/Result	References
Coated before fried or heating				
Potato French fries	Potato French fries	CaCl$_2$	Reduction of oil intake and water evaporation during deep-fat frying. Double-coated French fries had higher moisture contents and firmer structures than the single coated.	Khalil (1999)
Doughnut mix and potatoes discs	Soy protein	Gellan gum or glycerin	Soy protein/gellan gum coating reduced fat intake in doughnut mix and coated fried potatoes.	Rayner et al. (2000)
Mashed potato balls	Zein, hydroxypropyl methylcellulose or methylcellulose		All the coatings tested reduced moisture loss and fat uptake during deep-fat frying.	Mallikarjunan et al. (1997)
Mackerel nuggets	Hydroxypropyl methylcellulose coating between mackerel fish meat and batter		Inhibition of water diffusion from fish meat into the crust during microwave reheating.	Chen et al. (2008)

REFERENCES

Atarés, L., Pérez-Masia, R., and Chiralt, A. The role of some antioxidants in the HPMC film properties and lipid protection in coated toasted almonds. *Journal of Food Engineering* 104 (2011): 649–656.

Baldwin, E.A. and Wood, B. Use of edible coating to preserve pecans at room temperature. *HortScience* 41(1) (2006): 188–192.

Baysal, T., Bilek, S.E., and Apaydın, E. The effect of corn zein edible film coating on intermediate moisture apricot (*Prunus armeniaca* L.) quality. *GIDA* 35(4) (2010): 245–249.

Beverlya, R.L., Janes, M.E., Prinyawiwatkula, W., and No, H.K. Edible chitosan films on ready-to-eat roast beef for the control of *Listeria monocytogenes*. *Food Microbiology* 25(3) (2008): 534–537.

Bonilla, J., Atares, L., Vargas, M., and Chiralt, A. Edible films and coatings to prevent the detrimental effect of oxygen on food quality: Possibilities and limitations. *Journal of Food Engineering* 110 (2012): 208–213.

Cagri, A., Ustunol, Z., Osburn, W., and Ryser, E.T. Inhibition of *Listeria monocytogenes* on hot dogs using antimicrobial whey protein-based edible casings. *Journal of Food Science* 68(1) (2003): 291–299.

Cagri, A., Ustunol, Z., and Ryser, E.T. Inhibition of three pathogens on bologna and summer sausage using antimicrobial edible films. *Journal of Food Science* 67(6) (2002): 2317–2324.

Caprioli, I., O'Sullivan, M., and Monahan, F.J. Use of sodium caseinate/glycerol edible films to reduce lipid oxidation in sliced turkey meat. *European Food Research and Technology* 228 (2009): 433–440.

Chen, Ch.L., Lim, P.Y., Hu, W.H., Lan, M.H., Chen, M.J., and Chen, H.H. Using HPMC to improve crust crispness in microwave-reheated battered mackerel nuggets: Water barrier effect of HPMC. *Food Hydrocolloids* 22 (2008): 1337–1344.

Coskun, B.K., Calikoglu, E., Emiroglu, Z.K., and Candogan, K. Antioxidant active packaging with soy edible films and oregano or thyme essential oils for oxidative stability of ground beef patties. *Journal of Food Quality* 37(3) (2014): 203–212.

Datta, S., Janes, M.E., Xue, J., Losso, Q.G., and La Peyre, J.F. Control of *Listeria monocytogenes* and *Salmonella anatum* on the surface of smoked salmon coated with calcium alginate coating containing oyster lysozyme and nisin. *Journal of Food Science* 73(2) (2008): M67–M71.

Erkan, N., Doğruyol, H., Günlü, A., and Genç, I.Y. Use of natural preservatives in seafood: Plant extracts, edible film and coating. *Journal of Food and Health Science* 1(1) (2015): 33–49.

Falguera, V., Quintero, J.P., Jiménez, A., Muñoz, A.J., and Ibarz, A. Edible films and coatings: Structures, active functions and trends in their use. *Trends in Food Science and Technology* 22(6) (2011): 292–303.

Gadang, V.P., Hettiarachchy, N.S., Johnson, M.G., and Owens, C. Evaluation of antibacterial activity of whey protein isolate coating incorporated with nisin, grape seed extract, malic acid, and EDTA on a turkey frankfurter system. *Journal of Food Science* 73(8) (2008): 389–394.

García, M., Díaz, R., Martínez, Y., and Casariego, A. Effects of chitosan coating on mass transfer during osmotic dehydration of papaya. *Food Research International* 43 (2010): 1656–1660.

Guldas, M., Bayizit, A.A., Yilsay, T.O., and Yilmaz, L. Effects of edible film coatings on shelf-life of mustafakemalpasa sweet, a cheese based dessert. *Journal of Food Science and Technology* 47(5) (2010): 476–481.

Guo, M., Jin, T.Z., Wang, L., Scullen, O.J., and Sommers, C.H. Antimicrobial films and coatings for inactivation of *Listeria innocua* on ready-to-eat deli turkey meat. *Food Control* 40 (2014): 64–70.

Herald, T.J., Hachmeister, K.A., Huang, S., and Bowers, J.R. Corn zein packaging materials for cooked turkey. *Journal of Food Science* 61(2) (1996): 415–418.

Jalaee, F., Fazeli, A., Fatemian, H., and Tavakolipour, H. Mass transfer coefficient and the characteristics of coated apples in osmotic dehydrating. *Food and Bioproducts Processing* 89 (2011): 367–374.

Janes, M.E., Kooshesh, S., and Johnson, M.G. Control of *Listeria monocytogenes* on the surface of refrigerated, ready to eat chicken coated with edible zein film coatings containing nisin and/or calcium propionate. *Journal of Food Science* 67(7) (2002): 2754–2757.

Jiang, Z., Neetoo, H., and Chen, H. Efficacy of freezing, frozen storage and edible antimicrobial coatings used in combination for control of *Listeria monocytogenes* on roasted turkey stored at chilled temperatures. *Food Microbiology* 28(7) (2011): 1394–1401.

Jiménez, A., Fabra, M.J., Talens, P., and Chiralt, A. Effect of sodium caseinate on properties and ageing behavior of corn starch based. *Food Hydrocolloids* 29(2) (2012): 265–271.

Joerger, R.D. Antimicrobial films for food applications: A quantitative analysis of their effectiveness. *Packaging Technology and Science* 20 (2007): 231–273.

Juck, G., Neetoo, H., and Chen, H. Application of an active alginate coating to control the growth of *Listeria monocytogenes* on poached and deli turkey products. *International Journal of Food Microbiology* 142 (2010): 302–308.

Kang, H.J., Jo, C., Kwon, J.H., Kim, J.H., Chung, H.J., and Byun, M.W. Effect of a pectin-based edible coating containing green tea powder on the quality of irradiated pork patty. *Food Control* 18 (2007): 430–435.

Khalil, A.H. Quality of French fried potatoes as influenced by coating with hydrocolloids. *Food Chemistry* 66 (1999): 201–208.

Khin, M.M., Zhou, W., and Perera, C.O. A study of the mass transfer in osmotic dehydration of coated potato cubes. *Journal of Food Engineering* 77(2006): 84–95.

Khin, M.M., Zhou, W., and Yeo, S.Y. Mass transfer in the osmotic dehydration of coated apple cubes by using maltodextrin as the coating material and their textural properties. *Journal of Food Engineering* 81 (2007): 514–522.

Lago-Vanzela, E.S., do Nascimento, P., Fontes, E.A.F., Mauro, M.A., and Kimura, M. Edible coatings from native and modified starches retain carotenoids in pumpkin during drying. *LWT—Food Science and Technology* 50 (2013): 420–425.

Li, T., Zhou, P., and Labuza, T.P. Effects of sucrose crystallization and moisture migration on the structural changes of a coated intermediate moisture food. *Frontiers of Chemical Engineering in China* 3(4) (2009): 346–350.

Lim, G.O., Hong, Y.H., and Song, K.B. Application of *Gelidium corneum* edible films containing carvacrol for ham packages. *Journal of Food Science* 75(1) (2010): C90–C93.

Lin, D. and Zhao, Y. Innovations in the development and application of edible coatings for fresh and minimally processed fruits and vegetables. *Comprehensive Reviews in Food Science and Food Safety* 6 (2007): 60–75.

Mallikarjunan, P., Chinnan, M.S., Balasubramaniam, V.M., and Phillips, R.D. Edible coatings for deep-fat frying of starchy products. *LWT—Food Science and Technology* 30 (1997): 709–714.

Matuska, M., Lenart, A., and Lazarides, H.N. On the use of edible coatings to monitor osmotic dehydration kinetics for minimal solids uptake. *Journal of Food Engineering* 72 (2006): 85–91.

Min, B., Han, I.Y., and Dawson, P.L. Antimicrobial gelatin films reduce *Listeria monocytogenes* on turkey bologna. *Poultry Science* 89(6) (2010): 1307–1314.

Min, S., Harris, L.J., Han, J.H., and Krochta, J.M. *Listeria monocytogenes* inhibition by whey protein films and coatings incorporating lysozyme. *Journal of Food Protection* 68(11) (2005): 2317–2325.

Moradi, M., Tajik, H., Rohani, S.M.R., and Oromiehie, A.R. Effectiveness of *Zataria multiflora* Boiss essential oil and grape seed extract impregnated chitosan film on ready-to-eat mortadella-type sausages during refrigerated storage. *Journal of Science and Food Agriculture* 91 (2011): 2850–2857.

Moreira, M.R., Pereda, M., Marcovich, N.E., and Roura, S.I. Antimicrobial effectiveness of bioactive packaging materials from edible chitosan and casein polymers: Assessment on carrot, cheese, and salami. *Journal of Food Science* 76(1) (2011): M54–M63.

Morsy, M.K., Khalaf, H.H., Sharoba, A.M., El-Tanahi, H.H., and Cutter, C.N. Incorporation of essential oils and nanoparticles in pullulan films to control foodborne pathogens on meat and poultry products. *Journal of Food Science* 79 (2014): M675–M684.

Nguyen, V.T., Gidley, M.J., and Dykes, G.A. Potential of a nisin containing bacterial cellulose film to inhibit *Listeria monocytogenes* on processed meats. *Food Microbiology* 25(3) (2008): 471–478.

No, H.K., Meyers, S.P., Prinyawiwatkul, W., and Xu, Z. Applications of chitosan for improvement of quality and shelf life of foods: A review. *Journal of Food Science* 72(5) (2007): R87–R100.

Ouattara, B., Simard, R.E., Piette, G., Bégin, A., and Holley, R.A. Inhibition of surface spoilage bacteria in processed meats by application of antimicrobial films prepared with chitosan. *International Journal of Food Microbiology* 62 (2000): 139–148.

Pinto, A.M.B., Santos, T.M., Caceres, C.A., Lima, J.R., Ito, E.N., and Azeredo, H.M.C. Starch-cashew tree gum nanocomposite films and their application for coating cashew nuts. *LWT—Food Science and Technology* 62 (2015): 549–554.

Pranoto, Y., Rakshit, S.K., and Salokhe, V.M. Enhancing antimicrobial activity of chitosan films by incorporating garlic oil, potassium sorbate and nisin. *LWT—Food Science and Technology* 38 (2005): 859–865.

Rao, D.N. Intermediate moisture foods based on meats—A review. *Food Reviews International* 13(4) (1997): 519–551.

Rao, M.S., Chander, R., and Sharma, A. Development of shelf-stable intermediate moisture meat products using active edible chitosan coating and Irradiation. *Journal of Food Science* 70(7) (2005): M325–M331.

Ravishankar, S., Jaroni, D., Zhu, L., Olsen, C., McHugh, T., and Friedman, M. Inactivation of *Listeria monocytogenes* on ham and bologna using pectin-based apple, carrot, and hibiscus edible films containing carvacrol and cinnamaldehyde. *Journal of Food Science* 77(7) (2012): 377–382.

Rayner, M., Ciolfi, V., Maves, B., Stedman, P., and Mittal, G.S. Development and application of soy-protein films to reduce fat intake in deep-fried foods. *Journal of the Science of Food and Agriculture* 80 (2000): 777–782.

Santiago-Silva, P., Soares, N.F.F., Nobrega, J.E.N. et al. Antimicrobial efficiency of film incorporated with pediocin (ALTA 2351) on preservation of sliced ham. *Food Control* 20(1) (2009): 85–89.

Siripatrawan, U. and Noipha, S. Active film from chitosan incorporating green tea extract for shelf life extension of pork sausages. *Food Hydrocolloids* 27(1) (2012): 102108.

Soukoulis, C., Yonekura, L., Gan, H.H., Behboudi-Jobbehdar, S., Parmenter, C., and Fisk, I. Probiotic edible films as a new strategy for developing functional bakery products: The case of pan bread. *Food Hydrocolloids* 39 (2014): 231–242.

Talens, P., Pérez-Masía, R., Fabra, M.J., Vargas, M., and Chiralt, A. Application of edible coatings to partially dehydrated pineapple for use in fruit–cereal products. *Journal of Food Engineering* 112 (2012): 86–93.

Tauxe, R.V. Emerging foodborne pathogens. *International Journal of Food Microbiology* 78 (2002): 31–41.

Tavera-Quiroz, M.J., Romano, N., Mobili, P., Pinotti, A., Gómez-Zavaglia, A., and Bertola, N. Green apple baked snacks functionalized with edible coatings of methylcellulose containing *Lactobacillus plantarum*. *Journal of Functional Foods* 16 (2015): 164–173.

Tharanathan, R.N. Biodegradable films and composite coatings: Past, present and future. *Trends in Food Science and Technology* 14 (2003): 71–78.

Theinsathid, P., Visessanguan, W., Kruenate, J., Kingcha, Y., and Keeratipibul, S. Antimicrobial activity of lauric arginate-coated polylactic acid films against *Listeria monocytogenes and Salmonella typhimurium* on cooked sliced ham. *Journal of Food Science* 77(2) (2012): 142–149.

Velusamy, V., Arshak, K., Korostynska, O., Oliwa, K., and Adley, C. An overview of foodborne pathogen detection: In the perspective of biosensors. *Biotechnology Advances* 28 (2010): 232–254.

Wang, G.H. Inhibition and inactivation of five species of foodborne pathogens by chitosan. *Journal of Food Protection* 55(11) (1992): 916–919.

Wu, H., Tassou, S.A., Karayiannis, T.G., and Jouhara H. Analysis and simulation of continuous food frying processes. *Applied Thermal Engineering* 53 (2013): 332–339.

Applications of Films and Coatings for Special Missions

Michelle J. Richardson, Ann H. Barrett, and Lauren O'Conner

CONTENTS

23.1 Introduction

Foods for both the military and National Aeronautics and Space Administration (NASA) have a much longer shelf-life requirement when compared to civilian. The shelf-life requirements for most military fielded rations are 3 years at 27°C (80°F) or 6 months at 38°C (100°F). For the upcoming mission to Mars, NASA requires shelf-stable menu items with at least a 5-year shelf life (Cooper et al., 2011). When developing long shelf-life foods for military and space feeding, there will always be unremitting requirements to ensure food safety is not compromised, menu monotony does not occur, nutritional intake is not reduced, and performance is not diminished. To meet these requirements, the military has been exploiting films and coatings for over 70 years. An edible coating is defined as a thin, edible film that can be deposited onto the surface of a food that can provide protection to extend the shelf life of the coated food by acting as a barrier to moisture, oil, and vapor transmission (Yang, 1994). Films and coatings can be used to develop, optimize, and/or produce foods that are appropriate for military field feeding. Benefits include the following:

1. Extending shelf life by controlling microbial growth
2. Preserving quality by controlling moisture and fat migration
3. Increasing variety
4. Providing nutrient stability and delivery
5. Masking off-flavors
6. Controlling release of specialized ingredients

Because food stabilization is essential for warfighters and astronauts, the Department of Defense (DoD) and NASA have been exploiting edible films and barriers, in order to develop components that are safe, nutritious, acceptable, and stable, thus ensuring the completion of missions, even in austere environments.

This chapter is divided into five sections: the first provides an overview of early efforts, the second focuses on coatings to preserve quality and increase variety, the third describes films and coatings for moisture and fat control, the fourth discusses coating applications for microbial control, and the fifth focuses on coatings for nutrient stability.

23.1.1 Early Efforts

M&M's® are an example of one of the first documented uses of coated food items for military feeding during World War II (Figure 23.1). These pan-coated discs originally patented in 1941 were intended for sale to the general public but were sold exclusively to the military in World War II because, unlike other chocolate candies, these coated discs "melt in your mouth, not in your hands." It was not until after the war when various quotas and war material restrictions were lifted that the popular candy treats were sold commercially to the public. M&M's continue to be a part of the Meal, Ready-to-Eat (MRE™) today (in commercial packaging and overwrapping) and one of the most popular candies with soldiers now as they were in the 1940s when they first appeared in C rations.

In the early 1960s, a research program was initiated in support of NASA on the encapsulation of bite-sized solids to increase space food variety and to prevent crumbs from floating while in zero gravity (Schuetze et al., 1962; AMRL, 1963). A fat-based coating with a high melting point (58°C) was tested in an effort to control the release of free-floating crumbs during flight. However, the coatings proved to be unpalatable; moreover, digestibility trials demonstrated that they were poorly absorbed in the gut and could result in a steatorrhea or excess fat in the stool. Subsequently, components of a novel feeding system based on reversibly compressed, dehydrated food bars and cubes of concentrated sauces and seasonings were developed and demonstrated for space feeding (Durst, 1963, 1964, 1967) (Figure 23.2a and b).

These items were further optimized to increase acceptability under a contract with NASA, Natick Soldier Research Development and Engineering Center (NSRDEC), and the Pillsbury Company (Blodgett, 1967). One of the objectives of the contract was to control the release of various ingredients such as salts, acids, and flavors from dual function bars, that is, bars that can be eaten as is or rehydrated. Two controlled-release systems were developed or procured and evaluated. The first system consisted of a delayed solubility system (DSS), in which ingredients were partially encapsulated in a hydrocolloid component that was less soluble than the ingredients coated. In this case, gum arabic was employed to prevent the rapid solubility or dispersion of citric acid and vinegar. The second system consisted of a thermal-release system (TRS) with melting point between 46°C and 52°C in which lipids served as the encapsulating material. Once the melting point of the lipid was reached, the encapsulated ingredients (citric acid, salt, and barbecue flavor) were released, thus stimulating the olfactory and gustatory receptors.

The coating materials used for both systems were either commercially available or prepared in-house and incorporated into the dual functioning bars. For one of the DSSs, a solution containing 20% citric

FIGURE 23.1 Advertisement for M&M's during World War II.

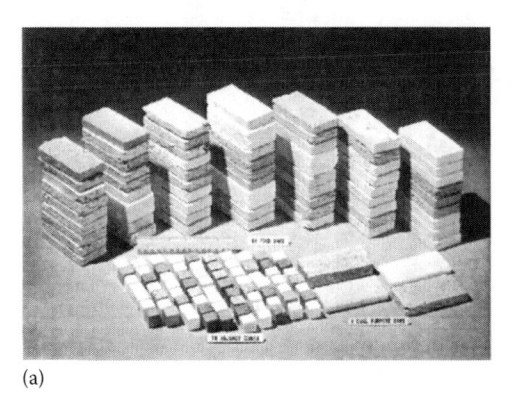
(a)

(b)

FIGURE 23.2 (a) Compressed dehydrated bars and cubes and (b) compressed and dehydrated meal components.

acid, 20% gum arabic, and 60% water was prepared by mixing at high speed in a commercial blender; then spray-dried, using a Bowen laboratory spray dryer; and sifted through a U.S. no. 12 stainless steel sieve. For the TRS, a molten mixture was prepared by heating 75% anhydrous citric acid with 25% Durkee, KLX flakes to 100°C using a commercial Waring-blender bowl. The mixture was poured on polyethylene, allowed to cool, broken into 1 in. pieces, chilled for 2 h in dry ice, ground using a Waring blender, and sifted using U.S. no. 12 stainless steel sieve. TRS salt was "Experimental Salt no. 7308" (85% salt encapsulated with hydrogenated vegetable oil), provided by Presco Food Products Inc., Flemington, New Jersey. The thermal-release barbecue flavor was supplied by Gentry Corporation, Paramus, New Jersey.

Prior to this encapsulation technology, the orange, lemon, beef/barley, barbecue beef, and chili bars were impossible to directly consume because of their high levels of acid, salt, and flavors. However, all these bars became acceptable in the dry form when these ingredients were encapsulated in the release systems. Table 23.1 summarizes these results.

TABLE 23.1

Bar Type, Bar Characteristics Prior to Use of Controlled-Release System, Controlled-Release Systems Employed, Ingredients Encapsulated, and Bar Characteristics after Controlled-Release System Addition

NASA Dual Purpose Bar Type	Characteristics Prior to Addition of Controlled-Release Systems Consumed as Is[a]	Controlled-Release System Employed and Ingredient Encapsulated	Characteristics after the Addition of Controlled-Release Systems	
			Consumed as Is[a]	Consumed after Hydration[b]
Lemon	Excessive acidity	DSS (citric acid)	Lemon drop candy	Good lemonade flavor
Orange	Excessive acidity	DSS (citric acid)	Slightly tart orange candy	Excellent orange-flavored drink
Chili	Excess saltiness, spice, and acidity	TRS (citric and salt)	Mild chili flavor	Good mild chili-type flavor
Beef barely soup	Excess saltiness	TRS (salt)	Bouillon, spice acceptable salt level	Rich beef barely soup
Barbecue beef	Excessive saltiness, acidity, and spice	TRS (barbecue flavor, vinegar, citric, and salt)	Mild barbecue flavor	Very good barbecue beef flavor

Source: Adapted from Blodgett, J. Maximum variety from feeding unit of low weight and bulk. Technical report no. 70-29-FL, contract no. DAAG 17-68-C-O148. U.S. Army Natick Laboratories, Natick, MA, 1967.

[a] Bars consumed were considered acceptable if 67% or more panelist members have rating of five or above a nine-point hedonic scale.

[b] Hydrated bars were considered acceptable if 67% or more panelist members scored the product six or above on a nine-point hedonic scale.

Another objective of this effort was to develop a coating material as well as application techniques to prevent fragmentation and fracturing of the components. Under this effort, a number of coating materials were subjectively evaluated and included zein, edible shellac, gelatin, fats, and acetylated monoglycerides; however, most were found unsuitable. Two coatings, however, were developed and applied to food bars and their effectiveness was validated using vibration, impact adhesion, and stickiness tests. Results showed that the coatings markedly reduced fragmentation and fracture of the bars, provided good adhesion, and were not sticky.

Cole (1966a) developed edible barriers for dehydrated foods, evaluated them after accelerated storage, and determined their effects on oxygen and moisture transport, fragmentation, and mold growth. He found effective coating material that included hot melts of acetoglyceride and ethylcellulose; mixtures of monoglyceride and polyglycolesters, protein films (soy and gelatin), fatty esters of amylase, monoglycerides, and hard fats; and combinations of these materials in the form of laminates or mixtures. Chemical preservatives, such as sorbic acid, potassium sorbate, methyl, and propyl p-hydroxybenzoates, were effective in inhibiting mold growth when they were incorporated into coating formulations. In Phase II of this effort, Cole (1966) focused on a commercially feasible procedure for the application of the previously developed films, evaluating the films after they were applied to representative foods stored in harsh temperature and humidity environments. His results showed a hot-melt curtain coating technique to be commercially feasible; however, the coatings did not withstand temperature cycling between −18°C and 40°C during storage.

In a collaborative effort with the U.S. Department of Interior, Fish and Wildlife Services, edible coatings were developed to retain the structural integrity of irradiated fish fillets during enzyme inactivation processing (microwave heating and deep fat frying) and during room temperature storage (Ronsivalli, 1972). The coating consisted of ground fish flesh, methylcellulose, and slow-browning breading material. Other coatings evaluated under this effort included methocel applied with and without a starch batter, acetylated monoglycerides (Myvacet) as a dip, and shellac applied as a dip or spray coated. Visual and organoleptic assessments showed that the methocel and methocel–starch coatings functioned well during processing but were unstable during storage: the Myvacet could not withstand the enzyme inactivation process, and the shellac became very brittle and cracked and had an extremely bitter aftertaste. The ground-blended fish muscle coating was initially considered highly acceptable by the taste panel, but after 2–3 months, the product became unacceptable. The coating did, however, minimize the physical damage during the heating process and through 9 months of storage at room temperature.

Under a cooperative agreement with NSRDEC, Brown and Swift demonstrated the feasibility of controlling the flavor intensity in compressed foods (Pavey, 1973, 1976). This investigation was conducted in two phases: Phase I involved evaluating and testing commercially available encapsulated flavoring materials. During this phase, chili with beans and barbecue pork bars were successfully developed. In Phase II, an extension of the previous study, four different dual-purpose bars were developed. Coatings were also developed and evaluated and included caseinate, caseinate with fat, and gelatin. All coatings achieved desired flavor inhibition when bars were consumed in the dry state and allowed release of the full flavor intensity when the bars were rehydrated. All items received an average rating of six or more on a nine-point hedonic scale.

Harris et al. (1974) used a powdered hydroxypropyl methylcellulose (HPMC), dextrose, low-DE hydrolyzed cereal solids, and pregelatinized tapioca starch mixture with water to produce a smooth, nonviscous, and uniform coating on the exterior surface of freeze-dried meat prior to cooking. This effort improved the texture and moisture retention properties of the cooked meat and other solid food. This author also evaluated hard domestic butters as a replacement for coconut fat in the enrobing material for three military ration candy types (Harris and Westcott, 1974). Candy produced with 100 hour A.O.M. coconut fat in the coating was more acceptable than that produced with 100 hour A.O.M. domestic fat after 6 months of storage. It was concluded that domestic hard butter was a satisfactory substitute for hydrogenated coconut fat in formulating candy centers, where hydrolytic rancidity can be a problem but not for formulating candy where oxidative rancidity is more likely to occur.

A limited investigation was undertaken to develop an edible coating that would retard moisture loss (freezer burn) and oxidative rancidity in frozen meat products. Shaw and Natress (1996) conducted a literature review, communicated with subject-matter experts, conducted laboratory screenings, carried out limited storage studies, and concluded that edible coating technology "has not advanced significantly since 1968." During the screening process, approximately 15 commercial coatings were evaluated on flaked, formed, and frozen lamb, which was grilled after thawing. Two promising coatings, calcium chloride alginate and a carrageenan mixture (iota and kappa types), were identified and evaluated. These two coatings produced a transparent or translucent film that evenly coated the lamb and were free of off-odors and off-flavors after a limited storage period.

23.1.2 Coatings to Preserve Quality and Increase Variety

Variety and acceptability are key considerations to ensure warfighter acceptance. Ration items and menus are designed to provide a variety of foods that will help prevent menu fatigue. Trail mix, composed of peanuts, walnuts, almonds, filberts, and raisins, is included in military rations and provides variety while supplying a good source of protein, vitamin A, potassium, fiber, and iron. However, obtaining shelf stability in a heterogeneous product with a relatively high fat content (30%) and ingredients with different moisture contents/water activities was difficult to accomplish. Prior to the use of coatings, the combined nut–raisin mix tended to aggregate or stick together, and the raisins became hard and sticky after storage, while the lipids of the nuts underwent oxidation. The stability of the raisin–nut trail mix was increased by using various coatings on individual components (PCR-N-003A; Defense Logistics Agency, 2013): the raisins were coated with an oil to reduce this agglomeration, and peanuts and walnuts were separately coated with 1.5% edible shellac or corn protein (zein) dissolved in alcohol to arrive at a 0.05% coating intended to control lipid oxidation.

Baked items are highly acceptable and one of the most consumed items in a field environment. To increase variety and acceptability, brownies, cookies, and bars were coated with chocolate. Besides providing variety and increasing acceptability, the chocolate coatings on brownies, cookies, chocolate fudge bars, and vanilla cream bars served as delivery systems for various vitamins. These items were coated with chocolate that was fortified with vitamin A, ascorbic acid, thiamin, and pyridoxine (Branagan and Pruskin, 1993).

23.1.3 Films and Coatings for Moisture or Fat Control

23.1.3.1 Past Applications

The control of moisture and fat migrations in foods is very important. Unwanted migration can result in undesirable microbial, sensory, and objective texture changes. One strategy that can effectively reduce moisture and fat migration is the incorporation of an edible barrier between layers. There are a variety of potential edible barriers that can be constructed from solutions containing combinations of different proteins, lipids, and carbohydrates (Kester and Fennema, 1986). Films can be applied to food items through spraying, wrapping, or dipping techniques (Guillard et al., 2004). Films can also be cast by drying the solution on a flat surface and then incorporated as a thin extra layer in the composite—a technique that is sometimes advocated due to the improved uniformity in film thickness of cast films compared to sprayed or dip-formed films.

Conca and Yang (1993) evaluated a shellac-based coated material as an edible film for intermediate-moisture beef sticks to preserve quality and to reduce packaging volume, weight, and waste. Samples were coated with a two-phase, shellac-based coating and stored for 4 weeks at room temperature. After 4 weeks, the coated beef sticks lost significantly less weight (moisture) than the noncoated beef sticks. Also noted were significant changes in color, texture, and size of the uncoated samples. Conca (1995) also evaluated a collagen-based film to reduce or eliminate the need for plastic bags or wraps for frozen meats. Microbial, sensory, oxidation, color changes, strength, moisture vapor transmission, and oxygen transmission were evaluated to determine the barrier effectiveness. Though there were several differences that were observed between the edible barrier and plastic-wrapped meat samples, the feasibility of using collagen-based films as an edible meat wrap was demonstrated.

Several studies were conducted to evaluate efficacy of moisture and fat migration barriers in sandwich systems. Edible intercomponent barriers were used to reduce moisture and fat migration in model mustard sandwich systems (Conca and Kensil, 2007, 2008; Conca et al., 2008). Fat migration was assessed in a thermally treated sandwich system. In this study, Conca tested zein/alcohol and sodium caseinate

(NaCas)/water film formulations for their ability to reduce migration of a higher fat tomato paste/olive oil/lycopene oil filling. Lycopene migration was reduced with the NaCas films and one formulation of zein and to a smaller degree for the other zein films. Conca (2008) also assessed zein, NaCas, collagen, and cornstarch-/cellulose-based films in a similar thermally treated sandwich system to prevent moisture and fat migration into the bread from the mustard/olive oil filling. Water activity, percent moisture, and pheophytin from olive oil contained in mustard filling were measured as indicators of migration and used to access barrier efficacy. The NaCas film was 12.69% more effective than the control and more effective than all other films at preventing migration of oil into the bread. NaCas film was also shown to have the best moisture barrier properties as demonstrated by the water activity and percent moisture data. Zein film was the second most effective film. One of the collagen films tested provided fairly good barrier properties. NaCas films effectively inhibited fat and moisture migration in pocket sandwich systems.

Though barriers have proven to be effective for moisture and fat migrations (Barrett 2006; Conca, 2002), they also have an effect on the sensory acceptability. Barrett et al. (2008) quantified the sensory attributes of zein films during storage. Results showed that zein barriers can be incorporated into selected sandwich systems without significant sensory consequences. The effects on perceived odor, flavor, and overall quality were significant only in bread/crust/roll systems containing no filling and were mitigated in meat-containing sandwiches. Conca (2008) showed sandwiches developed using NaCas or zein barrier films that may improve flavor and subjective texture qualities, and sandwiches with NaCas barriers were of better quality.

The U.S. Army expanded intermediated moisture food technology (Briggs et al., 1992; Taoukis and Richardson, 1995, 2007, 2012) to develop multicomponent shelf-stable products such as sandwiches (Figure 23.3). These multicomponent, intermediate-moisture foods (IMFs) are seemingly more familiar and "fresh like" than typical thermally processed, semiliquid ration items. However, despite careful formulation of crust and filling to create complementary water activities, in practice sandwich products will undergo interphase equilibration of moisture. This equilibration is largely due to the development of a relatively dry crust layer formed during baking, which draws moisture from the more moist bread crumb and, ultimately, from the even moister filling phase.

One of the most promising film types for moisture migration control is the corn protein, zein (Fu et al., 1999), which is among the most hydrophobic of food proteins. Zein has so little compatibility with water that it must be dissolved in alcohol in order to produce films.

"Model" experiments with zein consisted of investigations of barrier efficacy in three different systems. Zein-based edible barriers of different thicknesses and plasticizer concentrations were obtained commercially or developed, incorporated into three different model systems, and the moisture barrier properties of each were characterized. Model systems included (1) commercial snack bread or crackers

FIGURE 23.3 Photograph of a variety of shelf-stable sandwiches/sandwich-type items developed by the military.

interfaced with high-water-activity cheese, with and without barriers, (2) baked sandwich systems to assess effects of thermal treatments on films efficacy, and (3) ready-to-eat (unbaked) sandwich system to determine the film sensory properties. Moisture transfer was determined over time in this system.

23.1.3.1.1 Model Moisture Migration Experiments: Velveeta Cheese into Rubschlager Snack Bread

Results for efficacy of zein barriers inhibiting moisture transfer into a product of typical bread water activity are shown in Figure 23.4a and b. Each film/plasticizer type reduced moisture migration into Rubschlager snack bread. Results were significant for each system (Table 23.2). Increasing film thickness also generally progressively inhibited moisture sorption in each system ($p < 0.05$).

23.1.3.1.2 Model Moisture Migration Experiments: Velveeta Cheese into Melba Toast

Zein barriers also effectively slowed moisture migration into a very low water activity material (Figure 23.5), where water transfer is accelerated due to a larger a_w difference driving force; the timescale of these experiments is thus comparatively shorter than that for the Rubschlager bread tests. Again, film thickness was a significant factor in barrier efficacy ($p < 0.05$).

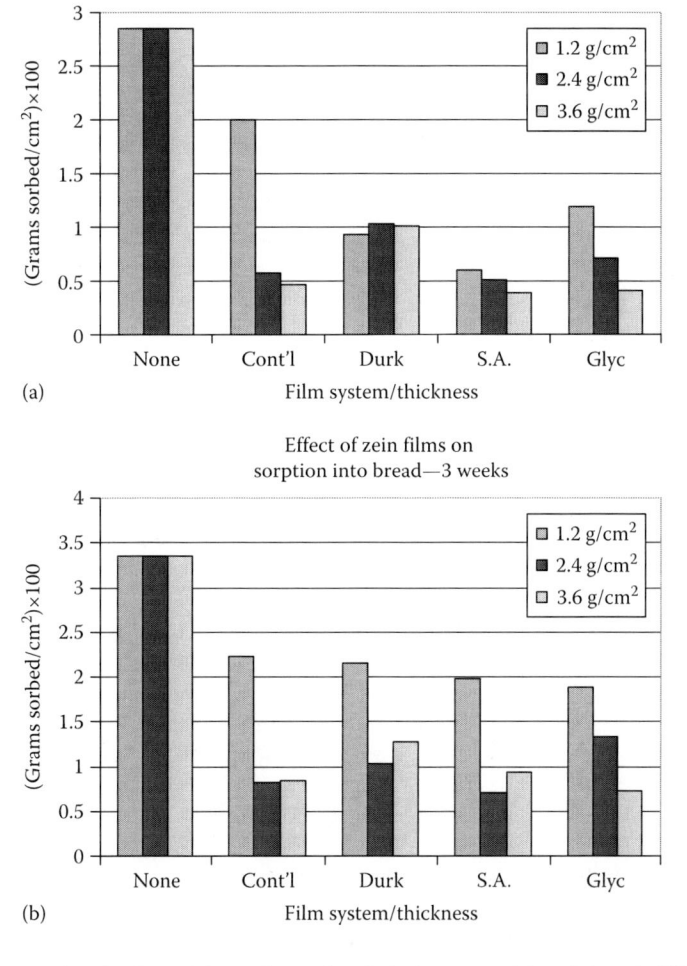

FIGURE 23.4 Moisture migration from Velveeta Cheese into Rubschlager snack bread through different thicknesses of zein film. (a) 1 week and (b) 3 weeks.

TABLE 23.2

Analysis of Variance for Film Efficacy: Unbaked Bilayer Systems

Film Type	Effect	F	p
Rubschlager cocktail bread			
No plasticizer	Film	41.1	>0.001
	Time	3.2	0.067
	Film × time		
Durkex—0.2%	Film	34.6	>0.001
	Time	7.7	0.035
	Film × time		
Stearic acid—0.2%	Film	9.3	0.002
	Time	15.8	>0.001
	Film × time		
Glycerol—2.0%	Film	45.3	>0.001
	Time	14.5	>0.001
	Film × time		
Melba toast			
No plasticizer	Film	8.9	0.002
	Time	21.3	>0.001
	Film × time		
Durkex—0.2%	Film	10.5	>0.002
	Time	22.8	>0.001
	Film × time		
Stearic acid—0.2%	Film	11.0	0.001
	Time	23.4	>0.001
	Film × time		
Glycerol—2.0%	Film	14.8	>0.001
	Time	20.7	>0.001
	Film × time		

23.1.3.1.3 Moisture Migration Experiments Employing Baked Systems

Zein baked into sandwiches maintained partial barrier effectiveness (Figure 23.6), but barrier properties were reduced relative to those for nonthermally treated sandwich systems (Barrett and Ndou, 2007): as can be seen in Table 23.3, one film layer did not produce a significant difference in sorption at 1 week, and three film layers were necessary to significantly inhibit sorption at 2 weeks. However, increasing film thickness (number of films) significantly decreased moisture penetration ($p < 0.05$). It is likely that thermal treatment, possibly by the evolution of localized steam pressure within the sandwich, affects the physical integrity of the barriers by introducing cracks in the films.

23.1.3.1.4 Sensory Characteristics and Scoring of Composites Incorporating Zein Films

Sensory scores of each sandwich system constructed from each bread/crust type are shown in Figures 23.7 through 23.9. The presence of the zein film in each case lowered sensory scores relative to bread/crust alone, most appreciably for flavor, odor, and overall quality (all $p \leq 0.05$). This effect was mitigated, however, by incorporating turkey that has a distinct and familiar flavor and texture into the sandwich. Table 23.4 compares average declines in specific sensory scores due to the zein films in composites with and without turkey filling, showing significant differences for flavor, odor, and overall quality only in sandwiches without meat.

The highest scores for the sandwich-with-film system, on average, were those that included the crescent rolls, largely because of high scores for appearance and texture; however, highest scores for the (problematic) flavor and odor attributes were for the unbaked sandwiches with turkey. Significance of effects due to zein film in specific sandwich systems (pooling data across the five sensory attributes) is listed in Table 23.5, which shows no significant differences in either baked product that contained turkey.

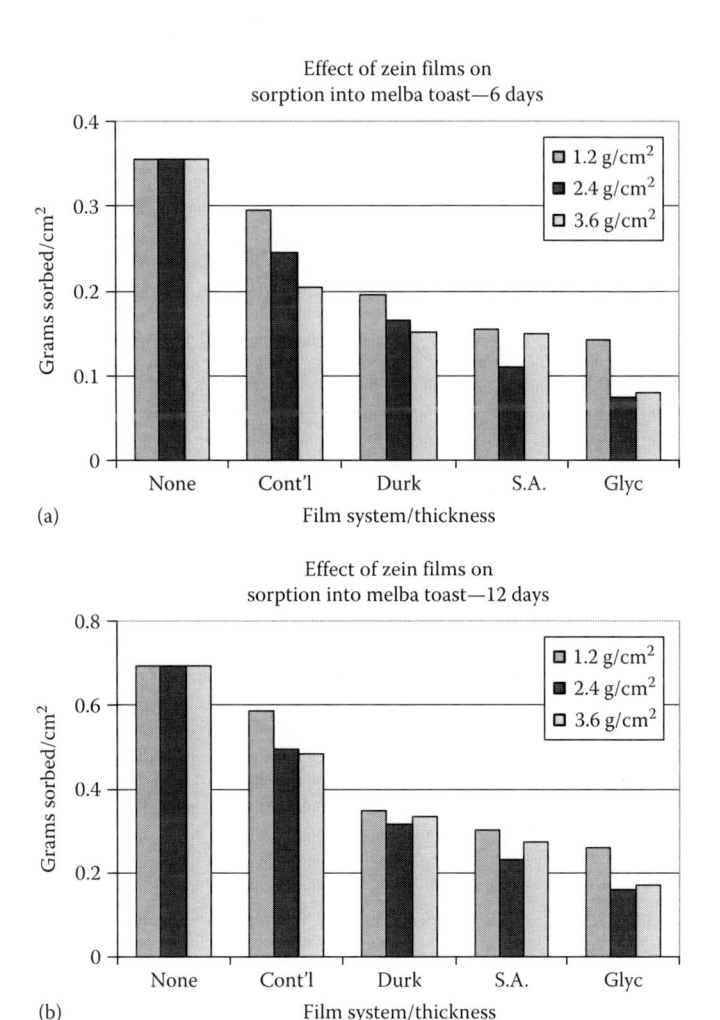

FIGURE 23.5 Moisture migration from Velveeta Cheese into Melba toast through different thicknesses of zein film. (a) 6 days and (b) 12 days.

It was concluded that edible barriers made from corn zein effectively inhibited moisture migration between sandwich layers that were unequally matched in water activity, while some long-term barrier properties, however, are better retained in nonthermally treated systems. Scores for specific sensory characteristic of the sandwiches were moderately reduced by the zein barriers, but less so in products containing a filling along with the zein. The sensory properties most affected were texture and odor. However, in baked sandwiches containing a meat filling, sensory scores grouped across attributes were not significantly (i.e., by $p < 0.05$) reduced. While the films have a characteristic "corn" aroma and flavor, the inherent sensory properties can be masked through the use of highly flavored fillings.

23.1.3.2 Current Applications

Laminate layer (LL), a technical-based effort, was aimed at exploring and modifying possible edible barriers that would work to prevent moisture and fat migration in military rations. Moisture and fat migration tend to negatively affect the ration components due to the extended shelf-life requirements of the military and the nature of the components. Laminate layers are edible barriers or films that would aid in the control of moisture and fat migration. Moisture migration itself often yields displeasing physical and chemical changes within the food system, affecting its safety, shelf life, and sensory quality.

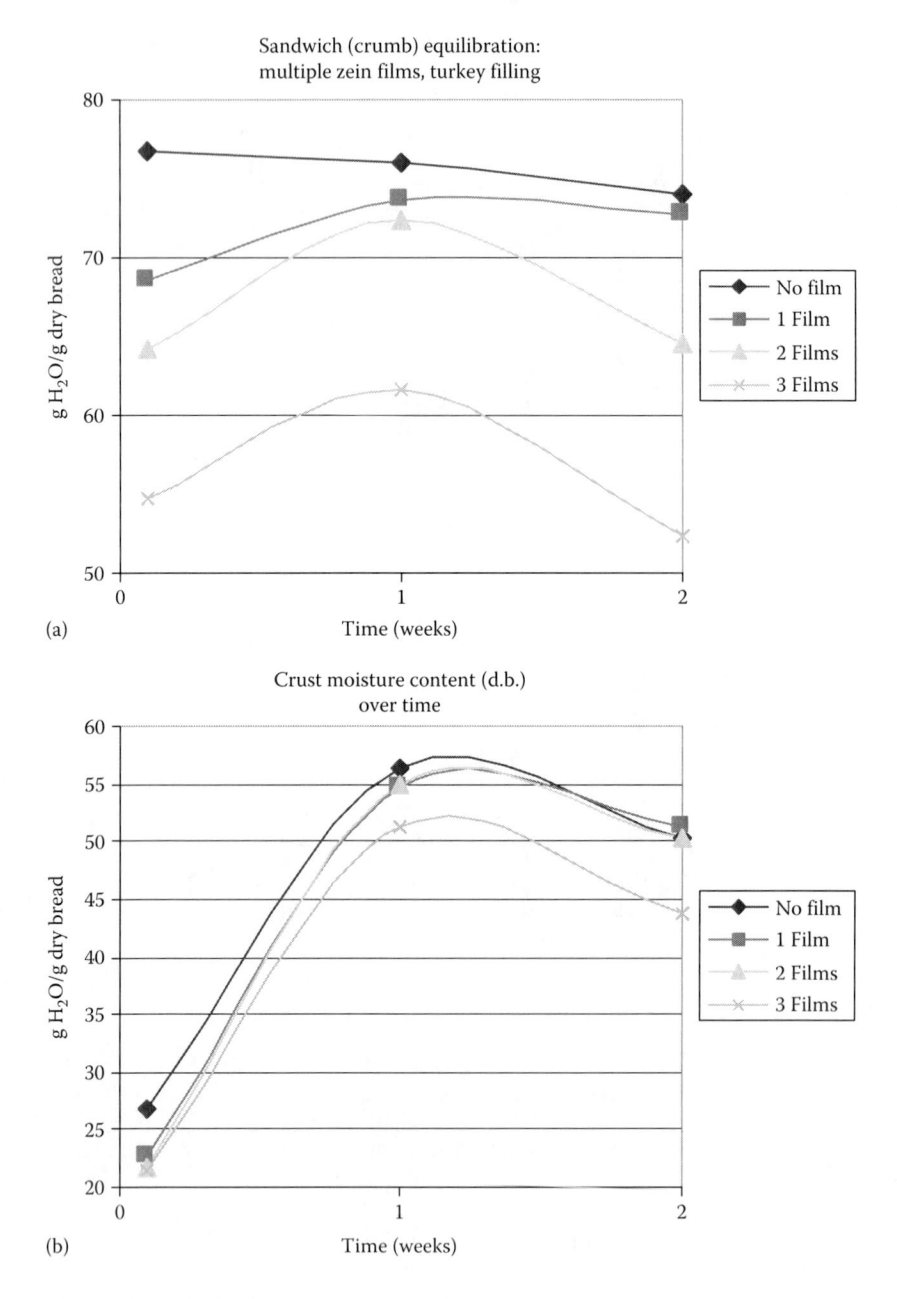

FIGURE 23.6 Equilibration of moisture with time in sliced turkey/modified Meal, Ready-to-Eat dough sandwich. (a) Migration into crumb; (b) migration into crust.

This migration can lead to lipid oxidation, nonenzymatic browning, crystallization, staling, and possibly microbial growth. This phenomenon occurs within a multicomponent food system in which the components have large differences in water activity. The component with a higher water activity will tend to lose moisture to the component with a lower water activity. The greater the difference in water activity that exists between the two components, the greater the detrimental effects on the product quality.

This large difference was very challenging, especially when trying to develop a shelf-stable peanut butter and jelly sandwich. Issues arise such as the oil migration from peanut butter to bread, color migration from jelly into bread, and moisture migration from the bread to the peanut butter. The moisture

TABLE 23.3

Two Sample Comparisons for Film Efficacy: Baked Meat Containing

Number of Films	Time	t-Ratio, Crust	p-Value, Crumb	t-Ratio, Crumb	p-Value, Crumb
1	After baking	2.2	0.04	3.7	0.01
	1 week	2.3	0.05	NS	NS
	2 weeks	NS	NS	NS	NS
2	After baking	5.1	0.001	5.1	0.001
	1 week	2.0	0.08	3.1	0.021
	2 weeks	NS	NS	NS	NS
3	After baking	4.2	0.002	7.7	0.001
	1 week	2.9	0.02	5.2	0.001
	2 weeks	3.0	0.02	4.3	0.008

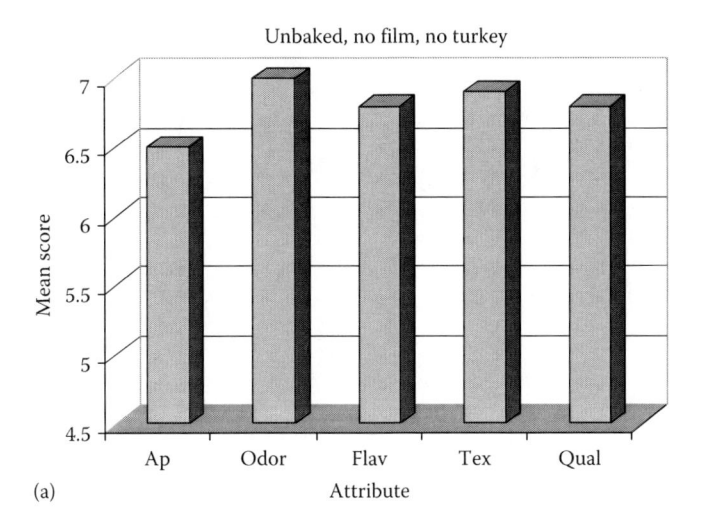

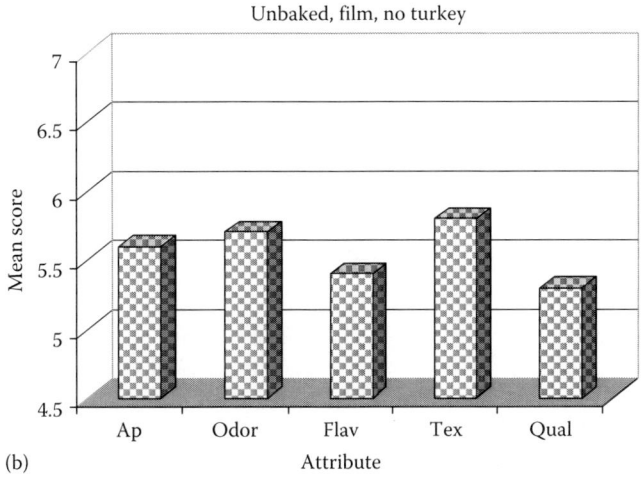

FIGURE 23.7 (a) Scores for unbaked, without film or meat. (b) Scores for unbaked, with film and without meat.

(Continued)

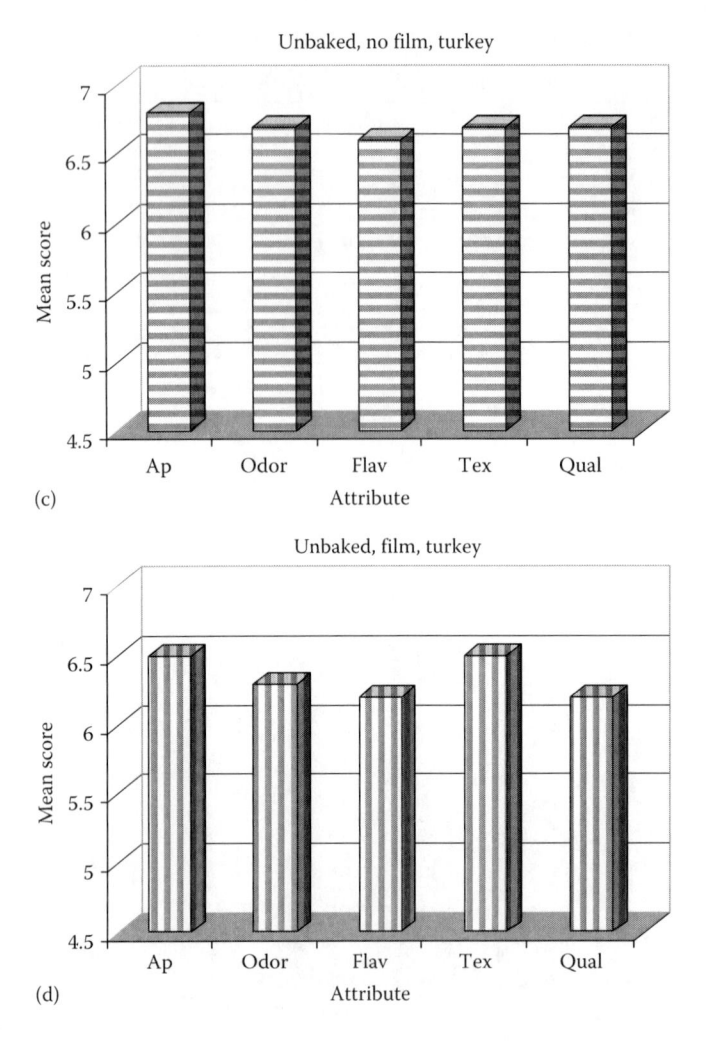

FIGURE 23.7 (*Continued*) (c) Scores for unbaked, without film and with meat. (d) Scores for unbaked, with film and with meat.

migration is due to the fact that the peanut butter has a lower water activity than the bread. Once the moisture migrates, a peanut butter with a higher water activity is now more prone to lipid oxidation than it was in its original state. Oil migration into the bread also causes it to be more prone to oxidative rancidity as well. Other examples common in the marketplace currently include ready-to-eat pies, in which the filling tends to cause a soggy texture to the pie crust, or ready-to-eat cakes where the moisture migrates from the cake to the frosting, causing a staling in the cake. The high-fat frosting would then be highly prone to oxidative rancidity.

The type of barrier to be used in a food system is largely dependent on the food system itself. For example, a barrier used to prevent lipid migration would have much different properties than a barrier used to prevent moisture migration. The properties in the film would vary mainly based on solubility. For this effort, moisture (water) migration was the main focus.

Various edible barriers were examined in model systems to determine their effectiveness at preventing moisture migration, and then feasible barriers were applied to a complex food system. However, it is important to note that once this technology is applied to a shelf-stable food system, the barrier itself must not be organoleptically displeasing. By incorporating laminated layers and/or edible barriers into the interface regions of complex ration components, the negative effects of moisture and fat migration can be mitigated.

To determine initial effectiveness of LL prototypes, a model system was designed. The model system used was an "open-faced" sandwich. MRE white wheat snack bread (a_w = 0.780) and Kraft American cheese singles (a_w = 0.955) were used. The height of the cheese portion was equal to the height of the bread and weighed approximately five times as much in order to supply a relatively large reservoir of moisture for migration. The water activity gradient between the two components was great enough to allow moisture migration to occur in a short period of time.

A wide variety of LLs were examined and tested. The layers that were evaluated include the following provided by Watson Inc.: pullulan beeswax moisture barrier (BWMB), pullulan moisture barrier film, alginate BWMB film, single-sided HPMC moisture barrier film, and dual-sided HPMC moisture barrier film. Films containing pullulan (a polysaccharide polymer produced from the fungus *Aureobasidium pullulans*) and sodium alginate were water soluble; films containing HPMC were water resistant. All films from Watson Inc. had a semiopaque appearance. Other films evaluated were provided by the NewGem Foods™, including Origami® Wrap and GemWrap®, which were made primarily from fruit

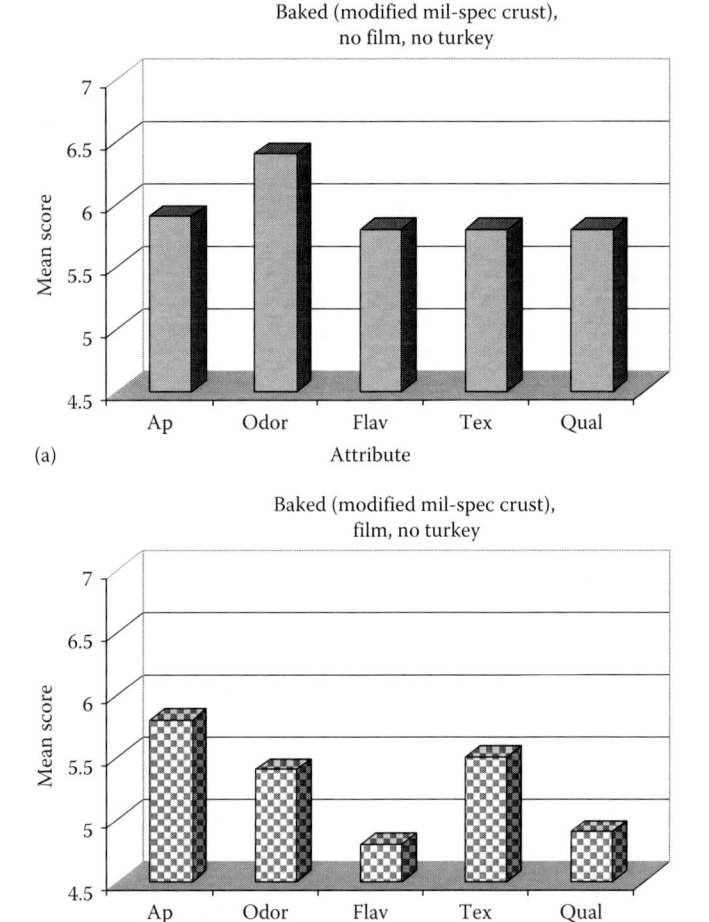

FIGURE 23.8 (a) Scores for baked model sandwich, without film or meat. (b) Scores for baked model sandwich, with film and without meat. *(Continued)*

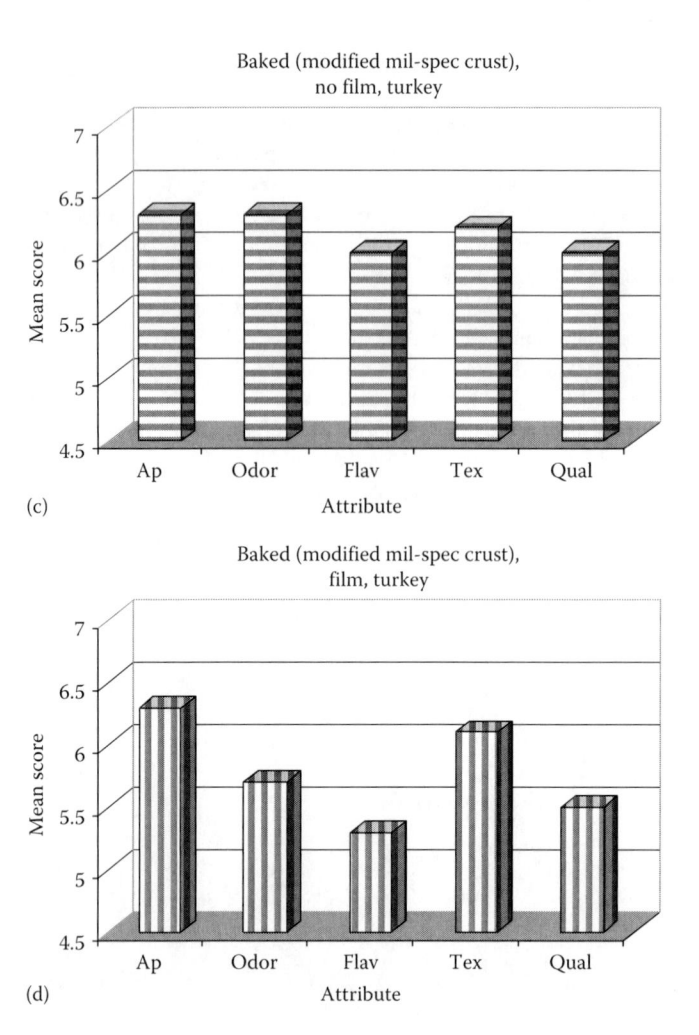

FIGURE 23.8 (*Continued*)   (c) Scores for baked model sandwich, without film and with meat. (d) Scores for baked model sandwich, with film and with meat.

and vegetable purees, soy protein isolate, and cellulose. While these barriers are available in a variety of flavors, the tomato flavor was used in our experiment.

Each of these LL prototypes was inserted between the bread and cheese components in the model system "sandwich," in either a single layer or a double layer (Figure 23.10). For the control, no barrier was inserted between the bread and cheese. It was ensured that the film fully covered the bread so that no cheese came into direct contact with the bread. The samples were made in triplicates for each storage time, wrapped in plastic wrap, sealed in military grade trilaminate foil pouches, and stored at 40°F for 3, 7, 14, and 28 days.

Initial analytical testing was performed on the bread, cheese, and moisture barriers to determine the water activity and initial moisture contents of each. All dry basis moisture contents were determined using a vacuum oven drying method. The components of the sandwich were carefully separated and individually weighed before drying. Analysis was performed on three replicates for each of the storage pulls. Texture and color analysis were also performed on each to determine sogginess of the bread and any off-color formation throughout storage. Percent moisture gain at each pull was calculated to determine which barriers were most effective at inhibiting moisture migration.

Once the most effective barriers were chosen, they were incorporated into a complex food system. These barriers included the Origami Wrap (tomato flavor), GemWrap (tomato flavor), and single-sided

HPMC film. Pizza was chosen as the test food due to its tendency for intercomponent moisture, color, and fat migration in addition to its desirability as a military ration component. The pizza was assembled in a rectangular pan using approximately 800 g of shelf-stable pizza dough, 215 g of shelf-stable pizza sauce, and 200 g of a cheese mix (50% shelf-stable cheese and 50% low-moisture mozzarella) (Figure 23.11). For each of the LL prototypes, the barrier was placed on top of raw dough, then topped with shelf-stable sauce, and finally topped with shelf-stable cheese. A control was constructed using no barrier between the sauce and dough. The specimens were baked at 375°F for 13 min in a Hobart convection oven, cooled to 80°F, cut into approximately 3 in.², placed in trilaminate pouches with an oxygen scavenger, and sealed. Samples analyzed were at time zero, 2 weeks at 120°F, 4 weeks at 120°F, 3 months at 100°F, and 6 months at 100°F. The pizza was tested for microbiological safety including APC, *Escherichia coli*, *Staphylococcus aureus*, yeast, and mold in addition to water activity and pH at each pull. Samples were then analytically evaluated by trained sensory panelists who scored the samples based on appearance, odor, flavor, texture, and overall quality using a nine-point hedonic scale. Moisture measurements were obtained by vacuum, drying the pizza crust only, due to the difficulty in separating the components of the pizza, particularly after storage. Interior sections of the crust were evaluated only. Figures 23.12 and 23.13 demonstrate the performance of the edible barriers in model systems at preventing moisture migration.

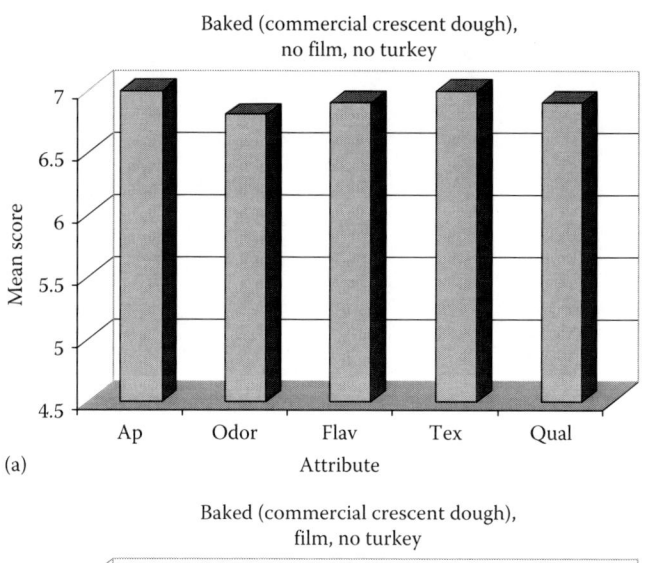

(a)

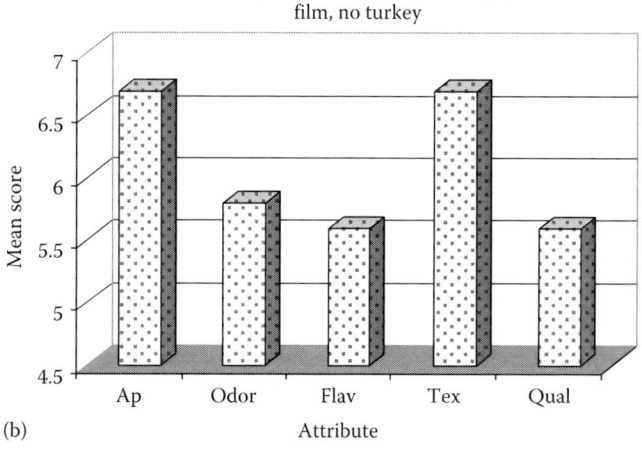

(b)

FIGURE 23.9 (a) Scores for baked crescent, without film or meat. (b) Scores for baked crescent, with film and no meat.

(Continued)

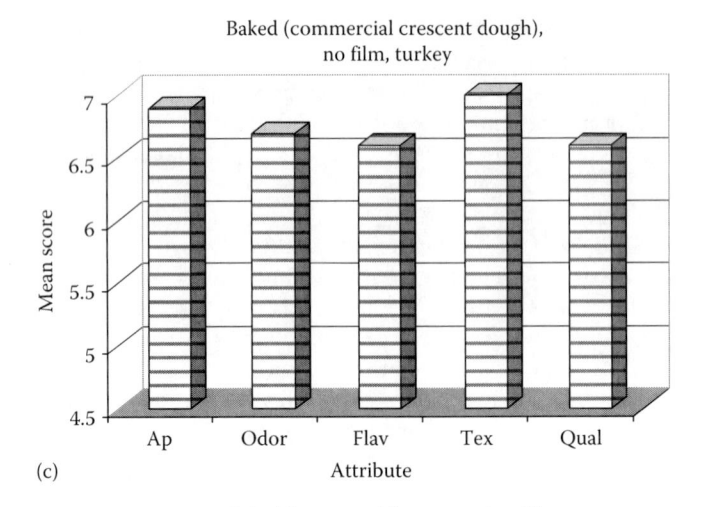

(c)

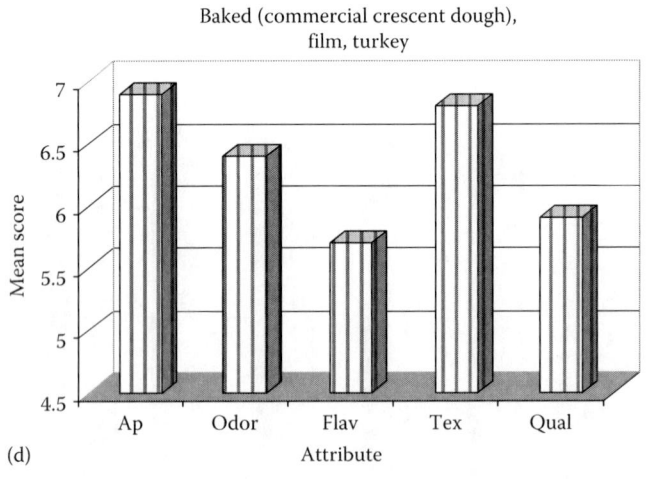

(d)

FIGURE 23.9 (*Continued*)   (c) Scores for baked crescent, without film and with meat. (d) Scores for baked crescent, with film and with meat.

TABLE 23.4

Significance of Film Effect on Specific Sensory Attributes (Pooled across Sandwich Type)

Attribute	t-Value, p	Significance
Appearance, without turkey	−0.93, 0.42	NS
Appearance, with turkey	−0.39, 0.72	NS
Odor, without turkey	−5.15, 0.014	*
Odor, with turkey	−1.69, 0.19	NS
Flavor, without turkey	2.90, 0.05	*
Flavor, with turkey	−2.03, 0.14	NS
Texture, without turkey	−1.14, 0.34	NS
Texture, with turkey	−0.54, 0.63	NS
Overall acceptance, without turkey	−3.04, 0.05	*
Overall acceptance, with turkey	−1.90, 0.15	NS

*$p < 0.05$.

TABLE 23.5

Significance of Film Effects on Sensory Scores for Different Sandwiches
(Pooled across Attributes)

Sandwich	t-Value, p	Significance
Unbaked, Pepperidge Farms bread, without turkey	9.93, <0.0001	***
Unbaked, Pepperidge Farms bread with turkey	4.81, 0.005	**
Baked, MRE dough, without turkey	2.98, 0.025	*
Baked, MRE dough, with turkey	1.92, 0.11	NS
Baked, Pillsbury crescent dough, without turkey	3.33, 0.029	*
Baked, Pillsbury crescent dough, with turkey	1.67, 0.17	NS

***$p < 0.001$, **$p < 0.01$, *$p < 0.05$.

FIGURE 23.10 The model system. From top to bottom: American cheese, Origami Wrap, and MRE white wheat snack bread.

FIGURE 23.11 Pizza prototype with Origami layer incorporated.

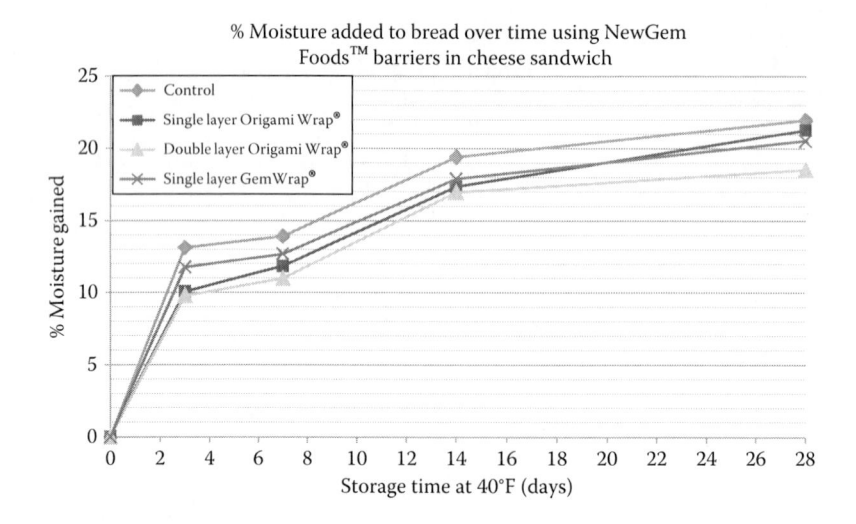

FIGURE 23.12 Three types of films provided by Watson, Inc., were tested against a control (no films) and stored at 40°F for 4 weeks. Results demonstrated that the most moisture migrated from the cheese ($a_w = 0.955$) to the MRE bread ($a_w = 0.780$) in the control. All three LLs demonstrated a high level of moisture migration retardation; however, the single-sided hydroxypropyl methylcellulose film was the most.

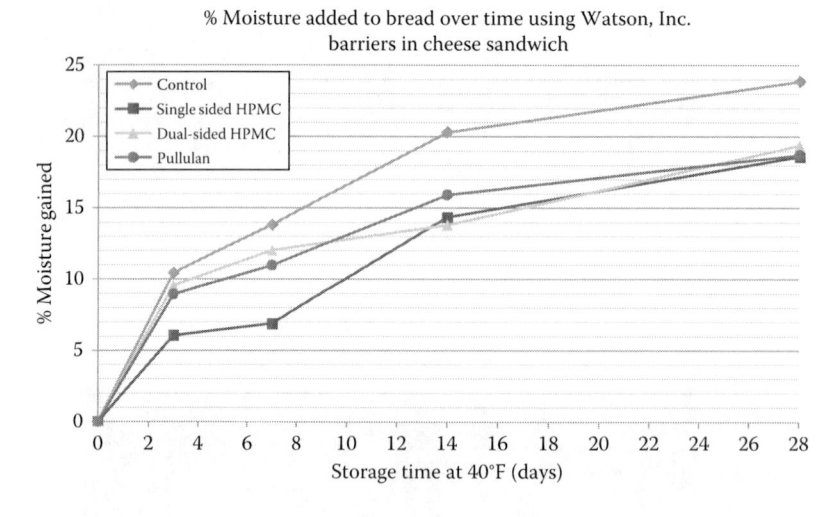

FIGURE 23.13 Three types of films provided by NewGem Foods were tested against a control (no films) and stored at 40°F for 4 weeks. Results demonstrated that the most moisture migrated from the cheese ($a_w = 0.955$) to the Meal, Ready-to-Eat bread ($a_w = 0.780$) in the control. The double-layer Origami Wrap demonstrated the highest level of effectiveness at reducing moisture migration by 15.6% in comparison with the control.

The most effective LLs were selected for incorporation into the shelf-stable pizza prototypes and evaluated for sensory quality. Although HPMC was an effective moisture barrier, it was not organoleptic acceptable in that it did not adhere to the bread and would separate during incisor bite down. However, after 3-month storage, the control sample scored the lowest of all variables, indicating the film's textural effects were less detrimental to sensory quality than was increased crust moisture. The most effective moisture barriers that also received high sensory ratings were the GemWrap and the Origami Wrap, both vegetable-based edible films.

While the efficacy of the commercially available barrier has thus been demonstrated in model food system and one popular food item, further research is needed to investigate the behavior of these layers in a wide variety of complex food matrices. These barriers may further serve as carriers for antimicrobial compounds, antioxidants, nutrients, flavors, and phytonutrients.

23.1.4 Coating Application for Microbial Control

The DoD Combat Feeding Directorate (CFD) has, for years, been developing and optimizing shelf-stable bread for the MRE (Leung, 1986). The shelf-stable bread was commercially produced in 1988 and then patented and successfully fielded in 1991 (patent number 5,059,432; Berkowitz and Oleksyk, 1991). Like all ration components, "pouched bread" was developed to have microbial and textural stability. Hurdle technology was used to prevent the growth of pathogenic bacteria. Organic acids and/or their salts (e.g., sorbic acid and potassium sorbate), have previously been used as a hurdle to prevent the growth of yeast and mold in baked items. However, in addition to inhibiting yeast and mold growth, organic acids can also adversely affect yeast activity during proofing and processing. Decreased yeast activity can result in inadequate leavening and lack of flavor formation during proofing. To prevent this decrease in yeast activity, several researchers have investigated incorporating encapsulated organic acids into the dough. In 1992, potassium sorbate was removed and replaced with encapsulated sorbic acid that consisted of 70% ± 2% sorbic acid and 30% ± 2% partially hydrogenated vegetable oil with a melting point of 141°F–147°F. In 1993, encapsulated potassium sorbate was approved for use in lieu of encapsulated sorbic acid that consisted of 50% ± 2% potassium sorbate and 50% ± 2% partially hydrogenated vegetable oil with a melting point of 152°F–158°F. In combination with the oxygen-scavenging sachet, the encapsulated sorbic acid and potassium sorbate were effective in preventing the growth of yeast and mold in military pouch bread and did not adversely affect dough strength, leavening, and furthermore maintained eating quality of the pouched bread.

The growth and survival of microorganisms in foods is influenced by several factors such as water activity (a_w), temperature, pH, oxidation–reduction potential, preservatives, and competitive microflora. Many of these factors are used synergistically to inhibit the growth of pathogenic bacteria in IMFs. To date, over 50 different IMFs are part of military rations. Table 23.6 summarizes IMF components in military field rations. Many of these IMF products contain encapsulated acids, which were evaluated in intermediate-moisture shelf-stable sandwiches developed in the early 1990s. Acidulants evaluated included glucono delta-lactone (GDL), citric acid, and lactic acid, which provide microbiological stability by increasing the acidity of the shelf-stable sandwiches. All the acid evaluated successfully reduced the sandwich pH from ~5.4 to ~4.8; however, GDL had the least effect on flavor and texture attributes.

Powers (1999) assessed the influence of water activity and pH on the growth of *S. aureus* in various multicomponent items in order to provide safety guidelines to manufacture shelf-stable sandwiches. Encapsulated GDL was added to a frankfurter bun, which was packaged with a shelf-stable frankfurter inoculated with a three-strain cocktail of *S. aureus* and stored at 35°C for 6 months. Growth that was expected at 0.89 water activity was prevented in the frankfurter and bun, most likely due to encapsulated GDL, which lowered the pH from 5.4 to 4.8. Figure 23.14 shows the effect of water activity and pH on the growth and survival of the *S. aureus* cocktail in frankfurters and buns. The items are considered microbiologically stable if the initial level of the challenge organism did not increase more than one log during the 6-month storage period, which it did not for the 0.89 pH GDL samples.

Increasing a_w and/or pH of IMF will improve the acceptability of these items but will also increase the possibility of pathogenic growth. To decrease this possibility, a first generation of controlled-release bacteriocins/antimicrobials (FGCRB/A) is currently being developed and validated under a small business innovative research proposal to maintain the microbial stability of intermediate-moisture ration components with increased a_w and pH levels, throughout a 3-year shelf life. So far, the ability to encapsulate and control the release of a nisin from microparticles produced using Orbis Bioscience Precision Particle Fabrication technology has been demonstrated. Orbis Bioscience fabricated degradable microspheres and measured the controlled release and bioactivity of nisin, which was not found to be negatively impacted by encapsulating; however, salt extraction noticeably increased the amount of nisin encapsulated and released from the microspheres. Orbis is currently demonstrating the efficacy of the FGCRB/A over a range of a_w (i.e., 0.88, 0.90, 0.95), pH (i.e., 5.0, 5.5, 6.0), and storage temperatures (80°F, 100°F, and 120°F) using microbial populations of 10^2, 10^4, and 10^6 colony-forming units/mL or gram. Maturity of this food additive technology will ensure the ability to biopreserve a safe, high-quality ration component and will support the development of future ration components.

TABLE 23.6

Intermediate-Moisture Foods in Military Rations

Product	Measured or Required Water Activity (a_w)	Required pH
Hamburger buns, shelf stable	>0.9	
Tortilla	NGT[a] 0.85	4.8–5.7
Biscuits	NGT 0.85	
Snack bread, plain or wheat	NGT 0.85	
Griddle breads (waffles and pancakes)	0.8–0.86	
Cakes, packaged in flexible pouch (vanilla, lemon, orange, pineapple, chocolate mint with drops, lemon poppy seed, spiced poppy seed, pumpkin, and carrot)	NGT 0.85	
Cakes packaged in polymeric trays (chocolate with vanilla-crumb topping, marble with toffee-crumb topping, devil's fudge with white icing, spice with vanilla-crumb topping, coffee with cinnamon-crumb topping, walnut tea, lemon crumb, dulce de leche with white icing, breakfast with maple-flavored syrup, yellow with chocolate icing, yellow with white icing, devil's fudge, spice with white icing, lemon with white icing)	NGT 0.90	
Brownies, packaged in flexible pouch fudge with chocolate drops	NGT 0.85	
Brownies fudge packaged in polymeric trays (fudge brownies with chocolate icing, brownies with pan-coated disc topping, brownies with butterfinger pieces®, and blonde brownies)	NGT 0.90	
Miniloafs packaged in polymeric trays (banana nut and banana nut frosted)	NGT 0.90	
Muffin tops, packaged in flexible pouch (chocolate banana nut)	NGT 0.85	
Sweet rolls packaged in polymeric trays (cinnamon swirl and raspberry swirl)	NGT 0.90	
Scones with icing (cinnamon, blueberry, and apple)	NGT 0.85	
Bacon, precooked sliced in flexible pouch	>0.86	
Dried and jerky products (beef and turkey):	0.750	
Moist cured/kipper product	0.850	NGT 5.6
Moist cured/lactate product	0.850	NGT 6.2
Filled French toast	0.85	
Turnovers (apple, cherry, and blueberry)	NGT 0.86	5.2
Toaster pastries	0.6	
Fruit infused and dried (cherries, apples–blueberries, strawberries, cranberries–peaches, raisins, red raspberries, apricots, pineapples, mangos, and papaya)	0.62	
Sandwiches, shelf stable (nacho-flavored beef, pepperoni, honey barbecue chicken, honey barbecue beef, Italian, and bacon cheddar)	0.83–0.89	4.8–5.2
Filled wraps, shelf stable (barbecue seasoned pork and Mexican style beef)	0.8–0.85	5.5

Source: Richardson, M., Intermediate moisture technologies for rations, In *Military Food Engineering and Ration Technology*, eds. A.H. Barrett and A. Cardello, pp. 225–255. DESteck Publications, Inc., Lancaster, PA, 2012.

[a] NGT, not greater than.

23.1.5 Coatings for Nutrient Stability

Both the military and NASA have been investigating and developing ways to mitigate the loss and increase retention and stability of vitamins (Eames, 1983; Morril et al., 1987; Zwart, 2009), proteins and amino acids (Anderson, 2012), and omega-3s (Barrett, 2011) in ration components. Combat rations are formulated to make available both macro- and micronutrients that meet the nutritional standards for operational rations (NSORs) and Army Regulation 40-25, which is mandated by the Office of the Surgeon General. The NSORs are based on the Food and Nutrition Board of the National Academy of Sciences, Institute of Medicine's Recommended Dietary Allowances.

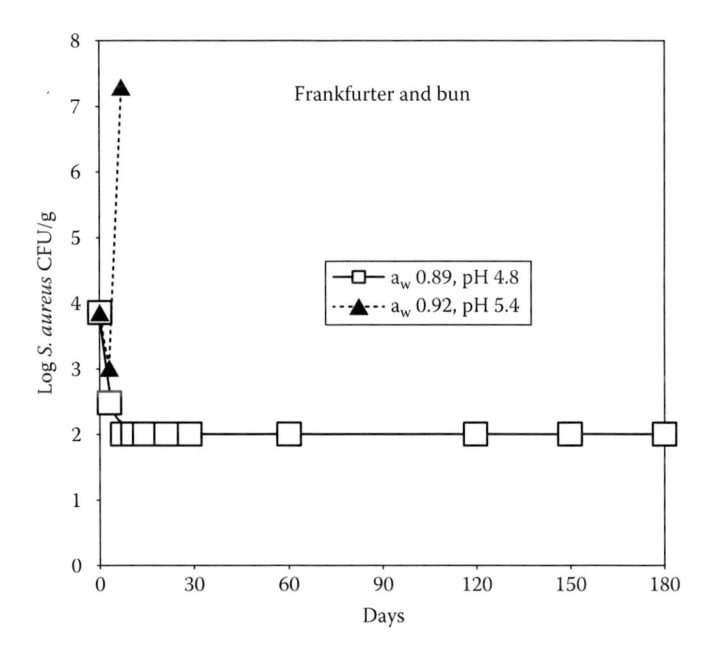

FIGURE 23.14 The effects of water activity and pH on the growth and survival of *Staphylococcus aureus* three-strain cocktail in frankfurter and buns stored at 35°C for 6 months.

23.1.5.1 Coatings for Vitamin Stability

Vitamin nutrient loss is inevitable during processing and storage. Like microorganism survival, macro- and microingredient stability is dependent on factors such as temperature, light, water activity, and pH. NSRDEC was tasked with developing nutrient-dense space foods with 5-year retention of vitamins and packaging materials compatible with alternative quality-preserving sterilization processes. Determining the effects of food matrix composition/polarity on vitamin stability is the primary goal of this effort. Of course, 5 years is well beyond the shelf life of commercial products. During this long period, even microbiologically safe foods can undergo significant losses in quality and nutrient content (Zwart et al., 2009), which in space may be accelerated by cosmic radiation. Astronauts require adequate vitamin intake because they are at risk from damaging radiation. Foods that vary in hydrophilic/hydrophobic character were developed and are being tested in a multiyear study funded by NASA. Effects of the matrix environment on vitamin retention and vitamin encapsulation techniques are currently being tested. Vitamins B_1, B_9, C (water soluble), A, and E (lipid soluble) were encapsulated in lipid-based or carbohydrate-based coatings, incorporated into low a_w polar vs. nonpolar food matrices, and either compacted into bars or maintained as loose powders. The encapsulated vitamins are also being tested in high-a_w products sterilized by retort and nonretort processes such as irradiation, high-pressure, and microwave-assisted thermal processing. For all samples produced, encapsulated vitamins were added at two times the space flight requirements, as listed in the NASA's Constellation Program (C × P) document 70024, "Human-Systems Integration Requirements," section 3.5.1.3.1 (Cooper et al., 2011). Vitamin content and sensory quality are being measured throughout storage at different temperatures. Best chemical environments and processing/packaging strategies for maintenance of nutrients and quality will be determined. Preliminary results show comparatively greater loss of vitamins A and B_9 with some benefits of opposite polarity environment on retention of B_9.

23.1.5.2 Coatings for Omega-3 Stability

Omega-3 fatty acids are long-chain polyunsaturated fatty acids that have many health benefits. High intake of omega-3 fats is reported to be negatively associated with incidence of poor coronary health

(Wang et al., 2009; Guerra et al., 2014), mental illness (Mais and Smith, 1998; Lewis et al., 2011; Lucas et al., 2011), cancer (Cavazos et al., 2013; Jiang et al., 201), and inflammation (Song, 2007; Dumancas et al., 2014). They are additionally reported to increase resilience after traumatic brain injury (Mills et al., 2011). Though nutritionally important, they are lacking in the military diet (Barrett et al., 2014). Fish and flaxseed oils are excellent sources of omega-3s, but because of their high level of unsaturation, they are susceptible to oxidation (references) and are not suitable for extended shelf life foods.

Barrett et al. (2011) produced and evaluated the oxidative stability of high-load fish and flaxseed oil powders with different concentrations of different antioxidants in performance bars to validate the feasibility of incorporating omega-3 lipids in ration components. Antioxidants and omega-3 lipids were emulsified, encapsulated in starch, and spray-dried. The antioxidants evaluated included mixed tocopherols, rosmarinic acid, citric acid, EDTA, carsonic, and ascorbyl palmitate, and each was added at concentrations from 100 to 2000 ppm.

Front-face fluorometry was used to assess effects of both encapsulation and antioxidant incorporation on oxidative stability. Encapsulation itself provided increased stability for both oils. The unencapsulated, unsupplemented flaxseed oil had a lower induction time (by a factor of ~6) when compared to the encapsulated, unsupplemented flaxseed oil. Also, the rapidity of oxidation, measured by initial fluorescence ratio vs. time slopes for the unencapsulated flaxseed oil, was much higher than that for encapsulated flaxseed oil. Encapsulation of fish oil was also protective, but to a smaller extent, increasing induction time by a factor of ~2. Regarding benefits of antioxidants, fish oil stability was improved by all antioxidants tested, whereas flaxseed oil stability was improved by rosmarinic acid only. Flaxseed oil is inherently much more stable than fish oil due to its native antioxidant.

Sensory assessment showed that performance bars produced with 20% encapsulated fish or flaxseed oils were initially not significantly different from the control performance bars for all sensory attributes. However, during accelerated storage, both encapsulated fats significantly increased off-flavor intensity, with encapsulated fish oil–containing samples having lower flavor quality scores than did encapsulated flaxseed oil–containing samples. Both of the encapsulated fats also decreased the sensory firmness.

Advances in encapsulation technology have more recently provided commercially available stable omega-3 fat-containing powders, resulting in the feasibility of producing a variety of shelf-stable omega-3-fortified products for military field feeding. These powders with encapsulated DHA and EPA were obtained from Ocean Nutrition Co. Barrett et al. (2014) evaluated the quality and shelf stability of omega-3 fat-supplemented ration components for military field feeding. In this study, representative low-moisture products (chocolate performance bar packaged without oxygen scavengers), intermediate-moisture products (orange cake packaged with oxygen scavengers), and high-moisture products (a thermostabilized beef lo mein) were produced with various targeted levels of fish gelatin encapsulated omega-3 fat (0, 100, 200, and 400 mg/serving) and stored for 4 weeks at 49°C and 3 and 6 months at 37°C. Sensory quality, physical properties, retained omega-3 fats (DHA and EPA methyl esters), and fluorescent volatile production were measured.

Table 23.7 provides omega-3 levels and overall sensory scores for all products. Sensory scores for all attributes did not significantly diminish after storage for the fortified orange cake and beef lo mein. However, there was a noticeable decrease in sensory scores for the chocolate performance bar after storage, which was proportional to omega-3 content. The bars with the higher omega-3 content had flavor and aroma scores that decreased by 23% ($p < 0.01$) and 43% after 6 months. The decrease in these scores resulted in a 38% ($p < 0.01$) decrease in overall quality scores. The bars containing 100 mg/serving had smaller but significant decreases in flavor, aroma, and overall quality scores after storage. The decreased flavor, odor, and overall acceptability scores were attributed to the difference in packaging. The cakes were packaged with an oxygen scavenger, and the beef lo mein was packaged with a very low headspace. It was concluded that it is feasible to produce shelf-stable omega-3-fortified products, provided an optimal packaging environment is available (i.e., limiting oxygen).

23.1.5.3 Coatings for Amino Acid and Protein Stability

Proteins and amino acids are vulnerable to chemical changes in unprocessed, minimally processed, and fully processed foods. Some of these chemical changes adversely affect bioavailability of amino acids

TABLE 23.7

Omega-3 Levels and Overall Sensory Scores for Performance Bar, Orange Cake, and Beef Lo Mein

	Storage Time Months							
	Overall Sensory Score				% EPA Levels		% DHA Levels	
Representative Products	Initial	1	3	6	3	6	3	6
Low moisture								
Chocolate performance bar, control	6.3	7.0	6.9	7.0				
Chocolate performance bar, 100 mg omega-3 fat/ serving	6.2	5.8	5.8	5.8	8.3 (NS)	20.5 ($p < 0.1$)	15.4 ($p < 0.1$)	19.1 ($p < 0.1$)
Chocolate performance bar, 200 mg omega-3 fat/ serving	6.0	3.7	5.0	4.2				
Intermediate moisture								
Orange cake, control	6.8	6.2	6.4	6.4	NS, $p > 0.05$		NS, $p > 0.05$	
Orange cake, 200 mg omega-3 fat/serving	6.7	6.3	6.4	6.5	NS, $p > 0.05$		NS, $p > 0.05$	
Orange cake, 400 mg omega-3 fat/serving	6.6	6.2	6.4	6.5	NS, $p > 0.05$		NS, $p > 0.05$	
High moisture								
Beef lo mein, control	6.8	6.6	6.5	6.5	NS, $p > 0.05$		NS, $p > 0.05$	
Beef lo mein, 200 mg omega-3 fat/serving	6.5	6.2	6.5	6.5	NS, $p > 0.05$		NS, $p > 0.05$	
Beef lo mein, 400 mg omega-3 fat/serving	6.2	6.2	6.5	6.5	NS, $p > 0.05$		NS, $p > 0.05$	

Source: Adapted from Barrett, A.H. et al., Quality and shelf stability of omega-3 fat supplemented foods, in: W. Khan (ed.), *Omega-3 Fatty Acids: Chemistry, Dietary Sources and Health Effects*, Nova Science Publishers, Inc., New York, 2014, pp. 213–231.

and proteins as well as food quality. Some of the more common reactions that take place during storage and processing of foods include proteolysis, which deteriorates the sensory quality of meat, seafood, and dairy products; cross-linking, which reduces protein bioavailability; and nonenzymatic browning, which negatively affects color, texture, digestibility, and reactive amino acid content. CFD is currently investigating encapsulation of both proteins and amino acids to increase their bioavailability.

Members of the U.S. Armed Forces in physically active positions have protein requirements similar to athletes, and any insufficient energy intake may result in an energy deficit (Pasiakos, 2013). To overcome this deficit, CFD has been exploiting the use of protein encapsulation. Using the rotating disc technology, 30–1000 microncapsule were created, and whey protein was spun with molten solutions of zein, stearine, or shellac (Anderson et al., 2012), resulting in an 80% payload of whey protein. Preliminary results show that an increase in stabilization of encapsulated whey protein was at higher temperatures. Microspheres produced using zein were not stable in the presence of high reducing sugars and showed decreased protein quality. Stearine microspheres were stable in the presence of high reducing sugars, but they were not stable at high temperatures.

Amino acids, the backbone of proteins, are essential for muscle protein synthesis (Wolf, 2000) and are needed to limit muscle loss during enhanced physical activity (Lemon, 2002). However, free amino acids are very unstable and also unpalatable. However, encapsulated amino acids are not currently commercially available. The CFD is evaluating various encapsulation technologies to stabilize these free amino acids in various food matrixes. Amino acids of interest include arginine, leucine, glutamine, and tyrosine. After encapsulation, increased digestibility of these amino acids will be verified, their improved sensory quality will be quantified, their objective color will be measured, and their off-flavor will be masked through formulation and product development research.

23.2 Conclusions/Future Applications of Films and Coatings

The application of innovative film and coating technology for special missions will provide the following: increased variety through items that could not be previously offered, extended shelf life through controlled microbial growth, preserved quality through control of moisture and fat migration, and stabilized nutrients through protection of nutrients during processing and storage. As the field of films and coatings continues to advance, both the military and NASA will continue to exploit innovations to address the challenges of providing safe, nutritious, high-quality products for both warfighters and astronauts.

REFERENCES

Aerospace Medical Research Laboratories (AMRL). Current research for space travel: Nutrition and food technology. AMRL Memorandum M33, Wright Patterson AFB, OH, 1963.

Anderson, D., Racicot, K., and Davis, B. 2012. Performance-optimizing ration components. In *Military Food Engineering and Ration Technology*, eds. A.H. Barrett and A. Cardello. pp. 275–299. DESteck Publications, Inc., Lancaster, PA.

Army Regulation 40-25. BLUMEDINST 10110.6, AF 44-141. Nutrition Standards and Education, July, 2001.

Barrett, A.H., Anderson, D., Ndou, T., Kensil, K., and Conca, K. 2014. Quality and shelf stability of omega-3 fat supplemented foods. In *Omega-3 Fatty Acids: Chemistry, Dietary Sources and Health Effects*, ed. W. Khan. pp. 213–231. Nova Science Publishers, Inc., New York.

Barrett, A.H. and Ndou, T. 2007. Efficacy of zein edible barriers in inhibiting moisture migration in baked sandwich systems. In *Annual Meeting of the Institute of Food Technologists*, New Orleans, LA.

Barrett, A.H., Ndou, T., and Conca, K. 2008. Sensory characteristics of zein moisture barriers in baked and unbaked sandwiches. In *Annual Meeting of the Institute of Food Technologists*, New Orleans, LA.

Barrett, A.H., Porter, W., Marando, G., and Chiachoti, P. Effects of various antioxidants, antioxidant levels, and encapsulation on the stability of fish and flaxseed oils: Assessment by fluorometric analysis. *Journal of Food Processing and Preservation* 35 (2011) 349–358.

Berkowitz, D. and Oleksyk, L.E. Leavened bread with an extended shelf life. U.S. Patent 5059432, October 22, 1991.

Blodgett, J. Maximum variety from feeding unit of low weight and bulk. Technical report no. 70-29-FL, contract no. DAAG 17-68-C-Ö148. U.S. Army Natick Laboratories, Natick, MA, 1967.

Branagan, M. and Pruskin, L. Effects of storage on sensory and nutritional quality of meal ready-to-eat, individual (MRE-1). Natick/TR-94/004, U.S. Army Natick RD&E Center, Natick, MA, 1993.

Briggs, J., Richardson, M.J., Yang, T., Berkowitz, D., and Tartarini, K. 1992. Shelf stable sandwich for military rations. Published abstract. In *Annual Meeting of the Institute of Food Technologists*, New Orleans, LA.

Cole, M. Edible coatings for dried and compacted foods: Part I. Technical report no. 66-37-FD for contract no. DA19-129-AMC-102(N), U.S. Army Natick Laboratories, Natick, MA, 1966a.

Cole, M. Edible coatings for dried and compacted foods: Phase II. Technical report no. 66-38-FD for contract no. DA19-129-AMC-102(N), U.S. Army Natick Laboratories, Natick, MA, 1966b.

Conca, K., Barrett, A., and Ndou, T. 2008. Effects of moisture migration during storage on the textural attributes of sandwich composites constructed from different oil content and crumb density in bread. In *Annual Meeting of the Institute of Food Technologists*, New Orleans, LA.

Conca, K.R. 1995. Evaluation of collagen-based film as an edible packaging material for frozen meats. In *Annual Meeting of the Institute of Food Technologists*, Anaheim, CA.

Conca, R.K. 2002. Protein-based films and coatings for military packaging applications. In *Protein Based Films and Coatings*, ed. Aristippos Gennadios. pp. 551–578, Boca Raton, FL: CRC Press.

Conca, K.R. and Kensil, K. 2007. Efficacy of fat migration barriers in baked pocket sandwich systems. In *Annual Meeting of the Institute of Food Technologists*, Chicago, IL.

Conca, K.R. and Kensil, K. 2008. Efficacy of moisture and fat migration barriers in sandwich systems. In *Annual Meeting of the Institute of Food Technologists*, New Orleans, LA.

Conca, K.R. and Yang, T.C.S. 1993. Evaluation of a shellac-based coating material as an edible film for intermediate moisture foods. Published abstract. In *Annual Meeting of the Institute of Food Technologists*, Chicago, IL.

Cooper, M., Douglas, G., and Perchonok, M. Developing the NASA food system for long-duration missions. *Journal of Food Science* 76(2) (2011) 40–48.

Defense Logistics Agency. Nut and fruit mix, packaged in a flexible pouch. Shelf stable PCR-N-003A, August 2013, http://www.dla.mil/Portals/104/Documents/TroopSupport/Subsistence/Rations/pcrs/mre/mre37/n003A.pdf.

Dumancas, G., Koralege, R., Mojica, E., Murdianti, B., and Pham-Bugayong, P. 2014. The Link between omega-3 fatty acids and rheumatoid arthritis: Properties, mechanisms and therapeutic efficacy. In *Omega-3 Fatty Acids; Chemistry Dietary Sources and Health Effects*, ed. W. Khan. pp. 1–20. Nova Science Publishers, Inc., New York.

Durst, J.R. Formulation and fabrication of food bars. Contract no. DA19-129-QM-1970(01-6063), ILS. Army Natick Laboratories, Natick, MA, 1963.

Durst, J.R. All purpose matrix for compressed food bars. Contract no. DA19-129-AMC-2103(X), U.S. Army Natick Laboratories, Natick, MA, 1964.

Durst, J.R. Compressed food components to minimize storage space. Technical report no. 68-22-FL for contract no. DA19-129-AMC~36Q(N), U.S. Army Natick Laboratories, Natick, MA, 1967.

Eames, C.M., Sherman, D.E., Atwood, B.M., and Branagan, M.T. Effects of storage on vitamin-fortified cheddar cheese spread. Natick Tech Report TR-84/012, Soldier Systems Center, Natick, MA, 1983.

Fu, D., Weller, C., and Wehling, R.L. Zein: Properties, preparations, and applications. *Food Science and Biotechnology* 8(1) (1999) 1–10.

Guerra, F., Piangerelli, L., Romandini, A., Maffei, S., and Capucci, A. 2014. Omega-3 fatty acids and cardiovascular disease. In *Omega-3 Fat Acids: Chemistry, Dietary Sources and Health Effects*, ed. W. Khan. pp. 37–62. Nova Science Publishers, Inc., New York.

Guillard, V., Guilbeli, S., BOnazzi, C., and Gontard, N. Edible acetylated monoglyceride films: Effect of film-forming technique on moisture barrier properties. *Journal of the American Oil Chemists Society* 81(11) (2004) 1053–1057.

Harris, E.N. and Westcott, E.D. Substitution of domestic fat for coconut (lauric) fat in coating military chocolate candies. Technical report no. AD 777535. Army Natick Laboratories, Natick, MA, 1974.

Harris, N. and Lee, F. 1974. Coating composition for foods and method of improving texture of cooked foods. U.S. Patent 3,794,742, filed March 21, 1972 and issued February 6, 1924.

Kester, J. and Fennema, J. Edible films and coatings: Review. *Journal of Food Technology* December 40 (1986) 47–59.

Lemon, P.W.R. The role of protein and amino acid supplements in the athlete's diet: Does type or timing of ingestion matter? *Current Sports Medicine Reports* August (4) (2002) 214–221.

Leung, H.K. 1986. Bread quality: Effects of water binding ingredients. In *Research and Development Association Annual Meeting Report*, U.S. Army Natick Laboratories, Natick, MA, pp. 43–48.

Lewis, M.D., Hibbein, J.R., Johnson, J.E., Lin, Y.H., Hyun, D.Y., and Loewke, J.D. Suicide deaths of active duct US military and omega-3 fatty acid status: A case control comparison. *Journal of Clinical Psychiatry* 72 (2011) 1565–1590.

Lucas, M., Mirzaie, F., O'Reielly, E.J., Willett, W.C., Kawachi, I., Koenen, K., and Ascherio, A. Dietary Intake of n-3 and n6 fatty acids and the risk of clinical depression in women: A 10 y prospective follow-up study. *American Journal of Clinical Nutrition* 93 (2011) 1337–1143.

Mais, M. and Smith, R. Fatty acids, cytokines and major depression. *Biological Psychiatry* 43 (1998) 313–314.

Morrill, A., Klicka, M.V., Sherman, D.E., Branagan, M.T., and Fossum, I. Effects of storage time and temperature on nutritional content of fortified fruitcake. Program Element Number BP(.14), 1987.

Pavey, R.L. 1973. Study techniques for controlling flavor intensity in compressed foods (Phase I). Contract no. DAAG 17-67-C-1021, NATICK/TR; 77/008, Natick, MA: Swift and Co.

Pavey, R.L. Study techniques for controlling flavor intensity in compressed foods (Phase II). Natick T75-49-FEL (FEL-6), 1975 (ADA 006031), Tech report TR-77/008, Natick Laboratories, Natick, MA, 1976.

Richardson, M. 2012. Intermediate moisture technologies for rations. In *Military Food Engineering and Ration Technology*, eds. A.H. Barrett and A. Cardello. pp. 225–255. DESteck Publications, Inc., Lancaster, PA.

Richardson, M., Briggs, J., Senecal, A., Dunne, P., and Lee, C. 1995. Development of intermediate moisture meats for use in military shelf-stable sandwiches. In *Proceedings of the 41st Annual International Congress of Meat Science and Technology*, San Antonia TX.

Schuetze, C.E., McMahon, E., Adams, L.M., and Barnes, M. Encapsulation of foods. Technical Documentary Report No. MRL-TDR-62-53, Wright Patterson Air Force Base, May 1962.

Shaw, C. and Natress, D. The effects of packaging conditions on the sensory quality of trail mix after storage. Natick/TR-96/044, Natick, MA: Soldier Systems Center, 1996.

Taoukis, P.S. and Richardson, M. 2007. Principles of intermediate-moisture foods and related technology. In *Water Activity in Foods*, eds. G. Barbosa-Cánovas, A.J. Fontana Jr., S.J. Schmidt, and T.P. Labuza. pp. 273–312. Blackwell Publishing, Ames, IA.

Wang, S., Wu, L., Matthan, N.R., Lamon-Fava, S., Lecker, J.L., and Lichtenstein, A.H. Reduction in dietary omega-6 polyunsaturated fatty acids; eicosapentaenoic acid plus docosahexaenoic acid ratio minimizes atherosclerotic lesion formation and inflammatory response in the LDL receptor null mouse. *Atherosclerosis* 204 (2009) 147–155.

Wolf, R.R. Protein supplements and exercise. *American Journal of Clinical Nutrition* 72(2 Suppl.) (August 2000) 551S–557S. Review.

Yang, T. The use of films as suitable packaging materials for minimally processed foods—A review. Natick/TR-94/029, Army Natick Laboratories, Natick, MA, 1994.

Zwart, S.R., Kloeris, V.L., Perchonok, M.H., Braby, L., and Smith, S.M. Assessment of nutrient stability in foods from the space food system after long-duration spaceflight in the ISS. *Journal of Food Science* 74(7) (2009) H209–H217.

Section VI

Coatings and Films: Drawbacks and Challenges

24

Films and Coatings: Migration of Ingredients

Lia Noemi Gerschenson, Ana María Rojas, and Silvia Karina Flores

CONTENTS

Abstract

Edible films and coatings can be used as an emergent technology to lengthen the shelf life of different food products because they can retard the migration of moisture or fat/oil and the gas (O_2, CO_2) or solute transport, improve the mechanical properties or structural integrity of food, improve food flavor retention, and support food additives or nutrients. For their formulation, the use of hydrocolloids is required (i.e., polysaccharides or proteins), which allow to obtain a continuous biopolymeric tridimensional structure and a plasticizer (i.e., glycerol, sorbitol), which provides flexibility to that structure.

In this chapter, the migration of antimicrobials, antioxidants, and plasticizers is analyzed with the purpose of the evaluation of the effect of this phenomenon on the performance of these edible matrices.

24.1 Introduction

Biodegradable edible films and coatings constitute a technological hurdle for food preservation because they can act as selective gas barriers (i.e., oxygen, aroma), while their microstructure is applied to carry and localize the activity and to control the release of food additives at interfaces. Compartmentalization of preservatives into these edible matrices can also overcome negative interactions with other components of the food system. By localization of the activity of additives at film interfaces where they are necessary, lower amounts of them would be necessary in the whole formulation in order to extend the shelf life of foods.

In general, these structures are named as films when they are constituted to stand alone and as coatings when they are constituted on the surface of the food products.

In the last 10 years, a great deal of effort has been applied to the study of formulation and production of these matrices and to their characterization. Anyhow, the answer to many challenges faced has not been found so far.

The objectives of this chapter are

- To evaluate the migration of ingredients from these edible matrices and its effects on the stability of the food and the usefulness of these films and coatings
- To conclude about the drawbacks and challenges that this migration produces in relation to the development and use of edible matrices

24.2 Results and Discussion

24.2.1 Antimicrobial or Antioxidant Migration

The application of an antimicrobial or antioxidant solution by dipping or spraying on the surface of a food may produce an uncontrolled migration into the product, and partial inactivation of the active compounds might happen due to the interaction with other food components. This inactivation can also occur when the additives are incorporated in the bulk of the food. A new approach to overcome these problems is the use of edible packaging techniques where active agents are incorporated into biopolymeric matrices that cover the food in order to control the diffusion into food matrix and/or to localize their activity and/or to protect them from the undesirable interactions that could promote their inactivation or degradation. These actions contribute to the decrease of the needed amount of food additives for extending the shelf life, contributing to the development of healthier foods.

The release of active compounds from polymeric matrices is influenced mainly by the properties of both the polymer and the active compound as well as the characteristics of the food product (López de Dicastillo et al. 2013).

Other aspects to consider when evaluating migration are the initial load of the active substance in the carrier matrix, which determines the driving force as well as the environmental conditions (i.e., temperature, moisture, etc.).

Depending on the objectives of the film or coating application, a fast release of the additive may be desirable to act on the food bulk or, conversely, a slow release rate could be necessary to maintain a critical concentration at the surface to avoid the food deterioration.

24.2.1.1 Antimicrobial Migration

Many studies have been carried out in order to describe antimicrobial diffusion from biopolymeric matrices to the food. In general, these studies have been performed using liquid or solid food models and different operating conditions.

There are two main types of assays that have been used to evaluate the migration of antimicrobials: (1) The first type includes the test of diffusion in agar or the inhibition zone (halo) test used to analyze the amount of active agent released from films and its antimicrobial effectiveness against a microorganism in controlled conditions. This test depends not only on the rate of release of the agent but also on the antimicrobial capacity to inhibit the target microorganism. (2) The second type includes studies focused on the description of the mechanism of the release, giving an approach to the type of diffusion observed, which have given origin to several models that describe the movement of antimicrobials.

24.2.1.1.1 Inhibition Zone Test

The technique further explored for testing the availability of the antimicrobial agent present in the edible matrices is the inhibition zone (halo) test, which is based on the observation of the growth of a selected microorganism on an appropriate culture medium. With this objective, disks of films containing preservative are applied on the inoculated surface and incubated at the appropriate temperature for the development of the target microorganism. The observation of light areas of nongrowth around and/or under the disks of films depends on the diffusion of preservative from the film into the agar and on the antimicrobial activity of the active agent. That is, nonvisualization of a clear zone of inhibition

TABLE 24.1

Some Results of the Inhibition Zone Test Performed with Antimicrobial Edible Films

Antimicrobial Amount[a]	Biopolymer	Target Microorganism	Inhibition Zone (Diameter, mm)	References
Oregano or thyme essential oils 1%, 2%, 3%, 4%, and 5% (v/v)	Soy protein	*E. coli, S. aureus, E. coli* O157:H7, *P. aeruginosa, L. plantarum*	20.5–50.5	Emiroğlu et al. (2010)
Potassium sorbate 15% (w/w)	Sweet potato starch	*E. coli, S. aureus*	5.6–11.3	Shen et al. (2010)
Oregano, thyme essential oils, and citrus extract 1% (v/v)	Gelatin, methylcellulose, and their blend 50:50 (w/w)	*Pseudomonas fluorescens, Aeromonas hydrophila/ caviae,* and *Listeria innocua*	11.5–20.5	Iturriaga et al. (2012)
Green tea extract 50/50 (v/v)	Agar–gelatin	*Listeria monocytogenes,* (and other 25 microbial strains)	99–178	Giménez et al. (2013a)
Potassium sorbate 0.5%, 1.0%, or 1.5% (w/w)	Whey protein	Eight *E. coli* non-O157	3–12.7	Pérez et al. (2014)
ZEO and *MEO* 1%, 2%, and 3% (v/v)	κ-Carrageenan	*S. aureus, B. cereus, E. coli, P. aeruginosa, S. typhimurium*	7.0–27	Shojaee-Aliabadi et al. (2014)

Sources: Data reprinted from *Meat Sci.*, 86, Emiroğlu, Z., Yemiş, G., Coşkun, B., and Candoğan, K., Antimicrobial activity of soy edible films incorporated with thyme and oregano essential oils on fresh ground beef patties, 283–288. Copyright 2010; *Food Hydrocoll.*, 24, Shen, X.L., Wu, J.M., Chen, Y., and Zhao, G., Antimicrobial and physical properties of sweet potato starch films incorporated with potassium sorbate or chitosan, 285–290. Copyright 2010; *Int. J. Food Microbiol.*, 158, Iturriaga, L., Olabarrieta, I., and Martínez de Marañón, I., Antimicrobial assays of natural extracts and their inhibitory effect against *Listeria innocua* and fish spoilage bacteria, after incorporation into biopolymer edible films, 58–64. Copyright 2012; *Food Hydrocoll.*, 30, Giménez, B., López de Lacey, A., Pérez-Santín, E., López-Caballero, M., and Montero, P., Release of active compounds from agar and agar–gelatin films with green tea extract, 264–271. Copyright 2013a; *Food Control*, 37, Pérez, L., Soazo, M.d.V., Balagué, C., Rubiolo, A., and Verdini, R., Effect of pH on the effectiveness of whey protein/glycerol edible films containing potassium sorbate to control non-O157 shiga toxin-producing *Escherichia coli* in ready-to-eat foods, 298–304. Copyright 2014; *Carbohydrate Polym.*, 101, Shojaee-Aliabadi, S., Hosseini, H., Mohammadifar, M.A., Mohammadi, A., Ghasemlou, M., Hosseinia, S. M., and Khaksar, R., Characterization of k-carrageenan films incorporated plant essential oils with improved antimicrobial activity, 582–591. Copyright 2014, with permission form Elsevier.

[a] Concentration in the film-forming solution.

might indicate that the antimicrobial studied is not effective against the test organism, at least in the concentration used, and/or that the antimicrobial do not have appropriate diffusional properties for this assay. Table 24.1 summarizes some results reported in bibliography for the last 5 years concerning the inhibition zone test.

To explain the procedure generally followed, two cases are explained in more detail. Emiroğlu et al. (2010) tested the antibacterial activity of oregano or thyme essential oils incorporated into soy protein edible films at levels of 1%, 2%, 3%, 4%, and 5% (v/v), against *Escherichia coli, Staphylococcus aureus, E. coli* O157:H7, *Pseudomonas aeruginosa,* and *Lactobacillus plantarum.* Disks cut from the films were placed on nutrient agar plates, previously surface spread with the inoculum containing indicator microorganisms in the range of 10^5–10^6 CFU/mL. The plates were then incubated at 37°C for 24 h. The diameter of inhibitory zone surrounding film disks and contact area of edible films with agar surface were then measured in millimeters. The authors reported that oregano and thyme essential oils exhibited similar antibacterial activity against all bacteria. While *E. coli, E. coli* O157:H7, and *S. aureus* were significantly inhibited by antimicrobial films, *L. plantarum* and *P. aeruginosa* appeared to be the more resistant bacteria.

Shojaee-Aliabadi et al. (2014) increased films' antimicrobial activity by incorporating essential oils, particularly *Zataria multiflora* Boiss (ZEO) and *Mentha pulegium* essential oils (MEO). *S. aureus* was found to be the most sensitive bacterium to either ZEO or MEO, followed by *Bacillus cereus* and *E. coli.* The highest inhibition halo zone of 544.05 mm² was observed for *S. aureus* around the films incorporated with 3% (v/v) ZEO. The total inhibitory zone of 3% (v/v) MEO-formulated films was 20.43 for *Salmonella typhimurium* and 10.15 mm^2 for *P. aeruginosa*.

In general, these tests allow to conclude about the adequacy of biopolymers used to constitute the edible matrices, which means that they can support the antimicrobial compound that is released in an adequate amount as to inhibit target microorganism growth.

24.2.1.1.2 Diffusional Tests

The effectiveness of edible films and coatings to ensure the microbiological quality of a food product depends on the mobility properties and release rate of the antimicrobial supported.

The mathematical theory of diffusion in isotropic media considers that the transfer speed of the substance that diffuses through a section of unit area is proportional to the concentration gradient that exists, perpendicular to that section. This is expressed through the Fick's first law:

$$F = -D\frac{\partial C}{\partial x}$$

where
  F is the flow or the rate of transfer through a section of unit area in the x direction
  C is the concentration of the diffusing substance
  x is the spatial coordinate perpendicular to the section
  D is the diffusion coefficient, which depends on the diffusing species and test conditions

In some instances, D may be assumed constant, while in others, it depends strongly on the concentration. Cases where transfer of a compound occurs, but its concentration does not change with time at any position of the system (steady-state diffusion), may be described using the aforementioned equation.

Conversely, for the case of diffusion in nonsteady state, where the local concentration of the compound that diffuses changes over time (accumulation appears), the first law of Fick should be derived taking into account the mass balance for an element of volume where the diffusion is carried out and the rate of change of the compound amount in such volume. For concentration gradients along the x axis, the following differential equation known as Fick's second law is obtained:

$$\frac{\partial C}{\partial t} = D\frac{\partial^2 C}{\partial x^2}$$

The solutions to this equation can be obtained by analytical, numerical, and graphical methods. Analytical solutions for diffusion in one dimension in an infinite plate, infinite cylinder, and a sphere have been found assuming a constant value of D and appropriate boundary conditions (Crank 1975).

Typically, the solution to Fick's second law is expressed in the form of a trigonometric series that converges for large values of time, or it may be expressed as a series of error functions or integrals related that are more suitable for numerical evaluation when time values are small (Welty et al. 1999). In general, it is assumed that the diffusion coefficient is constant. Anyhow, a mean value or effective diffusivity "D_{eff}" can be used when that assumption is not adequate.

Table 24.2 summarizes some results published in the last 5 years that were obtained through the study of the diffusion and/or the description of the kinetic of release of antimicrobials supported in edible film matrices. Such information is useful for the rational design of antimicrobial films and coatings. As can be observed, the receptor media is, in general, a liquid or a solid model system that resembles real food products because of the need for adequate geometric characteristics for modeling the phenomena.

Guillard et al. (2009) analyzed the sorbic acid mass transfer phenomena in a simulated coated food. The system was assumed to be composed of two finite plane sheets placed side by side (Figure 24.1).

For the characterization of the diffusion, it was mathematically described by considering the surface layer and the model food as separate regions that were linked through boundary conditions

TABLE 24.2

Effective Diffusion Coefficients (D_{eff}) of Some Antimicrobial/Edible Films Systems

Antimicrobial	Biopolymer	Receptor Medium	D_{eff} (cm²/s)	References
Potassium sorbate	Cornstarch	Liquid media	1.4×10^{-7} to 9×10^{-8}	Flores et al. (2007)
Sorbic acid	Wheat gluten	Agar gel	7.5×10^{-8}	Guillard et al. (2009)
	Beeswax		2.4×10^{-12}	
Natamycin	Chitosan	Liquid media	3.60×10^{-10}	Fajardo et al. (2010)
		Cheese	1.29×10^{-12}	
Natamycin	Alginate/pectin	Liquid media	3.2×10^{-9} to 9×10^{-12}	Krause et al. (2012)
Potassium sorbate	Cornstarch	Agar gel	3×10^{-10}	López et al. (2013)
OSCN⁻ LPOS[a]	Soybean meal	Ham disk	$1.2–24 \times 10^{-11}$	Lee and Min (2013)

Sources: Data reprinted from *J. Food Eng.*, 81, Flores, S., Conte, A., Campos, C., Gerschenson, L., and Del Nobile, M., Mass transport properties of tapioca-based active edible films, 580–586. Copyright 2007; *Innov. Food Sci. Emerg. Technol.*, 10, Guillard, V., Issoupov, V., Redl, A., and Gontard, N., Food preservative content reduction by controlling sorbic acid release from a superficial coating, 108–115. Copyright 2009; *J. Food Eng.*, 101, Fajardo, P., Martins, J., Fuciños, C., Pastrana, L., Teixeira, J., and Vicente, A., Evaluation of a chitosan-based edible film as carrier of natamycin to improve the storability of Saloio cheese, 349–356. Copyright 2010; *J. Food Eng.*, 110, Krause Bierhalz, A., Altenhofen da Silva, M., and Guenter Kieckbusch, T., Natamycin release from alginate/pectin films for food packaging applications, 18–25. Copyright 2012; *Mater. Sci. Eng. C*, 33, López, O., Giannuzzi, L., Zaritzky, N., and García, A., Potassium sorbate controlled release from corn starch films, 1583–1591. Copyright 2013; *LWT Food Sci. Technol.*, 54, Lee, H. and Min, S.C., Antimicrobial edible defatted soybean meal-based films incorporating the lactoperoxidase system, 42–50. Copyright 2013, with permission from Elsevier.

[a] OSCN⁻ LPOS: hypothiocyanite in lactoperoxidase system. D_{eff} refers to OSCN⁻ migration.

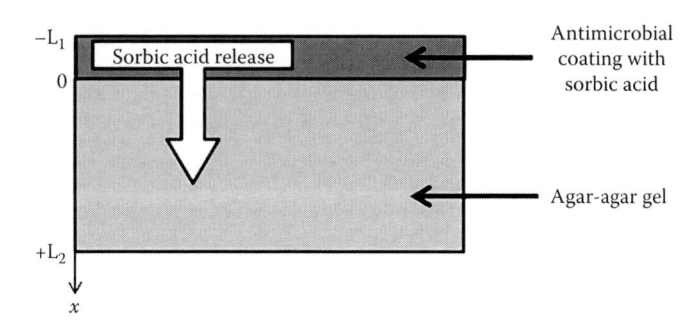

FIGURE 24.1 Simulated coated food.

at the film/model food interface. Variations of sorbic acid concentration within each region were given by

$$\frac{\partial C_1(x,t)}{\partial t} = D_1 \frac{\partial^2 C_1(x,t)}{\partial x^2}, \quad -L_1 < x < 0$$

$$\frac{\partial C_2(x,t)}{\partial t} = D_2 \frac{\partial^2 C_2(x,t)}{\partial x^2}, \quad 0 < x < L_2$$

The initial conditions were

$$C_1(x,0) = C_0, \quad -L_1 < x < 0$$

$$C_2(x,0) = 0, \quad 0 < x < L_2$$

The boundary conditions were the following:

$$\frac{\partial C_1(-L_1,t)}{\partial x} = 0$$
There is no interchange of antimicrobial agent
with surrounding and considering the coating or the gel.

$$\frac{\partial C_2(L_2,t)}{\partial x} = 0$$

At the coating/model food interface,
$$D_1 \frac{\partial C_1(0,t)}{\partial x} = D_2 \frac{\partial C_2(0,t)}{\partial x}$$
the flux leaving the surface layer is
equal to the flux entering the model food.

Across the interface, the concentrations
$$C_1(0,t) = k_{1/2}^{SA} C_2(0,t)$$
in the two regions are related by partition
equilibrium $\left(k_{1/2}^{SA}\right)$.

Guillard et al. (2009) used this model to evaluate the benefit of using an antimicrobial beeswax or wheat gluten coating instead of adding the preservative directly in relation to maintain a high preservative concentration of sorbic acid in the food surface. They concluded that the beeswax coating was appropriate because the calculated amount of sorbic acid required to maintain a 0.2% surface concentration during 23 days was 100 times lower when supported in this matrix than when introduced directly in the food. However, sorbic acid diffused very fast through wheat gluten edible coating displaying an inadequate behavior.

In the case of hydrophilic polymers, such as proteins and carbohydrates, a high polarity of the chains was observed. Due to the presence of groups such as $-OH$ and $-NH_2$, which have the potential of forming hydrogen bonds, they can easily interact with water, which is directly sorbed at high relative humidity (RH). This water sorption affects the intermolecular interactions between chains, greatly increasing the diffusivity of many substances. Since classical models were developed for nonpolar polymer matrices, it is necessary to develop alternative models for biopolymers because big deviations are expected due to their particular characteristics. Flores et al. (2007) mathematically modeled the release of potassium sorbate from edible films constituted by tapioca starch to a liquid media. The model was developed by taking into account the phenomena involved during the release of the antimicrobial compound from the hydrophilic polymeric matrix: water diffusion to the film, macromolecular matrix relaxation, and sorbate diffusion through the film. This model was based on diffusion and on structural characteristics of the polymer matrix while also considering its swelling. Diffusion parameters for the film at a fixed temperature and film composition were calculated with the following equation:

$$M(t) = M_F(t) + M_R(t)$$

where
- $M(t)$ is the total amount of water sorbed or total amount of active compound released at time t
- $M_F(t)$ is the contribution of stochastic phenomena to either the amount of water absorbed or the amount of active compound released at time t
- $M_R(t)$ is the contribution of polymer relaxation phenomena to either the amount of water sorbed or the amount of active compound released at time t

By defining X_F as the ratio between the M_F^{eq} (amount of low-molecular-weight compound sorbed or released at equilibrium as a consequence of stochastic phenomena) and M^{eq} being $M^{eq} = M_F^{eq} + M^{eq}$, the aforementioned equation can be rearranged in the following form:

$$M(t) = M^{eq} \cdot \left\{ X_F \cdot \left\{ 1 - \frac{8}{\pi^2} \cdot \sum_{n=0}^{n=\infty} \frac{1}{(2 \cdot n + 1)^2} \cdot \exp\left\{ -D \cdot (2 \cdot n + 1)^2 \cdot \pi^2 \cdot \frac{t}{\ell^2} \right\} \right\} \right\}$$
$$+ M^{eq}(1 - X_F) \cdot \left[1 - \exp\left(-\frac{t}{\tau} \right) \right]$$

By definition, X_F ranges from 0 to 1; for X_F equal to 1, the aforementioned equation is the solution of Fick's second law, whereas for X_F equal to 0, anomalous diffusion is obtained; D is the diffusivity coefficient; ℓ is the film thickness; τ is the relaxation time associated to polymer relaxation.

From the obtained results, it could be inferred that the proposed model satisfactorily fits the experimental data ($M(t)/M^{eq}$ plotted as a function of time, t), and it was concluded that casting technique produces films with high amorphous degree determining a great contribution of matrix relaxation to sorbate release. This trend can be attributed to the increased influence of matrix relaxation on the water sorption kinetic.

According to data obtained from diffusional assays, it can be concluded that the effective coefficients of diffusion are lower when the receptor media is semisolid than when it is liquid. Some examples can be observed in Table 24.2. Fajardo et al. (2010) reported that chitosan-based coating/films containing natamycin can be used as controlled-release systems creating an additional hurdle for mold or yeast growth contributing to the extension of cheese shelf life. In this case, the diffusion coefficient values of natamycin from the film to phosphate-buffered saline solution and to cheese were 3.60×10^{-10} and 1.29×10^{-12} cm²/s trend that can be ascribed to the lack of swelling effect in the release phenomenon in the case of cheese. It is important to state that a low diffusion coefficient is desirable in food applications if the target is to maintain a critical surface concentration of natamycin.

The antimicrobial release rate from edible films can be modulated using different edible matrix formulations for similar external conditions. Lee and Min (2013) developed edible films using defatted soybean meal and lactoperoxidase system that were added with the antimicrobial hypothiocyanite (OSCN⁻). The authors reported that when the concentration of glycerol increased from 20% to 50% (w/w), the D values of OSCN⁻ at 5°C (1.5×10^{-11} cm²/s) increased 3.3- and 7.2-fold at 96% RH and at 10°C and 22°C, respectively, probably due to the increase of free volume in the film matrix available for OSCN⁻ diffusion. On the other hand, faster diffusion at a higher RH was also observed, and it can be ascribed to a higher polymer relaxation induced by water molecules. In general, the D value increased with the storage temperature due to higher kinetic energy and molecular motion. Therefore, process or storage conditions (temperature, humidity) and formulation must be considered in the development of antimicrobial films and coatings to assure the obtention of the adequate release rates that optimize the antimicrobial activity.

The controlled release can help to assure a continuous provision of the antimicrobial along storage while precluding the interaction of the compound with other food ingredients contributing to preservative functionality. However, if the surface is protected, the movement of the active from the biopolymeric matrix to the food must be slow enough to maintain a superficial antimicrobial concentration higher than the minimum inhibitory concentration for the target microorganism to avoid the microbial growth during storage (Guillard et al. 2009). If both effects are important, an intermediate release must be achieved.

According to actual information, the general consensus is that the use of antimicrobial edible matrices helps to increase the effectiveness of preservatives, preventing microbial growth on food surfaces more effectively than the direct antimicrobial application. They can also help to decrease the amount of antimicrobials used. However, it is highly recommended that the results of studies performed with model systems are confirmed using real food systems.

24.2.1.2 Antioxidant Migration

Although edible coatings and films may not provide a good water vapor barrier, at low RH, these materials *per se* are very good barriers to gases (oxygen) and consequently can preserve food interfaces from lipid oxidation (Pérez et al. 2013). Moreover, edible films and coatings can be used to support active compounds such as natural antioxidants localizing their activity and controlling their release at interfaces.

There are certainly not many studies about migration of antioxidants from edible films and coatings. Gómez-Estaca et al. (2007) developed films carrying oregano or rosemary aqueous extract (250 and 4.7 μg/mL of total phenolic content, respectively) and based on gelatin (4% in the film-forming solution). The free-standing films of uniform thickness (100 μm) equilibrated at 57.7% RH before testing were casted, using sorbitol and glycerol for plasticization (15% on gelatin basis). The combined use of each of these films and of a high-pressure fish processing (300 MPa/20°C/15 min) showed the best performance for improving the shelf life of previously salted and smoked sardine (*Sardina pilchardus*). They observed that phenolics present in the extracts migrated from gelatin producing an increased phenolic content and antioxidant power in the muscle. This combination lowered lipid oxidation levels measured through the peroxide and TBARS indices.

Walnut oil is a valuable functional ingredient highly susceptible to oxidation due to its high content of polyunsaturated fatty acids. Tocopherols are the natural antioxidants of walnut oil. Pérez et al. (2013) developed edible films made with high methoxyl (72%—degree of methylation) pectin or a blend of methylcellulose and high methoxyl pectin (50:50, w/w) for carrying L-(+)-ascorbic acid (AA). This natural antioxidant showed a half-life time of 83 and 100 days, in each of the films previously cited, when the study was performed in the dark at 57.7% RH and 25°C. These films were applied as antioxidant interfaces on walnut oil in a 50-day storage study performed at the same conditions used for the kinetics studies. Fluorescence intensity was evaluated at the 308 nm peak, attributed to the oil tocopherols. It was determined that films carrying AA acted like sacrificing materials or oxygen scavengers reducing tocopherol destruction along storage as can be observed in Figure 24.2. The tocopherols were probably related to the delay of the early signs of oxidative oil spoilage (development of lipid peroxides). On the other hand, when AA was absent in the high methoxyl pectin film, the tocopherol levels of the covered walnut oil samples were also significantly higher ($p < 0.05$) than the ones of the uncovered control samples. The trials performed on films without AA demonstrated their ability to act *per se* as effective barriers to oxygen.

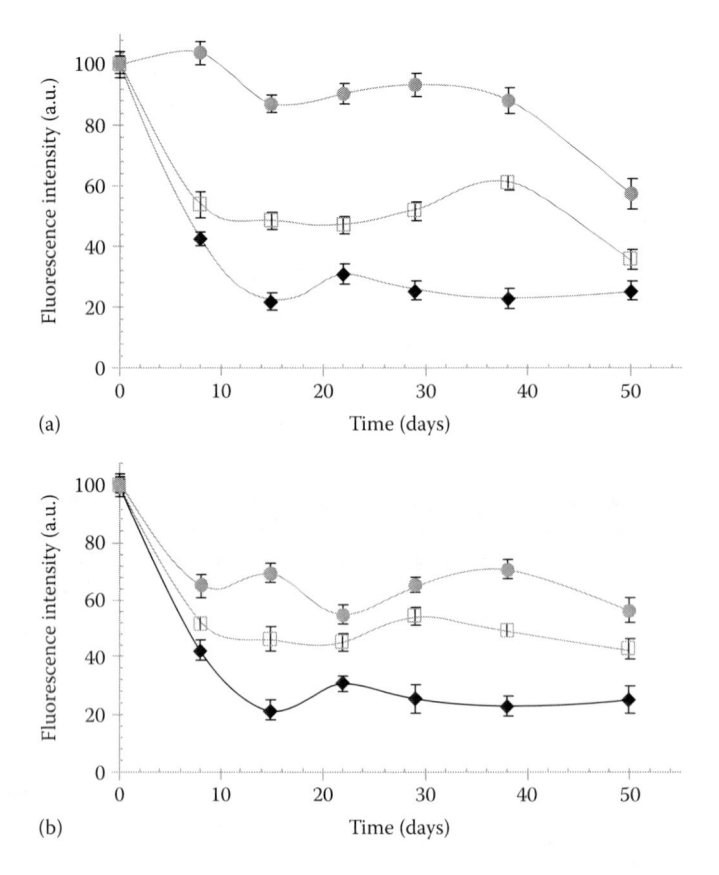

FIGURE 24.2 Fluorescence intensity collected at the 308 nm peak, attributed to the oil tocopherols, is plotted against the storage time (57% relative humidity, 25°C) of walnut oil covered with high methoxyl pectin (a) and 50:50 (w/w) high methoxyl pectin–methylcellulose (b) films made either with (●) or without (□) ascorbic acid. Control (uncovered) sample (◆) is also shown in both panels. The fluorescence intensity was normalized with respect to the initial value. Error bars (SD, *n* = 4). (Data reprinted from *J. Food Eng.*, 116(1), Pérez, C., De'Nobili, M., Rizzo, S., Gerschenson, L., Descalzo, A., and Rojas, A., High methoxyl pectin-methyl cellulose films with antioxidant activity at a functional food interface, 162–169. Copyright 2013, with permission from Elsevier.)

On the other hand, high methoxyl pectin–methylcellulose film carrying AA maintained the tocopherol content of walnut oil significantly above ($p < 0.05$) the ones shown by the uncovered control samples and films without AA, during a 50-day storage period (Figure 24.2b). The half-life times of AA in both film systems assayed were not very different, assuring similar concentrations of AA remaining into the films during the 50-day storage of covered walnut oil samples; film thickness showed similar values. Anyhow, the film containing AA and based on high methoxyl pectin showed a higher antioxidant capability up to 40 days of film storage (Figure 24.2a) than the film with AA based on high methoxyl pectin/methylcellulose. It was then observed that the polymeric microstructure determined a different antioxidant capability, which might be attributed to different AA availability. Hence, the network characteristics seemed to affect not only the stability of the compartmentalized AA but also the performance of the edible films as antioxidant interfaces for food preservation. The AA presence in the edible films was decisive for higher antioxidant protection of the tocopherols in the walnut oil. Studies of migration were not performed and AA was supposed to remain compartmentalized in the edible film during the study, whereas natural tocopherols persisted dissolved in the walnut oil, and, hence, migration was suggested to be negligible, though the antioxidant protection was successfully exerted by the active barrier interface.

Mathew and Abraham (2008) developed films where ferulic acid was esterified to the active $-NH_2$ groups of chitosan in starch–chitosan blend films with the object of obtaining a non-migration-type material, which can be useful for antioxidant preservation of food surfaces.

Concerning the studies of antioxidant migration from edible films and coatings performed on simulants, Giménez et al. (2013a) studied active biodegradable films based on agar and agar–fish gelatin (33% w/w) plasticized by glycerol (66% w/w of agar powder) and developed by casting that included green tea aqueous extracts. The effects of the partial replacement (33% of the agar) by fishskin gelatin and of the addition of green tea extract on the physical properties of the developed films were evaluated. To accomplish this, films were previously equilibrated at 58% RH, at 22°C. Special attention was given to the release of antioxidant compounds (phenolics) from the agar film matrices, with and without gelatin, in water simulant for 16 h at 22°C. The antioxidant power was determined through Ferric reducing antioxidant power (FRAP) and 2,2'-azinobis-(3-ethylbenzothiazoline-6-sulfonic acid) (ABTS) assays and the phenolic composition through reverse phase HPLC-MS. Agar–gelatin films were less resistant and more deformable than agar films. The presence of gelatin in the agar–green tea matrix hindered the release of total phenolic compounds, catechins, and flavonols in water. As a consequence, the antioxidant power of the simulant was lower in the case of films containing gelatin. A protein–phenolic interaction can be involved. The release of polyphenolics in water from both agar–green tea and agar–fish gelatin–green tea film matrices occurred mainly during the first 15 min period, showing both films similar values after this time (50 mg gallic acid equivalents/g film) (Giménez et al. 2013a). In the case of agar–green tea films, an increase of 1.5 times was observed in the release of polyphenols from 15 to 90 min, and only a slight but significant increase after 90 min, up to 16 h. However, in the case of agar–fish gelatin–green tea films, no increase was detected in the level of phenolics released from the matrix in the period between 15 and 90 min, whereas an increase of 1.2 times was observed between 90 min and 16 h of the delivery study, reaching values of 60 mg gallic acid equivalents/g film after this time. As reported by Giménez et al. (2013a), the most abundant polyphenols found in the green tea extract, epigallocatechin gallate and epigallocatechin, were also released in greater amounts from the films, showing values that were two orders of magnitude higher than the rest of catechins and flavonols, both in agar–green tea films and agar–fish gelatin–green tea films. After 16 h in water, all the polyphenols studied were released in higher quantities by agar–green tea films, with the exception of epicatechin and epicatechin-3-gallate. Therefore, the replacement of 33% of agar by fishskin gelatin in the film formulation seemed to hinder the release of the polyphenols from the film network. Furthermore, gelatin solubilization has been reported from films containing gelatin when these films are placed in water (Giménez et al. 2012), which involves the possible release of protein–polyphenol complexes or the formation of complexes between polyphenols and protein once released from the matrix. It is a well-known ability of polyphenols to interact with proteins. Giménez et al. (2013a) also found that the release of epigallocatechin gallate and epigallocatechin in water from agar–green tea films showed a highly significant correlation with FRAP and ABTS. However, in the case of agar–fish gelatin–green tea films, the release of epigallocatechin and epigallocatechin gallate showed no significant correlation with antioxidant activity determined through FRAP and ABTS.

In the literature, it has been also reported that a study of an *in vitro* enzymatic digestion mimicked the gastrointestinal digestion of fish gelatin antioxidant films carrying green tea extract (Giménez et al. 2013b). The residual antioxidant activity of the tea polyphenols as well as the recovery of the total polyphenols and major catechins after the *in vitro* digestion of films was determined. The antioxidant activity of the green tea extract supported in the gelatin films was maintained beyond the casting processing and during the enzymatic digestion of films. The authors concluded that the gelatin films could be a vehicle to deliver antioxidant compounds with potential benefits when these edible films are consumed. Also, it was determined that the polyphenols were effectively released from the phenolic–fish gelatin interaction in the intestinal tract.

Blanco-Fernández et al. (2013) developed antioxidant chitosan (1.0% and 1.5% w/w) films carrying α-tocopherol through casting. The effect of increasing concentration of α-tocopherol (0%–40% with respect to chitosan) emulsified with acetylated monoglyceride in 1:1 or 4:1 weight ratios on the film properties was studied. The films developed were able to produce the controlled release of the α-tocopherol content in 50% ethanol/water for more than 5 days; the antioxidant activity of α-tocopherol in the release medium was, after 20 days, as higher as that of a freshly prepared solution. Interestingly, it was observed that the rate of the α-tocopherol release to the medium was inversely related to the α-tocopherol content. Conversely, it was reported that the rapid release of tocopherols from gelatin films to a hydrophobic substrate like margarine was controlled by cross-linking of the protein matrix (Guilbert et al. 1996).

It can be concluded that studies on migration of antioxidants are scarce and that their deepening would be important to contribute to the adequate design of suitable films and coatings to protect surfaces or to control antioxidant release to food products.

24.2.2 Plasticizer Migration

Oses et al. (2009) studied edible films obtained through casting technique from 10% (w/w) aqueous solutions of whey protein isolate (WPI) and containing glycerol or sorbitol (30%, 40%, 50%, or 60% of the dry weight of the film). Rectangular strips (25.4 mm in width and 75.0 mm in length) were stored 1 week at ambient temperature and 50% or 75% RH observing that equilibrium moisture content increased with glycerol content for both RH and that sorbitol content showed a smaller effect on moisture content. The mechanical properties were evaluated after equilibration observing that, for the same plasticizer content and RH, films with glycerol presented significantly lower stresses at break and significantly greater deformations at break than the ones containing sorbitol, due to the more effective plasticizer action of the former. This trend was ascribed by the authors to the glycerol's greater moisture content due to its hygroscopicity, being water an excellent plasticizer.

The evolution of mechanical properties with storage was also studied observing that films plasticized with glycerol showed a constant value along 30 weeks and at both RH. On the contrary, films containing sorbitol turned to be less flexible along 30 weeks of storage at 75% RH and were rigid and impossible to be evaluated after storage at 50% RH. The authors attributed this trend to the migration of sorbitol to the surface of the films where it could crystallize according to previous reports of Krogars et al. (2003) for films based on cornstarch. The sorbitol crystallization weakened its ability to act as plasticizer allowing the increase of the interaction between proteins of the WPI and the increase in film rigidity. This phenomenon was more intense when storage was performed at 50% RH because the decrease in water content determined a decrease in sorbitol solubilization and an increase in crystallization rate.

Nilsson (2004) obtained edible films based on a fraction enriched in kafirin (ethanol-soluble protein obtained from kafir–sorghum with a protein content of 88.6%, w/w). The film was formulated with 0.64 g of defatted kafirin, 0.228 g of a mixture of 1:1:1 (w/w/w) lactic acid, glycerol, and polyethylene glycol 400 and 5.5 mL of ethanol 80% (v/v). The obtention was performed through casting and the procedure involved drying in an oven at 50°C (air velocity 2 m/s) for 2 h. Films were stored at 23°C and 50% RH until analysis. Release studies were performed from the films to a food model (a_w 0.95) constituted by gelatin, sucrose, and water, and the concentration of glycerol was evaluated along storage and at different depths of the food model. It was observed that glycerol was released rapidly to the food model and after 2 h, the highest concentration was found at a depth of 1 mm. After 24 h, the mechanical properties of the films have changed, losing its flexibility due to the loss of plasticizer, and the glycerol concentration was evaluated as homogeneous throughout the whole food revealing its migration from the film.

Kuutti et al. (1998) obtained edible films using oat starch, glycerol (30% w/w, dry basis), and water (12% w/w, dry basis), and the procedure involved the use of a twin-screw extruder, with a barrel temperature of 170°C and a die temperature of 120°C. The storage of the films was performed under 50% RH and at 20°C during 1–5 weeks. Small pieces of the films (4 mm × 4 mm) were cut and placed on sample stubs and studied through atomic force microscopy in air in the constant force mode (constant deflection). The films were also characterized through friction force microscopy that can be used to resolve different phases in phase-separated systems; for this technique, the tip was scanned sideways allowing to evaluate the torsion caused by the friction force. Figure 24.3 shows 10 μm × 10 μm micrographs of a 2-week-old oat film. The friction image revealed a partly heterogeneous surface structure (Figure 24.3b) featuring high (dark) and low (light) friction areas. The areas of low friction dominated the surface. The differences in friction were not due to changes in topography, as detected by comparing Figure 24.3a and b. At 5 weeks of storage, the initial higher friction phase disappeared almost completely from the surface of the oat film, although local differences were significant. The mean roughness increased. They attributed this trend to a starch–glycerol–water phase separation on the surface of oat starch films due to glycerol diffusion making those low friction areas and increasing the surface energy to values similar to those of glycerol.

Fontes et al. (2011) characterized through scanning electron microscopy (SEM) edible films casted from aqueous systems containing 1% (w/w) of methylcellulose (METHOCEL A15 FG) and 0.5% (w/w) of glycerol. The films were characterized after metallization through SEM at 5 kV. Images were obtained from the surface and from the cross section. The fractured film samples were fixed onto aluminum stubs with carbon adhesive tape. Figure 24.4 shows the structure of the surface (a) and cross section (b) of the films. According to the researchers, the cross section showed small white stripes (represented by letter e) inserted into the polymeric matrix, which could be related to the exudation of the glycerol from the matrix.

It can be concluded that there is scarce information about the migration of plasticizers in edible films and coatings along processing and storage. This phenomenon is important and must be studied because it can influence the mechanical properties of the films, as well as its surface tension and other physicochemical properties, potentially affecting film functionality.

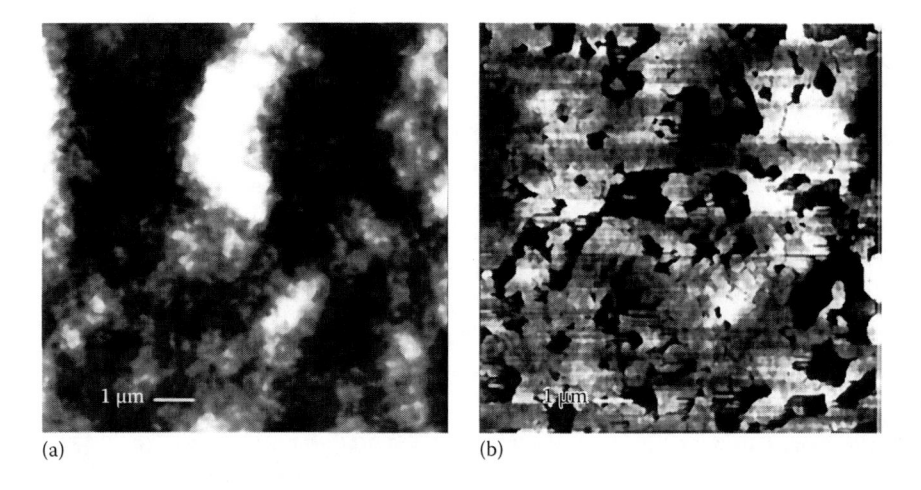

(a)   (b)

FIGURE 24.3 (a) A 10 μm × 10 μm top-view atomic force microscopy image of the 2-week-old oat film and (b) the respective lateral force image. (Data reprinted from *Carbohydr. Polym.*, 37(1), Kuutti, L., Peltonen, J., Myllarinen, P., Teleman, O., and Forssell, P., AFM in studies of thermoplastic starches during ageing, 7–12. Copyright 1998, with permission from Elsevier.)

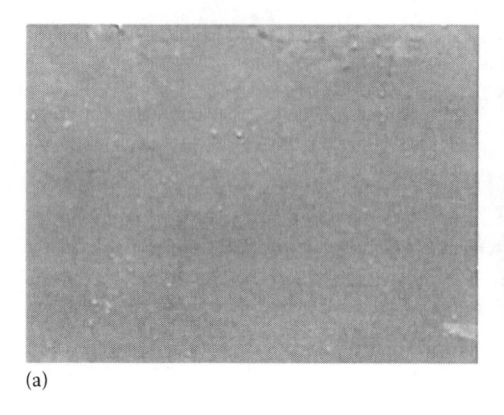
(a)

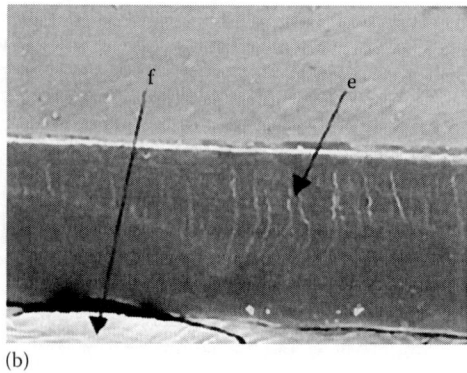
(b)

FIGURE 24.4 Scanning electron microscopy of methylcellulose films. (a) Surface; (b) cross section. Arrow with letter "e" is pointing to stripes. Arrow with letter "f" is pointing to carbon tape. (Data reprinted from Fontes, L. et al., *Am. J. Food Technol.*, 6(7), 555, 2011.)

24.3 Conclusions

The development of active edible films and coatings supporting antimicrobials or antioxidants still present a lot of challenges to be answered. The study of the release of these additives to food products in realistic external conditions will help to evaluate the efficiency of this emergent technology for lengthening shelf life and improving food quality. But the migration of other components, like plasticizers, needs also to be addressed because it can compromise film functionality. The scarce available information on some of these topics indicates that the path has been cleared but there is still far to go. Especially, considering the restrictions imposed in the formulations by the fact that being edible, the film and coating formulations must comply with food regulations.

ACKNOWLEDGMENTS

The authors acknowledge the financial assistance from Buenos Aires University (UBACyT 726, 070, 550BA); National Research Council of Argentina, CONICET (PIP 531, 349, 507); National Agency of Scientific and Technological Promotion of Argentina (PICT 2008 Number 2131 and PICT 2012 Number 0183).

REFERENCES

Blanco-Fernández, B., Rial-Hermida, M., Álvarez-Lorenzo, C., and Concheiro, A. Edible chitosan/acetylated monoglyceride films for prolonged release of vitamin E and antioxidant activity. *Journal of Applied Polymer Science* 129(2) (2013): 626–635.

Crank, J. *Mathematics of Diffusion*. London, U.K.: Oxford University Press, 1975.

Emiroğlu, Z., Yemiş, G., Coşkun, B., and Candoğan, K. Antimicrobial activity of soy edible films incorporated with thyme and oregano essential oils on fresh ground beef patties. *Meat Science* 86 (2010): 283–288.

Fajardo, P., Martins, J., Fuciños, C., Pastrana, L., Teixeira, J., and Vicente, A. Evaluation of a chitosan-based edible film as carrier of natamycin to improve the storability of Saloio cheese. *Journal of Food Engineering* 101 (2010): 349–356.

Flores, S., Conte, A., Campos, C., Gerschenson, L., and Del Nobile, M. Mass transport properties of tapioca-based active edible films. *Journal of Food Engineering* 81 (2007): 580–586.

Fontes, L., Ramos, K., Sivi, T., and Queiroz, F. Biodegradable edible films from renewable sources. Potential for their application in fried foods. *American Journal of Food Technology* 6(7) (2011): 555–567.

Giménez, B., Gómez-Guillén, M., López-Caballero, M., Gómez-Estaca, J., and Montero, P. Role of sepiolite in the release of active compounds from gelatin-egg white films. *Food Hydrocolloids* 27 (2012): 475–486.

Giménez, B., López de Lacey, A., Pérez-Santín, E., López-Caballero, M., and Montero, P. Release of active compounds from agar and agar-gelatin films with green tea extract. *Food Hydrocolloids* 30 (2013a): 264–271.

Giménez, B., Moreno, S., López-Caballero, M., Montero, P., and Gómez-Guillén, M. Antioxidant properties of green tea extract incorporated to fish gelatin films after simulated gastrointestinal enzymatic digestion. *LWT—Food Science and Technology* 53 (2013b): 445–451.

Gómez-Estaca, J., Montero, P., Giménez, B., and Gómez-Guillén, M. Effect of functional edible films and high pressure processing on microbial and oxidative spoilage in cold-smoked sardine (*Sardina pilchardus*). *Food Chemistry* 105 (2007): 511–520.

Guilbert, S., Gontard, N., and Gorris, L. Prolongation of the shelf-life of perishable food products using biodegradable films and coatings. *LWT—Food Science and Technology* 29(1–2) (1996): 10–17.

Guillard, V., Issoupov, V., Redl, A., and Gontard, N. Food preservative content reduction by controlling sorbic acid release from a superficial coating. *Innovative Food Science and Emerging Technologies* 10 (2009): 108–115.

Iturriaga, L., Olabarrieta, I., and Martínez de Marañón, I. Antimicrobial assays of natural extracts and their inhibitory effect against *Listeria innocua* and fish spoilage bacteria, after incorporation into biopolymer edible films. *International Journal of Food Microbiology* 158 (2012): 58–64.

Krause Bierhalz, A., Altenhofen da Silva, M., and Guenter Kieckbusch, T. Natamycin release from alginate/pectin films for food packaging applications. *Journal of Food Engineering* 110 (2012): 18–25.

Krogars, K., Heinamaki, J., Karjalainen, M., Niskanen, A., Leskela, M., and Yliruusi, J. Enhanced stability of rubbery amylose-rich maize starch films plasticized with a combination of sorbitol and glycerol. *International Journal of Pharmaceutics* 251(1–2) (2003): 205–208.

Kuutti, L., Peltonen, J., Myllarinen, P., Teleman, O., and Forssell, P. AFM in studies of thermoplastic starches during ageing. *Carbohydrate Polymers* 37(1) (1998): 7–12.

Lee, H. and Min, S.C. Antimicrobial edible defatted soybean meal-based films incorporating the lactoperoxidase system. *LWT—Food Science and Technology* 54 (2013): 42–50.

López, O., Giannuzzi, L., Zaritzky, N., and García, A. Potassium sorbate controlled release from corn starch films. *Materials Science and Engineering C* 33 (2013): 1583–1591.

López de Dicastillo, C., Ares Pernas, A., Castro López, M., López Vilariño, J., and González Rodríguez, M. Enhancing the release of the antioxidant tocopherol from polypropylene films by incorporating the natural plasticizers lecithin, olive oil, or sunflower oil. *Journal of Agricultural and Food Chemistry* 61(48) (2013): 11848–11857.

Mathew, S. and Abraham, E. Characterisation of ferulic acid incorporated starch-chitosan blend films. *Food Hydrocolloids* 22 (2008): 826–835.

Nilsson, K. Migration of substances from coating to food. ENVIROPAK. Environment-friendly packaging solutions for enhanced storage and quality of southern Africa's fruit and nut exports. ICA4-CT-2001-10062. Technical report. Deliverable 14 (2004). http://www.sik.se/enviropak.

Oses, J., Fernández-Pan, I., Mendoza, M., and Mate, J. Stability of the mechanical properties of edible films based on whey protein isolate during storage at different relative humidity. *Food Hydrocolloids* 23(1) (2009): 125–131.

Pérez, C., De'Nobili, M., Rizzo, S., Gerschenson, L., Descalzo, A., and Rojas, A. High methoxyl pectin-methyl cellulose films with antioxidant activity at a functional food interface. *Journal of Food Engineering* 116(1) (2013): 162–169.

Pérez, L., Soazo, M.d.V., Balagué, C., Rubiolo, A., and Verdini, R. Effect of pH on the effectiveness of whey protein/glycerol edible films containing potassium sorbate to control non-O157 shiga toxin-producing *Escherichia coli* in ready-to-eat foods. *Food Control* 37 (2014): 298–304.

Shen, X.L., Wu, J.M., Chen, Y., and Zhao, G. Antimicrobial and physical properties of sweet potato starch films incorporated with potassium sorbate or chitosan. *Food Hydrocolloids* 24 (2010): 285–290.

Shojaee-Aliabadi, S., Hosseini, H., Mohammadifar, M.A., Mohammadi, A., Ghasemlou, M., Hosseinia, S.M., and Khaksar, R. Characterization of κ-carrageenan films incorporated plant essential oils with improved antimicrobial activity. *Carbohydrate Polymers* 101 (2014): 582–591.

Welty, J., Wicks, C., and Wilson, R. *Fundamentos de transferencia de momento, calor y masa*. México, Mexico: Ed. Limusa, 1999.

25

Migration Analysis of Compounds in Food Packaging

Cristina Nerín

CONTENTS

25.1 Introduction

One of the risks involved in the use of food contact materials is the contamination of food from the compounds transferred from the packaging materials or articles to the food in contact with them. This transference of matter is called "migration." Migration process is governed by two main phenomena: (a) the partition coefficient of the migrant between the packaging material and the food and (b) the diffusion of migrants in the packaging material and in the food. There are a series of parameters that affect the migration process: temperature that influences the diffusion and also the partition coefficients; characteristics of the packaging material or article, such as composition, polarity, and crystallinity; and characteristics of the food in contact with them, such as polarity, fat matter, alcohol content, and composition.

Food contact materials have been studied since 1980s, and concentration limits of many substances have been established to ensure the safety in the use of such materials and articles. There are specific legislations in FDA, in Europe, Japan, China, Mercosur, and Australia, and in general around the world concerning the conditions that the materials and articles in contact with food have to comply. The Frame Regulation 1935/2004/CEE established the criteria that any material in contact with food has to fulfill, and the Regulation 10/2011/EU contains the list of monomers, additives, and starting substances that can be used to produce the packaging materials and articles of plastic in contact with food. Other materials, such as ceramics and glass, also contain limits of concentration of different compounds (metals and others). However, there are several materials in contact with food such as coatings and additional materials, which are present in the packaging but are not specifically regulated yet. In this context, adhesives, printing inks, biomaterials, and paper and board, among others can be mentioned. This is the case of adhesives, for example, which are present in most of the laminates to form multilayer structures, manufactured either from conventional materials or from biopolymers, and consequently, they supply a lot of chemical compounds to the migration process.

Many migrants such as antioxidants, plasticizers, monomers, and many additives can be common to several materials, but they do not behave in the same way as they play different roles. In some cases, the final result is different because of the mentioned interactions with other additives present in the matrix or even with the food or food simulant. That is why, when applying the migration studies, it is important to link the study to the material or application on which they have been applied to.

It is obvious that there is still a lot of work to do, as food contact materials are in continuous evolution. In the last 10 years, emerging technologies, such as active packaging and intelligent packaging, burst into the market. More recently, the interest for sustainable materials also increased, and new biopolymers and edible materials for being used in the food sector appeared. Although it is clear that all the materials have to be safe and have to comply with the existing legislation, specific migration tests should be applied to all materials before launching them into the market. Although biopolymers and edible packaging materials are not included in the plastics regulation, they are under the Frame Regulation 1935/2004, which implies that any material or article in contact with food have to be safe and cannot transfer any component to the food at the level that it could endanger the consumer health or change the properties of the packaged food. In this context, specific migration analysis is shown as the best way to demonstrate the safety in the use of these materials.

Migration can be measured in two different ways: global (overall) migration and specific migration. Overall migration is measured by gravimetric analysis, and it does not take into account the chemical nature of the migrants because only the total mass transferred from the packaging to the food is measured. This measure does not provide indication of safety in the use of the material and, in general, is very rare that overall migration is surpassed, as the maximum limit is 60 mg/kg of food. Specific migration means the identification and quantitation of every compound coming from the material, which means any migrant.

Analysis of specific migration is a very difficult task that requires a sophisticated and top-level analytical laboratory and a great experience in this frame of work. The main reasons for these requirements are as follows: (a) it is usually a blind analysis, where the composition of the material is unknown; (b) there are more than 1000 substances in the positive list, many of them with very low specific migration limits; and (c) impurities, degradation compounds, by-products, and in general nonintentionally added substances (NIAS) have to be identified at 10 ppb level in food, which is the maximum migration level accepted for the nonlisted substances in the EU legislation.

For all these reasons, the industry, the administration, and the control laboratories are concerned about the real risk assessment of the materials in contact with food. However, to apply the risk assessment the identification of compounds is required. Once identified, the toxicity of each compound can be classified according to Cramer list (Cramer et al. 1978; Ideaconsult Ltd. 2011) when no toxicity data exist. Based on the Cramer proposal the specific migration limit can be estimated for those non-listed compounds and NIAS. This chapter describes the main features of specific migration analysis for plastics, including both monolayer, multilayer multicomponent materials and biomaterials.

25.2 Migration Tests

The first step to face is the decision about the migration test. The European legislation (Regulation 10/2011/EU) gives the table of temperature and exposure time values, which should be applied in each case, depending on the intended use and the shelf life of the packaged product. To cover all possibilities, three migration tests are required. Food simulants are also defined in the Regulation, and it is a decision of the analyst to choose the most appropriate migration tests. However, there are multiple variables that remain undefined and left to the criterium of the analyst. The following ones can be pointed out:

1. Total immersion or one-layer tests. Total immersion can be used when the monolayer is tested. If the thickness of the specimen is higher than 300 μm, the total surface including all layers will be taken into account for calculating the migration values. One-layer test should be used when multilayer and multicomponent have to be tested. In this case, special migration cells or thermosealed pouches made from the material to be tested can be used.

2. Simulants to select when screening of migrants are required. Nontarget analysis demands a deep screening of migrants. Aqueous simulants can be extracted and processed by different analytical procedures, involving microextraction techniques and further chromatographic analysis with mass spectrometers as detectors. However, vegetable oil, which simulates the fatty food, cannot be analytically screened for searching unknown organic migrants at trace level. Most of the procedures used to eliminate the fatty acids from the oil also destroy the

migrants, and the screening is not possible. For this reason, screening of migrants and NIAS identification are always carried out in food simulants, in either Tenax (solid simulant) or liquid simulants such as isoctane or 95% ethanol for fatty food, 50% ethanol, 20% ethanol, or 10% ethanol.

3. When high temperature has to be tested, at 121°C, 131°C, or 175°C, to simulate the different processing conditions of food, liquid simulants are difficult to use, as they are not in liquid state at these conditions. Overpressure should be applied to maintain the simulant in liquid state. Otherwise, the vapor of the simulant could permeate through the material and extract the components from the internal layers where adhesives and other components are present. In this case, it is really an extraction more than migration test, and the concentration values of migrants would be overestimated. Retort or sterilization conditions can be simulated by applying 175°C for 2 h using Tenax as simulant or 131°C for 2 h using ethanol 95% under pressure.

4. Tenax (polyphenylene oxide) is recommended for high-temperature applications and to simulate dried foods, but for nonvolatile migrants, the migration can be underestimated. In these cases, the best solution is to confirm the migration values using the food intended to be in contact with this material (Canellas et al. 2014). But this is not always possible or feasible. Worst conditions can be simulated at 175°C for 2 h, as were mentioned earlier.

5. Migration test should be done to simulate the real exposure of the material to the food. This means that cutting the material in small pieces and placing the pieces in total immersion mode in the simulant will overestimate the migration. The border influence is very high, and although migration takes place in all directions, the surface exposed to the simulant is much higher than that in the real application. In these conditions, the migration values obtained cannot be considered as representative of the real application.

6. The exposure can be done using less volume of simulant than in the real application to get a more concentrated solution so that the detection limits of the analytical procedure further applied were easily surpassed. But this strategy is only to facilitate the analysis, and the final values should be referred to either the real application or the 6 dm^2 in contact with 1 kg of food, which are the criteria of surface to volume established by the EU legislation.

7. Biopolymers are often noncompatible with liquid simulants, and in this case, Tenax or the same food intended to be packed have to be used.

8. Edible packaging is an exception and additional considerations have to be taken into account. As an edible material, the food regulation should be applied, as the material is eaten together with the food. Then, migration issues do not make sense for these materials, as everything they contain has to be approved as a food or food additive, and consequently, the maximum concentration limits established by the food legislation have to be respected.

25.3 Sample Treatment

Once the exposure from the migration test is finished, the simulant or the food used in the test has to be analyzed. There are only a few examples in which the simulant can be directly analyzed, as, often, a concentration step to reach the detection limits of the migrants and to screen the simulant at 10 µg/kg of food (10 ppb) level is required. Solid phase extraction (SPE) (Aznar et al. 2009) microextraction techniques, such as solid phase microextraction (SPME) (Canellas et al. 2010), liquid-phase microextraction (LPME) (Pezo et al. 2007), and stir bar, are more and more applied in this case. Volatile compounds can be analyzed by headspace (HS) either static or dynamic (Purge & Trap, P&T) (Nerín et al. 1995, 1998) modes coupled to gas chromatography. Mass spectrometry (MS) is required to identify the compounds when screening the migrants, but other detectors such as FID can be used for target analysis and quantitative purposes. SPME is also applied to aqueous simulants when elimination of interferences and simultaneous concentration of the sample are required. A good example of SPE application is the determination of specific migration of primary aromatic amines (PAAs), where their specific migration limit (SML) is 10 ppb in food as a sum of all the PAAs present in the sample (Aznar et al. 2009; Pezo et al. 2012).

In general, working with biopolymers or with conventional polymers is not different in terms of sample treatment, as the simulants, either aqueous or organic solvents, such as ethanol 95%, are similarly treated. It is true that the lack of compatibility of organic solvents with some biopolymers makes them unsuitable for being used as food simulants, and in this case, the solid simulant Tenax can be employed. After the exposure, Tenax is extracted using an organic solvent, usually with acetone or methanol. This extraction has been already optimized for a wide series of migrants (Vera et al. 2011), and the best procedure resulted in applying two consecutive extractions with acetone. But migration into Tenax is not efficient for polar compounds, and migration of nonvolatile compounds can be underestimated, as was mentioned earlier. In this case, the best solution is to take the real food for migration analysis with target migrants. Screening of NIAS and potential migrants is not possible with real food, as the complex matrix overlaps the signals of many compounds at trace level, and it is not possible to confirm the identification and quantitation.

25.4 Analysis of Inorganic Migrants

Biopolymers and oxobiodegradable polymers appear in the market as sustainable food packaging materials. All of them may contain low concentration of toxic metals, such as arsenic, cadmium, lead, and mercury, and other metals such as cobalt or iron as an example. The presence of toxic metals is usually nonintentional, and the likely origin is the contaminated soil where the vegetables from which the biopolymer is produced are grown. For example, sugar can be contaminated by arsenic coming from India or China, where the soil in some geographic areas contains a high amount of arsenic. The same can happen with polysaccharides obtained from sugarcane or sugar beet. Cobalt and iron derivatives are commonly used as catalysts in oxobiodegradable polymers to facilitate the degradation of conventional polymers, although these are not biopolymers.

The analysis of metals in these materials requires the previous acidic digestion to dissolve the material, as one extraction is not applicable in this case. Further analysis by inductively coupled plasma–mass spectrometry (ICP-MS) is recommended, to do the qualitative and quantitative analysis. When applying the migration tests, the simulant that represents the worst scenario possible, which means the worst case of migration, is the simulant B, 3% acetic acid in distilled water. After the exposure, the simulant can be directly analyzed by ICP-MS without additional sample treatment, for all the metals except for arsenic and mercury. Arsenic requires the hydride generation step before ICP or atomic absorption spectrometry, and for mercury, the cold-vapor technique and atomic fluorescence are recommended.

25.5 Analysis of Volatile Organic Migrants

Most of the monomers are either gases or very volatile compounds. They can be easily analyzed by HS, directly applied to the simulant. In the case the oil used as simulant for fatty foodstuffs and for Tenax used as simulant for dried food, the simulant after the migration test can be directly analyzed by HS–GC–MS. However, the SML values established for some of the monomers are as low as 1 ppb in food or food simulant. In this case, direct HS is not enough to surpass this low quantification limit, and additional concentration has to be applied. SPME has shown a very good performance in these cases and can be recommended for this kind of analysis. SPME can be used in either HS or total immersion modes. SPME is HS mode is recommended for analysis of volatile compounds in oil, Tenax or viscous food after the migration tests. For aqueous simulants, total immersion mode is recommended. This is a clean technique, as the solvent is not injected into the chromatographic system; no solvent delay is required when using MS as detector and full chromatogram can be obtained. However, careful optimization of SPME conditions, such as the type of microfiber appropriate for the analysis, temperature, and time conditions for extraction and desorption as well as the chromatographic separation conditions and the MS parameters, is required.

Semivolatile migrants in liquid simulants can be also extracted and concentrated using the LPME techniques (Pezo et al. 2007; Rodríguez et al. 2008; Salafranca et al. 2009; Costa Oliveira et al. 2014). The small volume of a few microliters resulting from the microextraction device of solvent extraction can be easily injected and analyzed by GC–MS.

25.6 Analysis of Nonvolatile Compounds

The main problem to point out with the analysis of nonvolatile migrants is the lack of sensitivity of most of the analytical techniques. Concentration of the sample by either solid phase extraction (SPE), SPME, LPME, or just evaporation of the solvent is required. Target analysis is easier because we have to focus on the analytical effort on specific compounds to be analyzed. However, untarget analysis and the identification of unknowns and NIAS are really difficult. Several proposals have been published. A recent review dealing with the analysis of NIAS (Nerin et al. 2013) proposes a scheme (Figure 25.1) of the analytical options for this purpose. Of course, the identification of migrants demands the use of high-resolution MS detectors, and it can be emphasized as well that a great experience in working with packaging materials is also appreciated for being successful. Otherwise, the identification of the molecular structure and then the specific migrant among the great number of candidates for each molecular structure found in the simulants can be a nightmare. There is not a library available, as happens in GC–MS, and it is the analyst who makes his or her own library. For this reason, the experience here, more than in other areas, is very valuable. Software tools are also a great help and play a critical role for identification purposes in UPLC–MS–Q–TOF.

Most of the specific migration tests provide a series of unknown compounds coming from the packaging material. Of course, the unknowns and NIAS, as were mentioned earlier, have to be analyzed in food simulants as the food is a very complex matrix where organic or inorganic traces cannot be easily distinguished from the food and later analyzed. The "forest of peaks" name is usually applied to the series of chromatographic peaks that appear when specific migration analysis is carried out. A strategy for the identification of unknowns has been recently published and discussed in different forms. Figure 25.2 shows the strategy.

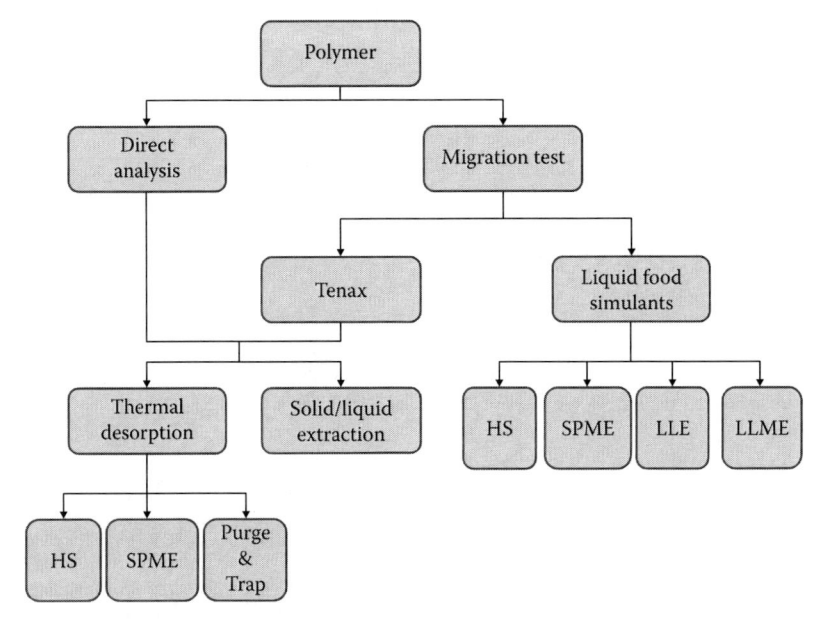

FIGURE 25.1 Scheme of analytical procedures. HS, Headspace; SPME, solid phase microextraction; LLE, liquid–liquid extraction; LLME, liquid–liquid microextraction. (From Nerin, C. et al., *Anal. Chim. Acta*, 775, 14, May 2, 2013.)

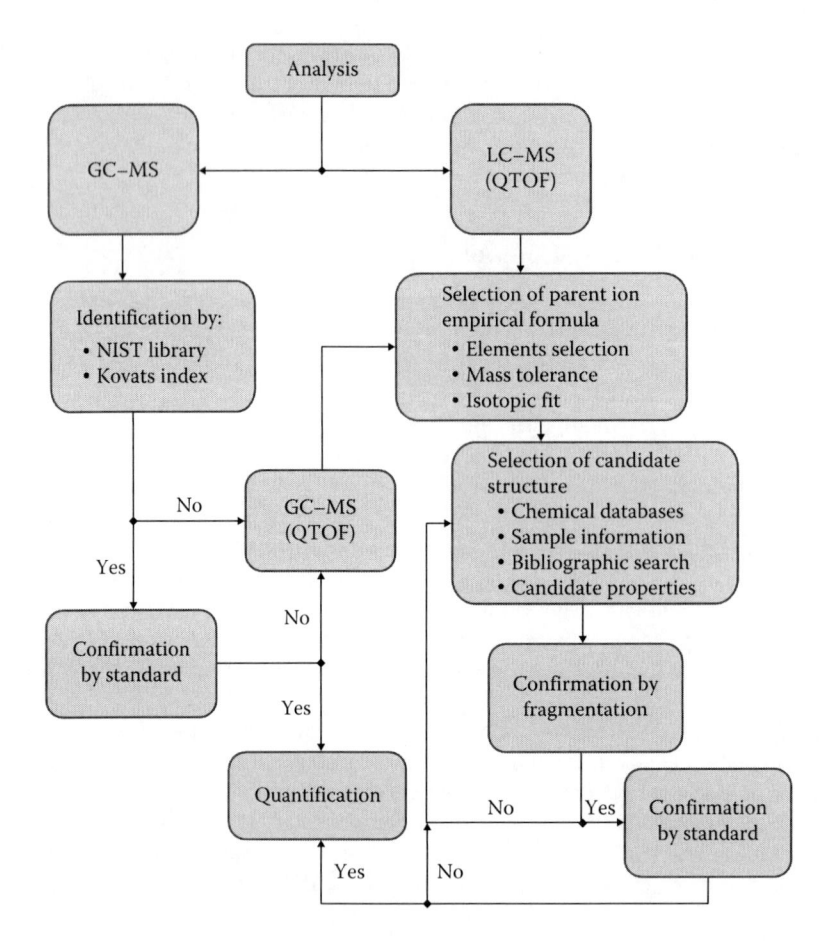

FIGURE 25.2 Analytical strategy for identification of unknowns. (From Nerin, C. et al., *Anal. Chim. Acta*, 775, 14, May 2, 2013.)

25.7 Challenge of Biopolymers in Migration Tests

Among the biopolymers, we can distinguish between those produced from sustainable sources, such as bioethylene, where the ethylene used as monomer comes from the ethanol obtained from sugarcane or other vegetable or residue, and those biopolymers produced by fermentation of residues or cereals, such as polylactic acid (PLA), polyhydroxyalkanoates (PHA), and starch or cellulose (Mahalik and Nambiar 2010; Peelman et al. 2013).

Most of the publications dealing with biopolymers deal with their barrier and mechanical properties, as these are the most important properties in terms of functionality. But there are no publications dealing with specific migration analysis. Two reasons can be pointed out to explain this situation. The first one is the impression that as biomaterials coming from natural sources, migration of substances from the packaging material is not an issue. The second one is that most of the additives used to enhance their properties are permitted without specific restriction. The legislation concerning food contact materials and more specifically that for plastics exclude the biopolymers, although the frame Regulation 1935/2004 affects all materials and articles in contact with food.

Some biopolymers are not compatible with liquid simulants, as the mutual interaction makes the simulant unsuitable for the large exposure. This is the case of starch, cellulose, and their derivatives, where Tenax is the appropriate simulant for migration tests. Other biopolymers such as PLA or PHA behave similarly as conventional polymers and liquid simulants can be used, but in any case, the compatibility has to be tested before the exposure. After the migration tests, there are no differences between the analytical procedures applied to conventional polymers for specific migration analysis.

REFERENCES

Aznar, M., Canellas, E., and Nerín, C. Quantitative determination of primary aromatic amines by cationic exchange solid phase extraction and UPLC-MS. *J. Chromatogr. A*, 1216 (2009): 5176–5182.

Canellas, E., Aznar, M., Mercea, P., and Nerín, C. Partition and diffusion of volatile compounds from acrylic adhesives used for food packaging multilayers manufacturing. *J. Mater. Chem.*, 20 (2010): 5100–5109.

Canellas, E., Vera, P., Domeño, C., Alfaro, A.P., and Nerín, C. Atmospheric pressure gas chromatography coupled to quadrupole-time of flight mass spectrometry as a powerful tool for identification of non intentionally added substances in acrylic adhesives used in food packaging materials. *J. Chromatogr. A*, 1235 (2012): 141–148.

Canellas, E., Vera, P., and Nerín, C. Atmospheric pressure gas chromatography coupled to quadrupole-time of flight mass spectrometry as a tool for identification of volatile migrants from autoadhesive labels used for direct food contact. *J. Mass Spectrom.*, 49 (2014): 1181–1190.

Commission Regulation (EU) No. 10/2011 of 14 January 2011 on plastic materials and articles intended to come into contact with food (2011).

Costa Oliveira, É., Echegoyen, Y., Cruz, S.A., and Nerin, C. Comparison between solid phase microextraction (SPME) and hollow fiber liquid phase microextraction (HFLPME) for determination of extractables from post-consumer recycled PET into food stimulants. *Talanta*, 127 (2014): 59–67.

Cramer, G.M., Ford, R.A., and Hall, R.L. Estimation of toxic hazard—A decision tree approach. *J. Cosmet. Toxicol.*, 16 (1978): 255–276.

Ideaconsult Ltd. Toxtree—Toxic Hazard Estimation by decision tree approach. Ideaconsult Ltd., Sofia, Bulgaria (2011). http://toxtree.sourceforge.net/, Accessed March 24, 2011.

Mahalik, N.P. and Nambiar, A.N. Trends in food packaging and manufacturing systems and technology. *Trends Food Sci. Technol.*, 21(3) (2010): 117–128.

Nerin, C., Alfaro, P., Aznar, M., and Domeño, C. The challenge of identifying the non intentionally added substances (NIAS) from food packaging materials. A review. *Anal. Chim. Acta*, 775 (2013): 14–24.

Nerín, C., Rubio, C., Cacho, J., and Salafranca, J. Determination of styrene in olive oil by an automatic purge-and-trap system coupled to gas chromatography–mass spectrometry. *Chromatographia*, 41(3/4) (1995): 216–220.

Nerín, C., Rubio, C., Cacho, J., and Salafranca, J. Parts-per-trillion determination of styrene in yoghurt by purge-and-trap gas chromatography with mass spectrometry detection. *Food Addit. Contam.*, 15(3) (1998): 346–354.

Peelman, N., Ragaert, P., De Meulenaer, B. et al. Application of bioplastics for food packaging. *Trends Food Sci. Technol.*, 32(2) (2013): 128–141.

Pezo, D., Fedeli, M., Bosetti, O., and Nerín, C. Aromatic amines from polyurethane adhesives in food packaging: The challenge of identification and pattern recognition using Q-TOF/MSE. *Anal. Chim. Acta*, 756 (2012): 49–59.

Pezo, D., Salafranca, J., and Nerín, C. Development of an automatic multiple dynamic hollow fiber liquid-phase microextraction procedure for specific migration analysis of new active food packagings containing essential oils. *J. Chromatogr. A*, 1174 (2007): 85–94.

Regulation (EC) no. 1935/2004 of the European Parliament and of the Council of 27, October 26, 2004.

Rodríguez, A., Pedersen-bjergaard, S., Rasmussen, K.E., and Nerín, C. Selective three phase liquid phase microextraction of acidic compounds from foodstuff simulants. *J. Chromatogr. A*, 38–440 (2008): 1198–1199.

Salafranca, J., Pezo, D., and Nerín, C. Assessment of specific migration to aqueous simulants of a new active food packaging containing essential oils by means of an automatic multiple dynamic hollow fiber liquid phase microextraction system. *J. Chromatogr. A*, 1216 (2009): 3731–3739.

Vera, P., Aznar, M., Mercea, P., and Nerin, C. Study of hotmelt adhesives used in food packaging multilayer laminates. Evaluation of the main factors affecting migration to food. *J. Mater. Chem.*, 21(2) (2011): 420–431.

Vera, P., Canellas, E., and Nerín, C. Migration of odorous compounds from adhesives used in market samples of food packaging materials by chromatography olfactometry and mass spectrometry (GC–O–MS). *Food Chem.*, 145 (2014): 237–244.

26

Edible Films and Coatings: Sensory Aspects

Kezban Candoğan, Gustavo V. Barbosa-Cánovas, and Emine Çarkcıoğlu

CONTENTS

26.1 Introduction

The food industry is looking for environmentally friendly technologies and sustainable food production to address expectations from consumers and regulatory agencies and, at the same time, project the best possible image in a very competitive environment. In this context, new packaging techniques have been introduced to replace or complement existing techniques. Edible films and coatings offer an alternative to commercial packaging practices, and they are receiving attention from the food industry because they offer some appealing advantages. Edible films and coatings are applied to food mainly to improve quality and to extend shelf life by protecting against contamination and thus, providing biochemical and microbial surface stability by acting as a barrier to moisture, oxygen and other gases, fats, and oils and as a carrier for active compounds such as antioxidants, antimicrobial and flavoring agents, pigments, and nutrients. While maintaining product freshness and its quality attributes, physical and sensory properties are also expected to be improved by the use of these covers (Pascall and Lin 2013, Pavlath and Orts 2009).

Scientific and technological innovations targeted at improving quality of life are not always easily adopted by consumers because of misperception of new technologies (Barrena-Figueroa and Garcia-Lopez-de-Meneses 2013, Deliza et al. 1999). In order to facilitate widespread adoption, the new technology should meet consumers' expectations related to, for example, price, convenience, nutritional quality, microbiological safety, methods of production, agrochemical residues, and environmental impact, as well as sensory quality attributes such as appearance, taste, flavor, and texture (Carel et al. 2012, Deliza et al. 2003, Nychas et al. 2008). Sensory aspects are crucial to guarantee the success of a given product, even if the scientific or technological innovation applied to the food conforms to nutritional and safety requirements (Grunert 2005, Ronteltap et al. 2007). To this end, when developing a new technology, process, or product, possible effects on sensory characteristics should be taken into account and evaluated to make sure consumer acceptability is not compromised (Valentin et al. 2012).

The use of edible films and coatings has drawn great attention over the last few decades as an innovative packaging strategy offering new and alternative ways of protecting foods and improving product quality. Although some types of edible films have been used for commercial purposes, most attempts on this topic have still been limited as research applications.

Edible films could be applied to the foods in two ways: (1) coating of the food by dipping, spraying, brushing, or panning and then by drying or (2) wrapping the food with the casted and dried films (Dhall 2013). Active compounds such as antimicrobials, antioxidants, antibrowning agents, and vitamins might be included in these films to enhance their functionality (Dhall 2013, Silva-Weiss et al. 2013). Whether in the form of dried freestanding film or in liquid form, with or without active agents, since it constitutes a part of the food, the film should be compatible with the product it encloses and should not negatively alter sensory properties of the food (Kim and Ustunol 2001). At the same time, sensory properties of the film (or coating) itself are important to show its level of consumer acceptance (Ozdemir and Floros 2008). In this chapter, edible films and coatings are discussed in terms of consumer acceptance by analyzing their sensory properties and how they influence final product to which they are applied.

26.2 Consumer Attitudes Related to Sensory Properties of Edible Films/Coatings

Sensory evaluation acquires the information of certain organoleptic properties of food products under consideration by using perceptions from different senses. This information is gathered by measuring, analyzing, and interpreting data from the reaction of consumers to the tested products (Lawless and Heymann 2010, Valentin et al. 2012). Organoleptic properties that are considered in sensory analysis of food products for characterization of consumer choice and acceptability are mainly texture, taste, flavor, appearance, color, odor, and aroma that can be evaluated by different methods such as hedonic, discriminative, and descriptive tests. By using these tests, not only are the effects of the new formulations, processes, or technologies on sensory attributes of the food products determined, but changes in sensory quality during storage and due to packaging processes are also of interest to food technologists.

Recent research has focused on evaluating consumers' opinions, beliefs, attitudes, and purchase intentions toward nonconventional technologies, as well as innovations on food products (Chen et al. 2013, Deliza et al. 2003, Wan et al. 2007). As an innovative packaging system, application of edible films and coatings on the surface of food products needs to be tested in terms of sensory properties in addition to their effects on microbiological and chemical properties, as well as functionality. However, not many detailed studies have been performed on the influence of edible films and coatings on sensory properties of food products. A focus group study was conducted by Wan et al. (2007) to investigate consumer attitudes, opinions, and concerns about novel food processing or packaging technologies, mainly, edible films and coatings. It was suggested that testing basic chemical and physical properties of edible films is not enough; the analysis of sensory properties of edible film applied food products is necessary. Evaluating sensory properties, proper labeling, and advertising these products with an emphasis on marketing the natural ingredients added were determined as steps that should be taken into account for commercializing edible films and coatings for food applications. It was also noted that sensory attributes of the coated products, together with safety, the types of products that are coated, cost, and perceived benefits were important factors affecting purchase intent for coated products (Figure 26.1).

It is obvious that knowledge of sensory properties and consumer demand for foods is crucial in order for a new technology or product to become successful and to survive in the market (Valentin et al. 2012). In this regard, important sensory attributes for food that should be considered when developing an edible film or coating becomes a crucially relevant issue. Although flavor is considered as the most important

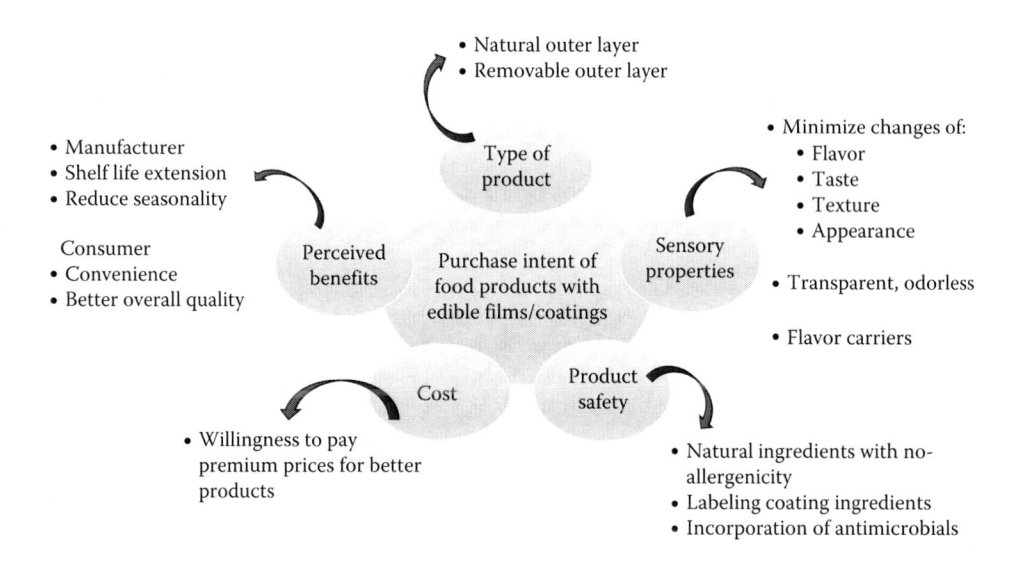

FIGURE 26.1 Factors affecting the purchase intent of foods covered with edible films/coatings and desirable characteristics of those factors identified by consumers groups. (Adapted from Wan, V.C.H. et al., *J. Sens. Stud.*, 22(3), 353, 2007.)

sensory attribute for many foods, other sensory properties such as appearance and texture become, for some specific products, particularly important when edible films or coatings are applied. Indicative organoleptic characteristics for foods that are supposed to change when an edible film or coating is applied are listed in Table 26.1.

Food structure and texture are affected by the characteristics and distribution of water in products (Tajner-Czopek et al. 2008). One of the main purposes of edible films is to control moisture transfer, in the form of either moisture uptake or moisture loss, depending on the specific product's characteristics. For instance, by using edible films, moisture loss is controlled to maintain juiciness in meat products and firmness in fruits (Fernández-Pan et al. 2011, Gennadios et al. 1997). On the other hand, it is possible to retard hydration of dried foods with edible films and thus to retain desirable textural characteristics such as crispness and crunchiness in the final product (Fernández-Pan et al. 2011). This could be ensured by the proper selection of moisture permeability properties of the edible films.

In fresh-cut fruits and vegetables, appearance and color attributes, as well as texture, are important sensory attributes to be considered. Enzymatic browning is one of the major problems in some fruits that could be minimized with the use of edible films. Also, foods with high-fat content, particularly those containing high polyunsaturated fatty acids, are subject to lipid oxidation, which might impart rancid flavor and also to negative change in the textural characteristics of the products. Furthermore, in heterogeneous products, that is, foods containing different ingredients with different water activities, internal water transmission between the moist and dry components leads to deleterious changes in sensory properties. Examples are some baked food products (i.e., cakes with dried fruits, frozen pizza), cereal with raisins, ice cream containing cookies, and/or candies. This problem also applies to composite confectionery products with nut-containing centers, such as coated biscuits and nut inclusions where oil migration might result in negative effects including deterioration of sensory quality (Debeaufort and Voilley 2009, Pavlath and Orts 2009, Ziegler et al. 2004). Application of edible films is a way to prevent these negative changes due to oil or moisture transmission through the various layers of the product (Guilbert 1986, Quezada-Gallo 2009). Selection of edible films considering the barrier requirements (appropriate water and gas transmission rates) for the particular products could provide protection and help maintain the desirable sensory characteristics of the product for longer periods of time.

TABLE 26.1

Sensory Attributes[a] That Should Be Considered When Developing Edible Films and Coatings for Food Applications

Sensory Attribute	Specific Attribute	Relevant Food	References
Appearance	Color	All foods	Castricini et al. (2012), Chidanandaiah et al. (2009), Chien et al. (2007), Gao et al. (2013), Kanmani and Rhim (2014), Oms-Oliu et al. (2008), Soukoulis et al. (2014), and Wu et al. (2013)
	Brightness	Fresh fruits and vegetables	Del Valle et al. (2005), Moreira et al. (2011), and Tanada-Palmu and Grosso (2005)
	Browning	Fresh-cut fruits and vegetables	Lee et al. (2003) and Moreira et al. (2011)
	Glossiness	Fresh fruits and vegetables, sliced cured meat products	Aloui et al. (2014), Han et al. (2005), and Lee and Min (2013)
	Shriveled	Citrus fruits	Machado et al. (2012)
	Wrinkling, ruggedness, flaccidity	Dried nuts, fruits and vegetables	Krasniewska et al. (2014)
Texture	Mechanical properties Crunchiness, crispness, hardness, firmness, cohesiveness, bite strength, tenderness, gumminess	Dried foods, muscle foods, fresh fruits and vegetables	Fernández-Pan et al. (2011), Kim et al. (2012), Mehyar et al. (2012), Lee et al. (2003), Lee and Min (2013), and Oms-Oliu et al. (2008)
	Geometrical properties Smoothness	Shelled eggs, dried nuts	Caner (2005) and Mehyar et al. (2012)
	Moisture properties Oiliness, juiciness	Muscle foods, fresh fruits	Fernández-Pan et al. (2011) and Kim et al. (2012)
Flavor	Aromatics Odor, aroma	All foods	Aloui et al. (2014), Garcia et al. (2010), Gniewosz and Synowiec (2011), Krasniewska et al. (2014), Lee and Min (2013), Lu et al. (2009), Moreira et al. (2011), and Vargas et al. (2006)
	Taste related Overall taste, acidity, sourness, sweetness, aftertaste	Fermented foods, muscle foods, fruits and vegetables	Aloui et al. (2014), Chien et al. (2007), Han et al. (2005), Gao et al. (2013), Kim et al. (2012), Tanada-Palmu and Grosso (2005), and Rodríguez et al. (2003)
	Chemical feelings related Astringency, tartness	Fermented foods, fruits and vegetables	Han et al. (2005) and Rodríguez et al. (2003)
	Overall flavor, flavor strength/ intensity	All foods	Ali et al. (2010), Aloui et al. (2014), Chidanandaiah et al. (2009), Contreras-Oliva et al. (2012), Gao et al. (2013), Garcia et al. (2010), Kim et al. (2012), Lee and Min (2013), Tanada-Palmu and Grosso (2005), and Viña et al. (2007)
	Off-flavor, rancid flavor	All foods	Bugnicourt et al. (2013), Contreras-Oliva et al. (2012), Rodríguez et al. (2003), Valencia-Chamorro et al. (2010), and Weerasinghe et al. (2013)

[a] General classification of the sensory attributes is based on Meilgaard et al. (2006).

26.3 Sensory Properties of Edible Films

Edible films or coatings have been defined as "any type of material used for enrobing i.e., coating or wrapping, selected foods to extend their shelf life where the cover might or might not be eaten" (Pavlath and Orts 2009). According to this definition, because edible films and coatings are eventually going to be a part of the food sensory properties of the film in addition to its technical characteristics become an important issue (Kim and Ustunol 2001, Ozdemir and Floros 2008). A summary of the expected sensory characteristics of the edible films or coatings to be taken into consideration for food applications is presented in Table 26.2.

In a study by Wan et al. (2007), sensory quality of coated food products was considered as one of the most important factors toward purchase intent of the consumer. In the focus group study performed by these researchers to address consumer attitudes, opinions, and concerns toward edible films and coatings, the majority of the panel stated that coating should be transparent and should not possess taste or odor affecting sensory quality of the product where it is applied with the exception of new food product development ideas for applying coatings as flavor carriers.

Film thickness is also a primary factor influencing sensory properties of the film itself and the food product, which could alter appearance and taste depending on the rate of moisture and gas transmission (Pavlath and Orts 2009). When the film is supposed to be consumed as a part of the food, a thinner film would be preferred from the sensory point of view because in this way, usefulness of the film is improved due to its less perceptible nature when the thickness is decreased (Ciannamea et al. 2014, Longares et al. 2004).

Transparency or opacity, which is as a result of morphology of a polymer, is related to the edible film appearance and plays a significant role in consumer acceptability (Kim and Ustunol 2001). Opacity of the film influences the appearance of the wrapped product. Fan et al. (2014) developed composite films with kidney bean protein isolate and chitosan using ultrasonic pretreatment to produce flexible films as controlled release carriers for nisin. Film opacity increased with increasing level of chitosan in the composite films. Kim et al. (2014) noted that in composite films with tapioca starch and pullulan, with increasing levels of pullulan, thickness was increased while transparency decreased; indicating a significant negative correlation between transparency and thickness, in that starch addition resulted in thinner and more transparent films. The most important factor affecting film thickness and transparency is the amount of the solids in the film formulation (Kim et al. 2014, Mali et al. 2005).

TABLE 26.2

Expected Sensory Characteristics for an Edible Film to Be Applied onto Foods

Sensory Attribute	Expected Sensory Attribute for Edible Films
Color	Colorless
Odor	Odorless
Taste	Tasteless
Stickiness	Not sticky
Adhesiveness	Adhesive
Appearance	
Smoothness, surface roughness	Smooth surface
Gloss	Glossy
Opacity	Transparent
Thickness	Thin film
Homogeneity	Homogenous

Sources: Modified from Kim, S.J. and Ustunol, Z., *J. Food Sci.*, 66(6), 909, 2001; Longares, A. et al., *Food Sci. Technol.*, 37(5), 545, 2004; Wan, V.C.H. et al., *J. Sens. Stud.*, 22(3), 353, 2007; Ozdemir, M. and Floros, J.D., *J. Food Eng.*, 86(2), 215, 2008; Park, J.W. et al., *Food Sci. Technol.*, 41(4), 692, 2008; Wu, J. et al., *Food Hydrocoll.*, 30(1), 82, 2013.

At the same time, Ciannamea et al. (2014) stated that physical properties of films such as opacity, as well as some tensile and barrier properties, are thickness dependent. These researchers also evaluated the effects of processing methods such as compression molding and solution casting during soybean protein concentrate–based film manufacturing plasticized by glycerol depending on molding temperature and pressure, drying conditions, and glycerol level as important factors affecting physical and mechanical properties and thus visual properties. They reported that films produced by compression molding were visually homogenous, transparent, and slightly yellowish, possessing a smooth surface. Films manufactured by solution casting were more opaque, having rougher surfaces and displaying bubbles in some cases. Similar results were previously noted by Park et al. (2008) who investigated differences in the properties of gelatin-based edible films produced by different methods (extrusion, casting by pouring, and extrusion combined with heat press) and plasticized with glycerol, sorbitol, or the mixture of both. They noted that the heat press method improved surface roughness and opacity of the films, thus smoothing the surface and increasing transparency.

Color is another relevant attribute influencing general appearance, and thus the sensory appeal and consumer acceptance of an edible film when it is used to enclose a food product. The method of manufacturing the edible films is an important factor affecting its ultimate color. Soybean and whey protein films manufactured with dry methods exhibited greater coloration (higher b^* values) as compared with those produced by casting (Kanmani and Rhim 2014, Wu et al. 2013). However, in a study by Ciannamea et al. (2014), it was postulated that compression molding and solution casting gave rise to similar b^* values close to 13, while films produced by compression molding had lower L^* value and higher a^* value than those produced by solution casting.

Optical properties of edible films are influenced by the film's manufacturing conditions. For instance, Gómez-Estaca et al. (2014) evaluated effects of pH (pH 2 and 11), thermal treatment (heating at 80°C), and the addition of a natural cross-linker (cinnamaldehyde) on the optical properties of shrimp protein isolate edible films. The results showed that films produced at pH 11 were more yellowish and had more intense color than those produced at pH 2. Application of heat treatment and higher pH values resulted in higher opacity. Incorporation of active agents might also alter optical characteristics of the films. For instance, the addition of antimicrobial zinc oxide nanoparticles into film formulations of agar, carrageenan, and carboxymethylcellulose resulted in lower lightness and higher greenness and yellowness of the films (Kanmani and Rhim 2014). Bitencourt et al. (2014) reported that when curcuma ethanol extract was incorporated into gelatin edible films in different concentrations to make an antioxidant active film, changes in color parameters were observed where the L^* value decreased and a^* and b^* values increased with increasing level of curcuminoid pigments. Wang et al. (2013) confirmed that chitosan edible films incorporated with tea polyphenols possessing natural pigments as coloring compounds could change the color parameters as compared with the control film.

The level of silver addition and molecular weight of the active agent incorporated into the formulation of the films could affect the appearance of the edible films. Regiel et al. (2013) tested some variables to optimize the bactericidal properties of chitosan films incorporated with silver nanoparticles. The results of this study indicated that the appearance of the films was changed depending on different levels of silver incorporation, as well as on the molecular weight of chitosan. It was found that higher concentrations of silver incorporation yielded darker color in the final film. Higher molecular weight of parent chitosan exhibited lighter color in the resulting films (Figure 26.2). Bitencourt et al. (2014) noted that with increasing curcuma extract concentrations in gelatin edible films, there was a decrease in gloss values; thus, surface roughness in comparison with control film could result from the increase in irregularities in the film surface due to the extract distribution in the matrix.

Wu et al. (2013) investigated the properties of nine different kinds of film solutions prepared with different ratios of pullulan–chitosan and pullulan–carboxymethyl chitosan–blended films by the casting/solvent evaporation technique and showed that chitosan or carboxymethyl chitosan incorporation resulted in more yellowishness in the surface of the blended films. They attributed this color change to increases in the number of the free amino groups with increases in chitosan to pullulan ratio in the film formulation. This formulation enhanced nonenzymatic browning reactions, resulting in more yellowness. In this study, pullulan film, when it is used alone, possessed low transparency due to the visible ripples on the surface of the film, which was attributed to the low viscosity of pullulan film solution

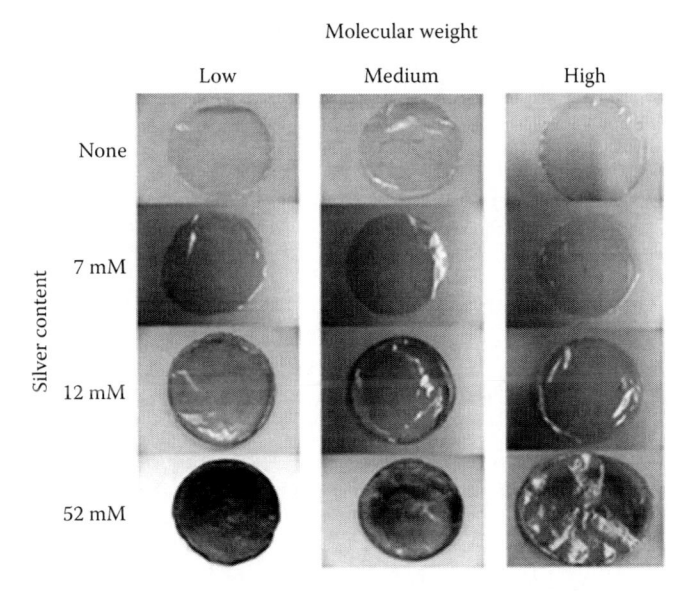

FIGURE 26.2 Appearance changes of chitosan and silver–chitosan films as a function of molecular weight of chitosan and silver content. (From Regiel, A. et al., *Nanotechnology*, 24(1), 2013, doi: 10.1088/0957-4484/24/1/015101.)

readily affected by the wind in the drying oven, whereas blended films possessed similar transparency values, indicating that the level of chitosan or carboxymethyl chitosan in the blends did not exhibit a significant effect on the transparency of the films.

Most studies that provide relevant information on the sensory attributes of edible films were based on instrumental methods. However, a few studies have been conducted on sensory properties of edible films using sensory panels. Kim and Ustunol (2001) evaluated sensory characteristics of emulsion edible films fabricated by using whey protein isolate (WPI) plasticized with sorbitol or glycerol and candelilla wax in a trained sensory panel to evaluate for milk odor, transparency/opaqueness, sweetness, and adhesiveness. While edible films without candelilla wax were clear and transparent, candelilla wax containing films was perceived to be opaque, which can be explained by the visible light scattering effect of candelilla wax particles through the film. While films plasticized with sorbitol were perceived more adhesive than those plasticized with glycerol, films containing candelilla wax were found to be less adhesive. Although whey protein isolate edible films did not exhibit distinctive milk odor, they were slightly sweet, which was attributed to the contribution of glycerol and sorbitol addition into the film formulations.

In another study by Ozdemir and Floros (2008), sensory properties of whey protein films as affected by protein, sorbitol, beeswax, and potassium sorbate concentrations were determined. In terms of film stickiness, the most important factor was beeswax, while the effects of potassium sorbate and sorbitol on stickiness was observed to a lesser extent, and the presence of protein was not an important factor influencing stickiness of the films. Beeswax also had a negative effect on appearance, followed by potassium sorbate, whereas protein and sorbitol did not exhibit any effects on the appearance of the films.

26.4 Sensory Properties of Foods as Affected by Application of Edible Films and Coatings

Edible films and coatings are generally classified based on the main materials used in their manufacture, that is, polysaccharides, lipids, proteins, and their composites. Each material used in the production imparts specific characteristics to the films and to the foods they are enclosing. These effects on the sensory properties of the foods might occur in two ways, that is, effects due to the prevention of deterioration of quality attributes and thus of possible deleterious changes in sensory attributes and negative effects resulting from the main materials used in their production. In the following sections, various types of edible films and coatings are reviewed in terms of their influence on the sensory attributes of food products.

26.4.1 Polysaccharide Films and Coatings

Polysaccharides produced from plants, algae, and microorganisms are capable of forming gels in water and have been utilized for edible film and coating production for application on the surfaces of a variety of food products including fruits, vegetables, meat products, and dairy products (Del Valle et al. 2005, Gennadios et al. 1997). Polysaccharide coatings, being hydrophilic, can act as a perfect barrier for gases and water transfer, thus maintaining high quality of the product for long periods (Del Valle et al. 2005, Krasniewska et al. 2014).

Chitosan has been widely used because it forms good quality films and coatings and it has significant antimicrobial and antioxidant activity. Both water- and acid-soluble chitosans can be used in the production of edible films and coatings. However, since water-soluble chitosan has poor film-forming ability, the acid-soluble form is generally preferred. When concentrations of acid-soluble chitosan higher than 1.0% are used, a high concentration of acid is required for better dissolution, resulting in astringent and bitter notes in the solutions that might limit the practical application of these coatings in real food systems. Furthermore, it has been noted that astringency decreased significantly when the pH of the solution reaches 6.3, which could be achieved by decreasing concentration of chitosan and neutralizing the pH of the solution (Han et al. 2005, Rodríguez et al. 2003). Han et al. (2005) developed three acid-soluble 1% chitosan-based coatings in 0.6% acetic acid solution, in 0.6% lactic acid solution, and in 0.6% lactic acid solution plus 0.2% vitamin E and tested these coatings on fresh strawberries in terms of sensory quality with consumer testing and trained panel descriptive analysis. No significant differences were detected in sourness and sweetness between groups. At the same time, adjustment of the pH of the solution at around 5 significantly prevented the undesirable astringency of strawberries coated with chitosan solutions. Another approach to reduce the astringent flavor due to chitosan film utilization is the addition of oleic acid to the coating formulation. Vargas et al. (2006) investigated the effect of edible coatings prepared with high-molecular-weight chitosan in combination with oleic acid on the quality characteristics of strawberries during cold storage and found that oleic acid improved some characteristics of the coatings; however, the coatings negatively affected the aroma and flavor of the strawberry fruit (less intense aroma and flavor than uncoated control samples), particularly when oleic acid concentration was increased in the film formulation. In addition, an oily aroma was detected in samples with higher content of oleic acid.

In a study conducted by Caner (2005), shell eggs were treated with chitosan coatings and sensory quality was evaluated during storage for 4 weeks. In comparison with uncoated control, samples treated with chitosan coatings possessed higher surface glossiness, odor, and stickiness scores while they exhibited similar surface smoothness scores to the uncoated control.

Chien et al. (2007) investigated the possibility of using 0.5%, 1%, or 2% chitosan coatings to maintain quality and extend shelf life of freshly sliced mangoes for a 7-day storage. It was noted that coating with chitosan maintained sensory quality of the mango slices after 7 days without affecting natural taste of the sliced mango. Moreira et al. (2011) noted that chitosan coating applied on broccoli improved sensory quality, while controlling growth of microorganisms present in broccoli in comparison with uncoated samples during refrigerated storage for 20 days. In a qualitative sensory test, color, texture, brightness, floret opening, aroma, and browning attributes were evaluated. In terms of color and brightness, samples coated with chitosan received higher scores than control samples, with no differences in texture and flavor properties. The application of chitosan coating maintained the typical green color of the florets and inhibited florets opening and also enzymatic browning over the refrigerated storage without the presence of undesirable odors. In another study, Han et al. (2014) reported that coatings formulated in aqueous solutions of 0.5% and 1.0% chitosan applied onto sponge gourd improved sensory properties evaluated in terms of appearance, taste, flavor, and acceptability during storage at 25°C with the greater effect when 1.0% chitosan coating was used.

Leceta et al. (2015) conducted a study to assess the efficacy of chitosan-based edible coatings on baby carrots. The film was applied by dipping or spraying the product and keeping it under modified atmosphere packaging during storage for 14 days at 4°C. Sensory evaluation results revealed that product texture after storage was better preserved using chitosan-based films while preventing the whitening of the surface. On the other hand, the intensity of sourness and off-flavor increased during storage where chitosan-coated samples resulted in slightly higher scores than the uncoated control, which was attributed to the astringency increasing effect of chitosan coatings prepared by dissolving in an acid medium as explained by Rodríguez et al. (2003). Coated carrots received similar flavor, sweetness, and overall acceptability scores to the control samples in this study (Leceta et al. 2015).

Gao et al. (2013) studied the effects of coatings prepared with 1% chitosan, 1% glucose, or chitosan–glucose mixture on the postharvest quality of fresh grapes and noted that coated grape samples had significantly higher flavor scores than uncoated samples during a 60-day cold storage. Grapes coated with chitosan–glucose complex received the highest flavor scores at the end of the storage and were perceived as being the most acceptable in terms of appearance, color, flavor, taste, and overall preference.

Petriccione et al. (2015) treated three sweet cherry cultivars with 0.5% chitosan edible coatings and evaluated quality changes during cold storage at 2°C for 14 days and noted that chitosan edible coating application extended postharvest life, improved storability, and enhanced the nutraceutical value of sweet cherry with better sensory attributes as compared to uncoated controls.

Moreira et al. (2011) investigated the effects of chitosan and carboxymethylcellulose with or without previous application of mild heat shock on minimally processed broccoli during storage for 20 days. In qualitative sensory analysis, broccoli samples coated with chitosan edible coating possessed higher sensory scores for all attributes. It was reported that edible coatings with or without mild heat shock treatment exhibited an inhibition effect on floret opening during storage, which is a very important quality indicator for broccoli.

Because edible films and coatings produced from starches are reported to be isotropic, odorless, tasteless, colorless, nontoxic, and biologically absorbable (Nisperos-Carriedo 1994), they would not exhibit adverse effects on the sensory quality of a product. For instance, there was no significant difference in sensory characteristics such as flavor, aroma, and texture between uncoated mango fruits and those coated with cassava starch–based coatings (Chiumarelli et al. 2011). Castricini et al. (2012) evaluated the sensory quality of papaya fruits coated with edible coatings of cassava starch and carboxymethyl starch (CMS) at 1%, 3%, and 5% concentrations during storage for 14 days and found that coatings with 3% and 5% concentrations were effective in maintaining green color of the product. In general, when the concentration of the coatings is high, there is a compromising effect on visual appearance, specifically, denser coating resulting in intense peeling and loss of skin integrity. This result is due to inadequate adhesion and flexibility because of the irregular shape of the product (Viña et al. 2007). Viña et al. (2007) further suggested that in order to overcome this adverse effect, incorporation of plasticizers and lipids to the formulations of the starch coatings should decrease apparent viscosity, which improves flexibility. The attributes of appearance were more affected by cassava starch and CMS coatings than by flavor attributes. Coatings at the concentrations of 3% and 5% maintained green color. CMS at higher concentrations resulted in bitter taste. The coatings, when applied to the fruits, deferred respirations and thus natural ripening, which might be effective for longer storage periods since coated fruits might mature later than uncoated ones.

Food gums are also widely used to produce edible coatings due to good gel-forming abilities. Oms-Oliu et al. (2008) evaluated the effects of alginate-based, pectin-based, and gellan-based edible coatings containing antibrowning agents such as *N*-acetylcysteine and glutathione on maintaining sensory quality and antioxidant properties of fresh-cut pears for 14 days at 4°C. It was observed that water vapor resistance increased and ethylene production was reduced with application of polysaccharide-based edible coatings on fresh-cut pears. In sensory evaluation, coated and uncoated fresh-cut pears received similar scores for odor and firmness at day 1 of storage, while coated pears had higher odor scores after 14 days, which was explained by good retention of volatile compounds in coated fruits. Incorporation of *N*-acetylcysteine and glutathione into coating formulations prevented enzymatic browning and thus decreased color changes during storage.

Alginate-based coatings could be of particular interest when applied to muscle foods to improve quality characteristics including sensory properties. Chidanandaiah et al. (2009) applied alginate coatings incorporated with preservatives on the surface of buffalo meat patties and evaluated quality changes during refrigerated storage at 4°C for 21 days. Sodium alginate coating at 2% level appeared to improve the appearance, overall texture, overall palatability, and color of the patties in comparison with other treatments and controls, while alginate concentration higher than 2% resulted in lower flavor scores. The improvement effect of alginate coating on sensory attributes of meat patties was due, most likely, to the good oxygen barrier properties and lipid oxidation retarding effects. Prevention of warmed-over flavor by application of alginate coatings on ground pork patties was also reported by Wanstedt et al. (1981).

Ali et al. (2010) applied 5%, 10%, 15%, and 20% gum arabic coatings on tomato surfaces during storage at 20°C for 20 days and reported that coating application improved pulp color, texture, flavor, and overall acceptability as compared to uncoated samples, obtaining the best results with 10% coating application. Coatings of 15% and 20% gum arabic resulted in poor pulp color and inferior texture and exhibited off-flavors, but still gave rise to higher scores than the uncoated control. As another example of using gum coatings on tomato, Mahfoudhi et al. (2014) evaluated the effectiveness of coatings from almond gum exudate or gum arabic on some quality characteristics of tomato during storage for 20 days. Coatings delayed changes in physical and chemical properties. At the same time, in sensory analysis, tomatoes with the application of both almond gum exudate and gum arabic edible coatings possessed higher scores for pulp color, texture, flavor, and overall acceptability attributes than for uncoated samples.

Pullulan, a water-soluble polysaccharide, which is produced by the *Aureobasidium pullulans* fungus, exhibits very good film-forming properties as a biodegradable packaging material due to its ability to be mixed with other biopolymers, high oxygen and carbon dioxide barrier properties, thermal stability, elasticity, glossiness, and transparency, in addition to being colorless and tasteless (Farris et al. 2014, Krasniewska et al. 2014, Singh and Saini 2008). Several recent studies have been conducted on the use of pullulan edible films and coatings, particularly for extending shelf life of nonclimacteric and climacteric fresh fruits and vegetables.

Del Valle et al. (2005) applied edible coatings produced from prickly pear cactus mucilage (*Opuntia ficus indica*) to strawberries to prolong shelf life during a 9-day storage. Effects of edible films on the sensory quality of the fruit, including attributes such as visual appearance, color, brightness, texture, taste, and overall acceptability, were tested during 9 days of storage at 5°C. The edible coating application did not have a negative effect on the natural taste of strawberries, with a preference for coated samples at the end of storage period.

26.4.2 Protein Films and Coatings

Proteins possess good film-forming abilities similar to polysaccharides. It has been noted that they exhibit an intermediate oxygen barrier and relatively high water vapor permeability. The hydrophilic nature of protein films constrains their resistance to water vapor permeability (Contreras-Oliva et al. 2012, Cuq et al. 1998). Protein edible films and coatings have been produced using collagen, wheat gluten, corn zein, soy protein, and whey protein (Dangaran et al. 2009, Ustunol 2009); however, examples of their individual use in real food systems are limited. They are usually produced with hydrophobic compounds to improve some quality properties by protecting the food from chemical and microbial deteriorations, thus, by extending the shelf life and also by retarding oil absorption during frying.

Due to their good oxygen barrier properties, protein films and coatings are convenient to use for high-fat foods to protect them from lipid oxidation and thus from rancid off-flavor generation. Whey protein isolate edible coatings delayed oxidative rancidity in peanuts as compared to uncoated samples in sensory rancidity rating and instrumental hexanal amounts. It was also noted that whey protein coated peanuts rated darker, glossier, more burnt, less roasted, more dry cardboard-like, less brittle, chewier, more rubbery, and sweeter (Lee et al. 2002, Lee and Krochta 2002). In another study by Lee et al. (2003), apples coated with edible coatings from whey protein concentrate (WPC) did not exhibit off-flavor in apple slices over storage for 2 weeks, while the coatings produced a milky smell.

Although there is research on the utilization of zein and collagen protein coatings in real food applications, these studies are limited and there are few in which effects of the coatings on sensory properties of the product were evaluated. Collagen films and coatings that are tasteless (Conca 2002) are generally used to improve the quality of meat products. In this case, when the product coated with collagen film was cooked or smoked, the film became an integral constituent of the product, resulting in an attractive appearance

(Ustunol 2009). A commercial edible collagen film, Coffi®, was studied in beef round steaks and demonstrated that there was a reduction in exudation with no significant effect on color and lipid oxidation (Farouk et al. 1990). When applied to improve the quality of nuts, zein protein coatings imparted more favorable sensory ratings over storage at 38°C in terms of appearance, odor, texture, and flavor (Conca 2002).

Rodriguez-Turienzo et al. (2012) reported that when Atlantic salmon was coated with 15 or 60 min ultrasound-treated whey protein coating, lipid oxidation was delayed during frozen storage at −10°C for 4 months without negatively affecting sensory properties determined in a quantitative affective consumer test.

Bugnicourt et al. (2013) packaged butter cheese with a whey protein–coated film produced at semi-industrial scale and conducted sensory evaluation in a triangle test and consensus sensory profiling to compare sensory characteristics of the fresh cheese, cheese packaged with this newly produced film and fully synthetic reference multilayer films of polyamide/polyethylene after storage for 42 days. Although consensus sensory profiling indicated complicated outcomes, there was no off-flavor and no negative effect of sensory changes on overall product characteristics. In the butter cheese stored in protein-coated film material, a slightly more buttery taste was detected, while the butter cheese with reference packaging showed slightly creamier and softer mouthfeel in triangle test.

Weerasinghe et al. (2013) used edible coatings produced from thermized (heated at 70°C for 0, 5, 10, and 15 min) cheddar WPC with or without enzymatically hydrolyzed casein to improve product characteristics of beefsteak for 8 days. Data obtained from the descriptive sensory test showed that edible coatings did not result in changes in general sensory properties of the steaks without imparting any undesirable flavors or objectionable odors to the products during storage. From a sensory point of view, the most positive effects in protecting meat against storage-induced off-flavor development was observed with thermization for 5 min.

Marquez et al. (2014) evaluated the effects of whey protein/pectin edible films with the addition of transglutaminase in deep-fried foods such as doughnuts and French fries and in "taralli" biscuits as a baked food in comparison with an uncoated control, and the samples coated with soy protein/whey protein. In the triangle test, when whey protein/pectin edible films with the addition of transglutaminase was used as coating material, there was no difference between coated and uncoated doughnuts and French fries in terms of texture which was also confirmed by sensory data. However, with "taralli," the majority of the panelists distinguished uncoated and whey protein/pectin/transglutaminase–coated samples after 50 days of storage. In the preference test, in both types of deep-fried foods, more than 46% of panelists preferred whey protein/pectin/transglutaminase–coated samples, while whey protein/pectin/transglutaminase–coated "taralli" as baked food was preferred by more than 60% of the panelists. Therefore, in addition to the oil content decreasing effect in deep-fried foods and the water absorption preventing effect during storage in baked foods, the edible whey protein/pectin film was also favorable in terms of sensory properties.

26.4.3 Lipid-Based Edible Films and Coatings

Lipid origin materials such as acylglycerols, waxes, and fatty acids are convenient in the production of edible films and coatings (Kester and Fennema 1986). Some lipids possess permeability values similar to plastic films. One common practice using edible films and coatings is to apply waxes or some other lipid coating by dipping or spraying not only to provide a glossy appearance on the surface of fruits but also to provide a good moisture barrier because of their hydrophobic nature, thus decreasing moisture loss, shriveling, and shrinkage of coated fruit. However, lipid- or wax-based edible coatings might result in opaque, slippery, and waxy tasting products and off-flavor development due to their sensitivity to rancidity (Contreras-Oliva et al. 2012, Debeaufort and Voilley 2009, Ustunol 2009).

Among lipid-based edible films and coatings, waxing is one of the foremost postharvest applications for fruits to avoid detrimental changes and extend shelf life. Waxes from animal sources such as beeswax, shellac wax, or films derived from plant sources such as carnauba wax and candelilla wax, mainly in emulsion form, have been commonly used as edible coatings, to impart a shiny appearance as well as to maintain gaseous exchange and moisture retention, while preserving quality over the storage period (Njombolwana et al. 2013).

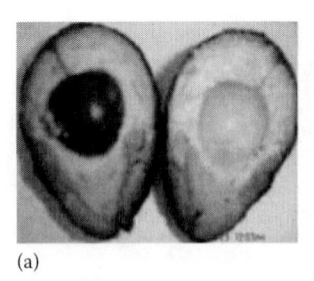 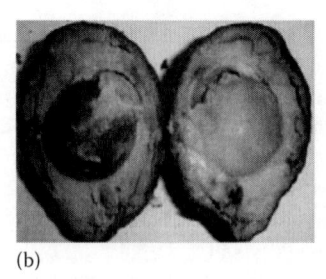 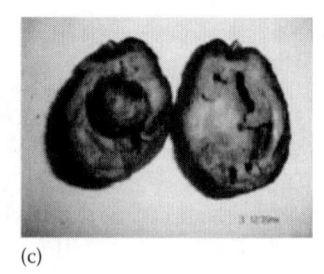

(a)   (b)   (c)

FIGURE 26.3 Internal appearance of avocados: (a) coated with edible candelilla wax with ellagic acid, (b) coated with edible candelilla wax, and (c) uncoated control. (From Saucedo-Pompa, S. et al., *Food Res. Int.*, 42(4), 511, 2009.)

Hagenmaier and Baker (1996) developed an edible coating with candelilla wax for fruit applications to improve glossiness and barrier properties by the use of gelatin and hydroxypropyl methylcellulose. For this purpose, glossy candelilla wax coatings were produced from ammonia-based microemulsions after incorporation of protein or hydroxypropyl methylcellulose.

Machado et al. (2012) evaluated quality changes in "Ortanique" tangor coated with two commercial carnauba-based waxes, Aruá Tropical® (18% soluble solids) or Star Light® (11% soluble solids), during storage at 22°C for 15 days. Based on the sensory evaluation, it was noted that Aruá Tropical coating application prolonged shelf life of "Ortanique" for at least 6 days. The tangors coated with the two commercial waxes maintained initial green color over storage, while uncoated fruits turned to yellow at the sixth day of storage. The fruits treated with Aruá Tropical coating possessed moistened peel toward the end of storage, whereas the peel of the tangors coated with Star Light was shriveled and wrinkled, suggesting that Aruá Tropical be a better commercial wax coating.

Candelilla wax recognized as GRAS by the FDA was applied as edible coating alone or with the inclusion of ellagic acid on the surface of avocado to minimize detrimental quality changes during 6 weeks of storage at 5°C (Saucedo-Pompa et al. 2009). Candelilla wax edible film provided better external and internal visual appearance (Figure 26.3) and delayed changes in brightness compared to the uncoated control, but the most efficient coating was when the edible film included ellagic acid.

Sreenivas et al. (2011) applied ash gourd peel wax onto the surface of strawberries and noted that wax coating prolonged the shelf life of the fruits to 7 days rendering good texture and color characteristics, while fruits without wax coating spoiled in less than 2 days.

26.4.4 Composite Films and Coatings

Proteins, polysaccharides, and lipids, the main polymeric ingredients used in the production of edible films and coatings, show advantages and disadvantages. In addition to their individual usage to provide better mechanical, physical, functional, and sensory characteristics, blends of two or all of them are generally utilized to make edible composite films (Contreras-Oliva et al. 2012, Pascall and Lin 2013). For instance, the hydrophilic nature of polysaccharides and proteins restricts desirable functions of edible films. Addition of hydrophobic compounds improves water barrier and mechanical properties of these films (Del Valle et al. 2005).

Conforti and Totty (2007) coated Golden Delicious apples with three individually developed lipid/hydrocolloid coatings containing different levels of maltodextrin, locust bean gum, gum arabic, algin, and sodium carboxymethylcellulose prepared with a given concentration of vegetable oil, parafilm, and surfactant in each treatment formulation. Data obtained from quantitative descriptive analysis for sensory testing showed that there were no significant differences in sensory properties of apples between the three coating applications. Coated apples were juicier than uncoated ones, while uncoated and coated samples did not differ in sweetness and tartness attributes. In terms of texture, the coated apples were crispy (rated based on initial bite outside) and firmer (rated depending on inside texture), whereas uncoated control was mushy (rated based on initial bite outside) and mealy (rated depending on inside texture). Results from this study show that the lipid/hydrocolloid edible coatings had a positive effect on maintaining quality.

Valencia-Chamorro et al. (2010) applied edible composite coatings based on hydroxypropyl methylcellulose together with beeswax and shellac as hydrophobic components on "Ortanique" mandarins. These composites included antifungal-specific preservatives: potassium sorbate, sodium benzoate, sodium propionate, or their mixtures to control citrus postharvest green molds (GM) and blue molds (BM). Quality changes were evaluated during storage at 5°C for 8 weeks. All the selected coatings were effective in reducing damage by GM and BM. These coatings were effective to control weight losses and preserve firmness. It was also reported that sensory flavor, off-flavor, and fruit appearance as well as external gas concentration, juice ethanol, and acetaldehyde contents, which have significant impact on sensory properties, were not adversely affected by the application of these coatings. Additional work was suggested to improve some physical characteristics of the coatings such as glossiness and appearance of the coated mandarins.

Contreras-Oliva et al. (2012) evaluated the effects of solid content and composition of coating produced from hydroxypropyl methylcellulose, beeswax, and shellac on some quality characteristics of "Oronules" mandarins. Hydroxypropyl methylcellulose–based coatings with high solid content, due to their higher ethanol concentration, resulted in a decrease in flavor and an increase in off-flavor when compared with those with low solid content after storage for 4 weeks at 5°C + 1 week at 20°C. All mandarins had acceptable appearance over the storage periods. Higher solid content in the edible coating formulation increased coating thickness and reduced transparency and gloss. Hydroxypropyl methylcellulose–based coatings did not improve glossiness of the fruit in comparison with uncoated mandarins. Among the edible coatings, the coating containing a beeswax/shellac ratio of 1:3 with low solid content was the most effective in terms of providing improvements in mandarin glossiness, likely due to the higher shellac concentration, which contributed higher gloss to the fruit than waxes. The effectiveness of edible composite coatings on the gloss was not as much as provided by commercial waxes, which was attributed to the difference in lipid particle size. Wax coatings produced by making microemulsions provided high gloss due to smaller lipid particle size because the higher the lipid particle size, the more the emulsions deplete transparency.

Mehyar et al. (2012) made edible coatings from whey protein isolate, pea starch, and a mixture of both with carnauba wax and applied the coatings onto the walnuts and pine nuts. In general, all coatings improved sensory properties of the nuts as compared to uncoated samples. Whey protein isolate in the coatings resulted in increased glossiness. Whey protein isolate/pea starch/carnauba wax (at the ratio of 1:1:1) coating received the highest scores in smoothness, taste, smell, bite strength, and overall acceptance of the nuts, whereas it caused unacceptable yellowish color on walnuts when it was used at 1:1:2 ratio.

Tanada-Palmu and Grosso (2005) evaluated the effect of wheat gluten–based coatings and films on strawberry quality and shelf life during refrigerated storage for 16 days. In sensory tests, appearance, color, brightness, and intention to buy after 1, 6, 12, and 16 days of refrigerated storage and taste and flavor after 5 days were evaluated in an uncoated control and fruits coated with gluten coating, composite coating (dipping the fruit in composite coating formulation of gluten, beeswax, stearic and palmitic acids), and bilayer coating (dipping the fruit first in gluten and then a mixture of lipids containing beeswax, stearic and palmitic acids). Despite the positive influence in maintaining the quality of strawberries, bilayer-coated strawberries were rejected by the panelists at all storage periods because the strawberries were found to be "artificial," with no characteristic color and an opaque, waxy appearance. They were also disliked because the coating was hard and difficult to bite and the taste was not pleasant. On the other hand, fruits applied with gluten and composite coating received higher scores for all evaluated attributes because they helped to improve shelf life and maintained the visual quality of the fruit. At the end of the storage (after 16 days), uncoated fruits received the lowest scores. Even though a residual taste in gluten and composite-coated samples was detected by the panelists, in general, no negative effect of gluten coating application was determined on the flavor and taste of the strawberries.

Kim et al. (2012) applied a composite edible coating produced from defatted mustard meal and xanthan gum on smoked salmon. In addition to decreasing the effect on lipid oxidation and volatile changes during storage, the composite film did not impart a negative sensory quality to the salmon. The coated and uncoated salmon samples did not differ in flavor strength, aftertaste, cohesiveness, gumminess, and oiliness attributes, indicating that the composite coating did not affect the texture of the product and the mustard flavor was not noticeable. No significant change was observed in color intensity, while coated salmon samples had the greatest glossiness during refrigerated storage for 7 days. The intensity of rancidity and fishy smell attributes did not exhibit significant change in coated salmon during storage, whereas uncoated control samples at day 7 of storage possessed the highest rancidity and fishy smell.

26.5 Effects of Incorporating Active Agents into Edible Films or Coatings on the Sensory Quality of Foods

Recent research on emerging active packaging technologies has focused on the inclusion of bioactive compounds, that is, antimicrobials, antioxidants, vitamins, and probiotics into formulations of edible films and coatings as good carriers for the delivery of these compounds, providing fresher high-quality products with longer shelf life. Inclusion of bioactive compounds has been demonstrated to be more effective than adding those chemicals directly onto the product because it offers better control of micro-biological and oxidative deteriorations, thereby providing health benefits to the consumer by preventing formation of particular harmful compounds during storage (Kanmani and Lim 2013, López de Lacey et al. 2012). When an active compound is incorporated into edible film formulation to enhance effective-ness of the films, it should be compatible with the sensory properties of food. There are a number of sub-stances being used for this purpose, some of which do not result in negative change in the food products, whereas other active agents, such as essential oils, might result in negative impact on sensory quality.

Lee et al. (2003) conducted a study on the shelf life extension of apple slices with antibrowning agents and carrageenan- or WPC-based edible coatings. The two films based on carrageenan had either ascor-bic acid and oxalic acid or ascorbic acid and citric acid as antibrowning agents, whereas the two using WPC had either ascorbic acid and oxalic acid or ascorbic acid and $CaCl_2$. All coated slices received higher scores than those noncoated. The antibrowning agents effectively retarded the enzymatic brown-ing and the main sensory attributes tested were firmness and color, where flavor did not play a role in all formulations tested. The carrageenan-based coatings received good scores but after a 2-week storage presented some problems, the one with oxalic acid in terms of color and the one with citric acid in terms of firmness. Apples treated with ascorbic acid plus $CaCl_2$-added WPI were free of browning with the greatest overall preference in terms of sensory color and firmness after 2-week storage.

Lu et al. (2009) evaluated the effects of alginate-calcium coatings incorporated with nisin and EDTA on the quality of fresh northern snakehead fillets during refrigerated storage at 4°C for 7 days. Three treatments were applied to fillets: (1) Nisin + EDTA, (2) alginate-calcium edible coating, and (3) algi-nate-calcium edible coating incorporating EDTA and nisin. In sensory hedonic test, fish fillets coated using alginate-calcium edible coatings, with or without nisin and EDTA, had the highest scores for appearance, odor, and texture during the storage period. Edible alginate-calcium coatings received lower scores for taste than when only nisin and EDTA were applied to the samples, but higher scores than the untreated samples. These low scores were attributed to the presence of calcium chloride in the coating formulation because it imparts a bitter taste. Based on their findings, the authors are recommending the use of these coatings because they adequately control microbial and chemical spoilage, water loss, and overall sensory attributes.

García et al. (2010) studied the potential of shelf life extension of strawberries by using cassava starch edible coatings at different concentrations (1%, 2%, and 3%) with or without the addition of potassium sorbate (0.05% and 0.1%) as antimicrobial agent. It was found that all coated strawberries did not differ from the uncoated ones in terms of appearance, aroma, flavor, texture, and overall impression. In addi-tion, the purchase intention was above 60% for all the studied conditions, which showed these coatings had great acceptability. The major differences detected among coated samples were on coating integrity, respiration rate, and water vapor resistance. There was no significant difference in instrumental color values between starch coated, with or without potassium sorbate, and uncoated strawberries, which could also be visually noticed in Figure 26.4. One of the outcomes of the aforementioned study indicated that potassium sorbate did not negatively affect sensory attributes of strawberries at the concentrations studied and both concentrations were equally effective.

(a) (b) (c)

FIGURE 26.4 Minimally processed strawberries treated with cassava starch edible coatings with or without potassium sorbate. (a) Uncoated, (b) coated with 3% cassava starch, (c) coated with 3% cassava starch with the inclusion of 0.1% potassium sorbate. (From Garcia, L.C. et al., *Food Bioprocess Technol.*, 3(6), 834, 2010.)

Gniewosz and Synowiec (2011) applied on the surface of mandarins and apples edible pullulan films incorporated with thymol in different proportions to determine antimicrobial activity of these films against selected Gram-positive and Gram-negative bacteria as well as sensory changes. The films were very transparent and shiny and had a strong antimicrobial activity during storage where the colors of the fruits were not affected. The characteristic aroma of thymol was perceptible on the surface of the coated fruits, while the aroma of thymol was not perceptible within the fruit flesh. These formulations showed a good balance between sensorial permissible changes and antimicrobial/antioxidant levels and promoted significant shelf life extension of the fruits.

Lee and Min (2013) developed antimicrobial films with defatted soybean meal and the lactoperoxidase system, applied them to ham slices, and evaluated sensory attributes such as glossiness, odor, hardness, cohesiveness, and flavor. They noted that there were no significant differences in all attributes between coated and uncoated samples, and the soybean flavor was not noticeable, suggesting that these film coatings can be used without altering sensory properties of sliced ham, and thus it could have important commercial applications for ham and other meat products.

Krasniewska et al. (2014) evaluated the impact of pullulan coatings with or without incorporation of summer savory *Satureja hortensis* (SH) extract on sensory properties of pepper and apple at 16°C for 14 days of storage among others. It is important to mention that SH has an important bactericidal/ fungicidal effect and for this reason is added to the formulation of the coatings. In hedonic tests, sensory attributes such as color, aroma, degree of coating wrinkling, adhesion, and general appearance were assessed with the coating and after the removal of the coating. SH extract incorporation into the pullulan coating resulted in lower color scores in comparison with uncoated samples and samples with only pullulan coating. While there were no differences within the coated and uncoated peppers in terms of aroma, coated apples had lower aroma scores than the control. Pullulan coating with or without the addition of SH extract gave rise to higher scores for wrinkling, suggesting that the coatings provided protection against excessive flaccidity and wrinkling of the surface, which ensures fresher and more attractive products. For both produce, there was no problem with coating adhesion to the surface of the material. All groups had similar general acceptability scores, with the exception of those with SH extract, which received lower scores than the other groups.

Essential oils or their active components generally have strong flavor notes, which might result in negative impact on the food product when they are used in the production of active edible films and coatings. The effects of 0.5% and 0.75% carvacrol incorporation into apple- and tomato-based edible film formulations and 0.5% and 0.75% cinnamaldehyde to apple-based edible film formulations on sensory characteristics of baked chicken pieces wrapped with these films were evaluated by Du et al. (2012). While 0.75% carvacrol incorporated into a tomato-based edible film reduced preference of the cooked chicken meat, 0.5% addition of carvacrol into the film solution did not have significant effect on the preference in the paired preference tests. Because of flavor incompatibility of the apple and carvacrol in comparison with the tomato and carvacrol, baked chicken pieces wrapped with carvacrol–apple films received negative responses from the panelists even at 0.5% added carvacrol level. In the preference test, chickens wrapped with apple films incorporated with 0.75% cinnamaldehyde were less preferred than those with 0.75% carvacrol; however, those with 0.5% cinnamaldehyde level were preferred over those with carvacrol at the same concentration. Therefore, 0.5% carvacrol was recommended for tomato edible films and 0.5% cinnamaldehyde for apple edible films. Positive and negative descriptors determined for sensory characteristics of cooked chicken meats wrapped with carvacrol- or cinnamaldehyde-added tomato- or apple-based edible films are summarized in Table 26.3. It is interesting to note that active

TABLE 26.3

Descriptors for Cooked Chicken Wrapped with Carvacrol- or Cinnamaldehyde-Added Tomato- or Apple-Based Edible Films

Film Type	Positive Descriptors	Negative Descriptors
0.5% carvacrol incorporated tomato-based edible film	Taste: "Nice, herbal seasoning, tangy, spicy and saltier" Odor: "Orange-like and herbal" Texture: "Firmer, juicy, and less slimy"	Taste: "Unpleasant aftertaste, bitter, metallic, paperlike, chemical, woody, less chicken-like, too salty, less sweet, and pungent" Odor: "Repulsive, medicinal and overly herbal smell" Texture: "Less tender, soft, moist, and driest"
0.75% carvacrol incorporated tomato-based edible film	Taste: "Natural flavor, good taste, spicy, herbal, like dried oregano, salty, and less sour" Texture: "Soft and tender" for texture	Taste: "Strong oregano flavor, bad, funny unpleasant aftertaste, smoky, less chicken taste, unpleasant artificial and strange flavor, bitter, astringent, less sweet, stale, inedible, metallic, mint, and herbal" Odor: "Funny, weird, rubbery and strange odor" Texture: "More dry"
0.5% carvacrol incorporated apple-based edible film	Taste: "Very tasty, herbal, oregano flavor, more flavorful, cleaner aftertaste, spice flavor and right saltiness" Texture: "Juicy, moist, tender, less granulated, and better consistency"	Taste: "Slight herbal bitter aftertaste, more salty, odd, strange, funny, plastic, medicine taste and flavor, oregano-like, taco seasoning, and not chicken-like" Odor: "Unpleasant smell, herbs scent and oregano odor" for odor Texture: "Less juicy and creamy"
0.75% carvacrol incorporated apple-based edible film	Taste: "Tangy, herbal, oregano-like, spicy, and saltier" for taste Odor: "Flavorful" for odor Texture: "Firmer and juicy"	Taste: "Less chicken taste, unpleasant aftertaste, strong herbal taste, bitter weird, chemical, metallic, paperlike, smoky, too salty, bad taste, overpowering plastic or medicine flavor" Odor: "Bad smell" Texture: "Dry, less tender, soft and moist"
0.5% cinnamaldehyde incorporated apple-based edible film	Taste: "Right, good flavoring; more complex, slightly better flavor, sweeter, salty, more spicy, less chicken aftertaste and stronger chicken flavor" Texture: "More tender, moist, easier to chew and juicy"	Taste: "Slight bitter aftertaste, altered, not as chicken, too salty, and slight sweeter" Odor: "Cinnamon smell, not as chicken and slight unpleasant odor" Texture: "Less tender, slightly drier and crumble after chewing"
0.75% cinnamaldehyde incorporated apple-based edible film	Taste: "Pleasant, distinct, salty enough, less bitter, sour, plain and raw" for taste Texture: "Firmer, moist, softer, chewy, juicy, less dry, and hard"	Taste: "Unusual, funny, strange, cinnamon-like, apple-like, like candy and spice (nutmeg, mint, anise, clove), maple, musty/earthy aftertaste, licorice off-taste, bitter, less sweet, chemical, unpleasant lingering after several bites and not chicken-like" Odor: "Cinnamon-like, bad smell and not as chicken smell" Texture: "Dry, flaky, less tender, smooth and juicy"

Source: Adapted from Du, W.X. et al., *J. Agric. Food Chem.*, 60(32), 7799, 2012.

agents added to the film formulations might have resulted in an increase in some sensations. For instance, in this study by Du et al. (2012), it was noted that carvacrol increased saltiness, while cinnamaldehyde increased sweetness of baked chicken. This ability of carvacrol to increase saltiness intensity could be used to compensate for sodium chloride reduction in foods.

Aloui et al. (2014) conducted a study to control postharvest decay in dates inoculated with *Aspergillus flavus* by using combined treatments of chitosan or locust bean gum edible coatings incorporated with different citrus essential oils at various concentrations. They also evaluated the impact of pure and combined essential oil–based coatings on sensory properties by using the sensory profile method to conduct a quantitative estimation of changes in sensory characteristics. They found significant differences in color, gloss, citrus odor, and flavor attributes. After coating application, color and glossiness intensity of dates was reduced, which was more pronounced in the groups coated with formulations containing citrus essential oils. The decreases in color and glossiness were attributed to the increase in the opacity of the film due to oil droplet aggregation during the drying process, resulting in reduction of light absorption by the surface of the fruits. In dates treated with either chitosan or locust bean gum edible coatings incorporated with citrus essential oils, citrus odor and flavor intensity were high, which might influence typical aroma and flavor of the samples.

Similar changes in glossiness on the surface of the product were observed with the addition of vitamin E and lemon essential oil into chitosan coating formulations. Han et al. (2005) noted that chitosan coating with the inclusion of vitamin E resulted in a loss of surface glossiness in strawberries. Perdones et al. (2012) investigated the impact of 1% chitosan edible coatings incorporated with 3% lemon essential oil and prepared using different homogenization treatments on some quality characteristics of strawberries during cold storage at 4°C for 10 days. In sensory evaluation, lemon essential oil incorporated samples were rated as being less glossy in comparison with samples coated with only chitosan (Figure 26.5). Strawberries treated with lemon essential oil–added coatings exhibited a decreased typical strawberry aroma and flavor due to the strong masking effect of lemon oil aroma compounds, which also gave rise to lower sample preference.

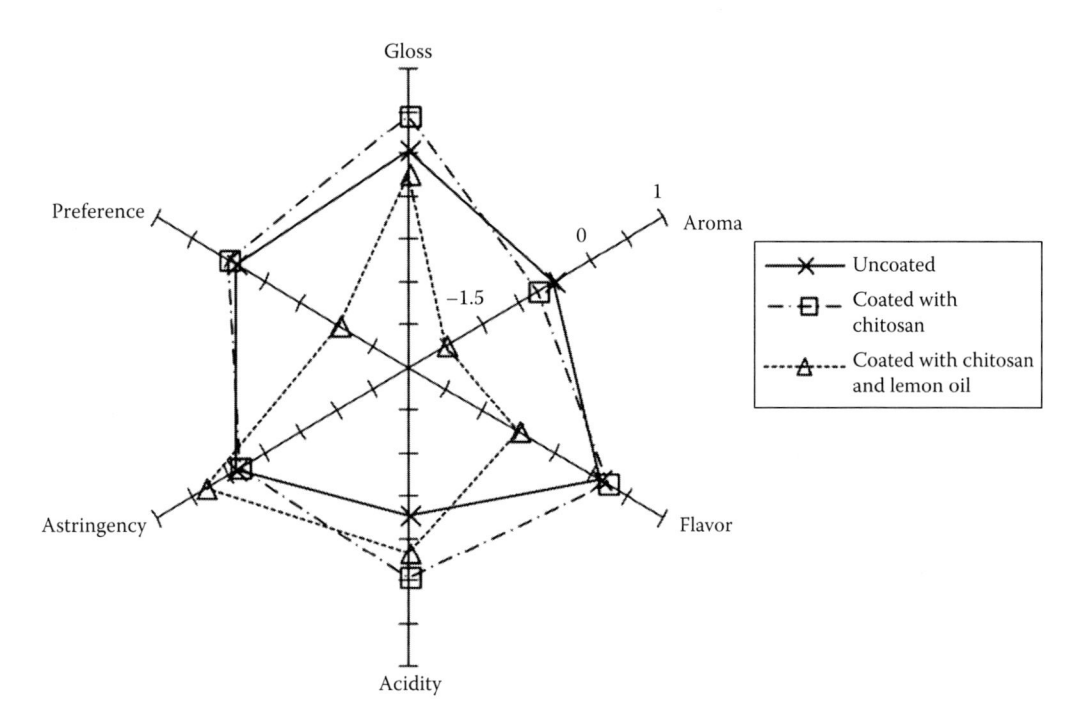

FIGURE 26.5 Sensory profile of strawberries coated with chitosan or chitosan and lemon essential oil. (From Perdones, A. et al., *Postharvest Biol. Technol.*, 70, 32, 2012.)

26.6 Final Remarks

- Sensory attributes should always be a significant factor while developing edible films and coatings before and after they are applied to foods. If sensory properties become compromised, it will not be easy to market these modified food products in the current highly competitive and consumer-oriented global market. At the same time, consumers' perception toward the utilization of these covers in food products should be properly analyzed before launching them.

- Aiming to have films or coatings with significant amount of physical, mechanical, and other desirable properties, a number of ingredients are included in their formulations, but the more ingredients are used, the more difficult is to control the sensory attributes of the final food product.

- The proper selection of the ingredients for a given formulation might make the covered product more appealing and accepted than the original one. If this is coupled with shelf life extension and the addition of functional ingredients, the final product would have good chances to become a successful one.

- Edible films and coatings are excellent carriers to host functional ingredients such as nutrients, antimicrobials, and antioxidants. This represents another reason why these covers are very useful in addition to shelf life extension. Some of the active agents used to enhance functionality might affect at least one sensory property of the film or the food itself. Such is the case of essential oils with excellent antioxidant and antimicrobial characteristics but with strong flavors. Therefore, a good compromise between the effectiveness of the active agents and their sensory effects on the edible films or coatings, as well as on the food products, should be reached in order to warrant acceptability.

- It is important to identify concentrations thresholds of the ingredients of a given edible film or coating formulation in terms of acceptability. Below a certain concentration, an ingredient might not negatively impact the sensory characteristics of the cover, but if it is used in excess, it might trigger undesirable properties. These thresholds very much depend on the type of food the cover is applied. At the same time, it is appropriate to carefully analyze interactions among ingredients and their concentrations. In some cases, we might have ingredients that neutralize the undesirable characteristics of another one, but the opposite might as well happen; that is, the combination of two or more ingredients might result in a cover with low sensory attributes.

- Sensorial analyses should be conducted throughout the formulations of the edible films and coatings to warrant the development of a quality, well-balanced final product. Those ingredients that might bring negative connotations to the formulation, such as off-flavors, strong flavors, different palatability, different colors, and different appearance, might be substituted by others with similar properties but not significantly affecting sensory attributes. This exercise of conducting sensory evaluation while developing the covers will result in better products while saving time and resources.

- There are no universal films or coatings, and this is, in part, related to sensory aspects resulting from the interaction of the original food product and the application of a given cover. Some edible films might not affect much certain sensory characteristics in a given type of foods but might be significant in others. At the same time, it is important to select ingredients for the covers compatible, in terms of sensory attributes, to the original food product.

REFERENCES

Ali, A., M. Maqbool, S. Ramachandran, and P.G. Alderson. Gum Arabic as a novel edible coating for enhancing shelf-life and improving postharvest quality of tomato (*Solanum lycopersicum* L.) fruit. *Postharvest Biology and Technology* 58(1) (2010): 42–47.

Aloui, H., K. Khwaldia, F. Licciardello et al. Efficacy of the combined application of chitosan and locust bean gum with different citrus essential oils to control postharvest spoilage caused by *Aspergillus flavus* in dates. *International Journal of Food Microbiology* 170 (2014): 21–28.

Barrena-Figueroa, R. and T. Garcia-Lopez-de-Meneses. The effect of consumer innovativeness in the acceptance of a new food product. An application for the coffee market in Spain. *Spanish Journal of Agricultural Research* 11(3) (2013): 578.

Bitencourt, C.M., C.S. Fávaro-Trindade, P.J.A. Sobral, and R.A. Carvalho. Gelatin-based films additivated with curcuma ethanol extract: Antioxidant activity and physical properties of films. *Food Hydrocolloids* 40 (2014): 145–152.

Bugnicourt, E., M. Schmid, O. McNerney et al. Processing and validation of whey-protein-coated films and laminates at semi-industrial scale as novel recyclable food packaging materials with excellent barrier properties. *Advances in Materials Science and Engineering* 11(10) (2013), http://www.hindawi.com/journals/amse/2013/496207/.

Caner, C. The effect of edible eggshell coatings on egg quality and consumer perception. *Journal of the Science of Food and Agriculture* 85(11) (2005): 1897–1902.

Carel, J.A., J.V. Garcia-Perez, J. Benedito, and A. Mulet. Food process innovation through new technologies: Use of ultrasound. *Journal of Food Engineering* 110(2) (2012): 200–207.

Castricini, A., R.C.C. Coneglian, and R. Deliza. Starch edible coating of papaya: Effect on sensory characteristics. *Ciencia E Tecnologia De Alimentos* 32(1) (2012): 84–92.

Chen, Q., S. Anders, and H. An. Measuring consumer resistance to a new food technology: A choice experiment in meat packaging. *Food Quality and Preference* 28(2) (2013): 419–428.

Chidanandaiah, S.M.K., R.C. Keshri, and M.K. Sanyal. Effect of sodium alginate coating with preservatives on the quality of meat patties during refrigerated (4 ± 1C) storage. *Journal of Muscle Foods* 20(3) (2009): 275–292.

Chien, P.J., F. Sheu, and F.H. Yang. Effects of edible chitosan coating on quality and shelf life of sliced mango fruit. *Journal of Food Engineering* 78(1) (2007): 225–229.

Chiumarelli, M., C.C. Ferrari, C.I.G.L. Sarantopoulos, and M.D. Hubinger. Fresh cut 'Tommy Atkins' mango pre-treated with citric acid and coated with cassava (*Manihot esculenta* Crantz) starch or sodium alginate. *Innovative Food Science & Emerging Technologies* 12(3) (2011): 381–387.

Ciannamea, E.M., P.M. Stefani, and R.A. Ruseckaite. Physical and mechanical properties of compression molded and solution casting soybean protein concentrate based films. *Food Hydrocolloids* 38 (2014): 193–194.

Conca, K.R. Protein-based films and coating for military packaging applications. In *Protein-Based Films and Coatings*, ed. A. Gennadios, pp. 551–577. Boca Raton, FL: CRC Press, 2002.

Conforti, F.D. and J.A. Totty. Effect of three lipid/hydrocolloid coatings on shelf life stability of golden delicious apples. *International Journal of Food Science and Technology* 42(9) (2007): 1101–1106.

Contreras-Oliva, A., C. Rojas-Argudo, and M.B. Perez-Gago. Effect of solid content and composition of hydroxypropyl methylcellulose-lipid edible coatings on physico-chemical and nutritional quality of 'Oronules' Mandarins. *Journal of the Science of Food and Agriculture* 92(4) (2012): 794–792.

Cuq, B., N. Gontard, and S. Guilbert. Proteins as agricultural polymers for packaging production. *Cereal Chemistry* 75(1) (1998): 1–9.

Dangaran, K., P.M. Tomasula, and P. Qi. Structure and function of protein-based edible films and coatings. In *Edible Films and Coatings for Food Applications*, ed. M. E. Embuscado and K. C. Huber, pp. 25–56. New York: Springer, 2009.

Debeaufort, F. and A. Voilley. Lipid-based edible films and coatings. In *Edible Films and Coatings for Food Applications*, ed. M. E. Embuscado and K. C. Huber, pp. 135–168. New York: Springer, 2009.

Deliza, R., A. Rosenthal, D. Hedderley, H.J.H. MacFie, and L. Frewer. The importance of brand, product information and manufacturing process in the development of novel environmentally friendly vegetable oils. *Journal of International Food and Agribusiness Marketing* 10(3) (1999): 67–78.

Deliza, R., A. Rosenthal, and A.L.S. Silva. Consumer attitude towards information on non conventional technology. *Trends in Food Science & Technology* 14(2) (2003): 43–49.

Del Valle, V., P. Hernandez-Munoz, A. Guarda, and M.J. Galotto. Development of a cactus-mucilage edible coating (*Opuntia ficus indica*) and its application to extend strawberry (*Fragaria ananassa*) shelf-life. *Food Chemistry* 91(4) (2005): 751–756.

Dhall, R.K. Advances in edible coatings for fresh fruits and vegetables: A review. *Critical Reviews in Food Science and Nutrition* 53(5) (2013): 435–450.

Du, W.X., R.J. Avena-Bustillos, R. Woods et al. Sensory evaluation of baked chicken wrapped with anti-microbial apple and tomato edible films formulated with cinnamaldehyde and carvacrol. *Journal of Agricultural and Food Chemistry* 60(32) (2012): 7799–7804.

Fan, J.M., W. Ma, G.Q. Liu, S.W. Yin, C.H. Tang, and X.Q. Yang. Preparation and characterization of kidney bean protein isolate (Kpi)–chitosan (Ch) composite films prepared by ultrasonic pretreatment. *Food Hydrocolloids* 36 (2014): 60–69.

Farouk, M.M., J.F. Price, and A.M. Salih. Effect of an edible collagen film overwrap on exudation and lipid oxidation in beef round steak. *Journal of Food Science* 55(6) (1990): 1510–1512.

Farris, S., I. Uysal Unalan, L. Introzzi, J.M. Fuentes-Alventosa, and C.A. Cozzolino. Pullulan-based films and coatings for food packaging: Present applications, emerging opportunities, and future challenges. *Journal of Applied Polymer Science* 131(13) (2014) DOI: 10.1002/app.40539.

Fernández-Pan, I., J. Ignacio, and M. Caballero. Biopolymers for edible films and coatings in food applications. In *Biopolymers—New Materials for Sustainable Films and Coatings*, ed. D. Plackett. pp. 233–254. Chichester, U.K.: John Wiley & Sons Inc., 2011.

Gao, P., Z. Zhu, and P. Zhang. Effects of chitosan-glucose complex coating on postharvest quality and shelf life of table grapes. *Carbohydrate Polymers* 95(1) (2013): 371–378.

Garcia, L.C., L.M. Pereira, C.I.G. de Luca Sarantópoulos, and M.D. Hubinger. Selection of an edible starch coating for minimally processed strawberry. *Food and Bioprocess Technology* 3(6) (2010): 834–842.

Gennadios, A., M.A. Hanna, and L.B. Kurth. Application of edible coatings on meats, poultry and seafoods: A review. *Food Science and Technology* 30(4) (1997): 337–350.

Gniewosz, M. and A. Synowiec. Antibacterial activity of pullulan films containing thymol. *Flavour and Fragrance Journal* 26(6) (2011): 389–395.

Gómez-Estaca, J., P. Montero, and M.C. Gómez-Guillén. Shrimp (*Litopenaeus vannamei*) muscle proteins as source to develop edible films. *Food Hydrocolloids* 41 (2014): 86–94.

Grunert, K.G. Food quality and safety: Consumer perception and demand. *European Review of Agricultural Economics* 32(3) (2005): 369–391.

Guilbert, S. Technology and application of edible protective films. In *Food Packaging and Preservation: Theory and Practice*, pp. 371–399. London, U.K.: Elsevier Applied Science Publishing Co., 1986.

Hagenmaier, R.D. and R.A. Baker. Edible coatings from candelilla wax microemulsions. *Journal of Food Science* 61(3) (1996): 562–565.

Han, C., J. Zuo, Q. Wang et al. Effects of chitosan coating on postharvest quality and shelf life of sponge gourd (*Luffa cylindrica*) during storage. *Scientia Horticulturae* 166 (2014): 1–8.

Han, C.R., C. Lederer, M. McDaniel, and Y.Y. Zhao. Sensory evaluation of fresh strawberries (*Fragaria ananassa*) coated with chitosan-based edible coatings. *Journal of Food Science* 70(3) (2005): 172–178.

Kanmani, P. and S.T. Lim. Development and characterization of novel probiotic-residing pullulan/starch edible films. *Food Chemistry* 141(2) (2013): 1041–1049.

Kanmani, P. and J.W. Rhim. Properties and characterization of bionanocomposite films prepared with various biopolymers and ZnO nanoparticles. *Carbohydrate Polymers* 106 (2014): 190–199.

Kester, J.J. and O.R. Fennema. Edible films and coatings—A review. *Food Technology* 40(12) (1986): 47–59.

Kim, I.H., H.J. Yang, B.S. Noh, S.J. Chung, and S.C. Min. Development of a defatted mustard meal-based composite film and its application to smoked salmon to retard lipid oxidation. *Food Chemistry* 133(4) (2012): 1501–1509.

Kim, J., Y. Choi, S.R.B. Kim, and S. Lim. Humidity stability of tapioca starch-pullulan composite films. *Food Hydrocolloids* 41 (2014): 140–145.

Kim, S.J. and Z. Ustunol. Sensory attributes of whey protein isolate and candelilla wax emulsion edible films. *Journal of Food Science* 66(6) (2001): 909–911.

Krasniewska, K., M. Gniewosz, A. Synowiec, J.L. Przybyl, K. Baczek, and Z. Weglarz. The use of pullulan coating enriched with plant extracts from *Satureja hortensis* L. to maintain pepper and apple quality and safety. *Postharvest Biology and Technology* 90 (2014): 63–72.

Lawless, H.T. and H. Heymann. *Sensory Evaluation of Food: Principles and Practice*. New York: Springer, 2010.

Leceta, I., S. Molinaro, P. Guerrero, J.P. Kerry, and K. de la Caba. Quality attributes of MAP packaged ready-to-eat baby carrots by using chitosan-based coatings. *Postharvest Biology and Technology* 100 (2015): 142–150.

Lee, H. and S.C. Min. Antimicrobial edible defatted soybean meal-based films incorporating the lactoperoxidase system. *LWT—Food Science and Technology* 54(1) (2013): 42–50.

Lee, J.Y., H.J. Park, C.Y. Lee, and W.Y. Choi. Extending shelf-life of minimally processed apples with edible coatings and antibrowning agents. *Food Science and Technology* 36(3) (2003): 323–329.

Lee, S.Y. and J.M. Krochta. Accelerated shelf life testing of whey-protein-coated peanuts analyzed by static headspace gas chromatography. *Journal of Agricultural and Food Chemistry* 50(7) (2002): 2022–2028.

Lee, S.Y., T.A. Trezza, J.X. Guinard, and J.M. Krochta. Whey-protein-coated peanuts assessed by sensory evaluation and static headspace gas chromatography. *Journal of Food Science* 67(3) (2002): 1212–1218.

Longares, A., F.J. Monahan, E.D. O'Riordan, and M. O'Sullivan. Physical properties and sensory evaluation of WPI films of varying thickness. *Food Science and Technology* 37(5) (2004): 545–550.

Lopez de Lacey, A.M., M.E. Lopez-Caballero, J. Gomez-Estaca, M.C. Gomez-Guillen, and P. Montero. Functionality of *Lactobacillus acidophilus* and *Bifidobacterium bifidum* incorporated to edible coatings and films. *Innovative Food Science & Emerging Technologies* 16 (2012): 277–282.

Lu, F., D. Liu, X. Ye, Y. Wei, and F. Liu. Alginate–calcium coating incorporating nisin and EDTA maintains the quality of fresh northern snakehead (*Channa argus*) fillets stored at 4°C. *Journal of the Science of Food and Agriculture* 89(5) (2009): 848–854.

Machado, F.L.C., F. Ligia, J.M. Correia Costa, and E.N. Batista. Application of carnauba-based wax maintains postharvest quality of 'Ortanique' Tangor. *Ciencia E Tecnologia De Alimentos* 32(2) (2012): 261–266.

Mahfoudhi, N., M. Chouaibi, and S. Hamdi. Effectiveness of almond gum trees exudate as a novel edible coating for improving postharvest quality of tomato (*Solanum lycopersicum* L.) fruits. *Food Science and Technology International* 20(1) (2014): 33–43.

Mali, S., M.V.E. Grossmann, M.A. Garcia, M.N. Martino, and N.E. Zaritzky. Mechanical and thermal properties of yam starch films. *Food Hydrocolloids* 19(1) (2005): 157–164.

Marquez, G.R., P. Di Pierro, M. Esposito, L. Mariniello, and R. Porta. Application of transglutaminase-crosslinked whey protein/pectin films as water barrier coatings in fried and baked foods. *Food and Bioprocess Technology* 7(2) (2014): 447–455.

Mehyar, G.F., K. Al-Ismail, J.H. Han, and G.W. Chee. Characterization of edible coatings consisting of pea starch, whey protein isolate, and carnauba wax and their effects on oil rancidity and sensory properties of walnuts and pine nuts. *Journal of Food Science* 77(2) (2012): 52–59.

Meilgaard, M.C., B.T. Carr, and G.V. Civille. Sensory attributes and the way we perceive them. In *Sensory Evaluation Techniques*, 4th ed., pp. 7–24. Boca Raton, FL: CRC Press, 2006.

Moreira, M.D., S.I. Roura, and A. Ponce. Effectiveness of chitosan edible coatings to improve microbiological and sensory quality of fresh cut broccoli. *Food Science and Technology* 44(10) (2011): 2335–2341.

Nisperos-Carriedo, M. Edible coatings and films based on polysaccharides. In *Edible Coatings and Films to Improve Food Quality*, pp. 305–335. Lancaster, U.K.: Technomic, 1994.

Njombolwana, N.S., A. Erasmus, J.G. Van Zyl, W. du Plooy, P.J.R. Cronje, and P.H. Fourie. Effects of citrus wax coating and brush type on imazalil residue loading, green mould control and fruit quality retention of sweet oranges. *Postharvest Biology and Technology* 86 (2013): 362–371.

Nychas, G.J.E., P.N. Skandamis, C.C. Tassou, and K.P. Koutsoumanis. Meat spoilage during distribution. *Meat Science* 78(2) (2008): 77–89.

Oms-Oliu, G., R. Soliva-Fortuny, and O. Martin-Belloso. Edible coatings with antibrowning agents to maintain sensory quality and antioxidant properties of fresh-cut pears. *Postharvest Biology and Technology* 50(1) (2008): 87–94.

Ozdemir, M. and J.D. Floros. Optimization of edible whey protein films containing preservatives for water vapor permeability, water solubility and sensory characteristics. *Journal of Food Engineering* 86(2) (2008): 215–224.

Park, J.W., W.S. Whiteside, and S.Y. Cho. Mechanical and water vapor barrier properties of extruded and heat-pressed gelatin films. *Food Science and Technology* 41(4) (2008): 692–700.

Pascall, M.A. and S. Lin. The application of edible polymeric films and coatings in the food industry. *Journal of Food Processing & Technology* 4(2) (2013): 116–117.

Pavlath, A.E. and W. Orts. Edible films and coatings: Why, what, and how? In *Edible Films and Coatings for Food Applications*, pp. 1–23. New York: Springer, 2009.

Perdones, A., L. Sanchez-Gonzalez, A. Chiralt, and M. Vargas. Effect of chitosan-lemon essential oil coatings on storage-keeping quality of strawberry. *Postharvest Biology and Technology* 70 (2012): 32–41.

Petriccione, M., F. De Sanctis, M.S. Pasquariello et al. The effect of chitosan coating on the quality and nutraceutical traits of sweet cherry during postharvest life. *Food and Bioprocess Technology* 8(2) (2015): 394–408.

Quezada-Gallo, J.A. Delivery of food additives and antimicrobials using edible films and coatings. In *Edible Films and Coatings for Food Applications*, pp. 315–333. New York: Springer, 2009.

Regiel, A., S. Irusta, A. Kyziol, M. Arruebo, and J. Santamaria. Preparation and characterization of chitosan-silver nanocomposite films and their antibacterial activity against *Staphylococcus aureus*. *Nanotechnology* 24(1) (2013), doi:10.1088/0957-4484/24/1/015101.

Rodríguez, M.S., L.A. Albertengo, I. Vitale, and E. Agulló. Relationship between astringency and chitosan–saliva solutions turbidity at different pH. *Journal of Food Science* 68(2) (2003): 665–667.

Rodriguez-Turienzo, L., A. Cobos, and O. Diaz. Effects of edible coatings based on ultrasound-treated whey proteins in quality attributes of frozen atlantic salmon (*Salmo salar*). *Innovative Food Science & Emerging Technologies* 14 (2012): 92–98.

Ronteltap, A., J.C.M. Van Trijp, R.J. Renes, and L.J. Frewer. Consumer acceptance of technology-based food innovations: Lessons for the future of nutrigenomics. *Appetite* 49(1) (2007): 1–17.

Saucedo-Pompa, S., R. Rojas-Molina, A.F. Aguilera-Carbó et al. Edible film based on candelilla wax to improve the shelf life and quality of avocado. *Food Research International* 42(4) (2009): 511–515.

Silva-Weiss, A., M. Ihl, P.J.A. Sobral, M.C. Gómez-Guillén, and V. Bifani. Natural additives in bioactive edible films and coatings: Functionality and applications in foods. *Food Engineering Reviews* 5(4) (2013): 200–216.

Singh, R.S. and G.K. Saini. Production, purification and characterization of pullulan from a novel strain of *Aureobasidium pullulans Fb-1*. *Journal of Biotechnology* 136 (2008): 506–507.

Soukoulis, C., L. Yonekura, H.H. Gan, S. Behboudi-Jobbehdar, C. Parmenter, and I. Fisk. Probiotic edible films as a new strategy for developing functional bakery products: The case of pan bread. *Food Hydrocolloids* 39(100) (2014): 231–242.

Sreenivas, K.M., K. Chaudhari, and S.S. Lele. Ash gourd peel wax: Extraction, characterization, and application as an edible coat for fruits. *Food Science and Biotechnology* 20(2) (2011): 383–387.

Tajner-Czopek, A., A. Figiel, and A.A. Carbonell-Barrachina. Effects of potato strip size and pre-drying method on french fries quality. *European Food Research and Technology* 227(3) (2008): 757–766.

Tanada-Palmu, P.S. and C.R.F. Grosso. Effect of edible wheat gluten-based films and coatings on refrigerated strawberry (*Fragaria ananassa*) quality. *Postharvest Biology and Technology* 36(2) (2005): 199–208.

Ustunol, Z. Edible films and coatings for meat and poultry. In *Edible Films and Coatings for Food Applications*, pp. 245–268. New York: Springer, 2009.

Valencia-Chamorro, S.A., M.B. Perez-Gago, M.A. Del Rio, and L. Palou. Effect of antifungal hydroxypropyl methylcellulose-lipid edible composite coatings on penicillium decay development and postharvest quality of cold-stored ortanique mandarins. *Journal of Food Science* 75(8) (2010): 418–426.

Valentin, D., S. Chollet, M. Lelievre, and H. Abdi. Quick and dirty but still pretty good: A review of new descriptive methods in food science. *International Journal of Food Science and Technology* 47(8) (2012): 1563–1578.

Vargas, M., A. Albors, A. Chiralt, and C. Gonzalez-Martinez. Quality of cold-stored strawberries as affected by chitosan-oleic acid edible coatings. *Postharvest Biology and Technology* 41(2) (2006): 164–171.

Viña, S.Z., A. Mugridge, M.A. García, R.M. Ferreyra, M.N. Martino, and A.R. Cheves. Effects of polyvinylchloride coating and edible starch coatings on quality aspects of refrigerated brussels sprouts. *Food Chemistry* 103(3) (2007): 701–709.

Wan, V.C.H., C.M. Lee, and S.Y. Lee. Understanding consumer attitudes on edible films and coatings: Focus group findings. *Journal of Sensory Studies* 22(3) (2007): 353–366.

Wang, L., Y. Dong, H. Men, J. Tong, and J. Zhou. Preparation and characterization of active films based on chitosan incorporated tea polyphenols. *Food Hydrocolloids* 32(1) (2013): 35–41.

Wanstedt, K.G., S.C. Seideman, L.S. Donnelly, and N.M. Quenzer. Sensory attributes of precooked, calcium alginate-coated pork patties. *Journal of Food Protection* 44(10) (1981): 732–735.

Weerasinghe, S., J.B. Williams, D. Mukherjee, D.K. Tidwell, S. Chang, and Z.U. Haque. Quality and sensory characteristics of cubed beefsteak dipped in edible protective solutions of thermized cheddar whey. *Journal of Food Quality* 36(2) (2013): 77–90.

Wu, J., F. Zhong, Y. Li, C.F. Shoemaker, and W. Xia. Preparation and characterization of pullulan–chitosan and pullulan–carboxymethyl chitosan blended films. *Food Hydrocolloids* 30(1) (2013): 82–91.

Ziegler, G.R., A. Shetty, and R.C. Anantheswaran. Nut oil migration through chocolate. Pennsylvania State University, State College, PA, 2004.

27

Digestibility and Toxicology of Edible Films and Coatings

Silvia Moreno and Begoña Giménez

CONTENTS

Abstract

The research area of edible films and coatings has received a great interest in recent years due to an increased consumer concern in health, nutrition, food safety, and environmental issues. The main advantage of edible coatings and films over traditional synthetics is that they can be consumed with food. A variety of active compounds can be incorporated in edible films, and therefore, they can be used both as an active packaging and as a vehicle to deliver compounds, which may exert potential beneficial effects in the organism when they are consumed. In this case, the active compounds should be released during digestion to be available for absorption. Although the edible coatings and films are developed by selecting diverse natural-based substances that are expected to be safe, it is essential to know if the stability and bioactivity of the functional ingredients will be affected by the gastrointestinal conditions. However, only a few reports deal with their digestibility and the residual activity of the active compounds incorporated to them after digestion. In addition, possible transformations, during or after manufacture and digestion, may convert them into toxic substances for the human. Furthermore, nanomaterials used for creating new coatings and films with improved physical, chemical, and biological properties have encouraged global concern regarding their effect in biological system, at the same time demanding parallel risk-assessment studies.

Edible films and coatings prepared from substances with different structures and functions have a complex nature; for this reason, no single experimental studies can predict accurately their physiologic activity and toxicity in humans. Therefore, for the coatings and films' toxicity assessment, there is a dilemma in the selection and validation of the test methods to be used. Moreover, their toxicological evaluation may vary from one formulation to another, whereas the toxicological evaluation of well-known components may pose much less problems. Multidisciplinary efforts from biologist, toxicologist, and physicist are necessary to facilitate the interlinking of different features assisting in the understanding of cellular responses to coatings and films' exposure. This chapter introduces some methods to evaluate how well coatings and films are in connection with their compatibility in mammalian and remarks on some toxicological studies of biodegradable films, coatings, and active compounds. At the end, major issues/challenges remaining to be done in the future in this emerging field are disclosed.

27.1 Introduction

Edible films and coatings are thin layers of edible materials applied on food products that play an important role on their conservation, distribution and marketing (Falguera et al., 2011). Different biopolymers can be used in the formulation of edible films, such as proteins, polysaccharides, lipids, or the combination of them (Gontard and Guilbert, 1994). Recent reviews have focused on edible films based on lipids (Debeaufort and Voilley, 2009), proteins (Ramos et al., 2012), and polysaccharides such as chitosan (Dutta et al., 2009), hemicelluloses (Hansen and Plackett, 2008), starch (Jiménez et al., 2012), and pectin (Pérez-Espitia et al., 2014).

In addition, to act as a barrier to moisture, gases, and solutes, one of the most important aspects of edible films is their ability to carry and release a variety of active compounds, including antioxidants, flavorings, antibrowning, and antimicrobial compounds, among others (Falguera et al., 2011). Therefore, edible films can be used as an active packaging to control the location or rate of release of these active compounds in a food but they can also be used as a vehicle to deliver compounds that may have potential beneficial effects in the organism when these edible films are consumed. Although films generally represent only a minor portion of the food they cover, the study of their digestibility properties may provide useful information for extending potential uses for this type of material in the design of functional foods. Extensive publications have shown wide reviews regarding the physicochemical and technological features of edible coatings and films as well as their application (Cagri et al., 2004; Campos et al., 2011; Falguera et al., 2011; Gennadios et al., 1997; Guilbert et al., 1995; Janjarasskul and Krochta, 2010). However, research on their digestibility and the residual activity of the active compounds incorporated to them after digestion is scarce.

The main advantage of edible coatings and films over traditional synthetics is that they can be consumed with food. As described earlier, these covers are developed by selecting diverse natural-based substances that are expected to be safe. However, possible transformations may convert them into toxic substances for the human. In addition, intended functional ingredients and leachable substances such as nanomaterials used for developing new coatings and films with improved physical, chemical, and biological properties have also encouraged global concern regarding their effect in biological system and resulting in a demand for parallel risk assessment.

The toxicity of coatings and films may occur for different reasons; for example, after their manufacture, the components may interact with each other or undergo degradation, changing its chemical structure and eventually leading to a series of unexpected compounds that displayed some level of physiologic activity, as well as show signs of cell toxicity.

Therefore, toxicological evaluation should be performed for a safe use of edible coatings and films in food, and many key questions should be answered:

- *Which is the tissue response of each coatings and films?*
- *Are the constituents of coatings and films innocuous after human digestion?*
- *Which are the in vivo/in vitro suitable methods to predict whether coatings or films present potential harm to the consumer?*

The responses to these questions have remarkable importance from research and development perspectives and are essential for performance, safety, and regulatory reasons.

27.2 Digestibility of Edible Films and Coatings

Up to date, little research on digestibility of edible films and coatings can be found in literature. Taking into account that the biopolymers used for the elaboration of edible films are supposed to provide a nutritional value, it would be interesting to study the changes in the nutritional quality that these biopolymers may suffer during and after the film preparation. Furthermore, in the case of edible coatings and films used as vehicle for active compounds that provide a beneficial effect on consumer health, it is necessary to know if these active compounds are released from the matrix film and if they are released in the right place for absorption, which can vary depending on the active compound that it is incorporated.

27.2.1 Digestibility of the Matrix Film

The biopolymers used for edible films' preparation are often subjected to different treatments during and after film preparation in order to improve the mechanical and barrier properties of the resulting films. These treatments can imply changes in the nutritional properties and bioavailability of the biopolymers that could be evaluated by *in vitro* digestibility studies. For example, in the case of protein-based films, which show high gas-barrier properties but low mechanical and vapor-barrier properties for practical application, some cross-linking agents such as gossypol, formaldehyde, glutaraldehyde, calcium salts, carbodiimide, ferulic acid, and transglutaminase (Kwok and Ou, 2002; Mariniello et al., 2003; Marquie et al., 1995; Morel et al., 2000; Park et al., 2001; Takahashi et al., 1999) are often added into the film-forming solution to improve their mechanical properties (Ou et al., 2004). Moreover, protein-based films are sometimes treated by ultrasound or γ-irradiation with the same aim (Banerjee et al., 1996; Sabato et al., 2000). Ou et al. (2004) studied the effect of different pHs and addition of tannin, ferulic acid, and corn starch on the *in vitro* digestibility and available lysine of soy protein isolate after formation of films. They concluded that, except for the addition of corn starch, the more the mechanical and water-barrier properties improved, the more the protein digestibility and content of available lysine decreased. Hernández et al. (2008) evaluated *in vitro* digestibility changes of starches from different botanical sources when incorporated into a film, concluding that film manufacture increased rates of amylolysis as a consequence of granule disruption during production of the film-forming starch paste.

Incorporation of plasticizers may also cause significant changes in the digestibility of the biopolymers used for the elaboration of edible films. Plasticizers are generally small molecules of low volatility that modify the three-dimensional organization when added to polymeric materials. They decrease intermolecular forces between polymer chains, increasing free volume and chain mobility. As a consequence, extensibility, dispensability, and flexibility of the developed film are increased, while cohesion and rigidity of the film are decreased (Kokoszka et al., 2010). Plasticizers are required for the elaboration of edible films, especially for polysaccharide and protein-based films, since the resultant structure is brittle and stiff due to extensive polymer–polymer interactions (Gennadios, 2004; Krochta, 2002). Food-grade plasticizers incorporated into edible films include polyols (glycerol, sorbitol, polyethylene glycol), sugars (glucose, honey), and lipids (monoglycerides, phospholipids, surfactants). Glycerol is the most common plasticizer used in film-making techniques, due to stability and compatibility with hydrophilic biopolymeric packaging chain (Chillo et al., 2008). Hernández et al. (2008) observed that the addition of glycerol into starch films affected the kinetics of starch digestion, giving lower starch digestion rates. This feature should be considered when aiming to produce low-glycemic-index foods.

27.2.2 Digestibility of Edible Films and Coatings as Vehicle for Active Compounds

In order to retain the potential health benefits of the active compounds incorporated into edible films, it is essential to know if they are released during digestion. Only the active compounds released from solid matrices become bioaccessible, potentially available for absorption by the gastrointestinal tract and, therefore, able to exert their beneficial effects in the human body (Tagliazucchi et al., 2010). The release of active compounds from the edible films does not necessarily imply the digestion of the matrix film. López de Lacey et al. (2012) demonstrated that green tea polyphenols contained in agar films were bioaccessible and therefore susceptible for absorption during simulated human digestion. The recovery of polyphenols took place mainly in the stomach, and the green tea compounds recovered showed both reducing powder and radical scavenging ability. After the gastrointestinal digestion, the polymeric matrix of agar with a certain amount of green tea extract retained in it remained as indigestible residue, which could be available in the colon. Okello et al. (2011) observed protective effects of colon-available extract of green tea against hydrogen peroxide and beta-amyloid-induced cytotoxicity in PC12 cells used as a model of neuronal cells. According to these authors, this colon-available green tea extract was depleted in flavan-3-ol and relatively enriched in certain flavonols, hydroxycinnamates, caffeine, theobromine, and a range of unidentified phenolic components. In the case of active edible films made of fish gelatin and green tea extract, high percentage of total polyphenols were recovered from the films after a simulated gastrointestinal digestion, although a significant degradation of the major catechins of the green tea epigallocatechin gallate and epigallocatechin was observed. Furthermore, the gelatin matrix of the films was efficiently hydrolyzed during the simulated digestion, resulting in gelatin hydrolysates composed of low-molecular-weight peptides (Giménez et al., 2013).

Besides the possible degradation that the active compounds released from the edible films and coatings may suffer during digestion, other factors involved in the bioavailability of these active compounds are the possible interactions with other food components, which may affect or even cancel the biological activity of the active compounds. However, only the films are usually subjected to digestion in most of the studies dealing with digestibility of edible films, and the possible interactions that may occur with other food components are not considered. López de Lacey et al. (2012) used gelatin in order to simulate the presence of diet protein during the digestion of agar films with green tea extract and to assess the interactions between protein and polyphenols, concluding that the presence of gelatin affected both the bioaccessibility and efficacy of green tea polyphenols.

On the other hand, it should be taken into account that sometimes the residual activity obtained for a certain active compound after *in vitro* edible film digestion cannot be extrapolated to the results that would be obtained *in vivo*, since the simulated bile and pancreatic juice used in the study of digestibility may interfere with the analytical technique used to measure the residual activity of the active compound.

27.2.3 Digestibility of Films and Coatings as Drug-Coating Materials

In the pharmaceutical field, polymeric materials that can be degraded by colonic microbiota have been investigated as new drug-coating materials. These polymeric materials should resist digestion in the stomach and duodenum, protecting the drug from the gastrointestinal environment and the brush border proteolytic enzymes and therefore providing colon-specific drug delivery systems. Because of the slow transit time, relatively low proteolytic activity, and pH values near to neutral, the colon is a preferred absorption site for oral administration of drugs, mainly proteins and peptides (Meneguin et al., 2014). Furthermore, colon-specific delivery systems can provide local treatment of several inflammatory bowel diseases such as ulcerative colitis, Crohn's disease, and colon carcinoma, allowing them to reduce the effective dose as well as the side effects (Souto Maior et al., 2008). For example, when pectin–ethylcellulose film-coated and uncoated pellets containing 5-fluorouracil were orally administered to rats, differences in biodistribution and pharmacokinetics were observed. The release of 5-fluorouracil from coated pellets took place mainly in the cecum and colon, and the mean residence time was 10-fold higher compared to the uncoated pellets (He et al., 2008). Based on their microbial biodegradability, polysaccharides have been proposed as promising drug-coating materials for the development of colon-specific delivery systems. Among the polysaccharides, pectin (Ahrabi et al., 2000; Ashrod et al., 1993; Fernández-Hervás and Fell, 1998; He et al., 2007, 2008), amylose (Basit et al., 2004; Thompson et al., 2002), guar gum (Krishnaiah et al., 2998, 2003; Tugcu-Demiroz et al., 2004), and chitosan (Shimono et al., 2002) have been studied for their potential for colonic delivery. The major inconvenient of native polysaccharides is their high solubility in aqueous media, and consequently, film coatings made of these polymers, alone or combinations of them, allow the release of drugs during the transit through the stomach and the small intestine (Ghaffari et al., 2007). To overcome this problem, one of the strategies used may be the combination of these polysaccharides with water-insoluble polymers such as ethylcellulose and acrylic polymers (Eudragit RS30D, Eudragit NE30D). The use of mixed films in drug delivery allows the optimization of the physicochemical and permeability properties of the resultant films (Ofori-Kwakye and Fell, 2003). Ghaffari et al. (2007) described that films made of pectin and chitosan were unable to protect premature swelling and theophylline release in simulated small intestine medium; but the incorporation of the water-insoluble polymer Eudragit RS drastically reduced the swelling ratio (Ghaffari et al., 2007). Resistant starch also may be a promising material to be used in the design of colon-specific drug delivery systems, since it resists digestion in the stomach and small intestine but it is degraded by colonic microbiota (Htoon et al., 2010; Meneguin et al., 2014).

27.3 Toxicology of Edible Coatings and Films' Constituents

Edible coatings and films have expanded their use to the field of human nutrition and health by the inclusion of nutrients or other bioactive compounds to be transported to the intestine. In consequence, a major part of the potential toxicity of the edible coatings and films begins after translocation from the lumen of the intestinal tract into different organs.

As mentioned in other chapters, edible coatings and films are based on biodegradable and biocompatible polymers from natural sources such as polysaccharides as pectin, proteins as gelatin, and some lipids listed as generally recognized as safe by the Food and Drug Administration and are used in food products (Doi et al., 2011; Falguera et al., 2011; Pérez-Espitia et al., 2014). Also, food-grade plasticizers are preferred as glycerol and sorbitol, due to their biocompatibility (Bjoervell et al., 1982; Chillo et al., 2008). Therefore, toxicity evaluation of the original constituents of edible coatings and films may not be in general necessary when data are available on the safety use of them. However, toxicity studies are essential in order to rule out potential transformations of the starting materials during or after manufacture. These processes may convert some edible coatings and films' components into toxic substances for the human. Moreover, digestion process or interaction with different food systems is also another cause in the synthesis of variants of greater toxicity of the initial edible coatings and films' components. As a result, the toxicological evaluation of the active packaging material after or before digestion as well as the cytotoxic level of the active compounds after the gastrointestinal digestion will demonstrate the potential degree of toxicity.

At present, there is scarce information on toxicological evaluation and awareness related to health of edible films and coatings (see Section 27.2.2) giving rise to a gap in crucial information that should be accomplished in the near future in order to meet the safety demands of the edible films and coatings as active packaging material.

27.3.1 Test for Toxicological Evaluation of Edible Coatings and Films

Helpful information about several useful testing procedures for the toxicological evaluation of these complex materials is addressed in this chapter. In a general way, as edible coatings and films can be part of foods, the toxicity evaluation could be performed in the same way as those made in any design of a food. In this sense, several toxicological tests for predicting the safety of these materials can be conducted using internationally agreed state-of-the-art protocols, such as the Organization for Economic Cooperation and Development (OECD) (2001, 2010) protocols and standard methods utilized in the past for safety evaluation of certain food additives as WHO (1974) guidelines using animals. Between them, acute toxicity after oral administration of edible coatings and films can be carried out to determine the effect on a whole organism on two different animal species, one rodent and one nonrodent (Stallard and Whitehead, 1995). The biomaterial to be tested can be administrated as small pieces mixed with the food, in the water solution, or in edible oil at 5, 50, 500, or 2000 mg/kg (OECD, 2001). Histopathological studies of several organs will be useful to rule out potential hepatic damage and nephropathy after ingestion of each formulation of edible coatings and films developed. Also, as applicable to any chemical compounds added to foods, genotoxic potential of edible coatings and films must be also assessed by induction of gene mutations using *in vitro* bacterial mutagenicity (Maron and Ames, 1983) and structural and numerical chromosomal alterations (Collins, 2004; EFSA, 2011; OECD, 2010; Shaposhnikov et al., 2011; Singh et al., 2009).

27.3.2 Cytotoxicity Evaluation of Edible Coatings and Films

To extend the breadth of toxicological analysis of edible coatings and films, mammalian cell culture or immortalized cells are an extremely useful *in vitro* technology. Actually, there have been a number of works published on *in vitro* cell biocompatibility and cytotoxicity assay to assess the toxicity of individual active compounds, which probably can display the most negative side effects after addition in the edible coatings and films (Chen et al., 2011; Gaya et al., 2013).

There are several tests to determine cell toxicity, and the widely employed assays are based on the detection of the mitochondrial dehydrogenase activity and uptake/release of a number of dyes. The activity of mitochondrial dehydrogenase can be measured by a tetrazolium dye called 3-(4,5-dimethylthiazol-2-yl)-2,5-diphenyltetrazolium bromide (MTT) dye or its soluble derivative, the MTS dye (Roehm, 1991). The mitochondrial dehydrogenase activity of viable cells cleaves the tetrazolium ring, yielding colored crystals of formazan, and the absorbance is measured by spectrophotometer. Other classic methods use exclusion dyes such as trypan blue, eosin, or propidium and neutral red (Borenfreund and Puerner, 1985; Strober, 2001).

There are only a few studies dealing with the cytotoxic effects of edible coatings and films as a whole material or around possible active compounds released to a greater or lesser degree after they are digested. The cytotoxicity of fish gelatin films containing Chinese green tea known as Lung Ching (*Camellia sinensis* L.) extract after gastrointestinal digestion on fibroblast cell line 3T3-L1 was evaluated by us (Giménez et al., 2013). The viability effect of gelatin film containing 2% green tea extract, 4% green tea extract, and 8% green tea extract after digestions or without green tea extract as control on 3T3-L1 cells was determined by the MTS assay (Figure 27.1). Results showed that only gelatin film with 2% green tea extract gave cell viabilities over 50%, while the cytotoxicity of films with 4% and 8% green tea extract, despite showing cell viabilities below 50%, may not be considered physiologically significant, because the high dilution of the green tea extract inside the human stomach would avoid the possible side effects. Further, it can be deduced that the cytotoxic level of the green tea extract was maintained after the gastrointestinal digestion (Giménez et al., 2013).

In another study, a cell biocompatibility assay using fibroblast NIH 3T3 cells was used to investigate the effect of a biodegradable polymers: poly(butylene succinate-co-cyclic carbonate) on cell morphology, adhesion properties, and cell viability determined by the MTT assay (Yang et al., 2004). Results obtained showed that pieces of the biodegradable polymer placed into the 24-well culture plates after 48 h at 37°C exhibited similar cytotoxicity than the commercial aliphatic polyesters (PLLA) used as reference.

27.3.3 Cytotoxicity of Edible Coatings and Films Containing Nanomaterials

The incorporation of nanomaterials in edible coatings and films opens up many possibilities for the generation of materials with enhanced properties. Nanostructured materials may allow a controlled release of bioactive compounds such as antioxidants, nutraceuticals and natural antimicrobial agents, as well as to protect micronutrients during manufacture and storage (Ishikawa et al., 2012). Also, nanofibers of cellulose incorporated to edible films based on pectin and mango were effective in increasing tensile strength of the matrix (Azeredo et al., 2009). However, researches related to the application of nanotechnology to edible film are scarce. This limitation is probably due to the current gap for completely understanding the toxicological effects of nanoparticles in human system after ingestion, which allows for determining the overall impact on consumers.

Nowadays, several studies have been initiated to investigate the cytotoxicity of nanocomplex and individual polymeric nanoparticles (Kumar et al., 2013; Sharma et al., 2012; von Staszewski et al., 2012). Nanocomplex formation between b-lactoglobulin or caseinomacropeptide and green tea polyphenols was developed, and the impact on protein gelation and the polyphenols' antiproliferative activity was reported (von Staszewski et al., 2012). Gelatin nanospheres were developed from gelatin aqueous solution containing luciferase small-interfering RNA coacelvated by acetone addition followed by the glutaraldehyde cross-linking of gelatin (Ishikawa et al., 2012). Results showed that nanospheres were internalized into the colon 26 cells and the small-interfering RNA was released inside the cell. Other authors encapsulated nanocurcumin with *N*-isopropylacrylamide, with *N*-vinyl-2-pyrrolidone, and with

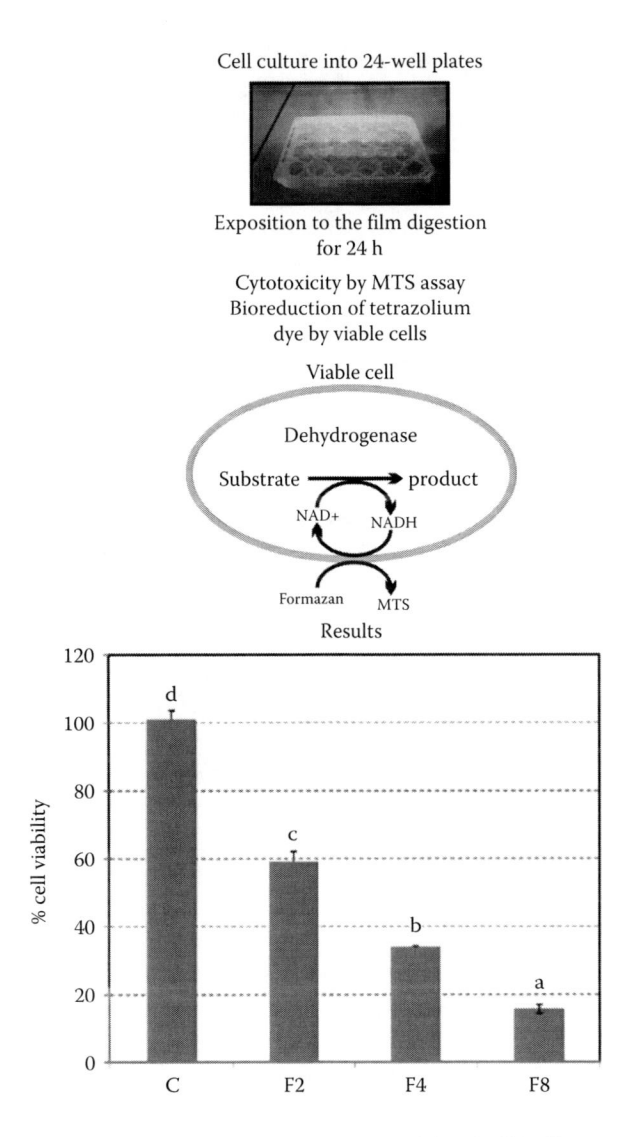

FIGURE 27.1 Schematic illustration of the method to evaluate cytotoxicity of edible films containing green tea extract after digestion. The 3T3-L1 cells cultured in a 24-well microplate were treated with films containing 2% green tea extract (F2), 4% green tea extract (F4), and 8% green tea extract (F4) after digestion or without green tea extract as control (C). Viability was determined after 24 h by the MTS assay measuring the absorbance in a microplate reader at 485 nm.

poly(ethylene glycol) monoacrylate that demonstrates comparable *in vitro* therapeutic efficacy to free curcumin against a panel of human pancreatic cancer cell lines as assessed by cell viability and clonogenicity assays in soft agar including induction of cellular apoptosis, blockade of nuclear factor kappa B activation, and downregulation of steady-state levels of multiple proinflammatory cytokines (IL-6, IL-8, and TNFα) (Bisht et al., 2007).

Due to the fact that the smaller size of nanoparticles imparts a different biokinetic behavior and ability to reach more distal regions of the body, it gives them unexpected and unanticipated consequences on interaction with biological systems. Indeed, many *in vivo* and *in vitro* investigations established accumulation of nanoparticles in organs like liver and adverse effects on DNA, immune system, and neurobehavioral patterns (Zhang et al., 2011). Another problematic issue is that there are no yet existing standard methodologies for risk assessment of nanomaterials. In this sense, recently modified cytotoxicity protocols have been reported (Dhawan and Sharma, 2010; Kumar et al., 2013).

In resume, when more toxicological studies are published in the coming years, the application of nanotechnology may lead the research toward the incorporation of diverse nanomaterials in edible films meeting the safety requirements for human use.

27.4 Conclusions

- Edible films and coatings represent a stimulating route for creating new materials with a wide range of properties that can be used as an active packaging, as well as a vehicle for active compounds with beneficial biological effect in the organism when they are consumed. However, potential functions and applications of coatings and films deserve increased considerations: studies of digestibility and safety/toxicity topics of coatings and films have remained far behind the rate at which they are being produced.

- The nutritional value of the biopolymers used in the elaboration of edible coatings and films may be modified during the film preparation.

- Depending on the proposed target, the edible films and coatings should be developed to release the active molecules incorporated into them in the upper gastrointestinal tract or in the colon. A more realistic assessment of the possible interactions between the film components and other food components during digestion should be done, in order to evaluate their bioavailability more accurately.

- Toxicological analysis of coatings and films should allow answering the question whether unintended adverse effects have been produced following their use in human.

- These assessments should be performed on a case-by-case basis and the coatings and films tested should be in a similar form that would be used by human (e.g., processed foods).

- Preclinical test in animals and/or cellular assays should be performed and complete end points (including biochemical, hematological, and histological end points) according to the OECD guidelines for toxicity testing are requested.

27.5 Future and Prospect

- Many aspects remain to be done in the area of films and coatings that should be developed in the near future. There is a need for more comprehensive studies to fully recognize and address the potential risks of engineered coatings and films. Further experimental data are required to make a full understanding of the potential toxicity of the edible biomaterials to arrive at the decision about the acceptability of them. This will help in creating biologically safe coatings and films for application in human use.

- The major concern of the scientific community when incorporating nanomaterials into edible coatings or food is still unresolved: the absence of studies into their possible toxicity. Unified toxicity evaluation of coatings and films using multitasking approaches is recommended. A multidisciplinary team work is necessary to facilitate the interlinking of different facets of the toxicological assessment of coatings and films, thus aiding in the understanding of cellular responses of these materials.

- A future prospect is that all approaches established are properly validated. Validation of the studies performed has to be conducted to assess how well its results are reproduced within and across laboratories and to measure the ability to correctly predict or measure the biological effect of interest. The guidelines of OECD and ICCVAM (http://iccvam.niehs.nih.gov/about/accept.htm) would be very useful to perform this task to validate protocols and test methods used and critical procedural details. Also, the risk assessment of substances to be used as food additives can be useful for edible coating and film evaluation (Gilsenan, 2011).

- As a final point, once toxicological requirements are met, it remains improved in the processes of regulation of these new materials for a successful implementation and commercialization of coatings and films generating a niche market that will be our future.

REFERENCES

Ahrabi SF, Madsen G, Dyrstad K, Sande SA, and Graffner C. Development of pectin matrix tablets for colonic delivery of model drug ropivacaine. *Eur J Pharm Sci* 10 (2000): 43–52.

Ashford M, Fell J, Attwood D, Sharma H, and Woodhead. An evaluation of pectin as a carrier for drug targeting to the colon. *J Control Release* 26 (1993): 213–220.

Azeredo HM, Mattoso LH, Wood D, Williams TG, Avena-Bustillos RJ, and McHugh TH. Nanocomposite edible films from mango puree reinforced with cellulose nanofibers. *J Food Sci* 74(5), (2009): 31–35.

Banerjee R, Chen H, and Wu J. Milk protein-based edible film mechanical strength changes due to ultrasound process. *J Food Sci* 61 (1996): 824–828.

Basit AW, Podczeck F, Newton JM, Waddington WA, Ell PJ, and Lacey LF. The use of formulation technology to assess regional gastrointestinal drug absorption in humans. *Eur J Pharm Sci* 21 (2004): 179–189.

Bisht S, Feldmann G, Soni S et al. Polymeric nanoparticle-encapsulated curcumin (nanocurcumin): A novel strategy for human cancer therapy. *J Nanobiotechnol* 5 (2007): 3.

Bjoervell H and Rössner S. Effects of oral glycerol on food intake in man. *Am J Clin Nutr* 36 (1982): 262–265.

Borenfreund E and Puerner JA. Toxicity determined in vitro by morphological alterations and neutral red absorption. *Toxicol Lett* 24 (1985): 119–124.

Cagri A, Ustunol Z, and Ryser LT. Antimicrobial edible films and coatings. *J Food Prot* 4 (2004): 636–848.

Campos CA, Gerschenson LN, and Flores SK. Development of edible films and coatings with antimicrobial activity. *Food Bioprocess Tech* 4 (2011): 849–875.

Chen D, Wan SB, Yang H et al. EGCG, green tea polyphenols and their synthetic analogs and prodrugs for human cancer prevention and treatment. *Adv Clin Chem* 53 (2011): 155–177.

Chillo S, Flores S, Mastromatteo M, Conte A, Gerschenson L, and Del Nobile MA. Influence of glycerol and chitosan on tapioca starch-based edible film properties. *J Food Eng* 88(2) (2008): 159–168.

Collins AR. The comet assay for DNA damage and repair: Principles, applications, and limitations. *Mol Biotechnol* 26(3) (2004): 249–261.

Debeaufort F, and Voilley A. Lipid-based edible films and coatings. In M.E. Embuscado, and K.C. Huber (eds.), *Edible Films and Coatings for Food Applications*, Springer, New York, 2009, pp. 246–268.

Dhawan A and Sharma V. Toxicity assessment of nanomaterials: Methods and challenges. *Anal Bioanal Chem* 398 (2010): 589–605.

Doi N, Jo JI, and Tabata Y. Preparation of biodegradable gelatin nanospheres with a narrow size distribution for carrier of cellular internalization of plasmid DNA. *J Biomater Sci Polym* 23 (2011): 991–1004.

Dutta PK, Tripathi S, Mehrotra GK, and Dutta J. *Food Chem* 114(4) (2009): 1173–1182.

EFSA Scientific Committee. Scientific opinion on genotoxicity testing strategies applicable to food and feed safety assessment. *EFSA J* 9(9) (2011): 2379.

Falguera V, Quintero JP, Jiménez A, AMuñoz J, and Ibarza A. *Trends Food Sci Technol* 22 (2011): 292–303.

Fernández-Hervás MJ and Fell JT. Pectin/chitosan mixtures as coatings for colon-specific drug delivery: An in vitro evaluation. *Int J Pharm* 169 (1998): 115–119.

Gaya M, Repetto V, Toneatto J et al. Antiadipogenic effect of carnosic acid, a natural compound present in *Rosmarinus officinalis*, is exerted through the C/EBPs and PPARγ pathways at the onset of the differentiation program. *Biochim Biophys Acta* 830 (2013): 3796–3806.

Gennadios A, Hanna MA, and Kurth LB. Application of edible coatings on meats, poultry and seafoods: A review. *LWT-Food Sci Technol* 30(4) (1997): 337–350.

Gennadios A. Edible films and coatings from proteins. In R. Yada (ed.), *Proteins in Food Processing*. Woodhead, Cambridge, U.K., 2004, pp. 442–467.

Ghaffari A, Navaee N, Oskoui M, Bayati K, and Rafiee-Tehrani M. Preparation and characterization of free mixed-film of pectin/chitosan/Eudragit RS intended for sigmoidal drug delivery. *Eur J Phram Biopharm* 67 (2007): 175–186.

Gilsenan MB. Additives dairy foods safety. In J.W. Fuquay (ed.), *Encyclopedia of Dairy Sciences*, 2nd edn. Leatherhead Food Research, Leatherhead, U.K., 2011, pp. 55–60.

Giménez B, Moreno S, López-Caballero ME, Montero P, and Gómez-Guillén MC. Antioxidant properties of green tea extract incorporated to fish gelatin films after simulated gastrointestinal enzymatic digestion LWT. *Food Sci Technol* 53 (2013): 445–451.

Gontard N, and Guilbert S. Biopackaging: Technology and properties of edible and/or biodegradable material of agricultural origin. In M. Mathlothi (ed.), *Food Packaging and Preservation*. Blackie Academic and Professional, New York, 1994, pp. 159–181.

Guilbert S, Gontard N, and Cuk B. Technology and applications of edible protective films. *Pack Technol Sci* 8(6) (1995): 339–346.

Hansen NML, and Plackett D. Sustainable films and coatings from hemicelluloses: a review. *Biomacromolecules* 9(6) (2008): 1493–1505.

He W, Du Q, Cao DY, Xiang B, and Fan LF. Pectin/ethylcellulose as film-coatings for colon-specific drug delivery: Preparation and *in vitro* evaluation using 5-fluorouracil pellets. *PDA J Pharm Sci Technol* 61 (2007): 121–130.

He W, Du Q, Cao DY, Xiang B, and Fan LF. Study on colon-specific pectin/ethylcellulose film-coated 5-fluorouracil pellets in rats. *Int J Pharm* 348 (2008): 35–45.

Hernández O, Emaldi U, and Tovar J. *In vitro* digestibility of edible films from various starch sources. *Carbohyd Polym* 71(4) (2008): 648–655.

Htoon AK, Uthayakumaran S, Piyasiri U, Appelqvist AM, López-Rubio A, Gilbert EP, and Mulder RJ. The effect of acid dextrinisation on enzyme-resistant starch content in extruded maize starch. *Food Chem* 120 (2010): 140–149.

Ishikawa H, Nakamura Y, Jo J et al. Gelatin nanospheres incorporating siRNA for controlled intracellular release. *Biomaterials* 33 (2012): 9097–9104.

Janjarasskul T, and Krochta JM. Edible packaging materials. *Annu Rev Food Sci Technol* 1 (2010): 415–448.

Jiménez A, Fabra MJ, Talens P, and Chiralt A. Edible and biodegradable starch films: a review. *Food Bioprocess Tech* 5(6), (2012): 2058–2076.

Kokoszka S, Debeaufort F, Lenart A, and Voilley A. Water vapour permeability, thermal and wetting properties of whey protein isolate based edible films. *Int Dairy J* 20 (2010): 53–60.

Krishnaiah YS, Satyanarayana S, Rama Prasad YV, and Narasimha Rao S. Gamma scintigraphic studies on guar gum matrix tablets for colonic drug delivery in healthy human volunteers. *J Control Release* 55 (1998): 245–252.

Krishnaiah YSR, Indira Muzib Y, Bhaskar P, Satyanarayana V, and Latha K. Pharmacokinetic evaluation of guar gum-based colon-targeted drug delivery systems of tinidazole in healthy human volunteers. *Drug Deliv* 10 (2003): 263–268.

Krochta JM. Proteins as raw materials for films and coatings: definitions, current status and opportunities. In A. Gennadios (ed.), *Protein Based Films and Coatings*. CRC Press, Boca Raton, FL, 2002, pp. 1–41.

Kumar A, Sharma V, and Dhawan A. Methods for detection of oxidative stress and genotoxicity of engineered nanoparticles. *Methods Mol Biol* 1028 (2013): 231–246.

Kwok KC, and Ou SY. Application of ferulic acid in preparation of edible films based on soy protein isolate. *Food Sci Technol* 129 (2002): 24–26

López de Lacey AM, Giménez B, Pérez-Santín E, Faulks R, Mandalari G, López-Caballero ME, and Montero P. Bioaccessibility of green tea polyphenols incorporated into an edible agar film during simulated human digestion. *Food Res Int* 48 (2012): 462–469.

Mariniello L, Di Pierro P, Esposito C, Sorrentino A, Masi P, and Porta R. Preparation and mechanical properties of edible pectin-/soy flour films obtained in the absence or presence of transglutaminase. *J Biotechnol* 102 (2003): 191–198.

Maron DM and Ames BN. Revised methods for the *Salmonella* mutagenicity test. *Mutat Res* 113 (1983): 173–215.

Marquie C, Aymard C, Cuq JL, and Guilbert S. Biodegradable packaging made from cottonseed flour: Formation and improvement by chemical treatments with gossypol, formaldehyde, and glutaraldehyde. *J Agric Food Chem* 43 (1995): 2762–2767.

Meneguin AB, Ferreira Cury BS, and Evangelista RC. Films from resistant starch-pectin dispersions intended for colonic drug delivery. *Carbohyd Polym* 99(2) (2014): 140–149.

Morel MH, Bonicel J, Micard V, and Guilbert S. Protein insolubilization and thiol oxidation in sulfide-treated wheat gluten films during aging at temperature and humidities. *J Agric Food Chem* 48 (2000): 186–190.

OECD. Organization for economic co-operation and development series on testing and assessment No. 24: Guidance document on acute oral toxicity testing (ENV7JM/MONO), 2001.

OECD. Guideline for the testing of chemicals. Section 4: Health effects. Proposal for updating guideline 487. *In Vitro* Mammalian Chromosome Aberration Test, 2010.

Ofori-Kwakye K, and Fell JT. Leaching of pectin from mixed films containing pectin, chitosan and HPMC intended for biphasic drug delivery. *Int J Pharm* 250(1) (2003): 251–257.

Okello EJ, McDougall GJ, Kumar S, and Seal CJ. *In vitro* protective effects of colon-available extract of *Camellia sinensis* (tea) against hydrogen peroxide and beta-amyloid ($A\beta_{(1-42)}$) induced cytotoxicity in differentiated PC12 cells. *Phytomedicine* 18 (2011): 691–696.

Ou S, Kwok KC, and Kang Y. Changes in in vitro digestibility and available lysine of soy protein isolate after formation of film. *J Food Eng* 64 (2004): 301–305.

Park SK, Rhee CO, Bae DH, and Hettiarachchy NS. Mechanical properties and water-vapor permeability of soy protein films affected by calcium salts and glucono-d-lactone. *J Agric Food Chem* 49 (2001): 2308–2312.

Pérez-Espitia PJ, Du W-X, Avena-Bustillos RJ, Ferreira Soares NF, and McHugh, TH. Edible films from pectin: Physical-mechanical and antimicrobial properties—A review. *Food Hydrocol* 35 (2014): 287–296.

Ramos OL, Fernandes JC, Silva SI, Pintado ME, and Malcata FX. Edible films and coatings from whey proteins: a review on formulation, and on mechanical and bioactive properties. *Crit Rev Food Sci Nutr* 52 (2012): 533–552.

Roehm NW. An improved colorimetric assay for cell proliferation and viability utilizing the tetrazolium salt XTT. *J Immunol Methods* 142 (1991): 257–265.

Sabato SF, Ouattara B, Yu H, D'Aprano G, Le Tien C, Mateescu MA, and Lacroix M. Mechanical and barrier properties of cross-linked soy and whey protein based films. *J Agric Food Chem* 49 (2000): 1397–1403.

Shaposhnikov S, Thomsen PD, and Collins AR. Combining fluorescent in situ hybridization with the comet assay for targeted examination of DNA damage and repair. *Methods Mol Biol* 682 (2011): 115–132.

Sharma V, Kumar A, and Dhawan A. Nanomaterials: Exposure, effects and toxicity assessment. *Proc Natl Acad Sci India Sect B Biol Sci* 82 (2012): 3–11.

Shimono N, Takatori T, Ueda M, Mori M, Higashi Y, and Nakamura Y. Chitosan dispersed system for colon-specific drug delivery. *Int J Pharm* 245(1–2) (2002): 45–54.

Singh N, Manshian B, Jenkins GJ et al. Nanogenotoxicology: The DNA damaging potential of engineered nanomaterials. *Biomaterials* 30(23–24) (2009): 3891–3914.

Souto Maior JFA, Reis AV, Muniz EC, and Cavalcanti OA. Reaction of pectin and glycidyl methacrylate and ulterior formation of free films by reticulation. *Int J Pharm* 355 (2008): 184–194.

Stallard N and Whitehead A. Reducing animal numbers in the fixed-dose procedure. *Hum Exp Toxicol* 14 (1995): 315–323.

Strober W. Trypan blue exclusion test of cell viability. *Curr Protoc Immunol* Appendix 3 (2001): Appendix 3B.

Tagliazucchi D, Verzelloni E, Bertolini D, and Conte A. *In vitro* bioaccessibility and antioxidant activity of grape polyphenols. *Food Chem* 120 (2010): 599–606.

Takahashi K, Nakata Y, Someya K, and Hattori M. Improvement of the physical properties of pepsin-solubilized elastin–collagen film by cross-linking. *Biosci Biotechnol Biochem* 63 (1999): 2144–2149.

Thompson RPH, Bloor JR, Ede RJ, Hawkey C, Hawthorne B, Muller FA, Palmer RMJ. Preserved endogenous cortisol levels during treatment of ulcerative colitis with COLAL-PRED, a novel oral system consistently delivering prednisolone metasulphobenzoate to the colon. *Gastroenterology* 122 (2002): 207.

Tugcu-Demiroz F, Acarturk F, Takka S, and Konus-Boyunaga O. *In vitro* and *in vivo* evaluation of mesalazine-guar gum matrix tablets for colonic drug delivery. *J Drug Target* 12 (2004): 105–112.

von Staszewski M, Jara FL, Ruiz LTG, Jagus RJ, Carvalho JE, and Pilosof AMR. Nanocomplex formation between b-lactoglobulin or caseinomacropeptide and green tea polyphenols: Impact on protein gelation and polyphenols antiproliferative activity. *J Funct Foods* 4(4) (2012): 800–809.

WHO. Toxicological evaluation of certain food additives with a review of general principles and of specifications. Technical report series no. 5, FAO nutrition meeting report series no. 53A, Published by FAO and WHO, 1974.

Yang J, Tian W, Li Q et al. Novel biodegradable aliphatic poly(butylene succinate-co-cyclic carbonate)s bearing functionalizable carbonate building blocks: II. Enzymatic biodegradation and *in vitro* biocompatibility assay. *Biomacromolecules* 5 (2004): 2258–2268.

Zhang XD, Wu D, Shen X et al. Size-dependent in vivo toxicity of PEG-coated gold nanoparticles. *Int J Nanomed* 6 (2011): 2071–2081.

28

Biodegradable Polymer for Food Packaging: Degradation and Waste Management

Almudena Ochoa-Mendoza, Carmen Fonseca-Valero, Jessica Acosta-García, and Teresa Agüinaco-Castro

CONTENTS

28.1 Introduction

Nowadays, there is a continuous increase in the volume of manufactured synthetic polymers used for various applications, which is one of the most important functions in food packaging. The disposal of the used materials is becoming a serious problem. Some fractions of plastic waste generated from food packaging materials may be recycled, but most of these residues are disposed in landfills due to technical and/or economical reasons.

Currently, plastic waste represents around 11% of the total municipal waste in the EU. In 2008, the total generation of post-consumer plastic waste in EU-27, Norway, and Switzerland was 24.9 Mt. Packaging is by far the largest contributor to plastic waste at 63% (European Commission DG ENV, 2011).

At this moment, the concern about the environmental impact of packaging wastes are increasing due to their high consumption, short shelf life, and nonbiodegradability that contribute to overflowing landfills, among other problems (Huang, 1995, Siracusa et al., 2008). The difficulties in collecting, identifying, sorting, transporting, cleaning, and reprocessing plastic packaging materials often render the attempt of recycling, which is noneconomical, thus making disposal to landfills a more convenient and easy alternative (Davis and Song, 2006).

Consequently, there is a growing interest and demand for biodegradable plastics, especially in the food packaging field, to make a more sustainable society and to solve global environmental and waste management problems. In the last decade, there has been an increased interest toward the development and application of biodegradable plastics in the food packaging field (Bastioli, 2005, Iwata, 2015, Murphy and Bartle, 2004, Pillai, 2010, Song et al., 2009, Vroman and Tighzert, 2009).

Thus, Davis et al., and other researchers, have explained that biodegradable packaging materials are most suitable for single-use disposable packaging applications where post-consumer package can be locally composted as a means of recycling the materials (Davis and Song, 2006).

The use of biodegradable plastics has been promoted positively in Europe, particularly in Italy, where it was decided in 2010 that all disposable shopping bags should be either reusable or produced from biodegradable plastics (Iwata, 2015).

Moreover, it is anticipated that as the materials are from renewable resources and biodegradable, they would contribute to sustainable development and, if properly managed, would reduce their environmental impact upon disposal. But to reach this situation, it will be necessary to study the impact of the biodegradable plastics on waste management in order to establish adequate waste management systems and legislation (Davis and Song, 2006).

Biodegradable plastics are materials that can be degraded by microorganism such as bacteria, fungi, and algae, via enzymatic action, under appropriate conditions of moisture, temperature, and oxygen availability, leading to fragmentation of macromolecular chains, producing no toxic residues for the environment (Fritz et al., 2001). The end products are CO_2, new biomass, and water (in the presence of oxygen, i.e., aerobic conditions) or methane (in the absence of oxygen, i.e., anaerobic conditions) (Avérous and Pollet, 2012, Chandra and Rustgi, 1998, Luckachan and Pillai, 2011, Vroman and Tighzert, 2009). These plastics are biodegradable and/or compostable according to the standards EN 13432:2000, EN 14995:2007, ISO 17088:2012, and ASTM D-6400/2004, having either a renewable or a fossil origin.

Compostable plastics must be biodegradable and disintegrable in a brief period of time in order to yield, during composting, water, CO_2, inorganic compounds, and fertile compost with no visible or toxic residue, without altering the quality of the produced compost (Luckachan and Pillai, 2011, Siracusa et al., 2008).

The speed of biodegradation depends on temperature ($50°C–70°C$), humidity, and the number and type of microbes. Degradation is fast only if all the three requirements are present. Generally, at home or in a supermarket, biodegradation occurs very low in comparison to composting. In industrial composting, bioplastics are converted into biomass, water, and CO_2 in about 6–12 weeks (www.european-bioplastic.org) (Siracusa et al., 2008). Depending on the type of standard to follow (ASTM or EN), different composting conditions (humidity and temperature cycle) must be realized to determine the compostability level. Therefore, the comparison of the results obtained from different standards is very difficult or sometimes impossible to establish (Avérous and Pollet, 2012).

On the other hand, biodegradable polymers from renewable resources, however, present some disadvantages such as high costs and poor barrier and mechanical properties, limiting their industrial applications. The problems associated with them are mainly those of performance, processing, and cost (Rhima et al., 2013), that is, in relation to performance are brittleness, low heat distortion temperature, and high gas and vapor permeability, while in relation to process are poor rheological properties and a higher tendency to degradation in comparison with conventional polymers (Bharadwaj, 2001, Koh et al., 2008, Krochta and De Mulder-Johnston, 1997, Neilsen, 1967, Sorrentino et al., 2006, Steinbüchel, 1995). These drawbacks have been limited either by adding nanofillers, by using them as bionanocomposites, or by coating or blending with other polymers.

According to the European Bioplastics, the term "bioplastics" can be defined as plastics that are biodegradable and/or compostable or as plastics based on renewable resources (bio-based), such as starch, cellulose, and others obtained by fermentation of renewable resources, for example, polylactic acid (PLA). The most important materials of bioplastics are summarized in Figure 28.1.

28.2 Biodegradable Polymers Used in Film Packaging Field

The performance expected from bioplastic materials used in food packaging application is their capability to contain the food and protect it from the environment, as well as to maintain good food quality (Arvanitoyannis, 1999). Also, it is important to study the change that can affect the characteristics of

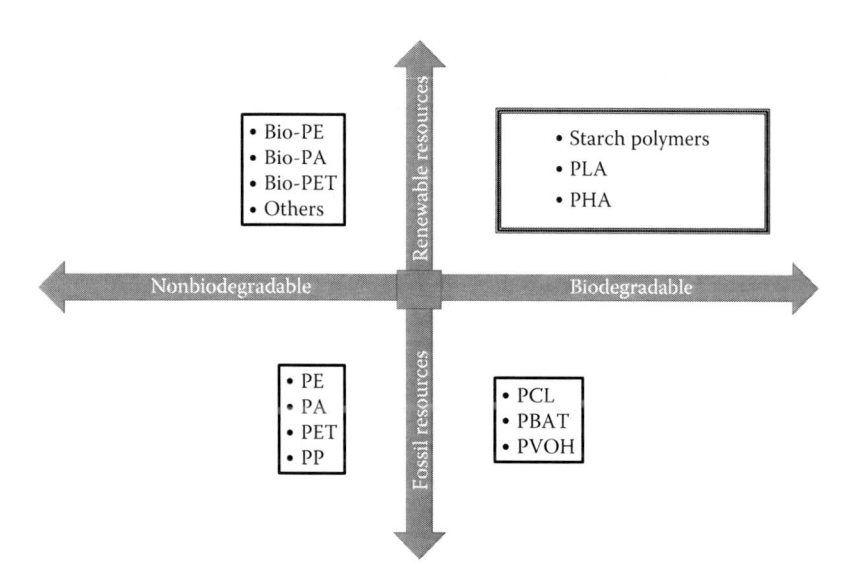

FIGURE 28.1 Bio-based plastic classification.

bioplastics during their interaction with food (Scott, 2000). Only a limited amount of biopolymers are able to maintain the food quality without important changes in their properties (Siracusa et al., 2008), and consequently, they are used for food packaging application (Arrieta et al., 2015).

The main renewable resource–based biodegradable polymers used as films for food packaging are starch, PLA, and polyhydroxyalkanoates (PHAs). Polycaprolactone (PCL) and poly(butylene adipate-co-terephthalate) (PBAT) are based on fossil resource, although biodegradables. Many researchers are developing bionanocomposites for these applications. Nanotechnology improves the thermal, barrier, and mechanical properties of the final material by the addition of a compound of nanometer size to the polymer matrix.

28.2.1 Biodegradable Plastics from Renewable Resources

Concerns about the potential damaging effects of nondegradable plastics on the environment and the increasing cost of petrochemical feedstocks have promoted new challenges in polymer research. In this sense, biodegradable polymers such as thermoplastic starch (TPS), PLA, or PHAs are an alternative to petroleum-based thermoplastics, and they depict an interesting and growing market. Both are thermoplastic bio-based polyesters with high promising perspectives in packaging applications. Currently, PLA is the most used bio-based material in the food packaging industry, for example, in the production of disposable cutlery, plates, lids, and cups and in postharvest packaging of fresh vegetables and fast-food containers. Polyhydroxybutyrate (PHB) is the most common PHAs, hence being recommended for short-term food packaging applications (Arrieta et al., 2014b).

28.2.1.1 Thermoplastic Starch

Starch is a cheap and widely available bioplastic. It is composed of amylose, a linear polysaccharide, and amylopectin, a branched one. Depending on the starch origin, it can contain 85%–70% amylopectin and 15%–23% amylose (Reddy et al., 2013).

Starch contains many intermolecular hydrogen bonds, and its degradation process is produced before melting (Halley, 2005); because of that, it cannot be processed as a plastic material by conventional techniques such as extrusion or injection. By addition of plasticizers, like water and alcohols prepared by heating, starch molecular structure is disrupted, a process named as "gelatinization" (Avérous, 2004). The resulting material has thermoplastic characteristics and it is hence known as "thermoplastic starch."

Starch, due to its proper structure, has high moisture sensitivity, thus affecting its performance in the packaging field. Therefore, the real application of starch-based bioplastics involves either the addition of fillers or the blending of TPS with other polymers and additives (Peelman et al., 2013). The addition of cellulose fibers reduces the water vapor permeability of starch, improving at the same time the barrier and mechanical properties of starch-based films (Dias et al., 2011, Muller et al., 2009). Ghanbarzadeh et al. (2011) reported better barrier properties of starch-based films to water vapor by the addition of carboxymethyl cellulose (CMC). The modification of CMC with ZnO added to plasticized starch improved water permeability but produced more brittle materials (Yu et al., 2009).

With respect to blends, those of two bio-based polymers seem to have a higher interest for their application in the food packaging field. It is noteworthy that many efforts are oriented in obtaining completely biodegradable materials from starch and biodegradable polyesters, including blends of TPS with PLA, PHB, PCL, polybutylenesuccinate, and PBAT (Reddy et al., 2013). The main objective is to enhance the compatibility between them, either by the introduction of a reactive functional group in the polymer or by chemical modification, for example, the esterification of starch hydroxyl groups (Thiebaud et al., 1997).

Blends of TPS with biodegradable polymer based on fossil resource, such as PCL or polyvinyl alcohol (PVOH), have been successfully applied on industrial level for foaming, film blowing, injection molding, blow molding, and extrusion applications (Peelman et al., 2013). On the other hand, the improvement of the barrier properties and the hydrolytic and UV stability of starch-based films was obtained by blending TPS with PHA, resulting in less degradation of TPS by processing (Shen et al., 2009).

These materials and their blends are widely developed in Chapter 9.

28.2.1.2 Polylactic Acid

PLA is a linear aliphatic–thermoplastic polyester produced either by the polycondensation of lactic acid or by the ring-opening polymerization of lactide. Lactide is a cyclic dimmer prepared by the controlled depolymerization of lactic acid, which is obtained from the fermentation of renewable sugar feedstock, such as corn or sugar beets. Because of the chirality of the lactic unit, the lactic acid cyclic dimers exist in three stereoisomeric forms: L-lactide, D-lactide, and DL-lactide. The equimolar mixture of L- and D-lactide is denominated as racemic DL-lactide (Vroman and Tighzert, 2009).

Lactic acid (2-hydroxypropanoic acid) is produced either through a fermentation process or a chemical synthesis. Industrial lactic acid production uses the lactic fermentation process rather than synthesis because the synthetic routes have some limitations, such as low capacity to produce the DL-lactide stereoisomer (Jamshidian et al., 2010).

The thermal, mechanical, and biodegradable properties of PLA are known to depend on the choice and distribution of stereoisomers within the polymer chains. The semicrystalline polymers poly(L-lactide) (PLLA) and poly(D-lactide) (PDLA) are obtained from L-lactide and D-lactide, respectively, whereas poly(DL-lactide) (PDLLA) is an amorphous polymer obtained from DL-isomer (Södergar and Stolt, 2002). PLLA and PDLA exhibit high tensile strength and low elongation at break, and they have also a high Young's modulus, making them more suitable for load-bearing applications. On the other hand, PDLLA has lower tensile strength, higher elongation at break, and, comparatively, a much more fast degradation, so it can be used for the production of transparent films and glues (Luckachan and Pillai, 2011).

Due to the high production costs, in the early stages of its development, PLA was used in limited fields, such as in the preparation of medical devices (bone surgery, suture, chemotherapy, etc.). Since the PLA production cost has been decreased by the development of new technologies and large-scale production, the application of PLA has been extended to other commodity areas such as packaging, textiles, and composite materials. Particularly, PLA has received considerable attention in food packaging applications with focus on films and coatings that are suitable for short shelf life and ready-to-eat food products (Kulinski and Piorkowska, 2005).

In food packaging and other consumer products, the use of PLA has increased considerably over the past few years due to their sustainable feedstock, compostability, and similar processing characteristics of conventional thermoplastics. In comparison to polyolefins, PLA has less melt strength, and therefore the formation of a stable bubble during extrusion blowing is more difficult. As a result, extrusion blowing of PLA film often requires the use of additives, such as melt viscosity enhancers, to increase its melt strength (Lim et al., 2010). Another way to improve the processability and the final properties of PLA is adjusting the proportion and sequence of L- and D-lactic acid unit in the synthesis to obtain high-molecular-weight PLA (Avérous and Pollet, 2012).

However, PLA shows a limitation for the application as gas barrier films to be used for the food packaging materials, as it has relatively low oxygen and water vapor permeation compared with conventional nondegradable polymer resins. It is a brittle and hard polymer with a very low elongation at break, with long degradation time in many cases. Therefore, much work has been performed to avoid gas permeation through biodegradable PLA resins by combining it with other materials (Bang and Kim, 2012). Blending is the easiest and, in general, the most cost-effective way to prepare polymeric materials with desirable properties. PLA can be blended with other polymers obtained from renewable resources, such as starch and PHB, or with petroleum-based polymers, such as PCL, in order to improve its properties. Blending is an effective approach to reduce PLA cost without diminishing its excellent biodegradability, while maintaining suitable mechanical and thermal properties. In other cases, PLA can be used as an *additive* to improve physical, mechanical, or barrier properties of other polymers from renewable sources (Almenar and Auras, 2010).

Otherwise, the basic properties that have influence on the degradation and biodegradation mechanisms of PLA are their physical and chemical structures (Luckachan and Pillai, 2011). Intrinsic polymer characteristics such as molecular weight, crystallinity, purity, the content of carboxyl or hydroxyl final groups, and the addition of additives acting as biodegradation catalyzers, as well as other factors such as pH, temperature, and water permeability, can influence the biodegradation process. For example, degradation is faster in the amorphous polymer region than in the crystalline region, and even the polymer hydrophilicity can enhance this process (Albertsson et al., 2010).

The first step to composting and/or biodegradation of PLA is the hydrolysis (Huang, 2005), and then it must be decomposed into carbon dioxide and water in a "controlled composting environment" in a time less than 90 days. Some researchers showed that PLA products can be degraded in less time, for instance, Pikon et al. (2013) studied the biodegradation of PLA film and demonstrated their decomposition in 3 weeks.

Microbial and enzymatic degradation of PLA was recently studied by many researchers because these types of degradations do not usually need high temperatures to take place (Jamshidian et al., 2010). Proteinase K is the only reported enzyme that would degrade amorphous regions of low molecular weight of PLA.

NatureWorks® LLC, the actual leader in PLA technology, has a 50–50 joint venture between Cargill Inc. and Dow Chemical Co. and was founded in November 1997. In 2002, they started the world's first full-scale PLA plant in Blair, NE, which is capable of producing 140,000 metric tons per year. NatureWorks entered into a joint venture between Cargill and Teijin Limited of Japan in December 2007 (http://www.natureworksllc.com).

In order to improve the biodegradability and the mechanical and processing properties of biodegradable polymers, they are blended or copolymerized with other monomers to achieve the desired properties.

PLA/starch: Starch, which is known as one of the cheapest functional eco-friendly materials, is an attractive option as biocompatible filler for PLA due to its abundant availability, compostability, and nontoxicity properties. Blending PLA with starch can greatly reduce its production cost. However, PLA and starch are immiscible due to the hydrophobicity of PLA, and starch is hydrophilic, so that PLA/starch blends show poor interfacial adhesion between phases. The use of compatibilizers and other additives are being studied in order to improve the interfacial interactions of these blends (Nuona et al., 2015). Maleic anhydride can be used for this purpose, which also entails an increase in the mechanical properties of these blends (Vroman and Tighzert, 2009).

PLA/PHB: Adding PHB to PLA improves the mechanical properties of the final blend compared to the neat polymers.

In this sense, Zhang and Thomas (2011) studied different blends of PLA and PHB and concluded that blending PLA with 25 wt% of PHB produced a reinforcement effect on PLA matrix due to the nucleating effect of PHB.

Other studies on PLA/PHB blends showed that the miscibility between these two polymers depended on the molecular weight of PHB, which has melt compatibility with low-molecular-weight PLA (M_w < 18,000) for the whole composition range, whereas the PHB blends with high-molecular-weight PLA (M_w > 18,000) show a biphasic separation. Similarly, the PLA component has melt compatibility with low-molecular-weight PHB (M_w = 9400) considering PHB contents up to 50 wt%, but it is immiscible with high-molecular-weight PHB. PLA/PHB blend properties also depend on composition, crosslinks, and processing conditions (Abdelwahab et al., 2012).

On the other hand, PLA/PHB films are very rigid and need to be modified for packaging applications. In order to increase the molecular mobility and consequently enhance the flexibility and ductile properties of the transparent blend films, the addition of plasticizers is usually done. Moreover, the processing performance and breaking properties increase with incorporation of these additives (Arrieta et al., 2014).

PLA/PCL: Blending PLA with PCL reduces the brittleness and slightly increases the thermal stability of PLA films, but decreases their barrier properties depending on the PCL percentage added. The barrier properties lost can be recovered by adding kaolinite nanoclays. Also, the combination of PLA with starch, plasticizers (glycerol/sorbitol), or other degradable polyesters can diminish the brittleness of the film (Peelman et al., 2013).

Some studies have revealed that PLA/PCL blends were modified by the addition of maleic anhydride and peroxide, which are new materials compatible with other polar polymers, by developing free radical generation via hydrogen abstraction from PLA and PCL molecules. Peroxide addition produces chemical cross-linking, generating at the same time a high-performance material (Semba et al., 2006).

PLA/PBAT: As it was mentioned earlier, PLA has high mechanical strength and Young's modulus, but it is too brittle, while PBAT is a flexible and tough polymer (with a strain at break around 700%). As consequence, blending of PLA with PBAT becomes a good choice to improve PLA mechanical properties without compromising its biodegradability (Long et al., 2006).

Several authors studied the compatibility, crystallization behavior, and tensile properties of different PLA/PBAT blends produced by melt blending. The PBAT carbonyl group increases the compatibility with PLA during the melt-blending process, finally obtaining a material tougher than PLA (Kanzawa and Tokumitsu, 2011). Other results showed that the toughness of PBAT and PLA blends was improved by the addition of 3%–5% GMA (glycidyl methacrylate) as a reactive compatibilizing agent (Jiang et al., 2006).

28.2.1.3 Polyhydroxyalkanoates

PHAs are a family of intracellular biopolymers, synthesized by many bacteria as intracellular carbon and energy storage granules. In most cases, they are produced and accumulated under stressed conditions such as nitrogen, phosphorous, or oxygen limitation with excess of carbon sources (Christopher et al., 2013). These polymers are classified structurally according to the number of carbon atoms, from 4 to 14, and the type of initial monomeric units, producing either homopolymers or heteropolymers (Keshavarz and Roy, 2010).

The mechanical properties of PHAs differ considerably depending on its chemical structure, obtaining either flexible or rigid plastics, according to the size of alkyd substituent (Avérous and Pollet, 2012).

PHB and polyhydroxybutyrate-co-hydroxyvalerate (PHBV) are the most well-known biopolymers of the PHA family. They are synthesized by exposing the bacteria to carbon source, while the other necessary nutrient becomes limited (Doi, 1990).

PHB is obtained by the polymerization of 3-hydroxybutyrate monomer, with crystallinity above 50%, due to exceptional stereochemical regularity, which leads to progressive crystallization by aging, thus making it a very brittle polymer (Tharanathan, 2003). Recently, it has been probed that switchgrass is capable of producing PHB (Somleva et al., 2008). Furthermore, PHB has a high melting point, T_m = 173°C–180°C, compared to other biodegradable polyesters, with some mechanical properties comparable to synthetic biodegradable polymers such as PLA. PHB is thermally unstable during conventional melt processing and it is only suitable for processing in a narrow temperature window. To overcome this drawback, it is blended with other biopolymers or plasticized with citrate ester (Lim et al., 2013).

Regarding the biodegradable behavior, PHAs can be degraded by abiotic process, that is, simple hydrolysis of the ester bond without requiring the presence of enzymes to catalyze this hydrolysis, or on the other hand, in a biotic degradation, the enzymes degrade the residual products till final mineralization (Avérous and Pollet, 2012).

The biodegradability PHAs depends mainly on their crystallinity and the polymer type; copolymers degrade faster than the homopolymers (Reddy et al., 2013). PHAs, unlike TPSs and PLAs, have hydrophobic nature, presenting good water vapor barrier properties (Petersen et al., 1999), which made them suitable for coating and film application in food packaging.

In particular, PHB is degraded by numerous microorganisms (bacteria, fungi, and algae) in various environments. Some external parameters can influence the biodegradation rate, including the type of environment (soil, lake, water, seawater, compost, etc.), microbial population, temperature, and the PHB design product such as porosity and texture, among others (Lee and Choi, 1999).

The copolymers of PHAs, such as poly(3-hydroxybutyrate-co-3-hydroxyhexanoate), have shown better barrier properties for gases such as O_2, water vapor, and CO_2 than a fossil resource–based ethylene vinyl alcohol (EVOH) copolymer, widely applied in multilayer packaging (Vandewijngaarden et al., 2014). Multilayer films for food packaging applications composed of PVOH as the core layer and PHA as the outer skin layers were produced by the coextrusion process, providing increased barrier performance over monolayer PVOH (Thellen et al., 2013).

Starch/PHB: PHB/starch blends have suitable physical properties for food packaging with important reduction in cost (Reis et al., 2008). Godbole et al. (2003) studied the thermal and mechanical properties of blends of PHB and starch. They reported that blends containing up to 30% starch could be advantageous for cost reduction with improved properties with respect to the virgin PHB.

28.2.2 Biodegradable Plastics from Fossil Resources

The main fossil resource–based biodegradable polymers used in the food packaging field are PCL and PBAT.

28.2.2.1 Polycaprolactone

PCL is obtained by two polymerization methods, the polycondensation of a hydroxycarboxylic acid (6-hydroxyhexanoic acid) and the ring-opening polymerization (ROP) of a lactone, for instance, ε-caprolactone (ε-CL), using stannous octoate as a process catalyzer. A wide range of catalyzers for the caprolactone ring-opening polymerization have been reviewed (Labet and Thielemans, 2009).

PCL is a thermoplastic polymer with a very high elongation at break and tensile strength (Matzinos et al., 2002), and it has also good chemical resistance, for example, to water, oil, solvent, and chloride.

On the other hand, PCL shows a very low glass transition temperature ($T_g \approx -61°C$) and a low melting point of 55°C–65°C, so that it is difficult to be processed by the conventional techniques used for thermoplastic materials. Moreover, these low transition temperatures involve very short degradation time, affecting their processing. For instance, PCL films produced by film blowing are especially sticky in extrusion and they are characterized by low melt strength at temperatures higher than 130°C (Bastioli, 1998).

For all those reasons, several blends of PCL with other biodegradable and nonbiodegradable polymers are being investigated in order to improve the processability, stability, and mechanical properties of these materials.

PCL biodegradation was studied by Lefebvre et al. (1995), which shows that its biodegradation rate depends on the polymer molecular weight and that the degradation process started in the vicinity of the chain macromolecular ends. Other authors discussed the PCL biodegradation by fungi, showing that PCL can be easily enzymatically degraded (Tokiwa and Suzuki, 1977).

The main blends used in the food packaging field are the following:

Starch/PCL: TPS is very much used in blends with other biodegradable polymers due to, comparatively, its lower production cost.

The use of PCL blends with starch reduces the cost of the produced final material due to the replacement of expensive synthetic PCL matrix by starch. It was shown that blending starch with PCL improved its processability and furthermore promoted its biodegradation (Matzinos et al., 2002). These blends were introduced into the market at the beginning of the 1990s as packaging materials. Nowadays, the most important PCL/starch blend is marketed by Novamont under the trade name MasterBi-Z.

In addition, it has been widely reported that the use of a compatibilizer in these blends can improve the material performance without changing their biodegradability (Avella et al., 2000). Moreover, it was shown that by adjusting the proportion of starch and compatibilizer, it was possible to obtain the suitable properties for specific applications (Yahiaoui et al., 2015).

In food application, as an intelligent and active packaging system, PCL was blended with starch to improve the suitability of packaging when being in contact with food and so that a modulated water-scavenger effect is obtained (Sebastien et al., 2013). On the other hand, bioactive trilayer films were prepared using one layer of methylcellulose with antimicrobial additives and two layers of PCL (Takala et al., 2013).

PCL/PH: The high crystallinity of PHB makes this polymer very brittle; this is the reason why it is usually blended with semicrystalline PCL, which is thermoplastic of low thermal transition temperatures, thus balancing the properties of these polyesters in order to obtain the most suitable final blend properties.

As mentioned earlier, a remarkable characteristic of polyesters like PHB and PCL is their biodegradability in the environment. Films and fibers made from those polymers can be degraded in soil, sludge, or seawater (Immirzi et al., 1999). Moreover, under optimum conditions the degradation can be extremely fast (La Cara et al., 2003).

28.2.2.2 Poly(Butylene Adipate-co-Terephthalate)

PBAT is an aliphatic–aromatic copolyester, a random copolymer of butylene adipate and butylene terephthalate prepared by melt polycondensation of 1,4-butanediol, dimethylterephthalate, and adipic acid, which is being catalyzed by tetrabutylorthotitanate (Shahlari and Lee, 2012). The terephthalate group contributes chemical stability and good mechanical properties, while the butylene adipate group provides the biodegradability of the material; so the proportion of these groups determines the performance of the polymers, of which the butylene adipate group affects the most.

This thermoplastic polymer has a relatively low tensile strength and low Young's modulus with a high elongation at break (Raquez et al., 2010). The polymer biodegradation process takes place with the use of lipases from *Pseudomonas cepacia* and *Candida cylindracea*, and it can be degraded in a few weeks once it is set in contact with these natural enzymes (Bastarrachea et al., 2010).

Since PBAT has high flexibility and low Young's modulus, this polymer is blended with other biodegradable polymers with a very high Young's modulus and tensile strength. PLA/PBAT blend is the most used in the packaging field; PLA in PLA/PBAT blends yields good mechanical properties such as high modulus and elongation at break, thus improving its performance.

PBAT is commercially marketed as a fully biodegradable plastic by BASF under the trade name Ecoflex®, showing 90% biodegradation after 80 days testing. It is also commercialized as a blend with PLA, which is called Ecovio®.

Specific PBAT applications, highlighted by the manufacturers, include cling wrap for food packaging and compostable plastic bags for gardening and agricultural field; and it is also used in waterproof materials and in mechanical strength coatings for other materials, such as paper cups.

PBAT is generally marketed as a fully biodegradable polymer for low-density polyethylene (LDPE) replacement since it has similar mechanical properties with that of pure LDPE, such as flexibility and resilience, and it is also used, in a similar way, as plastic bags and wraps.

PBAT/PHB: Considering that PHB has more toughness than PBAT, studies have been carried out to analyze the mechanical performance of different PBAT/PHB blends. These showed results that the specific toughness of PBAT was improved up to two orders of magnitude and that also the PBAT strain at break was increased by more than 500%. Moreover, Javadi et al. (2010) added recycled wood fiber and synthetic fillers as nanoclays in the these PHBV/PBAT blends to produce hybrid biocomposites (Nagarajan et al., 2013).

28.2.3 Bionanocomposites

Biodegradable polymers can be modified by nanoparticles, changing their characteristic properties and forming new bionanocomposite materials, which is necessary in order to attain good interaction between the polymer matrix (continuous phase) and the nanoparticles (discontinuous phase), thus getting an optimum behavior during the bionanocomposite application (Lagaron and Lopez-Rubio, 2011).

Recently, several research groups have prepared and characterized various types of bionanocomposites that showed interesting properties suitable for a wide range of applications, including food packaging (Sinha and Bousmina, 2005). Thus, biodegradable polymers, either renewable or fossil resource based, have been filled with nanoclays in order to improve their performance properties without losing their biodegradable character according to the standards. Nanoclay layers decrease the gas and water vapor permeabilities across the polymer wall, hence improving biopolymer barrier properties. Polymer nanocomposites with fully exfoliated and large aspect ratio clay minerals have the best gas barrier properties (Choudalakis and Gotsis, 2009).

Starch, PLA, PHAs, and PCL are some of the polymers most widely used to prepare bionanocomposites for their application in the food packaging field. As a general consideration, they have appropriate mechanical properties and good barrier properties against O_2, CO_2, water vapor, and flavor compounds, which are all required in order to extend the shelf life of fresh and processed foods (Rhima et al., 2013).

PLA-ZnO nanocomposite films were produced by melt compounding, evidencing multifunctional characteristics such as mechanical, anti-UV, antibacterial, electrical, and gas barrier properties, which are potentially of high interest as packaging biomaterials (Pantani et al., 2013).

In addition, low-weight packages with antimicrobial and antioxidant properties for food preservation can be manufactured with the application of recyclable bionanocomposites, with subsequent advantages as regards cost and waste reductions. The major potential food applications for antimicrobial bionanocomposite films are evident in the manufacture of foods that include meat, fish, poultry, bread, cheese, fruits, and vegetables (de Oliveira et al., 2007, Kerry et al., 2006, Moreira et al., 2011).

On the other hand, interesting bionanocomposite packaging applications include antifogging and antisticking films, light absorbing/regulation systems, microwave heating suitable films, gas-permeable/breathable films, bioactive agents for controlled release, and insect repellant packages (Ozdemir and Floros, 2004).

Nanocomposites have already led to several innovations with potential applications in the food packaging sector (Rhima et al., 2013). Innovative food packaging technologies such as nanosensors, freshness indicators, intelligent packaging, nanocoating, and high-barrier packaging are implemented by the use of specific types of nanoparticles, such as silver, silver zeolite, and nanoclays.

Functional materials, such as antioxidant nanobiocomposites based on PCL, were prepared by incorporating hydroxytyrosol (HT) and a commercial montmorillonite, obtaining a reduction in oxygen permeability and an increase in the hardness for the new films (Beltrán et al., 2014). Multilayer structures based on a commercial PHB and a polyhydroxybutyrate-co-valerate copolymer), containing a nanostructured zein interlayer, showed good microbial safety storing at less than 100% RH (Farida et al., 2015). Enhanced antibacterial properties were developed by biodegradable PCL-based nanocomposites for packaging applications, prepared by melt mixing of the poly(epsilon-caprolactone) with two organo-modified montmorillonites (Fabra et al., 2014).

According to an early report (Ratto et al., 1999), PCL/clay nanocomposites showed better biodegradability than pure PCL; this effect is actually explained by the catalytic effect of the organoclay on the biodegradation process. Similar results were obtained for other biopolymers such as PLA (Paul et al., 2005) and PHB (Maiti et al., 2007) on the basis of nanoclay hydrophilicity.

On the other hand, PLA/organoclay nanocomposites were disintegrated by composting faster than pure PLA due to the hydroxylated terminal groups of nanoclay layers (Sinha et al., 2003a,b, Sinha and Okamoto, 2003). Other authors found that the biodegradability of nanocomposites depends on the type and intercalation degree into the PLA matrix of nanoclays (Nieddu et al., 2009).

Active packaging and coatings were analyzed for food packaging applications in the form of either coatings or wrappings, by incorporation of antimicrobials such as bacteriocides or silver and copper nanoparticles in PHA and PLA polymer films, thus obtaining controlled release of active compounds, bioactive protection, and resistance to water (Vijayendra and Shamala, 2014).

Antioxidant agents have been incorporated into active packaging systems in different forms, mainly including independent sachet packages, adhesive-bonded labels, and physical adsorption/coating on packaging material surface, as well as being incorporated into packaging polymer matrix and multilayer films. Moreover, covalent immobilization technique is applied onto the food-contact packaging surface in antioxidant active packaging with the highlight of the development and application of nonmigratory active packaging systems with the focus on maintaining safety, quality, and nutrition of packaged foods (Tian et al., 2013).

28.3 Disposal and Biodegradable Plastic Waste Management

Waste management is a key contribution in building a resource-efficient society, where ambitious and realistic objectives must be arranged to find solutions at the end-of-life of products.

Plastic packaging wastes were traditionally sent to landfills due mainly to the difficulties in collecting, identifying, sorting, transporting, cleaning, and reprocessing large volumes of lightweight materials in mixed and contaminated forms. Biodegradable packaging materials are potentially suitable for inclusion in the composting process and open new routes for waste treatment (Davis and Song, 2006).

Sometimes, biodegradable plastics are blended with another polymer to obtain desired properties, but when the blend is with a nonbiodegradable plastic, only the biodegradable components will be degraded in the environment. Therefore, these materials should never be used as biodegradable plastics.

The different processes that can be used for biodegradable plastic waste management are recycling, incineration, and biodegradation (Huang, 1995). As these materials enable a potential option for waste treatment through composting as a way to recover the materials and to produce compost, this process must have special attention in the biodegradable plastic waste management.

28.3.1 Incineration with Recovery Energy

Incineration is a common form of general waste disposal. It can reduce the volume of disposed waste by up to 90% in waste streams with very high amounts of packaging materials, paper, cardboard, plastics, and horticultural waste.

Incineration without energy recovery (or nonautogenic combustion, the need to regularly add fuel) is not a preferred option due to high costs and pollution. Incineration with recovery of energy is an option after removing all recyclable elements.

Biodegradable polymers have gross calorific values (GCVs) in the same range as coal. Biopolymers such as natural fiber and starch have GCVs (Jacquinet, 1985, Ren, 2003) close to those of wood and thus still have considerable value for incineration. In addition, the production of these materials consumes significantly less energy than fossil resource polymers (Patel et al., 2003) and thus would possibly result in an overall positive energy balance. By these reasons, energy recovery by incineration is regarded as a suitable option for bioplastic polymers because it can contribute to the production of renewable energy (www.european-bioplastics.org). Nevertheless, incineration is not the better destination of biodegradable plastics and can only be considered a valid alternative to landfilling.

28.3.2 Landfilling

We might think that the entry of biodegradable plastics, except for the increase in volume, should not cause any problem in a standard landfill that is designed to be inert with wastes inside degrading very little. However, the entry of biodegradable plastics will enhance the degradation, producing more leachates and gases, and thus worsen the contamination of groundwater and surface water and the ambient environment. To avoid this, it is necessary to have a clear labeling and segregation of biodegradable plastics from other waste streams (Ren, 2003).

Article 5 of the Landfill Directive (99/31/EC) seeks to reduce the amount of biodegradable municipal waste in landfills, including biodegradable packaging waste, until their total elimination in 2020, due to the negative environmental impacts associated with leachate and methane production (Davis and Song, 2006).

Compostable plastics allow to divert organic waste from landfill and incineration to organic recycling, thus increasing the added value of these residues.

28.3.3 Recycling

Through mechanical recycling, materials are obtained with similar characteristics to that of the originals, but not with equal properties, as it can produce thermomechanical degradation. In general, biodegradable plastics degrade more than traditional plastics in the reprocessing, during which they decompose and complicate the later processing.

Although the mechanical recycling of some bioplastics such as PLA is possible several times without significant reduction in PLA properties, the lack of continuous and reliable supply of bioplastic polymer waste in large quantities, nowadays, makes bioplastic mechanical recycling less economically attractive than the conventional ones (Carrasco et al., 2010, Fonseca et al., 2014, Pillin et al., 2008).

Finally, a plastic packaging material may be a multilayer of different biopolymers that leads to better performance (e.g., in modified atmosphere packaging of meat products) or instead the production of heterogeneous mixing of biodegradable plastics in the feedstock of recycling, which might compromise the recyclability of these products. Therefore, for all these reasons, mechanical recycling is not the best option for the management of biodegradable plastic wastes (Ren, 2003, Song et al., 2009).

When the physical properties of the materials have deteriorated significantly, it is then possible to recover the monomers using chemical recycling that is potentially useful for certain polymers. The plastic materials are chemically broken down into smaller molecules, which can then be used as the building blocks for new materials.

PLA is chemically recycled by hydrolysis. In this process, PLA is broken down into its monomer, that is, lactic acid, and is then converted back again into a new resin (www.natureworksllc.com) (Piamonte et al., 2013).

There are many technologies developed in chemical recycling, but currently not all are economically attractive, as with the tightening of legislation and restrictions on oil supplies.

28.3.4 Composting

Composting may be defined as the aerobic biodegradation of organic material to form primarily carbon dioxide, water, and humus. At the end of the process, the initial waste is transformed into a substance called compost, which can be used in agriculture. The treatment of biodegradable plastics by composting is classified as material recovery.

In composting plants, this phenomenon is controlled and optimized in order to achieve a high conversion speed, control of the effluent, control of the quality of the final compost, etc. (Pikoń and Czop, 2014). Those control processes are very important because when the amount of partially biodegraded products increases rapidly, the soil becomes acidic owing to their high concentration. Accordingly, it is necessary to monitor the long-term influence of water-soluble biodegradable products on plant growth and soils (Itawa, 2015).

Biodegradable polymers can be brought along with the organic wastes to an industrial composting plant to be transformed into compost, but this requires the collected wastes already separated from non-biodegradable impurities. Therefore, in any case, products manufactured from biodegradable polymers should be deposited in landfills.

Composting is considered the most attractive route for the treatment of biodegradable plastic packaging waste (Kaplan et al., 1993), but this requires intensive research and development and, at the end, policies that facilitate their separation and management as waste before it becomes practical and real.

In contrast to incineration and chemical or mechanical recycling, composting can be achieved at comparably low capital costs, and it does not require extensive transportation. In addition, biodegradable materials disappear rather than being litter (Steinbüchel, 1995).

28.4 Unification of the Certification System and Social Acceptance

The establishment of a certification system and standard experimental methods is necessary to promote the use of biodegradable and bio-based plastics. For biodegradable plastics, the ISO standard has been established, and each country carries out biodegradation tests according to this standard; then, the certification bodies validate these materials like compostable plastics (Itawa, 2015).

TABLE 28.1

Currently Used Logos for Biodegradable Plastics

Certification Body	Reference Standard	Logo
Din Certco (Germany) AFOR (UK) Keurmerinstitute (the Netherlands) COBRO (Poland) ABA (Australia and New Zealand)	EN 13432-2000	
Vinçotte (Belgium)	EN 13432-2000	
Jatelaito-syhdisttys (Finland)	EN 13432-2000	
Certiquality/CIC (Italy)	EN 13432-2000	
Avfall Norge (Norway)	EN 13432-2000	
BPI (USA)	ASTM D 6400-04	
BNQ (Canada)	BNQ 9011-911/2007	
JBPA (Japan)	Green Pla	

In Europe, the products made of biodegradable plastics can be identified with logos that certify that an article is biodegradable/compostable according to the European Standard EN 13432, but each certification body can use its own logo. This situation is confused and damages the biodegradable plastic waste management, so a large number of organizations in EU are promoting the use of a unique logo to identify biodegradable/compostable products.

Table 28.1 shows the main logos, the reference standard, and the issuing certification bodies. Thus, products made with these materials should be labeled in that way so that they can be sorted out from recyclables.

In most cases, biodegradable plastics are eliminated as nonbiodegradable plastics because consumers do not differentiate them for the lack of labeling or the use of a common labeling. For this and other reasons, the biodegradable plastics go into the recycle streams of nonbiodegradable polymers and contaminate the recycled materials and probably lead to the degradation of their properties. For example, the additions of starch or natural fibers to traditional polymers have a deleterious effect on the mechanical properties of subsequent films (Hartmann and Rolin, 2002, Scott, 1995).

It is evident that clear labeling and identification of materials, which are the processes necessary for an effective sorting, will become key steps in the wide adaptation of biodegradable plastics. It would be necessary for policy makers and the industry itself to prepare themselves for the change in advance (Ren, 2003).

Japan Bioplastic Association labels biodegradable plastics as "green plastics." Furthermore, an international agreement has been established between Japan, Germany, and the United States. For example, when a plastic that is certificated as a "green plastic" in Japan is exported to Germany or the United States, the company can apply for this plastic to receive the certification and logo of "kompostierbar" in Germany or "compostable" in the United States without having to undergo necessary further testing of the biodegradability of the plastic in either country (Itawa, 2015).

REFERENCES

Abdelwahab, M.A., Flynn, A., Chiou, B., Imam, S., Orts, W., and Chiellini, E. Thermal, mechanical and morphological characterization of plasticized PLA-PHB blends. *Polymer Degradation and Stability* 97 (2012): 1822–1828.

Albertsson, A.-C., Kumari, I., Lochab, B., Finne-Wistrand, A., and Kumar, K. Design and synthesis of different types of poly(lactic acid). In: *Poly(lactic acid): Synthesis, Structure, Properties, Processing, and Application*, Auras, R., Lim, L.-T., Selke, E.M., Tsuji, H. (eds.), Vol. 4. 2010, John Wiley & Sons, Inc., pp. 43–55.

Almenar, E. and Auras, R. Permeation, sorption, and diffusion in poly(lactic acid). In: *Poly(lactic acid): Synthesis, Structure, Properties, Processing, and Application*, Auras, R., Lim, L.-T., Selke, E.M., Tsuji, H. (eds.), Vol. 12. 2010, John Wiley & Sons, Inc., pp. 155–176.

Arrieta, M.P., López, J., Hernández, A., and Rayón, E. Ternary PLA-PHB-Limonene blends intended for biodegradable food packaging applications. *European Polymer Journal* 50 (2014a): 255–270.

Arrieta, M.P., Peponi, L., López-Martínez, J., and Kenny, J.M. Nuevas tendencias en envases alimentarios plásticos. *Revista de Plásticos Modernos* 109(699) (2015): 21–23.

Arrieta, M.P., Samper, M.D., and López, J. Combined effect of poly(hydroxybutyrate) and plasticizers on polylactic acid properties for film intended for food packaging. *Journal of Polymers and the Environment* 22(4) (2014b): 460–470.

Arvanitoyannis, I.S. Totally and partially biodegradable polymer blends based on natural synthetic macromolecules: Preparation, physical properties, and potential as food packaging materials. *Journal of Macromolecular Science, Reviews in Macromolecular Chemistry and Physics* C39(2) (1999): 205–271.

Avella, M., Errico, M.E., Laurienzo, P., Martuscelli, E., Raimo, M., and Rimedio, R. Preparation and characterisation of compatibilised polycaprolactone/starch composites. *Polymer* 41 (2000): 3875–3881.

Avérous, L. Biodegradable multiphase systems based on plasticized starch: A review. *Journal of Macromolecular Science: Part C: Polymer Reviews* 44 (2004): 231–274.

Avérous, L. and Pollet, E (eds.). Chapter 2: Biodegradable polymers. In: *Environmental Silicate Nano-Biocomposites*, 2012, pp. 13–39.

Bang, G. and Kim, S.W. Biodegradable poly(lactic acid)-based hybrid coating materials for food packaging films with gas barrier properties. *Journal of Industrial and Engineering Chemistry* 18 (2012): 1063–1068.

Bastarrachea, L., Dhawan, S., Sablani, S., Mah, J.-H., Kang, D.-H., Zhang, J., and Tang, J. Biodegradable poly(butylene adipate-*co*-terephthalate) films incorporated with nisin: Characterization and effectiveness against *Listeria innocua*. *Journal of Food Science* 75 (2010): 215–224.

Bastioli, C. Properties and applications of Mater-Bi starch-based materials. *Polymer Degradation and Stability* 59 (1998): 263–272.

Bastioli, C. *Handbook of Biodegradable Polymers*. Shawbury, U.K.: Rapra Technology Ltd., 2005.

Beltran, A., Valente, A.J.M., and Jimenez, A. Characterization of poly(e-caprolactone)-based nanocomposites containing hydroxytyrosol for active food packaging. *Journal of Agricultural and Food Chemistry* 62(10) (2014): 2244–2252.

Bharadwaj, R.K. Modeling the barrier properties of polymer-layered silicate nanocomposites. *Macromolecules* 34 (2001): 189–192.

Carrasco, F., Pagès, P., Gámez-Pérez, P., Santana, O.O., and Maspoch, M.L. Processing of poly(lactic acid): Characterization of chemical structure, thermal stability and mechanical properties. *Polymer Degradation and Stability* 95(2) (2010): 116–125.

Chandra, I. and Rustgi, R. Biodegradable polymers. *Progress in Polymer Science* 23 (1998): 1273–1335.

Choudalakis, G. and Gotsis, A.D. Permeability of polymer/clay nanocomposites: A review. *European Polymer Journal* 45 (2009): 967–984.

Christopher, T., Cheney, S., and Ratto, J.A. Melt processing and characterization of polyvinyl alcohol and polyhydroxyalkanoate multilayer films. *Journal of Applied Polymer Science* 127(3) (2013): 2314–2324.

Davis, G. and Song, J.H. Biodegradable packaging based on raw materials from crops and their impact on waste management. *Industrial Crops and Products* 23 (2006): 147–161.

De Oliveira, T.M., Soares, N.F.F., Pereira, R.M., and Fraga, K.F. Development and evaluation of antimicrobial natamycin-incorporated film in Gorgonzola cheese conservation. *Packaging Technology and Science* 20 (2007): 147–153.

Dias, A.B., Muller, C.M.O., Larotonda, F.D.S., and Laurindo, J.B. Mechanical and barrier properties of composite films based on rice flour and cellulose fibers. *Food Science and Technology* 44 (2011): 535–542.

Doi, Y. *Microbial Polyesters*. New York: Wiley-VCH, 1990, p. 166.

European Commission DG ENV. Plastic Waste in the environment. BIO Intelligence Service, 2011. http://ec.europa.eu/environment/waste/pdf/2011_CDW_Report.pdf.

Fabra, M.J., Sanchez, G., and Lopez-Rubio, A. Microbiological and ageing performance of polyhydroxyalkanoate-based multilayer structures of interest in food packaging. *LWT—Food Science and Technology* 59(2) (2014): 760–767.

Farida, Y., Benhacine, F., and Ferfera-Harrar, H. Development of antimicrobial PCL/nanoclay nanocomposite films with enhanced mechanical and water vapor barrier properties for packaging applications. *Polymer Bulletin* 72(2) (2015): 235–254.

Fonseca, C., García, B., Martín, L., Ochoa, A., Flores, A., and Ania, F. Thermal and mechanical properties of recycled poly(lactic) acid. XIII Reunión del grupo especializado de polímeros (GEP) de la REEQ Y RSEF, 2014.

Fritz, J., Link, U., and Braun, R. Environmental impacts of biobased/biodegradable packaging. *Starch* 53(3–4) (2001): 105–109.

Ghanbarzadeh, B., Almasi, H., and Entezami, A.A. Improving the barrier and mechanical properties of corn starch-based edible films: Effect of citric acid and carboxymethyl cellulose. *Industrial Crops and Products* 33 (2011): 229–235.

Godbole, S., Gote, S., Latkar, M., and Chakrabarti, T. Preparation and characterization of biodegradable poly-3-hydroxybutyrate-starch blend films. *Bioresource Technology* 86 (2003): 33–37.

Halley, P.J. Thermoplastic starch biodegradable polymers. In: *Biodegradable Polymers for Industrial Applications*, Smith, R. (ed.). Cambridge, U.K.: CRC Press/Woodhead, 2005, pp. 140–162.

Hartmann, L. and Rolin, A. Post-consumer plastic recycling as a sustainable development tool: A case study. In: *GPEC 2002: Plastics Impact on the Environment*, February 13–14, 2002, Detroit, MI, pp. 431–438.

Huang, S.J. Polymer waste management: Biodegradation, incineration and recycling. *Journal of Macromolecular Science: Pure and Applied Chemistry* A32(4) (1995): 593–597.

Huang, S.J. Chapter 9: Poly(lactic acid) and copolyesters. In: *Handbook of Biodegradable Polymers*, Bastioli, C. (ed.). 2005, Shawbury, UK: Rapra Technology Limited, pp. 287–302.

Immirzi, B., Malinconico, M., Orsello, G., Portofino, S., and Volpe, M.G. Blends of biodegradable polyesters by reactive blending: Preparation characterization and properties. *Journal of Material Science* 34 (1999): 1625–1639.

Iwata, T. Biodegradable and bio-based polymers: Future prospects of eco-friendly plastics. *Angewandte Chemie International* 54 (2015): 3210–3215.

Jacquinet, P. Calorific value of composites. *Composites Plastiques Renforces Fibres de Verre Textile* 25 (1985): 14–16.

Jamshidian, M., Arab Tehrany, E., Imran, M., Jacquot, M., and Desobry, S. Poly-lactic acid: Production, applications, nanocomposites and release studies. *Comprehensive Reviews in Food Science and Food Safety* 9 (2010): 552–571.

Javadi, A., Kramschuster, A.J., Jungjoo, S.P., Gong, L., and Turng, L.-S. Processing and characterization of microcellular PHBV/PBAT blends. *Polymer Engineering and Science* 50 (2010): 1440–1448.

Jiang, L., Wolcott, M.P., and Zhang, J. Study of biodegradable polylactide/poly(butyleneadipate-co-terephthalate)blends. *Biomacromolecules* 7 (2006): 199–207.

Kanzawa, T. and Tokumitsu, K. Mechanical properties and morphological changes of poly(lactic acid)/polycarbonate/poly(butylene adipate-*co*-terephthalate) blend through reactive processing. *Journal of Applied Polymer Science* 121 (2011): 2908–2918.

Kaplan, D.L., Mayer, J.M., Ball, D., McCassie, J., Allen, A.L., and Sterhouse, P. *Biodegradable Polymers and Packaging.* Basel, Switzerland: Technomic Publishing, 1993, pp. 1–44.

Kerry, J.P., O'Grady, M.N., and Hogan, S.A. Past, current and potential utilization of active and intelligent packaging systems for meat and muscle-based products: A review. *Meat Science* 74 (2006): 113–130.

Keshavarz, T. and Roy, I. Polyhydroxyalkanoates: Bioplastics with a green agenda. *Current Opinion in Microbiology* 13 (2010): 321–326.

Koh, H.C., Parka, J.S., Jeong, M.A., Hwang, H.Y., Hong, Y.T., Ha, S.Y., and Nam, S.Y. Preparation and gas permeation properties of biodegradable polymer/layered silicate nanocomposite membranes. *Desalination* 233 (2008): 201–209.

Krochta, J.M. and De Mulder-Johnston, C. Challenges and opportunities. *Food Technology* 51(2) (1997): 61–74.

Kulinski, Z. and Piorkowska, E. Crystallization, structure and properties of plasticized poly(L-lactide). *Polymers* 46 (2005): 10290–10300.

La Cara, F., Immirzi, B., Ionata, E., Mazzella, A., Portofino, S., Orsello, G., and De Prisco, P.P. Biodegradation of poly-e-caprolactone/poly-b-hydroxybutyrate blend. *Polymer Degradation and Stability* 79 (2003): 37–43.

Labet, M. and Thielemans, W. Synthesis of polycaprolactone: A review. *Chemical Society Review* 38 (2009): 3484–3504.

Lagaron, J.M. and Lopez-Rubio, A. Nanotechnology for bioplastics: Opportunities, challenges and strategies. *Trends in Food Science and Technology* 22 (2011): 611–617.

Lee, S.Y. and Choi, J. Production and degradation of polyhydroxyalkanoates in waste environment. *Waste Management* 19 (1999): 133–139.

Lefebvre, F., Daro, A., and David, C. Biodegradation of polycaprolactone by microorganisms from an industrial compost of household refuse. Part II: *J. Macromol. Sci.*, Part A: *Pure Appl. Chem.* 32(4) (1995): 867–873.

Lim, J.S., Park, K., Chung, G.S., and Kim, J.H. Effect of composition ratio on the thermal and physical properties of semicrystalline PLA/PHB-HHx composites. *Material Science and Engineering: C* 33 (2013): 2131–2137.

Lim, L., Cink, K., and Vanyo, T. Processing of poly(lactic acid). In: *Poly(lactic acid): Synthesis, Structure, Properties, Processing, and Application*, Auras, R., Lim, L.-T., Selke, E.M., Tsuji, H. (eds.), Vol. 14. 2010, John Wiley & Sons, Inc., pp. 191–213.

Luckachan, G.E. and Pillai, C.K.C. Biodegradable polymers—A review on recent trends and emerging perspectives. *Journal Polymer Environmental* 19 (2011): 637–676.

Maiti, P., Batt, C.A., and Giannelis, E.P. New biodegradable polyhydroxybutyrate/layered silicate nanocomposites. *Biomacromolecules* 8 (2007): 3393–3400.

Matzinos, P., Tserki, V., Kontoyiannis, A., and Panayiotou, C. Processing and characterization of starch/polycaprolactone products. *Polymer Degradation and Stability* 77 (2002): 17–24.

Moreira, M.R., Pereda, M., Marcovich, N.E., and Roura, S.I. Antimicrobial effectiveness of bioactive packaging materials from edible chitosan and casein polymers: Assessment on carrot, cheese, and salami. *Journal of Food Science* 76(1) (2011): 54–63.

Muller, C.M.O., Laurindo, J.B., and Yamashita, F. Effect of cellulose fibers addition on the mechanical properties and water vapor barrier of starch-based films. *Food Hydrocolloids* 23(5) (2009): 1328–1333.

Murphy, R. and Bartle, I. Summary report, biodegradable polymers and sustainability: Insight from life cycle assessment. National Non Food Centre, York, U.K., 2004.

Nagarajan, V., Misra, M., and Mohanty, A.K. New engineered biocomposites from poly(3-hydroxybutyrate-*co*-3-hydroxyvalerate) (PHBV)/poly(butylene adipate-*co*-terephthalate) (PBAT) blends and switchgrass: Fabrication and performance evaluation. *Industrial Crops and Products* 42 (2013): 461–468.

Neilsen, L.E. Models for the permeability of filled polymer systems. *Journal of Macromolecular Science, Part A* 1 (1967): 929–942.

Nieddu, E., Mazzucco, L., Gentile, P., Benko, T., Balbo, V., Mandrile, R., and Ciardelli, G. Preparation and biodegradation of clay composite of PLA. *Reactive & Functional Polymers* 69 (2009): 371–379.

Nuona, A., Li, X., Zhu, X., Xiao, Y., and Che, J. Starch/polylactide sustainable composites: Interface tailoring with graphene oxide. *Composites: Part A* 69 (2015): 247–254.

Ozdemir, M. and Floros, J.D. Active food packaging technology. *Critical Reviews in Food Science and Nutrition* 44 (2004): 185–193.

Pantani, R., Gorrasi, G., Vigliotta, G. et al. PLA-ZnO nanocomposite films: Water vapor barrier properties and specific end-use characteristics. *European Polymer Journal* 49(11) (2013): 3471–3482.

Patel, M., Bastioli, C., Marini, L., and Würdinger, E. Life-cycle assessment of bio-based polymers and natural fibre composites. In: *Biopolymers*, Steinbüchel, A. (ed.), Vol. 10. London, UK: John Wiley & Sons, 2003.

Paul, M.A., Delcourt, C., Alexandre, M., Degée, Ph., Monteverde, F., and Dubois, Ph. Polylactide/montmorillonite nanocomposites: Study of the hydrolytic degradation. *Polymer Degradation and Stability* 87 (2005): 535–542.

Peelman, N., Ragaert, P., De Meulenaer, B., Adons, D., Peeters, R., Cardon, L., Van Impe, F., and Devlieghere, F. Application of bioplastics for food packaging. *Trends in Food Science and Technology* 32 (2013): 128–141.

Petersen, K., Nielsen, P.V., Bertelsen, G., Lawther, M., Olsen, M.B., Nilsson, N.H., and Mortensen, G. Potential of biobased materials for food packaging. *Trends in Food Science and Technology* 10 (1999): 52–68.

Piemonte, V., Sabatini, S., and Gironi, F. Chemical recycling of PLA: A great opportunity towards the sustainable development? *Journal of Polymers and the Environment* 21(3) (2013): 640–647.

Pikoń, K. and Czop, M. Environmental impact of biodegradable packaging waste utilization. *Polish Journal of Environmental Studies* 23(3) (2014): 969–973.

Pillai, C.K.S. Challenges for natural monomers and polymers: Novel design strategies and engineering to develop advanced polymers. *Designed Monomers and Polymers* 13 (2010): 87–121.

Pillin, I., Montrelay, N., Bourmaud, A., and Grohens, Y. Effect of thermo-mechanical cycles on the physicochemical properties of poly(lactic acid). *Polymer Degradation and Stability* 93(2) (2008): 321–328.

Raquez, J.M., Deléglise, M., Lacrampe, M.F., and Krawczak, P. Thermosetting (bio)materials derived from renewable resources: A critical review. *Progress in Polymer Science* 35 (2010): 487–509.

Ratto, J.A., Steeves, D.M., Welsh, E.A. et al. A study of polymer clay nanocomposites for biodegradable applications. In: *57th Annual Technical Conference of the SPE on Plastics Bridging the Millennia, ANTEC '99: Plastics Bridging the Millennia, Conference Proceedings*, Vols. I–III, Publisher Soc Plastics Engineers, Brookfield Center, CT, 1999, pp. 1628–1632.

Reddy, M.M., Vivekanandhana, S., Misraa, M., Bhatiac, S.K., and Mohantya, A.K. Biobased plastics and bionanocomposites: Current status and future opportunities. *Progress in Polymer Science* 38 (2013): 1653–1689.

Reis, K.C., Pereira, J., Smith, A.C., Carvalho, C.W.P., Wellner, N., and Yakimets, I. Characterization of polyhydroxybutyrate-hydroxyvalerate (PHB-HV)/maize starch blend films. *Journal of Food Engineering* 89 (2008): 361–369.

Ren, X. Biodegradable plastics: A solution or a challenge? *Journal of Cleaner Production* 11 (2003): 27–40.

Rhima, J.-W., Parkb, H.-M., and Hac, Ch.-S. Bio-nanocomposites for food packaging applications. *Progress in Polymer Science* 38 (2013): 1629–1652.

Scott, G. Photo-biodegradable plastics. In: *Degradable Polymers: Principles and Applications*, 1st ed., Scott, G., Gilead, D. (eds.). 1995, London, UK: Chapman & Hall, pp. 169–184.

Scott, G. Green polymer. *Polymer Degradation and Stability* 68 (2000): 1–7.

Sebastien, A., Angelique, M., Caroline, T. et al. Active pseudo-multilayered films from polycaprolactone and starch based matrix for food-packaging applications. *European Polymer Journal* 49(6) (2013): 1234–1242.

Semba, T., Kitagawa, K., Ishiaku, U.S., and Hamada, H. The effect of crosslinking on the mechanical properties of polylactic acid/polycaprolactone blends. *Journal of Applied Polymer Science* 101 (2006): 1816–1825.

Shahlari, M. and Lee, S. Mechanical and morphological properties of poly(butylene adipate-co-terephthalate) and poly(lactic acid) blended with organically modified silicate layers. *Polymer Engineering and Science* 52 (2012): 1420–1428.

Shen, L., Haufe, J., and Patel, M.K. Product overview and market projection of emerging bio-based plastics, 2009. http://en. european-bioplastics.org/wp-content/uploads/2011/03/publications/PROBIP2009_Final_June_2009.pdf (Accessed October, 2012).

Sinha, R.S. and Bousmina, M. Biodegradable polymers and their layered silicate nanocomposites: In greening the 21st century materials world. *Progress in Materials Science* 50 (2005): 962–1079.

Sinha, R.S. and Okamoto, M. New polylactide/layered silicate nanocomposites: Open a new dimension for plastics and composites. *Macromolecular Rapid Communications* 24 (2003): 815–840.

Sinha Ray, S., Yamada, K., Okamoto, M., and Ueda, K. Biodegradable polylactide/montmorillonite nanocomposites. *Journal for Nanoscience and Nanotechnology* 3 (2003a): 503–510.

Sinha Ray, S., Yamada, K., Okamoto, M., and Ueda, K. New polylactide layered silicate nanocomposites. 2. Concurrent improvements of materials properties, biodegradability and melt rheology. *Polymer* 44 (2003b): 857–866.

Siracusa, V., Rocculi, P., Romani, S., and Dalla Rosa, M. Biodegradable polymers for food packaging: A review. *Trends in Food Science and Technology* 19 (2008): 634–643.

Södergar, A. and Stolt, M. Properties of lactic acid based polymers and their correlation with composition. *Progress in Polymers Science* 27 (2002): 1123–1163.

Somleva, M.N., Snell, K.D., Beaulieu, J.J., Peoples, O.P., Garrison, B.R., and Patterson, N.A. Production of polyhydroxybutyrate in switchgrass, a value-added co-product in an important lignocellulosic biomass crop. *Plant Biotechnology Journal* 6 (2008): 663–678.

Song, J.H., Murphy, R.J., Narayan, R., and Davies, G.B.H. Biodegradable and compostable alternatives to conventional plastics. *Philosophical Transactions of the Royal Society B* 364 (2009): 2127–2139.

Sorrentino, A., Tortora, M., and Vittoria, V. Diffusion behavior in polymer–clay nanocomposites. *Journal of Polymer Science Part B: Polymer Physics* 44 (2006): 265–274.

Steinbüchel, A. Use of biosynthetic, biodegradable thermoplastics and elastomers form renewable resources: The pros and cons. In: *Degradable Polymers, Recycling and Plastics Waste Management*, Albertsson, A.C., Huang, S.J. (eds.). 1995, Boca Raton, FL: CRC Press, pp. 61–67.

Takala, P.N., Salmieri, S., and Boumail, A. Antimicrobial effect and physicochemical properties of bioactive trilayer polycaprolactone/methylcellulose-based films on the growth of foodborne pathogens and total microbiota in fresh broccoli. *Journal of Food Engineering* 116(3) (2013): 648–655.

Tharanathan, R.N. Biodegradable films and composite coating: Past, present and future. *Trends in Food Science and Technology* 14 (2003): 71–78.

Thellen, C., Cheney, S., and Ratto, J.A. Melt processing and characterization of polyvinyl alcohol and polyhydroxyalkanoate multilayer films. *Journal of Applied Polymer Science* 127(3) (2013): 2314–2324.

Thiebaud, S., Aburto, J., Alric, I. et al. Properties of fatty-acid esters of starch and their blends with LDPE. *Journal of Applied Polymer Science* 65 (1997): 705–721.

Tian, F., Decker, E.A., and Goddard, J.M. Controlling lipid oxidation of food by active packaging technologies. *Food & Function* 4(5) (2013): 669–680.

Tokiwa, Y. and Suzuki, T. Hydrolysis of polyesters by lipases. *Nature* 270(5632) (1977): 76–78.

Vandewijngaarden, J., Murariu, M., and Dubois, P. Gas permeability properties of poly (3-hydroxybutyrate-co-3-hydroxyhexanoate). *Journal of Polymers and the Environment* 22(4) (2014): 501–507.

Vijayendra, S.V.N. and Shamala, T.R. Film forming microbial biopolymers for commercial applications—A review. *Critical Reviews in Biotechnology* 34(4) (2014): 338–357.

Vroman, I. and Tighzert, L. Biodegradable polymers. *Materials* 2 (2009): 307–344.

Yahiaoui, F., Benhacine, F., Ferfera-Harrar, H. et al. Development of antimicrobial PCL/nanoclay nanocomposite films with enhanced mechanical and water vapor barrier properties for packaging applications. *Polymer Bulletin* 72(2) (2015): 235–254.

Yu, J., Yang, J., Liu, B., and Ma, X. Preparation and characterization of glycerol plasticized-pea starch/ZnO carboxymethylcellulose sodium nanocomposites. *Bioresource Technology* 100(11) (2009): 2832–2841.

Zhang, M. and Thomas, N.L. Blending polylactic acid with polyhydroxybutyrate: The effect on thermal, mechanical, and biodegradation properties. *Advances in Polymer Technology* 30 (2011): 67–79.

29

Agricultural Applications of Biodegradable Films

Hande Kaya-Celiker and P. Kumar Mallikarjunan

CONTENTS

29.1 Introduction

The use of polymers in agriculture dates back to, as early as, the 1950s when they were first introduced as greenhouse covers, made of cellophane in the United States and polyvinyl chloride (PVC) in Japan (Scarascia-Mugnozza et al., 2011). With a dramatic rise in demand for plastics in a wide variety of markets, total production of plastics has increased approximately 8.7% every year on global basis from 1.3 million tons back in the early 1950s to 288 million tons in 2012 (PlasticsEurope, 2013). Today, the leading country in plastic market is China, in terms of production and consumption. According to the report from 2013, China was holding 23.9% of the world's plastic materials production capacity, followed by Europe with 20.4% and North America with 19.9% (PlasticsEurope, 2013).

The preference of using plastics in a variety of conventional applications due to their longevity, stability, availability, and low cost has attracted the agricultural sectors, too. Plastic films have been a pioneer in agricultural applications by converting infertile areas into productive lands. Farmers have benefitted from plastics as they provide enhanced cultivation under adverse weather conditions, improved quality

TABLE 29.1

Application Areas and Estimated Production Level of Most Common Plastic Films in Agriculture Industry in the World

Application	Polymer	Volume of Production (Tons/Year)
Greenhouse and walk-in tunnel covers	LDPE, EVA, EBA, PVC, LLDPE	1,000,000
Small tunnel covers	EVA, EBA	170,000
Mulching	LLDPE, EVA, EBA	700,000

Source: Espi, E. et al., *J. Plast. Film Sheet.*, 22(2), 85, 2006.

and quantity of crops, and increased yield of harvest in a shorter time at a lower cost. Plastics found an imperative place in agricultural applications most popularly as greenhouse, walk-in tunnel and low tunnel covers, mulching reservoirs and irrigation systems, and silage. The preferred materials are low-density polyethylene (LDPE), PVC, and ethylene-vinyl acetate (EVA) or ethylene-butyl acrylate (EBA) copolymers for covers and linear LDPE (LLDPE) for mulching (Espi et al., 2006). The summary of the existing polymers used in agriculture is given in Table 29.1. Apart from common usage for greenhouse, tunnel covers, and mulching, other usual applications of agri-plastics include floating covers, covering films for vineyard orchards, silage films, bale twines, bale wraps, nets, windbreaks, shading, hydroponic sacks, protecting bags, nursery pots, strings and ropes, tapes and plastic parts for piping, and irrigation/drainage (Lamont, 2009; Martín-Closas and Pelacho, 2011).

Plastic has many advantages and disadvantages in agriculture. Plastic use in agricultural applications has helped many farmers increase the crop production yield and quality along with allowing growing season extension for crops, such as earlier or later crop production, or multicropping in one season. Not only do plastics allow for many crops, vegetables, and fruits to be produced in any season, but these products are usually in better quality compared to crops grown in open fields. Many benefits of using plastic mulch include, but not limited to, increase in soil temperature (Kasirajan and Ngouajio, 2012; Lamont, 2005), retention of soil structure loose and well aerated, access to adequate oxygen for the roots, and support of beneficial microbial activity (Sanders, 2001). Besides, plastic use in agriculture also provides weed control on mulching beds and conserves water. Plastic irrigation pipes prevent excess use of water and provide efficient use of water resources. Plastic reservoirs keep rain water and minimize the amount of water that may accumulate and cause erosion (Lamont, 2005). Plastic, resistant mulch provides maximum utilization of fertilizers by preventing leaching (Lamont, 2005; Sanders, 2001) and provides cleaner crop by elimination of soil splashing on the plants or fruits (Sanders, 2001). Moreover, the emissions of pesticides in the atmosphere will be reduced as they will remain fixed on the plastic cover and plastic mulches potentially decrease the incidence of diseases, insects, and vertebrate pests (Lamont, 2005).

The current demand for the plastic films used in agricultural applications sums up to 3.6 million tons, of which 41% is used for mulching, 40% is for greenhouse covers, and 19% is for silage (Martín-Closas and Pelacho, 2011). One big disadvantage of using plastic in agriculture is the disposal at the end of the life cycle. Majority of the mulches, greenhouse covers, and drip tapes are made of nonbiodegradable plastics. Heavy usage of plastics in agricultural applications created a widespread concern of public and governmental authorities because of their low degradability in nature. At the end of growing season, they are usually taken to the landfills. The recent weight of plastic waste generated after agricultural applications has been estimated to be 10% of the total weight of plastics in landfills (Rapa et al., 2011). One of the strategic standpoints of the present state of plastic waste disposal is utilization of the waste through recycling, which is a worthy way for source reduction and energy recovery (Figure 29.1). Considering the fact, plastics industry, as individual companies and through organizations such as American Plastics Council, has invested more than $1 billion to support increased recycling and educate communities (Kyrikou and Briassoulis, 2007). In Europe, 25.2 million tons of plastic waste is generated in 2012, of which agricultural plastics represent almost 5% (PlasticsEurope, 2013).

FIGURE 29.1 Picture of contemporary management of inhomogeneous and dirty agricultural plastic wastes delivered to a recycling industry in France before launching the *LabelAgriWaste* trials. (From Briassoulis, D. et al., *Waste Manag.*, 33(6), 1516, 2013.)

Until 2012, there was no common European legislation on agricultural plastic waste recovery, and still today, the majority of the plastic waste is left on the field or burnt uncontrollably by the farmers releasing harmful substances affecting human health, the safety of the farming products, and the environment (Briassoulis et al., 2013). In order to provide farmers a guideline on how to collect and sort the agricultural plastic waste, a project supported by the European Commission through the sixth framework program has been carried out (LabelAgriWaste, 2006–2009), and the system was field tested (Figure 29.1) (Briassoulis et al., 2013). Even though the recovery and recycling rates among the European countries differ greatly, overall recovery rate for agricultural plastics was reported as 49.5%, and mechanical recycling rate was around 23% in 2011 (PlasticsEurope, 2011). The problem is that sources vary greatly according to farm types or agricultural application from large sacks for fertilizers and seeds of vegetable farms to stretch films of silage bags from cattle farms and to greenhouse films. Many of these waste fractions need to be cleaned before material recovery as they are in touch with the ground and organic material, which makes recovery and utilization of agri-plastic waste fractions harder and costly. Thus, the majority of the farmers, especially in developing countries, do not find value in recycling the agricultural plastic. Volume of the certain plastic waste types used in farming practices affects the material collection and recovery efficiency. Some plastic wastes have been too small to cover cost or too much to collect and separate the plastic material for utilization purposes (Horttanainen et al., 2007).

In the last two decades, the resistance of synthetic polymers toward biological attack and subsequent deprivation through the soil created a need for development and study of biodegradable polymers. Numerous existing polymer industries, together with the newly formed biodegradable polymer producing facilities, are currently developing biodegradable polymers for ultimate use in the agricultural practices. A major problem in these areas is the improvement of competitive biodegradable polymers in terms of mechanical and physical properties and life-span.

29.2 Biodegradable Polymers

The renewable polymers are green alternatives to petroleum-derived polymers and also naturally biodegradable. Recent reviews divide biodegradable polymers used to manufacture agricultural mulches into three different categories: biodegradable polymers from renewable resources (including starch, poly(lactic acid) or polylactide [PLA], and polyhydroxyalkanoate [PHA] family of polymers), other compostable polymers from renewable resources (including, cellulose chitosan), and biodegradable polymers from petrochemical sources (including poly(butylene succinate terephthalate), poly(ε-caprolactone) [PCL], poly(butylene succinate) [PBS], poly(vinyl alcohol) [PVOH], and poly(vinyl acetate) [PVA]) (Rudnik, 2008).

29.2.1 Biodegradable Polymers from Renewable Resources

29.2.1.1 Starch

Plastic films were first introduced into agriculture in an effort to make cheaper greenhouses as a glass substitute for protected cropping as demonstrated by Dr. Emery Myers Emmert (Emmert, 1957). By then, plastics, especially polyethylene, were used extensively by the majority of the growers all over the world, which in turn created the environmental concerns and, hence, the need for green alternatives where starch comes the foremost.

Starch is one of the most abundant renewable resources known that is synthesized by various plants such as potatoes, wheat, cassava, rice, and maize. In plants, starch occurs in the form of granules. In its granular form, starch is composed of glucose units linked by glycosidic bonds forming amylose and amylopectin polymers (Mohammadi Nafchi et al., 2013). Starch granules vary in size (2–150 mm) and composition depending on the plant source. In general, the linear and helical amylose constitutes 20%–25% of the total starch granules, and the rest is the branched structured amylopectin. Pure starch is a white, tasteless, and odorless powder that is insoluble in cold water or organic solvents (Kerr, 1950; Radley, 1953).

Starch has many advantages for plastic production as it is renewable, abundant, cheap, and biodegradable and is a good oxygen barrier in dry state (Fabunmi et al., 2007). Yet, starch, by itself, is a poor substitute for petroleum-based plastics because of its hydrophilic nature and inferior mechanical properties like getting brittle by age. The history of developing starch-based biodegradable plastics is abound with scientific studies to improve the properties of starch-based composites through blending, grafting, reinforcing, or optimizing processing conditions, all with a purpose of redound of renewable biodegradable substitutes for commercial usage in agriculture and other industries (Fabunmi et al., 2007) (Table 29.2).

Incorporating starch derivatives into plastics to form rigid polyurethane foam dates back to the 1960s. In these early applications, the use of starch-derived polyols for the urethane market was the main concern (Otey, 1976) for improved properties of the polymer rather than the biodegradability. In the following years, the addition of starch as filler to various resin systems to increase the biodegradation and renewable content became more important, especially when environmental treaties started to restrict the use of nondegradable polymers. A remarkable success was, later, reported when starch was incorporated into LDPE as an inert filler (Griffin, 1974) using film extrusion process with an intent of improved biodegradability. In the same year, Otey and his coworkers developed starch–PVOH cast films, composed of starch, PVA, glycerol, surfactant, and formaldehyde. Plasticizers and PVA provided a more flexible starch film with higher elongation (10%–150%). Starch–PVA casting films were further coated with a solution of PVC or a copolymer of vinylidene chloride and acrylonitrile to deposit a water-repellent property (Otey et al., 1974). Agricultural mulch was thought as an application that might have allowed starch to attain larger market, which, indeed, was not produced in any significant amounts in the United States in those days (Otey, 1976). Water-resistant, ethylene acrylic acid copolymer (EAA)–starch cast films were also developed for possible application of starch films as agricultural mulch. Attempts of using water-insoluble plasticizers did not succeed; however, blending starch with EAA resulted in water-resistant cast films with acceptable physical properties for agricultural mulch applications (Otey et al., 1977). Back in the 1970s, starch was commonly thought as a filling agent for commercial polymer matrixes to improve the biodegradability, which was later proved that only starch components could biodegrade due to fungal attack while polymer matrix remains unaffected. Thus, blending starch and starch derivatives with polymers that can readily undergo microbial biodegradation was proposed as a solution for waste problem. Candidates of PCL and poly(3-hyroxybutyrate-*co*-3-hydroxyvalerate) were evaluated for their mechanical properties when blended with starch, and it was shown that starch could be added into cross-linked PCL at a level of 25 wt.% with a small decrease (20%) in strength (Koenig and Huang, 1995). Different blends of starch (thermoplastic starch [TPS]) and PCL were tested for their mechanical and thermomechanical properties and hydrophobicity, and problems associated with TPS such as low resilience, high moisture sensitivity, and high shrinkage were improved when TPS/PCL blends were prepared even with a level of PCL incorporation as low as 10 wt.%. Blending of TPS

TABLE 29.2

Biodegradable Plastics Available in the Market

Trade Name	Material[a]	Company	Application	Location
Mater-Bi® Y	TPS/cellulose derivatives	Novamont	Injection-molded items: flower pots, seedling, plant tray	Italy
Mater-Bi® V	TPS		Loose filler and packaging	
Mater-Bi® Z	Starch/PCL		Mulch film twines, wrapping films, nets	
Eastar Bio®	Starch/PBAT			
Origo-Bi® (ex Eastar Bio®)	TPS/PBAT		Mulch film, lawn and garden bags	
Vegemat®	Corn fiber/TPS/TPP	Vegeplast S.A.S.	Vineyard fastener, clip, jar	France
BioCore®	TPS	BiologiQ Inc.	Mulch film	United States
Cardia Biohybrid®	TPS/copolyester	Cardia Bioplastics	Injection-molded items: containers, caps, trays, tags, pipes	Australia
Cardia Compostable®	TPS/PE, TPS/PP		Horticultural products: flower pots, stakes	
Terraloy®	TPS/copolyester TPS/HIPS TPS/LLDPE	Teknor Apex Company	Blown film application: mulch films, tying film, sheet	United States
Biopar®	TPS/PP TPS/PBAT TPS/PVOH	Biop Biopolymer Technologies AG	Injection-molded items: pots, sacks, seed tape Foils and hoses for landscaping, mulching films	Germany
Cereplast® Hybrid Resin	Starch/PP, PE, HIPS	Trellis Earth Products Inc. (ex Cereplast Inc.)	Films	United States
Cereplast® Compostables	HDPE/PLA, PHA, PBS			
Plantic	TPS/PVOH	Plantic Technologies Ltd.	Seedling, planter pots	Australia
BioSafe®	PBAT/starch, PBS, PBSA	Zhejiang Hangzhou Xinfu Pharmaceutical Co. Ltd.	Cloth and net, planting tape, mulching film Delayed release of pesticide and fertilizer	China
Ingeo®	PLA/starch, PLA/PBS	NatureWorks LLC	Films, sheets	United States
Bio-Flex®	PLA/PBAT	Fkur Kunststoff GmbH	Sheets, films, bags	Germany
Biograde®	CA			
Fibrolon®-F	Wood fiber/PLA			
Fibrolon®-P	Wood fiber/PP			
Ecovio®	PLA/Ecoflex	BASF Corporation	Agricultural films	United States
Ecoflex	PBAT			
Biomax®	PLA/PBAT (resin modifier	DuPont	Mulch film, seed mats, plant pots	United States
Biomax Strong®				
Terramac®	PLA	Unitika Plastics Division	Flower pot, clip, tray, blow molding containers, soft ground stabilization	Japan
Mirel®	P3HB4HB	Metabolix/ADM joint venture, Telles	Agricultural film, sod netting, plant pots	United States
GreenBio® (ex Sogreen)	PHB, P3HB4HB	Tianjin Materials Co. Ltd.	Mulching films	China
Enmat®	PHBV	Tianan Biological Materials Co. Ltd.		China
Biomer®	PHB	Biomer Inc.		Germany

(Continued)

TABLE 29.2 (*Continued*)

Biodegradable Plastics Available in the Market

Trade Name	Material[a]	Company	Application	Location
Biocycle®	PHB, PHBV	PHB Industrials S.A.		Brazil
Nodax®	PHBA	MGH Meridian (ex P&G/Kaneka)	Weed barrier film, greenhouse film, hay bale wrap, pots, trays, insect trap	United States
Biomatera®	PHA	Biomatera	Agricultural films	Canada
reSound®	PLA/PHB, PHBV	PolyOne	Agricultural films	United States
GS Pla®	PBS	Mitsubishi Chemicals	Mulching film	Japan
EnPol®	PBAT/PBS	Ire Chemical	Agricultural films, plant pot, rope or string, clip, fishing gear	Korea
Bionelle®	PBS, PBSA, PBS/PLA	Showa Denko K.K.	Mulch films, plant pots, rope or string, clip	Korea
Bionelle Starcla®	PBS/Starch			
BioAmber®	Modified PBS/succinic acid	BioAmber	Sheets, films, fibers	France
NatureFlex®	Cellulose	Innovia Films	Mulching films, packaging	United Kingdom
Tenite®	CA, CAB, CAP	Eastman Chemical Company	Films	United States
NaturePlast®	Cellulose ester/PLA, PHA, TPS	NaturePlast	Horticultural films	France
Zelfo®	Cellulose fibers	Zelfo	Films	Germany
Daichitosan®	Chitosan	Dainichiseika Color & Chemicals Mfg. Co. Ltd.	Seed coating for yield enhancement	Japan
Sol-Actif®	Chitin	France Chitine	Plant growth enhancer, soil disease suppressive	France
—	Chitin, chitosan	India Sea Foods	Biofertilizer	India
—	Chitosan oligosaccharides	Qingdao Honghai-Biotech Co. Ltd.	Plant growth nutrient	China
—	Chitosan oligosaccharides	Quingdao Yuanzhou Biotechnology Co. Ltd.	Agricultural Chitosan[b]	China
Pratasan®	Chitosan salts	NovaMatrix (FMC Biopolymer)	Agricultural Chitosan[b]	Norway
Chitosan-Newsun	Chitosan	Chengdu Newsun Crop Science Co. Ltd.	Fungicide	China
Plant Nurse Seawinner Biofertilizer	Chitosan, chitosan oligosaccharides, seaweed extract	China Ocean University Organism Project Development Co. Ltd.	Biofertilizer, protect crops from baby plant period	China

[a] TPS, thermoplastic starch; TPP, thermoplastic protein; HDPE, high-density polyethylene; LDPE, low-density polyethylene; LLDPE, linear low-density polyethylene; PE, polyethylene; PP, polypropylene; HIPS, high-impact polystyrene; PLA, poly(lactic acid); PHA, polyhydroxyalkanoate; PHB, polyhydroxybutyrate; PHBA, poly(3-hydroxybutyrate-*co*-hydroxyalkanoate); PHBV, poly(3-hydroxybutyrate-*co*-valerate); P3HB4HB, poly(3-hydroxybutyrate-*co*-4 hydroxybutyrate); PBAT, poly(butylene adipate-*co*-terephthalate); PCL, poly(ε-caprolactone); PBS, poly(butylene succinate); PBSA, poly(butylene succinate)-*co*-adipate; PVOH, poly(vinyl alcohol); PVA, poly(vinyl acetate); CA, cellulose acetate; CAB, cellulose acetate butyrate; CAP, cellulose acetate propionate.

[b] Agricultural grade chitosan can be used as plant growth regulator, biofertilizer, biopesticide, seed and leaf coating for yield enhancement, soil disease suppressive, and antimicrobial agent.

with PCL resulted in decreased modulus with an increased impact resistance, when starch has a glassy behavior, and an increased modulus of material was observed when starch has a rubbery behavior (Averous et al., 2000).

The development of thermoplastic technology has played an important role in achieving mass production of any plastic product, so does in bioplastic production. Native starch is not a thermoplastic material, but when mixed with plasticizer (water, glycerin, sorbitol, etc.) and processed with heat and pressure, starch undergoes spontaneous destructurization. Extrusion technology is one of the fundamental processes that allows "destructurization" of starch completely and forms a homogeneous melt, known as TPS. Destructurization of starch means disordering the granules as molecular dispersion at which native crystallinity and granular structure disappear (Bastioli, 2012). TPS is a renewable and flexible material that can be further processed using injection molding, compression molding, or film blowing (Nafchi et al., 2013; Zdrahala, 1997). Blend of starch and commercial synthetic polymers treated with extrusion technology has resulted in improved mechanical properties to be utilized as agricultural mulch (Bastioli, 1998).

The Novamont group from Italy uses this approach successively for years in which starch is alloyed with various synthetic and natural polymers and additives (Bastioli, 2001). Since founded in 1990, Novamont's Mater-Bi starch-based technology was suited for agricultural applications such as mulch films, pots for the flower industry, bindings, and devices for the controlled release (CR) of clips and pheromones (Novamont, 2015). Mater-Bi starch-based technology utilizes the destructurized starch and specific polyesters to form amylose complexes. Starch with an amylose/amylopectin ratio above 20/80 (w/w) has a tendency to form droplet-like structure, where core is composed of amorphous amylopectin surrounded by complexed amylose molecules in the form of microdispersion. In film form, droplet-like structure gradually turns into layered form where amylopectin molecules intercalated by layers of complexed amylose. Amylose forms V-type structures when complexed by low-molecular-weight molecules such as butanol and fatty acids. The amount of the V-type amylose structure determines the biodegradation rate, as it increases the biodegradation rate decreases (Figure 29.2). Destructurized starch, having an amylose/amylopectin ratio above 20/80 (w/w), behaves like pseudoplastic at high-sheer stress in the presence of ethylene vinyl alcohol copolymer. Playing with the starch type, complexing agents, additives, and processing conditions allows to engineer a wide range of mechanical, physical–chemical, and rheological properties and the different biodegradation rates of Mater-Bi® products (Bastioli et al., 2012).

Mater-Bi mulching film provides an efficient, green alternative to traditional mulching films with similar mechanical properties and recently was awarded with "OK Biodegradable Soil" standard issued by one of the leading international certification bodies, Vinçotte. The OK Biodegradable Soil

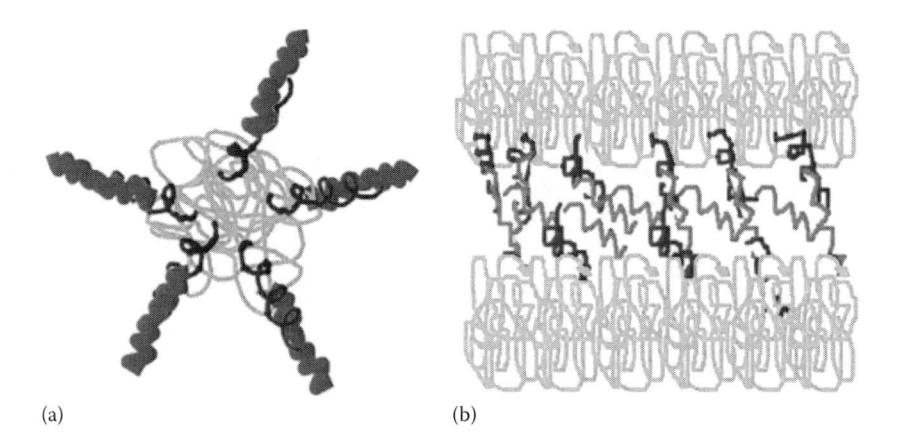

(a) (b)

FIGURE 29.2 Mater-Bi technology: (a) droplet-like structure and (b) layered structure. (Adapted from Bastioli, C. et al., Starch in polymers technology, in K. Khemani and C. Scholz, eds., *Degradable Polymers and Materials: Principles and Practice*, Vol. 1114, American Chemical Society, Washington, DC, 2012, pp. 87–112.)

label guarantees that products will completely biodegrade in soil without adversely affecting the environment (Vincotte, 2015). The production capacity of Novamont today is 120,000 tons/year and holds a patent portfolio that includes more than 90 patent families and 800 internationally registered patents (Bastioli et al., 2012). The patent portfolio of Novamont includes the complexed starch technology developed by Novamont, destructurized starch technology by Warner–Lambert (acquired by Novamont in 1997), film technology by Biotec GmbH & Co. KG (acquired by Novamont in 2001), and Easter Bio Technology of Eastman Chemical Company (acquired by Novamont in 2004) (Bastioli et al., 2012).

Vegeplast is an independent French company (Vegeplast, 2015) producing biosourced, biodegradable, and compostable plastic pieces through injection molding under the trade name Vegemat®. Vegemat films are composed of 40%–50% TPS and 30%–40% thermoplastic proteins to form film matrix, plant originated fibers to reinforce the structure and improve mechanical properties, and lubricants and plasticizers to ensure the technical feasibility. The final material is rigid as wood and is not water resistant. Vegemat products are in the market named as "vineyard fasteners," "clip tomato," and "horticultural jar." It usually biodegrades in 8 weeks (Flieger et al., 2003).

Established in 2002 as Biograde Limited, Cardia Bioplastics today produces cornstarch-based, renewable resins under the trade name Cardia Compostable B-F. Flexible films made of or based on blends of TPS, aliphatic polyesters, and natural plasticizers offer a high level of mechanical strength, elongation properties, and toughness, which makes blown films suitable for agricultural mulching (CardiaBioplastics, 2015).

Teknor Apex Company is a privately held company founded in 1924 and headquartered in Pawtucket, Rhode Island, United States. In December of 2008, Teknor Apex announced a new bioplastics division initially marketing compounds based on thermoplastic starch. TPS from corn, wheat, or potato is combined with biodegradable copolyester (poly(butylene adipate-*co*-terephthalate) [PBAT]), PLA, or PHAs or with petrochemical-based polymers such as polyolefins and polystyrene (PS) to create a product line under the trade name Terroloy™. Terroloy series are produced for blown film–extruded and injection-molded applications including horticulture such as plant pots, underlays, sacks, seed/fertilizer tape, covering film, mulching film, and tying film. The content of TPS ranges from 30% to 50% for masterbatch (TeknorApex, 2015).

Biop Biopolymer Technologies is in the market with Biopar® products since its launch in 2002. Biopar is a bioplastic resin consisting mainly of thermoplastic starch, aliphatic copolyesters, and additives suitable for film blowing. The compatibilizer allows the starch polymer to be in a bicontinuous phase with the polyester, which advances the resin to hold more starch (BiopBiopolymerTechnologies, 2015). Unlike commonly used polymers that are disperse, Biopar is multilayered and biphasic, owing to its formula having patented compatibilizer. Its high water vapor rate and gas barrier properties enable the use of polymer for agricultural applications and landscape gardening (BiobasedNews, 2011).

BiologiQ, founded in 2011, is another company that commercializes TPS- and PLA-based technology (eco-starch resin) to produce agricultural mulch film under the BioCore™ trademark. Biocore supplies different formulations, which enable the tailor-made mulch films with different thickness, width, length, and biodegradation rate of 50–60 to 120–150 days (BiologiQ, 2015). This new formula enables to reach biodegradable mulch film at a price that is comparable to fossil fuel–based counterparts.

Having the same claim, the California-based company Cereplast produces a family of Bio-polyolefins®, which are exhibiting similar properties of modulus and impact strength compared to traditional polyolefins at a pricing similar to the traditional plastics. Cereplast, founded in 1996, has started business by marketing the Novamont products in North America. In 2001, they produced their own TPS technology, Cereplast Hybrid Resin®, replacing 50% of synthetic polymers with starches from corn, tapioca, wheat, and potatoes (Cereplast, 2015). The proprietary process developed by Cereplast requires less energy as of using lower process temperatures, making the thermoplastic elastomer starch hybrid compounds suitable for injection molding applications. The blend of selected biopolymers are polymerized and treated with nanocomposites for surface optimization and reinforcement. These hybrids are not biodegradable but they increased the use of renewable resources and overcame the problems of heat resistance and stability that other bioresins are facing with (Marcin et al., 2009). Earth Products Inc. has acquired the assets of Cereplast, which filed for bankruptcy in 2014 (Voegele, 2014). According to Bill Collins, founder,

FIGURE 29.3 Process of biodegradation of Plantic® plant pots in compost. (From Plantic, 2015, www.plantic.com.au/.)

chairman, and president of Trellis Earth, his company is ready to restart Cereplast's manufacturing facility with an improved business plan (Voegele, 2014).

Plantic Technologies Ltd. is based in Australia. Plantic Technology Ltd.'s primary feedstock is a naturally high-amylose starch (up to 70% amylose), derived from corn, which has been hybridized over a number of generations. When starch is heated, the crystalline structure is disrupted and upon cooling, typically recrystallizes in a process called retrogradation. This is commonly known as staling, similar to a loaf of bread hardening several days after being baked. To prevent this from occurring in Plantic® products, the high-amylose starch currently used by Plantic Technologies Ltd. undergoes a chemical modification process called hydroxypropylation, which retards retrogradation and effectively plasticizes the starch, making it behave like a thermoplastic and providing a shelf life of many years. This hydroxy-propylated, high-amylose starch forms the base of all of Plantic Technology Ltd. products, and its use in packaging applications is protected through a family of patents. Water-resistant Plantic products have been molded into several sizes of planter pots. When placed in a typical domestic compost environment, these shelf-stable pots completely biodegrade in less than 1 month (Figure 29.3). Plantic also developed a customized resin with Freitec Kunststoffe GmbH in the form of plant seeds packaged in a compostable blister as a green marketing initiative (Plantic, 2015). In 2007, DuPont and Plantic announced an alliance for starch-based biomaterials, in which DuPont marketed and distributed Plantic's starch-based resins and sheet products under the DuPont™ Biomax family of products. In 2014, a Japanese biobased barrier film manufacturer Kuraray Co. Ltd. concluded a contract with Plantic Technologies to promote and distribute the biobased barrier material Plantic film in Japan, and a year after, the company announced the acquisition of all of the shares in Plantic Technologies Ltd. (Kuraray, 2015).

A joint venture company (Ever Corn, Inc.) was established between Japan Corn Starch Co. Ltd. and Grand River Technologies, Inc., a subsidiary of Michigan Biotechnology Institute in Michigan in 1993 (EverCorn, 2015). They are commercializing starch ester–based thermoplastic technology tailored with appropriate plasticizers and additives to give good mechanical properties, under the brand name Cornpole®. Cornpole resin has also water-repellent properties and is used for a wide range of applications, including agricultural mulching films (NihonCornStarch, 2015).

29.2.1.2 Poly(lactic Acid) or Polylactide

Among the family of biodegradable aliphatic polyesters, PLAs have been the focus of much attention because they are produced from annually renewable resources. They are considered biodegradable and biocompostable, and they have very low or no toxicity and high mechanical performance, comparable to those of commercial polymers (Yu et al., 2006). Poly(lactic acid) is one of the few polymers that can be modified using the L- or D-isomers to aspire to design high-molecular-weight amorphous or crystalline polymers that can be used for food contact and are generally recognized as safe (GRAS) (Garlotta, 2001).

PLA polymers are generally derived by fermenting the sugar from carbohydrate crops such as corn, wheat, barley, cassava, and sugarcane through catalytic polymerization (Briassoulis, 2004). The lactic acid (2-hydroxy propionic acid) is chiral molecule with an asymmetric carbon atom, which leads to optically active configurations (Figure 29.4). The L(+)-isomer is produced in humans and animals, whereas both D(−)- and L(+)-isomers are synthesized by bacterial fermentation. In industry, majority of lactic acid is produced homolactic organisms such as optimized or genetically modified strains of *Lactobacilli*. Among many *lactobacilli* strains, *Lactobacillus amylophilus*, *L. bavaricus*, *L. casei*, *L. maltaromicus*, and *L. salivarius* predominantly produce L(+)-isomer, while *L. delbrueckii*, *L. jensenii*, and *L. acidophilus* yield D(−)-lactic acid from the fermentation of carbohydrates (Garlotta, 2001).

PLA is a hard material (Rockwell H Scale of more than 60). Because of the hardness, the PLA fractures along the edges, creating a product that cannot be used (Briassoulis, 2004). The macroscopic morphology depends very much on the percentage and distribution of chiral repeating units, which can be very different, depending of the synthesis route and on the chiral monomer ratio. D-units in a poly-L chain behave like impurities. The percentage of L(+)-LA units corresponds to the polymer being more or less stereoregular. The melting temperature and crystallinity decrease rapidly with increasing percentage of L-LA unit, making the material strong but brittle. Less stereoregular forms of the polymers have better mechanical characteristics due to their lower crystallinity. Interestingly enough, the ratio of oligomers can be used to adjust the mechanical characteristics to create commercially interesting PLA from waxy amorphous oligomeric compounds to very strong crystalline devices (Vert et al., 1995).

Degradation of PLA consists of hydrolysis ester bonds and degradation rate depends on the size, shape, isomer ratio of polymer, and the temperature of the hydrolysis. Thermal degradation temperature for PLA is above 200°C, with a chain of reactions of hydrolysis, lactide reformation, oxidative chain scission, and inter- or intramolecular transesterification (Garlotta, 2001). PLA is hydrophobic and expensive and degradation process is rather slow when accumulation of waste of plastics in landfills is considered. It was, therefore, blended with other polymers or plasticizers to suit PLA for various applications. It can therefore be blended with starch to decrease the cost and improve the biodegradability; however, some causality in mechanical properties such as tensile strength and elongation and increase in water sensitivity due to the hydrophilic nature of native starch should be expected (Fabunmi et al., 2007). One common blend of PLA and starch is the use of plasticizers and/or grafting onto PLA (e.g., using maleic anhydride) to increase miscibility. Other polymers employed in blends with PLA include PBAT, polyethylene oxide, chitosan, epoxidized soybean oil, PCL, and PVOH with common plasticizers such as lactic acid, glycerol, and citrate esters (Hayes et al., 2012).

There are other manufacturers based in the United States and Japan, but Cargill Dow LLC presently manufactures approximately 95% of the world's production of PLA (Plackett and Vazquez, 2010). Nature Works LLC (Cargill Dow LLC–owned company) has developed a low-cost continuous production for

FIGURE 29.4 L(+)- and D(−)-isomers of PLA. (Adapted from Tsuji, H. et. al., *Macromol. Biosci.*, 5(7), 569, 2005.)

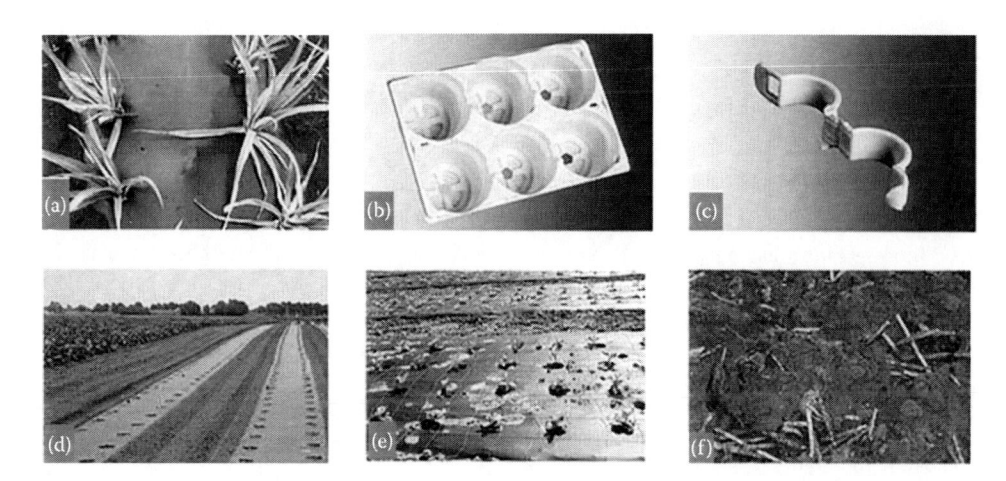

FIGURE 29.5 Horticultural and agricultural applications of Bio-Flex®: (a) mulch film for pineapple, (b) plant ray, (c) plant clips, (d) mulch film laid out, (e) mulch film in use, (f) mulch film biodegraded. (From Fkur, 2015, http://www.fkur.com.)

PLA, consisting of prepolymer production, which is catalytically converted into the cyclic dimer lactide and vaporized, purification via distillation and melt crystallization, and high-molecular-weight PLA production using solvent-free ring-opening lactide polymerization. The process is a continuous process, where meso and L fractions were combined to have a wide portfolio for use in different applications. Currently, Nature Works LLC, United States, is the major supplier of PLA sold under the brand name Ingeo, with a production capacity of 150,000 ton/year (Vink and Steve, 2015). There are many manufacturers using the Nature Works PLA for the production of blend to be used in agricultural applications. Fkur Kunststoff GmbH is one of them, selling PLA copolyester blends under the name of Bio-Flex®. Bio-Flex mulch films do not contain any starch or derivatives that decrease the sensitivity to moisture and improve the durability, so they do not biodegrade too quickly during their protective function on the surface of the field. However, they do biodegrade steadily once ploughed into the soil after use (Fkur, 2015). Bio-Flex is also a good barrier for vapor and oxygen, which makes it suitable to be used in the same way as conventional polymers, in agricultural applications (Figure 29.5) (Lörcks, 1998). Bio-Flex polymer has been tested for its mechanical properties and biodegradability in soil as mulching film on a variety of crops in different environmental conditions (Barragán et al., 2012; Mostafa and Sourell, 2011; Yang and Wu, 1999). Polymers from Bio-Flex series can also be used in horticultural applications such as ties and clips to support the plant shoots (Figure 29.5). Higher elongation together with rigidity of Bio-Flex polymers represents a practical alternative to standard plastic ties or clips, which may occasionally fall off during plant growth and may be trodden into the soil (Fkur, 2015).

Another company using Ingeo PLA is BASF AG, Germany. BASF offers Ecoflex®, an aliphatic–aromatic copolyester based on the monomers 1,4-butanediol, adipic acid, and terephthalic acid in the polymer chain. The content of terephthalic acid in the polymer is approximately 42–45 mol%. Modifications of the basic copolyester lead to a flexible material, which is especially suitable for film applications (Muller, 2005). Ecovio® is made of Ecoflex® plus renewable raw materials, including 45% by weight Ingeo™ bioresin (Worldwide Developments in More Sustainable Materials, 2009). Depending on the mixing ratio of Ecovio with Ecoflex® or PLA, more flexible or more rigid formulations are possible. Thus, among other things, it can be used to produce mulch film in agriculture (BASFNews, 2007).

Like Ecoflex®, the Eastman product, Eastar Bio®, is based on a copolyester composed of terephthalic acid, adipic acid, and 1,4-butanediol, but due to some special modifications, the material properties are different (Muller, 2005). In September 2004, Novamont acquired the "Eastar Bio Technology" of Eastman Chemical Company and integrated the line of renewably sourced polyesters technology into Mater-Bi bioplastics. Novamont markets this new technology under the trade name Origo-Bi (Esposito, 2014; Novamont, 2015). DuPont has also developed a PBAT-related copolymeric product referred to as Biomax. The product, Biomax Strong®, originally presented as a toughening modifier for PLA, is designed to improve the thermal stability during processing, to decrease the brittleness (DuPont, 2015).

Futerro, a joint venture between Galactic and Total Petrochemicals, produces PLA from renewable vegetable resources such as sugar beet, sugarcane, wheat, maize, and cellulose since 2007 (http://www.futerro.com). Another company, Plasthill Ltd., is a manufacturer of Kareline®, cellulose-PLA composite (Plasthill, 2015). The fiber is Nordic softwood pulp and content is changing (10–50 wt.%) for tailor-made products. Biolloy®, introduced by JSR Corporation, is a polymer alloy of PLA with thermoplastic resin. PLA combination can be altered (5%–80%) for specific applications (JSRCorporation, 2015). Unitika's biomass material, Terramac®, is a PLA-based GreenPla- and BiomassPla-certified biopolymer. The technology behind Terramac PLA makes a uniform, flexible, and high-strength polymer matrix, which enables the material ideal for agricultural areas such as weed control sheets, antiweed sheets, or soil stabilizer (Unitika, 2015). In 2011, Dow and PolyOne introduced an additive concentrate product, OnCap™ Bio, to improve the impact resistance of opaque, injection-molded PLA products with minimal effect on heat distortion temperature and stiffness (PolyOne, 2015).

29.2.1.3 Polyhydroxyalkanoates

The microbial polyesters of PHAs are a family of intracellular biopolymers produced by bacterial fermentation to provide a reserve of carbon and energy (Griffin, 1995) and can be utilized in various industrial applications because they are biodegradable and highly biocompatible thermoplastics (Figure 29.6). These polymers contain repeating units of 3-hydroxyacids (Anderson and Dawes, 1990) and can be produced by a variety of microorganisms such as *Bacillus* spp. (Full et al., 2006; Lstrok et al., 2001; Valappil et al., 2007), *Alcaligenes* spp. (Brandl et al., 1989; Cavalheiro et al., 2009; Peoples and Sinskey, 1989), *Pseudomonas* spp. (Durner et al., 2000; Fernández et al., 2005; Lemoigne, 1926; Pachence et al., 2007; Tobin and O'Connor, 2005; Ward et al., 2005; Ward and O'Connor, 2005), *Aeromonas hydrophila* (Chen et al., 2001; Qiu et al., 2006), recombinant *Escherichia coli* (Kahar et al., 2005; Mahishi et al., 2003; Nikel et al., 2006), and many others. *Pseudomonas* species (specifically *Pseudomonas oleovorans* and *Pseudomonas aeruginosa*, *Pseudomonas putida*, *Pseudomonas fluorescens*, *Pseudomonas testosterone*) accumulate PHA having 3-hydroxyacids monomers with $2n$ carbon shorter than the substrate that is utilized as sole carbon source (Anderson and Dawes, 1990). *Alcaligenes eutrophus* (*Ralstonia eutropha* or *Cupriavidus necator*) is a gram-negative soil bacterium, which synthesizes poly(3-hydroxybutyrate) (PHB) using glucose and three enzymes: 3-ketothiolase, acetoacetyl-CoA reductase, and PHA synthase (Ojumu et al., 2004). When glucose is substituted with propionic acid or valeric acid, *A. eutrophus* synthesizes polymers containing 3-hydroxybutyrate a (3HB) and 3-hydroxyvalerate (3HV) (Ojumu et al., 2004). Photosynthetic bacterium *Rhodospirillum rubrum* is, also, capable of producing PHAs β-hydroxybutyrate (HB), β-hydroxyvalerate (HV), and β-hydroxyhexanoate (HC) monomer units (Brandl et al., 1989).

Over 250 different bacteria reported to accumulate various PHAs (Ojumu et al., 2004), PHB is the best known member of short-chain-length PHA with monomers containing 3–5 carbon atoms (Smith, 2005).

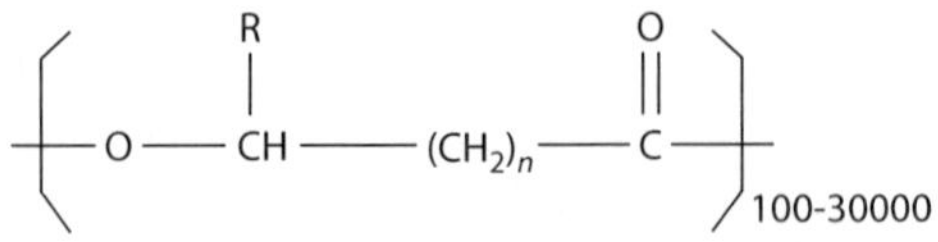

$n = 1$ R=   Hydrogen   Poly (3-hydroxypropionate)
             Methyl   Poly (3-hydroxybutyrate)
             Ethyl   Poly (3-hydroxyvalerate)
             Propyl   Poly (3-hydroxyhexanoate)
$n = 2$ R=   Hydrogen   Poly (4-hydroxybutyrate)

FIGURE 29.6 Structures of polyhydroxyalkanoates. (Adapted from Ojumu, T. et al., *African Journal of Biotechnology*, 3(1), 18, 2004.)

Since its discovery by Lemoigne as a constituent of bacterium *Bacillus megaterium* (Lemoigne, 1926), most studies of the physical properties of bacteria PHAs have been with PHB and poly(3-hydroxybutyrate-*co*-3-hydroxyvalerate) (Ojumu et al., 2004). The homopolymer PHB is stiff, which makes it quite brittle when melt crystallized and solvent cast (Pachence et al., 2007). Efforts have been made to develop PHB copolymers by altering the growth medium, PHB synthesizing microorganism, and carbon source. There have been many reports searching for PHB with better thermal and mechanical properties through copolymerization or blending with other biodegradable or nonbiodegradable polymers. Most common amendment is increasing the fraction of 3HV units that, in turns, creates tougher (higher impact strength), more flexible (lower Young's modulus), and more readily processible PHB copolyesters. However, they are still relatively hydrophobic, which makes the biodegradation time longer and consequently unsuitable for short-term applications (Pachence et al., 2007).

Potential agricultural applications of PHA biopolymers can include encapsulation of seeds, encapsulation of fertilizers for slow release, and biodegradable containers for hothouse facilities (Verlinden et al., 2007). The barrier properties of PHA biopolymers are comparable with polyethylene but film's moisture vapor barrier properties are not as good. This can be advantageous for plants as they can breathe while the soil temperature is kept lower especially for summer production. However, homopolymer PHB is a brittle, crystalline thermoplastic and undergoes thermal decomposition just at its melting point, thus making processing difficult and limiting its commercial usefulness. Therefore, extensive efforts have been directed toward synthesis of copolymers that have better properties than PHB (Rudnik, 2008).

The first commercial production of PHB and PHA dates back to 1982, when Imperial Chemical Industries (ICI/Zeneca BioProducts, Bellingham, United Kingdom) marketed the PHBV (poly(3-hydroxybutyrate-*co*-valerate)) copolymer from *A. eutrophus* under the trade name Biopol®, which is less stiff and less brittle than homopolymer PHB. The ratio of HB to HV monomer can be varied by changing the glucose to propionic acid ratio. By increasing the ratio of HV to HB, the melting temperatures are lower and mechanical properties are improved. In 1996, Zeneca sold its Biopol business to Monsanto, and then in 2001, Metabolix Inc (United States) acquired Monsanto Biopol patents (Rudnik, 2008). Cambridge-based company Metabolix owns patents that are built around PHA technologies and market PHA-based biopolymers with a brand name of Mirel™. Mirel polymers (3HB-4HB copolymers) are produced by recombinant *E. coli* harboring the *A. eutrophus* biosynthesis genes from corn syrup as carbon source (Chen, 2009). Mirel was commercialized by joint venture between Metabolix and Archer Daniels Midland Co. called Telles from 2006 to until the disclosure of Telles in 2012. Potential agricultural applications of Mirel biopolymer include agricultural film, sod netting, and plant pots. Mirel™ is certified compostable by the Biodegradable Products Institute and carries the "OK compost" certification from Vinçotte (EN 13432) as well as "OK biodegradable SOIL" certification (Metabolix, 2015). Likewise Metabolix, Tianjin GreenBio Materials Co. Ltd. (China) is utilizing the recombinant *E. coli* and *A. eutrophus* for PHB and P3HB4HB (poly(3-hydroxybutyrate-*co*-4 hydroxybutyrate)) production (Chen, 2009). GreenBio® polymers, in pellet or sheet form, are suitable for biodegradable mulching applications (Tianjin, 2015); because they biodegrade in soil, home compost, and industrial compost sites; they also biodegrade in freshwater and seawater environments. Another Chinese company, Tianan Biologic Material Co., is manufacturing PHBV-based polymers produced by *A. eutrophus* grown on dextrose or glucose derived from corn or cassava as sole carbon source and propionic acid for enhanced production of polymer. The brand name of the product is Enmat® (Tianan, 2015).

A German company, Biomer Inc., produces PHB on a commercial scale from *Alcaligenes latus*, which is one of the strains that satisfy the requirement for industrial PHB biosynthesis (Chen, 2009). This strain grows on sucrose, glucose, and molasses rapidly and can accumulate PHB as high as over 90% of cell dry weight (Chen, 2009). Biomer™ acquired the microorganisms from Austrian company, Chemie Linz (later btf Austria), and registered the trade name Biomer in 1995 (Chanprateep, 2010). In Brazil, PHB Industrial S.A., owner of the Biocycle® brand, uses sugarcane to produce PHB and PHBV biopolymers from *Bacillus* spp. in commercial scale (Biocycle, 2015). Biocycle is manufactured in a joint venture started in 1992 between a sugar producer (*Irmãos Biagi*) and an alcohol producer (the Balbo Group) (Valappil et al., 2007). The large-scale production of PHB in sugar mills has many advantages that energy needed to produce PHB is provided by biomass. Simultaneously, carbon dioxide emission can be assimilated through photosynthesis of sugarcane crop, and waste can be recycled via cane field, and that will lead a cheaper polymer production under low-price carbon source and energy (Nonato et al., 2001). Integrated pilot-scale plant is being prospected to increase the production capacity up to 3000 ton/year in 2015 (Biocycle, 2015).

Another company, Procter and Gamble, in partnership with Kaneka Corporation, directed efforts into the development of a variety of PHB biopolymers under the Nodax® brand name, such as fibers, nonwoven hygiene products (Chanprateep, 2010; Rudnik, 2008). The Nodax family of copolymers are PHBAs (poly(3-hydroxybutyrate-*co*-hydroxyalkanoate)s), with a copolymer content varying from 3 to 15 mol% and chain length from C7 up to C19 (Rudnik, 2008). In 2007, Meredian Holdings Group, also known as MHG (Georgia, United States), purchased the patent for Nodax PHA technology from Procter and Gamble to expand their product range from solely PLA to MHG PHA. MHG Nodax PHA technology is based on locally grown canola oil as the food stock for the microorganisms. The PHA produced at MHG is also Vinçotte certified, which can biodegrade within 12–18 weeks in six different mediums including anaerobic, soil, freshwater, marine, industrial, and home composting (Meredian, 2015). MGH Meredian (sister company of MGH-Danimer Scientific, United States) has the largest production capacity of medium-chain-length branched PHA (mcl-PHA) with a value of 272,000 tons/year (Chanprateep, 2010). MGH Nodax can be manufactured as landscaping weed barrier film, greenhouse film, agricultural mulch film, and hay bale wrap for agricultural applications or as horticultural pots, trays, tree-growing containers, landscaping, erosion control or tent stakes, and insect traps for horticultural applications (MGHNodax, 2015).

There are many other PHA-based biopolymers marketed for agricultural applications such as BioGreen® by Mitsubishi Gas Chemical, AmBio® Bioresin by Ecomann Biotechnology Co., Biomatera® by Biomatera Inc., and reSound® by PolyOne. Apart from the PHB resin producers, there are sectors integrating the PHA resins into specific products. A French company, NaturePlast and her daughter company Biopolynov, created a R&D department to utilize the PHA resins in final product for special applications (Biopolynov, 2015).

29.2.2　Other Compostable Polymers from Renewable Resources

29.2.2.1　Cellulose

Cellulose is the most abundant material worldwide: it is the main constituent of plants, serving to maintain their structure, and is also present in bacteria, fungi, algae, and even in animals (O'Sullivan, 1997). It is representing about 1.5×10^{12} tons of the annual biomass production and is considered as an almost inexhaustible source of raw material for the increasing demand biocompatible products (Klemm et al., 2005; O'Sullivan, 1997). Native plant cellulose is composed of a β(*1→4*)-*linked* D-*glucan* backbone with 3000–8000 polymerized glucose units, and the number of glucose units depends on the plant source (Nawrath et al., 1995). Wood pulp is the main source to produce cellulose regenerate fibers, films, cellulose esters, and ethers (approximately 3.2 million tons of wood pulp is used to process cellulose in 2003) (Klemm et al., 2005). Cellulose fibers isolated from wood pulp typically have chain lengths of 500–2000 glucose units. Because of regularity of repeating units of cellulose and the hydrogen bonding of hydroxyl groups in adjacent chains makes cellulose tightly packed crystalline material, which is insoluble in water and in most organic solvents (Nawrath et al., 1995). The insolubility of cellulose caused by its so-called supramolecular structure is the reason why esterification or etherification of the hydroxyl groups is performed to produce cellulose derivatives (Rudnik and Briassoulis, 2011; Sannino et al., 2009). Another reason is that cellulose is not melt processible, because it decomposes before it undergoes melt flow (Rudnik and Briassoulis, 2011).

Commercially available products are currently produced from cellulose esters and ethers and find a place in a variety of applications mostly in coating, medical, and CR applications. The most common cellulose esters are cellulose acetate (CA), cellulose acetate propionate (CAP), and cellulose acetate butyrate (CAB). They are thermoplastic materials produced through esterification of cellulose (Rudnik and Briassoulis, 2011). In the field of CR systems, most water-soluble cellulose derivatives are obtained via etherification of cellulose, which involves the reaction of the hydroxyl groups of cellulose with organic species, such as methyl and ethyl units. Cellulose-based hydrogels, either reversible or stable, can be formed by properly cross-linking aqueous solutions of cellulose ethers, such as methylcellulose (MC), hydroxypropyl methylcellulose (HPMC), ethyl cellulose (EC), hydroxyethyl cellulose (HEC), and sodium carboxymethylcellulose (NaCMC), which are among the most widely used cellulose derivatives (Rudnik and Briassoulis, 2011).

Innovia Films offers two wood pulp–derived cellulose product lines, NatureFlex® and Cellophane® (InnoviaFilms, 2015). Eastman (United States) manufactures several plasticizers to aid in the production of agricultural films exhibiting good weathering and heat stability properties. The Tenite® cellulosics is made from softwood material (Eastman, 2015). Fkur Kunststoff GmbH has a product line based on CA for injection molding, sheet, and profile, under the trade name Biograde® (Fkur, 2015). This thermoplastic was developed especially for injection molding and extrusion plants. It contains a high proportion of cellulose, exhibits excellent shape retention under heat up to a temperature of 122°C, and has similar properties to PS (Peters, 2011). Another product Fibrilon® by Fkur Kunststoff is a wood–plastic composite (wood fiber/PLA and wood fiber/PP) suitable for injection molding and extrusion (Fkur, 2015). The French manufacturer, NaturePlast, specialized in the production of bioplastics such as PLA, PHA, TPS, and cellulose derivatives (wood pulp). NaturePlast® ACI 001 is a thermoplastic resin of cellulose ester made from renewable vegetable resources and is specifically developed for injection molding (NaturePlast, 2015). Zelfo Technology GmbH (Germany) manufactures wheat straw–based micro- and nanofibrillated cellulose (M/NFC) under the trade name Zelfo® (Zelfo, 2015). This material is made completely from cellulose fibers of plant origin (e.g., hemp, flax, waste paper). It is transformed into a pliable mass without the addition of water or adhesives and can then be injection molded, extruded, or compression molded (Peters, 2011). Biowert Industrie GmbH (Germany) introduced AgriPlast BW plastic granules consisting of 40%–75% cellulose fibers, obtained from grasses, and 25%–60% PE or PP (Biowert, 2015). Biowert Industrie GmbH operates a grass-refining plant, which is based on the principles of "green biorefinery" and transforms moist biomass containing fibers to a composite granulate, without the use of chemical additives or solvents (Peters, 2011). Tecnaro GmbH (Germany) is a producer of thermoplastics from lignin (waste from pulp industry) and cellulose under the trade name Arboform® (Teknaro, 2015). Blending lignin with natural fibers (flax, hemp, or other fiber plants) and some natural additive produces a fiber composite, which can be worked using injection molding or extrusion and can be recycled (Peters, 2011).

29.2.2.2 *Chitin and Chitosan*

Chitin is a long-chain polymer of 2-acetamide-2-deoxy β-D-glucose bound through the β-(1→4)-glycoside linkage and is usually found in the shells of crabs, lobsters, shrimps, and insects (Chandra and Rustgi, 1998). The structure of chitin is comparable to cellulose, both form crystalline nanofibrils (Sharp, 2013), but unlike cellulose, chitin has an innate rigidity. Chitin is insoluble in its native form but chitosan, the partly deacetylated form, is water soluble. The materials are biocompatible and have antimicrobial activities as well as the ability to absorb heavy metal ions (Chandra and Rustgi, 1998). Chitosan can be obtained by: (1) the removal of acetyl group of chitin in the presence of strong alkali, (2) enzymatic hydrolysis in the presence of a chitin deacetylase, and (3) natural means of finding it in certain fungi as part of their structure. In chitosan, the amino groups are protonated in acid, which determines the positive charge of chitosan (Castro and Paulín, 2012). The positive charge of chitosan redounds to chitosan unique properties to be utilized in agriculture as plant elicitor (Lee et al., 1999), plant immunity regulator (Yin et al., 2010), insecticide or fungicide (Benhamou and Thériault, 1992; Ohta et al., 2004; Rabea et al., 2005; Rodríguez et al., 2007; Zhang et al., 2003), fertilizer (Ohta et al., 2004), CR and water-retention agent (Wu and Liu, 2008), and growth promoter (Nge et al., 2006; Xuan Tham et al., 2001), to mention a few. The uniqueness of biopolymer chitin and chitosan and their biodegradability, biocompatibility, and nontoxicity properties attracted many researchers and allow many applications in agriculture. Enormous data have emerged for chitin and chitosan uses from not less than 20 books, over 300 reviews, over 12,000 publications, and several patents (Sastry et al., 2015).

Some of the chitin and chitosan producers are Primex Ehf (Iceland), France Chitin (France), Qingdao Heppe Biotechnology Ltd. (China/Germany), FMC BioPolymer (United States), NovaMatrix (Norway), Dalwoo-Chitosan BLS (Korea), Meron Biopolymers (India), Golden-Shell Biochemical Co. Ltd. (China) (Niekraszewicz and Niekraszewicz, 2009), India Sea Foods (India), Qingdao Yunzhou Biochemistry Co. Ltd. (China), Dainichiseika Color & Chemicals Mfg. Co. Ltd. (Japan), and G.T.C. Bio Corporation (China).

29.2.3 Synthetic Biodegradable Polymers

29.2.3.1 *Poly(butylene Adipate-co-Terephthalate)*

Poly(butylene adipate-*co*-terephthalate) (PBAT) is a biodegradable copolyester and obtained by poly-condensation between 1,4-butanediol and a mixture of adipic acid and terephthalic acid (Figure 29.7). It shows good mechanical and thermal properties at a concentration in terephthalic acid higher than 35 mol %. The biodegradation rate decreases rapidly when the concentration became higher than 55% (Vroman and Tighzert, 2009). PBAT readily forms composites with several biopolymers, but among them PBAT/PLA blends are the most studied ones, as both being thermoplastic allowing to process using most conventional polymer-processing methods. Changing the PBAT content, blends with different mechanical properties are possible. For example, with the increase in PBAT content (5–20 wt.%), the blend showed decreased tensile strength and modulus; however, elongation and toughness were dramatically increased. With the addition of PBAT, the failure mode changed from brittle fracture of the neat PLA to ductile fracture (Jiang et al., 2006). The linear and nonlinear shear rheological behaviors of PLA/PBAT melts were investigated, and the results showed that PLA/PBAT blend is a kind of immiscible, two-phase system where PBAT disperses evenly in PLA matrix (Gu et al., 2008). PBAT is produced by many different manufacturers and known with brand names of Ecovio (Ecoflex®, Envoi®) prepared by BASF (Germany), Origo-Bi® (Eastar Bio) from Novamont (Italy), and Biosafe® prepared by Xinfu Pharmaceutical Co. (China) (Hayes et al., 2012). DuPont's Biomax is another example for PBAT-related copolymeric products.

29.2.3.2 *Poly(ε-Caprolactone)s*

PCL is an aliphatic polyester produced by the ring-opening polymerization of ε-caprolactone monomeric unit (Figure 29.7). This polyester is highly processible as it is soluble in a wide range of organic solvents and fully biodegradable (Briassoulis, 2004). It is nontoxic material and a good candidate for CR applications. Poor impact and tear strength behavior seen in PCL extrusion films (Briassoulis, 2004) can be solved by preparing blends or by copolymerizing with other biodegradable polymers such as PHB, PLA, and starch (Rudnik, 2008).

PCL is predominantly used as a component in polyester/starch blends (e.g., Mater-Bi by Novamont), as it reduces the moisture sensitivity and improves mechanical properties. In films, the fine dispersion of polycaprolactone phase in the polyethylene/starch matrix was reported to improve mechanical properties, while in injection specimens there was property decrease due to phase coalescence (Matzinos et al., 2002). In another study, starch-Cl/PCL blends were prepared by the mixing between starch-Cl and PCL solutions, and the miscibility was improved (Kweon et al., 2004). PCL was melt blended by extrusion with wheat TPS to improve the weakness of pure TPS such as low resilience, high moisture sensitivity, and high shrinkage, even at low PCL concentration, for example, 10 wt.% (Averous et al.,

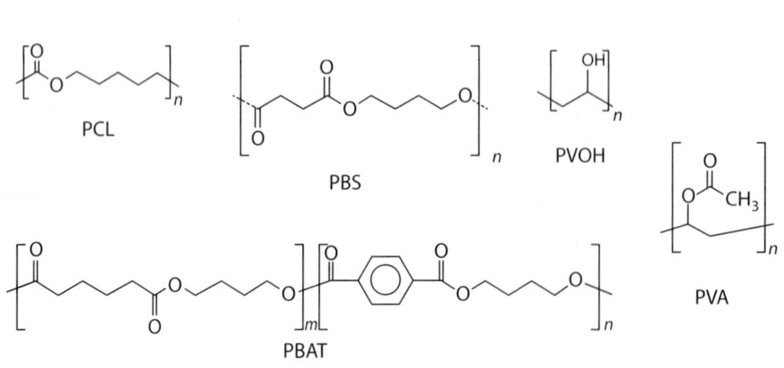

FIGURE 29.7 Chemical structures of poly(butylene adipate-*co*-terephthalate) (PBAT), poly-ε-caprolactones (PCL), poly(vinyl alcohol) (PVOH), poly(vinyl acetate) (PVA), and poly(butylene Succinate (PBS).

2000). Blending PCL with PLA has been studied widely to prepare degradable polymers with tailored properties. For example, the glassy PLA with high degradation rate shows better tensile strength, while the rubbery PCL with much slower degradation rate shows better toughness. The PLA/PCL blends compatibilized with the block copolymer (PEG–PPG) were investigated to improve phase miscibility, and it was found that the compatibilized blends exhibited a significant enhancement in tensile strain at break compared to that of the neat PLA/PCL blend (Chavalitpanya and Phattanarudee, 2013). Ternary blends of PLA/PCL and TPS were also reported that addition of PCL to form a ternary blend results in a substantial number of fine-dispersed particles present in the system (Sarazin et al., 2008).

PCL polyesters have long been available from companies such as Solvay and Union Carbide (acquired by Dow Performance Chemicals in 2001) for use in adhesives, compatibilizers, modifiers, and films. Union Carbide produces PCL, under the name Tone® for use in coatings and elastomers. Recently, the total biodegradability of tone has led to the polymer finding such applications as agricultural films, tree-planting containers, and matrices for the CR of pesticides, herbicides, and fertilizers. Solvay Interox Ltd. (Belgium) produces Capa®, a high-molecular-weight PCL. It has found applications as an additive, improving the strength of many conventional adhesives, bioplastics, coatings, elastomers, and resins (Moore and Saunders, 1997). Daicel Chemistry (Japan) markets caprolactone monomers to utilize in specialized polymers and also markets PCL polyesters, under the trade name Placcel® (Daicel, 2015).

29.2.3.3 Poly(butylene Succinate)

PBS belongs to the poly(alkenedicarboxylate) family. They are obtained by polycondensation reactions of glycols and 1,4-butanediol with succinic acid (Figure 29.7) (Vroman and Tighzert, 2009). The first attempt for PBS synthesis dates back to 84 years when Carothers et al. (1931) worked on synthesis of fibers of aliphatic polyesters by polycondensation reaction (Carothers, 1931). The PBS produced was of low molecular weight (less than 5000); hence, reaction products made of them were weak and brittle (Fujimaki, 1998). Therefore, researchers and companies attempted to manufacture high-molecular-weight PBS with the use of new catalysts and coupling reaction. In 1991, Showa Denko K.K. introduced high-molecular-weight PBS, produced via polycondensation reaction of glycols such as ethylene glycol and butanediol-1,4 and aliphatic dicarboxylic acids such as succinic acid and adipic acid. Polycondensation reaction is followed by chain extension to control the number of average molecular weight when higher weight is needed (Fujimaki, 1998). In 1994, Showa High Polymers commercialized PBS under the trade name Bionolle®, which is the first commercialized PBS resin (Xu and Guo, 2010). White crystalline aliphatic polymer family of Bionolle is biodegradable in compost, in moist soil, in freshwater with activated sludge, and in seawater. It has received numerous certification from association all over the world including "GreenPla," labeled by the Japanese BioPlastic Association (JBPA), and EN 13432 by DIN CERTCO and OK Compost by Vincotte (ShowaDenko, 2015). In 2005, Showa developed a new biodegradable formulation of PBS derived from biobased raw materials, for example, corn or other starches. Using more than 50% of starch as a raw material, the end product Bionolle Starcla® is a hybrid compound of Bionolle, starch, and PLA. Due to its biodegradable nature, Starcla® is the ideal resin for use in mulching films.

GS Pla® trademark is another aliphatic copolyester commercialized by Mitsubishi Chemical. Mitsubishi Chemical built a 3000 ton/year PBS production line and began its practical market introduction of GS Pla in April 2003 (Xu and Guo, 2010). According to Mitsubishi, GS Pla is a flexible polymer with high seal strength and is oxygen and moisture permeable. It is first introduced in Japan to be utilized as agricultural mulching films (MitsubishiNews, 2015). Now, the company produces succinic acid from vegetation-derived glucose, and biomass product named BioPBS is commercialized by a Thailand-based company, PTT MCC BioChem (joint venture between Mitsubishi Chemical and PTT PCL, focusing on the delivery of BioPBS™) (PTT-MCC, 2015). Therefore, there is a great demand for biobased succinic acid even just for PBS production. Succinic acid can also be obtained from fermentation of microorganisms on renewable feedstocks, such as glucose and xylose. Recently, some companies have built pilot, demonstration, or commercial scale factories to manufacture succinic acid via biofermentation (Xu and Guo, 2010). BioAmber is now the supplier of biobased succinic acid for PTT MCC BioChem. BioAmber is producing succinic acid using sugar from cane, beets, sorghum, wheat, and tapioca and transforms this biobased succinic acid into 1,4-butanediol and tetrahydrofuran in a single catalytic step. BioAmber produced succinic acid in a large-scale demonstration facility in France for 4 years. Recently, they announced a construction of the world's largest succinic acid plant in Sarnia, Canada, with a capacity of 30,000 tons/year. Beyond Sarnia, BioAmber is planning to build a second world-scale plant in North America that will produce both 1,4-butanediol (BDO) and succinic acid. The nameplate capacity is expected to be 100,000 tons of BDO and 70,000 tons of succinic acid per year and is expected to start up in late 2017 or early 2018 (Bioamber, 2015).

In 2006, Hexing Chemical Anhui (China) started producing PBS by direct melt polycondensation. In 2009, a facility with a production capacity of 10,000 tons/year was constructed by Hexing Chemical to manufacture PBS and its copolymers. The average molecular weight of the products can reach up to 200,000 (Xu and Guo, 2010). Xinfu Pharmaceutical (China) was established in 1994 as a fine chemical producer, and in 2007, the company built a PBS production line using one-step polymerization technology (Xu and Guo, 2010). Under the brand name Biosafe, Xinfu Pharmaceutical Co. Ltd. offers certified fully biodegradable PBAT, poly(butylene succinate)-*co*-adipate (PBSA), and PBS resins (Xinfu, 2015). A South Korean company SK Chemicals produces SkyGreen® PBS thermoplastics that is prepared by condensation of 1,2-ethylenediol and 1,4-butanediol with succinic and adipic acids (Vroman and Tighzert, 2009). This aliphatic polyester and aliphatic/aromatic copolyesters can be degraded into water and carbon dioxide by microorganisms (SkyChemicals, 2015). Ire Chemical (Korea) is another big company producing PBS with the trademark EnPol®. These polymers are produced through the polycondensation reaction of glycols and dicarboxylic acids with Ire Chemical's self-developed catalyst and process technology. Also, Ire Chemical is registered on the list of JBPA and has a certificate of Green Pla (IreChemicals, 2015).

29.2.3.4 Poly(vinyl Alcohol)

PVOH is the largest synthetic, water-soluble polymer produced in the world (Figure 29.7). Vinyl polymers are generally not susceptible to hydrolysis and require a catalyst to promote the oxidation or photooxidation process for biodegradation (Vroman and Tighzert, 2009). Thus, biodegradation mechanism involves the enzymatic oxidation of the secondary alcohol groups in PVA to ketone groups, and this is followed by hydrolysis that results in chain cleavage (Chandra and Rustgi, 1998). PVOH is not produced from direct polymerization of the corresponding monomer, but attained from the PVA. The polymerization of vinyl acetate occurs via a free-radical mechanism, usually in an alcoholic solution (methanol, ethanol), although for some specific applications, a suspension polymerization can be used. PVOH is produced on an industrial scale by hydrolysis (methanolysis) of stiff homopolymer PVA, and depending on the degree of hydrolysis, different grades of PVOH are attained (Ewa Rudnik, 2008). Some of the PVOH producers and their product trade names are Mowiol (Clariant GmbH, Germany), Erkol (ErkolSA, Spain), Sloviol (Novacky, Slovakia), Polyvinol (VinavilSpA, Italy), Elvanol (DuPont, United States), Cevol (Celanep, United States), Airvol (Air Products, United States), Kuraray Poval (Kuraray Co. Ltd., Japan), Unitica Poval (Unitica Ltd., Japan), Gohsenol (Nippon Gohsei-The Nippon Synthetic Chemical Industry Co. Ltd., Japan), and Hapol (Hap Heng, China) (Rudnik, 2008).

29.3 Agricultural Applications

29.3.1 Mulching

As stated earlier, the history of use of plastics in agriculture dates back to 1948 when polyethylene was first used as a greenhouse cover by professor Emmert at the University of Kentucky. Emmert demonstrated the use of black and clear plastic as mulches and row covers in the field culture of vegetables in his research (Emmert, 1957). Since then, plastic mulch has been used all over the world to protect crops from unfavorable climatic conditions, such as protecting soil from cold, or to protect from weed infestation or to allow season extension, multiple cropping, and higher yield of higher-quality produce. One tremendous benefit was attributed to peanut production in China. The use of plastic mulch was a revolution in the history of Chinese agriculture. Because of its contribution in increasing the yield, polyethylene mulching is called "White Revolution" (Kasirajan and Ngouajio, 2012). It helped farmers make massive strides toward increasing grain production.

Reasons for better plant growth under mulch include higher soil temperatures with little variations and more conserved soil moisture, which conducts accelerated biological activity that results in a greater liberation of nutrients and consequently to better nutrient uptake and greater plant growth (Magistad et al., 1935). As such, the temperature at root zone of a plant is an important parameter influencing the uptake of nutrients by different crop species (Tindall et al., 1990). It is well documented that mulching increases microbial activities and maintenance of organic matter in soil, which leads to better aeration and increased infiltration. For example, earthworms continually carry undecomposed plant materials downward thus favoring microbial activity. As mulch decomposes, some soluble carbohydrates are leached into the soil. The use of farmyard manure increased the infiltration capacity when applied to herbicide plots, compared to the plots treated with herbicide alone (Haynes, 1980).

Contributions of plastic-mulching applications to agriculture cannot be underestimated. However, there is a growing recognition and concern in the agricultural community that there is a great need for environmentally responsible disposal solutions for nondegradable plastic waste. China has emerged as a leading country where petroleum-based plastic mulches popularly are in use and consumes 1.42 million tons of agriculture mulch films each year (BASFNews, 2015). The dream of "White Revolution" is about to turn out a nightmare, as the residual pollution left in the soil was increased dramatically (Figure 29.8) such that the remnants cause lower yield by time as it decreases the water absorption of plant seeds from soil, which affects the sustainability. Recycling is an option but in China, lack of technology to clean soil residues makes residue recovery and recycling of mulch film cost intensive. Many mulching films are thin and usually buried deep in the soil, some of them are shred into small pieces during their useful lifetime and this makes hand picking unmanageable and time consuming. Thicker films are easier to collect but they can be expensive and unfavorable for farmers. In this case, it can be more cost-effective to use biodegradable mulch film, which can enable farmers to cultivate the film into the soil along with plant remaining, which indeed helps to preserve the soil quality.

Over the past years, BASF has worked with various stakeholders in China to promote the use of biodegradable mulch films of Ecovio in corn, tomato, and potato fields. Ecovio that consists of the certified compostable PBAT Ecoflex® and PLA is a compostable-certified polymer marketed by BASF. Ecovio mulching films have been used in Europe since 2012 with great success and introduced and tested in China since 2012. Recently, BASF has announced the starting of the largest trial project in cotton field

FIGURE 29.8 Landscape view of terraced hillside where plastic mulching is used, Guanghe County, Gansu province, China. (From Gansu PPMO team, 2014, http://ifad-un.blogspot.com/2014/10/growing-more-food-and-using-less-water.html.)

of Xinjiang Uyghur Autonomous Region with a promise of providing sustainable agriculture in China (BASFNews, 2015).

Depending on the crop and specific requirements and environmental conditions of the region, different polymer mulches have to be developed to achieve the critical design requests like retaining the mechanical properties for full-crop season, degrading in soil at the end of useful time, and being a cost-efficient alternative to conventional plastics that are in use (Riggi et al., 2011). Many researchers have carried out several trials using different bio- or photodegradable mulching films, which are greenhouse covers to prove that green alternatives are easy to adapt and economically feasible to replace the synthetic counterparts from petrochemical sources. The major concern is the useful lifetime of biodegradable polymers during growing cycle for long-lasting applications. It has been reported that premature breaking during growing cycle is mainly reasoned from long-term UV exposure rather than the water or high-temperature exposure (Briassoulis, 2007). From 2004 to 2007, under the EU project LIFE ENV/IT/463 (BIOMASS), Mater-Bi (Novamont, Italy) was tested to control the weed growth and adapted to a large number of crops mainly vegetables, (sweet basil, lettuce, tomato, artichoke, garlic, onion, sweet pepper, watermelon, zucchini, Brussels sprout, eggplant, strawberry), fruit trees (hazel tree, grapevine), and ornamental trees (Christmas tree). Researchers concluded that for short-crop duration, starch-based biodegradable mulch films can substitute PE alternatives; however, for long-lasting crops (strawberries, perennial), these films barely delayed the development of weed with limited effects on crop yield (Minuto et al., 2008).

There are typically three steps in biodegradation process: (1) temporary flexible chain formation to break crystalline structure of polymer, (2) breaking down of the polymer into small fragments of oligomers or monomers through microbial, enzymatic cleavage of polymer chain, and (3) degradation of the oligomers and monomers to form water and CO_2 or methane by microorganisms, to metabolize them within their cells (Siegenthaler et al., 2012). In parallel to enzymatic degradation, degradation of polymer through chemical hydrolysis, UV light, or heat is very common. Actually, biodegradation is an erosion process that takes place at the surface of the plastic article (Siegenthaler et al., 2012). The process of biodegradation has been studied for many commercially available polymers (Bilck et al., 2010; Cowan et al., 2013; Kapanen et al., 2008; Lopez-Marin et al., 2012; Miles et al., 2012; Ngouajio et al., 2008; Romic et al., 2003) on various crops including melon, broccoli, tomato, strawberry, lettuce (Table 29.3). Deterioration of Mater-Bi-based black film (BioAgri), along with other biodegradable polymers of PHA and experimental spunbonded PLA fabric (SP-PLA), has been studied in broccoli field; and deterioration rate was found to be greatest in PHA film followed by BioAgri and SP-PLA (Cowan et al., 2013). Postharvest soil quality index (SQI) showed a negligible effect of cellulose-based mulch (WeedGuardPlus) and relatively higher yet minor effects of starch based (BioAgri) and PLA based (SP-PLA) mulches on the soil's biological and chemical properties in tomato production systems (Li et al., 2014).

PBAT-based Ecoflex® (BASF, United States) was designed to be a strong and flexible material with mechanical properties similar to PE. Blends of Ecoflex® also shows very beneficial performances (e.g., the addition of 20% Ecoflex® in PLA reduces the stiffness of PLA by 25%) (Siegenthaler et al., 2012). Selection of the mulching film type, additives, color, and thickness should be considered and designed well to take specific environmental requirements of the region into account. For example, at Michigan State University, the field performance of Ecoflex® has been tested using "Mountain Fresh Plus" tomato (*Solanum lycopersicum*) as a model crop (Ngouajio et al., 2008). Soil temperature under mulches was higher compared to LDPE alternative, only for the first week, which gradually decreased as premature breakdown of the white mulches was observed (Ngouajio et al., 2008). PBAT (Ecoflex®)/TPS blends have been tested for strawberry plantation in the city of Londrina, Paraná, Brazil. The PBAT film showed small cracks in the structure 5 weeks after being laid onto the ground and, 8 weeks afterward, the maximum tensile strength, elongation at break, and water sorption were reduced (Figure 29.9). However, these changes in the film structure did not influence the quality and amount of the fresh produce (Bilck et al., 2010).

A full-scale experiment on biodegradation behavior in soil of PLA (EarthFirst PLA-Based on Ingeo resin of NatureWorks LLC and NatureWorks™), TPS (Cliper-net; Mater-Bi), and PHA (Mirel) having different thicknesses was conducted in Spata, Greece, to simulate agricultural mulch film use and fate in soil after use. Composting studies indicated that PHA films (Mirel) are more readily attached by microorganisms both in soil environment and under composting conditions. PLA films, on the other hand,

TABLE 29.3

Mulching Applications in Horticulture Crops

Product	Film Type	Uses	Reference
Melon	Black LDPE, PE with photodegradation additives, starch-based biodegradable mulches	Yield of melon production, solar radiation efficiency	Candido et al. (2001)
	Black PE, clear photodegradable PE, paper biodegradable cellulose mulches	Yield of watermelon production, soil temperature during deterioration of films	Romic et al. (2003)
	PE, biodegradable TPS mulch films	Yield, fruit size, and quality of winter melon	Vetrano et al. (2009)
Cucumber	Kraft paper, covered or uncovered with maize starch–based biodegradable film	Yield of production, color effect on soil temperature changes	Haapala et al. (2015)
Broccoli	Mater-Bi®-based black film (BioAgri), PHB, experimental spunbonded PLA fabric mulches	Deterioration rate before and after soil incorporation	Cowan et al. (2013)
	Black/transparent PE, green/black/transparent biodegradable mulch films	Growth rate feasibility of substituting biodegradable polymer with PE	Lopez-Marin et al. (2012)
Tomato	Mater-Bi® dyed with masterbatch black, wheat straw mulches	Yield of rain-fed fresh market tomato and weed control	Fontanelli et al. (2013)
	Green/brown/black biodegradable, PE mulch films	Yield of production, color effect on soil temperature	Moreno and Moreno (2008)
	BioAgri, BioTelo, WeedGuardPlus, SB-PLA-10, black PE mulch films	Yield of tomato production, deterioration rate, weed control	Miles et al. (2012)
	Black PE, black biodegradable, oxo-degradable plastic, paper, brown kraft paper mulches, barley straw	Weed control, yield, and economic aspects	Cirujeda et al. (2012)
Tomato	Black PE, white/black biodegradable film, silver film, fume film, green FSL film, green translucent film	Corky rot/virus control, color effect	Lops et al. (2011)
	Black PE, black biodegradable corn starch plastic, aluminized photodegradable plastic	Deterioration rate, soil temperature, tomato yield, and fruit quality	Moreno et al. (2009)
	Rice straw, barley straw, maize harvest residue, absinth wormwood plants, black biodegradable plastic (Mater-Bi®), brown kraft paper, PE	Tomato yield and weed control	Anzalone et al. (2010)
	Modified Ecoflex®, PBAT, LDPE mulch films	Yield of tomato, weed control, color effect	Kijchavengkul et al. (2008)
	Modified PBAT, Ecoflex® (BASF), LDPE (titanium dioxide and carbon black were used as additives to produce the white and black films)	Deterioration rate, soil temperature, tomato growth, weed density, and biomass	Ngouajio et al. (2008)
Strawberry	Mater-Bi® (black, silver white with different thicknesses)	Yield of production, quality	Costa et al. (2014)
	Black/white biodegradable films from cassava starch PBAT blends, Ecoflex, PE	Yield of strawberry production, weed growth, degradation profile	Bilck et al. (2010)
	Black Mater-Bi®-mulching films with different thicknesses, black Mater-Bi® low tunnel films with different UV stabilizers, LDPE mulch and low tunnel films	Degradation rate, effect of mulching and low tunnel films, radiometric properties	Kapanen et al. (2008)

(Continued)

TABLE 29.3 (*Continued*)

Mulching Applications in Horticulture Crops

Product	Film Type	Uses	Reference
	Clear/black biodegradable polymer covering brown kraft paper, Planters Paper	Fruit yield, degradation, weed growth	Weber (2003)
Raspberry, Blueberry	Mater-Bi	Soil fertility, growth rate of berry fruits	Girgenti et al. (2011)
Sunflower	Black/white-on-black coextruded LDPE mulch films, black biodegradable mulch spray with chitosan, black biodegradable mulch spray with galactomannans and agarose	Yield and quality, soil thermal behavior, irrigation effects	Anifantis et al. (2012)
Lettuce	Transparent/black LDPE, transparent/black oxo-biodegradable, transparent/black biodegradable mulch films	Degradation time, plant growth rate, color effect	Lopez-Marin et al. (2012)
Bell pepper	Black PE, biodegradable cellulose mulch	Nitrate leaching rate in bell pepper cultivation	Romic et al. (2003)
Zucchini squash	Silver spray mulch, silver PE mulch films	Color effect on virus onset	Summers et al. (1995)
Capsicum	White PE, black biodegradable film, burlap, hardwood, sawdust, sugarcane trash, paper film	Weed growth, yield of marketable capsicum fruit	Olsen and Gounder (2001)

FIGURE 29.9   (a) White biodegradable film, (b) black biodegradable film, (c) control polyethylene film; after 5 weeks in contact with soil. (From Bilck, A.P. et al., *Polym. Test.*, 29(4), 471, 2010.)

required longer time and/or higher temperature (at least 58°C) to degrade (Rudnik and Briassoulis, 2011). PLA in soil environment undergoes degradation by a combination of hydrolysis of polymer structure followed by biotic activity, which indeed confirms that soil temperature is a critical factor and enhances the biodegradation of PLA (Ho and Pometto III, 1999; Rudnik and Briassoulis, 2011).

The color of mulch is important as it determines solar radiation behavior and its influence on the microenvironments around plants. Black mulches can absorb most of solar radiation, which is transferred to the soil by conduction. By contrast, clear plastic mulch absorbs little solar radiation but transmits 85%–95%, with the relative transmission depending on the thickness and degree of opacity. Clear plastic mulches generally are used in the cooler regions. White, white-on-black, or silver-reflective mulches may result in a slight decrease in soil temperature (1.1°C at 2.5 cm depth or 0.4°C at 10 cm depth), because they reflect most of the incoming solar radiation back into the plant canopy (Lament, 1993). Depending on the plant type and environmental conditions, colored/or clear mulches have been studied to increase the yield and quality of the product.

Black biodegradable films seem to be a better alternative as a mulch film for vegetable production systems. For example, in tomato field study in the state of Michigan, black color in biodegradable mulching films (PBAT/carbon black for black color) used to cover the beds of tomato plots increased the degradation time, while white films (PBAT based) started to degrade after 2 weeks, because the white color catalyzed the photodegradation from solar radiation (Kijchavengkul et al., 2008). Similarly, the differences in soil temperature observed under the mulching films in tomato field in Ciudad Real, Spain, were attributed to film material (biodegradable or PE) rather than the color (green, brown, and black biodegradable films and black and blue/yellow PE films) as it was always the lowest under biodegradable mulches compared to PE, because of higher gas permeability (Moreno and Moreno, 2008). However, in both studies, the marketable fruit yield did not change significantly. The earlier degradation of material having white (Kijchavengkul et al., 2008) or green (Moreno and Moreno, 2008) color encouraged the weed growth, most probably because the premature cracks provided a moist, warm environment under the mulch for weed seed germination with sufficient light for growth (Weber, 2003). White or reflecting mulches are used to establish crops like cauliflower or tomatoes in midsummer, when soil temperatures are high and any reduction in soil temperatures is beneficial. And clear plastic mulches generally are used in the cooler regions (Lament, 1993).

Weeds usually cannot survive under mulch (Lament, 1993). Black mulch film always has an ability to suppress weeds better, because black color stabilizes the photodegradation (Kijchavengkul et al., 2008), which consequently increases the degradation time. Suppression of weed growth in matted-row strawberry planting was investigated for three different biodegradable mulch films, and it was concluded that both the color (black or white) and material of films (biodegradable polymer covering brown Kraft paper and Planters Paper) did not change the weed biomass; all types of mulching applications were effective at weed control compared to standard-control practices (preemergent herbicide application before or shortly after planting with supplemental hand and/or mechanical cultivation). However, it was reported that materials used had definite disadvantages, for example, Planters Paper was not durable and was torn and blown away by wind from the plot. Clear biodegradable film was not as effective on weed control as black biodegradable one; in fact, weeds were emerged only at the holes on black biodegradable film where the strawberries were planted. However, black biodegradable film did not degrade soon enough for rooting of runners and had to be slit so that sufficient runners could fill in the matted row (Weber, 2003).

29.3.1.1 Paper Mulching

Indeed, paper-based mulches are the earliest mulching systems developed for fruit and vegetable production. The use of paper for mulches definitely reduces the weed growth and in this manner reduces cultivation and weeding costs (Magistad et al., 1935). However, they have many inherent disadvantages such as being susceptible to tearing from the wind and losing mechanical strength when wetted (Hayes et al., 2012). They are light weight, which easies the long-distance transportation. They can be applied from a roll onto the field, but it is difficult to lay them down on the field (Hayes et al., 2012). Kraft paper mulch was improved in terms of rate of biodegradation in soil and the ability to inhibit weed growth, when cured oils (oxidatively polymerized linseed oil and polyester formed by the reaction of epoxidized soybean oil and citric acid) were incorporated. The polymerized oils acted as a barrier and delayed the microbial cleavage of cellulosic fibers. Resin-coated papers inhibited weed growth and hold the mulch intact for more than 10 weeks, while uncoated paper underwent total degradation in 6–9 weeks (Shogren, 1999). Black colored, paper–oil mulches were also effective on blocking the nutsedge growth in watermelon field, and still the handling and laying the oily paper are the main disadvantages (Shogren and Hochmuth, 2004). In recent studies, the water vapor barrier property of Kraft paper was improved by coating with pregelatinized starch (Zhang et al., 2008), and wet strength of paper sheet was improved by chitosan (Kamel et al., 2004).

29.3.1.2 Hydromulching

Hydromulching (or hydraulic mulch seeding, hydroseeding, hydraseeding, liquid mulching) is a method of using the slurry of seed and mulch to form a biodegradable mat *in situ*. Straw virgin and recycled wood fiber and recycled waste paper are all used in hydromulching applications (English, 1997). Hydromulching helps to maintain the moisture level of the seed and seedlings (Chou et al., 2011). Other ingredients in slurry of seed and mulch blends may include fertilizers, pH adjusters, moisture retention agents, and herbicides and can be applied simultaneously with the seed and the mulch (English, 1997). The ingredients should have effects on the adhesion of the hydrocolloids to increase erosion resistance and promote quick germination of seeds (Chou et al., 2011). The spray technology is applied similar to the applications of fertilizers or pesticides on soil, in which the polymeric water solution is dispersed onto soil, instead. Technology is familiar to farmers but application as mulching is innovative (Immirzi et al., 2009).

Hydrophilic polymers, such as poly(acrylamide) (PAM), PVOH, carboxymethyl cellulose, and hydrolyzed acrylonitrile-grafted starch (H-SPAN) are some examples of biodegradable polymers, which are commonly used as soil conditioner in these hydromulching techniques (Chiellini et al., 2003). PVOH is known with its positive effect on the soil structure and on the stability and porosity of loamy and clay soil (Chiellini et al., 2008). Mixtures of PVOH as polymer material supported with proteins (by-products of leather industry) and natural fillers (lignocellulose from by-products of wood industry, starch from ethanol biorefinery of maize) were proven to be suitable for application, in the form of water suspensions by conventional spraying equipment (Chiellini et al., 2003). The fillers added (sugarcane bagasse, wheat flour, sawdust, and wheat straw) improved the degradation of PVOH on soil surface and consequently the structuring effect in lettuce and corn field (Chiellini et al., 2008). Similarly, cross-linking with various dialdehyde agents in the polymer blends and composites was proven to reinforce the polymeric structure based on waste gelatin, PVOH, and sugarcane bagasse and consequently delay biodegradation on soil (Kerouani et al., 2013). PAM was used as a soil aggregate agent of silty clay loams and effects of PAM on the germination and growth of Bermuda and Bahia were observed (Chou et al., 2011). In the field, during the cultivation of strawberries, the spray coating of sodium alginate (NaAlg) and hydroxyethylcellulose (HeCell) solutions was compared with an LDPE mulching film and a straw mulch (Immirzi et al., 2009). Indeed, there is a strong interaction between NaAlg and the available calcium naturally present in the soil. NaAlg forms stiff polymer, and HeCell was introduced to provide long-lasting plasticizing by producing physically entangled three-dimensional networks of calcium alginate. Sixty-five percent of spray coating samples was biodegraded after 6 months in soil, and the coating maintained its capacity to suppress weeds during this period (Immirzi et al., 2009). The polymeric formulations including arabic gums, agarose, cellulose (Immirzi et al., 2008; Schettini et al., 2008), collagen hydrolyzate, and PEG derivatives (Schettini et al., 2012) have also been studied.

Spray biodegradable mulches (silver-pigmented and white "Styrofan" synthetic latex spray) developed by direct coating on soil of zucchini squash plantation field were investigated to study the effect of reflective mulches on repelling alate aphids and on delaying the onset of various viruses. These mulches reflect short-wave light, which claimed to reduce the incidence of alate aphids on plants. Plants grown on silver-pigmented spray mulch produced yield of marketable fruit equivalent to or more than those of the silver PE alternatives (Summers et al., 1995). Unmulched beds were 100% virus infected, while only 10% of plants grown over silver mulches were diseased (Summers et al., 1995). Similar results were observed for the onset of symptoms of cucumber mosaic cucumovirus and watermelon mosaic and zucchini yellow mosaic potyviruses, which were delayed 3–6 weeks in plants growing over the reflective mulches, which was critical for initiation of normal flowering and fruiting (Stapleton and Summers, 2002).

29.3.2 Other Applications

The usage of biodegradable films is definitely proven to eliminate the need for plastic waste management when utilized as mulching films. Biodegradable polymers can also be utilized for other horticultural applications, such as flower pots, seedling, plant tray, wrapping films, nets, lawn, and garden bags, caps, tags, pipes, sacks, and seed tape. One example of using commercially available biodegradable films is the attempt to produce biodegradable irrigation thin-wall pipes and drippers from Mater-Bi (Novamont SPA) and Bio-Flex (FKUR Kunststoff GmbH) to study biodegradation time under real soil conditions, mechanical behavior, and manufacturing process of biodegradable drip irrigation systems (Hiskakis et al., 2011). The mechanical behavior of the biodegradable thin-wall pipes during the irrigation period was more unstable when compared to the corresponding behavior of the rigid pipes. Despite the significant drop of the elongation at break, all biodegradable rigid pipes generally retained their tensile strength as well as a satisfactory hydraulic performance during almost the whole duration of their exposure (Briassoulis et al., 2011).

29.3.3 Controlled Release

CR technology has emerged as a method where active agricultural ingredients are enclosed in release carriers and permeated to be transferred to the targeted surface at designed rate and period of time (Kenawy, 1998). The polymer matrix serves as an encapsulation membrane for active reagents like a carrier in which release takes place through diffusion or after biological or chemical breakdown of the polymer. Or polymers are employed as a backbone support where active reagents are chemically bound and released by chemical or biological cleavage of the bond (Mitrus et al., 2010).

Delivery of agrochemicals (pesticides, herbicides, nutrients) through CR devices is important to improve the production of crops. An ideal pesticide formulation, for instance, should maintain the local concentration at desired level for pest control. As well, encapsulating polymers containing the pesticide moieties should minimize the chemical run off, and reduce leaching, volatilization and toxicity (Roy et al., 2014). CR technology has, therefore, many applications where crop protection is desired for prolonged time periods. The most commonly used modified natural biopolymers include starch (Cao et al., 2005; Doane et al., 1977; Riley, 1983; Shasha et al., 1976; Zhu and Zhuo, 2001), cellulose (Bote et al., 1993; Perez-Martinez et al., 2001; Sopeña et al., 2007), chitin and chitosan (Elbahri and Taverdet, 2005; Fan et al., 2012; Hussain et al., 2012), alginates (Fernández-Pérez et al., 1999; Işiklan, 2006; Johnson and Pepperman, 1995), and lignin (Fernández-Pérez et al., 2011; Garrido-Herrera et al., 2009). They all have common properties of being abundant and relatively inexpensive but insoluble in solvents necessary for encapsulation, dispersion formulations, and chemical reactions; and that results in limited amount of active material per weight of carrier polymer.

29.3.3.1 Use of Hydrogels in Controlled Release

Over the past decade, hydrogel-based active ingredient release devices became popular as formulations include cross-linking of the polymeric matrix in the presence of active material or emulsification and separation of microspheres. Hydrogels can also act as superabsorbent that can absorb and retain huge amounts of water or aqueous solutions. Hydrogels can be synthesized by methods, used commonly in polymer chemistry, such as solution polymerization, suspension polymerization, graft polymerization, and microencapsulation, depending on the type of starting material and type of end polymer desired (Rudzinski et al., 2002).

Superabsorbent hydrogels, already existing in the market, are generally acrylate-based products; hence, they are not biodegradable, and more importantly there are concerns about their toxicity when applied in agricultural practices. Hydrogels of natural polymers, alternatively, were developed to overcome these problems, as they are produced mostly from chemically modified starch moieties via vinyl graft copolymerization. The starch graft copolymer has been produced by generating free radicals on the surface of the starch granules followed by copolymerization of these free radicals with the respective vinyl monomers (Ahmed, 2015). In the 1970s, the U.S. Department of Agriculture research laboratory of the National Center for Agricultural Utilization Research announced a revolutionary "Super Slurper" superabsorbent hydrogel, formally hydrolyzed acrylonitrile-grafted starch (H-SPAN) for its chemical makeup, which is able to absorb 5000 times of its weight in water. Research group headed by George F. Fanta combined synthetic compounds with cornstarch to create Super Slurper, and since then, a number of patents were issued on the preparation and uses of H-SPAN (Fanta et al., 1979, 1984; Weaver et al., 1976a,b). Today, Super Slurper is mainly used for seed coating to accelerate the germination and used to coat roots of plants in dry areas to improve its water-holding ability. Besides H-SPAN is insoluble in water and when gets wet, forms a gel with high surface area, which is used as soil conditioner in agriculture (Ahmed, 2015). The addition of this material to the soil delays the moisture stress, reduces the water retention (Miller, 1979), and enhances soil permeability and infiltration rates (Al-Darby et al., 1990). When dry hydrogel is mixed with the soil, a layer of substrate/gel mixture might form, which further limits the flows of air and water within the soil (Figure 29.10). When water is added, large-granule gel is formed, which leads better airflow through the soil and higher oxygenation to plant roots (Sannino et al., 2009).

Alternative to polysaccharides, cellulose, and cellulose, xanthates are promising because of their hydrophilicity, biodegradability, biocompatibility, transparency, low cost, and nontoxicity (Chang

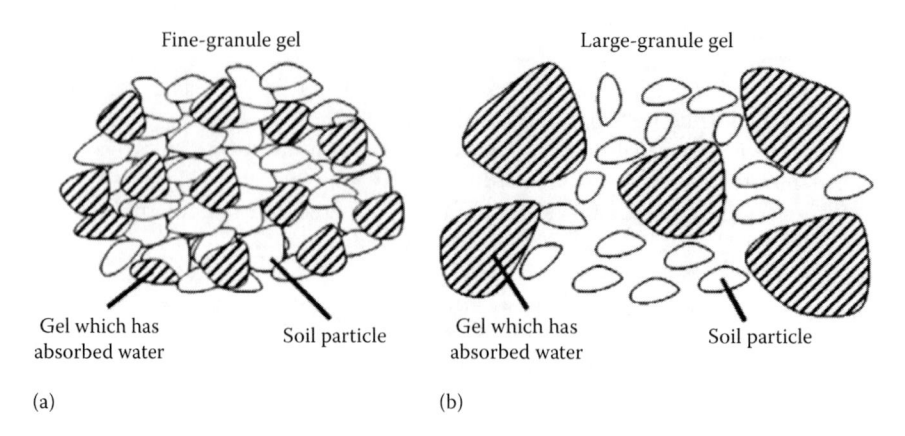

FIGURE 29.10   Effect of hydrogel swelling on soil porosity: (a) dry hydrogel and (b) swollen hydrogel. (From Sannino, A. et al., *Materials*, 2(2), 353, 2009.)

and Zhang, 2011). Mechanical properties of highly water-swollen cellulose-based hydrogels were described by Westman and Lindström (1981a). Cellulose was regenerated by removing the xanthate groups in epichlorohydrin (0%–24% w/w on cellulose) solutions (Westman and Lindström, 1981a,b). More recently, preparation of superabsorbent hydrogels based on cross-linked carboxymethyl cellulose polymer and acrylamide monomer through electron-beam irradiation was carried with intent of use as water-managing material for agriculture and horticulture in desert and drought-prone areas. Water-retention property of developed hydrogels was investigated on growth of rice (Ibrahim et al., 2007). Sannino and coworkers recently developed a novel class of microporous cellulose-based superabsorbent hydrogels (Demitri et al., 2008, 2013; Sannino et al., 2009, 2010; Sannino and Nicolais, 2005). Cellulose-based superabsorbent hydrogels were proposed to absorb up to one liter of water per gram of dry material. Active ingredients can be loaded within the gel for CR in which the rate depends on the swelling/deswelling changes. The swelling capacity was tested for the cultivation of cherry tomatoes, and hydrogel potential as water reservoir in soil was evaluated. Carbodiimide cross-linked hydrogels were applied in powder form to the soil and soil was watered to allow the hydrogels swell to their optimum capacity. Hydrogel amended soil was proven to retain water and sustain the water release for prolonged time with respect to the soil watered without the presence of the hydrogel. Dry matters of hydrogels have a similar granular size with substrate. When applied to the soil in powder form, then watered, they expand as they swell and increase the soil porosity, consequently the oxygenation through plant roots (Demitri et al., 2013).

The application of hydrogels as soil conditioner can be extended into reduced soil erosion and general stabilization. Hydrogels can aggregate to soil by surface treatment and hence prevent soil erosion and runoff by stabilizing the surface layers (Ahmed, 2015). Soil stabilization by cellulose xanthate dates back to the 1950s (Meadows, 1956, 1957). Cellulose xanthate solutions, made from fibrous cellulose from computer cards, at a concentration of about 0.1% in the soil were found to form tick paste when water was mixed and improve the mechanical strength of soil enough to reduce the possible erosion (Menefee and Hautala, 1978). When used as soil conditioner on soils conductive to capping, as described by Menefee and Hautala (1978), cellulose xanthate increased the water stability of soil aggregates without affecting the growth or taste of tested crops (lettuce, onions, carrots, leeks, sugar beet) or its interaction with herbicides (Page, 1980). Suspensions of chitosan, starch xanthate, cellulose xanthate, and acid-hydrolyzed cellulose microfibrils were compared for their efficacy as erosion retardants in irrigation water. Even cellulose and starch xanthates were found to be effective as soil stabilizers (Meadows, 1956; Menefee and Hautala, 1978); larger quantities of biopolymers were needed to conserve the soil measures compared to PAM copolymers in flowing water (Orts et al., 2000). Cellulose xanthate was also evaluated as water infiltration at exchangeable sodium fractions of 0, 0.05, and 0.10 (Wood and Oster, 1985).

Polysaccharide (guar) derived from guar bean was tested in soil flocculation, and when fully dissolved in acid solutions, guar tended to be more effective; otherwise, it formed weak water-stable aggregate (Wallace, 1986).

29.4 Conclusion

With the demand for reducing the use of synthetic polymers, mainly due to their environmental impact, the use of biodegradable polymers in agriculture is increasing. The growth is accelerating with increase in agricultural production in countries like China. Future development in the application of biodegradable polymers in agriculture depends on the development of economically viable production and utilization processes and also the use of diverse array of feedstock resources ranging from crop residues to forest resources. In addition, adoption of biodegradable polymers in agriculture will be driven by sustainability for energy, water, and waste reduction. Current trends in synthetic polymers due to volatile pricing of crude oil dampen the research in biodegradable polymers, but with the goal of reducing the dependence of fossil fuel resources, furthering the utilization of biodegradable polymers in agriculture is here to stay.

REFERENCES

Ahmed, E. M. (2015). Hydrogel: Preparation, characterization, and applications: A review. *Journal of Advanced Research*, 6(2), 105–121.

Al-Darby, A. M., Mustafa, M. A., and Al-Omran, A. M. (1990). Effect of water quality on infiltration of loamy sand soil treated with three gel-conditioners. *Soil Technology*, 3(1), 83–90.

Anderson, A. J. and Dawes, E. A. (1990). Occurrence, metabolism, metabolic role, and industrial uses of bacterial polyhydroxyalkanoates. *Microbiological Reviews*, 54(4), 450–472.

Anifantis, A., Canzio, G., Cristiano, G. et al. (2012). Influence of the use of drip irrigation systems and different mulching materials on ornamental sunflowers in greenhouse cultivation. *International Symposium on Advanced Technologies and Management Towards Sustainable Greenhouse Ecosystems: Greensys2011*, Halkidiki, Greece, Vol. 952, pp. 385–392.

Anzalone, A., Cirujeda, A., Aibar, J., Pardo, G., and Zaragoza, C. (2010). Effect of biodegradable mulch materials on weed control in processing tomatoes. *Weed Technology*, 24(3), 369–377.

Averous, L., Moro, L., Dole, P., and Fringant, C. (2000). Properties of thermoplastic blends: Starch–polycaprolactone. *Polymer*, 41(11), 4157–4167.

Barragán, D. H., Pelacho, A. M., and Martín-Closas, L. (2012). A respirometric test for assessing the biodegradability of mulch films in the soil. *XXVIII International Horticultural Congress on Science and Horticulture for People (IHC2010): International Symposium on Environmental, Edaphic, and Genetic Factors Affecting Plants, Seeds and Turfgrass*, Lisbon, Portugal.

BASFNews. (2007). Biodegradable BASF Plastics Quite Outstanding-Ecoflex and Ecovio receive iF material award, 2015. Retrieved from http://www.plasticsportal.net/wa/plasticsEU/portal/show/common/plasticsportal_news/2007/07_292.

BASFNews. (September 24, 2015). BASF contributes to soil protection in China with biodegradable mulch films made from ecovio®. Retrieved from https://www.basf.com/hk/en/company/news-and-media/news-releases/cn/2015/08/ecovio.html.

Bastioli, C. (1998). Biodegradable materials: Present situation and future perspectives. *Macromolecular Symposia*, 135, 193–204.

Bastioli, C. (2001). Global status of the production of biobased packaging materials. *Starch/Starke*, 53(8), 351–355.

Bastioli, C., Magistrali, P., and Garcia, S. G. (2012). Starch in polymers technology. In K. Khemani and C. Scholz (eds.), *Degradable Polymers and Materials: Principles and Practice*, Vol. 1114, pp. 87–112. Washington, DC: American Chemical Society.

Benhamou, N. and Thériault, G. (1992). Treatment with chitosan enhances resistance of tomato plants to the crown and root rot pathogen Fusarium oxysporum f. sp. radicis-lycopersici. *Physiological and Molecular Plant Pathology*, 41(1), 33–52.

Bilck, A. P., Grossmann, M. V. E., and Yamashita, F. (2010). Biodegradable mulch films for strawberry production. *Polymer Testing*, 29(4), 471–476.

Bioamber. (2015). Retrieved from http://www.bio-amber.com.

BiobasedNews. (April 12, 2011). BIOPAR® was presented with product Differentiation Excellence Award. Retrieved from www.bio-based.eu/news.

Biocycle. (2015). Retrieved from http://www.biocycle.com.br.

BiologiQ. (2015). Retrieved from http://www.biologiq.com.

BiopBiopolymerTechnologies. (2015). Retrieved from www.biop.eu.

Biopolynov. (2015). Retrieved from www.biopolynov.com.

Biowert. (2015). Retrieved from http://www.biowert.de.

Bote, A. N., Nadkarni, V. M., and Rajagopalan, N. (1993). Cellulose xanthide (CellX) as an encapsulating matrix. I. Comparison with starch xanthide (StX) on swelling and release properties. *Journal of Controlled Release*, 27(3), 207–217.

Brandl, H., Knee, E. J., Jr., Fuller, R. C., Gross, R. A., and Lenz, R. W. (1989). Ability of the phototrophic bacterium Rhodospirillum rubrum to produce various poly (beta-hydroxyalkanoates): Potential sources for biodegradable polyesters. *International Journal of Biological Macromolecules*, 11(1), 49–55.

Briassoulis, D. (2004). An overview on the mechanical behaviour of biodegradable agricultural films. *Journal of Polymers and the Environment*, 12(2), 65–81.

Briassoulis, D. (2007). Analysis of the mechanical and degradation performances of optimised agricultural biodegradable films. *Polymer Degradation and Stability*, 92(6), 1115–1132.

Briassoulis, D., Babou, E., and Hiskakis, M. (2011). Degradation behaviour and field performance of experimental biodegradable drip irrigation systems. *Journal of Polymers and the Environment*, 19(2), 341–361.

Briassoulis, D., Hiskakis, M., and Babou, E. (2013). Technical specifications for mechanical recycling of agricultural plastic waste. *Waste Management*, 33(6), 1516–1530.

Candido, V., Miccolis, V., Gatta, G., Margiotta, S., Picuno, P., and Manera, C. (2001). The effect of soil solarization and protection techniques on yield traits of melon in unheated greenhouse. In J. A. Fernandez, P. F. Martinez, and N. Castilla (eds.), *Proceedings of the Fifth International Symposium on Protected Cultivation in Mild Winter Climates: Current Trends for Sustainable Technologies*, Vols. I and II, pp. 705–712. Leuven, Belgium: International Society Horticultural Science.

Cao, Y., Huang, L., Chen, J., Liang, J., Long, S., and Lu, Y. (2005). Development of a controlled release formulation based on a starch matrix system. *International Journal of Pharmaceutics*, 298(1), 108–116.

CardiaBioplastics. (2015). Retrieved from http://www.cardiabioplastics.com.

Carothers, W. H. (1931). Polymerization. *Chemical Reviews*, 8(3), 353–426.

Castro, S. P. M. and Paulín, E. G. L. (2012). *Is Chitosan a New Panacea? Areas of Application*. Rijeka, Croatia: InTech.

Cavalheiro, J., de Almeida, M., Grandfils, C., and da Fonseca, M. M. R. (2009). Poly(3-hydroxybutyrate) production by *Cupriavidus necator* using waste glycerol. *Process Biochemistry*, 44(5), 509–515.

Cereplast. (2015). Retrieved from www.cereplast.com.

Chandra, R. and Rustgi, R. (1998). Biodegradable polymers. *Progress in Polymer Science*, 23(7), 1273–1335.

Chang, C. and Zhang, L. (2011). Cellulose-based hydrogels: Present status and application prospects. *Carbohydrate Polymers*, 84(1), 40–53.

Chanprateep, S. (2010). Current trends in biodegradable polyhydroxyalkanoates. *Journal of Bioscience and Bioengineering*, 110(6), 621–632.

Chavalitpanya, K. and Phattanarudee, S. (2013). Poly(lactic acid)/polycaprolactone blends compatibilized with block copolymer. *Energy Procedia*, 34, 542–548.

Chen, G.-Q. (2009). A microbial polyhydroxyalkanoates (PHA) based bio-and materials industry. *Chemical Society Reviews*, 38(8), 2434–2446.

Chen, G. Q., Zhang, G., Park, S. J., and Lee, S. Y. (2001). Industrial scale production of poly(3-hydroxybutyrate-co-3-hydroxyhexanoate). *Applied Microbiology and Biotechnology*, 57(1–2), 50–55.

Chiellini, E., Cinelli, P., D'Antone, S., Ilieva, V. I., Magni, S., Miele, S., and Pampana, S. (2003). Liquid mulch based on poly(vinyl alcohol). PVA-soil interaction. *Macromolecular Symposia*, 197, 133–142.

Chiellini, E., Cinelli, P., Magni, S., Miele, S., and Palla, C. (2008). Fluid biomulching based on poly(vinyl alcohol) and fillers from renewable resources. *Journal of Applied Polymer Science*, 108(1), 295–301.

Chou, L. H., Lu, C. K., and Chou, T. M. (eds.). (2011). Application of biodegradable polymers on hydroseeding. *Advanced Materials Research*, 415–417, 1257–1260.

Cirujeda, A., Aibar, J., Anzalone, A. et al. (2012). Biodegradable mulch instead of polyethylene for weed control of processing tomato production. *Agronomy for Sustainable Development*, 32(4), 889–897.

Costa, R., Saraiva, A., Carvalho, L., and Duarte, E. (2014). The use of biodegradable mulch films on strawberry crop in Portugal. *Scientia Horticulturae*, 173, 65–70.

Cowan, J. S., Inglis, D. A., and Miles, C. A. (2013). Deterioration of three potentially biodegradable plastic mulches before and after soil incorporation in a broccoli field production system in Northwestern Washington. *HortTechnology*, 23(6), 849–858.

Daicel. (2015). Retrieved from http://www.daicel.com.

Demitri, C., Del Sole, R., Scalera, F. et al. (2008). Novel superabsorbent cellulose-based hydrogels crosslinked with citric acid. *Journal of Applied Polymer Science*, 110(4), 2453–2460.

Demitri, C., Scalera, F., Madaghiele, M., Sannino, A., and Maffezzoli, A. (2013). Potential of cellulose-based superabsorbent hydrogels as water reservoir in agriculture. *International Journal of Polymer Science*, 2013, 6.

Doane, W. M., Shasha, B. S., and Russell, C. R. (1977). Encapsulation of pesticides within a starch matrix. In H. B Scher (ed.), *Controlled Release Pesticides*, Vol. 53, pp. 74–83. Washington, DC: American Chemical Society.

DuPont. (2015). Retrieved from http://www.dupont.com.

Durner, R., Witholt, B., and Egli, T. (2000). Accumulation of poly[(R)-3-hydroxyalkanoates] in *Pseudomonas oleovorans* during growth with octanoate in continuous culture at different dilution rates. *Applied and Environmental Microbiology*, 66(8), 3408–3414.

Eastman. (2015). Retrieved from http://www.eastman.com.

Elbahri, Z. and Taverdet, J. L. (2005). Optimization of an herbicide release from ethylcellulose microspheres. *Polymer Bulletin*, 54(4–5), 353–363.

Emmert, E. (1957). Black polyethylene for mulching vegetables. *American Society for Horticultural Science*, 69, 464–469.

English, B. (ed.). (1997). *Filters, Sorbents, and Geotextiles*. New York: CRC Press, Inc.

Espi, E., Salmeron, A., Fontecha, A., Garcia, Y., and Real, A. I. (2006). Plastic films for agricultural applications. *Journal of Plastic Film and Sheeting*, 22(2), 85–102.

Esposito, F. (March 3, 2014). Novamont buys majority stake in mater biopolymer venture. *Plastic News*. Retrieved from http://www.plasticsnews.com/article/20140303/NEWS/140309988/novamont-buys-majority-stake-in-mater-biopolymer-venture.

EverCorn. (2015). Retrieved from http://ever-corn.com.

Fabunmi, O. O., Tabil, L., Chang, P., and Panigrahi, S. (2007). Developing biodegradable plastics from starch. Paper presented at the *ASABE North Central Intersectional Meeting*, Fargo, ND.

Fan, L., Jin, R., Le, X., Zhou, X., Chen, S., Liu, H., and Xiong, Y. (2012). Chitosan microspheres for controlled delivery of auxins as agrochemicals. *Microchimica Acta*, 176(3–4), 381–387.

Fanta, G. F., Doane, W. M., and Stout, E. I. (1984). Modified starches as extenders for absorbent polymers. US Patent 4,483,950.

Fanta, G. F., Stout, E. I., and Doane, W. M. (1979). Highly absorbent graft copolymers of polyhydroxy polymers, acrylonitrile, and acrylic comonomers. US Patent 4,134,863.

Fernández, D., Rodríguez, E., Bassas, M. et al. (2005). Agro-industrial oily wastes as substrates for PHA production by the new strain Pseudomonas aeruginosa NCIB 40045: Effect of culture conditions. *Biochemical Engineering Journal*, 26(2–3), 159–167.

Fernández-Pérez, M., Villafranca-Sánchez, M., Flores-Céspedes, F., and Daza-Fernández, I. (2011). Ethylcellulose and lignin as bearer polymers in controlled release formulations of chloridazon. *Carbohydrate Polymers*, 83(4), 1672–1679.

Fernández-Pérez, M., Villafranca-Sánchez, M., González-Pradas, E., and Flores-Céspedes, F. (1999). Controlled release of diuron from an alginate-bentonite formulation: Water release kinetics and soil mobility study. *Journal of Agricultural and Food Chemistry*, 47(2), 791–798.

Fkur. (2015). Retrieved from http://www.fkur.com.

Flieger, M., Kantorova, M., Prell, A., Rezanka, T., and Votruba, J. (2003). Biodegradable plastics from renewable sources. *Folia Microbiologica*, 48(1), 27–44.

Fontanelli, M., Raffaelli, M., Martelloni, L., Frasconi, C., Ginanni, M., and Peruzzi, A. (2013). The influence of non-living mulch, mechanical and thermal treatments on weed population and yield of rainfed fresh-market tomato (*Solanum lycopersicum* L.). *Spanish Journal of Agricultural Research*, 11(3), 593–602.

Fujimaki, T. (1998). Processability and properties of aliphatic polyesters, 'BIONOLLE', synthesized by poly-condensation reaction. *Polymer Degradation and Stability*, 59(1–3), 209–214.

Full, T. D., Jung, D. O., and Madigan, M. T. (2006). Production of poly-beta-hydroxyalkanoates from soy molasses oligosaccharides by new, rapidly growing *Bacillus* species. *Letters in Applied Microbiology*, 43(4), 377–384.

Gansu PPMO team. (2014). Growing more food and using less water while preserving the natural environment in Guanghe County, China. Retrieved from http://ifad-un.blogspot.com/2014/10/growing-more-food-and-using-less-water.html.

Garlotta, D. (2001). A literature review of poly(lactic acid). *Journal of Polymers and the Environment*, 9(2), 63–84.

Garrido-Herrera, F. J., Daza-Fernández, I., González-Pradas, E., and Fernández-Pérez, M. (2009). Lignin-based formulations to prevent pesticides pollution. *Journal of Hazardous Materials*, 168(1), 220–225.

Girgenti, V., Peano, C., Giuggioli, N. R., Giraudo, E., and Guerrini, S. (2011). First results of biodegradable mulching on small berry fruits. In B. Mezzetti and P. B. DeOliveira (eds.), *XXVIII International Horticultural Congress on Science and Horticulture for People*, Vol. 926, pp. 571–576. Leuven, Belgium: International Society for Horticultural Science.

Griffin, G. J. L. (1974). Biodegradable fillers in thermoplastics. In R. D. Deanin and N. R. Schott (eds.), *Fillers and Reinforcements for Plastics*, Vol. 134, pp. 159–170. Washington, DC: American Chemical Society.

Griffin, G. J. L. (1995). Chemistry and technology of biodegradable polymers. *Journal of Chemical Education*, 72(3), A73.

Gu, S.-Y., Zhang, K., Ren, J., and Zhan, H. (2008). Melt rheology of polylactide/poly(butylene adipate-co-terephthalate) blends. *Carbohydrate Polymers*, 74(1), 79–85.

Haapala, T., Palonen, P., Tamminen, A., and Ahokas, J. (2015). Effects of different paper mulches on soil temperature and yield of cucumber (*Cucumis sativus* L.) in the temperate zone. *Agricultural and Food Science*, 24(1), 52–58.

Hayes, D. G., Dharmalingam, S., Wadsworth, L. C., Leonas, K. K., Miles, C., and Inglis, D. A. (2012). Biodegradable agricultural mulches derived from biopolymers. In K. Khemani and C. Scholz (eds.), *Degradable Polymers and Materials: Principles and Practice*, Vol. 1114, pp. 201–223. Washington, DC: American Chemical Society.

Haynes, R. J. (1980). Influence of soil management practice on the orchard agro-ecosystem. *Agro-Ecosystems*, 6(1), 3–32.

Hiskakis, M., Babou, E., and Briassoulis, D. (2011). Experimental processing of biodegradable drip irrigation systems-possibilities and limitations. *Journal of Polymers and the Environment*, 19(4), 887–907.

Ho, K.-L. and Pometto III, A. (1999). Temperature effects on soil mineralization of polylactic acid plastic in laboratory respirometers. *Journal of Environmental Polymer Degradation*, 7(2), 101–108.

Horttanainen, M., Friari, P., Honkanen, H., Luoranen, M., and Marttila, M. (2007). Recycling of the plastic waste of farms—Effects of high oil price and changes in waste management. *ISWA Conference Paper*, Lappeenranta University of Technology, Lappeenranta, Finland.

Hussain, M. R., Devi, R., and Maji, T. (2012). Controlled release of urea from chitosan microspheres prepared by emulsification and cross-linking method. *Iranian Polymer Journal*, 21(8), 473–479.

Ibrahim, S. M., El Salmawi, K. M., and Zahran, A. H. (2007). Synthesis of crosslinked superabsorbent carboxymethyl cellulose/acrylamide hydrogels through electron-beam irradiation. *Journal of Applied Polymer Science*, 104(3), 2003–2008.

Immirzi, B., Malinconico, M., Santagata, G., and Trautz, D. (2008). Characterization of galactomannans and cellulose fibres based composites for new mulching spray technology. *Acta Horticulturae*, 801(1), 195.

Immirzi, B., Santagata, G., Vox, G., and Schettini, E. (2009). Preparation, characterisation and field-testing of a biodegradable sodium alginate-based spray mulch. *Biosystems Engineering*, 102(4), 461–472.

InnoviaFilms. (2015). Retrieved from http://www.innoviafilms.com.

IreChemicals. (2015). Retrieved from http://irechem.en.ecplaza.net/.

Işiklan, N. (2006). Controlled release of insecticide carbaryl from sodium alginate, sodium alginate/gelatin, and sodium alginate/sodium carboxymethyl cellulose blend beads crosslinked with glutaraldehyde. *Journal of Applied Polymer Science*, 99(4), 1310–1319.

Jiang, L., Wolcott, M. P., and Zhang, J. (2006). Study of biodegradable polylactide/poly(butylene adipate-co-terephthalate) blends. *Biomacromolecules*, 7(1), 199–207.

Johnson, R. M. and Pepperman, A. B. (1995). Soil column mobility of metribuzin from alginate-encapsulated controlled release formulations. *Journal of Agricultural and Food Chemistry*, 43(1), 241–246.

JSRCorporation. (2015). Retrieved from http://www.jsr.co.jp/jsr_e/pdf/pd/pla/biolloy.pdf.

Kahar, P., Agus, J., Kikkawa, Y., Taguchi, K., Doi, Y., and Tsuge, T. (2005). Effective production and kinetic characterization of ultra-high-molecular-weight poly[(R)-3-hydroxybutyrate] in recombinant *Escherichia coli*. *Polymer Degradation and Stability*, 87(1), 161–169.

Kamel, S., El-Sakhawy, M., and Nada, A. M. A. (2004). Mechanical properties of the paper sheets treated with different polymers. *Thermochimica Acta*, 421(1–2), 81–85.

Kapanen, A., Schettini, E., Vox, G., and Itavaara, M. (2008). Performance and environmental impact of biodegradable films in agriculture: A field study on protected cultivation. *Journal of Polymers and the Environment*, 16(2), 109–122.

Kasirajan, S. and Ngouajio, M. (2012). Polyethylene and biodegradable mulches for agricultural applications: A review. *Agronomy for Sustainable Development*, 32(2), 501–529.

Kenawy, E.-R. (1998). Recent advances in controlled release of agrochemicals. *Journal of Macromolecular Science, Part C*, 38(3), 365–390.

Kerouani, S., Sadoun, T., and Azzouz, N. (2013). Infrared spectroscopic study of the thermo-oxidative aging of polyethylene containing pro-oxidant. In K. M. Gupta (ed.), *Advanced Materials Research III*, Vol. 685, pp. 316–323. Zurich, Switzerland: Trans Tech Publications Limited.

Kerr, R. W. (1950). Occurrence and varieties of starch. In R. W. E. Kerr (ed.), *Chemistry and Industry of Starch*, 2nd ed., pp. 3–25. New York: Academic Press Inc.

Kijchavengkul, T., Auras, R., Rubino, M., Ngouajio, M., and Fernandez, R. T. (2008). Assessment of aliphatic-aromatic copolyester biodegradable mulch films. Part I: Field study. *Chemosphere*, 71(5), 942–953.

Klemm, D., Klemm, D., Heublein, B., Fink, H. P., and Bohn, A. (2005). Cellulose: Fascinating biopolymer and sustainable raw material. *Angewandte Chemie (International ed.)*, 44(22), 3358–3393.

Koenig, M. F. and Huang, S. J. (1995). Biodegradable blends and composites of polycaprolactone and starch derivatives. *Polymer*, 36(9), 1877–1882.

Kuraray. (2015). Retrieved from www.kuraray.co.jp.

Kweon, D.-K., Kawasaki, N., Nakayama, A., and Aiba, S. (2004). Preparation and characterization of starch/polycaprolactone blend. *Journal of Applied Polymer Science*, 92(3), 1716–1723.

Kyrikou, I. and Briassoulis, D. (2007). Biodegradation of agricultural plastic films: A critical review. *Journal of Polymers and the Environment*, 15(2), 125–150.

LabelAgriWaste. (2006–2009). Labelling agricultural plastic waste for valorising the waste stream. Retrieved from https://labelagriwaste.aua.gr.

Lament, W. J. (1993). Plastic mulches for the production of vegetable crops. *HortTechnology*, 3(1), 35–39.

Lamont, W. J. Jr. (2005). Plastics: Modifying the microclimate for the production of vegetable crops. *HortTechnology*, 15(3), 477–481.

Lamont, W. J. (2009). Overview of the use of high tunnels worldwide. *HortTechnology*, 19(1), 25–29.

Lee, S., Choi, H., Suh, S. et al. (1999). Oligogalacturonic acid and chitosan reduce stomatal aperture by inducing the evolution of reactive oxygen species from guard cells of tomato and commelina communis. *Plant Physiology*, 121(1), 147–152.

Lemoigne, M. (1926). Produits de deshydration et de polymerisation de lacide β-oxybutyrique. *Bulletin de la Societe de Chimie Biologique*, 8, 770–782.

Li, C., Moore-Kucera, J., Lee, J., Corbin, A., Brodhagen, M., Miles, C., and Inglis, D. (2014). Effects of biodegradable mulch on soil quality. *Applied Soil Ecology*, 79, 59–69.

Lopez-Marin, J., Abrusci, C., Gonzalez, A., and Fernandez, J. A. (2012). Study of degradable materials for soil mulching in greenhouse-grown lettuce. *International Symposium on Advanced Technologies and Management Towards Sustainable Greenhouse Ecosystems: Greensys2011*, Halkidiki, Greece, Vol. 952, pp. 393–398.

Lopez-Marin, J., Gonzalez, A., Fernandez, J. A., Pablos, J. L., and Abrusci, C. (2012). Biodegradable Mulch Film in a Broccoli Production System. *XXVIII International Horticultural Congress on Science and Horticulture for People (Ihc2010): International Symposium on Organic Horticulture: Productivity and Sustainability*, Lisbon, Portugal, Vol. 933, pp. 439–444.

Lops, F., Carlucci, A., Camele, I., Raimondo, M. L., and Frisullo, S. (2011). Effectiveness of mulching plastic film to control corky rot and some viruses of tomato. In A. Crescenzi (ed.), *III International Symposium on Tomato Diseases*, Vol. 914, pp. 113–115. Leuven, Belgium: International Society for Horticultural Science.

Lörcks, J. (1998). Properties and applications of compostable starch-based plastic material. *Polymer Degradation and Stability*, 59(1–3), 245–249.

Labużek, S. and Radecka, I. (2001). Biosynthesis of PHB tercopolymer by *Bacillus* cereus UW85. *Journal of Applied Microbiology*, 90(3), 353–357.

Magistad, C., Farden, C. A., and Baldwin, W. A. (1935). Bagasse and paper mulches. *Journal of the American Society of Agronomy*, 27(10), 813–825.

Mahishi, L. H., Tripathi, G., and Rawal, S. K. (2003). Poly(3-hydroxybutyrate) (PHB) synthesis by recombinant *Escherichia coli* harbouring Streptomyces aureofaciens PHB biosynthesis genes: effect of various carbon and nitrogen sources. *Microbiological Research*, 158(1), 19–27.

Martín-Closas, L. and Pelacho, A. M. (2011). Agronomic potential of biopolymer films. In D. V. Plackett (ed.), *Biopolymers—New Materials for Sustainable Films and Coatings*, pp. 277–299. Chichester, U.K.: John Wiley & Sons, Ltd.

Matzinos, P., Tserki, V., Gianikouris, C., Pavlidou, E., and Panayiotou, C. (2002). Processing and characterization of LDPE/starch/PCL blends. *European Polymer Journal*, 38(9), 1713–1720.

Meadows, G. W. (1956). Process for conditioning soil with polysaccharide xanthates. US Patent 2,761,247.

Meadows, G. W. (1957). Stabilized soils. US Patent 2,801,933.

Menefee, E. and Hautala, E. (1978). Soil stabilisation by cellulose xanthate. *Nature*, 275(5680), 530–532.

Meredian. (2015). Retrieved from http://meredianinc.com/.

Metabolix. (2015). Retrieved from http://www.metabolix.com.

MGHNodax. (2015). Retrieved from http://www.mhgbio.com/.

Miles, C., Wallace, R., Wszelaki, A., Martin, J., Cowan, J., Walters, T., and Inglis, D. (2012). Deterioration of potentially biodegradable alternatives to black plastic mulch in three tomato production regions. *Hortscience*, 47(9), 1270–1277.

Miller, D. E. (1979). Effect of H-SPAN on water retained by soils after irrigation. *Soil Science Society of America Journal*, 43(3), 628–629.

Minuto, G., Pisi, L., Tinivella, F. et al. (2008). Weed control with biodegradable mulch in vegetable crops. *Acta Horticulturae*, 801(1), 291.

Mitrus, M., Wojtowicz, A., and Moscicki, L. (eds.). (2009). *Biodegradable Polymers and Their Practical Utility*. Weinheim, Germany: Wiley-VCH Verlag GmbH & Co. KGaA.

Mitrus, M., Wojtowicz, A., and Moscicki, L. (2010). Biodegradable polymers and their practical utility. In L. P. B. M. Janssen and L. Moscicki (eds.), *Thermoplastic Starch*, pp. 1–33. Weinheim, Germany: Wiley-VCH Verlag GmbH & Co. KGaA.

MitsubishiNews. (2015). Retrieved from http://www.m-kagaku.co.jp/index_en.htm.

Mohammadi Nafchi, A., Moradpour, M., Saeidi, M., and Alias, A. K. (2013). Thermoplastic starches: Properties, challenges, and prospects. *Starch/Stärke*, 65(1–2), 61–72.

Moore, G. F. and Saunders, S. M. (1997). *Advances in Biodegradable Polymers*, Vol. 9. Shrewsbury, U.K.: Rapra Technology Limited.

Moreno, M. M. and Moreno, A. (2008). Effect of different biodegradable and polyethylene mulches on soil properties and production in a tomato crop. *Scientia Horticulturae*, 116(3), 256–263.

Moreno, M. M., Moreno, A., and Mancebo, I. (2009). Comparison of different mulch materials in a tomato (*Solanum lycopersicum* L.) crop. *Spanish Journal of Agricultural Research*, 7(2), 454–464.

Mostafa, H. M. and Sourell, H. (2011). Characteristics of some biochemical materials used for agricultural foil mulch. *Journal of Agricultural Research*, 49(4), 539–550.

Muller, R.-J. (ed.). (2005). *Aliphatic-Aromatic Polyesters*. Shrewsbury, U.K.: Rapra Technology Limited.

Nafchi, A. M., Moradpour, M., Saeidi, M., and Alias, A. (2013). Thermoplastic starches: Properties, challenges, and prospects. *Starch-Starke, 65*(1–2), 61–72.

NaturePlast. (2015). Retrieved from http://www.natureplast.eu/en/.

Nawrath, C., Poirier, Y., and Somerville, C. (1995). Plant polymers for biodegradable plastics: Cellulose, starch and polyhydroxyalkanoates. *Molecular Breeding*, 1(2), 105–122.

Nge, K. L., Nwe, N., Chandrkrachang, S., and Stevens, W. F. (2006). Chitosan as a growth stimulator in orchid tissue culture. *Plant Science*, 170(6), 1185–1190.

Ngouajio, M., Auras, R., Fernandez, R. T., Rubino, M., Counts, J. W., and Kijchavengkul, T. (2008). Field performance of aliphatic-aromatic copolyester biodegradable mulch films in a fresh market tomato production system. *HortTechnology*, 18(4), 605–610.

Niekraszewicz, B. and Niekraszewicz, A. (eds.). (2009). Chapter 8: The structure of alginate, chitin and chitosan fibres. In S. Eichhorn (ed.), *Handbook of Textile Fibre Structure*, Vol. 2: Natural, regenerated, inorganic and specialist fibres. Cambridge, U.K.: Woodhead Publishing Limited.

NihonCornStarch. (2015). Retrieved from http://www.nihon-cornstarch.com.

Nikel, P. I., Pettinari, M. J., Galvagno, M. A., and Mendez, B. S. (2006). Poly(3-hydroxybutyrate) synthesis by recombinant *Escherichia coli* arcA mutants in microaerobiosis. *Applied and Environmental Microbiology*, 72(4), 2614–2620.

Nonato, R., Mantelatto, P., and Rossell, C. (2001). Integrated production of biodegradable plastic, sugar and ethanol. *Applied Microbiology and Biotechnology*, 57(1–2), 1–5.

Novamont. (2015). Retrieved from http://www.novamont.com.

O'Sullivan, A. (1997). Cellulose: The structure slowly unravels. *Cellulose*, 4(3), 173–207.

Ohta, K., Morishita, S., Suda, K., Kobayashi, N., and Hosoki, T. (2004). Effects of chitosan soil mixture treatment in the seedling stage on the growth and flowering of several ornamental plants. *Journal of the Japanese Society for Horticultural Science*, 73(1), 66–68.

Ojumu, T., Yu, J., and Solomon, B. (2004). Production of polyhydroxyalkanoates, a bacterial biodegradable polymers. *African Journal of Biotechnology*, 3(1), 18–24.

Olsen, J. K. and Gounder, R. K. (2001). Alternatives to polyethylene mulch film—A field assessment of transported materials in capsicum (*Capsicum annuum* L.). *Australian Journal of Experimental Agriculture*, 41(1), 93–103.

Orts, W. J., Sojka, R. E., and Glenn, G. M. (2000). Biopolymer additives to reduce erosion-induced soil losses during irrigation. *Industrial Crops and Products*, 11(1), 19–29.

Otey, F. H. (1976). Current and potential uses of starch products in plastics. *Polymer-Plastics Technology and Engineering*, 7(2), 221–234.

Otey, F. H., Mark, A. M., Mehltretter, C. L., and Russell, C. R. (1974). Starch-based film for degradable agricultural mulch. *Product R&D*, 13(1), 90–92.

Otey, F. H., Westhoff, R. P., and Russell, C. R. (1977). Biodegradable films from starch and ethylene-acrylic acid copolymer. *Product R&D*, 16(4), 305–308.

Pachence, J. M., Bohrer, M. P., and Kohn, J. (2007). Chapter twenty-three—Biodegradable polymers. In R. L. L. Vacanti (ed.), *Principles of Tissue Engineering*, 3rd ed., pp. 323–339. Burlington, MA: Academic Press.

Page, E. R. (1980). Cellulose xanthate as a soil conditioner: Field trials. *Journal of the Science of Food and Agriculture*, 31(7), 718–723.

Peoples, O. P. and Sinskey, A. J. (1989). Poly-beta-hydroxybutyrate (PHB) biosynthesis in Alcaligenes eutrophus H16. Identification and characterization of the PHB polymerase gene (phbC). *Journal of Biological Chemistry*, 264(26), 15298–15303.

Perez-Martinez, J. I., Morillo, E., Maqueda, C., and Gines, J. M. (2001). Ethyl cellulose polymer microspheres for controlled release of norflurazon. *Pest Management Science*, 57(8), 688–694.

Peters, S. (2011). Bioplastics based on cellulose. In S. Peters (ed.), *Material Revolution Sustainable and Multi-Purpose Materials for Design and Architecture*, pp. 38–40. Basel, Switzerland: Birkhauser GmbH.

Plackett, D. and Vazquez, A. (eds.). (2010). *Natural Polymer Sources*. Cambridge, U.K.: Woodhead Publishing.

Plantic. (2015). Retrieved from www.plantic.com.au/.

Plasthill. (2015). Retrieved from http://plasthill.fi/en/kareline.

PlasticsEurope. (2011). An analysis of European plastics production, demand and recovery for 2010. Retrieved from http://www.plasticseurope.org/.

PlasticsEurope. (2013). An analysis of European latest plastics production, demand and waste data. Retrieved from http://www.plasticseurope.org.

PolyOne. (2015). Retrieved from http://www.polyone.com.

PTT-MCC. (2015). Retrieved from http://www.pttmcc.com/new/faq.php.

Qiu, Y. Z., Han, J., and Chen, G. Q. (2006). Metabolic engineering of Aeromonas hydrophila for the enhanced production of poly(3-hydroxybutyrate-co-3-hydroxyhexanoate). *Applied Microbiology and Biotechnology*, 69(5), 537–542.

Rabea, E. I., Badawy, M. E. I., Rogge, T. M., Stevens, C. V., Höfte, M., Steurbaut, W., and Smagghe, G. (2005). Insecticidal and fungicidal activity of new synthesized chitosan derivatives. *Pest Management Science*, 61(10), 951–960.

Radley, J. A. (1953). Some physical properties of starch. In A. G. Gray (ed.), *Starch and Its Derivatives*, 3rd ed., pp. 58–80. London, U.K.: Chapman & Hall.

Rapa, M., Popa, M. E., Cinelli, P., Lazzeri, A., Burnichi, R., Mitelut, A., and Grosu, E. (2011). Biodegradable alternative to plastics for agriculture application. *Romanian Biotechnological Letters*, 16(6), 59–64.

Riggi, E., Santagata, G., and Malinconico, M. (2011). Bio-based and biodegradable plastics for use in crop production. *Recent Patents on Food, Nutrition & Agriculture*, 3(1), 49–63.

Riley, R. T. (1983). Starch-xanthate-encapsulated pesticides: A preliminary toxicological evaluation. *Journal of Agricultural and Food Chemistry*, 31(2), 202–206.

Rodríguez, A. T., Ramírez, M. A., Cárdenas, R. M., Hernández, A. N., Velázquez, M. G., and Bautista, S. (2007). Induction of defense response of *Oryza sativa* L. against Pyricularia grisea (Cooke) Sacc. by treating seeds with chitosan and hydrolyzed chitosan. *Pesticide Biochemistry and Physiology*, 89(3), 206–215.

Romic, D., Borosic, J., Poljak, M., and Romic, M. (2003). Polyethylene mulches and drip irrigation increase growth and yield in watermelon (*Citrullus lanatus* L.). *European Journal of Horticultural Science*, 68(4), 192–198.

Romic, D., Romic, M., Borosic, J., and Poljak, M. (2003). Mulching decreases nitrate leaching in bell pepper (*Capsicum annuum* L.) cultivation. *Agricultural Water Management*, 60(2), 87–97.

Roy, A., Singh, S., Bajpai, J., and Bajpai, A. (2014). Controlled pesticide release from biodegradable polymers. *Central European Journal of Chemistry*, 12(4), 453–469.

Rudnik, E. (2008). *Compostable Polymer Materials*. Oxford, U.K.: Elsevier.

Rudnik, E. and Briassoulis, D. (2011). Comparative biodegradation in soil behaviour of two biodegradable polymers based on renewable resources. *Journal of Polymers and the Environment*, 19(1), 18–39.

Rudzinski, W. E., Dave, A. M., Vaishnav, U. H., Kumbar, S. G., Kulkarni, A. R., and Aminabhavi, T. M. (2002). Hydrogels as controlled release devices in agriculture. *Designed Monomers and Polymers*, 5(1), 39–65.

Sanders, D. C. (January 31, 2001). Using plastic mulches and drip irrigation for vegetable production. Horticulture Information Leaflet, NC Cooperative Extension Resources, Raleigh, NC.

Sannino, A., Demitri, C., and Madaghiele, M. (2009). Biodegradable cellulose-based hydrogels: Design and applications. *Materials*, 2(2), 353–373.

Sannino, A., Madaghiele, M., Demitri, C., Scalera, F., Esposito, A., Esposito, V., and Maffezzoli, A. (2010). Development and characterization of cellulose-based hydrogels for use as dietary bulking agents. *Journal of Applied Polymer Science*, 115(3), 1438–1444.

Sannino, A. and Nicolais, L. (2005). Concurrent effect of microporosity and chemical structure on the equilibrium sorption properties of cellulose-based hydrogels. *Polymer*, 46(13), 4676–4685.

Sarazin, P., Li, G., Orts, W. J., and Favis, B. D. (2008). Binary and ternary blends of polylactide, polycaprolactone and thermoplastic starch. *Polymer*, 49(2), 599–609.

Sastry, K. R., Shrivastava, A., and Venkateshwarlu, G. (2015). Assessment of current trends in R&D of chitin-based technologies in agricultural production-consumption systems using patent analytics. *Journal of Intellectual Property Rights*, 20(1), 19–38.

Scarascia-Mugnozza, G., Sica, C., and Russo, G. (2011). Plastic materials in European agriculture: Actual use and perspectives. *Journal of Agricultural Engineering*, 42(3), 15–28.

Schettini, E., Sartore, L., Barbaglio, M., and Vox, G. (2012). Hydrolyzed protein based materials for biodegradable spray mulching coatings. *International Symposium on Advanced Technologies and Management Towards Sustainable Greenhouse Ecosystems: Greensys2011*, Halkidiki, Greece, Vol. 952, pp. 359–366.

Schettini, E., Vox, G., Candura, A., Malinconico, M., Immirzi, B., and Santagata, G. (2008). Starch-based films and spray coatings as biodegradable alternatives to LDPE mulching films. In S. DePascale, G. S. Mugnozza, A. Maggio, and E. Schettini (eds.), *Proceedings of the International Symposium on High Technology for Greenhouse System Management*, Vols 1 and 2, pp. 171–179. Leuven, Belgium: International Society Horticultural Science.

Sharp, R. (2013). A review of the applications of chitin and its derivatives in agriculture to modify plant-microbial interactions and improve crop yields. *Agronomy*, 3(4), 757.

Shasha, B. S., Doane, W. M., and Russell, C. R. (1976). Starch-encapsulated pesticides for slow release. *Journal of Polymer Science: Polymer Letters Edition*, 14(7), 417–420.

Shogren, R. L. (1999). Preparation and characterization of a biodegradable mulch: Paper coated with polymerized vegetable oils. *Journal of Applied Polymer Science*, 73(11), 2159–2167.

Shogren, R. L. and Hochmuth, R. C. (2004). Field evaluation of watermelon grown on paper-polymerized vegetable oil mulches. *Hortscience*, 39(7), 1588–1591.

ShowaDenko. (2015). Retrieved from http://www.showa-denko.com.

Siegenthaler, K. O., Kunkel, A., Skupin, G., and Yamamoto, M. (2012). Ecoflex (R) and Ecovio (R): Biodegradable, performance-enabling plastics. In B. Rieger, A. Kunkel, G. W. Coates, R. Reichardt, E. Dinjus, and T. A. Zevaco (eds.), *Synthetic Biodegradable Polymers*, Vol. 245, pp. 91–136. Berlin, Germany: Springer.

SkyChemicals. (2015). Retrieved from http://www.skchemicals.com/en/.

Smith, R. (2005). *Biodegradable Polymers for Industrial Applications*. Cambridge, U.K.: CRC Press.

Sopeña, F., Cabrera, A., Maqueda, C., and Morillo, E. (2007). Ethylcellulose formulations for controlled release of the herbicide alachlor in a sandy soil. *Journal of Agricultural and Food Chemistry*, 55(20), 8200–8205.

Stapleton, J. J. and Summers, C. G. (2002). Reflective mulches for management of aphids and aphid-borne virus diseases in late-season cantaloupe (*Cucumis melo* L. var. cantalupensis). *Crop Protection*, 21(10), 891–898.

Summers, C. G., Stapleton, J. J., Newton, A. S., Duncan, R. A., and Hart, D. (1995). Comparison of sprayable and film mulches in delaying the onset of aphid-transmitted virus diseases in zucchini squash. *Plant Disease*, 79(11), 1126–1131.

Teknaro. (2015). Retrieved from http://www.tecnaro.de/.

TeknorApex. (2015). Retrieved from http://www.teknorapex.com.

Tianan. (2015). Retrieved from http://www.tianan-enmat.com.

Tianjin. (2015), Retrieved from www.tjgreenbio.com.

Tindall, J. A., Mills, H. A., and Radcliffe, D. E. (1990). The effect of root zone temperature on nutrient uptake of tomato. *Journal of Plant Nutrition*, 13(8), 939–956.

Tobin, K. M. and O'Connor, K. E. (2005). Polyhydroxyalkanoate accumulating diversity of Pseudomonas species utilising aromatic hydrocarbons. *FEMS Microbiology Letters*, 253(1), 111–118.

Tsuji, H. (2005). Poly(lactide) stereocomplexes: formation, structure, properties, degradation, and applications. [Review]. *Macromolecular Bioscience*, 5(7), 569–597.

Unitika. (2015). Retrieved from https://www.unitika.co.jp/terramac/e/.

Valappil, S. P., Peiris, D., Langley, G. J., Herniman, J. M., Boccaccini, A. R., Bucke, C., and Roy, I. (2007). Polyhydroxyalkanoate (PHA) biosynthesis from structurally unrelated carbon sources by a newly characterized *Bacillus* spp. *Journal of Biotechnology*, 127(3), 475–487.

Vegeplast. (2015). Retrieved from www.vegeplast.com.

Verlinden, R. A. J., Hill, D. J., Kenward, M. A., Williams, C. D., and Radecka, I. (2007). Bacterial synthesis of biodegradable polyhydroxyalkanoates. *Journal of Applied Microbiology*, 102(6), 1437–1449.

Vert, M., Schwarch, G., and Coudane, J. (1995). Present and future of PLA polymers. *Journal of Macromolecular Science. Part A, Pure and Applied Chemistry*, A32(4), 787–796.

Vetrano, F., Fascella, S., Iapichino, G., Incalcaterra, G., Girgenti, P., Sutera, P., and Buscemi, G. (2009). Response of melon genotypes to polyethylene and biodegradable starch-based mulching films used for fruit production in the western coast of sicily. In Y. Tuzel, G. B. Oztekin, and M. K. Meric (eds.), *International Symposium on Strategies towards Sustainability of Protected Cultivation in Mild Winter Climate*, Vol. 807, pp. 109–113. Leuven, Belgium: International Society Horticultural Science.

Vincotte. (2015). Retrieved from http://www.vincotte.com.

Vink, E. T. H. and Steve, D. (2015). Life cycle inventory and impact assessment data for 2014 Ingeo™ polylactide production. *Industrial Biotechnology*, 11(3), 167–180.

Voegele, E. (2014). Trellis Earth restarts Cereplast plant with fresh business plan. *Biomass Magazine*. Retrieved from http://biomassmagazine.com/articles/10639/trellis-earth-restarts-cereplast-plant-with-fresh-business-plan.

Vroman, I. and Tighzert, L. (2009). Biodegradable polymers. *Materials*, 2(2), 307.

Wallace, A. (1986). A polysaccharide (guar) as a soil conditioner. *Soil Science*, 141(5), 371–373.

Ward, P. G., de Roo, G., and O'Connor, K. E. (2005). Accumulation of polyhydroxyalkanoate from styrene and phenylacetic acid by *Pseudomonas putida* CA-3. *Applied and Environmental Microbiology*, 71(4), 2046–2052.

Ward, P. G. and O'Connor, K. E. (2005). Bacterial synthesis of polyhydroxyalkanoates containing aromatic and aliphatic monomers by Pseudomonas putida CA-3. *International Journal of Biological Macromolecules*, 35(3–4), 127–133.

Weaver, M. O., Bagley, E. B., Fanta, G. F., and Doane, W. M. (1976a). Method of reducing water content of emulsions, suspensions, and dispersions with highly absorbent starch-containing polymeric compositions. US Patent 3,935,099.

Weaver, M. O., Bagley, E. B., Fanta, G. F., and Doane, W. M. (1976b). Highly absorbent starch-containing polymeric compositions. US Patent 3,981,100.

Weber, C. A. (2003). Biodegradable mulch films for weed suppression in the establishment year of matted-row strawberries. *HortTechnology*, 13(4), 665–668.

Westman, L. and Lindström, T. (1981a). Swelling and mechanical properties of cellulose hydrogels. I. Preparation, characterization, and swelling behavior. *Journal of Applied Polymer Science*, 26(8), 2519–2532.

Westman, L. and Lindström, T. (1981b). Swelling and mechanical properties of cellulose hydrogels. II. The relation between the degree of swelling and the creep compliance. *Journal of Applied Polymer Science*, 26(8), 2533–2544.

Wood, J. D. and Oster, J. D. (1985). The effect of cellulose xanthate and polyvinyl alcohol on infiltration, erosion, and crusting at different sodium levels. *Soil Science*, 139(3), 243–249.

Worldwide Developments in More Sustainable Materials. (Janaury 28 2009). *Ingeonews*, 6.

Wu, L. and Liu, M. (2008). Preparation and properties of chitosan-coated NPK compound fertilizer with controlled-release and water-retention. *Carbohydrate Polymers*, 72(2), 240–247.

Xinfu. (2015). Retrieved from http://www.xinfuchina.com/.

Xu, J. and Guo, B. H. (2010). Poly(butylene succinate) and its copolymers: Research, development and industrialization. *Biotechnology Journal*, 5(11), 1149–1163.

Xuan Tham, L., Nagasawa, N., Matsuhashi, S., Ishioka, N. S., Ito, T., and Kume, T. (2001). Effect of radiation-degraded chitosan on plants stressed with vanadium. *Radiation Physics and Chemistry*, 61(2), 171–175.

Yang, S.-R. and Wu, C.-H. (1999). Degradable plastic films for agricultural applications in Taiwan. *Macromolecular Symposia*, 144(1), 101–112.

Yin, H., Zhao, X., and Du, Y. (2010). Oligochitosan: A plant diseases vaccine—A review. *Carbohydrate Polymers*, 82(1), 1–8.

Yu, L., Dean, K., and Li, L. (2006). Polymer blends and composites from renewable resources. *Progress in Polymer Science*, 31(6), 576–602.

Zdrahala, R. J. (1997). Thermoplastic starch revisited. Structure/property relationship for "dialed-in" biodegradability. *Macromolecular Symposia*, 123, 113–121.

Zelfo. (2015). Retrieved from http://www.zelfo-technology.com.

Zhang, M., Tan, T., Yuan, H., and Rui, C. (2003). Insecticidal and fungicidal activities of chitosan and oligochitosan. *Journal of Bioactive and Compatible Polymers*, 18(5), 391–400.

Zhang, Y. C., Han, J. H., and Kim, G. N. (2008). Biodegradable mulch film made of starch-coated paper and its effectiveness on temperature and moisture content of soil. *Communications in Soil Science and Plant Analysis*, 39(7–8), 1026–1040.

Zhu, Z. and Zhuo, R. (2001). Controlled release of carboxylic-containing herbicides by starch-g-poly(butyl acrylate). *Journal of Applied Polymer Science*, 81(6), 1535–1543.

Index